# RNA Delivery Function for Anticancer Therapeutics

## Nanotechnology for Drugs, Vaccines and Smart Delivery Systems

Series Editors: Ram K. Gupta, Tuan Anh Nguyen

This book series aims to provide an overview of the recent development in vaccine and drug design, smart delivery systems, and characterizations. Many topics related to the applications of nanotechnology in the advancement of drugs, vaccines, and delivery systems will be discussed. Each book in this series will be authored or contributed by experts and global research teams will participate. The level of presentation is intended for students, scientists, researchers, and industries dealing with nanotechnology for advanced drugs, vaccines, and targeted delivery systems.

**RNA Delivery Function for Anticancer Therapeutics**
*Loutfy H. Madkour*

**Nanoparticle-Based Drug Delivery in Cancer Treatment**
*Loutfy H. Madkour*

For more information on this book series, please visit https://www.routledge.com/Nanotechnology-for-Drugs-Vaccines-and-Smart-Delivery-Systems/book-series/CRCNANDRUVAC

# RNA Delivery Function for Anticancer Therapeutics

*Loutfy H. Madkour*

**CRC Press**
Taylor & Francis Group
Boca Raton London New York

CRC Press is an imprint of the
Taylor & Francis Group, an **informa** business

First edition published 2022
by CRC Press
6000 Broken Sound Parkway NW, Suite 300, Boca Raton, FL 33487-2742

and by CRC Press
2 Park Square, Milton Park, Abingdon, Oxon, OX14 4RN

CRC Press is an imprint of Taylor & Francis Group, LLC

---

**Library of Congress Cataloging-in-Publication Data**

Names: Madkour, Loutfy H. (Loutfy Hamid), author.
Title: RNA delivery function for anticancer therapeutics / Loutfy H. Madkour.
Description: First edition. | Boca Raton : CRC Press, 2022. | Series: Nanotechnology for drugs, vaccines and smart delivery systems | Includes bibliographical references and index.
Identifiers: LCCN 2021039463 (print) | LCCN 2021039464 (ebook) | ISBN 9781032135168 (hardback) | ISBN 9781032135182 (paperback) | ISBN 9781003229650 (ebook)
Subjects: LCSH: RNA--Therapeutic use. | Cancer--Gene therapy. | Small interfering RNA. | MicroRNA.
Classification: LCC RC271.G45 M33 2022 (print) | LCC RC271.G45 (ebook) | DDC 616.99/4042--dc23/eng/20211108
LC record available at https://lccn.loc.gov/2021039463
LC ebook record available at https://lccn.loc.gov/2021039464

---

ISBN: 9781032135168 (hbk)
ISBN: 9781032135182 (pbk)
ISBN: 9781003229650 (ebk)

DOI: 10.1201/9781003229650

Typeset in Times
by KnowledgeWorks Global Ltd.

# Contents

# Preface

RNA interference can be considered a promising alternative for cancer therapy as it is less toxic than classical chemotherapy. Recent progress in RNA biology has broadened the scope of therapeutic targets of RNA drugs for cancer therapy. In the last decade, increasing evidence has proved the biological importance of miRNAs in physiological contexts and a huge number of studies have pointed to the fact that dysregulation of miRNAs plays a fundamental role in several pathological conditions, including cancer. Within a few years, we will know if miRNA-based therapeutics, alone or in combination with other modalities, will be clinically useful treatments for various diseases. Then it could be that blood-based circulating miRNA analysis has imminent clinical utility as disease markers. However, if this concept is to translate readily from bench to bedside, then supporting data demonstrating validity of this novel approach must stem from carefully planned studies. If the current momentum in translational research can be maintained, then an era of noninvasive rapid diagnostics is rapidly forthcoming. We hope that this book provides the readers great insight into the current advances and hurdles of RNAi therapeutics and accelerates RNAi drug development from bench to bedside.

Cancer is a multifactorial and epigenetic disease. Despite our ever-increasing knowledge of cancer, cancer is still the second leading cause of death, overall, and is predicted to become the leading killer as heart disease therapies improve. Cancer is caused by various mutations in hundreds of genes including both proto-oncogenes and tumor suppressor genes. There are many studies reporting the involvement of noncoding RNAs in cancer progression and development such as siRNAs, and miRNAs, but piRNAs have only recently been identified as new prognostic and diagnostic tools. In order to develop siRNAs as therapeutic agents for cancer treatment, delivery strategies for siRNAs must be carefully designed and potential gene targets carefully selected for optimal anticancer effects. Small interfering RNA (siRNA) has gained attention as a potential therapeutic reagent due to its ability to inhibit specific genes in many genetic diseases. Because RNAi is a highly conserved mechanism across mammals with the same siRNA sequences, it can also allow the animal results to be quickly translated into clinical design. In order to develop siRNAs as therapeutic agents for cancer treatment, delivery strategies for siRNA must be carefully designed and potential gene targets carefully selected for optimal anticancer effects.

RNA interference (RNAi) is a genetic regulatory system that functions to silence the activity of specific genes. RNAi occurs naturally, through the production of nuclear-encoded pre-microRNA (pre-miRNA), and can be induced experimentally, using short segments of synthetic double-stranded RNA (dsRNA). The synthetic dsRNA employed is typically either a small hairpin RNA (shRNA) or a short interfering RNA (siRNA). In both the natural and the experimental pathways, an enzyme known as Dicer is necessary for the formation of miRNA from pre-miRNA or of siRNA from shRNA. The miRNA or siRNA then binds to an enzyme-containing molecule known as RNA-induced silencing complex (RISC). The miRNA–RISC or siRNA–RISC complex binds to target, or complementary, messenger RNA (mRNA) sequences, resulting in the enzymatic cleavage of the target mRNA. The cleaved mRNA is rendered nonfunctional and hence is "silenced."

The ability of RNAi to silence genes was discovered in the 1990s by American scientists Andrew Z. Fire and Craig C. Mello, who shared the 2006 Nobel Prize in Physiology or Medicine for their work. Fire and Mello successfully inhibited the expression of specific genes by introducing short double-stranded RNA (dsRNA) segments into the cells of nematodes (*Caenorhabditis elegans*). The dsRNA segments underwent enzymatic processing that enabled them to attach to molecules of mRNA possessing complementary nucleotide sequences. The attachment of the two RNAs inhibited the translation of the mRNA molecules into proteins.

The fundamental principles for the regulation of gene expression were identified more than 40 years ago by the French Nobel Laureates François Jacob and Jacques Monod. Today, we know that similar principles operate throughout evolution, from bacteria to humans. They also form the basis for gene technology, in which a DNA sequence is introduced into a cell to produce new protein.

Fire and Mello published their findings in the journal *Nature* on February 19, 1998. Their discovery clarified many confusing and contradictory experimental observations and revealed a natural mechanism for controlling the flow of genetic information. This heralded the start of a new research field.

After a series of simple but elegant experiments, Fire and Mello deduced that dsRNA can silence genes, that this RNAi is specific for the gene whose code matches that of the injected RNA molecule, and that RNAi can spread between cells and even be inherited. Injecting tiny amounts of dsRNA was enough to achieve an effect, and Fire and Mello therefore proposed that RNA interference (now commonly abbreviated to RNAi) is a catalytic process.

RNAi is used to regulate gene expression in the cells of humans as well as worms. Hundreds of genes in our genome encode small RNA molecules called microRNAs. They contain pieces of the code of other genes. Such a microRNA molecule can form a double-stranded structure and activate the RNAi machinery to block protein synthesis. The expression of that particular gene is silenced. We now understand that genetic regulation by microRNAs plays an important role in the development of the organism and the control of cellular functions.

Our genome operates by sending instructions for the manufacture of proteins from DNA in the nucleus of the cell to the protein synthesizing machinery in the cytoplasm. These instructions are conveyed by mRNA. This mechanism, RNAi, is activated when RNA molecules occur as double-stranded pairs in the cell. The dsRNA activates biochemical machinery, which degrades those mRNA molecules that carry a genetic code identical to that of the dsRNA. When such mRNA molecules disappear, the corresponding gene is silenced and no protein of the encoded type is made.

Our genome consists of approximately 30,000 genes. However, only a fraction of them is used in each cell. Which genes are expressed (i.e., govern the synthesis of new proteins) is controlled by the machinery that copies DNA to mRNA in a process called transcription. This process can be modulated by various factors.

The genetic code in DNA determines how proteins are built. The instructions contained in the DNA are copied to mRNA and subsequently used to synthesize proteins. This flow of genetic information from DNA via mRNA to protein has been termed the central dogma of molecular biology by the British Nobel Laureate Francis Crick. Proteins are involved in all processes of life, for instance as enzymes digesting our food, as receptors receiving signals in the brain, and as antibodies defending us against bacteria.

Although Fire and Mello's work involved the experimental introduction of RNAi into cells, gene silencing by RNAi is a natural genetic mechanism in eukaryotes, which takes place following transcription (the synthesis of mRNA from DNA). Special microRNA (miRNA) segments, each of which is approximately 20 nucleotides (nt) in length, are encoded by the genomes of eukaryotic organisms. Each miRNA is produced from a precursor transcript (pre-miRNA). After the pre-miRNA migrates from the nucleus into the cytoplasm, it is cleaved into a mature miRNA by an enzyme known as Dicer. The mature miRNA molecule then binds to an RNA-induced silencing complex (RISC), which contains multiple proteins, including a ribonuclease enzyme. The miRNA nucleotide sequence directs the protein complex to bind to a complementary sequence of mRNA. Once bound to the mRNA, the miRNA–RISC complex then enzymatically cleaves targeted sites on the mRNA molecule, thereby inhibiting the translation of the gene into a protein, which effectively silences the gene.

RNAi plays an important role not only in regulating genes but also in mediating cellular defense against infection by RNA viruses, including influenza viruses and rhabdoviruses, a group that contains the causative agent of rabies. In fact, a number of plants and animals have evolved antiviral RNAi genes that encode short segments of RNA molecules with sequences that are complementary to viral sequences. This complementarity enables RNAi produced by the cell to bind to and inactivate specific RNA viruses.

RNAi also is an innate mechanism by which cells can suppress the activity of transposons, or "jumping genes." Certain types of transposable elements are able to produce mobile copies of them, which subsequently are inserted into various regions of the genome, giving rise to repetitive sequences of DNA. These insertions generally are of little concern. However, some insertions lead to increased or decreased gene activity and can give rise to disease in humans. For example, certain types of cancer and Duchenne muscular dystrophy, a hereditary muscle-wasting disorder, are associated with insertions of transposons.

The discovery that genes can be silenced by segments of dsRNA that are introduced into cells in tissue culture revolutionized the study of gene function. Gene silencing by dsRNA makes use of the naturally occurring cell machinery that is involved in the processing of miRNA in eukaryotic cells. For example, each dsRNA is cleaved into small pieces by the Dicer enzyme. These pieces are called short interfering RNAs (siRNAs) and are about 20–25 nt in length. Similar to miRNA, siRNA binds to RISC and cleaves targeted sequences of mRNA.

RNAi is a posttranscriptional mechanism that involves the inhibition of gene expression through promoting cleavage on a specific area of a target mRNA. This technology has shown promising therapeutic results for a good number of diseases, especially in cancer. MicroRNA molecules are small, single-stranded RNA molecules that function to regulate networks of genes. The discovery of miRNAs in general and particularly of circulating miRNAs is one of the major scientific breakthroughs in recent years and has revolutionized current cell biology and medical science. Whilst significant progress has been made in our understanding of the biological role of miRNAs, the complexity of miRNA regulation, including specific cell/tissue type and temporal expression, target multiplicity, and miRNA redundancy, presents many challenges to the miRNA researcher. Therapeutic strategies based on modulation of miRNA activity hold great promise due to the ability of these small RNAs to significantly alter cellular behavior.

Different types of synthetic dsRNAs can be employed to disrupt gene function. Commonly used molecules include siRNA, which bypasses Dicer cleavage, and small hairpin RNA (shRNA), which actually is one RNA strand containing two unique siRNA segments that is folded into a double strand, with the adjacent nucleotides joined through heating (annealing) rather than complementary base pairing. This creates a structure that resembles a hairpin because it has a tight loop at one end. Inside a cell, Dicer cleaves the shRNA into its two component siRNAs.

RNAi is an exceptionally powerful research tool. Synthetic dsRNAs are designed to prevent the expression of specific genes, thereby enabling geneticists to manipulate the activity of genes in order to better understand their functions. In addition, abnormally overactive genes contribute to certain human diseases, and silencing this activity using RNAi has become an important area of investigation. RNAi is being explored as a form of treatment for a variety of diseases, including macular degeneration, hepatitis, AIDS, Huntington disease, and cancer.

In macular degeneration, RNA sequences that block the production of a protein called vascular endothelial growth factor (VEGF) in cells of the retina can inhibit the excess growth of retinal blood vessels, which leak and lead to vision loss. RNAi treatments for macular degeneration involve the injection of naked RNA into the eye. The term *naked RNA* is used to distinguish this approach from those that employ viral vectors to introduce dsRNA into diseased cells. Interfering RNAs incorporated into vectors are being studied for their effectiveness in slowing tumor growth. For example, mRNA transcripts of genes known to be overactive in certain forms of cancer serve as useful targets for RNAi-based treatments, which can silence overactive genes and slow disease progression.

Factors such as ensuring that interfering RNAs reach the cells and that the viral vectors themselves do not give rise to dangerous side effects have complicated the development of RNAi therapies. Furthermore, sequence similarities between genes can result in the binding of dsRNAs to otherwise properly functioning genes. This can result in the silencing of healthy genes that are vital to normal cell function. Nevertheless, the technique remains promising for applications in medicine.

Figure 1 represents the gene silencing by RNAi.

When comparing RNAi therapeutics with small-molecule drugs and biologics, the RNAi therapeutics have the distinct advantage of being able to wipe out the disease-causing proteins from being translated, thus avoiding the need to attack other downstream components in a molecular cascade, as small-molecule drugs often do [1]. In addition, RNAi therapeutics can target even intracellular proteins hard to be reached by monoclonal antibodies. Because RNAi is a highly conserved mechanism across mammals with the same siRNA sequences, it can also allow the animal results to be quickly translated into clinical design.

Nevertheless, the journey of RNAi therapies to the clinic didn't go as smoothly as expected. In the early stage, both biotechs and major pharmaceutical companies competitively jumped into RNAi therapeutics regardless of technical challenges such as delivery. Consequently, a staggering number of early clinical studies failed to deliver patient benefit [2]. These initial failures made Big Pharma players such as Roche, Pfizer, and Merck halt their own RNAi programs. But thankfully, the past few years have seen considerable technical refinement in RNA chemical structure, targeting, and delivery. Recently the clinical success rate of RNAi therapeutics has increased and now some are on the right track to gain US Food and Drug Administration (FDA) approval in the next year or two [3].

Although there still remain some challenges in translating RNAi therapeutics to the clinic and commercialization,

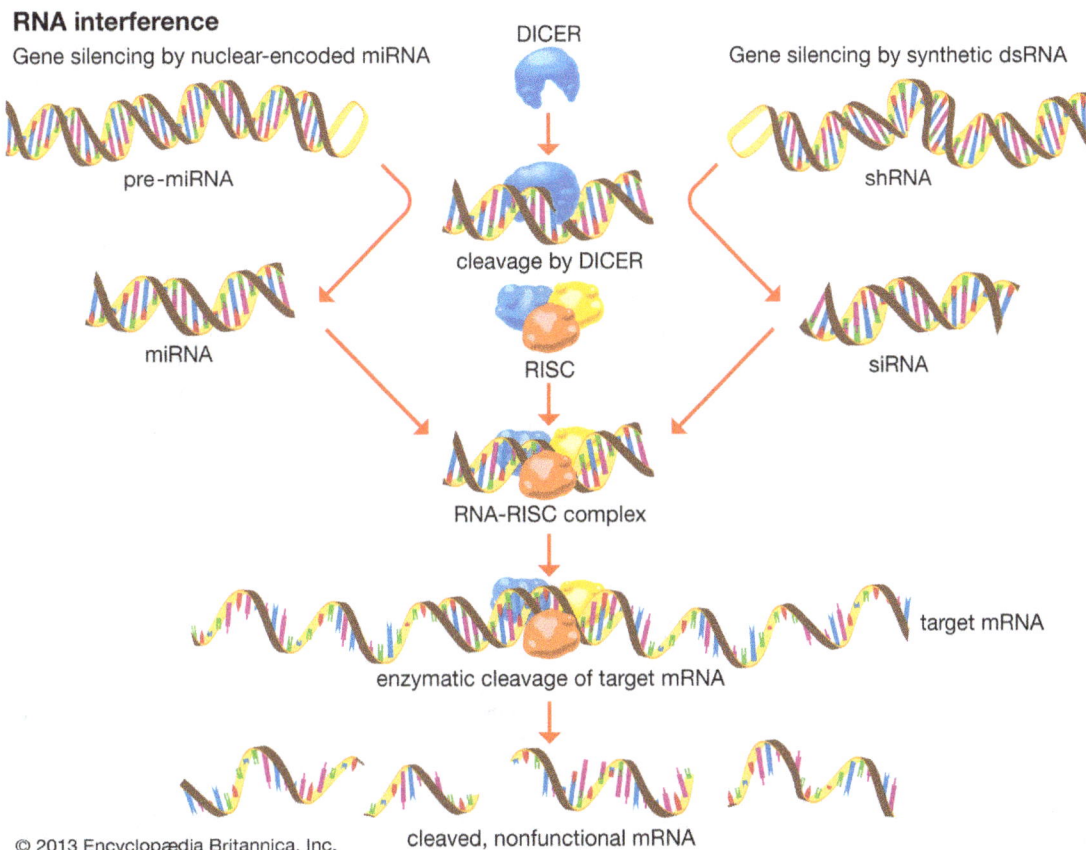

© 2013 Encyclopædia Britannica, Inc.

**FIGURE 1** Gene silencing by RNA interference.

researchers certainly wouldn't have been able to proceed this far in developing RNAi drugs without the efforts of many academic and industry leaders to design RNAi therapeutics. These efforts are delineated throughout the book, where we highlight the unforeseen hurdles and the potential pathways to translating RNAi therapeutics to the clinic. This book covers the discovery of valuable disease targets for RNAi therapeutics (Thomas Roberts et al.) [4], the chemical and structural modification of siRNA molecules (Sun Hwa Kim et al. and Hyukjin Lee et al.) [4], cancer-targeted delivery systems for RNAi therapeutics (Xiaoyuan Chen et al. and Kanjiro Miyata et al.) [4], and preclinical and clinical issues in RNAi therapeutics (Hyejung Mok et al. and Yoon Yeo et al.) [4].

## SUMMARY

The RNA interference (RNAi) technique is a new modality for cancer therapy, and several candidates are being tested clinically. In the development of RNAi-based therapeutics, imaging methods can provide a visible and quantitative way to investigate the therapeutic effect at the anatomical, cellular, and molecular levels; to noninvasively trace the distribution; and to study the biological processes in the preclinical and clinical stages. Their abilities are important not only for therapeutic optimization and evaluation but also for shortening of the time of drug development to market. Typically, imaging-functionalized RNAi therapeutic delivery that combines nanovehicles, and imaging techniques to study and improve their biodistribution and accumulation in tumor sites, has been progressively integrated into the anticancer drug discovery and development processes.

The discovery of RNAi in the late 1990s unlocked a new realm of therapeutic possibilities by enabling potent and specific silencing of theoretically any desired genetic target. The discovery of RNAi has opened doors that might introduce a novel therapeutic tool to the clinical setting. For many decades, small molecules have been developed and utilized in cancer therapy; however, critical problems, such as undesirable toxicity against normal tissues due to a lack of selectivity, still remain today. Using RNAi as a therapeutic tool will allow targeting previously unreachable targets with its potential to silence the function of any cancer-causing gene.

Small interfering RNA (siRNA) has gained attention as a potential therapeutic reagent due to its ability to inhibit specific genes in many genetic diseases. For many years, studies of siRNA have progressively advanced toward novel treatment strategies against cancer.

RNAi-based gene therapy has drawn tremendous attention due to its highly specific gene regulation by selective degradation of any target mRNA. There have been multiple reports regarding the development of various cationic materials for efficient siRNA delivery; however, many studies still suffer from conventional delivery problems such as suboptimal transfection performance, a lack of tissue specificity, and potential cytotoxicity. Despite the huge therapeutic potential of siRNAs, conventional gene carriers have failed to guarantee successful gene silencing in vivo, thus not warranting clinical trials. The relatively short double-stranded structure of siRNAs has resulted in uncompromising delivery formulations, as well as low transfection efficiency, compared with the conventional nucleic acid drugs such as plasmid DNAs. Recent developments in structural siRNA and RNAi nanotechnology have enabled more refined and reliable in vivo gene silencing with multiple advantages over naked siRNAs. This book focuses on recent progress in the development of structural DNA/RNA-based RNAi systems and their potential therapeutic applications. In addition, we provide an extensive list of prior reports on various RNAi systems, categorized by their distinctive molecular characters.

When sense and antisense RNA molecules meet, they bind to each other and form dsRNA. Could it be that such a dsRNA molecule silences the gene carrying the same code as this particular RNA? Fire and Mello tested this hypothesis by injecting dsRNA molecules containing the genetic codes for several other worm proteins. In every experiment, injection of dsRNA carrying a genetic code led to silencing of the gene containing that particular code. The protein encoded by that gene was no longer formed.

The components of the RNAi machinery were identified during the following years. DsRNA binds to a protein complex, Dicer, which cleaves it into fragments. Another protein complex, RISC, binds these fragments. One of the RNA strands is eliminated but the other remains bound to the RISC complex and serves as a probe to detect mRNA molecules. When an mRNA molecule can pair with the RNA fragment on RISC, it is bound to the RISC complex, cleaved, and degraded. The gene served by this particular mRNA has been silenced.

Jumping genes, also known as transposons, are DNA sequences that can move around in the genome. They are present in all organisms and can cause damage if they end up in the wrong place. Many transposons operate by copying their DNA to RNA, which is then reverse-transcribed back to DNA and inserted at another site in the genome. Part of this RNA molecule is often double-stranded and can be targeted by RNA interference. In this way, RNA interference protects the genome against transposons. This method has already become an important research tool in biology and biomedicine. In the future, it is hoped that it will be used in many disciplines including clinical medicine and agriculture. Several recent publications show successful gene silencing in human cells and experimental animals. For instance, a gene causing high blood cholesterol levels was recently shown to be silenced by treating animals with silencing RNA. Plans are underway to develop silencing RNA as a treatment for virus infections, cardiovascular diseases, cancer, endocrine disorders, and several other conditions.

This book presents an overview of the current status of translating the RNAi cancer therapeutics in the clinic, a brief description of the biological barriers in drug delivery,

# RNA interference
## — gene silencing by double-stranded RNA

### 1. The central dogma

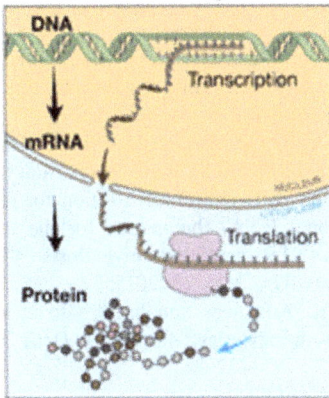

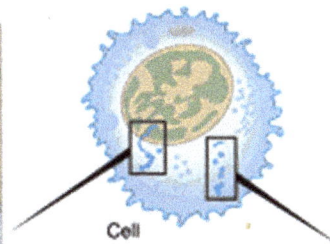

Our genome operates by sending information from double-stranded DNA in the nucleus, via single-stranded mRNA, to guide the synthesis of proteins in the cytoplasm.

### 2. The experiment

RNA carrying the code for a muscle protein is injected into the worm *C. elegans*. Single-stranded RNA has no effect. But when double-stranded RNA is injected, the worm starts twitching in a similar way to worms carrying a defective gene for the muscle protein.

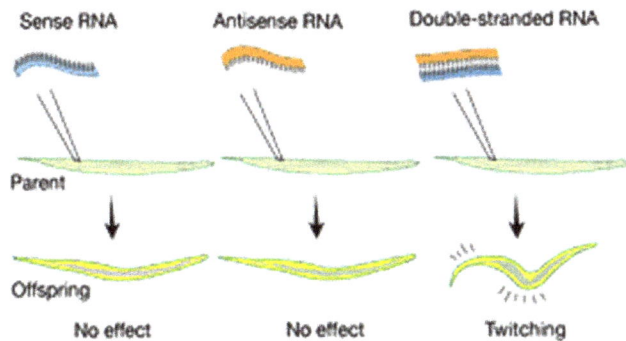

### 3. The RNAi mechanism

RNA interference (RNAi) is an important biological mechanism in the regulation of gene expression.

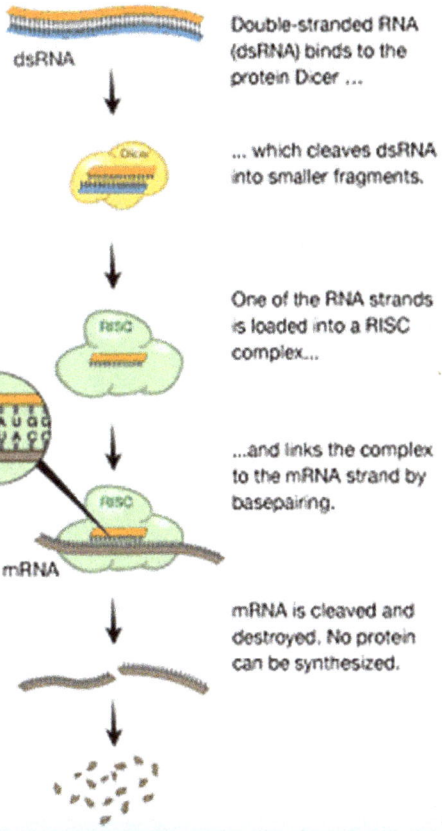

Double-stranded RNA (dsRNA) binds to the protein Dicer ...

... which cleaves dsRNA into smaller fragments.

One of the RNA strands is loaded into a RISC complex...

...and links the complex to the mRNA strand by basepairing.

mRNA is cleaved and destroyed. No protein can be synthesized.

### 4. Several processes in the cell use RNAi

A. When an RNA virus infects the cell, it injects its genome consisting of double-stranded RNA. RNA interference destroys the viral RNA, preventing the formation of new viruses.

B. Synthesis of many proteins is controlled by genes encoding microRNA. After processing, microRNA prevents the translation of mRNA to protein.

C. In the research laboratory, dsRNA molecules are tailor-made to activate the RISC complex to degrade mRNA for a specific gene.

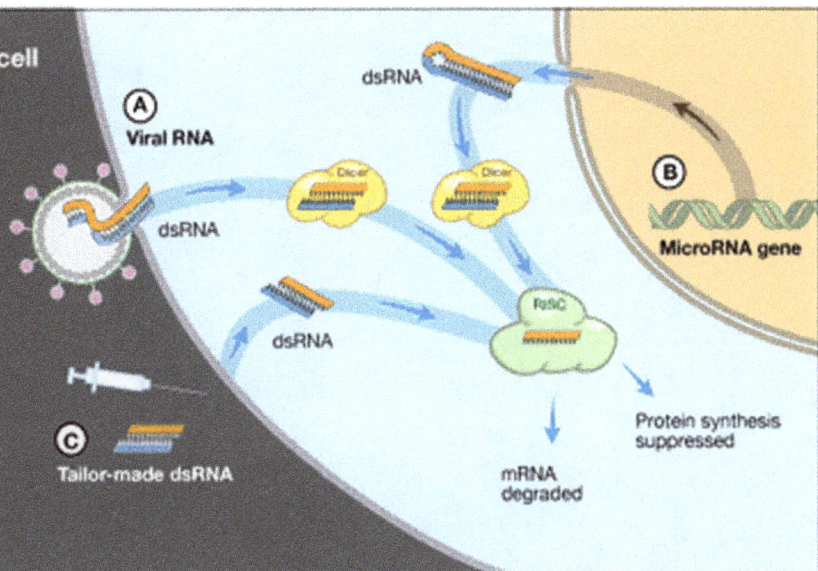

© The Nobel Committee for Physiology or Medicine      Illustration: Annika Röhl

**FIGURE 2**   Several mechanism processes in the cell use RNAi.

and the roles of imaging in aspects of administration route, systemic circulation, and cellular barriers for the clinical translation of RNAi cancer therapeutics, with an overview of the safety concerns. It then focuses on imaging-guided delivery of RNAi therapeutics in preclinical development, including the basic principles of different imaging modalities, and their advantages and limitations for biological imaging. With the growing number of RNAi therapeutics entering the clinic, various imaging methods will play an important role in facilitating the translation of RNAi cancer therapeutics from bench to bedside. RNAi technique has become a powerful tool for basic research to selectively knock down gene expression in vitro and in vivo. Our scientific and industrial communities have started to develop RNAi therapeutics as the next class of drugs for treating a variety of genetic disorders, such as cancer and other diseases that are particularly difficult to address with current treatment strategies.

The central dogma of RNA interference and several mechanism processes in the cell use RNAi are shown in Figure 2.

## REFERENCES

[1] Bender E. The second coming of RNAi. *Scientist*. September 1, 2014;28. LabX Media Group. https://www.the-scientist.com/features/the-second-coming-of-rnai-36936
[2] Krieg AM. Is RNAi dead? *Mol Ther*. 2011;19:1001–1002.
[3] Mason M. Alnylam receives orphan drug designation from the United States Food & Drug Administration for revusiran, an investigational RNAi therapeutic for the treatment of transthyretin (TTR)-mediated amyloidosis (ATTR Amyloidosis). *Business Wire*. May 20, 2015.
[4] Kwon IC, Kataoka K. Advances and hurdles to clinical translation of RNAi therapeutics. *Adv Drug Deliv Rev*. 2016;104:1.

# Author

**Loutfy H. Madkour** is a Professor of Physical Chemistry and Nanoscience at Tanta University, Egypt. He received his BSc, MSc, and PhD in physical chemistry from the Cairo, Minia, and Tanta universities in Egypt, respectively. He has published 200 peer-reviewed original research articles, 20 review articles, and 8 books. He is an editorial board member of several international journals, including *International Journal of Industrial Chemistry* (IJIC-Springer), *International Journal of Ground Sediment & Water, Global Drugs and Therapeutics, Chronicles of Pharmaceutical Science, Journal of Targeted Drug Delivery, UPI Journal of Pharmaceutical, Medical and Health Sciences, Global Journal of Nanomedicine, Clinical Pharmacology and Toxicology Research, Journal of Pharmacology & Pharmaceutical Research, LOJ Pharmacology & Clinical Research, CPQ Medicine, Pharmaceutical Sciences & Analytical Research Journal, Japan Journal of Research, Organic & Medicinal Chemistry International Journal, Nanotechnology & Applications, Materials Science Journal, Journal of Chemical Science and Chemical Engineering, United Journal of Nanotechnology and Medicine, Clinical Practice (Therapy), Journal of Materials New Horizons, Journal of Radiology and Medical Imaging* (MedDocs Publishers), *World Journal of Pharmacy and Pharmaceutical Sciences, Journal of Material Science and Technology Research, Ecronicon: EC Clinical & Medical Case Reports (ECCMC), Medical Research and Health Sciences,* and *Acta Scientific Women's Health Journal, PHARMACOGNOSY Journal and Journal of Community Medicine and Health Care Management.*

# 1 Cancer Epigenetic Mechanisms
## DNA Methylomes, Histone Codes, and MiRNAs

## 1.1 BACKGROUND

The term *epigenetics* is made of two parts: the Greek prefix *epi*, which means upon or over, and *genetics*, which is the science of genes, heredity, and variations in living organisms. Epigenetics defines what is occurring in the physical state of the genes and chromatin. This word was first defined by Conrad Hal Waddington (1905–1975) as the interaction between genes and their environment that creates the phenotype emphasizing that epigenetic mechanisms are different in response to a given environment. Waddington later pointed out that one of the main characteristics of epigenetic changes will occur in gene expression without any mutations. Nongenetic manifestation of traits in morphology had been introduced by Jean-Baptiste Lamarck (1744–1829) many years before Waddington propounded this idea. In this new definition, epigenetics is referred to as those changes in the genes' functions that are transmitted through both mitosis and meiosis without causing any alterations in the DNA sequence [1].

Our current knowledge of the deregulation that occurs during the onset and progression of cancer and other diseases leads us to recognize both genetic and epigenetic alterations as being at the core of the pathological state. Genetics alone cannot explain disease: in fact, people sharing the same DNA sequence (monozygotic twins) often present different levels of disease penetrance. The term *epigenetics* partially explains this phenomenon. Originally introduced to name the causal interactions between genes and their products, bringing the phenotype into being, it was subsequently used to define the heritable gene expression changes not related to any alteration in DNA sequence.

Interest in epigenetics has grown over the past decades, especially since it was found to play a major role in physiologic phenomena such as embryogenesis, imprinting, and X chromosome inactivation, and in disease states such as cancer. Cancer had been previously thought of as a disease with an exclusive genetic etiology. However, recent data have demonstrated that the complexity of human carcinogenesis cannot be accounted for by genetic alterations alone, but also involves epigenetic changes in processes such as DNA methylation, histone modifications, and microRNA (miRNAs) expression. In turn, these molecular alterations lead to permanent changes in the expression of genes that regulate the neoplastic phenotype, such as cellular growth and invasiveness. Targeting epigenetic modifiers has been referred to as epigenetic therapy. The success

of this approach in hematopoietic malignancies validates the importance of epigenetic alterations in cancer, not only at the therapeutic level but also with regard to prevention, diagnosis, risk stratification, and prognosis.

MiRNAs are small noncoding RNAs that regulate the expression of complementary messenger RNAs and function as key controllers in a myriad of cellular processes, including proliferation, differentiation, and apoptosis. In the last few years, increasing evidence has indicated that a substantial number of miRNA genes are subjected to epigenetic alterations, resulting in aberrant patterns of expression upon the occurrence of cancer.

Since its initial characterization of methylation in human tumors, epigenetic research has greatly expanded, recently introducing preliminary descriptions of epigenomes of human cells. MiRNAs are a class of small noncoding RNAs in many diseases including cancer. This chapter focuses on the link between epigenetics and miRNAs in cancer.

## 1.2 EPIGENETIC CRITERION LANDSCAPE

The term *epigenetics* refers to variability in gene expression, heritable through mitosis and potentially meiosis, without any underlying modification in the actual genetic sequence. This alteration in gene expression plays a fundamental role in several aspects of natural development, from embryogenesis, in which a resetting of the "epigenetic code" takes place in the very early moments after conception [2], to the determination of cellular fate and its commitment to a particular lineage. Epigenetics also plays a fundamental role in biological diversity, such as phenotypic variation among genetically identical individuals [3]. Indeed, epigenetic processes account fully for the differences between queen bees and worker bees in *Apis mellifera* species [4]. Several mechanisms fall under the banner of the epigenetic machinery, the most studied of which are DNA methylations, histone modifications, and small, noncoding RNAs.

All somatic cells possess the same genotype, because they have originated from the growth and division of a common progenitor cell. However, during the differentiation process cells become specialized and obtain a variety of functions and features by expressing and suppressing different sets of genes. Normally these settings are controlled by epigenetic processes. The genetics of changes and cell division is heritable. Epigenetic features are changed during tumor induction and cancer development with different patterns and characteristics [5]. The concept of epigenesis

DOI: 10.1201/9781003229650-1

is ancient: it can be attributed to the theory of development by Aristotle in his book *On the Generation of Animals*. In its traditional understanding it represents the concept that heterogeneous complex structures during development arise from less complex structures or even a homogeneous state. Today, in light of more complex studies, this concept has more molecular aspects. In fact, with the advent of molecular genetics, this concept has new meaning. The term *epigenesis* can now be considered as the science about what stands above the genes, and in this context the term is substituted with *epigenetics*. Epigenetics is the study of changes in the hereditary material not involving a change in the DNA sequence or the sequence of the proteins associated with DNA. Epigenetic regulation includes DNA methylation and histone modifications. DNA methylation is a reversible reaction, primarily occurring by the covalent modification of cytosine residues in CpG dinucleotide. These nucleotides are concentrated in short CpG-rich DNA regions called CpG islands (present in >50% of human gene promoters) and regions of large repetitive sequences. DNA methylation is catalyzed by DNA methyltransferases (DNMTs), known to catalyze the transfer of a methyl group from the methyl donor *S*-adenosyl methionine onto the $5'$ position on the cytosine ring. Today, three DNMTs are known: DNMT1, DNMT3A, and DNMT3B [6]. DNMT1 acts during replication showing preference for hemimethylated DNA sequences, whereas DNMT3A and DNMT3B act independently of replication, methylating both unmethylated and hemimethylated DNA sequences [7, 8]. Histones, the main protein components of chromatin and comprising the nucleosome core, are proteins with a globular C-terminal domain and an unstructured protruding N-terminal tail that can undergo a variety of chemical reactions (such as acetylation, methylation, phosphorylation, SUMOylation, and ubiquitylation), favoring the switch versus the accessible euchromatin or the inaccessible heterochromatin. Histone modifications can lead to either transcriptional activation or repression. For example, lysine acetylation correlates with transcriptional activation [9] while trimethylation of lysine 4 on histone H3 (H3K4me3) is present at gene promoters that are transcriptionally active [10] and in euchromatin [11]; on the other hand, trimethylation of H3K9 (H3K9me3) and H3K27 (H3K27me3) is present at transcriptionally repressed gene promoters [9]. Histone modification patterns are regulated by enzymes that add and remove covalent modifications such as histone acetyltransferases (HATs), histone methyltransferases (HMTs), histone deacetylases (HDACs), and histone demethylases (HDMs).

The concept of epigenetics includes those heritable changes that do not involve an alteration of the genome at the level of nucleotide sequences. Recent progress in the field has highlighted the fundamental role of epigenetic mechanisms in ensuring the proper control of key biological processes, such as imprinting, X chromosome inactivation, or the establishment and maintenance of cell identity. The functional significance of epigenetic control becomes apparent in the deregulated state: we now know that alterations of both genetic and epigenetic mechanisms are responsible for the establishment and progression of cancer, amongst other diseases (for recent reviews, see [12, 13]).

The mammalian genome is organized and packaged into chromatin, a highly compact and structured complex of DNA and proteins that can adopt different tridimensional conformations depending on the nuclear context and biochemical modifications present in both DNA and histones [14]. A simplified view of chromatin recognizes two basic states: an open, transcriptionally competent euchromatin and a more condensed, transcriptionally silent heterochromatin. Besides those structural regions of the genome that present constitutive heterochromatin (essentially found at centromeres), other genomic regions can undergo transitions between a more open conformation and a more compact, facultative heterochromatin. These transitions are vital to set the different transcription patterns required during embryonic life and development, or in the adult stage, and rely largely on the epigenetic control of histone and DNA modifications.

The main covalent chemical modification in DNA is the methylation of cytosine bases found in CpG dinucleotides, often located in enriched regions called CpG islands (CGIs). Around 60% of the promoters of protein-coding genes in the human genome seem to contain CGIs, and most of them appear methylated in differentiated tissues [15, 16]. Other promoters that lack bona fide CGIs nonetheless show some tissue-specific methylation patterns that can be correlated with their transcriptional status. Methylation of cytosines represents a stable, heritable, and reversible mark that is generally associated with transcriptional repression, either because it inhibits the binding of factors to their cognate DNA recognition sequences or because it recruits methyl-CpG-binding proteins (MeCPs and MBDs) together with co-repressor molecules. DNA methylation is controlled by DNMTs, which catalyze the transfer of a methyl group from the methyl donor, *S*-adenosyl methionine, onto the $5'$ position of the cytosine ring. In mammals, three catalytically active DNMTs have been identified: DNMT1, DNMT3A, and DNMT3B. The latter two can add methyl groups onto CG sites that were unmethylated on the parental DNA strand and are efficient in de novo methylation, whereas DNMT1 is responsible for maintaining the pattern of DNA methylation during DNA replication [8, 17].

Histones are the protein moiety around which DNA is packaged within the chromatin, and they can suffer a variety of posttranslational modifications of their N-terminal tails, including acetylation, methylation, phosphorylation, SUMOylation, ubiquitylation, and ADP ribosylation (for a review, see [9, 18]). These modifications can alter DNA–histone interactions and have an impact on higher-order chromatin structure. For example, acetylation of lysine residues is catalyzed by HATs and (with some exceptions) tends to create a more open chromatin, whereas deacetylation (carried out by HDACs) is associated with transcriptional repression. Methylation of lysine and arginine residues can occur in histones H3 and H4, in the mono-, di-, or trimethylated form, with specific HMTs acting on particular residues. Depending on the site and type of histone, the methylation pattern will result in a different transcriptional outcome. Hence, methylation of H3K9, H3K27, and H4K20 is generally linked to heterochromatin formation and the presence of the transcriptional repressor HP1, whereas methylation of H3K4 and H3K36 is associated with transcriptionally active regions [19]. Histone modifications can influence each other and also interact with DNA methylation, and it is the combined presence of a myriad of modifications in a spatial and temporal-dependent controlled manner (known as the "histone code") that ultimately programs the appropriate genome expression profile for each cellular identity. Two main types of complexes can be found accompanying these epigenetic modifications of histones, containing members of the trithorax group (TrxG) and polycomb group (PcG) of proteins. Some components of the PcG and TrxG complexes display histone methyltransferase activity, whereas other members of the complexes interpret the histone marks, thereby playing central roles in epigenetic gene regulation: they coordinate DNA accessibility throughout development and during the establishment of cell lineages, essentially by switching the balance between transcriptionally silenced heterochromatin (PcG-bound) and transcriptionally competent euchromatin (TrxG-bound; [20]).

It often happens that DNMTs, MeCPs, and MBDs associate with other chromatin remodeling activities, such as HDACs or nucleosomal remodeling complexes. In fact, it has been recently shown that the link between DNA methylation and gene chromatin silencing can be facilitated by the coupling action of covalent modifications of histones [21, 22], demonstrating the tight connection established between DNA and histone covalent modifications and the nucleosomal remodeling machinery. It is the concerted functioning of these three mechanisms that finally regulates the epigenetic silencing of discrete genomic regions.

The aberrant functions of chromatin modifier enzymes, which can result in altered histone modification or DNA methylation patterns, are hallmarks of human diseases. For instance, a common feature of human cancer is the loss of acetylation at lysine 16 and trimethylation at lysine 20 of histone H4 [23]. Likewise, hypermethylation of DNA promoter and coding sequences is a major epigenetic mechanism that contributes to cancer progression by causing the inactivation of a number of tumor suppressor genes in a wide range of tumor types [24]. Recent models also suggest that the epigenetic disruption of the stem cell population is the origin of cancer [25].

In this chapter, we will first describe the general mechanisms through which the epigenetic code is established and then focus on the alterations of the epigenome taking place in cancer, with an emphasis on how these aberrations can potentially be used in the clinical setting. We will focus on the epigenetic changes that occur in miRNA genes with reported tumor suppressor or oncogenic activities.

## 1.3 EPIGENETIC MECHANISMS

Epigenetic regulations are derived from the fact that the DNA packaging in the nucleus directly affects gene expression [26]. In general, the increased condensation of DNA enhances the probability of gene silencing. In return, decreasing compression of DNA leads to its accessibility for transcription machinery and increased expression of genes. Physically, the genome in the eukaryotic cells is packed in chromatin structure that determines its accessibility for functions such as transcription, replication, and DNA repair [27]. In general, three common biochemical mechanisms occur in the cell for epigenetic changes: DNA methylation, histone changes, and association of nonhistone proteins such as PcG and TrxG complexes.

## 1.4 DNA METHYLATION

In mammals, DNA methylation is a common epigenetic change in DNA. After DNA synthesis, cytosines within the dinucleotide CpGs are methylated at their carbon 5 by DNMT (Figure 1.1). CpGs that undergo methylation could be found either in isolation or in clusters, so-called CGIs [28]. But if the methylation happens in the promoter region of the genes, it would likely lead to gene silencing [29]. Normally, long-term silencing of genes occurs only in X-linked, imprinted, and germ-cell specific genes. CGIs of DNA sequence that contain plenty of C and G nucleotides are commonly hypermethylated in tumor cells, which could result in silencing of tumor suppressor genes [30]. See **Figure 1.1**.

An important step toward understanding the role of DNA methylation is to specify its location in the genome. Nowadays, this can be achieved by utilizing methods developed for genome-wide mapping of 5-methylcytosine, such as microarrays or high-throughput sequencing [31].

Data obtained from methylation studies show that cytosine methylation is available throughout the genome of mammals. Moreover, in most of the genomes in which DNA has lower CpG content, there is a high degree of cytosine methylation, while CGIs often remain nonmethylated [32, 33]. DNA methylation is a covalent modification of the cytosine ring at the 5′ position of a CpG dinucleotide, whereby a methyl group is deposited on the carbon 5 of that ring using S-adenosyl methionine as a methyl donor. This transfer of methyl group is a replication-dependent reaction catalyzed by DNMTs, present at the replication fork during the S-phase [34]. CpG dinucleotides, the usual targets

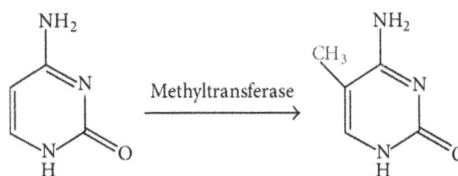

**FIGURE 1.1** Methylation of cytosine in carbon 5.

of DNA methylation in mammals, are scattered throughout the genome and present in a lower-than-expected abundance. This has been explained over evolution by the spontaneous deamination of the cytosine in the CpG site into a thymine [35].

However, in certain areas of the genome, a high concentration of these CpG dinucleotides is found, and these are referred to as CpG islands, or CGIs [36]. These CGIs average 1000 base pairs and can be found at the 5′ promoter region of approximately 50% of genes. In a normal differentiated cell, CpG loci disseminated across the genome are highly methylated, whereas most promoter CGIs are protected from the spreading of methylation inside their boundaries [36].

DNA methylation at gene promoter CGIs has been correlated with permanent expression silencing such as that noted in the inactive X chromosome in women [34]. DNA methylation leads to silencing by direct inhibition of transcription factor binding to their relative sites and by recruitment of methyl-binding domain proteins (MBDs) [34]. These MBDs are present in transcription co-repressor complexes involving several other members of the epigenetic machinery such as HDACs and HMTs, resulting in chromatin reconfiguration and gene silencing [37].

One such MBD is MeCP2, the deletion of which causes the neurodevelopmental disorder called Rett syndrome [38]. Throughout evolution, DNA methylation has been used to silence the expression of endogenous repeats and infecting retrotransposons, keeping them from disrupting normal gene expression [39]. An overview of epigenetic regulation in eukaryotic cells is presented in Figure 1.2. Other physiological phenomena in which DNA methylation in CGIs plays a fundamental role are X chromosome inactivation [40], genomic imprinting in which one allele is expressed depending on its paternal or maternal origin [41], and somatic tissue–specific repression of a set of germ cell–specific genes [42].

Although DNA methylation patterns in adult cells are relatively stable, important changes have been described in aging tissues. A global decrease in 5-methylcytosine content was reported in cultured human fibroblasts [43] and promoter-specific hypermethylation was observed in epithelial tissues [44, 45]. Global profiling using methylated CGI amplification in combination with microarray analysis demonstrated several hundreds of gene promoters to acquire methylation in aging mice intestinal mucosae whereas hundreds of others were found to have a parallel loss of DNA methylation [46]. This linear change of 5-methylcytosine

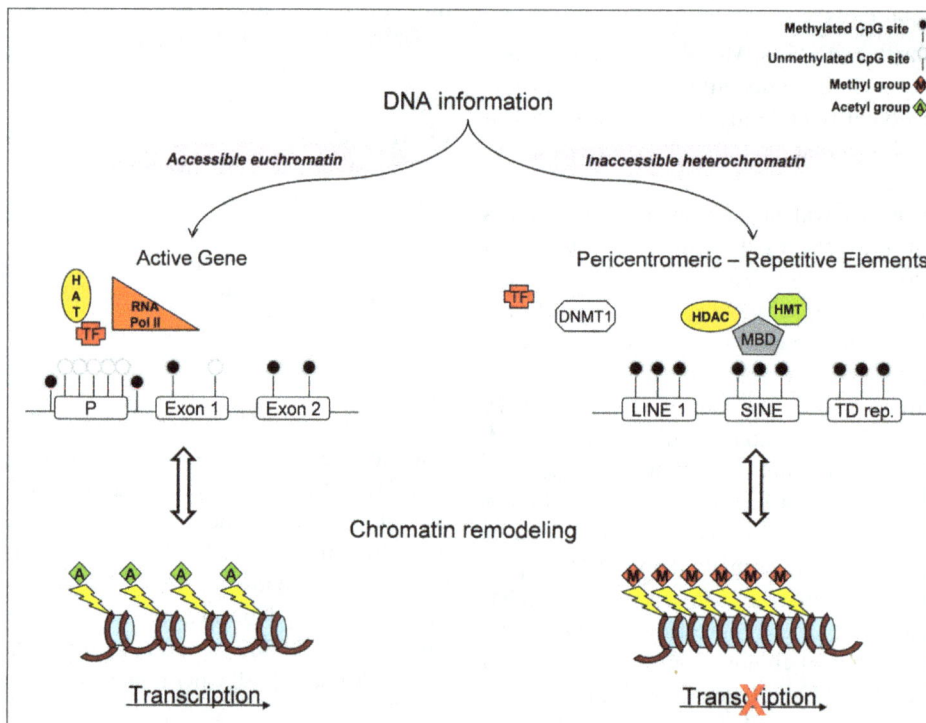

**FIGURE 1.2**  Normal transcriptional regulation in higher eukaryotes. DNA is packaged in nucleosomal building blocks in a way that determines its accessibility to the nuclear environment and transcriptional status. (*Left*) Transcriptionally active genes are marked by methylation-free promoters and an open, highly acetylated chromatin configuration that allows access to transcription factors and polymerase II. (*Right*) Repetitive elements are silenced by high levels of DNA methylation, specific histone lysine methylation, and a closed chromatin state. A switch from active to inactive chromatin characterizes some genes in cancer cells. DNMT1, DNA methyltransferase 1; HAT, histone acetyltransferase; HDAC, histone deacetylase; HMT, histone methyltransferase; LINE 1, long interspersed nuclear element 1; MBD, methyl-CpG binding protein; P, gene promoter; RNA Pol II, RNA polymerase II; SINE, short interspersed nuclear element; TD rep, tandem repeats; TF, transcription factor.

content with aging has a strong tissue specificity and has been shown to be common across mammals. Indeed, both the amount and pattern of DNA methylation have been found to diverge between human monozygotic twins as they age [47]. It is still not clear whether the accumulation of these DNA methylation defects with time is of a random or rather programmed nature, and although their pathophysiologic consequences are unknown, they have been proposed to play a role in aging disorders, including cancer.

DNA methylation is catalyzed by a group of enzymes in mammals called DNMT1, DNMT3A, and DNMT3B. DNMT1, known as the maintenance methyltransferase, has been shown to have a ten-fold preference for hemimethylated DNA (only one of the two DNA strands is methylated) compared with a methylated strand, and is used mostly by the cell to maintain the DNA methylation status in a stable fashion through cell division [48]. DNMT3A and DNMT3B, known as de novo methyltransferases, are used by the mammalian cell to methylate previously unmethylated DNA. It is worth mentioning that DNMT1 demonstrates far higher catalytic activity than DNMT3a and DNMT3b [49], and all three are involved in important cellular functions such as differentiation [50]. The functional importance of these enzymes is highlighted by the fact that DNMT deletion is embryonically lethal in mice [51].

## 1.5 HISTONE CHANGES

Histone changes include posttranslation modifications in the histone proteins of nucleosomes. The long N-terminal tail of histones, which makes the interaction between neighbor nucleosomes, may be affected and undergo a variety of modifications such as lysine and arginine methylation, lysine acetylation, and serine phosphorylation (Figure 1.3). Histone changes affect the organization of the nucleosomes in higher-order DNA packaging [9]. According to Turner, histone modifications are used for scheduling the activity of genes during the stages of differentiation [52].

Nonhistone proteins may affect the chromatin configuration through interacting with DNA and histones. For instance, ATP-dependent complexes that rearrange the chromatin structure directly move and relocate nucleosomes along the DNA [54]. The second group of protein complexes propounded in epigenetics includes HP1, PcG, and TrxG groups. These proteins attach to the DNA or especially modified histones and catalyze DNA methylation or other changes in histones; for example, PcG protein complex arrests histone methylation by interacting with DNMT and recruits DNA methylation [20]. Transcription factors can also affect the chromatin structure and be involved in the heredity of epigenetic changes during cell division [55].

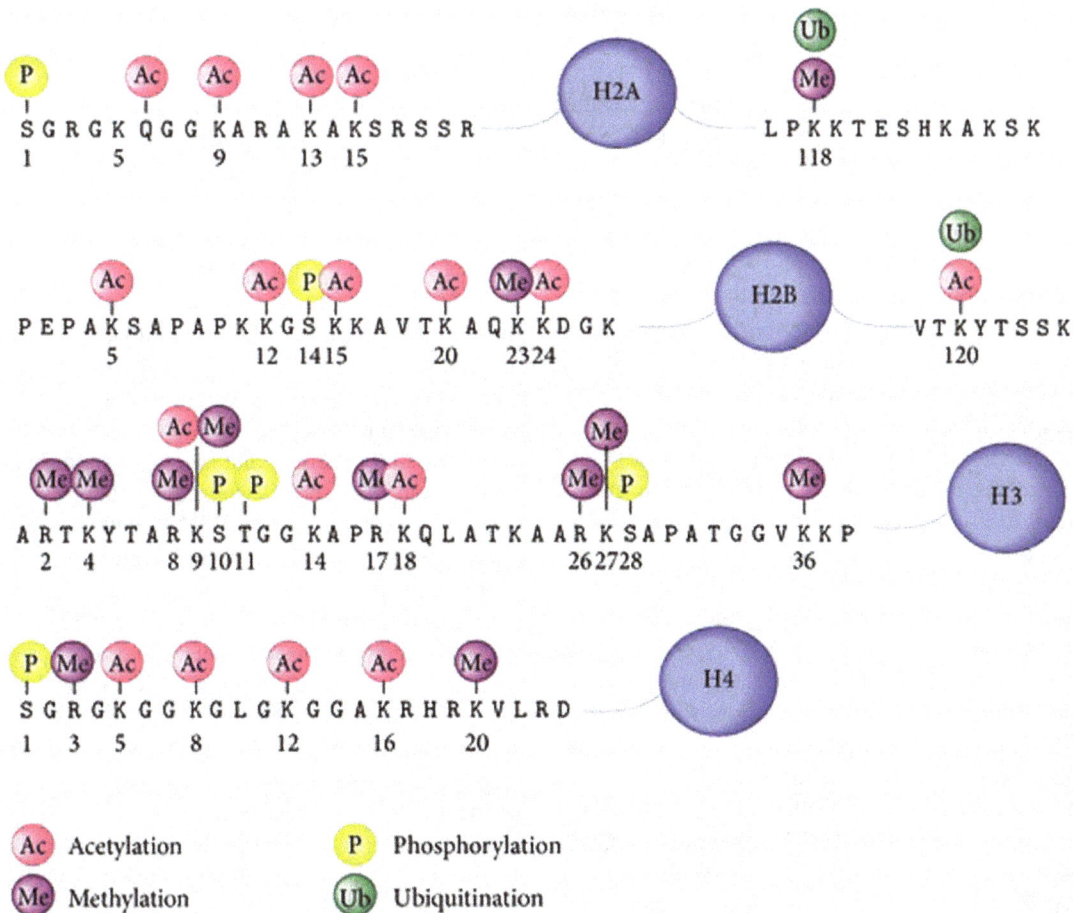

**FIGURE 1.3**  Epigenetic modifications of histone amino acids [53].

In brief, the interaction between different epigenetic mechanisms controls the accessibility of genes by the transcription machinery. Epigenetic mechanisms are completely intertwined in controlling one another and the function of the target genes in an intensifying or attenuating manner during which the activities of these genes can be aligned or nonaligned [56].

## 1.6 POSTTRANSLATIONAL HISTONE MODIFICATIONS

DNA is wrapped around histone proteins to form nucleosomes, in a way that regulates accessibility of the genetic sequence to the nuclear environment [57]. Each nucleosome is comprised of a tetramer of two histone 2A (H2A) and two histone 2B (H2B) molecules, flanked by H3 and H4 dimers. H3 and H4 have N-terminal tails that, in their deacetylated form, are positively charged, leading to a closed and tight chromatin configuration around the negatively charged deoxyribonucleic acid. The addition of an acetyl group neutralizes the positive charge of the lysine residues in these N-terminal tails, loosening up this tight bond between DNA and histones, resulting in a more open chromatin configuration accessible to being successfully transcribed [58]. Two consecutive nucleosomes are tied together by linker histone H1. Recent studies have shown that the abundance of these linker histones is tightly related to chromatin configuration and might be altered in cancer cells [59].

Histone modifications comprise a multitude of covalent reactions affecting the histone N-terminal tails and form a code that fine-tunes the way DNA is wrapped around these proteins. These posttranslational modifications include acetylation, methylation, phosphorylation, ubiquitination, SUMOylation, and ADP ribosylation [57]. These reactions occur in a very targeted and amino acid–specific way, the most studied of which are acetylation and methylation of specific lysine residues on histones H3 and H4. Several enzymes catalyzing these reactions, namely HATs, HDAC, HMTs, and HDMTs, have been identified. These enzymes exert their function in the setting of either transcriptional activator or repressor complexes, depending on the specific substrate residue.

Histone acetylation status results from an intricate cross-talk between HATs and HDACs. HATs are separated according to their cellular location and function into two distinct groups: the cytoplasmic B-type HATs and the nuclear A-type HATs [60]. The latter are presumed to have more impact on gene transcription, whereas cytoplasmic HATs can catalyze acetylation of nonhistone proteins. The most-studied HAT families are GCN5-related N-acetyltransferase (GNAT), MYST (MOZ, Ybf2/Sas3, Sas2, and Tip60), and p300/CREB-binding protein (CBP), all of which are associated with complexes such as GCN5, PCAF, MOF, and p300/CBP, respectively. These complexes interact with each other and, through both targeted promoter-specific and nontargeted general acetylation

reactions, play significant roles in development, differentiation, and cell cycle progression [61].

HDACs are a class of enzymes catalyzing the opposite action to HATs. They influence a myriad of cellular processes including signal transduction, apoptosis, cell cycle regulation, and cell growth [62]. HDACs catalyze deacetylation of both histone and nonhistone proteins and, similar to HATs, can be either nuclear or cytoplasmic. This cytoplasmic deacetylase activity can lead to posttranslational modifications of transcription factors and chaperone proteins, and can have major effects on several important pathways, such as the NF-κB (nuclear factor kappa-light-chain enhancer of activated B cells) pathway [63], the APE1-Ref1 oxidative stress response pathway [64], and the phosphatase and tensin homolog (PTEN) phosphatase gene [65]. Similarly to HATs, HDACs exert their catalytic activity through an association with protein complexes, such as the sirtuin (silent mating type information regulation 2 homolog 1 [SIRT1]) protein deacetylase complex [66].

Histone methylation also plays a major role in gene expression regulation [67]. Histone methylation is associated with transcriptional repression or activation depending on the specific amino acid affected. For example, methylation of histone H3 lysines 4 and 36 is associated with active gene expression, whereas methylation of histone H3 lysines 9 and 27 is associated with gene silencing. Histone methylation is catalyzed by a large number of enzymes, the majority of which contain a specific protein module called SET (su(var)3-9, enhancer-of-zeste, trithorax) domain [68]. Similar to acetylation/deacetylation, histone methylation is reversible and catalyzed by two families of HDMTs, namely the lysine-specific demethylase 1 (LSD1) and the Jumonji domain-containing enzymes [69, 70]. Histone methylases and HDMTs are usually part of large protein complexes that regulate gene transcription.

Histones can also be targeted by other posttranslational modifications such as phosphorylation, ADP-ribosylation, and ubiquitination. These affect a limited number of residues but could play an important role in gene regulation. For example, serine 10 phosphorylation is inversely correlated with lysine methylation, and this methylation/phosphorylation module is conserved across different proteins [71].

Histone modifications (and DNA methylation) ultimately affect gene expression in part by influencing nucleosome positioning. Active genes demonstrate a lack of nucleosomes at their transcription start site, whereas epigenetically silenced genes have a nucleosome positioned critically at the start of transcription [72]. Thus, nucleosome positioning can be involved in either the activation or repression of gene transcription [73]. The SWI/SNF protein complexes play a major role in this process [74]. Through their targeting to specific gene promoters, these complexes can activate or repress transcription via three biochemical processes: nucleosome remodeling, nucleosome sliding, and octamer transfer [75]. It is still unknown whether nucleosome formation and positioning are mainly determined by underlying proximal genetic sequences ("cis effect") or by other mechanisms operated by ATP-dependent nucleosome remodeling complexes in a sequence-independent manner ("trans effect"). Recent studies have suggested that the answer is more likely to be a mixture of the two, in some type of a nucleosome positioning code governing histone–DNA interactions [72].

## 1.7 NONCODING RNAs

Small noncoding RNAs refer to a family of RNAs that, by complementarity to the 3′ untranslated region of messenger RNAs, lead to their degradation and subsequent inhibition of gene expression [76]. Part of this family of noncoding RNAs are 20- to 22-nucleotide miRNAs, resulting from the sequential splicing of primary then pre-RNAs. These oligonucleotides are first synthesized as long, noncoding RNAs that are processed by the RNA cleaving enzyme DROSHA in the nucleus, transported into the cytoplasm in the form of short hairpin RNAs, and further cleaved by the enzyme Dicer into their final configuration of double-stranded miRNAs [76]. MiRNAs are then incorporated in the RNA-induced silencing complex and transported back in the nucleus, where they exert their biological effect. Through Watson-Crick base pairing, miRNAs bind to complementary sequences of mRNAs and induce either degradation or translational silencing of the target mRNAs [76]. It is interesting to note that miRNAs are also themselves epigenetically regulated at their promoter level, and target many genes that play important roles in such processes as cell cycle progression, apoptosis, and differentiation [77]. A single miRNA can have hundreds of target mRNAs, highlighting the implication of this gene regulation system in cellular functions [78]. The study of miRNAs has become the subject of intense interest, especially after the discovery of the fundamental role of these small, noncoding RNAs in a myriad of cellular and biological processes ranging from development to disease states [79].

## 1.8 DNA METHYLTRANSFERASE ENZYMES

In mammals, DNA methyltransferase enzymes include four members in two families, which are distinct from each other both structurally and functionally. The DNMT3 family provides the CpG methylation through the de novo path, whereas the DNMT1 family maintains the methylation models during replication [80].

The DNMT3 family includes two enzymes, DNMT3A and DNMT3B. A regulatory factor similar to DNMT3 enzyme is placed beside these enzymes, which is referred to as DNMT3L [81]. DNMT3A and DNMT3B have similar domains. They both have a variable region in the N-terminus and a PWWP domain that probably affects the nonproprietary connection to DNA [82]. Despite possessing homologous structures, they have distinctly proprietary targets [83]. The rate of DNMT3B is very high at the early embryonic period, and it is responsible for DNA methylation during implantation [84], whereas DNMT3A is active in the late phases of the embryonic period and cell differentiation. DNMT3A also is responsible for DNA methylation in matured gametes [85, 86].

DNMT1 enzyme causes methylated models of CpG regions to be copied on new synthesized DNA. Therefore, these changes deemed as epigenetic modifications are inherited during cell division and are hereditary. Recently, data obtained from the DNMT1 structure indicate how DNMT1 connection to the hemimethylated CpG region provides DNA accessibility to enzyme catalytic domain while links to nonmethylated CpGs protect the newly synthesized fiber by CXXC domain from its methylation. The DNMT1 activity is performed through interaction with proliferating cell nuclear antigen (PCNA) and Np95 during phase S of the cell cycle [87, 88].

## 1.9 THE ROLE OF DNA METHYLATION AND HISTONE ACETYLATION IN THE REGULATION OF GENE EXPRESSION

Epigenetic mechanisms and modifications play a crucial role in the regulation of gene expression during growth periods and cell differentiation. Epigenetic mechanisms may result in gene silencing. This fact has been revealed in many types of human cancers. Despite much research that has been done in the field of epigenetics, the role of DNA methylation, histone acetylation, and deacetylation in controlling gene silencing is still not fully understood.

Today, lines of evidence demonstrate the importance of DNA methylation and histone acetylation in transcription. However, the mechanisms that cause changes in the histone acetylation and methylation of CpG sites are not completely identified. Similarly, the rise of epigenetic marks in human cancer cannot be explained. Although it appears that a correct understanding is achieved for maintenance and reserving the DNA methylation patterns after replication by DNMT1 enzyme, the causes of methylated cytosine in CpG sites remain largely unknown.

It is believed that some DNA sequences are probably identified by DNA methyltransferase enzyme as target. However, there is some evidence proving that the initial DNA sequence causes cytosine methylation in CGIs [8, 89]. Notwithstanding that the DNMT3A and DNMT3B enzymes are specific to the CGI methylation, they seem not to be able to differentiate between the DNA sequences [90].

According to some research, DNA methylation is led by histone modifications [91, 92]. Today, we know with certainty that transcriptional silencing of genes is directly connected to the amount of histone acetylation and DNA methylation. Nevertheless, the propounded question is whether DNA hypermethylation or histone hypoacetylation leads to gene silencing or epigenetic modifications are consequences of gene silencing. Identifying events that cause inactivation of genes are critical and very important for understanding the mechanisms that lead to cancer; thus, the correct perception of these events ultimately can be helpful in cancer prevention and treatment.

Recent studies show a significant correlation between histone acetylation, DNA methylation, and gene silencing. Since then, interactions between these mechanisms are thought to be likely. In fact, implicit and explicit lines of evidence indicate the connections between histone acetylation and DNA methylation [89, 90, 93, 94]. However, there is still no general consensus that epigenetic modifications can lead to cancer inception.

## 1.10  RELATIONSHIP BETWEEN GENE SILENCING AND DISEASE

Any changes made in gene expression play an important role in the incidence of disease (or pathogenesis). In some cases, changes in gene expression can result from genetic defects in genes and nucleotide sequences. In some groups of patients, the aberrant epigenetic changes may alter the gene expression. As mentioned earlier, two major epigenetic mechanisms that alter and modify the gene expression are DNA methylation and histone changes [95]. Several changes in the pattern of DNA methylation and silencing the genes that cause various diseases are illustrated in Table 1.1.

Change in methylation pattern may lead to congenital defects. For example, downregulation of some genes affects the growth rate and changes in the DNMT3B catalytic domain result in immunodeficiency [95, 96].

## 1.11  EPIGENETICS IN NORMAL CELLS

The well-defined gene functions in eukaryotic cells can be affected by factors such as growth stage, cell periphery, or other circumstances surrounding the cell. The sequence of genes and their regulatory factors are conserved in initial DNA except for a few variables such as B and T lymphocytes. DNA and its associated proteins are subjected to covalent changes, and the location and pattern of these changes determine the growth phase and cell condition.

### 1.11.1  DNA METHYLATION

DNA methylation plays an important role in controlling cellular processes such as embryonic growth, transcription, inactivation of chromosome X, and determination of gene mapping [98].

In humans, methylation occurs in carbon 5 of cytosine, which is positioned before guanine [89]. In normal cells, the CpG methylation is mainly done in the repetitive regions of genome including satellite regions, endogenous retroviruses, LINEs, and SINEs [99]. CpGs do not exist randomly in the genome, but they are located in CGIs. These islands are usually found at the end of 5′ regulatory regions of many genes [100]. In normal cells, a great portion (about 94%) of CGIs is nonmethylated.

**TABLE 1.1**
**Gene Silencing Proteins and Diseases [97]**

| Protein | Cellular Defect/Disease |
| --- | --- |
| DNMT1 | Developmental abnormalities |
|  | *Igf2* imprinting |
|  | Colon cancer |
|  | Lymphoma |
|  | Pancreatic cancer |
| DNMT3B | Developmental abnormalities |
|  | ICF |
|  | Bladder cancer |
|  | Breast cancer |
|  | Colon cancer |
|  | Hepatocellular carcinoma |
|  | Lung cancer |
| MeCP2 | Chromosome instability/cell cycle defects |
|  | Breast cancer |
|  | Rett syndrome |
| EZH2 | Cell cycle defects |
|  | Barrett's esophagus |
|  | Bladder cancer |
|  | Breast cancer |
|  | Colorectal cancer |
|  | Melanoma |
|  | Myeloma/lymphoma |
|  | Hepatocellular carcinoma |
|  | Prostate cancer |
|  | Wilms tumor |
| Suv39h1 | Blood cell defects (RBC and WBC) |
|  | Chromosome instability |
|  | Chromosome instability/cell cycle defects |
| HP1 | Breast cancer |
|  | Medulloblastoma |
|  | Papillary thyroid carcinoma |
|  | Viral latency |

*Abbreviations:* ICF, immunodeficiency, centromeric instability, and facial anomalies; RBC, red blood cell; WBC, white blood cell.

Altogether, CGI methylation is associated with gene silencing; it has also been shown that methylation of these islands is connected with the histones' deacetylation in addition to other factors being involved in gene transcription [101]. The methylation can play a major role in reducing the gene expression as well as inactivation of the X chromosome in females and chromosome stability through hypermethylation of repetitive sequences [102].

### 1.11.2  HISTONE CHANGES

Currently, more than 60 histone modifications have been identified, including acetylation, methylation, phosphorylation, and ubiquitylation, either by specific antibodies or spectrophotometric techniques [9]. Among all known changes, the lysine methylation shows the highest degree of compression where each methylated region might independently affect the gene activity. In general, the main function of these changes is still unknown. Transcription initiation and elongation are associated with changes of the N-terminal region of nucleosomal H3 and H4 histone proteins in the promoter region [103, 104].

Lysine acetylation is generally associated with increased accessibility of chromatin and consequently enhanced activity of transcription enzymes. But the effect of the lysine methylation on histones varies depending on the location where the histone modification has been detected [105]. In addition to the regulatory role of histone changes in transcription, lysine methylation also participates in replication, repair, and organization of DNA [9].

## 1.12 EPIGENETIC CHANGES IN CANCERS

Cancer cells have genome-wide aberrations at the epigenetic level, including global hypomethylation, promoter-specific hypermethylation, histone deacetylation, global downregulation of miRNAs, and upregulation of certain

actors of the epigenetic machinery such as EZH2. These aberrations confer a selective growth advantage to neoplastic cells, leading to apoptotic deficiency, uninhibited cellular proliferation, and tumorigenicity (Figure 1.4). In the following sections, we will describe these different layers of epigenetic regulation and their aberrant functioning in cancer cells.

Today, there are various techniques that reveal the correlation between cancers and epigenetics. They illustrate the connection between the inactivation of miRNAs, genes, and epigenetic changes in cancer progression.

### 1.12.1 DNA METHYLATION IN CANCER

Change of DNA methylation pattern in CGIs was the first and most significant abnormal epigenetic change identified in cancerous cells. There is a lot of evidence representing the DNA methylation as a change toward the beginning of the carcinogenesis processes. In particular, in age-dependent changes, the DNA methylation pattern may be altered [106]. Age-related modifications can be closely related to

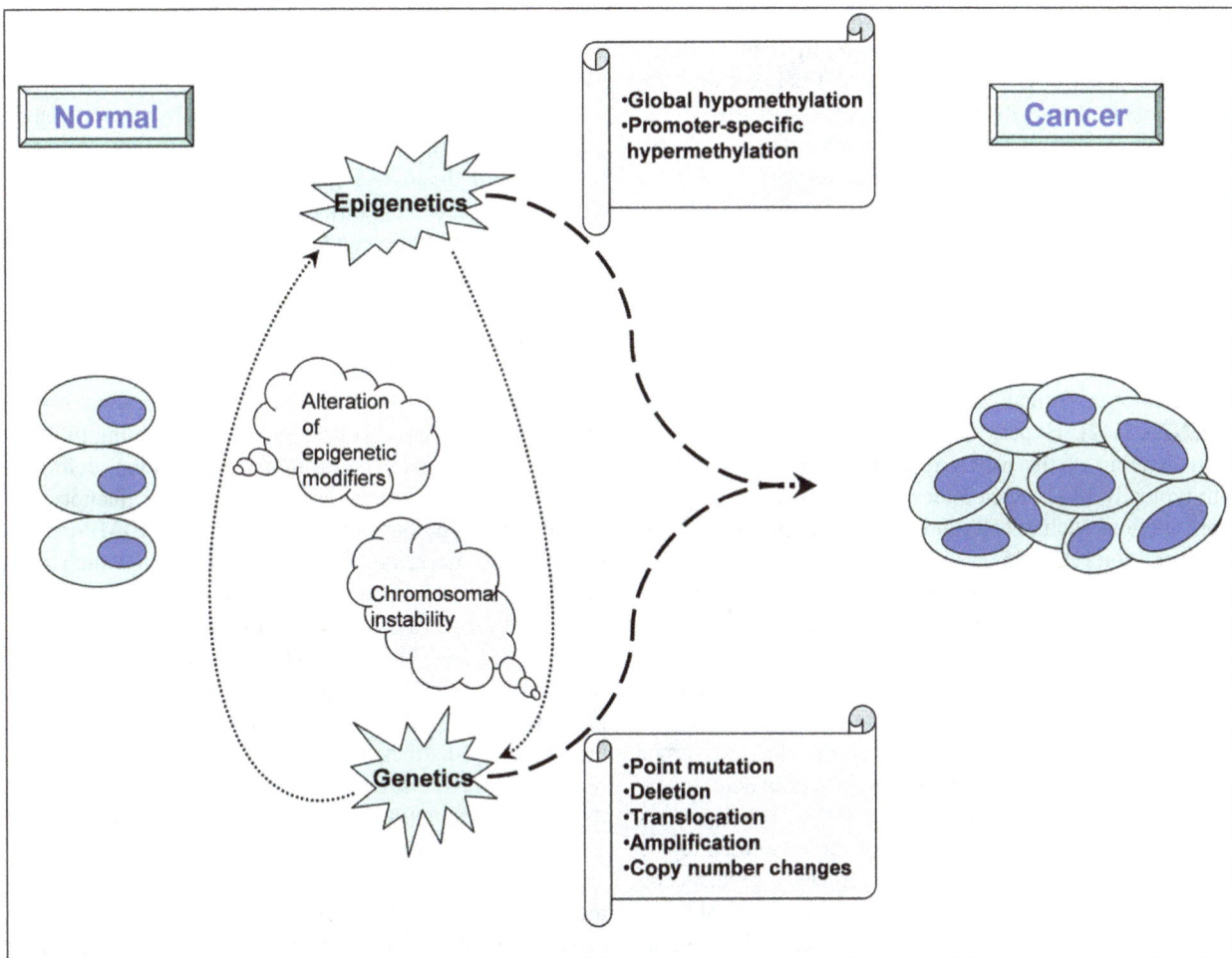

**FIGURE 1.4** Tumorigenic mechanisms in mammalian cells. Both genetic and epigenetic aberrations are involved in neoplastic transformation. These two alternate pathways of tumorigenesis are linked by an intricate cross-talk and can, either individually or in synergy, lead to the development of the malignant phenotype.

the changes in the methylation pattern and the occurrence of cancer in a specific region of the body. Furthermore, in case of the relationship between the environment and the underlying DNA methylation, there is a strong correlation between environmental factors and development of cancer [107]. The DNA methylation pattern in cancers varies. Hypomethylation is an extremely common change associated with many cancers such as stomach, kidney, colon, pancreas, liver, and lung [108–110].

Low-level methylation in cancers is substantially due to the loss of methylation at repetitive sequences as well as the intron demethylation [111]. During the development of cancer, hypomethylation degree may increase in the DNA, and the progressive lesion could participate in the reproduction and metastasis of cancer cells [111].

The following three mechanisms have been suggested for DNA hypomethylation: (1) the increased instability of the genome, (2) the reactivation of factors that are capable of movement on DNA, and (3) the functional defect in elements related to the genome [112].

Transcription of the regions with a modified methylation pattern that has resulted in hypomethylation would cause the genome impairment. Hypomethylation of HPV genome in cervical cancer leads to progression of cancer [110]. The reverse can also occur. In such a mode, hypermethylation of tumor suppressor genes in CGIs as well as in the miRNA genes leads to inactivation of tumor suppressor genes. The hypermethylation of CGIs in the promoter region is a crucial incident in the initiation of carcinogenesis.

Aberrant methylation of CGIs is also an important factor in the proliferation of tumor cells, which leads to silencing of tumor suppressor genes and molecules involved in cell differentiation. For instance, increased methylation (hypermethylation) of CGIs within the promoter of tumor suppressor genes is associated with various cancers such as E-cadherin, human mutL homolog 1 (*MLH1*), and CDKN2A [112]. Hypermethylation of CGIs is associated with silencing of the miRNA. miRNAs are small RNAs with typically 18–22 nucleotides, regulating many intracellular functions such as cell reproduction, apoptosis, and cell differentiation [113–115].

Tumorigenesis is a result of the activation of oncogenic and/or inactivation of proapoptotic or tumor suppressor pathways. Initially, these were believed to result exclusively from genetic events such as mutations, amplifications, gene rearrangements, or deletions [116]. We now understand that DNA methylation is an alternate way of silencing tumor suppressor genes, in a manner equivalent to genetic mutations [13]. Examples of this mechanism of tumorigenesis are numerous, notably methylation of the mismatch repair gene *MLH1* in colorectal cancer, the DNA repair gene O-6-methylguanine-DNA methyltransferase (MGMT) in gliomas and colorectal cancer, and the cell cycle regulator *p16* (cyclin dependent kinase inhibitor 2A [*CDKN2A*]) in colorectal and other malignancies [117]. Cross-talk has been shown to exist between these mechanisms and genetic ones in a cell. This is exemplified in colorectal cancer, in which CGI promoter hypermethylation has been shown to be present only in the wild-type allele of silenced genes [118]. In addition, aberrant DNA methylation was more frequent than copy number changes when studied on a whole-genome level in malignant gliomas [119]. This is the case in colorectal cancer as well, in which individual tumors are found to harbor more hypermethylated genes than genetic mutations, and within individual genes, hypermethylation was found to be more frequent than genetic changes [120]. DNA methylation effects on pathway alterations can be either direct, by affecting promoters of tumor suppressor genes, or indirect, by silencing known inhibitors of oncogenes, such as the silencing of the secreted frizzled-related protein (SFRP) family of genes, leading to the activation of the Wnt pathway in colorectal carcinogenesis [121]. Similar to mutations, silencing of tumor suppressor genes confers a selective proliferative advantage to corresponding cells, mediates invasiveness, and facilitates metastasis.

DNA hypermethylation is an early event in tumorigenesis, most likely playing a major role in tumor initiation and progression, and creating a fertile ground for the accumulation of a multitude of simultaneous genetic and epigenetic aberrations [122]. This is supported by the finding of a "field defect," in which normal tissue adjacent to a tumor is found to harbor several epimutations as well, most notably in colorectal cancers [123] but also in gastric cancer and liver cancer. Another example is *MGMT* hypermethylation, which plays a direct role in the accumulation of G-to-A mutations in the *KRAS* gene in colorectal tumors [124]. These data led to new thinking regarding the mechanisms behind tumor initiation and progression, even at the earliest stages of carcinogenesis.

Aberrant patterns of DNA methylation in cancer have significant interneoplastic and interindividual variability, accounting not only for tumor type specificity but also personal variability [125]. The latter is best represented by the presence of a subgroup of patients demonstrating high levels of simultaneous gene promoter methylation, defining a phenomenon now known as CGI methylator phenotype or CIMP [126]. The best-studied subgroup of CIMP-positive patients was described in colon cancer, in which these tumors were reported to comprise 20% to 40% of cases and were found to be associated with microsatellite instability (MSI), a defective human *MLH1* function, a location mostly in the ascending colon, an older patient age, and female predominance [127]. These CIMP-positive tumors often are clinically distinct from those in the rest of the patient population for the tumor type in question, which suggests that DNA methylation could be used for personalized cancer treatment in the clinical oncology setting.

On the other end of the spectrum is global DNA hypomethylation, the first epigenetic alteration noted in cancer cells [128]. In various cancers, 5-methylcytosine content was found to decrease by an average of 10% [129]. This affects both repetitive elements such as LINE1 and Alu [130] and specific gene promoters [131]. One potential consequence of profound hypomethylation is genomic instability, predisposing patients to mutations, deletions, amplifications, inversions, and translocations [132]. This may occur in part through reactivation of mobile elements. Indeed, hypomethylation correlates with a higher rate of chromosomal changes in patients with colon cancer [133] and is associated with a poor prognosis [134]. Another potential consequence of DNA hypomethylation is the reactivation of normally silenced genes [135]. This could lead to the disruption of normal gene expression and potential activation of growth-promoting and antiapoptotic pathways. Furthermore, promoter hypomethylation can lead to reactivation of miRNAs embedded in the coding regions of certain genes, resulting in silencing or aberrant expression of the corresponding protein [77]. Hypomethylation by genetic disruption of DNMT1 is protective against carcinogenesis in some models [136] but can promote tumor formation in others [137].

### 1.12.2 Histone Modifications in Cancer

The relationship between histone changes and cancers is specified after identifying the correlation between cancer and the DNA methylation [109]. Changes in the methylation of CGIs could result in various histone modifications and consequently different types of cancers [138]. Independent of CGI methylation, the histone changes can also be directly associated with cancer progression. Activity of many histone acetylase enzymes would result in deacetylation of histone H3 in the tumor suppressor gene [139]. Various pathways lead to histone modification and cancer development. In some cases, the links between cancers and histone modifications are clearly marked and suggest a direct relationship [140, 141].

There is limited information regarding global histone modification profiling in cancer cell lines and primary tumors. Recent studies have demonstrated a global loss of histone H4 lysine 16 monoacetylation and histone H4 lysine 20 trimethylation in cancer [12]. These modifications were found to occur throughout the genome, specifically overlapping with areas of DNA hypomethylation in repetitive sequences. Conversely, loss of histone H3 lysine 9 acetylation and lysine 4 dimethylation or trimethylation and gain of histone H3 lysine 9 dimethylation or trimethylation and lysine 27 trimethylation can be found at specific gene promoters and can contribute to tumorigenesis by silencing critical tumor suppressor genes [142]. One interesting observation is the correlation between genes that are marked by DNA methylation in cancer and those found to be bound to the repressive PcG proteins in embryonic cells [143]. These two groups appear to overlap, implying that certain genes are poised for silencing and predetermined to be the target of specific repressive histone marks in cancer.

Unlike DNA methylation, in which a bona fide DNA demethylase has not yet been identified, posttranslational histone modifications are well characterized as a two-way street governed by a balance of catalytic enzymes [144]. Shifting of this balance in cancer can occur through altered expression or function of epigenetic modifiers, and this has been found to play a role in both murine and human neoplasia. For example, the histone methyltransferase PcG protein EZH2 catalyzes H3K27 trimethylation [145]. Its overexpression was found to promote tumor growth both in vitro and in vivo [146], and it is present in several cancers in the clinical setting, such as melanomas, lymphomas, and prostate and breast cancers [147]. EZH2 has also been found to be useful as a potential biomarker to distinguish aggressive prostate and breast tumors from more indolent ones [146, 148]. In prostate cancer, EZH2 expression has been correlated with aberrant H3K27 trimethylation affecting potential tumor suppressor genes [149]. Recently, mutations of EZH2 were found in lymphomas [150], but their functional significance there remains to be clarified. The H3K27 repressive methylation mark can also be overrepresented in cancer through an alternative mechanism, inactivation of a specific H3K27 demethylase, *UTX* [151]. The latter has been shown to be somatically mutated in several tumor types, such as multiple myeloma, esophageal squamous cell carcinoma, and renal cell carcinoma [151]. Reintroduction of *UTX* in cancer cells presenting with an inactivating mutation of this gene led to a reversion of the malignant phenotype [151]. Another histone methylase, multiple myeloma SET domain (*MMSET*), is genetically altered by a common chromosomal translocation in multiple myeloma, resulting in altered expression of target genes [152]. In addition, the histone H3 lysine 9 methyltransferase *SUV39H* may play a role in carcinogenic initiation and progression [153]. Its deletion in mice was found to lead to chromosomal instability and increased tumor formation [153]. Perhaps one of the most relevant clinical entities highlighting the importance of HMTs in cancer is the 11q23 translocation in leukemias [154]. These have rearrangements giving rise to a multitude of fusion proteins involving the mixed lineage leukemia (MLL1) H3 lysine 4 HMT. MLL1 fusion proteins act as constitutively active chimeric transcription factors and lead to upregulation of downstream homeobox (*HOX*) genes and activation of several leukemogenic pathways such as RAS and fms-related tyrosine kinase 3 (FLT3). MLL leukemias appear to have a unique transcriptional signature [155] and a poor prognosis overall [154].

In addition to alterations in histone methylases/HDMTs in cancer, numerous changes in gene-specific histone acetylation have also been described. These can be primary or secondary to aberrant recruitment. For example, the chimeric oncoprotein promyelocytic leukemia-retinoic acid receptor α (PML-RARα) produced by the t(15:17) translocation in acute promyelocytic leukemia targets specific promoters through the aberrant recruitment of HDACs and HMTs, leading to silencing of gene expression [156, 157]. In addition, DNA hypermethylation can lead to aberrant HDAC and HMT recruitment to specific promoters [34]. Conversely, direct primary changes in HATs/HDACs can also occur in cancer. Several studies have demonstrated a direct effect of p300/CBP HAT on cellular proliferation [158]. There is an interesting interaction reported between p300/CBP and the viral oncogenic protein E1A [159]. This association disrupts the interaction between the p300/CBP complex and other HATs, in turn leading to increased tumorigenesis. This mimics the effect of E1A on the retinoblastoma (Rb) tumor suppressor gene [160]. Mutations of p300/CBP are also found in Rubinstein-Taybi syndrome, a developmental disorder associated with an increased risk of solid tumors, leukemias, and lymphomas [159]. p300 mutations have also been noted in several human malignancies, including glioblastomas and breast and colorectal cancers [159].

One of the limitations of studying histone modifications in cancer is the requirement for a relatively large number of fresh or fresh frozen cells. This has limited the study of these 9 modifications in clinical tissue samples, although some data are beginning to accumulate in leukemias [161]. Advances in technology to analyze histone modifications are needed to improve our understanding of various tumors.

## 1.13 IMPACTS OF EPIGENETIC CHANGES ON MiRNAs

MiRNAs are a class of small, noncoding, and endogenous RNAs that play a crucial role in gene expression through transcription prevention by induction of a group of regulatory molecules. The miRNAs can also be important in controlling DNA methylation and histone modifications (Figure 1.5). Epigenetic mechanisms such as promoter methylation and histone acetylation may be adjusted by miRNA expression. In some pathologic cases such as in cancers, the balance between miRNA and epigenetic processes becomes disordered. The lack of appropriate function of miRNAs and their aberrant expression are associated with the development and progression of human cancers. This process follows cell proliferation and impairment of the apoptosis [162].

Epigenetic changes on gene promoters and miRNAs dependent to the 3-UTR are considered as two important regulatory mechanisms in eukaryotics. Both DNA methylation and miRNAs can suppress the gene expression where miRNAs tend to target the genes with low DNA methylation in their promoter [163]. Recently, it has been shown that

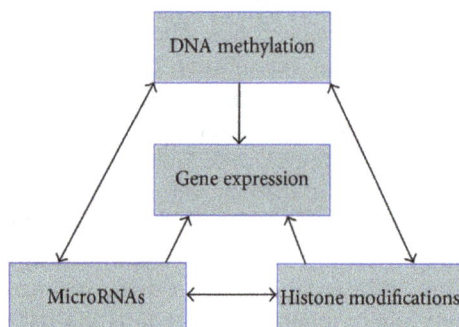

**FIGURE 1.5** Cross-talk between genomic methylation, histone modifications, and the effects of microRNAs on gene expression.

epigenetic mechanisms including DNA methylation and histone modifications affect not only gene expression but also the miRNAs. In fact, miRNAs like any other involved genes are affected by similar epigenetic regulations.

Epigenetic abnormalities can increase the rate of miRNA expression in tumor cells, and this increase may lead to the formation of these tumor cells. Almost half of miRNA genes are associated with CGIs and their expression can be adjusted by methylation [142]. Scientific lines of evidence suggest that epigenetic mechanisms are responsible for regulating more than 100 miRNAs and more than half of them have been identified in the form of methylated specific cancers and about 20 kinds of tumors [164]. MiRNAs can be impaired due to abnormal expression of other epigenetic regulatory mechanisms, including the HDACs (histone deacetylases) or PcG suppressor groups [165].

Incorporation of three methyl groups on lysine 27 of histone H3 (H3k27me3) and two methyl groups on lysine 79 in histone H3 (H3k79me2) has been identified in 47 miRNAs (out of 174 miRNAs) in colorectal cancer cells, which is the result of epigenetic mechanisms [166].

Tumor cells often not only show the change in the DNA methylation pattern but also face a lot of histone changes. HDACs are overexpressed in some specific cancers such as chronic lymphocytic leukemia (CLL) and can intervene in the silencing of several miRNA genes, including miR-15a, miR-16, and miR-29b.

## 1.14 EPIGENETIC BIOMARKERS IN CANCER

It has been shown that DNA proprietary methylation especially in CGIs can be used as a marker for early cancer detection, disease progression, and prediction of the response to cancer treatment [167].

There are reasons to suggest the methylated tumor suppressor genes as good diagnostic markers. Some of these reasons include the following: (1) DNA is a stable molecule that can be easily extracted from body fluids and tissues and includes the methylation data; (2) detection of methylation change signals can be considered as an important marker to diagnose gene expression change; however, identifying them in normal cells is very difficult; (3) the operating protocols for analyzing the protein or the cDNA expression level are not difficult. Epigenetic biomarkers can be easily extracted from bodily fluids such as blood, saliva, or urine and used to identify and diagnose tumors at the initial phase of disease [168].

Detection of methylated DNA, which is derived from tumors in serum and plasma, has shown a similar methylation pattern in all kinds of tumors; therefore, by tracing these markers, the beginning phase of tumor genesis processes could be detected [167]. Promoter methylation of gene P16 is a good marker for early detection of lung cancer. This marker is detectable and identifiable 3 years prior to the clinical symptoms of cancer [169]. The value of disease prognosis by means of DNA methylation is different depending on the cancer. For instance, in myelodysplastic syndrome, while the prognosis is very weak in the early stage of the disease, promoter methylation of genes HIC1, CDH1, ER, and P15 is diagnosable and traceable [170].

*MGMT* gene silencing, which is due to methylation in the promoter region, is an appropriate diagnostic marker in glioblastoma patients [171]. If the functions of DNMT and HDAC suppressors lead to the activation of tumor suppressor genes, then the DNA methylation would be a favorable marker for predicting response to treatment by epidrugs. Further identification of these markers is required for the production of epidrugs in the future, but the overall response to the treatment in patients who are being treated by DNMT and HDAC suppressors is not significantly considerable at the present. It should be noted that the focus on DNA methylation as a marker is essentially applied for early detection of disease and very few studies have been done on the assessment of DNA methylation as a marker for examination of response to treatments by DNMT and HDAC suppressors [172]. It is also known that biomarkers such as different HDAC isoforms and histone acetylation rate are very valuable for prediction of patient response to the treatment by the HDAC suppressors [173].

## 1.15 EPIGENETIC TREATMENTS

Upon revealing the aberrant epigenetic changes in malignant cells that lead to tumor suppressor gene silencing, the research has been expanded for achieving new drugs that are able to reactivate the silenced genes due to these changes [174].

Both genetic and epigenetic changes are factors that contribute to cancer development and progression; while genetic changes are not reversible, epigenetic changes can revert to their normal state. Reversibility of epigenetic changes caused the epigenetic treatments to be prioritized among the suggested treatments. The DNMT and HDAC enzymes have been assumed to be the most important targets of these treatments. Today, some DNMT and HDAC suppressors have been approved by the US Food and Drug Administration (FDA) as an anticancer treatment. The use of epigenetic targets for cancer treatment shows promise [175]. Azacitidine and decitabine were the first drugs synthesized for this purpose that suppress the DNMT enzyme. In 2004, the FDA approved decitabine (5-aza-2-deoxycytidine) as an effective medication for myelodysplastic syndrome. The use of these drugs has remained problematic due to their volatility in water, disruption in growth, differentiation of myeloid blood cells, and side effects such as poisoning [174].

Epigenetic changes allow cancer cells to adapt to their surrounding environment. However, the tumor suppressor genes that are hypermethylated can be reactivated by drugs. Reexpression of hypermethylated genes in the cancer cells is conducted by demethylating agents [176]. DNA demethylating drugs in low doses have antitumor activities. For example, drug 5 azacitidine (Vidaza) has received approval from the FDA for the treatment of leukemia [177–179]. The

HDAC enzyme-suppressing drugs arrest the cell cycle in the differentiation stage. Although they have demonstrated apoptotic activity in vitro, the specific mechanism for this has not yet been discovered [180].

All together, the HDAC enzyme suppressors are insignificant in tumor treatment. Sometimes, DNA demethylating drugs and HDAC suppressors can result in unintended and contradictory consequences because of the increased gene expression. However, achieving epigenetic specific treatments, which directly target transcription factors and gene promoters, has provided a lot of hope for cancer treatment.

## 1.16 MiRNAs' ROLE IN CANCER

MiRNAs are small noncoding molecules, 18–25 nucleotides long, functioning as negative regulators of protein-encoded genes. MiRNA genes are often located at an intronic region of a protein-coding gene, but they can also be present in exons or between genes. MiRNA is derived from a complicated process of maturation of its primary transcript named pri-miRNA [181] (Figure 1.6). Then, the pri-miRNA is endonucleolytically processed into a ~70-nucleotide hairpin-like precursor miRNA (pre-miRNA) by the RNAse III Drosha [182]. Then, the pre-miRNA is transported from nucleus to cytoplasm where the RNAse III endonuclease Dicer processes pre-miRNAs into miRNA duplexes. Next, the miRNA duplex is unwound and the released mature miRNA binds to an Argonaute protein (Ago) forming a core effector complex (miRNP/RISC) able now to interact with their mRNA targets typically in the 3' untranslated region. Since miRNAs can target different and various genes, the modulation of a single miRNA might affect many pathways at the same time. Today, different studies suggest that miRNAs can act as switches, turning cell signaling pathways on/off. The deregulation (up/down) of specific miRNAs can trigger the switch of specific cellular pathways such as apoptosis, cell proliferation, development, differentiation, metabolism, and cancer. At least half of the known miRNAs are located close to or inside of fragile sites and common breakpoints associated with cancer. Different studies have shown that deregulated levels of selected miRNAs are related to human cancer development and/or progression. In fact, different findings have linked miRNAs with specific type of cancers such as CLL [183], Burkitt's lymphoma [184], colorectal cancer [185], glioblastoma [186], hepatocellular carcinoma [187], lung cancer [188], papillary thyroid cancer [189], pancreas cancer [190], prostate cancer [191], and renal carcinoma [192]. See Figure 1.6.

The first studies that suggested a link between miRNA deregulation and cancer focused on observations made in *Caenorhabditis elegans* and later in *Drosophila*, with the discovery of *lin-4* and *let-7* miRNAs in the former [193, 194] and the *Bantam* miRNA in the latter [195]. Knockout of *lin-4* or *let-7* in *C. elegans* led to abnormal differentiation [196], whereas *Bantam* upregulation in *Drosophila* led to cellular growth and the inhibition of apoptosis [197]. Mice studies confirmed the previous findings, and Dicer knockout led to

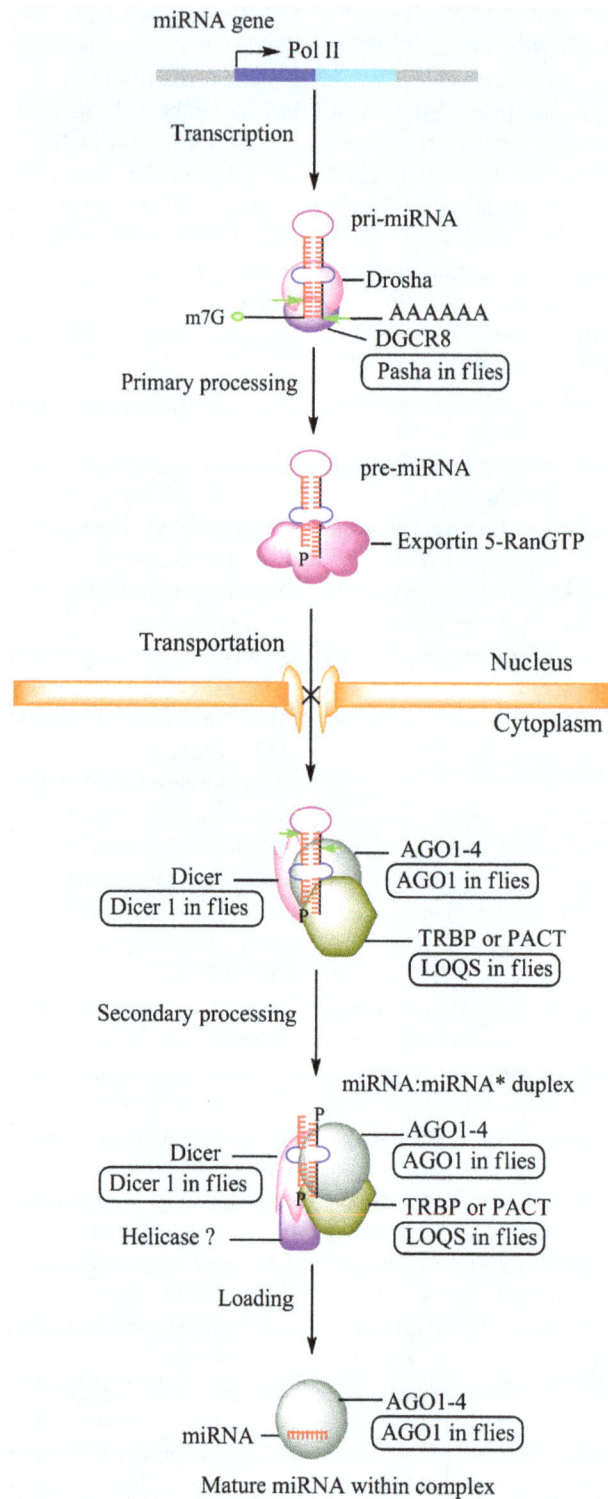

**FIGURE 1.6**   The animal miRNA synthesis pathway. The microRNA (miRNA) genes are transcribed by RNA polymerase II (Pol II), which results in the production of a pri-miRNA. Drosha, along with DiGeorge syndrome critical region gene-8 (DGCR-8; Pasha in flies), mediates the initial processing step (primary processing) that produces a ~65 nucleotide (nt) pre-miRNA. The pre-miRNA has a short stem of 2–3 nt 3′ overhangs, which is recognized by exportin 5 (EXP5) that mediates transport to cytoplasm. In the cytoplasm, RNase III Dicer is thought to catalyze the second processing step (secondary processing), which generates the miRNA/miRNA* duplex. Dicer, TRBP or PACT (LOQS in flies), and Argonaute 1–4 (Ago 1–4) (Argonaute 1 in flies) are responsible for pre-miRNA processing and RISC (RNA-induced silencing complex) assembly. An unknown helicase is thought to mediate unwinding of the duplex. One strand of the duplex remains the mature miRNA (miRNA) on Ago, whereas the miRNA* or passenger strand is degraded. The figure shows the mammalian miRNA synthesis pathway and fly factors are in the squares.

a defective miRNA production and impaired cellular differentiation [198]. These observations suggested that miRNAs might play a role in human neoplasia. Indeed, microarray studies have shown that there are global alterations in miRNA expression in cancer [199], with many miRNAs that are downregulated by genetic or epigenetic events, and some that are upregulated. For example, the *let-7* family of miRNAs is aberrantly downregulated in breast and lung tumors, leading to RAS pathway oncogenic activation [200]. Another example is the downregulation of miR-15 and miR-16 in CLL and resultant activation of the BCL2 proto-oncogene [201]. Overexpressed miRNAs include the miR-17-92 cluster, which plays a role in the development of lung and breast cancers as well as chronic myeloid leukemia through targeting of the transcription factor E2F1, a major cell cycle regulator [202]. MiR-17-92 cluster amplification has also been shown to frequently play a role in the development of B-cell lymphoma [203]. Its overexpression led to increased disease aggressiveness in mouse models [204]. This cluster of miRNAs has also been shown to be activated by the oncogene *c-myc* [205], highlighting its importance in tumorigenesis.

An interesting question relates to mechanisms of miRNA deregulation in cancer. Many miRNAs are transcriptionally regulated in a similar way as protein-coding genes and can be overexpressed by genetic mechanisms (eg, amplification) or suppressed by genetic (eg, deletion) or epigenetic (eg, hypermethylation) ones. Recently, Dicer and *DROSHA* expressions were also found to be altered in some cancers [206].

Molecular biologists have long been aware of the existence of noncoding RNAs (ncRNAs) in different organisms, but the 21st century has witnessed a genuine revolution in the field. Nevertheless, in spite of the growing amount of information concerning the function and types of ncRNAs in different species, we are still far from fully comprehending the role of the large fraction of the transcriptome that does not encode for proteins. Some estimates indicate that as many as 50% of all nucleotides in the genome might be transcribed [207, 208]. This includes long ncRNAs (hundreds to thousands of nucleotides) as well as the more abundant small ncRNAs.

A variety of small ncRNA pathways operate in yeast, plants, and animals to regulate gene expression. As a general rule, the effector complex is an RNA-induced silencing complex (RISC) that consists of the mature 20–30 nt RNA fragment bound to members of the Argonaute family of proteins [209]. Argonaute proteins can have endonuclease activity directed against messenger RNAs (mRNAs) with total complementarity to the small interfering RNA (siRNA). In other instances, such as with miRNAs, Argonaute acts with cofactors to induce destabilization or translational repression of the target mRNAs. Besides their well-characterized cytoplasmic role as posttranscriptional repressors, siRNAs from fungi and plants also display prominent functions controlling chromatin structure and transcriptional output. In this regard, much has been learned from the study of the

fission yeast *Schizosaccharomyces pombe*: in this model organism, the interaction between siRNAs originated from centromeric repeats and Argonaute proteins to form the RNA-induced initiation of transcriptional gene silencing complex results in the guiding of HMT activity to specific chromatic regions and the establishment of silenced heterochromatin [210–212]. The inheritance of heterochromatin throughout the cell cycle in *S. pombe* has also recently been explained by the action of the RNAi machinery: during S phase, a number of heterochromatic transcripts preferentially transcribed from centromeric repeats are transiently expressed and rapidly processed into siRNAs. This promotes the restoration of H3K9me2 marks and Swi6 (the yeast homolog of HP1) once replication is completed [213, 214]. Also in plants, Ago4 directs DNA methylation through interaction with RNA polymerase IVb, which has previously been implicated in RNA-directed DNA methylation [215]. In mammals, however, evidence of an equivalent role for ncRNAs has so far remained elusive.

MiRNAs/MiRs are a special type of small ncRNAs that also regulate gene expression of complementary mRNAs. The first miRNA to be identified was *lin-4* in *Caenorhabditis elegans* [216, 217]. Since then, thousands of miRNAs have been identified in a wide range of viruses, plants, and animal genomes, mainly by using cloning or bioinformatic approaches. The latest release of the Sanger miRNA Registry (http://microrna.sanger.ac.uk) annotates more than 700 human miRNAs (release 11.0), but it has been predicted that the human genome may contain more than 1000 different miRNAs [218]. It is estimated that at least 30% of all human genes are regulated by miRNAs [219], and although a reliable prediction of miRNA target genes remains a bioinformatic challenge, it is becoming increasingly clear that multiple miRNAs may cooperate to coordinate the regulation of the targeted transcript [78, 220].

MiRNAs are initially synthesized in the nucleus by RNA Pol II as long-capped and polyadenylated transcripts, called primary miRNAs (or pri-miRNAs). They can be expressed from their own transcriptional units or, most frequently, can be embedded within other host genes, very often as part of their introns or, more rarely, included within exonic regions [221]. It is not uncommon to find them clustered as a group of several miRNAs that are transcribed together and later processed into independent, mature molecules. Their posttranscriptional biogenesis pathway involves, firstly, cropping of pri-miRNA transcripts and release of hairpin-shaped precursors (or pre-miRNAs, 70–90 nucleotides long) by the Drosha/DGCR8 complex [222, 223]. This step depends on the secondary structure of the pri-miRNA transcript and may require additional cofactors [224]. Subsequent events include export from the nucleus, further trimming by the type III ribonuclease Dicer, and incorporation into the Argonaute-containing RISC complex, which delivers mature miRNAs to their mRNA targets through imperfect Watson-Crick base pairing along the 3′ untranslated regions (3′ UTRs), resulting in either translational arrest or degradation of the transcripts [225] (see Figure 1.7). Although miRNAs are generally considered to play a negative role in target gene expression, a recent study suggests that some miRNAs may switch from causing repression in proliferating cells to inducing translational activation upon cell cycle arrest [226].

Most miRNAs are expressed in a temporal- and tissue-specific manner and play crucial roles in cell proliferation, apoptosis, and differentiation during mammalian development [219, 227], but also in stem cell biology [228]. If we consider the number of cellular processes that require the involvement of miRNA-mediated regulation at one level or another, it is not surprising to observe that their expression profiles are significantly altered in tumors. Even though the general picture in cancer shows a widespread reduction in miRNA levels [229], it has been known for some years that both overexpression and depletion of certain miRNA loci play pathogenic roles in tumor progression [230]. One of the first instances of an miRNA characterized as oncogenic was the miR-17–92 cluster, which is amplified in human B-cell lymphomas and can enhance tumor growth in mouse models [231]. Other examples of miRNAs that might have oncogenic features are BIC/miR-155, which is overexpressed in

B-cell lymphomas [232] and miR-21, upregulated in breast cancer [233]. In contrast, the *let-7* family of genes (which includes at least 11 homologous miRNAs) is also directly associated with cancer but as an antiproliferative agent. Let-7 miRNAs downregulate the expression of all *ras* oncogenes in *C. elegans* and in human cells [234]. In lung and colon cancers, depletion of the let-7 family causes enhanced tumorigenicity [234, 235]. Another important mRNA target of the let-7 family of miRNAs is the high-mobility group AT-hook protein HMGA2. Disruption of the interaction between let-7 family members and their multiple target sites across HMGA2 3′ UTR results in enhanced expression of HGMA2 and neoplastic transformation [236–238]. Other well-characterized miRNAs with antiproliferative function include miR-15 and miR-16, which regulate the antiapoptotic factor BCL2 and are downregulated in CLLs [239, 240].

MiRNA microarray platforms, together with quantitative reverse transcription-polymerase chain reaction (qRT-PCR) designed for the detection of mature miRNAs, have facilitated the analysis of global miRNA expression profiles [241, 242]. Given the involvement of miRNAs in the etiology of cancer, it is not surprising that miRNA profiling is

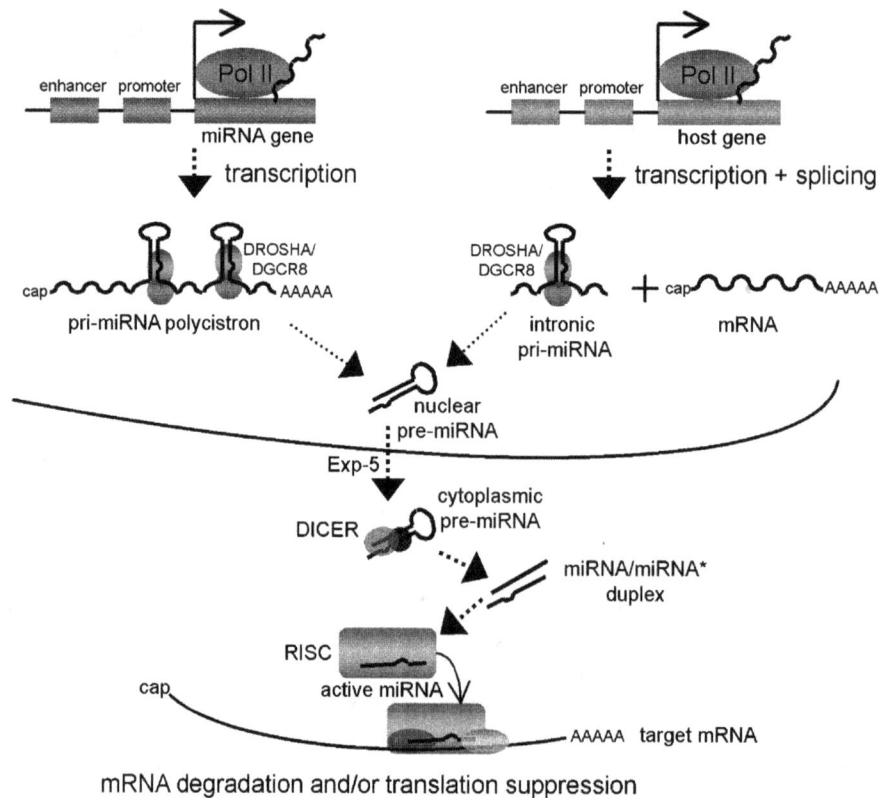

**FIGURE 1.7**   The biogenesis of microRNAs. Primary miRNAs are either transcribed by RNA Pol II as independent transcriptional units or co-expressed from the intronic regions of host gene transcripts. Concomitant with, or after splicing, the pri-miRNAs are recognized and cropped in the nucleus by the microprocessor complex, which contains the RNAse III type Drosha, to release a shorter hairpin named pre-miRNA. Pre-miRNAs are then actively exported to the cytoplasm by exportin-5 and further processed by Dicer to generate a short, ~21 nts duplex from which the mature miRNA will be incorporated into RISC. The miRNA-loaded RISC complex mediates negative posttranscriptional regulation, promoting either mRNA degradation or suppression of translation.

emerging as a useful tool in the characterization of a variety of human cancers, potentially being of even greater prognostic value than the analysis of the expression of protein-coding genes [243], and references therein [229, 244]. For example, the expression profile of more than 300 miRNAs in primary human breast cancer was compared with normal tissue and suggested to have prognostic value for the classification into different molecular subtypes [245].

## 1.17   EPIGENETIC ALTERATIONS AND MiRNAs

Even though different studies have contributed to better information regarding the biological importance of miR-NAs, the regulation of miRNA expression is still not fully understood. Different studies suggest that tumor suppressor miRNAs (the ones targeting oncogene transcripts) can be activated by chromatin-modifying drugs. In human cancer cells, DNA hypermethylation and chromatin structure can silence tumor suppressor miRNAs since they are present around their promoter regions. Chromatin-modifying drugs (DNA methylation and HDAC inhibitors) can activate transcription of pri-miRNAs that can be processed in pre-miRNAs and then in mature miRNAs. Then, the mature tumor suppressor miRNAs can translationally repress the target genes. An example of activation of a miRNA by chromatin-modifying drugs in human cancer cells is miR-127. Recently, it was demonstrated that miR-127 is located within a CGI and highly induced by DNA demethylation agent 5-5-aza-2'- deoxycytidine (decreasing its expression) and HDAC inhibitor 4-phenylbutyric acid (increasing its expression) in bladder cancer cells. MiR-127 is usually expressed as part of a cluster (containing miR-136, -431, -432, and -433) in normal cells but not in cancer cells; all these findings suggest an epigenetic regulation. The silencing of miR-127 was noted only when the drugs were used in combination suggesting a role of both epigenetic processes in controlling the expression of miR-127. This epigenetic silencing of miR-127 unlocks the expression of BCL6 oncogene contributing to bladder carcinogenesis [246]. In another study, Luiambio et al. analyzed miRNA expression profiling of HCT116 colon cancer cells and DNMT1–/–DNMT3B–/– HCT116 cells [247]. More than 5% of the 320 analyzed miRNAs were upregulated in DNMT1–/– DNMT3B–/– HCT116 cells. The authors found five miRNAs embedded in canonical CGIs and methylated in HCT116 cells but only miR-124a was unmethylated in normal colon tissue and hypermethylated in most primary colorectal tumors. The silencing of miR-124a leads to upregulation of CDK6 oncogene, known to regulate the tumor suppressor protein Rb. Very recently, it was demonstrated that miR-124a is methylated also in more than 50% of acute lymphoblastic leukemia (ALL) patients; its epigenetic silencing confers a poor prognosis to ALL patients [248, 249] and its promoter hypermethylation is an early event in gastric cancer [250]. The oncogene CDK6 is also targeted by miR-107 that seems to be epigenetically silenced in pancreatic cancer [251]. In fact, miR-107 is upregulated in pancreatic cancer cell lines when treated with a combination of the demethylating agent 5-aza-2'-deoxycytidine and the HDAC inhibitor trichostatin A.

Some researchers believe that transcriptional regulation of miRNA expression might be achieved by epigenetic alterations of target gene regulatory elements distant from the miRNA locus. In fact, it is known that many miRNAs are encoded in introns of host genes [252, 253] explaining why they might be susceptible to transcriptional repression by aberrant methylation of CGIs located in the 5'UTR of the target gene. An example of human intronic miRNA is miR-342. This miRNA is embedded, on the plus strand, in the center of a 25.9 kb intron between the third and fourth exons of the EVL gene on chromosome 14. A recent study reported that the expression of hsa-miR-342 is commonly suppressed in human colorectal cancer, the expression of EVL and hsa-miR-342 is coordinately suppressed in colorectal cancer, and the repression of hsa-miR-342 and EVL is associated with CGI methylation upstream of EVL [254].

Another example of intronic miRNA epigenetically regulated is miR-126. The tumor suppressor miR-126 is located within intron 7 of EGFL7, an epidermal growth factor-domain gene frequently downregulated in several cancer cell lines. Saito et al. demonstrated that miR-126 is downregulated in human cancer cell lines and bladder and prostate tumors, but is upregulated together with gene EGFL7 by epigenetic treatment. In fact, miR-126 is activated by inhibitors of DNA methylation (5-aza-2'-deoxycytidine) and histone deacetylation (4-phenylbutyric acid). Interestingly, treatment of cancer cell lines with the 4-phenylbutyric acid alone was not able to activate miR-126 expression [255].

Another miRNA epigenetically regulated is miR-1. Datta et al. analyzed the miRNA expression profile in hepatocarcinoma (HCC) cell lines HepG2 and Hep3B treated with a DNA hypomethylating agent (5-azacytidine) and/or a HDAC inhibitor (trichostatin A) [256]. Among the analyzed miRNAs, miR-1 was found significantly upregulated ($p \le 0.0001$) in both cell lines upon treatment with 5-azacytidine alone or in combination with trichostatin A. Furthermore, miR-1, coded by an intron 1 of the putative ORF166 is embedded in CGIs of which the one located upstream of miR-1 is methylated in both HCC cell lines and primary hepatocellular carcinomas.

Recently, Lujambio et al. identified miRNAs undergoing transcriptional silencing in lymph node metastatic cancer cells from colon, melanoma, and head and neck by miRNA expression microarray analysis upon DNA-demethylating agent 5-aza-2'-deoxycytidine treatment [257]. The authors [258] identified miR-148a, miR-34b/c, and miR-9 undergoing specific hypermethylation-associated silencing in cancer cells compared with normal tissues. The epigenetic inactivation of these three miRNAs contributed to in vitro and in vivo tumor dissemination, and the epigenetic silencing of miR-148a and miR-34b/c mediated the activation of oncogenic and metastasis target genes such as c-MYC, E2F3, CDK6, and TGIF2.

Epigenetic regulation is also a mechanism for miRNA inactivation in human breast cancer. In fact, an aberrant hypermethylation was shown for miR-9-1, miR-124a3, miR-148, miR-152, and miR-663 in 34–86% of cases in a series of 71 primary human breast cancer specimens [259]. Also, the authors demonstrated a reactivation of miR-9-1 in breast cancer cell lines treated with 5-aza-2'-deoxycytidine and hypermethylation of miRNA genes in human breast cancer, suggesting that miRNA gene methylation might serve as a sensitive marker for epigenetic instability. Another known CGI-embedded miRNA is miR-370. Meng et al. reported that miR-370 showed IL-6-driven methylation regulation in cholangiocarcinoma cells. In this study, IL-6 was found to enhance the growth of cholangiocarcinoma cells by repressing the expression of this miRNA epigenetically [260].

MiR-137 is closely associated with a large CGI and together with miR-124 might be activated in glioblastoma multiforme cell lines following treatment with a DNA methylation inhibitor (5-aza-2'-deoxycytidine) and/or a histone deacetylase inhibitor (trichostatin A). Expression of both miRNAs did not relatively change in cells treated with trichostatin A alone [261]. In epithelial ovarian cancer cell lines, miR-34b, miR-372, miR-516, miR-518a, miR-519d, miR-519e, and miR-520e are reported upregulated by treatment with the DNA demethylating agent 5-aza-2'-deoxycytidine and the HDAC inhibitor 4-phenylbutyric acid [262]. Interestingly, miR-34b is reported to be epigenetically regulated also in other cancers [257, 263, 264]. In 2008, Kozaki et al. demonstrated that the expression of miR-34b, miR-137, miR-193a, and miR-203, four miRNAs located close to CGIs, was restored by treatment with 5-aza-2'-deoxycytidine in the oral squamous cell carcinoma cells lacking their expression [265]. The expression levels of the four miRNAs were inversely correlated with their DNA methylation status in the oral squamous cell carcinoma cells. MiR-137 and miR-193a are most likely miRNAs frequently silenced in oral squamous cell carcinoma and they both have tumor suppressive effects on the growth of oral squamous cell carcinoma cell lines.

Brueckner et al. noticed that the human let-7a-3 miRNA gene on chromosome 22q13.31 was associated with a CGI heavily methylated in normal human tissues but hypomethylated in some human lung adenocarcinomas [266]. Lung cancer cells combinatorially treated with 5-aza-2'- deoxycytidine and the histone deacetylase inhibitor valproic acid showed a clear demethylation and transcriptional upregulation of let-7a-3. The epigenetic reactivation of let-7a-3 by hypomethylation induced tumor phenotypes and oncogenic changes in transcription profiles. These results suggest that let-7a-3 is an onco-miRNA promoting human lung carcinogenesis. In certain cases, the histone modification alone may regulate the miRNA expression. Scott et al. reported rapid alteration of miRNA levels by the potent hydroxamic acid HDAC inhibitor LAQ824 in the breast cancer cell line SKBr3 [267]. The miRNA profiling by miRNA microarray

analysis revealed significant changes in 40% of the 67 different miRNAs expressed in SKBr3 cells, with 5 miRNAs upregulated and 22 miRNAs downregulated.

The epigenetic regulation of miRNAs might be also cell-type specific. For example, miR-127 expression can be significantly upregulated by 5-aza-deoxycytidine and phenylbutyrate treatment in several cell lines, including CFPAC-1 pancreatic carcinoma, HCT116, HeLa, NCCIT embryonal carcinoma, and Ramos lymphoma [246] but not in CALU-1 lung carcinoma cells and MCF7 breast carcinoma. DNA demethylation or HDAC inhibition can also have no effect on miRNA in lung cancer cells [188, 268]. MiRNA methylation patterns between different cell lines can be distinct and sometimes variable, which might be explained at least in part due to their differences in tissue origins and differentiation states. An example is human miR-200c, recently described unmethylated in HCT116 colon carcinoma cells and HES7 embryonic stem cells but partially methylated in PHF primary fibroblast cells and HeLa cervical carcinoma cells [269].

At least 29,000 CGIs are predicted to be present in the human genome [270, 271], with many located at the 5' of the genes. A recent study reported that 70% of promoters in the human genome are associated with CGIs [272]. Also, several miRNAs linked to epigenetic regulation are closely associated with CGIs. In fact, it was reported that at least >40% of human miRNAs genes are associated with CGIs [269], indicating that several miRNAs can be considered candidate DNA methylation targets.

## 1.18   MiRNA GENES TARGETED BY EPIGENETIC MODIFICATIONS

If we consider the importance of miRNAs in controlling all stages of cell physiology, it is perhaps not surprising that their own expression needs tight regulation and that this also takes place at the level of epigenetic control. In the last few years, a number of studies have assessed the relevance of changes in chromatin modifications and their impact on the transcriptional control of a number of miRNA genes [269]. As happens with protein-coding genes, an aberrant pattern of methylation of CGIs near or within miRNA genes could result in a misregulated expression of key miRNAs and ultimately in pathogenic alterations, including tumorigenesis. In support of this, a recent study analyzing miRNAs that are aberrantly expressed in ovarian cancer identified a number of hypomethylated miRNAs genes (including miR-21, miR-203, and miR-205) with the encoded miRNAs displaying upmodulated expression [273]. In another report, Lujambio and coworkers [247, 274] used the colon cancer cell line HCT-116 as a model. When the miRNA expression profile in this cell line was compared with the corresponding line in which both the DNMTs 1 and 3b had been knocked out (double knockout, DKO), approximately 6% of the 320 miRNAs analyzed were upregulated more than three-fold in the DKO line. Of these, only miR-124a is embedded in a CGI that is densely methylated in the cancer cell line but not in the normal tissue. Subsequent analysis proved that miR-124a is also frequently methylated in other colon, breast, and lung carcinoma cell lines, as well as in leukemias and lymphomas. Most importantly, CGI hypermethylation around miR-124 is concomitant with an absence of activating histone modification marks, and with occupancy by methyl–CpG binding domain proteins. Furthermore, the investigators showed that methylation-mediated silencing of miR-124a in cancer cells results in an increased expression of its target CDK6, and as a consequence in the phosphorylation status of the downstream CDK6-regulated Rb protein. Thus, the loss of epigenetic control of methylation levels in miRNA genes in cancer cells contributes to the malignant phenotype by altering regulatory networks triggered by miRNA function. Another recent analysis of methylation of miRNA genes has shown hypermethylation of miR-9-1, miR-124a3, miR-148a, miR-152, and miR-663 in 34–86% of all primary human breast cancers analyzed [259]. This hypermethylation occurs rather early in cancer development and is associated with a marked decrease of miRNA expression in tumor tissue.

As stated before, a general downregulation of the let-7 family of miRNAs has been linked to lung tumorigenesis, and a tumor suppressing function had been proposed for members of this family [234, 275]. However, detailed analysis of let-7a-3 gene methylation and expression patterns reveals that this particular miRNA is often hypomethylated and highly expressed in different lung tumors, its overexpression correlating with increased oncogenic characteristics in A549 lung adenocarcinoma cells [266]. It is therefore clear that complex, differential epigenetic control can take place even among closely related members of the same family of miRNAs. The same let-7a-3 gene was found to be methylated in a clinical study of 214 human epithelial ovarian cancer samples, by using a real-time methylation-specific PCR (qMSP) approach [276]. The average level of methylation was 56.3%. Even though the expression of let-7a seemed to be only weakly affected (which could actually be due to the masking effect of let-7a-1 and let-7a-2 on let-7a-3 levels), this methylation correlated with low levels of insulin-like growth factor 2 (IGF2), strong expression of insulin-like growth factor-binding protein-3 (IGFBP-3), and better survival rates for patients [276].

In oral squamous cell carcinomas, comparison of tumor cell lines with normal lines revealed four miRNAs (miR-34b, miR-137, miR-193a, and miR-203) located around CGIs that were hypermethylated and whose expression was downregulated in the cancer state [265]. Of these, miR-137 and miR-193a were more consistently hypermethylated in 11 primary tumors, their ectopic expression resulting in reduced cell growth and downregulation of key proliferation factors, indicating that these two miRNAs possess tumor suppressor characteristics that are epigenetically silenced during oral cancer progression. MiR-34b/c, which is a crucial component of the p53 tumor suppressor network [277], is also a target of epigenetic silencing in most colorectal cancer primary tumors and derived cell lines. Notably, downregulation of miR-34b/c expression correlates with hypermethylation of its neighboring CGI [257, 263], which displays bidirectional promoter activity and also controls transcription of another tumor suppressor gene, B-cell translocation gene 4 (BTG4; [257, 263]). Interestingly, miR-34b/c, miR-148, and miR-9 CGI hypermethylation are associated with the development of human cancer metastasis [257].

The silencing of miR-223 in leukemias illustrates in detail how individual miRNAs can suffer altered expression in cancer through deregulation of chromatin modifiers. MiR-223 is highly specific for hematopoietic cells and constitutes a regulator of myelopoiesis [278, 279], but its transcription is inhibited in distinct subtypes of primary leukemias and this repression may underlie the block in myeloid differentiation that occurs in the cancer state. The oncogenic fusion protein AML1/ETO is the product of the most frequent chromosomal rearrangement in leukemias. AML1/ETO is able to bind DNA through its AML1 moiety but acts as a dominant-negative repressor of AML1 target genes, functioning as a docking site for several transcriptional regulators (such as the repressors SMRT, N-CoR; the HDACs HDAC1, HDAC2 and HDAC3; or DNMTs [280, 281]). Transcription of miR-223 is under direct control of AML1/ETO, and expression of the fusion protein in cancer drives histone deacetylation and DNA methylation of the miR-223 gene, resulting in heterochromatic silencing of miR-223 [282]. AML1/ETO binds to the natural target site for AML1 present at the 5′ end of miR-223 gene core promoter, and ChIP analysis reveals a concomitant occupancy of the region by DNMT1, DNMT3a, DNMT3b, and HDAC1. As a consequence, H3 and H4 histones become deacetylated and a small CGI present on the core promoter region of miR-223 close to the DNA region around the AML1-binding site is hypermethylated. Newly methylated CpGs act as binding sites for the DNA–methyl CpG-binding protein MeCP2. All these changes in chromatin modifications depend critically on the presence of AML1/ETO, and thus illustrate how an aberrantly formed chromatin remodeling complex may control the transcriptional silencing of a differentiation-associated miRNA gene upon the onset of cancer.

Taken together, these data show that miRNA genes are subject to hypermethylation and hypomethylation in a tumor- and tissue-specific manner. The enhanced definition of downstream mRNA targets for these miRNAs will undoubtedly shed light on the functional consequences of their altered epigenetic regulation and how this contributes to human tumorigenesis. Furthermore, if we take into account the functional importance of miRNAs in maintaining cell fate, and the fact that the PRC2 component Suz12 binds to several miRNA gene clusters (see supplementary data in [283]), it is clear that additional layers of control by miRNA function linked to the epigenetic regulators of the PcG type may exist during development and disease.

## 1.19 EPI-MiRNAs: A NEW GROUP OF MiRNAs

Today, miRNA can be considered an indirect mechanism through which epigenetic mechanisms regulate expression genes involved in human cancer development and/or progression. There is increasing evidence that miRNA-encoding genes are not only targets but also regulators of methylation and acetylation processes; in other words, miRNAs might act as epigenetic players. In fact, miRNAs can target genes coding for enzymes responsible for histone modifications (EZH2) and DNA methylation (DNMT3A and DNMT3B) [284, 285]. A perfect example is miR-101 and EZH2 gene (PcG enhancer of zest homolog 2). EZH2 is the catalytic subunit of the polycomb repressive complex 2 (PRC2), and it is able to trimethylate lysine 27 of histone H3 (H3K27me3). This trimethylation acts as a molecular mark and is recognized and bound by the polycomb repressive complex 1, causing gene repression by still-unclear mechanisms in which histone modifications, recruitment of chromatin-binding proteins such as heterochromatin binding protein 1, and chromatin compaction are involved [286]. Abnormally high levels of EZH2 lead to de novo silencing of genes, contributing to epigenetic reprogramming in cancer [287]. In normal cells, miR-101 is expressed and targets EZH2 3′ UTR mRNA in a sequence-dependent manner causing gene expression repression and/or transcript destabilization of this gene. Normal levels of EZH2 create normal epigenetic modifications and gene expression. In cancer cells, miR-101 expression is decreased, causing abnormal high levels of EZH2 and aberrant tumor suppressor and pro differentiation gene silencing via H3K27me3. EZH2 is also downregulated by miR-26a during myogenesis [288].

HDACs are also targeted from epi-miRNAs: HDAC1 3′ UTR is targeted from miR- 449a and re-expression leads to its reduction [289]; miR-1 and miR-140 directly target HDAC4 gene [290, 291].

Another example is the miR-29 family and DNMT3A and DNMT3B. In a recent study, it was demonstrated that the expression of miRNA-29 family (29a, 29b, and 29c) is inversely correlated to DNMT3A and DNMT3B in lung cancer and that miRNA-29 family directly targets the 3′ UTR of DNMT3A and DNMT3B genes 45. MiR-29b is not only able to target both DNMT3A and DNMT3B genes but also can indirectly target DNMT1 gene via SP1 [292].

Certain splice variants of DNMT3B gene are targeted by miR-148a and miR-148b. This miRNA family binds with high homology within the coding sequence of DNMT3B gene, but the mechanism of DNMT3B repression is still unknown [293]. The fact that miR-148a is also epigenetically regulated [257], suggests a self-epigenetic regulation loop for this epi-miRNA, but more studies are needed.

In 2008, two independent studies demonstrated that miR-290 cluster directly targets *Rb2/p130* gene in mouse embryonic stem cells [294, 295]. In Dicer–/– mouse ES cells, miR-290 cluster is not expressed leading to downregulation of DNMT3 genes and disruption of DNA methylation pattern. These effects were reversed by reintroduction of miR-290 cluster [294, 295].

## 1.20 CONTROLLING MiRNA EXPRESSION WITH EPIGENETIC DRUGS

Even though the global decrease in the methylation levels of genomic DNA was the first epigenetic change described in human cancers, a consistent hypermethylation of CGIs has been found in tumors. Indeed, for several years hypermethylation-induced silencing of tumor suppressor genes has been the focus of much epigenetic research [142] (and references therein). More recently, DNMT and HDAC inhibitors have both been used in cancer cells to unmask tumor suppressor genes that become epigenetically silenced in human cancers [296, 297]. The same strategy has been applied in a number of studies to unravel miRNA genes whose expression is altered by chromatin modifications in cancer samples when compared with normal tissues.

Pharmacologic inhibitors of class I and II HDACs induce drastic changes in cellular programming, promoting growth arrest, differentiation, and apoptosis, and can act as antitumoral agents [298]. Although their detailed molecular mechanism of action is still unknown, it is estimated that between 5% and 10% of transcribed genes become upregulated or downregulated following HDAC treatment. Why only genes of this subset become transcriptionally challenged by HDACs remains a mystery, but a report on the breast cancer cell line SKBr3 indicates that miRNA genes may also be rapidly altered and highly sensitive to HDAC action [267]. Using proapoptotic doses of the histone deacetylating agent LAQ824, the analysis of 200 miRNA expression profiles showed that around 40% of miRNAs with significant expression signals in non-treated cells display altered expression after 5 h of treatment. Of these, the majority (80%) of miRNAs underwent downregulation, and remarkably, some of them (miR-27a and miR-27b) arose as bona fide repressors of mRNAs known to be upregulated upon LAQ824 treatment, indicating that HDAC can alter cellular mRNA levels by targeting miRNA genes.

Chromatin-modifying drugs can alter the expression patterns of some miRNAs as elegantly illustrated by Saito et al. [246], who combined inhibitors of DNA methylation and histone deacetylation to induce the expression of particular miRNAs. Simultaneous treatment of T24 human bladder cancer cells with chromatin-modifying drugs 5-aza-2′deoxycytidine (5-Aza-CdR, a DNA-demethylating agent) and 4-phenylbutyric acid (PBA, a HDACi) resulted in more than three-fold upregulation in ~5% of 313 miRNAs analyzed [246]. Among the upregulated miRNAs, the strongest effect (~50-fold upregulation) was seen for miR-127, which is embedded in a CGI. It is normally silenced in cancer cells and the proto-oncogene BCL6 is a potential target. Thus, the combination of DNA demethylation and HDAC agents results in the activation of miRNAs that are usually subjected to epigenetic silencing in cancer and can act, when activated, as tumor suppressors. MiR-127 is expressed together with a cluster of four other miRNAs (miR-136, miR-431, miR-432, and miR-433) in normal tissues. Interestingly, the whole cluster is silenced in tumors, and miR-127 is the only miRNA from the cluster whose expression is markedly recovered in cancer cells upon treatment with 5-Aza-CdR + PBA, suggesting that epigenetic control can differentially affect miRNAs that are clustered at the same genomic loci and are normally processed from a large single transcript. The induction of miR-127 after treatment is associated with a decrease in DNA methylation levels and an increase in acetylated histone H3 and trimethylated histone H3K4 around specific transcription sites within the miR-127 CGI. These marks are typical of open chromatin and active gene expression, and indicate that CGIs can become active promoters of transcription for individual miRNAs in cancer cells. Despite this compelling evidence, the miR-127 gene was seen to be methylated in many human tissues and no methylation changes were found when four of each type of normal and primary human tumor samples (from prostate, bladder, and colon) were compared, pointing to the existence of a very complex transcriptional regulation. An unresolved question is why the subset of miRNAs whose expression is affected by the combination of drugs is different in normal fibroblasts and cancer cell lines.

In another study, the use of 5-Aza-CdR alone proved potent enough to reactivate the expression of miRNAs downregulated in breast cancer, such as miR-9-1 [259]. The use of demethylating drugs is also sufficient to restore the functionality of an endogenous mature miRNA, as in the aforementioned case of miR-223 and myeloid differentiation: its epigenetic silencing in leukemias was reverted with 5-Aza-CdR, the expression of the miRNA increased two-fold to three-fold, and granulocytic maturation was restored [282].

## 1.21 MiRNAs: REGULATORS OF CHROMATIN STRUCTURE?

In addition to being subjected to epigenetic regulation through chromatin modifications of their corresponding genes, miRNAs may also play a more decisive role in chromatin structure control by directly targeting the posttranscriptional regulation of key chromatin-modifying enzymes. Figure 1.8 depicts the influence that epigenetic control can have on miRNA expression but also how miRNA activity may regulate the synthesis of chromatin remodelers. In support of this, a study predicting miRNA target genes in humans has listed a number of HMTs, methyl CpG-binding proteins, chromo-domain-containing proteins, and HDACs as potential targets of miRNA regulation [299].

Along these lines, several recent reports have highlighted the central roles that particular miRNAs display as regulators of DNMT activity. Because gene hypermethylation is responsible for the silencing of many tumor suppressor genes in a variety of human cancers, the identification of

miRNAs as upstream regulatory factors of DNMT enzymes represents an exciting avenue for innovative therapeutic strategies.

The first study to show a direct link between miRNA function and DNMT regulation concerned lung cancer tissues [284]. In non–small cell lung cancer, the miR-29 family (miR-29a, miR-29b, and miR-29c) is downregulated, whereas DNMT3A and DNMT3B show increased expression. The complementarity between miR-29s and the 3′ UTR in DNMT3A and DNMT3B suggested that these mRNAs are bona fide targets of miR-29 regulation. Indeed, the use of luciferase reporters containing the predicted complementary sites from DNMT3 3′ UTRs as readouts shows an inverse correlation between luciferase activity and miR-29a, miR-29b, or miR-29c expression. Enforced expression of miR-29s in the A549 lung cancer cell line recovers normal levels of DNA methylation, with an effect (30% reduction) similar to the use of decitabine, a DNMT1 inhibitor, indicating that both de novo and maintenance of methylation are important for regulating methylation patterns in these systems. Most importantly, overexpression

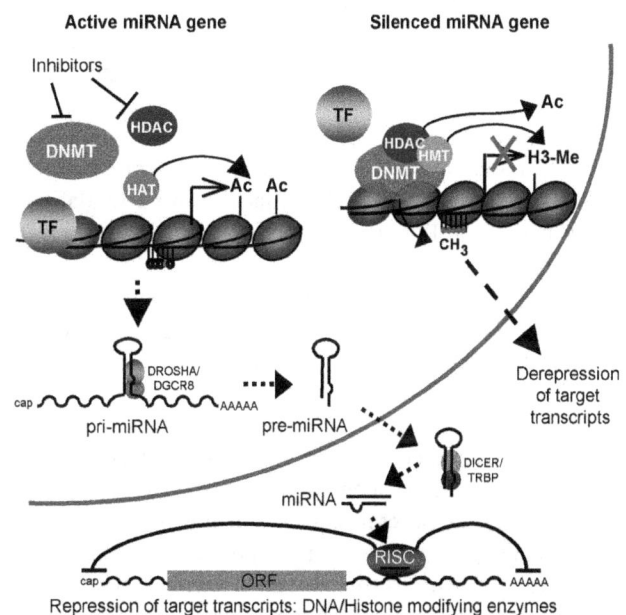

FIGURE 1.8 Epigenetic changes may affect miRNA expression and vice versa. Primary miRNA transcripts are generated from transcriptionally active miRNA genes and undergo a two-step biogenesis processing to produce the mature, functional 21-nucleotide-long repressor that associates with the RISC complex to downregulate target mRNAs. This regulation can involve a direct or indirect repression of DNA and histone-modifying enzymes, as well as chromatin-remodeling factors. Chemical inhibitors may be used to maintain the activation of miRNA genes. Under physiological or disease-caused deregulation, a number of repressive epigenetic marks act cooperatively to silence discrete genomic loci, among which, silencing of miRNA genes might result in the altered expression pattern of a subset of downstream targets. DNMT, DNA methyltransferase; HAT, histone acetyltransferase; HDAC, histone deacetylase; HMT, histone methyltransferase; ORF, open reading frame; TF, transcription factor.

of miR-29s increased the levels of tumor suppressor genes, such as FHIT and WWOX (frequently silenced by promoter methylation in lung cancer), ultimately suppressing tumorigenicity both in vitro and in vivo.

Two other recent studies that connect the miRNA processing machinery and DNMT activity show the intricate way in which different cellular pathways converge in a coordinated fashion to dictate the epigenetic control of cell fates and homeostasis. In mouse Dicer1-null ES cells, the levels of global DNA methylation are decreased to ratios similar to those seen in DNMT1−/− cells [294]. This results in a reduction in the levels of de novo methylation in differentiating ES cells (for example, in the Oct4 core pluripotency factor) accompanied by an abnormal upregulation of pluripotency markers and the inability of Dicer−/− ES cells to differentiate [295]. In fact, Dicer deficiency causes downregulation of DNMT1, DNMT3a2, and DNMT3b both at the level of mRNA and protein in a miR-290 cluster-dependent manner. MiR-290 is a member of a family of miRNAs restricted to the placenta in mammals that targets Rbl2, which in turn is a posttranscriptional repressor of DNMT3a and DNMT3b. Ectopic expression of de novo DNMTs or transfection of the miR-290 cluster rescues the defect in DNA methylation. Thus, downregulation of Dicer results in lack of miRNA-mediated repression of Rbl2, which as a consequence decreases DNMT activity through posttranscriptional regulation of DNMT3a and DNMT3b. These results illustrate how the RNAi machinery and miRNA function in particular can exert their control on the epigenetic regulation of key cellular events.

An even more striking link between Dicer and the global pattern of DNA methylation has been recently reported by Ting et al. [300]. In HCT116 colon cancer cells, a hypomorphic Dicer does not alter the overall level of genomic DNA methylation but causes a number of epigenetically silenced genes to become activated, this reactivation being concomitant with the demethylation of their promoters. Among them are several members of the WNT signaling pathway antagonists, which are repressed by CGI hypermethylation in various tumors and cancer cell lines [120]. How Dicer controls demethylation at these loci in particular is not known, but the effect does not seem to be mediated by miRNAs targeting DNMTs, since the levels of DNMT activity remain unaffected. It is conceivable that other types of small noncoding RNAs might have a role in nuclear chromatin silencing, mirroring mechanisms already described in yeast, plants, and *Drosophila* [301] (and references therein), but supporting evidence is still scarce. Previous studies had reported the ability of promoter-directed, exogenously introduced, short double-stranded RNA to induce transcriptional silencing of targeted genes in mammalian cells, hinting at the existence of an evolutionarily conserved mechanism [302, 303]. However, even though RNA-dependent transcriptional silencing in mammals causes chromatin structural changes near the targeted genes and is associated with altered histone modification patterns, this silencing is not always linked to changes in the DNA methylation state.

At this point, the role of a special kind of small noncoding RNA on DNA methylation deserves particular mention. The Piwi-associated RNAs (piRNAs) are molecules of 25–30 nucleotides in length that, like miRNAs, are processed from long single-stranded precursors, although they are not produced by the same biogenesis pathway. They are present in all animals, with a restricted pattern of expression in mammalian testes, and are essential for germ cell development. The piRNAs associate with the Piwi subfamily of Argonaute proteins and are involved in the silencing of transposable elements during gametogenesis [304, 305]. Remarkably, the recent finding that loss of piRNA-interacting proteins MILI and MIWI2 results in defective de novo DNA methylation of retrotransposons in fetal male germ cells [306] supports a mechanism for RNA-guided DNA methylation in the mammalian germline and opens up exciting new avenues in the field. In addition to piRNAs, a related second class of small RNAs, the endogenous siRNAs, have been recently identified in female germ cells [307, 308]. Both classes of small RNAs seem to be involved in the repression of mobile elements, although they may also regulate additional targets, including protein-coding mRNAs. Production of the oocyte-specific siRNAs is Dicer- and Ago2-dependent, and their biogenesis may involve the pairing of antisense, pseudogene-derived transcripts with the sense, spliced protein-coding mRNAs. The building of a comprehensive list of specific targets for this new subset of small RNAs will require further investigation. Furthermore, we currently know little about their ability to guide chromatin modifications directly at discrete genomic loci.

## 1.22 CLINICAL APPLICATIONS: EPIGENETIC TUMOR MARKERS

The rationale for the use of aberrant DNA methylation of a particular gene or a set of selected genes for clinical assessment comes from its frequency, stability, and variability between patients, which may indicate clinical usefulness. As mentioned earlier, DNA methylation is a stable and clonally propagated mark. Furthermore, DNA is less prone to degradation than RNA. Highly sensitive and/or quantitative methylation detection techniques are available, such as bisulfite pyrosequencing [309], methylation-specific polymerase chain reaction [310], or bisulfite treatment combined with high-throughput deep sequencing [311]. Moreover, aberrant methylation of some gene promoters is more common and easier to detect than the presence of mutations. This is especially valuable if the cancer cell or the cancer cell–derived free DNA is embedded in nonneoplastic cells or normal DNA molecules. Examples illustrating the potential use of epigenetic biomarkers in a clinical setting are described below and in Table 1.2 [127, 133, 312–330].

**TABLE 1.2**

**Examples of Clinically Relevant Epigenetic Biomarkers**

| Epigenetic Biomarker | Clinical Relevance | Supporting Literature | Sensitivity/Specificity/OR/HR |
|---|---|---|---|
| Hypermethylation of GSTP1 | Diagnosis/early detection of prostate cancer | Cairns 2001 [312] Lee 1994 [313] Eilers 2007 [32] | Sensitivity/specificity: 92%/86% |
| Hypermethylation of DAPK | Association with early recurrence and pathological stage in bladder cancer | Jarmalaite 2008 [314] Catto 2005 [315] | OR, 2.2 (95% CI, 1.04–4.5) |
| Hypermethylation of MGMT | Predictor of response to carmustine and temozolomide in gliomas | Hegi 2005 [316] Esteller 2000 [318] | HR for death associated with nonmethylation, 9.5 (95% CI, 3.0–42.7) HR for progression of disease associated with nonmethylation, 10.8 (95% CI, 4.4–30.8) |
| CIMP | Subtype classification, risk stratification, and prognostic relevance in colorectal cancer, leukemias, MDS, etc. | Issa 2008 [133] Shen 2007 [127] Issa 2005 [318] Issa 2004 [319] Shen 2002 [320] Shen 2010 [329] | HR for overall survival in MDS patients, 1.68 (95% CI, 1.0–2.81) HR for progression-free survival in MDS patients, 1.95 (95% CI, 1.18–3.21) |
| CIMP | Correlation with favorable prognosis in gliomas | Noushmehr 2010 [321] | G-CIMP status as an independent predictor of survival (p<0.1) |
| CIMP | Determinant of poor prognosis in neuroblastomas | Abe 2005 [322] | HR, 22.1 (95% CI, 5.3–93.4) |
| Promoter methylation of p16, CDH13, RASSF1A, and APC | Association with early recurrence in stage I NSCLC | Brock 2008 [323] | OR of recurrent cancer, 25.25 |
| Promoter methylation of p16 and of MGMT RASSF1A-DAPK-PAX5α in plasma and sputum, respectively | Association with smoking and lung cancer risk | Belinsky 2005 [324] Belinsky 2006 [330] | OR for cancer development, 6.5 Sensitivity/specificity: 65%/65% |
| Quantitation of promoter methylation of p16, p14ARF, MGMT, and GSTP1 | Detection of bladder cancer in urine sediment DNA | Hoque 2006 [325] | Sensitivity/specificity: 82%/96% |
| Global histone modification profiling in primary prostatectomy tissue samples | Correlation with prognosis and risk of recurrence in low-grade prostate cancer | Seligson 2005 [326] | HR, 9.2 (95% CI, 1.02–82.2) |
| microRNA signature | Association with clinical outcome (event-free survival) in cytogenetically normal AML patients with high-risk molecular features | Marcucci 2008 [327] | HR for an event, 1.8 (95% CI, 1.0–3.0) |

*Abbreviations:*   AML, acute myeloid leukemia; APC, adenomatous polyposis coli; CDH13, cadherin 13, H-cadherin (heart); CI, confidence interval; CIMP, CpG island methylator phenotype; DAPK, death-associated protein kinase; G-CIMP, glioma CpG island methylator phenotype; GSTP1, glutathione S-transferase-π; HR, hazard ratio; OR, odds ratio; MDS, myelodysplastic syndrome; MGMT, O-6-methylguanine-DNA methyltransferase; NSCLC, non–small cell lung cancer; PAX5α, paired box gene 5α;.RASSF1A, RAS association family 1A.

## 1.23   ABERRANT DNA METHYLATION IN CANCER RISK ASSESSMENT AND PREVENTION

There are two potential ways by which DNA methylation can be used for risk assessment: the detection of constitutional aberrant DNA methylation and the detection of acquired abnormalities that are harbingers of cancer development. The first relates to the transgenerational transmissibility of epigenetic alterations. Although a resetting of epigenetic marks takes place in the germline [2], making

the heritability of epigenetic modifications between parents and their offspring highly improbable, constitutional epigenetic alterations are noted in certain individuals [331], which could be either inherited or an acquired germline defect. The clinical entity that illustrates this clearly is the autosomal dominant hereditary nonpolyposis colorectal cancer (HNPCC) syndrome, in which affected individuals are highly predisposed to developing colorectal and endometrial cancers at a relatively young age [332]. This syndrome is caused by defects in mismatch repair, leading to MSI. Genes potentially involved are *MLH1*, human mutS homolog 2 (*MSH2*), *MSH6*, and postmeiotic segregation increased 2 (S.cerevisiae) (*PMS2*). It is interesting to note that a few individuals with HNPCC were described in whom no sequence mutation was detected in any of these genes, whereas *MLH1* or *MSH2* promoter methylation was found to be present in normal tissues, including circulating white blood cells [333]. In the case of *MSH2*, this has been traced to a mutation in the tumor-associated calcium signal transducer 1 (*TACSTD1*) gene immediately adjacent to *MSH2*, leading to aberrant transcription through its promoter and associated DNA hypermethylation [334]. No such mutation was detected for *MLH1*, which therefore appears to be a rare germline defect that is occasionally inherited. Constitutional epigenetic changes (epimutations) can also result from genetic variations in the form of single-nucleotide polymorphisms occasionally occurring in close proximity to a promoter and resulting in a predisposition toward acquired DNA methylation. This likely occurs via disruption of binding of transacting protective proteins such as Sp1 [335]. Thus, the transgenerational heritability of epigenetic modifications can result either from cis-acting events or from epigenetic transmission per se, but a familial cancer predisposition related exclusively to epigenetic phenomena appears to be relatively rare.

The second approach to cancer risk assessment is based on methylation studies of normal or preneoplastic tissues to detect acquired epimutations. For example, in lung cancer, methylation of the p16 gene was found to be present in preneoplastic lesions in smokers whereas no methylation was detected in never-smokers. Hence, p16 methylation in conjunction with other genes (such as p14, p15, E-cadherin, and RAS association family 1A [*RASSF1A*]) has been proposed as a biomarker to assess a patient's risk for developing lung cancer, and this is being tested by detecting methylation in sputum [324]. Indeed, in one prospective study of 98 cases and 92 matched controls, promoter methylation of 14 genes in sputum was evaluated for lung cancer risk assessment. Promoter hypermethylation of 6 genes was found to be associated with a >50% risk for subsequently developing lung cancer. The concomitant hypermethylation of 3 or more of these 6 genes was associated with an odds ratio of 6.5 for developing lung cancer, with sensitivity and specificity in the range of 65% [330]. It is interesting to note that hypermethylation of p16 and MGMT was detectable in sputum years before the clinical occurrence of lung cancer [336]. Another example is found in colorectal cancer patients, in whom loss of imprinting (LOI) of insulin-like growth factor 2 (IGF2) was found concurrently in cancer and adjacent normal colorectal tissue. LOI of IGF2 was also found in peripheral blood lymphocytes and its measurement was found to be predictive of the risk of developing colon cancer [337]. Also in the colon, age-related methylation in normal tissues has been proposed to mark a field defect associated with cancer risk, and measurement of this field could be a useful biomarker [122]. These data are relevant to cancer prevention because DNA methylation can be reversed by drug intervention. Therefore, its detection at a preneoplastic

stage would open the door to cancer prevention strategies, either passively through close monitoring of the investigated tissue (serial colonoscopies/bronchoscopies, imaging studies, etc.) or actively by the use of hypomethylating drugs and/or chromatin-remodeling agents to try to revert the premalignant phenotype.

## 1.24   ABERRANT DNA METHYLATION AS A DIAGNOSTIC TOOL

Aberrant methylation has been tested in the clinical setting as a diagnostic biomarker in biopsy specimens or in bodily fluids such as serum, sputum, bronchoalveolar lavage, saliva, urine, pleural or peritoneal effusions, and stool. For example, glutathione S-transferase-π (*GSTP1*) promoter hypermethylation was found in 100% of human prostatic carcinoma tissue specimens in one study [313] and was able to detect the presence of malignancy in biopsy samples in a study of 86 patients in whom prostate cancer was suspected, with a sensitivity and specificity of 92% and 86%, respectively, and positive and negative predictive values of 82% and 94%, respectively [328]. Similarly, the presence of vimentin methylation in stool samples was found to have a 46% sensitivity (95% confidence interval [CI], 35–56) and a 90% specificity (95% CI, 85–94) in diagnosing colon cancer [338]. A potential lack of specificity of single markers can be remedied by the use of a panel of several aberrantly methylated genes. For example, methylation of a panel of nine genes in urine sediment DNA from 175 patients and 94 controls was able to predict the presence of bladder cancer with a sensitivity of 82% (95% CI, 75–87) and a specificity of 96% (95% CI, 90–99) [325]. One limitation to the use of DNA methylation as a biomarker for disease diagnosis and assessment is the possibility that aberrant methylation could originate from a precancerous lesion or reflect an age-related phenomenon [44]. Indeed, most studies published to date have suggested that this approach has a low positive predictive value despite relatively good sensitivity and specificity. More sensitive methods are being developed to address this issue [311].

## 1.25   ABERRANT DNA METHYLATION AND ASSESSMENT OF PROGNOSIS/ RESPONSE TO THERAPEUTICS

Methylation patterns can be useful to assess clinical outcomes or response to chemotherapeutic agents. In general, high levels of DNA methylation are associated with a poor prognosis such as in lung cancer [323] or myelodysplastic syndrome (MDS) [329]. In a study of 51 cases with stage I non–small cell lung cancer (NSCLC) who developed an early recurrence after curative surgical resection and 116 controls who were free of disease recurrence after surgery, the promoter methylation status of seven genes was investigated in tumor and lymph node samples for its association with NSCLC recurrence. Methylation of four of those genes (p16, cadherin 13 [*CDH13*], *RASSF1A*, and adenomatous polyposis coli [*APC*]) demonstrated an independent association with tumor recurrence, with methylation of p16 and *CDH13* found to have an odds ratio of recurrent cancer of 15.5 and 25.25, respectively, in the training and combined training-validation cohorts. Similarly, MDS patients with higher levels of methylation, as assessed by studying a panel of ten genes, were found to have a shorter median overall survival (12.3 months vs 17.5 months, respectively; p=0.04) and a shorter median progression-free survival (6.4 months vs 14.9 months, respectively; p=0.009) when compared with patients with lower levels of methylation. However, in some instances, intense hypermethylation defines a distinct subgroup of cancers that may have a favorable prognosis. This is the case in colon cancer, in which simultaneous methylation of multiple genes termed CIMP is associated with *MLH1* methylation, which results in a favorable prognosis [126]. CIMP has also been described recently in glioblastoma multiforme, in which it also was found to be associated with a better outcome; CIMP-positive cases were significantly younger at the time of diagnosis (median age of 36 years vs 59 years, respectively), closely associated with *IDH1* somatic mutations, and had a significantly better survival (median survival of 150 weeks vs 42 weeks, respectively) compared with CIMP-negative cases (p=0.0165) [321].

Methylation can also be useful as a predictive biomarker. For example, methylation of the MGMT DNA repair gene reportedly correlates with a good response to temozolomide and better overall clinical outcome in patients with glioblastoma multiforme [316]. Indeed, MGMT promoter methylation, present in approximately 45% of cases, was found to be correlated with a significant benefit from temozolomide therapy (median survival of 21.7 months vs 15.3 months without temozolomide therapy; p=0.007). In patients without MGMT methylation, the effects of temozolomide were less clear (median survival of 12.7 months with vs 11.8 months without temozolomide therapy; p=0.06). These data suggest that tumor methylation profiling could be useful for risk stratification and making therapeutic decisions.

## 1.26 MiRNAs IN CANCER DIAGNOSIS, CLASSIFICATION, AND PROGNOSIS

MiRNA profiling has been shown to be informative both as a diagnostic tool and as a potential prognostic biomarker [200]. For example, miRNA profiling was shown to be useful in a series of tissue samples derived from metastatic sites of unknown primary origins [339]. The prognostic significance of miRNAs in cancer is currently being extensively studied. In CLL, the expression of a panel of 13 miRNAs was shown to correlate with disease aggressiveness as reflected by the time elapsed between diagnosis and first treatment [340]. However, this predictive ability has not been shown to be independent from other CLL prognostic markers [341]. In lung cancer, miR-155 and let-7 miRNA levels were found to be correlated with disease aggressiveness and clinical outcome [341]. Higher let-7 levels were associated with a more indolent disease and better survival after surgical resection. MiR-155 has also been shown to be of prognostic value in patients with diffuse large B-cell lymphoma [342], in whom it is present at significantly higher levels in the activated B-cell phenotype than in the germinal center phenotype. MiRNA profiling could also be useful in the future as part of a model integrating multiple prognostic information [327].

## 1.27 CLINICAL APPLICATIONS: EPIGENETIC THERAPY

With the understanding of the mechanisms underlying the silencing of tumor suppressor genes in cancer came the idea of pharmacologically relieving the inhibitory effects of DNA methylation and chromatin remodeling on gene expression. Two classes of drugs that modify epigenetics have been approved by the US Food and Drug Administration (FDA) for the treatment of cancer: DNA methylation inhibitors and HDAC inhibitors (Table 1.3) [343–347].

## TABLE 1.3
## Epigenetic-Acting Drugs Approved by FDA

| Epigenetic-Acting Drug | Clinical Indication | Major Data | Supporting Literature |
|---|---|---|---|
| **DNA Methyltransferase Inhibitors** | | | |
| 5-azacytidine (azacitidine) | Symptomatic MDS | 16% overall response rate; 66% hematologic improvement/transfusion independence | Kaminskas 2005 [343] Fenaux 2009 [344] |
| 5-aza-2′-deoxycytidine (decitabine) | Intermediate and high-risk MDS | 73% objective response rate; 34% complete response rate | Kantarjian 2007 [345] |
| **Histone Deacetylase Inhibitors** | | | |
| Suberoylanilide hydroxamic acid (vorinostat) | Progressive, persistent, or recurrent cutaneous T-cell lymphoma | 30% objective response rate | Mann 2007 [346] |
| Romidepsin (depsipeptide) | Progressive, persistent, or recurrent cutaneous T-cell lymphoma | 34% overall response rate; 6% complete response rate | Piekarz 2009 [347] |

*Abbreviation:* MDS, myelodysplastic syndrome.

The available DNA methylation inhibitors are nucleoside analogues that exert their demethylating activity through the establishment of an irreversible covalent bond with DNMTs after their incorporation into DNA [348]. Hypomethylation requires that the cells be proliferating after DNMT inhibition. These DNA methylation inhibitors were first introduced in the clinic several decades ago. At that time, they were used as cytotoxic chemotherapy at relatively high doses [349] and were found to be toxic (at these doses) without great antitumor activity. In the past 10 to 15 years, these drugs were reintroduced at lower doses that promoted the hypomethylating effect, and results of clinical trials indicated that repeated exposure induced DNA demethylation accompanied by a better antineoplastic effect than when used at higher doses [350]. This led to the FDA's approval of 5-azacytidine (azacitidine) in 2004 [351] and 5-aza-2′-deoxycytidine (decitabine) in 2006 [352] for the treatment of patients with MDS. Azacitidine induced an overall response rate in the range of 20–60% and significantly improved survival compared with standard of care [344]. Decitabine induced a high response rate at optimal doses [345] (complete response [CR]/pathologic CR rate of 40%) and has been shown to prolong survival when compared with historical controls [353]. The major side effect with these drugs is myelosuppression and the regimens used currently are well tolerated [350]. Some shortcomings of these drugs are their relatively short half-lives, their instability in aqueous solutions, a lack of specificity inherent to their mechanism of action, and the fact that acquired resistance is nearly universal, without a clear mechanism. This has led to a search for potentially different DNA methylation inhibitors; several were identified such as the cytidine analogue zebularine, the antiarrhythmic procainamide (a weak inhibitor), and SGI-1027, a drug that may inhibit DNA methylation without requiring incorporation [352].

Another interesting class of epigenetically targeted drugs is HDAC inhibitors [355]. HDAC inhibitors were initially identified through differentiation screens. These drugs target the catalytic domain of HDACs, thus interfering with their substrate recognition. Most HDAC inhibitors affect zinc-dependent HDACs and are divided into several classes depending on their chemical nature. The ones described to date comprise the short-chain fatty acids (such as sodium phenylbutyrate, sodium butyrate, and valproic acid); the hydroxamic acids (such as trichostatin A, vorinostat, and panobinostat); the cyclic peptides (such as romidepsin); and the benzamides, comprised of MGCD-0103 and entinostat. In 2006, suberoylanilide hydroxamic acid (vorinostat) was approved by the FDA for the treatment of patients with progressive, persistent, or recurrent cutaneous T-cell lymphoma [346]. Recently, depsipeptide (romidepsin) received FDA approval for use in the refractory form of the same disease. Clinical trials of these and other HDAC inhibitors in other malignancies are currently ongoing. Early results suggest activity in other lymphoid malignancies such as Hodgkin lymphoma, but limited activity in solid tumors [356].

Similar to DNA hypomethylating agents, HDAC inhibitors suffer from nongene selectivity. The exact mechanism by which these drugs exert their gene expression reactivating effect is still unclear. One straightforward mechanism proposed is the induced hyperacetylation of histone proteins, leading to an open chromatin configuration and transcriptional activation [355]. However, the mechanism of action of these drugs is more complex because they are active both in the nucleus and the cytoplasm, and HDACs catalyze the deacetylation of both histone and nonhistone proteins. In fact, HDAC inhibitors might very well be exerting their antitumor activity through apoptosis or cellular differentiation induction by affecting multiple cellular pathways, some transcriptionally and some posttranscriptionally. Some of these pathways, along with the biological effects epigenetically targeted drugs have on tumor cells, are shown in Figure 1.9. There is currently interest in developing drugs that target other epigenetic pathways such as histone methylases/HDMT, MBDs, and histone readers.

The lack of specificity of epigenetically targeted drugs raises concerns about their use in clinical practice. Some of these concerns would be the reactivation of normally silenced sequences (such as repetitive elements) or imprinted genes. This reactivation could theoretically lead to allelic imbalance or genomic instability, and other deleterious effects of retrotransposon activation. To date, there are no data supporting these concerns clinically, but it is possible that problems will emerge after several years of therapy. This has led researchers in the field to try to develop new compounds selectively targeting specific genes. One example is the development of a methylated oligonucleotide directed toward the 5′ promoter region of the IGF2 growth-promoting gene, subsequently leading to the methylation of this promoter and transient silencing of the gene [357]. This line of research is still in its infancy and could face significant problems in drug delivery.

It is important to mention that epigenetic drugs are promising not only as single agents but also in combination with other epigenetically targeted drugs or with conventional chemotherapy. Several studies, both in vitro and in vivo, have demonstrated the synergistic effect of sequentially administering DNMT inhibitors (such as decitabine) and HDAC inhibitors (such as vorinostat) [348], and this approach is currently being tested in clinical trials. Furthermore, a synergistic effect was also found when combining epigenetic drugs with conventional chemotherapy [358], and trials are currently testing these combinations in the clinical setting for several tumor types.

Several studies have tried to link DNA methylation profiles at study entry with response to therapy. To date, these studies of a limited number of genes have been negative [329]. Entire epigenome studies of this issue are currently ongoing. Studies also have tried to correlate global

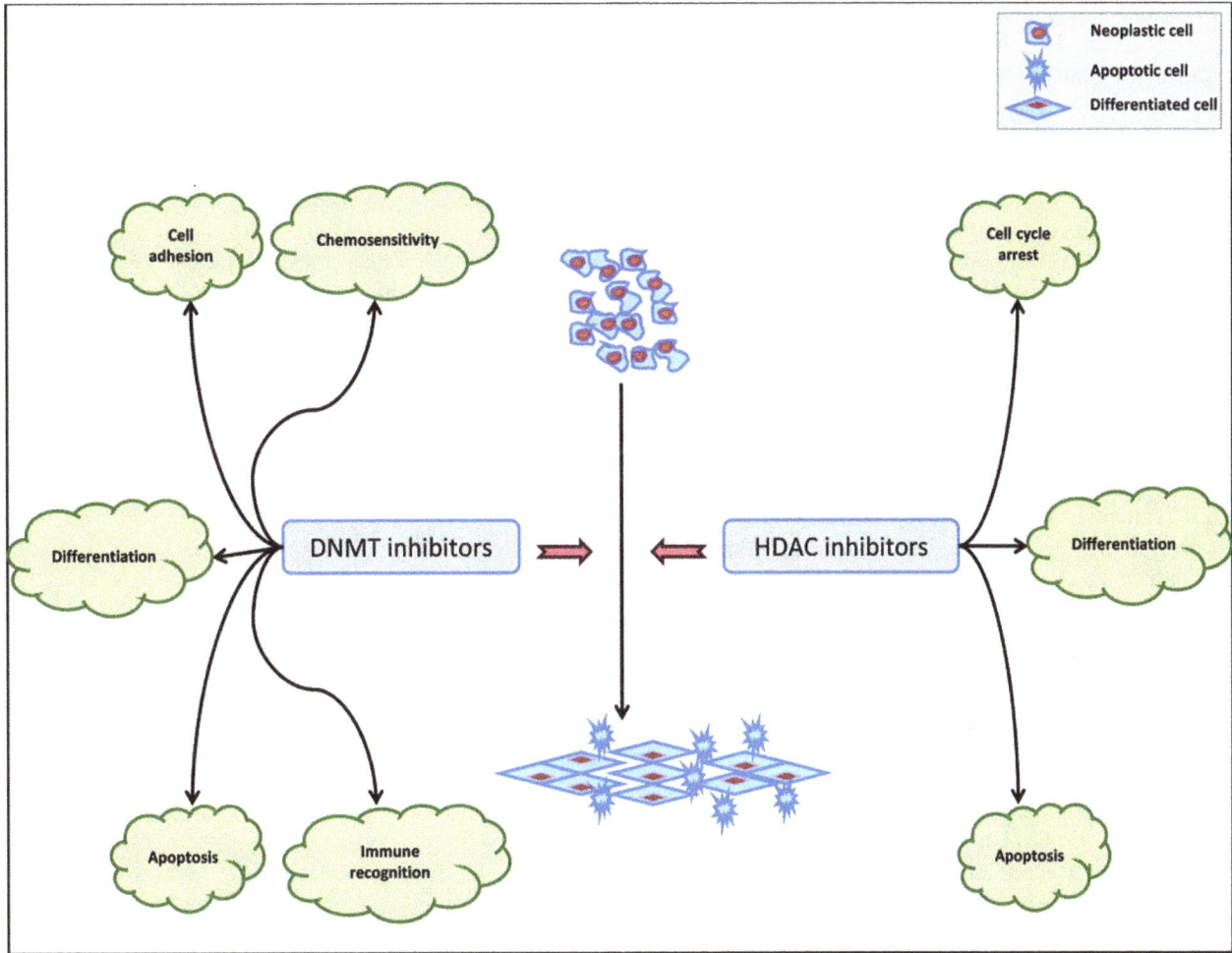

**FIGURE 1.9 Epigenetic therapy.** The two main families of epigenetically acting drugs, DNA methyltransferase (DNMT) inhibitors and histone deacetylase (HDAC) inhibitors, exert their antineoplastic effect via several mechanisms such as cell cycle arrest, apoptosis induction, and immune recognition. These effects eventually result in differentiation or cancer cell death.

hypomethylation, as assessed by the methylation levels of LINE1 and Alus repetitive elements, at days 5 and 12 after decitabine therapy with clinical response. Results were controversial. Indeed, some studies found a trend toward a positive correlation between global hypomethylation at day 5 and clinical response [350] in patients with leukemia, whereas other studies found an inverse correlation between levels of hypomethylation at day 12 and achievement of CR [359] in patients with chronic myelogenous leukemia, that was resistant to imatinib mesylate. The latter finding was hypothesized to be due to a cell death mechanism of response, with the resistant cells capable of sustaining more hypomethylation.

In contrast to DNA methylation markers, gene expression induction has been consistently linked to subsequent response to decitabine. This has been demonstrated for *P15* [345], *ER* [360], *P53R2*/ribonucleosidediphosphate reductase subunit M2 B (*RRM2B*) [361], and miR-29b [362].

Perhaps one of the major drawbacks of epigenetic therapy is the presence of spontaneous and/or acquired resistance to

these drugs, both in vitro and in vivo. Indeed, in a panel of cancer cell lines, resistance to the hypomethylating agent decitabine was manifested by a 1000-fold difference in the half maximal (50%) inhibitory concentration ($IC_{50}$) of this drug among the cell lines tested [363]. Resistance mechanisms were hypothesized to be related to variations in the parameters affecting nucleoside analogue metabolism, starting with transport inside the cell (human equilibrative nucleoside transporter 1 [hENT1] and hENT2), initial phosphorylation (deoxycytidine kinase [DCK] for decitabine and uridine-cytidine kinase [UCK] for azacitidine), and finally catabolism by the enzyme cytidine deaminase (CDA). Indeed, in vitro studies demonstrated that low levels of DCK and hENT1 and high levels of CDA were correlated with resistance to hypomethylating agents. In fact, the observed cross-resistance between decitabine and cytarabine (two nucleoside analogues sharing the same need for phosphorylation by DCK for incorporation into the DNA) and the lack of cross-resistance between decitabine and azacytidine indicate that incorporation into the DNA plays a major role in cancer cell resistance to nucleoside analogues, including decitabine. These observations were found to be relevant in vivo as well, because low levels of DCK/low DCK activities were correlated with poor response to nucleoside analogues in, for example, childhood acute lymphoblastic leukemias [364] and pancreatic cancer [365].

The full-scale implementation of these molecular markers of sensitivity to hypomethylating agents in the clinical setting faces several challenges. One is the less-than-perfect correlation between the clinical activity of these drugs and their hypomethylating effect. One possibility is that beyond a certain threshold, more hypomethylation does not correlate with a better clinical outcome. In fact, there may be molecular barriers downstream of hypomethylation that prevent adequate gene reactivation. Another possibility is the hypomethylation-independent mechanisms of antineoplastic activity of decitabine and azacitidine. Both drugs can induce DNA damage at relatively high doses, and azacitidine also affects RNA methylation [366, 367]. Both of these effects may also be involved in clinical responses.

## 1.28 CONCLUSIONS

Epigenetics is a hereditable process that affects the gene expression without changing the DNA sequence. Epigenetic processes include DNA methylation and histone and chromatin changes. The importance of epigenetic modifications in cancer is well known. Hence, plenty of research has been conducted in recent years on mechanisms involved in epigenetic changes in cells. Recent advancements in the field provide a precise pattern of methylation, acetylation, and miRNA level. These findings have led to the identification of biomarkers in various diseases bringing optimism and hope for curing these diseases in the near future.

Understanding the complexity of the epigenome and of all the actors involved in modulating its interactions with genomic sequences is of fundamental importance in health and disease. This understanding will allow us to reach newer horizons in our search for the mechanisms governing cellular fate. On the tumorigenic spectrum, the time when we switch from untargeted cytotoxicity to reversion of the malignant phenotype is drawing near.

## REFERENCES

[1] Gräff J, Mansuy IM. Epigenetic codes in cognition and behaviour. Behav Brain Res. 2008;192(1):70–87.

[2] Reik W, Dean W, Walter J. Epigenetic reprogramming in mammalian development. Science. 2001;293:1089-1093.

[3] Morgan HD, Sutherland HG, Martin DI, et al. Epigenetic inheritance at the agouti locus in the mouse. Nat Genet. 1999;23:314–318.

[4] Wang Y, Jorda M, Jones PL, et al. Functional CpG methylation system in a social insect. Science. 2006;314:645–647.

[5] Gal-Yam EN, Saito Y, Egger G, et al. Cancer epigenetics: modifications, screening, and therapy. Annu Rev Med. 2008;59:267–280.

[6] Jeltsch A. Beyond Watson and Crick: DNA methylation and molecular enzymology of DNA methyltransferases. Chembiochem. 2002;3:274–293.

[7] Kim GD, Ni J, Kelesoglu N, Roberts RJ, et al. Co-operation and communication between the human maintenance and de novo DNA (cytosine-5) methyltransferases. EMBO J. 2002;21:4183–4195.

[8] Okano M, Bell DW, Haber DA, et al. DNA methyltransferases Dnmt3a and Dnmt3b are essential for de novo methylation and mammalian development. Cell. 1999;99:247–257.

[9] Kouzarides T. Chromatin modifications and their function. Cell. 2007;128:693–705.

[10] Liang G, Lin JCY, Wei V, et al. Distinct localization of histone H3 acetylation and H3-K4 methylation to the transcription start sites in the human genome. Proc Natl Acad Sci USA. 2004;101:7357–7362.

[11] Bernstein BE, Meissner A, Lander ES. The mammalian epigenome. Cell. 2007;128:669–681.

[12] Esteller M. Epigenetics in cancer. N Engl J Med. 2008; 358:1148–1159.

[13] Jones PA, Baylin SB. The epigenomics of cancer. Cell. 2007;128:683–692.

[14] Espada J, Esteller M. Epigenetic control of nuclear architecture. Cell Mol Life Sci. 2007;64:449–457.

[15] Ehrlich M. Amount and distribution of 5-methylcytosine in human DNA from different types of tissues or cells. Nucleic Acids Res. 1982;10:2709–2721.

[16] Bird A. DNA methylation patterns and epigenetic memory. Genes Dev. 2002;16:6–21.

[17] Pradhan S, Talbot D, Sha M, et al. Baculovirus-mediated expression and characterization of the full-length murine DNA methyltransferase. Nucleic Acids Res. 1997;25: 4666–4673.

[18] Bhaumik SR, Smith E, Shilatifard A. Covalent modifications of histones during development and disease pathogenesis. Nat Struct Mol Biol. 2007;14:1008–1016.

[19] Barski A, Cuddapah S, Cui K, et al. High-resolution profiling of histone methylations in the human genome. Cell. 2007;129:823–837.

[20] Schuettengruber B, Chourrout D, Vervoort M, et al. Genome regulation by polycomb and trithorax proteins. Cell. 2007;128(4):735–745.

[21] Li H, Ilin S, Wang W, et al. Molecular basis for site specific read-out of histone H3K4me3 by the BPTF PHD finger of NURF. Nature. 2006;442:91–95.

[22] Wysocka J, Swigut T, Xiao H, et al. A PHD finger of NURF couples histone H3 lysine 4 trimethylation with chromatin remodelling. Nature. 2006;442:86–90.

[23] Fraga MF, Ballestar E, Villar-Garea A, et al. Loss of acetylation at Lys16 and trimethylation at Lys20 of histone H4 is a common hallmark of human cancer. Nat Genet. 2005;37:391–400.

[24] Jones PA, Baylin SB. The fundamental role of epigenetic events in cancer. Nat Rev Genet. 2002;3:415–428.

[25] Feinberg AP, Ohlsson R, Henikoff S. The epigenetic progenitor origin of human cancer. Nat Rev Genet. 2006;7:21–33.

[26] Dillon N. Gene regulation and large-scale chromatin organization in the nucleus. Chromosome Res. 2006;14(1): 117–126.

[27] Woodcock CL. Chromatin architecture. Curr Opin Struct Biol. 2006;16(2):213–220.

[28] de Groote ML, Verschure PJ, Rots MG. Epigenetic Editing: targeted rewriting of epigenetic marks to modulate expression of selected target genes. Nucleic Acids Res. 2012;40(21):10596–10613.

[29] Maunakea AK, Nagarajan RP, Bilenky M, et al. Conserved role of intragenic DNA methylation in regulating alternative promoters. Nature. 2010;466(7303):253–257.

[30] De Smet C, Lurquin C, Lethé B, et al. DNA methylation is the primary silencing mechanism for a set of germ line- and tumor-specific genes with a CpG-rich promoter. Mol Cell Biol. 1999;19(11):7327–7335.

[31] Laird PW. Principles and challenges of genome-wide DNA methylation analysis. Nat Rev Genet. 2010;11(3):191–203.

[32] Lister R, Pelizzola M, Dowen RH, et al. Human DNA methylomes at base resolution show widespread epigenomic differences. Nature. 2009;462(7271):315–322.

[33] Meissner A, Mikkelsen TS, Gu H, et al. Genome-scale DNA methylation maps of pluripotent and differentiated cells. Nature. 2008;454(7205):766–770.

[34] Klose RJ, Bird AP. Genomic DNA methylation: the mark and its mediators. Trends Biochem Sci. 2006;31:89–97.

[35] Glass JL, Thompson RF, Khulan B, et al. CG dinucleotide clustering is a species specific property of the genome. Nucleic Acids Res. 2007;35:6798–6807.

[36] Illingworth RS, Bird AP. CpG islands–'a rough guide.' FEBS Lett. 2009;583:1713–1720.

[37] Nan X, Ng HH, Johnson CA, et al. Transcriptional repression by the methyl-CpG binding protein MeCP2 involves a histone deacetylase complex. Nature. 1998;393:386–389.

[38] Zoghbi HY. Rett syndrome: what do we know for sure? Nat Neurosci. 2009;12:239–240.

[39] Bestor TH, Tycko B. Creation of genomic methylation patterns. Nat Genet. 1996;12:363–367.

[40] Mohandas T, Sparkes RS, Shapiro LJ. Reactivation of an inactive human X chromosome: evidence for X inactivation by DNA methylation. Science. 1981;211:393–396.

[41] Swain JL, Stewart TA, Leder P. Parental legacy determines methylation and expression of an autosomal transgene: a molecular mechanism for parental imprinting. Cell. 1987;50:719–727.

[42] Shen L, Kondo Y, Guo Y, et al. Genomewide profiling of DNA methylation reveals a class of normally methylated CpG island promoters. PLoS Genet. 2007;3:2023–2036.

[43] Wilson VL, Jones PA. DNA methylation decreases in aging but not in immortal cells. Science. 1983;220:1055–1057.

[44] Issa JP, Ottaviano YL, Celano P, et al. Methylation of the oestrogen receptor CpG island links ageing and neoplasia in human colon. Nat Genet. 1994;7:536–540.

[45] Ahuja N, Li Q, Mohan AL, et al. Aging and DNA methylation in colorectal mucosa and cancer. Cancer Res. 1998;58:5489–5494.

[46] Maegawa S, Hinkal G, Kim HS, et al. Widespread and tissue specific age-related DNA methylation changes in mice. Genome Res. 2010;20:332–340.

[47] Fraga MF, Ballestar E, Paz MF, et al. Epigenetic differences arise during the lifetime of monozygotic twins. Proc Natl Acad Sci USA. 2005;102:10604–10609.

[48] Jones PA, Liang G. Rethinking how DNA methylation patterns are maintained. Nat Rev Genet. 2009;10:805–811.

[49] Jair KW, Bachman KE, Suzuki H, et al. De novo CpG island methylation in human cancer cells. Cancer Res. 2006;66:682–692.

[50] Bestor TH, Verdine GL. DNA methyltransferases. Curr Opin Cell Biol. 1994;6:380–389.

[51] Li E, Bestor TH, Jaenisch R. Targeted mutation of the DNA methyltransferase gene results in embryonic lethality. Cell. 1992;69:915–926.

[52] Nightingale KP, O'Neill LP, Turner BM. Histone modifications: signaling receptors and potential elements of a heritable epigenetic code. Curr Opin Genet Dev. 2006; 16(2):125–136.

[53] Kato S, Inoue K, Youn MY. Emergence of the osteoepigenome in bone biology. IBMS BoneKEy. 2010;7(9):314–324.

[54] Gangaraju VK, Bartholomew B. Mechanisms of ATP dependent chromatin remodeling. Mutat Res. 2007;618(1-2):3–17.

[55] Zhou GL, Liu DP, Liang CC. Memory mechanisms of active transcription during cell division. BioEssays. 2005;27(12):1239–1245.

[56] Li Y, Zhu J, Tian G, et al. The DNA methylome of human peripheral blood mononuclear cells. PLOS Biology. 2010;8(11):e1000533.

[57] Strahl BD, Allis CD. The language of covalent histone modifications. Nature. 2000;403:41–45.

[58] Struhl K. Histone acetylation and transcriptional regulatory mechanisms. Genes Dev. 1998;12:599–606.

[59] Clapier CR, Cairns BR. The biology of chromatin remodeling complexes. Annu Rev Biochem. 2009;78:273–304.

[60] Yang XJ. The diverse superfamily of lysine acetyltransferases and their roles in leukemia and other diseases. Nucleic Acids Res. 2004;32:959–976.

[61] Utley RT, Ikeda K, Grant PA, et al. Transcriptional activators direct histone acetyltransferase complexes to nucleosomes. Nature. 1998;394:498–502.

[62] Yang XJ, Seto E. HATs and HDACs: from structure, function and regulation to novel strategies for therapy and prevention. Oncogene. 2007;26:5310–5318.

[63] Ashburner BP, Westerheide SD, Baldwin AS Jr. The p65 (RelA) subunit of NF-kappaB interacts with the histone deacetylase (HDAC) corepressors HDAC1 and HDAC2 to negatively regulate gene expression. Mol Cell Biol. 2001;21:7065–7077.

[64] Tell G, Quadrifoglio F, Tiribelli C, et al. The many functions of APE1/Ref-1: not only a DNA repair enzyme. Antioxid Redox Signal. 2009;11:601–620.

[65] Ikenoue T, Inoki K, Zhao B, et al. PTEN acetylation modulates its interaction with PDZ domain. Cancer Res. 2008;68:6908–6912.

[66] Vaziri H, Dessain SK, Ng Eaton E, et al. hSIR2 (SIRT1) functions as an NAD-dependent p53 deacetylase. Cell. 2001;107:149–159.

[67] Jenuwein T, Allis CD. Translating the histone code. Science. 2001;293:1074–1080.

[68] Lachner M, Jenuwein T. The many faces of histone lysine methylation. Curr Opin Cell Biol. 2002;14:286–298.

[69] Shi Y, Lan F, Matson C, et al. Histone demethylation mediated by the nuclear amine oxidase homolog LSD1. Cell. 2004;119:941–953.

[70] Tsukada Y, Fang J, Erdjument-Bromage H, et al. Histone demethylation by a family of JmjC domain-containing proteins. Nature. 2006;439:811–816.

[71] Zhang K, Lin W, Latham JA, et al. The Set1 methyltransferase opposes Ipl1 aurora kinase functions in chromosome segregation. Cell. 2005;122:723–734.

[72] Segal E, Fondufe-Mittendorf Y, Chen L, et al. A genomic code for nucleosome positioning. Nature. 2006;442:772–778.

[73] Schones DE, Cui K, Cuddapah S, et al. Dynamic regulation of nucleosome positioning in the human genome. Cell. 2008;132:887–898.

[74] Weissman B, Knudsen KE. Hijacking the chromatin remodeling machinery: impact of SWI/SNF perturbations in cancer. Cancer Res. 2009;69:8223–8230.

[75] Langst G, Becker PB. Nucleosome remodeling: one mechanism, many phenomena? Biochim Biophys Acta. 2004; 1677:58–63.

[76] Ghildiyal M, Zamore PD. Small silencing RNAs: an expanding universe. Nat Rev Genet. 2009;10:94–108.

[77] Davalos V, Esteller M. MicroRNAs and cancer epigenetics: a macrorevolution. Curr Opin Oncol. 2010;22:35–45.

[78] Lim LP, Lau NC, Garrett-Engele P, et al. Microarray analysis shows that some microRNAs downregulate large numbers of target mRNAs. Nature. 2005;433:769–773.

[79] Schickel R, Boyerinas B, Park SM, et al. MicroRNAs: key players in the immune system, differentiation, tumorigenesis and cell death. Oncogene. 2008;27:5959–5974.

[80] Chen T, Li E. Establishment and maintenance of DNA methylation patterns in mammals. Curr Top Microbiol Immunol. 2006;301:179–201.

[81] Bestor TH. The DNA methyltransferases of mammals. Hum Mol Genet. 2000;9(16):2395–2402.

[82] Lukasik SM, Cierpicki T, Borloz M, et al. High resolution structure of the HDGF PWWP domain: a potential DNA binding domain. Protein Sci. 2006;15(2):314–323.

[83] Borgel J, Guibert S, Li Y, et al. Targets and dynamics of promoter DNA methylation during early mouse development. Nat Genet. 2010;42(12):1093–1100.

[84] Kato Y, Kaneda M, Hata K, et al. Role of the Dnmt3 family in de novo methylation of imprinted and repetitive sequences during male germ cell development in the mouse. Hum Mol Genet. 2007;16(19):2272–2280.

[85] Kaneda M, Okano M, Hata K, et al. Essential role for de novo DNA methyltransferase Dnmt3a in paternal and maternal imprinting. Nature. 2004;429(6994):900–903.

[86] Smallwood SA, Tomizawa SI, Krueger F, et al. Dynamic CpG island methylation landscape in oocytes and preimplantation embryos. Nat Genet. 2011;43(8):811–814.

[87] Song J, Rechkoblit O, Bestor TH, et al. Structure of DNMT1-DNA complex reveals a role for autoinhibition in maintenance DNA methylation. Science. 2011; 331(6020):1036–1040.

[88] Song J, Teplova M, Ishibe-Murakami S, et al. Structure-based mechanistic insights into DNMT1-mediated maintenance DNA methylation. Science. 2012;335(6069):709–712.

[89] Bird A. DNA methylation patterns and epigenetic memory. Genes Dev. 2002;16(1):6–21.

[90] Richards EJ, Elgin SCR. Epigenetic codes for heterochromatin formation and silencing: rounding up the usual suspects. Cell. 2002;108(4):489–500.

[91] Bartee L, Malagnac F, Bender J. Arabidopsis cmt3 chromomethylase mutations block non-CG methylation and silencing of an endogenous gene. Genes Dev. 2001;15(14): 1753–1758.

[92] Papa CM, Springer NM, Muszynski MG, et al. Maize chromomethylase Zea methyltransferase2 is required for CpNpG methylation. Plant Cell. 2001;13(8):1919–1928.

[93] Mutskov V, Felsenfeld G. Silencing of transgene transcription precedes methylation of promoter DNA and histone H3 lysine 9. EMBO J. 2004;23(1):138–149.

[94] Mutskov VJ, Farrell CM, Wade PA, et al. The barrier function of an insulator couples high histone acetylation levels with specific protection of promoter DNA from methylation. Genes Dev. 2002;16(12):1540–1554.

[95] Ehrlich M, Buchanan KL, Tsien F, et al. DNA methyltransferase 3B mutations linked to the ICF syndrome cause dysregulation of lymphogenesis genes. Hum Mol Genet. 2001;10(25):2917–2931.

[96] Delaval K, Wagschal A, Feil R. Epigenetic deregulation of imprinting in congenital diseases of aberrant growth. BioEssays. 2006;28(5):453–459.

[97] Moss TJ, Wallrath LL. Connections between epigenetic gene silencing and human disease. Mutat Res. 2007;618(1–2): 163–174.

[98] Robertson KD. DNA methylation and human disease. Nat Rev Genet. 2005;6(8):597–610.

[99] Yoder JA, Walsh CP, Bestor TH. Cytosinemethylation and the ecology of intragenomic parasites. Trends Genet. 1997;13(8):335–340.

[100] Antequera F, Bird A. Number of CpG islands and genes in human and mouse. PNAS. 1993;90(24):11995–11999.

[101] Lopez-Serra L, Esteller M. Proteins that bind methylated DNA and human cancer: reading the wrong words. Br J Cancer. 2008;98(12):1881–1885.

[102] Schulz WA, Steinhoff C, Florl AR. Methylation of endogenous human retroelements in health and disease. Curr Top Microbiol Immunol. 2006;310:211–250.

[103] Sims RJ III, Belotserkovskaya R, Reinberg D. Elongation by RNA polymerase II: the short and long of it. Genes Dev. 2004;18(20):2437–2468.

[104] Sims RJ III, Reinberg D. Histone H3 Lys 4 methylation: caught in a bind? Genes Dev. 2006;20(20):2779–2786.

[105] Laribee RN, Krogan NJ, Xiao T, et al. BURkinase selectively regulatesH3K4 trimethylation and H2B ubiquitylation through recruitment of the PAF elongation complex. Curr Biol. 2005;15(16):1487–1493.

[106] Piyathilake CJ, Henao O, Frost AR, et al. Race- and age-dependent alterations in global methylation of DNA in squamous cell carcinoma of the lung (United States). Cancer Causes Control. 2003;14(1):37–42.

[107] Edwards TM, Myers JP. Environmental exposures and gene regulation in disease etiology. Environ Health Perspect. 2007;115(9):1264–1270.

[108] Akiyama Y, Maesawa C, Ogasawara S, et al. Cell-type-specific repression of the maspin gene is disrupted frequently by demethylation at the promoter region in gastric intestinal metaplasia and cancer cells. Am J Pathol. 2003;163(5):1911–1919.

[109] Feinberg AP, Tycko B. The history of cancer epigenetics. Nat Rev Cancer. 2004;4(2):143–153.

[110] Badal V, Chuang LSH, Tan EHH, et al. CpG methylation of human papillomavirus type 16 DNA in cervical cancer cell lines and in clinical specimens: genomic hypomethylation correlates with carcinogenic progression. J Virol. 2003;77(11):6227–6234.

[111] Fraga MF, Herranz M, Espada J, et al. A mouse skin multistage carcinogenesis model reflects the aberrant DNA methylation patterns of human tumors. Cancer Res. 2004;64(16):5527–5534.

[112] Esteller M. Molecular origins of cancer: epigenetics in cancer. N Engl J Med. 2008;358(11):1148–1159.

[113] Calin GA, Croce CM. MicroRNA signatures in human cancers. Nat Rev Cancer. 2006;6(11):857–866.

[114] Chen CZ. MicroRNAs as oncogenes and tumor suppressors. N Engl J Med. 2005;353(17):1768–1771.

[115] Zhang B, Pan X, Cobb GP, et al. microRNAs as oncogenes and tumor suppressors. Dev Biol. 2007;302(1):1–12.

[116] Vogelstein B, Kinzler KW. Cancer genes and the pathways they control. Nat Med. 2004;10:789–799.

[117] Herman JG, Baylin SB. Gene silencing in cancer in association with promoter hypermethylation. N Engl J Med. 2003;349:2042–2054.

[118] Myohanen SK, Baylin SB, Herman JG. Hypermethylation can selectively silence individual p16ink4A alleles in neoplasia. Cancer Res. 1998;58:591–593.

[119] Zardo G, Tiirikainen MI, Hong C, et al. Integrated genomic and epigenomic analyses pinpoint biallelic gene inactivation in tumors. Nat Genet. 2002;32:453–458.

[120] Schuebel KE, Chen W, Cope L, et al. Comparing the DNA hypermethylome with gene mutations in human colorectal cancer. PLoS Genet. 2007;3:1709–1723.

[121] Suzuki H, Watkins DN, Jair KW, et al. Epigenetic inactivation of SFRP genes allows constitutive WNT signaling in colorectal cancer. Nat Genet. 2004;36:417–422.

[122] Issa JP. Cancer prevention: epigenetics steps up to the plate. Cancer Prev Res (Phila Pa). 2008;1:219–222.

[123] Ushijima T. Detection and interpretation of altered methylation patterns in cancer cells. Nat Rev Cancer. 2005;5:223–231.

[124] Shen L, Kondo Y, Rosner GL, et al. MGMT promoter methylation and field defect in sporadic colorectal cancer. J Natl Cancer Inst. 2005;97:1330–1338.

[125] Aggerholm A, Guldberg P, Hokland M, et al. Extensive intra- and interindividual heterogeneity of p15INK4B methylation in acute myeloid leukemia. Cancer Res. 1999;59:436–441.

[126] Toyota M, Ahuja N, Ohe-Toyota M, et al. CpG island methylator phenotype in colorectal cancer. Proc Natl Acad Sci USA. 1999;96:8681–8686.

[127] Shen L, Toyota M, Kondo Y, et al. Integrated genetic and epigenetic analysis identifies three different subclasses of colon cancer. Proc Natl Acad Sci USA. 2007;104:18654–18659.

[128] Lapeyre JN, Becker FF. 5-Methylcytosine content of nuclear DNA during chemical hepatocarcinogenesis and in carcinomas which result. Biochem Biophys Res Commun. 1979; 87:698–705.

[129] Feinberg AP, Gehrke CW, Kuo KC, et al. Reduced genomic 5-methylcytosine content in human colonic neoplasia. Cancer Res. 1988;48:1159–1161.

[130] Estecio MR, Gharibyan V, Shen L, et al. LINE-1 hypomethylation in cancer is highly variable and inversely correlated with microsatellite instability. PLoS One. 2007;2:e399.

[131] Dunn BK. Hypomethylation: one side of a larger picture. Ann NY Acad Sci. 2003;983:28–42.

[132] Chen RZ, Pettersson U, Beard C, et al. DNA hypomethylation leads to elevated mutation rates. Nature. 1998;395: 89–93.

[133] Issa JP. Colon cancer: it's CIN or CIMP. Clin Cancer Res. 2008;14:5939–5940.

[134] Ogino S, Nosho K, Kirkner GJ, et al. A cohort study of tumoral LINE-1 hypomethylation and prognosis in colon cancer. J Natl Cancer Inst. 2008;100:1734–1738.

[135] Ehrlich M. DNA methylation in cancer: too much, but also too little. Oncogene. 2002;21:5400–5413.

[136] Trinh BN, Long TI, Nickel AE, et al. DNA methyltransferase deficiency modifies cancer susceptibility in mice lacking DNA mismatch repair. Mol Cell Biol. 2002;22:2906–2917.

[137] Eden A, Gaudet F, Waghmare A, et al. Chromosomal instability and tumors promoted by DNA hypomethylation. Science. 2003;300:455.

[138] Fraga MF, Ballestar E, Villar-Garea A, et al. Loss of acetylation at Lys16 and trimethylation at Lys20 of histone H4 is a common hallmark of human cancer. Nat Genet. 2005;37(4):391–400.

[139] Gibbons RJ. Histone modifying and chromatin remodeling enzymes in cancer and dysplastic syndromes. Hum Mol Genet. 2005;14(1):R85–R92.

[140] Shi Y. Histone lysine demethylases: emerging roles in development, physiology and disease. Nat Rev Genet. 2007;8(11):829–833.

[141] Klose RJ, Yan Q, Tothova Z, et al. The retinoblastoma binding protein RBP2 is an H3K4 demethylase. Cell. 2007;128(5):889–900.

[142] Esteller M. Cancer epigenomics: DNA methylomes and histone-modification maps. Nat Rev Genet. 2007;8:286–298.

[143] Widschwendter M, Fiegl H, Egle D, et al. Epigenetic stem cell signature in cancer. Nat Genet. 2007;39:157–158.

[144] Bannister AJ, Schneider R, Kouzarides T. Histone methylation: dynamic or static? Cell. 2002;109:801–806.

[145] Cao R, Wang L, Wang H, et al. Role of histone H3 lysine 27 methylation in Polycomb-group silencing. Science. 2002;298:1039–1043.

[146] Kleer CG, Cao Q, Varambally S, et al. EZH2 is a marker of aggressive breast cancer and promotes neoplastic transformation of breast epithelial cells. Proc Natl Acad Sci USA. 2003;100:11606–11611.

[147] Martinez-Garcia E, Licht JD. Deregulation of H3K27 methylation in cancer. Nat Genet. 2010;42:100–101.

[148] Yu J, Rhodes DR, Tomlins SA, et al. A polycomb repression signature in metastatic prostate cancer predicts cancer outcome. Cancer Res. 2007;67:10657–10663.

[149] Kondo Y, Shen L, Cheng AS, et al. Gene silencing in cancer by histone H3 lysine 27 trimethylation independent of promoter DNA methylation. Nat Genet. 2008;40:741–750.

[150] Morin RD, Johnson NA, Severson TM, et al. Somatic mutations altering EZH2 (Tyr641) in follicular and diffuse large B-cell lymphomas of germinal-center origin. Nat Genet. 2010;42:181–185.

[151] van Haaften G, Dalgliesh GL, Davies H, et al. Somatic mutations of the histone H3K27 demethylase gene UTX in human cancer. Nat Genet. 2009;41:521–523.

[152] Keats JJ, Maxwell CA, Taylor BJ, et al. Overexpression of transcripts originating from the MMSET locus characterizes all t(4;14)(p16;q32)-positive multiple myeloma patients. Blood. 2005;105:4060–4069.

[153] Peters AH, O'Carroll D, Scherthan H, et al. Loss of the Suv39h histone methyltransferases impairs mammalian heterochromatin and genome stability. Cell. 2001;107:323–337.

[154] Schoch C, Schnittger S, Klaus M, et al. AML with 11q23/MLL abnormalities as defined by the WHO classification: incidence, partner chromosomes, FAB subtype, age distribution, and prognostic impact in an unselected series of 1897 cytogenetically analyzed AML cases. Blood. 2003; 102:2395–2402.

[155] Armstrong SA, Staunton JE, Silverman LB, et al. MLL translocations specify a distinct gene expression profile that distinguishes a unique leukemia. Nat Genet. 2002;30:41–47.

[156] Carbone R, Botrugno OA, Ronzoni S, et al. Recruitment of the histone methyltransferase SUV39H1 and its role in the oncogenic properties of the leukemia-associated PML-retinoic acid receptor fusion protein. Mol Cell Biol. 2006;26:1288–1296.

[157] Segalla S, Rinaldi L, Kilstrup-Nielsen C, et al. Retinoic acid receptor alpha fusion to PML affects its transcriptional and chromatin-remodeling properties. Mol Cell Biol. 2003;23:8795–8808.

[158] Goodman RH, Smolik S. CBP/p300 in cell growth, transformation, and development. Genes Dev. 2000;14:1553–1577.

[159] Iyer NG, Ozdag H, Caldas C. p300/CBP and cancer. Oncogene. 2004;23:4225–4231.

[160] Deng Q, Li Y, Tedesco D, et al. The ability of E1A to rescue ras-induced premature senescence and confer transformation relies on inactivation of both p300/CBP and Rb family proteins. Cancer Res. 2005;65:8298–8307.

[161] Neff T, Armstrong SA. Chromatin maps, histone modifications and leukemia. Leukemia. 2009;23:1243–1251.

[162] Iorio MV, Piovan C, Croce CM. Interplay between microRNAs and the epigenetic machinery: an intricate network. Biochimica et Biophysica Acta. 2010;1799(10–12):694–701.

[163] Su Z, Xia J, Zhao Z. Functional complementation between transcriptional methylation regulation and posttranscriptional microRNA regulation in the human genome. BMC Genomics. 2011;12(supplement 5):S15.

[164] Kunej T, Godnic I, Ferdin J, et al. Epigenetic regulation of microRNAs in cancer: an integrated review of literature. Mutat Res. 2011;717(1–2):77–84.

[165] Richly H, Aloia L, Di Croce L. Roles of the Polycomb group proteins in stem cells and cancer. Cell Death Dis. 2011;2(9):e204.

[166] Suzuki H, Takatsuka S, Akashi H, et al. Genome-wide profiling of chromatin signatures reveals epigenetic regulation of microRNA genes in colorectal cancer. Cancer Res. 2011;71(17):5646–5658.

[167] Shi H, Wang MX, Caldwell CW. CpG islands: their potential as biomarkers for cancer. Expert Rev Mol Diagn. 2007;7(5):519–531.

[168] Cottrell SE, Laird PW. Sensitive detection of DNA methylation. Ann NY Acad Sci. 2003;983:120–130.

[169] Palmisano WA, Divine KK, Saccomanno G, et al. Predicting lung cancer by detecting aberrant promoter methylation in sputum. Cancer Res. 2000;60(21):5954–5958.

[170] Aggerholm A, Holm MS, Guldberg P, et al. Promoter hypermethylation of p15INK4B, HIC1, CDH1, and ER is frequent in myelodysplastic syndrome and predicts poor prognosis in early-stage patients. Eur J Haematol. 2006;76(1):23–32.

[171] Hegi ME, Diserens AC, Gorlia T, et al. MGMT gene silencing and benefit from temozolomide in glioblastoma. N Engl J Med. 2005;352(10):997–1003.

[172] Raj K, John A, Ho A, et al. CDKN2B methylation status and isolated chromosome 7 abnormalities predict responses to treatment with 5-azacytidine. Leukemia. 2007;21(9):1937–1944.

[173] Stimson L, La Thangue NB. Biomarkers for predicting clinical responses to HDAC inhibitors. Cancer Lett. 2009;280(2):177–183.

[174] Kantarjian HM, O'Brien S, Cortes J, et al. Results of decitabine (5-aza-2′deoxycytidine) therapy in 130 patients with chronic myelogenous leukemia. Cancer. 2003;98(3):522–528.

[175] Cai FF, Kohler C, Zhang B, et al. Epigenetic therapy for breast cancer. Int J Mol. 2011;12(7):4465–4487.

[176] Yoo CB, Jones PA. Epigenetic therapy of cancer: past, present and future. Nat Rev Drug Discov. 2006;5(1):37–50.

[177] Müller CI, Rüter B, Koeffler HP, et al. DNA hypermethylation of myeloid cells, a novel therapeutic target in MDS and AML. Curr Pharm Biotechnol. 2006;7(5):315–321.

[178] Oki Y, Aoki E, Issa J-PJ. Decitabine-Bedside to bench. Crit Rev Oncol Hematol. 2007;61(2):140–152.

[179] Mack GS. Epigenetic cancer therapy makes headway. J Natl Cancer Inst. 2006;98(20):1443–1444.

[180] Bolden JE, Peart MJ, Johnstone RW. Anticancer activities of histone deacetylase inhibitors. Nat Rev Drug Discov. 2006;5(9):769–784.

[181] Denli AM, Tops BB, Plasterk RH, et al. Processing of primary microRNAs by the Microprocessor complex. Nature. 2004;432:231–235.

[182] Lee Y, Ahn C, Han J, et al. The nuclear RNase III Drosha initiates microRNA processing. Nature. 2003;425:415–419.

[183] Calin GA, Dumitru CD, Shimizu M, et al. Frequent deletions and down-regulation of micro-RNA genes miR15 and miR16 at 13q14 in chronic lymphocytic leukemia. Proc Natl Acad Sci USA. 2002;99:15524–15529.

[184] Metzler M, Wilda M, Busch K, et al. High expression of precursor microRNA-155/BIC RNA in children with Burkitt lymphoma. Genes Chromosomes Cancer. 2004;39:167–169.

[185] Cummins JM, He Y, Leary RJ, et al. The colorectal microRNAome. Proc Natl Acad Sci USA. 2006;103:3687–3692.

[186] Ciafre SA, Galardi S, Mangiola A, et al. Extensive modulation of a set of microRNAs in primary glioblastoma. Biochem Biophys Res Commun. 2005;334:1351–1358.

[187] Murakami Y, Yasuda T, Saigo K, et al. Comprehensive analysis of microRNA expression patterns in hepatocellular carcinoma and non-tumorous tissues. Oncogene. 2006;25:2537–2545.

[188] Yanaihara N, Caplen N, Bowman E, et al. Unique microRNA molecular profiles in lung cancer diagnosis and prognosis. Cancer Cell. 2006;9:189–198.

[189] He H, Jazdzewski K, Li W, et al. The role of microRNA genes in papillary thyroid carcinoma. Proc Natl Acad Sci USA. 2005;102:19075–19080.

[190] Roldo C, Missiaglia E, Hagan JP, et al. MicroRNA expression abnormalities in pancreatic endocrine and acinar tumors are associated with distinctive pathologic features and clinical behavior. J Clin Oncol. 2006;24:4677–4684.

[191] Musiyenko A, Bitko V, Barik S. Ectopic expression of miR-126*, an intronic product of the vascular endothelial EGF-like 7 gene, regulates protein translation and invasiveness of prostate cancer LNCaP cells. J Mol Med. 2008;86:313–322.

[192] Gottardo F, Liu CG, Ferracin M, et al. Micro-RNA profiling in kidney and bladder cancers. Urol Oncol. 2007;25:387–392.

[193] Lau NC, Lim LP, Weinstein EG, et al. An abundant class of tiny RNAs with probable regulatory roles in *Caenorhabditis elegans*. Science. 2001;294:858–862.

[194] Lagos-Quintana M, Rauhut R, Lendeckel W, et al. Identification of novel genes coding for small expressed RNAs. Science. 2001;294:853–858.

[195] Brennecke J, Hipfner DR, Stark A, et al. Bantam encodes a developmentally regulated microRNA that controls cell proliferation and regulates the proapoptotic gene hid in Drosophila. Cell. 2003;113:25–36.

[196] Reinhart BJ, Slack FJ, Basson M, et al. The 21-nucleotide let-7 RNA regulates developmental timing in *Caenorhabditis elegans*. Nature. 2000;403:901–906.

[197] Hipfner DR, Weigmann K, Cohen SM. The bantam gene regulates Drosophila growth. Genetics. 2002;161:1527–1537.

[198] Kanellopoulou C, Muljo SA, Kung AL, et al. Dicer-deficient mouse embryonic stem cells are defective in differentiation and centromeric silencing. Genes Dev. 2005;19:489–501.

[199] Calin GA, Croce CM. MicroRNA signatures in human cancers. Nat Rev Cancer. 2006;6:857–866.

[200] Peter ME. Let-7 and miR-200 microRNAs: guardians against pluripotency and cancer progression. Cell Cycle. 2009;8:843–852.

[201] Cimmino A, Calin GA, Fabbri M, et al. miR-15 and miR-16 induce apoptosis by targeting BCL2. Proc Natl Acad Sci USA. 2005;102:13944–13949.

[202] Olive V, Jiang I, He L. mir-17-92, a cluster of miRNAs in the midst of the cancer network. Int J Biochem Cell Biol. 2010;42:1348–1354.

[203] Inomata M, Tagawa H, Guo YM, et al. MicroRNA-17-92 down-regulates expression of distinct targets in different B-cell lymphoma subtypes. Blood. 2009;113:396–402.

[204] Xiao C, Srinivasan L, Calado DP, et al. Lymphoproliferative disease and autoimmunity in mice with increased miR-

17-92 expression in lymphocytes. Nat Immunol. 2008;9: 405–414.

[205] Mu P, Han YC, Betel D, et al. Genetic dissection of the miR-17-92 cluster of microRNAs in Myc-induced B-cell lymphomas. Genes Dev. 2009;23:2806–2811.

[206] Merritt WM, Lin YG, Han LY, et al. Dicer, Drosha, and outcomes in patients with ovarian cancer. N Engl J Med. 2008;359:2641–2650.

[207] Carninci P, Kasukawa T, Katayama S, et al. The transcriptional landscape of the mammalian genome. Science. 2005;309:1559–1563.

[208] Katayama S, Tomaru Y, Kasukawa T, et al. Antisense transcription in the mammalian transcriptome. Science. 2005;309:1564–1566.

[209] Peters L, Meister G. Argonaute proteins: mediators of RNA silencing. Mol Cell. 2007;26:611–623.

[210] Noma K, Sugiyama T, Cam H, et al. RITS acts in cis to promote RNA interference-mediated transcriptional and post-transcriptional silencing. Nat Genet. 2004;36:1174–1180.

[211] Lippman Z, Martienssen R. The role of RNA interference in heterochromatic silencing. Nature. 2004;431:364–370.

[212] Verdel A, Moazed D. RNAi-directed assembly of heterochromatin in fission yeast. FEBS Lett. 2005;579:5872–5878.

[213] Chen ES, Zhang K, Nicolas E, et al. Cell cycle control of centromeric repeat transcription and heterochromatin assembly. Nature. 2008;451:734–737.

[214] Kloc A, Zaratiegui M, Nora E, et al. RNA interference guides histone modification during the S phase of chromosomal replication. Curr Biol. 2008;18:490–495.

[215] El-Shami M, Pontier D, Lahmy S, et al. Reiterated WG/GW motifs form functionally and evolutionarily conserved ARGONAUTE-binding platforms in RNAi-related components. Genes Dev. 2007;21:2539–2544.

[216] Wightman B, Ha I, Ruvkun G. Posttranscriptional regulation of the heterochronic gene lin-14 by lin-4 mediates temporal pattern formation in C. elegans. Cell. 1993;75:855–862.

[217] Lee RC, Feinbaum RL, Ambros V. The C. elegans heterochronic gene lin-4 encodes small RNAs with antisense complementarity to lin-14. Cell. 1993;75:843–854.

[218] Berezikov E, Guryev V, van de Belt J, et al. Phylogenetic shadowing and computational identification of human microRNA genes. Cell. 2005;120:21–24.

[219] Bartel DP. MicroRNAs: genomics, biogenesis, mechanism, and function. Cell. 2004;116:281–297.

[220] Krek A, Grün D, Poy MN, et al. Combinatorial microRNA target predictions. Nat Genet. 2005;37:495–500.

[221] Kim VN, Nam JW. Genomics of microRNA. Trends Genet. 2006;22:165–173.

[222] Lee Y, Jeon K, Lee JT, et al. MicroRNA maturation: stepwise processing and subcellular localization. EMBO J. 2002;21:4663–4670.

[223] Lee Y, Kim M, Han J, et al. MicroRNA genes are transcribed by RNA polymerase II. EMBO J. 2004;23:4051–4060.

[224] Guil S, Cáceres JF. The multifunctional RNA-binding protein hnRNP A1 is required for processing of miR-18a. Nat Struct Mol Biol. 2007;14:591–596.

[225] He L, Hannon GJ. MicroRNAs: small RNAs with a big role in gene regulation. Nat Rev Genet. 2004;5:522–531.

[226] Vasudevan S, Tong Y, Steitz JA. Switching from repression to activation: microRNAs can up-regulate translation. Science. 2007;318:1931–1934.

[227] Ambros V. The functions of animal microRNAs. Nature. 2004;431:350–355.

[228] Cheng LC, Tavazoie M, Doetsch F. Stem cells: from epigenetics to microRNAs. Neuron. 2005;46:363–367.

[229] Lu J, Getz G, Miska EA, et al. MicroRNA expression profiles classify human cancers. Nature. 2005;435:834–838.

[230] Hammond SM. MicroRNAs as oncogenes. Curr Opin Genet Dev. 2006;16:4–9.

[231] He L, Thomson JM, Hemann MT, et al. A microRNA polycistron as a potential human oncogene. Nature. 2005;435:828–833.

[232] Eis PS, Tam W, Sun L, et al. Accumulation of miR-155 and BIC RNA in human B cell lymphomas. Proc Natl Acad Sci USA. 2005;102:3627–3632.

[233] Iorio MV, Ferracin M, Liu CG, et al. MicroRNA gene expression deregulation in human breast cancer. Cancer Res. 2005;65:7065–7070.

[234] Johnson SM, Grosshans H, Shingara J, et al. RAS is regulated by the let-7 microRNA family. Cell. 2005;120:635–647.

[235] Akao Y, Nakagawa Y, Naoe T. Let-7 microRNA functions as a potential growth suppressor in human colon cancer cells. Biol Pharm Bull. 2006;29:903–906.

[236] Lee YS, Dutta A. The tumor suppressor microRNA let-7 represses the HMGA2 oncogene. Genes Dev. 2007;21: 1025–1030.

[237] Mayr C, Hemann MT, Bartel DP. Disrupting the pairing between let-7 and Hmga2 enhances oncogenic transformation. Science. 2007;315:1576–1579.

[238] Kumar MS, Erkeland SJ, Pester RE, et al. Suppression of non-small cell lung tumor development by the let-7 microRNA family. Proc Natl Acad Sci USA. 2008;105:3903–3908.

[239] Calin GA, Dumitru CD, Shimizu M, et al. Frequent deletions and down-regulation of micro-RNA genes miR15 and miR16 at 13q14 in chronic lymphocytic leukemia. Proc Natl Acad Sci USA. 2002;99:15524–15529.

[240] Cimmino A, Calin GA, Fabbri M, et al. miR-15 and miR-16 induce apoptosis by targeting BCL2. Proc Natl Acad Sci USA. 2005;102:13944–13949.

[241] Shi R, Chiang VL. Facile means for quantifying microRNA expression by real-time PCR. Biotechniques. 2005;39:519–525.

[242] Tang F, Hajkova P, Barton SC, et al. MicroRNA expression profiling of single whole embryonic stem cells. Nucleic Acids Res. 2006;34:e9.

[243] Esquela-Kerscher A, Slack FJ. Oncomirs-microRNAs with a role in cancer. Nat Rev Cancer. 2006;6:259–269.

[244] Yanaihara N, Caplen N, Bowman E, et al. Unique microRNA molecular profiles in lung cancer diagnosis and prognosis. Cancer Cell. 2006;9:189–198.

[245] Blenkiron C, Goldstein LD, Thorne NP, et al. MicroRNA expression profiling of human breast cancer identifies new markers of tumor subtype. Genome Biol. 2007;8:R214.

[246] Saito Y, Liang G, Egger G, et al. Specific activation of microRNA-127 with downregulation of the proto-oncogene BCL6 by chromatin modifying drugs in human cancer cells. Cancer Cell. 2006;9:435–443.

[247] Lujambio A, Ropero S, Ballestar E, et al. Genetic unmasking of an epigenetically silenced microRNA in human cancer cells. Cancer Res. 2007;67:1424–1429.

[248] Roman-Gomez J, Agirre X, Jiménez-Velasco A, et al. Epigenetic regulation of microRNAs in acute lymphoblastic leukemia. J Clin Oncol. 2009;27:1316–1322.

[249] Agirre X, Vilas-Zornoza A, Jiménez-Velasco A, et al. Epigenetic silencing of the tumor suppressor microRNA Hsa-miR-124a regulates CDK6 expression and confers a poor prognosis in acute lymphoblastic leukemia. Cancer Res. 2009;69:4443–4453.

[250] Ando T, Yoshida T, Enomoto S, et al. DNA methylation of microRNA genes in gastric mucosae of gastric cancer

patients: it's possible involvement in the formation of epigenetic field defect. Int J Cancer. 2009;124:2367–2374.

[251] Lee KH, Lotterman C, Karikari C, et al. Epigenetic silencing of MicroRNA miR-107 regulates cyclin-dependent kinase 6 expression in pancreatic cancer. Pancreatology. 2009;9:293–301.

[252] Kim YK, Kim VN. Processing of intronic microRNAs. EMBO J. 2007;26:775–783.

[253] Saini HK, Griffiths-Jones S, Enright AJ. Genomic analysis of human microRNA transcripts. Proc Natl Acad Sci USA. 2007;104:17719–17724.

[254] Grady WM, Parkin RK, Mitchell PS, et al. Epigenetic silencing of the intronic microRNA hsa-miR-342 and its host gene EVL in colorectal cancer. Oncogene. 2008;27:3880–3888.

[255] Saito Y, Friedman JM, Chihara Y, et al. Epigenetic therapy upregulates the tumor suppressor microRNA-126 and its host gene EGFL7 in human cancer cells. Biochem Biophys Res Commun. 2009;379:726–731.

[256] Datta J, Kutay H, Nasser MW, et al. Methylation mediated silencing of MicroRNA-1 gene and its role in hepatocellular carcinogenesis. Cancer Res. 2008;68:5049–5058.

[257] Lujambio A, Calin GA, Villanueva A, et al. A microRNA DNA methylation signature for human cancer metastasis. Proc Natl Acad Sci USA. 2008;105:13556–13561.

[258] Russo G, Puca A, Masulli F, et al. Epigenetics, microRNAs and cancer: an update. In: Cancer Epigenetics: Biomolecular Therapeutics for Human Cancer. John Wiley & Sons, Inc; 2011:101–112.

[259] Lehmann U, Hasemeier B, Christgen M, et al. Epigenetic inactivation of microRNA gene hsa-mir-9-1 in human breast cancer. J Pathol. 2008;214:17–24.

[260] Meng F, Wehbe-Janek H, Henson R, et al. Epigenetic regulation of microRNA-370 by interleukin-6 in malignant human cholangiocytes. Oncogene. 2008;27:378–386.

[261] Silber J, Lim DA, Petritsch C, et al. miR-124 and miR-137 inhibit proliferation of glioblastoma multiforme cells and induce differentiation of brain tumor stem cells. BMC Med. 2008;6:14.

[262] Zhang L, Volinia S, Bonome T, et al. Genomic and epigenetic alterations deregulate microRNA expression in human epithelial ovarian cancer. Proc Natl Acad Sci USA. 2008;105:7004–7009.

[263] Toyota M, Suzuki H, Sasaki Y, et al. Epigenetic silencing of microRNA-34b/c and B-cell translocation gene 4 is associated with CpG island methylation in colorectal cancer. Cancer Res. 2008;68:4123–4132.

[264] Herman JG, Baylin SB. Gene silencing in cancer in association with promoter hypermethylation. N Engl J Med. 2003;349:2042–2054.

[265] Kozaki K, Imoto I, Mogi S, et al. Exploration of tumor suppressive microRNAs silenced by DNA hypermethylation in oral cancer. Cancer Res. 2008;68:2094–2105.

[266] Brueckner B, Stresemann C, Kuner R, et al. The human let-7a-3 locus contains an epigenetically regulated microRNA gene with oncogenic function. Cancer Res. 2007;67:1419–1423.

[267] Scott GK, Mattie MD, Berger CE, et al. Rapid alteration of microRNA levels by histone deacetylase inhibition. Cancer Res. 2006;66:1277–1281.

[268] Diederichs S, Haber DA. Sequence variations of microRNAs in human cancer: alterations in predicted secondary structure do not affect processing. Cancer Res. 2006; 66:6097–6104.

[269] Weber B, Stresemann C, Brueckner B, et al. Methylation of human microRNA genes in normal and neoplastic cells. Cell Cycle. 2007;6:1001–1005.

[270] Lander ES, Waterston RH, Sulston J, et al. Initial sequencing and analysis of the human genome. Nature. 2001;409:860–921.

[271] Venter JC, Adams MD, Myers EW, et al. The sequence of the human genome. Science. 2001;291:1304–1351.

[272] Saxonov S, Berg P, Brutlag DL. A genome-wide analysis of CpG dinucleotides in the human genome distinguishes two distinct classes of promoters. Proc Natl Acad Sci USA. 2006;103:1412–1417.

[273] Iorio MV, Visone R, Di Leva G, et al. MicroRNA signatures in human ovarian cancer. Cancer Res. 2007;67:8699–8707.

[274] Lujambio A, Esteller M. CpG island hypermethylation of tumor suppressor microRNAs in human cancer. Cell Cycle. 2007;6:e1–4.

[275] Takamizawa J, Konishi H, Yanagisawa K, et al. Reduced expression of the let-7 microRNAs in human lung cancers in association with shortened postoperative survival. Cancer Res. 2004;64:3753–3756.

[276] Lu L, Katsaros D, Rigault de la Longrais IA, et al. Hypermethylation of let-7a-3 in epithelial ovarian cancer is associated with low insulin-growth factor-II expression and favourable prognosis. Cancer Res. 2007;67:10117–10122.

[277] Corney DC, Flesken-Nitkin A, Godwin AK, et al. MicroRNA-34c are targets of p53 and cooperate in control of cell proliferation and adhesion-independent growth. Cancer Res. 2007;67:8433–8438.

[278] Chen CZ, Li L, Lodish HF, et al. MicroRNAs modulate hematopoietic lineage differentiation. Science. 2004; 30:83–86.

[279] Fazi F, Rosa A, Fatica A, et al. A minicircuitry comprised of microRNA-223 and transcription factors NFI-A and C/EBPalpha regulates human granulopoiesis. Cell. 2005;123:819–831.

[280] Hess JL, Hug BA. Fusion-protein truncation provides new insights into leukemogenesis. Proc Natl Acad Sci USA. 2004;101:16985–16986.

[281] Nimer SD, Moore MA. Effects of the leukaemia-associated AML1-ETO protein on hematopoietic stem and progenitor cells. Oncogene. 2004;23:4249–4254.

[282] Fazi F, Racanicchi S, Zardo G, et al. Epigenetic silencing of the myelopoiesis regulator microRNA-223 by the AML1/ETO oncoprotein. Cancer Cell. 2007;12:457–466.

[283] Lee TI, Jenner RG, Boyer LA, et al. Control of developmental regulators by polycomb in human embryonic stem cells. Cell. 2006;125:301–313.

[284] Fabbri M, Garzon R, Cimmino A, et al. MicroRNA-29 family reverts aberrant methylation in lung cancer by targeting DNA methyltransferases 3A and 3B. Proc Natl Acad Sci USA. 2007;104:15805–15810.

[285] Friedman JM, Liang G, Liu C-C, et al. The putative tumor suppressor microRNA-101 modulates the cancer epigenome by repressing the polycomb group protein EZH2. Cancer Res. 2009;69:2623–2629.

[286] Sparmann A, van Lohuizen M. Polycomb silencers control cell fate, development and cancer. Nat Rev Cancer. 2006;6: 846–856.

[287] Gal-Yam EN, Egger G, Iniguez L, et al. Frequent switching of Polycomb repressive marks and DNA hypermethylation in the PC3 prostate cancer cell line. Proc Natl Acad Sci USA. 2008;105:12979–12984.

[288] Wong CF, Tellam RL. MicroRNA-26a targets the histone methyltransferase Enhancer of Zeste homolog 2 during myogenesis. J Biol Chem. 2008;283:9836–9843.

[289] Noonan EJ, Place RF, Pookot D, Basak S, et al. miR-449a targets HDAC-1 and induces growth arrest in prostate cancer. Oncogene. 2009;28:1714–1724.

[290] Chen JF, Mandel EM, Thomson JM, et al. The role of microRNA-1 and microRNA-133 in skeletal muscle proliferation and differentiation. Nat Genet. 2006;38:228–233.

[291] Tuddenham L, Wheeler G, Ntounia-Fousara S, et al. The cartilage specific microRNA-140 targets histone deacetylase 4 in mouse cells. FEBS Lett. 2006;580:4214–4217.

[292] Garzon R, Liu S, Fabbri M, et al. MicroRNA-29b induces global DNA hypomethylation and tumor suppressor gene reexpression in acute myeloid leukemia by targeting directly DNMT3A and 3B and indirectly DNMT1. Blood. 2009;113:6411–6418.

[293] Duursma AM, Kedde M, Schrier M, et al. miR-148 targets human DNMT3b protein coding region. RNA. 2008;14:872–877.

[294] Benetti R, Gonzalo S, Jaco I, et al. A mammalian microRNA cluster controls DNA methylation and telomere recombination via Rbl2-dependent regulation of DNA methyltransferases. Nat Struct Mol Biol. 2008;15:268–279.

[295] Sinkkonen L, Hugenschmidt T, Berninger P, et al. MicroRNAs control de novo DNA methylation through regulation of transcriptional repressors in mouse embryonic stem cells. Nat Struct Mol Biol. 2008;15:259–267.

[296] Yamashita K, Upadhyay S, Osada M, et al. Pharmacological unmasking of epigenetically silenced tumor suppressor genes in esophageal squamous cell carcinoma. Cancer Cell. 2002;2:485–495.

[297] Suzuki H, Gabrielson E, Chen W, et al. A genomic screen for genes upregulated by demethylation and histone deacetylase inhibition in human colorectal cancer. Nat Genet. 2002;31:141–149.

[298] Drummond DC, Noble CO, Kirpotin DB, et al. Clinical development of histone deacetylase inhibitors as anticancer agents. Annu Rev Pharmacol Toxicol. 2005;45:495–528.

[299] Lewis BP, Burge CB, Bartel DP. Conserved seed pairing, often flanked by adenosines, indicates that thousands of human genes are microRNA targets. Cell. 2005;120:15–20.

[300] Ting AH, Suzuki H, Cope L, et al. A requirement for DICER to maintain full promoter CpG island hypermethylation in human cancer cells. Cancer Res. 2008;68:2570–2575.

[301] Grewal SIS, Elgin SCR. Transcription and RNA interference in the formation of heterochromatin. Nature. 2007;447:399–406.

[302] Morris KV, Chan SW, Jacobsen SE, et al. Small interfering RNA-induced transcriptional gene silencing in human cells. Science. 2004;305:1289–1292.

[303] Ting AH, Schuebel KE, Herman JG, et al. Short double-stranded RNA induces transcriptional gene silencing in human cancer cells in the absence of DNA methylation. Nat Genet. 2005;37:906–910.

[304] Aravin AA, Sachidanandam R, Girard A, et al. Developmentally regulated piRNA clusters implicate MILI in transposon control. Science. 2007;316:744–747.

[305] Carmell MA, Girard A, van de Kant HJ, et al. MIWI2 is essential for spermatogenesis and repression of transposons in the mouse male germline. Dev Cell. 2007;12:503–514.

[306] Kuramochi-Miyagawa S, Watanabe T, Gotoh K, et al. DNA methylation of retrotransposon genes is regulated by Piwi family members MILI and MIWI2 in murine fetal testes. Genes Dev. 2008;22:908–917.

[307] Tam OH, Aravin AA, Stein P, et al. Pseudogene-derived small interfering RNAs regulate gene expression in mouse oocytes. Nature. 2008;453:534–538.

[308] Watanabe T, Totoki Y, Toyoda A, et al. Endogenous siRNAs from naturally formed dsRNAs regulate transcripts in mouse oocytes. Nature. 2008;453:539–543.

[309] Colella S, Shen L, Baggerly KA, et al. Sensitive and quantitative universal Pyrosequencing methylation analysis of CpG sites. Biotechniques. 2003;35:146–150.

[310] Cottrell SE, Laird PW. Sensitive detection of DNA methylation. Ann NY Acad Sci. 2003;983:120–130.

[311] Li M, Chen WD, Papadopoulos N, et al. Sensitive digital quantification of DNA methylation in clinical samples. Nat Biotechnol. 2009;27:858–863.

[312] Cairns P, Esteller M, Herman JG, et al. Molecular detection of prostate cancer in urine by GSTP1 hypermethylation. Clin Cancer Res. 2001;7:2727–2730.

[313] Lee WH, Morton RA, Epstein JI, et al. Cytidine methylation of regulatory sequences near the pi-class glutathione S transferase gene accompanies human prostatic carcinogenesis. Proc Natl Acad Sci USA. 1994;91:11733–11737.

[314] Jarmalaite S, Jankevicius F, Kurgonaite K, et al. Promoter hypermethylation in tumour suppressor genes shows association with stage, grade and invasiveness of bladder cancer. Oncology. 2008;75:145–151.

[315] Catto JW, Azzouzi AR, Rehman I, et al. Promoter hypermethylation is associated with tumor location, stage, and subsequent progression in transitional cell carcinoma. J Clin Oncol. 2005;23:2903–2910.

[316] Hegi ME, Diserens AC, Gorlia T, et al. MGMT gene silencing and benefit from temozolomide in glioblastoma. N Engl J Med. 2005;352:997–1003.

[317] Esteller M, Garcia-Foncillas J, Andion E, et al. Inactivation of the DNA-repair gene MGMT and the clinical response of gliomas to alkylating agents. N Engl J Med. 2000;343:1350–1354.

[318] Issa JP, Shen L, Toyota M. CIMP, at last. Gastroenterology. 2005;129:1121–1124.

[319] Issa JP. CpG island methylator phenotype in cancer. Nat Rev Cancer. 2004;4:988–993.

[320] Shen L, Issa JP. Epigenetics in colorectal cancer. Curr Opin Gastroenterol. 2002;18: 68–73.

[321] Noushmehr H, Weisenberger DJ, Diefes K, et al. Identification of a CpG island methylator phenotype that defines a distinct subgroup of glioma. Cancer Cell. 2010;17:510–522.

[322] Abe M, Ohira M, Kaneda A, et al. CpG island methylator phenotype is a strong determinant of poor prognosis in neuroblastomas. Cancer Res. 2005;65:828–834.

[323] Brock MV, Hooker CM, Ota-Machida E, et al. DNA methylation markers and early recurrence in stage I lung cancer. N Engl J Med. 2008;358:1118–1128.

[324] Belinsky SA, Klinge DM, Dekker JD, et al. Gene promoter methylation in plasma and sputum increases with lung cancer risk. Clin Cancer Res. 2005;11:6505–6511.

[325] Hoque MO, Begum S, Topaloglu O, et al. Quantitation of promoter methylation of multiple genes in urine DNA and bladder cancer detection. J Natl Cancer Inst. 2006;98:996–1004.

[326] Seligson DB, Horvath S, Shi T, et al. Global histone modification patterns predict risk of prostate cancer recurrence. Nature. 2005;435:1262–1266.

[327] Marcucci G, Radmacher MD, Maharry K, et al. MicroRNA expression in cytogenetically normal acute myeloid leukemia. N Engl J Med. 2008;358:1919–1928.

[328] Eilers T, Machtens S, Tezval H, et al. Prospective diagnostic efficiency of biopsy washing DNA GSTP1 island hypermethylation for detection of adenocarcinoma of the prostate. Prostate. 2007;67:757–763.

[329] Shen L, Kantarjian H, Guo Y, et al. DNA methylation predicts survival and response to therapy in patients with myelodysplastic syndromes. J Clin Oncol. 2010;28:605–613.

[330] Belinsky SA, Liechty KC, Gentry FD, et al. Promoter hypermethylation of multiple genes in sputum precedes lung cancer incidence in a high-risk cohort. Cancer Res. 2006;66:3338–3344.

[331] Dobrovic A, Kristensen LS. DNA methylation, epimutations and cancer predisposition. Int J Biochem Cell Biol. 2009;41:34–39.

[332] Lynch HT, Lynch PM, Lanspa SJ, et al. Review of the Lynch syndrome: history, molecular genetics, screening, differential diagnosis, and medicolegal ramifications. Clin Genet. 2009;76:1–18.

[333] Hitchins MP, Ward RL. Constitutional (germline) MLH1 epimutation as an aetiological mechanism for hereditary non-polyposis colorectal cancer. J Med Genet. 2009;46:793–802.

[334] Ligtenberg MJ, Kuiper RP, Chan TL, et al. Heritable somatic methylation and inactivation of MSH2 in families with Lynch syndrome due to deletion of the 3′ exons of TACSTD1. Nat Genet. 2009;41:112–117.

[335] Boumber YA, Kondo Y, Chen X, et al. An Sp1/Sp3 binding polymorphism confers methylation protection. PLoS Genet. 2008;4:e1000162.

[336] Palmisano WA, Divine KK, Saccomanno G, et al. Predicting lung cancer by detecting aberrant promoter methylation in sputum. Cancer Res. 2000;60:5954–5958.

[337] Cui H, Cruz-Correa M, Giardiello FM, et al. Loss of IGF2 imprinting: a potential marker of colorectal cancer risk. Science. 2003;299:1753–1755.

[338] Chen WD, Han ZJ, Skoletsky J, et al. Detection in fecal DNA of colon cancer specific methylation of the nonexpressed vimentin gene. J Natl Cancer Inst. 2005;97:1124–1132.

[339] Rosenfeld N, Aharonov R, Meiri E, et al. MicroRNAs accurately identify cancer tissue origin. Nat Biotechnol. 2008;26:462–469.

[340] Calin GA, Ferracin M, Cimmino A, et al. A MicroRNA signature associated with prognosis and progression in chronic lymphocytic leukemia. N Engl J Med. 2005;353:1793–1801.

[341] Raponi M, Dossey L, Jatkoe T, et al. MicroRNA classifiers for predicting prognosis of squamous cell lung cancer. Cancer Res. 2009;69:5776–5783.

[342] Eis PS, Tam W, Sun L, et al. Accumulation of miR-155 and BIC RNA in human B cell lymphomas. Proc Natl Acad Sci USA. 2005;102:3627–3632.

[343] Kaminskas E, Farrell AT, Wang YC, et al. FDA drug approval summary: azacitidine (5-azacytidine, Vidaza) for injectable suspension. Oncologist. 2005;10:176–182.

[344] Fenaux P, Mufti GJ, Hellstrom-Lindberg E, et al. Efficacy of azacitidine compared with that of conventional care regimens in the treatment of higher-risk myelodysplastic syndromes: a randomised, open-label, phase III study. Lancet Oncol. 2009;10:223–232.

[345] Kantarjian H, Oki Y, Garcia-Manero G, et al. Results of a randomized study of 3 schedules of low-dose decitabine in higher risk myelodysplastic syndrome and chronic myelomonocytic leukemia. Blood. 2007;109:52–57.

[346] Mann BS, Johnson JR, Cohen MH, et al. FDA approval summary: vorinostat for treatment of advanced primary cutaneous T-cell lymphoma. Oncologist. 2007;12:1247–1252.

[347] Piekarz RL, Frye R, Turner M, et al. Phase II multi-institutional trial of the histone deacetylase inhibitor romidepsin as monotherapy for patients with cutaneous T-cell lymphoma. J Clin Oncol. 2009; 27:5410–5417.

[348] Issa JP, Kantarjian HM. Targeting DNA methylation. Clin Cancer Res. 2009;15:3938–3946.

[349] Von Hoff DD, Slavik M, Muggia FM. 5-Azacytidine. A new anticancer drug with effectiveness in acute myelogenous leukemia. Ann Intern Med. 1976;85:237–245.

[350] Issa JP, Garcia-Manero G, Giles FJ, et al. Phase 1 study of low-dose prolonged exposure schedules of the hypomethylating agent 5-aza-2′-deoxycytidine (decitabine) in hematopoietic malignancies. Blood. 2004;103:1635–1640.

[351] Kaminskas E, Farrell A, Abraham S, et al. Approval summary: azacitidine for treatment of myelodysplastic syndrome subtypes. Clin Cancer Res. 2005;11:3604–3608.

[352] Gore SD, Jones C, Kirkpatrick P. Decitabine. Nat Rev Drug Discov. 2006;5:891–892.

[353] Kantarjian HM, O'Brien S, Huang X, et al. Survival advantage with decitabine versus intensive chemotherapy in patients with higher risk myelodysplastic syndrome: comparison with historical experience. Cancer. 2007;109: 1133–1137.

[354] Datta J, Ghoshal K, Denny WA, et al. A new class of quinoline-based DNA hypomethylating agents reactivates tumor suppressor genes by blocking DNA methyltransferase 1 activity and inducing its degradation. Cancer Res. 2009;69:4277–4285.

[355] Xu WS, Parmigiani RB, Marks PA. Histone deacetylase inhibitors: molecular mechanisms of action. Oncogene. 2007;26:5541–5552.

[356] Prince HM, Bishton MJ, Harrison SJ. Clinical studies of histone deacetylase inhibitors. Clin Cancer Res. 2009;15:3958–3969.

[357] Yao X, Hu JF, Daniels M, et al. A methylated oligonucleotide inhibits IGF2 expression and enhances survival in a model of hepatocellular carcinoma. J Clin Invest. 2003;111:265–273.

[358] Kim MS, Blake M, Baek JH, et al. Inhibition of histone deacetylase increases cytotoxicity to anticancer drugs targeting DNA. Cancer Res. 2003;63:7291–7300.

[359] Issa JP, Gharibyan V, Cortes J, et al. Phase II study of low-dose decitabine in patients with chronic myelogenous leukemia resistant to imatinib mesylate. J Clin Oncol. 2005;23:3948–3956.

[360] Blum W, Klisovic RB, Hackanson B, et al. Phase I study of decitabine alone or in combination with valproic acid in acute myeloid leukemia. J Clin Oncol. 2007;25: 3884–3891.

[361] Link PA, Baer MR, James SR, et al. p53-inducible ribonucleotide reductase (p53R2/RRM2B) is a DNA hypomethylation- independent decitabine gene target that correlates with clinical response in myelodysplastic syndrome/acute myelogenous leukemia. Cancer Res. 2008;68:9358–9366.

[362] Blum W, Garzon R, Klisovic RB, et al. Clinical response and miR-29b predictive significance in older AML patients treated with a 10-day schedule of decitabine. Proc Natl Acad Sci USA. 2010;107:7473–7478.

[363] Qin T, Jelinek J, Si J, et al. Mechanisms of resistance to 5-aza-2′-deoxycytidine in human cancer cell lines. Blood. 2009;113:659–667.

[364] Kakihara T, Fukuda T, Tanaka A, et al. Expression of deoxycytidine kinase (dCK) gene in leukemic cells in childhood: decreased expression of dCK gene in relapsed leukemia. Leuk Lymphoma. 1998;31:405–409.

[365] Sebastiani V, Ricci F, Rubio-Viqueira B, et al. Immunohistochemical and genetic evaluation of deoxycytidine kinase in pancreatic cancer: relationship to molecular mechanisms of gemcitabine resistance and survival. Clin Cancer Res. 2006;12:2492–2497.

[366] Schaefer M, Hagemann S, Hanna K, et al. Azacytidine inhibits RNA methylation at DNMT2 target sites in human cancer cell lines. Cancer Res. 2009;69:8127–8132. Cancer Epigenetics 392 CA.

[367] Madkour LH. Nucleic Acids as Gene Anticancer Drug Delivery Therapy. Elsevier. 2020. https://www.elsevier.com/books/nucleic-acids-as-gene-anticancer-drug-delivery-therapy/madkour/978-0-12-819777-6

# 2 Circulating MiRNAs in Human Cancer
## *Cancer Biomarkers and Epigenetic Modifications*

Lung cancer stands among the leading causes of cancer-related death in the world. Although the molecular network implicated in lung cancer development has been extensively revealed, the mortality rate has improved only slightly. MiRNAs are small, endogenous, single-stranded, evolutionarily conserved, noncoding RNAs that contribute to a wide variety of biological processes including cell growth, proliferation, metabolism, and differentiation.

## 2.1 BLOOD CELL ORIGIN OF CIRCULATING MiRNAs

Since the initial description of circulating miRNAs in 2008 [1–4], more than 200 papers have reported circulating miRNAs as biomarkers for detection of a range of cancer types and other diseases [5, 6]. At least 79 miRNAs have been reported as plasma or serum miRNA biomarkers of solid tumors (i.e., nonhematopoietic malignancies), including prostate, lung, breast, colon, ovarian, esophageal, melanoma, and gastric cancer. However, little attention has been given to the cellular origin of circulating miRNAs and what impact this has on biomarker specificity. We hypothesized that blood cells may contribute significantly to circulating miRNA, and that this could have important implications for interpretation of results from circulating miRNA cancer biomarker studies. In this study we show that a majority of solid tumor–associated circulating miRNA biomarkers reported to date are highly expressed in blood cells, and that plasma levels of these biomarkers are correlated to blood cell counts. We discuss the implications of these findings on the interpretation of circulating miRNA tumor biomarker results reported to date.

Of 79 solid tumors circulating miRNA biomarkers reported in the literature, it has been found that 58% (47/79) are highly expressed in one or more blood cell type. Plasma levels of miRNA biomarkers expressed by myeloid (e.g., miR-223, miR-197, miR-574-3p, let-7a) and lymphoid (e.g., miR-150) blood cells tightly correlated with corresponding white blood cell (WBC) counts. Plasma miRNA biomarkers expressed by red blood cells (RBCs) (e.g., miR-486-5p, miR-451, miR-92a, miR-16) could not be correlated to RBC counts due to limited variation in hematocrit in the cohort studied, but were significantly increased in hemolyzed specimens (20- to 30-fold plasma increase; $p<0.0000001$).

Finally, in a patient undergoing autologous hematopoietic cell transplantation, plasma levels of myeloid- and lymphoid-expressed miRNAs (miR-223 and miR-150, respectively) tracked closely with changes in corresponding blood counts. We present evidence that blood cells are a major contributor to circulating miRNA, and that perturbations in blood cell counts and hemolysis can alter plasma miRNA biomarker levels by up to 50-fold.

In the future, better knowledge of the reciprocal connection between miRNAs and the epigenome will help to develop novel miRNA-orientated diagnostic, prognostic, and therapeutic strategies related to human lung cancer.

## 2.2 OVERVIEW OF LUNG CANCER

Lung cancer is one of the important leading causes of cancer-related death around the world. Histologically, lung cancer is divided into two main groups: non–small cell lung cancer (NSCLC; about 85%) and small cell lung cancer (SCLC; about 15%). NSCLC, the most common epithelial cancer, can be further subclassified into three more subtypes, including (1) adenocarcinoma (AD), (2) squamous cell carcinoma (SCC), and (3) large cell carcinoma (LCC) [7]. Despite years of research, the overall prognosis of lung cancer remains unclear. Currently, routine traditional diagnostics include chest imaging and sputum cytology, which can discover only 15%–20% of lung cancer prior to spread of the disease; therefore, most lung cancer patients are diagnosed at a late stage, so they do not have a notable chance of survival [7, 8]. Since the last century, the high mortality rate of patients has not been considerably improved. The two major reasons are as follows: first, the lack of effective methods to diagnose at an early stage, and second, the ineffectiveness of therapies for the advanced stage of the disease. Hence, a wide understanding of the molecular network of lung carcinogenesis is urgently needed [8].

Genetic alterations could occur at three genomic levels: at the chromosomal level, at the nucleotide level, and at the epigenetic level. These changes could result in the regulation of key genes. So, there are different mutations and expression profiles, which are not only associated with biological processes but also altered cancer development [8].

Epigenetic regulations refer to a series of biological processes that control gene expression, including DNA methylation and histone modification, which are closely associated [9]. Epigenetic patterns connect genetic information to visible phenotypes [9]. These mechanisms not only regulate the protein-coding genes but also affect non-protein-coding genes. Therefore, frequently aberrant methylation of CpG islands (CGIs) on the promoters of oncogenes and/or tumor suppressor genes may lead to lung cancer development [10].

Non-protein-coding RNAs (ncRNAs) have a critical role in all biological processes. MiRNAs are small noncoding

DOI: 10.1201/9781003229650-2

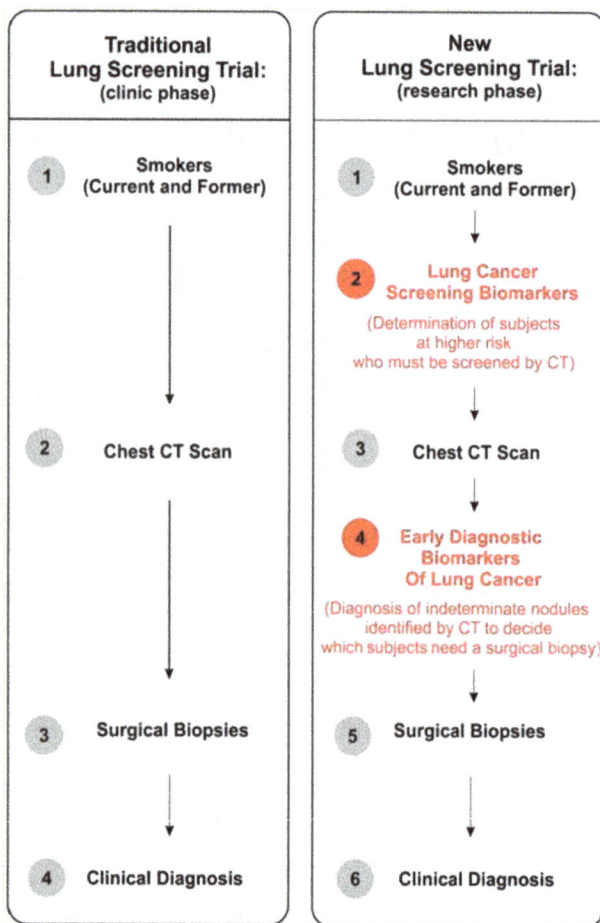

**FIGURE 2.1** An overview of lung cancer diagnosis traditional steps according to the National Lung Screening Trial and new molecular-based method [13].

RNAs that are generated from hairpin transcripts [11]. The expression of a single miRNA or a few miRNAs in the same cluster is altered in tumor cells by genetic or epigenetic mechanisms. So, the miRNAs have the ability to modulate the epigenetic patterns of multiple different keys [12]. Called *epi-miRNAs*, these miRNAs target epigenetic machinery effectors. Therefore, a reciprocal relationship exists between miRNA and epigenetic modifications in carcinogenesis [11].

The effective role of miRNAs in tumorigenesis has prompted the search for and evaluation of new strategies to make use of these molecules in cancer diagnosis (Figure 2.1).

In this chapter, we will summarize the recent studies about miRNAs and also the recent findings about miRNAs and epigenetic machinery.

## 2.3 MiRNAs

RNA, as a genetic material, synthesizes all groups of proteins through the translation process. RNAs have various types of coding and noncoding molecules such as miRNAs in animals [14]. Noncoding RNAs are produced by RNA polymerase enzyme. Moreover, more than 90% of the human genome is transcribed with no protein-coding capacity. NcRNAs are involved in different biological processes, including protein synthesis, gene regulation, and inactivation of X chromosome [15]. MiRNAs are small endogenous RNAs, 19–25 nucleotides (nt) in length, which control messenger RNA (mRNA) translation into proteins. Additionally, miRNAs not only regulate the translation of approximately 30%–60% of protein-coding genes but also cooperate in the regulation of all biological processes [16].

### 2.3.1 MiRNAs: Biogenesis

MiRNAs are transcribed by RNA polymerase II (RNA Pol II), generating long primary transcripts (pri-miRNA) with poly-(A)-tailed and -capped species. Then, pri-miRNAs are processed by Drosha (an RNAse type-III protein) complexes in cooperation with the DiGeorge syndrome critical region 8 (DGCR8), giving rise to long (70–100 nt) pre-miRNAs that are transferred from the nucleus to the cytoplasm via the nuclear export protein Exportin-5 (XPO5) [17]. Then, pre-miRNAs are processed by Dicer (a ribonuclease-III protein) in collaboration with transactivator RNA binding protein (TRBP) to generate a functional double-stranded (ds) RNA approximately 22 nt long [18]. This dsRNA includes the mature miRNA (miRNA-5p) and the complementary strand, which is termed *miRNA\** (miRNA-3p). Then, miRNA\* is normally degraded and the mature miRNA can regulate the gene expression by base-pairing to partially complementary specific sequences at the 3′-untranslated regions (3′-UTRs) of the target mRNAs [19]. Cis-regulatory RNA elements located at the 3′-UTR position control gene regulation [14]. Mature miRNAs recruit Argonaute protein subfamily (Ago), the RNA-induced silencing complex (RISC) catalytic components, as well as the other additional factors such as the TRBPs, which form a complex known as RISC. The responsibility of this complex is the regulation of the target mRNA and inhibition of translation [20]. The Ago protein family contains four types. In mammals, Ago2 creates endonucleolytic cleavage and catalyzes degradation of target mRNAs [14].

Over the past decade, researchers have paid considerable attention to miRNAs as a new generation of biomarkers and therapeutic targets [21] mainly due to their unique features such as petty size, supernatural gene repression role, relative stability [11], and degree of sequence heterogeneity [14].

Up to now, about 1000 miRNAs have been recognized. In 2008, the Sanger Center contained entries of 8619 pre-miRNAs and 8273 mature miRNAs in plants, primates, rodents, birds, fishes, worms, flies, and viruses (http://microrna.sanger.ac.uk/) [8]. A helpful summary of miRNA databases with structural and functional annotations was collected by Nagendra Kumar Singh [14].

### 2.3.2  MiRNAs: The Regulatory Role

In 1993, the first miRNA was discovered, which was associated with the development of the nematode *Caenorhabditis elegans* by regulating the lin-14 protein. Actually, it took more years to recognize miRNAs in several other organisms, such as *Homo sapiens*. With more studies, it was found that miRNAs are specific to tissues, are greatly conserved across different species, and have important functions in all biological processes of organisms [22].

In the last decade, the crucial role of miRNAs helped establish their place in biology. MiRNAs regulate target genes through two mechanisms according to the degree of complementarity between target mRNA and miRNA. In the first mechanism, mature miRNA binds to the 3'-UTR sequence of the target mRNA with absolute complementarity, which induces cleavage and degradation of the target molecule by deadenylation and decapping [23]. In this pathway, the miRNAs act as small interfering RNAs (siRNAs), which are a type of ncRNA and produced by Dicer, usually of exogenous origin. Through the RNA interference pathway, siRNAs direct RNA cleavage [24]. Actually, most of the characterized miRNAs were unable to completely interact with their targeted sequence of mRNA, which prevented miRNA entrance into the siRNA pathway [25]. Therefore, in many cases, the mechanisms of the disruption of protein synthesis in ribosome are considered a second pathway [20].

The second mechanism is most common at the posttranscriptional level involving suppression of translation in the initiation phase because of the imperfect sequence complementarity between the miRNA and target mRNA [26]. In cytoplasm, the RISC-bound mRNAs are packed in structures called processing bodies (P bodies) and are degraded [27]. Unstable mRNA degradation is dedicated by adenylate uridylate (AU)-rich elements (AREs) in the 3'-UTR of mRNAs [28]. In this mechanism of mRNA turnover, miRNAs and Dicer cooperate. The complex of miRNAs, Dicer enzyme, Ago proteins, and ARE-binding proteins is needed for target mRNA decay [29].

These mechanisms could be important in several cancer pathways such as invasiveness, metastasis, tumor recurrence, and resistance to therapy. Based on these mechanisms, miRNAs in high abundance are able to target their available mRNA target sites and significantly affect mRNA stability in several diseases, and also less abundant miRNAs can synergistically regulate target expression [30].

### 2.3.3  MiRNAs: Targets

Gene silencing is the most thoroughly studied trait of miRNAs, as miRNAs have the ability to modulate gene transcription during the cell cycle [20]. For example, the ability of let-7 and miRNA-369-3 swings between repression and activation, depending on the cell cycle progression. When cells are in the proliferation step, these miRNAs repress the expression of their targets, but in the cell arrest level or in the differentiation step, they trigger activation [31].

Taking into account all presented explanations, some miRNAs bind selectively to the 5'-UTR sequence of mRNA and induce gene expression [32]. MiR-10a is a good example of miRNA-mediated activation of mRNA expression by binding to the 5'-UTR sequence of the target mRNAs [33].

Furthermore, researchers have reported four rather unusual processing mechanisms [34]. First, after Drosha processing, miR-451 enters the RISC complex by loading its precursor, and the activation phase with Dicer is skipped during the replacement of the activity of Dicer by Ago2 protein [35]. Second, Drosha-independent mechanisms include mirtrons. In initial maturation, a mirtron uses a splicing machinery system to bypass Drosha cleavage, for example, miR-320 [36]. Third, another mechanism is conducted by staphylococcal nuclease homology domain containing 1 (SND1), one of the components of let-7-directed RISC regulation in RAS signaling [37]. Finally, there is significant similarity between a motif of Ago2 protein with a domain of an essential translation initiation factor (eIF4E). These two molecules compete together to bind to the domain and promote or repress translation [38].

### 2.3.4  MiRNAs: Transports

According to recent findings, miRNAs may circulate via two pathways: the first, free attachments to selected proteins (Ago) or lipids, and second, packaged within extracellular vesicles (exosomes, apoptotic bodies, and microvesicles) [39]. Under physiological conditions, many cell types produce exosomes, releasing them into the extracellular matrix. The exosomes contain proteins, mRNA, and miRNA, which move from a donor cell to the recipient cell using the "exosomal shuttle RNA mechanism." The aim of this mechanism is to regulate protein production in other cells [40]. MiRNA regulatory pathways are summarized in Figure 2.2.

Further explorations about the mechanisms of miRNA biogenesis will be crucial to our knowledge of the regulatory role of miRNAs, thus aiding in the development of diagnostic methods and targeted therapeutic strategies.

## 2.4  MiRNAs AND CANCER

Cancer involves genetic and epigenetic alterations, which trigger uncontrolled cell growth. In recent decades, scientists have focused on coding genes and their proteins and have classified them into two main groups in lung cancer: (1) oncogenes, and (2) tumor suppressor genes (TSGs; [37, 41–69]). The ncRNAs as well as coding genes can be affected by genetic and epigenetic alterations.

Genome-wide transcriptional analysis revealed that an aberrant miRNA profiling (miRNoma) was useful to classify a variety of human tumors, suggesting a putative diagnostic, prognostic, and therapeutic value of these molecules [22]. MiRNA profiling can be used to discriminate

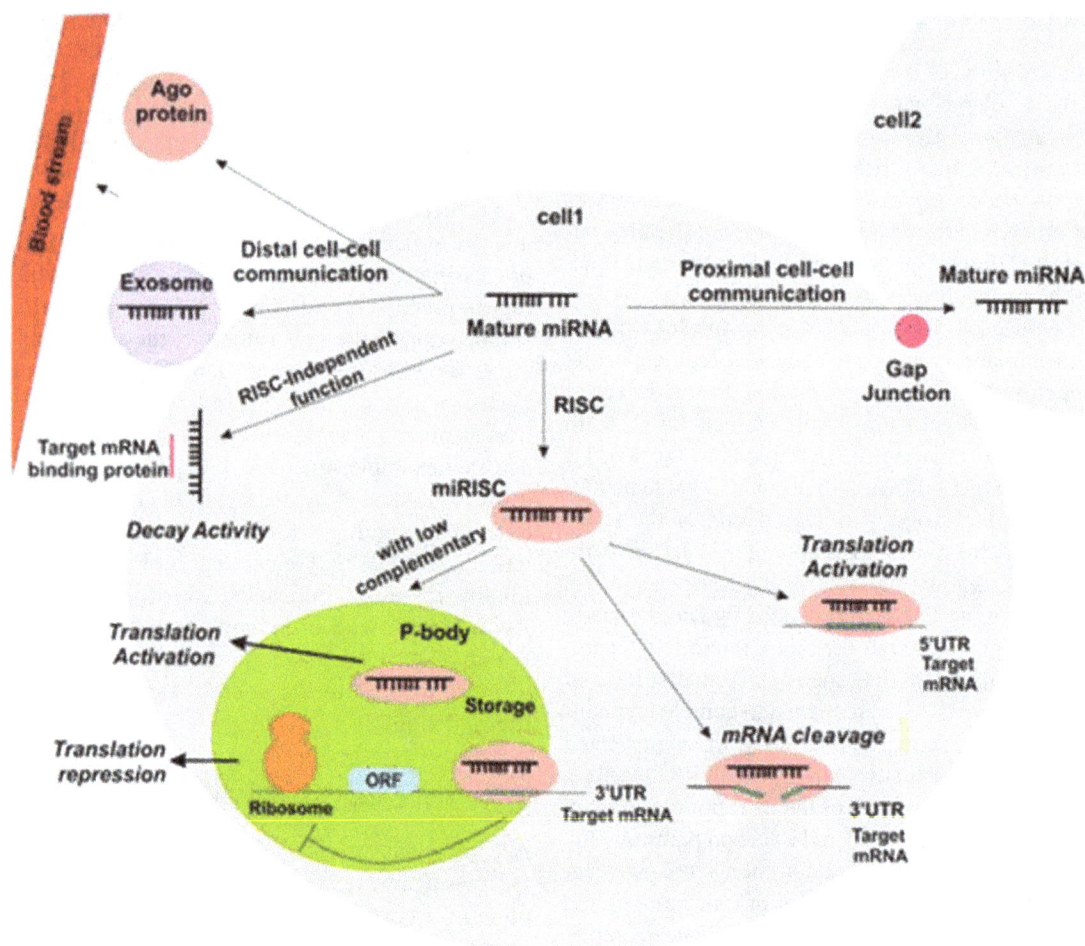

**FIGURE 2.2**    MicroRNA regulatory/transport pathways [20, 34].

different subtypes of a particular cancer and also is applied as a novel biomarker [12]. In 2002, the first documentation of a link between miRNAs and cancer was found in chronic lymphocytic leukemia (CLL) with deletion of chromosome 13q14, which leads to dysregulation of miR-15 and miR-16 encoded by this locus [70]. Human miRNA mapping revealed that many miRNAs were usually located at fragile sites of the chromosomal instability regions that were involved in cancer development [56]. In lung tissue, the miRNA expression pattern varies from the fetal stage to adulthood and from normal to cancerous tissue. MiRNAs as oncogenes and tumor suppressors may have key roles in lung tissue development, and ectopic expression profiles of them could induce and promote lung cancer development [8] (Figure 2.3).

Nowadays, miRNAs can be extracted from tissue, serum, cells, and bodily fluids. The optimum quality of miRNAs is identified by (1) low-throughput experimental technology (northern blotting, expressed sequence tag [EST], serial analysis of gene expression [SAGE], and Southern blotting), and (2) high-throughput techniques (deep sequencing, microarray, and real-time quantitative reverse transcription polymerase chain reaction [RT-qPCR]) [14].

## 2.5    EPIGENETICS

Epigenetics is currently defined as the evaluation of heritable changes in downstream phenotypes. However, epigenetics includes structural modifications in chromosomal regions, which are independent of changes in DNA sequence. Two major epigenetic events are involved: DNA methylation and histone modification, both of which contribute to gene regulation and carcinogenesis (Figure 2.4) [71].

### 2.5.1    DNA METHYLATION

DNA methylation is characterized by the joining of a methyl group to cytosine residues and 5-methylcytosine formation [72]. Often, these modifications happen in the CGIs, which are observed in about 60% of human gene promoters [72]. Methylation of CpG dinucleotides to a high degree leads to downregulation of gene expression. In normal cells, this is an important way to correct gene expression patterns, cell differentiation, and development [13]. Furthermore, this mechanism cooperates with the establishment of imprinting phenomena and inactivation of X chromosomes. In this way, high-degree methylation of repetitive sequences helps to maintain chromosomal integrity [73].

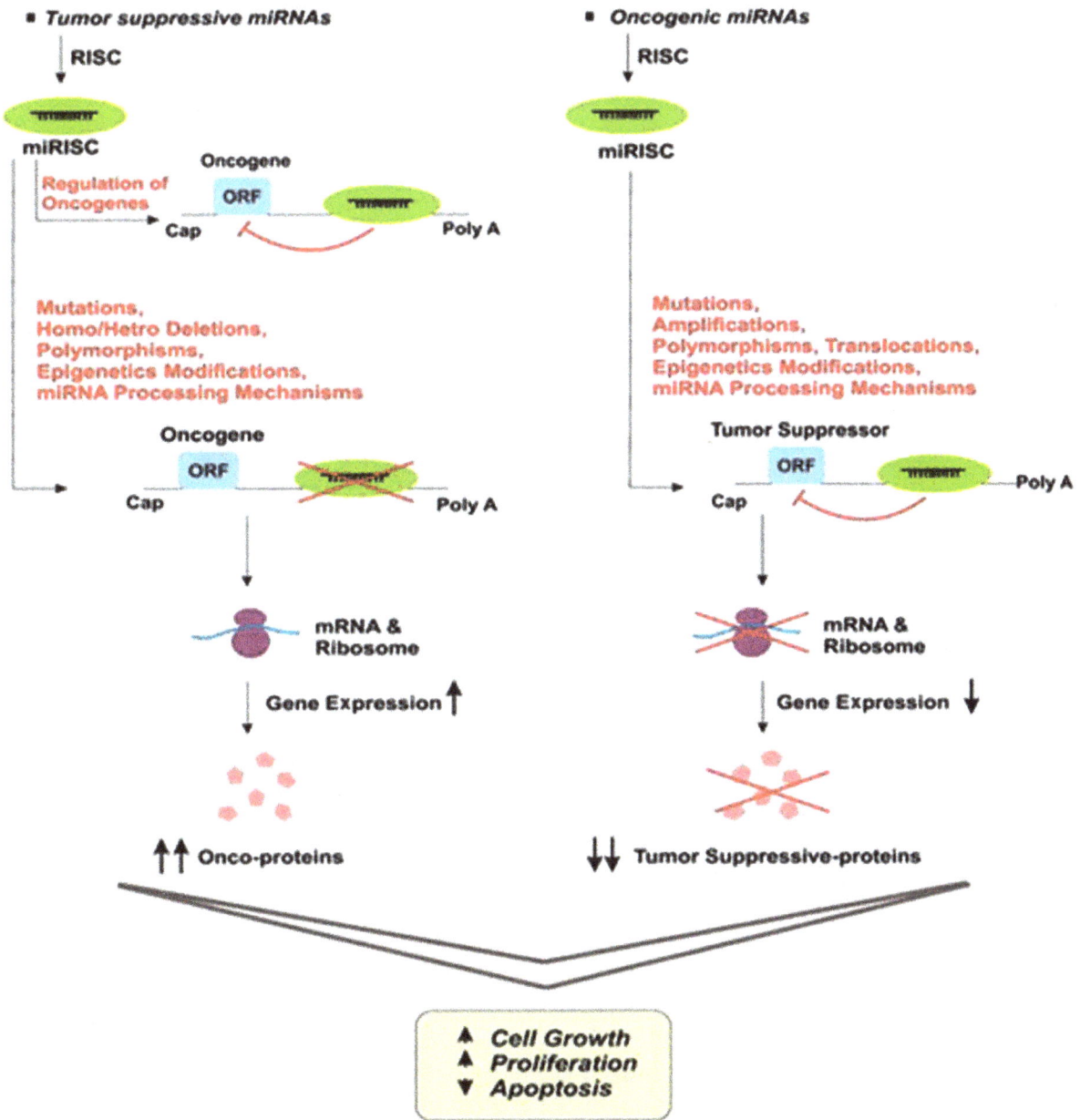

**FIGURE 2.3** MicroRNAs affecting oncogenic and tumor suppressor pathways [34].

There are three enzymes in human DNA methylation: DNMT1, which acts in parental methylation patterns, and DNMT3A and DNMT3B, which act in de novo methylation. Expression patterns of these enzymes are different in certain cancerous cells but not always associated with hypermethylation/hypomethylation status. Chedin [74] partially demonstrated that expression of DNA methyltransferase (DNMT) enzymes may sometimes be regulated by miRNA molecules.

Another mechanism observed during the development process is the loss of DNA methylation that is catalyzed by DNA demethylases [75]. In particular, one key mechanism is carried out by hydroxylases such as ten-eleven translocation 1–three proteins (TET1-3), which convert 5-methylcytosine (5mC) to the other different intermediates in the DNA demethylation process [76]. Hypermethylation is usually correlated with gene silencing, while hypomethylation induces gene activation [75].

### 2.5.2 HISTONE MODIFICATION

Another way that gene expression is epigenetically regulated is through alterations in histone tails, which are dynamic regulators of gene expression [77]. The most common histone posttranslational modifications are acetylation, methylation, ubiquitylation, phosphorylation, and SUMOylation [77]. Histone modifications through two mechanisms regulate gene expression; first, histone protein

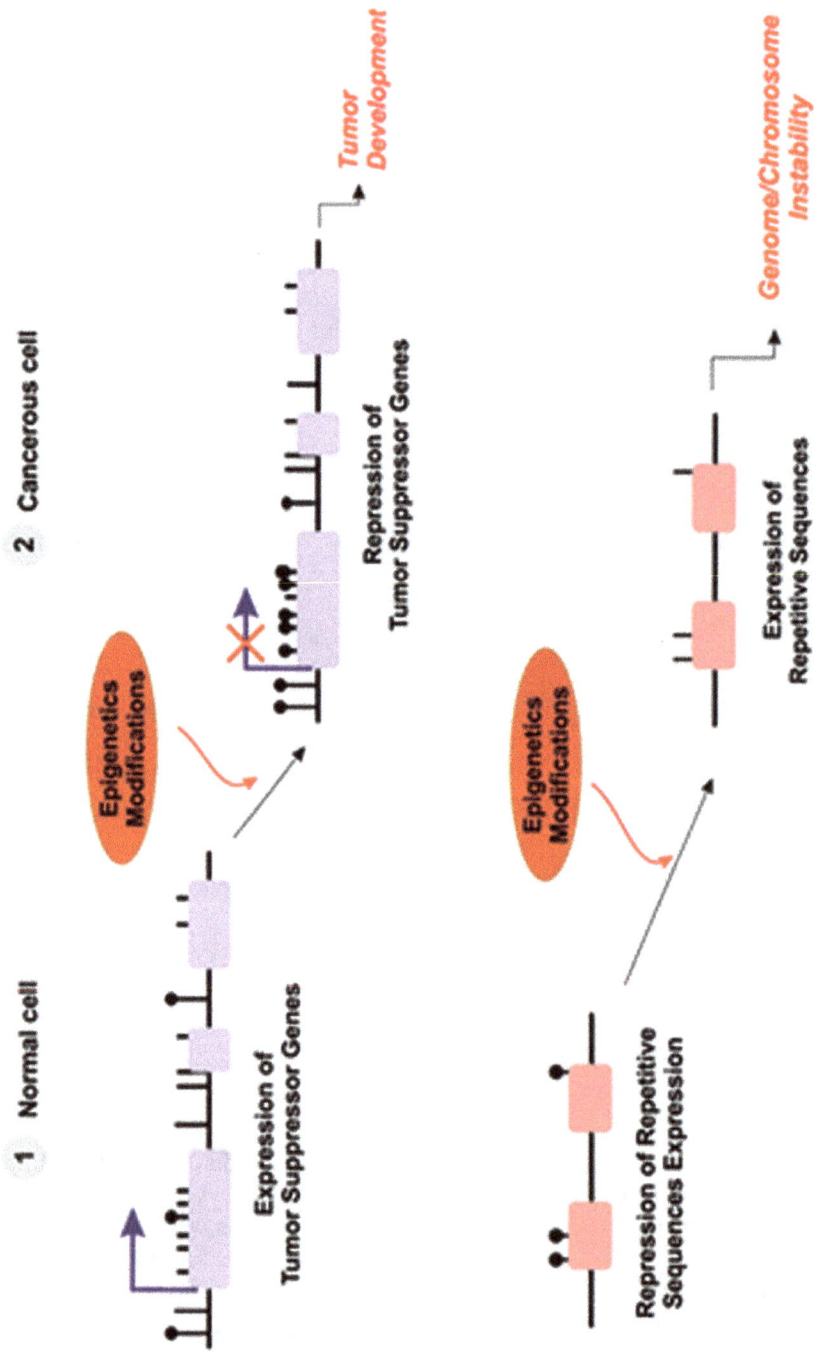

**FIGURE 2.4**   Aberrant epigenetic patterns in cancer [20].

acetylations are correlated with transcriptional activity, and second, histone methylations lead to condensed and inactive chromatin formation [78]. The enzymes that are responsible for adding and removing these groups have been expressed differentially in many cancerous cells. For example, (1) acetyl groups by histone acetyltransferase (HAT) function were added to lysine residues and removed by histone deacetylases (HDACs), and (2) methyl groups by histone methyltransferase (HMT) activity were added to lysine and arginine residues and removed by histone demethylase (HDM) enzyme [75].

Until now, the roles of these enzymes in cancer have not been completely understood. The histone modification hypothesis postulates that the expression regulation of a specific region of DNA depends upon the combination of the histone protein's modifications.

## 2.6 EPIGENETICS AND CANCER

Cancer can arise from a combination of genetic and epigenetic alterations, which lead to dysregulated gene expression and function. So, altered epigenetic patterns, along with genetic abnormalities, can give us a great understanding of the tumorigenesis process [79]. The most well-known epigenetic regulations existing in cancer are aberrant methylations of CGIs, especially in promoter regions, and deacetylation/methylation alterations of histones [79]. Compared with normal cells, cancerous cells are widely hypomethylated, which can mediate the overexpression of oncogenes; most tumor suppressor genes are often silenced because of hypermethylation [75]. Indeed, the epigenetic machinery controls gene expression and cooperates in different cellular/physiological processes; dysregulation of this mechanism in malignancies can be a preliminary transforming event and initiate genetic instability. In the case of mutation in a single allele of a gene, the epigenetic modifications could affect the other allele of the same gene via a "second hit" mechanism, consequently leading to loss of heterozygosity (LOH). Finally, LOH phenomena induce inactivation of the key tumor suppressor genes and activation of oncogenes to promote cell overgrowth and tumor development [22].

The discovery of the cooperation of epigenetic modifications in tumor development has been of considerable importance, and it may usher in a new era of cancer treatment. Furthermore, epigenetic markers could be applied as biomarkers for many different reasons: first, isolation of genomic DNA for evaluation of methylation profiling is easier, and second, DNA molecules are more stable than mRNA for analysis of gene expression [75].

In recent years, the overlapped network of gene expression and epigenetic modification of tumor development has been complicated by the discovery of the key role of miRNAs, which are able to directly regulate target genes. Furthermore, a straight connection exists between the epigenetic mechanisms and gene regulation [22].

## 2.7 EPIGENETIC MODIFICATIONS AND MiRNA EXPRESSION

The number of miRNA studies related to lung cancer has increased rapidly in recent years, as a result of understanding the crucial role of miRNAs in lung cancer development and their capability as diagnostic, prognostic, predictive, and therapeutic markers [8]. In human cancers, CGI promoter hypermethylation is one of the most common causes of the silencing of tumor suppressor miRNAs [80]. Furthermore, cancerous cells could be affected by DNA hypomethylation. The results of epigenetic modifications in tumor cells are genome instability, chromosomal rearrangements, and genomic DNA disruption (Figure 2.5) [20].

### 2.7.1 EPIGENETIC MODIFICATIONS AFFECT MiRNA EXPRESSION

The DNA methylation pattern of tumor cells could affect miRNA expression and can be used as a novel hallmark biomarker to classify tumor type, clinical diagnosis, prognosis, and response to therapy. Also, histone modifications are able to affect miRNA expression profiles associated with silencing [81].

Some methods can detect the miRNA epigenetic status. Most of the studies for this used a pharmacologic unmasking strategy for the identification of epigenetically altered miRNAs based on microarray approach and RT-qPCR. The other studies used the double knockout (DKO) method

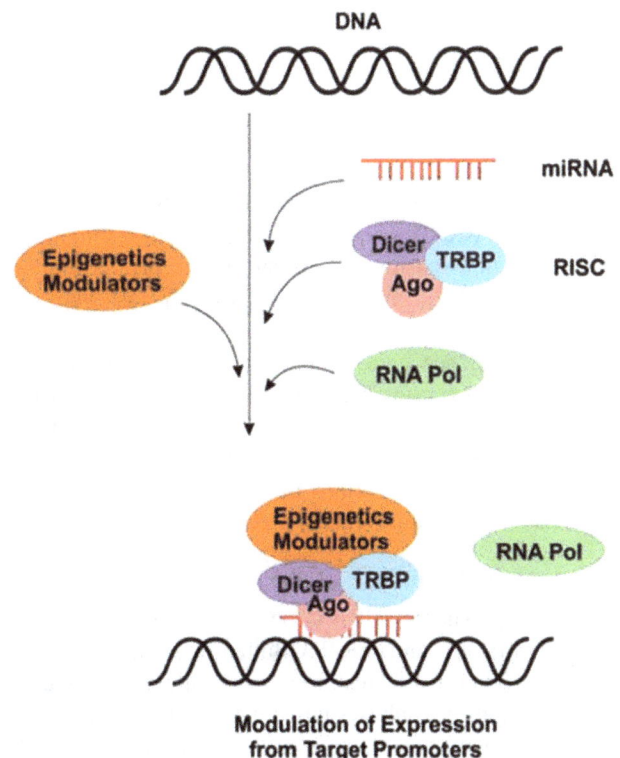

**FIGURE 2.5** Interaction of microRNAs and epigenetic regulators [19].

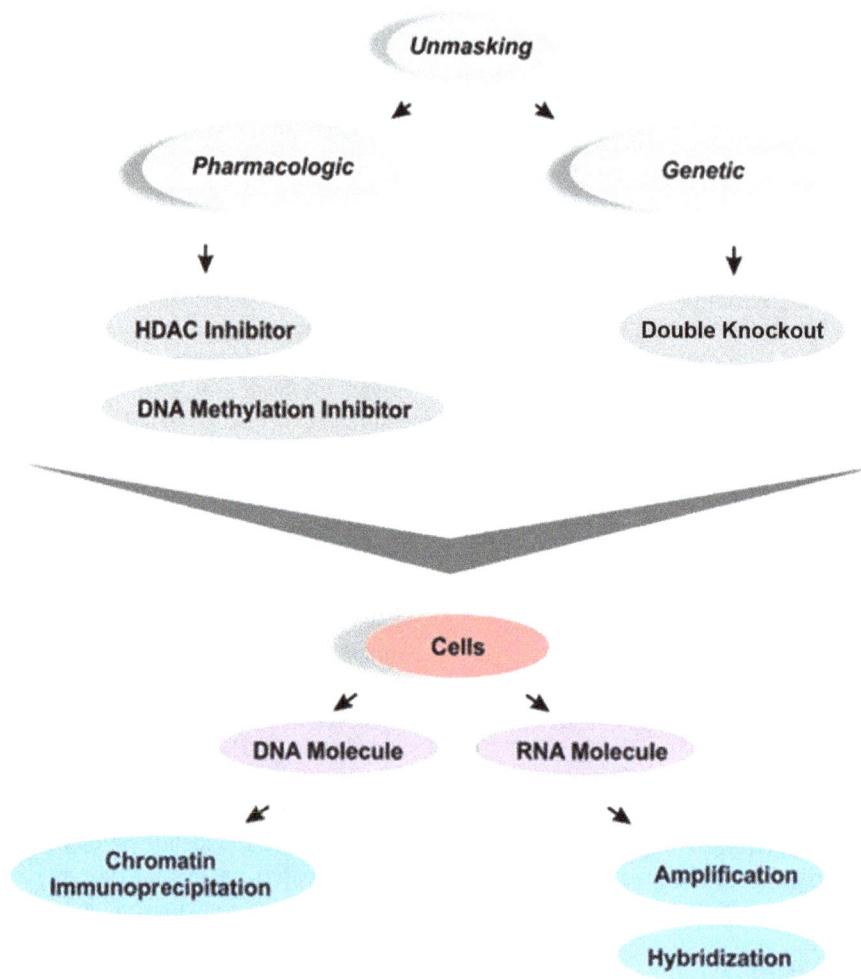

**FIGURE 2.6** Unmasking strategy for detection of miRNA epigenetic regulation [82].

for DNMT1/3B. So, DNA methylation was evaluated by (1) methylation-specific PCR (MSP), (2) bisulfite genomic sequencing PCR (BSP), (3) combined bisulfate restriction analysis (COBRA), (4) bisulfite pyrosequencing, and (5) methylation-sensitive single-nucleotide primer extension (Ms-SNuPE), and (6) MassARRAY assays [82] (Figure 2.6).

Recent evidence suggests that cigarette smoking, as the leading risk factor for lung cancer, affects human bronchial epithelial cells (HBECs) and induces the epigenetic silencing of specific miRNAs, leading to the decrease in the expressions of certain miRNAs and epithelial-to-mesenchymal transition (EMT) [83, 84].

### 2.7.1.1 MiR-9 Family

In the human genome, miR-9 is located at three distinct loci: miR-9-1, miR-9-2, and miR-9-3. Both miR-9-1 and miR-9-3 have CGIs, but miR-9-2 does not [85]. In all normal tissues, the miR-9 family is always unmethylated. Lujambio et al. [80] treated metastatic cancer cell lines with the DNMT inhibitor 5-aza-2′-deoxycytidine and measured the miRNA expression. They reported that the DNA methylation of miR-9-3 CGI was associated with the metastasis of human

lung cancer. Kitano et al. found that the DNA methylation of miR-9-3 was associated with an advanced T factor. They found a similar proportion of methylated miR-9-3; however, they did not observe any correlation between the methylation status of miR-9-3 and lymph metastasis [85]. Heller et al. [86] evaluated the miRNA expression profile in A549 cells, which had been treated with both tricostatin A (TSA) and 5-aza-2′-deoxycytidine. They found that miR-9-3 could be a target for DNA methylation in NSCLC. Wang et al. evaluated the role of miR-9-3 demethylation in NSCLCs and reported that methylation silencing of miR-9-3 may affect epigenetic mechanisms in tumor cell growth, survival, and cancer progression. So, they proposed that demethylation of miR-9-3 cooperates with the anticancer properties of 5-AzaC and can soon apply as a new therapeutic strategy in NSCLC treatment [87]. Tan et al. observed that methylation levels of miR-9 were higher in NSCLC. Furthermore, miR-9-1 and miR-9-3 were found to be correlated with age, implying that miRNA hypermethylation could further affect the risk in aged NSCLC patients. Additionally, their study is the first to propose that miR-9-2 hypermethylation pattern plays an important role in survival of lung cancer patients [88].

## 2.7.1.2 MiR-34 Family

The miR-34 family consists of three miRNAs: miR-34a, miR-34b, and miR-34c, which are located on chromosomes 1 and 11 [89]. In mice, miR-34a, miR-34b, and miR-34c are mainly expressed through the brain and lung [90]. The miR-34 family by the downregulation of multiple targets like Bcl-2, CyclinD1/E2, CDK4/6, and Myc [91] induces cell cycle arrest and apoptosis. Moreover, this family has been demonstrated as targets of the p53 transcription factor [90]. According to these results, the miR-34 family plays a crucial role as a tumor suppressor [92], and most importantly, miR-34b and miR-34c were always unmethylated in all normal tissues [80]. Through the epigenetic alterations in lung cancer, miR-34a, miR-34b, and miR-34c are silenced by DNA methylation [66, 93, 94]. Furthermore, miR-34b and miR-34c are methylated in 41% of primary NSCLC [94] and 67% of primary SCLC patients [95] and are associated with a poorer prognosis [93]. Lujambio et al. [80] treated metastatic cancer cell lines with the DNMT inhibitor 5-aza-2′-deoxycytidine and reported that the DNA methylation of miR-34b/c was correlated with human lung cancer metastasis. Watanabe et al. [94] measured the expression level of 55 in silico selected miRNAs treated with 5-aza-2′-deoxycytidine and proposed the DNA methylation–silenced miR-34b/c in NSCLC. Tanaka et al. transected miR-34b/c to SCLC, which resulted in the significant inhibition of growth and invasion compared with control transfectants. Their results demonstrate that the ectopic methylation of miR-34b/c has an important role in the pathogenesis of SCLC. Actually, in SCLC, miR-34b/c can be a useful therapeutic target [95]. Furthermore, Tan et al. [88] reported that miR-34b and miR-34c are often hypermethylated in cancer tumors compared with normal tissues.

## 2.7.1.3 MiR-148a

It is reported that miR-148a expression is important in tumor development. Studies have demonstrated that miR-148a is usually downregulated in different types of human cancers and has some crucial functions, including suppression of tumor growth, invasion, angiogenesis, and induction of apoptosis [96]. MiR-148a was always unmethylated in all normal cells. Lujambio et al. used the DNMT inhibitor 5-aza-2′-deoxycytidine for treating metastatic lung cancer, and they evaluated the miRNA expression and reported that there was a correlation between the DNA methylation of miR-148a and human lung cancer metastasis. By using computational prediction, they found that the TGIF2 gene was one of the best targets for miR-148a [80].

## 2.7.1.4 MiR-193a

It has been found that miR-193a modulates expression of specific oncogenic factors. Heller et al. [86] treated A549 cells with 5-aza-2′-deoxycytidine and TSA. Then, they measured the miRNA expression profile and identified miR-193a to be a good target for DNA methylation in NSCLC. Wang et al. evaluated the effect of miR-193a demethylation on proliferation and apoptosis in NSCLCs and found that methylation silencing of miR-193a may affect epigenetic mechanisms in NSCLC cell growth, survival, and progression. So, demethylation of miR-193a contributes, at least, to the anticancer properties of 5-AzaC and can be a new therapeutic strategy in NSCLC treatment [87].

## 2.7.1.5 MiR-126

MiR-126 is located within the intron of EGFL7 gene and silenced by the DNA methylation of its host gene in NSCLC [94]. MiR-126 inhibition of EGFL7 is proposed to reduce proliferation of cells in NSCLC [88] by acting as a tumor-suppressive miRNA and through targeting the Crk gene inhibiting the invasion of the NSCLC [97]. In addition, miR-126 downregulation is significantly correlated with a shorter survival of NSCLC patients [98]. In contrast, Donnem et al. [99] reported that high expression of miR-126 is correlated with a shorter survival period and increase in vascular endothelial growth factor A (VEGF-A) expression in NSCLC patients. Watanabe et al. analyzed the expression profiles of 55 in silico candidate miRNAs treated with 5-aza-2′-deoxycytidine. They proposed that miR-126 was silenced by DNA methylation in NSCLC [94]. Tan et al. [88] showed that miR-126 is hypermethylated in tumor tissues compared to normal tissues and would result in loss of EGFL7 inhibition causing increased cellular proliferation in lung cancer.

## 2.7.1.6 MiR-124 Family

In the human genome, miR-124 is located at three distinct loci: miR-124-1/2/3 embedded within CGIs. The methylation of miR-124 is previously reported in lung cancer. Exhibited silencing of miR-124a conducted CDK6 activation and phosphorylation of the retinoblastoma gene. Lujambio et al. [80, 100] found that CDK6 and BCL6 are the important targets for miR-124a, and epigenetic downregulation of miR-124a in cancerous cells leads to CDK6 upregulation, which is involved in cell cycle progression and differentiation. Kitano et al. revealed that miR-124-2 and miR-124-3 showed similar methylation profiles in NSCLC specimen that were distinct from miR-124-1. The methylation statuses of miR-124-2 and miR-124-3 were individually related to recurrence of NSCLC [85]. Tan et al. [88] conducted research on the NSCLC tumor samples and found higher amounts of methylated miR-124a, especially miR-124a-2 and miR-124a-3, in tumors compared with normal lung tissues.

## 2.7.1.7 MiR-152

MiR-152 is located at a single locus of DNA, which is embedded within CGIs. Kitano et al. conducted the first study that focused on miR-152 methylation in lung cancer. They showed that the CGI methylation of miR-152 was common in NSCLC. However, no considerable correlation was observed between the methylation of miR-152 and clinicopathological features. Therefore, miR-152 may not have a dominant role in lung cancer [85].

### 2.7.1.8 MiR-200 Family

The miR-200 family consists of five members: miR-200a, -200b, -200c, -141, and -429, which are expressed in two genomic clusters: chromosome 1p36.33 and chromosome 12p12.31 [101]. The regulatory region of miR-200b and miR-200c consists of CpG-rich sequences. So, miR-200b and miR-200c were implicated in the dedifferentiation of HBECs and also in primary lung tumors [83]. The miR-200 family, by targeting ZEB1 and ZEB2 and consequently conserving E-cadherin junctions, act as negative regulators of the EMT and prevent tumor progression [102]. Tellez et al. [83] found that the exposure of HBECs to tobacco smoking decreased the expression of miR-200b and miR-200c through DNA methylation mechanisms and finally induced EMT. As a result of epigenetic modification of the promoter region, miR-200c expression is decreased, and it has been associated with a poor grade of differentiation and induced an aggressive and invasive phenotype of NSCLC [101].

### 2.7.1.9 Let-7a-3

Let-7a-3 has an unusual oncogenic role unlike most other let-7 miRNAs, which act as tumor suppressors [103]. Let-7a-3 gene, on chromosome 22q13.31, is correlated with CGIs. According to the study of Brueckner et al. [104], let-7a-3 is methylated in normal lung but hypomethylated in AD. Let-7a-3 can be affected by DNMT1 and DNMT3B, and the high expression of let-7a-3 in human lung cancer cell lines is able to induce enhanced tumor phenotypes.

### 2.7.1.10 MiR-127

MiR-127 can affect BCL6 and modulates DNA damage–induced apoptotic responses. Epigenetic silencing of miR-127 induces the activation of BCL6 [105]. Tan et al. [88] reported that the hypermethylation of miR-127 is correlated with an increased risk of death and shorter survival in NSCLC, and they suggested that the epigenetic status of miR-127 could be a biomarker for the prediction of NSCLC patients' outcomes.

### 2.7.1.11 MiR-487b

Xi et al. examined the effects of cigarette smoking on miR-487b, a novel tumor suppressor miRNA, in cultured respiratory epithelial and lung tumor cells of two groups: smokers and nonsmokers, and they reported that cigarette smoke induces downregulation of miR-487b, thereby upregulating its targets like SUZ12, BMI1, WNT5A, MYC, and KRAS. This mechanism promotes proliferation, invasion, tumorigenicity, and metastasis of lung cancer [84]. Subsequent studies confirmed notable downregulation of miR-487b in primary lung cancer, especially in smokers, compared to adjacent normal lung tissues. In cultured cells, repression of miR-487b via cigarette smoke coincided with epigenetic modification and recruitment of polycomb repressor proteins to the regulatory region of miR-487b [106].

### 2.7.1.12 MiR-205

MiR-205, located at the region of chromosome 1, is amplified in lung cancer [8]. Relative quantification of miR-205 can be a promising diagnostic tool and the distinction between AD and SCC [107]. Actually, miR-205 affects the epithelial cells of lung by targeting ZEB1 and ZEB2 directly and E-cadherin indirectly [102]. Furthermore, miR-205 was implicated in HBEC dedifferentiation and also in primary lung tumors. Tellez et al. [83] studied the role of EMT and epigenetic silencing of miR-205, a tumor-suppressive miRNA, which affected the primary lung tumor dedifferentiation after exposure to tobacco smoking.

## 2.7.2 MiRNAs that Target Epigenetic Machinery

MiRNA expression is regulated by epigenetic machinery, and it is now clear that the epigenetic modifications are targeted by miRNAs [89]. MiRNAs have the capability to regulate the expression of important epigenetic regulators, including DNMTs and HDACs, and to create an exact controlled feedback mechanism. These types of miRNA are called *epi-miRNAs*, and their dysregulated expression has been usually related to the development of human cancers [19, 22].

### 2.7.2.1 MiR-29 Family

The first observed epi-miRNA in lung cancer is the miR-29 family [59]. The miR-29 family, including miR-29a (on chromosome 7), miR-29b, and miR-29c (on chromosome 1 and chromosome 7), is upregulated in normal cells and downregulated in human lung cancer [108, 109]. So, miR-29a and miR-29b through the regulation of the Src-ID1 pathway act as antiproliferative and antimetastatic miRNAs in lung cancer [110]. The miR-29 family targets DNMT3A/3B directly [59] and DNMT1 by targeting Sp1 indirectly [111]. The induction of miR-29 expression in lung cancer cell lines has some effects: (1) restoration of the normal pattern of epigenetic modifications, (2) reactivation of methylated silenced tumor suppressors, and (3) repression of the tumor development process [59].

### 2.7.2.2 MiR-148a

MiR-148a is a dysregulated miRNA in lung cancer [88]. Duursma et al. [112] showed that human miR-148 inhibits expression of the DNMT3B gene through a region in its coding sequence. Chen et al. studied the epigenetic regulation of miR-148a in NSCLC. Due to hypermethylation of the miR-148a encoding region, the expression levels of miR-148a were decreased in NSCLC, which is associated with lymph node metastasis, advanced clinical stage, and shortened overall survival in NSCLC. So, enforced expression of miR-148a in lung cancer cells resulted in a considerable decline in the expression of DNMT1 and led to reduction in the DNA methylation of E-cadherin as a tumor suppressor. MiRNA-148a can be a potential therapeutic target to suppress lung cancer metastasis [96].

### 2.7.2.3 MiR-101

MiR-101 is frequently downregulated in lung cancer. Overexpression of miR-101 importantly induces antitumorigenic properties. MiR-101 inactivation causes the overexpression of EZH2, which leads to hypermethylation and aggressive tumorigenesis [113]. Yan et al. demonstrated that miR-101 downregulation correlated with DNMT3A overexpression in lung cancer. Ectopic expression of miR-101 significantly targeted the luciferase activity of DNMT3A and induced downregulation of endogenous DNMT3A, which finally led to the reexpression of CDH1, a tumor suppressor, via its promoter DNA hypomethylation, suppression of cell clonability, and migration of lung cancer cell lines and tissues [114].

These results clearly show a potential interplay between the miRNA expression and the epigenetic machinery and provide new insights into the molecular mechanisms of aberrant epigenetic modifications in lung cancer.

## 2.8 CLINICAL SAMPLES AND PLASMA PREPARATION

Individual miRNAs were detected by qRT-PCR as previously described [115]. TaqMan assays for human miRNAs hsa-miR-451, hsa-miR-16, hsa-miR-92a, hsa-miR-486-5p, hsa-let-7a, hsa-miR-223, hsa-miR-150, hsa-miR-574-3p, hsa-miR-197, and hsa-miR-122 and *C. elegans* miRNAs cel-miR-39, cel-miR-54, and cel-miR-238 were obtained from Applied Biosystems. Oligonucleotides corresponding to the mature sequence of each miRNA were synthesized (Integrated DNA Technologies) and diluted for standard curves. These studies were performed in absolute quantitation for each miRNA both in blood cells [116] and in plasma.

The researchers used qRT-PCR to examine expression of 79 miRNAs reported as circulating solid tumor biomarkers in purified subpopulations of blood cells and in matched healthy donor plasma [117–121]. Of these literature-reported cancer biomarkers, 58% (46/79) are highly expressed in one or more blood cell type. Most of the blood cell–expressed biomarkers (42/46, 91%) were also present at considerable basal levels in healthy donor plasma (>50th percentile among all miRNAs detected), suggesting that blood cells could be a major source for these plasma miRNAs (Figure 2.7). Comparing expression profiles of miRNA detectable in plasma with miRNA profiles of blood cells demonstrated that the distribution of miRNA expression values in plasma mirrors that of blood cells (Figure 2.7).

If circulating miRNA is derived from blood cells, researchers expected levels of these miRNAs to vary as a function of blood cell counts. In order to test this hypothesis, they collected a cohort of 42 plasma samples from a hospital clinical hematology lab, in which a complete differential blood count (CBC with differential) was performed prior to preparing plasma. The cohort consisted of consecutive residual patient plasma samples collected at a single academic medical center, including both inpatients and outpatients with a wide variety of underlying diseases. They selected ten miRNAs to measure in plasma and correlate with blood cell counts [116]. Eight of these were selected by virtue of being published circulating miRNA cancer biomarkers, which researchers found in their blood cell miRNA profiling studies to be expressed in blood cells in a cell-type-enriched manner (for RBCs: miR-451, miR-92a, miR-16, and miR-486-5p; for myeloid blood cells: miR-223, let-7a, miR-197, and miR-574-3p) (Figure 2.8a). One additional miRNA, miR-150, was chosen that demonstrated strong lymphoid cell–enriched expression (Figure 2.8a) but has not yet been reported as a circulating cancer biomarker. The liver-specific miRNA, miR-122, that is not

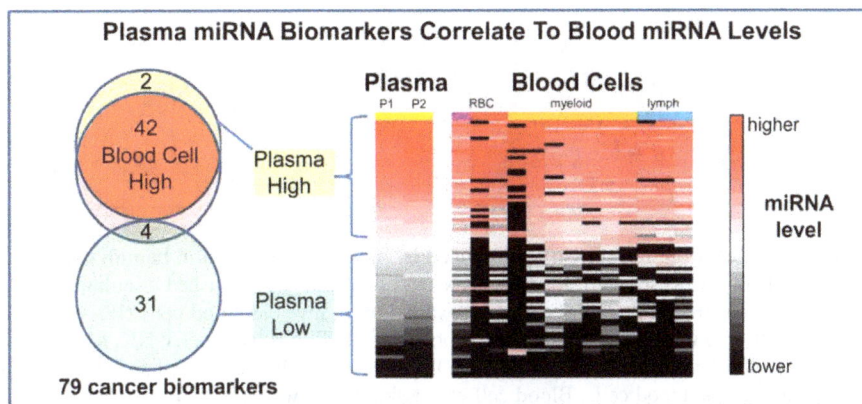

**FIGURE 2.7** **Relationship between blood cell and plasma microRNA expression among published circulating cancer biomarkers.** (*Left*) The Venn diagram depicts the distribution of 79 miRNAs published as biomarkers of nonhematopoietic cancers in healthy donor plasma and matched blood cells. "Plasma high" (*yellow circle*) refers to expression above the 50th percentile among 292 reliably detectable miRNAs by qPCR; "Plasma low" (*green circle*) refers to miRNAs detected below the 50th percentile, or not detected. "Blood cell high" (*red circle*) refers to miRNAs detected in the top 50th percentile in whole blood and at least one blood cell class as described in detail in the supplemental methods. (*Right*) A heat map showing side-by-side comparisons of plasma and blood cell expression of the 79 reported biomarkers demonstrates the close correlation between plasma and blood cell miRNA expression. P1 and P2 represent two independent plasma specimens drawn on different days. For blood cells, the columns represent multiple replicates and/or cell types as detailed in the methods. Circulating miRNA biomarkers are sorted in descending order of expression in healthy donor plasma and are not clustered.

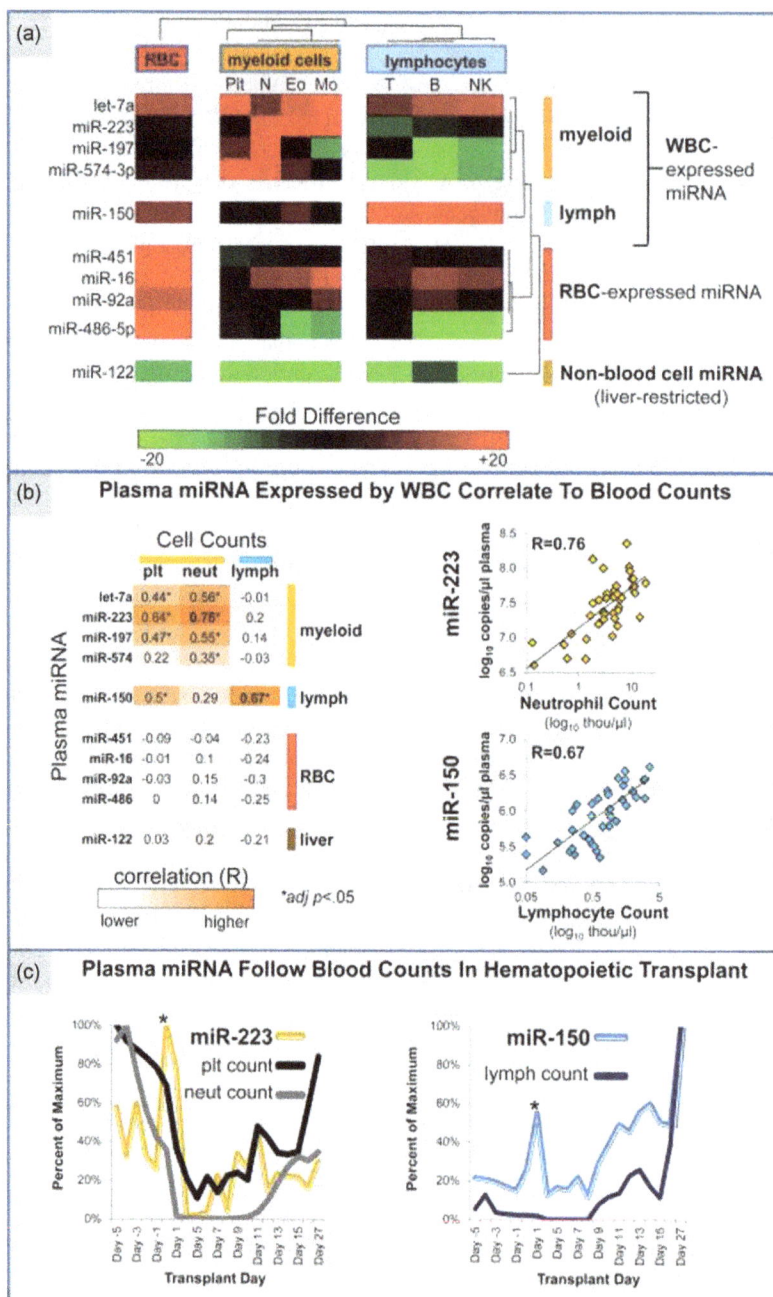

**FIGURE 2.8    Circulating microRNA biomarkers are influenced by blood cell counts and hemolysis.** (a) A heat map depicts the relative expression in blood cells of the 10 selected plasma miRNAs, 8 of which are published circulating miRNA cancer biomarkers. Let-7a, miR-223, miR-197, and miR-574-3p had the highest expression in myeloid blood cells (Plt, platelets; N, neutrophils; Eo, eosinophils; Mo, monocytes). MiR-150 was most abundant in lymphocytes (T, T cells; B, B cells; NK, natural killer cells), while miR-451, miR-16, miR-92a, and miR-486-5p were enriched in RBC. A liver-specific miRNA selected as a negative control (miR-122) was not appreciably expressed in any of the blood cells. Blood cell expression levels were determined using Exiqon v1 qRT-PCR arrays and confirmed in independent samples with TaqMan qRT-PCR assays as described in the supplemental methods. (b) (*Left*) Results of plasma miRNA correlation to blood cell counts in 42 consecutive plasma samples from an academic hospital clinical laboratory. Blood cell expression levels of the 10 miRNAs were inferred based on qRT-PCR Ct values as described in the supplemental methods. Pearson correlation coefficients of plasma miRNA levels and blood cell counts in the 42 clinical samples are shown in the table. Statistically significant correlations after correcting for multiple comparisons (i.e., adjusted $p<0.05$) are highlighted by *deep color boxes*. (*Right*) Data corresponding to correlations for miR-223 with neutrophil count and miR-150 with lymphocyte count are plotted. (c) In a patient undergoing myeloablative chemotherapy and autologous hematopoietic stem cell transplant, plasma miR-223 tracked with changes in myeloid blood counts (neutrophils and platelets), and plasma miR-150 correlated with changes in lymphocyte counts. Notably, spikes in plasma miR-223 and miR-150 were observed following infusion of hematopoietic stem cells (*asterisks*). (d) Shown are the differences in mean expression of plasma miRNAs in hemolyzed specimens (n=3) compared to nonhemolyzed specimens (n=39). Error bars represent the standard error of the difference of means. The biomarkers that were most highly expressed in RBC (miR-451, miR-16, miR-92a, miR-486-5p) were 20- to 30-fold higher in hemolyzed specimens, which was highly statistically significant ($p<10^{-7}$, two-tailed t-test).

*(Continued)*

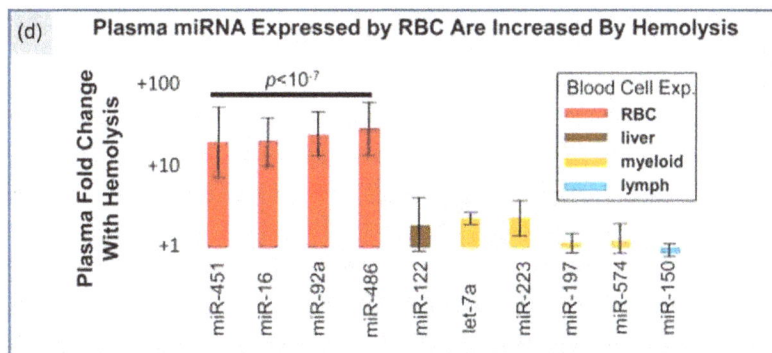

**FIGURE 2.8** *(Continued)*

expressed in blood cells, was selected as a negative control (Figure 2.8a) [122].

Plasma levels of the myeloid-expressed miRNAs let-7a, miR-223, miR-197, and miR-574-3p showed significant positive correlations with myeloid blood cell counts (granulocytes and platelets) (Figure 2.8b, left), with >50-fold differences in plasma miRNA biomarker levels between patients with the highest and lowest overall cell counts (Figure 2.8b, right). Also consistent with the hypothesis, levels of the lymphoid-enriched miRNA, miR-150, were most highly correlated with lymphocyte count (Figure 2.8b). Importantly, plasma levels of a non–blood cell expressed miRNA (miR-122) did not strongly correlate with blood cell counts, and the selected RBC-expressed miRNAs did not show correlations with WBC or platelet counts (Figure 2.8b, left). Furthermore, when these ten miRNAs were subjected to unsupervised hierarchical clustering based on their expression in plasma across the 42 patient samples, they clustered into separate groups corresponding to the four RBC-expressed miRNAs, the five WBC-expressed miRNAs, and the non–blood cell expressed miRNA, consistent with the notion that plasma expression patterns for many miRNAs are reflective of their blood cell origin.

To test whether there is a causal relationship underlying correlations observed between blood cell count and plasma miRNA biomarkers, researchers examined miR-223 (myeloid cell-expressed) and miR-150 (lymphoid cell-expressed) miRNAs in the plasma of a patient over a time-course of myeloablative chemotherapy and hematopoietic stem cell transplant engraftment. They collected serial plasma samples and corresponding blood count data at 25 time points over a 32-day period (Figure 2.8c). They found that plasma levels of the myeloid-enriched miRNA, miR-223, closely tracked with changes in myeloid blood counts (platelets and neutrophils), while plasma levels of the lymphoid-enriched miRNA, miR-150, mirrored lymphoid counts (Figure 2.8c), demonstrating that WBC counts can significantly influence plasma levels of blood cell–expressed miRNAs.

Thus, the study could not assess correlations between RBC-expressed miRNAs and RBC counts because there was limited variation in RBC count among individuals in the analyzed cohort. However, 3 of the 42 plasma specimens

were noted to be hemolyzed. Comparing expression of the ten miRNAs above in hemolyzed vs nonhemolyzed specimens, researchers found that all four RBC-associated miRNAs were increased by 20- to 30-fold in hemolyzed plasma, including the published colon cancer plasma biomarker miR-92a [123, 124]. In contrast, none of the six non-RBC-associated miRNAs were significantly increased with hemolysis (Figure 2.8d). Also, plasma levels of RBC-expressed miRNAs were tightly intercorrelated even among nonhemolyzed specimens, but not substantially correlated to six non-RBC-expressed miRNAs [116]. The results suggest that RBCs can contribute significant levels of reported cancer biomarkers into plasma and have important implications for biomarker interpretation. For example, a greater propensity for RBC hemolysis in colon cancer patients [125] could explain the relatively modest (<5-fold) increase in plasma miR-92a that has been reported in colon cancer patients [123, 124].

Taken together, all results [116] indicate that a majority of miRNAs reported in the literature as circulating cancer biomarkers may originate in large part from blood cells. This finding is supported by two recent studies in this area [126, 127]. Importantly, the studies in the setting of myeloablative chemotherapy and hematopoietic stem cell transplant engraftment provide the first direct, in vivo evidence that blood cell abundance can influence circulating miRNA levels. Among the examined miRNA biomarkers, the team observed variation attributable to blood cell effects that was greater in magnitude than many of the differences reported between cancer patients and controls [121, 123, 128]. This raises the concern that many miRNAs reported as circulating cancer biomarkers reflect a secondary effect on blood cells rather than a tumor cell–specific origin. For example, elevated neutrophil counts are associated with shortened progression-free and overall survival in several cancers including NSCLC and breast cancer [129, 130]. A recent study published in *PNAS* found that increased ratios of neutrophil-expressed plasma miRNAs (miR-197, miR-142-3p, miR-140-5p, miR-17, miR-21) compared with RBC-expressed miRNAs (miR-92a, miR-486-5p, miR-16, and miR-451) were associated with poorer outcomes in NSCLC patients [121]. Although blood cell counts were not available in that report, the findings suggest the possibility that the reported plasma miRNA ratios primarily reflect differences in blood counts that could be more readily measured with a routine CBC. A similar concern may be raised for many other studies where circulating miRNAs that are expressed highly in blood cells have been reported as cancer biomarkers [117–119, 123, 124, 128, 131–140].

In contrast to blood cell–expressed miRNAs, plasma levels of liver-restricted miR-122 that was selected as a negative control were not significantly correlated to blood cell counts or influenced by RBC hemolysis. This suggests that organ-restricted miRNAs may escape the problem of blood cell interference.

The blood cell findings should be interpreted in the context of other preanalytical and analytical sources of variation that may influence plasma miRNA levels. For example, differences in collection procedures and specimen processing conditions have been shown to contribute to plasma miRNA variability [126, 127]. Imprecision attributable to specific methods of RNA extraction, miRNA measurement, data acquisition, and data normalization is also likely to have a meaningful impact on plasma miRNA biomarker assessments, especially when fixed biases exist.

The study [116] does not directly assess the relative contribution of different blood cell types to plasma miRNA, or examine mechanisms of miRNA release into plasma. Here, we provide a framework and rationale for future investigations into these questions. A more detailed understanding of the cellular origin of circulating miRNA will inform the appropriate use of this exciting new class of analyte as a cancer biomarker.

## 2.9 MiRNA LIMITATIONS IN CLINICAL APPLICATION AND PLANS FOR INTRODUCING MiRNA AS A BIOMARKER IN CLINIC

Computed tomography (CT) screening programs, in the post–National Lung Screening Trial era, have an urgent need to discover, develop, and validate new biomarkers that can both help identify high-risk subjects and distinguish the benign mass from malignant lesions and also be applied in therapy. While new genomic, transcriptomic, and epigenomic biomarkers have been explained above, we have yet to see progression from biomarker discovery in the research phase to clinical application; reviewed biomarkers should be validated in several trials to confirm their validity. These biomarkers need to provide information about cancer risk, diagnosis, prognosis, and response to therapy that is independent of the clinical and radiographic factors that have been well established for disease. Even after recent significant progress in miRNA discovery, miRNAs have not reached clinical application in human lung cancer. Ongoing investigation in this field suggests that miRNA signatures will be of value in the settings of indeterminate lung nodules, early detection, histologic classification, and response to therapy [39].

So, better understanding of the interaction of miRNAs and epigenomics will improve our knowledge of the tumorigenic processes and will help us to discover effective strategies that can be used to conquer lung cancer in the near future.

## 2.10 CONCLUSION

According to evidence reviewed here, miRNAs can be considered as a part of multilevel regulatory machinery; however, miRNAs not only are able to target certain molecules at the posttranscriptional level [141–143], like the epigenetic machinery members, but they also are strictly targeted by epigenetic mechanisms. According to these findings, aberrant balance among the members of this complicated network leads to lung cancer.

The discovery of miRNAs will probably change the landscape of lung cancer genetics. Additional studies using bioinformatic methods are required to understand the mechanisms by which miRNAs cooperate with cancer origin and development. Data from several different sources will enable scientists to evaluate the effects of epigenetic modifications on miRNA expression and their aberrant role in lung cancer.

In conclusion, we demonstrate for the first time that blood cell counts can substantially influence plasma miRNA biomarker levels. For studies of circulating miRNA biomarkers that are expressed in blood cells, we propose that CBC data be collected and that miRNA expression levels be interpreted in light of blood cell counts. Acceptable ranges for blood cell counts might be established for specific miRNA biomarkers that are particularly vulnerable to blood cell effects. In the future, deeper quantitative understanding of the contributions of specific blood cell types to circulating miRNAs may enable correction for variation in blood cell number in some cases. That said, it is important to note that a significant minority of literature-reported solid tumor biomarkers were *not* highly expressed in blood cells. In clinical contexts where highly specific circulating miRNA biomarkers of cancer are sought, efforts may be most effective if focused on such miRNAs that are not blood cell expressed.

All these recent findings open the possibility of new diagnostic and therapeutic strategies for lung cancer patients. This integrative project is ongoing and there is a need to update the database with upcoming studies.

## REFERENCES

[1] Mitchell PS, Parkin RK, Kroh EM, et al. Circulating microRNAs as stable blood-based markers for cancer detection. Proc Natl Acad Sci USA. 2008;105:10513–10518.

[2] Chim SS, Shing TK, Hung EC, et al. Detection and characterization of placental microRNAs in maternal plasma. Clin Chem. 2008;54:482–490.

[3] Lawrie CH, Gal S, Dunlop HM, et al. Detection of elevated levels of tumour-associated microRNAs in serum of patients with diffuse large B-cell lymphoma. Br J Haematol. 2008;141:672–675.

[4] Chen X, Ba Y, Ma L, et al. Characterization of microRNAs in serum: a novel class of biomarkers for diagnosis of cancer and other diseases. Cell Res. 2008;18:997–1006.

[5] Kosaka N, Iguchi H, Ochiya T. Circulating microRNA in body fluid: a new potential biomarker for cancer diagnosis and prognosis. Cancer Sci. 2010;101:2087–2092.

[6] Wittmann J, Jack HM. Serum microRNAs as powerful cancer biomarkers. Biochim Biophys Acta. 2010;1806:200–207.

[7] Travis WD, Travis LB, Devesa SS. Lung cancer. Cancer. 1995;75:191–202.

[8] Wang QZ, Xu W, Habib N, et al. Potential uses of microRNA in lung cancer diagnosis, prognosis, and therapy. Curr Cancer Drug Targets. 2009;9(4):572–594.

[9] Portela A, Esteller M. Epigenetic modifications and human disease. Nat Biotechnol. 2010;28(10):1057–1068.

[10] Yoshino M, Suzuki M, Tian L, et al. Promoter hypermethylation of the p16 and Wif-1 genes as an independent prognostic marker in stage IA non-small cell lung cancers. Int J Oncol. 2009;35(5):1201–1209.

[11] Kumar R, Xi Y. MicroRNA, epigenetic machinery and lung cancer. Thorac Cancer. 2011;2(2):35–44.

[12] Iorio MV, Croce CM. MicroRNA dysregulation in cancer: diagnostics, monitoring and therapeutics. A comprehensive review. EMBO Mol Med. 2012;4(3):143–159.

[13] Brothers JF, Hijazi K, Mascaux C, et al. Bridging the clinical gaps: genetic, epigenetic and transcriptomic biomarkers for the early detection of lung cancer in the post-National Lung Screening Trial era. BMC Med. 2013;11(1):168.

[14] Singh NK. MicroRNAs databases: developmental methodologies, structural and functional annotations. Interdiscip Sci. Epub ahead of print March 28, 2016. doi:10.1007/s12539-016-0166-7.

[15] Xu N, Donohoe ME, Silva SS, et al. Evidence that homologous X-chromosome pairing requires transcription and Ctcf protein. Nat Genet. 2007;39(11):1390–1396.

[16] Friedman RC, Farh KK-H, Burge CB, et al. Most mammalian mRNAs are conserved targets of microRNAs. Genome Res. 2009;19(1):92–105.

[17] Yang J-S, Lai EC. Alternative miRNA biogenesis pathways and the interpretation of core miRNA pathway mutants. Mol Cell. 2011;43(6):892–903.

[18] Hutvagner G, McLachlan J, Pasquinelli AE, et al. A cellular function for the RNA interference enzyme Dicer in the maturation of the let-7 small temporal RNA. Science. 2001;293(5531):834–838.

[19] Malumbres M. MiRNAs and cancer: an epigenetics view. Mol Aspects Med. 2013;34(4):863–874.

[20] Lopez-Serra P, Esteller M. DNA methylation-associated silencing of tumor-suppressor microRNAs in cancer. Oncogene. 2012;31(13):1609–1622.

[21] Sheervalilou R, Ansarin K, Fekri Aval S, et al. An update on sputum MicroRNAs in lung cancer diagnosis. Diagn Cytopathol. 2016;44:442–449.

[22] Iorio MV, Piovan C, Croce CM. Interplay between microRNAs and the epigenetic machinery: an intricate network. Biochim Biophys Acta. 2010;1799(10):694–701.

[23] Zhu S, Si ML, Wu H, et al. MicroRNA-21 targets the tumor suppressor gene tropomyosin 1 (TPM1). J Biol Chem. 2007;282(19):14328–14336.

[24] Plasterk RH. RNA silencing: the genome's immune system. Science. 2002;296(5571):1263–1265.

[25] Elbashir SM, Lendeckel W, Tuschl T. RNA interference is mediated by 21- and 22-nucleotide RNAs. Genes Dev. 2001;15(2):188–200.

[26] Frankel LB, Christoffersen NR, Jacobsen A, et al. Programmed cell death 4 (PDCD4) is an important functional target of the microRNA miR-21 in breast cancer cells. J Biol Chem. 2008;283(2):1026–1033.

[27] Pillai RS, Bhattacharyya SN, Artus CG, et al. Inhibition of translational initiation by Let-7 MicroRNA in human cells. Science. 2005;309(5740):1573–1576.

[28] Shaw G, Kamen R. A conserved AU sequence from the 3′ untranslated region of GM-CSF mRNA mediates selective mRNA degradation. Cell. 1986;46(5):659–667.

[29] Jing Q, Huang S, Guth S, et al. Involvement of microRNA in AU-rich element-mediated mRNA instability. Cell. 2005;120(5):623–634.

[30] Hu Z, Chen J, Tian T, et al. Genetic variants of miRNA sequences and non–small cell lung cancer survival. J Clin Invest. 2008;118(7):2600–2608.

[31] Vasudevan S, Tong Y, Steitz JA. Switching from repression to activation: microRNAs can up-regulate translation. Science. 2007;318(5858):1931–1934.

[32] Stark A, Lin MF, Kheradpour P, et al. Discovery of functional elements in 12 Drosophila genomes using evolutionary signatures. Nature. 2007;450(7167):219–232.

[33] Orom UA, Nielsen FC, Lund AH. MicroRNA-10a binds the 5′ UTR of ribosomal protein mRNAs and enhances their translation. Mol Cell. 2008;30(4):460–471.

[34] Parasramka MA, Ho E, Williams DE, et al. MicroRNAs, diet, and cancer: new mechanistic insights on the epigenetic actions of phytochemicals. Mol Carcinog. 2012;51(3): 213–230.

[35] Cheloufi S, Dos Santos CO, Chong MMW, et al. A dicer independent miRNA biogenesis pathway that requires Ago catalysis. Nature. 2010;465(7298):584–589.

[36] Babiarz JE, Ruby JG, Wang Y, et al. Mouse ES cells express endogenous shRNAs, siRNAs, and other Microprocessor independent, Dicer-dependent small RNAs. Genes Dev. 2008;22(20):2773–2785.

[37] Johnson SM, Grosshans H, Shingara J, et al. RAS is regulated by the let-7 microRNA family. Cell. 2005;120(5):635–647.

[38] Mathonnet G, Fabian MR, Svitkin YV, et al. MicroRNA inhibition of translation initiation in vitro by targeting the cap-binding complex eIF4F. Science. 2007;317(5845):1764–1767.

[39] DeVita VT, Lawrence TS, Rosenberg SA. Cancer: Principles & Practice of Oncology. 10th ed. Wolters Kluwer Publications; 2016:490–492.

[40] Valadi H, Ekstrom K, Bossios A, et al. Exosome-mediated transfer of mRNAs and microRNAs is a novel mechanism of genetic exchange between cells. Nat Cell Biol. 2007;9(6):654–659.

[41] Chang T-C, Yu D, Lee YS, et al. Widespread microRNA repression by Myc contributes to tumorigenesis. Nat Genet. 2008;40(1):43–50.

[42] Chou Y-T, Lin HH, Lien YC, et al. EGFR promotes lung tumorigenesis by activating miR-7 through a Ras/ERK/Myc pathway that targets the Ets2 transcriptional repressor ERF. Cancer Res. 2010;70(21):8822–8831.

[43] Hong L, Lai M, Chen M, et al. The miR-17-92 cluster of microRNAs confers tumorigenicity by inhibiting oncogene induced senescence. Cancer Res. 2010;70(21):8547–8557.

[44] Zhu S, Wu H, Wu F, et al. MicroRNA-21 targets tumor suppressor genes in invasion and metastasis. Cell Res. 2008;18(3):350–359.

[45] Hatley ME, Patrick DM, Garcia MR, et al. Modulation of K-Ras-dependent lung tumorigenesis by MicroRNA-21. Cancer Cell. 2010;18(3):282–293.

[46] Zhang J-g, Wang JJ, Zhao F, et al. MicroRNA-21 (miR-21) represses tumor suppressor PTEN and promotes growth and invasion in non-small cell lung cancer (NSCLC). Clinica Chimica Acta. 2010;411(11):846–852.

[47] Puissegur M, Mazure NM, Bertero T, et al. MiR-210 is overexpressed in late stages of lung cancer and mediates mitochondrial alterations associated with modulation of HIF-1 activity. Cell Death Differ. 2011;18(3):465–478.

[48] Garofalo M, Di Leva G, Romano G, et al. MiR-221 & 222 regulate TRAIL resistance and enhance tumorigenicity through PTEN and TIMP3 downregulation. Cancer Cell. 2009;16(6):498–509.

[49] Liu X, Sempere LF, Ouyang H, et al. MicroRNA-31 functions as an oncogenic microRNA in mouse and human lung cancer cells by repressing specific tumor suppressors. J Clin Invest. 2010;120(4):1298–1309.

[50] Yang M, Shen H, Qiu C, et al. High expression of miR-21 and miR-155 predicts recurrence and unfavourable survival in nonsmall cell lung cancer. Eur J Cancer. 2013;49(3):604–615.

[51] Raponi M, Dossey L, Jatkoe T, et al. MicroRNA classifiers for predicting prognosis of squamous cell lung cancer. Cancer Res. 2009;69(14):5776–5783.

[52] Liu B, Wu X, Liu B, et al. MiR-26a enhances metastasis potential of lung cancer cells via AKT pathway by targeting PTEN. Biochim Biophys Acta. 2012;1822(11):1692–1704.

[53] Liu Y, Li M, Zhang G, et al. MicroRNA-10b overexpression promotes nonsmall cell lung cancer cell proliferation and invasion. Eur J Med Res. 2013;18(41):2047–2783.

[54] Zhang N, Wei X, Xu L. MiR-150 promote the proliferation of lung cancer cells by targeting P53. FEBS Lett. 2013;587(15):2346–2351.

[55] Cai J, Fang L, Huang Y, et al. MiR-205 targets PTEN and PHLPP2 to augment AKT signaling and drive malignant phenotypes in non–small cell lung cancer. Cancer Res. 2013;73(17):5402–5415.

[56] Calin GA, Sevignani C, Dumitru CD, et al. Human micro-RNA genes are frequently located at fragile sites and genomic regions involved in cancers. Proc Natl Acad Sci USA. 2004;101(9):2999–3004.

[57] Takamizawa J, Konishi H, Yanagisawa K, et al. Reduced expression of the let-7 micro-RNAs in human lung cancers in association with shortened postoperative survival. Cancer Res. 2004;64(11):3753–3756.

[58] He XY, Chen JX, Zhang Z, et al. The let-7a microRNA protects from growth of lung carcinoma by suppression of k-Ras and c-Myc in nude mice. J Cancer Res Clin Oncol. 2010;136:1023–1028.

[59] Fabbri M, Garzon R, Cimmino A, et al. MicroRNA-29 family reverts aberrant methylation in lung cancer by targeting DNA methyltransferases 3A and 3B. Proc Natl Acad Sci USA. 2007;104(40):15805–15810.

[60] Zhang J-g, Guo JF, Liu DL, et al. MicroRNA-101 exerts tumor-suppressive functions in non-small cell lung cancer through directly targeting enhancer of zeste homolog 2. J Thorac Oncol. 2011;6(4):671–678.

[61] Sun Y, Fang R, Li C, et al. Hsa-mir-182 suppresses lung tumorigenesis through down regulation of RGS17 expression in vitro. Biochem Biophys Res Commun. 2010;396(2):501–507.

[62] Chen Z, Zeng H, Guo Y, et al. MiRNA-145 inhibits non-small cell lung cancer cell proliferation by targeting c-Myc. J Exp Clin Cancer Res. 2010;29(1):151.

[63] Incoronato M, Garofalo M, Urso L, et al. MiR-212 increases tumor necrosis factor–related apoptosis-inducing ligand sensitivity in non–small cell lung cancer by targeting the antiapoptotic protein PED. Cancer Res. 2010;70(9):3638–3646.

[64] Nasser MW, Datta J, Nuovo G, et al. Down-regulation of micro-RNA-1 (miR-1) in lung cancer suppression of tumorigenic property of lung cancer cells and their sensitization to doxorubicin induced apoptosis by miR-1. J Biol Chem. 2008;283(48):33394–33405.

[65] Kumarswamy R, Mudduluru G, Ceppi P, et al. MicroRNA-30a inhibits epithelial to mesenchymal transition by targeting Snail and is downregulated in non-small cell lung cancer. Int J Cancer. 2012;130(9):2044–2053.

[66] Gallardo E, Navarro A, Vinolas N, et al. MiR-34a as a prognostic marker of relapse in surgically resected non-small cell lung cancer. Carcinogenesis. 2009;30(11):1903–1909.

[67] Wang G, Mao W, Zheng S, et al. Epidermal growth factor receptor-regulated miR-125a-5p–a metastatic inhibitor of lung cancer. FEBS J. 2009;276(19):5571–5578.

[68] Sun Y, Bai Y, Zhang F, et al. MiR-126 inhibits non-small cell lung cancer cells proliferation by targeting EGFL7. Biochem Biophys Res Commun. 2010;391(3):1483–1489.

[69] Seike M, Goto A, Okano T, et al. MiR-21 is an EGFR regulated anti-apoptotic factor in lung cancer in never-smokers. Proc Natl Acad Sci USA. 2009;106(29):12085–12090.

[70] Calin GA, Dumitru CD, Shimizu M, et al. Frequent deletions and down-regulation of micro-RNA genes miR15 and miR16 at 13q14 in chronic lymphocytic leukemia. Proc Natl Acad Sci USA. 2002;99(24):15524–15529.

[71] Bird A. Perceptions of epigenetics. Nature. 2007;447(7143):396–398.

[72] Bird A. DNA methylation patterns and epigenetic memory. Genes Dev. 2002;16(1):6–21.

[73] Kaneda M, Okano M, Hata K, et al. Essential role for de novo DNA methyltransferase Dnmt3a in paternal and maternal imprinting. Nature. 2004;429(6994):900–903.

[74] Chedin F. The DNMT3 family of mammalian de novo DNA methyltransferases. Prog Mol Biol Transl Sci. 2011;101:255–285.

[75] Blair LP, Yan Q. Epigenetic mechanisms in commonly occurring cancers. DNA Cell Biol. 2012;31(S1):S-49–S-61.

[76] He Y-F, Li BZ, Li Z, et al. Tet-mediated formation of 5-carboxylcytosine and its excision by TDG in mammalian DNA. Science. 2011;333(6047):1303–1307.

[77] Egger G, Liang G, Aparicio A, et al. Epigenetics in human disease and prospects for epigenetic therapy. Nature. 2004;429(6990):457–463.

[78] Peters AH, Mermoud JE, O'Carroll D, et al. Histone H3 lysine 9 methylation is an epigenetic imprint of facultative heterochromatin. Nat Genet. 2002;30(1):77–80.

[79] Jones PA, Baylin SB. The epigenomics of cancer. Cell. 2007;128(4):683–692.

[80] Lujambio A, Calin GA, Villanueva A, et al. A microRNA DNA methylation signature for human cancer metastasis. Proc Natl Acad Sci USA. 2008;105(36):13556–13561.

[81] Esteller M. Epigenetics in cancer. N Engl J Med. 2008;358(11):1148–1159.

[82] Kunej T, Godnic I, Ferdin J, et al. Epigenetic regulation of microRNAs in cancer: an integrated review of literature. Mutat Res. 2011;717(1):77–84.

[83] Tellez CS, Juri DE, Do K, et al. EMT and stem cell–like properties associated with miR-205 and miR-200 epigenetic silencing are early manifestations during carcinogen-induced transformation of human lung epithelial cells. Cancer Res. 2011;71(8):3087–3097.

[84] Xi S, Xu H, Shan J, et al. Cigarette smoke mediates epigenetic repression of miR-487b during pulmonary carcinogenesis. J Clin Invest. 2013;123(3):1241–1261.

[85] Kitano K, Watanabe K, Emoto N, et al. CpG island methylation of microRNAs is associated with tumor size and recurrence of non-small-cell lung cancer. Cancer Sci. 2011;102(12):2126–2131.

[86] Heller G, Weinzierl M, Noll C, et al. Genome-wide miRNA expression profiling identifies miR-9-3 and miR-193a as targets for DNA Methylation in non–small cell lung cancers. Clin Cancer Res. 2012;18(6):1619–1629.

[87] Wang J, Yang B, Han L, et al. Demethylation of miR-9-3 and miR-193a genes suppresses proliferation and promotes apoptosis in non-small cell lung cancer cell lines. Cell Physiol Biochem. 2013;32(6):1707–1719.

[88] Tan W, Gu J, Huang M, et al. Epigenetic analysis of microRNA genes in tumors from surgically resected lung cancer patients and association with survival. Mol Carcinog. Epub ahead of print March 24, 2014. doi:10.1002/mc.22149.

[89] Watanabe K, Takai D. Disruption of the expression and function of microRNAs in lung cancer as a result of epigenetic changes. Front Genet. 2013;4:275.

[90] Bommer GT, Gerin I, Feng Y, et al. p53-mediated activation of miRNA34 candidate tumor-suppressor genes. Curr Biol. 2007;17(15):1298–1307.

[91] Hermeking H. The miR-34 family in cancer and apoptosis. Cell Death Differ. 2010;17(2):193–199.

[92] Cole KA, Attiyeh EF, Mosse YP, et al. A functional screen identifies miR-34a as a candidate neuroblastoma tumor suppressor gene. Mol Cancer Res. 2008;6(5):735–742.

[93] Wang Z, Chen Z, Gao Y, et al. DNA hypermethylation of microRNA-34b/c has prognostic value for stage I non-small cell lung cancer. Cancer Biol Ther. 2011;11(5):490–496.

[94] Watanabe K, Emoto N, Hamano E, et al. Genome structure based screening identified epigenetically silenced microRNA associated with invasiveness in non-small-cell lung cancer. Int J Cancer. 2012;130(11):2580–2590.

[95] Tanaka N, Toyooka S, Soh J, et al. Frequent methylation and oncogenic role of microRNA-34b/c in small-cell lung cancer. Lung Cancer. 2012;76(1):32–38.

[96] Chen Y, Min L, Zhang X, et al. Decreased miRNA-148a is associated with lymph node metastasis and poor clinical outcomes and functions as a suppressor of tumor metastasis in non-small cell lung cancer. Oncol Rep. 2013;30(4):1832–1840.

[97] Crawford M, Brawner E, Batte K, et al. MicroRNA-126 inhibits invasion in nonsmall cell lung carcinoma cell lines. Biochem Biophys Res Commun. 2008;373(4):607–612.

[98] Jusufović E, Rijavec M, Keser D, et al. let-7b and miR-126 are down-regulated in tumor tissue and correlate with microvessel density and survival outcomes in non–small–cell lung cancer. PLoS ONE. 2012;7:e45577.

[99] Donnem T, Lonvik K, Eklo K, et al. Independent and tissue specific prognostic impact of miR-126 in nonsmall cell lung cancer. Cancer. 2011;117(14):3193–3200.

[100] Lujambio A, Ropero S, Ballestar E, et al. Genetic unmasking of an epigenetically silenced microRNA in human cancer cells. Cancer Res. 2007;67(4):1424–1429.

[101] Ceppi P, Mudduluru G, Kumarswamy R, et al. Loss of miR-200c expression induces an aggressive, invasive, and chemoresistant phenotype in non–small cell lung cancer. Mol Cancer Res. 2010;8(9):1207–216.

[102] Gregory PA, Bert AG, Paterson EL, et al. The miR-200 family and miR-205 regulate epithelial to mesenchymal transition by targeting ZEB1 and SIP1. Nat Cell Biol. 2008;10(5):593–601.

[103] Roush S, Slack FJ. The let-7 family of microRNAs. Trends Cell Biol. 2008;18(10):505–516.

[104] Brueckner B, Stresemann C, Kuner R, et al. The human let-7a-3 locus contains an epigenetically regulated microRNA gene with oncogenic function. Cancer Res. 2007;67(4):1419–1423.

[105] Saito Y, Liang G, Egger G, et al. Specific activation of microRNA-127 with downregulation of the proto-oncogene BCL6 by chromatin modifying drugs in human cancer cells. Cancer Cell. 2006;9(6):435–443.

[106] Jakopovic M, Thomas A, Balasubramaniam S, et al. Targeting the epigenome in lung cancer: expanding approaches to epigenetic therapy. Front Oncol. 2013;3:261.

[107] Del Vescovo V, Cantaloni C, Cucino A, et al. MiR-205 expression levels in nonsmall cell lung cancer do not always distinguish adenocarcinomas from squamous cell carcinomas. Am J Surg Pathol. 2011;35(2):268–275.

[108] Yanaihara N, Caplen N, Bowman E, et al. Unique microRNA molecular profiles in lung cancer diagnosis and prognosis. Cancer Cell. 2006;9(3):189–198.

[109] Xu H, Cheung IY, Guo H, et al. MicroRNA miR-29 modulates expression of immunoinhibitory molecule B7-H3: potential implications for immune based therapy of human solid tumors. Cancer Res. 2009;69(15):6275–6281.

[110] Rothschild SI, Tschan MP, Federzoni EA, et al. MicroRNA-29b is involved in the Src-ID1 signaling pathway and is dysregulated in human lung adenocarcinoma. Oncogene. 2012;31(38):4221–4232.

[111] Garzon R, Liu S, Fabbri M, et al. MicroRNA-29b induces global DNA hypomethylation and tumor suppressor gene reexpression in acute myeloid leukemia by targeting directly DNMT3A and 3B and indirectly DNMT1. Blood. 2009;113(25):6411–6418.

[112] Duursma AM, Kedde M, Schrier M, et al. MiR-148 targets human DNMT3b protein coding region. RNA. 2008;14(5):872–877.

[113] Cho HM, Jeon HS, Lee SY, et al. MicroRNA-101 inhibits lung cancer invasion through the regulation of enhancer of zeste homolog 2. Exp Ther Med. 2011;2(5):963–967.

[114] Yan F, Shen N, Pang J, et al. Restoration of miR-101 suppresses lung tumorigenesis through inhibition of DNMT3a-dependent DNA methylation. Cell Death Dis. 2014;5(9):e1413.

[115] Kroh EM, Parkin RK, Mitchell PS, et al. Analysis of circulating microRNA biomarkers in plasma and serum using quantitative reverse transcription-PCR (qRT-PCR). Methods. 2010;50:298–301.

[116] Pritchard CC, Kroh E, Wood B, et al. Blood cell origin of circulating microRNAs: a cautionary note for cancer biomarker studies. Cancer Prev Res (Phila). 2012;5(3):492–497. doi: 10.1158/1940-6207.CAPR-11-0370

[117] Brase JC, Johannes M, Schlomm T, et al. Circulating miRNAs are correlated with tumor progression in prostate cancer. Int J Cancer. 2011;128:608–616.

[118] Heneghan HM, Miller N, Lowery AJ, et al. Circulating microRNAs as novel minimally invasive biomarkers for breast cancer. Ann Surg. 2010;251:499–505.

[119] Taylor DD, Gercel-Taylor C. MicroRNA signatures of tumor-derived exosomes as diagnostic biomarkers of ovarian cancer. Gynecol Oncol. 2008;110:13–21.

[120] Zhang C, Wang C, Chen X, et al. Expression profile of microRNAs in serum: a fingerprint for esophageal squamous cell carcinoma. Clin Chem. 2010;56:1871–1879.

[121] Boeri M, Verri C, Conte D, et al. MicroRNA signatures in tissues and plasma predict development and prognosis of computed tomography detected lung cancer. Proc Natl Acad Sci USA. 2011;108:3713–3718.

[122] Zhang Y, Jia Y, Zheng R, et al. Plasma microRNA-122 as a biomarker for viral-, alcohol-, and chemical-related hepatic diseases. Clin Chem. 2010;56:1830–1838.

[123] Ng EK, Chong WW, Jin H, et al. Differential expression of microRNAs in plasma of patients with colorectal cancer: a potential marker for colorectal cancer screening. Gut. 2009;58:1375–1381.

[124] Huang Z, Huang D, Ni S, et al. Plasma microRNAs are promising novel biomarkers for early detection of colorectal cancer. Int J Cancer. 2010;127:118–126.

[125] Avinash SS, Anitha M, Vinodchandran, et al. Advanced oxidation protein products and total antioxidant activity in colorectal carcinoma. Indian J Physiol Pharmacol. 2009;53:370–374.

[126] Duttagupta R, Jiang R, Gollub J, et al. Impact of cellular miRNAs on circulating miRNA biomarker signatures. PLoS One. 2011;6:e20769.

[127] McDonald JS, Milosevic D, Reddi HV, et al. Analysis of circulating microRNA: preanalytical and analytical challenges. Clin Chem. 2011;57:833–840.

[128] Hu Z, Chen X, Zhao Y, et al. Serum microRNA signatures identified in a genome-wide serum microRNA expression profiling predict survival of non-small-cell lung cancer. J Clin Oncol. 2010;28:1721–1726.

[129] Azab B, Bhatt VR, Phookan J, et al. Usefulness of the neutrophil-to-lymphocyte ratio in predicting short- and long-term mortality in breast cancer patients. Ann Surg Oncol. 2012;19(1):217–224.

[130] Teramukai S, Kitano T, Kishida Y, et al. Pretreatment neutrophil count as an independent prognostic factor in advanced non-small-cell lung cancer: an analysis of Japan Multinational Trial Organisation LC00-03. Eur J Cancer. 2009;45:1950–1958.

[131] Heegaard NH, Schetter AJ, Welsh JA, et al. Circulating microRNA expression profiles in early stage non-small cell lung cancer. Int J Cancer. 2012;130(6):1378–1386.

[132] Heneghan HM, Miller N, Kerin MJ. Circulating miRNA signatures: promising prognostic tools for cancer. J Clin Oncol. 2010;28:e573–e574; author reply e5–e6.

[133] Lodes MJ, Caraballo M, Suciu D, et al. Detection of cancer with serum miRNAs on an oligonucleotide microarray. PLoS One. 2009;4:e6229.

[134] Tsujiura M, Ichikawa D, Komatsu S, et al. Circulating microRNAs in plasma of patients with gastric cancers. Br J Cancer. 2010;102:1174–1179.

[135] Wang J, Chen J, Chang P, et al. MicroRNAs in plasma of pancreatic ductal adenocarcinoma patients as novel blood-based biomarkers of disease. Cancer Prev Res (Phila). 2009;2:807–813.

[136] Xu J, Wu C, Che X, et al. Circulating MicroRNAs, miR-21, miR-122, and miR-223, in patients with hepatocellular carcinoma or chronic hepatitis. Mol Carcinog. 2011;50:136–142.

[137] Zhu W, Qin W, Atasoy U, et al. Circulating microRNAs in breast cancer and healthy subjects. BMC Res Notes. 2009;2:89.

[138] Ali S, Almhanna K, Chen W, et al. Differentially expressed miRNAs in the plasma may provide a molecular signature for aggressive pancreatic cancer. Am J Transl Res. 2010;3:28–47.

[139] Moltzahn F, Olshen AB, Baehner L, et al. Microfluidic-based multiplex qRT-PCR identifies diagnostic and prognostic microRNA signatures in the sera of prostate cancer patients. Cancer Res. 2011;71:550–560.

[140] Zhao H, Shen J, Medico L, et al. A Pilot Study of Circulating miRNAs as Potential Biomarkers of Early Stage Breast Cancer. PLoS One. 2010;5:e13735.

[141] Sheervalilou R, Khamaneh AM, Sharifi A, et al. Using miR-10b, miR-1 and miR-30a expression profiles of bronchoalveolar lavage and sputum for early detection of non-small cell lung cancer. Biomed Pharmacother. 2017;88:1173–1182.

[142] Madkour LH. Nucleic Acids as Gene Anticancer Drug Delivery Therapy. Elsevier; 2020. https://www.elsevier.com/books/nucleic-acids-as-gene-anticancer-drug-delivery-therapy/madkour/978-0-12-819777-6

[143] Sheervalilou R, Shirvaliloo S, Fekri Aval S, et al. A new insight on reciprocal relationship between microRNA expression and epigenetic modifications in human lung cancer. Tumor Biol. May 2017;39(5):1010428317695032.

# 3 The Role of Circulating MiRNAs in Diagnosis, Prognosis, and Treatment Targets of Cancer and Diseases

MicroRNAs (miRNAs) are small (~22 nucleotide [nt]), noncoding RNAs that regulate a myriad of biological processes and are frequently dysregulated in cancer. MiRNAs have been implicated in different areas such as the immune response, neural development, DNA repair, apoptosis, oxidative stress response, and cancer. The majority of miRNAs are found intracellularly, but a significant number of miRNAs have been observed outside of cells, including in various bodily fluids.

MiRNAs are regulatory RNAs that are frequently dysregulated in cancer and have shown promise as tissue-based markers for cancer classification and prognostication. We show here that miRNAs are present in human plasma in a remarkably stable form that is protected from endogenous RNase activity. MiRNAs originating from human prostate cancer xenografts enter the circulation, are readily measured in plasma, and can robustly distinguish xenografted mice from controls.

Since the discovery of microRNAs and their involvement in hepatocarcinogenesis, the literature has described their usefulness as potential new biomarkers and treatment targets. Some of these miRNAs can also be found in the systemic circulation. With advances in detection and sequencing technologies, an increasing amount of data demonstrate the possibility of using circulating miRNAs as biomarkers to improve our current management of hepatocellular carcinoma (HCC) in a less invasive manner. Additionally, the role of circulating miRNAs could be particularly relevant in the context of neoplastic diseases. At least 79 miRNAs have been reported as plasma or serum miRNA biomarkers of solid and hematologic tumors.

This chapter will review circulating miRNAs with a known function in HCC, describing their role and function in tumorigenesis. This chapter discusses their potential use as biomarkers in conjunction with emerging treatments in the diagnosis and targeting of this disease.

## 3.1 GENERAL CONSIDERATIONS ON CELLULAR MiRNAS

MiRNAs are small, typically 22 nt, noncoding (nc), endogenous, single-stranded RNAs. Until recently, the central dogma of genetics was that RNA played the role of messenger between the gene and the final proteins encoded by the gene, and ncRNAs were ignored in the field of genome sequencing. Before the discovery of miRNAs, it was known that a large part of the genome is not translated into proteins. This so-called junk DNA was thought to be evolutionary debris with no real function. Recently, the explosion in research in this area has established miRNAs as powerful regulators of gene expression. Genome-wide transcriptional analyses have estimated that most transcribed mammalian genomic sequences are ncRNAs [1–6]. While only about 1424 human miRNA sequences have been identified so far, genomic computational analysis indicates that as many as 50,000 miRNAs may exist in the human genome, and each may have multiple targets based on similar sequences in the 3'-UTR of mRNA. In fact, it has been estimated that more than 60% of mammalian mRNAs are targeted by at least one miRNA [7].

MiRNAs were first identified in *Caenorhabditis elegans* in the early 1990s [8] but have since been reported in a wide variety of organisms ranging from single-cell algae to humans, suggesting that miRNA-mediated biological function is an ancient and critical cellular regulatory system [9, 10]. The importance of miRNA function is further suggested by the extreme evolutionary conservation of both individual miRNA sequences and the miRNA processing machinery. MiRNA genes are evolutionarily conserved and are located within the introns or exons of protein-coding genes, as well as in intergenic areas. In addition, the number of miRNAs in the genome appears to be correlated with the complexity of the developmental program, with mammals having the largest number of miRNAs.

MiRNAs are transcribed in much the same way as protein-coding genes. The majority of miRNAs are transcribed by RNA polymerase II [11], though a minor fraction of miRNAs are transcribed by RNA polymerase III [12]. These primary miRNA transcripts (pri-miRNAs) are often several hundred nucleotides long and are modified similarly to protein-coding transcripts by the addition of a 5' cap and a 3' poly-A tail. Then, pri-miRNAs are processed first in the nucleus and later in the cytosol by the RNase III enzymes Drosha and Dicer, respectively. This sequential processing of pri-miRNAs first yields a miRNA precursor (pre-miRNA) of ~70 nt in length, and eventually a mature double-stranded miRNA of 19–24 nucleotides. Typically, one strand of this mature miRNA duplex, termed the *guide strand*, associates with the RNA-induced silencing complex (RISC). While it is generally believed that upon incorporation into the RISC complex, the other strand (the passenger strand) is unwound from the guide strand and

DOI: 10.1201/9781003229650-3

degraded, there is evidence that in some cases, both strands of the miRNA duplex are functional [13]. MiRNA-RISC complexes interact with mRNA targets through partial sequence complementation, typically within the 3' untranslated region of target mRNAs. It is thought that the extent of base pairing between the miRNA and its mRNA target determines whether the mRNA is degraded or translationally repressed [14].

MiRNAs have been implicated in different areas such as the immune response, neural development, DNA repair, apoptosis, and oxidative stress response. It is surprising the list of diseases that have been found to be associated with abnormal miRNA expression. Moreover, several authors have focused their attention on the importance of cancer regulator miRNAs. They are divided into oncomiRs and anti-oncomiRs that negatively regulate tumor suppressor genes and oncogenes, respectively. Importantly, the association of miRNAs with cancer has prompted their potential relevance in cancer diagnosis, prognosis, and treatment [15].

## 3.2 CIRCULATING MiRNAs: THE NEW FRONTIER OF INTERCELLULAR COMMUNICATION

If the discovery of miRNAs by Ambros and colleagues in 1993 introduced another level of intricacy in the regulation of the genome [8], the discovery of circulating miRNAs highlighted the possibility of new mediators of gene regulation. In fact, while the majority of miRNAs are found intracellularly, a significant number of miRNAs have been observed outside of cells, including in various bodily fluids [16–20]. These miRNAs are stable and show distinct expression profiles in different fluid types. Given the instability of most RNA molecules in the extracellular environment, the presence and apparent stability of miRNAs here is surprising. Serum and other bodily fluids are known to contain ribonuclease [21], which suggests that secreted miRNAs are likely packaged in some manner in order to protect them against RNase digestion. MiRNAs could be shielded from degradation by packaging in lipid vesicles, in complexes with RNA-binding proteins, or both [22, 23]. Despite accumulating evidence for the presence of miRNAs in bodily fluids, the origin and the function of these circulating extracellular miRNAs remain poorly understood. One of the more interesting ideas is that extracellular miR-NAs are used as mediators of cell–cell communication [24–26]. If this is the case, then certain miRNAs are presumably targeted for export in one cell, and can be recognized, taken up, and utilized by another. The presence of miRNAs in bodily fluids may represent an infinite resource of noninvasive biomarkers in cancer [27, 28]. Recent studies have also identified miRNAs in two types of cell-derived lipid vesicles; microvesicles and exosomes. Microvesicles are relatively large (100 nm$^{-1}$ μm) vesicles released from the cell through blebbing. Exosomes, on the other hand, are smaller vesicles (30–100 nm) released when endosomally derived multivesicular bodies fuse with the plasma membrane. MiRNAs have been identified in both exosomes and microvesicles derived from a variety of sources, including human and mouse mast cells, glioblastoma tumors [29], plasma [30], saliva [31] and urine [32].

Studies have suggested that exosomes can be secreted by many cells, including T cells, B cells, mast cells, dendritic cells, cancer cells, and macrophages [33–37]. Probably most of the circulating miRNAs are found in exosomes, which protect them from degradation and are responsible for their excellent stability [23, 30, 38]. Exosomes can be transferred from one cell to another, and their components can function in the new environment. Much work has focused on understanding the mechanisms of exosome-mediated cell–cell interactions [39]. Some of these interactions involve communication between different cell types and convey regulatory effectors [40, 41], whereas others occur among cells of the same type [42]. It was found that endogenous plasma miRNAs exist in a form that is resistant to 4°C or 37°C incubation, freeze-thaw cycles, and even to RNase activity [43].

Circulating miRNAs are surprisingly tractable. Some of the key molecular properties of these species include high stability in circulation and the ability to survive unfavorable physiological conditions such as extreme variations in pH and multiple freeze-thaw cycles [18, 23, 44, 45]. However, although exosomal miRNA has been hypothesized to be involved in intercellular communication [18, 23, 46], it remains unclear whether all extracellular miR-NAs are associated with exosomes and whether extracellular miRNA are present in physiologically relevant amounts for cell-to-cell signaling. A different study has indicated that cells in culture predominantly exported miRNA in exosome-independent form [30]. In fact, Turchinovich et al. have shown that the overwhelming majority of the nuclease-resistant extracellular miRNA in plasma and cell culture media is floating outside exosomes and is bound to Ago2 protein—a part of the RNA-induced silencing complex. They also found indications that extracellular miRNA can be bound to other Ago proteins (especially Ago1). However, it is possible that parts of extracellular circulating miRNAs are by-products of dead or dying cells that persist due to the high stability of the miRNA/Ago2 complex [47]. As has already been said, circulating miRNAs are not only abundant in blood but are also very stable; these traits are important prerequisites for clinical biomarkers. Chen et al. [48] treated miRNAs with RNase A digestion. Surprisingly, more than half of the miRNAs remained intact after 3 h of exposure to RNase A. Furthermore, circulating miRNAs remain stable after being subjected to harsh conditions including boiling, low/high pH, and extended storage [49]. Circulating mature miRNA in contrast to mRNA is strikingly stable in blood plasma and cell culture media. The high stability of miRNAs constitutes an enormous advantage from a clinical diagnostic point of view as it allows an efficient isolation from clinical specimens including sputum [50], plasma [44, 51, 52], and serum [48, 53]. Finally, miRNA signatures in blood are similar in men and women, as well as individuals of different ages [30]. Together, these results indicate that circulating miRNAs have many characteristics of ideal biomarkers, most notably their inherent stability and resilience.

## 3.3 CIRCULATING MiRNAs IN HEALTH AND DISEASE

Several studies have compared the expression profiles of hundreds of blood-borne miRNAs across a variety of nonmalignant and malignant diseases to identify disease-specific expression patterns. The resulting miRNA expression data could be used to discriminate disease samples with a high level of accuracy, demonstrating the potential for using microRNA signatures for the blood-based diagnosis of disease [54]. The ability to profile miRNAs in circulation thus provides a noninvasive opportunity to investigate disease-specific miRNAs and represents an exciting alternative approach to current strategies for cancer surveillance.

The field of study of circulating miRNAs is relatively new but it has been developing rapidly. Taken together, the data suggest many circulating miRNAs are promising biomarkers that are superior to AFP and can allow an earlier diagnosis and monitoring of HCC in a noninvasive manner. Many circulating miRNAs have been shown to be able to distinguish HCC patients from healthy individuals and patients with chronic cirrhosis with extremely high accuracy, especially in patients with HBV infection. Undoubtedly, the future of HCC management needs to focus on earlier diagnosis, better indicators of prognosis, and improved treatments.

Numerous serum tumor markers have been proposed over the last decade in the hope to increase diagnostic accuracy. These include DCP, AFP-L3, VEGF, and IGF-1 [55, 56]. This is a reflection of the heterogeneous nature of the HCC tumors as a result of their diverse etiology, which leads to wide variability of marker expressions [57, 58]. Furthermore, independently arising HCCs within the same patient can also show distinctive gene expression [59]. There is a clear need for a novel tumor marker that can assist with diagnosis, prognosis, and treatment.

MiRNAs are a group of small noncoding RNAs, approximately 22 nt in length that can accurately regulate gene expression by complementary base pairing with the target messenger RNAs (mRNAs) [60]. Since their discovery in nematode *C. elegans* in 1993 [8], significant research has led to a better understanding of their function. It has been estimated by computational analysis that over 50% of human protein–coding genes are regulated by miRNAs and that each miRNA may target hundreds of different targets [61]. Some miRNAs can work together as a cluster to accomplish a common function; they often have a similar sequence and each cluster usually consists of two to three (but may exceed eight) members [62]. Through their ability to regulate gene expression, miRNAs can participate not only in normal cell homeostasis and development processes such as cell cycle, cell growth, proliferation, and apoptosis, but can also act as tumor suppressors and oncogenes, controlling tumor invasion, metastasis, and drug resistance [63, 64]. Dysregulated miRNA with oncogenic potential is often termed *oncomir* [65].

Numerous articles describe the association of miRNAs with hepatocarcinogenesis and their use as potential markers for all aspects of HCC management including diagnosis, differentiating etiology, monitoring disease progression, treatment response, prognosis, and as novel therapeutic targets [66, 67]. This is summarized in Table 3.1. Although

**TABLE 3.1**

**Summary of Aberrantly Expressed Circulating MiRNA in HCC [69]**

| MiRNA | Expression Changes [Reference] | Study Sample | Significance |
|---|---|---|---|
| miR-1 | Up [51] | 50 healthy liver (HL) 55 HBV-HCC | Hepatitis B related HCC (HBV-HCC) significantly upregulated from control (p<0.0001) |
| miR-15b | Up [91] | 30 HL 153 HCC (141 HBV) 30 HBV carriers | Distinguish HCC from NC with a sensitivity of 98.3%, specificity of 15.3%, and AUC[1] of 0.485 Level decrease post resection (p=0.0637) |
| miR-16 | Down [120] | 71 HL 105 HCC (20 HBV, 66 HCV) 107 chronic liver disease (CLD) (8 HBV, 59 HCV) | Significant association with HCC (HCC vs NC, HCC vs CLD, CLD vs NC, all p<0.01), combine with AFP improves accuracy |
| miR-18a | Up [50] | 60 HL 101 HBV-HCC 30 HBV chronic hepatitis or cirrhosis | Discriminate HBV-HCC from healthy controls with AUC 0.881, sensitivity 86.1%, specificity 75% HBV-HCC from HBV cirrhosis or chronic hepatitis AUC 0.775, sensitivity 77.2%, specificity 70% |
| miR-21 | Up [105] | 89 HL, 101 HCC (76 HBV-HCC), 48 CH (all HBV+ve) | Significantly elevated in HCC (p<0.0001) |
| miR-21 | Up [86] | 20 HL, 46 HCC (30 HBV-HCC) | Elevated in HCC |

*(Continued)*

**TABLE 3.1** *(Continued)*
**Summary of Aberrantly Expressed Circulating MiRNA in HCC [69]**

| MiRNA | Expression Changes [Reference] | Study Sample | Significance |
|---|---|---|---|
| miR-21 | Up [70] | 68 HL<br>204 HBV-HCC<br>75 chronic HBV<br>60 HBV-cirrhosis | Significantly upregulated HBV-HCC vs healthy+CBH+HBV-cirrhotic group (p<0.001)<br>AUC to distinguish HBV-HCC from all those other groups combined is 0.626 |
| miR-21 | Up [94] | 30 HL<br>29 HBV<br>57 HBV-HCC | Distinguish HBV-HCC from healthy+HBV group with sensitivity 89.47%, specificity 71.19%, AUC 0.865<br>Level decrease post resection (p=0.0648) |
| miR-25 | Up [51] | 50 HL<br>55 HBV-HCC | Significantly upregulated from control (p<0.0001)<br>When miR-25 and let-7f were tested together with miR-375, it increased the AUC to 0.9967+/− 0.015 (specificity 99.1%; sensitivity 99.1%) |
| miR-26a | Down [70] | 68 HL<br>204 HBV-HCC<br>75 chronic HBV<br>60 HBV-cirrhosis | Significantly downregulated when compared to healthy+CBH+HBV-cirrhotic group (p<0.001)<br>AUC to distinguish HBV-HCC from all those other groups combined is 0.665 |
| miR-27a | Down [70] | 68 HL<br>204 HBV-HCC<br>75 chronic HBV<br>60 HBV-cirrhosis | Significantly downregulated when compared to healthy+CBH+HBV-cirrhotic group (p<0.001)<br>AUC to distinguish HBV-HCC from all those other groups combined is 0.638 |
| miR-92a | Down [52] | 10 HL, 10 HCC (3HBV, 5 HCV, 2 non-b, non-c) | Decreased compared to HL (p=0.022)<br>Decreases post resection (p=0.082) |
| miR-92a | Up [51] | 50 HL<br>55 HBV-HCC | Significantly elevated in HBV-HCC (p<0.0001) |
| miR-122 | Up [105] | 89 HL, 101 HCC (76 HBV-HCC), 48 CH (all HBV+ve) | Significantly elevated in HCC (p<0.0001)<br>AUC 0.93 to diagnose CH from control, AUC 0.79 to diagnose HBV-HCC from CH, if other cause for liver injury has been excluded |
| miR-122 | Up [115] | 24 HL<br>38 chronic HCV with normal ALT<br>64 chronic HCV with elevated ALT | Significantly upregulated in chronic HCV hepatitis (p<0.001)<br>Discriminate chronic HCV infection from HL (p=0.026) AUC 0.97 |
| miR-122 | Down [70] | 68 HL<br>204 HBV-HCC<br>75 chronic HBV<br>60 HBV-cirrhosis | Significantly downregulated when compared to healthy+CBH+HBV-cirrhotic group (p<0.062)<br>AUC to distinguish HBV-HCC from all those other groups combined is 0.553 |
| miR-122 | Down [87] | 85 HL<br>85 HCC (75 are HBV-HCC) | Significantly upregulated compared to healthy (p<0.001)<br>Diagnose HCC, AUC O.707, sensitivity 70.6%, specificity 67.1% (not superior to AFP)<br>When combined with AFP, AUC 0.943, sensitivity 87.1%, specificity 98.8% |
| miR-130b | Up [94] | 30 HL<br>153 HCC patients (141 HBV+ve)<br>30 HBV carriers | Distinguish HCC from healthy AUC 0.913, sensitivity of 87.7%, specificity of 81.4%<br>When combined with miR-15b, AUC 0.981, sensitivity of 98.3%, specificity of 91.5%<br>Level decrease post resection (p=0.0158) |
| miR-183 | Up [94] | 30 HL<br>153 HCC patients (141 HBV+ve)<br>30 HBV carriers | Significantly elevated in HBV-HCC. Distinguish HBV-HCC from healthy+HBV group with sensitivity 57.89%, specificity 69.49%, AUC 0.661<br>Level decrease post resection (p=0.0084) |
| miR-192 | Up [115] | 24 HL<br>38 chronic HCV with normal ALT<br>64 chronic HCV with elevated ALT | Significantly upregulated in chronic HCV hepatitis (p<0.001) |

*(Continued)*

**TABLE 3.1** *(Continued)*
**Summary of Aberrantly Expressed Circulating MiRNA in HCC [69]**

| MiRNA | Expression Changes [Reference] | Study Sample | Significance |
|---|---|---|---|
| miR-192 | Up [70] | 68 HL<br>204 HBV-HCC<br>75 chronic HBV<br>60 HBV-cirrhosis | Significantly upregulated in HBV-HCC when compared to healthy+CBH+HBV-cirrhotic group (p=0.016)<br>Distinguish HBV-HCC from all those groups combined, AUC 0.569 |
| miR-195 | Down [120] | 71 HL, 105 HCC (20 HBV, 66 HCV), 107 CLD (8 HBV, 59 HCV) | Significantly downregulated in HCC compared to HL (p<0.01)<br>HCC vs CLD (p=0.04)<br>CLD vs normal (p<0.01) |
| miR-199a | Down [120] | 71 HL, 105 HCC (20HBV, 66HCV), 107 CLD (8HBV, 59HCV) | Significantly downregulated in HCC compared to healthy control (p<0.01), HCC vs CLD (p<0.01), CLD vs normal (p<0.01) |
| miR-206 | Up [51] | 50 HL<br>55 HBV-HCC | Significantly upregulated in HBV-HCC compared to control (p<0.0001) |
| miR-221 | Up [86] | 20 HL, 46 HCC (30 HBV-HCC, 16 others) | Elevated in 35/46 HCC, correlates with HCC stage and prognosis (5 year overall survival rate is significantly lower in patients with high miR-221 expression (p<0.05) |
| miR-221 | Up [87] | 85 HL<br>85 HCC (75 are HBV-HCC) | Level is higher in the HCC group but it is not statistically significant (p=0.225) |
| miR-222 | Up [86] | 20 HL, 46 HCC (30 HBV-HCC, 16 others) | Elevated in HCC |
| miR-223 | Up [51] | 160 HL<br>135 HBV (both chronic and asymptomatic)<br>48 HCV<br>65 HBV-HCC | Significantly elevated in all HBV related liver groups compared to HL (p<0.0001)<br>HCV vs control (p=0.0043) |
| miR-223 | Up [105] | 89 HL, 101 HCC (76 HBV-HCC), 48 CH (all HBV+ve) | Distinguish HCC from HL (p<0.0001) |
| miR-223 | Down [70] | 68 HL<br>204 HBV-HCC<br>75 chronic HBV<br>60 HBV-cirrhosis | Significantly downregulated in HBV-HCC when compared to HL+CBH+HBV-cirrhotic group (p<0.001)<br>AUC to distinguish HBV-HCC from all those groups combined is 0.643 |
| miR-224 | Up [86] | 20 HL, 46 HCC (30 HBV-HCC, 16 others) | Elevated in HCC |
| miR-375 | Up [51] | 50 HL<br>55 HBV-HCC | HBV-HCC significantly upregulated from HL (p<0.0001)<br>Distinguish HBV-HCC from HL<br>AUC 0.96 (specificity of 96%; sensitivity of 100%)<br>When miR-25 and let-7f were tested together with miR-375, it increased the AUC to 0.9967+/- 0.015 (specificity 99.1%; sensitivity 99.1%) |
| miR-500 | Up [95] | 10 HCC (unspecified etiology) | Levels increased in 3/10 HCC patients, significantly reduced within 6 months post HCC resection |
| miR-801 | Up [70] | 68 HL<br>204 HBV-HCC<br>75 chronic HBV<br>60 HBV-cirrhosis | Significantly upregulated when compared to HL+CBH+HBV-cirrhotic group (p<0.001)<br>AUC to distinguish HBV-HCC from all those other groups combined is 0.629 |
| miR-885 (-5p) | Up [93] | 24 HL, 46 HCC (33 HBV-HCC), 26 liver cirrhosis (LC) (20 HBV), 26 chronic HBV | The data demonstrated that patients with HCC, LC, or CHB had significantly (p<0.0001) higher serum levels of miR-885-5p than HL |
| Let-7f | Up [51] | 50 HL<br>55 HBV-HCC | HBV-HCC significantly upregulated from HL. (p<0.0001)<br>Distinguish HBV-HCC from HL when combined with miR-25 and -375, AUC=0.9967 (specificity 99.1%, sensitivity 97.9%) |
| miR-21, -26a, -27a, -122, -192, -223 | Up (for miR-21, miR-192), down for the rest [70] | 68 HL<br>204 HBV-HCC<br>75 chronic HBV<br>60 HBV-cirrhosis | Panel consists of miR-21, -26a, -27a, -122, -192, -223, can diagnose HBV-HCC with an AUC of 0.888, regardless of the BCLC stages<br>Can also differentiate HBV-HCC from HL (AUC=0.941), chronic hepatitis B (AUC=0.842), and cirrhosis (AUC=0.884) |

[1] AUC, area under the operating characteristic curve, where 1 is a perfect test.

most of the information came from HCC tissue studies, increasing evidence suggests alterations were also reflected in patients' serum miRNA level [68]. Enumerating circulating miRNAs would allow for the possibility of gathering information during treatment planning prior to surgery, which would make circulating miRNAs a novel marker that can improve our current HCC management. This section will focus on this subgroup of miRNAs.

The literature search involved all relevant published peer-reviewed articles from PubMed, Ovid, and Medline in English language until July 2013. The search terms were combinations of "hepatocellular carcinoma/hepatoma/liver cancer/liver neoplasm" plus one of the following: *microRNA, serum microRNA, circulating microRNA, diagnosis, prognosis,* or *recurrence*. Only articles that have discussed the use of miRNA as markers of HCC were included. We subsequently found 104 articles that fit these criteria.

Circulating miRNA in liver disease was first demonstrated by Wang et al., using an acetaminophen overdose–induced liver injury mouse model. The authors found a group of miRNAs (e.g., miR-122, miR-192, among many others) that showed dose-dependent and exposure duration–dependent changes in their plasma level, which paralleled changes in serum aminotransferase (AST) level. Zhang et al. [70] subsequently found that circulating miR-122 would rise before AST in patients developing liver injury from HBV infection, and, furthermore, circulating miR-122 level was disease severity–dependent and correlated with histological stages [71].

MiRNAs are remarkably stable in the systemic circulation as they can avoid ribonucleases by forming lipid or lipoprotein complexes (such as apoptotic bodies, microvesicles, and exosomes), and these complexes can withstand extreme pH and temperature, remaining relatively stable for detection ex vivo (i.e., formalin-fixed tissue) [48, 68, 72]. Circulating miRNA can be detected with RT-PCR, microarray, and next-generation sequencing. RT-PCR is the most common technique that utilizes a stem-loop primer binding to the mature miRNA during the reverse transcription step to amplify the desired miRNA. Chen et al. found that miRNA RT-PCR can distinguish single nucleotide differences between related miRNAs [73]. The stability of circulating miRNAs allows reliable enumeration, which is vital as a novel diagnostic biomarker.

Study of circulating miRNAs is not complete without understanding their role in liver tissue. Numerous studies have attempted to correlate aberrant miRNA expression with various roles in hepatocarcinogenesis. Many used microarrays to compare miRNA expression between HCC tissues to their adjacent normal tissues; any aberrantly expressed miRNAs would be confirmed with qRT-PCR and analyzed for statistical significance. Various in silico and in vitro methods would then be employed to identify possible targets of those miRNAs. As a result, several hundred tissue miRNAs and their potential targets have been shown to play a role in hepatocarcinogenesis [74, 75]. These tissue-study findings will be summarized as we study their

corresponding alterations in the circulation (serum) in more than 20 important miRNAs below.

This chapter has highlighted that in order to identify and compare aberrantly expressed circulating miRNAs, a large patient sample with an age-, gender-, and race-matched control group, along with knowledge of hepatitis serology and underlying liver disease is essential [69]. As HCC is a heterogeneous disease with various etiologies, a small sample size could easily obscure true associations and might explain the inconsistencies in the various reports. Furthermore, HBV- and HCV-related HCC often express a different miRNA profile. The majority of the studies were conducted on HBV-related mainland Chinese patients. Although the study reflects the demographics of the disease, patients with other etiology and ethnicities need to be recruited in the future to generate a more coherent picture of the role of miRNAs in hepatocarcinogenesis. It is likely that different panels of miRNAs would need to be employed in the future for patients with different ethnic backgrounds and HCC etiologies.

Another obstacle for comparing miRNA expression across different studies is the lack of a standardized internal reference. Many studies used U6 RNA, which is normally highly expressed. However, RNA is unstable in the plasma and may be subject to degradation if the sample handling is not appropriate and timely [69]. Other studies have used stable and highly expressed miRNAs as their internal control, such as miR-16, -181a, -181c, -638, and -1228. These need to be validated to reach a consensus as to which one is most appropriate and should be used by all researchers in the future.

Future studies will also need to correlate the circulating miRNA level to other aspects of HCC management beyond diagnosis and monitoring, such as prognosis, likelihood of recurrence, and response to chemotherapy. Ji et al. discovered patients with low tissue miR-26 expressing tumors respond better to interferon therapy [76]. This supports the rationale that miRNAs may have a potential to distinguish a subpopulation of patients who may respond favorably to a systemic therapy. The ability to predict tumor recurrence and to prognosticate a patient's disease also has other implications such as decision for liver transplantation or other treatments. The result will be a more targeted treatment, which can improve patient care and distribution of resources.

## 3.4 CIRCULATING MiRNAs IN AUTOIMMUNE DISEASES, INFLAMMATORY AND METABOLIC DISORDERS

Before we cover extensively the role of miRNAs in the pathogenesis and diagnosis of cancer, we can briefly summarize the action of miRNAs in noncancer diseases, in order to fully understand the extension of their role in pathophysiological mechanisms. A promising field of research for the use of miRNAs as markers of disease is definitely one constituted by autoimmune diseases. Wang et al. studied the serum and urinary level of several circulating miRNA species (miR-200 family, miR-205, and miR-192) in patients with systemic lupus erythematosus (SLE) and compared them with that of healthy controls. The serum levels of miR-200a, miR-200b, miR-200c, miR-429, miR-205, and miR-192, and urinary miR-200a, miR-200c, miR-141, miR-429, and miR-192 of SLE patients were lower than those of controls. Glomerular filtration rate (GFR) correlated with serum miR-200b, miR-200c, miR-429, miR-205, and miR-192; proteinuria inversely correlated with serum miR-200a and miR-200c. SLE disease activity index (SLEDAI) inversely correlated with serum miR-200a. Serum miR-200b and miR-192 correlated with platelet count, while serum miR-205 correlated with red cell count and hematocrit [77]. But a prognostic significance can be assigned to the circulating microRNAs also in autoimmune pathologies of a different type. Michael et al. extracted miRNAs from exosomes obtained from the saliva of patients with Sjögren's syndrome [31]; although many cellular miRNA expressions appear to be involved in the development of less serious immunologic diseases such as allergic rhinitis [78], no study has confirmed a different pattern of circulating miRNAs in these types of patients.

Several studies have instead pointed out an altered expression of circulating miRNA in inflammatory diseases such as Crohn's disease (CD) and sarcoidosis. A survey of miRNA levels in the sera of control and patients with CD detected significant elevation of 24 miRNAs, 11 of which were chosen for further validation. All of the candidate biomarker miRNAs were confirmed in an independent CD sample set. To explore the specificity of the CD-associated miRNAs, they were measured in the sera of patients with celiac disease; none was changed compared with healthy controls. Receiver operating characteristic (ROC) analyses revealed that serum miRNAs have promising diagnostic utility, with sensitivities for CD above 80%. Of particular interest is the data that indicate significant decreases in serum miRNAs observed in patients with pediatric CD after 6 months of treatment [79]. Moreover, peripheral blood miRNAs could be used to distinguish active CD and ulcerative colitis (UC) from healthy controls. Five miRNAs were significantly increased and two miRNAs (149* and miRplus-F1065) were significantly decreased in the blood of active CD patients as compared with healthy controls. Twelve miRNAs were significantly increased and miRNA-505* was significantly decreased in the blood of active UC patients as compared with healthy controls. Ten miRNAs were significantly increased and one miRNA was significantly decreased in the blood of active UC patients as compared to active CD patients [80]. However, peripheral blood miRNAs can be developed as unique biomarkers that are reflective and predictive of metabolic health and disorder. It has been reported that miRNAs circulating in the blood can potentially serve as novel noninvasive biomarkers of diseases including diabetes [81]. Karolina et al. have, in fact, identified signature miRNAs that could possibly explain the pathogenesis of T2D and the significance of miR-144 in insulin signaling. Increased circulating levels of miR-144

have been found to correlate with downregulation of its predicted target, insulin receptor substrate 1 (IRS1) at both the mRNA and protein levels. They also experimentally demonstrated that IRS1 is indeed the target of miR-144 [82].

Circulating microRNA expression profiles may be promising biomarkers for diagnosis and assessment of the prognosis of neurologic diseases as well. Cogswell et al. recovered miRNAs from cerebrospinal fluid and discovered Alzheimer's disease (AD)-specific miRNA changes consistently with their role as potential biomarkers for the disease [83], while it was identified that miR-137, miR-181c, miR-9, and miR-29a/b miRNAs were downregulated in the blood serum of probable AD patients. The levels of these miRNAs were also reduced in the serum of AD risk factor models. Although the ability of these miRNAs to conclusively diagnose AD is currently unknown, it was suggested as a potential use for circulating miRNAs, along with other markers, as noninvasive and relatively inexpensive biomarkers for the early diagnosis of AD [84]. However, circulating mRNAs could be useful in a complex neuropsychiatric disorder that involves disturbances in neural circuitry and synaptic function as in schizophrenia. The capacity for discrete postsynaptic remodeling of neurons requires coordination by an elaborate intracellular network of molecular signal transduction systems. The redundancy of these networks means that many combinations of gene variants have the potential to cause system dysfunction that manifests as related neurobehavioral syndromes. Recent investigation has revealed that posttranscriptional gene regulation and associated small noncoding miRNA are likely to be important factors that shape the topography of these networks. MiRNAs display complex temporal spatial expression patterns in the mammalian brain and have the potential to regulate thousands of target genes by functioning as the specificity factor for intracellular gene-silencing machinery. They are emerging as key regulators of many neurodevelopmental and neurological processes such as in the pathophysiology of schizophrenia, as their dysregulation could lead to pervasive changes in the network structure during development and in the mature brain. Furthermore, modification of miRNAs could be found not only in the cerebral cortex but at a peripheral level as well [85]. Finally, circulating mRNAs could be used as biomarker in Duchenne muscular dystrophy (DMD), a lethal X-linked disorder caused by mutations in the dystrophin gene, which encodes a cytoskeletal protein, dystrophin. Creatine kinase (CK) is generally used as a blood-based biomarker for muscular disease including DMD, but it is not always reliable because it is easily affected by stress to the body, such as exercise. The expression levels of specific serum-circulating miRNAs may be useful to monitor the pathological progression of muscular diseases. Mizuno et al. found that the serum levels of several muscle-specific miRNAs (miR-1, miR-133a, and miR-206) are increased in both dystrophin-deficient muscular dystrophy mouse models and canine X-linked muscular dystrophy in Japan dog models. Interestingly, unlike CK levels, expression levels of

**TABLE 3.2**
**Circulating MiRNAs in Nonneoplastic Diseases**

| Disease | MiRNA | Variation | Significance |
|---|---|---|---|
| Systemic lupus erythematosus | miR-200a, 200b, 200c, 205, 429, 192 (serum), miR-200a, 200c, 141, 429, 192 (urinary) | Reduced | Correlation with GFR, proteinuria, and SLE disease activity index |
| Crohn's disease | miR-149, plus F1065 | Reduced | |
| Ulcerative colitis | miR-505 | Reduced | |
| Diabetes | miR-144 | Increased | Correlation with insulin receptor substrate 1 |
| Alzheimer's disease | miR-137, 181c, 9, 29a/b | Reduced | Markers of early disease |
| Duchenne muscular dystrophy | miR-1, 133a, 206 | Increased | Experimental models |
| Preeclampsia | miR-210 | Increased | |
| | miR-152 | Reduced | |

*Abbreviations:*   GFR, glomerular filtration rate; SLE, systemic lupus erythematosus.

these miRNAs in mdx serum are only slightly influenced by exercise when using a treadmill. These results suggest that serum miRNAs are useful and reliable biomarkers for muscular dystrophy [86]. Laterza et al. have demonstrated that aberrant expression of circulating miRNAs corresponds to the origin of tissue injury [87]. Circulating miRNAs are in fact biomarkers for diverse cardiovascular diseases, including acute myocardial infarction, heart failure, coronary artery disease, stroke, essential hypertension, and acute pulmonary embolism [88]. But circulating miRNAs seem to be able to regulate various physiological processes such as pregnancy. It is in fact well known that maternal plasma miRNA expression profiles dynamically change during pregnancy. The results of function analysis suggested that miRNAs may play an important role in regulating the pregnancy process, which can help us understand the refined regulatory mechanism in pregnancy [89]. It has also been shown that placenta-associated circulating miRNAs correlate with pregnancy progression [90]. Even more relevant could be the circulating mRNA profile, in that it could be useful to predict pregnancy-related disease such as preeclampsia, which continues to be a fatal disease among pregnant women. Gunel et al. found upregulated miR-210 levels as well as downregulated miR-152 levels in preeclampsia patients (Table 3.2) [91].

## 3.5 CELLULAR AND CIRCULATING MiRNAs IN NEOPLASTIC DISEASES

### 3.5.1 CELLULAR MiRNAs IN ONCOLOGIC PATIENTS

Because cellular miRNAs play a fundamental role in regulation of a variety of cellular, physiological, and developmental processes, it is easy to understand that their aberrant expression can lead to a variety of human diseases including cancer. A number of cellular miRNAs were shown in fact to be located in fragile regions of the human genome that are associated with cancer [92], and dysregulated miRNAs have been shown to play a crucial role in tumor initiation, progression, and metastasis, and are often associated with diagnosis, prognosis, and response to therapy (Table 3.3) [93–95]. Lu et al. developed a method for bead-based miRNA profiling. Employing this technique on 20 different cancers, they found that each cancer had a specific miRNA profile and that most poorly differentiated tumors could be classified to their tissues of origin based on their miRNA expression levels [96]. Among their functional roles, cellular miRNAs regulate in fact development, organogenesis, hematopoiesis, and cell proliferation, and they can intervene with various mechanisms in the genesis of neoplastic disease acting on apoptosis, angiogenesis, and neoplastic microenvironments.

Several studies have pointed out the action of cellular miRNA on the onset of cancer, and miRNA-21 is one of the most prominent miRNAs implicated in the genesis and progression of human cancer. The earliest study showed that miRNA-21 is commonly and markedly upregulated in human glioblastoma, and inhibition of miRNA-21 expression leads to caspase activation and associated apoptotic cell death in multiple glioblastoma cell lines [97]. Subsequently, there is a growing body of evidence to prove that miRNA-21 is overexpressed in a variety of tumors such as breast cancer [97], lung cancer [98], colon cancer, and HCC, with proproliferative and antiapoptotic function [99–102]. Upregulation of miRNA-31 was found in colon cancer and squamous cell carcinomas of the tongue. Interestingly, in colon cancer studies not only was miRNA-31 upregulated but this upregulation was found to be associated with poorer clinical outcomes and increased invasiveness [103–106]. However, there is some support for downregulation of miRNA-31 in gastric cancer and urothelial cancers [107–110]. Moreover, a comparison of miRNAs and mRNA profiles of primary and metastatic cancer lesions showed that miRNAs provided a more reliable and distinctive signature than mRNAs and found that miRNA signatures were superior to mRNAs in identifying the organ source of metastases of unknown origin [96, 111]. As for the action of cellular miRNAs on apoptosis, it is well known that miRNA-195 is an important member of the micro-15/16/195/424/497 family, which is

## TABLE 3.3
## Cellular MiRNAs in Cancer Diseases

| Tumor Type | Expression | MiRNAs |
| --- | --- | --- |
| Glioblastoma | Reduced | 21 |
| | Increased | — |
| Breast cancer | Reduced | 193b, Let-7, 29, 200/141 |
| | Increased | 21, 155, 17-92 |
| Lung cancer | Reduced | 486-5p, let-7, 29 |
| | Increased | 21, 155, 17-92 |
| Colon cancer | Reduced | 34 |
| | Increased | 21, 31, 155 |
| HCC | Reduced | 21 |
| | Increased | — |
| Tongue | Reduced | 31 |
| | Increased | — |
| Gastric cancer | Reduced | 31, 192, 215, 200/141 |
| | Increased | 21 |
| Urothelial cancer | Reduced | 31 |
| | Increased | — |
| Bladder cancer | Reduced | 143, 145 (CAFs), 200/141 |
| | Increased | 16, 320 (CAFs) |
| Endometrial cancer | Reduced | 503, 424, 29b, 146a |
| | Increased | 31 (CAFs) |
| Pancreatic cancer | Reduced | 18a, 21 |
| | Increased | — |
| Testicular cancer | Reduced | 372, 373 |
| | Increased | — |
| Lymphomas | Reduced | 155, 17-92 |
| | Increased | — |
| CLL | Reduced | 15a, 16-1, 29 |
| | Increased | 155, 21 |
| AML | Reduced | 29 |
| | Increased | 155, 21, let-7, 126, 196b |
| ALL | Reduced | 128 |
| | Increased | — |

*Abbreviations:* ALL, acute lymphoblastic leukemia; AML, acute myeloid leukemia; CLL, chronic lympho-cytic leukemia; HCC, hepatocellular carcinoma.

activated in multiple diseases, such as cancers. MiRNA-195 regulates a plethora of target proteins, which are involved in the cell cycle, apoptosis, and proliferation. WEE1, CDK6, and Bcl-2 are confirmed target genes of miRNA-195 that are involved in miRNA-195-mediated cell cycle and apoptosis process effects [112]. Moreover, Donzelli et al. showed that p53R175H, a hotspot p53 mutant, induces miRNA-128-2 expression. Mutant p53 binds to the putative promoter of miR-128-2 host gene, ARPP21, determining a concomitant induction of ARPP21 mRNA and miRNA-128-2. MiRNA-128-2 expression in lung cancer cells inhibits apoptosis and confers increased resistance to cisplatin, doxorubicin, and 5-fluorouracil treatments. At the molecular level, miRNA-128-2 posttranscriptionally targets E2F5 and leads to the abrogation of its repressive activity on p21(waf1) transcript. The p21(waf1) protein localizes to the cytoplasmic compartment, where it exerts an antiapoptotic effect by preventing

procaspase-3 cleavage [113]. It is also well known that a tumor needs to grow its own vasculature. The miRNA-17-92 cluster is highly expressed in human endothelial cells that participate in angiogenesis. Yin et al. showed that miRNA-19b-1, a component of this cluster, controls the intrinsic angiogenic activity of human umbilical vein endothelial cells (HUVECs) in vitro. In silico and in vitro analyses have suggested that miRNA-19b-1 targets mRNA corresponding to the proangiogenic protein, FGFR2, and blocks the cell cycle from the S phase to the G2/M phase transition by controlling the expression of cyclin D1. Thus, miRNA-19b-1 may serve as a valuable therapeutic agent in the context of tumor angiogenesis [114]. Finally, it is possible that cellular miRNAs can create better conditions for the growth of the tumor. A recent bladder cancer study has identified specific miRNA profiles in cancer-associated fibroblasts (CAFs) [115]. These studies observed an increase in miRNA-16 and miRNA-320 in CAFs in comparison with normal bladder fibroblasts. It is possible that the upregulated miRNAs in CAFs facilitate tumor survival or progression. In contrast to the upregulated miRNAs, miRNA-143 and miRNA-145 are downregulated in CAFs compared with foreskin fibroblasts. There was no significant decrease in miRNA-143 and miRNA-145 between CAFs and normal bladder fibroblasts, however, although a trend was reported. Previous studies in bladder cancer have shown decreased levels of miRNA-145 correlated with a decrease in apoptotic ability suggesting that a decrease in miRNA-145 in CAFs may also contribute to evasion of apoptosis in the micro-environment [116–119]. Another study looking at different CAF cell lines from endometrial cancers identified several differentially expressed miRNAs in the CAFs in comparison to normal fibroblasts [18]. These included increased expression of miRNA-503, miRNA-424, miRNA-29b, miRNA-146a, and miRNA-31 [120].

A number of target genes for miRNA-31 are involved in cellular movement, cytoskeletal organization, phagocytosis, transport, cellular transformation, anchorage-independence, and chromatin remodeling, such as CCNJ, ELAVL1, ENY2, RHOBTB1, CLASP2, VAMP4, STX12, TACC2, and SATB2. In any case it has been confirmed that cellular miRNAs are involved in the development of a variety of tumors, such as leukemia, neuroblastoma, pituitary adenoma, breast cancer, thyroid cancer, hepatocarcinoma, colorectal cancer, and lung cancer. The upregulation or downregulation of different miRNAs in these tumor tissues were found with most of the miRNA targets located in regions of tumor-related genome such as fragile sites, loss of heterozygosity, and amplified regions, thereby exhibiting the same effect as oncogenes or tumor suppressor genes [92, 121]. Bullock et al. demonstrated their importance in the early stages of the metastatic cascade. An important event in the metastatic cascade is epithelial to mesenchymal transition (EMT), a reversible phenotypic switchover, which endows malignant epithelial cells with the capacity to break free from one another and invade the surrounding stroma. Their understanding of EMT has been significantly improved by the characterization of miRNAs that influence the signaling pathways and downstream events that define EMT on a molecular level [122]. In addition, miRNAs were prognostic indicators in colon cancer patients [52], while miRNA expression differentiated histology and predicted survival and relapse of lung cancer (Table 3.3) [123–125].

### 3.5.2 Circulating MiRNAs and Cancer

Circulating miRNAs have been implicated in regulation of stem cells as well as cancer stem cells. Given that cancer stem cells are believed to be responsible for cancer initiation, metastasis, and chemotherapy resistance, a better understanding of how circulating miRNAs mediate gene expression in cancer stem cells will help identify novel cancer biomarkers and therapeutic targets (Table 3.4). As a result, it will aid in the development of better strategy for cancer treatment [126]. Mitchell et al. demonstrated the presence of circulating tumor-derived miRNAs in blood by using a mouse prostate cancer xenograft model system and showed that measurements obtained from plasma were strongly correlated with those obtained from sera, suggesting that both serum and plasma samples would be adequate for measuring specific miRNA levels [44]. In another study, Chen et al. demonstrated that by using serum directly or by extracting RNA from the serum they could identify unique miRNA expression profiles for lung cancer, and colorectal cancer patients compared with healthy subjects [48]. Circulating miRNAs have also been postulated as novel biomarkers for ovarian cancer, pancreatic cancer, and colorectal cancer [38, 52, 127–129]. Although the clinical significance of these findings has not been elucidated in detail, those findings demonstrated that circulating miRNAs could be noninvasive diagnostic or prognostic markers for cancer. We might wonder what could be the mechanism by which circulating miRNAs can exert their action. It was pointed out that there must be some method of communication between the cancer cells and their microenvironment. The dynamic cross-talk between cancer cells and normal cells in the microenvironment is a crucial point in the progression of disease. One manner of cell–cell communication is through the secretion of molecules and paracrine signaling. Molecules of secretion are no longer limited to cytokines, chemokines, growth factors, and other protein molecules, but now include miRNA species [130].

Observations that miRNAs act as mediators of heterotypic signaling in the tumor stroma and ECM could provide researchers with a novel target for such therapies. Given their biological importance, it is not surprising that circulating miRNA expression is frequently dysregulated in human cancer. Accumulating evidence suggests that circulating miRNAs can contribute to tumorigenesis either by directly modulating oncogenic or tumor suppressor pathways and/or being regulated by oncogenes or tumor suppressor genes. To date dozens of cancer types have been investigated in which expression profiling of circulating miRNAs have revealed both diagnostic and prognostic utility for this class of biomarkers [131, 132], and several authors recently showed that aberrant plasma

### TABLE 3.4
### Circulating MiRNAs in Cancer Diseases

| Tumor Type | Expression | MiRNAs |
| --- | --- | --- |
| Prostate | Reduced | — |
|  | Increased | 16, 92a, 103, 107, 197, 485-3p 486-5p, 26a, 92b, 574-3p, 636, 640, 766, 885-5p, 141, 195, leti, 375, 298, 346 |
| Breast cancer | Reduced | — |
|  | Increased | 10b, 34a, 195, let-7, 223 |
| Ovarian cancer | Reduced | 155, 127, 99b |
|  | Increased | 21, 141, 200a, 200b, 200c, 203, 205, 214 |
| Lung cancer | Reduced | 486-5p, 146b, 221, 17-5p, 106a, let-7b, 223 |
|  | Increased | 7i, 206, 21, 30d, 1, 491, 210, 10b, 34a, 141, 155, 197, 182, 29c |
| Colorectal cancer | Reduced | — |
|  | Increased | 17-3p, 92a, 29a, 141 |
| Oral cancer | Reduced | 125a, 200a (saliva) |
|  | Increased | 31, 24 |
| Hepatocellular carcinoma | Reduced | — |
|  | Increased | 21, 122, 34a, 16 |
| Pancreatic cancer | Reduced | — |
|  | Increased | 155 (pancreatic juice), 18 |
| Biliary tract cancer | Reduced | — |
|  | Increased | 9, 145 (bile) |
| Esophageal carcinoma | Reduced | 375 |
|  | Increased | 21 |
| Gastric cancer | Reduced | — |
|  | Increased | 187, 371-5p, 378 |
| Renal cancer | Reduced | — |
|  | Increased | 1233 |
| Lymphoma | Reduced | — |
|  | Increased | 21, 155, 210 |
| Chronic lymphocytic leukemia | Reduced | — |
|  | Increased | 195, 29a, 222, 20a, 150, 451, 135a, 486-5p |
| Acute myeloid leukemia | Reduced | — |
|  | Increased | 92 |
| Myelodisplastic syndrome | Reduced | Let7a, 16 |
|  | Increased | — |
| Multiple myeloma | Reduced | — |
|  | Increased | 148a, 181a, 20a, 221, 625, 99b |

expressions of miRNAs could distinguish cancer patients from healthy individuals (Table 3.4) [43, 50, 133, 134].

In order to efficiently discover serum miRNAs that are differentially abundant between cancer cases vs controls [135], real-time PCR-based miRNA TaqMan low-density arrays (TLDA) have been used to screen for differential abundance of 365 miRNAs in serum RNA pooled from mCRPC patients (n=25) vs age-matched controls (n=25; all controls had normal PSA and normal digital rectal exam findings). After normalization of data using spike-in control miRNAs and comparison of pooled serum RNA from mCRPC cases with that of controls (Figure 3.1a and inset),

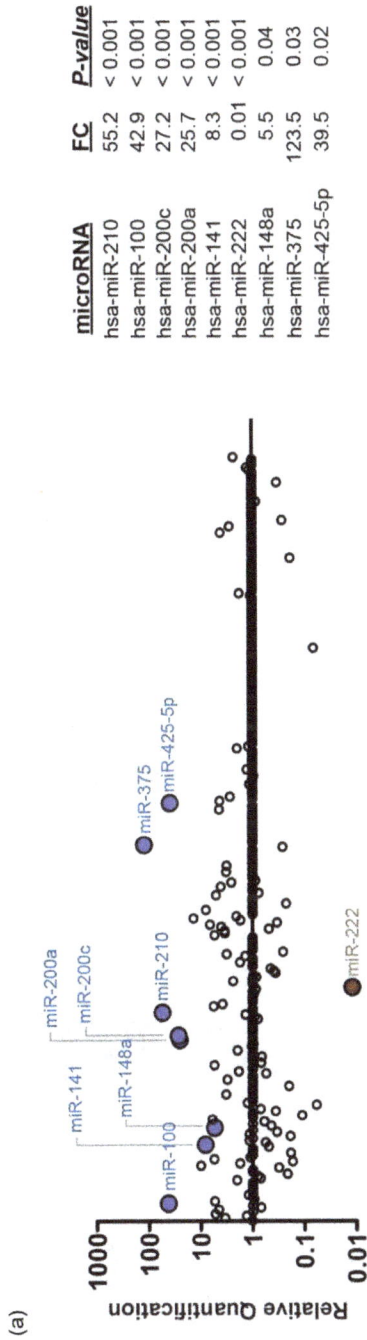

| microRNA | FC | P-value |
|----------|-----|---------|
| hsa-miR-210 | 55.2 | < 0.001 |
| hsa-miR-100 | 42.9 | < 0.001 |
| hsa-miR-200c | 27.2 | < 0.001 |
| hsa-miR-200a | 25.7 | < 0.001 |
| hsa-miR-141 | 8.3 | < 0.001 |
| hsa-miR-222 | 0.01 | < 0.001 |
| hsa-miR-148a | 5.5 | 0.04 |
| hsa-miR-375 | 123.5 | 0.03 |
| hsa-miR-425-5p | 39.5 | 0.02 |

**FIGURE 3.1    Serum miRNA profiling and validation.** (a) Measurement of circulating miRNAs in sera pooled from patients with advanced prostate cancer as compared to healthy donors (comprising a Discovery Set) by TLDA profiling. *Blue-* and *brown-filled circles* represent serum miRNAs increased or decreased (with unadjusted p-value<0.05), respectively, in mCRPC patients compared with healthy controls. Inset: Nine miRNAs demonstrated 0.5-fold change (unadjusted p<0.05, student's t-test). FC, fold-change. (b) Confirmation of mCRPC-associated serum miRNAs in individual samples from the Discovery Set from the University of Washington samples. Upper: miRNA biomarker candidates were measured in individual samples by TaqMan miRNA qRT-PCR (p-value assigned by Wilcoxon signed-rank test), where miRNA abundance is given in terms of miRNA copies/mL serum. *Red bars*, mean +/− SEM of miRNA copies/mL serum for each group. Lower: Receiver operating characteristic (ROC) curves plot sensitivity vs (1 − specificity) to assess the ability of each miRNA biomarker to distinguish cases from controls. (c) Validation of mCRPC-associated serum miRNAs in an independent Validation Set. Upper: Serum concentration (copies/μl) of miR-141, miR-200a, miR-200c, miR-210, and miR-375 was measured by TaqMan miRNA qRT-PCR. Dot plot–associated p-values were assigned by Wilcoxon signed-rank test. Dot plots and ROC curves were generated as described for **Figure 3.1**. Lower: *Red*, results from the validation sample set obtained from the University of Michigan. *Black*, results from the primary sample set obtained from the University of Washington reproduced from **Figure 3.1b**, lower. AUC, area under the curve; mCRPC, prostate cancer patient sera; FC, fold-change; CTL, control sera (from age-matched male individuals with normal PSA and negative digital rectal exam).

*(Continued)*

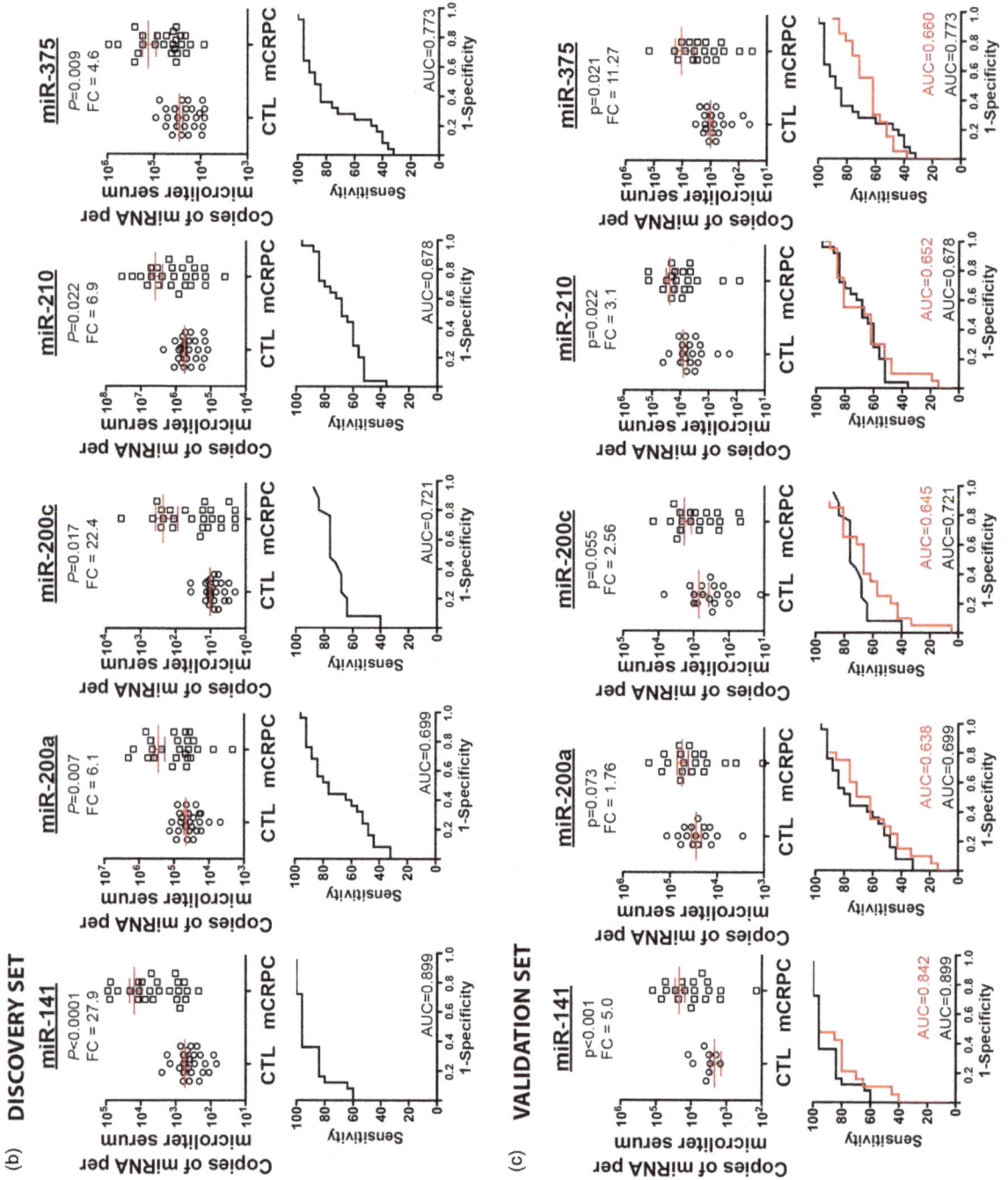

**FIGURE 3.1** *(Continued)*

nine miRNAs have demonstrated a greater than five-fold change in abundance (unadjusted P,0.05, student's t-test). Next, these nine miRNAs from the serum RNA of individual mCRPC cases and controls were individually measured using miRNA-specific TaqMan qRT-PCR assays [135]. Serum levels of five miRNAs were confirmed to be significantly elevated in mCRPC cases compared with controls (miR-141: p,0.0001, miR-200a:

p=0.007, miR-200c: p=0.017, miR-375: p=0.009, and miR-210: p=0.022, Wilcoxon signed-rank analysis). The average fold difference between cases and controls ranged from 4.6 (miR-375) to 27.9 (miR-141) (Figure 3.1b, upper). In addition, ROC plots demonstrate the capacity of these miRNAs to discriminate between the two groups (miR-141 area under the curve [AUC]=0.899; miR-200a AUC=0.699; miR-375 AUC=0.773; miR-200c AUC=0.721; and miR-210 AUC=0.678) (Figure 3.1b, lower). Importantly, they [135] verified that control miRNAs were not differentially expressed between the two populations.

To validate these findings in an independent specimen set collected at a different institution, we measured miR-141, miR-200a, miR-200c, miR-375, and miR-210 from the sera of an additional 21 mCRPC patients and 20 age-matched healthy controls collected at the University of Michigan. All five miRNAs were elevated in sera from mCRPC cases relative to controls in this independent validation set. MiR-141, miR-375, and miR-210 were significant at a p-value threshold of 0.01 in the second cohort (p=0.001, p=0.021, p=0.022, respectively) and miR-200a and miR-200c tended toward significance (p=0.073, p=0.055, respectively) (Figure 3.1c, upper). ROC curves were generally concordant between the specimen sets from the two institutions (Figure 3.1c, lower). Analysis of serum miRNA markers in various combinations demonstrated that adding more miRNAs to serum miR-141 (which had the best performance alone) did not improve the ability to distinguish between cases and controls. Consistent with this observation, it has been found that among cancer cases in which expression of miR-141, miR-200a, miR-200c, and miR-375 was higher than all healthy controls, these miR-NAs were also significantly correlated with each other and with serum PSA. In contrast, miR-210 did not show significant correlation with any of these four miRNAs nor with serum PSA, suggesting that it provides distinct information about disease biology.

Three of the serum prostate cancer–associated miRNAs identified (miR-141, miR-200a, and miR-200c) are epithelial specific, highly related in sequence, and have known roles in maintaining the epithelial state by suppression of the epithelial-to-mesenchymal transition [136]. It has been hypothesized that elevated circulating levels of miR-141, miR-200a, and miR-200c reflect the epithelial origin of prostate cancer cells.

The presence of elevated circulating miR-141 and miR-375 in mCRPC patients has also been observed in recent mCRPC circulating miRNA biomarker studies [137–141].

Interestingly, elevated miR-210 was not reported in these other studies, despite the fact that we observed this in independent specimen sets from two different institutions. This could be due to different comparison groups used (e.g., localized prostate cancer rather than healthy controls as the comparator to mCRPC), the use of plasma rather than serum, differences in the data analytic approach used to identify differentially expressed miRNAs, as well as potential differences in the clinical characteristics of the mCRPC patients across different studies.

The elevated levels of miR-210 in serum from patients with mCRPC were particularly interesting because this miRNA is well known to be transcriptionally activated by the hypoxia-inducible factor 1 alpha (HIF-1α) [142, 143] and may contribute to adaptation to hypoxia in tumors [144, 145]. This raises the possibility that miR-210 is produced and released by hypoxic cells in prostate cancer (and/or by the tumor microenvironment), a potential explanation for elevated levels of miR-210 we observed in the serum of a subset of patients with mCRPC.

## 3.6 SPECIFIC PATTERNS OF CIRCULATING MiRNAs IN CANCER PATIENTS

### 3.6.1 LUNG CANCER

It is well known that upregulation of cellular miRNA-21 leads to tumor development and progression [97, 146, 147], but circulating miRNA-21 has also been described as a biomarker for different tumor entities [148, 149]. For instance, it was recently found that miRNA-21 was one of the plasma miRNAs that could differentiate early-stage lung cancer patients from healthy nonsmoking individuals [150]. In a recent study, miRNA-21 and miRNA-210 display higher plasma expression levels, whereas miRNA-486-5p has a lower expression level in patients with malignant solitary pulmonary nodules (SPNs), as compared to subjects with benign SPNs and healthy controls. A logistic regression model with the best prediction was built on the basis of miR-21, miR-210, and miR-486-5p. The three miRNAs used in combination produced the area under ROC curve at 0.86 in distinguishing lung tumors from benign SPNs with 75.00% sensitivity and 84.95% specificity. Validation of the miRNA panel in the testing set confirms their diagnostic value that yields significant improvement over any single one [150]. All these data were confirmed by other studies. Plasma miRNA-21 levels were in fact significantly higher in non–small cell lung cancer (NSCLC) patients than in age- and sex-matched controls and were related to Tumor, Node, Metastasis (TNM) classification stage, but not related to age, sex, smoking status, histological classification, lymph node status, and metastasis. Importantly, miRNA-21 plasma levels in partial response samples were several folds lower than that in stable disease plus progressive disease samples, and were close to that in healthy controls. In such a case, plasma miRNA-21 can serve not only as a circulating tumor biomarker for the early diagnosis of NSCLC, but also to the sensitivity of platinum-based chemotherapy [151].

A different promising miRNA is miRNA-486-5p, which was shown to regulate tumor progression and OLFM4 antiapoptotic factor [152]. Furthermore, cellular miRNA-486-5p was underexpressed in several types of solid tumors and in lung cancer [153, 154]. Several studies showed that plasma miRNA-486-5p expression in lung cancer patients is significantly lower compared to subjects with both benign solitary pulmonary nodules (SPNs) and healthy smokers. The findings in both surgical tissues and plasma specimens suggest that miRNA-486-5p downregulation might play a role as a tumor suppressor in lung tumorigenesis [150, 155]. The concentrations of four circulating microRNAs (miRNA-10b, miRNA-34a, miRNA-141, and miR-RNA155) significantly discriminated cancer patients from healthy individuals. The levels of miRNA-10b, miRNA-141, and miRNA-155 were significantly higher in lung cancer patients than those in patients with benign disease. In lung cancer patients, high serum miRNA-10b values associated with lymph node metastasis and elevated levels of tissue polypeptide antigen (TPA), whereas high serum miR-141 values associated with elevated levels of urokinase plasminogen activator (uPA) [156]. A significant correlation between circulating miRNA and stage disease was pointed out by Zheng et al., who found that the levels of miR-155, miR-197, and miR-182 in the plasma of lung cancer including stage I patients were significantly elevated compared with controls. The combination of these three miRNAs yielded 81.33% sensitivity and 86.76% specificity in discriminating lung cancer patients from controls. The levels of miR-155 and miR-197 were higher in the plasma from lung cancer patients with metastasis than in those without metastasis and were significantly decreased in responsive patients during chemotherapy [157]. Heegaard et al. had different findings, specifically that the expression of miRNA-146b, miRNA-221, let-7a, miRNA-155, miRNA-17-5p, miRNA-27a, and miRNA-106a were significantly reduced in the serum of NSCLC cases while miRNA-29c was significantly increased. No significant differences were observed in plasma of patients compared with controls. Overall, expression levels in serum did not correlate well with levels in plasma. In secondary analyses, reduced plasma expression of let-7b was modestly associated with worse cancer-specific mortality in all patients and reduced serum expression of miR-223 was modestly associated with cancer-specific mortality in stage IA/B patients [158]. Finally of extreme interest could be the datum that Bianchi et al. developed a test, based on the detection of 34 miRNAs from serum, that could identify patients with early-stage NSCLCs in a population of asymptomatic high-risk individuals with 80% accuracy. The signature could assign disease probability accurately either in asymptomatic or symptomatic patients, distinguish between benign and malignant lesions, and capture the onset of the malignant disease in individual patients over time [159].

### 3.6.2 Breast Cancer

Analyses of cellular miRNA profiles in breast cancer have determined that many miRNAs display expression patterns linked to molecular subtype [160–163] as well as ER status, tumor grade [161], and other tumor-related processes [164–166]. Experimental evidence has confirmed that miRNA levels can play a role in determining disease course; for example, re-expression of miRNA-193b, downregulated in highly metastatic derivatives of the MDA-MB-231 cell line, significantly inhibited tumor growth and dissemination in a mouse xenograft model. Thus, integrated analysis of miRNA in breast cancer constitutes an important new frontier [167]. Yan et al. found that overexpression of miR-21 in breast cancer is associated to lymph node metastasis and poor prognosis [107]. Roth and colleagues reporting the findings of altered tumor-specific miRNAs in sera of breast cancer patients, while Mitchell et al. have previously demonstrated that circulating miRNAs that are elevated in breast cancer patients when the tumor is in situ (miR-195 and let-7a) decrease to basal levels by 2 weeks post tumor resection [44, 132, 168]. Recent data have demonstrated that macrophages are able to produce microvesicles, which shuttle proteins or miRNAs into adjacent cells within the microenvironment [23, 25, 169–172]. Yang et al. demonstrated that exogenous miRNAs transfected into IL-4-activated M2 macrophages can be shuttled into cocultivated breast cancer cells in the absence of direct cell–cell contact with the macrophages. Exosomes containing miRNA-223 were released by M2 cells and were then internalized by cocultivated breast cancer cells that did not express this miRNA. The exosome-shuttled miR-223 promoted the invasiveness of breast cancer cells in vitro. This process of invasion could be inhibited by transfecting miRNA-223 antisense oligonucleotides (ASO) into the tumor cells. The above provides evidence for the delivery of invasion-potentiating miRNA-223 by IL-4-activated macrophages to breast cancer cells via exosomes and may highlight a novel communication mechanism between tumor-associated macrophages (TAMs) and cancer cells [173].

### 3.6.3 Patients with Liver Damage and Hepatocellular Carcinoma

In chronic hepatitis C (CHC) infection patients and in patients with nonalcoholic fatty liver disease (NAFLD), plasma levels of miR-122 were elevated compared to healthy controls. In CHC patients, miR-122 levels correlated with fibrosis stage and inflammation activity but did not correlate with HCV viral load [174–178]. In HCC patients, plasma miRNA-21 level was measured by qRT-PCR before and after curative resection of HCC. Plasma miRNA-21 was also compared in other groups of patients with chronic hepatitis, and healthy volunteers. Plasma miRNA-21 level in the patients with HCC was significantly higher than in patients with chronic hepatitis and healthy volunteers. In the patient group after resection of the carcinoma, plasma miRNA-21 levels significantly diminished after surgery compared with the preoperative values. ROC analysis of plasma miRNA-21 yielded an AUC of 0.773 with 61.1% sensitivity and 83.3% specificity when differentiating HCC from chronic hepatitis, and an AUC of 0.953 with 87.3% sensitivity and 92.0% specificity when differentiating HCC from healthy volunteers. Both sets of values were superior to α-fetoprotein and improved for the combination of miRNA-21 and α-fetoprotein [179]. Plasma miRNA-21 level remains as a promising biochemical marker for HCC and liver damage, although Cermelli et al. found that whereas miR-21 extracellular levels were unchanged, extracellular levels of miR-122, miR-34a, and to a lesser extent miR-16, steadily increased during the course of liver damage [180]. MiR-122 in serum was significantly higher in HCC patients than healthy controls. More importantly, it was found that the levels of miRNA-122 were significantly reduced in the postoperative serum samples when compared with the preoperative samples. Although serum miR-122 was also elevated in HBV patients with HCC comparing with those without HCC, the difference was at the border line (p=0.043) [181].

### 3.6.4 PANCREATIC CANCER

Aberrant expression of miRNAs is associated with phenotypes of various cancers, including pancreatic cancer (PaC). However, the mechanism of the aberrant expression is largely unknown. Activation of the mitogen-activated protein kinase (MAPK) signaling pathway plays a crucial role in gene expression related to the malignant phenotype of PaC. Ikeda et al. studied the role of MAPK in the aberrant expression of miRNAs in PaC cells. They found that the cellular miRNAs, miRNA-7-3, miRNA-34a, miRNA-181d, and miRNA-193b, were preferentially associated with MAPK activity. Among these miRNAs, miRNA-7-3 was upregulated by active MAPK, while the others were downregulated. Promoter assays indicated that the promoter activities of the host genes of miRNA-7-3 and miRNA-34a were both downregulated by alteration in MAPK activity. Exogenous overexpression of the MAPK-associated miRNAs had the effect of inhibition of the proliferation of cultured PaC cells; miRNA-193b was found to exhibit the most remarkable inhibition [182]. Regarding circulating miRNAs, the comparative analysis of miRNA expression during malignant progression in the mouse model points to some conclusions about relevant miRNAs in the circulation that can indicate the presence of precursor lesions. Habbe et al. [183] reported on the expression levels of miRNAs in human intraductal papillary mucinous neoplasm (IPMN) tissues and in pancreatic juice samples, and concluded that miRNA-155 is upregulated and a possible tissue biomarker of preinvasive disease. La Conti et al. reported that changes in miRNA expression patterns during progression of normal tissues to invasive pancreatic adenocarcinoma in the p48-Cre/LSL-KrasG12D mouse model mirrors the miR changes observed in human PaC tissues. MiRNA-148a/b and miRNA-375 expressions were found decreased whereas miRNA-10, miRNA-21, miRNA-100, and miRNA-155 were increased when comparing normal tissues, premalignant lesions, and invasive carcinoma in the mouse model. Predicted target mRNAs FGFR1 (miRNA-10) and MLH1 (miRNA-155) were found to be downregulated. Quantitation of nine microRNAs in plasma samples from patients distinguished pancreatic cancers from other cancers as well as noncancerous pancreatic disease. Finally, gemcitabine treatment of control animals and p48-Cre/LSL-KrasG12D animals with PaC caused distinct and up to 60-fold changes in circulating miRNAs that indicate differential drug effects on normal and cancer tissues [184]. MiRNA-18a is located in the miRNA-17-92 cluster and reported to be highly expressed in PaC tissues. The expression of miRNA-18a was significantly higher in PaC tissues and PaC cell lines than in normal tissues and fibroblasts. Moreover, plasma concentrations of miRNA-18a were significantly higher in PaC patients than in controls. Plasma levels of miRNA-18a were significantly lower in postoperative samples than in preoperative samples [185]. Several other studies confirmed the relevance of circulating miRNAs in PaC cancer patients. Seven miRNAs displayed significantly different expression levels in PaC compared with controls. This serum 7-miRNA-based biomarker had high sensitivity and specificity for distinguishing various stages of PaC from cancer-free controls and also accurately discriminated PaC patients from chronic pancreatitis (CP) patients. Among the seven miRNAs, miRNA-21 levels in serum were significantly associated with overall PaC survival. The diagnostic accuracy rate of the serum 7-miRNA profile was 83.6% in correctly classifying 55 cases with clinically suspected PaC [186]. Hwang et al. retrospectively studied the effect of miRNA-21 on the chemotherapy efficacy and prognosis of PaC: patients with low expression of miRNA-21 have significantly prolonged overall and disease-free survival after chemotherapy; antisense inhibition of miRNA-21 can enhance chemotherapeutic sensitivity. MiRNA-21 may be related to platinum sensitivity. The high expression of plasma miRNA-21 may also be involved in cell resistance to platinum chemotherapy. Inhibiting miRNA-21 expression is likely to restore cell sensitivity to platinum chemotherapy [187].

### 3.6.5 BILIARY TRACT CANCER: MiRNA IN BODILY FLUIDS

Besides the possibility of assessing the serum or plasma circulating concentrations of miRNAs, as has already been mentioned, the possibility to dose such markers in other bodily fluids exists. This is especially true in patients with biliary tract cancers (BTCs) in which the evaluation can be carried out in the bile. Shigehara et al. sampled bile from patients who underwent biliary drainage for biliary diseases such as BTC and choledocholithiasis. PCR-based miRNA detection and miRNA cloning were performed to identify bile miRNAs. Using high-throughput real-time PCR-based miRNA microarrays, the expression profiles of 667 miRNAs were compared in patients with malignant disease and age-matched patients with the benign disease choledocholithiasis. They subsequently characterized bile miRNAs in terms of stability and localization. Through cloning and using PCR methods, they confirmed that miRNAs exist in bile. Differential analysis of bile miRNAs demonstrated that 10 of the 667 miRNAs were significantly more highly expressed in the malignant group than in the benign group. Setting the specificity threshold to 100% showed that some miRNAs (miRNA-9, miRNA-302c*, miRNA-199a-3p, and miRNA-222*) had a sensitivity level of 88.9%, and ROC analysis demonstrated that miRNA-9 and miRNA-145* could be useful diagnostic markers for BTC. Moreover, they verified the long-term stability of miRNAs in bile, a characteristic that makes them suitable for diagnostic use in clinical settings. These findings suggest that bile miRNAs could be informative biomarkers for hepatobiliary disease and that some miRNAs, particularly miR-9, may be helpful in the diagnosis and clinical management of BTC [188].

### 3.6.6 Upper Digestive Tract: Esophageal Carcinoma and Gastric Cancer

The plasma level of miRNA-21 was significantly higher and that of miRNA-375 was significantly lower in esophageal squamous cell carcinoma (ESCC) patients than controls. The high plasma miRNA-21 levels reflected tumor levels in all cases (100%). The plasma level of miRNA-21 was significantly reduced in postoperative samples. The miRNA-21/miRNA-375 ratio was significantly higher in ESCC patients than in controls. Patients with a high plasma level of miRNA-21 tended to have greater vascular invasion and a high correlation with recurrence [189].

Several authors have studied expression levels of cellular miRNAs in patients with gastric cancer (GC). In fact, miRNA-192 and miRNA-215 were downregulated in MGC-803 cells, BGC-823 cells, and SGC-7901 cells. The downregulation of miRNA-192 and miRNA-215 was also demonstrated to be associated with increased tumor sizes. Moreover, the expression of miRNA-192 was significantly lower in the pT4 stage of gastric cancer than in the pT1, pT2, and pT3 stages. Furthermore, there was a strong correlation between miR-192 and miR-215 in GC tissues. MiRNA-192 and miRNA-215 might be related to the proliferation and invasion of GC [190]. Cellular miRNAs can modify invasion of cancer cells with a negative effect on metastatic phenomena [191]. Concerning the circulating miRNAs, Liu et al. showed that serum miRNA-378 could serve as a novel noninvasive biomarker in GC detection. Genome-wide miRNA expression profiles followed with real-time quantitative RT-PCR (qRT-PCR) assays revealed that miRNA-187*, miRNA-371-5p, and miRNA-378 were significantly elevated in GC patients.

Further validation indicated that miRNA-378 alone could yield a ROC curve area of 0.861 with 87.5% sensitivity and 70.73% specificity in discriminating GC patients from healthy controls [192].

### 3.6.7 Colorectal Cancer

The high expression of cellular miRNA-21 in patients with colorectal cancer (CRC) is correlated to clinical stage, lymph node metastasis, and distant metastasis [104]. MiRNA-34a, a transcriptional target of p53, is a well-known tumor suppressor gene. Wu et al. identified Fra-1 as a new target of miRNA-34a, and demonstrated that miRNA-34a inhibits Fra-1 expression at both protein and mRNA levels. In addition, they found that p53 indirectly regulates Fra-1 expression via a miRNA-34a-dependent manner in colon cancer cells. Overexpression of miR-34a strongly inhibited colon cancer cell migration and invasion, which can be partially rescued by forced expression of the Fra-1 transcript lacking the 3'-UTR. Moreover, they found that miRNA-34a was downregulated in 25 of 40 (62.5%) colon cancer tissues, as compared with the adjacent normal colon tissues, and that the expression of miRNA-34a was correlated with the DNA binding activity of p53 [193, 194]. Cheng et al. found that circulating miRNA-141 was significantly associated with stage IV colon cancer in a cohort of 102 plasma samples. ROC analysis was used to evaluate the sensitivity and specificity of candidate plasma miRNA markers. They observed that combination of miRNA-141 and carcinoembryonic antigen (CEA), a widely used marker for CRC, further improved the accuracy of detection. Furthermore, their analysis showed that high levels of plasma miRNA-141 predicted poor survival in both cohorts and that miR-141 was an independent prognostic factor for advanced colon cancer. They propose that plasma miRNA-141 may represent a novel biomarker that complements CEA in detecting colon cancer with distant metastasis and that high levels of miRNA-141 in plasma were associated with poor prognosis [195]. A different circulating miRNA was found altered in colon cancer patients. MiRNA-92 levels were in fact significantly higher in plasma samples from patients than in healthy controls and can be a potential marker for CRC detection [129].

### 3.6.8 Renal Cell Carcinoma and Prostate Cancer

Although 109 miRNAs were circulating at higher levels in cancer patients' serum, Wulfken et al. identified only few miRNAs with upregulation in renal cell carcinoma (RCC) tissue and serum of RCC patients. A multicenter cohort of RCC patients and healthy controls using quantitative real-time PCR (sensitivity 77.4%, specificity 37.6%) shows miRNA-1233 was increased in RCC patients. They also studied samples of patients with angiomyolipoma or oncocytoma, whose serum miRNA-1233 levels were similar to RCC patients. However, circulating miRNAs were not correlated with clinicopathological parameters [196].

Mahn et al. instead analyzed circulating miRNAs in serum as noninvasive biomarkers in patients with localized prostate cancer (PCa), with benign prostate hyperplasia (BPH), and in healthy individuals (HI). Circulating oncogenic miRNA levels were different, and especially the miRNA-26a level allowed sensitive (89%) discrimination of PCa and BPH patients at a moderate specificity (56%; AUC: 0.703); the analysis of oncogenic miRNAs in combination increased the diagnostic accuracy (sensitivity: 78.4%; specificity: 66.7%; AUC: 0.758). Despite the small number of patients limiting the statistical power of the study, they observed correlations with clinicopathologic parameters: miRNA-16, miRNA-195, and miRNA-26a were significantly correlated with surgical margin positivity; miRNA-195 and miRNA-let7i were significantly correlated with the Gleason score. Tissue miRNA levels were correlated with pre-prostatectomy miRNA levels in serum, and serum miRNA decreased after prostatectomy, thereby indicating tumor-associated release of miRNA [197]. Moreover, recent reports found that some circulating prostate cancer-associated miRNAs, such as miRNA-375 and miRNA-141, correlate with risk of disease progression and other predictors of disease outcome, such as Gleason score and lymph node status, highlighting the prognostic potential of this class of molecules [44, 149, 198–200]. Wach et al. used an autochthonous transgenic model of prostate cancer, transgenic adenocarcinoma of mouse prostate (TRAMP), to discover prostate cancer–associated miRNAs in serum. They showed that the levels of miR-141, miR-298, and miR-375 were increased in localized primary tumors and metastases. They proposed that these miRNAs are expressed in prostate cancer cells and released into surrounding blood vessels during disease progression. Interestingly, miR-346 was elevated in the serum of men with mCRPC but reduced in metastases in the MSKCC cohort. This tissue-associated loss appears to be a consequence of enhanced release of microvesicle-encapsulated miRNAs. Thus, hsa-miR-346 export from prostate cancer cells may be augmented without a change in expression of the hsa-miR-346 gene, leading to it being progressively lost in malignant tissue while it accumulates in the blood [201–203]. However, expression of both miR-141 and miR-375 was correlated with risk of biochemical relapse. Moreover, Cox proportional hazards regression and C-statistics revealed that miR-375 possesses prognostic potential that is independent from Gleason score. Therefore, the capacity of tumor and/or serum levels of miR-141 and miR-375 to predict prostate cancer recurrence is quite interesting. Given the association of hsa-miR-141 and hsa-miR-375 with biochemical relapse, it was hypothesized that they could have direct roles in prostate cancer pathophysiology by deregulating normal gene expression. Analysis of the predicted gene targets of these miRNAs revealed enrichment of many pathways likely to be important in prostate cancer. Furthermore, preliminary bioinformatic analyses revealed that four genes, SPAG9, SOCS5, MBNL1, and MTPN, were downregulated in prostate tumors and inversely correlated with levels of miR-141

and miR-375, suggesting that they are targets of these miR-NAs [204].

As circulating miRNA-141 is detected in plasma of patients with PCa, several authors compared the temporal changes of miR-141 with the levels of CTC, LDH, and PSA in patients with PCa, and longitudinally examined these markers alone or in combinations to determine the utility of miR-141 in predicting a patient's clinical course and response to therapy. A total of 35 intervals were assessed. Directional changes (increasing or decreasing) in PSA, CTC, and miR-141 had sensitivity in predicting clinical outcome (progression vs nonprogressing) of 78.9%. Logistic regression modeling of the probability of clinical progression demonstrates that miR-141 levels predicted clinical outcomes with an odds ratio of at least 8.3. MiR-141 also had the highest correlation with temporal changes of PSA. In this retrospective study, miR-141 demonstrated a similar ability to predict clinical progression when compared with other clinically validated biomarkers. Furthermore, miR-141 demonstrated high correlation with changes of the other biomarkers [205].

### 3.6.9 Salivary MiRNAs and Oral Cancer Detection

Upregulation of hsa-miR-24 and hsa-miR-31 has been found in squamous cell carcinoma of the oral cavity [206], while decreased levels of miR-125a and miR-200a in saliva are associated with the same disease [207].

### 3.6.10 Hematologic Neoplasias

One area of great interest for use in diagnostic, prognostic, and therapeutic research of circulating miRNAs is certainly hemopoiesis and hematologic diseases. Patients with diffuse large B-cell lymphoma (DLBCL) had high serum levels of miRNA-21, which was associated with improved relapse-free survival [30]. This result is consistent with previous findings in biopsy material from a different cohort of DLBCL patients, where high tumor miR-21 expression was also associated with a more favorable clinical outcome. Moreover, elevated levels of miR-155 and miRNA-210 were found in the serum of patients with DLBCL [208]. Why high levels of miRNA-21 should be associated with a more favorable clinical outcome for DLBCL patients remains to be determined. Although miRNA-21 has been found to have an antiproliferative effect in some cancers [209], the opposite appears to be true for other cancers [210]. Presumably it is the targeting of oncogenes and/or antiapoptotic molecules by miRNA-21 that is important in determining clinical outcome in DLBCL patients.

### 3.6.11 LLC

Calin et al. [92] were the first to show that their cellular miRNA microarray could differentiate between B-cell chronic lymphocyte leukemia (CLL) cells and normal cells. Furthermore, they classified CLL samples into two different groups based on their miRNA profiles, and these profiles corresponded to high or low levels of a protein that is associated with a positive prognosis at low levels. One commonly observed chromosomal aberration in CLL is the deletion of chromosomal 13q14.3, a region containing miR-15a and miR-16, which suggests, but does not prove, the involvement of miRNAs in the pathogenesis of CLL. In addition, an aberrant cellular miRNA expression profile in CLL B cells has been described and the changes correlate well with prognostic factors including ZAP-70 expression status and $IgV_H$ mutations in CLL patients. Recent studies also demonstrated the decrease of miR-29c and miR-223 levels in cells during the progression of the disease [211].

Moussay et al. showed that circulating miRNAs could also be sensitive biomarkers for CLL, because certain extracellular miRNAs are present in CLL patient plasma at levels significantly different from healthy controls and from patients affected by other hematologic malignancies. The levels of several of these circulating miRNAs also displayed significant differences between zeta-associated protein 70 (ZAP-70)+ and ZAP-70− CLL. The changes of circulating miR-195 (AUC=0.951) or miR-20a (AUC=0.920) levels were the best classifiers to separate CLL patients from healthy controls. They tested the possibility of improving performance by combining the changes of several miRNAs. When 14 miRNAs were combined, the AUC value derived from a standard principal component analysis and ROC analyses reached 0.950. Excellent separation between CLL patients and controls can be reached by using only three of several strongly affected miRNAs, miRNA-195, miRNA-29a, and miRNA-222, in CLL patients; the AUC value reached 0.982. Moreover in CLL patients there was a significant increase in the levels of miRNA-150 in ZAP-70 plasma samples. The level of miRNA-150 increased with the severity of the diseases in ZAP-70− samples, so there is staging information associated with this marker. They also determined that the level of circulating miRNA-20a correlates reliably with diagnosis-to-treatment time. Network analysis of their data suggests a regulatory network associated with BCL2 and ZAP-70 expression in CLL [212].

However, we have to bear in mind that because miRNAs are exported from cells under some circumstances [18] the changes of the most abundant plasma miRNA species from miRNA-223 to miRNA-150 may be the result of the changing composition of lymphoid cells in circulation. Bone marrow stromal cells also provide key influences and protection for CLL B cells, which suggests they may play a role in producing circulating miRNAs in B CLL patients. MiRNA-451, miRNA-135a*, and miRNA-486-5p are more abundant in plasma (compared with B cells), which suggests that a significant fraction of these miRNAs in circulation were released by other cell types, perhaps including bone marrow stromal cells [213].

## 3.6.12  ACUTE LEUKEMIA, MYELODYSPLASTIC SYNDROME, AND MULTIPLE MYELOMA

Leukemia cells can carry a small subset of poorly differentiated cells, which are considered to be precursors of lymphoblasts, myeloblasts, or monoblasts. Thus these cells are also called *leukemia stem cells* (LSCs) because they are capable of maintaining and propagating leukemia in vivo, while retaining the ability to differentiate into committed progeny that lack these properties. Like hematopoietic stem cells (HSCs), LSCs possess the ability of self-renewal under a complex regulatory system. The recent discovery of miRNAs may shed new light on regulation of LSCs and leukemogenesis. As master gene regulators, cellular and circulating miRNAs participate in these processes through coordinated work with key transcription factors required for hematopoiesis. Therefore, miRNAs could play a critical role in normal HSCs as well as LSCs, and in the onset of diseases such as acute leukemia and myelodysplastic syndrome [214]. Cellular miRNAs are in fact of great importance in the pathogenesis, diagnosis, and prognosis of acute leukemia (AL). Zhu et al. studied five AL-related miRNAs to confirm the significance of these miRNAs in AL. Samples tested included acute myeloid leukemia (AML) and acute lymphoblastic leukemia (ALL). Analysis showed that miRNA-128 expression was significantly higher in ALL. However, the let-7b and miRNA-223 expressions in ALL were significantly lower than in AML. Compared with normal controls, miR-128 expression was significantly higher in ALL, but there was no significant difference in AML. The expressions of let-7b and miRNA-223 in the AL group were higher than in normal controls. MiRNA-181a was quantitatively detected in AML patients, and they found that the expression of miRNA-181a in M1 or M2 patients was significantly higher compared with it in M4 or M5. It was found that the expression of miRNA-181a in the favorable-prognosis group was significantly lower than in the poor-prognosis group. In FLT3-ITD mutation positive patients, the miRNA-155 expression was significantly higher than in the negative group [215]. At this time very few studies have been performed on circulating miRNA in AL, but a recent study found that plasma levels of miRNA-92 might be a biomarker for AML [216].

Two miRNAs, let-7a and miR-16, are known to play important roles in myeloid leukemogenesis by regulating the cell cycle and apoptosis, both of which are important in myelodysplastic syndrome (MDS) pathogenesis [92, 217–219]. Let-7a is a tumor suppressor gene that regulates oncogenes such as RAS and HMGA2, and miR-16 targets multiple oncogenes, including BCL2, MCL1, CCND1, and WNT3A. Decreased miR-16 expression also has been found in blasts isolated from high-risk MDS patients [220, 221]. Zuo et al. examined plasma levels of let-7a and miR-16, in patients with MDS and healthy persons using quantitative real-time PCR. Circulating levels of both miRNAs were similar among healthy controls but were significantly lower in MDS patients. The distributions of these miRNA levels were bimodal in MDS patients, and these levels were significantly associated with their progression-free survival and overall survival. This association persisted even after patients were stratified according to the International Prognostic Scoring System. Multivariate analysis revealed that let-7a level was a strong independent predictor for overall survival in this patient cohort. These findings suggest that let-7a and miR-16 plasma levels can serve as noninvasive prognostic markers in MDS patients [222].

Finally, Huang et al. evaluated global miRNA expression profiles in the plasma of multiple myeloma (MM) patients and healthy controls. Six miRNAs (miRNA-148a, miRNA-181a, miRNA-20a, miRNA-221, miRNA-625, and miRNA-99b) that were significantly upregulated in MM were selected and further quantified independently by quantitative reverse transcription PCR in plasma from MM patients and healthy controls. Moreover, within the patient group, the expression levels of miRNA-99b and miRNA-221 were associated with chromosomal abnormalities t(4;14) and del(13q), respectively. High levels of miRNA-20a and miRNA-148a were related to shorter relapse-free survival [223].

## 3.7 ADVANTAGES AND POTENTIAL OF MiRNAs AS BLOOD-BASED CANCER BIOMARKERS

The availability of powerful approaches for global miRNA characterization and simple, universally applicable assays for quantitation (e.g., qRT-PCR) suggests that the discovery–validation pipeline for miRNA biomarkers will be more efficient than traditional proteomic biomarker discovery–validation pipelines, which typically encounter bottlenecks at the point of antibody generation and quantitative assay development for validation of biomarker candidates [224]. In addition, the inherent regulatory function of miRNAs makes it likely that many miRNAs expressed in tumor tissue influence the biological behavior and clinical phenotype of the tumor. As the functional roles of miRNAs in tumor biology are unraveled, we envision that blood-based miRNA biomarkers that predict clinical behavior and/or therapeutic response will be identified.

Although there is a long history of investigation of circulating mRNA molecules as potential biomarkers [225], blood-based miRNA studies are in their infancy. Recently, Chim et al. [226] reported the detection by qRT-PCR of miRNAs of presumed placental origin in the plasma of pregnant women, and Lawrie et al. [227] reported detecting elevations in miRNAs in serum from lymphoma patients. Beyond confirming the early reports, the study yielded (i) a more comprehensive view of plasma miRNAs by direct cloning and sequencing from a plasma small RNA library, (ii) unique results on miRNA stability that provide a firm grounding for further investigation of this class of

molecules as blood-based cancer biomarkers, and (iii) evidence that tumor-derived miRNAs can enter the circulation even when originating from an epithelial cancer type (as compared with hematopoietic malignancies like lymphoma). Most importantly, the study of miR-141 in prostate cancer patients demonstrates that serum levels of a tumor-expressed miRNA can distinguish, with significant specificity and sensitivity, patients with cancer from healthy controls.

The available data indicate that miR-141 is expressed in an epithelial cell type–specific manner in a range of common human cancers. Given this, we speculate that it could have value in the setting of detecting cancer recurrence for cancer types for which clinically validated blood biomarkers are lacking (e.g., lung cancer, breast cancer). We also anticipate that advances in miRNA qRT-PCR assay design and assay optimization, and the application of alternative miRNA quantitation strategies, will substantially improve the approach and will likely be needed to detect cancer at lower tumor burdens (i.e., early-stage disease). It is likely that other blood-based miRNA markers that are specific for particular cancer types will be discovered. The results presented [228] establish the foundation and rationale to motivate future global investigations of miRNAs as circulating cancer biomarkers for a variety of common human cancers.

The remarkable stability of miRNAs in clinical plasma samples raises important and intriguing questions regarding the mechanism by which miRNAs are protected from endogenous RNase activity. One tantalizing hypothesis is that they are packaged inside exosomes that are secreted from cells. Exosomes are 50- to 90-nm [229], membrane-bound particles that have been reported to be abundant in plasma [230] and that have recently been shown (in cell culture studies) to contain miRNAs [231]. The results [228] on filtration (through a 0.22-μm filter) and differential centrifugation of plasma containing tumor-derived miRNAs are certainly consistent with this hypothesis [228]. Alternative explanations include protection via association with other molecules (e.g., in a RNA–protein complex) or modifications of the miRNAs that make them resistant to RNase activity.

The high abundance of many circulating miRNAs also raises provocative questions regarding their potential biological role as extracellular messengers mediating short- and long-range cell–cell communication, reminiscent of the similar role played by small RNAs that spread within the organism in plants and C. elegans [232, 233]. Additional studies will be needed to explore these exciting hypotheses.

## 3.8 LIMITS AND CHALLENGES OF CIRCULATING MiRNAs

Conventional strategies for blood-based biomarker discovery have shown promise, but their clinical use has been limited due to lack of sufficient sensitivity, specificity, and stability. New approaches that can complement and improve on current strategies for cancer detection are needed. Circulating miRNAs have immense potential for refinement of the current processes for diagnosis, staging, and prognostic prediction, and they may also provide potential future therapeutic targets in the management of diseases [234]. However, many challenges regarding miRNAs in sera need to be confronted. Firstly, the specificity of miRNAs: one miRNA can distinguish different cancers, which have the same serum miRNA, e.g., miR-21 in DLBCL and pancreatic cancer. Secondly, the standardization of miRNAs: the preparation of serum/plasma will need to be standardized in order to generalize findings from different patients, groups, or labs. Thirdly, the quantification of miRNAs: what will be used as a standard for qRT-PCR for measuring circulating miRNAs (e.g., as "housekeeping" serum miRNA/small RNA), as other classes of RNAs or mRNAs are not stable in serum? Moreover, the origin of miRNAs is unclear and lacks experimental evidence, but current assumptions of the origin include cancer cells in peripheral blood [16] or tumor-associated bodies [38, 128].

Duttagupta et al. found considerable proportions of miRNAs derived from red and white blood cells, present as contaminants in plasma preparations with the potential to mask the intensities of truly circulating miRNA species [235]. Pritchard et al. hypothesized that blood cells may contribute significantly to circulating miRNA, and that could have important implications for interpretation of results from circulating miRNA cancer biomarker studies [236]. They pointed out that a majority of miRNAs reported in the literature as circulating cancer biomarkers may originate in large part from blood cells, and this finding is supported by other studies in this area [237]. Of 79 solid-tumor circulating miRNA biomarkers reported in the literature, it was found that 58% (47/79) are highly expressed in one or more blood cell type. Plasma levels of miRNA biomarkers expressed by myeloid and lymphoid blood cells tightly correlated with corresponding white blood cell counts. Plasma miRNA biomarkers expressed by red blood cells were significantly increased in hemolyzed specimens (20- to 30-fold plasma increase). Finally, in a patient undergoing autologous hematopoietic cell transplantation, plasma levels of myeloid- and lymphoid-expressed miRNAs tracked closely with changes in corresponding blood counts. There is evidence that blood cells are a major contributor to circulating miRNA, and that perturbations in blood cell counts and hemolysis can alter plasma miRNA biomarker levels by up to 50-fold. Given that a majority of reported circulating miRNA cancer biomarkers are highly expressed in blood cells, caution is needed in interpretation of such results as they may reflect a blood cell–based phenomenon rather than a cancer-specific origin. Stratification of circulating miRNAs based on detection in circulation and/or contaminant classes reveals an approximate loss of 66% of all detected miRNAs through the removal of cellular miRNA-derived signatures. The proportion of this reduction is comparable to similar extensive overlaps seen between miRNA profiles derived from plasma microvesicles and peripheral blood mononuclear cells [25]. A direct comparison of the 20 most common circulating miRNAs from healthy individuals over five different datasets reveals that at least 75% (15/20) of the reported circulating miRNA species can be mapped to cellular miRNA signatures [30, 238].

## 3.9 CONCLUSIONS AND FUTURE PERSPECTIVES FOR CIRCULATING MiRNAs

MiRNAs function as extracellular communication RNAs that play an important role in cell proliferation and differentiation [23, 239]. A more comprehensive study of circulating miRNAs and their association with various physiopathological conditions may lead to another dimension in the discovery of biomarkers in the blood for many physiological and pathological conditions. As circulating miRNAs regulate the expression of multiple genes in a disease pathway, circulating miRNAs and the genes they influence can be therapeutic targets. Currently, there are clinical trials evaluating therapy based on miRNA inhibition or overexpression. For instance, overexpression and downregulation of specific miRNAs are being evaluated as a novel approach to the treatment of myocardial infarction [240].

Technological applications [241–244] and novel biomarkers in the field of molecular diagnostics have never been evolving at a more rapid pace. Wang et al. showed that a nanopore sensor based on the α-haemolysin protein can selectively detect miRNAs at the single molecular level in plasma samples from lung cancer patients without the need for labels or amplification of the miRNA. The sensor, which uses a programmable oligonucleotide probe to generate a target-specific signature signal, can quantify subpicomolar levels of cancer-associated miRNAs and can distinguish single-nucleotide differences between miRNA family members. This approach is potentially useful for quantitative miRNA detection [245]. However, circulating miRNAs will find a space in several different fields of application such as transplantation, and recently Spiegel et al. pointed out the role of miRNAs in immunology and transplantation medicine and their role as potential biomarkers. They also focused on the molecular mechanisms and therapeutic implications of the use of miRNA-based therapeutic strategies to improve long-term allograft survival [246].

Within a few years, we will know if miRNA-based therapeutics, alone or in combination with other modalities, will be clinically useful treatments for various diseases [247]. Then it could be that blood-based circulating miRNA analyses have imminent clinical utility as disease markers. However, if this concept is to translate readily from bench to bedside, then supporting data demonstrating validity of this novel approach must stem from carefully planned studies. If the current momentum in translational research can be maintained, then an era of noninvasive rapid diagnostics is rapidly forthcoming.

# REFERENCES

[1] Setoyama T, Ling H, Natsugoe S, et al. Non-coding RNAs for medical practice in oncology. Keio J Med. 2011;60:106–113.

[2] Carninci P, Kasukawa T, Katayama S, et al. The transcriptional landscape of the mammalian genome. Science. 2005;309:1559–1563.

[3] Kapranov P, Drenkow J, Cheng J, et al. Examples of the complex architecture of the human transcriptome revealed by RACE and high-density tiling arrays. Genome Res. 2005;15:987–997.

[4] Mercer TR, Dinger ME, Mattick JS. Long non-coding RNAs: insights into functions. Nat Rev Genet. 2009;10:155–159.

[5] Taft RJ, Pang KC, Mercer TR, et al. Non-coding RNAs: regulators of disease. J Pathol. 2010;220:126–139.

[6] Gupta RA, Shah N, Wang KC, et al. Long non-coding RNA HOTAIR reprograms chromatin state to promote cancer metastasis. Nature. 2010;464:1071–1076.

[7] Friedman RC, Farh KK, Burge CB, et al. Most mammalian mRNAs are conserved targets of microRNAs. Genome Res. 2009;19:92–105.

[8] Lee RC, Feinbaum RL, Ambros V. The C. elegans heterochronic gene lin-4 encodes small RNAs with antisense complementarity to lin-14. Cell. 1993;75(5):843–854. doi:10.1016/0092-8674(93)90529-Y.

[9] Bartel DP, Chen CZ. Micromanagers of gene expression: the potentially widespread influence of metazoan microRNAs. Nat Rev Genet. 2004;5:396–400.

[10] Zhao T, Li G, Mi S, Li S, et al. A complex system of small RNAs in the unicellular green alga *Chlamydomonas reinhardtii*. Genes Dev. 2007;21:1190–1203.

[11] Lee Y, Kim M, Han J, et al. MicroRNA genes are transcribed by RNA polymerase II. EMBO J. 2004;23:4051–4060.

[12] Borchert GM, Lanier W, Davidson BL. RNA polymerase III transcribes human microRNAs. Nat Struct Mol Biol. 2006;13:1097–1101.

[13] Hu HY, Yan Z, Xu Y, et al. Sequence features associated with microRNA strand selection in humans and flies. BMC Genomics. 2009;10:413.

[14] Fabian MR, Sonenberg N, Filipowicz W. Regulation of mRNA translation and stability by microRNAs. Annu Rev Biochem. 2010;79:351–379.

[15] Fiorucci G, Chiantore MV, Mangino G, et al. Cancer regulator microRNA: potential relevance in diagnosis, prognosis and treatment of cancer. Curr Med Chem. 2012;19:461–474.

[16] Zen K, Zhang CY. Circulating microRNAs: a novel class of biomarkers to diagnose and monitor human cancers. Med Res Rev. 2012;32:326–348.

[17] Weber JA, Baxter DH, Zhang S, et al. The microRNA spectrum in 12 body fluids. Clin Chem. 2010;56:1733–1741.

[18] Wang K, Zhang S, Weber J, et al. Export of microRNAs and microRNA-protective protein by mammalian cells. Nucleic Acids Res. 2010;38:7248–7259.

[19] Zubakov D, Boersma AW, Choi Y, et al. MicroRNA markers for forensic body fluid identification obtained from microarray screening and quantitative RT-PCR confirmation. Int J Legal Med. 2010;124:217–226.

[20] Hanson EK, Lubenow H, Ballantyne J. Identification of forensically relevant body fluids using a panel of differentially expressed microRNAs. Anal Biochem. 2009;387:303–314.

[21] Weickmann JL, Glitz DG. Human ribonucleases. Quantitation of pancreatic-like enzymes in serum, urine, and organ preparations. J Biol Chem. 1982;257:8705–8710.

[22] Gibbings DJ, Ciaudo C, Erhardt M, et al. Multivesicular bodies associate with components of miRNA effector complexes and modulate miRNA activity. Nat Cell Biol. 2009;11:1143–1149.

[23] Valadi H, Ekstrom K, Bossios A, et al. Exosome-mediated transfer of mRNAs and microRNAs is a novel mechanism of genetic exchange between cells. Nat Cell Biol. 2007;9:654–659.

[24] Iguchi H, Kosaka N, Ochiya T. Secretory microRNAs as a versatile communication tool. Commun Integr Biol. 2010;3:478–481.

[25] Camussi G, Deregibus MC, Bruno S, et al. Exosomes/microvesicles as a mechanism of cell-to-cell communication. Kidney Int. 2010;78:838–848.

[26] Muralidharan-Chari V, Clancy JW, Sedgwick A, et al. Microvesicles: mediators of extracellular communication during cancer progression. J Cell Sci. 2010;123:1603–1611.

[27] Cortez MA, Bueso-Ramos C, Ferdin J, et al. MicroRNAs in body fluids—the mix of hormones and biomarkers. Nat Rev Clin Oncol. 2011;8:467–477.

[28] Etheridge A, Lee I, Hood L, et al. Extracellular microRNA: a new source of biomarkers. Mutat Res. 2011;717:85–90.

[29] Skog J, Wurdinger T, van Rijn S, et al. Glioblastoma microvesicles transport RNA and proteins that promote tumour growth and provide diagnostic biomarkers. Nat Cell Biol. 2008;10:1470–1476.

[30] Hunter MP, Ismail N, Zhang X, et al. Detection of microRNA expression in human peripheral blood microvesicles. PLoS One. 2008;3:e3694.

[31] Michael A, Bajracharya SD, Yuen PS, et al. Exosomes from human saliva as a source of microRNA biomarkers. Oral Dis. 2010;16:34–38.

[32] Dimov I, Velickovic L, Stefanovic V. Urinary exosomes. Sci World J. 2009;9:1107–1118.

[33] Blanchard N, Lankar D, Faure F, et al. TCR activation of human T cells induces the production of exosomes bearing the TCR/CD3/zeta complex. J Immunol. 2002;168:3235–3241.

[34] Raposo G, Nijman HW, Stoorvogel W, et al. B lymphocytes secrete antigen-presenting vesicles. J Exp Med. 1996;183:1161–1172.

[35] Thery C, Regnault A, Garin J, et al. Molecular characterization of dendritic cell-derived exosomes. Selective accumulation of the heat shock protein hsc73. J Cell Biol. 1999;147:599–610.

[36] Mears R, Craven RA, Hanrahan S, et al. Proteomic analysis of melanoma-derived exosomes by two-dimensional polyacrylamide gel electrophoresis and mass spectrometry. Proteomics. 2004;4:4019–4031.

[37] O'Neill HC, Quah BJ. Exosomes secreted by bacterially infected macrophages are proinflammatory. Sci Signal. 2008;1:8.

[38] Taylor DD, Gercel-Taylor C. MicroRNA signatures of tumor-derived exosomes as diagnostic biomarkers of ovarian cancer. Gynecol Oncol. 2008;110:13–21.

[39] Denzer K, Kleijmeer MJ, Heijnen HF, et al. Exosome: from internal vesicle of the multivesicular body to intercellular signaling device. J Cell Sci. 2000;113:3365–3374.

[40] Huber V, Filipazzi P, Iero M, et al. More insights into the immunosuppressive potential of tumor exosomes. J Transl Med. 2008;6:63.

[41] Xiang X, Poliakov A, Liu C, et al. Induction of myeloid-derived suppressor cells by tumor exosomes. Int J Cancer. 2009;124:2621–2633.

[42] Vallhov H, Gutzeit C, Johansson SM, et al. Exosomes containing glycoprotein 350 released by EBV-transformed B cells selectively target B cells through CD21 and block EBV infection in vitro. J Immunol. 2011;186:73–82.

[43] Shen J, Todd NW, Zhang H, et al. Plasma microRNAs as potential biomarkers for non-small-cell lung cancer. Lab Invest. 2011;91:579–587.

[44] Mitchell PS, Parkin RK, Kroh EM, et al. Circulating microRNAs as stable blood-based markers for cancer detection. Proc Natl Acad Sci USA. 2008;105:10513–10518.

[45] Gilad S, Meiri E, Yogev Y, et al. Serum microRNAs are promising novel biomarkers. PLoS One. 2008;3:e3148.

[46] Kosaka N, Iguchi H, Yoshioka Y, et al. Secretory mechanisms and intercellular transfer of microRNAs in living cells. J Biol Chem. 2010;285:17442–17452.

[47] Turchinovich A, Weiz L, Langheinz A, et al. Characterization of extracellular circulating microRNA. Nucleic Acids Res. 2011;39:7223–7233.

[48] Chen X, Ba Y, Ma L, et al. Characterization of microRNAs in serum: a novel class of biomarkers for diagnosis of cancer and other diseases. Cell Res. 2008;18(10):997–1006. doi:10.1038/cr.2008.282.

[49] Li J, Smyth P, Flavin R, et al. Comparison of miRNA expression patterns using total RNA extracted from matched samples of formalin-fixed paraffin-embedded (FFPE) cells and snap-frozen cells. BMC Biotechnol. 2007;7:36.

[50] Xie Y, Todd NW, Liu Z, et al. Altered miRNA expression in sputum for diagnosis of non-small cell lung cancer. Lung Cancer. 2010;67:170–176.

[51] Tsujiura M, Ichikawa D, Komatsu S, et al. Circulating microRNAs in plasma of patients with gastric cancers. Br J Cancer. 2010;102:1174–1179.

[52] Huang Z, Huang D, Ni S, et al. Plasma microRNAs are promising novel biomarkers for early detection of colorectal cancer. Int J Cancer. 2010;127:118–126.

[53] Xi Y, Nakajima G, Gavin E, et al. Systematic analysis of microRNA expression of RNA extracted from fresh frozen and formalin-fixed paraffin-embedded samples. RNA. 2007;13:1668–1674.

[54] Schöler N, Langer C, Kuchenbauer F. Circulating microRNAs as biomarkers—true blood? Genome Med. 2011;3:72.

[55] Behne T, Copur MS. Biomarkers for hepatocellular carcinoma. Int J Hepatol. 2012;859076. doi:10.1155/2012/859076.

[56] Shiota G, Miura N. Biomarkers for hepatocellular carcinoma. Clin J Gastroenterol. 2012;5(3):177–182.

[57] Iizuka N, Oka M, Yamada-Okabe H, et al. Comparison of gene expression profiles between hepatitis B virus- and hepatitis C virus infected hepatocellular carcinoma by oligonucleotide microarray data on the basis of a supervised learning method. Cancer Res. 2002;62(14):3939–3944.

[58] Bellodi-Privato M, Kubrusly MS, Stefano JT, et al. Differential gene expression profiles of hepatocellular carcinomas associated or not with viral infection. Braz J Med Biol Res. 2009;42(12):1119–1127.

[59] Chen X, Cheung ST, So S, et al. Gene expression patterns in human liver cancers. Mol Biol Cell. 2002;13(6):1929–1939. doi:10.1091/mbc.02-02-0023.

[60] Bartel DP. MicroRNAs: genomics, biogenesis, mechanism, and function. Cell. 2004;116(2):281–297. doi:10.1016/S0092-8674(04)00045-5.

[61] Wu W, Sun M, Zou GM, et al. MicroRNA and cancer: Current status and prospective. Int J Cancer. 2007;120(5):953–960. doi:10.1002/ijc.22454.

[62] Yu J, Wang F, Yang GH, et al. Human microRNA clusters: genomic organization and expression profile in leukemia cell lines. Biochem Biophys Res Commun. 2006;349(1):59–68. doi:10.1016/j.bbrc.2006.07.207.

[63] Aravalli RN, Steer CJ, Cressman EN. Molecular mechanisms of hepatocellular carcinoma. Hepatology. 2008;48(6):2047–2063. doi:10.1002/hep.22580.

[64] Negrini M, Ferracin M, Sabbioni S, et al. MicroRNAs in human cancer: from research to therapy. J Cell Sci. 2007;120(Pt 11):1833–1840. doi:10.1242/jcs.03450.

[65] He L, Thomson JM, Hemann MT, et al. A microRNA polycistron as a potential human oncogene. Nature. 2005;435(7043):828–833. doi:10.1038/nature03552.

[66] Gailhouste L, Gomez-Santos L, Ochiya T. Potential applications of miRNAs as diagnostic and prognostic markers in liver cancer. Front Biosci (Landmark Ed). 2013;18:199–223. doi:10.2741/4096.

[67] Borel F, Konstantinova P, Jansen PL. Diagnostic and therapeutic potential of miRNA signatures in patients with hepatocellular carcinoma. J Hepatol. 2012;56(6):1371–1383. doi:10.1016/j.jhep.2011.11.026.

[68] Kosaka N, Iguchi H, Ochiya T. Circulating microRNA in body fluid: a new potential biomarker for cancer diagnosis and prognosis. Cancer Sci. 2010;101(10):2087–2092. doi:10.1111/j.1349-7006.2010.01650.x.

[69] Tsao SC-Hao, Behren A, Cebon J, et al. The role of circulating microRNA in hepatocellular carcinoma. Front Biosci (Landmark Ed). 2015;20:78–104, January 1.

[70] Wang K, Zhang S, Marzolf B, et al. Circulating microRNAs, potential biomarkers for drug-induced liver injury. Proc Natl Acad Sci USA. 2009;106(11):4402–4407. doi:10.1073/pnas.0813371106.

[71] Zhang Y, Jia Y, Zheng R, et al. Plasma microRNA-122 as a biomarker for viral-, alcohol-, and chemical-related hepatic diseases. Clin Chem. 2010;56(12):1830–1838. doi:10.1373/clinchem.2010.147850.

[72] Mitchell PS, Parkin RK, Kroh EM, et al. Circulating microRNAs as stable blood-based markers for cancer detection. Proc Natl Acad Sci USA. 2008;105(30):10513–10518. doi:10.1073/pnas.0804549105.

[73] Chen C, Ridzon DA, Broomer AJ, et al. Real-time quantification of microRNAs by stem-loop RT-PCR. Nucleic Acids Res. 2005;33(20):e179. doi:10.1093/nar/gni178.

[74] Gailhouste L, Ochiya T. Cancer related microRNAs and their role as tumor suppressors and oncogenes in hepatocellular carcinoma. Histol Histopathol. 2013;28(4):437–451.

[75] Murakami Y, Yasuda T, Saigo K, et al. Comprehensive analysis of microRNA expression patterns in hepatocellular carcinoma and nontumorous tissues. Oncogene. 2006;25(17):2537–2545. doi:10.1038/sj.onc.1209283.

[76] Yang X, Liang L, Zhang XF, et al. MicroRNA-26a suppresses tumor growth and metastasis of human hepatocellular carcinoma by targeting interleukin-6-Stat3 pathway. Hepatology. 2013;58(1):158–170. doi:10.1002/hep.26305.

[77] Wang G, Tam LS, Li EK, et al. Serum and urinary free microRNA level in patients with systemic lupus erythematosus. Lupus. 2011;20:493–500.

[78] Shaoqing Y, Ruxin Z, Guojun L, et al. Microarray analysis of differentially expressed microRNAs in allergic rhinitis. Am J Rhinol Allergy. 2011;25:242–246.

[79] Zahm AM, Thayu M, Hand NJ, et al. Circulating microRNA is a biomarker of pediatric Crohn disease. J Pediatr Gastroenterol Nutr. 2011;53:26–33.

[80] Wu F, Guo NJ, Tian H, et al. Peripheral blood microRNAs distinguish active ulcerative colitis and Crohn's disease. Inflamm Bowel Dis. 2011;17:241–250.

[81] Zampetaki A, Kiechl S, Drozdov I, et al. Plasma microRNA profiling reveals loss of endothelial miR-126 and other microRNAs in type 2 diabetes. Circ Res. 2010;107: 810–817.

[82] Karolina DS, Armugam A, Tavintharan S, et al. MicroRNA 144 impairs insulin signaling by inhibiting the expression of insulin receptor substrate 1 in type 2 diabetes mellitus. PLoS One. 2011;6:e22839.

[83] Cogswell JP, Ward J, Taylor IA, et al. Identification of miRNA changes in Alzheimer's disease brain and CSF yields putative biomarkers and insights into disease pathways. J Alzheimers Dis. 2008;14:27–41.

[84] Geekiyanage H, Jicha GA, Nelson PT, et al. Blood serum miRNA: non-invasive biomarkers for Alzheimer's disease. Exp Neurol. 2012;235:491–496.

[85] Beveridge NJ, Cairns MJ. MicroRNA dysregulation in schizophrenia. Neurobiol Dis. 2012;46:263–271.

[86] Mizuno H, Nakamura A, Aoki Y, et al. Identification of muscle-specific microRNAs in serum of muscular dystrophy animal models: promising novel blood-based markers for muscular dystrophy. PLoS One. 2011;6:e18388.

[87] Laterza OF, Lim L, Garrett-Engele PW, et al. Plasma microRNAs as sensitive and specific biomarkers of tissue injury. Clin Chem. 2009;55:1977–1983.

[88] Xu J, Zhao J, Evan G, et al. Circulating microRNAs: novel biomarkers for cardiovascular diseases. J Mol Med (Berl). 2012;90:865–875.

[89] Li H, Guo L, Wu Q, et al. A comprehensive survey of maternal plasma miRNAs expression profiles using high-throughput sequencing. Clin Chim Acta. 2012;413:568–576.

[90] Chim SS, Shing TK, Hung EC, et al. Detection and characterization of placental microRNAs in maternal plasma. Clin Chem. 2008;54:482–490.

[91] Gunel T, Zeybek YG, Akçakaya P, et al. Serum microRNA expression in pregnancies with preeclampsia. Genet Mol Res. 2011;10:4034–4040.

[92] Calin GA, Sevignani C, Dumitru CD, et al. Human microRNA genes are frequently located at fragile sites and genomic regions involved in cancers. Proc Natl Acad Sci USA. 2004;101:2999–3004.

[93] Asaga S, Kuo C, Nguyen T, et al. Direct serum assay for microRNA-21 concentrations in early and advanced breast cancer. Clin Chem. 2011;57:84–91.

[94] Cho WC. MicroRNAs in cancer-from research to therapy. Biochim Biophys Acta. 2010;1805:209–217.

[95] Cho WC. MicroRNAs: potential biomarkers for cancer diagnosis, prognosis and targets for therapy. Int J Biochem Cell Biol. 2010;42:1273–1281.

[96] Lu J, Getz G, Miska EA, et al. MicroRNA expression profiles classify human cancers. Nature. 2005;435:834–838.

[97] Chan JA, Krichevsky AM, Kosik KS. MicroRNA-21 is an antiapoptotic factor in human glioblastoma cells. Cancer Res. 2005;65:6029–6033.

[98] Iorio MV, Ferracin M, Liu CG, et al. MicroRNA gene expression deregulation in human breast cancer. Cancer Res. 2005;65:7065–7070.

[99] Markou A, Tsaroucha EG, Kaklamanis L, et al. Prognostic value of mature microRNA-21 and microRNA-205 overexpression in nonsmall cell lung cancer by quantitative real-time RT-PCR. Clin Chem. 2008;54:1696–1704.

[100] Schetter AJ, Leung SY, Sohn JJ, et al. MicroRNA expression profiles associated with prognosis and therapeutic outcome in colon adenocarcinoma. JAMA. 2008;299:425–436.

[101] Huang YS, Dai Y, Yu XF, et al. Microarray analysis of microRNA expression in hepatocellular carcinoma and non-tumorous tissues without viral hepatitis. J Gastroenterol Hepatol. 2008;23:87–94.

[102] Ladeiro Y, Couchy G, Balabaud C, et al. MicroRNA profiling in hepatocellular tumors is associated with clinical features and oncogene/tumor suppressor gene mutations. Hepatology. 2008;47:1955–1963.

[103] Bandrés E, Cubedo E, Agirre X, et al. Identification by real-time PCR of 13 mature microRNAs differentially expressed in colorectal cancer and non-tumoral tissues. Mol Cancer. 2006;5:29.

[104] Slaby O, Svoboda M, Fabian P, et al. Altered expression of miR-21, miR-31, miR-143 and miR-145 is related to clinicopathologic features of colorectal cancer. Oncology. 2007;72:397–402.

[105] Motoyama K, Inoube H, Takatsuno Y, et al. Over- and under-expressed microRNAs in human colorectal cancer. Int J Oncol. 2009;34(4):1069–1075.

[106] Wang CJ, Zhou ZG, Wang L, et al. Clinicopathological significance of microRNA-31, -143 and -145 expression in colorectal cancer. Dis Markers. 2009;26:27–34.

[107] Yan LIX, Huang XF, Shao Q, et al. MicroRNA miR-21 overexpression in human breast cancer is associated with advanced clinical stage, lymph node metastasis and patient poor prognosis. RNA. 2008;14:2348–2360.

[108] Valastyan S, Reinhardt F, Benaich N, et al. A pleiotropically acting microRNA, miR-31, inhibits breast cancer metastasis. Cell. 2009;137:1032–1046.

[109] Guo J, Miao Y, Xiao B, et al. Differential expression of microRNA species in human gastric cancer versus non-tumorous tissues. J Gastroenterol Hepatol. 2009;24: 652–657.

[110] Veerla S, Lindgren D, Kvist A, et al. MiRNA expression in urothelial carcinomas: important roles of miR-10a, miR-222, miR-125b, miR-7 and miR-452 for tumor stage and metastasis, and frequent homozygous losses of miR-31. Int J Cancer. 2009;124:2236–2242.

[111] Rosenfeld N, Aharonov R, Meiri E, et al. MicroRNAs accurately identify cancer tissue origin. Nat Biotechnol. 2008;26:462–469.

[112] He JF, Luo YM, Wan XH, et al. Biogenesis of miRNA-195 and its role in biogenesis, the cell cycle, and apoptosis. J Biochem Mol Toxicol. 2011;25:404–408.

[113] Donzelli S, Fontemaggi G, Fazi F, et al. MicroRNA-128-2 targets the transcriptional repressor E2F5 enhancing mutant p53 gain of function. Cell Death Differ. 2012;19: 1038–1048.

[114] Yin R, Bao W, Xing Y, et al. MiR-19b-1 inhibits angiogenesis by blocking cell cycle progression of endothelial cells. Biochem Biophys Res Commun. 2012;417:771–776.

[115] Enkelmann A, Heinzelmann J, von Eggeling F, et al. Specific protein and miRNA patterns characterise tumour-associated fibroblasts in bladder cancer. J Cancer Res Clin Oncol. 2011;137:751–759.

[116] Schaar DG, Medina DJ, Moore DF, et al. miR-320 targets transferrin receptor 1 (CD71) and inhibits cell proliferation. Exp Hematol. 2009;37:245–255.

[117] Ichimi T, Enokida H, Okuno Y, et al. Identification of novel microRNA targets based on microRNA signatures in bladder cancer. Int J Cancer. 2009;125:345–352.

[118] Chiyomaru T, Enokida H, Tatarano S, et al. miR-145 and miR-133a function as tumour suppressors and directly regulate FSCN1 expression in bladder cancer. Br J Cancer. 2010;102:883–891.

[119] Ostenfeld MS, Bramsen JB, Lamy P, et al. miR-145 induces caspase-dependent and -independent cell death

in urothelial cancer cell lines with targeting of an expression signature present in Ta bladder tumors. Oncogene. 2010;29:1073–1084.

[120] Aprelikova O, Yu X, Palla J, et al. The role of miR-31 and its target gene SATB2 in cancer-associated fibroblasts. Cell Cycle. 2010;9:4387–4398.

[121] Rana TM. Illuminating the silence: understanding the structure and function of small RNAs. Nat Rev Mol Cell Biol. 2007;8:23–36.

[122] Bullock MD, Sayan AE, Packham GK, et al. microRNAs: critical regulators of epithelial to mesenchymal (EMT) and mesenchymal to epithelial transition (MET) in cancer progression. Biol Cell. 2012;104:3–12.

[123] Avila-Moreno F, Urrea F, Ortiz-Quintero B. MicroRNAs in diagnosis and prognosis in lung cancer. Rev Invest Clin. 2011;63:516–535.

[124] Segura MF, Belitskaya-Lévy I, Rose AE, et al. Melanoma microRNA signature predicts post-recurrence survival. Clin Cancer Res. 2010;16:1577–1586.

[125] Yu SL, Chen HY, Chang GC, et al. MicroRNA signature predicts survival and relapse in lung cancer. Cancer Cell. 2008;13:48–57.

[126] Zhou N, Mo YY. Roles of microRNAs in cancer stem cells. Front Biosci. 2012;4:810–818.

[127] Ho AS, Huang X, Cao H, et al. Circulating miR-210 as a novel hypoxia marker in pancreatic cancer. Transl Oncol. 2010;3:109–113.

[128] Resnick KE, Alder H, Hagan JP, et al. The detection of differentially expressed microRNAs from the serum of ovarian cancer patients using a novel real-time PCR platform. Gynecol Oncol. 2009;112:55–59.

[129] Ng EK, Chong WW, Jin H, et al. Differential expression of microRNAs in plasma of patients with colorectal cancer: a potential marker for colorectal cancer screening. Gut. 2009;58:1375–1381.

[130] Wentz-Hunter KK, Potashkin JA. The role of miRNAs as key regulators in the neoplastic microenvironment. Mol Biol Int. 2011:839872.

[131] Kosaka N, Iguchi H, Ochiya T. Circulating microRNA in body fluid: a new potential biomarker for cancer diagnosis and prognosis. Cancer Sci. 2010;101:2087–2092.

[132] Roth C, Rack B, Muller V, et al. Circulating microRNAs as blood-based markers for patients with primary and metastatic breast cancer. Breast Cancer Res. 2010;12:R90.

[133] Yu L, Todd NW, Xing L, et al. Early detection of lung adenocarcinoma in sputum by a panel of microRNA markers. Int J Cancer. 2010;127:2870–2878.

[134] Xing L, Todd NW, Yu L, et al. Early detection of squamous cell lung cancer in sputum by a panel of microRNA markers. Mod Pathol. 2010;8:1157–1164.

[135] Cheng HH, Mitchell PS, Kroh EM, et al. Circulating microRNA profiling identifies a subset of metastatic prostate cancer patients with evidence of cancer-associated hypoxia. PLoS ONE. July 2013;8(7):e69239. www.plosone.org

[136] Gregory PA, Bert AG, Paterson EL, et al. The miR-200 family and miR-205 regulate epithelial to mesenchymal transition by targeting ZEB1 and SIP1. Nat Cell Biol. 2008;10:593–601.

[137] Brase JC, Johannes M, Schlomm T, et al. Circulating miRNAs are correlated with tumor progression in prostate cancer. Int J Cancer. 2011;128:608–616.

[138] Zhang HL, Qin XJ, Cao DL, et al. An elevated serum miR-141 level in patients with bone-metastatic prostate cancer is correlated with more bone lesions. Asian J Androl. 2013;15:231–235.

[139] Nguyen HC, Xie W, Yang M, et al. Expression differences of circulating microRNAs in metastatic castration resistant prostate cancer and low-risk, localized prostate cancer. Prostate. 2013;73(4):346–354.

[140] Bryant RJ, Pawlowski T, Catto JW, et al. Changes in circulating microRNA levels associated with prostate cancer. Br J Cancer. 2012;106:768–774.

[141] Watahiki A, Macfarlane RJ, Gleave ME, et al. Plasma miRNAs as Biomarkers to Identify Patients with Castration-Resistant Metastatic Prostate Cancer. Int J Mol Sci. 2013;14:7757–7770.

[142] Ivan M, Harris AL, Martelli F, et al. Hypoxia response and microRNAs: no longer two separate worlds. J Cell Mol Med. 2008;12:1426–1431.

[143] Huang X, Le QT, Giaccia AJ. MiR-210–micromanager of the hypoxia pathway. Trends Mol Med. 2010;16:230–237.

[144] Huang X, Ding L, Bennewith KL, et al. Hypoxia-inducible mir-210 regulates normoxic gene expression involved in tumor initiation. Mol Cell. 2009;35:856–867.

[145] Mathew LK, Simon MC. mir-210: a sensor for hypoxic stress during tumorigenesis. Mol Cell. 2009;35:737–738.

[146] Pezzolesi MG, Platzer P, Waite KA, et al. Differential expression of PTEN-targeting microRNAs miR-19a and miR-21 in Cowden syndrome. Am J Hum Genet. 2008;82:1141–1149.

[147] Zhu S, Si ML, Wu H, et al. MicroRNA-21 targets the tumor suppressor gene tropomyosin 1 (TPM1). J Biol Chem. 2007;282:14328–14336.

[148] Du Rieu MC, Torrisani J, Selves J, et al. MicroRNA-21 is induced early in pancreatic ductal adenocarcinoma precursor lesions. Clin Chem. 2010;56:603–612.

[149] Zhang HL, Yang LF, Zhu Y, et al. Serum miRNA-21: elevated levels in patients with metastatic hormone-refractory prostate cancer and potential predictive factor for the efficacy of docetaxel-based chemotherapy. Prostate. 2011;71:326–331.

[150] Shen J, Liu Z, Todd NW, et al. Diagnosis of lung cancer in individuals with solitary pulmonary nodules by plasma microRNA biomarkers. BMC Cancer. 2011;11:374.

[151] Wei J, Gao W, Zhu CJ, et al. Identification of plasma microRNA-21 as a biomarker for early detection and chemosensitivity of non-small cell lung cancer. Chin J Cancer. 2011;30:407–414.

[152] Oh HK, Tan AL, Das K, et al. Genomic loss of miR-486 regulates tumor progression and the OLFM4 antiapoptotic factor in gastric cancer. Clin Cancer Res. 2011;17:2657–2667.

[153] Mees ST, Mardin WA, Sielker S, et al. Involvement of CD40 targeting miR-224 and miR-486 on the progression of pancreatic ductal adenocarcinomas. Ann Surg Oncol. 2009;16:2339–2350.

[154] Bansal A, Lee IH, Hong X, et al. Feasibility of microRNAs as biomarkers for Barrett's esophagus progression: a pilot cross-sectional, phase 2 biomarker study. Am J Gastroenterol. 2011;106:1055–1063.

[155] Huang X, Ding L, Bennewith KL, et al. Hypoxia-inducible mir-210 regulates normoxic gene expression involved in tumor initiation. Mol Cell. 2009;35:856–867.

[156] Roth C, Kasimir-Bauer S, Pantel K, et al. Screening for circulating nucleic acids and caspase activity in the peripheral blood as potential diagnostic tools in lung cancer. Mol Oncol. 2011;5:281–291.

[157] Zheng D, Haddadin S, Wang Y, et al. Plasma microRNAs as novel biomarkers for early detection of lung cancer. Int J Clin Exp Pathol. 2011;4:575–586.

[158] Heegaard NH, Schetter AJ, Welsh JA, et al. Circulating microRNA expression profiles in early stage non-small cell lung cancer. Int J Cancer. 2012;130:1378–1386.

[159] Bianchi F, Nicassio F, Marzi M, et al. A serum circulating miRNA diagnostic test to identify asymptomatic high-risk individuals with early stage lung cancer. EMBO Mol Med. 2011;3:495–503.

[160] Bockmeyer CL, Christgen M, Müller M, et al. MicroRNA profiles of healthy basal and luminal mammary epithelial cells are distinct and reflected in different breast cancer subtypes. Breast Cancer Res Treat. 2011;130:735–745.

[161] Blenkiron C, Goldstein LD, Thorne NP, et al. MicroRNA expression profiling of human breast cancer identifies new markers of tumor subtype. Genome Biol. 2007;8:214.

[162] Adams BD, Guttilla IK, White BA. Involvement of microRNAs in breast cancer. Semin Reprod Med. 2008;26:522–536.

[163] Enerly E, Steinfeld I, Kleivi K, et al. miRNA-mRNA integrated analysis reveals roles for miRNAs in primary breast tumors. PLoS One. 2011;6:e16915.

[164] Yu F, Yao H, Zhu P, et al. let-7 regulates self renewal and tumorigenicity of breast cancer cells. Cell. 2007;131:1109–1123.

[165] Tavazoie SF, Alarcón C, Oskarsson T, et al. Endogenous human microRNAs that suppress breast cancer metastasis. Nature. 2008;451:147–152.

[166] Hurteau GJ, Carlson JA, Spivack SD, et al. Overexpression of the microRNA hsa-miR-200c leads to reduced expression of transcription factor 8 and increased expression of E-cadherin. Cancer Res. 2007;67:7972–7976.

[167] Bertos NR, Park M. Breast cancer—one term, many entities? J Clin Invest. 2011;121:3789–3796.

[168] Heneghan HM, Miller N, Kerin MJ. Circulating microRNAs: promising breast cancer. Breast Cancer Res. 2011;13:402.

[169] Ratajczak J, Wysoczynski M, Hayek F, et al. Membrane-derived microvesicles: important and underappreciated mediators of cell-to-cell communication. Leukemia. 2006;20:1487–1495.

[170] Ohshima K, Inoue K, Fujiwara A, et al. Let-7 microRNA family is selectively secreted into the extracellular environment via exosomes in a metastatic gastric cancer cell line. PLoS One. 2010;5:e13247.

[171] Gourzones C, Gelin A, Bombik I, et al. Extra-cellular release and blood diffusion of BART viral micro-RNAs produced by EBV-infected nasopharyngeal carcinoma cells. Virol J. 2010;7:271.

[172] Luo SS, Ishibashi O, Ishikawa G, et al. Human villous trophoblasts express and secrete placenta-specific microRNAs into maternal circulation via exosomes. Biol Reprod. 2009;81:717–729.

[173] Yang M, Chen J, Su F, et al. Microvesicles secreted by macrophages shuttle invasion-potentiating microRNAs into breast cancer cells. Mol Cancer. 2011;10:117.

[174] Sarasin-Filipowicz M, Krol J, Markiewicz I, et al. Decreased levels of microRNA miR-122 in individuals with hepatitis C responding poorly to interferon therapy. Nat Med. 2009;15:31–33.

[175] Morita K, Taketomi A, Shirabe K, et al. Clinical significance and potential of hepatic microRNA-122 expression in hepatitis C. Liver Int. 2011;31:474–484.

[176] Wang K, Zhang S, Marzolf B, et al. Circulating microRNAs, potential biomarkers for drug-induced liver injury. Proc Natl Acad Sci USA. 2009;106:4402–4407.

[177] Zhang Y, Jia Y, Zheng R, et al. Plasma microRNA-122 as a biomarker for viral-, alcohol-, and chemical-related hepatic diseases. Clin Chem. 2010;56:1830–1838.

[178] Xu J, Wu C, Che X, et al. Circulating microRNAs, miR-21, miR-122, and miR-223, in patients with hepatocellular carcinoma or chronic hepatitis. Mol Carcinog. 2011;50:136–142.

[179] Tomimaru Y, Eguchi H, Nagano H, et al. Circulating microRNA-21 as a novel biomarker for hepatocellular carcinoma. J Hepatol. 2012;56:167–175.

[180] Cermelli S, Ruggieri A, Marrero JA, et al. Circulating microRNAs in patients with chronic hepatitis C and non-alcoholic fatty liver disease. PLoS One. 2011;6:e23937.

[181] Qi P, Cheng SQ, Wang H, et al. Serum microRNAs as biomarkers for hepatocellular carcinoma in Chinese patients with chronic hepatitis B virus infection. PLoS One. 2011;6:e28486.

[182] Ikeda Y, Tanji E, Makino N, et al. MicroRNAs associated with mitogen-activated protein kinase in human pancreatic cancer. Mol Cancer Res. 2012;10:259–269.

[183] Habbe N, Koorstra J, Mendell J, et al. MicroRNA miR-155 is a biomarker of early pancreatic neoplasia. Cancer Biol Ther. 2009;8:340–346.

[184] La Conti JL, Shivapurkar N, Preet A, et al. Tissue and serum microRNAs in the KrasG12D transgenic animal model and in patients with pancreatic cancer. PLoS One. 2011;6:e20687.

[185] Morimura R, Komatsu S, Ichikawa D, et al. Novel diagnostic value of circulating miR-18a in plasma of patients with pancreatic cancer. Br J Cancer. 2011;105:1733–1740.

[186] Liu R, Chen X, Du Y, et al. Serum microRNA expression profile as a biomarker in the diagnosis and prognosis of pancreatic cancer. Clin Chem. 2012;58:610–618.

[187] Hwang JH, Voortman J, Giovannetti E, et al. Identification of microRNA-21 as a biomarker for chemoresistance and clinical outcome following adjuvant therapy in resectable pancreatic cancer. PLoS One. 2010;5:e10630.

[188] Shigehara K, Yokomuro S, Ishibashi O, et al. Real-time PCR-based analysis of the human bile microRNAome identifies miR-9 as a potential diagnostic biomarker for biliary tract cancer. PLoS One. 2011;6:e23584.

[189] Komatsu S, Ichikawa D, Takeshita H, et al. Circulating microRNAs in plasma of patients with oesophageal squamous cell carcinoma. Br J Cancer. 2011;105:104–111.

[190] Chiang Y, Zhou X, Wang Z, et al. Expression levels of microRNA-192 and -215 in gastric carcinoma. Pathol Oncol Res. 2012;18:585–591.

[191] Wang J, Zhang J, Wu J, et al. MicroRNA-610 inhibits the migration and invasion of gastric cancer cells by suppressing the expression of vasodilator-stimulated phosphoprotein. Eur J Cancer. [Epub ahead of print] 2011.

[192] Liu H, Zhu L, Liu B, et al. Genome-wide microRNA profiles identify miR-378 as a serum biomarker for early detection of gastric cancer. Cancer Lett. 2012;316:196–203.

[193] Wu J, Wu G, Lv L, et al. MicroRNA-34a inhibits migration and invasion of colon cancer cells via targeting to Fra-1. Carcinogenesis. 2012;33:519–528.

[194] Nugent M, Miller N, Kerin MJ. MicroRNAs in colorectal cancer: function, dysregulation and potential as novel biomarkers. Eur J Surg Oncol. 2011;37:649–654.

[195] Cheng H, Zhang L, Cogdell DE, et al. Circulating plasma MiR-141 is a novel biomarker for metastatic colon cancer and predicts poor prognosis. PLoS One. 2011;6:e17745.

[196] Wulfken LM, Moritz R, Ohlmann C, et al. MicroRNAs in renal cell carcinoma: diagnostic implications of serum miR-1233 levels. PLoS One. 2011;6:e25787.

[197] Mahn R, Heukamp LC, Rogenhofer S, et al. Circulating microRNAs (miRNA) in serum of patients with prostate cancer. Urology. 2011;77:9–16.

[198] Lodes MJ, Caraballo M, Suciu D, et al. Detection of cancer with serum miRNAs on an oligonucleotide microarray. PLoS One. 2009;4:e6229.

[199] Brase JC, Johannes M, Schlomm T, et al. Circulating miRNAs are correlated with tumor progression in prostate cancer. Int J Cancer. 2010;128:608–616.

[200] Moltzahn F, Olshen AB, Baehner L, et al. Microfluidic-based multiplex qRT-PCR identifies diagnostic and prognostic microRNA signatures in the sera of prostate cancer patients. Cancer Res. 2011;71:550–560.

[201] Wach S, Nolte E, Szczyrba J, et al. MicroRNA profiles of prostate carcinoma detected by multiplatform microRNA screening. Int J Cancer. 2012;130:611–621.

[202] Kuwabara Y, Ono K, Horie T, et al. Increased microRNA-1 and microRNA-133a levels in serum of patients with cardiovascular disease indicate the existence of myocardial damage. Circ Cardiovasc Genet. 2011;4:446–454.

[203] Hanke M, Hoefig K, Merz H, et al. A robust methodology to study urine microRNA as tumor marker: microRNA-126 and microRNA-182 are related to urinary bladder cancer. Urol Oncol. 2010;28:655–661.

[204] Selth LA, Townley S, Gillis JL, et al. Discovery of circulating microRNAs associated with human prostate cancer using a mouse model of disease. Int J Cancer. 2012;131:652–661.

[205] Gonzales JC, Fink LM, Goodman OB Jr, et al. Comparison of circulating MicroRNA 141 to circulating tumor cells, lactate dehydrogenase, and prostate-specific antigen for determining treatment response in patients with metastatic prostate cancer. Clin Genitourin Cancer. 2011;9:39–45.

[206] Lin SC, Liu CJ, Lin JA, et al. miR-24 up-regulation in oral carcinoma: positive association from clinical and in vitro analysis. Oral Oncol. 2010;46:204–208.

[207] Park NJ, Zhou H, Elashoff D, et al. Salivary microRNA: discovery, characterization, and clinical utility for oral cancer detection. Clin Cancer Res. 2009;15:5473–5477.

[208] Lawrie CH, Gal S, Dunlop HM, et al. Detection of elevated levels of tumour-associated microRNAs in serum of patients with diffuse large B-cell lymphoma. Br J Haematol. 2008;141:672–675.

[209] Si ML, Zhu S, Wu H, et al. miR-21-mediated tumor growth. Oncogene. 2006;26:2799–2803.

[210] Cheng AM, Byrom MW, Shelton J, et al. Antisense inhibition of human miRNAs and indications for an involvement of miRNA in cell growth and apoptosis. Nucleic Acids Res. 2005;33:1290–1297.

[211] Stamatopoulos B, Meuleman N, Haibe-Kains B, et al. MicroRNA- 29c and microRNA-223 downregulation has in vivo significance in chronic lymphocytic leukemia and improves disease risk stratification. Blood. 2009;113:5237–5245.

[212] Moussay E, Wang K, Cho JH, et al. MicroRNA as biomarkers and regulators in B-cell chronic lymphocytic leukemia. Proc Natl Acad Sci USA. 2011;108:6573–6578.

[213] Fulci V, Chiaretti S, Goldoni M, et al. Quantitative technologies establish a novel microRNA profile of chronic lymphocytic leukemia. Blood. 2007;109:4944–4951.

[214] Huang J, Mo YY. Role of microRNAs in leukemia stem cells. Front Biosci (Schol Ed). 2012;4:799–809.

[215] Zhu YD, Wang L, Sun C, et al. Distinctive microRNA signature is associated with the diagnosis and prognosis of acute leukemia. Med Oncol. [Epub ahead of print] 2011.

[216] Tanaka M, Oikawa K, Takanashi M, et al. Down-regulation of miR-92 in human plasma is a novel marker for acute leukemia patients. PLoS One. 2009;4:e5532.

[217] Aqeilan RI, Calin GA, Croce CM. miR-15a and miR-16-1 in cancer: discovery, function and future perspectives. Cell Death Differ. 2010;17:215–220.

[218] Cammarata G, Augugliaro L, Salemi D, et al. Differential expression of specific microRNA and their targets in acute myeloid leukemia. Am J Hematol. 2010;85:331–339.

[219] Tsang WP, Kwok TT. Let-7a microRNA suppresses therapeutics-induced cancer cell death by targeting caspase-3. Apoptosis. 2008;13:1215–1222.

[220] Calin GA, Dumitru CD, Shimizu M, et al. Frequent deletions and down-regulation of micro-RNA genes miR15 and miR16 at 13q14 in chronic lymphocytic leukemia. Proc Natl Acad Sci USA. 2002;99:15524–15529.

[221] Pons A, Nomdedeu B, Navarro A, et al. Hematopoiesis-related microRNA expression in myelodysplastic syndromes. Leuk Lymphoma. 2009;50:1854–1859.

[222] Zuo Z, Calin GA, De Paula HM, et al. Circulating microRNAs let-7a and miR-16 predict progression-free survival and overall survival in patients with myelodysplastic syndrome. Blood. 2011;118:413–415.

[223] Huang JJ, Yu J, Li JY, et al. Circulating microRNA expression is associated with genetic subtype and survival of multiple myeloma. Med Oncol. [Epub ahead of print] 2012.

[224] Rifai N, Gillette MA, Carr SA. Protein biomarker discovery and validation: the long and uncertain path to clinical utility. Nat Biotechnol. 2006;24:971–983.

[225] Tsang JC, Lo YM. Circulating nucleic acids in plasma/serum. Pathology. 2007;39:197–207.

[226] Chim SS, Shing TKF, Hung ECW, et al. Detection and characterization of placental microRNAs in maternal plasma. Clin Chem. 2008;54:482–490.

[227] Lawrie CH, Gal S, Dunlop HM, et al. Detection of elevated levels of tumor-associated microRNAs in serum of patients with diffuse large B-cell lymphoma. Br J Haematol. 2008;141:672–675.

[228] Mitchell PS, Parkin RK, Kroh EM, et al. Circulating microRNAs as stable blood-based markers for cancer detection. PNAS. July 29, 2008;105(30):10513–10518. www.pnas.org_cgi_doi_10.1073_pnas.0804549105

[229] van Niel G, Porto-Carreiro I, Simoes S, et al. Exosomes: A common pathway for a specialized function. J Biochem. 2006;140:13–21.

[230] Caby MP, Lankar D, Vincendeau-Scherrer C, et al. Exosomal-like vesicles are present in human blood plasma. Int Immunol. 2005;17:879–887.

[231] Valadi H, Ekström K, Bossios A, et al. Exosome-mediated transfer of mRNAs and microRNAs is a novel mechanism of genetic exchange between cells. Nat Cell Biol. 2007;9:654–659.

[232] Hunter CP, Winston WM, Molodowitch C, et al. Systemic RNAi in Caenorhabditis elegans. Cold Spring Harb Symp Quant Biol. 2006;71:95–100.

[233] Voinnet O. Noncell autonomous RNA silencing. FEBS Lett. 2005;579:5858–5871.

[234] Yu DC, Li QG, Ding LXW, et al. Circulating microRNAs: potential biomarkers for cancer. Int J Mol Sci. 2011;12:2055–2063.

[235] Duttagupta R, Jiang R, Gollub J, et al. Impact of cellular miRNAs on circulating miRNA biomarker signatures. PLoS One. 2011;6:e20769.

[236] Pritchard CC, Kroh E, Wood B, et al. Blood cell origin of circulating microRNAs: a cautionary note for cancer biomarker studies. Cancer Prev Res (Phila). 2012;5:492–497.

[237] McDonald JS, Milosevic D, Reddi HV, et al. Analysis of circulating microRNA: preanalytical and analytical challenges. Clin Chem. 2011;57:833–840.

[238] Reid G, Kirschner MB, van Zandwijk N. Circulating microRNAs: association with disease and potential use as biomarkers. Crit Rev Oncol Hematol. 2011;80:193–208.

[239] Benner SA. Extracellular 'communicator RNA'. FEBS Lett. 1988;233:225–228.

[240] D'Alessandra Y, Pompilio G, Capogrossi MC. MicroRNAs and myocardial infarction. Curr Opin Cardiol. 2012;27: 228–235.

[241] Madkour LH. Reactive Oxygen Species (ROS), Nanoparticles, and Endoplasmic Reticulum (ER) Stress-Induced Cell Death Mechanisms. Academic Press; 2020.

[242] Madkour LH. Nanoparticles Induce Oxidative and Endoplasmic Reticulum Antioxidant Therapeutic Defenses. Springer International Publishing; 2020. doi:10.1007/978-3-030-37297-2.

[243] Madkour LH. Nucleic Acids as Gene Anticancer Drug Delivery Therapy. Elsevier; 2020.

[244] Madkour LH. Nanoelectronic Materials: Fundamentals and Applications (Advanced Structured Materials). Springer International Publishing; 2019. doi:10.1007/978-3-030-21621-4.

[245] Wang Y, Zheng D, Tan Q, et al. Nanopore-based detection of circulating microRNAs in lung cancer patients. Nat Nanotechnol. 2011;6:668–674.

[246] Spiegel JC, Lorenzen JM, Thum T. Role of microRNAs in immunity and organ transplantation. Expert Rev Mol Med. 2011;13:e37.

[247] Ha TY. The role of microRNAs in regulatory T cells and in the immune response. Immune Netw. 2011;11:11–41.

# 4 MicroRNA's Potential in Human Cancer as Therapeutic Targets and Novel Biomarkers

Mature miRNAs are single-stranded RNA molecules of 20- to 23-nucleotide (nt) length that control gene expression in many cellular processes. These molecules typically reduce the translation and stability of mRNAs, including those of genes that mediate processes in tumorigenesis, such as inflammation, cell cycle regulation, stress response, differentiation, apoptosis, and invasion. MiRNA targeting is initiated through specific base-pairing interactions between the 5′ end ("seed" region) of the miRNA and sites within coding and untranslated regions (UTRs) of mRNAs; target sites in the 3′ UTR lead to more effective mRNA destabilization. Because miRNAs frequently target hundreds of mRNAs, miRNA regulatory pathways are complex.

We are only beginning to comprehend the functional repercussions of the gain or loss of particular microRNAs on cancer. Nonetheless, although microRNAs were discovered in humans a mere 8 years ago, a host of promising potential applications in the diagnosis, prognoses, and therapy of cancer are emerging at a rapid pace.

## 4.1 CANCER AND MiRNA OVERVIEW

Cancer is a complex disease where a group of abnormal cells grows without control and becomes able to invade adjacent tissues and colonize other organs. Such invasive behavior results in organ dysfunction and ultimately organ failure. Several lines of evidence indicate that carcinogenesis is a multistep process where the malignant cells accumulate genetic and epigenetic alterations that drive the progressive transformation of normal cells into malignant ones [1]. In this way, cancer cells select alterations in genes that promote cancer progression (oncogenes) or genes that impair it (tumor suppressor genes). At the end of the transformation process, the malignant cells lose their cellular identity and acquire growth independence, invasiveness, and resistance to senescence and apoptosis.

MiRNAs are a class of small RNA molecules that regulate gene expression at the posttranscriptional level. Initially discovered in *Caenorhabditis elegans* [2, 3], they were considered a peculiarity of nematodes until it was realized that some of them were phylogenetically conserved in a wide variety of animals including humans [4–6]. Today, microRNAs are increasingly seen as important regulators of gene expression, ushering in a renewed appreciation of the regulative capabilities of noncoding RNA. At the cellular level, miRNAs are important in the establishment and maintenance of cell identity [7, 8]. Abnormal levels of miRNAs often result in loss of differentiation, a hallmark of cancer. Not surprisingly, therefore, dysfunctions of the miRNA pathway affect many cellular processes that are routinely altered in cancer, such as differentiation, proliferation, apoptosis, metastasis [9], and telomere maintenance [10, 11].

MiRNAs were originally shown to be important in the timing of larval development in *C. elegans,* leading to the identification of the miRNAs lin-4 and let-7 [3, 4]. Our initial understanding of miRNA–mRNA target recognition came from observations of sequence complementarity of the lin-4 RNA to multiple conserved sites within the lin-14 3′ UTR [2, 3]; molecular genetic analysis had shown that this complementarity was required for the repression of lin-14 by lin-4 [12]. Homologs of let-7 or lin-4/mir-125 were thereafter shown to have temporal expression patterns in other organisms, including mammals, and to regulate mammalian development [13–16]. Given their integral role in development, it was no surprise that miRNAs were soon found to be important in tumorigenesis, and since their discovery close to 5000 publications associate miRNAs to cancer, including more than 1000 reviews (recent examples include [17–19]). MiRNAs were initially linked to tumorigenesis due to their apparent proximity to chromosomal breakpoints [20] and their dysregulated expression levels in many malignancies [21, 22].

Given the wealth of rapidly accumulating information implicating miRNAs in cancer, to allow the reader to critically assess the reports exploring the function of miRNAs in malignancies, we first review the methods used to study the expression and role of miRNAs in tumors, and then review the evidence that relates miRNA genomic organization, biogenesis, target recognition, and function to tumorigenesis. An overview of miRNA cistronic expression and sequence similarity allows a better understanding of the regulation of miRNA expression and the factors contributing to technical limitations in accuracy of miRNA detection. Understanding the regulatory potential of miRNAs based on sequence similarity families and miRNA abundance allows evaluation of which miRNAs are important regulators of tumorigenesis pathways.

In hepatocellular carcinoma (HCC), miRNAs frequently present aberrant expression profiles, which make them potentially attractive for diagnostic or prognostic applications. Currently, accumulating evidence is indicating the

DOI: 10.1201/9781003229650-4

role of miRNAs as tumor suppressors or oncogenes in hepatic malignancies. In particular, comprehensive studies have made possible a better understanding of HCC behavior, such as tumor growth, response to therapies, metastatic potential, or recurrence, regarding the altered expression of cancer-related miRNAs. Based on these findings, efforts are underway to define new markers for liver cancer in both invasive (hepatic biopsy or tumor resection) and noninvasive (circulating miRNAs in blood serum) ways. Due to their implication in the control of various cell processes altered in HCC, cancer-related miRNAs also offer encouraging perspectives for the development of innovative cancer therapies. In this chapter, we review the importance of miRNA deregulation in HCC progression and the role of these small noncoding RNAs as tumor suppressors and oncogenes. The significance of miRNAs in HCC diagnosis and miRNA-based therapeutic strategies is then discussed.

## 4.2　MicroRNAs: GENOMICS, BIOGENESIS, AND MODE OF ACTION

MiRNAs are a large family of 19- to 25-nucleotide-long non-protein-coding RNAs (ncRNAs) that are currently still growing in number and in diversity of function [23, 24]. At the time of writing, more than 5500 miRNA loci from 58 species are listed in the miRNA registry [25]. Around 550 miRNAs have been documented in humans but computational models predict up to roughly 1000 human miRNAs [26].

About half of the miRNAs genes are organized in clusters that are transcribed as polycistronic primary transcripts from which the individual miRNAs are processed [13, 15]. The remaining miRNAs are encoded in single loci. More than two-thirds of the miRNAs share transcriptional units with protein-coding genes or longer ncRNAs. Almost all the miRNAs that arise from protein-coding genes are located in introns, while those that arise from longer ncRNAs can be located in either introns or exons [27].

The basic scheme of the microRNA pathway has been revealed (Figure 4.1). Most miRNAs are transcribed by RNA polymerase II as long primary RNA (pri-miRNAs) that contain a 5' CAP structure and a 3' polyadenylated tail [28]. These pri-miRNAs can be several kilobases long, and are processed in the nucleus by the RNAase III enzyme Drosha and its double-stranded binding protein DGCR8 [29–31], into one or more miRNA precursors (pre-miRNA). The posttranscriptional maturation of miRNAs is regulated in response to proliferative stimuli and cellular differentiation [32]. It has been recently shown that, at least in the case of the let-7 family of miRNAs, the RNA-binding protein Lin28 is necessary and sufficient for blocking the cleavage of the pri-let-7 miRNAs [33].

Pre-miRNAs are about 65–85 bp long and are folded in stem-loop structures that are exported to the cytoplasm by the nuclear export factor Exportin-5 and its cofactor RAN-GTP [35, 36]. In the cytoplasm, the pre-miRNAs are processed by another RNAse III enzyme (Dicer) into 22-bp double-stranded RNAs [37–39]. Argonaute 2 (hAgo2)

is subsequently recruited, completing the RNA-induced silencing complex (RISC) [40, 41]. Only one of the two strands (guide strand) will remain in the mature RISC. The main factor determining the selection of the guide strand seems to be the higher stability of the duplex in the 5' half [42–45]. The repression of the mRNA is achieved in two different ways depending on the degree of complementarity between the miRNA and the target. If the miRNA binds with near-perfect complementarity to protein-coding mRNA sequences, the messenger is cut by RISC and degraded. This mechanism commonly occurs in plants [46] and only occasionally in animals [47]. In animals, the miRNAs bind more frequently to their targets' 3' untranslated regions (UTR) with imperfect complementarity, resulting in translational inhibition, followed by a variable degree of mRNA degradation [48, 49]. Several mechanisms of translational silencing have been suggested, including blocking of the initiation or elongation steps of translation, or miRNA-mediated sequestration of the mRNA targets in the P-bodies, which are specialized cytoplasm compartments where translational repression and mRNA turnover takes place [48, 50–52].

## 4.3　MiRNA BIOGENESIS AND MECHANISM OF ACTION

MiRNAs are evolutionarily conserved, small, noncoding RNAs of approximately 22 nt that accurately regulate gene expression by complementary base pairing with the 3'-untranslated regions (3'-UTRs) of messenger RNAs (mRNAs) [28]. These fine posttranscriptional regulators were first evidenced in *C. elegans* by Ambros and coworkers, who discovered that lin-4, a gene known to control the timing of nematode larval development, did not code for a protein but produced small RNAs that can specifically bind to lin-14 mRNA and repress its translation [2, 3]. Then, miRNAs have been reported in a variety of organisms ranging from virus to mammalians. In order to facilitate miRNA-based investigations, a miRNA registry (miRBase) has been established and is currently maintained by the University of Manchester [53]. So far, 1921 mature miRNA sequences have been registered in the miRBase database (http://www.mirbase.org, release 18, November 2011). It was estimated that more than 25% of all mature human miRNAs belong to a family comprising two or more members (based on a seven-seed sequence homology). As attested by computational studies, more than 30% of the protein-coding regions may be directly targeted and modulated by miRNAs [54]. An essential feature of miRNAs is that a single miRNA can recognize numerous mRNAs, and, conversely, one mRNA can be recognized by several miRNAs. These pleiotropic properties enable miRNAs to exert a wide control on a plethora of targets, attesting to the complexity of this mechanism of gene expression regulation. In addition, recent studies have demonstrated that certain miRNAs exhibit a tissue-specific distribution in rodent endoderm–derived tissues [55], although other miRNAs can be expressed ubiquitously [14].

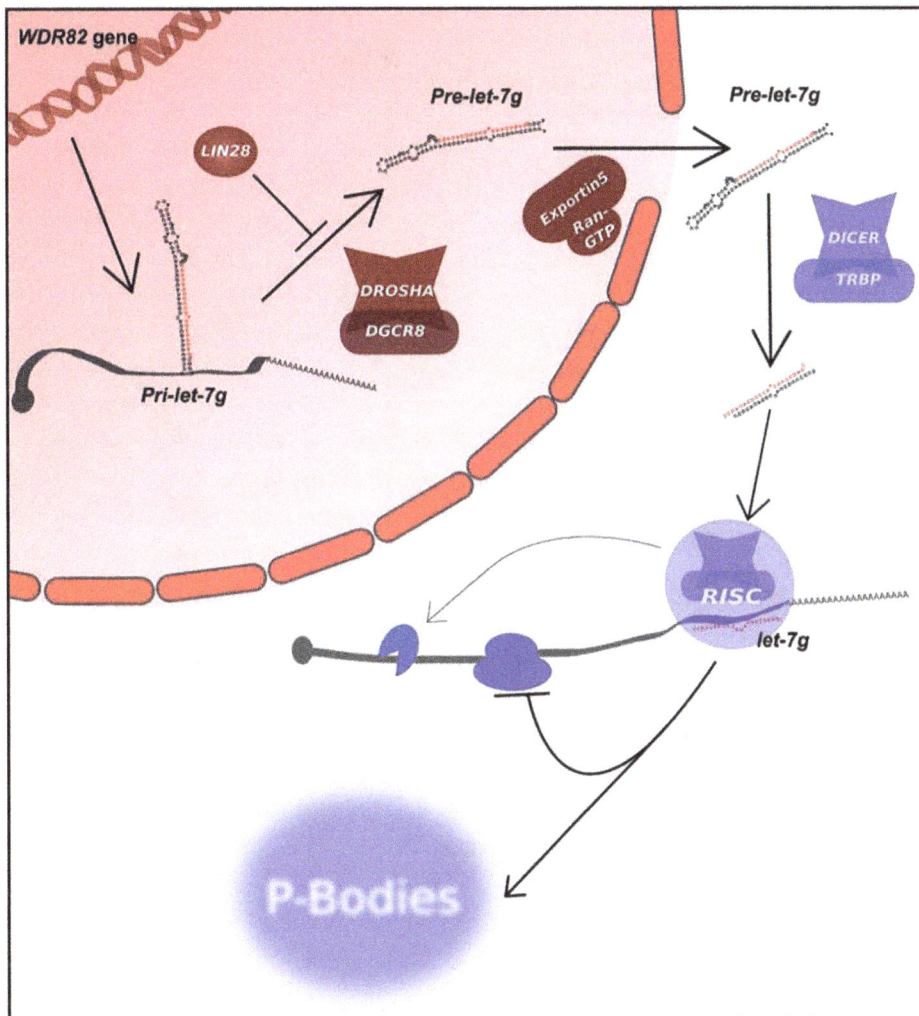

**FIGURE 4.1** Cartoon of the biogenesis of let-7g as a representative example of miRNAs involved in cancer. Let-7g is coded in the second intron of the WDR82 gene. The messenger RNA is transcribed by RNA polymerase II and it is capped and polyadenylated. This first precursor is called pri-let-7g and folds into a characteristic stem-loop structure from which mature miRNAs are generated in two sequential processing steps. First the pri-miRNA is recognized and cleaved by the microprocessor complex composed by a nuclear RNase III endonuclease called Drosha and its RNA binding partner DGCR8 resulting in the formation of an 84-nucleotide precursor miRNA (pre-miRNA). LIN28 inhibits this first processing step at least in the miRNA let-7 family. Exportin-5 and Ran-GTP are involved in the transport of the pre-miRNA to the cytoplasm where a second endonucleolytic cleavage occurs mediated by the RNAse III Dicer, in association with its RNA binding partner TRBP, generating a double-stranded RNA with 2-nt overhang that is recruited to the RNA-induced silencing complex (RISC). Subsequently one of the strands, the passenger strand, is removed and degraded and the mature let-7g (UGAGGUAGUAGUUUGUACAGUU) guides the RISC to target mRNAs whose expression can be inhibited by translational repression, although the direct degradation of the mRNA also has been reported [34]. This mRNA inhibition may be performed in specialized cytoplasm regions called P-bodies enriched in proteins involved in mRNA turnover. See text for further details.

Thus, endodermal-derived hepatic and pancreatic tissues display differential miRNA enrichment in comparison to other organs. Typically, miR-122, miR-21, miR-101, miR-192, and miR-221 expression progressively increases during liver morphogenesis to become predominant in the adult hepatic tissue. Among these miRNAs highly expressed in hepatocytes, the liver-specific miR-122 represents 70% of the total amount of miRNAs in the organ. Intriguingly, the pri- and pre-miR-122 are regulated in a circadian manner [56]. In addition, the turnover of mature miR-122 appears to be relatively long compared to other miRNAs, as its half-life may reach several weeks.

MiRNA biogenesis is a multistep process that has been reviewed extensively (Figure 4.2). Briefly, miRNAs are produced by the RNA polymerase II into transcriptional precursors of hundreds of nucleotides called *primary miRNAs* (pri-miRNAs). These long primary precursor transcripts exhibit several stem-loop structures of approximately 80 nt. In the nucleus, pri-miRNAs undergo processing by the nuclear endonuclease Drosha and the double-stranded RNA-binding protein Pasha to be cleaved into precursor miRNAs (pre-miRNAs). Pre-miRNAs are then exported to the cytoplasm by exportin-5, where they undergo further processing by the RNase III endonuclease

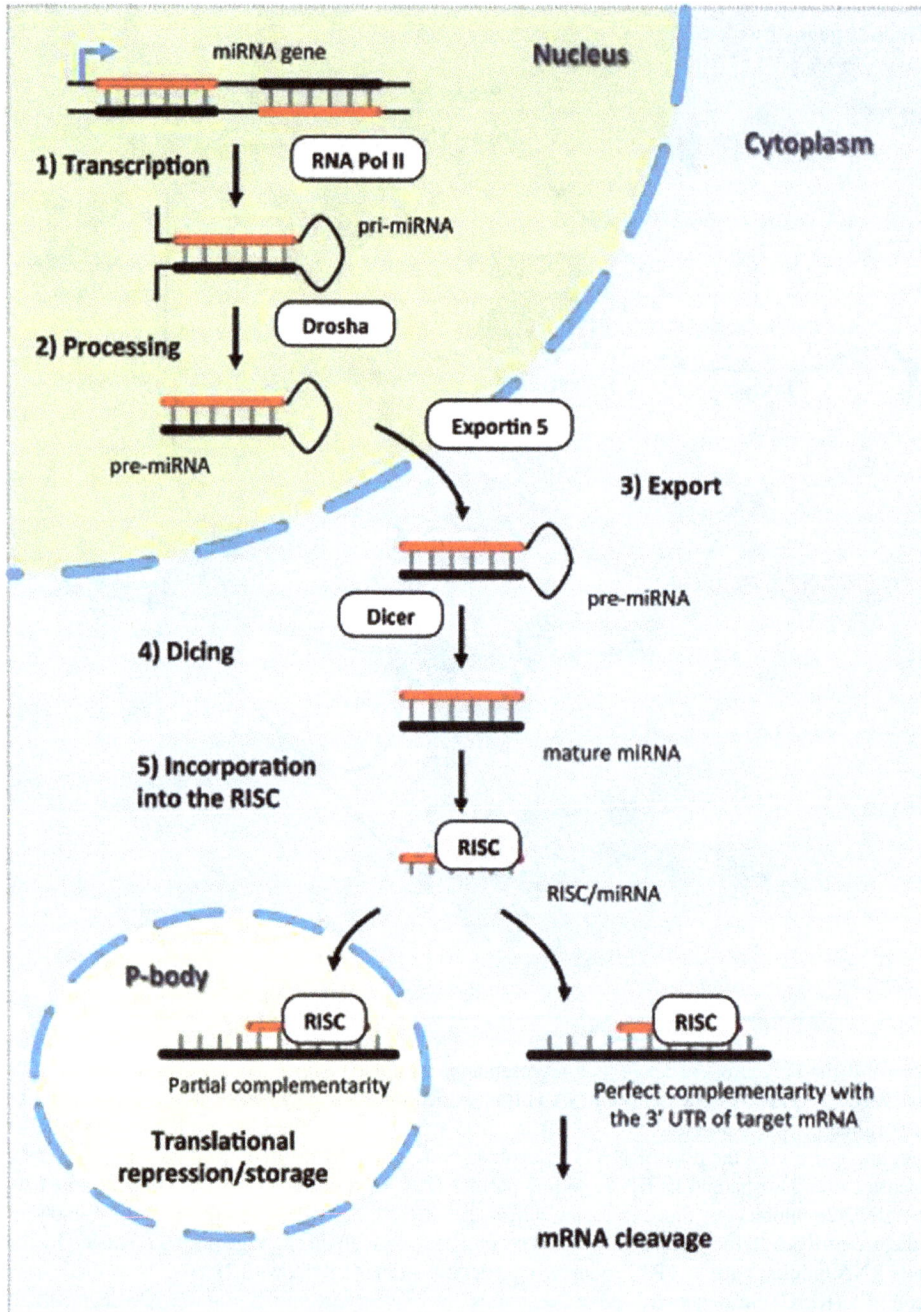

**FIGURE 4.2** **The RNA interference process: biogenesis and regulation of miRNAs.** Transcription from the miRNA genes by the RNA polymerase II occurs in the nucleus. The primary precursor miRNAs (pri-miRNAs) are then cleaved by the RNase III enzyme Drosha, producing precursor molecules (pre-miRNAs). With the help of Exportin-5, the pre-miRNAs are exported into the cytoplasm, where they undergo further processing by the ribonuclease Dicer to generate mature miRNAs. Mature single-stranded miRNAs are incorporated onto the RNA-induced silencing complex (RISC) to carry out their silencing function. The regulation mechanism is dependent on the degree of complementarity between the 3'-UTR region of the target mRNA and the seed region in the 5'-end of the miRNA. In the case of perfect complementarity, the mRNA is cleaved by RISC. If the complementarity is partial, the regulation is carried out by repression of the translation in the P-body.

Dicer. Dicer cleaves the pre-miRNA loop to produce an imperfect duplex consisting of a mature miRNA and a complementary fragment of a similar size (miRNAs*). A mature miRNA measures 20–23 nt in length, which can be incorporated into the RISC, whereas the complementary miRNA* separates from the duplex and is generally degraded. Functional target sites within the mRNA usually consist of a 6- to 7-nt-long sequence, the so-called miRNA seed sequence. The silencing complex binds complementarily the 3'-UTR of the target sequences and negatively regulates gene expression either through endonucleolytic cleavage of the mRNA or inhibition of its translation [28]. Lastly, miRNAs and their target mRNAs will be localized in the cytoplasmic processing bodies (P-bodies) where they will be degraded [52]. The turnover of miRNAs is still a largely unexplored area. However, RNA degradation enzymes might target not only mature miRNAs but also the precursor pri- and pre-miRNAs. In humans, the decrease of mRNA levels related to miRNA activity has been shown to precede protein diminution in 84% of cases [57].

## 4.4 METHODS FOR STUDYING MiRNA GENETICS AND EXPRESSION

### 4.4.1 MiRNA PROFILING

The main methods currently used for miRNA profiling are sequencing, microarray, and real-time RT-PCR based approaches (reviewed in [58–60]). The input material initially used for these studies comprised high-quality preserved fresh frozen samples, but recently it has been possible to obtain reproducible and comparable profiles using formalin-fixed paraffin-embedded tissues (FFPE), making these archived tumor collections accessible for study [61–63]. Microarrays generally provide fold-changes in miRNA expression between samples, with members of miRNA sequence families prone to cross-hybridization [64–67]. More recently, calibration cocktails of synthetic miRNAs were used in array experiments to derive absolute abundance of miRNAs [68]. RT-PCR methods are lower throughput and require normalization (i.e., candidate reference genes including other small noncoding RNAs [69, 70]). Mean expression normalization has been suggested as an alternative RT-PCR normalization method for reduction of technical variation to allow appreciation of biological changes [71]. If external miRNA standards are used for quantification (i.e., [72, 73]), the most abundant miRNA, which may vary in length due to 3′ end heterogeneity, should be used as a calibration standard. Sequencing methods, besides their obvious potential to identify new miRNAs, editing, and mutation events, estimate miRNA abundance based on frequency of sequence reads (e.g., [13, 15, 16, 74–77]). Given the dramatic increase in sequencing power, bar-coding samples can allow multiple specimens to be processed at the same time, reducing the cost and effort of profiling, and paving the way for large specimen studies [77–79]. Ligation biases between miRNAs and 5′ and 3′ adapters for RT-PCR amplification exist in sequencing methods, and miRNA read frequencies may not always reflect the absolute expression levels, but these variations are irrelevant when monitoring fold-changes between samples. A study with a synthetic pool of 770 miRNA sequences showed that overall, these biases did not prevent identification of miRNAs and allowed estimation of these biases [79]. For example, certain miRNAs could be overrepresented due to higher ligation efficiency (such as miR-21, which was ~two-fold overrepresented), while other miRNAs could be underrepresented (such as miR-31,

which was >five-fold underrepresented). However, given the increasing depth of sequencing, most underrepresented miRNAs are identified with sufficient sequence reads to allow for a statistically significant comparison across parallel processed samples.

Recent studies have compared the results obtained using multiple platforms [80]. A study of miRNA expression in liposarcoma revealed excellent agreement between barcoded next generation sequencing (NGS) and microarray profiles [81], while another study of miRNA expression in breast cancer showed good agreement between bar-coded NGS and another hybridization-based method, northern blotting [82].

Finally, choosing the appropriate statistical analysis to evaluate the data depends on the methodology used to obtain the profiles, ranging from established significance analysis of microarrays (SAM) analysis for microarray data [83], to newly developed techniques for sequencing data [77, 84, 85]. Recent in situ hybridization (ISH) advances allowed sensitive detection of miRNAs in heterogeneous tissues, defining miRNA cellular localization [86–88]. The potential of miRNA localization to suggest function for a subpopulation of cells was demonstrated early on, as in the case of lsy-6 expressed in less than ten neurons in *C. elegans* controlling left/right asymmetry [89].

### 4.4.2 MiRNA DATABASES AND VALIDATION

It is critical to know which miRNAs are validated and have the potential to regulate cellular functions, especially given the frequent revisions of the miRNA database, miRBase (www.mirbase.org) [25], and the dramatic increase in the number of novel and reannotated miRNAs through the use of deep-sequencing technologies. It is extremely challenging to establish the validity of novel miRNAs, particularly when their definition is based on a handful of sequence reads. The latest release (version 17) of miRBase includes 1424 human miRNA precursors. Compared with version 16, version 17 includes 385 novel human miRNA precursors, 45 name changes, 1 sequence revision, and the removal of 2 precursors. Given the recent explosion in acquisition of NGS profiles, miRBase has now added features to allow evaluation of microRNA annotation [90]. The database mapped reads from short RNA deep-sequencing experiments to miRNAs and developed web interfaces to view these mappings. This is an important step in characterizing the newly identified miRNAs as prototypical miRNAs (consisting of a hairpin structure and processing sites consistent with RNase III cleavage steps). The challenge of constantly revising and curating existing databases based on newly acquired sequencing data is illustrated in two recent studies re-evaluating mouse and human miRNAs. A recent study of 60 million small RNA sequence reads generated from a variety of adult and embryonic mouse tissues confirmed 398 annotated miRNA genes and identified 108 novel miRNA genes but was unable to find sequencing evidence for 150 previously annotated mouse miRNAs. Ectopic expression of the confirmed and newly identified miRNA hairpin sequences yielded small RNAs with the classical miRNA features but failed to support other previously annotated sequences (of the 17 tested miRNAs with no read evidence, only 1 yielded a single sequence read, while of 28 tested miRNAs with insufficient number of reads, only 4 were verified) [91]. A more recent study has reannotated human miRNAs based on read evidence from more than 1000 human samples [82]. MiRNAs were curated both on the basis of read counts, as well as patterns compatible with traditional miRNA processing, redefining prototypical miRNAs (557 precursors, corresponding to 1112 mature and star sequences [miRNA*, described in the following section], miR-451 and miR-618 being the only miRNAs without a star sequence). Also, 269 not-yet-reported star sequences were added (compared to miRBase 16); putative miRNAs from miRBase, for which read evidence was not obtained, were ignored; and specific miRNAs were renamed according to the read ratio between mature and star sequences. The importance of curated miRNA databases is especially evident in assessing the statistical significance of differentially expressed miRNAs to identify potential biomarkers based on microarray studies. Including miRNAs without strong read evidence in such comparisons could skew the results.

## 4.5 MECHANISMS OF ALTERATION OF MiRNA LEVELS IN MALIGNANCY

We review miRNA biogenesis (Figure 4.3) and illustrate which steps of the biogenesis pathway are linked to malignancy, starting from miRNA genomic localization, transcriptional regulation, processing steps and posttranscriptional modification. There is evidence supporting the association of the first three processes and/or the factors that control them with tumorigenesis, whereas evidence relating posttranscriptional miRNA modifications to cancer is not clear-cut.

### 4.5.1 GENERAL PRINCIPLES OF MiRNA GENOMIC ORGANIZATION

MiRNAs are frequently expressed as polycistronic transcripts. To date, 1424 human miRNA precursor sequences have been deposited in miRBase [25]. Approximately one-third (497) of these miRNAs are located in 156 clusters, each measuring ≤51 kb in the human genome (51 kb being the longest distance between miRNAs belonging to the same cluster, Figure 4.4). These miRNA clusters are co-expressed based on evidence from miRNA profiling data from a variety of tissues and cell lines [65, 76, 77, 91]. The genomic organization of representative oncogenic (miR-17 and miR-21) and tumor suppressor (let-7 and miR-141) sequence families (described in following section) is illustrated in Figure 4.4. Presentation of miRNA profiles in the

form of expression clusters provides a readily interpretable summary of expression data and stresses the importance of cistronic expression regulation; dysregulation of one member of the cluster should be accompanied by similar dysregulation of other cluster members [82]. Because miRNA genes are frequently multicopy, determining the relative contribution of each genomic location to mature miRNA expression is challenging.

### 4.5.2 ALTERATIONS IN GENOMIC MiRNA COPY NUMBERS AND LOCATION

Changes in miRNA expression between normal and tumor specimens are often attributed to the location of miRNAs in regions of chromosomal instability (amplification, translocation, or deletion), or nearby chromosomal breakpoints, initially locating 52.5% of miRNA genes in cancer-associated regions or fragile sites [20]. The miRNA cluster mir-15a/16-1 is located in a frequently deleted genomic locus containing a putative tumor suppressor–containing region in B-cell chronic lymphocytic leukemia (B-CLL) [92]. Other examples include deletion of let-7g/mir-135-1 in a variety of human malignancies [20], amplification of mir-17-92 cluster in lymphoma [93], translocation of mir-17-92 in T-cell acute lymphoblastic leukemia (T-ALL) [94], and amplification of mir-26a in glioblastoma [95].

### 4.5.3 ALTERATIONS IN MiRNA TRANSCRIPTIONAL REGULATION

Some autonomously expressed miRNA genes have promoter regions that allow miRNAs to be highly expressed in a cell-type-specific manner, and can even drive high levels of oncogenes in cases of chromosomal translocation. The mir-142 gene, strongly expressed in hematopoietic cells, is located on chromosome 17 and was found at the breakpoint junction of a t(8;17) translocation to MYC, which causes an aggressive B-cell leukemia [96]. The translocated MYC gene, which was also truncated at the first exon, was located only four nucleotides from the 3's end of the mir-142 precursor, placing it under the control of the upstream mir-142 promoter. In an animal model for HCC, a similar event placed C-MYC downstream of the mir-122a promoter, which is active only in hepatocytes [97].

Many transcription factors regulate miRNA expression in a tissue-specific and disease state–specific fashion, and some miRNAs are regulated by well-established tumor suppressor or oncogene pathways such as TP53, MYC, and RAS (reviewed in [98]). The miRNA and its transcriptional regulators can participate in complex feedback regulation loops. Examples include the TP53-regulated mir-34a [99, 100], the RAS-regulated mir-21 [76, 101, 102], and the MYC-regulated mir-17-92 gene cluster [103, 104].

MiRNA dysregulation has also been linked to changes in epigenetic regulation, such as the methylation status of miRNA genes, which results in alterations in their

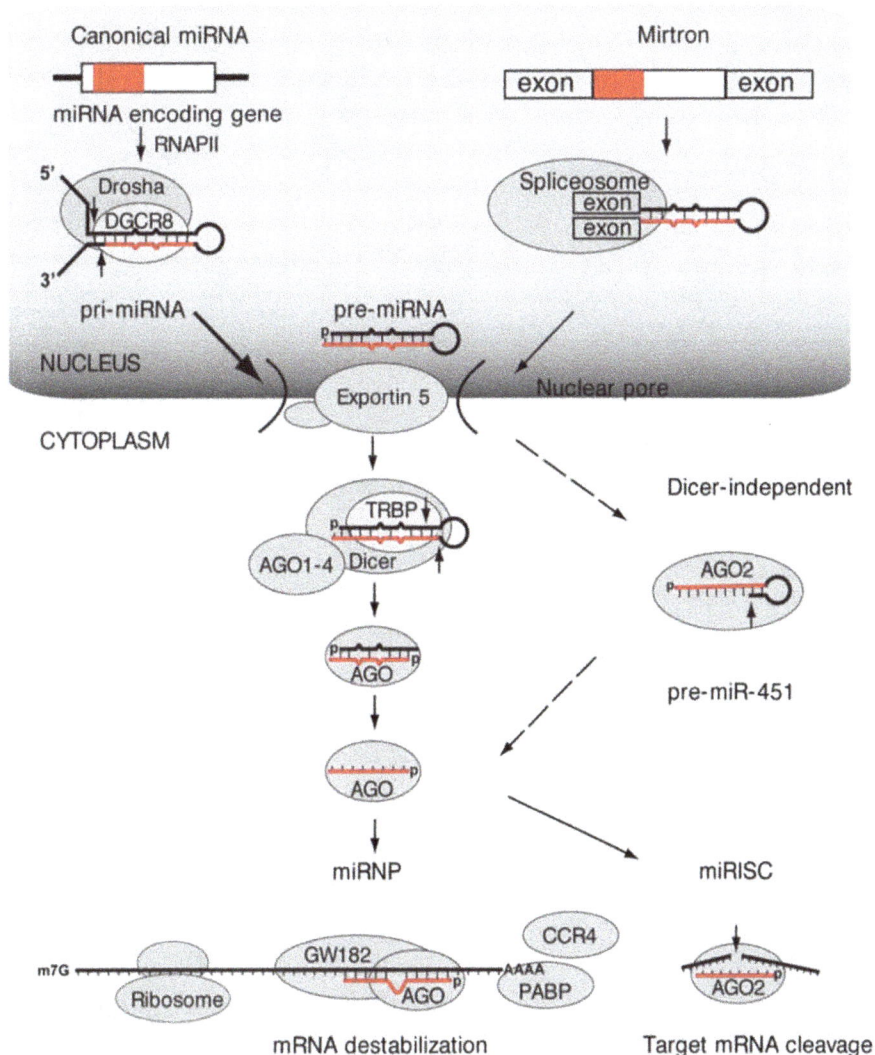

**FIGURE 4.3  MiRNA biogenesis pathway.** MiRNAs are transcribed by RNAPII to produce pri-miRNAs. Canonical miRNAs are processed by the endoribonuclease Drosha in partnership with its RBP partner DGCR8; mirtrons are instead processed by the spliceosome. The processed pre-miRNA is transported to the cytoplasm through an export complex consisting of exportin 5. The pre-miRNA is subsequently processed in the cytoplasm by another endoribonuclease Dicer in partnership with its RBP partner TRBP to form the final 21- to 23-nucleotide miRNA product. MiR-451 is not processed by Dicer, but is rather cleaved by AGO2. Mature miRNAs (indicated in *red*) are then incorporated into AGO 1 through 4, forming miRNPs, also known as miRISC. MiRNPs also incorporate other proteins, such as GW182. MiRNPs are thought to direct miRNA-mediated destabilization (i.e., through interaction with CCR4) or miRNA-mediated translational repression (i.e., through interaction with ribosomes) of miRNAs without perfectly complementary mRNA targets. MiRISC is thought to direct AGO2-catalyzed target mRNA cleavage of miRNA fully or nearly fully complementary mRNA targets.

expression levels [105, 106]. Examples of methylated miRNA genes include mir-127 in bladder cancer cells [107] and mir-9-1 in breast cancer [108].

### 4.5.4  MiRNA Biogenesis Pathway in Tumorigenesis

MiRNA biogenesis has been reviewed extensively (Figure 4.3) [28, 57, 98, 109–113]. MiRNA pathway components could either be misexpressed in tumors or mutated (reviewed in [114, 115]). Posttranscriptional regulation of miRNAs themselves through RNA editing or terminal modifications was shown to alter miRNA targeting, processing, and stability, but connection of these modifications to tumorigenesis has not yet been definitive (reviewed in [98, 115, 116]).

#### 4.5.4.1  Alterations in RNASEN/DGCR8 and DICER1/TARBP2

Inhibition of the miRNA biogenesis pathway leads to severe developmental defects and is lethal in many organisms (reviewed earlier in [8], recent examples include [117, 118]), and perturbations of this pathway predispose to tumorigenesis [119]. Initial miRNA expression profiling experiments suggested that miRNAs are less abundant in tumors

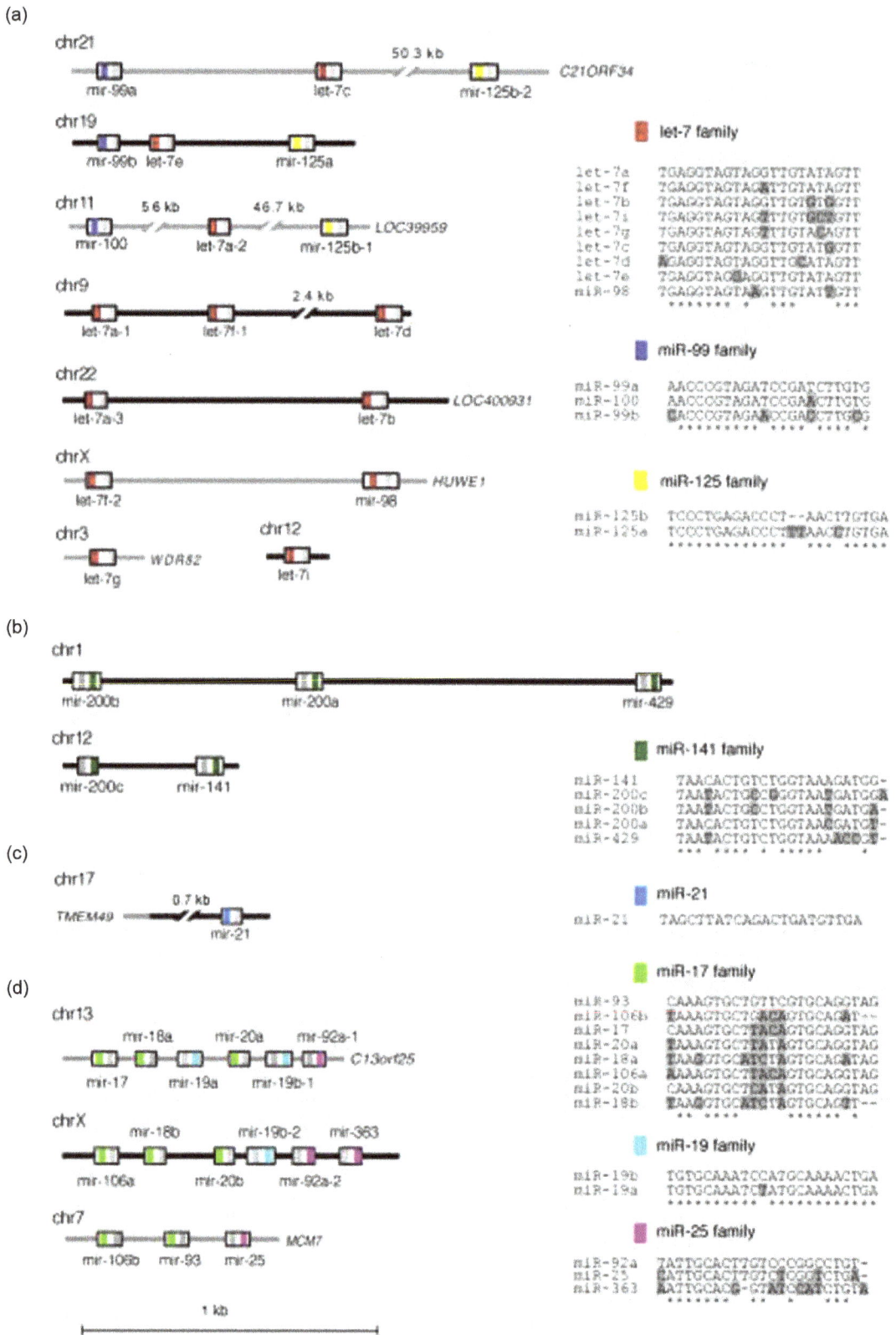

**FIGURE 4.4   MiRNA genomic and functional organization.** The genomic and functional organization of four miRNA clusters is clarified: (a) let-7/mir-98 cluster, (b) mir-141/mir-200a cluster, (c) mir-21 cluster, and (d) mir-17-92 cluster. The genomic locations for each of the miRNA members are defined. *Grey* lines denote intronic regions. MiRNA mature sequences are color coded according to the sequence family they belong to (i.e., in the let-7/mir-98 cluster *red* signifies the let-7 sequence family). The star sequence is defined with a *grey* bar. The sequence families are depicted as sequence alignments compared to the most highly expressed miRNA family member shown on *top,* based on profiles of more than 1000 human specimens [82]. *Shaded residues* denote differences from the most highly expressed miRNA family member.

compared with their normal tissue counterparts [22], leading to the proposal that miRNAs are predominantly tumor suppressors rather than oncogenes. Quantification of absolute miRNA levels, not only relative abundance, in miRNA profiling methods is necessary to clarify these observations. Of various tumors, 27% are found to have a hemizygous deletion of the gene that encodes DICER1 [120]. Global knockdown of mature miRNAs by targeting DICER1, RNASEN, and its cofactor DGCR8 increases the oncogenic potential of already transformed cancer cell lines and accelerates tumor formation [119]. Reductions in the amount of DICER1 resulting in impaired miRNA processing have also been shown to increase the rate of tumor formation in two different cancer mouse models, a K-RAS-driven lung cancer [120] and an Rb-driven retinoblastoma [121]. DICER1 is therefore considered a haplo-insufficient tumor suppressor, requiring partial deletion for its associated tumorigenesis phenotype [121]. The phosphorylation of the DICER1 cofactor TARBP2 by the mitogen-activated protein kinase Erk enhances pre-miRNA processing of oncogenic miRNAs, such as miR-21, and decreases production of tumor suppressor let-7a [122]. Moreover, TARBP2 is mutated in some colon and gastric cancers with microsatellite instability and TARBP2 frameshift mutations correlate with DICER1 destabilization; in cell lines and xenografts with TARBP2 mutations, reintroduction of wild-type TARBP2/DICER1 slowed tumor growth [123, 124]. Finally, DICER1 was also recently implicated as a metastasis suppressor (reviewed in [125]).

### 4.5.4.2 Alterations in Other Pathway-Related RBPs

Firstly, expression of LIN28A blocks the processing of tumor suppressor pri- and pre-let-7 [33, 126–129], thus maintaining expression of genes that drive self-renewal and proliferation (reviewed in [130]); tumors that express LIN28A were indeed shown to be poorly differentiated and more aggressive than LIN28A-negative tumors. Secondly, the helicases DDX5 and DDX17 are thought to stimulate processing of one-third of all murine miRNAs by acting as a scaffold and recruiting factors to the RNASEN complex and thereby promoting pri-miRNA processing [131]. Association of DDX17 and DDX5 RNA helicases through interactions mediated by the tumor suppressor TP53 with the RNASEN/DGCR8 complex facilitates the conversion of pri- to pre-miRNAs [132]. Specifically, the DDX5-mediated interaction of the RNASEN complex with the tumor suppressor TP53 was shown to have a stimulatory effect on the tumor suppressor pri-miR-16-1, pri-miR-143, and pri-miR-145 processing in response to DNA damage in cancer cells [132]. Thus, TP53 mutations, often observed in malignancies, led to a decrease in pre-miRNA production. Thirdly, oncogenic SMADs, downstream effectors of the TGF-β superfamily pathways, have been shown to control RNASEN-mediated miRNA maturation through interaction with DDX5, promoting expression of oncogenic miR-21 [133]. KSRP promotes the biogenesis of a subset of miRNAs, including let-7a, by serving as a component of

both DICER1 and RNASEN complexes affecting proliferation, apoptosis, and differentiation [134]. In a final example, inactivating mutations of XPO5 in tumors with microsatellite instability results in the nuclear retention of miRNAs [135]. Restoration of XPO5 function reverses the impaired export of pre-miRNAs and has tumor suppressor features.

## 4.6 DYSREGULATION OF MiRNA-MRNA TARGET RECOGNITION

### 4.6.1 MiRNA Function/Mechanism

As described above, miRNAs function through the AGO proteins, containing both RNA-binding domains and RNase H domains (reviewed in [136]). The four human Ago genes are co-expressed and bind to miRNAs irrespective of their sequence. AGO2, in contrast to the other members, retains an active RNase H domain and thus is able to directly cleave target RNAs with extensive complementarity to the bound miRNAs. The assembly of the miRNP complex involves multiple AGO conformational transitions captured in a series of crystal structures (reviewed in [137]). The mRNA target is recognized by pairing of the miRNA seed region (position 2–8) to complementary sequences located mainly in the target 3′ UTR, but also in the coding regions. Target mRNA recognition and regulation involve members of the GW182/TNRC6 family. TNRC6 proteins act at the effector step of silencing, downstream of AGO proteins, and play a crucial role in miRNA silencing in animals (reviewed in [138]). Proteomic approaches identified additional AGO-interacting proteins, some of which likely represent mRNA-interacting partners that co-purified with miRNA-targeted mRNPs; their function in RNA silencing processes and potentially tumorigenesis remains to be established.

In mammalian cells under steady-state conditions, miRNAs have been shown to destabilize targeted transcripts [34, 139–141] through a variety of mechanisms, including decapping and deadenylation; target mRNA and protein abundance changes track closely [139, 140, 142, 143]. These studies also showed that miRNAs destabilize mRNAs preferably through binding sites located in their 3′ UTRs [144–148]. Ribosome profiling studies demonstrated that the ribosome density of miRNA targets was unaltered, while changes in miRNA levels were inversely correlated to mRNA and protein abundance, emphasizing the role of miRNAs in regulation of mRNA stability but not translation [149]. Translational regulation by miRNA targeting is considered to predominantly act at the level of translation initiation. Identification of miRNA/mRNA ribonucleoprotein components in processing bodies (P-bodies) also implies their role in mRNA storage and RNA turnover. An excellent recent review describes the different mechanisms implicated in miRNA function, highlighting the different experiments supporting translational repression versus mRNA decay and the evolution in our current thinking [138].

### 4.6.2 Organization of MiRNAs into Sequence Families

Certain miRNAs share sequence similarity in regions that are critical for mRNA target recognition, specifically the seed region, and are best viewed as a family when considering mRNA target regulation and functional consequences of altered miRNA expression. MiRNAs can be grouped in sequence families, based not only on their seed sequence similarity but also overall sequence similarity given that the miRNA 3′ end also contributes to miRNA targeting, although to a lesser extent (Figure 4.4) (reviewed in [57]). Changes in the overall abundance of miRNA sequence families relate directly to target regulation. In a MYC-driven B-cell lymphoma mouse model, a conditional knockout of the oncogenic miR-17-92 gene cluster induces apoptosis, which can be reduced by reintroduction of only one of the four sequence families produced from the cluster [150].

### 4.6.3 MiRNA-MRNA Stoichiometry

The majority of miRNA profiling studies do not provide an estimate of miRNA abundance, which is critical in our understanding of the role of miRNA–mRNA mediated regulation in tumorigenesis.

Only the most abundantly expressed miRNAs occupy a substantial fraction of their available mRNA target sites and affect target mRNA stability [148]. Abundant miRNAs that behave as "switches," turned on or off during the tumorigenesis process, as shown in developmental processes, have the most significant regulatory potential, given that miRNAs usually only lead to modest 1.5- to 4-fold regulation of their target expression [142, 143, 145]. However, given that specific mRNAs are subject to regulation by multiple miRNAs of unrelated families, cumulative effects of lower expressed miRNAs may be relevant [28, 151, 152]. Furthermore, in the rare circumstance that miRNAs share near-perfect complementarity to mRNAs, they may act in an siRNA-like catalytic mode, cleaving mRNA targets even at low miRNA abundance. To conclude, the interplay between miRNAs expressed in particular tissues, the levels of their respective expressed targets, as well as other posttranscriptional gene regulatory mechanisms (such as regulation by RBPs or other competing interactions—see below) is likely responsible for balancing miRNA-conferred regulation.

## 4.7 MiRNA TARGET IDENTIFICATION

Not all protein-coding genes are regulated by miRNAs. Interestingly, some genes involved in basic cellular processes avoid microRNA regulation due to short 3′ UTRs that are specifically depleted of microRNA binding sites [153]. However, it has been estimated that miRNAs regulate the translation of about 30% of the protein-coding genes [54]. Conceivably, therefore, these regulators directly or indirectly affect most, if not all cellular pathways.

The accurate prediction of the target genes is made difficult by the imperfect complementarity between the miRNA and the regulated mRNAs. To add further complexity to the issue, some genes can have alternative 3′ UTRs, which could be regulated by a different set of miRNAs. Several bioinformatics approaches for the prediction of miRNAs targets have been developed [144, 152, 154]. Most search algorithms rely upon the search for perfect complementarity in the so-called seed region, that is, the seven nucleotides between positions 2–8 of the miRNA [54]. Other features of the target sequence, such as phylogenetic conservation, position within the 3′ UTR, and absence of stable secondary structures, are also accounted for in recent predictions [144]. It has been estimated that a single miRNA family can regulate as many as 200 different genes [54]. Thus the effects of the miRNAs are likely to be pleotropic, and their aberrant expression could conceivably unbalance the cell's homeostasis, contributing to diseases, including cancer.

The currently available target prediction databases (reviewed in [57]) do not easily allow prioritizing the involvement of reported targets in certain phenotypes, thus necessitating the selection of a few targets from a list of hundreds for further study and validation, based on a priori knowledge of potentially involved biological pathways. Since the prediction algorithms do not always produce identical target lists, use of multiple algorithms and comparison or intersection of their results narrows the list to higher confidence targets. Targets are only relevant to a specific phenotype if they are expressed in the studied tissue, an issue not addressed by most computational prediction algorithms. Recently, new algorithms are trying to prioritize computationally predicted targets using integrated miRNA and mRNA profiles [155]. Biochemical identification methods in cell lines and tissues are being established and further refine our understanding of miRNA–mRNA target binding recognition. These methods involve two approaches: overexpression or downregulation of studied miRNAs followed by assessment of transcriptome-wide mRNA levels by mRNA microarray analysis (e.g., [148]) or deep sequencing technology after immunoprecipitation of miRNAs and mRNAs complexed with AGO, the main component of the miRNA effector complex, to not only identify mRNA targets, but also localize their precise binding sites [156, 157].

### 4.7.1 Changes in MiRNA Targets

The binding sites of miRNAs in mRNAs can be altered through a variety of mechanisms, such as point mutations, translocations, shortening of the 3′ UTR, competition with other RBPs, or decoy molecules for mRNA binding. Point mutations in miRNA targets can both create or destroy a miRNA binding site [158–160]. Chromosomal translocations can remove miRNA binding sites from their regulated oncogenes, such as in the case of let-7 targeting of the 3′ UTR of the Hmga2 gene [161]. Shortening of the 3′ UTR through alternative polyadenylation can relax miRNA-mediated regulation of known oncogenes, such as IGF2BP1/IMP1, and lead to oncogenic transformation [162], as does use of decoy pseudogenes, as in the case of PTEN, by saturating miRNA binding sites [163]. Finally, cooperativity or competition of miRNAs for mRNA target site binding with other RBPs, such as ELAVL1 (HuR), DND1, and PUM1, can also de-repress target expression [164–167]. This topic is discussed in a recent review [168].

## 4.8 DISTINCT MiRNA PROFILES OF CANCER TISSUES

We will first discuss the state of current miRNA profile databases, and then explore the issue of tissue heterogeneity in the tissue profiles before summarizing the role of miRNA dysregulation in malignancies.

### 4.8.1 MicroRNAs and Cancer

Since miRNAs can alter gene expression, they are good candidates for maintaining balance in the growth kinetics of the cell. Pioneering studies showed a differential miRNA expression profile between tumors and normal tissue [22, 169, 170]. But this result does not mean that all of the perturbed miRNAs are directly involved with tumor progression. Many of them are bystanders indirectly altered by the genomic, epigenomic, or physiological changes that arise during the carcinogenesis and are not causative agents. MiRNAs that truly are involved in the cancer progress are called oncomirs [171], and their discovery is currently one of the major goals in the cancer research field.

Not only are specific miRNAs involved in cancer (as we will discuss later), but also components of the miRNA biogenesis machinery are involved. Recently it has been shown that global repression of miRNA maturation by mutation of Drosha, DGCR8, and Dicer1 promotes cellular transformation and tumorigenesis [119]. These findings are in agreement with previous observations of Dicer1 loss in some tumors [172], and a general downregulation of miRNAs in tumors compared to normal tissues [22]. Interestingly, conditional loss of Dicer1 [173, 174] or DGCR8 [175] in murine embryonic stem cells (ES) results in impaired proliferation and differentiation.

Other components of the miRNA machinery that have been implicated in cancer include Argonaute family members hAgo1/EIF2C1, hAgo3/EIF2C3, hAgo4, and Hiwi. hAgo1, hAgo3 and hAgo4 cluster on the 1p34-35 chromosomal region, which is often lost in human cancers such as Wilms tumors, neuroblastoma, and carcinomas of the breast, liver, and colon [176]. On the other hand, overexpression of Hiwi (that belongs to an Argonaute protein family clade involved in germline maintenance) [177] has been related with germinal tumors, among others [178–180].

In the following section, we will review, without pretension to completeness given the rapid pace of the field, some of the better characterized examples of oncomirs

**TABLE 4.1**

**Representative MiRNA Involved in Cancer, Proposed Role (TS, Tumor Suppressor; OG, Oncogen), and Representative Targets Tested Experimentally**

| MiRNA | Role | Representative Targets Tested Experimentally | References |
|---|---|---|---|
| miR-15a miR-16-1 | TS | BCL2 | [181] |
| let-7 family | TS | RAS family, HGMA2, MYC, CDK6, CDC25 | [161,182–185] |
| miR-34 family | TS | E2F3, CDK4, CDK6, CCNE2, BIR3, DCR3, BCL2 | [186–189] |
| miR-17-92-1 cluster | OG | E2F family | [103,190,191] |
| | OG | PTEN, TPM1, PDCD4 | [192–195] |
| | OG | TP53INP1 | [196] |
| | OG | LAST2 | [197] |

(Table 4.1). We have classified the oncomirs from a classic point of view considering them as tumor suppressors when they impair tumor progression or as oncogenes when they promote it. However, because the effects of the miRNAs are intrinsically pleotropic, this classification should be considered flexible.

### 4.8.2 MiRNA Cancer Database

The development of miRNA microarrays, RT-PCR platforms, and deep sequencing methodologies has resulted in an exponential acquisition of miRNA profiles. Some of the published miRNA profiles are available in the NCBI Gene Expression Omnibus, similarly to mRNA profiles (other resources include www.microrna.org, hittp://www.mirz.unibas.ch). Larger cancer and blood-borne disease collections have recently been published using various platforms [155, 198, 199]. However, there is no database or viewer that allows for cross-platform comparison of existing data.

### 4.8.3 Tissue Heterogeneity

Tissues are generally composed of multiple cell types, each with their distinct gene-expression program. Disease not only alters the expression programs of the affected cell type, but often also its cell type composition. To best separate these effects in the profiling of heterogeneous tumor samples, it may be useful to profile tumor cell lines and individual cell types that may be present in a tumor sample or define miRNA cellular localization by performing RNA ISH. Figure 4.5 compares miRNA abundance profiles of normal breast, an estrogen-receptor positive invasive ductal breast carcinoma, the estrogen receptor positive ductal cell line MCF7, human fat, and blood [81, 82]. Strikingly, we can model the profile of a human cancer by simply combining tumor cell line and human fat profiles at equal ratio.

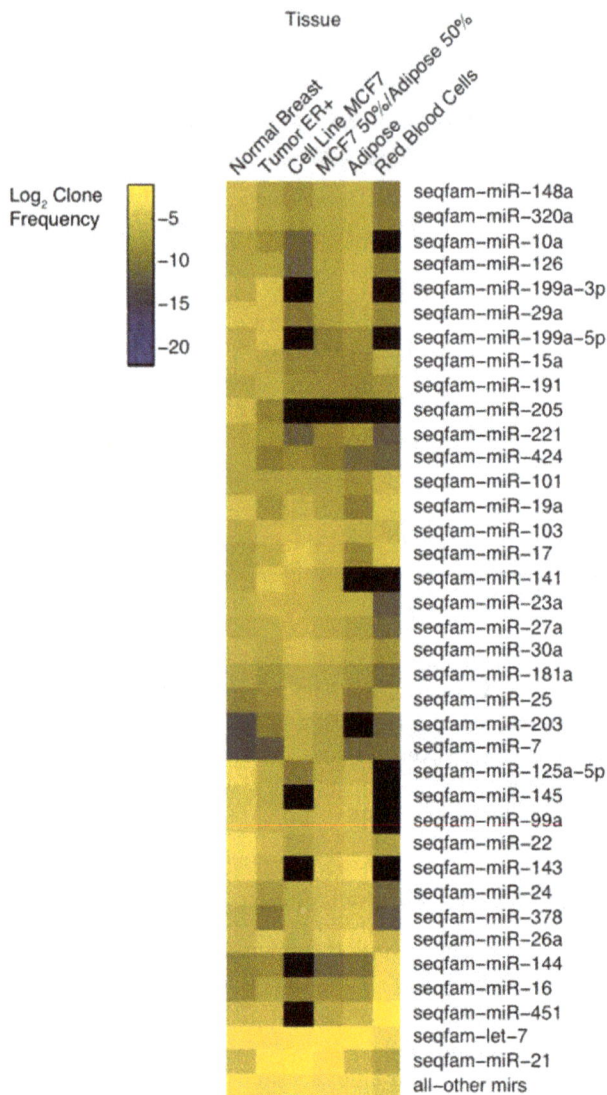

**FIGURE 4.5 MiRNA breast tumor and cell line profiles.** Comparison of abundance profiles of the top-expressed miRNA sequence families of normal breast, an estrogen receptor positive invasive ductal carcinoma breast tumor (ER+), the MCF7 ductal-derived cell line, human subcutaneous adipose tissue, and red blood cells.

This demonstrates that the MCF7 tumor cell line may be a good disease model for deciphering miRNA regulatory networks, as it expresses many of the miRNAs present in the predominant tumor-derived cell type and highlights the need for individual cell type miRNA profiles.

## 4.8.4 MicroRNAs as Tumor Suppressors

### 4.8.4.1 MicroRNA-15a and MicroRNA-16-1

Pioneering studies on chronic lymphocytic leukemia (CLL) revealed the first direct evidence that miRNAs can play a role in cancer. The frequent deletion of the 13q14 region in the majority of B-cell CLL cases suggested the presence of a tumor suppressor gene in that locus. For several years, the search for a candidate tumor suppressor in the

13q14 region had been frustrated by the absence of a suitable protein-coding gene. Deletion analyses showed a 30 kb region, which comprised a non-protein-coding gene called leukemia associated gene 2 (LEU2), as the minimal common region lost in CLL patients [200]. Calin and his collaborators became aware that LEU2 contains miR-15a and miR-16-1 in its first intron. Reduced abundance of these two miRNAs was then documented in 68% of CLL cases analyzed [92]. Accordingly, germline mutations associated with the downregulation of these miRNAs were found in CLL patients [201].

Further functional analysis described the anti-apoptotic oncogene BCL2 as one of the targets regulated by miR-15a and miR-16-1. The levels of these miRNAs were inversely correlated to BCL2 expression and reporter assays determined that both microRNAs negatively regulate BCL2 at a posttranscriptional level. Moreover, the Bcl2 repression by these microRNAs induces apoptosis in a CLL cell line model [181].

Because the 13q14 region is also lost in other kinds of cancer [202], future investigations are likely to uncover involvement of miR-15a and miR-16-1 in other neoplasias.

### 4.8.4.2 let-7 MicroRNA Family

let-7 is the first miRNA identified in humans and the let-7 family consists of 12 very closely related genes [4]. In *C. elegans*, let-7 is temporally regulated and controls the timing of terminal differentiation, acting as a master temporal regulator of multiple genes required for cell cycle exit in seam cells, a stem cell–like population [4, 182]. Many human let-7 genes map to regions altered or deleted in human tumors [20], indicating that these genes may function as tumor suppressors. In fact, let-7g maps to 3p21, which has been implicated in the initiation of lung cancers [169]. Several studies have shown that let-7 is expressed at lower levels in lung tumors than in normal tissue [20, 160, 169, 182]. Furthermore, let-7 levels have been correlated to lung cancer prognosis: patients with lower let-7 expression survived for a shorter time than those with higher let-7 expression in lung [160, 169, 203] and other tumors [204]. A possible mechanistic explanation for the let-7 biological role as tumor suppressor was provided by the discovery that the RAS oncogenes are let-7 targets [182]. The RAS family encodes 21-kDa protein kinases that bind guanine nucleotides (GTP and GDP) and are implicated in signal transduction processes, including mitogenesis. Mutations in the RAS family of proto-oncogenes (comprising H-RAS, N-RAS, and K-RAS) are found in 20%–30% of all human tumors [205]. About one-third of human lung adenocarcinomas (LACs) carry RAS oncogenic mutations, mainly in codon 12 of the KRAS gene [206]. Interestingly, the 3′ UTRs of human RAS genes contain multiple let-7-complementary elements, and let-7 represses the expression of KRAS and NRAS in tissue culture through their 3′ UTRs. Moreover, in lung squamous cell carcinoma, low let-7 levels correlate with high RAS expression, consistent with let-7-mediated regulation of RAS protein levels in vivo [182]. Recently, other oncogenes and proteins involved in the cell cycle have been reported to be regulated by the let-7 family, such as: HGMA2 [161, 183], myc [184], CDK6, and CDC25 [185]. Moreover, let-7 expression also slows down cell growth in tissue culture [185] and tumor growth both in xenograft and lung cancer mouse models [207, 208].

Recent evidence shows that the tumor suppressor activity of let-7 encompasses other tumors types. For example, let-7 represses self-renewal and tumorigenicity of breast cancer cells [209]. Increased expression of let-7 in self-renewing tumor-initiating cells (T-IC) decreased their proliferation and metastatic capacity.

### 4.8.4.3 MicroRNA-34 Family

In vertebrates, the mir-34 family is composed of three evolutionarily conserved members: miR-34a, miR-34b, and miR-34c. In humans, miR-34a resides in the 1p36 chromosomal region and mir-34b and mir-34c are derived from the same transcript in the 11q23.1 region. Initial links to cancer arose from the observation of low levels of miR-34 in neuroblastomas [186]. In this report, the reintroduction of this miRNA into neuroblastoma cell lines caused a dramatic reduction in cell proliferation through the induction of a caspase-dependent apoptotic pathway. Shortly after this initial observation, independent studies documented the involvement of miR-34 in the p53 pathway [99, 187, 188, 210, 211]. Interestingly, the expression of the mir-34 family members is regulated directly by p53, and consequently the expression of the mir-34 family reflects p53 activity. Accordingly, these miRNAs act as tumor suppressor genes, and their reintroduction in defective cells promotes cell cycle arrest and senescence or apoptosis depending on the genetic background. Experimental analysis and bioinformatics predictions have implicated the miR-34 family in the regulation of important genes implicated in the control of cell cycle and apoptosis, such as E2F3, CDK4, CDK6, CCNE2, BIR3, DCR3, and BCL2 [186–189].

### 4.8.5 Tumor Suppressive MiRNAs and Oncomirs

Under physiological conditions, the mechanisms for DNA repair, cell proliferation, motility, and programmed cell death are tightly regulated in order to maintain tissue homeostasis. Alteration of critical genes that modulate these cellular processes may tilt this balance, predisposing cells to transformation. MiRNAs have been closely associated with a number of these critical genes and found to exert an essential role in conditioning tumorigenesis and cancer progression. The current consensus is that cancer-related miRNAs function as oncogenes or tumor suppressors [212, 213]. As for other malignancies, two situations can occur in HCC: (1) tumor suppressor miRNAs can be downregulated in HCC and cause the upregulation of oncogenic target genes, repressed in normal hepatic tissues, increasing cell growth, migration, or invasion, and potentially leading to hepatocarcinogenesis; (2) oncogenic miRNAs can be upregulated in HCC and lead to the downregulation of

target tumor suppressor genes, participating in the development of the cancer phenotype. Importantly, the expression profile of certain miRNAs has been found to reflect the biological behavior of HCC tumors, such as aggressiveness, invasiveness, or drug resistance. As a consequence, miRNA investigations may offer the opportunity to determine miRNA signatures that would provide valuable information to stratify and refine HCC diagnosis in terms of prognosis, response to treatment, and disease relapse. However, the gap between bench and bedside has not been bridged, and a better understanding of the cellular mechanisms that are altered in HCC through miRNA deregulation will be required to identify the miRNAs that would serve as relevant diagnostic makers and therapeutic targets for clinical practice.

### 4.8.5.1 HCC Recurrence

Highly active drug-metabolizing pathways and multidrug resistance transporter proteins are known to diminish the efficiency of current chemotherapeutic treatments. In addition, HCC recurrence after surgical resection of the primary tumor represents one of the characteristics leading to the low survival rate associated with liver cancer. Specific miRNA signatures have been linked to the increased risk of tumor recurrence and poor prognosis. The expression profiling of apoptosis-associated and metastasis-related miRNAs may provide clues for individual patients to predict drug resistance and invasiveness of HCC that conditions the recurrence of their disease. Fornari and colleagues demonstrated that miR-199a-3p repression observed in HCC leads to the overexpression of mTOR and MET, whereas the experimental restoration of miR-199a-3p reduces the growth and invasive properties of HCC cells and increases the apoptosis induced by doxorubicin [214]. Thus, an inverse correlation was revealed between miR-199a-3p and mTOR, as well as a shorter time to recurrence after tumor resection, in the patients with lower miR-199a-3p. Another study showed that low expression levels of miR-26 are well correlated with a better response to IFN-based treatment in patients with HCC but are associated with short survival [215].

The accurate assessment of cancer-related miRNA expression may predict the risk of relapse and represent an attractive prognostic tool. In particular, the high expression of miR-15b is associated with a low risk of tumor recurrence following surgical resection, as shown by Chung and colleagues who reported a negative correlation between miR-15b expression and the reappearance of HCC [216]. Experimentally, targeting miR-15b with antagonists increased HCC cell proliferation and inhibited TRAIL-induced apoptosis in vitro, while the miR-15b precursor transfection decreased proliferation and enhanced apoptosis by repressing the anti-apoptotic Bcl-w. In addition to their prognostic significance, modulating the expression of specific drug resistance–related miRNAs may clearly represent a valuable method to improve apoptosis-sensitizing strategies for HCC treatment and avoid the recurrence of the tumor.

### 4.8.5.2 "MiRNA Perspective" in Liver Cancer

The discovery of miRNAs has considerably modified and complicated the conventional concepts regarding gene regulation. Concerning cancer biology, understanding the molecular mechanisms by which miRNAs promote carcinogenesis may lead to novel concepts in the diagnosis and treatment of a large number of malignancies. In addition to the deregulation of cancer-related miRNAs observed in HCC, an association has also been found between miRNA expression and the clinicopathological outcome of liver cancer (tumor growth, response to treatment, metastatic potential, and recurrence). Therefore, the use of a miRNA-based classification correlated with the etiology and the aggressiveness of the tumor could significantly enhance the molecular diagnosis accuracy of HCC and its classification, leading to the consideration of more appropriate therapeutic strategies. In this regard, several teams have reported particular miRNA expression profiles that could be considered as valuable HCC prognostic indicators [217]. Budhu and collaborators defined a combination of 20 miRNAs as an HCC metastasis signature and showed that this 20-miRNA-based profile was capable of predicting the survival and recurrence of HCC in patients with multinodular or single tumors, including those at an early stage of the disease [218]. Remarkably, the highlighted expression profile showed a similar accuracy regarding patient prognosis when compared to the conventional clinical parameters, suggesting the clinical relevance of this miRNA signature. Consequently, the profiling of aberrantly expressed cancer-related miRNAs might establish the basis for the development of a rational system of classification in order to refine the diagnosis and the prediction of HCC evolution.

The potential implication of miRNAs as oncogenes or tumor suppressors supports the interest paid to cancer-related miRNAs in the past decade for the development of new curative approaches. MiRNAs represent relevant candidates as therapeutic targets, and several strategies have been reported to amend the altered expression of cancer-related miRNAs in the liver [219]. First, miRNA replacement therapies use short RNA duplexes that mimic downregulated miRNAs. On the other hand, miRNA inhibitors are chemically modified single-stranded oligonucleotides that antagonize the miRNAs overexpressed in cancer. In combination with the latest developments, which render miRNA delivery safer and more efficient, the use of RNA interference (RNAi) therapeutic strategies will pave the way to innovative perspectives in the clinical management of HCC. Pertinent studies have already argued that miRNA-based therapy may represent an attractive approach to target hepatic primary tumors. For example, Kota and collaborators showed that a systemic administration of miR-26a in rodents led to a dramatic slowing of HCC progression without notification of toxicity [220]. Thus, the delivery of tumor suppressor miRNAs, which are typically highly expressed in the liver but altered in HCC, may provide a valuable curative approach. However, miRNA-based therapeutics are still in an early stage of development and

more work will be required to identify relevant cancer-related miRNAs and understand the complex implication of these small noncoding RNAs in early or late HCC. In addition, as one miRNA can substantially affect the expression of several downstream targets, precautions are necessary to avoid undesirable off-target effects. Finally, the safety of the reagents used to deliver miRNA mimics and antagomirs needs to be validated for future clinical applications.

### 4.8.6 MICRORNAS AS ONCOGENES

### 4.8.6.1 MicroRNA-17-92-1 Cluster

Mir-17-92-1 is a cluster of miRNAs (miR-17-5p, miR-17-3p, miR-18, miR-19a, miR-20a, miR-19b-1, and miR-92-1) that arises from the polycistronic transcript C13orf25 placed in 13q31.3. Interestingly, this chromosomal region is often amplified in B-cell lymphomas and other malignancies, leading to overexpression of the miRNAs of the mir-17-92-1 cluster [104, 221]. Recent studies have shown that the miR-17-92-1 cluster synergizes with the oncogenic properties of c-myc [103, 104, 222]: the overexpression of the miR-17-92 cluster acted cooperatively with c-myc to accelerate tumor development in a mouse B-cell lymphoma model increasing the tumor resistance to apoptosis [104].

On the other hand, it was shown that c-myc binds directly to the mir-17-92-1 cluster locus and activates its expression. Moreover, it was shown that E2F1 is negatively regulated by two miRNAs from the mir-17-92-1 cluster, miR-17-5p and miR-20a.91. E2F1 is a member of the E2F transcription factors family that promotes the transition from $G_1$ to S phase of the cell cycle. Interestingly, E2F1 is also a target of c-myc. Therefore, c-myc regulates E2F1 expression in a sophisticated way, by directly activating E2F1 transcription but indirectly limiting its translation through the miR-17-92-1 cluster, therefore achieving tight control of the proliferative signal. A further level of regulation seems to be in place, as E2F1, E2F2, and E2F3 directly bind the promoter of the miR-17-92-1 cluster and activate its transcription [103, 190, 191].

Although c-myc upregulates the miR-17-92-1 clusters, the predominant consequence of activation of c-myc is widespread repression of miRNA expression. Much of this repression is likely to be a direct result of c-myc binding to miRNA promoters. The reactivation of the miRNAs repressed by c-myc diminishes tumorigenicity in a lymphoma cell model [223].

### 4.8.6.2 MicroRNA-372 and MicroRNA-373

A genetic screen for miRNAs that cooperate with oncogenes in cellular transformation identified oncogenic proprieties in miR-372 and miR-373 in human testicular germ cell tumors. These miRNAs promote tumorigenesis of primary human cells that harbor both oncogenic RAS and active wild-type p53 [197]. Mechanistic assays determined that these miRNAs could disable the p53 pathway by inhibition of CDK2, possibly through direct inhibition of the expression of the large tumor suppressor homologue 2 (LATS2). Thus, these miRNAs render the cells insensitive to the suppression abilities of p53 and so overcoming senescence. The abilities of miR-372 and miR-373 to promote growth in cells with wild-type p53 could help to explain why mutations of the p53 gene are not frequently detected in testicular cancer [224].

### 4.8.6.3 MicroRNA-21

MicroRNA-21 is frequently found to be overexpressed in miRNA profiling of many types of tumors, including neuroblastoma, glioblastoma, colorectal, lung, breast, and pancreatic cancer [225–230]. Apparently mir-21 has anti-apoptotic abilities, since blocking it with antisense mir-21 results in increased apoptosis [225, 229]. Further studies have identified several tumor suppressors as mir-21 targets, such as PTEN [192], TPM1 [193], and PDCD4 [194, 195], which could help to explain its biological role in carcinogenesis. Measures of the level of mir-21 could be useful in the diagnosis and prognosis of cancer [230].

### 4.8.6.4 MicroRNA-155

MicroRNA-155 is coded in a conserved region of the noncoding gene B-cell integration cluster (BIC). BIC was initially identified as a common integration site for the avian leucosis virus, inducing B-cell lymphomas in collaboration with MYC [231–233]. Overexpression of miR-155 has been observed in both hematological [234, 235] and solid [169, 170] tumors.

Experiments conducted in transgenic mice have shed light on the miR-155 involvement in cancer and in the immune system. When miR-155 was overexpressed under control of $V_H$ promoter-Ig heavy chain Eμ, it initially exhibited a preleukemic pre-B-cell proliferation evident in spleen and bone marrow, followed by B-cell malignancy. These findings indicate that miR-155 can induce polyclonal expansion, favoring the capture of secondary genetic changes for full transformation [236]. Conversely, using genetic deletion and transgenic approaches, two groups independently showed that miR-155 has an important role in the mammalian immune system [196, 237, 238]. Defective miR-155 mice are immunodeficient and have abnormalities in the maturation of B and T lymphocytes. Mouse transcriptome analysis identified a wide spectrum of miR-155-regulated genes, including cytokines, chemokines, and transcription factors. Another miR-155 target, the tumor protein 53-induced nuclear protein 1 (TP53INP1), a pro-apoptotic factor, could explain, at least in part, the oncogenic role of the miR-155 gene [239].

Interestingly, it has been shown that viral miR-K12-11 encoded by Kaposi's sarcoma–associated herpes virus (KSHV) functions as an orthologue of miR-155. It has been observed that miR-K12-11 and miR-155 have an extensive set of common mRNA targets, including genes with known roles in cell growth regulation. This observation suggests that miR-K12-11 may contribute to the induction of B-cell tumors in infected patients [240].

MiRNA dysregulation could be used as a diagnostic tool even if the particular miRNAs do not serve any regulatory function. Alternatively, miRNA dysregulation could drive tumorigenesis through the roles miRNAs can adopt as tumor suppressors or oncogenes. MiRNAs that are up- or downregulated in malignancies are respectively referred to as oncogenic or tumor suppressor miRNAs, sometimes even if there is no evidence for their causative role in tumorigenesis.

Functional studies performed in cancer cell lines or mouse models of various malignancies through overexpression or knockdown of miRNAs have supported a role for some of these miRNAs in tumorigenesis. Overexpression of tumor suppressor miRNAs, such as let-7g, reduced tumor burden in a K-RAS murine lung cancer model [208]. Overexpression of the oncogenic mir-17-92 gene cluster led to a lymphoproliferative disorder, and higher-level expression of the cluster in MYC-driven B-cell lymphomas dramatically increased tumorigenicity [104, 241]. Overexpression of another oncogene, miR-21, frequently highly expressed in solid and hematologic malignancies, resulted in a pre-B malignant lymphoid-like phenotype whereas subsequent miR-21 inactivation in the same model led to apoptosis and tumor regression [242]. Transgenic mice models with loss and gain of function of miR-21 combined with a model of lung cancer confirmed the role of miR-21 as an enhancer of tumorigenesis when overexpressed, or a partial protector when genetically deleted [101]. Ectopic expression of miR-155 in bone marrow induced polyclonal pre–B cell proliferation progressing to B-cell leukemia or myeloproliferation in mice [243, 244].

Metastasis-related miRNAs have been identified in various malignancies mainly from cell line and xenograft experiments (reviewed in [245]). Examples include breast cancer–related miR-10b, miR-9, miR-31, and miR-335, among others. The interesting regulatory roles of these miRNAs cannot easily be validated in large clinical studies. Two clinical studies with long-term follow-up data instead identified miR-210 to be associated with tumor aggressiveness [246, 247], pointing to difficulties reconciling cell line, xenograft model, and patient materials, due to tissue heterogeneity discussed earlier, the heterogeneous nature of the malignancy, and timing of clinical specimen acquisition. Tumor miRNA profiles cannot dissect contributions from subpopulations of cells that may be important for tumor characteristics such as metastasis, while cell line miRNA profiles cannot capture the cellular interactions from supporting cell types in the tumor microenvironment. Patient samples are often collected at the time of diagnosis, by which time a tumor is already well established and cannot unravel early changes that may be critical in tumor initiation or later changes important in metastasis.

### 4.8.7 MiRNA-Regulated Pathways

The observed effects of miRNA misexpression on tumor initiation, maintenance, or metastasis can be explained by the mRNA targets and pathways they regulate, which include known tumor suppressors and oncogenes (reviewed in [19]). MiRNAs regulate a large number of genes, some estimates reporting miRNA regulation of up to 60% of the human genome, making it challenging to attribute a phenotype after misexpression of a particular miRNA through its action on only a subset of targets [141, 248]. If a few of these targets control rate-limiting steps in the studied tumorigenesis processes within the specified tissues and cell types, such as metastasis, then miRNA regulation of a handful of targets could potentially explain the phenotype resulting from miRNA misexpression [249].

Examples of miRNA-regulated cancer pathways include differentiation, apoptosis, proliferation, and stem cell maintenance, a process important for disease relapse and/or metastasis. The skeletal muscle–specific miR-206 blocks human rhabdomyosarcoma growth in mouse xenograft models by inducing myogenic differentiation [73], while the mir-141/200a cluster is critical in the epithelial-to-mesenchymal transition (EMT) in various malignancies (reviewed in [250]). Sustained expression of endogenous mir-17-92 cluster is required to suppress apoptosis in Myc-driven B-cell lymphomas in a conditional knockout allele of mir-17-92 cluster [150]. TP53-regulated, ectopically expressed miR-34 induced cell cycle arrest in both primary and tumor-derived cell lines, downregulating genes promoting cell cycle progression (reviewed in [100]). In a final example of miRNA-regulated cancer pathways, isolation of a subset of highly tumorigenic breast cancer cells that were thought to have stemness properties showed that these cells do not express let-7 family members and that expression of let-7 or its known target RAS leads to loss of self-renewal [209].

## 4.9 CANCER-RELATED MiRNAs AND THEIR ALTERED EXPRESSION IN HCC

HCC represents the third largest cause of death from cancer and the major form of liver malignancy worldwide, as it accounts for almost 90% of primitive hepatic tumors [251]. HCC is generally encountered in patients exhibiting an underlying chronic liver disease related to well-known risk factors, including hepatitis B virus (HBV) and/or C virus (HCV) infection, alcohol abuse, genetic diseases (e.g., hemochromatosis), genotoxic intoxication (e.g., aflatoxin B1), and liver steatosis. In the absence of diagnosis and clinical management, chronic hepatitis leads to fibrosis and gradually evolves into cirrhosis. Global studies estimate that approximately 80%–90% of all HCCs arise from cirrhotic livers [252]. To date, surgical resection and liver transplantation remain the only effective therapeutic options. However, a majority of HCC patients present an unresectable tumor due to late diagnosis; as such the disease tends to remain asymptomatic until late advancement or distant metastasis. In addition, chemotherapy resistance, high metastatic potential, and tumor recurrence are generally associated with liver cancer, leading to a low survival rate (less than 10% after 5 years). Consequently, the discovery of innovative and effective biomarkers ensuring an early diagnosis of the disease in order to maximize

the positive response of therapeutics before spreading and metastasizing remains an essential purpose in modern hepatology.

The critical role of miRNAs has been described in the control of various biological processes frequently altered in cancer. In addition, several reports indicate that the altered expression of specific sets of miRNAs can contribute to liver tumorigenesis. Remarkably, even slight changes in the amount of a few miRNAs can substantially modify cellular physiology and contribute to carcinogenesis. It has been shown that more than 50% of miRNA genes are located at fragile sites or in cancer-associated genome regions [20]. Therefore, following mutation, deletion, translocation, or amplification, miRNAs can be subjected to the same alterations as classic oncogenes or tumor suppressors [171]. In the last decade miRNA functions have begun to be elucidated, especially in the understanding of their major physiological implications. In mammals, miRNAs are predicted to participate in the regulation of almost all cellular processes, including development, cell differentiation, proliferation, and apoptosis [57]. As the abnormal expression of a number of miRNAs has been reported in a wide range of human cancers, a strong consensus has emerged that these cancer-related miRNAs may function as oncogenes or tumor suppressors [212, 253]. Regardless of cell origin, a plethora of studies has revealed the overall and recurrent downregulation of miRNAs in tumor tissues compared with normal tissues [22, 254]. More recently, the establishment of miRNA signatures is interesting with regard to the management of liver cancer patients from both diagnostic and therapeutic perspectives [213]. In the present chapter, we describe the miRNA biogenesis mechanism and focus on miRNA-altered expression in liver cancer. By considering well-defined cases, the role of miRNAs as tumor suppressors and oncogenes is then explored. Lastly, the potential applications of cancer-related miRNAs for diagnosis and their therapeutic value in human HCC are discussed.

MiRNA expression can be regulated at various levels from sequence identity, processing, stability, and mRNA binding. Thus, all these steps are susceptible to be altered in cancer cells, impacting the global production of miRNAs. In general, miRNAs display a globally repressed expression profile regardless of tumor origin [22]. However, the oncogenic properties of a number of miRNAs and their overexpression in several types of cancers have also been reported [255].

The progression of HCC generally involves various genetic and epigenetic aberrations [256]. Genetic alterations usually result from chromosomal abnormalities that can lead to the depletion, amplification, or translocation of miRNAs. Approximately 50% of all annotated human miRNA genes are located at fragile sites of the genome that are associated with cancer [20]. For instance, the gene that codes for miR-21 is located in a region at chromosome 17q23.2 frequently amplified in various types of solid tumors [170]. Epigenetic mechanisms and DNA methylation are also critical for miRNA regulation [257]. Thus, it has been demonstrated that the downregulation of several tumor suppressor miRNAs observed in cancer cells is mediated by the hypermethylation of their promoting sequences [258, 259]. For example, Furata and colleagues showed that the silencing of the tumor suppressive miR-124 and miR-203 through CpG island methylation represents a major event in hepatocarcinogenesis [260]. This aberrant methylation is frequently due to the increased expression of DNA methyltransferase (DNMT) enzymes and observed in a number of human cancers, including HCC [261]. The case of miR-148a and miR-152 is of interest as the inactivation of these two miRNAs by DNA methylation frequently occurs in malignancies, such as gastric tumors [262], pancreatic cancers [263], and cholangiocarcinoma [264].

In addition to genomic and epigenetic alterations, miRNA processing itself is frequently altered in liver cancer. First, the transcription of pri-miRNAs can be regulated and modified by several transcription factors. The oncogenic transcription factor c-Myc binds, for instance, the upstream of let-7 and miR-26a, repressing the transcription of these two miRNAs and contributing to tumorigenesis [217]. Another crucial point is the altered expression of Dicer. In most cancer types, the downregulation of Dicer has been shown to be related to the global deregulation of miRNA expression [265–268]. In the liver, Sekine and co-workers tested the consequence of DICER1 depletion by performing conditional knockout in hepatocytes [269]. Remarkably, the hepatocytes exhibiting DICER1-specific depletion displayed a gene expression profile indicative of cell growth and dedifferentiation into liver progenitors. At 1 year of age, approximately 60% of the mutant mice spontaneously developed HCC derived from the DICER1-deficient hepatocytes. In another study, TARBP2 (TAR RNA-binding protein 2), encoding an integral component of a DICER1-containing complex, has been described as important for maintaining DICER1 stability, its mutation leading to DICER1 alteration and the global downregulation of miRNAs [123]. Lastly, and on the Mrna side, oncogenes acquire mutations that remove miRNA binding sites in tumor cells. This phenomenon has been described in lipoma development, where the disruption of the pairing between let-7 and HMGA2 enhanced oncogenic transformation [161].

## 4.10 MiRNA AS A DIAGNOSTIC TOOL

In recent years, several groups have reported the overexpression or the downregulation of a number of miRNAs in a large variety of cancers. MiRNA signatures are believed to serve as accurate molecular biomarkers for the clinical classification of tumors as well as for the development of innovative therapeutic strategies. Therefore, the availability of consistent technologies that enable the detection of miRNAs has become of interest for both fundamental and clinical purposes. The current detection methods commonly used are microarray, real-time quantitative polymerase chain reaction (RT-qPCR), NGS, and, in a less routine way, northern blot or in situ hybridization. Microarray analyses present the advantage of offering a high speed of screening by employing various miRNA probes within a single microchip. However, the technique has lower sensitivity and specificity than RT-qPCR, which is still the gold standard for miRNA analyses. MiRNA RT-qPCR is based on the use of stem-loop primers, which can specifically bind to the mature miRNA during reverse transcription, conferring a high degree of accuracy to the method [270]. Analysis of miRNAs by RT-qPCR is a cost-effective technique and, due to its efficiency, a valuable method for the validation of miRNA signatures. Moreover, the development of RT-qPCR protocols has improved the sensitivity of miRNA detection down to a few nanograms of total RNAs. This amount can be easily and routinely obtained by extracting total RNAs from a small fragment of a hepatic percutaneous biopsy. However, the procedures employed for the normalization of miRNA data still remain a major point of discordance that requires attention. Indeed, distinct miRNA signatures have been emphasized despite the fact that miRNA profiling assays were carried out under similar conditions (the same type of tumor and the same screening technique). This dispersion might be caused by variations in patient population, such as ethnicity, gender, difference in diagnostic classification, as well as the standardization methods employed or the references used (adjacent nontumor tissue or hepatic samples from healthy donors). Obviously, the disparity observed between miRNA profiles established in the same type of cancer may also reflect different underlying causes of the diseases as well as diverse HCC outcomes.

The aberrant expression of cancer-related miRNAs in HCC frequently contributes to the deregulation of a tumor suppressor and/or oncogene pathways, indicating the direct and crucial role of miRNAs in liver carcinogenesis [213, 271, 272]. Different miRNA signatures have also been related to chronic hepatic infections [212], cirrhosis, and steatosis [273]. Using microarray technologies, Murakami and colleagues were one of the first groups to report a pattern of seven mature miRNAs that exhibit differential expression between HCC and adjacent nontumor samples [274]. In the 25 pairs analyzed, 5 miRNAs (miR-199a, miR-199a*, miR-195, miR-200a, and miR-125a) appeared to be significantly downregulated, whereas 2 miRNAs (miR-18 and miR-224) displayed a higher expression in HCC samples. Ladeiro and co-workers also identified specific miRNA expression profiles that can unambiguously differentiate between benign and malignant HCCs as well as between several subtypes of HCC tumors [275]. In this chapter, HCC tumors exhibited a redundant overexpression of miR-224 regardless of the underlying disease. Moreover, the case of the hepato-specific miR-122 may also be of prime interest. Krutzfeldt and colleagues demonstrated that the silencing of miR-122 by using antagomir resulted in an increased expression of hundreds of genes known to be putatively targeted by miR-122 and normally repressed in normal hepatocytes [147].

This argues for the involvement of miR-122 in maintaining the "adult-liver" phenotype by suppressing the expression of several nonhepatic genes. Furthermore, the replication of HCV is known to be related to the expression of miR-122 in infected cells. Thus, HCV viral RNA can replicate in the Huh-7 cell line, which expresses miR-122, but not in HepG2 cells, which do not express miR-122. In addition, the experimental sequestration of miR-122 in cells leads to a marked loss of HCV RNA replication [276]. Importantly, miR-122 was highlighted as downregulated in more than 70% of the samples from HCC patients with underlying cirrhosis, as well as in 100% of the HCC-derived cell lines analyzed [277]. Finally, miR-122 knockout mice display hepatosteatosis, fibrosis, and a high incidence of HCC, suggesting the tumor suppressor role of miR-122 in the liver [278].

A plethora of studies have reported various miRNA profiles potentially reflecting HCC initiation and progression, which could be employed as specific cancer biomarkers [278, 280]. To generalize, miR-21, miR-221, miR-222, and miR-224 were widely reported as upregulated in HCC, whereas miR-122, miR-199, miR-223, and the let-7 family members were frequently found to be downregulated in most studies. Considering these data, the establishment of accurate miRNA-based signatures could be of prime interest in the development of new tools for the diagnosis and advancement staging of liver cancer. However, elucidating the functional implication of the hepatic cancer–related miRNAs and the consequence of their deregulation in HCC progression also remains of prime interest from the perspective of miRNA-based curative management.

MiRNAs demonstrated their potential as diagnostic tumor markers early on when their profiles were shown to correlate with the tumor embryonic origin, thus defining tumors of unknown origin indistinguishable by histology and assigned based on clinical information [22]. MiRNA expression patterns have been linked to clinical outcomes given that miRNAs modulate tumor behavior such as tumor progression and metastasis. Expression of let-7 is downregulated in non–small cell lung cancer patients [185] and is associated with poor prognosis [160, 169], whereas a miRNA signature was identified to be associated with prognosis in CLL [201]. Advances in miRNA detection, such as ISH or RT-PCR, may allow miRNAs to be used as diagnostic and prognostic markers in the clinic.

Briefly, the mature 20- to 23-nt miRNA molecules are excised in a multistep process from primary transcripts (pri-miRNAs) that contain one or more 70-nt hairpin miRNA precursors (pre-miRNA) and have their own promoters or share promoters with coding genes. These hairpin structures are recognized in the nucleus by DGCR8, a double-stranded RNA-binding protein (dsRBP), and RNASEN, also known as RNase III Drosha, and excised to yield pre-miRNAs. These molecules are subsequently transported by XPO5 (exportin 5) to the cytoplasm where they are further processed by DICER1 (Dicer) in complex with dsRBPs TARBP2 (TRBP), and/or PRKRA to yield an RNA duplex processing intermediate composed of mature miRNA and miRNA* sequences. Some miRNAs bypass the general miRNA processing and their maturation can be independent of DGCR8 and RNASEN, such as miR-320 or miR-484 [117], or are DICER1 independent, such as erythropoiesis-related miR-451 [118, 281]. DGCR8- and RNASEN-independent miRNAs include mirtrons and tailed mirtrons, which release their pre-miRNA by splicing and exonuclease trimming [16, 282]. A recent review describes alternative processing pathways and enumerates settings in which alternative miRNA pathways contribute to distinct phenotypes among miRNA biogenesis mutants [283].

While the mature miRNA is loaded into the Argonaute/ EIF2C (AGO) proteins that are at the core of the miRNA-containing ribonucleoprotein complex (miRNP), sometimes also referred to as RNA-induced silencing complex (miRISC), the miRNA* is released and degraded. MiR-451 is generated from an unusual hairpin structure that is processed by AGO2 instead of DICER1 [118, 281]. The miRNPs contain a member of the AGO family (1–4), which binds the miRNA and mediates target mRNA recognition. Several other RBPs have been implicated in miRNA biogenesis, including DHX9, DDX6, MOV10, DDX5, DDX17, LIN28A, HNRNPA1, and KSRP [98, 168]. Following transcription, miRNAs can be modified by several enzymes, including deaminases, resulting in miRNA editing and terminal uridylyl transferases (TUTases), leading to pre-miRNA uridylylation, potentially affecting the amount and ratio of miRNA and miRNA* (e.g., [284]) or their sequences (e.g., [285]).

## 4.11   CIRCULATING MiRNAs

To date, the development of consistent and reproducible methods to improve the diagnosis of HCC at an early stage in a noninvasive way represents a real necessity in clinical hepatology. HCC tissues secrete various tumor-related compounds into the blood, and these may serve as circulating biomarkers for the diagnosis of liver cancer. It was recently proposed that miRNAs can be conveyed in blood serum, participating in intercellular communication or conditioning the tumor environment [286, 287]. The concept that miRNAs could serve as potential plasma markers for liver diseases is, thus, gaining attention. Moreover, the American Association for the Study of Liver Diseases (AASLD), in its practice guideline, recently discontinued the use of the blood tumor marker alpha-fetoprotein (AFP) for surveillance and diagnosis, increasing the need for novel HCC biomarkers in blood tests.

Tumor-derived miRNAs have been efficiently detected in the serum of liver disease patients and characterized as potential biomarkers for HCC [288]. In a relevant manner, the levels of three miRNAs (miR-21, miR-122, and miR-223) have been found to be significantly elevated in the serum of the patients exhibiting HBV infection or HBV-related HCC [289]. A study carried out by Li and collaborators highlighted a specific set of miRNAs significantly upregulated in HBV-positive HCC samples [289, 290]. Among them, miR-122 was highly increased in the serum of HBV patients but not in those of HCV patients. By employing a combination of characterized miRNAs, the authors could finally discriminate HCC cases from the controls or the infected non-HCC patients. Another study also emphasized the prognostic significance of serum miR-221, which represents another miRNA frequently overexpressed in HCC. In this chapter, the high expression of circulating miR-221 was correlated with the size of the tumor and the advancement of the disease [291, 292]. Furthermore, the overall survival rate of patients exhibiting high levels of serum miR-221 was significantly lower than that of patients with low miR-221 rates. Remarkably, miR-500 has also been detected in increased amounts in the serum of HCC patients and found to be significantly reduced after the surgical resection of their tumor [293]. Conversely, Shigoka and collaborators highlighted the low level of circulating miR-92a in HCC, whereas tumor resection was followed by the drastic augmentation of this miRNA in blood serum [294]. More recently, Tomimaru and collaborators evaluated the significance of plasma miRNAs as biochemical markers for HCC and demonstrated the relevance of assessing circulating miR-21 for the noninvasive diagnosis of liver cancer [295].

To validate the clinical relevance of serum miRNAs, further studies will be required, with a special focus on the standardization methods or the choice of the most appropriate miRNAs for internal references. Despite the availability of good endogenous normalizers in liver tissues, no circulating miRNAs have been identified and validated as a standard reference. Although the process of assessing serum miRNAs remains under improvement and will require procedures for validations, cancer-related circulating miRNAs represent an exciting and promising field of investigation in order to develop more accurate technologies for the noninvasive diagnosis of HCC.

## 4.12 ALTERATIONS OF MiRNA SEQUENCE

MiRNA dysregulation could be a result of mutations in miRNA genes in well-conserved regions in their mature sequence affecting mRNA targeting, or the remainder of the miRNA precursor potentially affecting processing and stability of the mature miRNA (reviewed in [115]). For example, a mutation in the seed region of mir-96 was shown to lead to hearing loss in a mouse model [296] and was identified in families with nonsyndromic progressive sensorineural hearing loss [297], while a point mutation in the viral mir-K5 precursor stem loop was shown to interfere with its processing and reduce mature miR-K5 accumulation [298]. Germline deletion of the mir-17-92 gene cluster was another recent example causing skeletal growth defects in humans [299]. If miRNAs are drivers of oncogenic and tumor suppressor pathways, we would expect to find miRNA mutations that can also be causative of the disease. So far the only mutation identified in a miRNA that could lead to malignancy is miR-16, where a germline mutation

potentially affects miR-16 biogenesis and abundance in a kindred with familial CLL [201] and New Zealand black mice that naturally develop CLL-like disease [300]. Single nucleotide polymorphisms (SNPs), located both in precursor and mature miRNA sequences, have been examined in the context of disease risk for various malignancies but have not been validated as causative (reviewed in [115]).

## 4.13 MiRNAs AS THERAPEUTICS

Because miRNAs affect the expression of multiple genes and thereby tune multiple points in disease pathways, miRNAs and their regulated genes represent interesting drug targets. Antisense oligonucleotide targeting experiments in human cell lines, mice [147, 301–304], and nonhuman primates [305] have demonstrated the feasibility of manipulating miRNA levels. MiR-143 was initially shown to promote adipocyte differentiation and could be a target for therapies in obesity and metabolic diseases [301]. Alternatively, "miRNA sponges" have been exploited to reduce miRNA expression in mammalian cells and mouse models by using RNA transcripts expressed from strong promoters containing miRNA-complementary binding sites (reviewed in [306]). Systemic administration of antisense oligonucleotide therapeutics to miR-122, a liver-enriched miRNA, in mice and primates was shown to alter lipid metabolism and hepatitis C viral load, resulting in reduced liver damage [147, 302–304, 307, 308]. At the same time, systemic delivery of a miRNA mimic for miR-26a in a murine model of HCC reduced tumor size [220]. The new and exciting advances in delivery of miRNA inhibitors and mimics hold the promise of quickly translating our knowledge of miRNAs into treating disease.

## 4.14 MicroRNAs AND THEIR FUTURE USE IN THE CLINIC: DIAGNOSIS, PROGNOSIS, AND THERAPY

Although the microRNA era started only a few years ago, it has brought great promises for diagnosis, prognosis, and therapy of cancer. The quick development of powerful techniques such as miRNA microarrays, bead-based miRNA profiling, specific quantitative-PCR of miRNAs, and antisense technologies are expected to have a significant impact on clinical oncology in the next decade.

Because miRNAs are key factors that define cell identity, they could be used as a valuable tool in cancer diagnosis. Pioneering studies using miRNA microarray analysis and bead-based miRNA profiling identified statistically unique profiles, which could easily discriminate cancers from noncancerous tissues [169, 170, 309]. Indeed, miRNA expression profiles seem to be more informative than traditional mRNA profiling for the classification of tumors with respect to their tissue of origin and differentiation. Thus, the profile of only 200 miRNAs was sufficient to classify poorly differentiated tumors in a recent study, with greater accuracy than a profile of 16,000 mRNAs [22]. Another study has achieved almost-perfect accuracy in classifying the tissue origin of 400 tumor samples from 22 different tumor tissues and metastases [310]. These findings demonstrate the effectiveness of miRNAs as biomarkers for tracing the tissue of origin of cancers of unknown primary origin, a major clinical problem [311].

But miRNA expression profiles also provide important information regarding the prognosis of cancer patients. For example, it was shown that miRNA expression profiles obtained by miRNA microarrays correlated with survival with lung adenocarcinomas, including those in precocious pathological stages. High levels of miR-155 and low let-7a-2 expression show a correlation with poor survival [169]. Another recent miRNA profiling effort in lung cancer identified five microRNAs as important for prognosis: high levels of miR-221 and let-7a appeared to be protective, whereas high levels of miR-137, miR-372, and miR-182 were correlated with worse clinical outcome. The levels of these miRNAs could also help in predicting relapse of the cancer [203]. A recent study performed in colorectal cancer showed that high miR-21 expression is associated with poor survival and poor therapeutic outcome [230].

On the therapeutic side, the gene therapy approaches to replace missing miRNAs have shown effectiveness in mouse models [207–209]. On the other hand, the discovery of miRNAs acting as oncogenes suggests that antisense oligonucleotides could specifically block their pathogenic activity. These anti-miRs are oligonucleotides complementary to the miRNAs that have a chemical modification to improve their stability and/or the delivery. Among the more promising chemical modifications being tested are the substitution of phosphate bounds by phosphorothioate bounds, the methylation of the oxygen at position 2 in the riboses (2'O-Methyl), or the addition of an extra bridge between the carbons of the ribose ring (locked nucleic acid, LNA) [311]. These antisense technologies are used frequently in cell culture in research settings. Some studies have provided evidence of their successful use in vivo. In this regard, a study has shown that adding a cholesterol molecule to the 3' of anti-miRs improved their effectiveness in vivo [147]. A vein tail injection of these anti-miRs (called antagomirs) resulted in a marked reduction of the corresponding miRNA with remarkable efficiency and specificity in all the tested tissues, except the brain. As in the case of many others drugs used in cancer therapies, problems related to the delivery, stability, or toxicity of these new compounds are to be expected as the experimentation moves more into the clinical phases. As each miRNA has multiple targets, its inhibition could cause side effects. Moreover, it is known that the RNA interference processing machinery that is shared by siRNAs and miRNAs can be oversaturated, producing fatal effects [34, 312]. Despite these possible pitfalls, the harnessing of miRNAs for clinical applications holds great promise, and the biotechnology community is embracing these new tiny regulators of gene expression with increasing interest.

## 4.15 CONCLUSION

We are only beginning to comprehend the functional repercussions of the gain or loss of particular microRNAs on cancer. A host of promising potential applications [313, 314] in the diagnosis, prognoses, and therapy of cancer [315, 316] are emerging at a rapid pace. Some miRNAs exhibit differential expression levels in cancer and have demonstrated capability to affect cellular transformation, carcinogenesis, and metastasis acting either as oncogenes or tumor suppressors. Increasing evidence has highlighted the frequent alteration of miRNA expression in liver cancer, as well as the critical role of these small RNAs in tumorigenesis. Collectively, the investigative studies performed to date have resulted in a better understanding of cancer-related miRNA functions and their role as tumor suppressors and oncogenes. Given the implication of a large number of miRNAs in the control of key tumor suppressors and oncogenes, the deregulation of specific miRNAs has been shown to greatly influence HCC growth, invasiveness, treatment response, and liver tumor curability. From a diagnostic point of view, miRNA profiling (from hepatic tissues and sera) may be beneficial, as it offers additional information that could be used in combination with the conventional methods available for the clinical assessment of liver cancer. In addition, a better understanding of the processes leading to the deregulation of miRNA expression in HCC will yield further insight into the molecular mechanisms of tumorigenesis and provide a promising perspective regarding the development of new curative approaches.

## REFERENCES

[1] Hanahan D, Weinberg RA. The hallmarks of cancer. Cell. 2000;100:57–70.

[2] Wightman B, Ha I, Ruvkun G. Posttranscriptional regulation of the heterochronic gene lin-14 by lin-4 mediates temporal pattern formation in *C. elegans*. Cell. 1993;75:855–862.

[3] Lee RC, Feinbaum RL, Ambros V. The *C. elegans* heterochronic gene lin-4 encodes small RNAs with antisense complementarity to lin-14. Cell. 1993;75:843–854.

[4] Reinhart BJ, Slack FJ, Basson M, et al. The 21-nucleotide let-7 RNA regulates developmental timing in *Caenorhabditis elegans*. Nature. 2000;403:901–906.

[5] Pasquinelli AE, Reinhart BJ, Slack F, et al. Conservation of the sequence and temporal expression of let-7 heterochronic regulatory RNA. Nature. 2000;408:86–89.

[6] Kato M, Slack FJ. microRNAs: small molecules with big roles—*C. elegans* to human cancer. Biol Cell. 2008;100:71–81.

[7] Stadler BM, Ruohola-Baker H. Small RNAs: keeping stem cells in line. Cell. 2008;132:563–566.

[8] Stefani G, Slack FJ. Small non-coding RNAs in animal development. Nat Rev Mol Cell Biol. 2008;9:219–230.

[9] Tavazoie SF, Alarcon C, Oskarsson T, et al. Endogenous human microRNAs that suppress breast cancer metastasis. Nature. 2008;451:147–152.

[10] Benetti R, Gonzalo S, Jaco I, et al. A mammalian microRNA cluster controls DNA methylation and telomere

recombination via Rbl2-dependent regulation of DNA methyltransferases. Nat Struct Mol Biol. 2008;15:268–279.

[11] Sinkkonen L, Hugenschmidt T, Berninger P, et al. MicroRNAs control de novo DNA methylation through regulation of transcriptional repressors in mouse embryonic stem cells. Nat Struct Mol Biol. 2008;15:259–267.

[12] Wightman B, Burglin TR, Gatto J, et al. Negative regulatory sequences in the lin-14 3′-untranslated region are necessary to generate a temporal switch during *Caenorhabditis elegans* development. Genes Dev. 1991;5:1813–1824.

[13] Lagos-Quintana M, Rauhut R, Lendeckel W, et al. Identification of novel genes coding for small expressed RNAs. Science. 2001;294:853–858.

[14] Lagos-Quintana M, Rauhut R, Yalcin A, et al. Identification of tissue-specific microRNAs from mouse. Curr Biol. 2002;12:735–739.

[15] Lau NC, Lim LP, Weinstein EG, et al. An abundant class of tiny RNAs with probable regulatory roles in *Caenorhabditis elegans*. Science. 2001;294:858–862.

[16] Lee RC, Ambros V. An extensive class of small RNAs in Caenorhabditis elegans. Science. 2001;294:862–864.

[17] Garofalo M, Croce CM. microRNAs: master regulators as potential therapeutics in cancer. Annu Rev Pharmacol Toxicol. 2010;51:25–43.

[18] Medina PP, Slack FJ. microRNAs and cancer: an overview. Cell Cycle. 2008;7:2485–2492.

[19] Ventura A, Jacks T. MicroRNAs and cancer: short RNAs go a long way. Cell. 2009;136:586–591.

[20] Calin GA, Sevignani C, Dumitru CD, et al. Human microRNA genes are frequently located at fragile sites and genomic regions involved in cancers. Proc Natl Acad Sci USA. 2004;101:2999–3004.

[21] Calin GA, Liu CG, Sevignani C, et al. MicroRNA profiling reveals distinct signatures in B cell chronic lymphocytic leukemias. Proc Natl Acad Sci USA. 2004;101:11755–11760.

[22] Lu J, Getz G, Miska EA, et al. MicroRNA expression profiles classify human cancers. Nature. 2005;435:834–838.

[23] Pang KC, Stephen S, Dinger ME, et al. RNAdb 2.0—an expanded database of mammalian non-coding RNAs. Nucleic Acids Res. 2007;35:178–182.

[24] He S, Liu C, Skogerbo G, et al. NONCODE v2.0: decoding the non-coding. Nucleic Acids Res. 2008;36:170–172.

[25] Griffiths-Jones S, Saini HK, van Dongen S, et al. miRBase: tools for microRNA genomics. Nucleic Acids Res. 2008;36:D154–D158.

[26] Berezikov E, van Tetering G, Verheul M, et al. Many novel mammalian microRNA candidates identified by extensive cloning and RAKE analysis. Genome Res. 2006;16:1289–1298.

[27] Rodriguez A, Griffiths-Jones S, Ashurst JL, et al. Identification of mammalian microRNA host genes and transcription units. Genome Res. 2004;14:1902–1910.

[28] Bartel DP. MicroRNAs: genomics, biogenesis, mechanism, and function. Cell. 2004;116:281–297.

[29] Gregory RI, Shiekhattar R. MicroRNA biogenesis and cancer. Cancer Res. 2005;65:3509–3512.

[30] Han J, Lee Y, Yeom KH, et al. The Drosha-DGCR8 complex in primary microRNA processing. Genes Dev. 2004;18:3016–3027.

[31] Denli AM, Tops BB, Plasterk RH, et al. Processing of primary microRNAs by the Microprocessor complex. Nature. 2004;432:231–235.

[32] Thomson JM, Newman M, Parker JS, et al. Extensive post-transcriptional regulation of microRNAs and its implications for cancer. Genes Dev. 2006;20:2202–2207.

[33] Viswanathan SR, Daley GQ, Gregory RI. Selective blockade of microRNA processing by Lin28. Science. 2008;320:97–100.

[34] Bagga S, Bracht J, Hunter S, et al. Regulation by let-7 and lin-4 miRNAs results in target mRNA degradation. Cell. 2005;122:553–563.

[35] Lund E, Guttinger S, Calado A, et al. Nuclear export of microRNA precursors. Science. 2004;303:95–98.

[36] Yi R, Qin Y, Macara IG, et al. Exportin-5 mediates the nuclear export of premicroRNAs and short hairpin RNAs. Genes Dev. 2003;17:3011–3016.

[37] Hutvagner G, Zamore PD. A microRNA in a multiple-turnover RNAi enzyme complex. Science. 2002;297:2056–2060.

[38] Ketting RF, Fischer SE, Bernstein E, et al. Dicer functions in RNA interference and in synthesis of small RNA involved in developmental timing in *C. elegans*. Genes Dev. 2001;15:2654–2659.

[39] Hammond SM. Dicing and slicing: the core machinery of the RNA interference pathway. FEBS Lett. 2005;579:5822–5829.

[40] Gregory RI, Chendrimada TP, Cooch N, et al. Human RISC couples microRNA biogenesis and posttranscriptional gene silencing. Cell. 2005;123:631–640.

[41] Chendrimada TP, Gregory RI, Kumaraswamy E, et al. TRBP recruits the Dicer complex to Ago2 for microRNA processing and gene silencing. Nature. 2005;436:740–744.

[42] Rand TA, Petersen S, Du F, et al. Argonaute2 cleaves the anti-guide strand of siRNA during RISC activation. Cell. 2005;123:621–629.

[43] Matranga C, Tomari Y, Shin C, et al. Passenger-strand cleavage facilitates assembly of siRNA into Ago2-containing RNAi enzyme complexes. Cell. 2005;123:607–620.

[44] Martinez J, Patkaniowska A, Urlaub H, et al. Single-stranded antisense siRNAs guide target RNA cleavage in RNAi. Cell. 2002;110:563–574.

[45] Schwarz DS, Hutvagner G, Du T, et al. Asymmetry in the assembly of the RNAi enzyme complex. Cell. 2003;115:199–208.

[46] Tang G, Reinhart BJ, Bartel DP, et al. A biochemical framework for RNA silencing in plants. Genes Dev. 2003;17:49–63.

[47] Yekta S, Shih IH, Bartel DP. MicroRNA-directed cleavage of HOXB8 mRNA. Science. 2004;304:594–596.

[48] Pillai RS, Bhattacharyya SN, Artus CG, et al. Inhibition of translational initiation by Let-7 MicroRNA in human cells. Science. 2005;309:1573–1576.

[49] Olsen PH, Ambros V. The lin-4 regulatory RNA controls developmental timing in *Caenorhabditis elegans* by blocking LIN-14 protein synthesis after the initiation of translation. Dev Biol. 1999;216:671–680.

[50] Pillai RS, Bhattacharyya SN, Filipowicz W. Repression of protein synthesis by miRNAs: how many mechanisms? Trends Cell Biol. 2007;17:118–126.

[51] Chan SP, Slack FJ. microRNA-mediated silencing inside P-bodies. RNA Biol. 2006;3:97–100.

[52] Liu J, Valencia-Sanchez MA, Hannon GJ, et al. MicroRNA-dependent localization of targeted mRNAs to mammalian P-bodies. Nat Cell Biol. 2005;7:719–723.

[53] Griffiths-Jones S. The microRNA Registry. Nucleic Acids Res. 2004;32:D109–111.

[54] Lewis BP, Burge CB, Bartel DP. Conserved seed pairing, often flanked by adenosines, indicates that thousands of human genes are microRNA targets. Cell. 2005;120:15–20.

[55] Gao Y, Schug J, McKenna LB, et al. Tissue-specific regulation of mouse microRNA genes in endoderm-derived tissues. Nucleic Acids Res. 2011;39:454–463.

[56] Gatfield D, Le Martelot G, Vejnar CE, et al. Integration of microRNA miR-122 in hepatic circadian gene expression. Genes Dev. 2009;23:1313–1326.

[57] Bartel DP. MicroRNAs: target recognition and regulatory functions. Cell. 2009;136:215–233.

[58] Aravin A, Tuschl T. Identification and characterization of small RNAs involved in RNA silencing. FEBS Lett. 2005;579:5830–5840.

[59] Creighton CJ, Reid JG, Gunaratne PH. Expression profiling of microRNAs by deep sequencing. Brief Bioinform. 2009;10:490–497.

[60] Meyer SU, Pfaffl MW, Ulbrich SE. Normalization strategies for microRNA profiling experiments: a 'normal' way to a hidden layer of complexity? Biotechnol Lett. 2010; 32(12):1777–1788.

[61] Lawrie CH, Soneji S, Marafioti T, et al. MicroRNA expression distinguishes between germinal center B cell-like and activated B cell-like subtypes of diffuse large B cell lymphoma. Int J Cancer. 2007;121:1156–1161.

[62] Weng L, Wu X, Gao H, et al. MicroRNA pro filing of clear cell renal cell carcinoma by whole-genome small RNA deep sequencing of paired frozen and formalin- fixed, paraffin-embedded tissue specimens. J Pathol. 2010;222: 41–51.

[63] Xi Y, Nakajima G, Gavin E, et al. Systematic analysis of microRNA expression of RNA extracted from fresh frozen and formalin-fixed paraffin-embedded samples. RNA. 2007;13:1668–1674.

[64] Barad O, Meiri E, Avniel A, et al. MicroRNA expression detected by oligonucleotide microarrays: system establishment and expression profiling in human tissues. Genome Res. 2004;14:2486–2494.

[65] Baskerville S, Bartel DP. Microarray profiling of microRNAs reveals frequent coexpression with neighboring miRNAs and host genes. RNA. 2005;11:241–247.

[66] Thomson JM, Parker JS, Hammond SM. Microarray analysis of miRNA gene expression. Methods Enzymol. 2007;427:107–122.

[67] Nelson PT, Baldwin DA, Scearce LM, et al. Microarray-based, high-throughput gene expression profiling of microRNAs. Nat Methods. 2004;1:155–161.

[68] Bissels U, Wild S, Tomiuk S, et al. Absolute quantification of microRNAs by using a universal reference. RNA. 2009;15:2375–2384.

[69] Peltier HJ, Latham GJ. Normalization of microRNA expression levels in quantitative RT-PCR assays: identification of suitable reference RNA targets in normal and cancerous human solid tissues. RNA. 2008;14:844–852.

[70] Fiedler SD, Carletti MZ, Christenson LK. Quantitative RT-PCR methods for mature microRNA expression analysis. Methods Mol Biol. 2010;630:49–64.

[71] Mestdagh P, Van Vlierberghe P, De Weer A, et al. A novel and universal method for microRNA RT-qPCR data normalization. Genome Biol. 2009;10:R64.

[72] Smith RD, Brown B, Ikonomi P, et al. Exogenous reference RNA for normalization of realtime quantitative PCR. Biotechniques. 2003;34:88–91.

[73] Taulli R, Bersani F, Foglizzo V, et al. The muscle-specific microRNA miR-206 blocks human rhabdomyosarcoma growth in xenotransplanted mice by promoting myogenic differentiation. J Clin Invest. 2009;119:2366–2378.

[74] Berezikov E, Thuemmler F, van Laake LW, et al. Diversity of microRNAs in human and chimpanzee brain. Nat Genet. 2006;38:1375–1377.

[75] Houbaviy HB, Murray MF, Sharp PA. Embryonic stem cell-specific microRNAs. Dev Cell. 2003;5:351–358.

[76] Landgraf P, Rusu M, Sheridan R, et al. A mammalian microRNA expression atlas based on small RNA library sequencing. Cell. 2007;129:1401–1414.

[77] Witten D, Tibshirani R, Gu SG, et al. Ultrahigh throughput sequencing-based small RNA discovery and discrete statistical biomarker analysis in a collection of cervical tumours and matched controls. BMC Biol. 2010;8:58.

[78] Vigneault F, Sismour AM, Church GM. Efficient microRNA capture and bar-coding via enzymatic oligonucleotide adenylation. Nat Methods. 2008;5:777–779.

[79] Hafner M, Renwick N, Brown M, et al. RNA ligase-dependent biases in miRNA representation in deep-sequenced small RNA cDNA libraries. RNA. 2011;17:1697–1712.

[80] Git A, Dvinge H, Salmon-Divon M, et al. Systematic comparison of microarray profiling, realtime PCR, and next-generation sequencing technologies for measuring differential microRNA expression. RNA. 2010;16: 991–1006.

[81] Ugras S, Brill E, Jacobsen A, et al. Small RNA sequencing and functional characterization reveals microRNA-143 tumor suppressor activity in liposarcoma. Cancer Res. 2011;71:5659–5669.

[82] Farazi TA, Horlings HM, Ten Hoeve JJ, et al. MicroRNA sequence and expression analysis in breast tumors by deep sequencing. Cancer Res. 2011;71:4443–4453.

[83] Tusher VG, Tibshirani R, Chu G. Significance analysis of microarrays applied to the ionizing radiation response. Proc Natl Acad Sci USA. 2001;98:5116–5121.

[84] Berninger P, Gaidatzis D, van Nimwegen E, et al. Computational analysis of small RNA cloning data. Methods. 2008;44:13–21.

[85] Robinson MD, McCarthy DJ, Smyth GK. EdgeR: a Bioconductor package for differential expression analysis of digital gene expression data. Bioinformatics. 2010;26:139–140.

[86] Nelson PT, Baldwin DA, Kloosterman WP, et al. RAKE and LNA-ISH reveal microRNA expression and localization in archival human brain. RNA. 2006;12:187–191.

[87] Pena JT, Sohn-Lee C, Rouhanifard SH, et al. miRNA in situ hybridization in formaldehyde and EDC-fixed tissues. Nat Methods. 2009;6:139–141.

[88] Sempere LF, Christensen M, Silahtaroglu A, et al. Altered MicroRNA expression confined to specific epithelial cell subpopulations in breast cancer. Cancer Res. 2007;67:11612–11620.

[89] Johnston RJ, Hobert O. A microRNA controlling left/right neuronal asymmetry in Caenorhabditis elegans. Nature. 2003;426:845–849.

[90] Kozomara A, Griffiths-Jones S. miRBase: integrating microRNA annotation and deep-sequencing data. Nucleic Acids Res. 2010;39:D152–D157.

[91] Chiang HR, Schoenfeld LW, Ruby JG, et al. Mammalian microRNAs: experimental evaluation of novel and previously annotated genes. Genes Dev. 2010;24:992–1009.

[92] Calin GA, Dumitru CD, Shimizu M, et al. Frequent deletions and downregulation of micro-RNA genes miR15 and miR16 at 13q14 in chronic lymphocytic leukemia. Proc Natl Acad Sci USA. 2002;99:15524–15529.

[93] Tagawa H, Seto M. A microRNA cluster as a target of genomic amplification in malignant lymphoma. Leukemia. 2005;19:2013–2016.

[94] Mavrakis KJ, Wolfe AL, Oricchio E, et al. Genome-wide RNA-mediated interference screen identifies miR-19 targets in Notch-induced T-cell acute lymphoblastic leukaemia. Nat Cell Biol. 2010;12:372–379.

[95] Huse JT, Brennan C, Hambardzumyan D, et al. The PTEN-regulating microRNA miR-26a is amplified in high-grade glioma and facilitates gliomagenesis in vivo. Genes Dev. 2009;23:1327–1337.

[96] Gauwerky CE, Huebner K, Isobe M, et al. Activation of MYC in a masked t(8;17) translocation results in an aggressive B-cell leukemia. Proc Natl Acad Sci USA. 1989;86:8867–8871.

[97] Etiemble J, Moroy T, Jacquemin E, et al. Fused transcripts of c-myc and a new cellular locus, hcr in a primary liver tumor. Oncogene. 1989;4:51–57.

[98] Krol J, Loedige I, Filipowicz W. The widespread regulation of microRNA biogenesis, function and decay. Nat Rev Genet. 2010;11:597–610.

[99] Chang TC, Wentzel EA, Kent OA, et al. Transactivation of miR-34a by p53 broadly influences gene expression and promotes apoptosis. Mol Cell. 2007;26:745–752.

[100] He L, He X, Lowe SW, et al. microRNAs join the p53 network – another piece in the tumour-suppression puzzle. Nat Rev Cancer. 2007;7:819–822.

[101] Hatley ME, Patrick DM, Garcia MR, et al. Modulation of K-Ras-dependent lung tumorigenesis by MicroRNA-21. Cancer Cell. 2010;18:282–293.

[102] Huang TH, Wu F, Loeb GB, et al. Up-regulation of miR-21 by HER2/neu signaling promotes cell invasion. J Biol Chem. 2009;284:18515–18524.

[103] O'Donnell KA, Wentzel EA, Zeller KI, et al. c-Myc-regulated microRNAs modulate E2F1 expression. Nature. 2005;435:839–843.

[104] He L, Thomson JM, Hemann MT, et al. A microRNA polycistron as a potential human oncogene. Nature. 2005;435:828–833.

[105] Han L, Witmer PD, Casey E, et al. DNA methylation regulates MicroRNA expression. Cancer Biol Ther. 2007;6:1284–1288.

[106] Saito Y, Jones PA. Epigenetic activation of tumor suppressor microRNAs in human cancer cells. Cell Cycle. 2006;5:2220–2222.

[107] Saito Y, Liang G, Egger G, et al. Specific activation of microRNA-127 with downregulation of the proto-oncogene BCL6 by chromatin-modifying drugs in human cancer cells. Cancer Cell. 2006;9:435–443.

[108] Lehmann U, Hasemeier B, Christgen M, et al. Epigenetic inactivation of microRNA gene hsamir-9-1 in human breast cancer. J Pathol. 2008;214:17–24.

[109] Brodersen P, Voinnet O. Revisiting the principles of microRNA target recognition and mode of action. Nat Rev Mol Cell Biol. 2009;10:141–148.

[110] Carthew RW, Sontheimer EJ. Origins and mechanisms of miRNAs and siRNAs. Cell. 2009;136:642–655.

[111] Ghildiyal M, Zamore PD. Small silencing RNAs: an expanding universe. Nat Rev Genet. 2009;10:94–108.

[112] Kim VN, Han J, Siomi MC. Biogenesis of small RNAs in animals. Nat Rev Mol Cell Biol. 2009;10:126–139.

[113] Winter J, Jung S, Keller S, et al. Many roads to maturity: microRNA biogenesis pathways and their regulation. Nat Cell Biol. 2009;11:228–234.

[114] Kwak PB, Iwasaki S, Tomari Y. The microRNA pathway and cancer. Cancer Sci. 2010;101(11):2309–2315.

[115] Ryan BM, Robles AI, Harris CC. Genetic variation in microRNA networks: the implications for cancer research. Nat Rev Cancer. 2010;10:389–402.

[116] Nishikura K. Functions and regulation of RNA editing by ADAR deaminases. Annu Rev Biochem. 2010;79:321–349.

[117] Yi R, Pasolli HA, Landthaler M, et al. DGCR8-dependent microRNA biogenesis is essential for skin development. Proc Natl Acad Sci USA. 2009;106:498–502.

[118] Cheloufi S, Dos Santos CO, Chong MM, et al. A dicer-independent miRNA biogenesis pathway that requires Ago catalysis. Nature. 2010;465:584–589.

[119] Kumar MS, Lu J, Mercer KL, et al. Impaired microRNA processing enhances cellular transformation and tumorigenesis. Nat Genet. 2007;39:673–677.

[120] Kumar MS, Pester RE, Chen CY, et al. Dicer1 functions as a haplo-insufficient tumor suppressor. Genes Dev. 2009;23:2700–2704.

[121] Lambertz I, Nittner D, Mestdagh P, et al. Monoallelic but not biallelic loss of Dicer1 promotes tumorigenesis in vivo. Cell Death Differ. 2010;17:633–641.

[122] Paroo Z, Ye X, Chen S, et al. Phosphorylation of the human microRNA-generating complex mediates MAPK/Erk signaling. Cell. 2009;139:112–122.

[123] Melo SA, Ropero S, Moutinho C, et al. A TARBP2 mutation in human cancer impairs microRNA processing and DICER1 function. Nat Genet. 2009;41:365–370.

[124] Garre P, Perez-Segura P, Diaz-Rubio E, et al. Reassessing the TARBP2 mutation rate in hereditary nonpolyposis colorectal cancer. Nat Genet. 2010;42:817–818;author reply 818.

[125] Valastyan S, Weinberg RA. Metastasis suppression: a role of the Dice(r). Genome Biol. 2010;11:141.

[126] Newman MA, Thomson JM, Hammond SM. Lin-28 interaction with the Let-7 precursor loop mediates regulated microRNA processing. RNA. 2008;14:1539–1549.

[127] Piskounova E, Viswanathan SR, Janas M, et al. Determinants of microRNA processing inhibition by the developmentally regulated RNA-binding protein Lin28. J Biol Chem. 2008;283:21310–21314.

[128] Rybak A, Fuchs H, Smirnova L, et al. A feedback loop comprising lin-28 and let-7 controls prelet-7 maturation during neural stem-cell commitment. Nat Cell Biol. 2008;10:987–993.

[129] Viswanathan SR, Powers JT, Einhorn W, et al. Lin28 promotes transformation and is associated with advanced human malignancies. Nat Genet. 2009;41:843–848.

[130] Viswanathan SR, Daley GQ. Lin28: a microRNA regulator with a macro role. Cell. 2010;140:445–449.

[131] Fukuda T, Yamagata K, Fujiyama S, et al. DEAD-box RNA helicase subunits of the Drosha complex are required for processing of rRNA and a subset of microRNAs. Nat Cell Biol. 2007;9:604–611.

[132] Suzuki HI, Yamagata K, Sugimoto K, et al. Modulation of microRNA processing by p53. Nature. 2009;460:529–533.

[133] Davis BN, Hilyard AC, Lagna G, et al. SMAD proteins control DROSHA-mediated microRNA maturation. Nature. 2008;454(7200):56–61.

[134] Trabucchi M, Briata P, Garcia-Mayoral M, et al. The RNA-binding protein KSRP promotes the biogenesis of a subset of microRNAs. Nature. 2009;459:1010–1014.

[135] Melo SA, Moutinho C, Ropero S, et al. A genetic defect in exportin-5 traps precursor microRNAs in the nucleus of cancer cells. Cancer Cell. 2010;18:303–315.

[136] Ender C, Meister G. Argonaute proteins at a glance. J Cell Sci. 2010;123:1819–1823.

[137] Parker JS. How to slice: snapshots of Argonaute in action. Silence. 2010;1:3.

[138] Huntzinger E, Izaurralde E. Gene silencing by microRNAs: contributions of translational repression and mRNA decay. Nat Rev Genet. 2011;12:99–110.

[139] Linsley PS, Schelter J, Burchard J, et al. Transcripts targeted by the microRNA-16 family cooperatively regulate cell cycle progression. Mol Cell Biol. 2007;27:2240–2252.

[140] Zhao Y, Ransom JF, Li A, et al. Dysregulation of cardiogenesis, cardiac conduction, and cell cycle in mice lacking miRNA-1-2. Cell. 2007;129:303–317.

[141] Lim LP, Lau NC, Garrett-Engele P, et al. Microarray analysis shows that some microRNAs downregulate large numbers of target mRNAs. Nature. 2005;433:769–773.

[142] Baek D, Villen J, Shin C, et al. The impact of microRNAs on protein output. Nature. 2008;455(7209):64–71.

[143] Selbach M, Schwanhausser B, Thierfelder N, et al. Widespread changes in protein synthesis induced by microRNAs. Nature. 2008;455(7209):58–63.

[144] Grimson A, Farh KK, Johnston WK, et al. MicroRNA targeting specificity in mammals: determinants beyond seed pairing. Mol Cell. 2007;27:91–105.

[145] Hausser J, Landthaler M, Jaskiewicz L, et al. Relative contribution of sequence and structure features to the mRNA binding of Argonaute/EIF2CmiRNA complexes and the degradation of miRNA targets. Genome Res. 2009;19:2009–2020.

[146] Karginov FV, Conaco C, Xuan Z, et al. A biochemical approach to identifying microRNA targets. Proc Natl Acad Sci USA. 2007;104:19291–19296.

[147] Krützfeldt J, Rajewsky N, Braich R, et al. Silencing of microRNAs in vivo with antagomirs'. Nature. 2005;438: 685–689.

[148] Landthaler M, Gaidatzis D, Rothballer A, et al. Molecular characterization of human Argonaute-containing ribonucleoprotein complexes and their bound target mRNAs. RNA. 2008;14:2580–2596.

[149] Guo H, Ingolia NT, Weissman JS, et al. Mammalian microRNAs predominantly act to decrease target mRNA levels. Nature. 2010;466:835–840.

[150] Mu P, Han YC, Betel D, et al. Genetic dissection of the miR-17 92 cluster of microRNAs in Myc induced B-cell lymphomas. Genes Dev. 2009;23:2806–2811.

[151] Wu S, Huang S, Ding J, et al. Multiple microRNAs modulate p21Cip1/Waf1 expression by directly targeting its 3′ untranslated region. Oncogene. 2010;29:2302–2308.

[152] Krek A, Grun D, Poy MN, et al. Combinatorial microRNA target predictions. Nat Genet. 2005;37:495–500.

[153] Stark A, Brennecke J, Bushati N, et al. Animal MicroRNAs confer robustness to gene expression and have a significant impact on 3'UTR evolution. Cell. 2005;123:1133–1146.

[154] John B, Enright AJ, Aravin A, et al. Human MicroRNA targets. PLoS Biol. 2004;2:363.

[155] Mestdagh P, Lefever S, Pattyn F, et al. The microRNA body map: dissecting microRNA function through integrative genomics. Nucleic Acids Res. 2011;39(20):e136.

[156] Chi SW, Zang JB, Mele A, et al. Argonaute HITS-CLIP decodes microRNA-mRNA interaction maps. Nature. 2009; 460:479–486.

[157] Hafner M, Landthaler M, Burger L, et al. Transcriptome-wide identification of RNA-binding protein and microRNA target sites by PAR-CLIP. Cell. 2010;141:129–141.

[158] Chin LJ, Ratner E, Leng S, et al. A SNP in a let-7 microRNA complementary site in the KRAS 3′ untranslated region increases non-small cell lung cancer risk. Cancer Res. 2008;68:8535–8540.

[159] Jiang S, Zhang HW, Lu MH, et al. MicroRNA-155 functions as an OncomiR in breast cancer by targeting the suppressor of cytokine signaling 1 gene. Cancer Res. 2010;70:3119–3127.

[160] Takamizawa J, Konishi H, Yanagisawa K, et al. Reduced expression of the let-7 microRNAs in human lung cancers in association with shortened postoperative survival. Cancer Res. 2004;64:3753–3756.

[161] Mayr C, Hemann MT, Bartel DP. Disrupting the pairing between let-7 and Hmga2 enhances oncogenic transformation. Science. 2007;315:1576–1579.

[162] Mayr C, Bartel DP. Widespread shortening of 3′ UTRs by alternative cleavage and polyadenylation activates oncogenes in cancer cells. Cell. 2009;138:673–684.

[163] Poliseno L, Salmena L, Zhang J, et al. A coding-independent function of gene and pseudogene mRNAs regulates tumour biology. Nature. 2010;465:1033–1038.

[164] Bhattacharyya SN, Habermacher R, Martine U, et al. Relief of microRNA-mediated translational repression in human cells subjected to stress. Cell. 2006;125:1111–1124.

[165] Kim HH, Kuwano Y, Srikantan S, et al. HuR recruits let-7/RISC to repress c-Myc expression. Genes Dev. 2009; 23:1743–1748.

[166] Kedde M, Strasser MJ, Boldajipour B, et al. RNA-binding protein Dnd1 inhibits microRNA access to target mRNA. Cell. 2007;131:1273–1286.

[167] Kedde M, van Kouwenhove M, Zwart W, et al. A Pumilio-induced RNA structure switch in p27-3′ UTR controls miR-221 and miR-222 accessibility. Nat Cell Biol. 2010;12:1014–1020.

[168] van Kouwenhove M, Kedde M, Agami R. MicroRNA regulation by RNA-binding proteins and its implications for cancer. Nat Rev Cancer. 2011;11:644–656.

[169] Yanaihara N, Caplen N, Bowman E, et al. Unique microRNA molecular profiles in lung cancer diagnosis and prognosis. Cancer Cell. 2006;9:189–98.

[170] Volinia S, Calin GA, Liu CG, et al. A microRNA expression signature of human solid tumors defines cancer gene targets. Proc Natl Acad Sci USA. 2006;103:2257–2261.

[171] Esquela-Kerscher A, Slack FJ. Oncomirs—microRNAs with a role in cancer. Nat Rev Cancer. 2006;6:259–269.

[172] Karube Y, Tanaka H, Osada H, et al. Reduced expression of Dicer associated with poor prognosis in lung cancer patients. Cancer Sci. 2005;96:111–115.

[173] Murchison EP, Partridge JF, Tam OH, et al. Characterization of Dicer-deficient murine embryonic stem cells. Proc Natl Acad Sci USA. 2005;102:12135–12140.

[174] Kanellopoulou C, Muljo SA, Kung AL, et al. Dicer-deficient mouse embryonic stem cells are defective in differentiation and centromeric silencing. Genes Dev. 2005;19:489–501.

[175] Wang Y, Medvid R, Melton C, et al. DGCR8 is essential for microRNA biogenesis and silencing of embryonic stem cell self-renewal. Nat Genet. 2007;39:380–385.

[176] Koesters R, Adams V, Betts D, et al. Human eukaryotic initiation factor EIF2C1 gene: cDNA sequence, genomic organization, localization to chromosomal bands 1p34-p35, and expression. Genomics. 1999;61:210–218.

[177] Aravin A, Gaidatzis D, Pfeffer S, et al. A novel class of small RNAs bind to MILI protein in mouse testes. Nature. 2006;442:203–207.

[178] Taubert H, Greither T, Kaushal D, et al. Expression of the stem cell self-renewal gene Hiwi and risk of tumour-related death in patients with soft-tissue sarcoma. Oncogene. 2007;26:1098–1100.

[179] Liu X, Sun Y, Guo J, et al. Expression of hiwi gene in human gastric cancer was associated with proliferation of cancer cells. Int J Cancer. 2006;118:1922–1929.

[180] Qiao D, Zeeman AM, Deng W, et al. Molecular characterization of hiwi, a human member of the piwi gene family whose overexpression is correlated to seminomas. Oncogene. 2002;21:3988–3999.

[181] Cimmino A, Calin GA, Fabbri M, et al. miR-15 and miR-16 induce apoptosis by targeting BCL2. Proc Natl Acad Sci USA. 2005;102:13944–13949.

[182] Johnson SM, Grosshans H, Shingara J, et al. RAS is regulated by the let-7 microRNA family. Cell. 2005;120:635–647.

[183] Lee YS, Dutta A. The tumor suppressor microRNA let-7 represses the HMGA2 oncogene. Genes Dev. 2007;21: 1025–1030.

[184] Sampson VB, Rong NH, Han J, et al. MicroRNA let-7a downregulates MYC and reverts MYC-induced growth in Burkitt lymphoma cells. Cancer Res. 2007;67: 9762–9770.

[185] Johnson CD, Esquela-Kerscher A, Stefani G, et al. The let-7 microRNA represses cell proliferation pathways in human cells. Cancer Res. 2007;67:7713–7722.

[186] Welch C, Chen Y, Stallings RL. MicroRNA-34a functions as a potential tumor suppressor by inducing apoptosis in neuroblastoma cells. Oncogene. 2007;26:5017–5022.

[187] He L, He X, Lim LP, et al. A microRNA component of the p53 tumour suppressor network. Nature. 2007;447:1130–1134.

[188] Bommer GT, Gerin I, Feng Y, et al. p53-mediated activation of miRNA34 candidate tumor-suppressor genes. Curr Biol. 2007;17:1298–1307.

[189] Tazawa H, Tsuchiya N, Izumiya M, et al. Tumor-suppressive miR-34a induces senescence-like growth arrest through modulation of the E2F pathway in human colon cancer cells. Proc Natl Acad Sci USA. 2007;104:15472–15477.

[190] Woods K, Thomson JM, Hammond SM. Direct regulation of an oncogenic micro-RNA cluster by E2F transcription factors. J Biol Chem. 2007;282:2130–2134.

[191] Sylvestre Y, De Guire V, Querido E, et al. An E2F/miR-20a autoregulatory feedback loop. J Biol Chem. 2007; 282:2135–2143.

[192] Meng F, Henson R, Wehbe-Janek H, et al. MicroRNA-21 regulates expression of the PTEN tumor suppressor gene in human hepatocellular cancer. Gastroenterology. 2007; 133:647–658.

[193] Zhu S, Si ML, Wu H, et al. MicroRNA-21 targets the tumor suppressor gene tropomyosin 1 (TPM1). J Biol Chem. 2007;282:14328–14336.

[194] Frankel LB, Christoffersen NR, Jacobsen A, et al. Programmed Cell Death 4 (PDCD4) Is an Important Functional Target of the MicroRNA miR-21 in Breast Cancer Cells. J Biol Chem. 2008;283:1026–1033.

[195] Asangani IA, Rasheed SA, Nikolova DA, et al. MicroRNA-21 (miR-21) post-transcriptionally downregulates tumor suppressor Pdcd4 and stimulates invasion, intravasation and metastasis in colorectal cancer. Oncogene. 2008;27(15):2128–2136.

[196] Vigorito E, Perks KL, Abreu-Goodger C, et al. microRNA-155 Regulates the Generation of Immunoglobulin Class-Switched Plasma Cells. Immunity. 2007;27:847–859.

[197] Voorhoeve PM, le Sage C, Schrier M, et al. A genetic screen implicates miRNA-372 and miRNA-373 as oncogenes in testicular germ cell tumors. Adv Exp Med Biol. 2007;604:17–46.

[198] Volinia S, Galasso M, Costinean S, et al. Reprogramming of miRNA networks in cancer and leukemia. Genome Res. 2010;20:589–599.

[199] Keller A, Leidinger P, Bauer A, et al. Toward the blood-borne miRNome of human diseases. Nat Methods. 2011; 8(10):841–843.

[200] Bullrich F, Fujii H, Calin G, et al. Characterization of the 13q14 tumor suppressor locus in CLL: identification of ALT1, an alternative splice variant of the LEU2 gene. Cancer Res. 2001;61:6640–6648.

[201] Calin GA, Ferracin M, Cimmino A, et al. A MicroRNA signature associated with prognosis and progression in chronic lymphocytic leukemias. N Engl J Med. 2005;353:1793–1801.

[202] Chen C, Frierson HF Jr, Haggerty PF, et al. An 800-kb region of deletion at 13q14 in human prostate and other carcinomas. Genomics. 2001;77:135–144.

[203] Yu SL, Chen HY, Chang GC, et al. MicroRNA signature predicts survival and relapse in lung cancer. Cancer Cell. 2008;13:48–57.

[204] Shell S, Park SM, Radjabi AR, et al. Let-7 expression defines two differentiation stages of cancer. Proc Natl Acad Sci USA. 2007;104:11400–11405.

[205] Malumbres M, Barbacid M. RAS oncogenes: the first 30 years. Nat Rev Cancer1140 2003;3:459–465.

[206] Sanchez-Cespedes M. Dissecting the genetic alterations involved in lung carcinogenesis. Lung Cancer. 2003;40: 111–121.

[207] Esquela-Kerscher A, Trang P, Wiggins JF, et al. The let-7 microRNA reduces tumor growth in mouse models of lung cancer. Cell Cycle. 2008;7(6):759–764.

[208] Kumar MS, Erkeland SJ, Pester RE, et al. Suppression of non-small cell lung tumor development by the let-7 microRNA family. Proc Natl Acad Sci USA. 2008;105:3903–3908.

[209] Yu F, Yao H, Zhu P, et al. let-7 regulates self renewal and tumorigenicity of breast cancer cells. Cell. 2007;131: 1109–1123.

[210] Tarasov V, Jung P, Verdoodt B, et al. Differential regulation of microRNAs by p53 revealed by massively parallel sequencing: miR-34a is a p53 target that induces apoptosis and G1-arrest. Cell Cycle. 2007;6:1586–1593.

[211] Corney DC, Flesken-Nikitin A, Godwin AK, et al. MicroRNA-34b and MicroRNA-34c Are Targets of p53 and Cooperate in Control of Cell Proliferation and Adhesion-Independent Growth. Cancer Res. 2007;67:8433–8438.

[212] Calin GA, Croce CM. MicroRNA signatures in human cancers. Nat. Rev. Cancer. 2006;6: 857–866.

[213] Gailhouste L, Gómez-Santos L, Ochiya T. Potential applications of miRNAs as diagnostic and prognostic markers in liver cancer. Front Biosci. 2013;18:199–223.

[214] Fornari F, Milazzo M, Chieco P, et al. MiR-199a-3p regulates mTOR and c-Met to influence the doxorubicin sensitivity of human hepatocarcinoma cells. Cancer Res. 2010;70:5184–5193.

[215] Ji J, Shi J, Budhu A, et al. MicroRNA expression, survival, and response to interferon in liver cancer. N Engl J Med. 2009;361:1437–1447.

[216] Chung GE, Yoon JH, Myung SJ, et al. High expression of microRNA-15b predicts a low risk of tumor recurrence following curative resection of hepatocellular carcinoma. Oncol Rep. 2010;23:113–119.

[217] Villanueva A, Hoshida Y, Toffanin S, et al. New strategies in hepatocellular carcinoma: genomic prognostic markers. Clin Cancer Res. 2010;16:4688–4694.

[218] Budhu A, Jia HL, Forgues M, et al. Identification of metastasis-related microRNAs in hepatocellular carcinoma. Hepatology. 2008;47:897–907.

[219] Wang XW, Heegaard NH, Orum H. MicroRNAs in liver disease. Gastroenterology 2012;142:1431–1443.

[220] Kota J, Chivukula RR, O'Donnell KA, et al. Therapeutic microRNA delivery suppresses tumorigenesis in a murine liver cancer model. Cell. 2009;137:1005–1017.

[221] Ota A, Tagawa H, Karnan S, et al. Identification and characterization of a novel gene, C13orf25, as a target for 13q31-q32 amplification in malignant lymphoma. Cancer Res. 2004;64:3087–3095.

[222] Dews M, Homayouni A, Yu D, et al. Augmentation of tumor angiogenesis by a Myc-activated microRNA cluster. Nat Genet. 2006;38:1060–1065.

[223] Chang TC, Yu D, Lee YS, et al. Widespread microRNA repression by Myc contributes to tumorigenesis. Nat Genet. 2008;40:43–50.

[224] Peng HQ, Hogg D, Malkin D, et al. Mutations of the p53 gene do not occur in testis cancer. Cancer Res. 1993;53:3574–3578.

[225] Si ML, Zhu S, Wu H, et al. miR-21-mediated tumor growth. Oncogene. 2007;26:2799–2803.

[226] Corsten MF, Miranda R, Kasmieh R, et al. MicroRNA-21 knockdown disrupts glioma growth in vivo and displays synergistic cytotoxicity with neural precursor cell delivered S-TRAIL in human gliomas. Cancer Res. 2007;67:8994–9000.

[227] Roldo C, Missiaglia E, Hagan JP, et al. MicroRNA expression abnormalities in pancreatic endocrine and acinar tumors are associated with distinctive pathologic features and clinical behavior. J Clin Oncol. 2006;24:4677–4684.

[228] Iorio MV, Ferracin M, Liu CG, et al. MicroRNA gene expression deregulation in human breast cancer. Cancer Res. 2005;65:7065–7070.

[229] Chan JA, Krichevsky AM, Kosik KS. MicroRNA-21 is an antiapoptotic factor in human glioblastoma cells. Cancer Res. 2005;65:6029–6033.

[230] Schetter AJ, Leung SY, Sohn JJ, et al. MicroRNA expression profiles associated with prognosis and therapeutic outcome in colon adenocarcinoma. JAMA. 2008;299:425–436.

[231] Tam W, Ben-Yehuda D, Hayward WS. bic, a novel gene activated by proviral insertions in avian leukosis virus-induced lymphomas, is likely to function through its noncoding RNA. Mol Cell Biol. 1997;17:1490–1502.

[232] Zhang T, Nie K, Tam W. BIC is processed efficiently to microRNA-155 in Burkitt lymphoma cells. Leukemia. 2008;22:1795–1797.

[233] Clurman BE, Hayward WS. Multiple proto-oncogene activations in avian leukosis virus-induced lymphomas: evidence for stage-specific events. Mol Cell Biol. 1989;9:2657–2664.

[234] Kluiver J, Poppema S, de Jong D, et al. BIC and miR-155 are highly expressed in Hodgkin, primary mediastinal and diffuse large B cell lymphomas. J Pathol. 2005;207:243–249.

[235] Eis PS, Tam W, Sun L, et al. Accumulation of miR-155 and BIC RNA in human B cell lymphomas. Proc Natl Acad Sci USA. 2005;102:3627–3632.

[236] Costinean S, Zanesi N, Pekarsky Y, et al. Pre-B cell proliferation and lymphoblastic leukemia/high-grade lymphoma in E(mu)-miR155 transgenic mice. Proc Natl Acad Sci USA. 2006;103:7024–7029.

[237] Thai TH, Calado DP, Casola S, et al. Regulation of the germinal center response by microRNA-155. Science. 2007;316:604–608.

[238] Rodriguez A, Vigorito E, Clare S, et al. Requirement of bic/microRNA-155 for normal immune function. Science. 2007;316:608–611.

[239] Gironella M, Seux M, Xie MJ, et al. Tumor protein 53-induced nuclear protein 1 expression is repressed by miR-155, and its restoration inhibits pancreatic tumor development. Proc Natl Acad Sci USA. 2007;104:16170–16175.

[240] Gottwein E, Mukherjee N, Sachse C, et al. A viral microRNA functions as an orthologue of cellular miR-155. Nature. 2007;450:1096–1099.

[241] Xiao C, Srinivasan L, Calado DP, et al. Lymphoproliferative disease and autoimmunity in mice with increased miR-17-92 expression in lymphocytes. Nat Immunol. 2008;9:405–414.

[242] Medina PP, Nolde M, Slack FJ. OncomiR addiction in an in vivo model of microRNA-21-induced pre-B-cell lymphoma. Nature. 2010;467:86–90.

[243] Costinean S, Zanesi N, Pekarsky Y, et al. Pre-B cell proliferation and lymphoblastic leukemia/high-grade lymphoma in E(mu)-miR155 transgenic mice. Proc Natl Acad Sci USA. 2006;103:7024–7029.

[244] O'Connell RM, Rao DS, Chaudhuri AA, et al. Sustained expression of microRNA-155 in hematopoietic stem cells causes a myeloproliferative disorder. J Exp Med. 2008;205:585–594.

[245] Hurst DR, Edmonds MD, Welch DR. Metastamir: the field of metastasis-regulatory microRNA is spreading. Cancer Res. 2009;69:7495–7498.

[246] Camps C, Buffa FM, Colella S, et al. hsamiR- 210 is induced by hypoxia and is an independent prognostic factor in breast cancer. Clin Cancer Res. 2008;14:1340–1348.

[247] Foekens JA, Sieuwerts AM, Smid M, et al. Four miRNAs associated with aggressiveness of lymph node-negative, estrogen receptor-positive human breast cancer. Proc Natl Acad Sci USA. 2008;105:13021–13026.

[248] Friedman RC, Farh KK, Burge CB, et al. Most mammalian mRNAs are conserved targets of microRNAs. Genome Res. 2009;19:92–105.

[249] Valastyan S, Benaich N, Chang A, et al. Concomitant suppression of three target genes can explain the impact of a microRNA on metastasis. Genes Dev. 2009;23:2592–2597.

[250] Cano A, Nieto MA. Non-coding RNAs take centre stage in epithelial-to-mesenchymal transition. Trends Cell Biol. 2008;18:357–359.

[251] Farazi PA, DePinho RA. Hepatocellular carcinoma pathogenesis: from genes to environment. Nat Rev Cancer. 2006;6:674–687.

[252] El-Serag HB, Rudolph KL. Hepatocellular carcinoma: epidemiology and molecular carcinogenesis. Gastroenterology. 2007;132:2557–2576.

[253] Kent OA, Mendell JT. A small piece in the cancer puzzle: microRNAs as tumor suppressors and oncogenes. Oncogene. 2006;25:6188–6196.

[254] Lujambio A, Lowe SW. The microcosmos of cancer. Nature. 2012;482: 347–355.

[255] Calin GA, Croce CM. Chromosomal rearrangements and microRNAs: a new cancer link with clinical implications. J Clin Invest. 2007;117:2059–2066.

[256] Aravalli RN, Steer CJ, Cressman EN. Molecular mechanisms of hepatocellular carcinoma. Hepatology. 2008;48:2047–2063.

[257] Sato F, Tsuchiya S, Meltzer SJ, et al. MicroRNAs and epigenetics. FEBS J. 2011;278:1598–1609.

[258] Lujambio A, Esteller M. CpG island hypermethylation of tumor suppressor microRNAs in human cancer. Cell Cycle. 2007;6:1455–1459.

[259] Datta J, Kutay H, Nasser MW, et al. Methylation mediated silencing of MicroRNA-1 gene and its role in hepatocellular carcinogenesis. Cancer Res. 2008;68:5049–5058.

[260] Furuta M, Kozaki KI, Tanaka S, et al. miR-124 and miR-203 are epigenetically silenced tumor-suppressive microRNAs in hepatocellular carcinoma. Carcinogenesis. 2010;31:766–776.

[261] Kondo Y, Kanai Y, Sakamoto M, et al. Genetic instability and aberrant DNA methylation in chronic hepatitis and cirrhosis–A comprehensive study of loss of heterozygosity and microsatellite instability at 39 loci and DNA hypermethylation on 8 CpG islands in microdissected specimens from patients with hepatocellular carcinoma. Hepatology. 2000;32:970–979.

[262] Zhu A, Xia J, Zuo J, et al. MicroRNA-148a is silenced by hypermethylation and interacts with DNA methyltransferase 1 in gastric cancer. Med Oncol. 2011;29:2701–2709.

[263] Hanoun N, Delpu Y, Suriawinata AA, et al. The silencing of microRNA 148a production by DNA hypermethylation is an early event in pancreatic carcinogenesis. Clin. Chem. 2010;56:1107–1118.

[264] Braconi C, Huang N, Patel T. MicroRNA-dependent regulation of DNA methyltransferase-1 and tumor suppressor gene expression by interleukin-6 in human malignant cholangiocytes. Hepatology. 2010;51:881–890.

[265] Faggad A, Budczies J, Tchernitsa O, et al. Prognostic significance of Dicer expression in ovarian cancer-link to global microRNA changes and oestrogen receptor expression. J Pathol. 2010;220:382–391.

[266] Wu JF, Shen W, Liu NZ, et al. Down-regulation of Dicer in hepatocellular carcinoma. Med Oncol. 2011a;28:804–809.

[267] Wu L, Cai C, Wang X, et al. MicroRNA-142-3p, a new regulator of RAC1, suppresses the migration and invasion of hepatocellular carcinoma cells. FEBS Lett. 2011b;585:1322–1330.

[268] Wu N, Liu X, Xu X, et al. MicroRNA-373, a new regulator of protein phosphatase 6, functions as an oncogene in hepatocellular carcinoma. FEBS J. 2011c;278:2044–2054.

[269] Sekine S, Ogawa R, Ito R, et al. Disruption of Dicer1 induces dysregulated fetal gene expression and promotes hepatocarcinogenesis. Gastroenterology. 2009;136:2304–2315, e2301–2304.

[270] Chen C, Ridzon DA, Broomer AJ, et al. Real-time quantification of microRNAs by stem-loop RT-PCR. Nucleic Acids Res. 2005;33:e179.

[271] Gramantieri L, Fornari F, Callegari E, et al. MicroRNA involvement in hepatocellular carcinoma. J Cell Mol Med. 2008;12:2189–2204.

[272] Mott JL. MicroRNAs involved in tumor suppressor and oncogene pathways: implications for hepatobiliary neoplasia. Hepatology. 2009;50:630–637.

[273] Cheung O, Puri P, Eicken C, et al. Nonalcoholic steatohepatitis is associated with altered hepatic MicroRNA expression. Hepatology. 2008;48:1810–1820.

[274] Murakami Y, Yasuda T, Saigo K, et al. Comprehensive analysis of microRNA expression patterns in hepatocellular carcinoma and non-tumorous tissues. Oncogene. 2006;25:2537–2545.

[275] Ladeiro Y, Couchy G, Balabaud C, et al. MicroRNA profiling in hepatocellular tumors is associated with clinical features and oncogene/tumor suppressor gene mutations. Hepatology. 2008;47:1955–1963.

[276] Jopling CL, Yi M, Lancaster AM, et al. Modulation of hepatitis C virus RNA abundance by a liver-specific MicroRNA. Science. 2005;309:1577–1581.

[277] Gramantieri L, Ferracin M, Fornari F, et al. Cyclin G1 is a target of miR-122a, a microRNA frequently down-regulated in human hepatocellular carcinoma. Cancer Res. 2007;67:6092–6099.

[278] Hsu SH, Wang B, Kota J, et al. Essential metabolic, anti-inflammatory, and anti-tumorigenic functions of miR-122 in liver. J Clin Invest. 2012;122:2871–2883.

[279] Chen XM. MicroRNA signatures in liver diseases. World J Gastroenterol. 2009;15:1665–1672.

[280] Ji J, Wang XW. New kids on the block: diagnostic and prognostic microRNAs in hepatocellular carcinoma. Cancer Biol Ther. 2009;8:1686–1693.

[281] Yang JS, Maurin T, Robine N, et al. Conserved vertebrate mir-451 provides a platform for Dicer-independent, Ago2-mediated microRNA biogenesis. Proc Natl Acad Sci USA. 2010;107:15163–15168.

[282] Babiarz JE, Ruby JG, Wang Y, et al. Mouse ES cells express endogenous shRNAs, siRNAs, and other Microprocessor-independent, Dicer-dependent small RNAs. Genes Dev. 2008;22:2773–2785.

[283] Yang JS, Lai EC. Alternative miRNA biogenesis pathways and the interpretation of core miRNA pathway mutants. Mol Cell. 2011;43:892–903.

[284] Hagan JP, Piskounova E, Gregory RI. Lin28 recruits the TUTase Zcchc11 to inhibit let-7 maturation in mouse embryonic stem cells. Nat Struct Mol Biol. 2009;16:1021–1025.

[285] Kawahara Y, Zinshteyn B, Chendrimada TP, et al. RNA editing of the microRNA-151 precursor blocks cleavage by the Dicer-TRBP complex. EMBO Rep. 2007;8:763–769.

[286] Kosaka N, Iguchi H, Ochiya T. Circulating microRNA in body fluid: a new potential biomarker for cancer diagnosis and prognosis. Cancer Sci. 2010;101:2087–2092.

[287] Kosaka N, Iguchi H, Yoshioka Y, et al. Competitive interactions of cancer cells and normal cells via secretory microRNAs. J. Biol. Chem. 2012;287:1397–1405.

[288] Borel F, Konstantinova P, Jansen PL. Diagnostic and therapeutic potential of miRNA signatures in patients with hepatocellular carcinoma. J Hepatol. 2012;56:1371–1383.

[289] Li J., Fu H., Xu C., et al. miR-183 inhibits TGF-beta1-induced apoptosis by downregulation of PDCD4 expression in human hepatocellular carcinoma cells. BMC Cancer 2010a;10:354.

[290] Li LM, Hu ZB, Zhou ZX, et al. Serum microRNA profiles serve as novel biomarkers for HBV infection and diagnosis of HBV-positive hepatocarcinoma. Cancer Res. 2010b;70:9798–9807.

[291] Li D, Liu X, Lin L, et al. MicroRNA-99a inhibits hepatocellular carcinoma growth and correlates with prognosis of patients with hepatocellular carcinoma. J Biol Chem. 2011a;286:36677–36685.

[292] Li J, Wang Y, Yu W, et al. Expression of serum miR-221 in human hepatocellular carcinoma and its prognostic significance. Biochem Biophys Res Commun. 2011b;406:70–73.

[293] Yamamoto Y, Kosaka N, Tanaka M, et al. MicroRNA-500 as a potential diagnostic marker for hepatocellular carcinoma. Biomarkers. 2009;14:529–538.

[294] Shigoka M, Tsuchida A, Matsudo T, et al. Deregulation of miR-92a expression is implicated in hepatocellular carcinoma development. Pathol Int. 2010;60:351–357.

[295] Tomimaru Y, Eguchi H, Nagano H, et al. Circulating microRNA-21 as a novel biomarker for hepatocellular carcinoma. J Hepatol. 2012;56:167–175.

[296] Lewis MA, Quint E, Glazier AM, et al. An ENU-induced mutation of miR-96 associated with progressive hearing loss in mice. Nat Genet. 2009;41:614–618.

[297] Mencia A, Modamio-Hoybjor S, Redshaw N, et al. Mutations in the seed region of human miR-96 are responsible for nonsyndromic progressive hearing loss. Nat Genet. 2009;41:609–613.

[298] Gottwein E, Cai X, Cullen BR. Expression and function of microRNAs encoded by Kaposi's sarcoma-associated herpesvirus. Cold Spring Harb Symp Quant Biol. 2006;71:357–364.

[299] de Pontual L, Yao E, Callier P. Germline deletion of the miR-17 approximately 92 cluster causes skeletal and growth defects in humans. Nat Genet. 2011;43(10):1026–1030.

[300] Raveche ES, Salerno E, Scaglione BJ, et al. Abnormal microRNA-16 locus with synteny to human 13q14 linked to CLL in NZB mice. Blood. 2007;109:5079–5086.

[301] Esau C, Kang X, Peralta E, et al. MicroRNA-143 regulates adipocyte differentiation. J Biol Chem. 2004;279: 52361–52365.

[302] Krutzfeldt J, Kuwajima S, Braich R, et al. Specificity, duplex degradation and subcellular localization of antagomirs. Nucleic Acids Res. 2007;35:2885–2892.

[303] Elmen J, Lindow M, Silahtaroglu A, et al. Antagonism of microRNA-122 in mice by systemically administered LNA-antimiR leads to up-regulation of a large set of predicted target mRNAs in the liver. Nucleic Acids Res. 2008;36:1153–1162.

[304] Esau C, Davis S, Murray SF, et al. miR-122 regulation of lipid metabolism revealed by in vivo antisense targeting. Cell Metab. 2006;3:87–98.

[305] Elmen J, Lindow M, Schutz S, et al. LNA-mediated microRNA silencing in non-human primates. Nature. 2008;452:896–899.

[306] Ebert MS, Sharp PA. MicroRNA sponges: progress and possibilities. RNA. 2010;16:2043–2050.

[307] Lanford RE, Hildebrandt-Eriksen ES, Petri A, et al. Therapeutic silencing of microRNA-122 in primates with chronic hepatitis C virus infection. Science. 2010;327: 198–201.

[308] Meister G, Tuschl T. Mechanisms of gene silencing by double-stranded RNA. Nature. 2004;431:343–349.

[309] Liu CG, Calin GA, Meloon B, et al. An oligonucleotide microchip for genome-wide microRNA profiling in human and mouse tissues. Proc Natl Acad Sci USA. 2004;101: 9740–9744.

[310] Rosenfeld N, Aharonov R, Meiri E, et al. MicroRNAs accurately identify cancer tissue origin. Nat Biotechnol. 2008;26:462–469.

[311] Gleave ME, Monia BP. Antisense therapy for cancer. Nat Rev Cancer. 2005;5:468–479.

[312] Grimm D, Streetz KL, Jopling CL, et al. Fatality in mice due to oversaturation of cellular microRNA/short hairpin RNA pathways. Nature. 2006;441:537–541.

[313] Madkour LH. Reactive Oxygen Species (ROS), Nanoparticles, and Endoplasmic Reticulum (ER) Stress-Induced Cell Death Mechanisms. Academic Press; 2020.

[314] Madkour LH. Nanoparticles Induce Oxidative and Endoplasmic Reticulum Antioxidant Therapeutic Defenses. Springer International Publishing; 2020.

[315] Madkour LH. Nucleic Acids as Gene Anticancer Drug Delivery Therapy. Elsevier; 2020.

[316] Madkour LH. Nanoelectronic Materials: Fundamentals and Applications (Advanced Structured Materials). Springer International Publishing; 2019. doi:10.1007/978-3-030-21621-4.

# 5 Biological Function of miRNA and piRNA Targets in Cancer Tissues

Several studies demonstrated that miRNAs take part in numerous biological processes, such as proliferation, apoptosis, and migration. Different mechanisms involved in regulating miR-193a-3p expression have been reported, including epigenetic modifications and transcription factors. In physiological contexts, miR-193a-3p seemed able to limit proliferation and cell cycle progression in normal cells. Remarkably, several publications demonstrated that miR-193a-3p acted as a tumor suppressor miRNA in cancer by targeting different genes involved in proliferation, apoptosis, migration, invasion, and metastasis. Furthermore, the downregulation of miR-193a-3p has been observed in many primary tumors and altered levels of circulating miR-193a-3p have been identified in serum or plasma of cancer patients and subjects affected by Parkinson's disease or by schizophrenia. We summarized current advances in our knowledge of the roles of piRNAs in cancer.

## 5.1 SncRNAs (SMALL NONCODING RNAS)

Irrespective of socioeconomic context, cancer is the second leading cause of death globally, because it is the third leading cause of death in low- and middle-income countries and the second leading cause of death in high-income countries [1–4].

SncRNAs (small noncoding RNAs) are part of noncoding oligonucleotide regulators with wide morphologic and physiologic functions. At the transcriptional and posttranscriptional levels, these molecules are primary mediators of gene regulation [5]. SncRNAs have a variety of family members, among which the most investigated are small nucleolar RNAs, small nuclear RNAs, siRNAs (small interfering RNAs) [6], miRNAs (microRNAs) [7], and piRNAs (PIWI-interacting RNAs) [8].

RNAi (RNA interference), also denoted as RNA silencing, in most eukaryotes has emerged as one of the key gene regulatory pathways [9, 10]. Central to RNAi pathways is the generation of small RNAs of 20–31 nt (nucleotides). These small RNAs, produced by a processing protein known as Dicer or by Dicer-independent processes, form complexes with RISC (RNA-induced silencing complex) carrying Argonaute (AGO) proteins [11, 12]. RISCs are directed to the target genes based on the complementarities between target gene transcripts and small RNAs and inhibit their expression by RNA instability, by inducing translational inhibition, or by cleaving the transcripts [13], and/or heterochromatinization [14, 15]. MiRNAs and endogenous short interfering RNAs (endo-siRNAs) associate with the AGO subfamily members, while piRNAs produced by a Dicer-independent process associate with PIWI subfamily members of proteins (Figure 5.1) [16].

PiRNAs are a class of sncRNA molecules that have been recently recognized to be relevant to cancer biology. More than 30,000 piRNAs have been found in humans [17, 18]. They are best characterized by their role in guiding associated chromatin-silencing machinery to transposon-encoding DNA sequences in the genome [18, 19]. Researchers have offered that a subset of piRNAs may also be capable of regulating protein-coding genes via DNA methylation, which, if the regulatory targets are cancer-relevant, may bear on cancer development [20–22].

PiRNAs independent of RNase III enzymes are generated from single-stranded precursors in a manner [23–26] and they are typically 26–31 nt long. The mechanisms underlying piRNA functions and biogenesis mainly remain unknown, mostly because the piRNA pathway has little in common with the miRNA and endo-siRNA pathways as well as the restriction of piRNA territories to the reproductive tissues [23].

The piRNAs are generally processed from their longer precursors that are transcribed from introns, 3'UTR regions, and repetitive elements [27, 28]. They are a distinct class of small RNAs that function in transposon silencing, epigenetic regulation, and germline development [29–31]. These molecules have remarkable diversity, with tens of thousands of unique sequences in mammals and more than 1.5 million unique sequences in Drosophila [32, 33].

The recent research on the piRNA in cancer biology has been analyzed [34], and some of the severe cancer that happens in the human body also has been summarized.

## 5.2 THE ROLE OF piRNAs IN CANCER

The functions of PIWI proteins and piRNAs have started to emerge in human cancers [35]. A growing amount of evidence has shown that PIWI proteins in mice and humans, such as PIWIL2-like proteins, HIWI, and PIWIL2, are expressed in various types of tumor cells [36, 37]. Furthermore, piRNAs were also detected in these cells [37]. The deregulated expression of piRNAs has been reported in human cancers, including gastric cancer, bladder cancer, breast cancer, colorectal cancer, and lung cancer. These findings indicate that the piRNA pathway may be linked to cancer development. Though the potential role of piRNAs in cancer has just emerged and remains to be investigated, the functional role of specific piRNAs is poorly understood in human cancer. These results highlight the importance of understanding the exact role of the piRNA pathway during tumorigenesis, which may offer new possibilities for tumor therapy (Table 5.1).

DOI: 10.1201/9781003229650-5

## Small RNA Loci

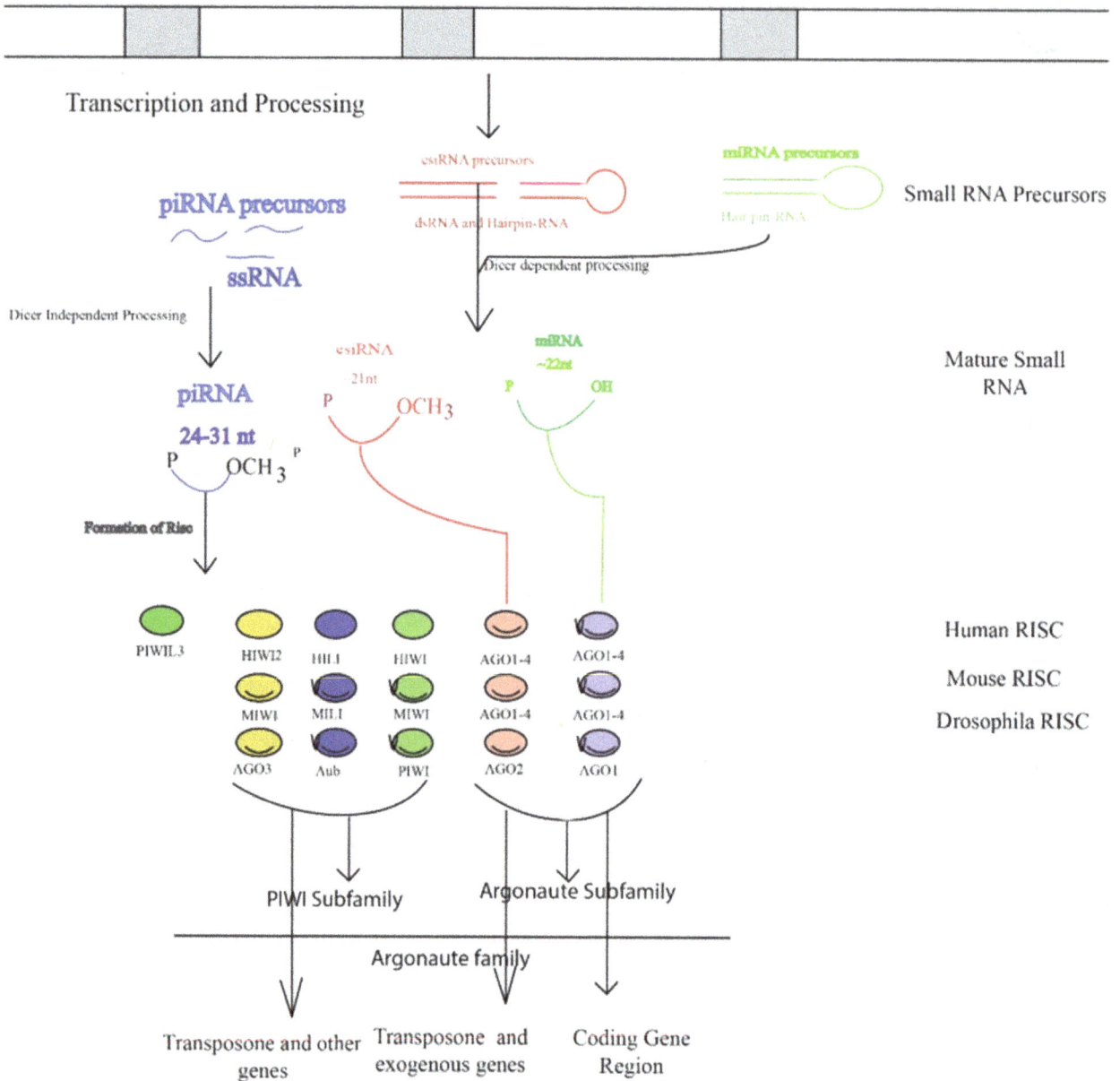

**FIGURE 5.1** RNA silencing by small RNAs and their partner Argonaute family proteins in Drosophila, human, and mouse RNA. The expression level of four PIWI proteins (PIWIL3) has been a discovery in humans. The main roles of the mature sequences, RISC formation, piRNA precursors, and target genes are summarized for miRNAs and piRNAs. A correlation between piRNA and human PIWI protein has not yet been detected. *Abbreviations:* dsRNA, double-stranded RNA; esiRNA, endogenous small interfering RNA (siRNA); miRNA, microRNA; nt, nucleotide; piRNA, PIWI-interacting RNA; RISC, RNA-induced silencing complex; ssRNA, single-stranded RNA.

### 5.2.1 GASTRIC CANCER

Gastric cancer is the second most significant cause of global cancer-related mortalities [45, 46] and the fifth most prevalent cancer in the world [47, 48]. In the United States, there are about 7.4 new cases of gastric cancer per 100,000 women and men per year [49]. Also in 2016, with 26,420 deaths and 42,280 new cases, esophageal and gastric cancers rank among the deadliest malignant diseases in the United States [50]. Based on research in Europe, four of five patients with gastric cancer die within the first 5 years after diagnosis [51].

PiR-823 was downregulated in gastric cancer tissue with no association between its clinicopathological features and expression levels [44]. Most importantly, the growth of gastric cancer cells was inhibited by piR-823 mimics in vitro, and tumor growth in vivo (in a xenograft model) was significantly suppressed, both in a dose-dependent manner. These

**TABLE 5.1**

**Role of the piRNA Pathway in Cancer**

| piRNA | Up | Down | Cancer | Function | Reference |
|---|---|---|---|---|---|
| piR-34736 | | * | Breast | Induced by cell cycle progression | [38] |
| piR-36249 | | * | Breast | Induced by cell cycle progression | [38] |
| piR-35407 | | * | Breast | Induced by cell cycle progression | [38] |
| piR-36318 | | * | Breast | Induced by cell cycle progression | [38] |
| piR-34377 | | * | Breast | Induced by cell cycle progression | [38] |
| piR-36743 | * | | Breast | Induced by cell cycle progression | [38] |
| piR-36026 | * | | Breast | Induced by cell cycle progression | [38] |
| piR-31106 | * | | Breast | Induced by cell cycle progression | [38] |
| piRABC | | * | Bladder | Increase the expression of TNFSF4 protein | [39] |
| PiR-823 | | * | Gastric | Inhibit cancer cell growth | [40] |
| piR-55490 | | * | Lung | Suppress the activation of Akt/mTOR pathway by binding 3/UTR of mTOR messenger RNA and induce its degradation | [41] |
| piR-L-163 | | * | Lung | Bind directly to phosphorylated ERM protein (p-ERM) | [42] |
| piR-Hep1 | * | | Liver | Deep sequencing of cell lines, validated in matched tumor–normal tissues via PCR | [43] |
| piR–015551 | | * | Colorectal | Associated with recurrence-free survival | [44] |

results suggest that piR-823 is a possible therapeutic target and tumor suppressor for gastric cancer. PiR-651 as an oncogene was overexpressed in gastric cancer, resulting in a positive correlation with Tumor, Node, Metastasis (TNM) stage. This observation was consistent with results in other cancer tissues and cell lines such as breast, lung, colon, and liver cancers. Of course, the research has not shown that the expression of piR-651 was associated with other clinicopathological findings such as sex, age, invasion, and tumor size. In addition, in the G2/M phase, a piR-651 inhibitor could inhibit cell growth, which was an important indication that piRNAs play a crucial role in tumorigenesis [52]. In addition, piR-651 and piR-823 were both reported to be at lower levels in circulating tumor cells (CTCs) compared with normal controls, in the peripheral blood of gastric cancer patients [53]. Both piR-823 and piR-651 were more sensitive than the commonly used biomarkers for gastric cancer such as CA19-9 (carbohydrate antigen 19–9) and CEA (serum carcinoembryonic antigen). Some reasons for this include: (1) as a short fragment, piRNAs are not as easily degraded, (2) levels of piR-651 as well as piR-823 in blood samples are relatively stable, (3) piRNAs can pass through the cell membrane, and (4) piRNAs can be detected and isolated easily from bodily fluids. It has also been suggested that gastric cancer patients can be distinguished from healthy controls, by way of measuring the levels of piR-823 and piR-651 in peripheral blood, making possible an early diagnosis of gastric cancer. These findings suggest that piRNAs could be novel therapeutic targets in gastric cancer [54].

### 5.2.2 Breast Cancer

The United States is home to around 3.1 million breast cancer survivors. The chance of any woman dying from breast cancer is around 2.7%, or 1 in 37. In 2021, an estimated 281,550 new cases of invasive breast cancer are expected to be diagnosed in women in the U.S., along with 49,290 new cases of non-invasive (in situ) breast cancer. About 2,650 new cases of invasive breast cancer are expected to be diagnosed in men in 2021. Deep sequencing was carried out in four matched nontumor tissues and four breast cancer tissues, to screen out differentially expressed piRNAs. Afterwards, by RT-PCR in 50 breast cancer tissues, 4 piRNAs (piR-20365, piR-20582, piR-20485, and piR-4987) were confirmed to be upregulated. The clinical pathology features of patients, such as estrogen receptor (ER) status, tumor size, Her2 status and, lymph node status, were recorded. The upregulation of piR-4987 was also positively associated with lymph node metastasis [55]. In a study it was shown that piR-932/PIWIL2 complex through promoting the methylation of latex may positively regulate the process of breast cancer stem cells, which in turn promotes EMT (epithelial-mesenchymal transition). It has been suggested that both PIWIL2 and piR-932 could be potential targets for blocking the metastasis of breast cancer [56]. Similarly, another study found that piR-021285 is involved in methylation at a number of known breast cancer–related genes, in particular, attenuated 5′ untranslated region (UTR)/first exon methylation at the proinvasive ARHGAP11A gene and invasiveness in an in vitro cell line model [57, 58].

### 5.2.3 Bladder Cancer

Bladder cancer is the world's ninth most common malignancy and the most common malignancy of the urinary tract [59, 60]. As per the American Cancer Society, 83,730 new cases (about 64,280 in men and 19,450 in women) and

17,200 deaths are expected (about 12,260 in men and 4,940 in women) from bladder cancer in the United States in the year 2021 [61]. Using the ArrayAtarHG19 piRNA array, for 23,677 human piRNAs, the researchers profiled three pairs of bladder cancer tissues and their adjacent normal tissues. They identified piRABC (also called DQ594040) as a relevant piRNA being downregulated in bladder cancer [62]. PiRABC showed very high differential expression levels between normal tissues and bladder cancer. Studies in vitro on human bladder cancer cell lines suggested that the overexpression of piRABC may inhibit the promotion of cell apoptosis, cell proliferation, and colony formation. Researchers showed a possible interaction with tumor necrosis factor superfamily member 4 (TNFSF4), hypothesizing that piRABC may promote bladder cancer cell apoptosis by upregulation of TNFSF4 [63].

### 5.2.4 LUNG CANCER

The leading cause of cancer-related death in the world [64], lung cancer is broadly divided into non–small cell lung cancer (NSCLC) (approximately 85% cases) and small cell lung cancer (approximately 15% cases). Researchers demonstrated that piRNAs are expressed in somatic human bronchial epithelial (HBE) cells, and the expression patterns are distinctive between lung cancer cells and normal bronchial epithelial cells [64]. Furthermore, they have shown that piR-L-163 could directly bind and regulate phosphorylated ERM and play a critical role in protein activation [65, 66]. The researchers identified that piR-L-138 was upregulated upon cisplatin (CDDP)-based chemotherapy both in vivo and in vitro, and that targeting it could be a potential strategy to overcome chemoresistance in patients with lung squamous cell carcinoma (LSCC) [67].

### 5.2.5 LIVER CANCER

Liver cancer is the second most common cause of cancer death for men and women combined worldwide [68]. Liver cancer is the fifth most prevalent cancer among men and the ninth most common cancer among women. It occurs more frequently in less developed regions of the world, but it is still a significant health burden in the United States [69]. By combining biological and bioinformatics analyses, a number of studies have identified and characterized less well-explored noncoding RNAs in human hepatocellular carcinoma (HCC) [70, 71]. Law et al. identified piR-Hep1 as being upregulated in nearly half (46.6%) of HCC tumors compared with their corresponding adjacent nontumoral liver. Silencing of this piRNA inhibited invasiveness, cell viability, and motility with a concomitant reduction in the level of active AKT phosphorylation [43, 72, 73].

### 5.2.6 COLORECTAL CANCER

Colorectal cancer is the third most common cancer in men and the second in women worldwide [74, 75]. Some

researchers have observed that PIWI contributes to the development of colorectal cancer [76, 77]. Chu et al. proposed that piRNAs through binding to PIWI may play an important role in the risk of colorectal cancer. They hypothesized that genetic variants in piRNAs could modulate colorectal cancer susceptibility [78]. Yin et al. proposed that piRNA-823 was one of the piRNAs that contributed to colorectal carcinogenesis. They observed that knocking down this piRNA could suppress G1 phase arrest and cellular apoptosis, the viability of colorectal cancer cells. They suggested that piRNA-823 could be a potential therapeutic target for colorectal cancer.

## 5.3 BIOLOGICAL FUNCTION OF MicroRNAs

MiRNAs constitute a biologically very important class of small, noncoding RNAs, about 18–22 nt long, that mainly act as negative regulators of gene expression at the posttranscriptional level by controlling the translation and stability of mRNA targets. It is known that a miRNA may target several mRNAs and that a mRNA can be under the control of several miRNAs. Most of the findings reported in the literature show clearly that miRNAs play an important role in several physiological and pathological processes exerting a highly precise regulation of most mRNA expression.

In this chapter, we focus on the human miR-193a-3p because an increasing amount of evidence has demonstrated its importance in several biological functions. Moreover, its role as an important tumor suppressor miRNA has recently emerged in both liquid and solid tumors. According to UCSC Genome Browser (Human Dec. 2013 Assembly - GRCh38/hg38) [79], the miR-193a coding gene, defined as MIR193a, is located on human chromosome 17q11.2 (chr17:31,558,803-31,560,358) (Figure 5.2(a)). By analyzing the region of 2000 bp spanning MIR193a, a typical CpG island (CGI) is identified (chr17:31558803-31560358) in which the miR-193a coding sequence is embedded. Interestingly, MIR193a is found internal to a sequence that displays a high level of enrichment of H3K27Ac, H3K4Me1, and H3K4Me3 histone marks. In detail, the acetylation of lysine 27 of the H3 histone protein, the monomethylation of lysine 4 of the H3 histone protein, and the trimethylation of lysine 4 of the H3 histone protein have been associated with enhanced transcription, enhancer, and active promoter, respectively. In addition, a regulatory region characterized by transcription factor binding sites is found upstream MIR193a (chr17:31558470-31559544), indicating that miR-193a coding sequence is localized in an active transcriptional region. The pre-miR-193a generates two mature miRNAs, miR-193a-3p and miR-193a-5p, depending on the arm that is processed during miRNA biogenesis (Figure 5.2(b)). Consequently, the different sequence that characterizes miR-193a-3p and miR-193a-5p determines distinct target sets for each miRNA. In this chapter, we focus on the miR-193a-3p biological and molecular mechanisms from both the physiological and the pathological perspectives.

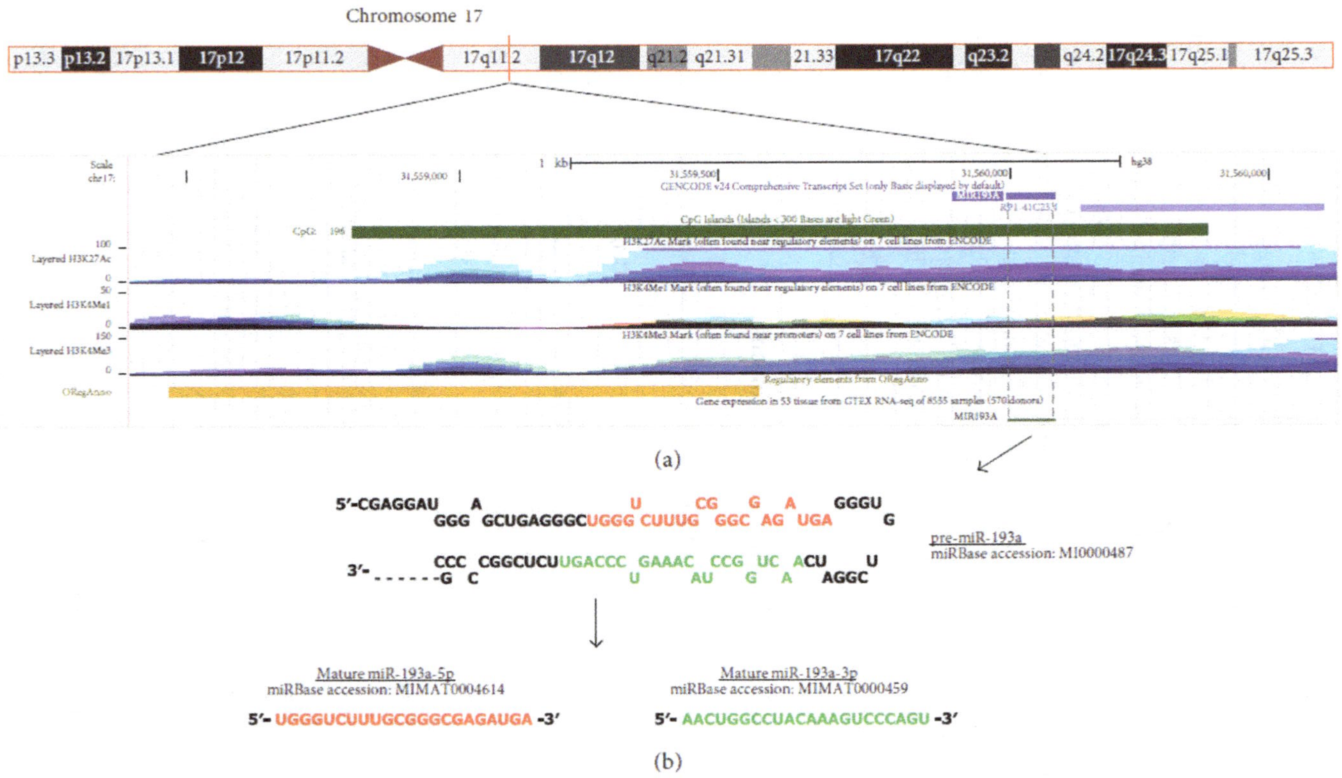

**FIGURE 5.2** Genomic location of miR-193a coding sequence, stem-loop hairpin structure of pre-miR-193a and miR-193a-3p/miR-193a-5p sequences. (a) The analysis of the genomic region coding miR-193a referred to Genome Browser (https://genome-euro.ucsc.edu/). MIR193a coding sequence is located on human chromosome 17q11.2 characterized by a typical CpG island (in *green*). The layered H3K27Ac, H3K4Me1, and H3K4Me3 show the levels of histone marks across the genome in seven cell lines (data obtain from ENCODE on the basis of ChiP-seq assay). By default, this track displays data from a number of cell lines in the same vertical space and each cell line is associated with a particular color. The regulatory element (in *orange*) is described as transcription factor binding sites by ORegAnno (open regulatory annotation). (b) The MIR193A gene is transcribed into a precursor (pre-miR-193a) with 88 nucleotides that is processed during miRNA biogenesis to yield mature miR-193a-5p (in *red*) and mature miR-193a-3p (in *green*) with 22 nucleotides in length.

Furthermore, the mechanisms involved in its expression regulation are also addressed. Finally, we highlight the aberrant expression of miR-193a-3p both at tissue and at circulating levels in several pathological conditions, including cancer, in order to offer novel insights in the role of miR-193a-3p as a diagnostic and prognostic biomarker.

## 5.4 THE REGULATION OF miR-193A-3P EXPRESSION

miR-193a-3p was upregulated in patients with glioma and could affect the invasion, migration, and MT of glioma by regulating BTRC [80]. miR-193a-3p serves as a promoter in the immune-inflammatory response via targeting LGR4 in LPS-primed endometritis [81]. According to this, we speculated that inhibition of miR-193a-3p could be a potential molecular target for endometritis treatment in the future (last update February 28, 2021). Several mechanisms, including transcription factors, DNA methylation, and competing endogenous RNAs (ceRNAs), have been reported to be involved in the dysregulation of miR-193a-3p

in pathological contexts (Table 5.2). These data unmistakably suggest a multifactorial regulation of miR-193a-3p at the transcriptional or posttranscriptional level with the possibility of a context-dependent activation of specific mechanisms.

### 5.4.1 TRANSCRIPTION FACTORS AND REGULATORY PROTEINS

Like protein-coding genes, the expression of miRNAs may be under the control of transcription factors (TFs) that bind to specific DNA sequences in the miR promoter and may act either as transcriptional activators or as repressors. The transcription silencing by specific TFs was reported to play a critical role in the inactivation of miR-193a-3p in different pathological contexts. Iliopoulos et al. showed that the downregulation of miR-193a-3p was driven by Max, the Myc-associated factor X, and RXR$\alpha$, a nuclear receptor, both involved in the processes causing cellular transformation of breast epithelial cells. Using chromatin immunoprecipitation (ChIP) and siRNA-mediated inhibition,

## TABLE 5.2
### Regulation of miR-193a-3p Expression by Different Mechanisms

| Mechanisms of Regulation | Effect on miR-193a-3p Expression | Sample Type | Experimental Procedures | Ref. |
|---|---|---|---|---|
| **Transcription Factors** | | | | |
| Max and RXR$\alpha$ | Downregulation | Transformed breast epithelial cells | ChIP; siRNA-mediated inhibition experiments | [82] |
| AML1/ETO | Downregulation | AML cell lines primary AML samples with t(8;12) | Luciferase reporter assay; ChIP | [83] |
| HNF4$\alpha$ | Upregulation | Liver from mice with liver-specific knockout of HNF4$\alpha$ | miRNA microarray and qPCR in Hnf4$\alpha$-LivKO mice | [84] |
| **DNA Methylation** | | | | |
| DNA hypermethylation | Downregulation | AML cell lines AML primary samples | MSP Bisulfite sequencing | [85] |
| DNA hypermethylation | Downregulation | OSC carcinoma cells and primary AML samples | COBRA Bisulfite sequencing | [86] |
| DNA hypermethylation | Downregulation | NSCL cancer cells NSCL specimens | MSP Bisulfite sequencing | [87, 88] |
| DNA hypermethylation | Downregulation | Highly metastatic osteosarcoma cells | Bisulfite sequencing | [89] |
| DNA hypomethylation | No effect | HCC cell lines HCC specimens | MSP | [90] |
| DNA hypomethylation | No effect | Mesothelioma | MSP | [91] |
| **Regulatory Protein** | | | | |
| XB130 | Downregulation | Thyroid carcinoma cells | shNA-mediated inhibition and ectopic expression experiments | [92] |
| **Competing Endogenous RNA (ceRNA) Network** | | | | |
| Linc00152 | Downregulation | Colon cancer cells | RIP; luciferase reporter assay | [93] |
| LncRNA-UCA1 | Downregulation | NSCL cells | RIP; luciferase reporter assay | [94] |

*Abbreviations:* AML, acute myeloid leukemia; COBRA, combined bisulfite restriction analysis; HCC, hepatocellular carcinoma; MSP, methylation-specific PCR; NSCL, non–small cell lung; OSC, oral squamous cell.

they demonstrated that Max and RXRα bound directly to the miR-193a regulatory region and repressed its transcription in ER-Src-transformed cells [82]. Similarly, Li et al. revealed that the downregulation of miR-193a-3p was strongly associated with fusion protein AML1/ETO expressed in hematopoietic cells isolated from patients affected by acute myeloid leukemia (AML) with t(8;21). In this pathological context, AML1/ETO acted as a transcriptional repressor by localizing the AML1 binding site on the MIR193a upstream region and recruiting histone deacetylases (HDACs) and DNA methyltransferases (DNMTs). The chromatin-remodeling complex formed by AML1/ETO and the DNA hypermethylation triggered the silencing of miR-193a-3p in t(8;21) AML [83].

Two other factors, hepatocyte nuclear factor α (HNF4α) and XB130, may also have a relevant role in the regulation of miR-193a-3p. HNF4α is a regulator of hepatic gene expression essential for liver development and function. The lacking of HNF4α expression in the liver of young adult mice (Hnf4α-LivKO) determined the downregulation of some miRs, including miR-193a that is in cluster with miR-365 on the chromosome 11 of *Mus musculus* [84]. XB130 is a member of the actin filament–associated protein (AFAP) family affecting the downstream signaling PI3k/Akt pathway by functioning as an adaptor protein and tyrosine kinase substrate [95]. In human thyroid carcinoma WRO cells and MRO cells, the gene silencing of XB130 by stable transfection of short hairpin increased both pri-miR-193a and its mature form (miR-193a-3p), while the ectopic expression of XB130 induced their downregulation [92]. These data indicated that the regulation of miR-193a-3p may be mediated by HNF4α and XB130 in a healthy liver and thyroid carcinoma, respectively. However, further studies based on ChIP and gene reporter assays are needed in order to examine the direct and specific mechanisms that link these factors to miR expression.

### 5.4.2 Epigenetic Regulation by DNA Methylation

The hypermethylation of CGIs located around miR genes is a key mechanism of epigenetic downmodulation of miRs that acts as a tumor suppressor in specific tumors. The MIR193A gene is embedded in a 1556 bp CGI that counts 196 CpG sites (Figure 5.2(a)). Several studies have found altered DNA methylation occurring in the CpG sites of the miR-193a promoter in different types of tumors. MiR-193a-3p was silenced in oral squamous cell carcinoma (OSCC) cell lines and in primary tumors through aberrant DNA methylation of the CpG sites near the miR coding sequence as verified by COBRA (combined bisulfite restriction analysis) assay and bisulfite sequencing [86]. Gao et al. demonstrated that the promoter hypermethylation repressed miR-193a-3p expression in acute myeloid leukemia (AML). They studied [96] the DNA methylation levels in several leukemia cell lines and bone marrow (BM) samples from AML patients and healthy donors by bisulfite sequencing and methylation-specific PCR (MSP). Treatment with the inhibitor of DNA

methylation 5-azacytidine (5-aza-dC) restored miR-193a-3p expression and reduced its target, the oncogene c-kit. In this situation, growth inhibition, the induction of apoptosis, and differentiation of AML cells were observed [85]. MiR-193a was also found tumor specifically methylated in patients with NSCLC [87]. Treatment with 5-aza-dC upregulated miR-193a-3p expression, impaired cell proliferation ability, and promoted apoptosis in NSCLC cells via downregulation of one of the miR-193a-3p targets, the anti-apoptotic myeloid leukemia cell sequence-1 (Mcl-1) [88]. Finally, Pu et al. found that both miR-193a-3p and miR-193a-5p were hypermethylated and downregulated in a metastatic osteosarcoma cell line [89].

Although altered DNA methylation levels have been associated with several types of tumor, this cannot be generalized. In malignant pleural mesothelioma (MPM), miR-193a-3p was inhibited when compared with normal pleura, but the DNA hypermethylation of miR-193a-associated CGI was not responsible for the inhibition of miR-193a-3p in MPM cells as verified by MSP [91]. Interestingly, the results obtained in human hepatocellular carcinoma (HCC) by Grossi et al. were in line with those in MPM. The authors demonstrated that the downmodulation of miR-193a-3p in HCC was not mediated by DNA methylation in a cohort of 30 matched peritumoral and HCC tissues from bioptic samples. However, the miR-193a-3p CpG sites methylated in the differentiated HepG2 cells and the treatment with 5-aza-dC led to miR-193a-3p increasing as observed also by Ma and colleagues [90, 97]. These results would point toward a variable miR-193a dependence on CpG DNA methylation in HCC. In conclusion, it has been proved that DNA hypermethylation of CGI associated to MIR193A gene was responsible for miR-193a-3p downmodulation in certain types of cancer, which in turn led to increased expression levels of miR-193a-3p targets involved in cell malignant behavior. In other types of cancer, DNA methylation may not contribute to the regulation of this miR.

### 5.4.3 Competing Endogenous RNA (ceRNA)

In recent years, the important results obtained through the application of high-throughput RNA-seq have shed a light on the complex landscape of long noncoding RNAs (lncRNAs). These are RNA, typically longer than 200 nt, without protein-coding potential. Several studies established the biological functions of lncRNAs comprising transcriptional and posttranscriptional regulation and chromatin modification. Furthermore, some lncRNAs have been characterized for their ability to regulate miRNA function by competing for miRNA binding and decreasing the negative effect of miRNAs on their targets. For this reason, these lncRNAs have been named as competing endogenous RNAs (ceRNAs) or miRNA sponges [98].

Consistent with the competing endogenous RNA role of lncRNAs, two different ceRNAs have been found to target miR-193a-3p, where the physical association with mature miR-193a-3p has been demonstrated by RNA immunoprecipitation (RIP) and luciferase reporter assays. In particular, oncogenic Linc00152 (long intergenic non-coding RNA 152) and LncRNA-UCA1 (urothelial carcinoma-associated 1) competitively bind miR-193a-3p in colon cancer and NSCLC cell lines, respectively [99, 100]. Interestingly, Linc00152 and UCA1 functioned as miRNA sponges and suppressed the endogenous effect of miR-193a-3p by silencing the miR target ERBB4. In addition, the overexpression of Linc00152 or UCA1 increased cell growth through modulation of ERBB4 while this effect was attenuated by transfection of miR-193a-3p mimics in both cell lines.

This evidence strongly suggests that the ceRNA regulatory network should be considered as a mechanism involved in the dysregulation of miR-193a-3p.

## 5.5 EXPRESSION PROFILE OF miR-193a IN NORMAL HUMAN TISSUE

To provide complete information on the global expression profile of miR-193a-3p in normal tissues, we referred to data deposited in Genome Browser. The data have been reported as the median gene expression levels in 51 tissues and 2 cell lines, based on RNA-seq data obtained from the NIH Genotype-Tissue Expression (GTEx) project [101, 102]. This release is based on data from 8555 tissue samples obtained postmortem of 570 adult individuals with no evidence of disease. As indicated in Figure 5.3, the expression of miR-193a (without discriminating between miR-193a-3p and miR-193a-5p) was detected in all tissues, with the exception of the bladder, some brain components (hippocampus, nucleus accumbens, and spinal cord), and the cervix (endocervix). Adipose and breast tissues displayed the highest miR-193a expression level.

## 5.6 BIOLOGICAL FUNCTION OF miR-193a-3p IN DEVELOPMENT AND CELL PHYSIOLOGY

Very little is known about the biological function of miR-193a-3p in cell physiology. To the best of our knowledge, there have not been reports on the role of this miR in development. Concerning its role in cell physiology, the main available data were obtained from studies on the following: (a) from cord blood and peripheral blood endothelial colony-forming cells (CB/PB-ECFC) derived from donations of healthy subjects; (b) from skeletal muscle specimens of control subjects (CTRL) with no sign of muscle pathology detectable by immunohistochemistry; and (c) from endometrial epithelial cells of healthy volunteer women aged 18 through 36 years old. In particular, it has been found that miR-193a-3p was one of the 25 miRNAs differentially regulated in CB-ECFC versus PB-ECFC [103]. It was highly expressed in the less proliferative PB-ECFCs where its inhibition using anti-miR molecules improved the in vitro proliferation, migration, and vascular tubule formation of

these cells. Conversely, miR-193a-3p was expressed in low amounts in the proliferative CB-ECFCs with its in vitro ectopic overexpression limiting the proliferation and the cell cycle progression of these cells and consequent reduction of their vascular tubule formation and cell migration. Altogether, the data obtained by Khoo et al., by using the miRnome studies, in silico miRNA target database analyses combined with proteome arrays and luciferase reporter assays in miR mimic-treated ECFCs, allowed the identification of the negative regulatory role of miR-193a-3p in the vascular function of these cells and in the proliferation and migration abilities of these cells via directly targeting HMGB1 expression. Thus, this miR has a regulatory role in cell physiology of ECFCs that are considered circulating endothelial lineage progenitors. Targeting the miR itself by anti-miR molecules may improve the abilities of PB-ECFC cells in proliferation and migration, in their angiogenic function thus contributing to a positive clinical outcome in ischemic diseases (stroke, myocardial infarction, and limb ischemia).

For skeletal muscle, a miRNA profiling approach combined with bioinformatics analyses and qPCR experimental validation has identified 11 miRNAs including miR-193a-3p involved in the homeostasis of normal myofibers. In particular, downregulation of miR-193a-3p has been associated with events contributing to myofiber alterations of patients with myotonic dystrophy type 2 (DM2, OMIM 602688) [104]. DM2 is an autosomal dominant multisystemic disorder affecting the skeletal muscles, the heart, the eye, the central nervous system, and the endocrine system. The findings reported by Greco et al. clearly showed that the level of miR-193a-3p downmodulation contributed to the DM2 miRNA score, allowing distinguishing the muscle specimens of DM2 patients from those of controls. However, these results do not lead to a hypothesis on the functional role in normal myofibers of miR-193a-3p as well as those of the other 10 miRNAs deregulated in DM2 muscle biopsies.

Finally regarding the human endometrium, several data suggested a regulatory function of miRNAs during the physiological cycle phases. In particular, Kuokkanen et al. provided strong evidence of the hormonal regulation of miR-193a-3p expression in isolated uterine epithelial cells derived from mid-reproductive-aged women [105]. The study examined [96] miRNAs at two stages: (a) in uterine epithelial cells derived from late proliferative phase biopsies (cycle day, CD, 12 ± 1) to target the time of maximal endometrial response to female steroid hormone estradiol-17 beta (E2), and (b) from secretory biopsy specimens from midluteal phase on CDs 19 through 23 to target the endometrial window of receptivity and maximum P4 (progesterone) action. The findings obtained using a genomic profiling of miRNAs and mRNAs clearly showed that miR-193a-3p is one of the 12 miRs found to be upregulated in the midsecretory phase samples. This expression is suggestive of a role in downregulating some cell cycle genes in the secretory phase thereby suppressing proliferation of the endometrial epithelial cells in this specific physiological

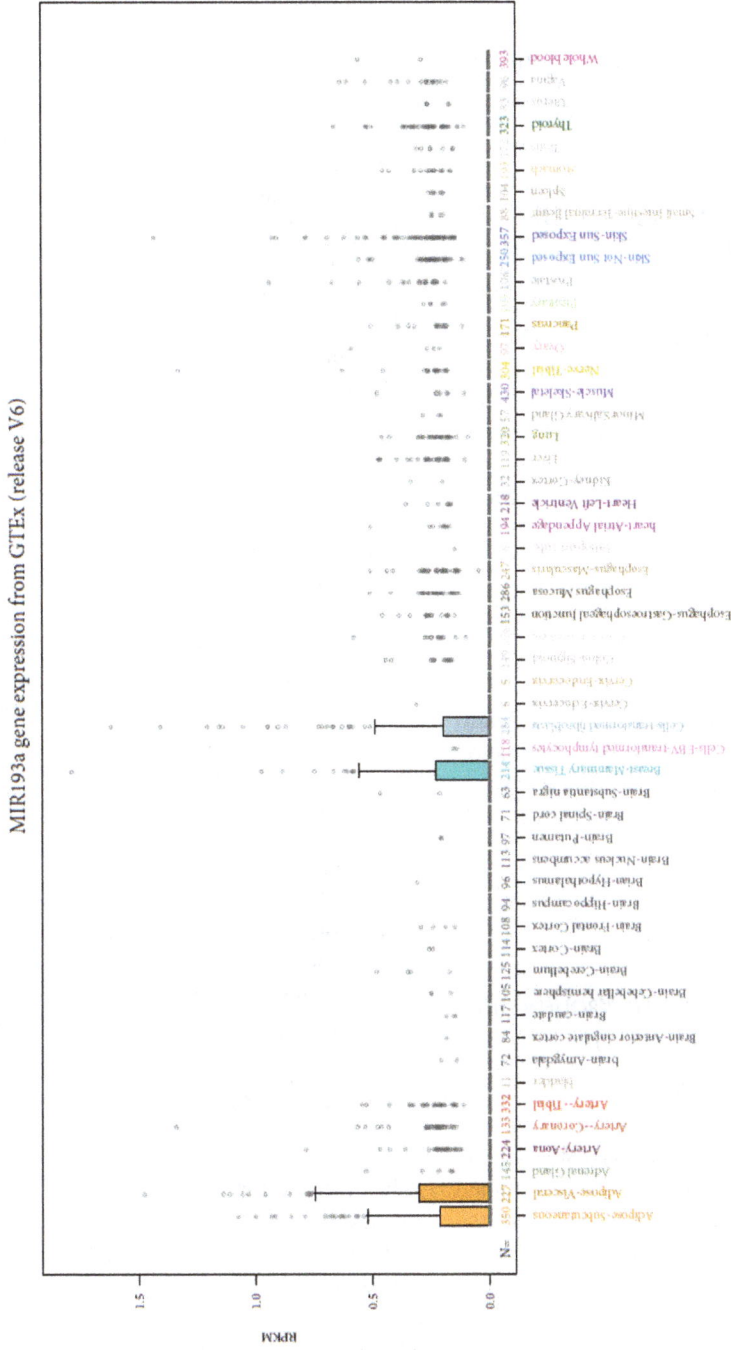

**FIGURE 5.3** Expression profile of miR-193a in normal tissues. The expression level of miR-193a is reported for 51 tissues and two cell lines (EBV-transformed lymphocytes and transformed fibroblasts) and is referred to a GTEx project collected in Genome Browser. Expression values are shown in RPKM (reads per kilobase of transcript per million mapped reads). The height of each bar represents the median expression level across all samples for a tissue, and points are outliers if they are above or below 1.5 times the interquartile range. Each color represents a specific tissue, conformed to GTEx consortium publication convention.

context. Further, these data demonstrate hormonal regulation in miRNA (i.e., miR-193a-3p) expression in a human endometrium.

In summary, the overexpression of miR-193a-3p in cultured normal cells derived from physiological contexts, in particular in PB-ECFC cells and in midsecretory uterine epithelial cells, seemed to limit cell proliferation and cell cycle progression. However, the lack of data (KO, KI, and conditional KD) in the development of an organism (i.e., *Mus musculus* and *Danio rerio*) carrying the ortholog MIR193A (Gene Card data) has prevented the proposal of a hypothesis on the role of miR-193a-3p during development [106].

## 5.7 miR-193a-3p FUNCTIONS AS TUMOR SUPPRESSOR IN CANCER

It is widely documented that the aberrant expression of miRNAs has a critical impact on cell biological processes and contributes to a number of pathological conditions, such as cancer. To the best of our knowledge, all published data point toward a role of miR-193a-3p as tumor suppressor miRNA (ts-miRNA) in both solid and liquid cancers since it impaired tumor cell–aggressive properties by targeting oncogenes. In addition, miR-193a-3p is found downregulated in transformed cells and its downregulation seemed to be required for cellular transformation in two isogenic models (breast epithelial cells and fibroblasts) [82]. By considering a small cohort of cancer patients (only 36 cases), Yi et al. found that miR-193a-3p is upregulated in 24/36 esophageal squamous cell carcinoma (ESCC) tissues compared with adjacent normal tissues and the downregulation of miR-193a-3p by a synthesized inhibitor decreases migration and proliferation and promotes apoptosis in ESCC cells [107]. For these reasons, they described miR-193a-3p as an oncogenic miRNA in ESCC and suggested further studies to define the controversial role of miR-193a-3p in ESCC.

### 5.7.1 MiR-193a-3p Limits Cancer Cell Proliferation and Impairs Cell Cycle Progression

Many studies have confirmed that miR-193a-3p has a significant role in the regulation of cancer cell growth. In particular, miR-193a-3p directly targeted JNK-1, a tyrosine kinase, because the ectopic expression of miR-193a-3p determined the dysregulation of cell cycle components including the decrease of CDK4, PIK3CA, and cyclin D1 and the overexpression of p27. In association with miR-124 and miR-147, miR-193a-3p has been shown to co-regulate and inhibit G1/S transition and proliferation in breast cancer and glioblastoma cell lines [108]. Moreover, miR-193a-3p repressed cell proliferation of AML cells through the inhibition expression of c-kit, an oncogene encoding a transmembrane glycoprotein belonging to the type III receptor tyrosine kinase family [85]. In the same clinical context, miR-193a-3p has also been found to directly regulate the expression of DNMT3a, HDAC3, and cyclin D1, consequently blocking

the cell cycle progression during granulopoiesis and inducing the differentiation of myeloid precursors [83]. It has also been shown that miR-193a-3p decreased the abilities of proliferation by inhibiting the expression of some TFs, including E2F6, and other genes involved in the growth of several cancer types, for example, K-Ras, ERBB4, and cyclin D1 [82, 86, 109, 110]. In particular, it has been demonstrated that miR-193a-3p negatively regulated K-Ras in lung cancer cells by binding two 3′UTR sites that have not been reported previously to be mutated in cancer. The overexpression of miR-193a-3p not only downregulated K-Ras but also reverted the whole protein signature associated with the signaling downstream of K-Ras identified by proteomic analysis of lung cancer samples. The authors [96] clearly determined the effects of miR-193a-3p on the cell-aggressive properties via the targeting of K-Ras. Interestingly, miR-193a-3p decreased cell cycle progression (G1-S) and cell proliferation in vitro and blocked colony formation in three-dimensional (3D) cultures. These findings have been translated into exciting ex vivo and in vivo experiments. For ex vivo experiments, the authors harvested lung–heart blocks from Sprague Dawley rats. To create a metastatic ex vivo four-dimensional (4D) lung model, the lungs were decellularized and placed in a bioreactor with an oxygenator and a pump with the right main stem ligated with silk suture, and A549 lung adenocarcinoma human epithelial cells were seeded in the left lung through the tracheal cannula. The use of 1,2-dioleoyl-sn-glycero-3-phosphocholine (DOPC) nanoliposomes to deliver miR-193a-3p reduced the number of viable cells and impaired the presence of cancer cells in the contralateral lobe, which is indicative of metastasis formation in this model. In addition, this ex vivo model allowed the collection of circulating tumor cells (CTCs) from the perfused cell media present in a bioreactor bottle. Interestingly, cells derived from a 4D model treated with miR-193a-3p showed less proliferation ability than those from an untreated model. In orthotopic xenograft K-Ras-mutated lung tumor models, miR-193a-3p encapsulated in DOPC nanoliposomes showed a reduction of tumor growth and metastasis at various sites [111].

### 5.7.2 MiR-193a-3p Induces Cell Death Mainly by Promoting Apoptosis

When considering the miRNA target genes validated in different cancers, the role of miR-193a-3p in affecting genes with the consequence of promoting apoptosis stood out. In fact, among the putative targets, Mcl-1 is the most validated one involved in programmed cell death. Mcl-1 is a multidomain protein belonging to the Bcl-2 family that binds and sequesters the BH3-only pro-apoptotic Bcl-2 family members (Bim, Bid, Bik, Noxa, and Puma), which in turn induce Bak and Bax homo-oligomerization and activation. Kwon et al. demonstrated that miR-193a-3p expression was induced by ionizing irradiation in U-251 glioma cells and HeLa cells. MiR-193a-3p negatively regulated Mcl-1 and promoted apoptosis by inducing ROS accumulation and DNA damage [114]. The direct binding of miR-193a-3p and Mcl-1 has also been demonstrated in human ovarian cancer cell lines where the overexpression of miR-193a-3p induced the activation of caspase 3/7 and resulted in apoptotic cell death [113]. Furthermore, the transfection with miR-193a-3p mimics in MPM cells reduced Mcl-1 protein level and increased the number of late apoptotic cells. In addition, the release of lactate dehydrogenase (LDH) from MPM cells transfected with miR-193a-3p could suggest that miR-193a-3p induced cell death at least in part by the induction of necrosis. The ability of miR-193a-3p to promote apoptosis was further demonstrated in MPM xenograft models when targeted by miR mimics delivered using EDV nanocells, bacterially derived minicells that can be packaged with a variety of cargoes and be delivered to tumors via bispecific antibodies attached on the surface [91]. Interestingly, Salvi et al. showed that the combination of miR-193a-3p mimics and sorafenib had additional effects on HCC inhibition of cell proliferation and induction of apoptosis suggesting that miR-193a-3p could also play an important role in promoting the sensitivity to sorafenib [114], the only innovative drug used for advanced HCC.

### 5.7.3 MiR-193a-3p Impairs Cancer Migration, Invasion, and Metastasis

Tumor cell invasion and metastasis are events of primary importance in the prognosis of cancer patients. Metastatic cells are able to invade the basal membrane (BM) and extracellular matrix (ECM) to penetrate and move into the lymphatic or vascular circulation and to produce a secondary tumor by the extravasation process and subsequent cell proliferation. Several studies have highlighted the roles of miRNAs in these complex processes.

Recently, it has been demonstrated that miR-193a-3p acted as a negative regulator of urokinase-type plasminogen activator (uPA) in breast cancer and HCC cell lines, and the high expression of miR-193a by mimics transfection strongly inhibited uPA expression and decreased cell aggressive properties [82, 114, 115]. uPA is a serine protease that converts the proenzyme plasminogen into the serine protease plasmin, thus making malignant cancer cells able to degrade BM and ECM. Furthermore, the interaction with its receptor, uPAR, leads to the activation of different intracellular signaling pathways, altering cell proliferation and migration abilities and expression of specific genes. The essential role played by uPA in migration has been well characterized in pathological contexts like cancer, and its overexpression was detected in various tumors, at both the mRNA and protein level, representing an unfavorable prognostic factor [116–118].

Recently, Yu et al. validated miR-193a-3p as a negative regulator of ErbB4 belonging to the ErbB family of tyrosine kinase receptors and the ribosomal protein S6K2, both playing a critical role in cell movement, growth, and development [119]. The expression of miR-193a-3p and miR-193a-5p was positively associated with cellular invasion and migration

by assessing human lung cancer cell with high metastatic potential (SPC-A-1sci) previously established from weakly metastatic cell (SPC-A-1) through in vivo selection in NOD/SCID mice [120]. Furthermore, the overexpression of miR-193a-3p inhibited migration, invasion, and epithelial mesenchymal transition in vitro and impaired the formation of metastasis in vivo. In addition, the protein profile of SPC-A-1sci cells stably transfected with miR-193a-3p has been determined by using a proteomic approach (iTRAQ and Nano LC-MS/MS) followed by DAVID (database for annotation, visualization, and integrated discovery; http://david.abcc.ncifcrf.gov) and STRING analysis. Interestingly, 112 proteins were differentially expressed (62 upregulated and 50 downregulated) compared with miR control transfected cells, and some of them have been associated to lung cancer metastasis and proliferation [121].

Similarly, Pu et al. reported that miR-193a-3p and miR-193a-5p were downregulated in osteosarcoma cells defined as highly tumorigenic and metastatic (MG63.2) with respect to the less metastatic parental MG63 cell line. The ability of MG63.2 cells to invade and migrate was decreased by restoring the miR-193a-3p expression level using transient transfection of miR mimics. Correspondingly, the inhibition of miR-193a-3p by antagomir transfection in MG63 cells induced invasive properties. Furthermore, the authors identified Rab27, a member of the RabGTPase family, as a direct target of miR-193a-3p. By Cignal reporter finder assay, they showed that the Rab27 knockdown repressed some pathways that were clearly implicated in metastasis, including TGFβ, Myc/Max, and ATF2/ATF3/ATF4. These results indicated the negative impact of miR-193a-3p on cancer invasion by repressing Rab27B and its downstream pathways in osteosarcoma cells [89].

Taken together, these data strongly suggest a potential role of miR-193a-3p as a metastasis-preventing miRNA. However, further work will be required to explore the exact molecular pathways by which this miRNA could exert its functions.

### 5.7.4 MiR-193a-3p Modulates drug Resistance in Cancer Cells

To date, chemotherapy still represents one of the most-used therapeutic options for the treatment of solid tumors worldwide. However, the clinical efficacy of these treatments is limited by the onset of drug resistance and the side effects of the drug both contributing to reduce cancer patients' positive outcomes. Even if the specific regulatory mechanism involved in chemoresistance remains very often unclear, increasing data show that miRNAs can have a crucial role in chemosensitivity by regulating cancer-related genes. As indicated in Figure 5.4(a), miR-193a-3p seems to be implicated in the activation of drug resistance pathways via repressing different targets. By a systematic analysis that compared the H-bc multidrug-resistant bladder cancer (BCa) cells versus 5637 sensitive ones, Lv et al. reported that miR-193a-3p was silenced by DNA hypermethylation

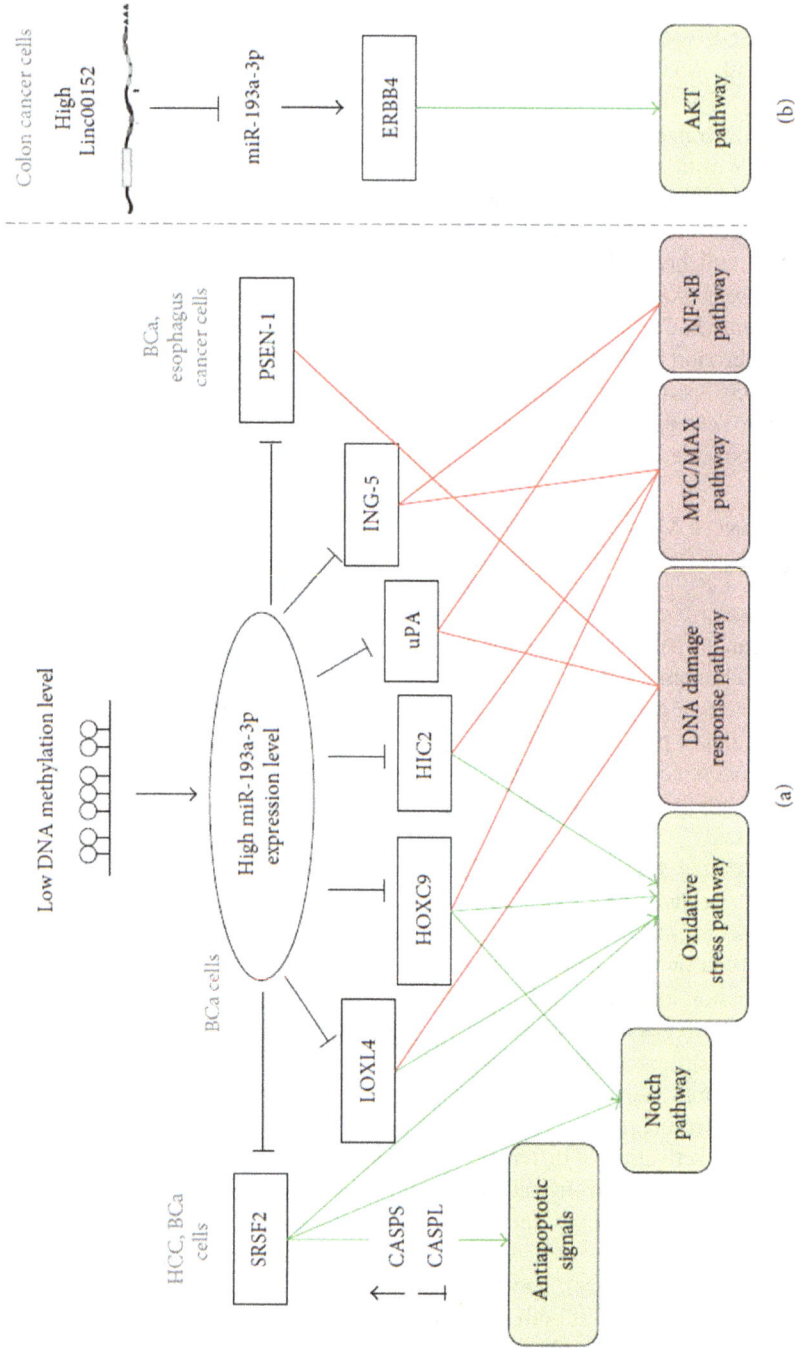

**FIGURE 5.4** Proposed model for the role of miR-193a-3p in the regulation of chemoresistance in different cancer cell lines. (a) Involvement of high miR-193a-3p expression level in chemoresistance in HCC cells, BCa cells, and esophageal cancer cells. (b) Low levels of miR-193a-3p are involved in resistance to oxaliplatin in colon cancer cells. Altered expression of miR-193a-3p will adversely affect immediate targets indicated inside *black boxes*. In turn, these targets will affect several downstream pathways with functional activation (*green arrows*) and repression (*red lines*).

in the sensitive cells, the cell line presenting the lowest $IC_{50}$ to different drugs. In chemoresistant BCa cells, miR-193a-3p decreased the expression of the following targets: SRSF2, PLAU, HIC2 [122], LOXL4 [123], ING-5 [124], HOXC9 [125], and PSEN-1 [126]. The last target has been considered relevant in chemoresistant and radioresistant esophageal cancer cells [127]. The results obtained through pathway reporter system assays revealed that the reduced level of these targets affected the activities of five signaling pathways in resistant cells. In particular, DNA damage, NF-κB, and Myc/Max pathways were found with lower activities, while Notch and oxidative stress pathways were activated in resistant cells compared with sensitive cells. In addition, the modulation of miR-193a-3p level by the injection of antagomir or agomir molecules in either resistant or sensitive BCa cells reversed the chemoresistance in tumor xenografts nude mice. Similarly, miR-193a-3p expression was found increased by DNA hypomethylation in HCC cells presenting resistance to 5-fluorouracil (5-Fu). MiR-193a-3p seemed to induce anti-apoptotic signals in 5-Fu resistant cells by suppressing SRSF2, a splicing factor that preferentially upregulates the pro-apoptotic form of caspase 2 (CASP2L) to the anti-apoptotic form CASP2S [97]. These intriguing data will require further investigation because it is not clear how the overexpression of miR-193a-3p can dictate chemoresistance even if its tumor suppressor functions have been well established in many primary cancers. In support of this notion, Yue et al. reported that the reduced activity of miR-193a-3p caused by the sponge effect of Linc00152 was related to oxaliplatin (L-OHP) resistance in colon cancer both in vitro and in vivo (Figure 5.4(b)). Linc00152 is usually overexpressed in human colon cancer tissues and is associated with poor prognosis in patients undergoing L-OHP treatment after surgery. Interestingly, Linc00152 competitively bound miR-193a-3p inducing the upregulation of its ERBB4 target and the consequent activation of AKT signaling pathway which, in turn, conferred resistance [99].

## 5.8 GENE ANNOTATION ANALYSIS ON PREDICTED AND EXPERIMENTALLY VALIDATED MIR-193a-3p TARGETS

It is well known that a given miR can regulate the expression of hundreds of targets, and conversely, dozens of miRs can target a single mRNA. For hsa-miR-193a-3p, TargetScan 7.1 predicted 293 putative target genes with 168 displaying the highest score (cumulative weighted context++ score < −0.24). The functional annotation analysis conducted using DAVID 6.8 on the candidate target genes of miR-193a-3p highlighted eight biological KEGG pathways overrepresented as statistically significant (p<0.05). They were the following: microRNA in cancer, ErbB signaling pathway, GnRH signaling pathway, acute myeloid leukemia, PI3K-Akt signaling pathway, Ras signaling pathway, pathways in cancer, and focal adhesion. The importance of these

pathways is indicated by the fact that they are highly relevant in the processes of onset, progression, and metastasis of several types of cancer. In addition, among the 24 experimentally validated and published miR-193a-3p targets, the KEGG pathway enrichment analysis outlines six terms (p<0.05): acute myeloid leukemia, chronic myeloid leukemia, microRNA in cancer, proteoglycans in cancer, Erb signaling pathway, pancreatic cancer, and pathways in cancer. This bioinformatics analysis underscored once again the key role of miR-193a-3p in cancer contexts.

## 5.9 miR-193a-3p AS DIAGNOSTIC AND PROGNOSTIC BIOMARKER

It is well known that aberrant miRNA expression has a critical impact on many cell biological processes and contributes to a number of pathological conditions. For this reason, the study of miRNA expression profile in pathological contexts is necessary to support the clinical significance of a specific miRNA and its possible role as a diagnostic and prognostic biomarker.

### 5.9.1 DYSREGULATION OF MIR-193A-3P IN CANCER TISSUES

Consistent with the role of ts-miRNA, miR-193a-3p was found downregulated in the majority of primary cancer tissues, such as HCC [90, 128, 129], NSCLC [119, 130], MPM [91], and AML [83, 85]. Furthermore, the low expression of miR-193a-3p was significantly related to reduced overall survival (OS) and disease-free survival (DFS) of HCC patients indicating its possible prognostic role in this cancer type [90]. In NSCLC, the expression of miR-193a-3p was negatively correlated to tumor size, lymph node metastasis (LNM), and TNM stages. Interestingly, miR-193a-3p was reported as downregulated in BRAF mutation with respect to wild-type melanoma [131], suggesting that miR-193a-3p may have a role in BRAF-associated events.

### 5.9.2 CIRCULATING MIR-193A-3P LEVELS IN PATHOLOGICAL CONDITIONS

Several data have determined that circulating microRNAs in human bodily fluids (i.e., serum and plasma) offer unique opportunities as biomarkers for early diagnosis of clinical conditions [132]. Indeed, some small ncRNAs have been found highly stable under extreme conditions (i.e., extreme pH and temperature and ribonuclease digestion) and numerous studies of circulating microRNA profiling have been conducted for several diseases. In regard to circulating miR-193a-3p, the use of next-generation sequencing and qPCR revealed different levels of this miR in patients with schizophrenia or with Parkinson's disease (PD) when compared with control subjects. Regarding PD, serum miRNA level obtained from a small number of patients revealed that miR-193a-3p was among the panel of four miRNAs significantly decreased in the PD patients compared to controls. Furthermore, miR-193a-3p could also be used to distinguish the HY-stage1 in PD patients from healthy controls [133]. In schizophrenia, the higher level of plasma miR-193a-3p (and miR-130b) in patients compared to controls was determined by global plasma miRNA profiling in a test cohort of 164 schizophrenia patients and 187 control subjects and subsequently validated by qRT-PCR in an independent cohort of 400 schizophrenia patients [134]. We think that these findings are extremely interesting, but further studies are needed in order to support the detection of circulating miR-193a-3p as a noninvasive biomarker for these diseases.

Regarding cancer, circulating miR-193a-3p levels were found increased in many malignancies. By comparing two independent miRNA microarrays, one in tissue and one in blood of colorectal cancer patients, Yong et al. identified higher levels of miR-193a-3p (in combination to miR-23a and miR-338-5p) in cancer patients, and the positive correlation was demonstrated between tissue and blood samples [135]. The high-throughput TaqMan low-density array (TLDA) combined with qPCR validation allowed the establishment of the high level of miR-193a-3p included in two different five-serum miRNA panels, either in renal cell carcinoma (RCC) or in NSCLC patients. It was clearly demonstrated by ROC analysis that the 5-miRNA-based panels (miR-193a-3p, miR-362, miR-572, miR-425-5p, and miR-543) had a high sensitivity and specificity in the discrimination of patients with early-stage RCC compared with healthy controls [136]. In NSCLC, the effectiveness of the 5-miRNA panel (miR-193a-3p, miR-483-5p, miR-214, miR-25, and miR-7) in discriminating cancer patients from normal subjects was confirmed in a multiethnic, multicenter study in which 438 participants from both China and America were enrolled (221 NSCLC patients, 161 normal controls, and 56 benign nodules) [137]. By using the same experimental approaches (TLDA followed by qPCR), Wu et al. identified significantly elevated levels of miR-193a-3p in sera from patients with esophageal squamous cell carcinoma (ESCC). The authors [96] indicated that miR-193a-3p may be used to discriminate between ESCC cases and healthy controls with high sensitivity and specificity in a cohort of 63 patients and 63 controls. The level of circulating miR-193a-3p was reduced after ESCC surgical removal, indicating that this miR may have been originally secreted by the tumor cells [138]. Interestingly, in a retrospective longitudinal phase 3 biomarker study, a set of five serum miRNAs (miR-193a-3p, miR-369-5p, miR-672, miR-429, and let-7i*) was identified as a specific biomarker for the surveillance and preclinical screening of HCC in a high-risk population of patients infected by hepatitis B virus (HBV). In particular, the different expression levels of these miRNAs (including the downregulation of miR-193a-3p) were identified in HBV patients who developed HCC (preclinical HCC patients) compared to the HBV group that did not develop HCC [139].

Although the origin of circulating miRNAs remains unclear, it has been reported that they may originate through different pathways including passive leakage from broken cells, active secretion via microvesicles, and active secretion through an RNA-binding protein-dependent pathway that has been suggested as the major source of circulating miRNAs [140, 141].

Extracellular vesicles (EVs) are mediators of intercellular communications during several physiopathological processes, such as differentiation, tissue repair, proliferation, and apoptosis, and they are released from both cancer cells and noncancer cells. Among the EVs, exosomes are small vesicles (50–150 nm) able to transport and deliver proteins, mRNAs, and ncRNAs including miRs from a donor to recipient cells [142]. Oh et al. demonstrated that the exosomes containing miR-193a-3p were able to induce differentiation of F11 cells (rat dorsal root ganglion and mouse neuroblastoma hybrid cells). By a microfluidic assay that collected real-time images of exosome migration, they verified that miR-193a-3p was a neurogenic miR that promoted the differentiation of recipient undifferentiated cells [143]. Teng et al. demonstrated that miR-193a-3p was present in exosomes obtained from tissue and cell culture media and serum, derived from primary mouse colon tumors and human liver metastasis of colon cancer. In severe disease, the high level of miR-193a-3p in the exosomes led to the reduction of cytoplasmatic miR-193a-3p, which in turn promoted the progression of premetastatic cells to metastatic ones. The authors found that overexpression of major vault protein (MVP) transported miR-193a-3p from the tumor cells to exosomes. On the contrary, MVP knockout determined miR-193a-3p accumulation in tumor cells triggering the inhibition of cell proliferation and cell cycle G1 arrest due to miR-193a-3p binding to its target, caprin-1 [144].

## 5.10  CONCLUSIONS AND PERSPECTIVES

In the last decade, an increasing number of studies have shown the biological importance of miRNAs in physiological contexts and a huge number of studies [145–148] have pointed to the fact that dysregulation of miRNAs plays a fundamental role in several pathological conditions, including cancer. There are many studies reporting the involvement of noncoding RNAs in cancer progression and development, such as siRNAs and miRNAs, but piRNAs have only recently been identified as new prognostic and diagnostic tools. Investigation into the role of piRNAs in the establishment of transcriptional patterns—and their involvement in key processes in germline and epigenetic, genetic, and, recently, somatic cells—is a rapidly growing field of research.

In the present chapter, we focused on miR-193a-3p because several findings support its role as a tumor suppressor miR both in solid and in liquid tumors leading researchers to believe that the detection of this miR at the tissue and/or circulating level may be employed as a diagnostic and prognostic biomarker for certain types of tumors. The functional role of a given miR can be tissue- or tumor-type dependent, and few data indicate a possible oncogenic role of miR-193a-3p in ESCC; nevertheless, we think that only strong clinical evidence, as well as biological studies of the miR mimics on the proliferation of ESCC cell lines, will elucidate this possible role of miR-193a-3p in the specific context of human cancer.

From a general point of view, miR-193a-3p has been studied in *Homo sapiens* and, to the best of our knowledge, no information is available in development. Regarding its biological function in physiological contexts, certain types of human normal cells seem to require high expression levels of miR-193a-3p when they do not need to proliferate. As a consequence, the cancer cells that usually need high proliferative capacity show low expression levels of this miR.

Different factors can contribute to the regulation of MIR193a expression, including TFs, DNA methylation, and, at the posttranscriptional level, ceRNAs. The alteration of these factors is context dependent and determines the aberrant expression of miR-193a-3p in cancer. The elucidation of these mechanisms may allow extending our knowledge on the level of miR-193a-3p dysregulation.

As already mentioned, the dysregulation of a given microRNA may alter the expression of hundreds of genes in cancer, affecting the entire network in which targets are involved. By considering all experimental validation studies in cancer, genes targeted by miR-193a-3p are involved in several biological processes, including proliferation, apoptosis, migration, and metastasis. To acquire major advancements in knowledge and comprehension of the canonical and noncanonical mRNA targets, more studies involving the use of proteomic profiling and RNA pull-down with biotinylated microRNA mimics are needed.

To date, many clinical trials have demonstrated the use of miRNA-based therapy as a promising strategy for the treatment of different diseases, making miRNA highly relevant for clinical use [149]. In this regard, the delivery of miR-193a-3p mimics by nano-sized particles could represent a novel therapeutic tool for the treatment of cancer because it may hamper tumor aggressive properties in tumor xenograft models by restoring original levels of the miR. On the other hand, the local delivery of anti-miR-193a-3p molecules could be an effective intervention for local ischemic diseases. These findings may pave the way to further studies aimed to elucidate the possible use of miR-193a-3p for experimental therapeutic procedures.

Understanding piRNA complex interactions will provide many novel interventions either in medical discovery, for clinical practice, or for biological uses, improving our understanding and management of cancer. The above-mentioned studies indicate that investigations focus on exploration of the importance of piRNAs in cancer. Therefore, this research in the future is likely to result in better treatment strategies based on these piRNAs interactions.

Finally, compelling evidence indicates that miR-193a-3p is detectable not only in primary cancer tissues but also at the circulating level (in exosomes or not) in cancer patients indicating the possible diagnostic and prognostic value of miR-193a-3p. In addition, altered circulating levels of this miR have been identified in subjects affected by PD or schizophrenia. Further studies are necessary to verify whether the detection of miR-193a-3p may be helpful for the characterization of these two diseases.

## REFERENCES

[1] American Cancer Society, Global Cancer Facts & Figs, 3rd ed, American Cancer Society, 2015.

[2] V. Marmari, H. Mahmoodzadeh, H. Dana, et al., "In silico analysis, cloning and expression of recombinant CD166 in E. coli BL21 (DE3) as a marker for detection and treatment of colorectal cancer," J Med Microb Diagn, vol. 6, p. 249, 2017.

[3] N. Mourouti, D. B. Panagiotakos, E. A. Kotteas, et al., "Optimizing diet and nutrition for cancer survivors: a review," Maturitas, vol. 105, pp. 33–36, 2017.

[4] H. Dana, A. Mazraeh, G. M. Chalbatani, et al., "Cloning and expression of C2 and V domains of ALCAM protein in E. coli BL21 (DE3)," Clin Microbiol, vol. 6, p. 271, 2017. http://dx.doi.org/10.4172/2327-5073.1000271

[5] J. C. van Wolfswinkel, R. F. Ketting, "The role of small non-coding RNAs in genome stability and chromatin organization," J Cell Sci, vol. 123, p. 1825e1839, 2010.

[6] H. Dana, G.M. Chalbatani, H. Mahmoodzadeh, et al., "Molecular mechanisms and biological functions of siRNA," Int J Biomed Sci: IJBS, vol. 13, no. 2, pp. 48–57, 2017.

[7] G. Mahmoodi Chalbatani, H. Mahmoodzadeh, E. Gharagozlou, et al., "Microrna a new gate in cancer and human disease," J Biol Sci, vol. 17, pp. 247–254, 2017.

[8] O. V. Klimenko, "Small non-coding RNAs as regulators of structural evolution and carcinogenesis," Non-coding RNA Res, vol. 2, p. 88e92, 2017.

[9] M. Ghildiyal, P. D. Zamore, "Small silencing RNAs: an expanding universe," Nat Rev Genet, vol. 10, pp. 94–108, 2009.

[10] H. Siomi, M. C. Siomi, "On the road to reading the RNA-interference code," Nature, vol. 457, pp. 396–404, 2009.

[11] E. M. Weick, E. A. Miska, "piRNAs: from biogenesis to function," Development, vol. 141, no. 18, pp. 3458–3471, 2014.

[12] M. A. Carmell, "The argonaute family: tentacles that reach into RNAi, developmental control, stem cell maintenance, and tumorigenesis," Genes Dev, vol. 16, pp. 2733–2742, 2002.

[13] K. Sato, M. C. Siomi, "Piwi-interacting RNAs: biological functions and biogenesis," Essays Biochem, vol. 54, pp. 39–52, 2013.

[14] D. Moazed, "Small RNAs in transcriptional gene silencing and genome defence," Nature, vol. 457, pp. 413–420, 2009.

[15] M. C. Siomi, K. Sato, D. Pezic, et al., "PIWI-interacting small RNAs: the vanguard of genome defence," Nat Rev Mol Cell Biol, vol. 12, pp. 246–258, 2011.

[16] Y. W. Iwasaki, M. C. Siomi, H. Siomi, "PIWI-interacting RNA: its biogenesis and functions," Annu Rev Biochem, vol. 84, pp. 405–433, 2015.

[17] H.-Y. Ku, H. Lin, "PIWI proteins and their interactors in piRNA biogenesis, germline development and gene expression," Natl Sci Rev, vol. 2, pp. 205–218, 2014. DOI: 10.1093/nsr/nwu014

[18] H. Yin, H. Lin, "An epigenetic activation role of Piwi and a Piwi-associated piRNA in Drosophila melanogaster," Nature, vol. 450, pp. 304–308, 2007.

[19] X. A. Huang, H. Yin, S. Sweeney, et al., "A major epigenetic programming mechanism guided by piRNAs," Dev Cell, vol. 24, pp. 502–516, 2013.

[20] P. Rajasethupathy, I. Antonov, R. Sheridan, et al., "A role for neuronal piRNAs in the epigenetic control of memory-related synaptic plasticity," Cell, vol. 149, pp. 693–707, 2012.

[21] T. Watanabe, S.-i. Tomizawa, K. Mitsuya, et al., "Role for piRNAs and noncoding RNA in de novo DNA methylation of the imprinted mouse Rasgrf1 locus," Science, vol. 332, pp. 848–852, 2011.

[22] D. I. Jacobs, Q. Qin, M. C. Lerro, et al., "PIWI-interacting RNAs in gliomagenesis: evidence from Post-GWAS and functional analyses," Cancer Epidemiol Biomark Prev, vol. 25, no. 7, pp. 1073–1080, 2016.

[23] H. Ishizu, H. Siomi, M.C. Siomi, "Biology of PIWI-interacting RNAs: new insights into biogenesis and function inside and outside of germlines," Genes Dev, vol. 26, no. 21, pp. 2361–2373, 2012.

[24] V. V. Vagin, A. Sigova, C. Li, et al., "A distinct small RNA pathway silences selfish genetic elements in the germline," Science, vol. 313, pp. 320–324, 2006.

[25] J. Brennecke, A. A. Aravin, A. Stark, et al., "Discrete small RNA-generating loci as master regulators of transposon activity in Drosophila," Cell, vol. 128, pp. 1089–1103, 2007.

[26] S. Houwing, L.M. Kamminga, E. Berezikov, et al., "A role for Piwi and piRNAs in germ cell maintenance and transposon silencing in zebrafish," Cell, vol. 129, pp. 69–82, 2007.

[27] M. C. Siomi, K. Sato, D. Pezic, et al., "PIWI-interacting small RNAs: the vanguard of genome defence," Nat Rev Mol Cell Biol, vol. 12, pp. 246–258, 2011.

[28] C. Kwon, H. Tak, M. Rho, et al., "Detection of PIWI and piRNAs in the mitochondria of mammalian cancer cells," Biochem Biophys Res Commun, vol. 446, no. 1, pp. 218–223, 2014.

[29] A. Ashe, A. Sapetschnig, E.-M. Weick, et al., "piRNAs can trigger a multigenerational epigenetic memory in the germline of C. elegans," Cell, vol. 150, pp. 88–99, 2012.

[30] P. Rajasethupathy, I. Antonov, R. Sheridan, et al., "A role for neuronal piRNAs in the epigenetic control of memory-related synaptic plasticity," Cell, vol. 149, pp. 693–707, 2012.

[31] C. Juliano, J. Wang, H. Lin, "Uniting germline and stem cells: the function of Piwi proteins and the piRNA pathway in diverse organisms," Annu Rev Genet, vol. 45, pp. 447–469, 2011.

[32] Z. Williams, P. Morozov, A. Mihailovic, et al., "Discovery and characterization of piRNAs in the human fetal ovary," Cell Rep, vol. 13, pp. 854–863, 2015.

[33] A. Aravin, D. Gaidatzis, S. Pfeffer, et al., "A novel class of small RNAs bind to MILI protein in mouse testes," Nature, vol. 442, pp. 203–207, 2006.

[34] G. M. Chalbatani, H. Dana, F. Memari, et al., "Biological function and molecular mechanism of piRNA in cancer," Pract Lab Med, 13, e00113, 2018. https://doi.org/10.1016/j.plabm.2018.e00113

[35] Y. Liu, "MicroRNAs and PIWI-interacting RNAs in oncology," Oncol Lett, vol. 12, no. 4, pp. 2289–2292, 2016.

[36] S. Siddiqi, I. Matushansky, "Piwis and piwi-interacting RNAs in the epigenetics of cancer," J Cell Biochem, vol. 113, pp. 373–380, 2012.

[37] M. Esteller, "Non-coding RNAs in human disease," Nat Rev Genet, vol. 12, pp. 861–874, 2011.

[38] A. S. Wilson, B. E. Power, P. L. Molloy, "DNA hypomethylation and human diseases," Biochim. Biophys Acta, vol. 1775, pp. 138–162, 2007.

[39] S. B. Baylin, "DNA methylation and gene silencing in cancer," Nat Clin Pract Oncol, vol. 2, pp. S4–S11, 2005.

[40] D. I. Jacobs, Q. Qin, M. C. Lerro, et al., "PIWI-interacting RNAs in gliomagenesis: evidence from Post-GWAS and functional analyses," Cancer Epidemiol Biomark Prev, vol. 25, pp. 1073–1080, 2016.

[41] S. Siddiqi, I. J. Matushansky, "Piwis and piwi-interacting RNAs in the epigenetics of cancer," Cell Biochem, vol. 113, pp. 373–380, 2012.

[42] M. Xu, Y. You, P. Hunsicker, et al., "Mice deficient for a small cluster of Piwi-interacting RNAs implicate Piwi-interacting RNAs in transposon control," Biol Reprod, vol. 79, pp. 51–57, 2008.

[43] P.T. Law, H. Qin, A. K. Ching, et al., "Deep sequencing of small RNA transcriptome reveals novel non-coding RNAs in hepatocellular carcinoma," J Hepatol, vol. 58, no. 6, pp. 1165–1173, 2013.

[44] J. Cheng, H. Deng, B. Xiao, et al., "piR-823, a novel non-coding small RNA, demonstrates in vitro and in vivo tumor suppressive activity in human gastric cancer cells," Cancer Lett, vol. 315, pp. 12–17, 2012.

[45] H. W. Pan, S. C. Li, K. W. Tsai, "MicroRNA dysregulation in gastric cancer," Curr Pharm Des, vol. 19, no. 7, pp. 1273–1284, 2013.

[46] N. Shomali, B. Mansoori, A. Mohammadi, et al., "MiR-146a functions as a small silent player in gastric cancer," Biomed Pharmacother, vol. 96, pp. 238–245, 2017.

[47] L. A. Torre, F. Bray, R. L. Siegel, et al., "Global cancer statistics, 2012," CA Cancer J Clin, vol. 65, no. 2, pp. 87–108, 2012.

[48] D. Jiang, L. Jiang, B. Liu, et al., "Clinicopathological and prognostic significance of FoxM1 in gastric cancer: a meta-analysis," Int J Surg, vol. 48, pp. 38–44, 2017.

[49] C. Marcus, R. M. Subramaniam, "PET/computed tomography and precision medicine: gastric cancer," PET Clin Oct, vol. 12, no. 4, pp. 437–447, 2017.

[50] R. L. Siegel, K. D. Miller, A. Jemal, "Cancer statistics, 2016," CA Cancer J Clin, vol. 66, pp. 7–30, 2016.

[51] R. De Angelis, M. Sant, M. P. Coleman, et al., "Cancer survival in Europe 1999–2007 by country and age: results of EUROCARE–5-a population-based study," Lancet Oncol, vol. 15, pp. 23–34, 2014.

[52] J. Cheng, J. M. Guo, B. X. Xiao, et al., "piRNA, the new non-coding RNA, is aberrantly expressed in human cancer cells," Clin Chim Acta, vol. 412, pp. 1621–1625, 2011.

[53] L. Cui, Y. Lou, X. Zhang, et al., "Detection of circulating tumor cells in peripheral blood from patients with gastric cancer using piRNAs as markers," Clin Biochem, vol. 44, pp. 1050–1057, 2011.

[54] P.-F. Li, S.-C. Chen, T. Xia, et al., "Non-coding RNAs and gastric cancer," World J Gastroenterol WJG, vol. 20, no. 18, pp. 5411–5419, 2014.

[55] G. Huang, H. Hu, X. Xue, et al., "Altered expression of piRNAs and relation with clinicopathologic features of breast cancer," Clin Transl Oncol, vol. 15, pp. 563–568, 2013.

[56] H. Zhang, Y. Ren, H. Xu, et al., "The expression of stem cell protein Piwil2 and piR-932 in breast cancer," Surg Oncol, vol. 22, pp. 217–223, 2013.

[57] A. Fu, D. I. Jacobs, A. E. Hoffman, et al., "PIWI-interacting RNA 021285 is involved in breast tumorigenesis possibly

by remodeling the cancer epigenome," Carcinogenesis, vol. 36, pp. 1094–1102, 2015.

[58] Y. N. Han, Y. Li, S. Q. Xia, et al., "PIWI proteins and PIWI-interacting RNA: emerging roles in cancer," Cell Physiol Biochem, vol. 44, no. 1, pp. 1–20, 2017.

[59] J. C. Park, D. E. Citrin, P. K. Agarwal, et al., "Multimodal management of muscle invasive bladder cancer," Curr Probl Cancer, vol. 38, no. 3, pp. 80–108, 2014.

[60] M. Ploeg, K. K. Aben, L. A. Kiemeney, "The present and future burden of urinary bladder cancer in the world," World J Urol, vol. 27, pp. 289–293, 2009.

[61] R. Siegel, D. Naishadham, A. Jemal, "Cancer statistics, 2013," CA Cancer J. Clin. vol. 63, pp. 11–30, 2013.

[62] H. Chu, G. Hui, L. Yuan, et al., "Identification of novel piRNAs in bladder cancer," Cancer Lett, vol. 356, pp. 561–567, 2015.

[63] B. Pardini, A. Naccarati, "Altered piRNA profiles in bladder cancer: a new challenge in the next-generation sequencing era?," J Genet Genomes, vol. 1, pp. 110–114, 2018.

[64] S. Blandin Knight, P. A. Crosbie, H. Balata, et al., "Progress and prospects of early detection in lung cancer," Open Biol, vol. 7, no. 9, p. 170070, 2017. http://dx.doi.org/10.1098/rsob.170070

[65] Y. Mei, Y. Wang, P. Kumari, et al., "A piRNA-like small RNA interacts with and modulates p-ERM proteins in human somatic cells," Nat Commun, no. 6, p. 7316, 2015. http://dx.doi.org/10.1038/ncomms8316

[66] A. L. Neisch, R. G. Fehon, "Radixin Ezrin, Moesin, key regulators of membrane-cortex interactions and signaling," Curr Opin Cell Biol, vol. 23, pp. 377–382, 2011.

[67] A. I. McClatchey, R. G. Fehon, "Merlin and the ERM proteins-regulators of receptor distribution and signaling at the cell cortex," Trends Cell Biol, no. 19, pp. 198–206, 2009.

[68] A. B. Ryerson, C. R. Eheman, S. F. Altekruse, et al., "Annual report to the nation on the status of cancer, 1975-2012, featuring the increasing incidence of liver cancer," Cancer, no. 122, vol. 9, 1312–1337, 2016.

[69] J. Ferlay, I. Soerjomataram, R. Dikshit, et al., "Cancer incidence and mortality worldwide: sources, methods and major patterns in GLOBOCAN 2012," Int J Cancer, vol. 136, pp. E359–E386, 2015.

[70] Y. Wang, T. Gable, M. Z. Ma, et al., "A piRNA-like small RNA induces chemoresistance to cisplatin-based therapy by inhibiting apoptosis in lung squamous cell carcinoma," Mol Ther Nucleic Acids, vol. 6, pp. 269–278, 2017.

[71] D. W. -H. Ho, R. C. -L. Lo, L. -K. Chan, et al., "Molecular pathogenesis of hepatocellular carcinoma," Liver Cancer, vol. 5, no. 4, pp. 290–302, 2016.

[72] Y. Mei, D. Clark, L. Mao, "Novel dimensions of piRNAs in cancer," Cancer Lett, vol. 336, no. 1, pp. 46–52, 2013.

[73] K. W. Ng, C. Anderson, E. A. Marshall, et al., "Piwi-interacting RNAs in cancer: emerging functions and clinical utility," Mol Cancer, vol. 15, no. 5, 2016.

[74] H. Dana, V. Marmari, G. Mahmoodi, et al., "CD166 as a stem cell marker? A potential target for therapy colorectal cancer?," J Stem Cell Res Ther, vol. 1, no. 6, p. 00041, 2016.

[75] H. Dana, V. Marmari, A. Mazraeh, et al., "Cloning and expression of the V-domain of the CD166 in prokaryotic host cell," Int J Cancer Ther Oncol, vol. 5, no. 1, p. 5110, 2017. http://dx.doi.org/10.14319/ijcto.51.10

[76] Y. Zeng, L.K. Qu, L. Meng, et al., "HIWI expression profile in cancer cells and its prognostic value for patients with colorectal cancer," Chin Med J, vol. 124, pp. 2144–2149, 2011.

[77] S. J. Oh, S. M. Kim, Y. O. Kim, et al., "Clinicopathologic implications of PIWIL2 expression in colorectal cancer," Korean J Pathol, vol. 46, pp. 318–323, 2012.

[78] H. Chu, L. Xia, X. Qiu, et al., "Genetic variants in non-coding PIWI-interacting RNA and colorectal cancer risk," Cancer, vol. 15, no. 12, pp. 2044–2052, 2015.

[79] W. J. Kent, C. W. Sugnet, T. S. Furey, et al., "The human genome browser at UCSC," Genome Res, vol. 12, no. 6, pp. 996–1006, 2002.

[80] Dan-Dan Zhou, Hong-Li Li, Wei Liu, et al., "miR-193a-3p Promotes the Invasion, Migration, and Mesenchymal Transition in Glioma through Regulating BTRC," Biomed Res Int, vol. 2021, p. 8928509, 2021. DOI: 10.1155/2021/8928509

[81] Baoyi Yin, Talha Umar, Xiaofei Ma, et al., "MiR-193a-3p targets LGR4 to promote the inflammatory response in endometritis," Int Immunopharmacol, vol. 98, p. 107718, 2021. https://doi.org/10.1016/j.intimp.2021.107718

[82] D. Iliopoulos, A. Rotem, K. Struhl, "Inhibition of miR-193a expression by Max and RXRα activates K-Ras and PLAU to mediate distinct aspects of cellular transformation," Cancer Res, vol. 71, no. 15, pp. 5144–5153, 2011.

[83] Y. Li, L. Gao, X. Luo, et al., "Epigenetic silencing of microRNA-193a contributes to leukemogenesis in t(8;21) acute myeloid leukemia by activating the PTEN/PI3K signal pathway," Blood, vol. 121, no. 3, pp. 499–509, 2013.

[84] H. Lu, X. Lei, J. Liu, et al., "Regulation of hepatic microRNA expression by hepatocyte nuclear factor 4 alpha," World J Hepatol, vol. 9, no. 4, pp. 191–208, 2017.

[85] X. N. Gao, J. Lin, Y. H. Li, et al., "MicroRNA-193a represses c-kit expression and functions as a methylation-silenced tumor suppressor in acute myeloid leukemia," Oncogene, vol. 30, no. 31, pp. 3416–3428, 2011.

[86] K. Kozaki, I. Imoto, S. Mogi, et al., "Exploration of tumor-suppressive microRNAs silenced by DNA hypermethylation in oral cancer," Cancer Res, vol. 68, no. 7, pp. 2094–2105, 2008.

[87] G. Heller, M. Weinzierl, C. Noll, et al., "Genome-wide miRNA expression profiling identifies miR-9-3 and miR-193a as targets for DNA methylation in non-small cell lung cancers," Clin Cancer Res, vol. 18, no. 6, pp. 1619–1629, 2012.

[88] J. Wang, B. Yang, L. Han, et al., "Demethylation of miR-9-3 and miR-193a genes suppresses proliferation and promotes apoptosis in non-small cell lung cancer cell lines," Cell Physiol Biochem, vol. 32, no. 6, pp. 1707–1719, 2013.

[89] Y. Pu, F. Zhao, W. Cai, et al., "MiR-193a-3p and miR-193a-5p suppress the metastasis of human osteosarcoma cells by down-regulating Rab27B and SRR, respectively," Clin Exp Metastasis, vol. 33, no. 4, pp. 359–372, 2016.

[90] I. Grossi, B. Arici, N. Portolani, et al., "Clinical and biological significance of miR-23b and miR-193a in human hepatocellular carcinoma," Oncotarget, vol. 8, no. 4, pp. 6955–6969, 2017.

[91] M. Williams, M. B. Kirschner, Y. Y. Cheng, et al., "miR-193a-3p is a potential tumor suppressor in malignant pleural mesothelioma," Oncotarget, vol. 6, no. 27, pp. 23480–23495, 2015.

[92] H. Takeshita, A. Shiozaki, X. H. Bai, et al., "XB130, a new adaptor protein, regulates expression of tumor suppressive microRNAs in cancer cells," PLoS One, vol. 8, no. 3, article e59057, 2013.

[93] B. Yue, D. Cai, C. Liu, et al., "Linc00152 functions as a competing endogenous RNA to confer oxaliplatin resistance and holds prognostic values in colon cancer," Mol Ther, vol. 24, no. 12, pp. 2064–2077, 2016.

[94] W. Nie, H. J. Ge, X. Q. Yang, et al., "LncRNA-UCA1 exerts oncogenic functions in non-small cell lung cancer

by targeting miR-193a-3p," Cancer Lett, vol. 371, no. 1, pp. 99–106, 2016.

[95] R. Zhang, J. Zhang, Q. Wu, et al., "XB130: a novel adaptor protein in cancer signal transduction," Biomed Rep, vol. 4, no. 3, pp. 300–306, 2016.

[96] I. Grossi, A. Salvi, E. Abeni, et al., "Biological Function of MicroRNA193a-3p in Health and Disease," Int J Genomics, vol. 2017, 5913195. https://doi.org/10.1155/2017/5913195

[97] K. Ma, Y. He, H. Zhang, et al., "DNA methylation-regulated miR-193a-3p dictates resistance of hepatocellular carcinoma to 5-fluorouracil via repression of SRSF2 expression," J Biol Chem, vol. 287, no. 8, pp. 5639–5649, 2012.

[98] J. Liz, M. Esteller, "lncRNAs and microRNAs with a role in cancer development," Biochim Biophys Acta (BBA)—Gene Regul Mech, vol. 1859, no. 1, pp. 169–176, 2016.

[99] B. Yue, D. Cai, C. Liu, et al., "Linc00152 functions as a competing endogenous RNA to confer oxaliplatin resistance and holds prognostic values in colon cancer," Molecular Therapy, vol. 24, no. 12, pp. 2064–2077, 2016.

[100] W. Nie, H. J. Ge, X. Q. Yang, et al., "LncRNA-UCA1 exerts oncogenic functions in non-small cell lung cancer by targeting miR-193a-3p," Cancer Lett, vol. 371, no. 1, pp. 99–106, 2016.

[101] GTEx Consortium, "The genotype-tissue expression (GTEx) project," Nat Genet, vol. 45, no. 6, pp. 580–585, 2013.

[102] L. J. Carithers, K. Ardlie, M. Barcus, et al., "A novel approach to high-quality postmortem tissue procurement: the GTEx project," Biopreserv Biobank, vol. 13, no. 5, pp. 311–319, 2015.

[103] C. P. Khoo, M. G. Roubelakis, J. B. Schrader, et al., "miR-193a-3p interaction with HMGB1 downregulates human endothelial cell proliferation and migration," Sci Rep, vol. 7, article 44137, 2017.

[104] S. Greco, A. Perfetti, P. Fasanaro, et al., "Deregulated micro-RNAs in myotonic dystrophy type 2," PLoS One, vol. 7, no. 6, article e39732, 2012.

[105] S. Kuokkanen, B. Chen, L. Ojalvo, et al., "Genomic profiling of microRNAs and messenger RNAs reveals hormonal regulation in microRNA expression in human endometrium," Biol Reprod, vol. 82, no. 4, pp. 791–801, 2010.

[106] B. Mandriani, S. Castellana, C. Rinaldi, et al., "Identification of p53-target genes in Danio rerio," Sci Rep, vol. 6, article 32474, 2016.

[107] Y. Yi, J. Chen, C. Jiao, et al., "Upregulated miR-193a-3p as an oncogene in esophageal squamous cell carcinoma regulating cellular proliferation, migration and apoptosis," Oncol Lett, vol. 12, no. 6, pp. 4779–4784, 2016.

[108] S. Uhlmann, H. Mannsperger, J. D. Zhang, et al., "Global microRNA level regulation of EGFR-driven cell-cycle protein network in breast cancer," Mol Syst Biol, vol. 8, p. 570, 2012.

[109] H. Liang, M. Liu, X. Yan, et al., "miR-193a-3p functions as a tumor suppressor in lung cancer by down-regulating ERBB4," J Biol Chem, vol. 290, no. 2, pp. 926–940, 2015.

[110] K. W. Tsai, C. M. Leung, Y. H. Lo, et al., "Arm selection preference of microRNA-193a varies in breast cancer," Sci Rep, vol. 6, article 28176, 2016.

[111] E. G. Seviour, V. Sehgal, D. Mishra, et al., "Targeting KRas-dependent tumour growth, circulating tumour cells and metastasis in vivo by clinically significant miR-193a-3p," Oncogene, vol. 36, no. 10, pp. 1339–1350, 2017.

[112] J. E. Kwon, B. Y. Kim, S. Y. Kwak, et al., "Ionizing radiation-inducible microRNA miR-193a-3p in2duces

apoptosis by directly targeting Mcl-1," Apoptosis, vol. 18, no. 7, pp. 896–909, 2013.

[113] H. Nakano, Y. Yamada, T. Miyazawa, et al., "Gain-of-function microRNA screens identify miR-193a regulating proliferation and apoptosis in epithelial ovarian cancer cells," Int J Oncol, vol. 42, no. 6, pp. 1875–1882, 2013.

[114] A. Salvi, I. Conde, E. Abeni, et al., "Effects of miR-193a and sorafenib on hepatocellular carcinoma cells," Mol Cancer, vol. 12, p. 162, 2013.

[115] H. Noh, S. Hong, Z. Dong, et al., "Impaired MicroRNA processing facilitates breast cancer cell invasion by upregulating urokinase-type plasminogen activator expression," Genes Cancer, vol. 2, no. 2, pp. 140–150, 2011.

[116] A. Salvi, B. Arici, A. Alghisi, et al., "RNA interference against urokinase in hepatocellular carcinoma xenografts in nude mice," Tumour Biol, vol. 28, no. 1, pp. 16–26, 2007.

[117] G. D. Petro, D. Tavian, A. Copeta, et al., "Expression of urokinase-type plasminogen activator (u-PA), u-PA receptor, and tissue-type PA messenger RNAs in human hepatocellular carcinoma," Cancer Res, vol. 58, no. 10, pp. 2234–2239, 1998.

[118] J. A. Foekens, H. A. Peters, M. P. Look, et al., "The urokinase system of plasminogen activation and prognosis in 2780 breast cancer patients," Cancer Res, vol. 60, no. 3, pp. 636–643, 2000.

[119] T. Yu, J. Li, M. Yan, et al., "MicroRNA-193a-3p and -5p suppress the metastasis of human non-small-cell lung cancer by downregulating the ERBB4/PIK3R3/mTOR/S6K2 signaling pathway," Oncogene, vol. 34, no. 4, pp. 413–423, 2015.

[120] D. Jia, M. Yan, X. Wang, et al., "Development of a highly metastatic model that reveals a crucial role of fibronectin in lung cancer cell migration and invasion," BMC Cancer, vol. 10, p. 364, 2010.

[121] W. Deng, M. Yan, T. Yu, et al., "Quantitative proteomic analysis of the metastasis-inhibitory mechanism of miR-193a-3p in non-small cell lung cancer," Cell Physiol Biochem, vol. 35, no. 5, pp. 1677–1688, 2015.

[122] L. Lv, H. Deng, Y. Li, et al., "The DNA methylation-regulated miR-193a-3p dictates the multi-chemoresistance of bladder cancer via repression of SRSF2/PLAU/HIC2 expression," Cell Death Dis, vol. 5, article e1402, 2014.

[123] H. Deng, L. Lv, Y. Li, et al., "miR-193a-3p regulates the multidrug resistance of bladder cancer by targeting the LOXL4 gene and the oxidative stress pathway," Mol Cancer, vol. 13, p. 234, 2014.

[124] Y. Li, H. Deng, L. Lv, et al., "The miR-193a-3p-regulated ING5 gene activates the DNA damage response pathway and inhibits multi-chemoresistance in bladder cancer," Oncotarget, vol. 6, no. 12, pp. 10195–10206, 2015.

[125] L. Lv, Y. Li, H. Deng, et al., "MiR-193a-3p promotes the multi-chemoresistance of bladder cancer by targeting the HOXC9 gene," Cancer Lett, vol. 357, no. 1, pp. 105–113, 2015.

[126] H. Deng, L. Lv, Y. Li, et al., "The miR-193a-3p regulated PSEN1 gene suppresses the multi-chemoresistance of bladder cancer," Biochim Biophys Acta Mol Basis Dis, vol. 1852, no. 3, pp. 520–528, 2015.

[127] F. Meng, L. Qian, L. Lv, et al., "miR-193a-3p regulation of chemoradiation resistance in oesophageal cancer cells via the PSEN1 gene," Gene, vol. 579, no. 2, pp. 139–145, 2016.

[128] X. Dai, X. Chen, Q. Chen, et al., "MicroRNA-193a-3p reduces intestinal inflammation in response to microbiota via downregulation of colonic PepT1," J Biol Chem, vol. 290, no. 26, pp. 16099–16115, 2015.

[129] Y. Liu, F. Ren, Y. Luo, et al., "Downregulation of MiR-193a-3p dictates deterioration of HCC: a clinical real-time qRT-PCR study," Med Sci Monit, vol. 21, pp. 2352–2360, 2015.

[130] F. Ren, H. Ding, S. Huang, et al., "Expression and clinicopathological significance of miR-193a-3p and its potential target astrocyte elevated gene-1 in non-small lung cancer tissues," Cancer Cell Int, vol. 15, p. 80, 2015.

[131] S. Caramuta, S. Egyhazi, M. Rodolfo, et al., "MicroRNA expression profiles associated with mutational status and survival in malignant melanoma," J Invest Dermatol, vol. 130, no. 8, pp. 2062–2070, 2010.

[132] S. Gilad, E. Meiri, Y. Yogev, et al., "Serum microRNAs are promising novel biomarkers," PLoS One, vol. 3, no. 9, article e3148, 2008.

[133] H. Dong, C. Wang, S. Lu, et al., "A panel of four decreased serum microRNAs as a novel biomarker for early Parkinson's disease," Biomarkers, vol. 21, no. 2, pp. 129–137, 2016.

[134] H. Wei, Y. Yuan, S. Liu, et al., "Detection of circulating miRNA levels in schizophrenia," Am J Psychiatry, vol. 172, no. 11, pp. 1141–1147, 2015.

[135] F. L. Yong, C. W. Law, C. W. Wang, "Potentiality of a triple microRNA classifier: miR-193a-3p, miR-23a and miR-338-5p for early detection of colorectal cancer," BMC Cancer, vol. 13, p. 280, 2013.

[136] C. Wang, J. Hu, M. Lu, et al., "A panel of five serum miR-NAs as a potential diagnostic tool for early-stage renal cell carcinoma," Sci Rep, vol. 5, p. 7610, 2015.

[137] C. Wang, M. Ding, M. Xia, et al., "A five-miRNA panel identified from a multicentric case-control study serves as a novel diagnostic tool for ethnically diverse non-small-cell lung cancer patients," eBioMedicine, vol. 2, no. 10, pp. 1377–1385, 2015.

[138] C. Wu, C. Wang, X. Guan, et al., "Diagnostic and prognostic implications of a serum miRNA panel in oesophageal squamous cell carcinoma," PLoS One, vol. 9, no. 3, article e92292, 2014.

[139] L. Li, J. G. Chen, X. Chen, et al., "Serum miRNAs as predictive and preventive biomarker for pre-clinical hepatocellular carcinoma," Cancer Lett, vol. 373, no. 2, pp. 234–240, 2016.

[140] K. Zen, C. Y. Zhang, "Circulating microRNAs: a novel class of biomarkers to diagnose and monitor human cancers," Med Res Rev, vol. 32, no. 2, pp. 326–348, 2012.

[141] H. W. Liang, F. Gong, S. Y. Zhang, et al., "The origin, function, and diagnostic potential of extracellular microRNAs in human body fluids," WIREs RNA, vol. 5, no. 2, pp. 285–300, 2014.

[142] H. Valadi, K. Ekstrom, A. Bossios, et al., "Exosome-mediated transfer of mRNAs and microRNAs is a novel mechanism of genetic exchange between cells," Nat Cell Biol, vol. 9, no. 6, pp. 654–U672, 2007.

[143] H. J. Oh, Y. Shin, S. Chung, et al., "Convective exosome-tracing microfluidics for analysis of cell-non-autonomous neurogenesis," Biomaterials, vol. 112, pp. 82–94, 2017.

[144] Y. Teng, Y. Ren, X. Hu, et al., "MVP-mediated exosomal sorting of miR-193a promotes colon cancer progression," Nat Commun, vol. 8, article 14448, 2017.

[145] L. H. Madkour, Reactive Oxygen Species (ROS), Nanoparticles, and Endoplasmic Reticulum (ER) Stress-Induced Cell Death Mechanisms, Academic Press, 2020.

[146] L.H. Madkour, Nanoparticles Induce Oxidative and Endoplasmic Reticulum Antioxidant Therapeutic Defenses, Springer International Publishing, 2020, doi:10.1007/978-3-030-37297-2.

[147] L. H. Madkour, Nucleic Acids as Gene Anticancer Drug Delivery Therapy, Elsevier, 2020.

[148] L. H. Madkour, Nanoelectronic Materials: Fundamentals and Applications (Advanced Structured Materials), Springer International Publishing, 2019, doi:10.1007/978-3-030-21621-4.

[149] R. Rupaimoole, F. J. Slack, MicroRNA therapeutics: towards a new era for the management of cancer and other diseases, Nat Rev Drug Discov, vol. 16, no. 3, pp. 203–222, 2017.

# 6 Delivery Strategies for siRNA and Modifications Process of RNAi Therapeutics for Cancer Treatment

The discovery of RNAi in the late 1990s unlocked a new realm of therapeutic possibilities by enabling potent and specific silencing of theoretically any desired genetic target. Better elucidation of the mechanism of action, the impact of chemical modifications that stabilize and reduce nonspecific effects of siRNA molecules, and the key design considerations for effective delivery systems has spurred progress toward developing clinically successful siRNA therapies. A logical aim for initial siRNA translation is local therapies, as delivering siRNA directly to its site of action helps to ensure that a sufficient dose reaches the target tissue, lessens the potential for off-target side effects, and circumvents the substantial systemic delivery barriers. While locally injected or topically applied siRNA has progressed into numerous clinical trials, an enormous opportunity exists to develop sustained-release, local delivery systems that enable both spatial and temporal control of gene silencing.

In 1993, pioneering observations on RNA-mediated gene silencing were first reported in plants by John Lindbo and Bill Dougherty [1]. Half a decade later, RNA-mediated gene silencing known as RNA interference (RNAi) was made famous by Andrew Fire and Craig Mello's breakthrough study, which has decisively proven the RNAi mechanism working as an antiviral defense mechanism in *Caenorhabditis elegans* [2].

## 6.1 SMALL INTERFERING RNA (siRNA) THERAPEUTICS FOR CANCER TREATMENT

The discovery of RNA interference (RNAi) has opened doors that might introduce a novel therapeutic tool to the clinical setting [3]. For many decades, small molecules have been developed and utilized in cancer therapy; however, critical problems, such as undesirable toxicity against normal tissues due to a lack of selectivity, still remain today. Using RNAi as a therapeutic tool will allow targeting previously unreachable targets with its potential to silence the function of any cancer-causing gene [4]. This unique advantage is made possible by utilizing the biological functions of double-stranded RNA (dsRNA) molecules. Endogenous dsRNA is recognized by a ribonuclease protein, termed Dicer, and cleaved into small double-stranded fragments of 21 to 23 base pairs in length with 2-nucleotide (nt) overhangs at the 3' ends.

The cleaved products are referred to as small interfering RNAs (siRNAs). The siRNAs consist of a passenger strand and a guide strand, and are bound by an active protein complex called the RNA-induced silencing complex (RISC). After binding to RISC, the guide strand is directed to the target mRNA, which is cleaved between bases 10 and 11 relative to the 5' end of the siRNA guide strand, by the cleavage enzyme Argonaute-2. Thus, the process of mRNA translation can be interrupted by siRNA [5–7].

The therapeutic application of siRNA has the potential to treat various diseases including cancer [8, 9]. Cancer is a genetic disease caused by the generation of mutated genes within tumor cells; multiple gene mutations both activate disease-driving oncogenes and inactivate tumor suppressor genes in cancer [10–12]. SiRNAs that can inactivate specific cancer-driving genes have shown great potential as novel cancer therapeutics. Several anticancer siRNA-based drugs have entered clinical trials, and many are actively sought after in preclinical research [13–15].

Even though the usage of siRNA as therapy has shown promise in the treatment of cancer, many obstacles that hinder the ultimate functionality of siRNAs in the clinic remain to be solved [16, 17]. In order to make this therapy effective, the first and most crucial step is to ensure the delivery of siRNA to the tumor cells from the injection site. In practice, siRNAs face physiological and biological barriers that prevent their delivery to the active site when administered systemically [18–20]. These barriers include, but are not limited to, intravascular degradation, recognition by the immune system, renal clearance, impediments to tumor tissue penetration and uptake into tumor cells, endosomal escape once in tumor cells, and off-target effects [21–23]. Delivery formulations as well as chemical modification of siRNA are required to overcome these challenges and facilitate siRNAs in reaching their target cells [24]. Furthermore, selection of gene targets in cancer is also crucial in designing siRNA therapeutic strategies. Discoveries of mechanisms in cancer provide innovative targets for siRNA therapy that in many cases cannot be targeted with conventional drugs. However, the particular gene pool that drives cancer varies depending on the origins and types of the tumors. Thus, careful selection of gene targets according to their cancer type is essential in siRNA therapeutic strategies.

DOI: 10.1201/9781003229650-6

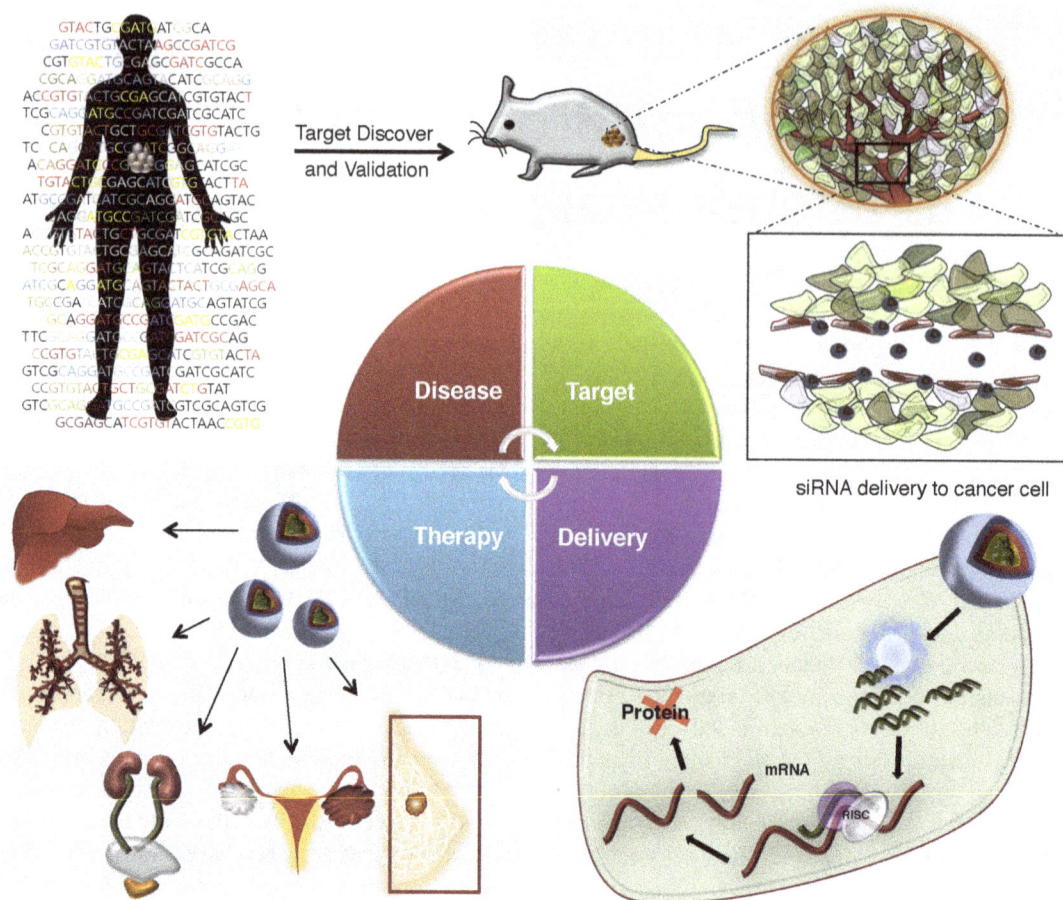

**FIGURE 6.1** Development process of siRNA therapeutics for cancer treatment.

To summarize, target discovery in cancer leads to the selection of siRNA gene targets, followed by their incorporation of the siRNAs into suitable delivery systems that allow access to the desired sites. Once therapeutic effect is observed, further application in varying organs and tissues can be anticipated as shown in Figure 6.1. This chapter examines current thoughts on the therapeutic potential of siRNA delivery strategies and the optimal targets for siRNA in major cancer types.

SiRNA, a 21- to 23-nt dsRNA responsible for posttranscriptional gene silencing, has attracted great interest as a promising genomic drug, due to its strong ability to silence target genes in a sequence-specific manner. Despite high silencing efficiency and on-target specificity, the clinical translation of siRNA has been hindered by its inherent features: poor intracellular delivery, limited blood stability, unpredictable immune responses, and unwanted off-targeting effects. To overcome these hindrances, researchers have made various advances to modify siRNA itself and to improve its delivery. In this chapter, first we briefly discuss the innate properties and delivery barriers of siRNA. Then, we describe recent progress in (1) chemically and structurally modified siRNAs to solve their intrinsic problems, and (2) siRNA delivery formulations including siRNA

conjugates, polymerized siRNA, and nucleic acid–based nanoparticles to improve in vivo delivery.

RNAi is a highly effective regulatory mechanism of gene expression at the posttranscriptional level [2, 25]. Small dsRNA, called small interfering RNA (siRNA), is responsible for RNAi-based gene silencing. When siRNA is generated via Dicer processing of long dsRNA or synthetic siRNA is delivered into the cytoplasmic region, it is incorporated into RISC and then removes the sense strand by the action of Argonaute-2. The activated RISC recognizes target mRNA with sequence homology and cleaves the mRNA at the opposite of position 10 from the 5′ end of the antisense strand [26]. SiRNA has been considered as a promising gene therapeutic because it can downregulate the expression of virtually all the genes, including previously undruggable targets. In past decades, the therapeutic potentials of siRNA have been proven in the treatment of genetic diseases, virus infections, and cancer [27, 28].

To date, several RNAi therapeutics, including naked siRNA and siRNA/carrier complexes, have undergone clinical trials for the treatment of ocular disorders, kidney disorders, and cancers (Table 6.1). However, most of them remain in Phase 1 (safety test), and very few have entered or are planning to enter Phase 2/3 (efficacy test). The first Phase

**TABLE 6.1**
**Clinical Trials of RNAi Therapeutics**

| Drug Name | Target Sequence | Target Disease | Phase | Status | Company |
|---|---|---|---|---|---|
| siG12D LODER | KRAS | Pancreatic tumor | 1 | Completed | Silenseed Ltd. |
| 15NP | P53 | Acute renal failure | 1 | Completed | Quark Pharmaceuticals |
| Atu027 | PKN3 | Advanced solid tumors | 1 | Completed | Silence Therapeutics GmbH |
| TD101 | Keratin6A | Pachyonychia congenita | 1 | Completed | Pachyonychia Congenita Project |
| AGN 211745 | VEGF | Age-related macular degeneration | 1/2 | Completed | Allergan |
| siRNA-EphA2-DOPC | EphA2 | Advanced cancers | 1/2 | Not yet recruiting | MD Anderson Cancer Center |
| CALAA-01 | RRM2 | Solid tumor | 1 | Terminated | Calando Pharmaceuticals |
| TKM-PLK1 | PKL1 | Hepatic metastases | 1 | Completed | Tekmira Pharmaceuticals Corporation |
| TKM-ApoB | ApoB | Hypercholesterolemia | 1 | Terminated | Tekmira Pharmaceuticals Corporation |
| ALN-VSP02 | VEGF, KSP | Solid tumor | 1 | Completed | Alnylam Pharmaceuticals |
| EZN-2968 | HIF-1 | Liver metastases | 1 | Completed | Santaris Pharma and Enzon Pharmaceuticals |
| PF-04523655 | PTP-801 | Choroidal neovascularization, diabetic macular edema | 2 | Completed | Quark Pharmaceuticals |
| Bevasiranib | VEGF | Age-related macular degeneration | 3 | Withdrawn | OPKO Health, Inc. |

3–entering RNAi drug, Bevasiranib by Opko Health, was withdrawn due to its low therapeutic efficacy. The development of siRNA drugs still suffers from practical problems, such as easy degradation of siRNA in vivo, unwanted off-target effects, and immunogenicity [3]. Delivery of the nucleic acids to desired tissues and cells is also one of the critical concerns [29, 30]. Modification of siRNA itself, both chemically and structurally, and development of an efficient delivery carrier can be considered as promising strategies to resolve such problems. Because the chemically or structurally modified siRNA can contribute to solve both intrinsic and delivery problems of siRNA, herein, we focus on recent advances of siRNA modification strategies. For a detailed description of the development of delivery carriers, please refer to recent review articles [31, 32].

In this chapter, we describe in detail the current challenges of siRNA for clinical applications; the intrinsic properties of siRNA itself and many concerns in siRNA delivery are discussed. Because siRNA modifications can be beneficial to achieve the clinical goal, several siRNA modification strategies are also summarized. First, we introduce chemically modified siRNAs or siRNA structural variants, which have been developed to overcome its inherent problems. Secondly, the development of siRNA conjugate systems or polymerized siRNA for enhancing delivery efficacy is described. It is generally known that gene carriers could increase the blood circulation time of siRNA, provide targeting moieties, and improve cellular uptake. The direct conjugation of siRNA and carriers or the use of polymerized siRNA can improve the loading efficiency of siRNA into gene carriers. Finally, recent advances in nucleic acid nanoparticle systems for efficient siRNA delivery are highlighted.

## 6.2 CHALLENGES IN CLINICAL APPLICATIONS OF siRNA

### 6.2.1 INHERENT PROPERTIES OF siRNA

The inherent properties of siRNA, which should be adjusted before the clinical applications, are mainly categorized into three groups: in vivo instability, off-target effects, and immunogenicity (Figure 6.2). These shortcomings may reduce the therapeutic efficacy of siRNA and cause unexpected toxicity. In this section, detailed mechanisms of easy degradation, unwanted gene silencing, and immune stimulation of siRNA are described, and strategies to overcome these weaknesses are proposed.

The easy degradation of siRNA under in vivo physiological conditions has been considered as one of the main problems for its clinical applications. Native siRNA has a short half-life (less than ~15 min) in serum due to its vulnerability toward nuclease activity. It has been reported that siRNA is susceptible to RNase A family enzyme in serum [33, 34]. Thus chemical modification of specific dinucleotide motifs, substrates for RNase A-like activity, can improve its serum stability. In addition, intracellular siRNA degradation occurs via the activity of the 3′ exonuclease Eri-1 [35], and chemically modified siRNA showed improved resistance to Eri-1 [36]. Alternatively, the use of gene carriers can prevent the access of nucleases to siRNA, subsequently leading to the enhancement of siRNA stability.

It has been known that synthetic siRNA induces unwanted gene silencing, called off-target effects, via two pathways: (1) miRNA-like pathway, and (2) sense strand-mediated pathway. MiRNA suppresses the translation of genes, which contain partial homology with 3′-untranslated

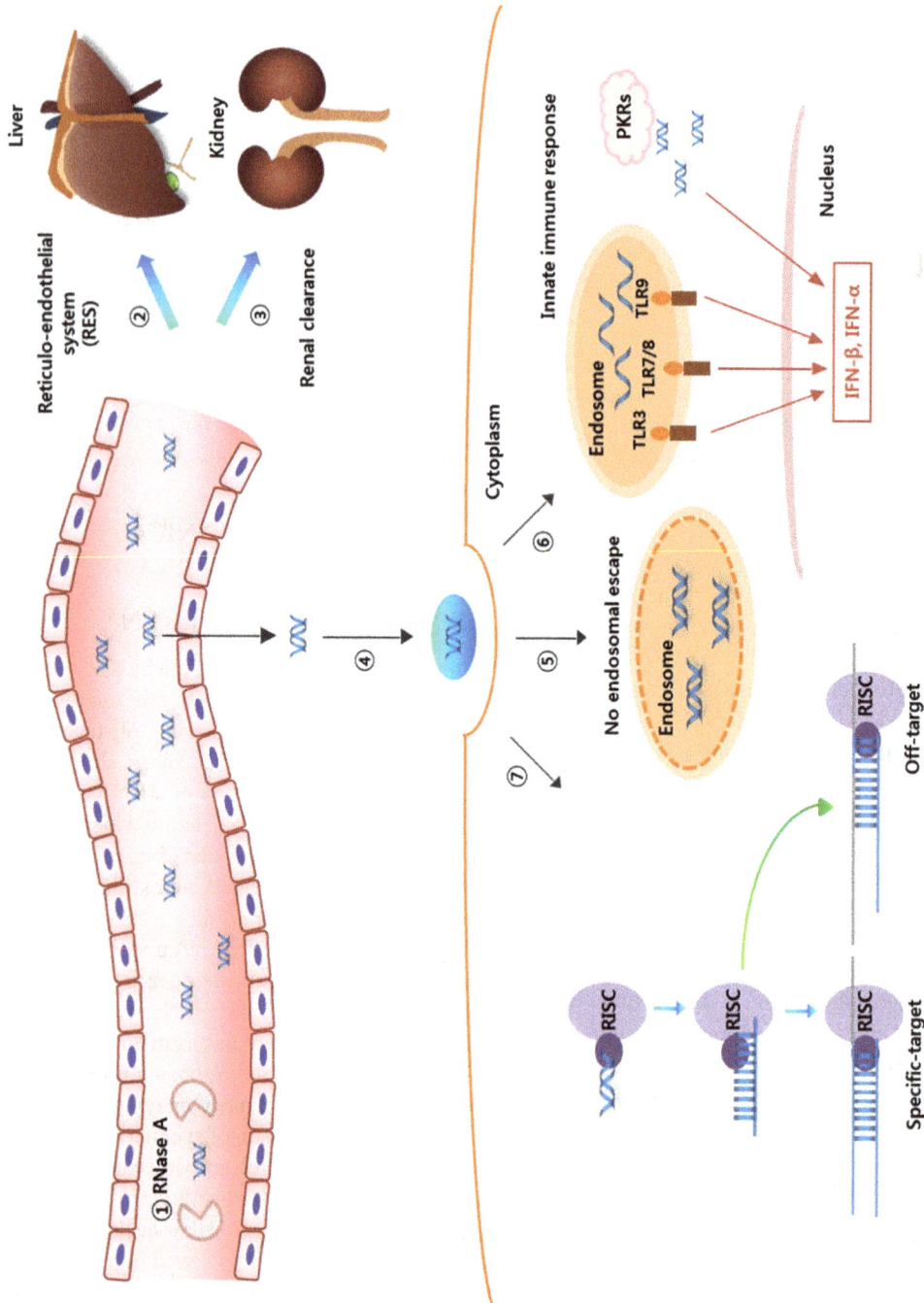

**FIGURE 6.2** Inherent problems and delivery barriers of synthetic siRNA in vivo. ① SiRNAs are influenced by enzymatic degradation in the blood system. Also, siRNAs can be rapidly eliminated from the blood circulation by the ② reticuloendothelial system (RES) and ③ renal clearance. In the siRNA delivery system, the current main drawbacks include ④ poor transport across cell membranes and ⑤ endosomal entrapment. ⑥ SiRNAs trapped in endosome can induce unwanted TLR-mediated immune responses, and external cytoplasmic RNAs activate the immune cells through PKRs. Finally, ⑦ siRNA can cause off-targeting mRNA degradation, leading to unintended transcription and translation suppression.

region (UTR) of the target mRNA, in nature [37]. Similarly, the seed region of siRNA antisense strand (position 2–8 from 5′ end) can interact with 3′-UTR of mRNA with partial homology. The translational suppression owing to imperfect matching with 3′-UTR, rather than mRNA degradation by Argonaute-2, consequently leads to unintended off-targeting [38, 39]. MiRNA target prediction, the use of the lowest dose of siRNAs, and multiple siRNA pools can minimize the unwanted gene silencing via miRNA-like pathways. The sense strand of siRNA can participate in the gene silencing mechanism [40]. Incorporation of sense strand in RISC may cause downregulation of nontarget gene expression. The improved selection of antisense strand in RISC via chemical modification at 5′-end of the sense strand can overcome these types of off-target effects.

The innate immune response to exogenous siRNA is categorized as Toll-like receptor (TLR)-mediated and non-TLR-mediated immune responses. Three types of TLR (TLR3, TLR7, and TLR8) among 13 TLRs are involved in the TLR-mediated immune response. TLR3 shows the length-dependent activity [41]; dsRNAs longer than 21–23 nt can stimulate the TLR-mediated immune response, though the length threshold is dependent on cell type [42]. TLR7 and TLR8 sense the nucleotide sequence; both of them are stimulated by GU-rich motifs, whereas AU-rich motifs primarily activate TLR8 [43, 44]. The latter group of immune responses includes the activation of dsRNA-dependent protein kinase R (PKR) and retinoic acid inducible protein (RIG-1). PKR is activated by dsRNA longer than 30 bp in a sequence-independent manner, though it can interact with short dsRNA containing 11 bp [45]. The activation of RIG-1 is not sequence specific, but length dependent [46]. Further the siRNA overhang can reduce the RIG-1-mediated immune response [47]. The modification of TLR-activatable motifs and the alteration of the interactions between RNA and immune-related proteins have been proposed as potential strategies to avoid siRNA-triggered immune reactions.

## 6.2.2 Barriers to siRNA Delivery

To play a role as a therapeutic agent, siRNA should be delivered to target tissue, be internalized into specific type of cells, and be placed at the site of action (cytosol). However, siRNA delivery is troubled by short blood circulation time, lack of target specificity, and difficulty to cellular uptake. Endosomal escape is also one of the main barriers in siRNA delivery (Figure 6.2).

Short blood circulation time has been considered as a critical barrier to clinical applications of siRNA therapeutics, and it may be caused by enzymatic degradation, renal clearance, and capture by the reticuloendothelial system (RES). As aforementioned, siRNA is rapidly degraded in the bloodstream by RNase A-like nucleases [33, 34]. The rapid renal clearance of naked siRNA occurs upon systemic administration because small molecules <50 kDa are excreted through the kidney [30]. Phagocytic cells in

RES also contribute to remove the foreign nucleic acids as well as gene carriers [48]. The pharmacokinetics of siRNA can be improved by chemical modification of siRNA itself, through inhibition of enzymatic degradation and incorporation with efficient delivery vehicles, through siRNA protection from nuclease attack, and prevention of renal clearance and phagocytosis in RES.

It is important to deliver the gene therapeutics into specific target tissue, but siRNA itself does not have any targeting moieties. To provide tissue-targeting efficacy, the introduction of a nanovehicle system, specific cell-targeting molecules, or both to siRNA has been widely studied. When siRNA is transported via nanocarriers, the resulting nanoparticles can be diffused into liver tissue through the fenestrated blood vessels or into tumor tissue through immature leaky endothelia. Inefficient lymphatic drainage in tumor tissue attributes to the retention of nanovehicles; this phenomenon is called enhanced permeability and retention (EPR) effects [49]. The incorporation of targeting molecules, such as antibodies, aptamers, and ligands for cell surface receptors, enables the recognition of specific types of cells [31].

The intracellular entrance of siRNA is hampered by large size (~15 kDa) and highly negative charge [30]. A positively charged carrier helps not only nanosized particle formation but also crossing of the negatively charged cell membrane. Receptor-binding ligands improve receptor-mediated endocytosis, and cell penetrating peptides (CPPs) have been also widely used to enhance cellular uptake of siRNA [50]. When siRNA is endocytosed, it should escape from the endosomes before the relocation to lysosomes, which contain nucleases. Considering the acidic environment of late endosomes, gene carriers having pH-responsive proton sponge effects or membrane disruption activities allow the endosomal escape of siRNA [51].

## 6.2.3 Development of Efficient siRNA Delivery System

For the use of siRNA in the clinic, the aforementioned intrinsic and delivery problems of siRNA must be overcome. The inherent properties of siRNA, including easy degradation against serum nucleases, unwanted off-target effects, and immunogenicity, can be somewhat conquered by chemical modification at the specific position or sequence of the nucleic acids and by structural alteration [52]. As will be described later, diverse chemical modification strategies and siRNA structural variants have been developed and are applicable to improve serum stability, to minimize off-target effects, and to reduce immune stimulation.

The weaknesses of siRNA in delivery issues, such as short blood circulation time, lack of targeting moieties, and difficulty of subcellular localization, can be overcome by using effective gene carriers. A delivery vehicle may protect siRNA from nuclease attack and from detection by macrophages. Nanoparticle formation itself by gene carriers and further introduction of targeting molecules can improve the delivery efficiency of siRNA into specific target tissues and cells. Gene carriers can be adopted by two strategies: direct conjugation of carrier with siRNA and complex formation between carrier and siRNA [32, 53]. The former generally includes covalent linkage between siRNA and carriers. Lipophilic molecules or polymers prolong the blood circulation of siRNA, and aptamers are used to provide targeting efficacy to siRNA. The latter groups of the siRNA delivery system have been extensively focused to use cationic carriers. Considering the stiff structure and low charge density of siRNA, however, the enhancement of physicochemical properties of siRNA has been required to improve the interaction between siRNA and gene carriers; thus polymerization strategies of siRNA have been proposed in the past decade.

## 6.3 CHEMICAL MODIFICATION OF siRNA

### 6.3.1 COMMON CHEMICAL MODIFICATION STRATEGIES

The most common backbone modification of siRNA is the substitution of nonbridging phosphate oxygen to sulfur (phosphorothioate, PS) (Figure 6.3). According to previous studies, PS modification of antisense oligonucleotides resulted in improved nuclease resistance and favorable pharmacokinetics [54]. Similarly, siRNA with PS modification exhibited high serum stability and high blood concentration in the period of time shortly post-injection [55]. Moderate PS modification of siRNA improved gene-silencing activity although the effects are highly position dependent; PS-modified siRNA at the position of 3, 5, and 17 from the 5′-end of the sense strand showed high silencing effects by improvement of RISC loading of the antisense strand [56]. However, a high degree of PS modification led to severe toxic effects, presumably attributing to the nonspecific binding to cellular membrane proteins; siRNA with 50% PS content (PS modification in every second nucleotide) showed cytotoxicity and reduced cell growth [57]. The substitution of two nonbridging oxygen atoms with sulfurs, called phosphodithioate (PS2), also resulted in enhanced serum stability and higher gene-silencing activity, which is position dependent [58]. Alternative backbone modifications include boranophosphate substitution, obtained by introduction of the BH3 group in place of nonbridging phosphate oxygen. This modification thermodynamically destabilizes siRNA with the decrease of

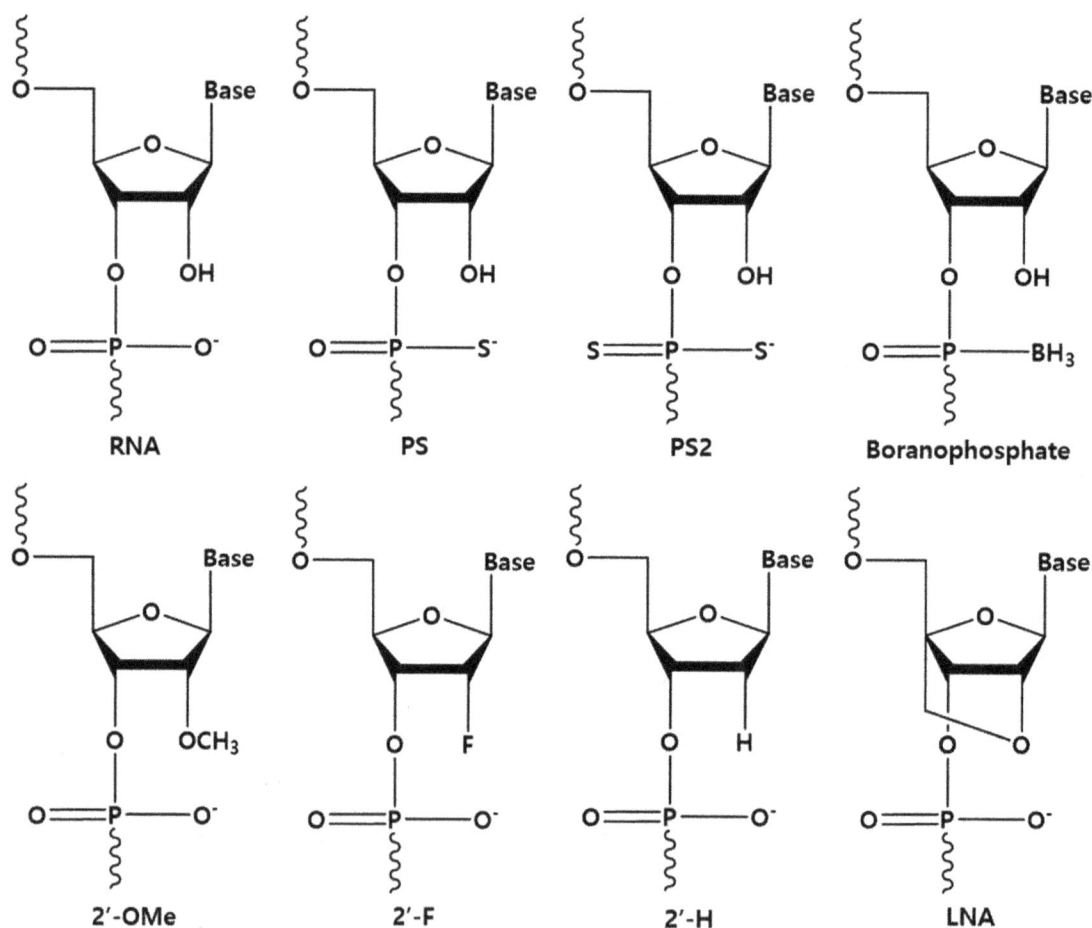

**FIGURE 6.3**  Popular chemical modifications of siRNA. LNA, locked nucleic acid; PS, phosphothioate; PS2, phosphodithioate; RNA, ribonucleic acid; 2′-OMe, 2′-O-methyl; 2′-F, 2′-fluoro; 2′-H, 2′-deoxy.

thermal stability ($T_m$) (0.5°C–0.8°C per modification) [59]. Boranophosphate modification resulted in siRNA potency when the seed region of the antisense strand was not modified, and serum stability of siRNA was enhanced after this modification [60].

Ribose 2'-OH is one of the most attractive modification sites because 2'-OH is not required for recognition by RNAi machinery or for the mRNA cleavage process by activated RISC [61]. Chemical modification of ribose 2'-OH involves the substitution of 2'-OH to other chemical groups, such as 2'-O-methyl (2'-OMe), 2'-F, and 2'-H (Figure 6.3). 2'-OMe modification improved the resistance to enzymatic digestion and thermal stability (0.5°C–0.7°C increase in $T_m$ per modification) [62]. When the antisense strand or both strands of siRNA were fully modified with 2'-OMe, the RNAi activity was completely abolished, whereas the same modification in the sense strand did not modulate the gene-silencing efficacy [63]. In contrast, the substitution of 2'-OH with fluorine (2'-F) can be accepted in both antisense and sense strands without loss of gene-silencing activity [61, 64]. 2'-F modification enhanced serum stability and the binding affinity of siRNA duplex (~1°C increase in $T_m$ per modification) [61, 65, 66]. 2'-H modification, DNA itself, is also well tolerated in siRNA duplex, particularly in the sense strand and at the end region (3'-overhangs or 5'-end of antisense strand) [67, 68].

Intramolecular linkage of 2'-oxygen to 4'-carbon is the alternative strategy for 2'-OH modification. The bridged nucleic acids contain the linkage between 2' and 4' positions of ribose ring via the methylene bridge (locked nucleic acid, LNA) or ethylene bridge (ethylene-bridged nucleic acid, ENA) (Figure 6.3). LNA modification locked the sugar ring in 3'-endo conformation, which increases in $T_m$ by 2°C–10°C per modification [69]. Further, this modification is highly position sensitive; the introduction of LNA modification at 10, 12, and 14 positions of the antisense strand abolishes RNAi activity due to the steric and conformational change near the cleavage site [70]. LNA modifications at 3'-overhangs protect siRNA from the 3' exonucleases, subsequently leading to improved serum stability [70].

With current bioorganic techniques, oligonucleotides can be synthesized and modified as single strands, then annealed into the desired double-stranded material. Customizable oligonucleotide synthesis incorporating artificial modifications enhances the potential of RNA therapeutics by overcoming problems associated with administration of naked siRNA. In particular, unmodified siRNA exposed in the bloodstream stimulates the innate immune response and is readily degraded by serum nucleases. One of the methods to increase stability in serum and potency of gene-silencing efficacy is to employ chemical modifications on the RNA-backbone of siRNA. A wide variety of chemical modifications, listed in Figure 6.4, have been proposed to overcome existing challenges of siRNA therapeutics.

One of the most common alterations of RNA is modification of the 2' position on the ribose backbone. These modifications include 2'-O-methyl, 2'-O-methoxyethyl, 2'-deoxy-2'-fluorouridine, locked nucleic acid (LNA), and many more [69, 71–73]. These chemical modifications increase stability against nucleases and improve thermal stability. As a naturally occurring RNA variant, 2'-O-methyl RNA has shown reduced potency or even inactivation in siRNA activity in the RNAi pathway upon heavy modification [64]. The 2'-fluoro modification is compatible with siRNA function and lends stability in the presence of nucleases. Combined modification with 2'-fluoro pyrimidines and 2'-O-methyl purines results in highly stable RNA duplexes in serum and improved in vivo activity [74]. The 2'-O-methoxyethyl RNA modification has also shown significant nuclease resistance as well as increased $T_m$. Nevertheless, this modification is not generally used as frequently as the 2'-O-methyl and 2'-fluoro RNAs. LNA contains a methylene bridge that connects the 2'-O with the 4'-C positions of the ribose backbone. This causes the siRNA to have "locked" sugar that results in higher stability with increased $T_m$. Though incorporation of LNA also interferes with the siRNA activity, limited modification retains the functionality [64].

In addition to the sugar modifications, variations in phosphate linkage of siRNA are also accepted as an alternative strategy to overcome functional limitations. The PS linkage, perhaps the most commonly modified linkage in siRNA, often displays cytotoxicity when used extensively; however, PS incorporation does not appear to have a major effect on biodistribution of siRNA [75].

Apart from modifications made on the backbone, chemical modifications are also made on other parts of siRNA to facilitate delivery to the target site. One of the hurdles in siRNA delivery is that weak negative charge and high molecular weight make the nucleic acid more prone to serum degradation and capture by the reticuloendothelial system (RES). In order to form more stable delivery complexes, polymerized siRNA can be synthesized, resulting in greater electrostatic interactions and facilitating incorporation into nanoparticles. Lee et al. developed polymerized siRNA using a thiol group to form a stable complex with glycol chitosan via not only electrostatic interaction but also disulfide bond crosslinking. Polymerized siRNA synthesized with thiol groups was also shown to form stable complexes with PEI, albumin, transferrin, hyaluronic acid, and other nanoparticles [76–81]. This delivery reagent was shown to have an antitumor effect in xenograft cancer models when systemically injected.

Other chemical alterations of siRNA include base modification, change in overhangs and termini of the RNA duplexes, and varying tertiary structure of the siRNA. In an attempt to develop siRNA for use in clinical trials as drugs, various chemical modifications are being investigated to improve qualities such as serum stability, siRNA potency, low immunostimulation, off-target effects, and target organ/cell delivery [82].

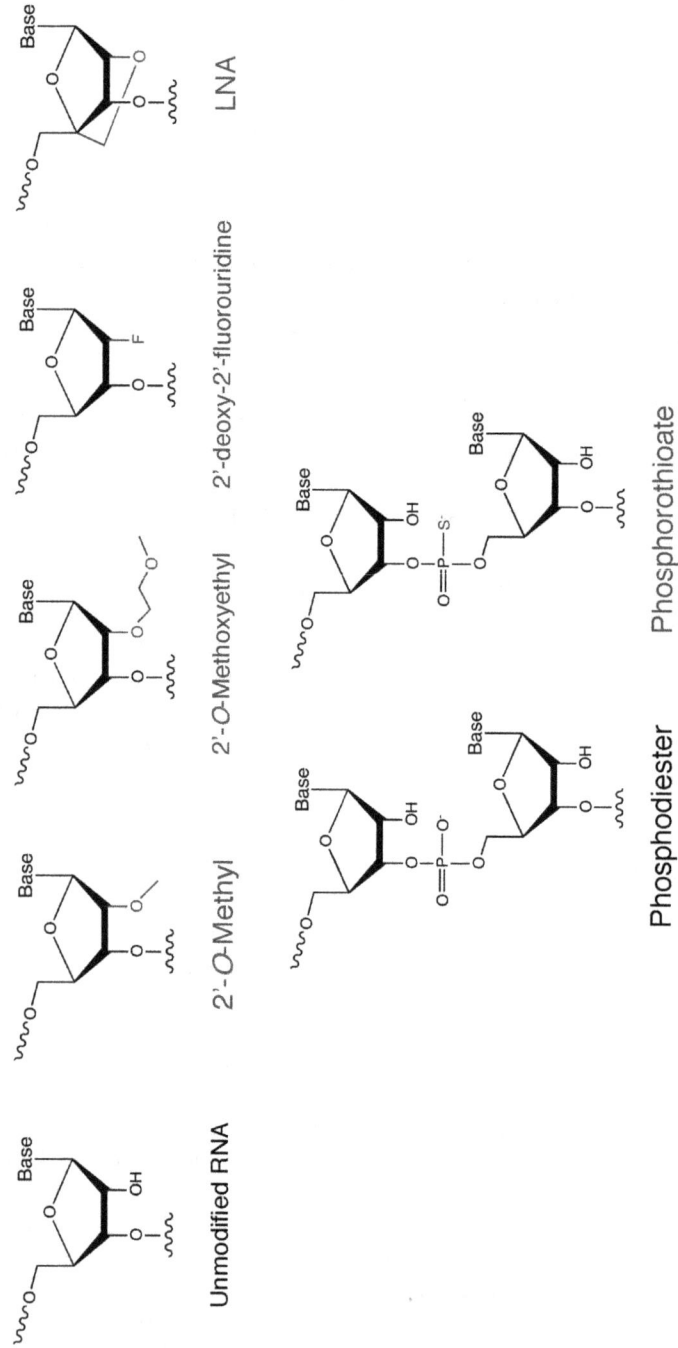

**FIGURE 6.4** Chemical modifications and siRNA.

## 6.3.2 Applications of Chemically Modified siRNA

As stated above, chemical modification can be used to overcome the inherent problems of siRNA. Considering that several nucleases catalyze the nucleophilic attack of 2′-OH and the hydrolysis of the interphosphate linkage in siRNA, modification of ribose 2′-OH position enables an improvement in serum stability. Particularly, 2′-OMe and 2′-F modification of nuclease-sensitive regions, such as UA and CA motifs, dramatically enhanced the resistance to nuclease digestion [33, 62, 83]. A combination of different modification strategies results in the highly stable siRNA in vivo; a successful example was the modified siRNA consisting of a sense strand with 2′-F on pyrimidine, 2′-H on purines, and 5′- and 3′-inverted a basic end caps and antisense strand with 2′-OMe on purines and PS at the 3′-terminus [84]. The 3′-overhangs are also susceptible to exonuclease attack; the chemical modified 2-nt 3′-overhangs, such as LNA modification, reduced the siRNA degradation in serum [70].

Off-target effects of siRNA via the miRNA-like pathway are dependent on the seed region homology with 3′-UTR of mRNA. Although in silico siRNA-mRNA sequence matching predictions may reduce these types of off-target effects, they cannot be fully avoided. According to the previous literature, chemical modification of the antisense strand modulated the undesired gene-silencing effects; 2′-OMe modification at position 2 of the antisense strand and introduction of eight DNA in the antisense strand seed region reduced the downregulation of nontarget gene expression [21, 85]. Increase in the incorporation selectivity of the antisense strand into RISC can modulate off-target effects, as a result of the contribution of sense strand to the gene-silencing process. Considering that phosphorylation of 5′-terminus is required for RISC activation [86], 5′-end modification of sense strand via 5′-OMe or LNA reduced its participating in the RNAi mechanism [87, 88].

Short RNA exhibits immunostimulatory properties; mediated by the TLR family or PKR, these exogenous siRNA-triggered immune responses can be decreased by applying several modification strategies. Concerning the incorporation of U-rich motif in TLR activation, the modification at ribose 2′-OH position of uridine residue enabled the minimization of the siRNA immunogenicity [89]. Introduction of alternating 2′-OMe modification in the sense strand reduced the cytokine induction without loss of gene-silencing activity [90]. LNA modification of the sense strand also blocked the TLR activation [91]. The activation of cytoplasmic PKR after intracellular delivery of siRNA can be abrogated by the reduction of hydrogen bonding between the RNA minor groove and PKR domain; chemical modification for hydrogen-bonding alteration, such as 2′-H or 2′-F modification, reduced PKR activation [92].

## 6.4 siRNA STRUCTURAL VARIANTS

A myriad of structural variants of RNAi-based therapeutics with their own advantages and disadvantages has been reported to improve the intrinsic properties of siRNAs. It

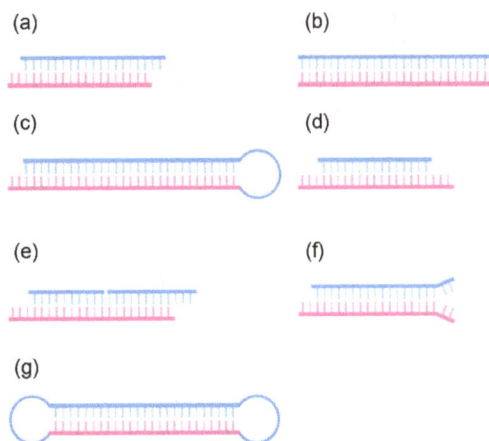

**FIGURE 6.5 Schematic diagrams of structural variants of siRNA.** (A) Conventional siRNA of 19 + 2. (B) 27-mer long dsRNA Dicer substrate. (C) 29-mer short-hairpin RNA. (D) Asymmetric siRNA. (E) Internally segmented siRNA. (F) Fork-shaped siRNA. (G) Dumbbell-shaped circular RNA.

suggests that nature's RNAi pathway machinery can tolerate various structurally different mediators of gene silencing. Tuschl and colleagues demonstrated the first successful sequence-specific gene silencing using chemically synthesized exogenous siRNA in mammalian cells without causing innate immune responses [93, 94]. The siRNA has a 19 base-paired duplex with 3′-end 2-nt overhangs at both the sense and antisense strand (19 + 2 traditional siRNAs) and is the most widely used siRNA (Figure 6.5(a)). The 19 + 2 siRNA has a structural similarity to nature's Dicer product.

### 6.4.1 RNAi Triggers with Increased Potency

Designing highly potent gene-suppressing RNAi triggers is one of the important goals for successful clinical application of RNAi therapeutics. In efforts to find RNAi triggers that work efficiently at a low concentration, Dicer substrates were found by two groups. Kim et al. found that long synthetic 27-mer duplex RNA without overhangs can be substantially more efficient in gene silencing than the corresponding traditional 21-mer siRNA (Figure 6.5(b)) [95]. In company with this report, Siolas et al. also identified a synthetic short-hairpin RNA (shRNA) as a potent mediator of RNAi (Figure 6.5(c)) [96]. The shRNAs composed of 29 base-paired stems with 2-nt 3′-overhangs and 4-nt loops. Both 27-mer RNA duplex and 29-mer shRNA were processed to 21- or 22-mer siRNA by Dicer (RNase III-family endonuclease) in vitro. It was reported that Dicer was involved in not only cleaving long dsRNAs but also RISC loading of processed RNA and RISC assembly [97, 98]. The improved potency of these RNAi triggers may be attributed to the fact that they are Dicer substrates. SiRNA of ~21 nt in length was produced when recombinant Dicer was treated to both 27-mer duplex RNA and 29-mer shRNA. Dicer processing may enhance the loading and incorporation of siRNA into RISC, thereby increasing gene-silencing efficiency. Furthermore, the Dicer substrate RNA did not induce innate immune responses, such as interferon production and PKR activation.

Exogenously introduced high concentration of siRNA can cause a saturation of cellular RNAi proteins, which would hinder the RNAi pathway and cause toxicity [99]. In addition, extents of unwanted off-target effects were proportional to the siRNA treatment concentration [100]. Therefore, these highly potent effectors of gene silencing will facilitate clinical translocation of RNAi therapeutics.

### 6.4.2  RNAi Triggers with Reduced Off-Target Effect

Incorporation of the sense strand into RISC is a main undesired off-target effect of RNAi-based therapeutics [101]. Novel designs of siRNA structural variants to reduce the off-target effects were suggested by Sun et al. [102]. They investigated whether asymmetric RNA duplexes with various lengths could induce gene silencing and found that asymmetric RNA duplexes with short 15-nt sense strand having both 3′ and 5′ antisense overhangs could mediate gene silencing (Figure 6.5(d)). The asymmetric interfering RNA (aiRNA) was incorporated into RISC more effectively than inhibited target gene expressions sequence specifically. More importantly, the sense strand–mediated off-target effects were reduced compared with conventional siRNAs, which may be attributed to the nature of structural asymmetry. They speculated that the asymmetric structure leads to preferential incorporation of antisense strand into RISC more than short sense strand, which resulted in reduced off-target effects.

Along with the aiRNA, a novel design of small internally segmented interfering RNA (sisiRNA) also showed decreased off-target effects caused by loading of sense strand into RISC (Figure 6.5(e)) [103]. The sisiRNA had an intact antisense strand and a sense strand that was divided into two segments. Because incorporation of the segmented sense strand into RISC was excluded and only the antisense strand could be loaded into RISC, this structural siRNA variant showed reduced off-target effects and increased target specificity.

Fork-shaped siRNA having 1- to 4-nt mismatch at the 3′-end of the sense strand was another structure that showed increased target specificity while maintaining gene-silencing activity (Figure 6.5(f)) [104]. The section suggested that the mismatch part may render the antisense strand more favored to be incorporated into RISC. It was reported that thermodynamically less stable 5′-end of siRNA was preferentially incorporated into RISC during the strand selection [105, 106]. Although chemical modifications to reduce off-target effects have been reported, structure-based asymmetry provides another way to overcome these problems [21].

### 6.4.3  RNAi Triggers with Increased Stability

Natural RNAs are rapidly degraded in biological fluid. Chemical modification can enhance siRNA stability,

although it often causes toxicity or decreases gene-silencing activity [107]. A method to increase RNA stability using natural RNA was proposed by Abe et al. [108]. One more loop was added into shRNA using T4 RNA ligase, which resulted in dumbbell-shaped circular RNA structure (Figure 6.5(g)). Due to the endless structure of the dumbbell-shaped RNA, it showed higher stability when treated with exonuclease, compared with the linear form of siRNA. In addition, the RNA dumbbell was processed more slowly and exhibited prolonged RNAi activity.

A variety of structural variants of siRNA have been reported with improved features including higher potency, reduced off-target effects, and increased stability. There is substantial structural flexibility of gene-silencing mediators in nature although more detailed structural, biochemical, and biological studies in RNAi mechanism are demanded. Optimization of the siRNA structure will provide safety and efficacy for clinical applications of RNAi therapeutics.

## 6.5  siRNA CONJUGATE SYSTEM

### 6.5.1  Lipophile–siRNA Conjugates

The introduction of lipophilic molecules, such as cholesterol and α-tocopherol, can improve the pharmacokinetic properties as well as cellular uptake of siRNA (Figure 6.6(a)) [109]. Cholesterol-conjugated siRNA (Chol-siRNA) exhibited prolonged blood-circulation time ($t_{1/2}$ of 95 min), compared with naked siRNA ($t_{1/2}$ of 6 min), after systemic administration, presumably attributing to the enhanced binding to serum proteins [24]. This enhanced pharmacokinetics resulted in the increase of specific gene-silencing efficacy; Chol-siRNA against apolipoprotein B (apoB) led to the downregulation of target mRNA in liver and jejunum and the decreased level of plasma apoB protein and serum cholesterol. Because lipoprotein particles, including both high-density lipoprotein (HDL) and low-density lipoprotein (LDL), play a critical role in cholesterol transport in vivo, the delivery of Chol-siRNA may be facilitated when Chol-siRNA is preassembled with HDL or LDL.

According to the previous results, Chol-siRNA with HDL was delivered into liver, gut, kidney, and steroidogenic organs, whereas LDL directed the Chol-siRNA primarily into the liver [110]. The binding with HDL or LDL further enhanced the cellular uptake of Chol-siRNA via HDL or LDL receptor, respectively.

The conjugation of lipophilic molecules can be used as a targeting moiety for siRNA delivery to specific tissue. Considering that α-tocopherol (vitamin E) is incorporated into lipoproteins and travels into liver for hepatic uptake, α-tocopherol-conjugated siRNA (Toc-siRNA) was delivered into the liver [111]. Toc-siRNA targeting apoB achieved the reduction of liver apoB mRNA expression and serum triglyceride/cholesterol level. Furthermore, neither interferon induction nor other side effects were observed after systemic administration of Toc-siRNA.

## 6.5.2 POLYMER–SIRNA CONJUGATES

The introduction of poly(ethylene glycol) (PEG) provided the stealth functionality, which avoids capture by RES, consequently resulting in prolonged blood circulation time. The half-life of siRNA was increased from 5 min to 1 h after conjugation with 20 kDa PEG, and the distribution of PEG-siRNA in liver, kidney, spleen, and lung was observed without significant degradation [112]. When PEG–siRNA conjugate was complexed with polyethylenimine (PEI), negatively charged siRNA and positively charged PEI formed polyelectrolyte core and hydrophilic PEG was present on the particle surface (Figure 6.6(b)) [113, 114]. The resultant PEG-siRNA/PEI nanocomplex exhibited enhanced serum stability and excellent tumor-targeting efficacy without any induction of interferon. When siRNA targeting vascular endothelial growth factor (VEGF) was used, the local or systemic administration of PEG-siRNA/PEI complex achieved the reduction of microvessel formation and the suppression of tumor growth [114]. When the polymer end was decorated with targeting moiety, such as folate and lactose, the delivery of PEG-siRNA into specific cell type was further improved [115, 116].

Hyaluronic acid (HA), a natural polymer having biocompatibility and binding affinity to CD44, has also been widely used as a siRNA carrier. Recently, HA–siRNA conjugates containing bioreducible disulfide bonds were developed [117, 118]. The HA conjugation led to the enhancement of serum stability and the compact nanocomplex formation with cationic carriers (Figure 6.6(b)). Further, HA-siRNA/PEI nanocomplex was successfully internalized into cells via receptor-mediated endocytosis and downregulated the target gene expression in CD44-expressing cancer cells. Attributing to the abundance of HA receptors in liver, HA–siRNA conjugate was delivered specifically into liver and silenced the target gene expression after systemic administration [118].

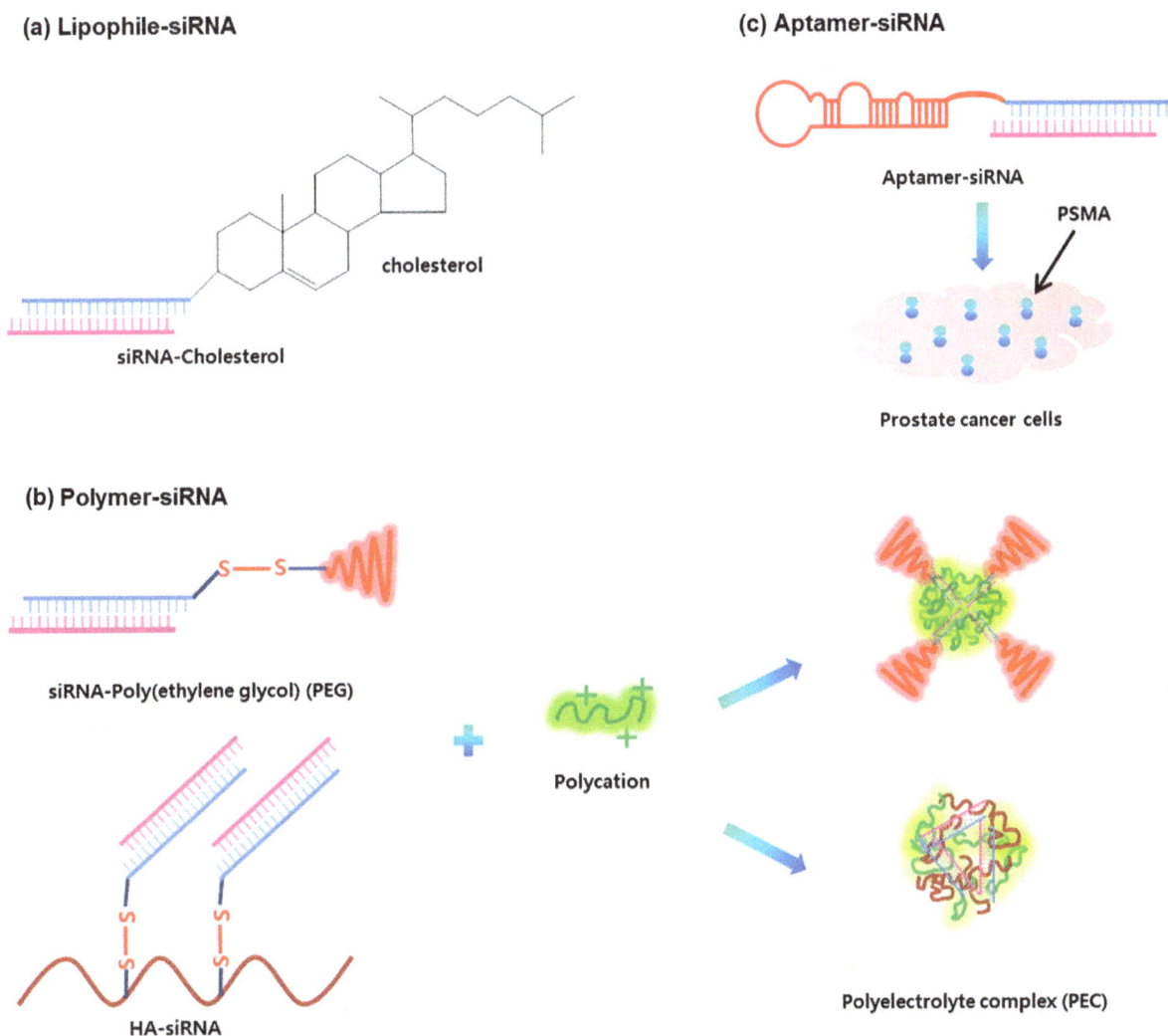

**(a) Lipophile-siRNA**

cholesterol

siRNA-Cholesterol

**(b) Polymer–siRNA**

siRNA-Poly(ethylene glycol) (PEG)

HA-siRNA

Polycation

**(c) Aptamer-siRNA**

Aptamer-siRNA

PSMA

Prostate cancer cells

Polyelectrolyte complex (PEC)

**FIGURE 6.6** **Schematic diagrams of siRNA conjugate system.** (a) Cholesterol–siRNA conjugates. (b) Polymer–siRNA conjugate, containing bioreducible disulfide linkage. The polymer-siRNA conjugate can form a stable polyelectrolyte complex with polycation. (c) Aptamer–siRNA chimera. PSMA-specific A10 aptamer is linked to siRNA.

### 6.5.3 APTAMER-siRNA CHIMERAS

Aptamer is single-stranded nucleic acids having high affinity to target molecules; thus it has been extensively studied as a targeting molecule in biomedical fields. Aptamer-siRNA chimera exhibited high specificity to adhere against target protein–expressing cells. For example, prostate-specific membrane antigen (PSMA)–targeting A10 aptamer specifically delivered siRNA into PSMA-expressing cell and tumor (Figure 6.6(c)) [119, 120]. A10 aptamer-siRNA chimera bound to PSMA on cell surfaces and mediated cellular uptake. After intracellular translocation, A10 aptamer-siRNA chimera was processed by Dicer and released the active siRNA. When A10 aptamer-siRNA targeting polo-like kinase 1 (plk1) was systemically administrated, the gene silencing and tumor growth inhibition were observed in PSMA-expressing tumor [120]. The incorporation of PEG into A10 aptamer-siRNA chimera further enhanced the pharmacokinetic and pharmacodynamic properties of the chimera [119]. Similarly, nucleolin-targeting aptamer has been considered as a promising tumor-targeting moiety due to the high expression of nucleolin on various cancer cells [121]. Combined treatment of nucleolin aptamer-siRNA chimeras against snail family zinc finger 2 (SLUG) and neuropilin 1 (NRP1) synergistically suppressed the invasion of lung cancer cells and tumor-induced angiogenesis.

## 6.6 siRNA POLYMERIZATION

Efficient and safe delivery of siRNA to target tissues and cells is one of the most critical goals for therapeutic application of siRNAs. A variety of siRNA delivery carriers based on polymers, lipids, and nanoparticles have been devised to improve delivery of siRNA [122]. The cationic polymers can form condensed polyelectrolyte complexes with anionic nucleic acid by electrostatic interactions. Furthermore, because synthetic cationic polymers have merits of the facile introduction of functional moieties and modification of its structure and molecular weight, they have been widely used as nucleic acid delivery carriers [123].

However, siRNA delivery is much more difficult than plasmid DNA delivery due to their different intrinsic physicochemical properties. The persistence length of dsRNA is approximately ~70 nm (corresponding to ~260 bp) [124]; therefore, siRNA of ~21 bp behaves like a rigid rod. Furthermore, having ~42 negative charges per molecule, siRNA has much lower spatial charge density than plasmid DNA, which has more than several thousand negative charges per molecule. The rigidity and low charge density of siRNA make it difficult to form small, stable, and condensed complexes through efficient electrostatic interactions with cationic polymers.

The shape, size, and surface properties of nanoparticles significantly affect the cellular uptake and in vivo biodistribution of nanoparticles [125]. It has been reported that stable, compact, and small nanoparticles were more efficiently taken up by cells [126]. Therefore, making stable, compact, and small nanocomplexes of siRNA/cationic polymers is a prerequisite of successful siRNA delivery systems. Although it is possible to make more stable and compact complexes by using large amounts of high-molecular-weight cationic polymeric carriers, the toxicity also depends on the concentration and structure of polymers [127]. Recently, structural modification of siRNA itself has been reported along with chemical modification of carriers to develop efficient delivery systems without causing toxicities and immune responses.

### 6.6.1 STICKY siRNA

A gene-like structure was constructed by connecting several siRNAs together to increase the charge density of the siRNAs [128]. Short additional complementary $A_{5-8}/T_{5-8}$ overhangs were introduced into 3′-ends of siRNA (called sticky siRNA) and siRNA concatemers were constructed by hybridizing the sticky siRNAs in aqueous solution (Figure 6.7(a)). When complexed with PEI, a widely used cationic polymer, the sticky siRNA concatemer showed increased complex stability and protection of siRNAs. The sticky siRNA/PEI complexes resulted in enhanced gene silencing in cultured A549 cells and in vivo mouse lung. The enhanced stability and delivery efficacy of the complexes were attributed to increased charge density of sticky siRNAs, which enabled more efficient electrostatic interactions with PEI. This study demonstrated the concept that increasing size of siRNA like pDNA could enhance complex stability and delivery efficacy even though there was stability concern with the sticky siRNA concatemers, with A8/T8 overhangs having a low melting point ($T_m < 10°C$).

### 6.6.2 MULTI-siRNA AND POLY-siRNA

Chemical crosslinking of several siRNAs together to increase the size was reported by two independent groups [129, 130]. Mok et al. synthesized multimerized siRNA (multi-siRNA) using a dithio-bismaleimidoethane (DTME), a cleavable chemical crosslinker (Figure 6.7(a)) [129]. Thiol-modified sense and antisense strands at 3′ end were reacted with DTME to construct dimeric sense and antisense strands. The prepared dimeric sense and antisense strands were hybridized by complement base pairing to produce multi-siRNAs. In company with this report, Lee et al. also prepared polymerized siRNA (poly-siRNA) without using chemical crosslinkers. Poly-siRNA was obtained by direct oxidation of siRNAs thiol-modified at 5′-end of both sense and antisense strands (Figure 6.7(a)) [130].

The synthesized multi-siRNA showed ladder-like band patterns on polyacrylamide electrophoresis analysis, which implied that mixtures of multi-siRNAs with various degrees of crosslinking were obtained. When incubated with reducing agents of dithiothreitol or glutathione, multi-siRNA was cleaved into monomeric siRNAs, which are active components of RNAi. It is well known that intracellular cytosol is more reductive than extracellular space [131]. Therefore, it

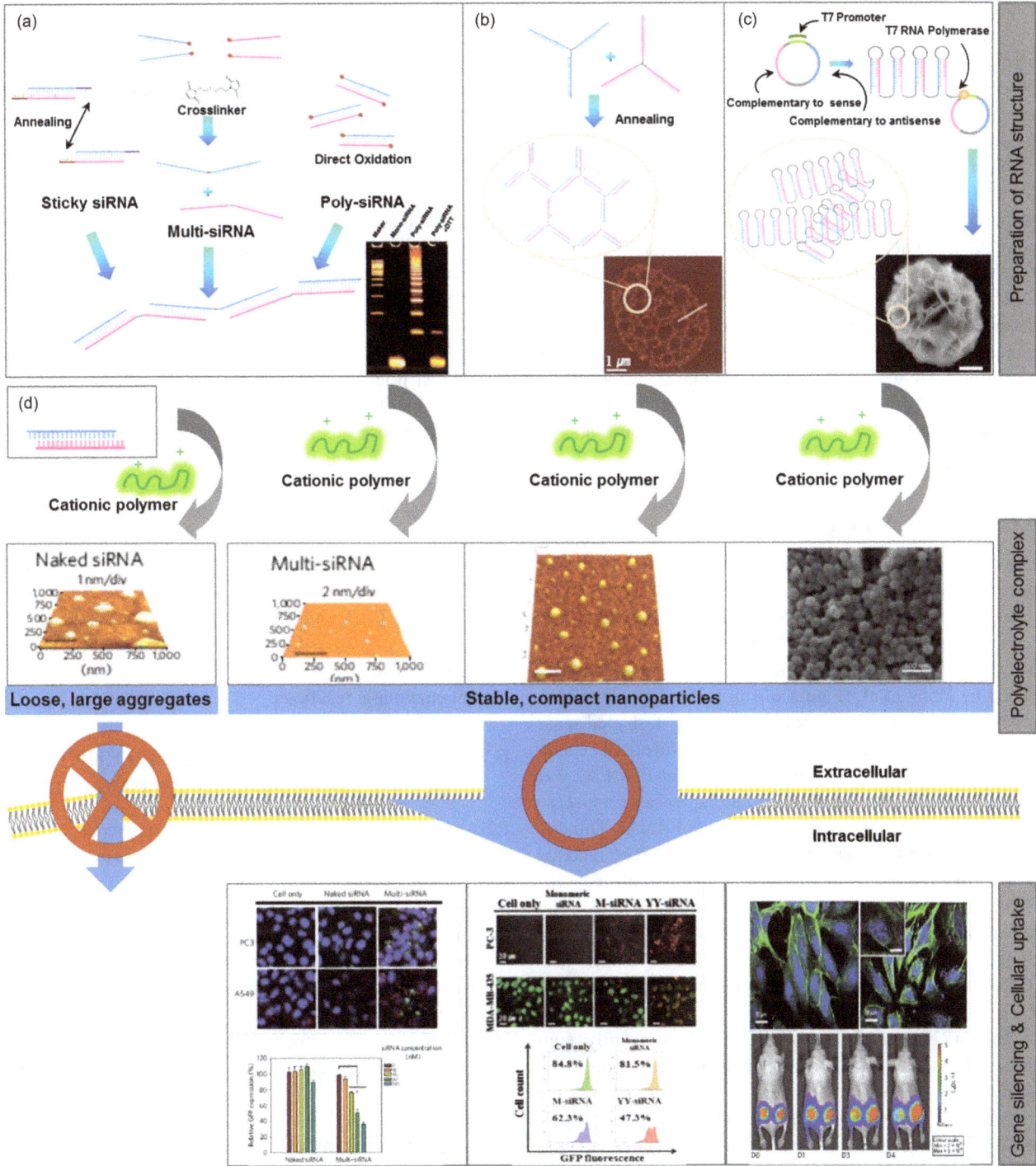

**FIGURE 6.7** Structured siRNAs for efficient condensation and delivery (preparation-*upper*, polyelectrolyte complex-*middle*, gene silencing/cellular uptake-*bottom*). (a) Polymerized siRNAs via complementary annealing, chemical crosslinker, or direct oxidation of thiol groups. (b) SiRNA microhydrogel. (c) RNAi microsponge. (d) Conventional siRNA. (Reprinted with permission from Ref. [129], Copyright 2010 Nature Publishing Group; Ref. [130], Copyright 2010 Elsevier; Ref. [135], Copyright 2011 American Chemical Society; and Ref. [136], Copyright 2012 Nature Publishing Group.)

was anticipated that monomeric siRNA would be generated by cleavage of disulfide bonds in the reductive cytosolic environments after cellular uptake.

According to the morphology analyses, multi-siRNA formed more stable and compact nanocomplexes with linear PEI (LPEI), more biocompatible cationic carrier than branched PEI (bPEI), whereas monomeric-siRNA/LPEI complexes showed unstable, large, and loose aggregates. The more stable and compact complex formation was largely attributed to increased charge density and introduction of flexible linkage of multi-siRNA, which enabled more efficient electrostatic interaction and entanglement with cationic polymers. Thus, much larger amounts of multi-siRNA/LPEI complexes were entered into cells, and thereby significantly enhanced gene silencing were observed, compared with tomonomeric-siRNA/LPEI complexes, in vitro PC3 cell and in vivo PC3 xenograft mouse model (Figure 6.7).

Nonspecific immune responses should be considered for clinical application of siRNAs because they could be induced by long dsRNAs [132]. The multi-siRNA/LPEI complexes did not elicit significant undesired INF-α induction when treated into peripheral blood monocyte cell or ICR mice. Furthermore, sequence-specific mRNA degradation was confirmed through reverse transcription-polymerase chain reaction (RT-PCR) and rapid amplification of cDNA ends (RACE). It was suggested that cleavable chemical linkage between siRNAs and regeneration of short monomeric-siRNA after cellular uptake of multi-siRNA prevented induction of immune responses.

### 6.6.3 siRNA Microhydrogel

Linear gene-like structural modification of siRNA further proceeded to construct three-dimensional (3D) siRNA structures. Although many 3D structures made of DNA have been reported, few of them have biological activities [133, 134]. Microhydrogels composed of networked siRNAs were introduced by Hong et al. (Figure 6.7(b)) [135]. The 3D siRNA microhydrogels were prepared by annealing Y-shaped sense strand with Y-shaped antisense strand. A trifunctional chemical crosslinker, tris-(2-maleimidoethyl)amine (TMEA), was reacted with thiol group at the 3′-end of sense or antisense strand to prepare Y-shaped single-stranded RNAs. Dimeric sense or antisense strands were also synthesized by reacting 3′ thiol-modified sense or antisense strand with bifunctional chemical crosslinker, 1, 8-bis(maleimidodiethylene) glycol (BM(PEG)$_2$). By controlling the ratios of Y-shaped and dimeric RNAs, several micrometered siRNA hydrogels with different pore size were obtained in aqueous solution through complement base pairing.

The siRNA-based microhydrogels were readily condensed to stable, ~100 nm of nanocomplexes upon interacting with a cationic polymer LPEI. Furthermore, the nanoscale complexes showed not only superior cellular uptake but also greatly enhanced gene-silencing activity in breast cancer cells (Figure 6.7). Significantly increased charges as well as flexibility of the siRNA microhydrogels enabled efficient condensation with cationic polymers. Even though no cleavable bond was introduced in siRNA microhydrogels, the construct was processed by Dicer and the processed product would participate in the sequence-specific gene inhibition.

### 6.6.4 RNAi Microsponge

Besides chemical reactions, biological enzymatic reactions were also used to produce condensed RNA 3D structures [136]. Single-stranded circular DNA encoding both strands

of siRNA was used as a template for rolling circle transcription (RCT) to produce hairpin RNA polymers (Figure 6.7(c)) [137]. Owing to the in vitro RCT process, a large amount of tandem repeats of hairpin RNA transcripts could be obtained efficiently. Interestingly, the RNA tandem repeats self-assembled into pleated sheets, which subsequently formed sponge-like microspheres (RNAi microsponge). The RNAi microsponge itself formed a highly dense structure without additional cationic materials.

Considering the molecular weight and concentration, approximately more than a half million copies of hairpin RNAs were included in a single RNAi microsponge. The tandem repeat of RNA was designed to generate ~21 nt siRNA under Dicer processing, and it was confirmed that RNAi microsponge was cleaved after treatment with recombinant Dicer. Due to its much higher negative charge of RNAi microsponge, cationic PEI readily interacted with the particle and formed ~200 nm condensed nanoparticles. The compact nanoparticle complexes exhibited superb cellular uptake and specific gene silencing (Figure 6.7). It is worthy of notice that extremely low numbers of RNAi/PEI particles were needed to induce similar gene-silencing efficiencies, compared with conventional nanoparticle delivery systems. The RNAi microsponge siRNA delivery system provides an easy method for high loading of siRNA and for production of large amounts of RNAi triggers using biological enzymatic reactions.

## 6.7 ADVANCES AND HURDLES TO CLINICAL TRANSLATION OF RNAi THERAPEUTICS

RNAi technique has recently become a powerful tool for basic research to selectively knock down gene expression in vitro and in vivo. At the same time, both the scientific and industrial communities have begun to develop RNAi therapeutics as the next class of drugs for treating a variety of genetic disorders, such as cancer and other diseases that are particularly hard to be addressed by current treatment strategies.

## 6.8 CONCLUSION

SiRNA has received much attention due to its sequence-specific gene-silencing efficacy and universality in therapeutic targeting. Despite its promising potential as a gene therapeutic [138–141], many limitations to clinical applications of siRNA remain to be overcome; not only the inherent properties of siRNA but also delivery barriers have been considered as serious challenges in clinical translation of siRNA drugs. There have been many efforts to develop safe and efficient gene carriers, but, recently, the improvement of the physicochemical properties of siRNA itself has been accomplished through chemical and structural modifications. The chemically or structurally modified siRNA exhibits enhanced stability, reduced off-target effects, and minimized immunogenicity. Furthermore, the development of siRNA conjugates, the increase in siRNA size, and the construction of nucleic acid nanostructures could achieve advancements in siRNA delivery properties. Thus, rational design of the modified siRNA and integrating it with efficient delivery carriers can overcome hurdles to clinical translation of siRNA therapeutics.

Finally, a combination approach using siRNA with a variety of cancer therapies, such as chemotherapy, immunotherapy, radiation therapy, or photodynamic therapy, may dramatically improve the efficacy of cancer therapy. In this strategy, each form of therapy can be used on targets particularly suited to the therapy type, such as small-molecule inhibitors for kinase targets and siRNAs for targets that are structurally unsuited to small-molecule attack. Moreover, because different therapeutic modalities may trigger different forms of resistance mechanisms such as P-glycoprotein (P-gp), multidrug resistance–associated proteins (MRP1, MRP2) for small-molecule drugs, and other yet-to-be-determined modes for siRNAs, such multimodal therapies may be more difficult for tumors to circumvent. It is our strong hope that by skillful delivery and careful target selection, siRNA nanoparticles may take a prominent place in the armamentarium that is being assembled to treat the many diseases that constitute cancer.40

## REFERENCES

[1] J.A. Lindbo, L. Silva-Rosales, W.M. Proebsting, et al., Induction of a highly specific antiviral state in transgenic plants: Implications for regulation of gene expression and virus resistance, Plant Cell 5 (1993) 1749–1759.

[2] A. Fire, S. Xu, M.K. Montgomery, et al., Potent and specific genetic interference by double-stranded RNA in *Caenorhabditis elegans*, Nature 391 (1998) 806–811.

[3] C.V. Pecot, G.A. Calin, R.L. Coleman, et al., RNA interference in the clinic: challenges and future directions, Nat Rev Cancer 11 (2011) 59–67.

[4] M.E. Davis, J.E. Zuckerman, C.H. Choi, et al., Evidence of RNAi in humans from systemically administered siRNA via targeted nanoparticles, Nature 464 (2010) 1067–1070.

[5] G. Meister, T. Tuschl, Mechanisms of gene silencing by double-stranded RNA, Nature 431 (2004) 343–349.

[6] D.G. Sashital, J.A. Doudna, Structural insights into RNA interference, Curr Opin Struct Biol 20 (2010) 90–97.

[7] J. Martinez, A. Patkaniowska, H. Urlaub, et al., Single-stranded antisense siRNAs guide target RNA cleavage in RNAi, Cell 110 (2002) 563–574.

[8] D. Bumcrot, M. Manoharan, V. Koteliansky, et al., RNAi therapeutics: a potential new class of pharmaceutical drugs, Nat Chem Biol 2 (2006) 711–719.

[9] E. Iorns, C.J. Lord, N. Turner, et al., Utilizing RNA interference to enhance cancer drug discovery, Nat Rev Drug Discov 6 (2007) 556–568.

[10] S. Van de Veire, I. Stalmans, F. Heindryckx, et al., Further pharmacological and genetic evidence for the efficacy of PlGF inhibition in cancer and eye disease, Cell 141 (2010) 178–190.

[11] M. Zhao, J. Sun, Z. Zhao, Synergetic regulatory networks mediated by oncogene driven microRNAs and transcription factors in serous ovarian cancer, Mol BioSyst 9 (2013) 3187–3198.

[12] R.C. Doebele, Targeted therapies: time to shift the burden of proof for oncogene-positive cancer? Nat Rev Clin Oncol 10 (2013) 492–493.

[13] I. Collins, P. Workman, New approaches to molecular cancer therapeutics, Nat Chem Biol 2 (2006) 689–700.

[14] P. Resnier, T. Montier, V. Mathieu, et al., A review of the current status of siRNA nanomedicines in the treatment of cancer, Biomaterials 34 (2013) 6429–6443.

[15] J. DeVincenzo, J.E. Cehelsky, R. Alvarez, et al., Evaluation of the safety, tolerability and pharmacokinetics of ALN-RSV01, a novel RNAi antiviral therapeutic directed against respiratory syncytial virus (RSV), Antivir Res 77 (2008) 225–231.

[16] A. de Fougerolles, H.P. Vornlocher, J. Maraganore, et al., Interfering with disease: a progress report on siRNA-based therapeutics, Nat Rev Drug Discov 6 (2007) 443–453.

[17] M.A. Behlke, Progress towards in vivo use of siRNAs, Mol Ther 13 (2006) 644–670.

[18] A.M. Gewirtz, On future's doorstep: RNA interference and the pharmacopeia of tomorrow, J Clin Invest 117 (2007) 3612–3614.

[19] D.H. Kim, J.J. Rossi, Strategies for silencing human disease using RNA interference, Nat Rev Genet 8 (2007) 173–184.

[20] A. Aigner, Nonviral in vivo delivery of therapeutic small interfering RNAs, Curr Opin Mol Ther 9 (2007) 345–352.

[21] A.L. Jackson, J. Burchard, D. Leake, et al., Position-specific chemical modification of siRNAs reduces "off-target" transcript silencing, RNA 12 (2006) 1197–1205.

[22] J.T. Marques, B.R. Williams, Activation of the mammalian immune system by siRNAs, Nat Biotechnol 23 (2005) 1399–1405.

[23] S.J. Lee, S. Son, J.Y. Yhee, et al., Structural modification of siRNA for efficient gene silencing, Biotechnol Adv 31 (2013) 491–503.

[24] J. Soutschek, A. Akinc, B. Bramlage, et al., Therapeutic silencing of an endogenous gene by systemic administration of modified siRNAs, Nature 432 (2004) 173–178.

[25] D.M. Dykxhoorn, C.D. Novina, P.A. Sharp, Killing the messenger: short RNAs that silence gene expression, Nat Rev Mol Cell Biol 4 (2003) 457–467.

[26] G.J. Hannon, RNA interference, Nature 418 (2002) 244–251.

[27] A. de Fougerolles, H.-P. Vornlocher, J. Maraganore, et al., Interfering with disease: a progress report on siRNA-based therapeutics, Nat Rev Drug Discov 6 (2007) 443–453.

[28] D.M. Dykxhoorn, D. Palliser, J. Lieberman, The silent treatment: siRNAs as small molecule drugs, Gene Ther 13 (2006) 541–552.

[29] J. Wang, Z. Lu, M.G. Wientjes, et al., Delivery of siRNA therapeutics: barriers and carriers, AAPS J 12 (2010) 492–503.

[30] K.A. Whitehead, R. Langer, D.G. Anderson, Knocking down barriers: advances in siRNA delivery, Nat Rev Drug Discov 8 (2009) 129–138.

[31] S.H. Ku, K. Kim, K. Choi, et al., Tumor-targeting multifunctional nanoparticles for siRNA delivery: recent advances in cancer therapy, Adv Healthcare Mater 3 (2014) 1182–1193.

[32] R. Kanasty, J.R. Dorkin, A. Vegas, et al., Delivery materials for siRNA therapeutics, Nat Mater 12 (2013) 967–977.

[33] J.J. Turner, S.W. Jones, S.A. Moschos, et al., MALDI-TOF mass spectral analysis of siRNA degradation in serum confirms an RNAse A-like activity, Mol BioSyst 3 (2007) 43–50.

[34] J. Haupenthal, C. Baehr, S. Kiermayer, et al., Inhibition of RNAse A family enzymes prevents degradation and loss of silencing activity of siRNAs in serum, Biochem Pharmacol 71 (2006) 702–710.

[35] S. Kennedy, D. Wang, G. Ruvkun, A conserved siRNA-degrading RNase negatively regulates RNA interference in C. elegans, Nature 427 (2004) 645–649.

[36] Y. Takabatake, Y. Isaka, M. Mizui, et al., Chemically modified siRNA prolonged RNA interference in renal disease, Biochem Biophys Res Commun 363 (2007) 432–437.

[37] E.C. Lai, Micro RNAs are complementary to 3′ UTR sequence motifs that mediate negative post-transcriptional regulation, Nat Genet 30 (2002) 363–364.

[38] A. Birmingham, E.M. Anderson, A. Reynolds, et al., 3′-UTR seed matches, but not overall identity, are associated with RNAi off-targets, Nat Methods 3 (2006) 199–204.

[39] A.L. Jackson, J. Burchard, J. Schelter, et al., Widespread siRNA "off-target" transcript silencing mediated by seed region sequence complementarity, RNA 12 (2006) 1179–1187.

[40] J.-X. Wei, J. Yang, J.-F. Sun, et al., Both strands of siRNA have potential to guide posttranscriptional gene silencing in mammalian cells, PLoS One 4 (2009) e5382.

[41] M.E. Kleinman, K. Yamada, A. Takeda, et al., Sequence- and target-independent angiogenesis suppression by siRNA via TLR3, Nature 452 (2008) 591–597.

[42] A. Reynolds, E.M. Anderson, A. Vermeulen, et al., Induction of the interferon response by siRNA is cell type- and duplex length-dependent, RNA 12 (2006) 988–993.

[43] A.D. Judge, V. Sood, J.R. Shaw, et al., Sequence-dependent stimulation of the mammalian innate immune response by synthetic siRNA, Nat Biotechnol 23 (2005) 457–462.

[44] A. Forsbach, J.-G. Nemorin, C. Montino, et al., Identification of RNA sequence motifs stimulating sequence-specific TLR8-dependent immune responses, J Immunol 180 (2008) 3729–3738.

[45] L. Manche, S.R. Green, C. Schmedt, et al., Interactions between double-stranded RNA regulators and the protein kinase DAI, Mol Cell Biol 12 (1992) 5238–5248.

[46] V. Hornung, J. Ellegast, S. Kim, et al., 5′-Triphosphate RNA is the ligand for RIG-I, Science 314 (2006) 994–997.

[47] J.T. Marques, T. Devosse, D. Wang, et al., A structural basis for discriminating between self and nonself double-stranded RNAs in mammalian cells, Nat Biotechnol 24 (2006) 559–565.

[48] R. Juliano, J. Bauman, H. Kang, et al., Biological barriers to therapy with antisense and siRNA oligonucleotides, Mol Pharm 6 (2009) 686–695.

[49] J. Fang, H. Nakamura, H. Maeda, The EPR effect: Unique features of tumor blood vessels for drug delivery, factors involved, and limitations and augmentation of the effect, Adv Drug Deliv Rev 63 (2011) 136–151.

[50] E. Vivès, J. Schmidt, A. Pèlegrin, Cell-penetrating and cell-targeting peptides in drug delivery, Biochim Biophys Acta 1786 (2008) 126–138.

[51] D.J. Gary, N. Puri, Y.-Y. Won, Polymer-based siRNA delivery: Perspectives on the fundamental and phenomenological distinctions from polymer-based DNA delivery, J Control Release 121 (2007) 64–73.

[52] J.B. Bramsen, J. Kjems, Development of therapeutic-grade small interfering RNAs by chemical engineering, Front Genet 3 (2012) 154.

[53] J.H. Jeong, H. Mok, Y.-K. Oh, et al., siRNA conjugate delivery systems, Bioconjug Chem 20 (2009) 5–14.

[54] S. Agrawal, J. Temsamani, J.Y. Tang, Pharmacokinetics, biodistribution, and stability of oligodeoxynucleotide phosphorothioates in mice, Proc Natl Acad Sci USA 88 (1991) 7595–7599.

[55] D.A. Braasch, Z. Paroo, A. Constantinescu, et al., Biodistribution of phosphodiester and phosphorothioate siRNA, Bioorg Med Chem Lett 14 (2004) 1139–1143.

[56] Z.-Y. Li, H. Mao, D.A. Kallick, et al., The effects of thiophosphate substitutions on native siRNA gene silencing, Biochem Biophys Res Commun 329 (2005) 1026–1030.

[57] J. Harborth, S.M. Elbashir, K. Vandenburgh, et al., Sequence, chemical, and structural variation of small interfering RNAs and short hairpin RNAs and the effect on mammalian gene silencing, Antisense Nucleic Acid Drug Dev 13 (2003) 83–105.

[58] X. Yang, M. Sierant, M. Janicka, et al., Gene silencing activity of siRNA molecules containing phosphorodithioate substitutions, ACS Chem Biol 7 (2012) 1214–1220.

[59] P. Li, Z.A. Sergueeva, M. Dobrikov, et al., Nucleoside and oligonucleoside boranophosphates: chemistry and properties, Chem Rev 107 (2007) 4746–4796.

[60] A.H.S. Hall, J. Wan, E.E. Shaughnessy, et al., RNA interference using boranophosphate siRNAs: structure–activity relationships, Nucleic Acids Res 32 (2004) 5991–6000.

[61] Y.-L. Chiu, T.M. Rana, siRNA function in RNAi: A chemical modification analysis, RNA 9 (2003) 1034–1048.

[62] S. Choung, Y.J. Kim, S. Kim, et al., Chemical modification of siRNAs to improve serum stability without loss of efficacy, Biochem Biophys Res Commun 342 (2006) 919–927.

[63] B.A. Kraynack, B.F. Baker, Small interfering RNAs containing full 2′-Omethylribonucleotide- modified sense strands display Argonaute2/eIF2C2-dependent activity, RNA 12 (2006) 163–176.

[64] D.A. Braasch, S. Jensen, Y. Liu, et al., RNA interference in mammalian cells by chemically-modified RNA, Biochemistry 42 (2003) 7967–7975.

[65] J.M. Layzer, A.P. McCaffrey, A.K. Tanner, et al., In vivo activity of nuclease-resistant siRNAs, RNA 10 (2004) 766–771.

[66] C.R. Allerson, N. Sioufi, R. Jarres, et al., Fully 2′-modified oligonucleotide duplexes with improved in vitro potency and stability compared to unmodified small interfering RNA, J Med Chem 48 (2005) 901–904.

[67] R.I. Hogrefe, A.V. Lebedev, G. Zon, et al., Chemically modified short interfering hybrids (siHYBRIDS): nanoimmuno-liposome delivery in vitro and in vivo for RNAi of HER-2, nucleosides, Nucleosides Nucleotides Nucleic Acids 25 (2006) 889–907.

[68] K.F. Pirollo, A. Rait, Q. Zhou, et al., Materializing the potential of small interfering RNA via a tumor targeting nanodelivery system, Cancer Res 67 (2007) 2938–2943.

[69] M. Petersen, J. Wengel, LNA: a versatile tool for therapeutics and genomics, Trends Biotechnol 21 (2003) 74–81.

[70] J. Elmén, H. Thonberg, K. Ljungberg, et al., Locked nucleic acid (LNA) mediated improvements in siRNA stability and functionality, Nucleic Acids Res 33 (2005) 439–447.

[71] R.A. Blidner, R.P. Hammer, M.J. Lopez, S et al., Fully 2′-deoxy-2′-fluoro substituted nucleic acids induce RNA interference in mammalian cell culture, Chem Biol Drug Des 70 (2007) 113–122.

[72] G. Dorn, S. Patel, G. Wotherspoon, et al., siRNA relieves chronic neuropathic pain, Nucleic Acids Res 32 (2004) e49.

[73] A.M. Kawasaki, M.D. Casper, S.M. Freier, et al., Uniformly modified 2′-deoxy-2′-fluoro phosphorothioate oligonucleotides as nuclease-resistant antisense compounds with high

affinity and specificity for RNA targets, J Med Chem 36 (1993) 831–841.

[74] D.V. Morrissey, J.A. Lockridge, L. Shaw, et al., Potent and persistent in vivo anti-HBV activity of chemically modified siRNAs, Nat Biotechnol 23 (2005) 1002–1007.

[75] D.A. Braasch, Z. Paroo, A. Constantinescu, et al., Biodistribution of phosphodiester and phosphorothioate siRNA, Bioorg Med Chem Lett 14 (2004) 1139–1143.

[76] S.J. Lee, M.S. Huh, S.Y. Lee, et al., Tumor-homing poly-siRNA/glycol chitosan self-cross-linked nanoparticles for systemic siRNA delivery in cancer treatment, Angew Chem 51 (2012) 7203–7207.

[77] S.J. Lee, J.Y. Yhee, S.H. Kim, et al., Biocompatible gelatin nanoparticles for tumor-targeted delivery of polymerized siRNA in tumor-bearing mice, J Control Release 172 (2013) 358–366.

[78] S.Y. Lee, M.S. Huh, S. Lee, et al., Stability and cellular uptake of polymerized siRNA (poly-siRNA)/polyethyleni-mine (PEI) complexes for efficient gene silencing, J Control Release 141 (2010) 339–346.

[79] S. Son, S. Song, S.J. Lee, et al., Self-crosslinked human serum albumin nanocarriers for systemic delivery of polymerized siRNA to tumors, Biomaterials 34 (2013) 9475–9485.

[80] J.Y. Yhee, S.J. Lee, S. Lee, et al., Tumor-targeting transferrin nanoparticles for systemic polymerized siRNA delivery in tumor-bearing mice, Bioconjug Chem 24 (2013) 1850–1860.

[81] H.Y. Yoon, H.R. Kim, G. Saravanakumar, et al., Bioreducible hyaluronic acid conjugates as siRNA carrier for tumor targeting, J Control Release 172 (2013) 653–661.

[82] J.K. Watts, G.F. Deleavey, M.J. Damha, Chemically modified siRNA: tools and applications, Drug Discov Today 13 (2008) 842–855.

[83] A.A. Volkov, N.Y.S. Kruglova, M.I. Meschaninova, et al., Selective protection of nuclease-sensitive sites in siRNA prolongs silencing effect, Oligonucleotides 19 (2009) 191–202.

[84] D.V. Morrissey, K. Blanchard, L. Shaw, et al., Activity of stabilized short interfering RNA in a mouse model of hepatitis B virus replication, Hepatology 41 (2005) 1349–1356.

[85] K. Ui-Tei, Y. Naito, S. Zenno, et al., Functional dissection of siRNA sequence by systematic DNA substitution: modified siRNA with a DNA seed arm is a powerful tool for mammalian gene silencing with significantly reduced off-target effect, Nucleic Acids Res 36 (2008) 2136–2151.

[86] A. Nykänen, B. Haley, P.D. Zamore, ATP requirements and small interfering RNA structure in the RNA interference pathway, Cell 107 (2001) 309–321.

[87] P.Y. Chen, L. Weinmann, D. Gaidatzis, et al., Strand-specific 5′-O-methylation of siRNA duplexes controls guide strand selection and targeting specificity, RNA 14 (2008) 263–274.

[88] O.R. Mook, F. Baas, M.B. de Wissel, et al., Evaluation of locked nucleic acid–modified small interfering RNA in vitro and in vivo, Mol Cancer Ther 6 (2007) 833–843.

[89] F. Eberle, K. Gießler, C. Deck, et al., Modifications in small interfering RNA that separate immunostimulation from RNA interference, J Immunol 180 (2008) 3229–3237.

[90] S. Hamm, E. Latz, D. Hangel, et al., Alternating 2′-O-ribose methylation is a universal approach for generating non-stimulatory siRNA by acting as TLR7 antagonist, Immunobiology 215 (2010) 559–569.

[91] V. Hornung, M. Guenthner-Biller, C. Bourquin, et al., Sequence-specific potent induction of IFN-α by short

interfering RNA in plasmacytoid dendritic cells through TLR7, Nat Med 11 (2005) 263–270.

[92] S.R. Nallagatla, P.C. Bevilacqua, Nucleoside modifications modulate activation of the protein kinase PKR in an RNA structure-specific manner, RNA 14 (2008) 1201–1213.

[93] S.M. Elbashir, J. Harborth, W. Lendeckel, et al., Duplexes of 21-nucleotide RNAs mediate RNA interference in cultured mammalian cells, Nature 411 (2001) 494–498.

[94] S.M. Elbashir, W. Lendeckel, T. Tuschl, RNA interference is mediated by 21-and 22-nucleotide RNAs, Genes Dev 15 (2001) 188–200.

[95] D.H. Kim, M.A. Behlke, S.D. Rose, et al., Synthetic dsRNA Dicer substrates enhance RNAi potency and efficacy, Nat Biotechnol 23 (2005) 222–226.

[96] D. Siolas, C. Lerner, J. Burchard, et al., Synthetic shRNAs as potent RNAi triggers, Nat Biotechnol 23 (2005) 227–231.

[97] Y.S. Lee, K. Nakahara, J.W. Pham, et al., Distinct roles for Drosophila Dicer-1 and Dicer-2 in the siRNA/miRNA silencing pathways, Cell 117 (2004) 69–81.

[98] E.J. Sontheimer, Assembly and function of RNA silencing complexes, Nat Rev Mol Cell Biol 6 (2005) 127–138.

[99] V. Bitko, A. Musiyenko, O. Shulyayeva, et al., Inhibition of respiratory viruses by nasally administered siRNA, Nat Med 11 (2005) 50–55.

[100] S.P. Persengiev, X.C. Zhu, M.R. Green, Nonspecific, concentration-dependent stimulation and repression of mammalian gene expression by small interfering RNAs (siRNAs), RNA Biol. 10 (2004) 12–18.

[101] A.L. Jackson, P.S. Linsley, Recognizing and avoiding siRNA off-target effects for target identification and therapeutic application, Nat Rev Drug Discov 9 (2010) 57–67.

[102] X.G. Sun, H.A. Rogoff, C.J. Li, Asymmetric RNA duplexes mediate RNA interference in mammalian cells, Nat Biotechnol 26 (2008) 1379–1382.

[103] J.B. Bramsen, M.B. Laursen, C.K. Damgaard, et al., Improved silencing properties using small internally segmented interfering RNAs, Nucleic Acids Res 35 (2007) 5886–5897.

[104] H. Hohjoh, Enhancement of RNAi activity by improved siRNA duplexes, FEBS Lett 557 (2004) 193–198.

[105] A. Khvorova, A. Reynolds, S.D. Jayasena, Functional siRNAs and rniRNAs exhibit strand bias, Cell 115 (2003) 209–216.

[106] D.S. Schwarz, G. Hutvagner, T. Du, et al., Asymmetry in the assembly of the RNAi enzyme complex, Cell 115 (2003) 199–208.

[107] M. Manoharan, RNA interference and chemically modified small interfering RNAs, Curr Opin Chem Biol 8 (2004) 570–579.

[108] N. Abe, H. Abe, Y. Ito, Dumbbell-shaped nanocircular RNAs for RNA interference, J Am Chem Soc 129 (2007) 15108–15109.

[109] M. Raouane, D. Desmaële, G. Urbinati, et al., Lipid conjugated oligonucleotides: a useful strategy for delivery, Bioconjug Chem 23 (2012) 1091–1104.

[110] C. Wolfrum, S. Shi, K.N. Jayaprakash, et al., Mechanisms and optimization of in vivo delivery of lipophilic siRNAs, Nat Biotechnol 25 (2007) 1149–1157.

[111] K. Nishina, T. Unno, Y. Uno, et al., Efficient in vivo delivery of siRNA to the liver by conjugation of a-tocopherol, Mol Ther 16 (2008) 734–740.

[112] F. Iversen, C. Yang, F. Dagnæs-Hansen, et al., Optimized siRNA-PEG conjugates for extended blood circulation and reduced urine excretion in mice, Theranostics 3 (2013) 201–209.

[113] S.H. Kim, J.H. Jeong, S.H. Lee, et al., PEG conjugated VEGF siRNA for anti angiogenic gene therapy, J Control Release 116 (2006) 123–129.

[114] S.H. Kim, J.H. Jeong, S.H. Lee, et al., Local and systemic delivery of VEGF siRNA using polyelectrolyte complex micelles for effective treatment of cancer, J Control Release 129 (2008) 107–116.

[115] M. Oishi, Y. Nagasaki, K. Itaka, et al., Lactosylated poly(ethylene glycol)-siRNA conjugate through acid-labile β-thiopropionate linkage to construct pH-sensitive polyion complex micelles achieving enhanced gene silencing in hepatoma cells, J Am Chem Soc 127 (2005) 1624–1625.

[116] C. Dohmen, T. Frohlich, U. Lachelt, et al., Defined folate-PEG-siRNA conjugates for receptor-specific gene silencing, Mol Ther–Nucleic Acids 1 (2012) e7.

[117] Y.L. Jang, S.H. Ku, S.J. Lee, et al., Hyaluronic acid-siRNA conjugate/reducible polyethylenimine complexes for targeted siRNA delivery, J Nanosci Nanotechnol 14 (2014) 7388–7394.

[118] K. Park, J.-A. Yang, M.-Y. Lee, et al., Reducible hyaluronic acid–siRNA conjugate for target specific gene silencing, Bioconjug Chem 24 (2013) 1201–1209.

[119] J.P. Dassie, X.-Y. Liu, G.S. Thomas, et al., Systemic administration of optimized aptamer-siRNA chimeras promotes regression of PSMA-expressing tumors, Nat Biotechnol 27 (2009) 839–846.

[120] J.O. McNamara, E.R. Andrechek, Y. Wang, et al., Cell type-specific delivery of siRNAs with aptamer siRNA chimeras, Nat Biotechnol 24 (2006) 1005–1015.

[121] W.-Y. Lai, W.-Y. Wang, Y.-C. Chang, et al., Synergistic inhibition of lung cancer cell invasion, tumor growth and angiogenesis using aptamer-siRNA chimeras, Biomaterials 35 (2014) 2905–2914.

[122] K.A. Whitehead, R. Langer, D.G. Anderson, Knocking down barriers: advances in siRNA delivery, Nat Rev Drug Discov 8 (2009) 129–138.

[123] H. Akita, H. Harashima, Advances in non-viral gene delivery: using multifunctional envelope-type nano-device, Expert Opin Drug Deliv 5 (2008) 847–859.

[124] P. Kebbekus, D.E. Draper, P. Hagerman, Persistence length of Rna, Biochemistry 34 (1995) 4354–4357.

[125] D.W. Bartlett, H. Su, I.J. Hildebrandt, et al., Impact of tumor-specific targeting on the biodistribution and efficacy of siRNA nanoparticles measured by multimodality in vivo imaging, Proc Natl Acad Sci USA 104 (2007) 15549–15554.

[126] H. Mok, T.G. Park, Self-crosslinked and reducible fusogenic peptides for intracellular delivery of siRNA, Biopolymers 89 (2008) 881–888.

[127] H.T. Lv, S.B. Zhang, B. Wang, et al., Toxicity of cationic lipids and cationic polymers in gene delivery, J Control Release 114 (2006) 100–109.

[128] A.L. Bolcato-Bellemin, M.E. Bonnet, G. Creusatt, et al., Sticky overhangs enhance siRNA-mediated gene silencing, Proc Natl Acad Sci USA 104 (2007) 16050–16055.

[129] H. Mok, S.H. Lee, J.W. Park, et al., Multimeric small interfering ribonucleic acid for highly efficient sequence-specific gene silencing, Nat Mater 9 (2010) 272–278.

[130] S.Y. Lee, M.S. Huh, S. Lee, et al., Stability and cellular uptake of polymerized siRNA (poly-siRNA)/polyethylenimine (PEI) complexes for efficient gene silencing, J Control Release 141 (2010) 339–346.

[131] F.H. Meng, W.E. Hennink, Z. Zhong, Reduction-sensitive polymers and bioconjugates for biomedical applications, Biomaterials 30 (2009) 2180–2198.

[132] M. Sioud, RNA interference and innate immunity, Adv Drug Deliv Rev 59 (2007) 153–163.

[133] P.W.K. Rothemund, Folding DNA to create nanoscale shapes and patterns, Nature 440 (2006) 297–302.

[134] A.V. Pinheiro, D.R. Han, W.M. Shih, et al., Challenges and opportunities for structural DNA nanotechnology, Nat Nanotechnol 6 (2011) 763–772.

[135] C.A. Hong, S.H. Lee, J.S. Kim, et al., Gene silencing by siRNA microhydrogels via polymeric nanoscale condensation, J Am Chem Soc 133 (2011) 13914–13917.

[136] J.B. Lee, J. Hong, D.K. Bonner, et al., Self-assembled RNA interference microsponges for efficient siRNA delivery, Nat Mater 11 (2012) 316–322.

[137] S.L. Daubendiek, K. Ryan, E.T. Kool, Rolling-circle RNA-synthesis-circular oligonucleotides as efficient substrates for T7 RNA-polymerase, J Am Chem Soc 117 (1995) 7818–7819.

[138] L.H. Madkour, Reactive Oxygen Species (ROS), Nanoparticles, and Endoplasmic Reticulum (ER) Stress-Induced Cell Death Mechanisms, Academic Press, 2020.

[139] L.H. Madkour, Nanoparticles Induce Oxidative and Endoplasmic Reticulum Antioxidant Therapeutic Defenses, Springer International Publishing, 2020. doi:10.1007/978-3-030-37297-2

[140] L.H. Madkour, Nucleic Acids as Gene Anticancer Drug Delivery Therapy, Elsevier, 2020.

[141] L.H. Madkour, Nanoelectronic Materials: Fundamentals and Applications (Advanced Structured Materials), Springer International Publishing, 2019. doi:10.1007/978-3-030-21621-4

# 7 Clinical siRNA-Based Conjugate Systems for RNAi Cancer Cell Therapy

Recent progress in RNA biology has broadened the scope of therapeutic targets of RNA drugs for cancer therapy. However, RNA drugs, typically small interfering RNAs (siRNAs), are rapidly degraded by RNases and filtrated in the kidney, thereby requiring a delivery vehicle for efficient transport to the target cells.

In this chapter, we introduce recent notable works regarding siRNA conjugates for clinical translations and focus on the considerations for the rational design of siRNA conjugates to improve biological benefits in preclinical and clinical studies. First, we describe the pros and cons of siRNA conjugates as therapeutics in terms of the physicochemical properties of siRNA, targeted delivery, therapeutic efficacy, and other biological benefits. Second, we introduce in detail various siRNA conjugates with aptamers, peptides, carbohydrates, lipids, polymers, and nanostructured materials in terms of in vitro and in vivo efficacy.

## 7.1 SMALL INTERFERING RNAs (siRNAs)

Small interfering RNAs (siRNAs), symmetric or asymmetric double-stranded RNAs with around 20 base pairs, have long been used as molecular tools to regulate the expression of genes of interest in basic research [1, 2]. After attachment to complementary target mRNAs, siRNAs allow any target genes to be suppressed specifically. Currently, many researchers have vigorously examined the feasibility of siRNAs as biotherapeutics to silence disease-causing genes that have not been regulated with conventional therapeutics [3–6]. Because RNA therapeutics are still at an early stage, many unexpected obstacles (e.g., off-target effect and immune response via activation of toll-like receptor [TLR]) have hampered preclinical and clinical studies in big pharmaceutical industries [3, 7, 8]. To overcome these obstacles, a variety of approaches, including modification of RNAs, development and optimization of carrier systems, and proper in vivo administration, have been explored intensively [3, 9–13]. Moreover, several recent promising clinical results still boost the development of RNAi-based therapeutics in biotech industries [14–16]. To derive more successful clinical translations, faced issues including poor delivery efficiency and off-target effects should be clearly understood and addressed. In addition, currently available carrier systems under clinical trials, including cationic liposomes, anionic liposomes, polymeric carriers (cyclodextrin-based nanoparticles), and siRNA conjugates, should be comparatively examined and evaluated in terms of feasibility in clinics [5, 17].

Bioconjugation techniques for a link between active molecules have been well established as delivery systems of biotherapeutics. A wide range of conjugate systems for drug delivery, such as antibody–drug conjugates and polymer–drug conjugates, have already been in clinics and lots of them are under clinical trials [18]. Since the first polyethylene glycol (PEG)–protein conjugate, Adagen (pegademase bovine), was approved by the US Food and Drug Administration (FDA) in 1990, around ten FDA-approved PEG–protein conjugates have become available [19, 20]. More recently, antibody–small molecule drug conjugates, Adcetris (brentuximab vedotin) and Kadcyla (ado-trastuzumab emtansine), were approved by the FDA in 2011 and in 2013, respectively, which demonstrated improved therapeutic effects by targeted drug delivery compared with unmodified small-molecule drugs [21]. In addition, one of the leading candidates in siRNA-based drugs under clinical trial is N-acetylgalactosamine-siRNA conjugates (GalNAc–siRNA conjugates), developed by Alnylam Pharmaceuticals [15, 22]. Previously, siRNA conjugate systems were discussed regarding type of chemical modifications and detailed synthetic schemes for conjugation (e.g., solid-phase synthesis, carbodiimide-mediated coupling reaction, and Michael addition reaction) [23, 24]. However, few studies have presented comparative evaluations of siRNA conjugates to other delivery systems, with considerations of the preclinical and clinical studies of siRNA conjugates and current status in their preclinical developments.

## 7.2 siRNA CONJUGATES: PROS AND CONS AS THERAPEUTICS COMPARED WITH OTHER DELIVERY STRATEGIES

Many research groups in industries and academia have paid attention to siRNA conjugates to endow favorable physicochemical properties and biological benefits for clinical translation. As shown in Figure 7.1, various functional molecules could be incorporated into siRNA conjugates to enhance their delivery efficiency. To date, a wide range of molecules have been attached to the ends of siRNAs to improve biological half-life and modify pharmacokinetics, which are crucial for in vivo therapeutic efficiency [25–27]. For example, PEG conjugation to siRNAs improved their physicochemical stability against enzymatic digestion and extended their blood half-life in vivo due to increase of hydrodynamic volume [26, 28]. In addition, conjugation of targeting ligands such as peptides and carbohydrates has greatly improved accumulation of siRNAs in target cells

DOI: 10.1201/9781003229650-7

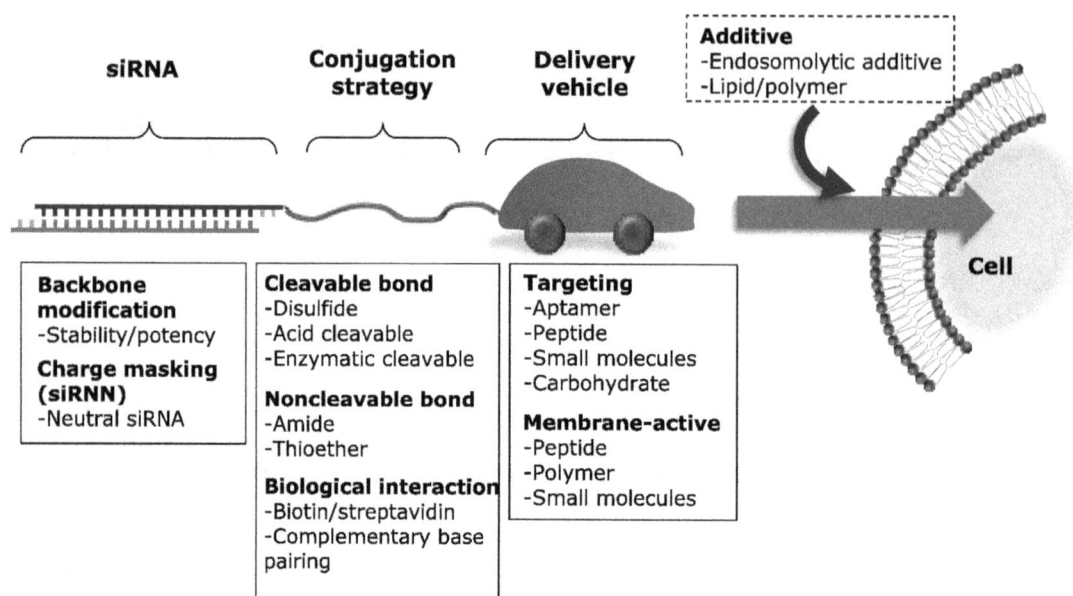

**FIGURE 7.1** Strategy of siRNA conjugate design for efficient delivery.

and tissues, which could reduce the siRNA dose required for in vivo therapeutic effects [29, 30]. These molecules could be linked to siRNA via various linkages, such as cleavable bonds, noncleavable bonds, and biological bonds, using different conjugation strategies. Cleavable bonds like reducible disulfide bonds or acid–labile hydrazine bonds can be cleaved in a reductive environment (e.g., cytosol) or in an acidic environment (e.g., endosomal lumen), respectively, and free siRNAs can be dissociated from conjugates. After bond cleavage, free siRNAs can be dissociated from conjugates without Dicer processing in cells [31]. SiRNA conjugates with noncleavable bonds such as amide bonds and thioether bonds need Dicer processing to release free siRNAs in cells [31]. Besides these covalent linkages, noncovalent interaction, such as biotin/streptavidin and complementary base pairing, can be used in siRNA conjugation [32, 33]. Backbone modification and charge masking of siRNA could be also considered for enhanced siRNA stability/potency and facile conjugation, respectively. After preparation of siRNA conjugates, they can be administered by themselves, or with additives like lipids and polymers in vitro and in vivo.

Conjugate systems have several advantages over other delivery strategies including particle-based carriers, as shown in Figure 7.2a. First, siRNA conjugate has a simple and well-defined molecular structure, and the formulation process could be well controlled in terms of homogeneity and reproducibility, which is a crucial factor in clinical translation. Second, covalent siRNA conjugation to hydrophobic molecules and delivery vehicles can modulate physicochemical properties that affect entrapment efficiency and pharmacokinetic properties. For example, free siRNAs noncovalently entrapped within polyelectrolyte complexes (PECs) were dissociated and released from PECs in the presence of exterior ions, while covalent conjugation of

siRNA to PECs resulted in the stable incorporation of siRNA within PECs during treatment to cells [34]. Besides direct conjugation of siRNAs to the delivery vehicles, conjugation of hydrophobic molecules like phospholipids and poly(lactic-co-glycolic acid) (PLGA) to siRNA can also lead to more favorable incorporation of siRNA within the particulate platforms due to modified physicochemical properties. For example, phospholipid–siRNA conjugates increase loading efficiency into hydrophobic nanoparticles and PLGA–siRNA conjugates form micelles in physiological condition, which can be readily internalized and induce target gene suppression in the presence of carriers like cationic polymers and lipids [35–37]. Third, the amount of additives in the conjugate systems is much lower than that in particle-based systems to cause the significant biological effect at the same dose of siRNA. When alpha-tocopherol (vitamin E)–siRNA conjugates and siRNAs within liposomes were injected at siRNA dose of 2 mg/kg and 2.5 mg/kg, vitamin E and lipids were administered at 46 µg/kg and 25 mg/kg, respectively [38, 39]. The administration of high amounts of additives can cause in vivo toxicity and immunostimulatory effects [40]. For example, liposome-based siRNA delivery via intravenous injection to monkey at a siRNA dose of 2.5 mg/kg caused the change of liver enzymes due to the liposome-associated liver toxicity [38]. However, another group reported that lipid-based particles elicited no severe adverse events except for infusion-related reaction in Phase 1 clinical trials, which suggests that toxicity depends on type of liposome [41]. The recent Phase 1 clinical study with targeted, polymer-based nanoparticle containing siRNA (CALAA-01) showed that the administration of cationic polymer-based nanoparticles caused major adverse events including acute immune responses, hypersensitivity, and dose-limiting toxic events [4]. Zuckerman et al. speculated that these toxicities are

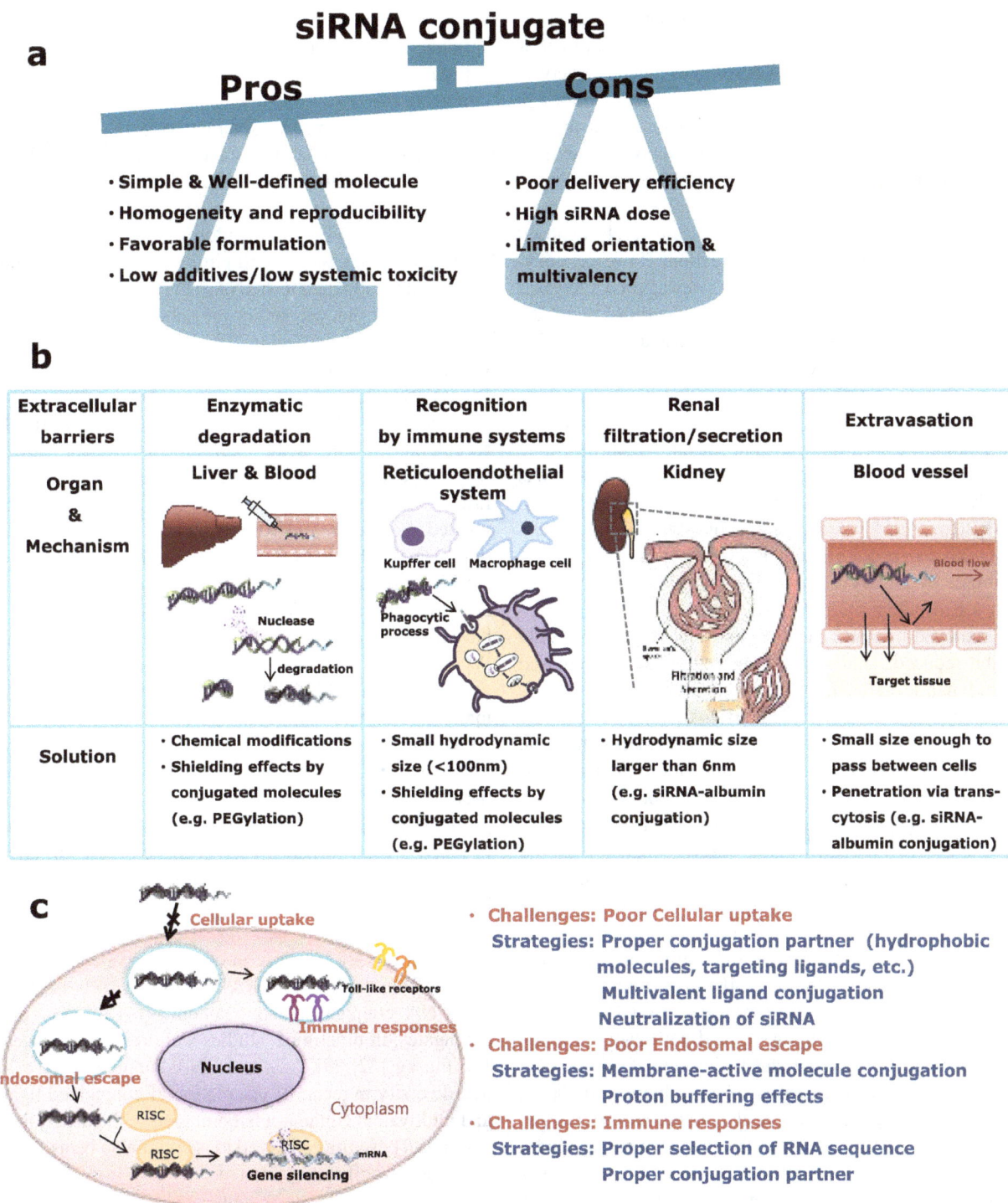

FIGURE 7.2 (a) Advantages and drawbacks of siRNA conjugates compared with other delivery strategies. (b) Extracellular and (c) intracellular barriers in vivo that are responsible for the poor delivery efficiency of the siRNA conjugates.

originated not only from siRNAs but also from polymeric components. Accordingly, minimal amounts of additives seem to be critical to be free from toxicity issues in pre-clinical and clinical studies. Lastly, diverse bioconjugate techniques, with their high stability and small molecular size, have been established not only for intravenous injection but also for other systemic administration routes. For example, triantennary GalNAc–siRNA conjugates can be

administered subcutaneously, which is more preferable than intravenous injection [15].

Despite many advantages of siRNA conjugates as therapeutics, the most important concern of siRNA conjugates for clinical translation is their poor delivery efficiency compared with particle-based systems. The poor delivery efficiency of siRNA conjugates is attributed to several extracellular and intracellular barriers in vivo that are responsible for blood circulation, extravasation, and accumulation in target tissues (Figure 7.2b). After administration via diverse routes, siRNA conjugates meet a variety of extracellular barriers including enzymatic degradations in blood, recognition by immune systems (monocytes, macrophages, Kupffer cells), renal filtration/secretion, and extravasation, which determine blood circulation of siRNAs [42]. While siRNAs within particles can be protected from nucleases in blood, siRNA conjugates are freely exposed and vulnerable to enzymatic degradation. However, enzymatic susceptibility of siRNA conjugates against nucleases has been improved by chemical modifications of RNAs and shielding effects by conjugated molecules like PEG, which have allowed long blood circulation [24, 28, 43]. After systemic injection, siRNA conjugates can be eliminated from the blood by immune cells like macrophage and Kupffer cells in the liver and spleen, which results in reduced blood circulation and accumulation of siRNAs mainly in the liver and spleen where reticuloendothelial systems are located [26, 44]. In addition, naked siRNAs with a size ~5.7 nm might be easily excreted via renal clearance by kidney. Considering that globular molecules with a hydrodynamic radius <6 nm were easily and rapidly removed by renal clearance, siRNA conjugates could exhibit reduced renal secretion probably due to increased hydrodynamic sizes [31, 45]. For example, cholesterol-siRNA conjugates and albumin-siRNA conjugates showed reduced renal secretion and elongated elimination half-life of ~95 min, while naked siRNAs had an elimination half-life of several minutes [44, 46]. Lastly, poor extravasation of siRNA conjugates is also a noteworthy extracellular barrier. For successful extravasation, siRNA conjugates should be small enough to pass through pores between endothelial cells or penetrate via transcytosis. For example, albumin-siRNA conjugate showed successful extravasation through albumin-transcytosis processing and significantly improved penetration into target tissues [44]. In addition, it should be noted that pharmacokinetic properties of siRNA-based carrier systems are closely associated with hydrodynamic radius in vivo. Foreign particles with a size of ~10–100 nm have been considered essential for long blood circulation and reduced recognition by reticuloendothelial systems [42, 47, 48]. Recent study also showed that particles around a size of 30 nm could penetrate into low vascularized tumor tissues while particles with bigger sizes failed to penetrate into them [47]. Thus, the hydrodynamic radius of siRNA conjugates should be precisely and properly designed to overcome those extracellular barriers. After extravasation of siRNA conjugates, they must overcome several intracellular barriers including intracellular uptake, endosome escape, and activation of immune-related receptors (Figure 7.2c). For efficient intracellular uptake via endocytosis and pinocytosis, strong and specific interactions of siRNA conjugates with target cells are necessary. Diverse factors including type of conjugated molecules (e.g., hydrophobic moieties and targeting ligands), density of conjugation, and orientation of conjugated molecules could affect the level of intracellular uptake. For example, multivalent aptamer–RNA conjugates showed significantly enhanced cellular uptake, compared with monovalent aptamer–RNA conjugate [49, 50]. Recently, charge of RNA was neutralized to reduce repulsive interaction with cellular membranes and result in facile intracellular uptake [29]. Within cells, siRNA conjugates should escape from endosome/lysosome to cytoplasm, which has been considered as a big challenge. To facilitate endosomal escape, membrane-active molecules with amphiphilic and fusogenic properties could be attached to siRNAs (Figure 7.1). Recent study demonstrated that secondary phosphate groups of nucleotides have proton-buffering effects and hemolytic activity, which can be also harnessed for endosome breaking [51]. Several cytoplasmic and endosomal RNA-binding proteins like retinoid-acid inducible gene 1 (RIG-1) and toll-like receptors can bind with internalized siRNAs, eliciting immune and inflammatory responses [23]. Interestingly, inflammatory responses were dependent on type of conjugated molecules and RNA sequences, which indicates that proper selection of conjugation partner could alleviate off-target effects by siRNA internalization [23, 52]. Due to poor in vivo delivery efficiency, high RNA doses have been needed for in vivo efficacy of siRNA conjugates. When the first systemic evaluation of cholesterol–siRNA conjugates via intravenous administration was reported, many researchers were worried about the feasibility of conjugate systems in clinics due to their high dose for in vivo gene silencing (50 mg/kg) compared with liposome-based systems (1–2.5 mg/kg) [38, 46]. Considering lipid particles have been administered at doses of 0.1–1.0 mg/kg in Phase 1 clinical trials, current in vivo doses of siRNA conjugates should be adjusted down for clinical feasibility [41]. Improvement of siRNA potency, stability by chemical modification, and the delivery efficiency have gradually allowed lower siRNA dose in conjugate systems. Through these efforts, doses of siRNA conjugates, such as vitamin E–siRNA conjugates and triantennary GalNAc–siRNA conjugates, in preclinical studies were recently reduced to 1–5 mg/kg [22, 39]. Lastly, siRNA conjugates have limited diversity in terms of orientation of conjugated ligands and multivalent conjugation, compared with particle-based systems. These limited diversities could also be associated with delivery efficiency in vivo.

## 7.3 siRNA-BASED CONJUGATE SYSTEMS

Various biomolecules as targeting ligands and membrane active molecules have been attached to siRNAs, as shown in Figure 7.3. Currently developed siRNA-based conjugate systems are presented especially regarding preclinical and clinical studies, which have been administered without and with additives for elevated delivery efficiency. We classify siRNA–conjugates into six categories: aptamer–siRNA, peptide–siRNA, carbohydrate–siRNA, lipid–siRNA, polymer–siRNA, and nanostructured materials–siRNA conjugates.

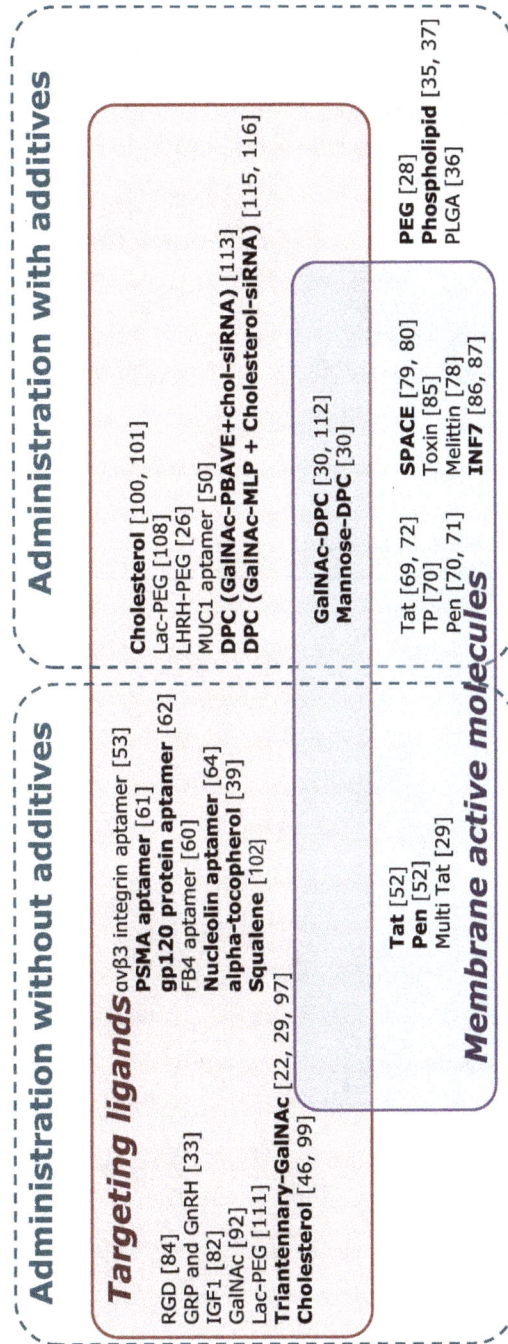

**FIGURE 7.3** Diverse delivery platforms of siRNA conjugates. Presented technologies are described in this review. Words in **bold type** represent the technology with in vivo study.

### 7.3.1 Aptamer–siRNA Conjugates

Aptamers have been noticed as good targeting ligands as well as therapeutic moieties due to their high affinity and specificity to the target molecules. Aptamer-mediated targeted drug delivery has several advantages including easy/fine production via chemical synthesis, prolonged shelf life, escape from multidrug resistance, reduced nonspecific toxicity, and enhanced efficacy on target site [53]. In addition, the physical stabilities of nucleic acid–based aptamers have been increased by diverse chemical modifications in phosphodiester bonds and 2′ ribose positions of nucleotides [54]. Many reviews have already described the feasibility of aptamers, aptamer-functionalized nanocarriers, and aptamer-screening methods including in vitro selection of aptamers in DNA/RNA libraries, named SELEX technology [54, 55]. Previously, aptamer–siRNA conjugates showed superior intracellular uptake via receptor-mediated endocytosis and efficient gene silencing in vitro and in vivo [56, 57]. In this section, we focus on aptamer–siRNA

conjugates and their in vivo applications as therapeutics for clinical translations. Both biologically derived aptamer–siRNA chimeras and chemically synthesized aptamer–siRNA conjugates are introduced.

#### 7.3.1.1 Biologically Generated Aptamer–siRNA Chimeras

Diverse aptamer–siRNA chimeras have been designed and generated via in vitro transcription using template DNAs (Table 7.1). Due to well-controlled enzyme reaction, finely tuned chimera can be produced, which is likely to be cleaved and release siRNAs with a size of 19–25 nucleotides (nt) via Dicer processing in cytoplasm [58]. However, Dicer processing within cells could also elicit immune responses such as small RNA-mediated viral immunity, which is one of the current challenges in biologically derived aptamer–siRNA chimera [59]. In addition, expensive enzyme reaction and laborious purification processing should be improved for mass production of chimera for clinical

**TABLE 7.1**

**Aptamer–siRNA Conjugates for Target Protein Silencing**

| Category | Aptamer | Conjugation strategy | Delivery method | dose in vitro in vivo | dose |
|---|---|---|---|---|---|
| Biologically generated chimeras | Alpha V beta3 integrin (αvβ3) aptamer | Apt / siRNA | Reagent free | $IC_{50}$ values of chimera - ~800 nM for U-87 MG cells - 1500 nM for PC-3 cells | N.D. |
| | Prostate-specific membrane antigen (PSMA) aptamer | Apt / siRNA | Reagent free | 400 nM | 200 pmol intratumoral injection |
| | HIV-1 envelope (gp120) protein aptamer | Apt / siRNA / Linker / dTdT | Reagent free | 400 nM | 250 pmol intravenous injection |
| | FB4 aptamer targeting to transferrin receptor | Apt / siRNA / hydrogen bond | Reagent free | 320 nM | N.D. |
| Chemically synthesized conjugates | Nucleolin aptamer (DNA aptamer) | Apt / siRNA / Linker / Spacer. Thiol-groups and poly(dT) spacers were adopted to aptamer at 5′ end. Amine-modified siRNAs and thiol-modified aptamers were reacted with SMPB crosslinker. | Reagent free | 50 nM | 100 pmol intratumoral injection |
| | Mucin1 aptamer (DNA aptamer) | Apt / siRNA / Linker. Thiol-modified, antisense siRNAs were crosslinked via DTME. Aptamer-sense siRNA conjugate was annealed to multimeric antisense siRNA. | Linear PEI(2.5K) | 144.6 nM | N.D. |

*Abbreviation:* N.D., not determined.

applications. Recently several aptamer–siRNA chimera showed successful gene knockdown as well as therapeutic effects in vitro and in vivo [53, 60–62]. An RNA aptamer targeting to transferrin receptor (FB4) facilitated siRNA delivery via biological adoption of chimeric RNA structures to RNA of bacterial virus phi29 (pRNA) [60]. Both siRNA against intercellular adhesion molecule-1 (ICAM-1) and aptamer FB4 were incorporated into the helical region of pRNA, which generated functional double-stranded RNAs (dsRNAs) such as aptamers and siRNAs after Dicer-mediated cleavage. The expression level of ICAM-1 in murine brain-derived endothelial cells was reduced by in vitro treatment of pRNA-based aptamer–siRNA chimera at a concentration of 160 nM. In another study, the integrin $\alpha v \beta 3$ recognizing RNA aptamer–siRNA chimera was produced by T7 RNA polymerase, and the silencing of target eukaryotic elongation factor 2 gene by aptamer–siRNA chimera caused significant apoptotic effects with integrin-expressing cells [53]. The induction of apoptosis by treatment of aptamer–siRNA chimera was observed on various integrin-expressing cancer cells (U-87MG, SiHa, and PC-3) at a chimera concentration of 500–2000 nM while HEK 293T with a low expression level of integrin did not show any apoptotic effects. Chimera of prostate-specific membrane antigen (PSMA) aptamer and siRNA against DNA-activated protein kinase, catalytic polypeptide (DNAPK), was generated from template DNA via in vitro transcription [61]. Interestingly, intratumoral injection of chimera (200 pmol) did not show any effects in mice tumor models but a combination of chimera and radiation resulted in great reduction of tumor size. The chimera of HIV-1 envelope (gp120) protein binding aptamer and siRNA against HIV-1 RNAs was produced via in vitro transcription, and the resulting chimera exerted potent antiviral activity in the humanized mouse model [62].

Instead of enzyme reaction using RNA polymerases, biological interactions such as biotin–streptavidin attachment could be harnessed for the production of aptamer–siRNA chimeras. Two types of aptamers against PSMA and epidermal growth factor receptor variant III (EGFRvIII) were labeled with biotin and simply connected to biotin-labeled siRNAs via biotin–streptavidin interaction [32, 63]. Chu et al. demonstrated that PSMA aptamer/siRNA structures suppressed efficiently the target gene expression in vitro in the absence of a transfection reagent [32].

### 7.3.1.2 Chemically Synthesized Aptamer–siRNA Conjugates

Chemically synthesized siRNAs and aptamers can be combined into one molecule by adoption of functional groups, like amines and thiols, and sequential conjugation reactions, like Michael addition, disulfide exchange, and carbodiimide-mediated coupling. Chemical conjugations have several advantages compared with biological enzyme reactions in order to produce aptamer–siRNA chimera. First, diverse crosslinkers could be incorporated into conjugates

to endow favorable characteristics, such as cleavability, stimuli-responsiveness, and structural flexibility, for efficient internalization of conjugates into cells and facile generation of free siRNAs at the site of action [24]. Another advantage of chemical conjugation is that only biologically functional RNAs/DNAs are incorporated within conjugates and additional nucleotides are not needed to form conjugates, decreasing the potential risk of nucleic acid–mediated immune responses. However, reaction yield and purity of chemical conjugations should be finely controlled to obtain reproducibility.

Recently, thiol groups of nucleolin aptamers at the 5'-end were chemically conjugated to the 5'-end of siRNAs using sulfo-succinimidyl 4-[p-maleimidophenyl]butyrate (SMPB) crosslinker [64]. For minimal steric hindrance, poly(dT) was also adopted to the 5'-end of aptamers. Without any additional transfection reagent, 50 nM of nucleolin aptamer–siRNA conjugate were treated in vitro for 48 h, which reduced target mRNA level to ~30%. After intratumoral injection of the aptamer–siRNA conjugate (100 pmol) on xenograft mice tumor model, the tumor growth rate decreased by four-fold compared with the PBS-injected control group. In addition, siRNAs could be chemically multimerized via crosslinkers, which could provide room for adoption of multivalent ligands [31, 65]. Yoo et al. synthesized multimeric antisense stranded siRNAs, which were complementary annealed with mucin1 aptamer-sense stranded siRNA chimeric nucleotides for adoption of multivalent aptamers to multimeric siRNAs [50]. Multivalent aptamer–siRNA conjugate showed higher internalization in the absence of transfection reagent via aptamer-mediated endocytosis by two-fold compared with monovalent aptamer–siRNA. However, gene suppression was negligible probably due to poor endosomal escape of multivalent aptamer–siRNA conjugates. Thus, in the presence of an endosome breaker, low-molecular-weight linear PEI, significant target gene suppression, and biological activity such as caspase-3 mediated apoptosis were observed by the treatment of multivalent aptamer–siRNA conjugate.

### 7.3.2 Peptide–siRNA Conjugates

Due to their well-defined structure, versatile biological function, and convenience of chemical conjugation, peptides have been considered as good candidates for siRNA delivery carriers by direct conjugation. Numerous synthetic strategies have been carried out to produce peptide–siRNA conjugates with a high yield via various linkages, such as disulfide bond, amide bond, thioether bond, and thiol–maleimide bond. Either carboxy terminus of peptide or additional cysteine residue in the peptide sequence has been harnessed for the conjugation of peptides to the functionalized siRNAs at their end terminal with amine, thiol, or maleimide group. Three types of peptides have been applied for peptide–siRNA conjugates: cell-penetrating peptides, targeting peptides, and lytic peptides.

### 7.3.2.1 Cell-Penetrating Peptides

Cell-penetrating peptides (CPPs), a class of short and basic amino acid-rich peptides, can facilitate intracellular uptake of various cargos. Over the past two decades, a number of CPPs have been applied for the intracellular delivery carriers of macromolecules [66]. Although their intracellular traffic is still controversial and their biological modalities are varied depending on delivery cargos and concentration, CPPs have been widely investigated for siRNA delivery when covalently conjugated to siRNA and when noncovalently complexed with siRNA via charge interaction [67, 68]. It is well known that CPPs (i.e., Tat, Transportan, and Penetratin) have been directly conjugated to siRNA and transfected to the cells in the absence of additives like transfection reagents [52, 69–72]. Despite their significant target gene silencing in vitro, these CPP–siRNA conjugates showed only limited success with in vivo results. For the formulation and application of CPP–siRNA conjugates, several issues should be carefully considered. Firstly, the CPP feasibility and toxicity can be significantly changed by chemical conjugation to siRNAs due to modification of CPP hydrophobicity or amphiphilicity [73, 74]. Additionally, charge interaction between CPP and siRNA could devastate CPP activity because this interaction neutralizes the basic residues of CPP, known to be crucial for biological activity of CPP [75]. In addition, it should be noted that most published studies have been performed with nonpurified CPP–siRNA conjugates, containing free peptides after synthesis [76]. With an excess amount of free peptides, the successful delivery was not likely associated to the CPP–siRNA conjugates, but to the polyelectrolyte complexes formed through charge interaction between positively charged CPP and negatively charged siRNA. To prevent this charge interaction, synthesis procedures and purification processes using HPLC should be precisely designed [77, 78]. Another problem is that the effective dose of CPPs that enables transport of the extracellular cargos into cytoplasm is at a submicromolar concentration, which is much higher than the dose of siRNA (at a subnanomolar concentration) in vitro. Thus, higher peptide–siRNA doses might be necessary for gene silencing, compared with conventional liposome-based or particle-based systems. In addition, the undesired immune responses and false-positive results by peptides themselves should be carefully considered for the clinical applications of CPP–siRNA conjugates [52].

For the topical treatment of siRNAs, Hsu et al. identified a peptide (SPACE peptide: skin penetrating and cell entering peptide) with high penetration efficiency across skin into the epidermis and dermis by a phage display technique [79]. Single SPACE was conjugated to siRNA via amide bond and SPACE–siRNA conjugates were topically administered on the mouse skin without the aid of transfection reagents. As a result, Interleukin-10 and GAPDH-specific SPACE–siRNA conjugates at 10 μM showed about 30% and 43% of target protein knockdown, respectively [79].

In a subsequent study, Chen et al. demonstrated that the combinational treatment of cationic liposome, supplementary free SPACE peptides, and SPACE–siRNA conjugates showed far enhanced in vivo results [80]. Supplementary peptides significantly improved the siRNA delivery efficiency, indicating that the amount of administered free peptide could exceed the dose needed to transport macromolecules into the target site. This noninvasive SPACE technology platform is currently being investigated by CONVOY Therapeutics to develop biological therapeutics for dermatological diseases.

More recently, Meade et al. demonstrated a novel self-delivering siRNA platform by conjugating multimeric Tat peptide domains on the charge-neutralized siRNA backbone, named as short interfering ribonucleic neutral (siRNN), to produce multi-Tat–siRNN conjugates [29]. This siRNN is composed of neutral phosphotriesters as internucleotide linkages, which could convert to phosphodiesters by cytoplasmic thioesterase, resulting in negatively charged siRNA. Hydrazine-containing peptide domains (hydrazine–Tat or hydrazine–Tat–PEG–Tat–PEG–Tat) were conjugated to the siRNNs containing two, three, and four aldehyde phosphotriester groups at their sense or antisense strands. Compared to the previous CPP–siRNA conjugate studies, this system could circumvent the loss of CPP activity due to charge interaction between siRNA and CPP. During in vitro experiments, these bioreversible multi-Tat–siRNNs conjugates were successfully delivered to the various cell lines in the absence of a transfection reagent, and inhibited target protein expressions, such as GFP, cMyc, and polo-like kinase 1 (Plk1). Interestingly, the conjugation site in siRNN backbone did not significantly affect the silencing efficiency because the cytoplasmic enzyme-mediated cleavage removes the conjugated peptide domains and the intrinsic structure of siRNAs could be recovered prior to their participation in the RNAi mechanism. It is worth noting that increasing the number of multi-Tat peptide domains in the siRNA structure enhanced the target protein silencing. This is consistent with previous results demonstrating that multiple CPP domains in a branched dendrimer or linear structure improved the delivery efficiencies of conjugated oligonucleotide [81]. Despite the promising in vitro results, this siRNN–peptide conjugate has not yet been applied for in vivo study.

### 7.3.2.2 Targeting Peptides

Several targeting peptides, which bind to the biomolecules expressed on the specific cell surface, have been conjugated to siRNA for cell- and tissue-specific siRNA delivery. Cearone et al. conjugated IGF1 (insulin growth factor 1 mimetic D-form peptide) to the end of IRS-1 (insulin receptor substrate 1) specific siRNA. This study showed that in vitro transfection with 200 nM of IGF1–siRNA conjugates to the IGF1 receptor overexpressing cells in the absence of a transfection reagent caused about 57% target mRNA knockdown [82]. Sioud et al. conjugated siRNA to gastrin-releasing peptide (GRP) and gonadotropin-releasing hormone (GnRH) peptides via disulfide bond to deliver siRNA to the target tissues, such as ovarian, colon, breast, prostate, and lung cancer, where GRP and GnRH receptors are overexpressed [33]. The biotinylated siRNAs, GRP, and GnRH were mixed with streptavidin to combine these three molecules in one single complex via biotin–streptavidin interaction. This bi-targeting system accelerated the intracellular uptake of siRNA in vitro compared with the mono-targeting system containing a single type of targeting peptide. Kim et al. conjugated the PEG-linked luteinizing hormone releasing hormone (LHRH) peptide, also known as GnRH, to siRNA. The LHRH–PEG–siRNA conjugates were complexed with cationic PEI and the targeting peptides were exposed outside of the electrolyte complex [83]. The in vitro study showed a high efficiency of target protein silencing in LHRH receptor overexpressing cells, A2780, but in vivo knockdown efficiency has not been evaluated.

To enhance binding affinity to the target cells, di-, tri-, and tetravalent cyclic Arg-Gly-Asp peptides (cRGD), binding to the $\alpha v\beta 3$ integrin, were conjugated to siRNA via thioether bond [84]. Although the intracellular uptake efficiency of cRGD–siRNA conjugates with different multivalencies did not show a significant difference, tri-/tetravalent cRGD–siRNA conjugates showed the enhanced luciferase silencing effect while bivalent cRGD–siRNA conjugates did not show any silencing effect. Despite promising in vitro results of several targeting peptide–siRNA conjugates in the absence of a transfection reagent, an intracellular trafficking mechanism to deliver siRNA to the site of action, such as cytosol, has not yet been clearly identified and the relevant in vivo studies have not been followed.

### 7.3.2.3 Lytic Peptides

Lytic peptides contain the membrane-active domains, facilitating the translocation of macromolecules into the cytosol. Detzer et al. showed that the conserved part of several bacterial toxin proteins, enabling the disruption of the cellular membrane, was conjugated to siRNA for its translocation into the cytosol [85]. The authors confirmed that conjugation of toxin peptide to siRNA promoted the delivery of siRNAs to the cytosol, resulting in efficient target gene silencing in vitro. One of the potential hurdles to using lytic peptides for the siRNA delivery system is toxicity. To overcome this hurdle, Meyer et al. investigated the acid-triggered activation of lytic peptide (melittin, derived from bee venom) by masking Lys residues in the peptide sequence with dimethylmaleic anhydride (DMMAn), which could be cleaved at endosomal acidic pH [78]. In complexation systems, siRNAs could be disassembled from the delivery carriers due to charge interaction of positively charged carriers with serum protein or cell surface proteoglycans. To prevent this dissociation problem, DMMAn-protected melittin, siRNA, and PEG

were conjugated to poly-L-lysine (PLL) via disulfide bond. The authors performed the erythrocyte lysis test by incubating PEG–PLL–melittin–siRNA conjugates with mouse erythrocytes, and confirmed that the immobilized melittin in PLL backbone still exhibits its intrinsic lytic ability. Even though in vitro results with this delivery platform were promising, both local and systemic administrations of PEG–PLL–melittin–siRNA conjugates caused severe toxicity in vivo. The authors [86] speculated that this toxicity might be attributed to the aggregation between positively charged polymer backbone and siRNA during injection because the control conjugates without siRNA did not cause any toxicity. The same group has developed another siRNA conjugate platform with negatively charged lytic peptides. INF7, derived from influenza virus hemagglutinin with pH-dependent lytic activity, was conjugated to siRNA via disulfide bond. The overall negatively charged INF7–siRNA conjugates were complexed with peptide-like polycations containing free thiol groups, PEG, and targeting moiety [87, 88]. During complexation of polycations and INF7–siRNA conjugates, thiol groups in cationic carriers are oxidized to form disulfide bonds, and this intermolecular crosslinking can increase the particle stability for clinical applications. Interestingly, the conjugation of siRNA and lytic peptide via noncleavable bond significantly decreased the target gene–silencing efficiency, indicating that the biological efficacy of either siRNA or peptide was severely damaged by conjugation via noncleavable bond. Despite the rapid renal clearance after intravenous administration in vivo due to the small particle size of polycations/INF7–siRNA complexes, the supplementary result demonstrated that the use of carrier containing a longer PEG spacer prolonged the circulation time.

### 7.3.3 Carbohydrate–siRNA Conjugates

Binding of specific carbohydrate in glycoprotein to the sugar-binding receptor on the cell surface plays an essential role in various biological processes such as cellular adhesion and signal transduction. Once glycoprotein binds to the cell surface, this glycoprotein could be internalized into the cell via receptor-mediated endocytosis [89]. Many studies demonstrated the efficient intracellular delivery of oligonucleotide by directly conjugating carbohydrate to the oligonucleotide or to polymer-based carriers [90–92]. Among various carbohydrates, galactose is the most popularly investigated carbohydrate by conjugating to siRNA for the hepatocyte targeting delivery. It is well known that systemically injected nano-sized particles can be easily taken up by liver due to the fenestrated structure of hepatic vessels [45]. For hepatocyte targeting, the uptake by Kupffer cells, specialized macrophages located in liver, should be avoided because the accumulation of delivered molecules in these immune cells can cause liver toxicity or immune response. To circumvent nonspecific uptake and to facilitate the hepatocyte targeting delivery, several research groups have addressed the hepatocyte-targeting siRNA delivery systems with galactose- or galactose derivative–siRNA conjugates [29, 93]. Avino et al. conjugated one, two, and four galactose molecules to siRNA in order to increase binding affinity to the asialoglycoprotein receptor (ASGP-R) [93]. SiRNA conjugates containing four galactose molecules at a concentration of 100 nM showed a decrease of target TNF-α release from HuH-7 cells in the absence of a transfection reagent in vitro, but the knockdown was not efficient (25% of inhibition). Recently Meade et al. introduced the multivalent N-acetylgalactosamine (GalNAc) molecules to the neutral siRNN system. GalNAc is a hepatocyte-targeting galactose derivative that has a highly enhanced binding affinity to the ASGP-R on hepatocytes compared with galactose [94, 95]. In this system, three GalNAc molecules linked via a triantennary spacer are conjugated to the 5′ end of the sense strand of siRNN. The optimal spacing and multivalency of GalNAc with high binding affinity to the ASGP-R on hepatocytes have been determined in previous studies [96, 97]. Triantennary GalNAc–siRNN conjugates at a concentration of 25 mg/kg were injected to mice via intravenous administration, resulting in robust target apolipoprotein B (apoB) mRNA silencing in liver (>60% knockdown from a single injection). The silencing effect lasted for N12 days [29]. Severgnini et al. presented the in vitro data of GalNAc–siRNA conjugates with primary mouse hepatocytes. Fifty percent of target apoB mRNA decreased after 4 h of incubation with 1 nM of conjugates [98]. This study showed that the expression level of ASGP-R in primary hepatocytes dramatically decreased by the incubation time in a culture medium after isolation, indicating that this dynamic change of receptor expression level during in vitro cell culture could influence the GalNAc-mediated intracellular uptake efficiency. Therefore this study was done with freshly isolated primary hepatocytes and minimal culture time (4 h) when the significant amount of ASGP-R still presented. More recently, the enhanced form of GalNAc–siRNA conjugates, composed of chemically modified siRNA with 2′-F, 2′-O-Me sugar modification and phosphorothioate linkage (enhanced stabilization chemistry, ESC), was treated to primary hepatocytes in vitro, showing higher potency and stability than the previous GalNAc–siRNA platform (standard template chemistry, STC) [22]. For in vivo experiments with nonhuman primates [22], the knockdown of target TTR (transthyretin) gene in liver was 80% after a single subcutaneous injection of GalNAc–siRNA conjugates at a dose of 5 mg/kg, and the weekly dosing at 2.5 mg/kg showed a long-lasting inhibition of serum TTR protein (80%) over 280 days without tachyphylaxis or sensitization. Interestingly, the subcutaneous administration of GalNAc–siRNA conjugates showed far higher uptake and target gene–silencing efficiency than the intravenous administration. This result suggests that slow release from the subcutaneous space to the liver through the systemic circulation could enhance the uptake efficiency.

Alnylam Pharmaceuticals has actively investigated the hepatocytes targeting siRNA delivery with triantennary GalNAc–siRNA conjugates, providing promising clinical results [22]. Alnylam Pharmaceuticals recently presented the positive initial Phase 2 data with ALN-TTRsc, composed of STC delivery platform. Multiple subcutaneous injections of ALN-TTRsc at doses of 5–7.5 mg/kg for 5 weeks resulted in maximum 98% knockdown of serum TTR. In addition, the advanced ESC delivery platform to target antithrombin (ALN-AT3) and proprotein convertase subtilisin/kexin type 9 (ALN-PCSsc) genes in hepatocytes entered Phase 1 clinical trials for the treatment of hemophilia/rare bleeding disorders and hypercholesterolemia, respectively (https://clinicaltrials.gov/show/NCT02035605). To date, most published data with GalNAc–siRNA conjugates are from in vivo experiments or in vitro experiments using primary hepatocytes. Thus, further studies are needed to elucidate the biological process of this delivery platform.

### 7.3.4 LIPID–SIRNA CONJUGATES

Hydrophobic and amphiphilic lipids including fatty acids, steroids (cholesterol and bile acids), phospholipids, and glycerolipids have been conjugated at terminal of siRNAs to improve half-life in blood and intracellular uptake [99]. Small molecular lipids have been attached to siRNAs via solid-phase synthesis or chemical conjugation, like carbidiimide-mediated coupling and Michael addition reaction. Previously, diverse conjugation schemes were described in detail [99]. In this chapter we focus on pharmacological efficacy and mode of mechanism of lipid–siRNA conjugates in vitro and in vivo.

Conjugation of cholesterol to siRNAs could facilitate their accumulation in liver tissue via lipid metabolic pathway. Soutschek et al. performed the first in vivo experiment with cholesterol–siRNA conjugates, demonstrating the efficient target ApoB silencing (65%) by three daily injections with 50 mg/kg of cholesterol–siRNA conjugates via intravenous administration [46]. Although the early study by Lorenz et al. showed that the in vitro transfection of cholesterol–siRNA conjugates without any additives allowed the target gene silencing [100], several studies demonstrated that serum lipoprotein particles play an important role in the efficient delivery of cholesterol–siRNA conjugates [101, 102]. As "lipoprotein vectors" for siRNA, these lipoprotein particles bind to cholesterol–siRNA conjugates and deliver them to the tissues, mainly the liver, overexpressing lipoprotein receptor and transmembrane receptor. Wolfrum et al. demonstrated that cholesterol–, stearoyl–, and docosanyl–siRNA conjugates showed a great target-gene knockdown in vivo after three daily intravenous administrations at a dose of 50 mg/kg [101]. Interestingly, high-density lipoprotein (HDL) and low-density lipoprotein (LDL) pre-associated cholesterol–siRNA conjugates showed more enhanced uptake efficiency in the liver than free cholesterol–siRNA conjugates or albumin-bound cholesterol–siRNA conjugates. Considering that cholesterol itself may not exhibit the sufficient affinity to be associated with systemically circulating lipoprotein particles in serum, this preassembly strategy gives an important guideline for tissue-targeting delivery of cholesterol–siRNA conjugates. It is worth noting that HDL- or LDL-bound cholesterol–siRNA conjugates showed different tissue distribution, indicating that the uptake of cholesterol–siRNA will be determined by the type of bound lipoproteins and the distribution of their receptors [101]. Apart from liver targeting, Kuwahara et al. reported that HDL pre-associated cholesterol–siRNA conjugates were successfully delivered to mouse brain capillary endothelial cells and reduced target mRNA expression levels in the brain by 50% after three intravenous injections in a dose of 10 mg/kg at 12-h intervals [102]. Nishina et al. demonstrated that siRNA conjugates with vitamin E at the 5′ end of antisense strand inhibited target gene expression in the liver without induction of interferons after an intravenous injection at a dose of 2 mg/kg [39]. Although the molecular mechanisms of vitamin E–siRNA conjugates, such as binding proteins for liver accumulations and intracellular processing for endosomal escape, were not clearly elucidated in this study, their extended blood half-life and liver-specific accumulation are likely due to the diverse vitamin E-binding proteins and vitamin E-transferring proteins in serum. Authors [86] observed dose-dependent reduction of target apoB Mrna level and phenotype changes like altered liver metabolisms, which evidently showed the efficacy of vitamin E–siRNA conjugates in vivo. Authors [86] also compared vitamin E–siRNA conjugates to cholesterol–siRNA conjugates, which showed that suppression of cholesterol–siRNA conjugates at a dose of 2 mg/kg was not statistically significant. Phosphothioethanol (PE-SH) was linked to the 2-pyridyl disulfide-activated siRNA via disulfide exchange to produce phospholipid–siRNA conjugates, which formed micelles in aqueous solution [35]. For the protection of siRNAs exposed on the surface of micelles, PEG–phosphoethanolamine (PEG–PE) was also incorporated into micelles. Phospholipid–siRNA/PEG–PE mixed micelles showed ~27% of target GFP gene knockdown in vitro without any additive transfection reagents at an siRNA concentration of 84 nM in the presence of serum. Liu et al. applied the phospholipid–siRNA conjugates for the formulation of PLGA nanoparticles [37]. Thiol-modified siRNA against Plk1 was conjugated to pyridyldithio propionate–PE (PDP–PE) via disulfide exchange. Phospholipid–siRNA conjugates dissolved in DMSO were mixed with PLGA and doxorubicin to form nanoparticles via coprecipitation methods, which allowed simultaneous delivery of siRNA and hydrophobic small-molecule drug, doxorubicin. $IC_{50}$ values of phospholipid–siRNA conjugates/doxorubicin co-delivery to cancer cells decreased by five-fold compared with a simple mixture composed of free doxorubicin and siRNA in vitro. Intravenous injection of PLGA particles containing phospholipid–siRNA conjugates/doxorubicin

at an siRNA dose of 1 mg/kg decreased the Plk1 expression level in tumor tissues and reduced tumor volume by 5.5-fold compared with PBS in vivo control study. Raouane et al. developed squalene–siRNA conjugates by attaching squalene to the 3′-end of the sense strand of siRNA via-maleimide–thiol coupling, and performed their feasibility studies in vivo [103]. Squalene–siRNA conjugates formed spherical nanoparticles with a size of 165 nm in aqueous solution, which improved serum stability of siRNAs in vitro. In vitro transfection of squalene–siRNA conjugates at a concentration of 50 nM to BHP cells showed ~80% knockdown of target genes. Intravenous injections of squalene–siRNA conjugates at a dose of 2.5 mg/kg five times during 19 days resulted in 70% tumor suppression. However, authors [86] noticed that control siRNAs, which did not cause the nonspecific gene knockdown in vitro, also slightly inhibited tumor growth in vivo.

### 7.3.5 POLYMER–siRNA CONJUGATES

Biocompatible polymers including hyaluronic acid and PEG have been conjugated to the ends of siRNAs in preclinical studies [28, 36, 104, 105]. Until now, diverse PEGs with different properties, in terms of polymer structure, linear and branched type, degree of conjugation, bond cleavability, and molecular weights, have been conjugated to siRNAs [28, 106–108]. PEGylations at both ends of siRNAs protected them more than 48 h in serum, which significantly enhanced serum stability compared with the single PEGylation [107]. Unlike protein drugs, which can therapeutically function at an extracellular location, siRNAs should be delivered to the cytosol, a site of action. However, degree of PEGylation can interfere with the intracellular translocation of siRNA due to steric effects with cellular membranes. To compensate for the steric effects at target cells, targeting ligands like lactose and folate have been conjugated to the terminal of PEG–siRNA conjugate [109, 110]. For example, galactose–PEG and mannose 6-phosphoate–PEG were conjugated to the 3′ end of sense strand of siRNA via disulfide bond to target hepatocytes and hepatic stellate cells (HSCs), respectively [111]. Four-hundred nanomoles of galactose–siRNA conjugates (Gal–PEG–siRNA) and mannose–siRNA conjugates (M6P–PEG–siRNA) were treated to the cells in the absence of transfection reagent, causing not robust but significant reduction of target TGF-β1 expression in vitro. Oishi et al. demonstrated the efficient intracellular delivery with lactosylated PEG–siRNA conjugates [112]. In vitro transfection with 100 nM of lactosylated PEG–siRNA conjugate without delivery carrier downregulated the target luciferase expression to 50% in HuH-7 cells, which were previously transfected with luciferase expressing DNA/lipofectamine complexes. A later study from the same group targeted the endogenous RecQL1 gene with HuH-7 cells, but the treatment of lactosylated PEG–siRNA in the absence of a transfection reagent did not show any inhibitive effect of target gene [109].

The advantage of polymer–siRNA conjugates is multivalent incorporation of different functional moieties into one molecule like PEG, targeting ligands, and membrane-active moieties, as shown in Dynamic Polyconjugates (DPCs), a technology currently developed by Arrowhead Pharmaceuticals [30]. The first generation of DPC was synthesized by conjugating siRNA to endosomolytic polymer (poly[butyl amino vinyl ether], PBAVE) via disulfide bond [30]. PEG and GalNAc molecules were reversibly conjugated to the amine groups of PBAVE via acid–labile bond to mask the membrane-active moieties of polymer. DPC–siRNA conjugates are taken into the cell via receptor-mediated endocytosis, and acidic pH in the endosome triggers the release of PEG and GalNAc, resulting in polymers recovering their endosomolytic activity. After endosomal escape, siRNA is released from the polymer via cleavage of disulfide bond in a reductive environment such as the cytosol, and the intact siRNA molecules can undergo the RNAi mechanism. Both in vitro and in vivo studies have demonstrated the efficient target gene knockdown with this multifunctional siRNA conjugate. In vitro transfection of DPC-containing siRNA against apoB (siRNA dose: 50 nM) to mouse primary hepatocytes resulted in 80% knockdown of target mRNA expression. For in vivo experiments, DPCs containing siRNAs against apoB and peroxisome proliferator-activated receptor alpha (ppra) (siRNA dose: 2.5 mg/kg) were injected to mice via intravenous administration, with the resulting levels of target apoB and ppra mRNA reduced by 76% and 64%, respectively. Looking into more details, this system contains an excess amount of polymer without siRNA (free DPC) and polymer–siRNA conjugates (DPC–siRNA) due to the lack of purification step after synthesis. The localization of DPC–siRNA and free DPC was monitored by in vivo PET/CT imaging [113]. Even though DPC–siRNA and free DPC showed slightly different tissue distribution and pharmacokinetics after intravenous injection in mice, two components were mostly co-localized in the mouse liver. For example, 70% of DPC–siRNA accumulated in liver, while 98% of free DPC accumulated in the liver, but at a slower rate. Accordingly, the residual polymers without siRNA (PBAVE masked with PEG and GalNAc) in the sample solution probably played an important role in the efficient delivery of siRNA. Wong et al. demonstrated the efficient hepatocyte-specific siRNA delivery by co-injection of liver targeting PBAVE (GalNAc–PBAVE) and cholesterol–siRNA conjugates [114]. Cholesterol–siRNA conjugates have shown their potent liver-targeting capacity but not elicited a sufficient biological effect due to the lack of endosomolytic function. The confocal microscopic data with mouse liver section revealed that more than 80% of intravenously injected cholesterol–siRNA conjugates were co-localized with GalNAc–PBAVE, intravenously injected 1 h before cholesterol–siRNA conjugates. One possible scenario is that the endosomes containing previously delivered siRNAs fuse with the endosomes containing GalNAc to release siRNA to the cytosol, in a phenomenon known as *endosomal fusion* [115]. This result provides a meaningful insight to design the siRNA delivery platform such that siRNA molecules do not have to be linked to the endosomolytic molecules when those two key components efficiently target the same cells, such as hepatocytes. The successful outcome has been achieved by co-injection of hepatocyte-targeting endosomolytic polymer and cholesterol–siRNA conjugates in nonhuman primates. This treatment increased the knockdown efficiency over 500-fold compared with the single injection of cholesterol–siRNA conjugates. The latest generation of DPC in clinical study, named ARC-520, is a co-injection system with two individual components, cholesterol–siRNA conjugates and hepatocyte-targeting membrane-active peptides. In this system, the previous endosomolytic PBAVE polymer was replaced with melittin-like peptide (MLP) with a well-defined structure [116, 117]. The preclinical studies with nonhuman primates showed that intravenous co-injection of GalNAc–MLP and cholesterol–siRNA conjugates elicited a highly efficient target-gene silencing in liver (N99% knockdown of target coagulation factor VII [F7] mRNA in serum after co-injection of 3 mg/kg of GalNAc–MLP and 2 mg/kg of cholesterol–siRNA conjugates). Considering that the intravenous injection with 6 mg/kg of GalNAc–MLP and 0.1 mg/kg of chol–siRNA showed N99% knockdown efficiency in mice, the dose of cholesterol–siRNA conjugates for nonhuman primates could be decreased to 0.1 mg/kg. The administration was well tolerated and around 80% knockdown of F7 activity lasted for nearly 1 month. Recently, Arrowhead announced that a Phase 1 single-dose study with ARC-520 has been completed and a Phase 2a single-dose study has begun for the treatment of chronic hepatitis B virus (HBV) infection. Arrowhead currently is developing another nucleic acid therapeutic with unlocked nucleobase analogs using the DPC delivery platform for the treatment of liver disease associated with alpha-1 antitrypsin deficiency (AATD).

## 7.3.6 NANOSTRUCTURED MATERIALS–siRNA CONJUGATES

Various nanostructured materials (e.g., gold nanoparticles [AuNPs], magnetic nanoparticles [MNPs], carbon nanotubes [CNTs], quantum dots [QDs], mesoporous silica nanoparticles [MSNPs]) have been utilized for siRNA conjugation either for theranostic (therapy and diagnostics) purposes or synergistic multidrug delivery. AuNPs have been explored as a promising siRNA delivery vehicle due to their biocompatibility and straightforward functionalization via strong pseudocovalent bonds between thiol group and Au [118–125]. In addition, their optical properties allow AuNP–siRNA conjugates to be applied for AuNP-induced hyperthermia or in vivo imaging [126]. VEGF siRNA–AuNP conjugates, functionalized with tumor-associated macrophages (TAMs) targeting peptide, were treated through intratracheal instillation at a dose of 0.05 mg/kg of siRNA three times during 21 days in a lung cancer murine model [118]. As a result, the level of target VEGF gene expression in the inflammatory tumor M2 macrophage and lung cancer cells was successfully inhibited and the tumor size decreased by 95%. MNPs are also an attractive metallic scaffold for siRNA conjugation due to their capacity to achieve magnetofection with an external magnetic field and high-resolution magnetic resonance imaging (MRI) [127, 128]. Medarova et al. developed near-infrared dye-labeled MNP–siRNA conjugates, modified with myristoylated polyarginine peptides as a membrane translocation module, not only for in vivo monitoring but also for siRNA-mediated cancer therapy [128]. Survivin siRNA–MNP conjugates were treated twice a week for 2 weeks at a dose of 5.7 mg/kg of siRNA to mice bearing subcutaneous adenocarcinoma intravenously, with the resulting survivin mRNA levels in tumors being 97% lower than in controls. Both MRI and optical imaging showed that conjugates were efficiently accumulated in tumors, resulting from an enhanced permeability and retention (EPR) effect [129]. Besides these, CNTs, QDs, and MSNPs are also promising candidates for siRNA delivery vehicles [130–134]. However, their evaluations as siRNA carriers have been demonstrated mainly in vitro. Accordingly, in vivo efficacy and toxicities of CNTs, QDs, and MSNPs, especially in terms of body clearance and side effects on specific organs, need to be studied before consideration of their translation to clinical applications [135].

## 7.4 RNAi AS GENE-SILENCING MECHANISM

RNA interference (RNAi) in mammalian cells has opened unprecedented research opportunities to take advantage of this mechanism for treatment of various diseases. RNAi is a posttranscriptional gene-silencing mechanism triggered by genome-origin microRNA, germline-specific Piwi-interacting RNA (piRNA), and small interfering RNA (siRNA) [136]. Whereas microRNA and piRNA are naturally produced from genome, siRNA describes exogenous synthetic molecules for biomedical research [136]. MicroRNA regulates the expression of multiple mRNA but siRNA inhibits the expression of one specific target mRNA [137]. Thus, the easier sequence design of siRNA allowed more scientists to focus on siRNA delivery in the early history of RNAi-based drugs. The siRNA is a double-stranded RNA molecule with typically 21–23 nt in each strand with two nucleotides overhung at both 3'-ends [138]. In the cytoplasm, siRNA forms an enzymatic machinery with several protein units, termed as the RNA-induced silencing complex (RISC), in which the sense (or passenger) strand of siRNA is degraded and ejected from RISC, and the remaining antisense (or guide) strand binds to its complementary mRNA to trigger mRNA degradation by RISC [2]. Considering that the sequence of siRNA can be designed to inhibit any target gene in theory [139, 140], its therapeutic potential covers various diseases ranging from viral infection to hereditary disorders and cancers [141]. In particular, target-gene candidates in cancer therapy are widely

managed through oncogenes, angiogenesis, proliferation, and the cell cycle. Polo-like kinase 1 (PLK1, cell cycle), EphA2 (oncoprotein) [142], ribonucleotide reductase M2 subunit (RRM2, DNA replication and repair) [143], kirsten rat sarcoma viral oncogene homolog (KRAS, cell proliferation) [144], protein kinase N3 (PKN3, cell cycle) [145], and vascular endothelial growth factor (VEGF, angiogenesis) [146] genes have been tested as target genes in early-phase clinical trials for RNAi-based cancer therapies.

The current major strategies to design vehicles for systemic siRNA delivery involve the construction of multimolecular assemblies from more than dozens of monomer components, including siRNA. In this way, siRNA-loaded nanoparticles with a larger size (or molecular weight) can avoid (or delay) renal filtration and allow more efficient accumulation in tumors [147–149]. Various materials, e.g., chitosan/biopolymers, cationic polymers, and inorganic materials, have been employed to improve RNAi-based therapeutics, where attempts have been made to lower the dose of siRNA, resulting in higher therapeutic effects on tumor growth inhibition. Nevertheless, much higher doses of siRNA are still required to elicit significant RNAi in cancer, compared with the liver. Lipid nanoparticle (LNP) targeting transthyretin (TTR) amyloidosis showed 80%–90% reduction in TTR protein levels in the liver at doses of 0.15–0.3 mg/kg [150], whereas gene silencing of similar LNPs in solid tumors was reported to be much less effective: 30%–70% reduction in VEGF mRNA level in 3 of 12 patients and no reduction in 8 patients at doses of 0.1–1.5 mg/kg [146]. These early clinical trial data indicate that the current delivery vehicles can be further improved for siRNA delivery to solid tumors [141]. This chapter (1) describes delivery vehicles, which have been evaluated in current clinical trials; (2) presents biological barriers in systemic siRNA delivery to solid tumors and considerations in designing multimolecular assemblies as delivery vehicles; (3) summarizes various techniques to develop delivery vehicles equipped with prolonged blood circulation property, selective release of siRNA, enhanced targetability, and more efficient endosomal escapability; and (4) discusses the next generation of delivery vehicles for cancer therapy.

## 7.5 CURRENT STATUS OF CLINICAL TRIALS IN RNAi-BASED CANCER THERAPY

To date, four RNAi-based drugs have been evaluated in early clinical trials for cancer therapy through intravenous administration (Table 7.2). Two stable nucleic acid lipid nanoparticle (SNALP) formulations, ALN-VSP02 and TKM-080301, have completed their Phase 1 trials targeting VEGF/KSP and PLK1 genes, respectively [146]. Another LNP formulation, Atu027, has also finished Phase 1 trials regarding its gene-silencing efficiency of PKN3 in solid tumors and safety in patients. A cyclodextrin-based polymer nanoparticle, CALAA-01, the first-in-human phase 1 clinical trial for cancer therapy using this drug, was

## TABLE 7.2

## RNAi-Based Drugs Tested in Clinical Trials for Cancer Therapy [157]

| Delivery Vehicle | Disease | Target | Phase | Status | Drug | Clinical Trial Government Identifier |
|---|---|---|---|---|---|---|
| LNP | Advanced solid tumor | PKN3 | 1 | Completed | Atu027 | NCT00938574 |
| LNP | Primary and secondary liver cancers | PLK1 | 1 | Completed | TKM-080301 | NCT01437007 |
| LNP | Liver cancer | VEGF, KSP | 1 | Completed | ALN-VSP02 | NCT00882180 |
| LNP | Liver cancer | MYC | 1b/2 | Recruiting | DCR-MYC | NCT02314052 |
| LNP | Advanced cancer | EphA2 | 1 | Not yet recruiting | siRNA-EphA2-DOPC | NCT01591356 |
| Polymer nanoparticle | Solid tumor | RRM2 | 1 | Terminated | CALAA-01 | NCT00689065 |
| Ex vivo transfection | Metastatic melanoma | LMP2, LMP7, MECL1 | 1 | Completed | — | NCT00672542 |
| Ex vivo transfection | Solid tumor | Cbl-b | 1 | Recruiting | APN401 | NCT02166255 |
| Polymer implant | Advanced pancreatic cancer | KRAS | 2 | Not yet recruiting | siG12D LODER | NCT01676259 |

evaluated for RRM2 silencing, but recently its trials have been terminated [143].

SNALP is composed of an ionizable cationic lipid, distearoylphosphatidylcholine, cholesterol, and poly(ethylene glycol) (PEG)-lipid [151]. ALN-VSP02 is a LNP formulation containing VEGF and KSP-targeted siRNAs at a 1:1 molar ratio and both siRNAs were chemically modified with partial phosphorothioate backbone and 2′-O-methylation [146]. The key ionizable cationic lipid (termed DLin-DMA) was selected from two important parameters, pKa of the amine in the head group of ionizable cationic lipids and bilayer-to-hexagonal HII phase transition temperature (TBH). These parameters are responsible for siRNA encapsulation/surface charge and endosome escapability of LNP formulations, respectively. TBH is defined as a temperature to transit lamellar structure into a non-bilayer structure at pH 4.8 in the measurement of differential scanning calorimetric analyses, where a protonated ionizable cationic lipid may form ion pairs with anionic lipids in the endosomal membranes and destabilize the membranes [151]. A detailed endosomal escape mechanism for lipid-based nanoparticles is explained later in this review. ALN-VSP02 has a particle diameter of 80–100 nm and is weakly charged with zeta potential of <6 mV at pH 7.4. The results obtained in the Phase 1 trial confirmed significant accumulation of siRNA in tumor biopsies, siRNA-mediated mRNA cleavage in the liver tumor, and antitumor activity that included complete regression of liver metastases in primary endometrial cancer [146]. The ALN-VSP02 was administered through a 15-min intravenous infusion via a controlled infusion device every 2 weeks (2 doses over 1 month). Tumor biopsies of 12 patients had detectable VEGF siRNA and 11 of 12 patients had detectable KSP siRNA. Although a substantial increase in the RNAi-mediated cleavage product of VEGF mRNA was observed in two patients, no patients were positive for KSP mRNA cleavage because the KSP

mRNA level in banked tumor and normal liver were substantially lower (1/50–1/100) than those of VEGF mRNA. Of note, the SNALP formulations comprising DLin-DMA and its next generation (DLin-MC3-DMA), ALN-TTR01, and ALN-TTR02, were also clinically tested in other diseases, including TTR-mediated amyloidosis. In their Phase 1 trial reports, ALN-TTR01 at a dose of 1.0 mg/kg suppressed TTR gene expression with a mean reduction of 38% at day 7 in 32 patients with TTR amyloidosis. For ALN-TTR02, the mean reduction in TTR levels at doses of 0.15 to 0.3 mg/kg ranged from 82% to 87%, with reductions of between 57% and 67% at 28 days, in 17 healthy volunteers [150].

Atu027 is composed of a cationic lipid (termed AtuFECT01) possessing a highly charged head group, a helper lipid, and a PEG-lipid. This vehicle is evaluated for PKN3 silencing to treat various primary tumors and metastatic lesions [145, 152]. Atu027 has a particle diameter of approximately 120 nm and a highly positive surface charge (approximately 50 mV), which allows more efficient siRNA binding to the vehicles. In the Phase 1 trial report, the plasma level of siRNA was increased during an initial 4-h infusion period, followed by a steep decline within the next 4 h. Atu027 was well tolerated up to 0.336 mg/kg, which approximately corresponds to twice the effective siRNA plasma level. The most frequent adverse events were mild fatigue (11 patients in a total of 66 patients) and increased lipase without clinical signs of pancreatitis (4 patients). After repeated treatment, 41% of patients experienced stabilization of advanced solid tumors.

CALAA-01 is composed of a cyclodextrin-based cationic polymer, a PEG-adamantane, and a transferrin-PEG-adamantane [153]. This nanoparticle has sizes from 60–150 nm, surface charges from 10–30 mV, and contains approximately 2000 siRNA molecules. The intracellularly localized nanoparticles were observed in tumor biopsies from melanoma patients and their amounts were correlated with the dose levels of the nanoparticles [143]. Furthermore, a reduction in both mRNA and protein of the target RRM2 was observed. The RNAi-mediated fragmentation of mRNA was also detected [143]. After treatment by infusion, the plasma concentration of siRNA in almost all patients rapidly declined and reached an undetectable level in 30 min [4]. When CALAA-01 was administered using four injections in a single patient, the decreasing rate of plasma siRNA concentration was similar among them. CALAA-01 was cleared mainly from the kidneys due to its interaction with the renal filtration barrier [147]. In addition, one or more of the delivery components of CALAA-01, rather than the siRNA, were primarily responsible for hypersensitivity and sequelae of acute immune responses (flushing, fever, and fatigue) [4]. Therefore, it is suggested that a toxicity profile may be alleviated by purifying any free polymer components of CALAA-01 after preparation. The scientific report on the causes of the termination of CALAA-01 in clinical trials has not been published. The relatively shorter blood circulation property of CALAA-01 (half life <30 min), compared to a clinically approved antibody drug conjugate (half-life 1–4 days) [154] and clinically tested anticancer drug–loaded polymeric micelles (half-life 16–80 h) [155], suggests that this unstable vehicle in the bloodstream may be difficult to take the advantage of delivery carrier (e.g., passive targeting by the nanoparticulate formation and active targeting by transferrin ligands), leading to inefficient accumulation of siRNA in solid tumors.

Two things can be learned from these clinical trial reports for development of the next generation of RNAi-based drugs. The dose of siRNA in cancer patients ranges between 0.1–1.5 mg/kg, providing a guideline for mouse model experiments. The pharmacokinetic data after infusion treatment of patients may be consistent with preclinical blood circulation properties in mice [4]. The clinically tested delivery vehicles have individual disadvantages when compared to one another. Polymer-based CALAA-01 is less stable because it exhibits shorter blood circulation properties (half-life <30 min) and approximately 60% nanoparticle of injected dose is entrapped with renal filtration barrier at 60 min after intravenous administration in mice [156]. Lipid-based ALN-VSP02 showed relatively longer blood circulation (half-life <2 h) but its hydrodynamic size is >50 nm, and nanoparticles of this size do not efficiently penetrate into stoma-rich and hypovascular tumor [157]. Approximately 50% nanoparticle of injected dose accumulates in the liver at 30 min after intravenous administration in mice [151].

## 7.6 IMMUNE CHECKPOINT INHIBITORS IN CLINICAL TRIALS

Clinicians have long sought to find combinations of drugs that might achieve synergy, defined as more than an additive effect. Because cancer drugs given in combination have the potential for increased tumor-cell killing, finding the best combination partners for programmed cell death 1 (PD-1) checkpoint inhibitors could improve clinical outcomes for patients with cancer. The success of immune checkpoint inhibitors (ICIs), notably anti-cytotoxic T lymphocyte associated antigen-4 (CTLA-4) as well as inhibitors of CTLA-4, programmed death 1 (PD-1), and programmed death ligand-1 (PD-L1), has revolutionized treatment options for solid tumors. However, the lack of response to treatment, in terms of de novo or acquired resistance, and immune-related adverse events (IRAE) remain as hurdles. One mechanism to overcome the limitations of ICIs is to target other immune checkpoints associated with tumor microenvironment. Immune checkpoints such as lymphocyte activation gene-3 (LAG-3), T cell immunoglobulin and ITIM domain (TIGIT), T cell immunoglobulin and mucin-domain containing-3 (TIM-3), V-domain immunoglobulin suppressor of T cell activation (VISTA), B7 homolog 3 protein (B7-H3), inducible T cell costimulatory (ICOS), and B and T lymphocyte attenuator (BTLA) are feasible and promising options for treating solid tumors, and clinical trials are currently under active investigation. This review aims to summarize the clinical aspects of the immune checkpoints and introduce novel agents targeting these checkpoints.

Cancer immunotherapy is one of the major pillars in the field of medical oncology, especially for the treatment of unrespectable, metastatic, and recurrent cancers. The success of ICIs, such as anti-CTLA-4 and anti-PD-1/PD-L1, in combination with chemotherapy, immunotherapy, and targeted agents, has changed the paradigm of cancer treatment. Nonetheless, the limited efficacy and IRAEs of ICIs have paved way for the discovery of novel checkpoints. Among the immune checkpoint inhibitors, anti-LAG-3 and anti-TIGIT are promising targets, and their efficacy in combination with anti-PD-1/PD-L1 may help overcome the limitations seen in prior treatments. More robust data are yet to follow on agents targeting TIM-3, B7-H3, VISTA, ICOS, and BTLA.

Advances in immunotherapy are changing treatment paradigms for most cancers. When given in combination, cancer drugs have the potential to increase tumor-cell killing. Therefore, following the success of programmed cell death 1 (PD-1) and programmed cell death ligand 1 (PD-L1) agents as monotherapies, research is turning to identifying effective combinations as shown in Table 7.3. Historically, the rationale for combining therapies has been that using two effective drugs with independent mechanisms of action would decrease the likelihood that resistant cancer cells could develop, a strategy that addresses clonal heterogeneity. DeVita and others have shown that agents active by any mechanism in shrinking tumors combine successfully. The activity of cancer treatments is defined by the US Food and Drug Administration (FDA) by using the overall response rate (ORR) in clinical trials, where the ORR is defined as the "proportion of patients with tumor size reduction of a predefined amount for a minimum time period."

**TABLE 7.3**

**Clinical Trial Landscape for Immune Checkpoint Inhibitors [158, 159]**

| Drug | Cancer Type | Clinical Trial ID |
|---|---|---|
| Pembrolizumab (Anti-PD-1) | NSCLC | NCT03134456, NCT02220894, NCT02142738, NCT02864394, NCT03302234, NCT01905657, NCT02504372, NCT02775435, NCT02578680 |
| | Small cell lung cancer | NCT03066778 |
| | Head and neck squamous cell carcinoma | NCT02252042, NCT03040999, NCT02358031 |
| | Renal cell carcinoma | NCT03142334, NCT02853331 |
| | Gastric adenocarcinoma | NCT02370498 |
| | Nasopharyngeal neoplasms | NCT02611960 |
| | Urothelial carcinoma | NCT02853305, NCT03244384, NCT02256436, NCT03374488, NCT03361865 |
| | Colorectal cancer | NCT02563002 |
| | Pleural mesothelioma | NCT02991482 |
| | TNBC | NCT02819518, NCT03036488, NCT02555657 |
| Nivolumab (Anti-PD-1) | NSCLC | NCT02041533, NCT01642004, NCT01673867 |
| | Mesothelioma | NCT03063450 |
| | Non-Hodgkin lymphoma | NCT03366272 |
| | Metastatic clear cell renal carcinoma | NCT01668784 |
| | Head and neck cancer | NCT02741570, NCT03342352 |
| | Lung cancer | NCT03348904 |
| | Melanoma | NCT03068455, NCT01844505 |
| Ipilimumab (Anti-CTLA-4) | NSCLC | NCT03469960, NCT03351361, NCT02785952, NCT03302234 |
| | Squamous cell lung carcinoma | NCT02785952 |
| | Mesothelioma | NCT02899299 |
| | Gastric cancer | NCT02872116 |
| | Gastroesophageal junction cancer | |
| | Metastatic melanoma | NCT03445533, NCT00636168, NCT01274338, NCT02339571, NCT02506153, NCT02224781, NCT00094653 |
| | Metastatic non-cutaneous melanoma | NCT02506153 |
| Atezolizumab (Anti-PD-L1) | Ovarian cancer, fallopian tube cancer, peritoneal neoplasms | NCT03038100, NCT02839707, NCT02891824 |
| | NSCLC | NCT02813785, NCT02008227, NCT02367781, NCT02366143, NCT02409342, NCT02486718, NCT02367794, NCT03191786, NCT02409355, NCT02657434, NCT03456063 |
| | Extensive stage small cell lung cancer | NCT02763579 |
| | TNBC | NCT03197935, NCT02425891, NCT03125902, NCT03281954 |
| | Renal cell carcinoma | NCT02420821, NCT03024996 |
| | Bladder cancer | NCT02302807 |
| | Squamous cell carcinoma of the head and neck | NCT03452137 |
| | Urothelial carcinoma | NCT02807636 |
| | Transitional cell carcinoma | NCT02450331 |
| | Prostatic neoplasms | NCT03016312 |
| Durvalumab (Anti-PD-L1) | NSCLC | NCT02352948, NCT03003962, NCT02453282, NCT02273375, NCT02542293, NCT03164616, NCT02125461, |
| | Squamous cell lung carcinoma | NCT02154490, NCT02551159 |
| | Recurrent or metastatic PD-L1 positive or negative SCCHN | NCT02369874 |

*(Continued)*

**TABLE 7.3** *(Continued)*
**Clinical Trial Landscape for Immune Checkpoint Inhibitors [158, 159]**

| Drug | Cancer Type | Clinical Trial ID |
|---|---|---|
| | Recurrent squamous cell lung caner | NCT02766335, NCT02154490 |
| | Urothelial cancer | NCT02516241 |
| | Advanced solid malignancies | NCT03084471 |
| | SCCHN, hypo pharyngeal squamous cell carcinoma, laryngeal squamous cell carcinoma | NCT02551159, NCT03258554 |
| BMS-936558 (Anti-PD-1) | Unresectable or metastatic melanoma | NCT01721746, NCT01721772 |
| SHR1210 (Anti-PD-1) | NSCLC | NCT03134872 |
| | Nasopharyngeal neoplasms | NCT03427827 |
| PDR001 (Anti-PD-1) | Melanoma | NCT02967692 |

*Abbreviations:*  NSCLC, non–small cell lung cancer; SCCHN, squamous cell carcinoma of the head and neck; TNBC, triple-negative breast cancer.

Exp Mol Med. 2018 Dec 13;50(12):165. doi: 10.1038/s12276-018-0191-1

There are 2250 active trials testing anti-PD1/ PDL1 agents as of September 2018, compared with 1502 trials in September 2017.

Nat Rev Drug Discov. 2018 Nov 28;17(12):854-855. doi:10.1038/nrd.2018.210

View this in web browser version.

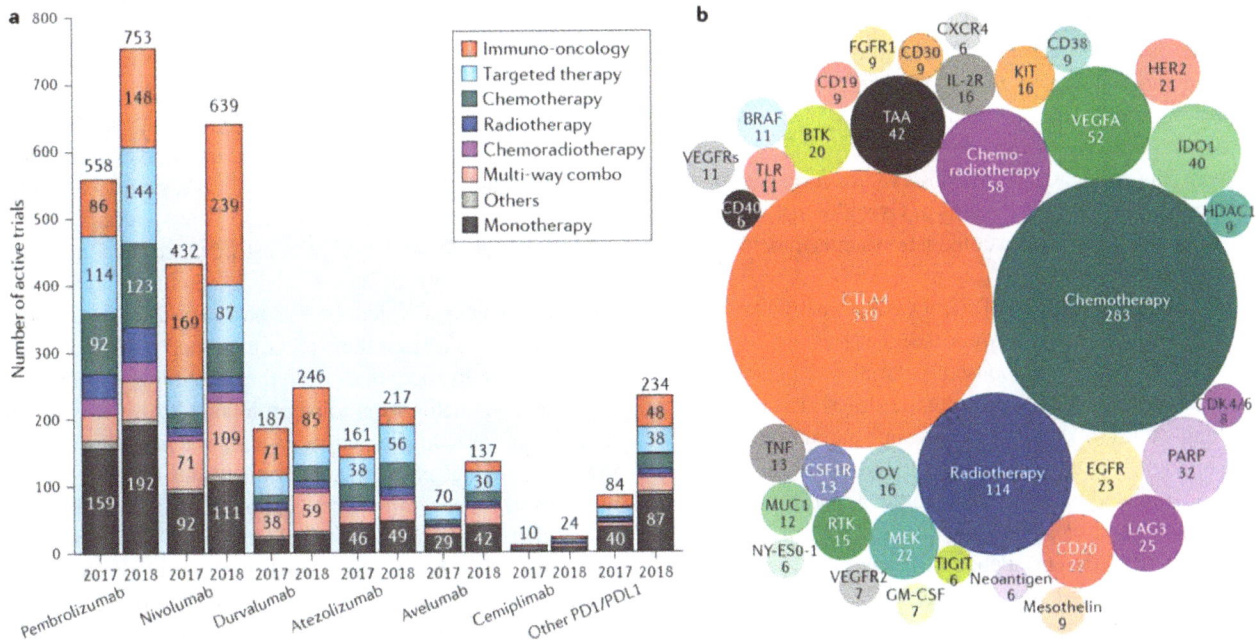

**FIGURE 7.4**  **(a)** There are 2,250 active trials testing anti-PD1/PDL1 agents as of September 2018, compared with 1,502 trials in September 2017. **(b)** The 1,332 trials evaluating anti-PD1/PDL1 agents in combination with the top 38 targets (among the 1,716 combination trials testing a total of 240 targets) are shown here. We have selected those targets being evaluated in at least 6 trials. The numbers of active clinical trials that are testing drugs against the target are indicated in each bubble.

## 7.7 LESSONS FROM PREVIOUS siRNA CONJUGATE STUDIES AND PERSPECTIVES OF CLINICAL APPLICATIONS

Currently, around nine naked siRNAs and nine lipid-based nanoparticles are in clinical trials for the treatment of eye-related diseases, viral infections, and cancers [160]. The latest clinical statuses of siRNA conjugates and particle-based carriers are shown in Table 7.4 in terms of type of carrier, conjugation/chemistry used, target genes, route of administration, and trial status. Three different siRNA conjugates including siRNA–GalNAc conjugate and DPC technology are under clinical status. While particle-based carriers have been mainly administered via intravenous injection, other formulations including siRNA conjugates and modified RNAi compounds allow different administration routes like intradermal and subcutaneous injection. Previous clinical trials have gradually resolved crucial issues, including identification of disease-causing genes, selection of target tissues and target mRNA sequences in genes, and chemical modifications. The closure of several trials with disappointing results, such as puzzling therapeutic outcomes caused by TLR activations and immune stimulations, has driven researchers to reconsider previous carrier systems. One promising alternative for clinical translation of siRNAs as therapeutics can be siRNA conjugates. However, siRNA conjugates usually have exhibited poorer knockdown efficiency for in vitro gene suppression than particle-based systems [52]. To allow favorable feasibility in preclinical and clinical studies, instead of the single conjugates, the combinational injection with cationic lipid/polymer could be a possible approach to obtain positive clinical outcomes. However, it should be also noticed that particle-based delivery systems can allow artifact effects by sedimentation during in vitro assay, which suggests that the superior in vitro efficiency of particle-based systems may not always be correlated with in vivo efficacy [161, 162]. Thus, comparable potency could be observed in conjugate systems to that in particle-based systems in vivo, although poorer efficacy has been observed in conjugate systems than in particle-based systems in vitro.

To improve feasibility of siRNA conjugates for preclinical studies, several factors in their design should be considered: cleavability, immune responses, and multivalent conjugation. Not many but a few studies demonstrated that cleavability of conjugation linkage and conjugation site on siRNA affect the biological activity of conjugates. For example, Detzer et al. conjugated bacterial toxin to the 3' end of the sense or antisense strand of siRNA via noncleavable thioether bond and showed different silencing efficiencies [85]. Dohmen et al. observed that noncleavable INF7–siRNA conjugates showed far less silencing activity than cleavable conjugates [87]. Aptamer–siRNA chimera via noncleavable bond should be cleaved by Dicer processing, which could elicit innate immune responses. Taken together, siRNA conjugates via cleavable bond for facile regeneration of free siRNAs could be much more favorable than siRNA conjugates via noncleavable bond. However, comparative studies with different synthetic strategies are still lacking with regards to estimating the modified biological activities of conjugates. Immunogenicity of biomolecules and their siRNA conjugates should be extensively monitored in vitro and in vivo not only to reduce toxicity but also to avoid false-positive therapeutic outcomes [52]. Although various chemical modifications on the siRNA backbone have been developed to prevent off-target effects and immune responses through TLR activations, these notorious side effects should be carefully monitored during administration of siRNA conjugates [163, 164]. Multivalent conjugations of biomolecules to siRNA could be a promising strategy to improve targeted delivery of siRNA conjugates in vivo. For example, conjugation of multiple peptides to siRNA could be an option to enhance the insufficient delivery potency of peptides. Multivalent targeting ligands have been designed to enhance binding affinity, and their functional benefits have been extensively investigated [19, 50]. Recently, drugs, targeting ligands, and functional molecules with desirable multivalency were tethered into one molecule, resulting in enhanced intracellular delivery efficiency or synergistic efficacy [165–167]. In particular, enhancement of delivery efficiency can reduce the doses of siRNA conjugates required for therapeutic outcomes in preclinical studies. Understanding the intracellular processing is crucial for the design of potent siRNA conjugates. In general, endosomal entrapment is one of the major challenges for siRNA cytosolic delivery when conjugates are taken up by the cells via endocytosis. To overcome this bottleneck, various additive reagents for endosomal escapes could be introduced to the system. For example, supplementary free peptides could be applied for topical administration of peptide–siRNA conjugates. Not by conventional proton sponge effects or membrane disruption but by TLR9-regulated mechanism, CpG oligonucleotide–siRNA conjugates also showed successful endosomal escape and consequent gene suppression [168, 169]. Although high concentration of siRNA (500 nM) was treated for in vitro gene knockdown, no additives were used for in vitro assay and in vivo tumor suppression was significant after local and systemic injection [169]. Among various siRNA conjugate systems, the most advanced technologies are Dynamic Polyconjugates and triantennary GalNAc–siRNA conjugates, which are on the Phase 2 clinical study. Even though data from their preclinical and clinical studies are still scant, the outcomes from the preclinical studies with nonhuman primates seem to be promising. Lastly it is worth noting that various evaluation systems for precise monitoring of biological effects in preclinical and clinical studies are needed for the clinical translation of siRNA conjugates. While diverse in vitro assays including 5'-RACE PCR, real-time PCR, ELISA, etc. have already been developed to assess gene knockdown quantitatively, in vivo monitoring systems are still few. Previously, liver-specific luciferase gene expression systems using adenovirus were developed to assess the pharmacodynamic studies of siRNAs in vivo [170]. Likewise, animal models can be designed to express reporter genes in a specific tissue, which could be harnessed for accurate, rapid, systemic, and real-time evaluation of siRNA efficacy noninvasively in vivo. The FDA novel drug approvals for 2019 are shown in Table 7.5.

**TABLE 7.4**

**Current Clinical Trials Using siRNA Conjugates and Particle-Based Carriers (Identification Numbers of the Clinical Trial From https://clinicaltrials.gov/)**

| Type of Carrier | Conjugation/ Chemistry Used | Target Genes | Route of Administration | Trial Status | Identification Number | Comments |
|---|---|---|---|---|---|---|
| siRNA conjugates | siRNA–GalNAc conjugate | AT | Subcutaneous injection | Phase 1: recruiting | NCT02035605 | Estimated study completion date: Nov. 2015 |
| | siRNA–GalNAc conjugate | TTR | Subcutaneous injection | Phase 1: ongoing, but not recruiting participants | NCT01814839 | Estimated study completion date: Sept. 2015 |
| | DPC technology (ARC-520) | HBV | Intravenous injection | Phase 2: recruiting | NCT02452528 | |
| Biodegradable polymeric matrix | Encapsulating SiRNA in LODER polymer (siG12D LODER) | KRAS-G12 | Intratumoral injection | Phase 2: not yet open for participant recruitment | NCT01676259 | Combination therapy with chemotherapy (gemcitabine or FOLFIRINOX) |
| Dendritic cell | In vitro transfection with siRNAs | LMP2, LMP7, and MECL1 | Intradermal injection | Phase 1: completed | NCT00672542 | Immunotherapy |
| Lipid nanoparticle (LNP) or liposome | Lipid nanoparticle | PLK1 | Intravenous injection | Phase 1/2: completed | NCT01262235 | |
| | Lipid nanoparticle | KSP/VEGF | Intravenous injection | Phase 1: completed | NCT01158079 | |
| | Lipid nanoparticle | PCSK9 | Intravenous injection | Phase 1: completed | NCT01437059 | |
| | siRNA–lipoplex | PKN3 | Intravenous injection | Phase 1: completed | NCT00938574 | Single treatment of Atu027 |
| | | | Intravenous injection | Phase 1b/2a: ongoing, but not recruiting participants | NCT01808638 | Combination therapy with gemcitabine and Atu027 |
| | TTR01 (first-generation of lipid nanoparticles) | TTR | Intravenous injection | Phase 1: completed | NCT01148953 | |
| | TTR02 (second-generation of lipid nanoparticles) | TTR | Intravenous injection | Phase 3: recruiting participants | NCT01960348 | |
| | TTR02 (second-generation of lipid nanoparticles) | TTR | Intravenous injection | Phase 3: enrolling participants by invitation only | NCT02510261 | Long-term safety and efficacy |
| | Neutral liposome | EphA2 | Intravenous injection | Phase 1: recruiting participants | NCT01591356 | Estimated study completion date: July 2020 |
| Hydrophobic, asymmetric RNAi compounds | sd-rxRNA compound | CTGF | Intradermal Injection | Phase 2: recruiting participants | NCT02246465 | Hypertrophic scar |
| | | | | Phase 2: ongoing, but not recruiting participants | NCT02079168 | Keloid excision surgery |
| Naked siRNA | | p53 | Intravenous injection | Phase 1: completed | NCT00554359 | Patients with major cardiovascular surgery |
| | | p53 | Intravenous injection | Phase 1/2: completed | NCT00802347 | Patients with decreased donor kidney transplantation |
| | | K6a | Injected into a callus on the bottom of one foot | Phase 1: completed | NCT00716014 | |

*(Continued)*

**TABLE 7.4** *(Continued)*

**Current Clinical Trials Using siRNA Conjugates and Particle-Based Carriers (Identification Numbers of the Clinical Trial From https://clinicaltrials.gov/)**

| Type of Carrier | Conjugation/ Chemistry Used | Target Genes | Route of Administration | Trial Status | Identification Number | Comments |
|---|---|---|---|---|---|---|
| | | Caspase 2 | Intravitreal injection | Phase 2/3: not yet open for participant recruitment | NCT02341560 | Multidose intravitreal injections |
| | | | | Phase 2: ongoing, but not recruiting participants | NCT01965106 | Single intravitreal injection |
| | | RTP801 | Intravitreal injection | Phase 2: completed | NCT01445899 | |
| | | TrpV1 | Eye drop | Phase 2: recruiting participants | NCT02455999 | |
| | | ADRB2 | Intraocular | Phase 2: recruiting participants | NCT02250612 | |
| | | RSV N | Nebulization once daily for 5 days | Phase 2b: completed | NCT01065935 | |

*Abbreviations:* ADRB2, β2 adrenoreceptor; AT, antithrombin; CTGF, connective tissue growth factor; DOPC, 1,2-dioleoyl-sn-glycero-3-phosphocholine; EphA2, ephrin type-A receptor 2; RTP801, Redd1 (for regulated in development and DNA damage responses); GalNAc, N-acetylgalactosamine; HBV, hepatitis B virus; KSP, kinesin spindle protein; LMP2, LMP7, low-molecular-mass polypeptides 2 and 7; MECL1, multicatalytic endopeptidase complex subunit; PCSK9, proprotein convertase subtilisin/kexin type 9; PKN3, protein kinase N3; PLK1, polo-like kinase 1; RSV N, respiratory syncytial virus nucleocapsid gene; sd-rxRNA, lipophilically enriched RNA-antisense hybrid; TrpV1, transient receptor potential cation channel subfamily V member 1; TTR, transthyretin; VEGF, vascular endothelial growth factor.

**TABLE 7.5**

**FDA Novel Drug Approvals for 2019**

| Active Ingredient | Drug Name | Approval Date | FDA-Approved Use on Approval Date |
|---|---|---|---|
| Trastuzumab | Enhertu | 12/20/2019 | To treat metastatic breast cancer |
| Voxelotor | Oxbryta | 11/25/2019 | To treat sickle cell disease |
| Zanubrutinib | Brukinsa | 11/14/2019 | To treat certain patients with mantle cell lymphoma, a form of blood cancer |
| Elexacaftor Ivacaftor Tezacaftor | Trikafta | 10/21/2019 | To treat patients 12 years of age and older with the most common gene mutation that causes cystic fibrosis |
| Lasmiditan | Reyvow | 10/11/2019 | For the acute treatment of migraine with or without aura, in adults |
| Istradefylline | Nourianz | 8/27/2019 | To treat adult patients with Parkinson's disease experiencing "off" episodes |
| Upadacitinib | Rinvoq | 8/16/2019 | To treat adults with moderately to severely active rheumatoid arthritis |
| Fedratinib | Inrebic | 8/16/2019 | To treat adult patients with intermediate-2 or high-risk myelofibrosis |
| Entrectinib | Rozlytrek | 8/15/2019 | To treat adult patients with metastatic non–small cell lung cancer (NSCLC) whose tumors are ROS1-positive |
| Pitolisant | Wakix | 8/14/2019 | To treat excessive daytime sleepiness (EDS) in adult patients with narcolepsy |
| Pretomanid | | 8/14/2019 | For treatment-resistant forms of tuberculosis that affects the lungs |
| Pexidartinib | Turalio | 8/2/2019 | To treat adult patients with symptomatic tenosynovial giant cell tumor |
| Darolutamide | Nubeqa | 7/30/2019 | To treat adult patients with non-metastatic castration resistant prostate cancer |
| Selinexor | Xpovio | 7/3/2019 | To treat adult patients with relapsed or refractory multiple myeloma (RRMM) |
| Alpelisib | Piqray | 5/24/2019 | To treat breast cancer |
| Tafamidis Meglumine | Vyndaqel | 5/3/2019 | To treat heart disease (cardiomyopathy) caused by transthyretin mediated amyloidosis (ATTR-CM) in adults |
| Erdafitinib | Balversa | 4/12/2019 | To treat adult patients with locally advanced or metastatic bladder cancer |
| Siponimod | Mayzent | 3/26/2019 | To treat adults with relapsing forms of multiple sclerosis |
| Triclabendazole | Egaten | 2/13/2019 | To treat fascioliasis, a parasitic infestation caused by two species of flatworms or trematodes |

*Note:* For research use only, not for human uses.

## 7.8 CONCLUSION AND PERSPECTIVES

Taking ongoing efforts and successful achievements into account, many researchers expect that therapeutic siRNAs will be translated into the clinics in the near future. Although most advanced carrier systems are lipid-based particles for siRNA delivery, diverse delivery systems have been developed to extend applications to target diseases and target tissues. One of the promising candidate systems is the siRNA-based conjugates. siRNA conjugates have major advantages, including their simple fabrication process, superior biocompatibility, and targeted delivery. However, several challenges, e.g., endosomal escape and high dose requirements, still remain. These challenges could be overcome by the design of potent siRNAs, multivalent conjugations, and combinational formulations with additive polymers and lipids. Currently, two types of siRNA conjugates, triantennary GalNAc–siRNA conjugates, and Dynamic Polyconjugates (ARC-520) are under clinical trial. Thus, it is expected that these front-running candidates will provide meaningful information and will lead to a better understanding of clinical applications of siRNA conjugates.

Several multimolecular delivery vehicles are under clinical trial for RNAi-based cancer therapy but the dose amounts of siRNA (0.1–1.5 mg/kg) are comparatively higher than levels observed in diseases of other organs (e.g., 0.15–0.3 mg/kg in the liver). This may indicate that the highest expected RNAi efficacy in tumor is similar to that in the liver. RNAi efficacy in rapid-growing cancer cells is not comparable to relatively slow-growing hepatocytes because siRNA concentration in cytoplasm will dilute in divided cells. But more efficient delivery vehicles for tumor may contribute to increase RNAi more than the present efficacies. Compared with a clinically approved trastuzumab emtansine (half-life 1–4 days) and clinically tested anticancer drug–loaded polymeric micelles (half-life 16–80 h), the current clinically tested vehicles showed shorter circulation properties (half-life <2 h) (Table 7.6). This indicated that the current delivery vehicle must perform better. The clinical trial results and new biological evidence provide clues for development of the next vehicle design. (1) The vehicle should exhibit long blood-circulation properties (half-life ≥2 h). The higher amounts of circulating delivery vehicle (containing siRNA) will increase the possibility that the vehicle diffuses/accumulates into the tumor microenvironment. Some vehicles introduced in this chapter showed a half-life longer than 2 h, but their doses for tumor growth inhibition in animal models were not significantly lower than other vehicles. These results indicate that other aspects in vehicle design should be considered. (2) The vehicle should be smaller than 30 nm diameter size to enhance diffusion/accumulation in tumor because nanoparticles of this diameter size penetrated in thick fibrotic stroma and hypovascular tumor in animal models. The research indicates that doxorubicin, with a diameter size of 90 nm, may be hampered in clinical tumor accumulation by this size limitation. Eventually, the behavior of the delivery vehicle inside the tumor is governed by

### TABLE 7.6
### Summary of Blood Circulation and Size of Delivery Vehicles in this Chapter

| Delivery Formulation | Half-Life in Mouse or Patient | Hydrodynamic Diameter (nm) |
|---|---|---|
| Naked siRNA [174, 175] | 3–10 min | 7 (length) × 2 (diameter) |
| ALN-VSP02 [146] | ≤2 h in patient | 80–100 |
| Atu027 [145, 152] | ≤2 h in patient | 120 |
| CALAA-01 [4, 153] | ≤30 min in patient | 60–150 |
| Hydrophobic interaction [177] | 10 min | 140 |
| Hydrophobic interaction [178] | 18 h | 100 |
| Hydrophobic interaction [179] | 3–4 h | 130 |
| Redox potential responsiveness [180] | 20 min | 40 |
| Extracellular pH responsiveness [181] | 5 h | 150 |
| MMP responsiveness [182] | ≤1 h | 80 |
| High quantity of siRNA [183] | 4 min in the first phase, 27 h in the second phase | 300 |
| Gold nanoparticle [184] | 1 min in the first phase, 8.5 h in the second phase | 31–34 |
| Gold nanoparticle [171] | 30 min | 40 |

diffusion, implicating that smaller particles with less than 30 nm size are preferred to reach cancer cells. Fabrication of these small nanoparticles has gradually been realized by various materials and techniques, e.g., unimer polyion complex/gold nanoparticles and polymers [171–174]. Repeatedly, we emphasize that size distribution of vehicle in buffer or fetal bovine serum does not guarantee the same size distribution in bloodstream. (3) Other functionalities (e.g., selective release of siRNA, high cell-specific recognition, and high endosome escapability) must endow the delivery vehicle, which is simultaneously satisfied with requirements (1) and (2). To date, it is not clear which functionality is the most critical factor to enhance RNAi in patients. Furthermore, a delivery vehicle that meets requirements (1) and (2), but not (3), does not expect to exhibit RNAi superior to the current vehicles in clinical trials. Ultimately, a simpler formulation of delivery vehicles can be more easily translated to their clinical use because of better quality control as well as lower possibility of unexpected adverse effects.

The success of RNAi-based cancer therapy [185–188] is closely associated with tumor biology as well as architecture of delivery vehicles. Tumor cell plasticity evokes a resistance mechanism against clinical treatments, and cancer stem cells are gradually being identified as the root of cancer recurrence. New target RNA genes should be discovered to increase apoptosis in cancer cells and simultaneously reduce side effects in normal and healthy cells. Multidisciplinary research studies will guide the development of highly effective and safer RNAi-based drugs in clinical trials.

# REFERENCES

[1] S.M. Elbashir, J. Harborth, W. Lendeckel, et al., Duplexes of 21-nucleotide RNAs mediate RNA interference in cultured mammalian cells, Nature 411 (2001) 494–498.

[2] M. Jinek, J.A. Doudna, A three-dimensional view of the molecular machinery of RNA interference, Nature 457 (2009) 405–412.

[3] S.H. Ku, K. Kim, K. Choi, et al., Tumor-targeting multifunctional nanoparticles for siRNA delivery: recent advances in cancer therapy, Adv Healthcare Mater 3 (2014) 1182–1193.

[4] J.E. Zuckerman, I. Gritli, A. Tolcher, et al., Correlating animal and human phase Ia/Ib clinical data with CALAA-01, a targeted, polymer-based nanoparticle containing siRNA, Proc Natl Acad Sci USA 111 (2014) 11449–11454.

[5] J.C. Burnett, J.J. Rossi, K. Tiemann, Current progress of siRNA/shRNA therapeutics in clinical trials, Biotechnol J 6 (2011) 1130–1146.

[6] K.A. Whitehead, J.R. Dorkin, A.J. Vegas, et al., Degradable lipid nanoparticles with predictable in vivo siRNA delivery activity, Nat Commun 5 (2014) 4277.

[7] J. Conde, N. Artzi, Are RNAi and miRNA therapeutics truly dead? Trends Biotechnol 33 (2015) 141–144.

[8] D. Haussecker, The business of RNAi therapeutics in 2012, Mol Ther Nucleic Acids 1 (2012) e8.

[9] A.D. Judge, G. Bola, A.C. Lee, et al., Design of noninflammatory synthetic siRNA mediating potent gene silencing in vivo, Mol Ther 13 (2006) 494–505.

[10] F.T. Vicentini, L.N. Borgheti-Cardoso, L.V. Depieri, et al., Delivery systems and local administration routes for therapeutic siRNA, Pharm Res 30 (2013) 915–931.

[11] R. Kanasty, J.R. Dorkin, A. Vegas, et al., Delivery materials for siRNA therapeutics, Nat Mater 12 (2013) 967–977.

[12] J. Conde, E.R. Edelman, N. Artzi, Target-responsive DNA/RNA nanomaterials for microRNA sensing and inhibition: the jack-of-all-trades in cancer nanotheranostics? Adv Drug Deliv Rev 81 (2015) 169–183.

[13] Y. Zhou, C. Zhang, W. Liang, Development of RNAi technology for targeted therapy — a track of siRNA based agents to RNAi therapeutics, J Control Release 193 (2014) 270–281.

[14] E.C. Hayden, RNA interference rebooted, Nature 508 (2014) 443.

[15] A. Bouchie, Markets, venture investors and big pharma interest in RNAi soars, Nat Biotechnol 32 (2014) 203–204.

[16] A. Marshall, Is this really the RNAissance? Nat Biotechnol 32 (2014) 201.

[17] S. Crunkhorn, Trial watch: success in amyloidosis trials supports potential of systemic RNAi, Nat Rev Drug Discov 12 (2013) 818.

[18] A.C. Anselmo, S. Mitragotri, An overview of clinical and commercial impact of drug delivery systems, J Control Release 190 (2014) 15–28.

[19] E.M. Pelegri-O'Day, E.W. Lin, H.D. Maynard, Therapeutic protein–polymer conjugates: advancing beyond PEGylation, J Am Chem Soc 136 (2014) 14323–14332.

[20] D. Pfister, M. Morbidelli, Process for protein PEGylation, J Control Release 180 (2014) 134–149.

[21] S. Panowksi, S. Bhakta, H. Raab, et al., Site-specific antibody drug conjugates for cancer therapy, MAbs 6 (2014) 34–45.

[22] J.K. Nair, J.L. Willoughby, A. Chan, et al., Multivalent N-acetylgalactosamine-conjugated siRNA localizes in hepatocytes and elicits robust RNAi-mediated gene silencing, J Am Chem Soc 136 (2014) 16958–16961.

[23] J.H. Jeong, H. Mok, Y.K. Oh, et al., siRNA conjugate delivery systems, Bioconjug Chem 20 (2009) 5–14.

[24] S.H. Lee, B.H. Chung, T.G. Park, et al., Small-interfering RNA (siRNA)-based functional micro- and nanostructures for efficient and selective gene silencing, Acc Chem Res 45 (2012) 1014–1025.

[25] C.J. Fee, Size comparison between proteins PEGylated with branched and linear poly(ethylene glycol) molecules, Biotechnol Bioeng 98 (2007) 725–731.

[26] S.H. Kim, J.H. Jeong, S.H. Lee, et al., Local and systemic delivery of VEGF siRNA using polyelectrolyte complex micelles for effective treatment of cancer, J Control Release 129 (2008) 107–116.

[27] D.R. Corey, Chemical modification: the key to clinical application of RNA interference? J Clin Invest 117 (2007) 3615–3622.

[28] S.H. Kim, J.H. Jeong, S.H. Lee, et al., PEG conjugated VEGF siRNA for anti-angiogenic gene therapy, J Control Release 116 (2006) 123–129.

[29] B.R. Meade, K. Gogoi, A.S. Hamil, et al., Efficient delivery of RNAi prodrugs containing reversible charge-neutralizing phosphotriester backbone modifications, Nat Biotechnol 32 (2014) 1256–1261.

[30] D.B. Rozema, D.L. Lewis, D.H. Wakefield, et al., Dynamic polyconjugates for targeted in vivo delivery of siRNA to hepatocytes, Proc Natl Acad Sci USA 104 (2007) 12982–12987.

[31] H. Mok, S.H. Lee, J.W. Park, et al., Multimeric small interfering ribonucleic acid for highly efficient sequence-specific gene silencing, Nat Mater 9 (2010) 272–278.

[32] T.C. Chu, K.Y. Twu, A.D. Ellington, et al., Aptamer mediated siRNA delivery, Nucleic Acids Res 34 (2006) e73.

[33] M. Sioud, A. Mobergslien, Efficient siRNA targeted delivery into cancer cells by gastrin-releasing peptides, Bioconjug Chem 23 (2012) 1040–1049.

[34] C.A. Hong, J.S. Kim, S.H. Lee, et al., Reductively dissociable siRNA–polymer hybrid nanogels for efficient targeted gene silencing, Adv Funct Mater 23 (2013) 316–322.

[35] T. Musacchio, O. Vaze, G. D'Souza, et al., Effective stabilization and delivery of siRNA: reversible siRNA–phospholipid conjugate in nanosized mixed polymeric micelles, Bioconjug Chem 21 (2010) 1530–1536.

[36] S.H. Lee, H. Mok, Y. Lee, et al., Self-assembled siRNA–PLGA conjugate micelles for gene silencing, J Control Release 152 (2011) 152–158.

[37] H.M. Liu, Y. Li, A. Mozhi, et al., SiRNA–phospholipid conjugates for gene and drug delivery in cancer treatment, Biomaterials 35 (2014) 6519–6533.

[38] T.S. Zimmermann, A.C. Lee, A. Akinc, et al., RNAi-mediated gene silencing in non-human primates, Nature 441 (2006) 111–114.

[39] K. Nishina, T. Unno, Y. Uno, et al., Efficient in vivo delivery of siRNA to the liver by conjugation of alpha-tocopherol, Mol Ther 16 (2008) 734–740.

[40] H. Lv, S. Zhang, B. Wang, et al., Toxicity of cationic lipids and cationic polymers in gene delivery, J Control Release 114 (2006) 100–109.

[41] T. Coelho, D. Adams, A. Silva, et al., Safety and efficacy of RNAi therapy for transthyretin amyloidosis, N Engl J Med 369 (2013) 819–829.

[42] F.M. Kievit, M. Zhang, Cancer nanotheranostics: improving imaging and therapy by targeted delivery across biological barriers, Adv Mater 23 (2011) H217–H247.

[43] H. Lee, A.K. Lytton-Jean, Y. Chen, et al., Molecularly self-assembled nucleic acid nanoparticles for targeted in vivo siRNA delivery, Nat Nanotechnol 7 (2012) 389–393.

[44] S. Lau, B. Graham, N. Cao, et al., Enhanced extravasation, stability and in vivo cardiac gene silencing via in situ siRNA–albumin conjugation, Mol Pharm 9 (2012) 71–80.

[45] H.S. Choi, W. Liu, P. Misra, et al., Renal clearance of quantum dots, Nat Biotechnol 25 (2007) 1165–1170.

[46] J. Soutschek, A. Akinc, B. Bramlage, et al., Therapeutic silencing of an endogenous gene by systemic administration of modified siRNAs, Nature 432 (2004) 173–178.

[47] H. Cabral, Y. Matsumoto, K. Mizuno, et al., Accumulation of sub-100 nm polymeric micelles in poorly permeable tumours depends on size, Nat Nanotechnol 6 (2011) 815–823.

[48] M.R. Dreher, W.G. Liu, C.R. Michelich, et al., Tumor vascular permeability, accumulation, and penetration of macromolecular drug carriers, J Natl Cancer Inst 98 (2006) 335–344.

[49] J. Kim, E. Lee, Y.Y. Kang, et al., Multivalent aptamer–RNA based fluorescent probes for carrier-free detection of cellular microRNA-34a in mucin1-expressing cancer cells, Chem Commun (Camb) 51 (2015) 9038–9041.

[50] H. Yoo, H. Jung, S.A. Kim, et al., Multivalent comb-type aptamer–siRNA conjugates for efficient and selective intracellular delivery, Chem Commun (Camb) 50 (2014) 6765–6767.

[51] H. Cho, Y.Y. Cho, Y.H. Bae, et al., Nucleotides as nontoxic endogenous endosomolytic agents in drug delivery, Adv Healthcare Mater 3 (2014) 1007–1014.

[52] S.A. Moschos, S.W. Jones, M.M. Perry, et al., Lung delivery studies using siRNA conjugated to TAT(48–60) and penetratin reveal peptide induced reduction in gene expression and induction of innate immunity, Bioconjug Chem 18 (2007) 1450–1459.

[53] A.F. Hussain, M.K. Tur, S. Barth, An aptamer–siRNA chimera silences the eukaryotic elongation factor 2 gene and induces apoptosis in cancers expressing alphavbeta3 integrin, Nucleic Acid Ther 23 (2013) 203–212.

[54] J. Zhou, J.J. Rossi, Cell-type-specific, aptamer-functionalized agents for targeted disease therapy, Mol Ther–Nucleic Acids 3 (2014) e169.

[55] V. Bagalkot, X. Gao, siRNA–aptamer chimeras on nanoparticles: preserving targeting functionality for effective gene silencing, ACS Nano 5 (2011) 8131–8139.

[56] J.O. McNamara, E.R. Andrechek, Y. Wang, et al., Cell type-specific delivery of siRNAs with aptamer–siRNA chimeras, Nat Biotechnol 24 (2006) 1005–1015.

[57] J.P. Dassie, X.Y. Liu, G.S. Thomas, et al., Systemic administration of optimized aptamer–siRNA chimeras promotes regression of PSMA-expressing tumors, Nat Biotechnol 27 (2009) 839–849.

[58] A. Fiszer, M. Olejniczak, P. Galka-Marciniak, et al., Self-duplexing CUG repeats selectively inhibit mutant huntingtin expression, Nucleic Acids Res 41 (2013) 10426–10437.

[59] R. Aliyari, S.W. Ding, RNA-based viral immunity initiated by the Dicer family of host immune receptors, Immunol Rev 227 (2009) 176–188.

[60] J. Hu, F. Xiao, X. Hao, et al., Inhibition of monocyte adhesion to brain-derived endothelial cells by dual functional RNA chimeras, Mol Ther–Nucleic Acids 3 (2014) e209.

[61] X. Ni, Y. Zhang, J. Ribas, et al., Prostate-targeted radiosensitization via aptamer–shRNA chimeras in human tumor xenografts, J Clin Invest 121 (2011) 2383–2390.

[62] C.P. Neff, J. Zhou, L. Remling, et al., An aptamer–siRNA chimera suppresses HIV-1 viral loads and protects from helper CD4(+) T cell decline in humanized mice, Sci Transl Med 3 (2011) 66ra66.

[63] X. Zhang, H. Liang, Y. Tan, et al., A U87-EGFRvIII cell-specific aptamer mediates small interfering RNA delivery, Biomed Rep 2 (2014) 495–499.

[64] W.Y. Lai, W.Y. Wang, Y.C. Chang, et al., Synergistic inhibition of lung cancer cell invasion, tumor growth and angiogenesis using aptamer–siRNA chimeras, Biomaterials 35 (2014) 2905–2914.

[65] S.Y. Lee, M.S. Huh, S. Lee, et al., Stability and cellular uptake of polymerized siRNA (poly-siRNA)/polyethylenimine (PEI) complexes for efficient gene silencing, J Control Release 141 (2010) 339–346.

[66] F. Heitz, M.C. Morris, G. Divita, Twenty years of cell-penetrating peptides: from molecular mechanisms to therapeutics, Br J Pharmacol 157 (2009) 195–206.

[67] S.H. Lee, B. Castagner, J.C. Leroux, Is there a future for cell-penetrating peptides in oligonucleotide delivery? Eur J Pharm Biopharm 85 (2013) 5–11.

[68] T. Endoh, T. Ohtsuki, Cellular siRNA delivery using cell-penetrating peptides modified for endosomal escape, Adv Drug Deliv Rev 61 (2009) 704–709.

[69] Y.L. Chiu, A. Ali, C.Y. Chu, et al., Visualizing a correlation between siRNA localization, cellular uptake, and RNAi in living cells, Chem Biol 11 (2004) 1165–1175.

[70] A. Muratovska, M.R. Eccles, Conjugate for efficient delivery of short interfering RNA (siRNA) into mammalian cells, FEBS Lett 558 (2004) 63–68.

[71] T.J. Davidson, S. Harel, V.A. Arboleda, et al., Highly efficient small interfering RNA delivery to primary mammalian neurons induces microRNA-like effects before mRNA degradation, J Neurosci 24 (2004) 10040–10046.

[72] H.Y. Nam, J. Kim, S.W. Kim, et al., Cell targeting peptide conjugation to siRNA polyplexes for effective gene silencing in cardiomyocytes, Mol Pharm 9 (2012) 1302–1309.

[73] S. Boeckle, E. Wagner, M. Ogris, C- versus N-terminally linked melittin–polyethylenimine conjugates: the site of linkage strongly influences activity of DNA polyplexes, J Gene Med 7 (2005) 1335–1347.

[74] S. El-Andaloussi, P. Jarver, H.J. Johansson, et al., Cargo-dependent cytotoxicity and delivery efficacy of cell-penetrating peptides: a comparative study, Biochem J 407 (2007) 285–292.

[75] B.R. Meade, S.F. Dowdy, Exogenous siRNA delivery using peptide transduction domains/cell penetrating peptides, Adv Drug Deliv Rev 59 (2007) 134–140.

[76] J.J. Turner, S. Jones, M.M. Fabani, et al., RNA targeting with peptide conjugates of oligonucleotides, siRNA and PNA, Blood Cells Mol Dis 38 (2007) 1–7.

[77] J.J. Turner, D. Williams, D. Owen, et al., Disulfide conjugation of peptides to oligonucleotides and their analogs, Curr Protoc Nucleic Acid Chem (2006) Chapter 4: Unit 4.28. doi 10.1002/0471142700.nc0428s24

[78] M. Meyer, C. Dohmen, A. Philipp, et al., Synthesis and biological evaluation of a bioresponsive and endosomolytic siRNA–polymer conjugate, Mol Pharm 6 (2009) 752–762.

[79] T. Hsu, S. Mitragotri, Delivery of siRNA and other macromolecules into skin and cells using a peptide enhancer, Proc Natl Acad Sci USA 108 (2011) 15816–15821.

[80] M. Chen, M. Zakrewsky, V. Gupta, et al., Topical delivery of siRNA into skin using SPACE–peptide carriers, J Control Release 179 (2014) 33–41.

[81] A.F. Saleh, A. Arzumanov, R. Abes, et al., Synthesis and splice-redirecting activity of branched, arginine-rich peptide dendrimer conjugates of peptide nucleic acid oligonucleotides, Bioconjug Chem 21 (2010) 1902–1911.

[82] G. Cesarone, O.P. Edupuganti, C.P. Chen, et al., Insulin receptor substrate 1 knockdown in human MCF7 ER+ breast cancer cells by nuclease-resistant IRS1 siRNA conjugated to a disulfide-bridged D-peptide analogue of insulin-like growth factor 1, Bioconjug Chem 18 (2007) 1831–1840.

[83] S.H. Kim, J.H. Jeong, S.H. Lee, et al., LHRH receptor-mediated delivery of siRNA using polyelectrolyte complex micelles self-assembled from siRNA–PEG–LHRH conjugate and PEI, Bioconjug Chem 19 (2008) 2156–2162.

[84] M.R. Alam, X. Ming, M. Fisher, et al., Multivalent cyclic RGD conjugates for targeted delivery of small interfering RNA, Bioconjug Chem 22 (2011) 1673–1681.

[85] A. Detzer, M. Overhoff, W. Wunsche, et al., Increased RNAi is related to intracellular release of siRNA via a covalently attached signal peptide, RNA 15 (2009) 627–636.

[86] S.H. Lee, Y.Y. Kang, H.-E. Jang, et al., Current preclinical small interfering RNA (siRNA)-based conjugate systems for RNA therapeutics, Adv Drug Deliv Rev 104 (2016) 78–92.

[87] C. Dohmen, D. Edinger, T. Frohlich, et al., Nanosized multifunctional polyplexes for receptor-mediated siRNA delivery, ACS Nano 6 (2012) 5198–5208.

[88] C.Y. Zhang, P. Kos, K. Muller, et al., Native chemical ligation for conversion of sequence-defined oligomers into targeted pDNA and siRNA carriers, J Control Release 180 (2014) 42–50.

[89] N. Yamazaki, S. Kojima, N.V. Bovin, et al., Endogenous lectins as targets for drug delivery, Adv Drug Deliv Rev 43 (2000) 225–244.

[90] M. Monsigny, P. Midoux, R. Mayer, et al., Glycotargeting: influence of the sugar moiety on both the uptake and the intracellular trafficking of nucleic acid carried by glycosylated polymers, Biosci Rep 19 (1999) 125–132.

[91] T.S. Zatsepin, T.S. Oretskaya, Synthesis and applications of oligonucleotide–carbohydrate conjugates, Chem Biodivers 1 (2004) 1401–1417.

[92] H. Yan, K. Tram, Glycotargeting to improve cellular delivery efficiency of nucleic acids, Glycoconj J 24 (2007) 107–123.

[93] A. Avino, S.M. Ocampo, R. Lucas, et al., Synthesis and in vitro inhibition properties of siRNA conjugates carrying glucose and galactose with different presentations, Mol Divers 15 (2011) 751–757.

[94] J.U. Baenziger, Y. Maynard, Human hepatic lectin. Physiochemical properties and specificity, J Biol Chem 255 (1980) 4607–4613.

[95] E.I. Park, Y. Mi, C. Unverzagt, et al., The asialoglycoprotein receptor clears glycoconjugates terminating with sialic acid alpha 2,6GalNAc, Proc Natl Acad Sci USA 102 (2005) 17125–17129.

[96] Y.C. Lee, R.R. Townsend, M.R. Hardy, et al., Binding of synthetic oligosaccharides to the hepatic Gal/GalNAc lectin. Dependence on fine structural features, J Biol Chem 258 (1983) 199–202.

[97] O. Khorev, D. Stokmaier, O. Schwardt, et al., Trivalent, Gal/GalNAc containing ligands designed for the asialoglycoprotein receptor, Bioorg Med Chem 16 (2008) 5216–5231.

[98] M. Severgnini, J. Sherman, A. Sehgal, et al., A rapid two-step method for isolation of functional primary mouse hepatocytes: cell characterization and asialoglycoprotein receptor based assay development, Cytotechnology 64 (2012) 187–195.

[99] M. Raouane, D. Desmaele, G. Urbinati, et al., Lipid conjugated oligonucleotides: a useful strategy for delivery, Bioconjug Chem 23 (2012) 1091–1104.

[100] C. Lorenz, P. Hadwiger, M. John, et al., Steroid and lipid conjugates of siRNAs to enhance cellular uptake and gene silencing in liver cells, Bioorg Med Chem Lett 14 (2004) 4975–4977.

[101] C. Wolfrum, S. Shi, K.N. Jayaprakash, et al., Mechanisms and optimization of in vivo delivery of lipophilic siRNAs, Nat Biotechnol 25 (2007) 1149–1157.

[102] H. Kuwahara, K. Nishina, K. Yoshida, et al., Efficient in vivo delivery of siRNA into brain capillary endothelial cells along with endogenous lipoprotein, Mol Ther 19 (2011) 2213–2221.

[103] M. Raouane, D. Desmaele, M. Gilbert-Sirieix, et al., Synthesis, characterization, and in vivo delivery of siRNA–squalene nanoparticles targeting fusion oncogene in papillary thyroid carcinoma, J Med Chem 54 (2011) 4067–4076.

[104] K. Park, J.A. Yang, M.Y. Lee, et al., Reducible hyaluronic acid–siRNA conjugate for target specific gene silencing, Bioconjug Chem 24 (2013) 1201–1209.

[105] H.Y. Yoon, H.R. Kim, G. Saravanakumar, et al., Bioreducible hyaluronic acid conjugates as siRNA carrier for tumor targeting, J Control Release 172 (2013) 653–661.

[106] S.W. Choi, S.H. Lee, H. Mok, et al., Multifunctional siRNA delivery system: polyelectrolyte complex micelles of six-arm PEG conjugate of siRNA and cell penetrating peptide with crosslinked fusogenic peptide, Biotechnol Prog 26 (2010) 57–63.

[107] S.H. Lee, H. Mok, T.G. Park, Di- and triblock siRNA–PEG copolymers: PEG density effect of polyelectrolyte complexes on cellular uptake and gene silencing efficiency, Macromol Biosci 11 (2011) 410–418.

[108] S. Jung, S.H. Lee, H. Mok, et al., Gene silencing efficiency of siRNA–PEG conjugates: effect of PEGylation site and PEG molecular weight, J Control Release 144 (2010) 306–313.

[109] M. Oishi, Y. Nagasaki, N. Nishiyama, et al., Enhanced growth inhibition of hepatic multicellular tumor spheroids by lactosylated poly(ethylene glycol)–siRNA conjugate formulated in PEGylated polyplexes, ChemMedChem 2 (2007) 1290–1297.

[110] C. Dohmen, T. Frohlich, U. Lachelt, et al., Defined folate–PEG–siRNA conjugates for receptor-specific gene silencing, Mol Ther–Nucleic Acids 1 (2012) e7.

[111] L. Zhu, R.I. Mahato, Targeted delivery of siRNA to hepatocytes and hepatic stellate cells by bioconjugation, Bioconjug Chem 21 (2010) 2119–2127.

[112] M. Oishi, Y. Nagasaki, K. Itaka, et al., Lactosylated poly(ethylene glycol)–siRNA conjugate through acid–labile beta-thiopropionate linkage to construct pH-sensitive poly-ion complex micelles achieving enhanced gene silencing in hepatoma cells, J Am Chem Soc 127 (2005) 1624–1625.

[113] S.R. Mudd, V.S. Trubetskoy, A.V. Blokhin, et al., Hybrid PET/CT for noninvasive pharmacokinetic evaluation of dynamic polyconjugates, a synthetic siRNA delivery system, Bioconjug Chem 21 (2010) 1183–1189.

[114] S.C. Wong, J.J. Klein, H.L. Hamilton, et al., Co-injection of a targeted, reversibly masked endosomolytic polymer dramatically improves the efficacy of cholesterol-conjugated small interfering RNAs in vivo, Nucleic Acid Ther 22 (2012) 380–390.

[115] F.M. Skjeldal, S. Strunze, T. Bergeland, et al., The fusion of early endosomes induces molecular-motor-driven tubule formation and fission, J Cell Sci 125 (2012) 1910–1919.

[116] C.I. Wooddell, D.B. Rozema, M. Hossbach, et al., Hepatocyte-targeted RNAi therapeutics for the treatment

of chronic hepatitis B virus infection, Mol Ther 21 (2013) 973–985.

[117] M.G. Sebestyen, S.C. Wong, V. Trubetskoy, et al., Targeted in vivo delivery of siRNA and an endosome-releasing agent to hepatocytes, Methods Mol Biol 1218 (2015) 163–186.

[118] J. Conde, C.C. Bao, Y.Q. Tan, et al., Dual targeted immunotherapy via in vivo delivery of biohybrid RNAi–peptide nanoparticles to tumor-associated macrophages and cancer cells, Adv Funct Mater 25 (2015) 4183–4194.

[119] Y.T. Chang, P.Y. Liao, H.S. Sheu, et al., Near-infrared light-responsive intracellular drug and siRNA release using Au nanoensembles with oligonucleotide-capped silica shell, Adv Mater 24 (2012) 3309–3314.

[120] J. Conde, F. Tian, Y. Hernandez, et al., In vivo tumor targeting via nanoparticle-mediated therapeutic siRNA coupled to inflammatory response in lung cancer mouse models, Biomaterials 34 (2013) 7744–7753.

[121] D.A. Giljohann, D.S. Seferos, A.E. Prigodich, et al., Gene regulation with polyvalent siRNA–nanoparticle conjugates, J Am Chem Soc 131 (2009) 2072–2073.

[122] J.S. Lee, J.J. Green, K.T. Love, et al., Gold, poly(betaamino ester) nanoparticles for small interfering RNA delivery, Nano Lett 9 (2009) 2402–2406.

[123] J. Conde, F. Tian, Y. Hernandez, et al., RNAi-based glyconanoparticles trigger apoptotic pathways for in vitro and in vivo enhanced cancer-cell killing, Nanoscale 7 (2015) 9083–9091.

[124] S. Guo, Y. Huang, Q. Jiang, et al., Enhanced gene delivery and siRNA silencing by gold nanoparticles coated with charge-reversal polyelectrolyte, ACS Nano 4 (2010) 5505–5511.

[125] J. Conde, A. Ambrosone, V. Sanz, et al., Design of multifunctional gold nanoparticles for in vitro and in vivo gene silencing, ACS Nano 6 (2012) 8316–8324.

[126] J. Lee, D.K. Chatterjee, M.H. Lee, et al., Gold nanoparticles in breast cancer treatment: promise and potential pitfalls, Cancer Lett 347 (2014) 46–53.

[127] S. Jiang, A.A. Eltoukhy, K.T. Love, et al., Lipidoid-coated iron oxide nanoparticles for efficient DNA and siRNA delivery, Nano Lett 13 (2013) 1059–1064.

[128] Z. Medarova, W. Pham, C. Farrar, et al., In vivo imaging of siRNA delivery and silencing in tumors, Nat Med 13 (2007) 372–377.

[129] H. Maeda, J. Fang, T. Inutsuka, et al., Vascular permeability enhancement in solid tumor: various factors, mechanisms involved and its implications, Int Immunopharmacol 3 (2003) 319–328.

[130] H.K. Na, M.H. Kim, K. Park, et al., Efficient functional delivery of siRNA using mesoporous silica nanoparticles with ultralarge pores, Small 8 (2012) 1752–1761.

[131] H. Meng, Y. Zhao, J. Dong, et al., Two-wave nanotherapy to target the stroma and optimize gemcitabine delivery to a human pancreatic cancer model in mice, ACS Nano 7 (2013) 10048–10065.

[132] Z. Liu, M. Winters, M. Holodniy, et al., siRNA delivery into human T cells and primary cells with carbon-nanotube transporters, Angew Chem Int Ed 46 (2007) 2023–2027.

[133] S. Li, Z. Liu, F. Ji, et al., Delivery of quantum dot–siRNA nanoplexes in SK–N–SH Cells for BACE1 gene silencing and intracellular imaging, Mol Ther Nucleic Acids 1 (2012) e20.

[134] A.M. Derfus, A.A. Chen, D.H. Min, et al., Targeted quantum dot conjugates for siRNA delivery, Bioconjug Chem 18 (2007) 1391–1396.

[135] M.S. Draz, B.A. Fang, P. Zhang, et al., Nanoparticle-mediated systemic delivery of siRNA for treatment of cancers and viral infections, Theranostics 4 (2014) 872–892.

[136] R.C. Wilson, J.A. Doudna, Molecular mechanism of RNA interference, Annu Rev Biophys 42 (2013) 217–239.

[137] J. Lam, M. Chow, Y. Zhang, et al., siRNA versus miRNA as therapeutics for gene silencing, Mol Ther Nucleic Acids 4 (2015) e252.

[138] A. Fire, S. Xu, M.K. Montgomery, et al., Potent and specific genetic interference by double-stranded RNA in Caenorhabditis elegans, Nature 391 (1998) 806–811.

[139] C. Fellmann, S.W. Lowe, Stable RNA interference rules for silencing, Nat Cell Biol 16 (2014) 10–18.

[140] B.L. Davidson, P.B. McCray Jr., Current prospects for RNA interference-based therapies, Nat Rev Genet 12 (2011) 329–340.

[141] S.Y. Wu, G. Lopez-Berestein, G.A. Calin, et al., RNAi therapies: drugging the undruggable, Sci Transl Med 6 (2014) 240ps7.

[142] C.N. Landen Jr., A. Chavez-Reyes, C. Bucana, et al., Therapeutic EphA2 gene targeting in vivo using neutral liposomal small interfering RNA delivery, Cancer Res 65 (2005) 6910–6918.

[143] M.E. Davis, J.E. Zuckerman, C.H.J. Choi, et al., Evidence of RNAi in humans from systemically administered siRNA via targeted nanoparticles, Nature 464 (2010) 1067–1071.

[144] E.Z. Khvalevsky, R. Gabai, I.H. Rachmut, et al., Mutant KRAS is a druggable target for pancreatic cancer, Proc Natl Acad Sci USA 110 (2013) 20723–20728.

[145] B. Schultheis, D. Strumberg, A.C. Vank, et al., First-in-human phase I study of the liposomal RNA interference therapeutic Atu027 in patients with advanced solid tumors, J Clin Oncol 32 (2014) 4141–4148.

[146] J. Tabernero, G.I. Shapiro, P.M. LoRusso, et al., First-in-humans trial of an RNA interference therapeutic targeting VEGF and KSP in cancer patients with liver involvement, Cancer Discov 3 (2013) 406–417.

[147] J.E. Zuckerman, C.H.J. Choi, H. Han, et al., Polycation-siRNA nanoparticles can disassemble at the kidney glomerular basement membrane, Proc Natl Acad Sci USA 109 (2012) 3137–3142.

[148] B. Naeye, H. Deschout, V. Caveliers, et al., In vivo disassembly of IV administered siRNA matrix nanoparticles at the renal filtration barrier, Biomaterials 34 (2013) 2350–2358.

[149] R.J. Christie, Y. Matsumoto, K. Miyata, et al., Targeted polymeric micelles for siRNA treatment of experimental cancer by intravenous injection, ACS Nano 6 (2012) 5174–5189.

[150] T. Coelho, D. Adams, A. Silva, et al., Safety and efficacy of RNAi therapy for transthyretin amyloidosis, N Engl J Med 369 (2013) 819–829.

[151] S.C. Semple, A. Akinc, J. Chen, et al., Rational design of cationic lipids for siRNA delivery, Nat Biotechnol 28 (2010) 172–176.

[152] A. Santel, M. Aleku, O. Keil, et al., A novel siRNA-lipoplex technology for RNA interference in the mouse vascular endothelium, Gene Ther 13 (2006) 1222–1234.

[153] D.W. Bartlett, M.E. Davis, Physiochemical and biological characterization of targeted, nucleic acid-containing nanoparticles, Bioconjug Chem 18 (2007) 456–468.

[154] S. Girish, M. Gupta, B. Wang, et al., Clinical pharmacology of trastuzumab emtansine (T-DM1): an antibody-drug conjugate in development for the treatment of HER2-positive cancer, Cancer Chemother Pharmacol 69 (2012) 1229–1240.

[155] R. Plummer, R.H. Wilson, H. Calvert, et al., A Phase I clinical study of cisplatin-incorporated polymeric micelles (NC-6004) in patients with solid tumours, Br J Cancer 104 (2011) 593–598.

[156] D.W. Bartlett, H. Su, I.J. Hildebrandt, et al., Impact of tumor-specific targeting on the biodistribution and efficacy of siRNA nanoparticles measured by multimodality in vivo imaging, Proc Natl Acad Sci USA 104 (2007) 15549–15554.

[157] H.J. Kim, A. Kim, K. Miyata, et al., Recent progress in development of siRNA delivery vehicles for cancer therapy, Adv Drug Deliv Rev 104 (2016) 61–77.

[158] Pramod Darvin, Salman M. Toor, Varun Sasidharan Nair, et al. Immune checkpoint inhibitors: recent progress and potential biomarkers, Exp Mol Med 50 (12) (2018) 1–11. doi:10.1038/s12276-018-0191-1

[159] Jun Tang, Jia Xin Yu, Vanessa M. Hubbard-Lucey, et al., The clinical trial landscape for PD1/PDL1 immune checkpoint inhibitors, Nat Rev Drug Discov 17 (2018), 854 -855. doi: https://doi.org/10.1038/nrd.2018.210

[160] S.Y. Wu, G. Lopez-Berestein, G.A. Calin, et al., RNAi therapies: drugging the undruggable, Sci Transl Med 6 (2014) 240ps247.

[161] E.C. Cho, Q. Zhang, Y. Xia, The effect of sedimentation and diffusion on cellular uptake of gold nanoparticles, Nat Nanotechnol 6 (2011) 385–391.

[162] S.H. Lee, E. Moroz, B. Castagner, et al., Activatable cell penetrating peptide– peptide nucleic acid conjugate via reduction of azobenzene PEG chains, J Am Chem Soc 136 (2014) 12868–12871.

[163] A.L. Jackson, P.S. Linsley, Noise amidst the silence: off-target effects of siRNAs? Trends Genet 20 (2004) 521–524.

[164] M. Robbins, A. Judge, I. MacLachlan, siRNA and innate immunity, Oligonucleotides 19 (2009) 89–102.

[165] Z. Poon, S. Chen, A.C. Engler, et al., Ligand-clustered "patchy" nanoparticles for modulated cellular uptake and in vivo tumor targeting, Angew Chem Int Ed 49 (2010) 7266–7270.

[166] P.R. Mallikaratchy, A. Ruggiero, J.R. Gardner, et al., A multivalent DNA aptamer specific for the B-cell receptor on human lymphoma and leukemia, Nucleic Acids Res 39 (2011) 2458–2469.

[167] H. Cai, Z.Y. Sun, M.S. Chen, et al., Synthetic multivalent glycopeptide–lipopeptide antitumor vaccines: impact of the cluster effect on the killing of tumor cells, Angew Chem Int Ed 53 (2014) 1699–1703.

[168] S. Nechaev, C. Gao, D. Moreira, et al., Intracellular processing of immunostimulatory CpG–siRNA: Toll-like receptor 9 facilitates siRNA dicing and endosomal escape, J Control Release 170 (2013) 307–315.

[169] M. Kortylewski, P. Swiderski, A. Herrmann, et al., In vivo delivery of siRNA to immune cells by conjugation to a TLR9 agonist enhances antitumor immune responses, Nat Biotechnol 27 (2009) 925–932.

[170] J. Wei, J. Jones, J. Kang, et al., RNA-induced silencing complex-bound small interfering RNA is a determinant of RNA interference-mediated gene silencing in mice, Mol Pharmacol 79 (2011) 953–963.

[171] H.J. Kim, H. Takemoto, Y. Yi, et al., Precise engineering of siRNA delivery vehicles to tumors using polyion complexes and gold nanoparticles, ACS Nano 8 (2014) 8979–8991.

[172] H. Shimizu, Y. Hori, S. Kaname, et al., siRNA-based therapy ameliorates glomerulonephritis, J Am Soc Nephrol 21 (2010) 622–633.

[173] A. Sizovs, X. Song, M.N. Waxham, et al., Precisely tunable engineering of sub-30 nm monodisperse oligonucleotide nanoparticles, J Am Chem Soc 136 (2014) 234–240.

[174] C. Dohmen, D. Edinger, T. Fröhlich, et al., Nanosized multifunctional polyplexes for receptor-mediated siRNA delivery, ACS Nano 6 (2012) 5198–5208.

[175] S. Gao, F. Dagnaes-Hansen, E.J.B. Nielsen, et al., The effect of chemical modification and nanoparticle formulation on stability and biodistribution of siRNA in mice, Mol Ther 17 (2009) 1225–1233.

[176] J.J. Turner, S.W. Jones, S.A. Moschos, et al., MALDI-TOF mass spectral analysis of siRNA degradation in serum confirms an RNAse A-like activity, Mol BioSyst 3 (2007) 43–50.

[177] H.J. Kim, M. Oba, F. Pittella, et al., PEG-detachable cationic polyaspartamide derivatives bearing stearoyl moieties for systemic siRNA delivery toward subcutaneous BxPC3 pancreatic tumor, J Drug Target 20 (2012) 33–42.

[178] N. Yagi, I. Manabe, T. Tottori, et al., A nanoparticle system specifically designed to deliver short interfering RNA inhibits tumor growth in vivo, Cancer Res 69 (2009) 6531–6538.

[179] Q. Sun, Z. Kang, L. Xue, et al., A collaborative assembly strategy for tumor-targeted siRNA delivery, J Am Chem Soc 137 (2015) 6000–6010.

[180] Y. Oe, R.J. Christie, M. Naito, et al., Actively-targeted polyion complex micelles stabilized by cholesterol and disulfide cross-linking for systemic delivery of siRNA to solid tumors, Biomaterials 35 (2014) 7887–7895.

[181] J. Li, X. Yu, Y. Wang, et al., A reduction and pH dual-sensitive polymeric vector for long-circulating and tumor-targeted siRNA delivery, Adv Mater 26 (2014) 8217–8224.

[182] H.-X. Wang, X.-Z. Yang, C.-Y. Sun, et al., Matrix metalloproteinase 2-responsive micelle for siRNA delivery, Biomaterials 35 (2014) 7622–7634.

[183] Z.J. Deng, S.W. Morton, E. Ben-Akiva, et al., Layer-by-layer nanoparticles for systemic codelivery of an anticancer drug and siRNA for potential triple-negative breast cancer treatment, ACS Nano 7 (2013) 9571–9584.

[184] S.A. Jensen, E.S. Day, C.H. Ko, et al., Spherical nucleic acid nanoparticle conjugates as an RNAi-based therapy for glioblastoma, Sci Transl Med 5 (2013), 209ra152.

[185] L.H. Madkour, Reactive Oxygen Species (ROS), Nanoparticles, and Endoplasmic Reticulum (ER) Stress-Induced Cell Death Mechanisms. Academic Press (2020).

[186] L.H. Madkour, Nanoparticles Induce Oxidative and Endoplasmic Reticulum Antioxidant Therapeutic Defenses. Springer International Publishing (2020). doi:10.1007/978-3-030-37297-2

[187] L.H. Madkour, Nucleic Acids as Gene Anticancer Drug Delivery Therapy. Elsevier (2020).

[188] L.H. Madkour, Nanoelectronic Materials: Fundamentals and Applications (Advanced Structured Materials) (2019). doi:10.1007/978-3-030-21621-4

# 8 Potential for siRNA in Types of Genetic Disease, Cancer Therapeutics

Cancer occurs as a result of a series of gene mutations in a cell. Generally, a combination of activating mutations in so-called oncogenes and the loss of tumor suppressor genes lead to uncontrolled cell growth and blockage of natural apoptotic processes [1, 2]. Because many key gene mutations involved in driving cancer, also known as driver genes, have been identified [3, 4], it is easy to see that siRNA therapeutics could be effective in cancer treatment [5, 6]. A major advantage of using siRNA in cancer treatment is its ability to specifically inhibit any of the large set of cancer-associated genes without regard to the druggability of their protein products [6]. This allows us to potentially drug the undruggable. Furthermore, a diverse set of therapeutic siRNA molecules can be developed to target genes associated with the multiple signaling pathways that are aberrantly activated in tumors [7]. Table 8.1 summarizes the current status of siRNA targets for major cancers, which are discussed in more detail below.

## 8.1 LUNG CANCER

Worldwide, lung cancer is the most frequently occurring tumor type with the highest incidence of cancer-related mortality. Among three main types of lung cancer, non–small cell lung cancer (NSCLC) is the most common, comprising approximately 85% of all lung cancers [70, 71]. Lung cancers often metastasize, leading to treatment failures. Despite the development of novel molecular therapies, aggressive surgery, and radiochemical therapy, prognosis for most of these patients is still very poor. It is possible to deliver siRNA-based therapeutics to the lung either systemically or via intrapulmonary administration [10, 72]. The latter route allows lower doses of siRNAs, reducing undesirable systemic side effects, and also improves the half-life of siRNAs in tumors. Hence, siRNA-based therapeutics should be considered for lung cancer treatment [73].

Because cancer is a genetic disease it is always worth asking which genes are most frequently altered in a given tumor type. Epidermal growth factor receptor (EGFR) nucleotide variants are found in various types of cancer including lung cancer [74]. NSCLC in particular displays frequent EGFR mutations, which occur on exons 18–21, encoding a portion of the EGFR kinase domain. The most common mutation class consists of exon 19 deletions, which activate the tyrosine kinase activity of EGFR, resulting in induction downstream of progrowth and survival signaling pathways [75]. Notably, it has been possible to construct allele-specific siRNAs against certain oncogenic EGFR mutants, which have displayed significant therapeutic

effects in lung tumor models with mutant EGFR alleles. The wet-weights of tumors treated with siRNAs targeting mutant EGFR were observed to be much lower than those of the control siRNA-treated group. Furthermore, caspase-3 activity was upregulated in EGFR-treated tumor tissue, indicating the induction of apoptosis. The allele-specific EGFR siRNA treatment inhibited specific oncogenic EGFR alleles without affecting the normal EGFR allele, leading to a safe and effective treatment [76].

Activating mutations in KRAS frequently occur in lung tumors. Notably, in EGFR mutant lung tumors KRAS mutations can render them resistant to treatment with EGFR-directed therapies. A seminal recent publication explored the anticancer therapeutic effect of both targeting KRAS activation and loss of p53 function, another common lesion in many tumor types including lung tumors. The authors used siRNA- and miRNA-loaded polymer-based nanoparticles in a genetically engineered mouse (GEM) model of lung cancer [77]. Nanoparticles loaded with siRNAs targeting KRAS and with a miRNA-34a mimic, which partially restores p53 downstream functions, were intravenously injected in a kras/p53 GEM model (KrasLSL-G12D/wt; p53flox/flox). Combination treatment with miR-34a and KRAS siRNA resulted in average lung tumor regression of 63% of its original volume and increased apoptotic cells. This form of targeted multigene combination therapy is perfectly suited for siRNA strategies and allows personalized cancer therapy adaptable to various mutations identified in a particular patient cancer type. The mice that were treated with both cisplatin and the nanoparticle formulation consisting of a combination of miR-34a and a KRAS siRNA survived significantly longer compared with mice treated with either treatment alone (combination of miR-34a/siKRAS and cisplatin; 159.5 ± 19.5d, siLuc; 93.7 ± 16.1 d, cisplatin; 127.4 ± 9.0d, miR-34a/siKRAS; 129.2 ± 16.2d).

Human-ribophorin II (RPN2) is part of an N-oligosaccharyltransferase complex. Generally, this protein is known to facilitate human cancer resistance against chemotherapy drugs such as docetaxel. In addition, it has been reported RPN2 can serve as an antiapoptotic protein that regulates tumor survival and cancer stem cell properties through the stabilization of mutant p53 [78]. To deliver naked RPN2 siRNA to the lung through inhalation, specifically structured siRNAs, termed as PnkRNATM and nkRNATM, were used [8]. The RPN2-targeting siRNAs inhibited the growth of a A549 luc-C8 lung xenograft and suppressed RPN2 expression.

The oncoprotein mouse double minute 2 (MDM2) is a negative regulator of the p53 tumor suppressor, which

**TABLE 8.1**

## siRNA Targets for Treatment of Major Cancers

| Cancer Type | siRNA/Target Gene | Route | Animal Model | siRNA Modification/Delivery System | Ref. |
|---|---|---|---|---|---|
| Lung | RPN2 | Inhalation | A549-luc-C8 cells via IV | Naked PnkRNA and nKRNAs | [8] |
| | C7orf24 | Jet injection | EBC-1, S.C. | Naked siRNA | [9] |
| | Mcl1 | Intratracheal | B16F10 or LLC, metastatic lung cancer | Liposome | [10] |
| | CD31 | IV | | Liposome | [11] |
| | Bcl2 | IV | LLC, metastatic lung cancer | Protein, cationic bovine serum albumin (CBSA) | [12] |
| | NPT2b | Inhalation | B16, metastatic lung cancer | | [13] |
| | MDM2 | IV | K-ras^L.A1 model | Polymer | [14] |
| | Survivin/cyclin B1 | IV | H2009, SC | Polymer | [15] |
| | HDM2/C-Myc/VEFG | IV | TSA-Luc, metastatic lung cancer | Sticky siRNA/polymer | [16] |
| | Survivin/Bcl-2 + cisplatin | IV | | Liposome | [17] |
| | | | H460 or A546 NSCLC, SC | Polymer | |
| | | | A549 resistant, SC | | |
| Liver | HDAC2 | IV | Huh7-luc+, orthotopic liver tumor | 2'-OMe modifications/LNPs | [18] |
| | Polo-like kinases (PLKs), ß1-integrin | IV | | Fusogenic liposome | [19] |
| | RRM2, adriamycin | IV | Induction by diethyl nitrosamine (DEN) | 2'-OMe-modification/LNPs | [20] |
| | Various siRNA (KRAS, NHP2L1, | IV | | Liposome | [21] |
| | BIRC5, CDCA1, PSMA2, Aurora B, | IV | MET/DN90-β-catenin-induced tumor | Liposome | [22] |
| | etc.) | IV | | RGD-polymer | [23] |
| | Survivin | IV | HepG2 orthotopic HCC | Cholesterol conjugation/liposome | [24] |
| | Pokemon | IT | HuH7 liver tumors | Cationic polymer | [25] |
| | hTERT | IV | Bel-7402, SC. | Anti-EGFR Fab'-liposomes | [26] |
| | RhoA | IV | HepG2, SC | 2'-Ome modifications/LNPs | [27] |
| | CSN5 | IV | HepG2, SC | 2'-Ome modifications/LNPs | [28] |
| | COP1 | | SMMC-7721, SC | | |
| | | | Huh7, orthotopic transplantation | | |
| | | | HepG2, Huh7, orthotopic | | |
| Prostate | HSP 27 | IT | PC-3, S | Amphiphilic dendrimer | [29] |
| | Survivin | IT | PC-3, SC | Lipid + PEI hybrid nanocarrier (LPN) | [30] |
| | Bcl2 | IV | PC-3, orthotopic | | [31] |
| | Myc | IP | PC-3, SC | cholesterol conjugation/lipid nanoplatform, | [32] |
| | plk-1 | IT | PC-3 and LNCaP, SC | | [33] |
| | YB-1 + rapamycin | Retro orbital | PC3, SC | In vivo-jet PEI | [34] |
| | Notch1 | IV | LNCaP, SC | Aptamer | [35] |
| | HIF-1α + Dox | IV | PC3, SC | Trilayer polymeric micelle | [36] |
| | VEGF | IV | PC3, SC | Anti-PSMA scFv fusion proteins | [37] |
| | AR | IV | PC3, SC | Micellar nanoparticle (MNP) | [38] |
| | Bcl-Xl + cisplatin | IV | TRAMP C1, SC | Thiolated siRNA/chitosan based polymer | [39] |
| | PKN3 | IV | LNCaP, SC | | [40] |
| | AURKB, and EGFR | IV | PC-3, SC | Cyclodextrin modified polymer | [41] |
| | Cyclin B1 | IV | DU1–45, S.C./PC-3, orthotopic | pH-triggered amphiphilic polymer | [42] |
| | EZH2, and p110a | IT | PC3, SC | | [43] |
| | plk-1 + paclitaxel | IV | PC3, SC | LNP | [44] |
| | Plk1 | IV | PC-3M-luc-C6, heart injection/ bone metastasis | Atelocollagen | [45] |
| | | | | Atu027 | |
| | | | | Hiperfect | |
| | | | | Amphipathic peptide carrier MPG-8 | |
| | | | | Atelocollagen | |

*(Continued)*

**TABLE 8.1** *(Continued)*
## siRNA Targets for Treatment of Major Cancers

| Cancer Type | siRNA/Target Gene | Route | Animal Model | siRNA Modification/Delivery System | Ref. |
|---|---|---|---|---|---|
| Breast | AURKB, and EGFR | IV | MDA-MB-435 s, SC | Polymer based micelleplex | [46] |
| | cyclin B1 | IV | MDA-MB-435 s, SC | Cationic lipid assisted PEG–PLA nanoparticles | [47] |
| | CDK1 | IV | SUM149 or BT549, SC | | [48] |
| | Erα | IV | MCF-7, SC | Cationic lipid assisted PEG–PLA nanoparticles | [49] |
| | DNMTs | IV | BT474, SC | | [50] |
| | RhoA | IT | MDA-MB-231, SC | Polymer nanocapsule | [51] |
| | Ataxia-telangiectasia mutated (ATM) | IV | MDA-MB-231, orthotopic | Anti-Her2 ScFv-protamine (F5-P) | [52] |
| | Osteopontin (OPN) | IT | MDA-MB-231, SC | | [53] |
| | Mcl-1 + RPS6KA5 | IT | MDA435_WT and MDA435_resistant (R), SC | Cytofectin transfection reagent | [54] |
| | P-gp + DOX | IV | | Porous silicon-based multistage vector (MSV) | [55] |
| | MnSOD | — | MCF7/A, SC | | [56] |
| | Raf-1 | IT | MCF7-BK-TR cells co-transplanted with siRNA/NPs | Glycerol propoxylate triacrylate-spermine copolymer | [57] |
| | VEGF | IV | | Lipid-substituted polymer | [58] |
| | | | MDA-MB-435, SC | RGD-Liposome | |
| | | | MCF-7 and HT1080, SC | PAMAM dendrimer based nanoparticles | |
| | | | | Histidine–lysine carrier | |
| | | | | Cholesterol conjugation/high-density lipoprotein | |
| Ovary | EZH2 + docetaxel | IV | HeyA8 or SKOV3ip1, orthotopic | PLGA-PRINT nanoparticles | [58] |
| | VEGF | IT | | Arginine-grafted polymer-microbubble | [59] |
| | EphA2 + paclitaxel | IP | A2780, SC | | [60] |
| | Akt + paclitaxel | IT | HeyA8, orthotopic | Neutral liposomes | [61] |
| | XIAP | IV | SKOV-3, SC | Dendrimer | [62] |
| | Vasohibin2 | IV | SKOV-3, SC | Ternary copolymer | [63] |
| | CD44 + paclitaxel | IP | DISS and SKOV-3, SC | Atelocollagen | [64] |
| | POSTN, FAK, and PLXDC1 | IV | Human ascitic cells, SC | Dendrimer | [65] |
| | EphA2 | IV | SKOV3ip1, HeyA8, and A2780, orthotopic | RGD-labeled chitosan nanoparticles | [66] |
| | c-Jun.-NH2-kinases (JNK) + docetaxel | IP | | | [67] |
| | KLF6-SV1 + cisplatin | IP | SKOV3ip1 or HeyA8, orthotopic | Mesoporous silicon particles | [68] |
| | FAK + docetaxel | IP | | Liposomes | [69] |
| | | | HeyA8 or SKOV3ip1, orthotopic | Accell chemically synthesized siRNAs (Dharmacon) | |
| | | | SKOV3-Luc, orthotopic | Neutral liposome | |
| | | | SKOV3ip1, A2780-CP20, and HeyA8MDR, orthotopic | | |

*Abbreviations:* IP, intraperitoneal; IT, intratumoral; IV, intravenous.

inhibits the transactivation activity of p53 via a direct interaction. A high level of MDM2 gene amplification has been detected in many human tumor types, including lung carcinomas [79, 80]. While inhibition of MDM2 function could obviously be beneficial in tumors with wild-type p53, it has also been reported that MDM2 can play a significant role in tumors featuring mutant p53. Thus MDM2 could be a potential therapeutic target for treatment of human NSCLC with either wild-type or mutant p53 [81]. In this regard it has been found that MDM2 siRNA modified with a pH-responsive diblock copolymer (termed MDM2 siRNA/PMPC-b-PDPA), showed significant downregulation of MDM2 expression in a xenograft of the p53

mutant H2009 cell line. MDM2 knockdown also resulted in impaired growth of H2009 tumors through cell cycle arrest and apoptosis. The tumor sizes of MDM2 siRNA/PMPC-b-PDPA treated H2009 tumor xenograft mice were significantly smaller than the tumors in control siRNA–treated mice [14].

## 8.2 LIVER CANCER

The total set of cancers in each organ type almost invariably constitutes a family of unique diseases. The major form of liver cancer is hepatocellular carcinoma (HCC), which is one of the most frequently occurring cancers worldwide.

HCC is a complicated disease due to its broad epidemiology. Its causes include not only mutagenic environmental insults common to many tumors but also viral insults including those arising from hepatitis viruses. Hepatitis B and C infections are major risk factors for the development of HCC. Currently the treatment of HCC is limited. Surgical resection and transplantation are available to a minority of HCC patients who have an organ donor and are free of preexisting liver conditions [82, 83]. In the search for alternative modes of liver cancer treatment, many nucleotide-based therapies, including siRNA therapeutics, that target signaling pathways specific to liver tumors, are now undergoing clinical trials [84].

Histone deacetylases (HDACs) are crucial enzymes that regulate gene expression by deleting acetyl groups from histone substrates [85]. HDACs interact with key cancer-associated transcription factors, such as ß-catenin, Myc, p53, Stat3, NF-Kb, and TFIIE, to regulate the expression of numerous proteins implicated in tumor formation and development. Thus, altered expression and pathological activity of HDACs can lead to tumor onset and progression. HDAC2, which is overexpressed in solid tumors, promotes cell cycle progression and prevents apoptosis by inhibiting action of the tumor suppressor p53 [86]. HDAC2-targeting siRNAs have demonstrated a reduction of liver cancer cell proliferation in vitro. The therapeutic effect of HDAC siRNA/lipid nanoparticles (LNP) was demonstrated in effectively reduced liver tumor growth in an orthotopic xenograft model [18].

Integrins are a superfamily of cell adhesion receptors that bind to and interact with the extracellular matrix (ECM). They are assembled from small families of α and ß glycoprotein subunits, generating 24 unique noncovalent α/ß heterodimers. The roles of integrins are both important and diverse; they take part in the regulation of cell motility, differentiation, survival, and proliferation. Among the integrin subunits, Itgb1 (ß1-integrin subunit) plays a critical role in liver formation and in the proliferation of liver tumor cells [87]. Overexpression of Itgb1 promotes survival and induces resistance to chemotherapy. It has been reported that activating mutations in Itgb1 affect tumorigenesis [88, 89]. Daniel G. Anderson's group reported that HCC progression can be delayed by targeting Itgb1 with siRNAs. In this study, human MET and δN90-ß-catenin were delivered to mice hepatocytes to construct a mouse model of spontaneous HCC [90, 91]. After the introduction of the oncogenes, injection of Itgb1 siRNA/LNP inhibited HCC progression [20]. Liver weights of the Itgb1 siRNA/LNP treated group were significantly reduced compared with those of control groups. HCC tissues also showed a significant reduction in the size and number of tumor foci and enlarged hepatocytes in the residual tumor nodules.

Survivin is a protein inhibitor of apoptosis that is strongly expressed in HCCs but is not expressed in differentiated adult tissues [92]. Overexpression of survivin protein can inhibit caspase activation, thereby leading to inhibition of apoptosis and stimulation of HCC cell proliferation [93]. A tumor growth inhibition study using surviving-targeting siRNA delivered by RGD-PEG-g-PEI-SPION nanoparticles was reported [23]. Mice bearing tumors arising from human HCC cell line, Bel-7402, were injected with survivin siRNA/RGD-PEG-g-PEI-SPION and exhibited delay in tumor growth. Inhibition of the survivin gene expression by siRNA resulted in an increase of cleaved caspase-3 expression.

Pokemon, a member of the POK family of transcriptional repressors, has been observed to have aberrant overexpression in multiple human cancers including liver tumors [94]. Pokemon plays a critical role in cellular transformation, as a central regulator of the ARF (alterative reading frame)/p53 pathway and the Rb (retinoblastoma)/E2F (early-region-2 transcription factor) pathway [95]. A reconstituted high-density lipoprotein (rHDL)–based delivery system was used for Pokemon siRNAs in a tumor model with HepG2 cells overexpressing scavenger receptor class B type I (SR-BI). SR-BI, which is expressed in the liver and most malignant cells, interacts with the major component of HDL, apoprotein A-I, to maintain cellular cholesterol homeostasis. The relative expression levels of the Pokemon and Bcl-2 protein were markedly reduced in tumor tissues of the mice treated with cholesterol-conjugated Pokemon siRNA/rHDL as compared with the control group, resulting in tumor growth inhibition [24].

Human telomerase reverse transcriptase (hTERT) is an essential component of human telomerase, which is often required to maintain stable telomere length in cancer cells. In liver and breast cancer cells, hTERT is highly expressed, which correlates with telomerase activity. Treatment with a siRNA-targeting hTERT conjugated with bioreducible polyethylenimine (SS-PEI) reduced telomerase activity levels and proliferation of HepG2 cells in vitro. Also, when hTERT siRNA/SS-PEI was injected into HepG2 tumor-bearing mice by intratumoral (IT) injection, tumor sizes were significantly smaller compared with the control siRNA–treated tumors [25].

The CSN5 protein is the catalytic center of the mammalian COP9 signalosome (CSN) complex and plays a crucial role in cell proliferation and senescence. In particular, CSN5 binds to transcription factors c-Jun and JunD. It also regulates the stability and controls the function of numerous intracellular regulators of cell proliferation and/ or apoptotic signaling such as MYC, c-Jun, JunD, NFkB, and p53 [96]. Generally, high expression of CSN5 has been found in numerous human cancers, suggesting that knockdown of CSN5 could be an effective anticancer strategy [97, 98]. Therapeutic efficacy of CSN5 gene knockdown was reported as early as 2011 [27]. In orthotopic mouse model of hepatocarcinoma, Huh-luc cells, were treated with either ß-galactosidase siRNA/SNALP or CSN5 siRNA/SNALP (2 mg/kg) by intravenous injection. The CSN5 siRNA/SNALP effectively inhibited the hepatic tumor growth whereas the ß-galactosidase siRNA/SNALP injected group showed partial liver parenchyma. Tumor growth inhibition effect arising from CSN5 knockdown was driven by induction of apoptotic cell death and delay of cell cycle progression.

## 8.3 PROSTATE CANCER

Prostate cancer is the most frequent malignant cancer in men. Several therapies are recommended for prostate cancers including prostatectomy, anti-androgenic hormone therapy, chemotherapy, and radiotherapy. However, these therapies often permanently lower the quality of life or result in negative impact to healthy organs. A frequent genetic lesion in prostate cancer is the loss of the PTEN tumor suppressor gene, the protein product that antagonizes the PI3K pathway. However in clinical trials, PI3K inhibitors have shown little effect as monotherapies. Thus the area is wide open for new targets and therapies directed toward them. Notably, oncogenes related to proliferation and metastasis of prostate cancer have been identified as potential targets for RNAi-based prostate cancer therapy [99].

Myc is a transcription factor associated with various biological processes, including replication, transcription, protein synthesis, cell division, and more [100]. Overexpression of Myc is observed frequently in primary and metastatic prostate cancers. A recent study by the Catapano lab provides insight into the efficacy of Myc inhibition by siRNA in prostate cancer stemlike cells using in vitro and in vivo models of human prostate cancer. Treatment with Myc-targeting siRNAs resulted in a reduction of stemlike properties such as self-renewal and tumor initiation, while further inducing cell senescence in monolayer cultures of human prostate cancer cells. Tumor masses of Myc siRNA/jet-PEI treated mice group were almost completely suppressed, whereas the control group had a sharp increase in tumor growth [32].

Casein kinase II (CK2) is a serine/threonine kinase, which exists as tetramer of two α catalytic subunits and two ß regulatory subunits. CK2 is amplified in various cancer types and can activate and regulate the stability of several tumor suppressor proteins, oncogenes, and even key transcription factors like c-Myc, c-Jun, and NFκB [101, 102]. To test whether CK2 might be a suitable target for therapy, mice bearing human prostate cancer cell xenografts were treated with CK2 siRNA using nanocapsules. The treated mice resulted in about 50% less tumor mass and reduced tumor nodules compared with the vehicle treated animals [103].

As noted above prostate cancer is largely promoted by altered signaling in the androgen receptor (AR) and the PI3K/PKB/mTOR pathways. Notably, signaling in both pathways is coordinated by KLK4 (Kallikrein-related peptidase 4). Thus KLK4 has an important role in PCa progression via its role in regulating cell-cycle gene expression. Although KLK4 is expressed in the normal prostate gland, it is significantly more expressed in malignant prostate cancer [104, 105]. Therefore, it is not surprising that LNCaP or VCaP prostate cancer cells treated with siRNA/liposome targeting KLK4 showed downregulated AR signaling and KLK4 gene expression. Furthermore, the KLK4 siRNA-treated mice showed a dramatic tumor regression of 90% [106].

Notch signaling is associated with multiple cellular processes including differentiation, proliferation, and apoptosis. Dysfunctional regulation of the Notch pathway has been demonstrated in a range of cancers. In prostate cancer cells, abundant Notch1 expression influences tumor invasion. Recent studies indicate that siRNA-mediated Notch1 knockdown inhibited invasion and proliferation of prostate cancer cells [107]. Downregulation of Notch1 expression by a Notch1-targeting siRNA/PSAM-protamine inhibited tumor growth and increased apoptosis in a LNCaP subcutaneous murine xenograft model [35].

## 8.4 BREAST CANCER

Breast cancer is the most frequently diagnosed type of cancer and the leading cause of cancer-related death in women [108]. Breast cancer can be divided into distinct subtypes by using either immunohistochemical (IHC) staining patterns for key receptors or, more recently, molecular profiling. The resulting subtypes display different clinical behaviors and responses to treatment [109]. There are many treatments for breast cancer including surgery, chemotherapy, and radiotherapy as well as the use of targeted therapies. Notably, inhibition of various genes that cause breast cancer through siRNA has been tested in animal models.

As one common way of subdividing breast cancers, tumors are classified by their expression of three receptor molecules: the estrogen receptor, the progesterone receptor, and the human epidermal growth factor receptor 2 (HER2). The estrogen receptor alpha (ER-α), a ligand-activated transcription factor, is one of two types of estrogen receptor. It plays a crucial role in regulation of the cell cycle progression of mammary epithelial cells. Approximately 70% of breast cancer cases are observed to be ER-α positive, often with overexpression of estrogen receptors [110]. To test siRNA targeting the estrogen receptor, MCF-7 cells, an ER-α positive cell line, were used to create xenografts. Intravenous injection with ER-α targeted siRNA-encapsulated stealth nanocapsules composed of PEG-co-poly (ε-caprolactone-co-dodecyl ßmalate) reduced tumor growth and downregulated the level of ER-α [49].

Approximately 15% of breast tumors are known as triple-negative breast cancer (TNBC), because they lack IHC expression in HER2, estrogen receptor, and progesterone receptor [111, 112]. Because the growth of TNBC is not dependent on hormones and/or HER2 receptors, common therapies such as antihormonal or HER2-targeting therapies are ineffective. Goga et al. provided a potential target for TNBC treatment by discovering that inhibition of cyclin-dependent kinase 1 (CDK1) induced synthetic lethality in TNBC overexpressing c-Myc. As a synthetic lethality-based TNBC treatment, CDK1-targeting siRNA was used with cationic lipid assisted poly(ethylene glycol)-b-poly(D,L-lactide) (PEG–PLA) nanoparticles (lipid NP) as the siRNA carrier. CDK siRNA/lipid NP delivered by systemic injection significantly reduced tumor growth in mice bearing c-Myc overexpressing SUM149 and BT549 xenografts without any systemic toxicity or innate immune response [48]. Obviously, siRNA directly targeting Myc, described earlier, could also be tried in this setting.

Another kinase target that has been examined in breast cancer is the ataxia-telangiectasia mutated (ATM) protein, a serine/threonine kinase that regulates DNA damage repair and cell cycle checkpoints and activates downstream signals including p53, CHK2, and BRCA following as DNA damage. For breast cancer therapy, an ATM targeting siRNA/porous silicon-based multistage vector (MSV) was administered in an orthotopic MDA-MB 231 mouse model. Biweekly treatment of ATM siRNA/MSV inhibited tumor growth and reduced ATM expression in IHC assay of tumor tissue [52].

Aberrant epigenetic regulation is frequently associated with the tumorigenic process. DNA methylation is a key epigenetic marker that serves as an important regulator of gene transcription. DNA methylation is organized and maintained by DNA methyltransferases (DNMTs). Aberrant patterns of DNA methylation have been identified in many types of human malignancies; for example, hypermethylation of tumor suppressor genes at CpG islands is associated with gene inactivation [113]. Wang et al. reported that a DNMT-targeted siRNA coupled with a fusion protein consisting of an anti-HER2 single-chain antibody fragment with a positively charged protamine carrier has successfully suppressed DNMTs in the HER2-expressing BT474 breast tumor model. Downregulation of DNMTs by siRNA induced re-expression of the RASSF1A tumor suppressor gene, which led to inhibition of tumor growth [50].

The osteopontin (OPN) protein binds to multiple cell surface receptors to induce cell adhesion and migration. OPN has been considered as a potential prognostic marker in breast cancer progression because elevated levels of OPN have been found in blood and plasma of patients with metastatic breast cancer. For this reason, suppression of OPN may be utilized as a therapeutic strategy. When mice bearing MDA-MB-231 xenografts were treated with an OPN targeting siRNA encapsulated in glycerol propoxylate triacrylate (GPT) and spermine (SPE) nanoparticles, significant inhibition of breast tumor growth with accompanying knockdown of OPN were observed [53].

## 8.5   OVARIAN CANCER

Ovarian cancer is one of the most common types of cancers in women and the leading cause of death in gynecologic cancers. Debulking surgery and chemotherapies with platinum-taxane drugs are generally used in the treatment of ovarian cancer. Despite treatment, the majority of ovarian cancer patients develop recurrent tumors that are resistant to chemotherapy. Consequently, there is an important need for alternative therapeutics for ovarian cancers [114,

115]. However, in contrast to the situation discussed above for TNBC, for ovarian cancers there is a relative dearth of genetically defined targets. One frequent genetic lesion is the loss of the tumor suppressor BRCA1, which has been attacked via PARP inhibitors. Genetic or epigenetic loss of expression of the tumor suppressor PTEN is also frequent, a lesion that can be attacked with PI3K inhibitors. However, there is a clear need for new targets and strategies in this disease.

Recently, Cheung et al. used genome-scale pooled shRNA screens in human cancer cell lines to define 54 overexpressed and essential genes in ovarian cancer that require further validation in vivo [116]. Among them, ID4, a helix-loop-helix (HLH) transcriptional regulator, is highly expressed in most primary ovarian cancers and is, moreover, overexpressed in 32% of high-grade serous ovarian cancers but not in normal tissues including ovary. In a flank xenograft tumor model using OVCAR-8 cells, both intravenous and intraperitoneal injection of ID4 siRNA in a tumor penetrating nanocomplex (TPN) lowered tumor burden compared with control groups. Histological analysis of tumor tissues from ID4 siRNA/TPN treated mice revealed suppression of ID4 expression levels and higher levels of apoptosis in tumor tissues [117].

Ephrin type-A receptor 2 (EphA2) is a tyrosine kinase receptor in the ephrin family that functions as an oncoprotein. EphA2 is highly expressed in many cancer types, but expressed at low levels in normal tissue in adults. This is particularly true for ovarian cancer, where approximately 70% of human ovarian tumors overexpress EphA2. Treatment of several orthotopic ovarian cancer models with paclitaxel and an EphA2 siRNA/DOPC liposome via intraperitoneal injection significantly reduced tumor growth by 81% in the case of a SKOV3ip1 model and by 48% in a HeyA8 model [60].

Enhancer of Zeste Homolog 2 (EZH2) functions as an epigenetic regulator of gene expression that works via histone methylation. Overexpression of EZH2 appears to be involved in cancer progression, functioning by silencing the expression of tumor suppressor genes via specific histone modifications [118]. Recently, EZH2 targeting siRNAs were incorporated into chitosan nanoparticles and utilized along with docetaxel-conjugated PLGA-PRINT nanoparticles as a combination treatment for ovarian cancer. The combination therapy resulted in 95% reduction in tumor weight and further reduced metastasis in orthotopic mice models of ovarian cancer [119].

CD44 is a cell-surface glycoprotein and a major receptor for hyaluronic acid. Hyaluronic acid is a main component of the peritoneum, where ovarian cancer metastases frequently occur. CD44 is involved in cancer progression and metastases but is not expressed in normal cells [120]. For ovarian cancer treatment, paclitaxel and CD44 targeting siRNA with a tumor-specific targeting peptide conjugated dendrimer were intraperitoneally administered in mice bearing human cancer cells directly isolated from malignant ascites of patients with advanced ovarian carcinoma.

The combination treatment group showed almost complete tumor inhibition of tumor growth compared with tumors in mice treated with paclitaxel and a control siRNA [64].

C-Jun-NH$_2$-kinases (JNKs) are serine/threonine kinases that bind and phosphorylate c-Jun. They are members of the mitogen-activated protein kinase family that regulates cell proliferation, apoptosis, and differentiation. Continuous activation of JNKs leads to cancer initiation and progression [121]. Of the two major JNK isoforms, JNK1 and JNK2, JNK1 specifically plays an important part in cell survival by controlling cell cycle arrest and apoptosis [122]. To determine whether JNK1 knockdown by siRNA is indeed capable of eliciting anti-ovarian tumor effects in vivo, JNK1 siRNA packaged in DOPC-liposomes was used to treat HeyA8-bearing and SKOV3ip1-bearing nude mice in combination with docetaxel, the current standard of care. Significant decreases in tumor weight and the number of metastatic nodules were observed in the combination treatment group in comparison to the groups treated with either JNK1 siRNA/DOPC or docetaxel alone [67].

X-linked inhibitor of apoptosis protein (XIAP) is an antiapoptotic protein that prevents apoptotic cell death in tumors. XIAP binds to caspase proteases that are primarily responsible for cell death and stops the proteolytic activation of caspases. Because deregulation of XIAP can contribute to cancer, a high level of XIAP expression has been identified as a tumor marker. In SKOV-3 tumor-bearing mice, administration of XIAP-siRNA/HER2-PLI (HER2 targeting PEI and PLL based copolymer) via tail vein injection significantly delayed tumor growth and increased survival time compared with the control groups [62].

## 8.6  DELIVERY CHALLENGE

Problems with pharmacokinetics and delivery are known to limit the usefulness of many of the small-molecule cancer therapeutics that have been developed in recent years. However, delivery of siRNA into target cells is even more challenging than delivery of conventional molecular drugs into target cells. Inherent characteristics of siRNAs, including negative charge, rigid structure, size, and stability, complicate their passive diffusion across the cell membrane; thus, endocytosis is the major mechanism for intracellular delivery of siRNA. Another challenge in siRNA therapy is the off-target silencing of unintended genes. While the design of a siRNA is meant to maximize knockdown of a specific gene target, nevertheless, marginal mismatch of the complementary sequence in mRNA is often tolerated, leading to nonspecific knockdown of genes [123].

Exogenous siRNA triggers Toll-like receptor (TLR)–mediated innate immune response in both sequence-dependent and independent manners. The TLR-3 pathway is activated by siRNA independent of sequence, whereas TLR-7 in dendritic cells and TLR-8 in monocytes are activated to produce proinflammatory cytokine in a sequence-specific manner [124–126]. Several modification strategies, including 2′-O-methyl chemical modification, have proved

to avoid stimulation of the innate immune response; however, the mechanism behind the activation of the immune system still needs further explanation to help avoid side effects in clinical trials.

The pharmacokinetics and pharmacodynamics of siRNA-based therapeutics in vivo are just beginning to be studied in the clinic. Optimal dosage levels, timing, and duration are expected to vary depending on delivery and targeting strategy, choice of target genes, and disease. Thus our understanding of siRNA-based therapies is in its infancy.

## 8.7  TARGETING CHALLENGE

As can be seen in the previous sections on specific diseases, cancer is a target-rich disease. A major question is which targets deserve the most attention in the target-rich environment that we have described. Here perhaps we can rely on a combination of our knowledge of the genetic basis of cancer and common sense to help make these difficult choices. Cancer is a genetic disease and many successful small-molecule therapeutics have directly targeted key genetic lesions. Thus, we should strongly consider potential genetic targets but perhaps concentrate our efforts on targets that are difficult to reach with small molecules. For example, oncoproteins such as c-Myc or KRAS are presented in mutated form in about 50% of human cancer types [127]. However, no drugs targeting either c-Myc or KRAS directly have been approved, mainly because of their chemically intractable characteristics. Therefore, RNAi therapeutics against these genes justifiably rank high on target lists.

Further targets are being identified that will allow marshaling of the immune system against tumors and blocking cancer support systems such as the vasculature that facilitate tumor growth. Immunotherapies targeting programmed cell death protein 1 (PD1, nivolumab), programmed cell death ligand 1 (PDL1), and cytotoxic T lymphocyte antigen 4 (CTLA4, ipilimumab) have shown exceptional success in melanoma, Hodgkin lymphoma, NSCLC, and bladder cancer, by blocking immune-inhibitory signals, thus enhancing the immune response against cancer [128]. SiRNA therapy is also suited for targeting immune cells because it can be modified to target a specific cell type and used for individual or multiple targets. Antitumor immunotherapies with siRNAs against immunosuppressive factors have been reported in dendritic cells [129–132], monocytes [133], and tumor-associated macrophage [134] and have exhibited significant results in cancer treatment. In a study by Dolina et al., Pdl1 was effectively silenced in Kupffer cells via in vivo administration of PD-L1 siRNA encapsulated in lipidoid nanoparticles (LNP) into mice. Silencing of Pdl1 effectively enhanced the NK cell and CD8+T cell intrahepatic accumulation, viral clearance, and CD8+T cell memory to hepatotropic viral infection. This study demonstrated that transient knockdown of PD-L1 using siRNA directly on the disease-causing cell type might even have benefits when compared with monoclonal antibody usage [135].

Alternatively, a high priority should be placed on investigating multitargeted gene therapy utilizing a combination of siRNAs to block several pathways. This approach parallels efforts to generate combinations of small-molecule therapies, but can be more easily achieved using siRNAs. As mentioned earlier, different types of small RNAs (e.g., siRNA and miRNA) can be partnered in combination therapy [77].

Combination therapy with siRNAs and chemotherapeutic drugs is a credible alternative method to combat tumor heterogeneity and chemoresistant tumors [136–138]. The combination therapy can overcome multidrug resistance and improve drug therapeutic response. Amiji group tested combination treatment with two antiapoptotic genes, bcl-2 and survivin, targeting siRNA and cisplatin to overcome drug resistance in NSCLC [139]. Overexpressed bcl-2 in NSCLC is involved in tumorigenesis and drug resistance as an activator of antiapoptotic cellular defenses. Using cisplatin-resistant tumor-bearing mice, combination therapy exhibited more effective tumor growth inhibition compared with single therapies as proved by the following percentages of inhibition: control siRNA+cisplatin 29%, bcl-2 siRNA+cisplatin 58%, and surviving siRNA+cisplatin 52%. Furthermore, the combination of bcl-2 and survivin siRNA and cisplatin treatment suppressed tumor growth by 62%.

## 8.8 CONCLUSION

Finally, a combination approach using siRNA with a variety of cancer therapies [140–143] such as chemotherapy, immunotherapy, radiation therapy, or photodynamic therapy may dramatically improve the efficacy of cancer therapy. In this strategy, each form of therapy can be used on targets particularly suited to the therapy type, such as small-molecule inhibitors for kinase targets and siRNAs for targets that are structurally unsuited to small-molecule attack. Moreover, because different therapeutic modalities may trigger different forms of resistance mechanisms such as P-glycoprotein (P-gp), multidrug resistance–associated proteins (MRP1, MRP2) for small-molecule drugs and other yet-to-be-determined modes for siRNAs, such multimodal therapies may be harder for tumors to circumvent. It is our strong hope that by skillful delivery and careful target selection, siRNA nanoparticles may take a prominent place in the armamentarium that is being assembled to treat the many diseases that constitute cancer.

## REFERENCES

[1] H.L. McLeod, Cancer pharmacogenomics: early promise, but concerted effort needed, Science 339 (2013) 1563–1566.

[2] M.L. Suva, N. Riggi, B.E. Bernstein, Epigenetic reprogramming in cancer, Science 339 (2013) 1567–1570.

[3] A.A. Farooqi, Z.U. Rehman, J. Muntane, Antisense therapeutics in oncology: current status, OncoTargets Ther 7 (2014) 2035–2042.

[4] W. Walther, P.M. Schlag, Current status of gene therapy for cancer, Curr Opin Oncol 25 (2013) 659–664.

[5] R.S. Bora, D. Gupta, T.K. Mukkur et al., RNA interference therapeutics for cancer: challenges and opportunities (review), Mol Med Rep 6 (2012) 9–15.

[6] S.M. Elbashir, J. Harborth, W. Lendeckel et al., Duplexes of 21-nucleotide RNAs mediate RNA interference in cultured mammalian cells, Nature 411 (2001) 494–498.

[7] E. Miele, G.P. Spinelli, E. Miele et al., Nanoparticle-based delivery of small interfering RNA: challenges for cancer therapy, Int J Nanomedicine 7 (2012) 3637–3657.

[8] Y. Fujita, F. Takeshita, T. Mizutani et al., A novel platform to enable inhaled naked RNAi medicine for lung cancer, Sci Rep 3 (2013) 3325.

[9] S. Hama, M. Arata, I. Nakamura et al., Prevention of tumor growth by needle-free jet injection of anti-C7orf24 siRNA, Cancer Gene Ther 19 (2012) 553–557.

[10] G. Shim, H.W. Choi, S. Lee et al., Enhanced intrapulmonary delivery of anticancer siRNA for lung cancer therapy using cationic ethylphosphocholine-based nanolipoplexes, Mol Ther 21 (2013) 816–824.

[11] V. Fehring, U. Schaeper, K. Ahrens et al., Delivery of therapeutic siRNA to the lung endothelium via novel lipoplex formulation DACC, Mol Ther 22 (2014) 811–820.

[12] J. Han, Q. Wang, Z. Zhang et al., Cationic bovine serum albumin based self-assembled nanoparticles as siRNA delivery vector for treating lung metastatic cancer, Small 10 (2014) 524–535.

[13] S.H. Hong, A. Minai-Tehrani, S.H. Chang et al., Knockdown of the sodium-dependent phosphate co-transporter 2b (NPT2b) suppresses lung tumorigenesis, PLoS One 8 (2013) e77121.

[14] H. Yu, Y. Zou, L. Jiang et al., Induction of apoptosis in non-small cell lung cancer by downregulation of MDM2 using pH responsive PMPC-b-PDPA/siRNA complex nanoparticles, Biomaterials 34 (2013) 2738–2747.

[15] M.E. Bonnet, J.B. Gossart, E. Benoit et al., Systemic delivery of sticky siRNAs targeting the cell cycle for lung tumor metastasis inhibition, J Control Release 170 (2013) 183–190.

[16] Y. Yang, Y. Hu, Y. Wang et al., Nanoparticle delivery of pooled siRNA for effective treatment of non-small cell lung cancer, Mol Pharm 9 (2012) 2280–2289.

[17] S. Ganesh, A.K. Iyer, J. Weiler et al., Combination of siRNA directed gene silencing with cisplatin reverses drug resistance in human nonsmall cell lung cancer, Mol Ther Nucleic Acids 2 (2013) e110.

[18] Y.H. Lee, D. Seo, K.J. Choi et al., Antitumor effects in hepatocarcinoma of isoform-selective inhibition of HDAC2, Cancer Res 74 (2014) 4752–4761.

[19] A. Chauhan, S. Zubair, A. Nadeem et al., Escheriosome-mediated cytosolic delivery of PLK1-specific siRNA: potential in treatment of liver cancer in BALB/c mice, Nanomedicine (London) 9 (2014) 407–420.

[20] R.L. Bogorad, H. Yin, A. Zeigerer et al., Nanoparticle-formulated siRNA targeting integrins inhibits hepatocellular carcinoma progression in mice, Nat Commun 5 (2014) 3869.

[21] J. Gao, H. Chen, Y. Yu et al., Inhibition of hepatocellular carcinoma growth using immunoliposomes for co-delivery of adriamycin and ribonucleotide reductaseM2 siRNA, Biomaterials 34 (2013) 10,084–10,098.

[22] L. Li, R. Wang, D. Wilcox et al., Developing lipid nanoparticle-based siRNA therapeutics for hepatocellular carcinoma using an integrated approach, Mol Cancer Ther 12 (2013) 2308–2318.

[23] C. Wu, F. Gong, P. Pang, et al., An RGD-modified MRI-visible polymeric vector for targeted siRNA delivery to hepatocellular carcinoma in nude mice, PLoS One 8 (2013) e66416.

[24] Y. Ding, W. Wang, M. Feng et al., A biomimetic nanovector-mediated targeted cholesterol-conjugated siRNA delivery for tumor gene therapy, Biomaterials 33 (2012) 8893–8905.

[25] W. Xia, P. Wang, C. Lin et al., Bioreducible polyethylenimine-delivered siRNA targeting human telomerase reverse transcriptase inhibits HepG2 cell growth in vitro and in vivo, J Control Release 157 (2012) 427–436.

[26] J. Gao, Y. Yu, Y. Zhang et al., EGFR-specific PEGylated immunoliposomes for active siRNA delivery in hepatocellular carcinoma, Biomaterials 33 (2012) 270–282.

[27] Y.H. Lee, A.D. Judge, D. Seo et al., Molecular targeting of CSN5 in human hepatocellular carcinoma: a mechanism of therapeutic response, Oncogene 30 (2011) 4175–4184.

[28] Y.H. Lee, J.B. Andersen, H.T. Song et al., Definition of ubiquitination modulator COP1 as a novel therapeutic target in human hepatocellular carcinoma, Cancer Res 70 (2010) 8264–8269.

[29] C. Liu, X. Liu, P. Rocchi et al., Arginine-terminated generation 4 PAMAM dendrimer as an effective nanovector for functional siRNA delivery in vitro and in vivo, Bioconjug Chem 25 (2014) 521–532.

[30] H.Y. Xue, H.L. Wong, Solid lipid-PEI hybrid nanocarrier: an integrated approach to provide extended, targeted, and safer siRNA therapy of prostate cancer in an all-in one manner, ACS Nano 5 (2011) 7034–7047.

[31] Q. Lin, C.S. Jin, H. Huang et al., Nanoparticle-enabled, image-guided treatment planning of target specific RNAi therapeutics in an orthotopic prostate cancer model, Small 10 (2014) 3072–3082.

[32] G. Civenni, A. Malek, D. Albino et al., RNAi-mediated silencing of Myc transcription inhibits stem-like cell maintenance and tumorigenicity in prostate cancer, Cancer Res 73 (2013) 6816–6827.

[33] J.O. McNamara II, E.R. Andrechek, Y. Wang et al., Cell type-specific delivery of siRNAs with aptamer siRNA chimeras, Nat Biotechnol 24 (2006) 1005–1015.

[34] S. Zeng, M.P. Xiong, Trilayer micelles for combination delivery of rapamycin and siRNA targeting Y-box binding protein-1 (siYB-1), Biomaterials 34 (2013) 6882–6892.

[35] Y. Su, L. Yu, N. Liu et al., PSMA specific single chain antibody-mediated targeted knockdown of Notch1 inhibits human prostate cancer cell proliferation and tumor growth, Cancer Lett 338 (2013) 282–291.

[36] X.Q. Liu, M.H. Xiong, X.T. Shu et al., Therapeutic delivery of siRNA silencing HIF-1 alpha with micellar nanoparticles inhibits hypoxic tumor growth, Mol Pharm 9 (2012) 2863–2874.

[37] S.J. Lee, M.S. Huh, S.Y. Lee et al., Tumor-homing poly-siRNA/glycol chitosan self-cross-linked nanoparticles for systemic siRNA delivery in cancer treatment, Angew Chem 51 (2012) 7203–7207.

[38] J. Guo, J.R. Ogier, S. Desgranges et al., Anisamide-targeted cyclodextrin nanoparticles for siRNA delivery to prostate tumours in mice, Biomaterials 33 (2012) 7775–7784.

[39] J. Guo, W.P. Cheng, J. Gu et al., Systemic delivery of therapeutic small interfering RNA using a pH-triggered amphiphilic poly-l-lysine nanocarrier to suppress prostate cancer growth in mice, Eur J Pharm Sci 45 (2012) 521–532.

[40] J.B. Lee, K. Zhang, Y.Y. Tam et al., Lipid nanoparticle siRNA systems for silencing the androgen receptor in human prostate cancer in vivo, Int J Cancer 131 (2012) E781–E790.

[41] P. Mu, S. Nagahara, N. Makita et al., Systemic delivery of siRNA specific to tumor mediated by atelocollagen: combined therapy using siRNA targeting Bcl-xL and cisplatin against prostate cancer, Int J Cancer 125 (2009) 2978–2990.

[42] M. Aleku, P. Schulz, O. Keil et al., Atu027, a liposomal small interfering RNA formulation targeting protein kinase N3, inhibits cancer progression, Cancer Res 68 (2008) 9788–9798.

[43] M.K. Addepalli, K.B. Ray, B. Kumar et al., RNAi-mediated knockdown of AURKB and EGFR shows enhanced therapeutic efficacy in prostate tumor regression, Gene Ther 17 (2010) 352–359.

[44] L. Crombez, M.C. Morris, S. Dufort et al., Targeting cyclin B1 through peptide-based delivery of siRNA prevents tumour growth, Nucleic Acids Res 37 (2009) 4559–4569.

[45] W. Tang, D.A. Weidner, B.Y. Hu et al., Efficient delivery of small interfering RNA to plant cells by a nanosecond pulsed laser-induced stress wave for posttranscriptional gene silencing, Plant Sci 171 (2006) 375–381.

[46] T.M. Sun, J.Z. Du, Y.D. Yao et al., Simultaneous delivery of siRNA and paclitaxel via a "two-in-one" micelleplex promotes synergistic tumor suppression, ACS Nano 5 (2011) 1483–1494.

[47] X.Z. Yang, S. Dou, T.M. Sun et al., Systemic delivery of siRNA with cationic lipid assisted PEG-PLA nanoparticles for cancer therapy, J Control Release 156 (2011) 203–211.

[48] Y. Liu, Y.H. Zhu, C.Q. Mao et al., Triple negative breast cancer therapy with CDK1 siRNA delivered by cationic lipid assisted PEG-PLA nanoparticles, J Control Release 192 (2014) 114–121.

[49] C. Bouclier, L. Moine, H. Hillaireau et al., Physicochemical characteristics and preliminary in vivo biological evaluation of nanocapsules loaded with siRNA targeting estrogen receptor alpha, Biomacromolecules 9 (2008) 2881–2890.

[50] S. Dou, Y.D. Yao, X.Z. Yang et al., Anti-Her2 single-chain antibody mediated DNMTs-siRNA delivery for targeted breast cancer therapy, J Control Release 161 (2012) 875–883.

[51] J.Y. Pille, C. Denoyelle, J. Varet et al., Anti-RhoA and anti-RhoC siRNAs inhibit the proliferation and invasiveness of MDA-MB-231 breast cancer cells in vitro and in vivo, Mol Ther 11 (2005) 267–274.

[52] R. Xu, Y. Huang, J. Mai et al., Multistage vectored siRNA targeting ataxia-telangiectasia mutated for breast cancer therapy, Small 9 (2013) 1799–1808.

[53] A. Minai-Tehrani, H.L. Jiang, Y.K. Kim et al., Suppression of tumor growth in xenograft model mice by small interfering RNA targeting osteopontin delivery using biocompatible poly(amino ester), Int J Pharm 431 (2012) 197–203.

[54] H.M. Aliabadi, R. Maranchuk, C. Kucharski et al., Effective response of doxorubicin-sensitive and -resistant breast cancer cells to combinational siRNA therapy, J Control Release 172 (2013) 219–228.

[55] J. Jiang, S.J. Yang, J.C. Wang et al., Sequential treatment of drug-resistant tumors with RGD-modified liposomes containing siRNA or doxorubicin, Eur J Pharm Biopharm 76 (2010) 170–178.

[56] S.K. Cho, A. Pedram, E.R. Levin et al., Acid-degradable core-shell nanoparticles for reversed tamoxifen-resistance in breast cancer by silencing manganese superoxide dismutase (MnSOD), Biomaterials 34 (2013) 10,228–10,237.

[57] Q. Leng, A.J. Mixson, Small interfering RNA targeting Raf-1 inhibits tumor growth in vitro and in vivo, Cancer Gene Ther 12 (2005) 682–690.

[58] Y. Ding, Y. Wang, J. Zhou et al., Direct cytosolic siRNA delivery by reconstituted high density lipoprotein for target-specific therapy of tumor angiogenesis, Biomaterials 35 (2014) 7214–7227.

[59] S. Florinas, J. Kim, K. Nam et al., Ultrasound-assisted siRNA delivery via arginine-grafted bioreducible polymer and microbubbles targeting VEGF for ovarian cancer treatment, J Control Release 183 (2014) 1–8.

[60] C.N. Landen, W.M. Merritt, L.S. Mangala et al., Intraperitoneal delivery of liposomal siRNA for therapy of advanced ovarian cancer, Cancer Biol Ther 5 (2006) 1708–1713.

[61] S. Kala, A.S. Mak, X. Liu et al., Combination of dendrimer-nanovector-mediated small interfering RNA delivery to target Akt with the clinical anticancer drug paclitaxel for effective and potent anticancer activity in treating ovarian cancer, J Med Chem 57 (2014) 2634–2642.

[62] J. Li, D. Cheng, T. Yin et al., Copolymer of poly(ethylene glycol) and poly(L-lysine) grafting polyethylenimine through a reducible disulfide linkage for siRNA delivery, Nanoscale 6 (2014) 1732–1740.

[63] T. Koyanagi, Y. Suzuki, Y. Saga et al., In vivo delivery of siRNA targeting vasohibin-2 decreases tumor angiogenesis and suppresses tumor growth in ovarian cancer, Cancer Sci 104 (2013) 1705–1710.

[64] V. Shah, O. Taratula, O.B. Garbuzenko et al., Targeted nanomedicine for suppression of CD44 and simultaneous cell death induction in ovarian cancer: an optimal delivery of siRNA and anticancer drug, Clin Cancer Res 19 (2013) 6193–6204.

[65] H.D. Han, L.S. Mangala, J.W. Lee et al., Targeted gene silencing using RGD-labeled chitosan nanoparticles, Clin Cancer Res 16 (2010) 3910–3922.

[66] T. Tanaka, L.S. Mangala, P.E. Vivas-Mejia et al., Sustained small interfering RNA delivery by mesoporous silicon particles, Cancer Res 70 (2010) 3687–3696.

[67] P. Vivas-Mejia, J.M. Benito, A. Fernandez et al., c-Jun.-NH2-kinase-1 inhibition leads to antitumor activity in ovarian cancer, Clin Cancer Res 16 (2010) 184–194.

[68] A. Difeo, F. Huang, J. Sangodkar et al., KLF6- SV1 is a novel antiapoptotic protein that targets the BH3-only protein NOXA for degradation and whose inhibition extends survival in an ovarian cancer model, Cancer Res 69 (2009) 4733–4741.

[69] J. Halder, A.A. Kamat, C.N. Landen Jr. et al., Focal adhesion kinase targeting using in vivo short interfering RNA delivery in neutral liposomes for ovarian carcinoma therapy, Clin Cancer Res 12 (2006) 4916–4924.

[70] J.R. Molina, P. Yang, S.D. Cassivi et al., Non-small cell lung cancer: epidemiology, risk factors, treatment, and survivorship, Mayo Clin Proc 83 (2008) 584–594.

[71] R.S. Herbst, J.V. Heymach, S.M. Lippman, Lung cancer, N Engl J Med 359 (2008) 1367–1380.

[72] J.K. Lam, W. Liang, H.K. Chan, Pulmonary delivery of therapeutic siRNA, Adv Drug Deliv Rev 64 (2012) 1–15.

[73] O.M. Merkel, I. Rubinstein, T. Kissel, siRNA delivery to the lung: what's new? Adv Drug Deliv Rev 75 (2014) 112–128.

[74] T. Mitsudomi, Y. Yatabe, Epidermal growth factor receptor in relation to tumor development: EGFR gene and cancer, FEBS J 277 (2010) 301–308.

[75] R. Sordella, D.W. Bell, D.A. Haber et al., Gefitinib-sensitizing EGFR mutations in lung cancer activate anti-apoptotic pathways, Science 305 (2004) 1163–1167.

[76] M. Takahashi, T. Chiyo, T. Okada et al., Specific inhibition of tumor cells by oncogenic EGFR specific silencing by RNA interference, PLoS One 8 (2013) e73214.

[77] W. Xue, J.E. Dahlman, T. Tammela et al., Small RNA combination therapy for lung cancer, Proc Natl Acad Sci USA 111 (2014) E3553–E3561.

[78] J. Kurashige, M. Watanabe, M. Iwatsuki et al., RPN2 expression predicts response to docetaxel in oesophageal squamous cell carcinoma, Br J Cancer 107 (2012) 1233–1238.

[79] Y. Haupt, R. Maya, A. Kazaz et al., Mdm2 promotes the rapid degradation of p53, Nature 387 (1997) 296–299.

[80] T. Iwakuma, G. Lozano, MDM2, an introduction, Mol Cancer Res 1 (2003) 993–1000.

[81] R. Honda, H. Tanaka, H. Yasuda, Oncoprotein MDM2 is a ubiquitin ligase E3 for tumor suppressor p53, FEBS Lett 420 (1997) 25–27.

[82] K. Schutte, J. Bornschein, P. Malfertheiner, Hepatocellular carcinoma—epidemiological trends and risk factors, Dig Dis 27 (2009) 80–92.

[83] B. Bartosch, R. Thimme, H.E. Blum et al., Hepatitis C virus-induced hepatocarcinogenesis, J Hepatol 51 (2009) 810–820.

[84] A. Sehgal, A. Vaishnaw, K. Fitzgerald, Liver as a target for oligonucleotide therapeutics, J Hepatol 59 (2013) 1354–1359.

[85] S. Ropero, M. Esteller, The role of histone deacetylases (HDACs) in human cancer, Mol Oncol 1 (2007) 19–25.

[86] O.H. Kramer, HDAC2: a critical factor in health and disease, Trends Pharmacol Sci 30 (2009) 647–655.

[87] C.C. Park, H. Zhang, M. Pallavicini et al., Beta1 integrin inhibitory antibody induces apoptosis of breast cancer cells, inhibits growth, and distinguishes malignant from normal phenotype in three dimensional cultures and in vivo, Cancer Res 66 (2006) 1526–1535.

[88] F. Aoudjit, K. Vuori, Integrin signaling inhibits paclitaxel-induced apoptosis in breast cancer cells, Oncogene 20 (2001) 4995–5004.

[89] T. Speicher, B. Siegenthaler, R.L. Bogorad et al., Knockdown and knockout of beta1-integrin in hepatocytes impairs liver regeneration through inhibition of growth factor signalling, Nat Commun 5 (2014) 3862.

[90] A.D. Tward, K.D. Jones, S. Yant et al., Distinct pathways of genomic progression to benign and malignant tumors of the liver, Proc Natl Acad Sci USA 104 (2007) 14,771–14,776.

[91] J.K. Stauffer, A.J. Scarzello, J.B. Andersen et al., Coactivation of AKT and beta-catenin in mice rapidly induces formation of lipogenic liver tumors, Cancer Res 71 (2011) 2718–2727.

[92] G. Ambrosini, C. Adida, D.C. Altieri, A novel anti-apoptosis gene, survivin, expressed in cancer and lymphoma, Nat Med 3 (1997) 917–921.

[93] T. Ito, K. Shiraki, K. Sugimoto et al., Survivin promotes cell proliferation in human hepatocellular carcinoma, Hepatology 31 (2000) 1080–1085.

[94] T. Maeda, R.M. Hobbs, P.P. Pandolfi, The transcription factor Pokemon: a new key player in cancer pathogenesis, Cancer Res 65 (2005) 8575–8578.

[95] K. Apostolopoulou, I.S. Pateras, K. Evangelou et al., Gene amplification is a relatively frequent event leading to ZBTB7A (Pokemon) overexpression in nonsmall cell lung cancer, J Pathol 213 (2007) 294–302.

[96] A. Yoshida, N. Yoneda-Kato, J.Y. Kato, CSN5 specifically interacts with CDK2 and controls senescence in a cytoplasmic cyclin E-mediated manner, Sci Rep 3 (2013) 1054.

[97] A. Fukumoto, N. Ikeda, M. Sho et al., Prognostic significance of localized p27Kip1 and potential role of Jab1/CSN5 in pancreatic cancer, Oncol Rep 11 (2004) 277–284.

[98] D. Ivan, A.H. Diwan, F.J. Esteva et al., Expression of cell cycle inhibitor p27Kip1 and its inactivator Jab1 in melanocytic lesions, Mod Pathol 17 (2004) 811–818.

[99] J. Guo, J.C. Evans, C.M. O'Driscoll, Delivering RNAi therapeutics with non-viral technology: a promising strategy for prostate cancer? Trends Mol Med 19 (2013) 250–261.

[100] C.V. Dang, MYC on the path to cancer, Cell 149 (2012) 22–35.

[101] D.W. Litchfield, Protein kinase CK2: structure, regulation and role in cellular decisions of life and death, Biochem J 369 (2003) 1–15.

[102] B. Guerra, O.G. Issinger, Protein kinase CK2 in human diseases, Curr Med Chem 15 (2008) 1870–1886.

[103] J.H. Trembley, G.M. Unger, V.L. Korman et al., Tenfibgen ligand nanoencapsulation delivers bifunctional anti-CK2 RNAi oligomer to key sites for prostate cancer targeting using human xenograft tumors in mice, PLoS One 9 (2014), e109970.

[104] Z. Xi, T.I. Klokk, K. Korkmaz et al., Kallikrein 4 is a predominantly nuclear protein and is overexpressed in prostate cancer, Cancer Res 64 (2004) 2365–2370.

[105] J. Lai, M.L. Lehman, M.E. Dinger et al., A variant of the KLK4 gene is expressed as a cis sense-antisense chimeric transcript in prostate cancer cells, RNA 16 (2010) 1156–1166.

[106] Y. Jin, S. Qu, M. Tesikova et al., Molecular circuit involving KLK4 integrates androgen and mTOR signaling in prostate cancer, Proc Natl Acad Sci USA 110 (2013) E2572–E2581.

[107] Z. Wang, Y. Li, S. Banerjee et al., Down-regulation of notch-1 and jagged-1 inhibits prostate cancer cell growth, migration and invasion, and induces apoptosis via inactivation of Akt, mTOR, and NF-kappaB signaling pathways, J Cell Biochem 109 (2010) 726–736.

[108] J. Zhang, X. Li, L. Huang, Non-viral nanocarriers for siRNA delivery in breast cancer, J Control Release 190 (2014) 440–450.

[109] H. Kennecke, R. Yerushalmi, R. Woods et al., Metastatic behavior of breast cancer subtypes, J Clin Oncol Off J Am Soc Clin Oncol 28 (2010) 3271–3277.

[110] A. Howell, M. Dowsett, Endocrinology and hormone therapy in breast cancer: aromatase inhibitors versus antioestrogens, Breast Cancer Res 6 (2004) 269–274.

[111] A. Bosch, P. Eroles, R. Zaragoza et al., Triple-negative breast cancer: molecular features, pathogenesis, treatment and current lines of research, Cancer Treat Rev 36 (2010) 206–215.

[112] J.S. Reis-Filho, A.N. Tutt, Triple negative tumours: a critical review, Histopathology 52 (2008) 108–118.

[113] S.B. Baylin, DNA methylation and gene silencing in cancer, Nat Clin Pract Oncol 2 (Suppl. 1) (2005) S4–11.

[114] R.C. Bast Jr., B. Hennessy, G.B. Mills, The biology of ovarian cancer: new opportunities for translation, Nat Rev Cancer 9 (2009) 415–428.

[115] B.A. Goff, L. Mandel, H.G. Muntz et al., Ovarian carcinoma diagnosis, Cancer 89 (2000) 2068–2075.

[116] H.W. Cheung, G.S. Cowley, B.A. Weir et al., Systematic investigation of genetic vulnerabilities across cancer cell lines reveals lineage-specific dependencies in ovarian cancer, Proc Natl Acad Sci USA 108 (2011) 12,372–12,377.

[117] Y. Ren, H.W. Cheung, G. von Maltzhan et al., Targeted tumor penetrating siRNA nanocomplexes for credentialing the ovarian cancer oncogene ID4, Sci. Transl. Med. 4 (2012) 147ra112 /DC1 (27 pages). DOI: 10.1126/scitranslmed.3003778

[118] M.L. Suva, N. Riggi, M. Janiszewska et al., EZH2 is essential for glioblastoma cancer stem cell maintenance, Cancer Res 69 (2009) 9211–9218.

[119] K.M. Gharpure, K.S. Chu, C.J. Bowerman et al., Metronomic docetaxel in PRINT nanoparticles and EZH2 silencing have synergistic antitumor effect in ovarian cancer, Mol Cancer Ther 13 (2014) 1750–1757.

[120] D. Naor, R.V. Sionov, D. Ish-Shalom, CD44: structure, function, and association with the malignant process, Adv Cancer Res 71 (1997) 241–319.

[121] R.J. Davis, Signal transduction by the JNK group of MAP kinases, Cell 103 (2000) 239–252.

[122] K. Sabapathy, K. Hochedlinger, S.Y. Nam et al., Distinct roles for JNK1 and JNK2 in regulating JNK activity and c-Jun.-dependent cell proliferation, Mol Cell 15 (2004) 713–725.

[123] O. Snove Jr., T. Holen, Many commonly used siRNAs risk off-target activity, Biochem Biophys Res Commun 319 (2004) 256–263.

[124] M.E. Kleinman, K. Yamada, A. Takeda et al., Sequence- and target-independent angiogenesis suppression by siRNA via TLR3, Nature 452 (2008) 591–597.

[125] A.D. Judge, V. Sood, J.R. Shaw et al., Sequence dependent stimulation of the mammalian innate immune response by synthetic siRNA, Nat Biotechnol 23 (2005) 457–462.

[126] A. Forsbach, J.G. Nemorin, C. Montino et al., Identification of RNA sequence motifs stimulating sequence-specific TLR8-dependent immune responses, J Immunol 180 (2008) 3729–3738.

[127] G.L. Verdine, L.D. Walensky, The challenge of drugging undruggable targets in cancer: lessons learned from targeting BCL-2 family members, Clin Cancer Res 13 (2007) 7264–7270.

[128] J.R. Cubillos-Ruiz, X. Engle, U.K. Scarlett et al., Polyethylenimine-based siRNA nanocomplexes reprogram tumor-associated dendritic cells via TLR5 to elicit therapeutic antitumor immunity, J Clin Invest 119 (2009) 2231–2244.

[129] A. Troegeler, C. Lastrucci, C. Duval et al., An efficient siRNA-mediated gene silencing in primary human monocytes, dendritic cells and macrophages, Immunol Cell Biol 92 (2014) 699–708.

[130] A. Alshamsan, S. Hamdy, A. Haddadi et al., STAT3 knockdown in B16 melanoma by siRNA lipopolyplexes induces bystander immune response in vitro and in vivo, Transl Oncol 4 (2011) 178–188.

[131] W. Hobo, T.I. Novobrantseva, H. Fredrix et al., Improving dendritic cell vaccine immunogenicity by silencing PD-1 ligands using siRNA-lipid nanoparticles combined with antigen mRNA electroporation, Cancer Immunol Immunother 62 (2013) 285–297.

[132] S. Warashina, T. Nakamura, H. Harashima, A20 silencing by lipid envelope-type nanoparticles enhances the efficiency of lipopolysaccharide-activated dendritic cells, Biol Pharm Bull 34 (2011) 1348–1351.

[133] F. Leuschner, P. Dutta, R. Gorbatov et al., Therapeutic siRNA silencing in inflammatory monocytes in mice, Nat Biotechnol 29 (2011) 1005–1010.

[134] J. Conde, C.C. Bao, Y.Q. Tan et al., Dual targeted immunotherapy via in vivo delivery of biohybrid RNAi-peptide nanoparticles to tumor-associated macrophages and cancer cells, Adv Funct Mater 25 (2015) 4183–4194.

[135] J.S. Dolina, S.S. Sung, T.I. Novobrantseva et al., Lipidoid nanoparticles containing PD-L1 siRNA delivered in vivo enter Kupffer cells and enhance NK and CD8(+) T cell-mediated hepatic antiviral immunity, Mol Ther Nucleic Acids 2 (2013) e72.

[136] H. Meng, W.X. Mai, H. Zhang et al., Codelivery of an optimal drug/siRNA combination using mesoporous silica nanoparticles to overcome drug resistance in breast cancer in vitro and in vivo, ACS Nano 7 (2013) 994–1005.

[137] J.A. MacDiarmid, N.B. Amaro-Mugridge, J. Madrid-Weiss et al., Sequential treatment of drug-resistant tumors with targeted minicells containing siRNA or a cytotoxic drug, Nat Biotechnol 27 (2009) 643–651.

[138] M. Abbasi, A. Lavasanifar, H. Uludag, Recent attempts at RNAi-mediated P-glycoprotein downregulation for reversal of multidrug resistance in cancer, Med Res Rev 33 (2013) 33–53.

[139] S. Ganesh, A.K. Iyer, J. Weiler et al., Combination of siRNA directed gene silencing with cisplatin reverses drug resistance in human nonsmall cell lung cancer, Mol Ther Nucleic Acids 2 (2013) e110.

[140] L.H. Madkour, Reactive Oxygen Species (ROS), Nanoparticles, and Endoplasmic Reticulum (ER) Stress-Induced Cell Death Mechanisms. Academic Press, 1st August 2020. https://www.elsevier.com/books/reactive-oxygen-species-ros-nanoparticles-and-endoplasmic-reticulum-er-stress-induced-cell-death-mechanisms/madkour/978-0-12-822481-6

[141] L.H. Madkour, Nanoparticles Induce Oxidative and Endoplasmic Reticulum Antioxidant Therapeutic Defenses, 1st Edition, 2020. Springer International Publishing, Switzerland. https://www.springer.com/gp/book/9783030372965?utm_campaign=3_pier05_buy_print&utm_content=en_08082017&utm_medium=referral&utm_source=google_books#otherversion=9783030372972

[142] L.H. Madkour, Nucleic Acids as Gene Anticancer Drug Delivery Therapy. 1st Edition, 2020. Elsevier. https://www.elsevier.com/books/nucleic-acids-as-gene-anticancer-drug-deliverytherapy/madkour/978-0-12-819777-6

[143] L.H. Madkour, Nanoelectronic Materials: Fundamentals and Applications (Advanced Structured Materials), 1st Edition, 2019. Springer International Publishing, Switzerland. https://books.google.com.eg/books/about/Nanoelectronic_Materials.html?id=YQXCxAEACAAJ&source=kp_book_description&redir_esc=y https://www.springer.com/gp/book/9783030216207

# 9 Recent Advances in miRNA Molecule Delivery as Anticancer Drugs

In this chapter, we provide insights into the development of nonviral synthetic microRNA (miRNA) vectors and the promise of miRNA-based anticancer therapies, including therapeutic applications of miRNAs, challenges of vector design to overcome delivery obstacles, and the development of miRNA delivery systems for cancer therapy. Understanding the regulation of miRNA molecule production and function may facilitate the development of novel diagnostic and therapeutic strategies to improve the prognosis of women with epithelial ovarian cancer. Additionally, understanding miRNA molecules and miRNA-regulatory machinery associations with clinical features may influence prevention and early detection efforts. This chapter catalogs the knowledge gained from collective studies, so as to assess the progress made so far. We discuss aberrant expression of miRNA molecules and miRNA-regulating machinery associated with clinical features of epithelial ovarian cancer. It is time to ponder the knowledge gained, so that more meaningful preclinical and translational studies can be designed to better realize the potential that miRNAs have to offer.

## 9.1 BRIEF INTRODUCTION TO miRNAs

miRNAs are naturally occurring, small, noncoding RNAs that serve as key regulators of gene expression [1]. Since the discovery of the first miRNA in 1993 [2, 3], more than 2500 human-sourced miRNAs have been identified [4]. Early studies showed that miRNAs regulate more than 30% of all protein-coding genes in fundamental cellular signaling or various other biological processes [5]. These include cell proliferation, differentiation, survival, motility, apoptosis, and death [6–8].

In mammalian cells, primary miRNA precursors (pri-miRNA) are initially produced in the nucleus by RNA polymerase II. They are then further processed by DiGeorge syndrome critical region 8 (DGCR8) and the ribonuclease (RNase) Drosha to form precursor miRNAs (pre-miRNA) with a long double-stranded hairpin. The pre-miRNA is transported by exportin-5 into the cytosol and finally processed into double-stranded duplexes by the RNAse Dicer [9]. The miRNA duplexes are then unwound and enter a miRNA-induced silencing complex (miRISC) causing an inhibition of translation and degradation of mRNA (Figure 9.1) [10, 11].

## 9.2 THERAPEUTIC APPLICATIONS AND CHALLENGES OF USING miRNAs

With increasing evidence that miRNAs function as global regulators in multiple biological systems (e.g., the endocrine system, hematopoiesis, and fat metabolism), the therapeutic and diagnostic values of miRNA are widely acknowledged [12]. There is thus dramatically growing interest in miRNA-based medicine.

### 9.2.1 BRIEF OVERVIEW OF miRNA THERAPEUTICS

miRNAs are biologically functional RNAs that cause gene silencing by targeting mRNA. Many pathological processes involve abnormal miRNA expression. For example, in rheumatoid arthritis, differential expression of miR-155, miR-124a, miR-146a, miR-15, and miR-16 has been identified [13]. In systemic lupus erythematosus, dysregulation of miR-21, miR-25, and miR-186 in peripheral blood mononuclear cells was reported recently [14]. In Sjögren's syndrome, increased miR-4524-b3p and an unidentified miR-5100 were found [15]. In multiple sclerosis patients, distinct expression of miR-20a, miR-326, and miR-17-5p was shown in CD4+ and CD8+ T cells [16].

It is well accepted that cancers often result from genetic mutations. In early studies comparing miRNA expression in normal tissues and primary carcinomas, approximately 50% of miRNA genes were located at fragile sites and genomic regions highly related to cancers [16–19]. Consequently, considering their important roles in tumorigenesis, miRNAs have been investigated as diagnostic biomarkers, oncogenic factors, or tumor suppressors with therapeutic potential (Figure 9.2) [20]. miRNAs are generally associated with various stages of cancer progression, such as unlimited proliferation, metastasis and invasion, sustained angiogenesis, apoptosis, and drug/radio-resistance.

Two approaches are used to modulate the function of miRNAs in cancers. The first approach is directed toward inhibiting the activity of oncogenic miRNA activity using miRNA antagonists, such as synthetic anti-miRNAs or locked nucleic acids (LNAs). For example, small antisense oligonucleotides with sequences complementary to endogenous mature miRNAs can be used as miRNA antagonists. Moreover, mRNAs involved with various targets can be applied as miRNA sponges (miRSP) to sequester the misexpressed miRNAs [21, 22].

OncomiRs (e.g., miR-125b, miR-20, miR-155, and the miR-17-92 family) that downregulate tumor suppressor genes are overexpressed in different types of cancers [23, 24]. In contrast, miRNAs (e.g., miR-34a, miR-15a, miR-16, miR-17, miR-29, miR-126, miR-143/145, and the let-7 family) that can function as tumor suppressors are underexpressed in many human cancers [25, 26]. Upregulation of these tumor suppressive miRNAs provides the second approach for miRNA-based cancer therapy. Specifically,

DOI: 10.1201/9781003229650-9

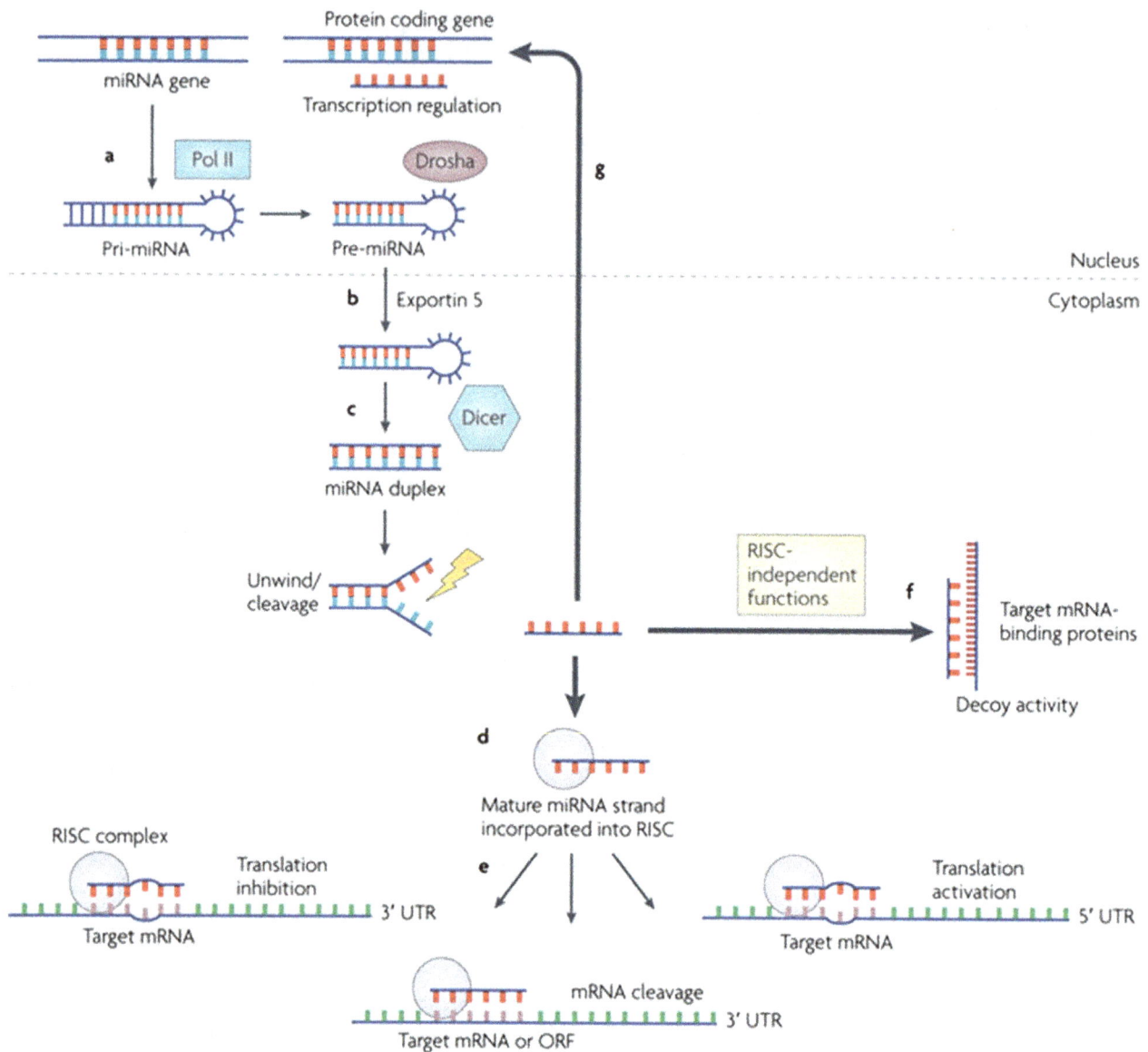

Nature Reviews | Drug Discovery

**FIGURE 9.1**  miRNA biogenesis and effector pathways. (Reprinted with permission from [11]. Copyright, 2010 by the Nature Publishing Group.)

miRNA mimics (functionally similarly to endogenous miRNAs) are employed to reprogram cancer cells by incorporation into the RNA-induced silencing complexes. Additionally, plasmid DNAs that encode hairpin pre-miRNA can be used to reactivate cellular pathways and induce therapeutic responses [27, 28].

Resistance to chemotherapy and radiotherapy is the major barrier against successful cancer therapy. There is growing evidence that the expression level of miRNAs is linked to the development of drug resistance [29]. Indeed, levels of various miRNAs correlate with both chemo- and radio-resistance through the regulation of drug efflux transporters, the cell cycle, and DNA repair [30]. For example, simultaneous upregulation of miR-27a and ABCB1 was found in paclitaxel-resistant ovarian carcinoma cells, while transfection with miR-27 inhibitors reduced the expression of P-gp [31]. Upregulation of miR-485-3p in human lymphoblastic leukemia CEM cells induced a decrease in nuclear factor-κB and an increase in the Top2α level, consequently increasing the sensitivity of these cells to topoisomerase II inhibitors [32]. In addition, overexpression of miR-9 and let-7g lowered the NFκB1 level and enhanced the sensitivity of lung tumors to radiation therapy [33].

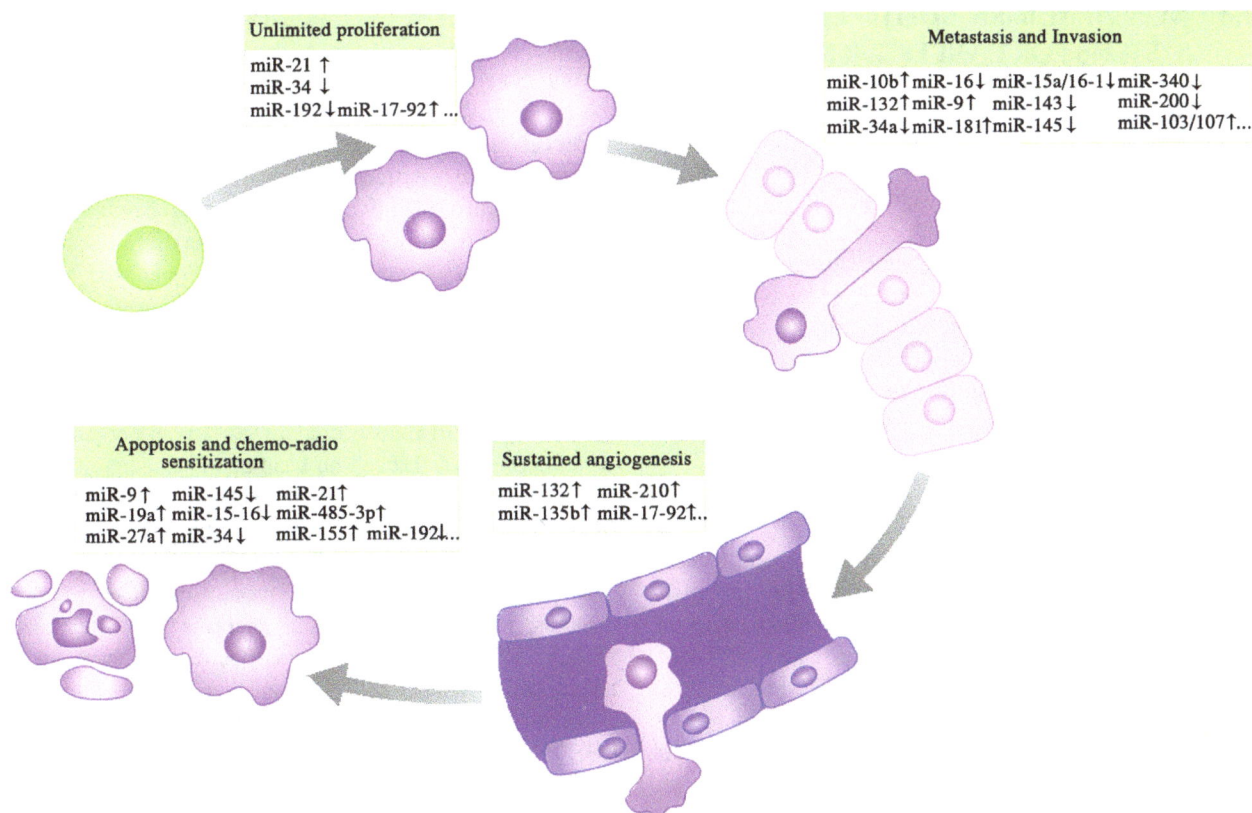

**FIGURE 9.2** miRNAs targeting the hallmarks of cancer at various progression stages including unlimited proliferation, metastasis and invasion, sustained angiogenesis, apoptosis, and drug/radio resistance. The *up arrows* represent oncomiRs and the *down arrows* indicate tumor suppressive miRNAs.

## 9.2.2 CHALLENGES OF miRNA THERAPY

Delivery of miRNAs for cancer therapy has received much attention [34]. Although there have been some remarkable advances in using miRNAs for inhibition and replacement therapies, formidable obstacles remain for their successful delivery. These obstacles include the instability of miRNAs in cellular environments [35], high rates of blood clearance, the risk of systemic toxicity, and an inability to deliver sufficient amounts of miRNAs to the targeted tissues and cells to achieve the desired therapeutic outcome [36, 37]. As a result, high doses are usually required for miRNA-based therapies, which increases the risk for unwanted toxicities [38], and significant effort has been expended in the development of effective nonviral carriers. Ideally, the carriers should be able to condense anionic miRNAs via charge interactions, protect the encapsulated miRNAs from degradation by nucleases, and selectively deliver miRNAs to the target cells inside diseased organs, eventually reaching the cytosolic targets [39–41].

We herein summarize the bio-barriers against the efficient delivery of miRNAs, including the biological and technical obstacles and their potential solutions [42–45], as shown in Table 9.1.

**TABLE 9.1**

**Barriers and Solutions to the Delivery of miRNAs**

| Barriers | Solutions |
|---|---|
| **Degradation and Elimination** | |
| 1. Degradation by nucleases | 1. Particle size |
| 2. Renal clearance | 2. Surface charge |
| 3. Removal by phagocytic immune cells | 3. Improved stability |
| | 4. Chemical modification |
| | 5. Local administration |
| **Poor Penetration through Tissues/Cells** | |
| 1. Failure to cross capillary endothelium in targeted tissues | 1. Targeting ligands |
| | 2. Nonspecific interactions |
| | 3. Cell-penetrating moieties |
| 2. Inefficient endocytosis in targeted cells | |
| **Intracellular Disposition** | |
| 1. Ineffective endosomal release | 1. Lytic lipids |
| 2. Requirement for intracellular localization | 2. Fusogenic peptides |
| | 3. Osmotic lysis |
| | 4. Targeting to RISC |

## 9.3 NONVIRAL miRNA DELIVERY SYSTEMS FOR CANCER THERAPY

miRNAs are highly susceptible to degradation in serum and to elimination by the reticuloendothelial system (RES). Moreover, interactions between miRNAs and the cell membrane are hindered by charge repulsion, leading to poor cellular uptake. miRNA-based treatments must, therefore, rely on vector-mediated delivery. Nonviral delivery systems have significant advantages including high biocompatibility, amenability to surface functionalization, and a loading capacity independent of the gene size (Figure 9.3). Wang et al. list the commonly used nonviral vectors [46].

### 9.3.1 LIPID-BASED NANOCARRIERS

Lipid-based nanocarriers are the most frequently used transfection reagents for delivering siRNAs and miRNAs in vitro. They have the advantages of easy chemical modification with the targeting moieties and the availability of fluorescent tracers [47, 48]. Often, specific lipids are strategically selected with positive charges to formulate the lipid-based vectors. This enhances endocytosis by their interaction with the negatively charged cell membrane. Several lipid-based vectors have been used for miRNA delivery, which are also summarized by Wang et al. [46].

Wu and coworkers used a cationic lipoplex-based vector to deliver miR-29b or miR-133b for the treatment of non–small cell lung cancer (NSCLC). This resulted in the downregulation of the key target proteins CDK6, DNMT3B, and MCL-1 (Figure 9.4) [49, 50]. These researchers also demonstrated that the lipid formulations were more effective in delivering miRNAs compared with the standard NeoFX transfection reagent. Another example of using a lipid-based vector is the local administration of a cationic lipoplex-based system carrying the miR-7-expressing plasmid. This treatment arrested tumor progression in an adenocarcinoma xenograft mouse model [51]. The effectiveness of systemic miRNA therapy on head and neck squamous cell carcinoma using lipoplexes of DDAB:cholesterol:TPGS/pre-miR-107 was also confirmed in a study showing a 45% decrease in the tumor growth rate [52].

Cationic lipids are normally prone to bind with serum proteins, which may lead to an increase in particle size, shorter half-lives in vivo, and nonspecific interactions with target cells. The resultant complexes readily stimulate immune recognition and trigger elimination by the RES [53, 54]. These characteristics of cationic lipid carriers often create a barrier against their systemic application. The potential toxicity of cationic liposomes is also a common

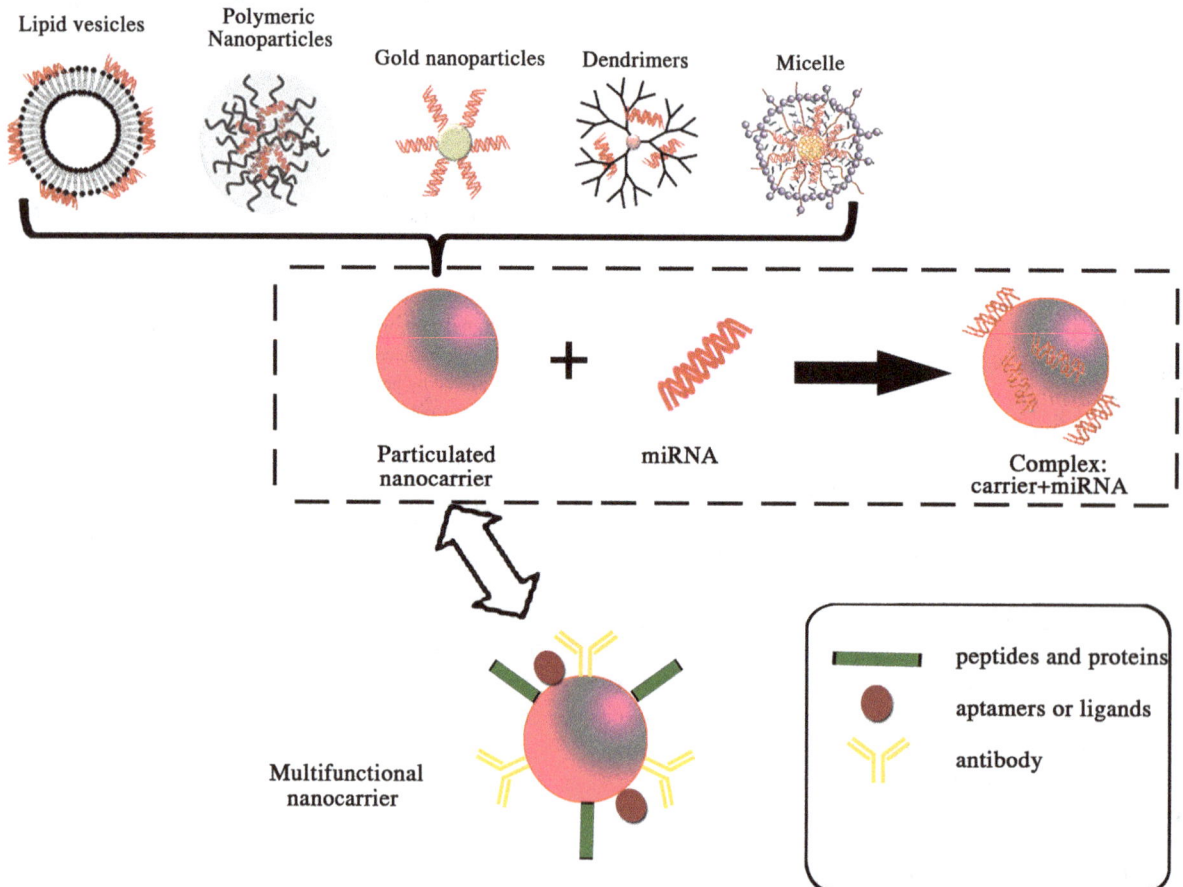

**FIGURE 9.3**  Nonviral delivery systems for miRNAs.

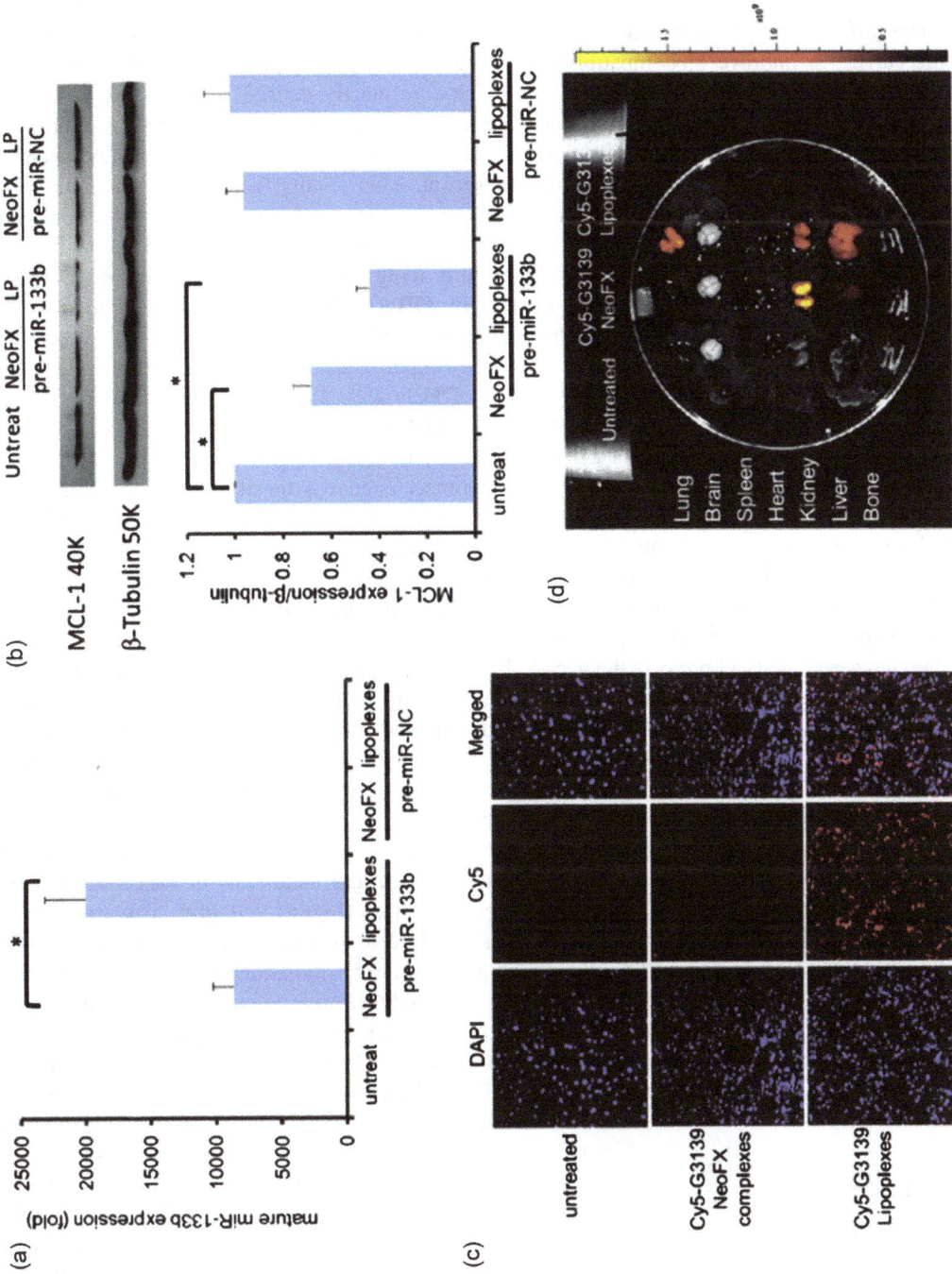

**FIGURE 9.4** Lipoplex-based nanocarriers DOTMA/cholesterol/TPGS. (a) Transfection efficiency of pre-miR-133b via commercial NeoFX and lipoplexes. (b) MCL-1 protein expression after transfection. (c) Lung tissue distribution of complexes after IV injection with imaging by confocal microscopy. (d) Biodistribution of fluorescence-labeled complexes in all tissues. (Reprinted with permission from [50]. Copyright, 2011 by the American Chemical Society.)

concern [55]. Generally, materials with an overall neutral or anionic surface charge are more biocompatible with the body compared with their cationic counterparts. However, neutral or anionic carriers typically contain cationic materials in their core for binding and condensing nucleic acid drugs. For example, the positively charged LMW-PEI is often selected as a core material to condense the negatively charged miR molecules to form complexes through electrostatic interactions. The miR-PEI core is then coated with a neutral or anionic carrier to improve biocompatibility [56]. The commercial MaxSuppressor (Bioo Scientific) is a proprietary structure formed from a neutral lipid that uses this approach. It has been used recently in several studies to inhibit the growth of NSCLC by systemically delivering tumor suppressor miRNAs (let-7 and miR-34a) [57, 58]. In a lung cancer mouse model, such a neutral lipid emulsion is preferentially accumulated in the lung rather than in other organs, and significantly upregulated the expression of let-7 and miR-34a in the lung, thereby decreasing the tumor burden.

PEGylation can not only increase the stability of a lipid carrier, but can also improve its pharmacokinetic profile. Surface modifications with hydrophilic and flexible PEGs can result in stable complexes that prevent degradation and yield a prolonged blood circulation time (e.g., up to 72 h) [59]. One example of this approach is the stable nucleic acid lipid particles consisting of lipids with an outer coating of PEG [60, 61]. The PEGylation neutralizes the positive surface charge and reduces interactions with serum proteins and erythrocytes. This suggests that PEG protects the vectors against opsonization, decreases complement activation, and promotes cargo stability. Furthermore, PEGylated polyplexes with a long circulation time may take advantage of the EPR effect in cancer therapy [62]. Complexes with a prolonged circulation time can extravasate and passively accumulate at tumor sites due to the leakiness of tumor vessels relative to normal vessels, and an ineffective lymphatic efflux.

Another example is the use of lipid nanoparticles that were structured using DOTAP:cholesterol:DSPE-PEG-OMe in a 1:1:0.2 molar ratio. Due to PEGylation-related prolonged circulation times and increased drug accumulation in tumors, the growth of pancreatic tumors was significantly inhibited via miR-34a and miR-143/145 PEGylated nanovectors [60].

### 9.3.2 Polymeric Vectors

#### 9.3.2.1 Polyethylenimine

Polyethyleneimines (PEIs) are a class of cationic synthetic polymers rich in amine groups with linear or branched structures. The protonatable amino groups on PEIs yield a high positive charge density even at neutral pH, rendering strong electrostatic interactions with the negatively charged nucleic acids, thus forming stable complexes [63]. Of note, PEIs are characterized by a unique property called the *proton sponge effect*. Once endocytosis

occurs, the protonated PEI will buffer or resist the acidification of the endosomal microenvironment, subsequently facilitating the release of polyplexes into the cytosol [64]. The successful delivery of nucleic acids by PEIs has been demonstrated in a wide variety of cells as well as a number of animal models. For example, PEI (25 kDa) is one of the most frequently used nonviral vectors and is considered as the gold standard for in vitro gene transfection. Despite the effectiveness of this vector, a major limitation to the clinical application of PEIs is their severe cytotoxicity, primarily caused by the high positive charge density and poor biodegradation [65]. Another challenge is their dramatically decreased efficiency when exposed to serum. This results from the nonspecific binding of PEIs with serum proteins and their consequent aggregation. To address these problems, researchers have focused on improving biocompatibility by using low-molecular-weight PEI (LMW-PEI) that showed reduced toxicity compared with high-molecular-weight PEI (HMW-PEI). For example, Bieber and Elsasser compared cytotoxicities of PEIs with different molecular weights and found that HMW-PEI caused a higher degree of membrane damage and greater cytotoxicity [66]. This may be due to differences in charge density and degradability. However, the smaller degree of branching in LMW-PEI is another possible reason for its low cytotoxicity compared with branched HMW-PEI [67].

Strong anticancer effects were found when LMW-PEI was employed both locally and by systemic injection to deliver miR-145 and miR-33a mimics in colon carcinoma xenograft mice [68]. In that study, miR-145 promoted apoptosis and inhibited tumor proliferation. Furthermore, the systemic delivery of LMW-PEI/miR-33a complexes was as potent as targeting the oncogenic kinase Pim-1. Polyurethane short-branched PEI(PU-PEI)/miR-145 nanoparticles (NPs) have been locally delivered to suppress tumor growth and metastasis in cancer stem cell (CSC) xenograft mice (Figure 9.5) [69]. This xenograft model was highly drug-resistant and did not respond to the co-administration of radiation and cisplatin. Nevertheless, PU-PEI mediated miR-145 expression significantly inhibited the CSC-like properties and prolonged the survival time of these mice. The mechanism underlying the effectiveness of PU-PEI/miR-145 NPs could be related to the repression of downstream targets including Oct4, Sox2, and Fascin1. The promising therapeutic effectiveness of PU-PEI/miR-145 NPs was further demonstrated in glioblastoma CSC-induced tumors [70]. The results of these studies suggested that the NPs could serve as a potential avenue to improve the treatment of chemo- and radio-resistant tumors.

#### 9.3.2.2 Atelocollagen

Naturally sourced cationic proteins or peptides are commonly utilized as drug carriers because they exhibit low cytotoxicity and immunogenicity, and are highly biodegradable. Among these proteins, atelocollagen has been increasingly used to deliver nucleic acids. Atelocollagen is

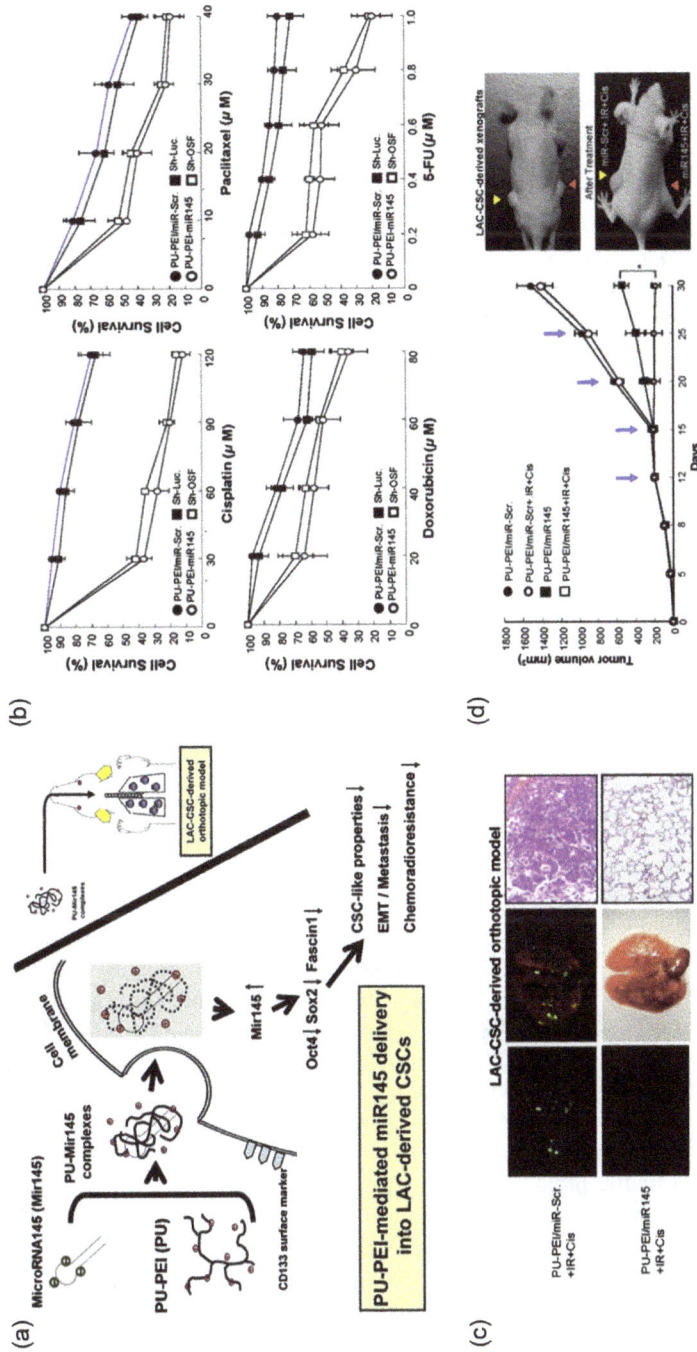

**FIGURE 9.5**  PU-PEI-based delivery system. (a) Construction of the PU-PEI delivery system. (b) Transfection with miR-145 retained the resistance to cisplatin, paclitaxel, DOX, and 5-FU. (c) Tumor nodules in lungs visible via a GFP image after treatment with PU-PEI/miR-145 complexes. (d) The tumor growth curve after treatment. (Reprinted with permission from [69]. Copyright, 2012 by Elsevier.)

prepared from pepsin-treated type I collagen. This treatment removes the immunogenic amino and carboxyl terminal peptides. Atelocollagen (0.5%) exists as a liquid at 4°C but turns into a gel at 37°C. The thermosensitive nature of atelocollagen assists in the prolonged release of nucleic acids and allows for intratumoral injection [71]. Atelocollagen can condense with nucleic acids through electrostatic interactions. The thus-formed nanocomplexes facilitate cellular uptake and perform sustained delivery.

Atelocollagen-mediated DNA or siRNA delivery is effective for gene therapy [72, 73]. As a result, its use for miRNA cancer therapy has been explored recently. miR-34a is a tumor suppressor with an ability to induce senescence, and is involved in the development of colon cancer. Importantly, a significant inhibition of tumor growth was found after intratumoral injection of miR-34a mimics/atelocollagen NPs in human colon tumor xenograft mice [74]. Using the same method, peritumoral injection of LNA miR-135b inhibitor/atelocollagen complexes in human lymphoma xenograft mice suppressed tumor growth [75]. In another study, miR-516a-3p/atelocollagen complexes were used to treat human gastric cancer. The results demonstrated a potential for using such complexes to suppress the metastasis of gastric tumors [76]. Interestingly, when atelocollagen was applied simultaneously with the systemic administration of miR-16 mimics in prostate tumor xenograft mice with potential bone tissue metastasis, the results showed suppression of tumor growth and metastasis without apparent toxicity [77]. These studies indicated the therapeutic potential of miRNA/atelocollagen complexes to treat cancer.

### 9.3.2.3 Poly (Lactide-Co-Glycolide)

Typically, poly(lactide-co-glycolide) (PLGA)-based carriers have excellent biocompatibility and biodegradability, and are widely used to deliver various therapeutic genes. Importantly, the delivery of nucleic acids with PLGA-based vectors permitted long-term gene transfection; e.g., greater than or equal to 4 weeks [78].

Oligonucleotide analogs of cell-penetrating peptide-modified PLGA NPs were efficiently transported into cancer cells and achieved miRNA suppression or alternative splicing. Systemic delivery of the NPs could knock down the function of oncogenic miRNAs as well as modulate splicing to downregulate the level of the proto-oncogene MCL-1. This effect slowed the progression of pre-B-cell lymphoma cancers in animal models [79].

Although PLGA is a US Food and Drug Administration (FDA)-approved biodegradable and biocompatible material used for drug delivery, there are still some disadvantages associated with the use of PLGA nanoparticles for nucleic acid therapeutics. These disadvantages include their hydrophobic nature and nonspecific interactions with target cells. Furthermore, the encapsulation techniques (e.g., spray-drying and modified double emulsion methods) enhance nucleic acid degradation due to the use of high-speed homogenization or sonochemical forces [80]. In addition, nucleic acids

can be rapidly degraded and damaged in the acidic and moist microenvironment that results from the hydrolysis of PLGA [81]. The application of physical or chemical surface modifications is a potential means to solve these problems. For example, cationic amine-modified branched polyesters (e.g., PVA-PLGA, DEAPA-PLGA, DMAPA, and other similar derivatives) usually stabilize nucleic acids within the polymer matrix [82].

### 9.3.3 Dendrimer-Based Vectors (Polyamidoamine Dendrimers)

Synthetic poly(amidoamine) dendrimers (PAMAMs) with positive charges can readily form nanocomplexes with negatively charged miRNAs through electrostatic attractions, and offer high transfection efficiency attributable to their "endosomal escape" capacity [40, 83]. Combining miRNAs with chemotherapy is often used to combat cancer. For example, a co-delivery system of the as-miR-21 oligonucleotide and 5-FU was prepared using G5-PAMAM. The nanocomplexes formed were smaller than 100 nm and simultaneously transferred the antisense miR-21 and 5-FU into human glioblastoma cells [84, 85]. This co-treatment enhanced the therapeutic effect [84].

When human glioblastoma cell lines containing wild-type or mutant PTEN were treated with a PAMAM dendrimer co-delivery system containing taxol and/or anti-sense-miR-21 [86], the anti-miR-21 reversed the resistance to taxol. Co-delivery of anti-miR-21 and taxol was an effective therapeutic approach against the growth of glioblastoma multiforme by suppressing the expression of STAT3 and p-STAT3.

Unfortunately, there is a major drawback to using PAMAM dendrimers as miRNA carriers: generation-dependent and surface charge–dependent toxicity. The toxicity of PAMAM increases with the generation number [87]. Nearly 90% of injected unmodified PAMAM is accumulated in the liver and may induce damage and the loss of glucose metabolism [88]. This toxicity can be reduced by masking the surface charge on PAMAM. For example, the hemolytic effect was reduced when using PEGylated dendrimers [89], and PAMAM with conjugated lauroyl chains exhibited reduced cytotoxicity [90].

### 9.3.4 Amphiphilic Star-Branched Copolymers

The amphiphilic star copolymer, which has a globular three-dimensional or branched structure, contains multiple arms connected to a central core and can serve as an effective carrier for nucleic acids because of its unique structure and physical characteristics. The molecular weight of star polymers is much higher (typically N300 kDa) compared with linear polymers, but its solubility and viscosity are similar [91].

Three amphiphilic star-branched copolymers were synthesized comprised of polylactic acid (PLA) as the hydrophobic branch and poly-dimethylaminoethyl methacrylate

(PDMAEMA) as the hydrophilic core. The molecular architectures of the three copolymers were PLA-PDMAEMA3, (PLA-PDMAEMA3)2, and (PLA-PDMAEMA3)3 [92]. The copolymers possessed a low critical micelle concentration and could form nano-sized micelles with positive surface charges. More importantly, their cytotoxicity was noticeably lower than PEI 25 kDa. The tumor suppressive effects of co-delivering doxorubicin (DOX) and a miR-21i inhibitor (miR-21i) using these polymers were investigated in LN229 glioma cells in vitro as well as in a subcutaneous tumor nude mouse model using colony formation and flow cytometry assays. The micelle-based co-delivery systems significantly decreased the expression of miR-21 and exhibited a greater antiproliferative efficiency compared with DOX or miR-21i treatment alone. The transfection efficiency of the micelle systems showed a remarkable dependency on their molecular architecture with the (PLA-PDMAEMA3)3 copolymer micelle displaying the highest transfection efficiency and tumor inhibition ability (Figure 9.6). These results demonstrate that copolymer micelles are promising vectors for cancer therapy.

### 9.3.5 INORGANIC MATERIALS

Inorganic materials, including gold NPs, nano-diamond [93], silica [94], and $Fe_3O_4$-based NPs [95], have wide applications in nanotechnologies and have been employed as miRNA vectors. For example, gold NPs have been used as nucleic acid carriers due to their easy surface functionalization such as adding thiol and amino groups [96].

miR-130b, a reported miRNA, was connected to gold NPs (miRNA-AuNPs) and the delivery efficiency was investigated [97]. Cy5-labeled miRNA-AuNPs efficiently entered multiple myeloma cells and downregulated gene expression. Ghosh and coworkers designed a different approach for transporting unmodified miRNAs into cancer cells by using multifunctional AuNPs that were modified with thiolpolyethylene glycol post miRNA packaging [98]. The miR1-AuNP10-SPEG0.5 had a high loading capacity, low toxicity, increased half-life, and enhanced miRNA transfection ability relative to other methods like AuNP polyelectrolyte complexes and lipofectamine. In addition, another type of functional AuNP with RNAi (cargo DNA) and an anti-miRNA antisense oligodeoxyribonucleotide was used to transport anti-miR-29b into HeLa cells. After treatment, the increase in MCL-1 protein levels was remarkable and the apoptosis induced by TRAIL (tumor necrosis factor-related apoptosis-inducing ligand) was suppressed. This resulted in prolonging cell survival and demonstrated the efficient intracellular delivery of anti-miR-29b by AuNPs [99].

### 9.4 TARGETED DELIVERY OF miRNA

miRNAs can be administered intratumorally in order to increase the local drug concentration, avoid nonspecific biodistribution, and mitigate the potential side effects that are often seen with systemic administration [57, 100, 101]. Thus, local injection of miRNA agents is primarily useful for topical tumors although a few studies have shown it could be applicable for interventional therapy with internal organ tumors. Targeting the delivery of miRNAs to tumors via systemic administration remains a challenge [102, 103]. However, a number of passive and active targeted delivery systems have been developed to transport miRNAs to internal tumors.

### 9.4.1 PASSIVE TARGETED DELIVERY

Passive targeted delivery relies mainly on the enhanced permeability and retention (EPR) effect in solid tumors [104–106]. miRNAs and other chemically modified oligonucleotides diffuse into the tumor mass in a concentration-driven manner, whereas nano-sized carriers can be entrapped and accumulate inside tumors due to their leaky vasculature and decreased lymphatic drainage. In the case of lymphoid tumors, the size-dependent effect of NPs displayed a more significant role in drug accumulation than did the active targeting effect [107].

Recently, Babar and coworkers designed an intricate intracorporeal vector for the delivery of anti-miR-155 to inhibit the expression of miR-155 in an animal model bearing a pre-B-cell tumor with a high miR-155 level [108]. Systemic application of anti-miR-155 involved using safe, biodegradable PLGA NPs containing a cell-penetrating peptide (penetratin) on the surface for efficacious delivery. The penetratin was connected to the NPs via a PEG spacer that improved passive target transport because the PEGylated NPs had a prolonged half-life of up to 5 days in the blood. The progression of a hypervascularized pre-B-cell tumor was inhibited by this treatment, providing a potential therapeutic approach for lymphoma/leukemia.

Solid lipid nanoparticles (SLNs) are an emerging approach for the passive delivery of miRNA. Dimethyldioctadecylammonium bromide (DDAB)-containing SLNs with an average size of 200 nm can be used for the systemic delivery of miR-34a via IV injection [109]. Fluorescence-labeling revealed a high accumulation of SLNs in the lung, avoiding unwanted liver capture by the RES. This is likely due to the PEG-like structure of DDAB covering the surface and the consequent EPR effect. When IV administered to B16F10-CD44+-bearing mice, CSC-like lung tumors were inhibited by the SLNs through the elevation of miR-34a concentrations in the lung [109].

### 9.4.2 ACTIVE TARGETED DELIVERY

miRNAs can target many different mRNAs. As a result, unwanted and potentially toxic off-target effects often occur. When administered systemically, it is important to specifically deliver miRNAs into the desired tissues or cells, thus reducing unwanted exposure to normal organs. Such targeting can be achieved by modifying the miRNA vectors with different moieties such as small molecules, peptides, antibodies,

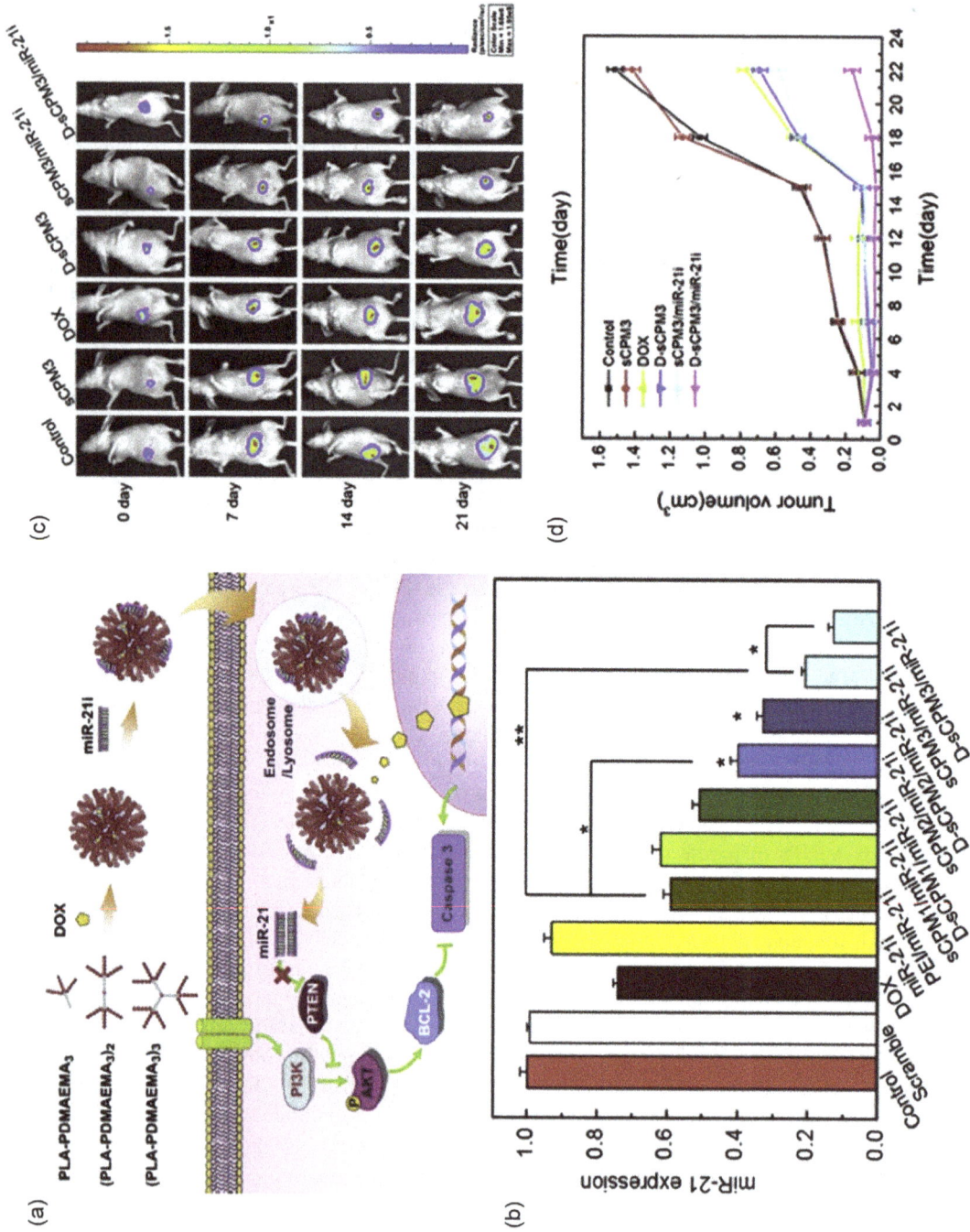

**FIGURE 9.6**   Star-branched amphiphilic copolymer micelles-based delivery system. (a) The mechanism for co-delivery of miR21i and DOX. (b) miR-21 expression after treatment with miR-21i micelles. (c) Representative images in mice. (d) Tumor growth curves after treatment. (Reprinted with permission from [92]. Copyright, 2014 by Elsevier.)

ligands, or aptamers that direct the miRNA to receptors specifically overexpressed in tumors. Active targeting will thus lower the miRNA dosage required for effective treatment. Because the solid tumor microenvironment is highly heterogeneous and the passive targeting of nanomedicine via EPR effect may not be sufficiently effective [110], many researchers have focused on multifunctional active targeting miRNA vectors (Table 9.2).

### 9.4.2.1 Peptides or Protein Ligands

The targeting efficacy of vectors can be enhanced by modification with functional peptides (e.g., specific tumor-homing peptides or integrin-binding peptides) or proteins. For example, $\alpha v\beta 3$ integrin is a crucial mediator of tumor progression and tumor metastasis. The integrin-targeting tripeptide, Arg-Gly-Asp (RGD), has been used to increase the specific delivery of chemotherapeutics and nucleic acids to the tumor neovasculature and cancer cells [123, 124]. Modified by the $\alpha v\beta 3$-binding peptide, cyclic RGD (cRGD), anti-miR-132-loaded liposomes have been effectively targeted to the tumor neovasculature with internalization of anti-miR-132 [48]. miR-132 is upregulated in neovascular endothelial cells and tumor cells with no measurable change in normal endothelial cells. miR-132 acts as an angiogenic switch by targeting endothelial p120RasGAP that is associated with vasoformation. After treatment, the anti-miR-132-loaded RGD-liposomal system exhibited a half-life of 2 days and, by the antagonism of miR-132, restored p120RasGAP expression, decreased Ras activity, and arrested tumor growth in human breast cancer xenograft animal models.

Liu and coworkers developed PEG-modified liposome-polycation hyaluronic acid (LPH) nanoparticles modified with cRGD (cRGD-LPHNP) for the targeted delivery of anti-miR-296 oligodeoxynucleotides to human umbilical vein endothelial cells (HUVECs) [111]. miR-296 belongs to a family of angio-miRs and promotes angiogenesis by reducing the hepatocyte growth factor-regulated tyrosine kinase substrate, and inducing the growth factor receptor VEGFR2 or platelet-derived growth factor receptor

## TABLE 9.2
## Active Targeted Moieties for miRNA-Based Cancer Therapy

| Targeted Moiety | Delivery System | Target miRNA | Administration | Therapeutic Approaches | Cancer | Ref. |
|---|---|---|---|---|---|---|
| **Peptides or Proteins** | | | | | | |
| Tumor-targeting peptide (cRGD) | Liposomes | miR-132 | Systemic, subcutaneous | Inhibition | Breast and pancreatic cancer | [48] |
| | PEGylated liposomepolycation-hyaluronic acid | miR-296 | Subcutaneous | Inhibition | Antiangiogenesis | [111] |
| Tumor-targeting peptide, ligand of EGFR, GE11 (YHWYGYTPQNVI) | Exosome | miR-let-7a | Systemic | Replacement | Breast cancer | [112] |
| Cell penetrating peptide DS 4-3 (CRIMRILRILKLAR) | b-PEI | miR-145 | In vitro | Replacement | Breast cancer | [113] |
| Specific tumor-homing and penetrating bifunctional peptide (CC9) | CD-PEI | miR-34a | Systemic | Replacement | Pancreas cancer | [114] |
| Peptide derived from the rabies virus glycoprotein (RVG) | SS-PEI | miR-124a | Systemic | Replacement | Neuronal | [115] |
| Transferrin | Lipopolyplex | miR-29b | Systemic, peritoneal | Replacement | Acute myeloid leukemia | [56] |
| **Antibodies** | | | | | | |
| scFv antibody | Liposome-polycation hyaluronic acid | miR-34a | Systemic | Replacement | Murine B16F10 melanoma | [116] |
| GD2 ch14.18 antibody | Silica nanoparticles | miR-34a | Systemic | Replacement | Neuroblastoma Tumor | [117] |
| **Aptamers** | | | | | | |
| A10-3,2 | PEGylated PAMAM | miR-15a/16-1 | In vitro | Replacement | Prostate cancer | [118] |
| AS1411 aptamer | Magnetic fluorescence (MF) nanoparticle | miR-221 | Intratumoral | Inhibition | Astrocytoma | [119] |
| **Other Molecule Ligands** | | | | | | |
| Hyaluronic acid | Chitosan NPs | miR-34a | Systemic | Replacement | Breast cancer | [120] |
| Folate acid | pRNA | AmiR-1 and AmiR-2 | In vitro | Replacement | HeLa cells | [121] |
| Magnetic nanoparticles | RGD-MNPs | miR-10b | Systemic | Inhibition | Breast cancer | [122] |

beta. cRGD-LPH-NP containing anti-miR-296 efficiently inhibited angiogenesis and endothelial cell migration. This approach caused a significant suppression of cell proliferation and microvessel formation by Matrigel plugs mixed with anti-miR-296 in LPH-NP or cRGD-LPH-NP via hypodermic injection in animal models whereas the NPs without cRGD showed poor cellular uptake and little miRNA silencing (Figure 9.7).

miRNA-related approaches are promising for the effective treatment of various tumors. However, targeted delivery in vivo and drug penetration into tumors and cells remain significant challenges. Tissue-specific peptides have been developed that can bind with the receptors that are overexpressed on some cells. For example, when the β-cyclodextrin-PEI (PEI-CD) gene carrier was conjugated with CC9 (CRGDKGPDC), a tumor-targeting and tumor-penetrating functional peptide was formed via its CRGDK motif that can bind to neuropilin-1 (NRP-1) [125]. When PEI-CD-CC9 was used for the delivery of a tumor suppressor miR-34a mimic, an effective therapy of pancreatic tumors was achieved. This delivery system significantly increased the miR-34a level in the PANC-1 cell line, induced cell cycle arrest and apoptosis, substantially inhibited the expression of related genes (e.g., E2F3, Bcl-2, c-myc, and cyclin D1), and suppressed cell migration. Importantly, the systemic administration of this gene carrier in an animal model of pancreatic cancer resulted in the successful delivery of miR-34, yielding significant inhibition of tumor growth (Figure 9.8) [114].

Another functional peptide, DS 4-3, was screened by a phage display technique and showed specificity toward metastatic breast cancer cells. The DS 4-3 peptide was conjugated to a polycationic bPEI miRNA carrier (DS-bPEI) for delivery of the tumor suppressor miR-145 into cells [113]. miR-145 significantly inhibited tumor cell growth and suppressed cell invasion. This DS-bPEI delivery system showed increased transfection in malignant murine breast cancer cells without causing any cytotoxicity. However, the increased transfection was not found in other non-breast cancer cells or normal cells, such as HeLa cervical cancer cells or NIH-3T3 murine fibroblast cells. These findings indicated that DS-bPEI/miR-145 NPs could be a potent delivery system targeting metastatic breast cancer.

The GE11 peptide (YHWYGYTPQNVI) can specifically bind to epidermal growth factor receptors (EGFR) that are overexpressed on many human tumor cells. The GE11 peptide thus can be used for active targeted delivery of miRNA for cancer treatment. GE11-conjugated NPs inhibited cancer cells or tumor xenografts that overexpressed EGFR [126]. In a representative research study, exosomes were modified with GE11 to prepare miRNA-loaded NPs. There was a greater uptake efficiency of the GE11-exosomes in breast cancer cells with higher levels of EGFR expression. After IV injection, GE11-exosomes showed a greater accumulation in tumor tissue than normal tissues, suggesting that they were selectively delivered

to EGFR-expressing breast cancer tissues in RAG2−/− mice [112].

Development of an effective nanovector that can transport therapeutic miRNAs for the treatment of brain diseases is extremely difficult because of the blood brain barrier (BBB). Recent research suggests that PEI-based delivery systems of miRNA using a small peptide derived from the rabies virus glycoprotein (RVG), with specific binding affinity to neuronal cells expressing the acetylcholine receptor, are effective in crossing the BBB [127, 128]. The RVG peptide was used with polyarginine to prepare miRNA nanocomplexes via electrostatic attraction [129]. In addition, RVG-modified noncytotoxic SS-PEI vectors were used to deliver miR-124a to neurons in vivo. As the major problem against clinical application of PEI is its severe cytotoxicity, primarily caused by the high positive charge density and poor biodegradation, studies have focused on improving the biocompatibility by using reducible linkages, e.g., disulfide (−S−S−) linkage. SS-PEI showed reduced toxicity compared with HMW-PEI, which might account for the degradability by endogenous enzymes such as glutathione reductase. As shown in Figure 9.9(a), the RVG-modified SS-PEI was synthesized by using an NHS-PEG-MAL linker. It is generally accepted that neuronal development is associated with a gradual increase in miR-124a expression levels. miR-124a accumulated in the brain more significantly when using the RVG-mediated SS-PEI system, compared to the vector without RVG (Figure 9.9) [127]. Together, these data suggest that RVG-SS-PEI is a potential delivery carrier of miRNA for the treatment of brain diseases. Other than using the RVG peptide, Kabanov and coworkers modified hydrophilic proteins with Pluronic P85 to penetrate the BBB [130, 131]. Modification of nanoparticles with Pluronic may be a novel approach to improve therapeutic miRNA delivery to the brain.

Transferrin (TF) is an iron transport glycoprotein that binds to transferring receptors (TfR) that are overexpressed on various cancer cells and at the BBB. TF-modified NPs are effective for active targeted delivery of nucleic acids to tumors and the BBB in vivo [132–134]. High levels of TfR expression are often found on cancer cells in patients with acute myeloid leukemia (AML). Thus, anionic lipopolyplex NPs with transferring (Tf-NP) were used to target miRNA-29b to AML cells. Tf-NP-miR-29b was twice as efficient as using non-Tf-modified NPs, and inhibited cancer cell growth by approximately 50% in vitro. DNA methylation-related genes (e.g., DNMTs, CDK6, and SP1) were significantly knocked down by Tf-NP-miR-29b. In an AML-bearing mouse model, Tf-NP-miR-29b prolonged the survival time. Finally, priming leukemia cells with Tf-NP-miR-29b resulted in an improved anticancer activity of decitabine [56].

### 9.4.2.2 Antibodies

Antibodies recognize their related cell surface receptors with high affinity and specificity. Antibody-based strategies for binding to specific receptors on tumor cells have been

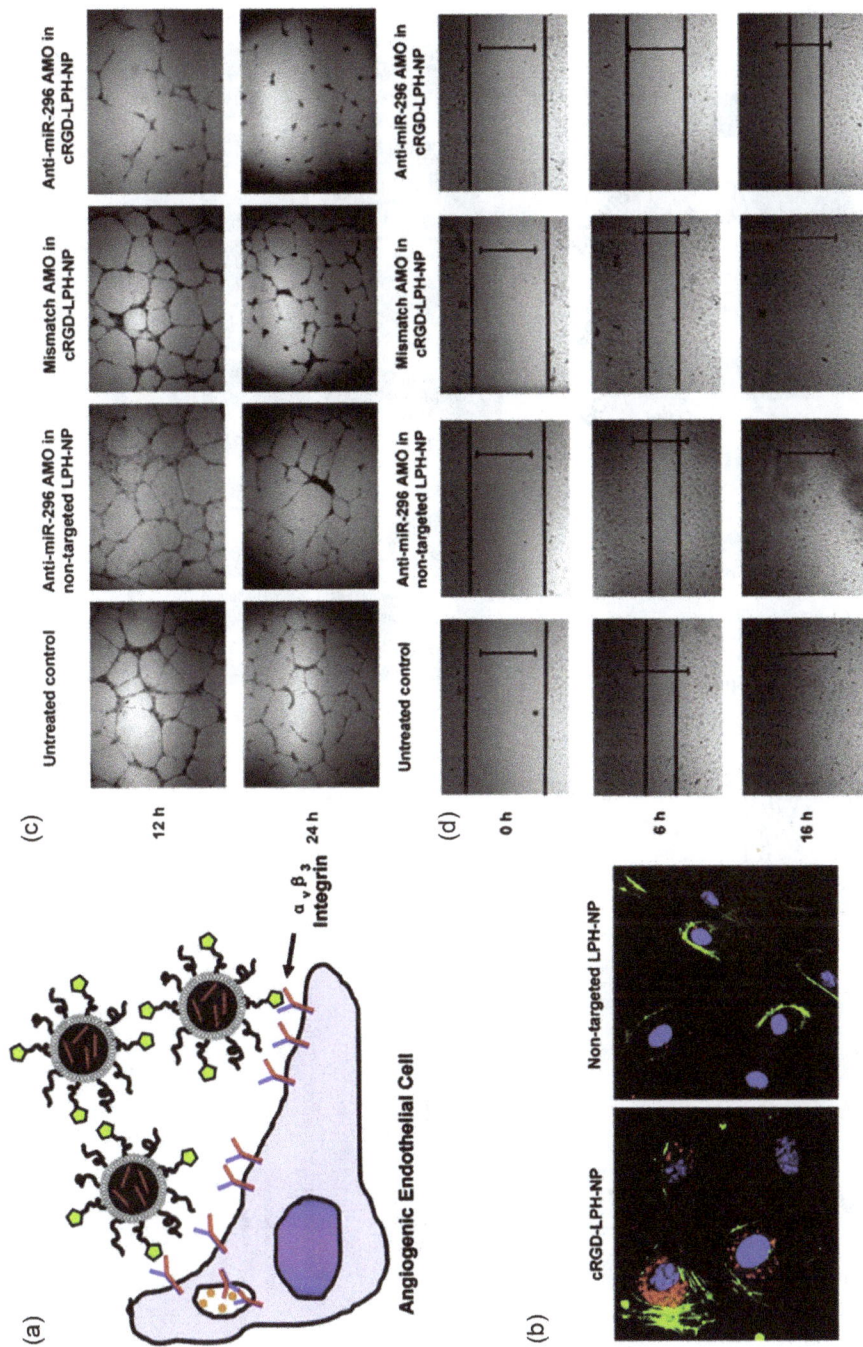

**FIGURE 9.7** cRGD-LPH-NP for miRNA delivery. (a) Illustration of a cyclic RGD-modified delivery system. (b) Cellular uptake of Cy3-labeled AMO via cRGD-LPH-NP. (c) Inhibition of tubule formation by the delivery of miR296. (d) Inhibition of migration by the delivery of miR296. (Reprinted with permission from [111]. Copyright, 2011 by the American Chemical Society.)

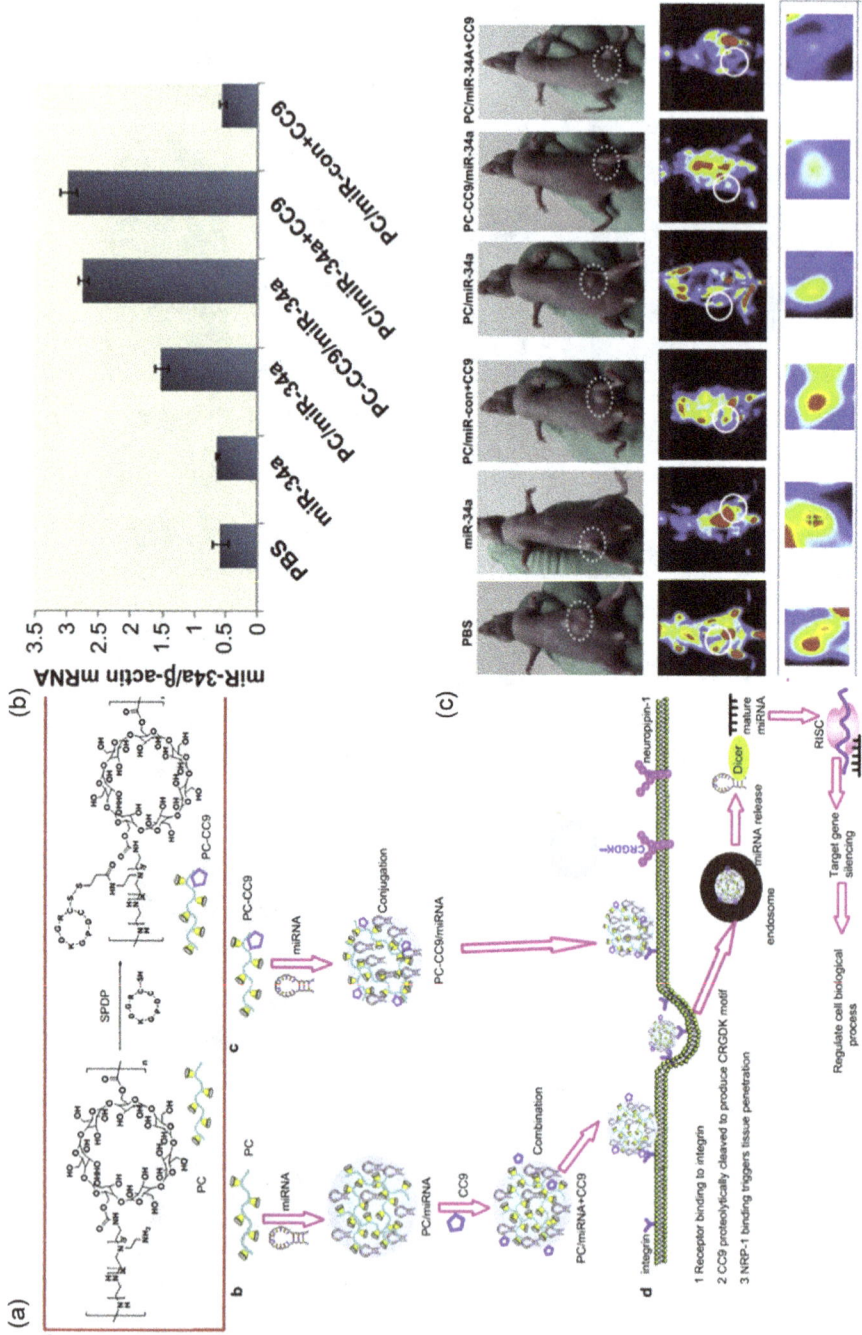

**FIGURE 9.8** The targeted peptide, modified bPEI, as a vector for miRNA delivery. (a) Illustration of targeted CC9-bPEI-mediated miRNA delivery. (b) miR-34a expression after treatment with PCCC9/miR-34a. (c) Representative images of tumor growth in mice after different treatments. (Reprinted with permission from [114]. Copyright, 2013 by Elsevier.)

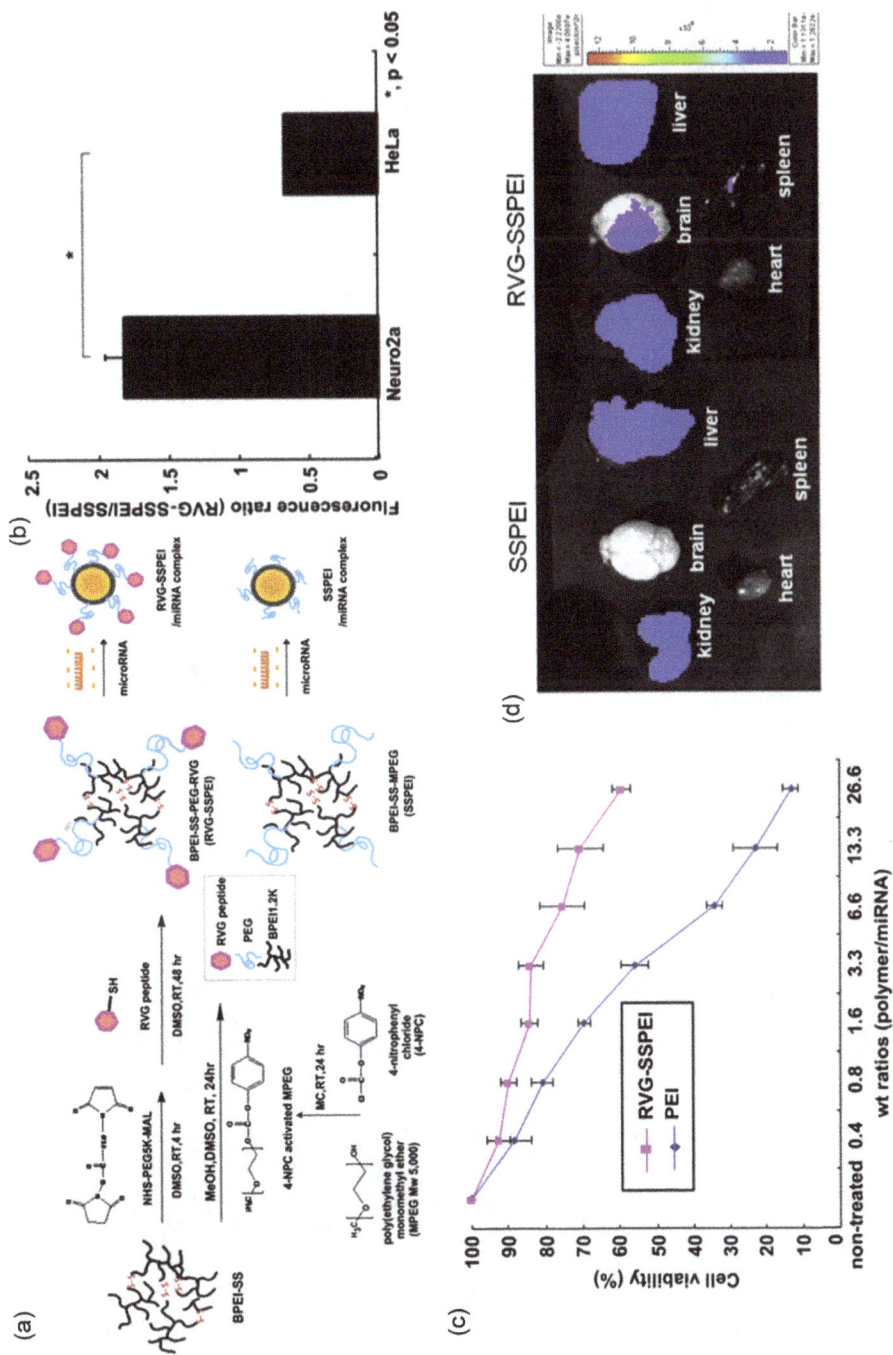

**FIGURE 9.9** RVG-modified SS-PEI vectors used for neuron-targeted miR-124a transport. (a) Illustration of RVG-SS-PEI synthesis. (b) Targeted cellular uptake of RVG-SS-PEI in receptor+ neuron cells. (c) Cytotoxicity of RVG-SS-PEI and PEI. (d) Biodistribution of cy5.5-labeled nanocomplexes in the major organs after IV injection. (Reprinted with permission from [127]. Copyright, 2011 by Elsevier.)

widely applied in the design of targeted delivery systems. For example, GD$_2$ and scFv are commonly used antibodies for active targeted miRNA delivery.

A scFv (ErbB2 single-chain variable fragment) antibody and protamine have been fused to produce a siRNA delivery system targeted to ErbB2-positive tumor cells [135]. Later, scFv-modified LPH (liposomepolycation-hyaluronic acid) NPs were developed as a carrier of miRNA to attack B16F10 lung metastasis in a murine model [116]. miR-34a was first packaged with protamine and then the electronegative cores were encapsulated in the cationic liposomes. To test their therapeutic utility, the NPs were injected IV into metastasis-bearing B16F10 mice. The NPs effectively delivered miR-34a, significantly inhibited surviving expression, downregulated the MAPK pathway, and induced apoptosis in the B16F10 cells, and consequently reduced the lung metastasis in the B16F10-bearing mice without noticeable side effects (Figure 9.10).

Tivnan and coworkers designed porous silica NPs with a disialoganglioside GD$_2$ ch14.18-antibody surface modification for the delivery of proapoptotic miR-34a into neuroblastoma cancers [117]. Neuroblastoma is a childhood cancer occurring mostly in those under the age of 5 years. Silica-based NPs were systemically administered to a well-characterized orthotopic xenograft mouse model with GD$_2$-overexpressing tumors. The tumor-specific delivery of miR-34a successfully inhibited the progression of the neuroblastoma tumor, increased apoptosis, and decreased vascularization. More specifically, the anti-GD$_2$/miR-34a NPs selectively induced an increase in miR-34a in the neuroblastoma cells but did not affect the negative control HEK 293 cell line. Furthermore, the existence of miR-34a in neuroblastoma cells decreased cell viability via a mechanism involving the activation of caspase-3 and caspase-7. The anti-GD$_2$-modified NPs containing miR-34a were further evaluated for their effect on NB1691$^{luc}$ and SKN-AS$^{luc}$ neuroblastoma tumors in vivo. There was a strong inhibition of tumor progression when anti-GD$_2$-miR-34a-NPs were administered systemically at various days post tumor transplantation. It was noteworthy that the systemic delivery of anti-GD2-miR-34a-NPs had no adverse effects on kidney or liver tissue.

### 9.4.2.3 Aptamers

Aptamers can fold into unique secondary or tertiary structures, and have been used successfully to deliver miRNA owing to their favorable characteristics including small size, high affinity and specificity with specific protein or receptors, low immunogenicity, and ease of synthesis.

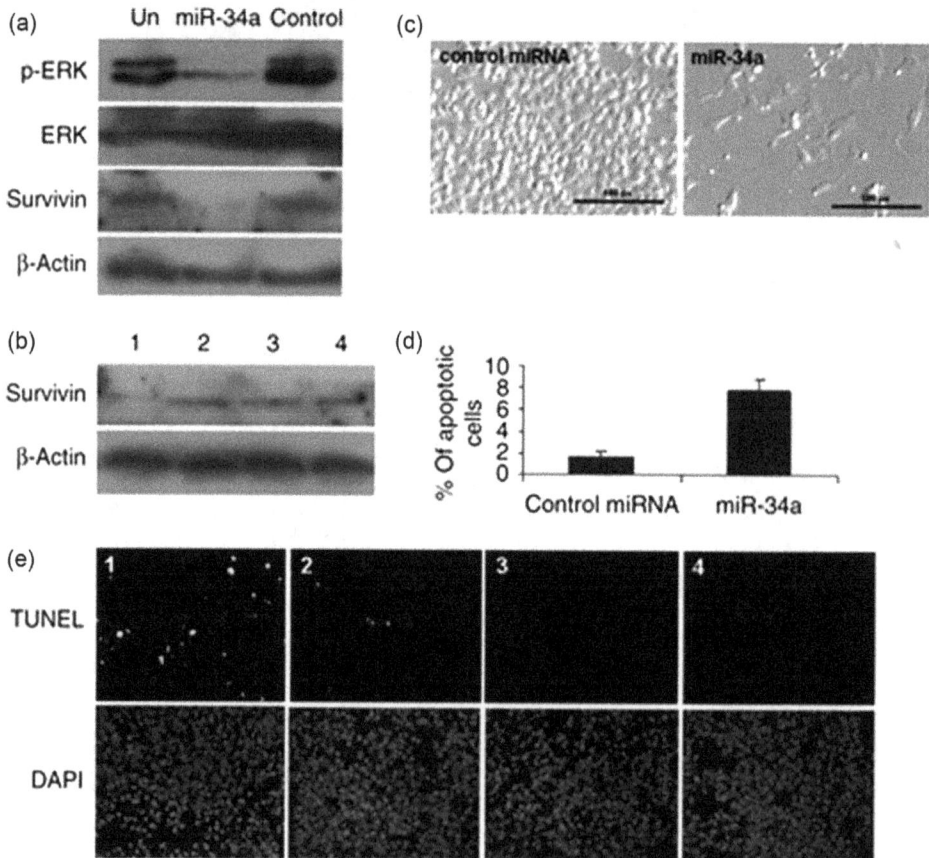

**FIGURE 9.10** scFv-modified LPH NPs. (a) Survivin protein expression after treatment with miR-34a. (b) Survivin protein expression in different formulations of miR-34a. (c) Cellular morphology after transfection of miR-34a. (d) The percentage of apoptotic cells after transfection with miR-34a. (e) TUNEL assay of B16F10 metastases after IV injection of miR-34a. (Reprinted with permission from [116]. Copyright, 2010 by the Nature Publishing Group.)

Wu and coworkers designed a PAMAM-aptamer nanocarrier for the delivery of miR-15a and miR-16-1 into prostate carcinoma cells [118]. PAMAM was conjugated to an enhanced second-generation aptamer (A10-3.2) for targeting to prostate tumor cells expressing the prostate-specific membrane antigen (PSMA). This polymer, which is rich in primary amino groups, potently encapsulated miR-15a and miR-16-1, and downregulated PSMA expression in prostate tumor cells. In addition, the PSMA-targeted system demonstrated effectiveness and selective cytotoxicity in prostate cancer.

Kim and coworkers designed a novel tumor-targeting vector containing the AS1411 aptamer and miR-221 beacon-conjugated magnetic fluorescence NPs to diagnose and treat tumors (Figure 9.11) [119]. The AS1411 aptamer can target the nucleolin protein, which is overexpressed on the cyto-membrane of cancer cells. The miR-221 molecular beacon (MB) was used to diagnose and suppress unregulated miR-221 in cancer cells. The miR-221 MB could detach from the theragnostic probe in cancer cells and specifically trace miR-221. Moreover, the NPs possessed anticancer activity by suppressing miRNA-221, which is overexpressed in papillary thyroid carcinoma, thus decreasing tumor invasion and side effects. The NPs displayed treatment selectivity to several other cell lines by using the respective, targeted miRNAs.

### 9.4.2.4 Other Ligands

Various ligands have been employed in the delivery of active targeted miRNA; for example, hyaluronic acid (HA) [120] and folate acid (FA) [121]. Chitosan NPs modified with hyaluronic acid (HA-CS NPs) are able to bind to the cancer cell biomarker CD44 (an integral membrane glycoprotein) via interactions between HA and CD44 [136]. Therefore, HA-CS NPs were used as a co-delivery vector of negatively charged miR-34a and positively charged DOX for targeting to breast cancer cells. The coating of HA on the chitosan NPs facilitated cellular uptake via interaction with CD44. Up to 80% of breast cancer cells were positive in a cellular uptake test. Furthermore, the intracellular accumulation of DOX and miR-34a was significantly decreased if the cells were pretreated with an excess of HA in culture media due to saturation of the binding to CD44. Cell migration was inhibited by the intracellular delivery of miR-34a with a decline in Notch-1 expression. In vivo studies further demonstrated that using HA-CS NPs to co-deliver miR-34a and DOX effectively inhibited tumor growth in an animal model of breast cancer (Figure 9.12). Co-delivery with miRNAs enhanced the antitumor effects of DOX by decreasing the expression of non-pump resistance proteins and the antiapoptosis proto-oncogene Bcl-2.

Folate receptors (FRs) are highly expressed on many cancer cells. FA-modified NPs have been used to deliver drugs to cancer cells that overexpress FR. AmiR-1 and AmiR-2 reduced coxsackievirus B3 replication nearly 100-fold [121]. To achieve targeted delivery, AmiRs were modified with the FA-conjugated bacteriophage Phi29 pRNA (FA-pRNA-AmiRs). These conjugates were able to target FR-positive HeLa cells via specific interactions with FRs.

### 9.4.2.5 Magnetic Nanoparticles

Magnetic-based nanoparticles (MNPs) can be targeted to tumor tissue if directed by an external magnetic field. MNPs have been successfully used for the effective and rapid delivery of nucleic acids [137–139]. MNP-based miRNA delivery is advantageous in vivo because of its selectivity and biocompatibility. For example, miR/PEI/MNPs could efficiently deliver miRNA into human mesenchymal stem cells with an external magnetic field. Over 60% of the cells were positive in tests for high miR uptake without any obvious cytotoxicity [140]. In addition, miR/PEI/MNPs provided a long-acting effect that could be beneficial for achieving successful transfection. This MNP system was further developed for cancer theragnostics [122]. The system (MNP-anti-miR-10b) was modified with a RGD targeting peptide and showed stability in the blood circulation with a half-life greater than 12 h. MRI and infrared fluorescence optical imaging revealed drug accumulation in the tumoral and lymph node interstitium. MNPs with anti-miR-10b showed the ability to inhibit breast cancer metastasis to lymph nodes but, because of miR-10b downregulation, were ineffective in inhibiting tumor growth in nude mice bearing breast tumors. This finding suggests the possibility of using MNP-anti-miR-10b for treating metastatic cancer by targeting to breast tumor tissue and adenocarcinomas.

## 9.5 miRNA MOLECULES AND miRNA-REGULATING MACHINERY ASSOCIATED WITH CLINICAL FEATURES

miRNA molecules are small, single-stranded RNA molecules that function to regulate networks of genes. They play important roles in normal female reproductive tract biology, as well as in the pathogenesis and progression of epithelial ovarian cancer. Drosha, Dicer, and Argonaute proteins are components of the miRNA-regulatory machinery and mediate miRNA production and function.

Ovarian cancer is the most lethal gynecologic malignancy with an estimated 21,410 new cases and 13,770 deaths expected for 2021 in the United States. According to the tissue of origin, ovarian tumors are classified into epithelial and non-epithelial types (https://www.frontiersin.org/articles/10.3389/fonc.2021.601512/full) [141]. Epithelial ovarian cancer (EOC) is the most prevalent type of ovarian cancer, accounting for 90% of all ovarian cancers. It is characterized by distinct histological phenotypes including serous, endometrioid, clear cell, and mucinous. Each histotype is thought to arise from distinct precursor lesions of the female reproductive tract [142]. Molecularly, the landscape of each individual EOC histotype is distinct at the gene expression and genomic DNA level, allowing novel means to classify tumors beyond traditional histology [143, 144].

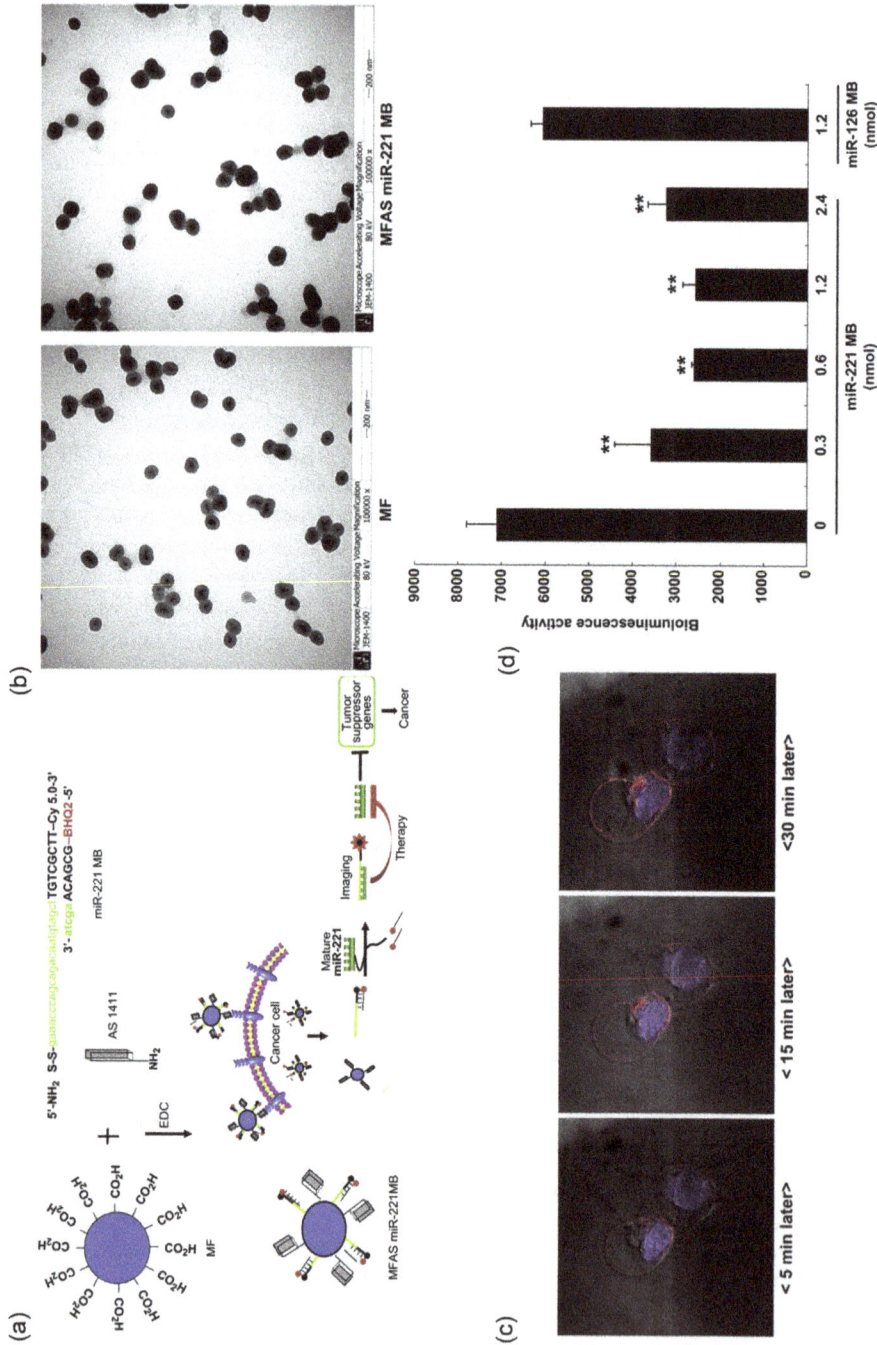

**FIGURE 9.11** AS1411 aptamer and miR-221 beacon-conjugated magnetic fluorescence NPs. (a) A detailed strategy of the targeting theragnostics. (b) TEM images of MF and MFAS miR-221 MB. (c) Cellular uptake of MFAS. (d) Anticancer effect of miR-221 MB. (Reprinted with permission from [119]. Copyright, 2012 by Elsevier.)

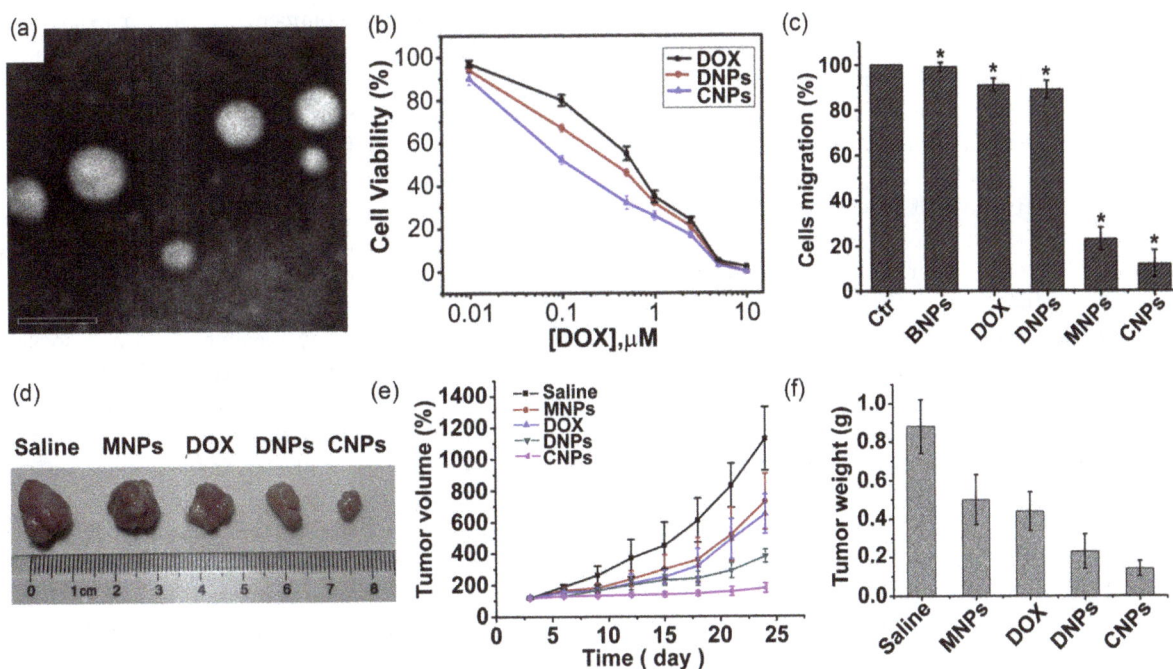

**FIGURE 9.12** HA-CS NPs for the co-delivery of DOX and miR-34a. (a) TEM images of DOX-miR-34a co-loaded HA-CS NPs. (b) Cell viability with co-delivery complexes. (c) The migration percentage of cancer cells. (d) Representative tumor images. (e) Tumor volume curves. (f) Tumor weights. (Reprinted with permission from [120]. Copyright, 2014 by Elsevier.)

Early-stage EOC has a 5-year survival of 92%, while late-stage disease has a 5-year survival of only 29%. Unfortunately, 79% of women with EOC have late-stage disease, defined as regional or distant metastasis, based on SEER data from 2006 to 2012 [141]. Improved screening approaches to detect early-stage disease and novel histotype- or molecular-marker-specific therapies for the treatment of late-stage disease are urgently needed. This review highlights the clinical associations of miRNA molecules and miRNA machinery, including Drosha, Dicer, and Argonaute proteins, in EOC identified since our last review [145]. The clinical relevance of these potential new biomarkers as prognostic, diagnostic, and therapeutic molecules is discussed.

## 9.6 GENESIS OF MATURE miRNA MOLECULES

### 9.6.1 miRNA-Regulating Machinery

RNA polymerase II transcribes miRNA molecules from genomic DNA into a primary miRNA molecule (pri-miRNA). Pri-miRNA molecules are typically greater than 200 nucleotides (nt) in length with a characteristic stem-loop structure. Furthermore, miRNA clusters containing multiple stem-loop structures, each coding for a mature miRNA molecule, can be in the kilobase size range. Pri-miRNA molecules are recognized by Drosha, an RNAse III, which cuts the double-stranded RNA into ~70-nt precursor miRNA (pre-miRNA) in the nucleus. Pre-miRNA molecules are exported to the cytoplasm and are processed by Dicer, an RNAse III, into two unique single-stranded

mature miRNA molecules representing each side of the stem-loop structure. Mature miRNA molecules are loaded onto the Argonaute-containing RNA-induced silencing complex (RISC). Within this structure, mature miRNA molecules function to repress gene expression by complementary binding of the 3' untranslated region (UTR) of the target gene to the miRNA seed sequence, nucleotides 2–8 of the mature miRNA molecule, leading to transcript degradation, and subsequent gene product loss [146]. Studies have shown that miRNA target genes play an important role in EOC cancer biology [147]. Thus, miRNA molecules and their biogenesis regulation as mediated by miRNA machinery are clinically important.

### 9.6.2 Primer on miRNA Nomenclature

Understanding how miRNA molecules are named is important for understanding how closely related miRNA molecules are to each other in the context of clinical associations and molecular functions. MiRNA molecules are sequentially named as they are discovered. For example, miR-21 was discovered and annotated in miRBase prior to miR-1307. Identical mature miRNA molecules of identical sequence may originate from different genomic loci with different primary miRNA molecule sequence due to the RNA processing. For example, miR-196a-1 and miR-196a-2 have identical mature miRNA sequence but originate from different genomic locations (i.e., chromosome 17 vs chromosome 12). This is different from miRNA molecules that have a closely related mature miRNA sequence such as miR-10a and miR-10b, which have a different mature sequence and are derived from different genomic locations. MiR-10a-5p and miR-10b-5p share the same seed sequence but differ in one nucleotide in the mature sequence [148].

Dicer processes each precursor miRNA into two mature molecules, with reverse complement sequence. Traditionally, the miRNA molecule with greater abundance was assigned the miRNA name (i.e., miR-29c), while the mature miRNA molecule on the other arm was called the * form (i.e., miR-29c*) [148]. This nomenclature, based on abundance, was phased out, and a new nomenclature, based on the location of the mature miRNA on the 5' or 3' strand, was phased in. Thus, the miRNA-3p forms are not necessarily less abundant or less functional. The nomenclature is now based on sequence location on the stem look. Mature miRNA molecules are grouped into families based on identical seed sequence, which are nucleotides 2–8 of the mature molecule. This sequence serves to function in complementary binding to the 3'UTR, leading to downstream effects of transcript repression [148]. MiRNA-5p and miRNA-3p molecules do not typically fall within the same miRNA family as they have a reverse complementary sequence. For example, the miR-10-5p family is comprised of miR-10a-5p and miR-10b-5p, whereas miR-10a-3p and miR-10b-3p are each members of their own family. Finally, isomiRs are mature miRNA molecules that differ from the mature sequence by one to two nucleotides. For example, the miR-21 + CA isomiR is formed from a unique tailing and trimming mechanism in proliferative diseases such as endometriosis and endometrial cancer. It contains the miR-21 mature sequence plus an additional two nucleotides [149]. While isomiRs have not yet been described in EOC, these unique molecules represent an opportunity to be utilized as biomarkers.

## 9.7  CLINICAL IMPLICATIONS

### 9.7.1  ABERRANT EXPRESSION OF MIRNA-BIOGENESIS MACHINERY COMPONENTS IN EOC

Multiple studies have examined the relative expression of Dicer and Drosha in EOC compared to control tissues. Choice of control tissue for comparison is critical in relative expression studies. Each histotype of ovarian cancer may arise from distinct precursor cells [142]. For example, high-grade serous EOC may arise from the fallopian tube or ovarian surface epithelium [150–152]. Endometrioid and clear-cell EOC may arise from a benign transformation of endometriosis [153, 154]. For these reasons, we will define histotype and control tissues used for each study.

Using 50 cases of high-grade serous ovarian cancer, Flavin et al. [155] revealed a significant upregulation of Drosha and a trend toward upregulation in Dicer by QPCR compared with normal ovary. Similarly, by immunohistochemistry in 37 samples, Dicer showed a significant upregulation in high-grade serous ovarian cancer. High Dicer expression was associated with an absence of lymph node metastasis and a low proliferation index. Dicer expression did not correlate with disease-free or overall survival [155].

The authors did not comment on Drosha expression by immunohistochemistry, nor did they comment on the association of Drosha with clinical factors. Additionally, only high-grade serous EOC tumor samples were examined.

Merritt et al. [156] examined Dicer and Drosha in 111 samples of EOC (2 endometrioid, 109 serous, 93 high-grade, 18 low-grade) by QPCR with validation by immunohistochemistry. Benign ovarian surface epithelium was used as a control. Using the bimodal expression of Dicer and Drosha in their dataset to classify tumors into low and high expression for Dicer and Drosha, they examined clinical associations. Low Dicer expression was associated with advanced-stage disease and reduced median survival. Low Drosha expression was associated with suboptimal cytoreduction and reduced median survival. Death from ovarian cancer was statistically associated with low levels of both Dicer and Drosha while high levels of both Dicer and Drosha were associated with increased median survival. Low Dicer was a predictor of poor prognosis (hazard ratio, 2.10; 95% CI, 1.15 to 3.85) but low Drosha was not. However, low Dicer and low Drosha was a predictor of death (hazard ratio, 4.00; 95% CI, 1.82 to 9.09). Cells with low Dicer expression could not process shRNA, an important potential therapeutic consideration [156]. A similar study using semi-quantitative RT PCR showed that Dicer expression was downregulated in both benign and malignant ovarian tumors compared with normal ovary. However, the specific histotype of malignant ovarian tumors was not defined [157].

Dicer expression was evaluated by immunohistochemistry in a large tissue microarray containing 87 serous and 39 nonserous ovarian tumors. Dicer expression negatively correlated with node status and tumor grade. Low Dicer expression in serous tumors was associated with poor overall survival. A similar trend was found for all tumors in the dataset, but the analysis was not performed for specific histotypes. miRNA profiling in tumors with low Dicer showed significant downregulation of many miRNA molecules compared with tumors with high Dicer expression. Additionally, tumors with low Dicer expression also had low estrogen-receptor expression [158]. Zhang et al. [159] did not find any significant difference in Dicer or Drosha expression by QPCR or immunohistochemistry staining between early- and late-stage disease without evaluation of normal control tissues, although there were differences in miRNA expression levels [159].

Flavin et al. [155] showed an association of high Dicer expression associated with better prognosis tumors (i.e., low metastatic lesions, low proliferative index tumors) but did not comment on low Dicer expression being associated with poor prognosis tumors per se [155]. Others have shown an association of poor survival with low expression of Dicer [156, 158]. However, Flavin et al. did not find a significant association of Dicer expression with survival rates [155]. Clinical associations for Dicer expression are intriguing for EOC. However, studies within the breast cancer literature have shown divergent expression results using different Dicer antibodies based on hormone receptor status of the tissues [160]. Additionally, each study in EOC determined a relative cutoff value for "low" and "high" Dicer expression. Prior to the use of Dicer expression in clinical care, these differences in cutoff values and antibody use must be standardized.

A meta-analysis showed an overall association of low Dicer expression with poor prognosis in multiple cancer types including ovarian cancer [161]. Lower expression of miRNA machinery is associated with poor prognosis tumors, suggesting that global loss of miRNA regulation may underlie the pathophysiology of poor prognosis EOC. These studies suggest that loss of Dicer may act as loss of a tumor suppressor, leading to disease that is more aggressive. Finally, these studies also suggest that lack of Dicer may influence attempts to treat tumors with small RNA molecules such as shRNA molecules.

Examination of expression of Dicer, Drosha, Argonaute 1 (AGO1), and Argonaute 2 (AGO2) in serous ovarian cancer revealed higher expression in metastatic lesions—either solid metastatic lesions or effusions—compared with the primary tumor [162]. Thus, high levels of miRNA machinery may be required for disease progression.

### 9.7.2 Aberrant miRNA-Regulating Machinery Expression in Mouse Models

MiRNA-regulating machinery is required for appropriate development of the female reproductive tract, as studies from mice have shown that full deletion of Dicer or Ago2 leads to embryonic lethality [163, 164]. Conditional deletion of Dicer or Ago2, allowing for postnatal studies with targeted deletion to the female reproductive tract using genetically engineered mouse models, revealed defects in female fertility [165]. However, deletion of Drosha in mice does not affect fertility or female reproductive tract development [165]. Importantly, none of these models results in a cancer phenotype, suggesting a critical role in normal female reproductive tract development and function, but not cancer initiation in the mouse, and thus, Dicer is not likely a true tumor suppressor.

Notably, loss of Dicer is critical to the development of high-grade serous EOC that begins in the oviduct of the mouse but also requires phosphatase and tensin homolog (Pten) deletion. In terms of translational importance, the histology of the ovarian tumors from anti-Müllerian hormone receptor type 2-Cre recombinase (Amhr2$^{Cre/+}$); Dicer$^{flox/flox}$; Ptenflox/flox mice recapitulated high-grade serous tumors from women. When the oviducts of Amhr2$^{Cre/+}$; Dicer$^{f/f}$; Pten$^{f/f}$ mice were removed, the mice failed to develop ovarian tumors. Thus, this model represents an important model to study the origins and mechanism of disease processes from the oviduct [150]. However, a majority (~96% of the Cancer Genome Atlas [TCGA] population) of high-grade serous EOC tumors from women contain mutations in tumor protein 53 (TP53) [166]. Addition of a TP53 mutation to the Amhr2$^{Cre/+}$; Dicer$^{f/f}$; Pten$^{f/f}$ mice (Amhr2$^{Cre/+}$; Dicer$^{f/f}$; Pten$^{f/f}$; TP53$^{R172H}$ mice) with removal of oviduct revealed metastatic tumors arising from the ovary.

However, loss of Dicer was not required as Amhr2Cre/+; Ptenf/f; TP53R172H mice developed similar metastatic lesions from the ovaries [152]. Thus, loss of Dicer seems to be important for high-grade serous EOC arising from the oviduct but not for disease arising from the ovary. The role of specific cell types and specific genetic hits in the origin of EOC requires further study prior to translational extrapolation to disease in women.

Other genetically engineered mouse models of Dicer deletion have revealed that loss of two alleles of Dicer is protective against cancer development, while loss of one allele gives a more aggressive cancer phenotype. Specifically, conditional loss of one allele of Dicer had a faster rate of lung cancer formation on an oncogenic KRAS proto-oncogene, GTPase (Kras$^{G12D}$) background, while loss of both Dicer alleles led to inhibition of tumor formation [167]. Similarly, Dicer haploinsufficiency on an oncogenic B-Raf proto-oncogene, serine/threonine kinase (Braf) (V600E) background led to increased metastasis in sarcomas [168]. Loss of one allele of Dicer in the retinoblasts of mice led to aggressive retinoblastoma with the inactivation of the retinoblastoma gene [169]. Similarly, loss of one allele of Dicer with Pten in the prostate led to more aggressive tumors [170]. These effects seem to be oncogene and tissue dependent, as conditional Dicer haploinsufficiency on an oncogenic MYC proto-oncogene, bHLH transcription factor (c-Myc) background does not facilitate cancers in B cells [171]. One hypothesis supported by in vitro studies is that loss of one allele of Dicer affects miRNA processing, or total expression of miRNA molecules, with loss of one allele of Dicer affecting the ratio of miRNA-5p to miRNA-3p from the precursor miRNA [172–174]. Further studies are needed to determine the effects of Dicer on miRNA processing in the initiation and progression of cancer in other female reproductive tract cancers, such as EOC.

### 9.7.3 Genetic Alterations of miRNA Machinery Genes

EOC is a genetically heterogeneous disease. Examination of copy number loss of Dicer in TCGA datasets revealed frequent loss of Dicer in high-grade serous EOC, consistent with low levels of expression and association with poor clinical outcome [150]. However, Dicer functional mutations are not recurrent in high-grade serous ovarian cancer [156, 175, 176]. Liang et al. [177] studied single-nucleotide polymorphisms from 8 miRNA-processing genes and 134 miRNA binding sites in genes in 339 EOC cases and 349 healthy controls. This work revealed that polymorphisms in specific miRNA binding sites in genes were associated with cancer risk, overall survival, and treatment response. In particular, a homozygous polymorphism in platelet-derived growth factor C (PDGFC) showed the most statistically significant effect on survival [177]. Single-nucleotide polymorphisms in Drosha were found to be associated with an increased risk of EOC diagnosis in a mostly Caucasian study population [177], but this association was not significant in a larger more ethnically diverse population [178]. Other studies have shown no association between polymorphisms in miRNA binding sites or mature miRNA molecules with EOC clinical outcomes [179–181]. Thus, use of single-nucleotide polymorphisms in miRNA machinery genes, mature miRNA molecules, or miRNA binding sites in genes to determine the risk of EOC diagnosis has been not been replicated clinically. Larger studies and grouping samples according to newly published genetic changes [144] instead of histotype or grade may allow for results that are more replicable.

## 9.8 miRNA MOLECULES AS CLINICAL BIOMARKERS FOR EOC

### 9.8.1 miRNA Molecules with Clinical Associations

In recent years, comprehensive profiling studies have revealed that miRNA molecules have distinct patterns of expression in EOC (Tables 9.3 and 9.4). A majority of these studies have used high-grade serous EOC or grouped all EOC histotypes together during analysis. The choice of

the control group for comparison has differed across many studies. Additionally, the miRNA-profiling platform has also differed. Thus, it is not surprising that the reproducibility of some miRNA-profiling datasets has been called into question, as the comparisons, samples, and technical platforms are not the same [182]. Attention to these details will be critical to deciphering important biomarkers.

Multiple studies have profiled miRNA molecules from clinical samples to determine whether miRNA molecules would be good biomarkers for EOC (Tables 9.3 and 9.4). Many of these large profiling studies aim to classify tumors into clinically important associations such as poor prognosis tumors, histotype, or chemotherapy responsive. Many of these studies do not limit samples to one particular histotype. Tables 9.3 and 9.4 list the studies with brief details of experimental design including sample number, histotype, miRNA profiling platform, and validation sample set if applicable. Fourteen studies use only serous histotype samples (Table 9.3), and 13 studies mix histotypes in the same study (Table 9.4). Since the two previous reviews [145, 165], the published studies for comprehensive miRNA profiling studies in EOC have expanded greatly in number and sophistication of experimental design. Clinical associations

**TABLE 9.3**

**MiRNAs as Biomarkers for Diagnosis of Epithelial Ovarian Cancer**

| miRNA | Analysis and Findings | ROC | Sensitivity | Specificity | Ref. |
|---|---|---|---|---|---|
| miR-141 | Serum samples: 74 EOC, 19 borderline, and 50 normal control | 0.75 | 69% | 72% | [183] |
| | High expression in EOC, especially late-stage disease with metastasis | | | | |
| miR-200c | Serum samples: 74 EOC, 19 borderline, and 50 normal control | 0.79 | 72% | 70% | [183] |
| | High expression in EOC without metastasis | | | | |
| | Expression variation across histotypes, with highest expression in serous | | | | |
| miR-125b | Serum samples: 54 controls with benign ovarian tumors and 135 patients with EOC | 0.737 | 75.6% | 68.5% | [184] |
| | High expression in EOC with no residual tumor after operation | | | | |
| | High expression in early-stage disease | | | | |
| miR-125b | Serum samples: 70 EOC and 70 age-matched controls | 0.728 | 62.3% | 77.1% | [185] |
| | High expression in early-stage disease | | | | |
| | High expression in no lymph node metastasis | | | | |
| miR-199a | Serum samples 70 EOC and 70 age-matched controls | 0.704 | 69.1% | 95.7% | [186] |
| | Low expression associated with aggressive tumor stage, lymph node metastasis, and distal metastasis | | | | |
| miR-200a | Serum samples: 70 EOC and 70 age-matched, cancer-free controls | 0.810 | 80.6% | 73.5% | [187] |
| | High expression in mucinous tumors | | | | |
| | High expression in metastatic tumors | | | | |
| miR-200c | Serum samples: 70 EOC and 70 age-matched, cancer-free controls | 0.833 | 83.3% | 73.1% | [187] |
| | High expression in advanced stage | | | | |
| | High expression in metastatic tumors | | | | |
| miR-200c | Serum samples: 70 EOC and 70 age-matched, cancer-free controls | 0.741 | 89.5% | 60.8% | [187] |
| | High expression in lymph node positive tumors | | | | |
| miR-1290 | Plasma samples: 42 serous EOC, 36 benign neoplasm, and 23 age-matched healthy controls | 0.87 | 63.3% | 100% | [188] |
| | High expression associated with long overall survival | | | | |
| miR-30a-5p | Urine samples: 34 serous EOC and 25 healthy controls | 0.862 | No number | | [189] |
| miR-6076 | Urine samples: 34 serous EOC and 25 healthy controls | 0.693 | | | [189] |

**TABLE 9.4**

**In Vivo Mouse Models of miRNA Therapy in Epithelial Ovarian Cancer**

| miRNA Treatment | Method | Effect | Ref. |
|---|---|---|---|
| miR-1307 | A2780 cells subcutaneous injection | Chemotherapy sensitivity (Taxol) | [190] |
| miR-192 | SKOV3ip1 cells intraperitoneal injection | Tumor growth | [191] |
| miR-506 | HeyA8-IP2 and SKOV3-IP1 cells intraperitoneal injection | Invasion Metastasis Chemotherapy sensitivity (cisplatin and olaparib) | [192,193] |
| miR-551b-3p | IGROV1 or HEYA8 cells intraperitoneal injection | Tumor growth Ascites formation | [194] |
| miR-6126 | HeyA8 cells intraperitoneal injection | Tumor growth Angiogenesis | [195] |

of specific miRNA molecules in EOC tissues are listed in Table 9.4. Thus, miRNA molecules hold promise for clinically useful biomarkers.

## 9.8.2 MiRNA MOLECULES IN RARE HISTOTYPES OF EOC

The molecular landscape of each histotype of EOC is distinct [143, 144]. However, all patients are treated similarly with standard-of-care debulking surgery and chemotherapy regimens, with dismal "cure" rates of 20% [141]. Notably, the presence of endometriosis, a benign pathologic growth of endometrium outside the uterus [196], leads to a 50% increase in the risk of ovarian cancer [197]. Further, women with endometriosis are at higher risk of developing rare EOC subtypes, including clear-cell or endometrioid histotypes, as opposed to the more common high-grade serous disease [197]. Therefore, a critical need exists to understand the genomic and pathophysiological differences between these histologic subtypes for early detection and for the development of histotype-specific therapies. Several studies have examined the role of miRNA molecules in these distinct EOC types.

miR-132, miR-9, miR-126, miR-34a, and miR-21 were found to differentiate clear-cell from serous tumors. miR-9, the most highly expressed of those five miRNA molecules in clear-cell samples, was found to be involved in cellular invasion in vitro [198]. In a similar study, miR-510, miR-129-3p, miR-483, and miR-449a were found to be differentially expressed between serous and clear-cell ovarian tumors. All were associated with advanced tumor stage. Low expression of miR-129-3p or low expression of miR-510 was associated with poor overall survival [199]. miR-30a*, miR-30e*, and miR-505* were the most upregulated miRNAs in clear-cell ovarian tumors compared with serous tumors [200]. Lower expression of miR-30a*, miR-30e*, and miR-505* was associated with poorer prognosis. Additionally, miR-134-3p was downregulated in CD44+/CD133+ ovarian cancer cells from clear-cell tumors. miR-134-3p directly targeted RAB27A, which is involved in downregulation of stem cell markers and adhesion proteins [201]. Overexpression of miR-21 was

associated with endometriosis in clear-cell ovarian cancer, and miR-21 targets the tumor suppressor PTEN [202]. Thus, miRNA molecules play a functional role in the regulation of gene expression in clear-cell ovarian cancer differently from high-grade serous ovarian cancer.

As a potential clinically relevant serum biomarker, miR-130a is elevated in the serum of women prior to surgery for clear-cell ovarian cancer and falls after surgery. Additionally, miR-130a levels increase prior to CA-125 levels prior to disease recurrence, making miR-130a a promising biomarker [203]. This was a well-normalized study in a specific histotype of ovarian cancer in a specific ethnic population. The results are promising and need further study in benign disease, normal control women, and other histotypes. A similarly important study showed a minimal signature of miRNA molecules detected in blood and the ability to determine whether a woman had endometriosis, clear-cell, endometrioid, or serous ovarian cancer [204]. Thus, the clinical usefulness of miRNA molecules as biomarkers for EOC, in particular, subtype-specific histotypes, has not been sufficiently studied.

Multiple studies have highlighted the function of miRNA molecules in terms of endometrioid histotypes [205–207]. Expression of miR-191was increased in ovarian endometriosis and endometrioid EOC. Importantly, miR-191 overexpression in vitro decreased tumor necrosis factor alpha-α (TNFα)-induced apoptosis [205]. Given the high expression of TNFα in the pelvic cavity of women with endometriosis [208], the failure of apoptosis due to high miR-191 may play a role in the over proliferation of endometrioid EOC associated with endometriosis. Additionally, miR-191 directly targets TIMP metallopeptidase inhibitor 3 (TIMP3) and loss of miR-191 leads to decreased cellular proliferation and decreased invasion in vitro [206]. Furthermore, miR-370 is downregulated in endometrioid EOC compared to the normal ovary or other histotypes of EOC. Overexpression of miR-370 in vitro showed decreased proliferation and increased sensitivity to platinum. Similar results were revealed in xenograft studies with miR-370 [207]. Thus, miRNA molecules may play significant roles in the underlying biology of endometrioid EOC.

### 9.8.3 miRNA MOLECULES IN BODILY FLUIDS AS BIOMARKERS

miRNA molecules offer promise as molecules for early detection. Urinary excretion of miR-30a-5p is found in serous ovarian cancer but not in other cancer types. After primary debulking surgery, less expression of miR-30a-5p is found in the urine [189]. While Zhou et al. [189] did not examine the expression of miR-30a-5p after recurrence of ovarian cancer, this miRNA may serve as a useful biomarker for determination of disease. Langhe et al. [209] used a discovery set of five malignant high-grade serous and five benign serous cystadenomas to examine the expression of miRNA levels in blood. They determined a significant downregulation of let-7i-5p, miR-122, miR-152-5p, and miR-25-3p in malignant disease compared with benign disease state [209]. Thus, miRNA molecules may serve as biomarkers for preoperative discrimination of benign and malignant disease. Table 9.3 lists the clinically relevant miRNA molecules as biomarkers for EOC and their associated clinical features.

### 9.8.4 miRNA MOLECULES AS MARKERS OF TREATMENT SPECIFICITY

A strategic priority of the NCI Gynecologic Cancer Steering Committee is identification of molecular and/or pathologic cancer subsets to drive therapy and improve outcomes. A large well-normalized study examined outcomes and expression of miRNA molecules in the blood of women with EOC in ICON7. This study aimed at assessing the safety and efficacy of adding the anti-angiogenesis immunologic therapy, specifically bevacizumab, to standard therapy. The isolation of miRNA molecules from blood was well described, and the experimental design contained a discovery and validation cohort. miR-1274a, miR-141, miR-200b, and miR-200c were associated with survival in the discovery studies. miR-141 and miR-200b were similarly associated in the validation cohort containing all histotypes. miR-200c was associated with better survival in women treated with bevacizumab compared to standard chemotherapy. Subclassification of serous tumors revealed low miR-1274a associated with prolonged survival in discovery and validation groups [210]. With the caveat that the study awaits independent replication, the results suggest the potential usefulness of miRNA profiles for the future choice of therapy regimens in EOC.

Petrillo et al. [211] analyzed miRNA profiles from matched tumor samples before and after neoadjuvant chemotherapy (NACT). They found that samples collected after NACT had distinct miRNA profiles and focused on miRNA molecules upregulated after NACT, hypothesizing that those miRNA molecules were necessary for chemotherapy resistance. They specifically studied miR-199, miR-29, miR-30, let-7, and miR-181 family members, as they were upregulated by QPCR in post NACT samples and associated with platinum resistance in other studies.

miR-199a-5p, miR-199a-3p, let-7a-5p, let-7g-5p, and miR-181a-5p were associated with worse progression-free survival. miR-199a-5p, miR-199a-3p, miR-199b-5p, let-7g-5p, and miR-181a-5p were associated with worse overall survival. Multivariate analysis showed only let-7g-5p associated with overall survival with a hazard ratio of 1.1 (1.04–1.23) [211]. Examination of P-SMAD2, an indirect target for miR-181a-5p, revealed a significant effect (hazard ratio 1.1, CI 1.0–1.2) on overall survival. Thus, these miRNA molecules and indirect targets may represent biomarkers of platinum resistance in a very specific group of patients. A potential concern, however, is represented by the rather modest hazard ratio, pointing to the need for large cohorts in any future validation study. Similar to studies in breast cancer, high expression of miR-622 in EOC was associated with loss of ku protein from the homologous recombination pathway and poor outcome. This suggests that miR-622 was associated with resistance to chemotherapy including poly ADP ribose polymerase (PARP) inhibitors [212]. Since PARP inhibitors are now FDA approved for ovarian cancer therapy [213], this miRNA marker may be an additional marker useful for selecting appropriate candidates for this particular therapy.

### 9.8.5 HYPOXIA-REGULATED miRNA MOLECULES IN EOC

Hypoxia plays an important role in miRNA molecule biogenesis and clinical oncology poor outcomes [214, 215]. Hypoxia is important for Dicer and Drosha downregulation in EOC [216]. Rupaimoole et al. [216] found that hypoxic conditions downregulated Dicer and Drosha expression in vitro and in vivo. The expression levels of Dicer and Drosha and a panel of hypoxia markers showed an inverse correlation in EOC. Additionally, these studies revealed global miRNA downregulation following exposure to hypoxia. Examination of survival with a hypoxia metagene signature in TCGA datasets revealed an association of poor survival with high levels of hypoxia [216]. miR-199a-3p, miR-216b, miR-548d-5p, and miR-579 were found downregulated and 19 miRNAs were upregulated in ovarian cancer under hypoxia [217]. Overexpression of miR-199a-3p reduces the invasion of ovarian cancer cells in response to hypoxia [217]. The role of hypoxia in miRNA molecule expression and function in EOC deserves additional study.

## 9.9 miRNA MOLECULES AS THERAPY FOR EOC

Given the ability of specific miRNA molecules to sensitize tumors to chemotherapy in vitro, studies have used miRNA molecules as adjuvant therapy with miRNA molecules in xenograft mouse models (Table 9.4). AntagomiR miR-1307 treatment of mice with xenograft tumors resulted in a decrease in tumor size and an even further decrease in tumor size with paclitaxel treatment [190]. Treatment of xenograft mouse models with miR-192-DOPC (1,2-dio-leoyl-sn-glycero-3-phosphocholine) leads to inhibition of angiogenesis and tumor growth [191]. miR-551b-3p is located in a genomic region frequently amplified in high-grade serous EOC [218]. miR-551b-3p overexpression in high-grade EOC is associated with decreased overall survival. Interestingly, miR-551-3p binds to the promoter of signal transducer and activator of transcription 3 (STAT3), leading to increased expression of STAT3. Treatment of mouse models with anti-miR-551b-3p via a liposomal delivery system decreased tumor burden in vivo [194]. These few preclinical models show promise for treatment of EOC with molecules that regulate miRNA expression.

## 9.10 CHALLENGES TO CLINICAL USE OF miRNA MOLECULES AND miRNA REGULATORY MACHINERY IN EOC

MiRNA molecules hold promise as biomarkers for clinical associations, early markers for disease, and adjuvant therapy. However, most of the corresponding studies have yet to be replicated. Studies using biological fluids such as urine, serum, or plasma need to consider details such as phase of the menstrual cycle, effect of exogenous hormones or bioidentical compounds, type of sample collection, and processing of samples. The reference range or control group needs to be considered carefully. EOC as a disease arising from the ovary has been challenged. Many studies have compared EOC to normal ovarian epithelium, but the cell of origin for high-grade serous EOC may be fallopian tube epithelium. Limited studies on the more rare subtypes of EOC, such as endometrioid, clear-cell, or mucinous, need to be addressed because understanding the differences may lead to improved therapy for specific histotypes or specific molecular characteristics—possibly based on miRNA molecule expression. Treatment of patients with miRNA molecules needs further study in drug delivery and miRNA effects. However, these preclinical studies hold promise for the future.

Supplementary data to these sections can be found online at http://dx.doi.org/10.1016/j.ygyno.2017.08.027.

## 9.11 CONCLUSIONS

Although miRNA-based therapies have received significant attention, most miRNA-related modulators remain in preclinical trials (Table 9.5). Among the candidate miRNA drugs that have entered clinical trials, LNA-modified

**TABLE 9.5**
**MiRNAs in R&D Pipeline**

| Biotech Company and Technology | Therapeutic Indication | Target miRNA | Status |
|---|---|---|---|
| Viridian Therapeutics | | | |
| | Heart failure | miR-208 | Preclinical |
| | Post-myocardial infarction remodeling | miR-15/195 | Preclinical |
| | Vascular disease | miR-145 | Preclinical |
| Chemical modifications[a] | Myeloproliferative disease | miR-451 | Preclinical |
| | Pathological fibrosis | miR-29 | Preclinical |
| | Cardiometabolic disease | miR-208 | Preclinical |
| | Peripheral arterial disease | miR-92 | Preclinical |
| | Cardiometabolic disease | miR-378 | Lead optimization |
| | Neuromuscular disease | miR-206 | Lead optimization |
| *Regulus Therapeutics* | | | |
| Phosphorothioate backbone to increase | Hepatocellular carcinoma | miR-122 | Preclinical |
| metabolic stability and tissue half-life[b] | Glioblastoma | miR-10b | Preclinical |
| | Kidney fibrosis | miR-21 | Preclinical |
| | Metabolic | miR-103/107 | Preclinical |
| | Oncology | miR-19 | Preclinical |
| *Santaris Pharma* | | | |
| Locked nucleic acids (Miravirsen)[c] | Liver transplant (HCV) | miR-122 | Phase 2 |
| *Mirna Therapeutics* | | | |
| Lipid-based vectors[d] | Cancer | miR34, let7 | Phase 1 |

[a]   https://www.ahajournals.org/doi/full/10.1161/CIRCRESAHA.111.247916

[b]   http://www.regulusrx.com/therapeutic-areas/#Oncology

[c]   http://www.santaris.com/sites/default/files/product-pipeline.png

[d]   https://www.ncbi.nlm.nih.gov/pmc/articles/PMC7378626/ OR https://onlinelibrary.wiley.com/doi/10.1002/jcp.28058

anti-miR-122 (SPC3649; http://www.santaris.com) has already completed the Phase 2 clinical stage for the treatment of HCV in liver transplants [219, 220]. Another successful case, the liposomal miR-34 mimic, MRX34 [221,222], for primary liver cancer or other solid cancers with liver involvement, has also entered a multicenter, open-label Phase 1 clinical trial (http://clinicaltrials.gov/ct2/show/NCT01829981).

MiRNA-based medicine is a relatively new field. To date, tumor therapies are the major application for miRNA-based medicine [223–227], and several biotech firms produce miRNA mimics, inhibitors, and vectors for the delivery of miRNAs for cancer therapy. However, within the past 20 years, miRNAs have undergone a rapid development from discovery to therapeutic application, and a new era of clinical translation is coming. Undoubtedly, the development of effective miRNA carriers would shorten the course from bench to bedside. Yet the roadmap to move from miRNA-related drug design, to preclinical studies, to clinical trials, to an approved product will be long because of the many challenges and risks associated with this novel approach to treatment. In the long term, future work should focus on identifying effective concentrations, dosage regimens, and combination with chemotherapy and radiotherapy, apart from the vectors. Once in use, the long-term results following miRNA therapy will need to be assessed.

Finally, we can conclude that:

- Expressions of miRNA-regulatory machinery may serve as unique biomarkers for prognosis in epithelial ovarian cancer.
- miRNA molecules are promising biomarkers for epithelial ovarian cancer.
- miRNA molecules are promising therapeutic adjuvants for epithelial ovarian cancer.
- Further work is needed to understand the molecular mechanisms of miRNA molecules in subtypes of epithelial ovarian cancer.

Supplementary data to these sections can be found online at http://dx.doi.org/10.1016/j.ygyno.2017.08.027.

## REFERENCES

[1] D.P. Bartel, MicroRNAs: genomics, biogenesis, mechanism, and function, Cell 116 (2004) 281–297.

[2] R.C. Lee, R.L. Feinbaum, V. Ambros, The C. elegans heterochronic gene lin-4 encodes small RNAs with antisense complementarity to lin-14, Cell 75 (1993) 843–854.

[3] B. Wightman, I. Ha, G. Ruvkun, Posttranscriptional regulation of the heterochronic gene lin-14 by lin-4 mediates temporal pattern formation in C. elegans, Cell 75 (1993) 855–862.

[4] A. Kozomara, S. Griffiths-Jones, miRBase: integrating microRNA annotation and deep-sequencing data, Nucleic Acids Res 39 (2011) 152–157.

[5] R.C. Friedman, K.K.H. Farh, C.B. Burge et al., Most mammalian mRNAs are conserved targets of microRNAs, Genome Res 19 (2009) 92–105.

[6] E. Berezikov, V. Guryev, J. van de Belt et al., Phylogenetic shadowing and computational identification of human microRNA genes, Cell 120 (2005) 21–24.

[7] P. Jakob, U. Landmesser, Role of microRNAs in stem/progenitor cells and cardiovascular repair, Cardiovasc Res 93 (2012) 614–622.

[8] V. Ambros, The functions of animal microRNAs, Nature 431 (2004) 350–355.

[9] Y. Lee, C. Ahn, J. Han et al., The nuclear RNase III Drosha initiates microRNA processing, Nature 425 (2003) 415–419.

[10] L.P. Lim, N.C. Lau, P. Garrett-Engele et al., Microarray analysis shows that some microRNAs downregulate large numbers of target mRNAs, Nature 433 (2005) 769–773.

[11] R. Garzon, G. Marcucci, C.M. Croce, Targeting microRNAs in cancer: rationale, strategies and challenges, Nat Rev Drug Discov 9 (2010) 775–789.

[12] C.C. Esau, B.P. Monia, Therapeutic potential for microRNAs, Adv Drug Deliv Rev 59 (2007) 101–114.

[13] R.P. Singh, I. Massachi, S. Manickavel et al., The role of miRNA in inflammation and autoimmunity, Autoimmun Rev 12 (2013) 1160–1165.

[14] E.A. Frangou, G.K. Bertsias, D.T. Boumpas, Gene expression and regulation in systemic lupus erythematosus, Eur J Clin Invest 43 (2013) 1084–1096.

[15] M. Tandon, A. Gallo, S.I. Jang et al., Deep sequencing of short RNAs reveals novel microRNAs in minor salivary glands of patients with Sjogren's syndrome, Oral Dis 18 (2012) 127–131.

[16] M. Thamilarasan, D. Koczan, M. Hecker et al., MicroRNAs in multiple sclerosis and experimental autoimmune encephalomyelitis, Autoimmun Rev 11 (2012) 174–179.

[17] A.G. Bader, D. Brown, J. Stoudemire et al., Developing therapeutic microRNAs for cancer, Gene Ther 18 (2011) 1121–1126.

[18] G.A. Calin, C. Sevignani, C. Dan Dumitru et al., Human microRNA genes are frequently located at fragile sites and genomic regions involved in cancers, Proc Natl Acad Sci USA 101 (2004) 2999–3004.

[19] S. Yang, Y. Li, MicroRNAs: novel factors in clinical diagnosis and prognosis for nasopharyngeal carcinoma, Acta Pharmacol Sin 33 (2012) 981–982.

[20] N. Rosenfeld, R. Aharonov, E. Meiri et al., MicroRNAs accurately identify cancer tissue origin, Nat Biotechnol 26 (2008) 462–469.

[21] J.A. Broderick, P.D. Zamore, MicroRNA therapeutics, Gene Ther 18 (2011) 1104–1110.

[22] Y.W. Kong, D. Ferland-McCollough, T.J. Jackson et al., MicroRNAs in cancer management, Lancet Oncol 13 (2012) 249–258.

[23] C.M. Croce, Causes and consequences of microRNA dysregulation in cancer, Nat Rev Genet 10 (2009) 704–714.

[24] A. Esquela-Kerscher, F.J. Slack, Oncomirs — microRNAs with a role in cancer, Nat Rev Cancer 6 (2006) 259–269.

[25] D. Bonci, V. Coppola, M. Musumeci et al., The miR-15a-miR-16-1 cluster controls prostate cancer by targeting multiple oncogenic activities, Nat Med 14 (2008) 1271–1277.

[26] M. Osaki, F. Takeshita, Y. Sugimoto et al., MicroRNA-143 regulates human osteosarcoma metastasis by regulating matrix metalloprotease-13 expression, Mol Ther 19 (2011) 1123–1130.

[27] A.G. Bader, D. Brown, M. Winkler, The promise of microRNA replacement therapy, Cancer Res 70 (2010) 7027–7030.

[28] J.C. Henry, A.C. Azevedo-Pouly, T.D. Schmittgen, MicroRNA replacement therapy for cancer, Pharm Res 28 (2011) 3030–3042.

[29] H.R. Li, B.B. Yang, Friend or foe: the role of microRNA in chemotherapy resistance, Acta Pharmacol Sin 34 (2013) 870–879.

[30] M. Garofalo, C.M. Croce, MicroRNAs as therapeutic targets in chemoresistance, Drug Resist Updat 16 (2013) 47–59.

[31] Z.M. Li, S. Hu, J. Wang et al., miR-27a modulates MDR1/P-glycoprotein expression by targeting HIPK2 in human ovarian cancer cells, Gynecol Oncol 119 (2010) 125–130.

[32] C.F. Chen, X. He, A.D. Arslan et al., Novel regulation of nuclear factor-YB by miR-485-3p affects the expression of DNA topoisomerase II alpha and drug responsiveness, Mol Pharmacol 79 (2011) 735–741.

[33] H. Arora, R. Qureshi, S. Jin et al., miR-9 and let-7g enhance the sensitivity to ionizing radiation by suppression of NFkappaB1, Exp Mol Med 43 (2011) 298–304.

[34] Y. Zhang, Z. Wang, R.A. Gemeinhart, Progress in microRNA delivery, J Control Release 172 (2013) 962–974.

[35] V. Bravo, S. Rosero, C. Ricordi et al., Instability of miRNA and cDNAs derivatives in RNA preparations, Biochem Biophys Res Commun 353 (2007) 1052–1055.

[36] W.C.S. Cho, Role of miRNAs in lung cancer, Expert Rev Mol Diagn 9 (2009) 773–776.

[37] C.V. Pecot, G.A. Calin, R.L. Coleman et al., RNA interference in the clinic: challenges and future directions, Nat Rev Cancer 11 (2011) 59–67.

[38] D. Grimm, K.L. Streetz, C.L. Jopling et al., Fatality in mice due to oversaturation of cellular microRNA/short hairpin RNA pathways, Nature 441 (2006) 537–541.

[39] X. Wang, B. Yu, W. Ren et al., Enhanced hepatic delivery of siRNA and microRNA using oleic acid based lipid nanoparticle formulations, J Control Release 172 (2013) 690–698.

[40] L. Aagaard, J.J. Rossi, RNAi therapeutics: principles, prospects and challenges, Adv Drug Deliv Rev 59 (2007) 75–86.

[41] H.W. Yang, C.Y. Huang, C.W. Lin et al., Gadolinium-functionalized nanographene oxide for combined drug and microRNA delivery and magnetic resonance imaging, Biomaterials 35 (2014) 6534–6542.

[42] J. Liu, M.A. Valencia-Sanchez, G.J. Hannon et al., MicroRNA-dependent localization of targeted mRNAs to mammalian P-bodies, Nat Cell Biol 7 (2005) 719–723.

[43] C. Scholz, E. Wagner, Therapeutic plasmid DNA versus siRNA delivery: common and different tasks for synthetic carriers, J Control Release 161 (2012) 554–565.

[44] K.A. Whitehead, R. Langer, D.G. Anderson, Knocking down barriers: advances in siRNA delivery, Nat Rev Drug Discov 8 (2009) 129–138.

[45] D.J. Gibbings, C. Ciaudo, M. Erhardt et al., Multivesicular bodies associate with components of miRNA effector complexes and modulate miRNA activity, Nat Cell Biol 11 (2009) 1143–1149.

[46] H. Wang, Y. Jiang, H. Peng et al., Recent progress in microRNA delivery for cancer therapy by non-viral synthetic vectors. Advanced Drug Delivery Reviews 81 (2015) 142–160.

[47] H. Hatakeyama, M. Murata, Y. Sato et al., The systemic administration of an anti-miRNA oligonucleotide encapsulated pH-sensitive liposome results in reduced level of hepaticmicroRNA-122 in mice, J Control Release 173 (2014) 43–50.

[48] S. Anand, B.K. Majeti, L.M. Acevedo et al., MicroRNA-132-mediated loss of p120RasGAP activates the endothelium to facilitate pathological angiogenesis, Nat Med 16 (2010) 909–914.

[49] Y. Wu, M. Crawford, Y. Mao et al., Therapeutic delivery of microRNA-29b by cationic lipoplexes for lung cancer, Mol Ther Nucleic Acids 2 (2013) e84.

[50] Y. Wu, M. Crawford, B. Yu et al., MicroRNA delivery by cationic lipoplexes for lung cancer therapy, Mol Pharm 8 (2011) 1381–1389.

[51] K. Rai, N. Takigawa, S. Ito et al., Liposomal delivery of microRNA-7-expressing plasmid overcomes epidermal growth factor receptor tyrosine kinase inhibitor-resistance in lung cancer cells, Mol Cancer Ther 10 (2011) 1720–1727.

[52] L. Piao, M. Zhang, J. Datta et al., Lipid-based nanoparticle delivery of pre-miR-107 inhibits the tumorigenicity of head and neck squamous cell carcinoma, Mol Ther 20 (2012) 1261–1269.

[53] P. de Antonellis, C. Medaglia, E. Cusanelli et al., miR-34a targeting of Notch ligand delta-like 1 impairs CD15+/CD133+ tumor propagating cells and supports neural differentiation in medulloblastoma, PLoS One 6 (2011) e24584.

[54] N. Kamaly, Z. Xiao, P.M. Valencia et al., Targeted polymeric therapeutic nanoparticles: design, development and clinical translation, Chem Soc Rev 41 (2012) 2971–3010.

[55] I.S. Zuhorn, J.B. Engberts, D. Hoekstra, Gene delivery by cationic lipid vectors: overcoming cellular barriers, Eur Biophys J 36 (2007) 349–362.

[56] X. Huang, S. Schwind, B. Yu et al., Targeted delivery of microRNA-29b by transferrin-conjugated anionic lipopolyplex nanoparticles: a novel therapeutic strategy in acute myeloid leukemia, Clin Cancer Res 19 (2013) 2355–2367.

[57] J.F. Wiggins, L. Ruffino, K. Kelnar et al., Development of a lung cancer therapeutic based on the tumor suppressor microRNA-34, Cancer Res 70 (2010) 5923–5930.

[58] P. Trang, J.F. Wiggins, C.L. Daige et al., Systemic delivery of tumor suppressor microRNA mimics using a neutral lipid emulsion inhibits lung tumors in mice, Mol Ther 19 (2011) 1116–1122.

[59] M. Bikram, M. Lee, C.W. Chang et al., Long circulating DNA-complexed biodegradable multiblock copolymers for gene delivery: degradation profiles and evidence of dysopsonization, J Control Release 103 (2005) 221–233.

[60] D. Pramanik, N.R. Campbell, C. Karikari et al., Restitution of tumor suppressor microRNAs using a systemic nanovector inhibits pancreatic cancer growth in mice, Mol Cancer Ther 10 (2011) 1470–1480.

[61] S.D. Li, L. Huang, Stealth nanoparticles: high density but sheddable PEG is a key for tumor targeting, J Control Release 145 (2010) 178–181.

[62] H. Maeda, The enhanced permeability and retention (EPR) effect in tumor vasculature: the key role of tumor-selective macromolecular drug targeting, Adv Enzym Regul 41 (2001) 189–207.

[63] O. Boussif, F. Lezoualch, M.A. Zanta et al., A versatile vector for gene and oligonucleotide transfer into cells in culture and in-vivo — polyethylenimine, Proc Natl Acad Sci USA 92 (1995) 7297–7301.

[64] W. Rodl, D. Schaffert, E. Wagner et al., Synthesis of polyethylenimine-based nanocarriers for systemic tumor targeting of nucleic acids, Methods Mol Biol 948 (2013) 105–120.

[65] K. Kunath, A. von Harpe, D. Fischer et al., Low molecular-weight polyethylenimine as a non-viral vector for DNA delivery: comparison of physicochemical properties, transfection efficiency and in vivo distribution with

high-molecular-weight polyethylenimine, J Control Release 89 (2003) 113–125.

[66] T. Bieber, H.P. Elsasser, Preparation of a low molecular weight polyethylenimine for efficient cell transfection, Biotechniques 30 (2001) 74–77.

[67] C.L. Gebhart, A.V. Kabanov, Evaluation of polyplexes as gene transfer agents, J Control Release 73 (2001) 401–416.

[68] A.F. Ibrahim, U. Weirauch, M. Thomas et al., MicroRNA replacement therapy for miR-145 and miR-33a is efficacious in a model of colon carcinoma, Cancer Res 71 (2011) 5214–5224.

[69] G.Y. Chiou, J.Y. Cherng, H.S. Hsu et al., Cationic polyurethanes-short branch PEI-mediated delivery of Mir145 inhibited epithelial–mesenchymal transdifferentiation and cancer stem-like properties and in lung adenocarcinoma, J Control Release 159 (2012) 240–250.

[70] Y.P. Yang, Y. Chien, G.Y. Chiou et al., Inhibition of cancer stem cell-like properties and reduced chemoradioresistance of glioblastoma using microRNA145 with cationic polyurethane-short branch PEI, Biomaterials 33 (2012) 1462–1476.

[71] T. Ochiya, S. Nagahara, A. Sano et al., Biomaterials for gene delivery: atelocollagen-mediated controlled release of molecular medicines, Curr Gene Ther 1 (2001) 31–52.

[72] Y. Minakuchi, F. Takeshita, N. Kosaka et al., Atelocollagen mediated synthetic small interfering RNA delivery for effective gene silencing in vitro and in vivo, Nucleic Acids Res 32 (2004) e109.

[73] F. Takeshita, Y. Minakuchi, S. Nagahara et al., Efficient delivery of small interfering RNA to bone-metastatic tumors by using atelocollagen in vivo, Proc Natl Acad Sci USA 102 (2005) 12177–12182.

[74] H. Tazawa, N. Tsuchiya, M. Izumiya et al., Tumor-suppressive miR-34a induces senescence-like growth arrest through modulation of the E2F pathway in human colon cancer cells, Proc Natl Acad Sci USA 104 (2007) 15472–15477.

[75] H. Matsuyama, H.I. Suzuki, H. Nishimori et al., miR-135b mediates NPM-ALK-driven oncogenicity and renders IL-17-producing immunophenotype to anaplastic large cell lymphoma, Blood 118 (2011) 6881–6892.

[76] Y. Takei, M. Takigahira, K. Mihara et al., The metastasis-associated microRNA miR-516a-3p is a novel therapeutic target for inhibiting peritoneal dissemination of human scirrhous gastric cancer, Cancer Res 71 (2011) 1442–1453.

[77] F. Takeshita, L. Patrawala, M. Osaki et al., Systemic delivery of synthetic microRNA-16 inhibits the growth of metastatic prostate tumors via downregulation of multiple cell-cycle genes, Mol Ther 18 (2010) 181–187.

[78] Y. Zhou, L. Zhang, W. Zhao et al., Nanoparticle-mediated delivery of TGF-beta1 miRNA plasmid for preventing flexor tendon adhesion formation, Biomaterials 34 (2013) 8269–8278.

[79] C.J. Cheng, W.M. Saltzman, Polymer nanoparticle-mediated delivery of microRNA inhibition and alternative splicing, Mol Pharm 9 (2012) 1481–1488.

[80] A.M. Tinsley-Bown, R. Fretwell, A.B. Dowsett et al., Formulation of poly(D,L-lactic-co-glycolic acid) microparticles for rapid plasmid DNA delivery, J Control Release 66 (2000) 229–241.

[81] G. Otten, M. Schaefer, C. Greer et al., Induction of broad and potent anti-human immunodeficiency virus immune responses in rhesus macaques by priming with a DNA vaccine and boosting with protein-adsorbed polylactide coglycolide microparticles, J Virol 77 (2003) 6087–6092.

[82] L.A. Dailey, M. Wittmar, T. Kissel, The role of branched polyesters and their modifications in the development of modern drug delivery vehicles, J Control Release 101 (2005) 137–149.

[83] R. Duncan, L. Izzo, Dendrimer biocompatibility and toxicity, Adv Drug Deliv Rev 57 (2005) 2215–2237.

[84] Y. Ren, C.S. Kang, X.B. Yuan et al., Co-delivery of as-miR-21 and 5-FU by poly(amidoamine) dendrimer attenuates human glioma cell growth in vitro, J Biomater Sci Polym Ed 21 (2010) 303–314.

[85] X. Li, S. Xin, Z. He et al., MicroRNA-21 (miR-21) post-transcriptionally downregulates tumor suppressor PDCD4 and promotes cell transformation, proliferation, and metastasis in renal cell carcinoma, Cell Physiol Biochem 33 (2014) 1631–1642.

[86] Y. Ren, X. Zhou, M. Mei et al., MicroRNA-21 inhibitor sensitizes human glioblastoma cells U251 (PTENmutant) and LN229 (PTEN-wild type) to taxol, BMC Cancer 10 (2010) 27.

[87] M. Labieniec-Watala, K. Karolczak, K. Siewiera et al., The janus face of PAMAM dendrimers used to potentially cure nonenzymatic modifications of biomacromolecules in metabolic disorders—a critical review of the pros and cons, Molecules 18 (2013) 13769–13811.

[88] Y.Y. Jiang, G.T. Tang, L.H. Zhang et al., PEGylated PAMAM dendrimers as a potential drug delivery carrier: in vitro and in vivo comparative evaluation of covalently conjugated drug and noncovalent drug inclusion complex, J Drug Target 18 (2010) 389–403.

[89] D. Bhadra, S. Bhadra, S. Jain et al., A PEGylated dendritic nanoparticulate carrier of fluorouracil, Int J Pharm 257 (2003) 111–124.

[90] R. Jevprasesphant, J. Penny, R. Jalal et al., The influence of surface modification on the cytotoxicity of PAMAM dendrimers, Int J Pharm 252 (2003) 263–266.

[91] H.Y. Cho, A. Srinivasan, J. Hong et al., Synthesis of biocompatible PEG-based star polymers with cationic and degradable core for siRNA delivery, Biomacromolecules 12 (2011) 3478–3486.

[92] X. Qian, L. Long, Z. Shi et al., Starbranched amphiphilic PLA-b-PDMAEMA copolymers for co-delivery of miR-21 inhibitor and doxorubicin to treat glioma, Biomaterials 35 (2014) 2322–2335.

[93] M.J. Cao, X.W. Deng, S.S. Su et al., Protamine sulfate-nanodiamond hybrid nanoparticles as a vector for MiR-203 restoration in esophageal carcinoma cells, Nanoscale 5 (2013) 12120–12125.

[94] A. Bitar, N.M. Ahmad, H. Fessi et al., Silica-based nanoparticles for biomedical applications, Drug Discov Today 17 (2012) 1147–1154.

[95] A. Schade, E. Delyagina, D. Scharfenberg et al., Innovative strategy for microRNA delivery in human mesenchymal stem cells via magnetic nanoparticles, Int J Mol Sci 14 (2013) 10710–10726.

[96] L. Vigderman, E.R. Zubarev, Therapeutic platforms based on gold nanoparticles and their covalent conjugates with drug molecules, Adv Drug Deliv Rev 65 (2013) 663–676.

[97] E. Crew, M.A. Tessel, S. Rahman et al., MicroRNA conjugated gold nanoparticles and cell transfection, Anal Chem 84 (2012) 10147.

[98] R. Ghosh, L.C. Singh, J.M. Shohet et al., A gold nanoparticle platform for the delivery of functional microRNAs into cancer cells, Biomaterials 34 (2013) 807–816.

[99] J.H. Kim, J.H. Yeom, J.J. Ko et al., Effective delivery of antimiRNA DNA oligonucleotides by functionalized gold nanoparticles, J Biotechnol 155 (2011) 287–292.

[100] P. Trang, P.P. Medina, J.F. Wiggins et al., Regression of murine lung tumors by the let-7microRNA, Oncogene 29 (2010) 1580–1587.

[101] S. Saini, S. Yamamura, S. Majid et al., MicroRNA-708 induces apoptosis and suppresses tumorigenicity in renal cancer cells, Cancer Res 71 (2011) 6208–6219.

[102] S. Shi, L. Han, T. Gong et al., Systemic delivery of microRNA-34a for cancer stem cell therapy, Angew Chem Int Engl 52 (2013) 3901–3905.

[103] V.J. Craig, A. Tzankov, M. Flori et al., Systemic microRNA-34a delivery induces apoptosis and abrogates growth of diffuse large B-cell lymphoma in vivo, Leukemia 26 (2012) 2421–2424.

[104] V. Torchilin, Tumor delivery of macromolecular drugs based on the EPR effect, Adv Drug Deliv Rev 63 (2011) 131–135.

[105] K. Maruyama, Intracellular targeting delivery of liposomal drugs to solid tumors based on EPR effects, Adv Drug Deliv Rev 63 (2011) 161–169.

[106] L. Brannon-Peppas, J.O. Blanchette, Nanoparticle and targeted systems for cancer therapy, Adv Drug Deliv Rev 56 (2004) 1649–1659.

[107] M.M. Schmidt, K.D. Wittrup, A modeling analysis of the effects of molecular size and binding affinity on tumor targeting, Mol Cancer Ther 8 (2009) 2861–2871.

[108] I.A. Babar, C.J. Cheng, C.J. Booth et al., Nanoparticle-based therapy in an in vivo microRNA-155 (miR-155)-dependent mouse model of lymphoma, Proc Natl Acad Sci USA 109 (2012) E1695–E1704.

[109] S.J. Shi, L. Han, T. Gong et al., Systemic delivery of microRNA-34a for cancer stem cell therapy, Angew Chem Int Ed Engl 52 (2013) 3901–3905.

[110] Y.H. Bae, Drug targeting and tumor heterogeneity, J Control Release 133 (2009) 2–3.

[111] X.Q. Liu, W.J. Song, T.M. Sun et al., Targeted delivery of antisense inhibitor of miRNA for antiangiogenesis therapy using cRGD-functionalized nanoparticles, Mol Pharm 8 (2011) 250–259.

[112] S. Ohno, M. Takanashi, K. Sudo et al., Systemically injected exosomes targeted to EGFR deliver antitumor microRNA to breast cancer cells, Mol Ther 21 (2013) 185–191.

[113] H.J. Lee, R. Namgung, W.J. Kim et al., Targeted delivery of microRNA-145 to metastatic breast cancer by peptide conjugated branched PEI gene carrier, Macromol Res 21 (2013) 1201–1209.

[114] Q.L. Hu, Q.Y. Jiang, X. Jin et al., Cationic microRNA-delivering nanovectors with bifunctional peptides for efficient treatment of PANC-1 xenograft model, Biomaterials 34 (2013) 2265–2276.

[115] D.W. Hwang, S. Son, J. Jang et al., A brain-targeted rabies virus glycoprotein-disulfide linked PEI nanocarrier for delivery of neurogenic microRNA, Biomaterials 32 (2011) 4968–4975.

[116] Y. Chen, X. Zhu, X. Zhang et al., Nanoparticles modified with tumor targeting scFv deliver siRNA and miRNA for cancer therapy, Mol Ther 18 (2010) 1650–1656.

[117] A. Tivnan, W.S. Orr, V. Gubala et al., Inhibition of neuroblastoma tumor growth by targeted delivery of microRNA-34a using anti-disialoganglioside GD2 coated nanoparticles, PLoS One 7 (2012) e38129.

[118] X. Wu, B. Ding, J. Gao et al., Second-generation aptamer-conjugated PSMA-targeted delivery system for prostate cancer therapy, Int J Nanomedicine 6 (2011) 1747–1756.

[119] J.K. Kim, K.J. Choi, M. Lee et al., Molecular imaging of a cancer-targeting theragnostics probe using a nucleolin aptamer- and microRNA-221 molecular beacon-conjugated nanoparticle, Biomaterials 33 (2012) 207–217.

[120] X. Deng, M. Cao, J. Zhang et al., Hyaluronic acid-chitosan nanoparticles for co-delivery of MiR-34a and doxorubicin in therapy against triple negative breast cancer, Biomaterials 35 (2014) 4333–4344.

[121] X. Ye, Z. Liu, M.G. Hemida et al., Targeted delivery of mutant tolerant anticoxsackievirus artificial microRNAs using folate conjugated bacteriophage Phi29 pRNA, PLoS One 6 (2011) e21215.

[122] M.V. Yigit, S.K. Ghosh, M. Kumar et al., Context-dependent differences in miR-10b breast oncogenesis can be targeted for the prevention and arrest of lymph node metastasis, Oncogene 32 (2013) 1530–1538.

[123] S.R. Singh, H.E. Grossniklaus, S.J. Kang et al., Intravenous transferrin, RGD peptide and dual-targeted nanoparticles enhance anti-VEGF intraceptor gene delivery to laser-induced CNV, Gene Ther 16 (2009) 645–659.

[124] N. Nasongkla, X. Shuai, H. Ai et al., cRGD functionalized polymer micelles for targeted doxorubicin delivery, Angew Chem Int Ed Engl 43 (2004) 6323–6327.

[125] K.N. Sugahara, T. Teesalu, P.P. Karmali et al., Coadministration of a tumor-penetrating peptide enhances the efficacy of cancer drugs, Science 328 (2010) 1031–1035.

[126] K. Klutz, D. Schaffert, M.J. Willhauck et al., Epidermal growth factor receptor-targeted I-131-therapy of liver cancer following systemic delivery of the sodium iodide symporter gene, Mol Ther 19 (2011) 676–685.

[127] D.W. Hwang, S. Son, J. Jang et al., A brain-targeted rabies virus glycoprotein-disulfide linked PEI nanocarrier for delivery of neurogenic microRNA, Biomaterials 32 (2011) 4968–4975.

[128] N.A. Maiorano, A. Mallamaci, Promotion of embryonic cortico-cerebral neuronogenesis by miR-124, Neural Dev 4 (2009) 40.

[129] P. Kumar, H. Wu, J.L. McBride et al., Transvascular delivery of small interfering RNA to the central nervous system, Nature 448 (2007) 39–43.

[130] X. Yi, D. Yuan, S.A. Farr et al., Pluronic modified leptin with increased systemic circulation, brain uptake and efficacy for treatment of obesity, J Control Release 191 (2014) 34–46.

[131] T.O. Price, S.A. Farr, X. Yi et al., Transport across the blood–brain barrier of pluronic leptin, J Pharmacol Exp Ther 333 (2010) 253–263.

[132] X. Zhang, C.G. Koh, B. Yu et al., Transferrin receptor targeted lipopolyplexes for delivery of antisense oligonucleotide g3139 in a murine k562 xenograft model, Pharm Res 26 (2009) 1516–1524.

[133] W.M. Pardridge, Blood–brain barrier drug delivery of IgG fusion proteins with a transferrin receptor monoclonal antibody, Expert Opin Drug Deliv (2014) 1–16.

[134] Q.H. Zhou, R.J. Boado, E.K. Hui et al., Chronic dosing of mice with a transferrin receptor monoclonal antibody-glial-derived neurotrophic factor fusion protein, Drug Metab Dispos 39 (2011) 1149–1154.

[135] Z.A. Ahmad, S.K. Yeap, A.M. Ali et al., scFv antibody: principles and clinical application, Clin. Dev. Immunol 2012 (2012) 980250.

[136] V. Orian-Rousseau, CD44, a therapeutic target for metastasising tumours, Eur J Cancer 46 (2010) 1271–1277.

[137] O. Veiseh, J.W. Gunn, M. Zhang, Design and fabrication of magnetic nanoparticles for targeted drug delivery and imaging, Adv Drug Deliv Rev 62 (2010) 284–304.

[138] W. Li, N. Ma, L.L. Ong et al., Enhanced thoracic gene delivery by magnetic nanobead-mediated vector, J Gene Med 10 (2008) 897–909.

[139] L.S. Lin, Z.X. Cong, J.B. Cao et al., Multifunctional Fe3O4@polydopamine core-shell nanocomposites for intracellular mRNA detection and imaging-guided photothermal therapy, ACS Nano 8 (2014) 3876–3883.

[140] A. Schade, P. Muller, E. Delyagina et al., Magnetic nanoparticle based nonviral microRNA delivery into freshly isolated CD105(+) hMSCs, Stem Cells Int 2014 (2014) 197154.

[141] R.L. Siegel, K.D. Miller, A. Jemal, Cancer statistics, 2017, CA Cancer J Clin 67 (1) (2017) 7–30.

[142] I. Romero, R.C. Bast Jr., Minireview: human ovarian cancer: biology, current management, and paths to personalizing therapy, Endocrinology 153 (4) (2012) 1593–1602.

[143] B. Winterhoff, H. Hamidi, C. Wang et al., Molecular classification of high grade endometrioid and clear cell ovarian cancer using TCGA gene expression signatures, Gynecol Oncol 141 (1) (2016) 95–100.

[144] Y.K. Wang, A. Bashashati, M.S. Anglesio et al., Genomic consequences of aberrant DNA repair mechanisms stratify ovarian cancer histotypes, Nat Genet 49 (6) (2017) 856–865.

[145] M. Logan, S.M. Hawkins, Role of microRNAs in cancers of the female reproductive tract: insights from recent clinical and experimental discovery studies, Clin Sci 128 (3) (2015) 153–180.

[146] D.P. Bartel, MicroRNAs: genomics, biogenesis, mechanism, and function, Cell 116 (2) (2004) 281–297.

[147] C.J. Creighton, A. Hernandez-Herrera, A. Jacobsen et al., Integrated analyses of microRNAs demonstrate their widespread influence on gene expression in high-grade serous ovarian carcinoma, PLoS One 7 (3) (2012) e34546.

[148] V. Ambros, B. Bartel, D.P. Bartel et al., A uniform system for microRNA annotation, RNA 9 (3) (2003) 277–279.

[149] J. Boele, H. Persson, J.W. Shin et al., PAPD5-mediated 3′ adenylation and subsequent degradation of miR-21 is disrupted in proliferative disease, Proc Natl Acad Sci USA 111 (31) (2014) 11,467–72.

[150] J. Kim, D.M. Coffey, C.J. Creighton et al., High-grade serous ovarian cancer arises from fallopian tube in a mouse model, Proc Natl Acad Sci USA 109 (10) (2012) 3921–3926.

[151] R.J. Kurman, Ie M. Shih, The origin and pathogenesis of epithelial ovarian cancer: a proposed unifying theory, Am J Surg Pathol 34 (3) (2010) 433–443.

[152] J. Kim, D.M. Coffey, L. Ma et al., The ovary is an alternative site of origin for high-grade serous ovarian cancer in mice, Endocrinology 156 (6) (2015) 1975–1981.

[153] C.M. Nagle, C.M. Olsen, P.M. Webb et al., Endometrioid and clear cell ovarian cancers: a comparative analysis of risk factors, Eur J Cancer 44 (16) (2008) 2477–2484.

[154] E. Somigliana, P. Vigano, F. Parazzini et al., Association between endometriosis and cancer: a comprehensive review and a critical analysis of clinical and epidemiological evidence, Gynecol Oncol 101 (2) (2006) 331–341.

[155] R.J. Flavin, P.C. Smyth, S.P. Finn et al., Altered eIF6 and Dicer expression is associated with clinicopathological features in ovarian serous carcinoma patients, Mod Pathol 21 (6) (2008) 676–684.

[156] W.M. Merritt, Y.G. Lin, L.Y. Han et al., Dicer, Drosha, and outcomes in patients with ovarian cancer, N Engl J Med 359 (25) (2008) 2641–2650.

[157] G. Pampalakis, E.P. Diamandis, D. Katsaros et al., Downregulation of dicer expression in ovarian cancer tissues, Clin Biochem 43 (3) (2010) 324–327.

[158] A. Faggad, J. Budczies, O. Tchernitsa et al., Prognostic significance of Dicer expression in ovarian cancer-link to global microRNA changes and oestrogen receptor expression, J Pathol 220 (3) (2010) 382–391.

[159] L. Zhang, S. Volinia, T. Bonome et al., Genomic and epigenetic alterations deregulate microRNA expression in human epithelial ovarian cancer, Proc Natl Acad Sci USA 105 (19) (2008) 7004–7009.

[160] N.S. Spoelstra, D.M. Cittelly, J.L. Christenson et al., Dicer expression in estrogen receptor-positive versus triple-negative breast cancer: an antibody comparison, Hum Pathol 56 (2016) 40–51.

[161] W. Shan, C. Sun, B. Zhou et al., Role of Dicer as a prognostic predictor for survival in cancer patients: a systematic review with a meta-analysis, Oncotarget 7 (45) (2016) 72,672–84.

[162] O. Vaksman, T.E. Hetland, C.G. Trope et al., Argonaute, Dicer, and Drosha are up-regulated along tumor progression in serous ovarian carcinoma, Hum Pathol 43 (11) (2012) 2062–2069.

[163] E. Bernstein, S.Y. Kim, M.A. Carmell et al., Dicer is essential for mouse development, Nat Genet 35 (3) (2003) 215–217.

[164] S. Morita, T. Horii, M. Kimura et al., One Argonaute family member, Eif2c2 (Ago2), is essential for development and appears not to be involved in DNA methylation, Genomics 89 (6) (2007) 687–696.

[165] S.M. Hawkins, G.M. Buchold, M.M. Matzuk, Minireview: the roles of small RNA pathways in reproductive medicine, Mol Endocrinol 25 (8) (2011) 1257–1279.

[166] Cancer Genome Atlas Research N, Integrated genomic analyses of ovarian carcinoma, Nature 474 (7353) (2011) 609–615.

[167] M.S. Kumar, R.E. Pester, C.Y. Chen et al., Dicer1 functions as a haploinsufficient tumor suppressor, Genes Dev 23 (23) (2009) 2700–2704.

[168] J.K. Mito, H.D. Min, Y. Ma et al., Oncogene-dependent control of miRNA biogenesis and metastatic progression in a model of undifferentiated pleomorphic sarcoma, J Pathol 229 (1) (2013) 132–140.

[169] I. Lambertz, D. Nittner, P. Mestdagh et al., Monoallelic but not biallelic loss of Dicer1 promotes tumorigenesis in vivo, Cell Death Differ 17 (4) (2010) 633–641.

[170] B. Zhang, H. Chen, L. Zhang et al., A dosage-dependent pleiotropic role of Dicer in prostate cancer growth and metastasis, Oncogene 33 (24) (2014) 3099–3108.

[171] M.P. Arrate, T. Vincent, J. Odvody et al., MicroRNA biogenesis is required for Myc-induced B-cell lymphoma development and survival, Cancer Res 70 (14) (2010) 6083–6092.

[172] M.S. Anglesio, Y. Wang, W. Yang et al., Cancer associated somatic DICER1 hotspot mutations cause defective miRNA processing and reverse-strand expression bias to predominantly mature 3p strands through loss of 5p strand cleavage, J Pathol 229 (3) (2013) 400–409.

[173] J. Chen, Y. Wang, M.K. McMonechy et al., Recurrent DICER1 hotspot mutations in endometrial tumours and their impact on microRNA biogenesis, J Pathol 237 (2) (2015) 215–225.

[174] A. Heravi-Moussavi, M.S. Anglesio, S.W. Cheng et al., Recurrent somatic DICER1 mutations in nonepithelial ovarian cancers, N Engl J Med 366 (3) (2012) 234–242.

[175] Y. Zou, M.Z. Huang, F.Y. Liu et al., Absence of DICER1, CTCF, RPL22, DNMT3A, TRRAP, IDH1 and IDH2

hotspot mutations in patients with various subtypes of ovarian carcinomas, Biomed Rep 3 (1) (2015) 33–37.

[176] M.S. Kim, S.H. Lee, N.J. Yoo et al., DICER1 exons 25 and 26 mutations are rare in common human tumours besides Sertoli-Leydig cell tumour, Histopathology 63 (3) (2013) 436–438.

[177] D. Liang, L. Meyer, D.W. Chang et al., Genetic variants in MicroRNA biosynthesis pathways and binding sites modify ovarian cancer risk, survival, and treatment response, Cancer Res 70 (23) (2010) 9765–9776.

[178] J. Permuth-Wey, Z. Chen, Y.Y. Tsai et al., MicroRNA processing and binding site polymorphisms are not replicated in the Ovarian Cancer Association Consortium, Cancer Epidemiol Biomark Prev 20 (8) (2011) 1793–1797.

[179] P.D. Pharoah, R.T. Palmieri, S.J. Ramus et al., The role of KRAS rs61764370 in invasive epithelial ovarian cancer: implications for clinical testing, Clin Cancer Res 17 (11) (2011) 3742–3750.

[180] E. Caiola, E. Rulli, R. Fruscio et al., KRas-LCS6 polymorphism does not impact on outcomes in ovarian cancer, Am J Cancer Res 2 (3) (2012) 298–308.

[181] G.L. Ryland, J.L. Bearfoot, M.A. Doyle et al., MicroRNA genes and their target 3′-untranslated regions are infrequently somatically mutated in ovarian cancers, PLoS One 7 (4) (2012) e35805.

[182] Y.W. Wan, C.M. Mach, G.I. Allen et al., On the reproducibility of TCGA ovarian cancer microRNA profiles, PLoS One 9 (1) (2014) e87782.

[183] Y.C. Gao, J. Wu, MicroRNA-200c and microRNA-141 as potential diagnostic and prognostic biomarkers for ovarian cancer, Tumour Biol 36 (6) (2015) 4843–4850.

[184] T. Zhu, W. Gao, X. Chen et al., A pilot study of circulating microRNA-125b as a diagnostic and prognostic biomarker for epithelial ovarian cancer, Int J Gynecol Cancer 27 (1) (2017) 3–10.

[185] M. Zuberi, I. Khan, R. Mir et al., Utility of serum miR-125b as a diagnostic and prognostic indicator and its alliance with a panel of tumor suppressor genes in epithelial ovarian cancer, PLoS One 11 (4) (2016) e0153902.

[186] M. Zuberi, I. Khan, G. Gandhi et al., The conglomeration of diagnostic, prognostic and therapeutic potential of serum miR-199a and its association with clinicopathological features in epithelial ovarian cancer, Tumor Biol 37 (8) (2016) 11,259–66.

[187] M. Zuberi, R. Mir, J. Das et al., Expression of serum miR-200a, miR-200b, and miR-200c as candidate biomarkers in epithelial ovarian cancer and their association with clinicopathological features, Clin Transl Oncol 17 (10) (2015) 779–787.

[188] I. Shapira, M. Oswald, J. Lovecchio et al., Circulating biomarkers for detection of ovarian cancer and predicting cancer outcomes, Br J Cancer 110 (4) (2014) 976–983.

[189] J. Zhou, G. Gong, H. Tan et al., Urinary microRNA-30a-5p is a potential biomarker for ovarian serous adenocarcinoma, Oncol Rep 33 (6) (2015) 2915–2923.

[190] W.T. Chen, Y.J. Yang, Z.D. Zhang et al., miR-1307 promotes ovarian cancer cell chemoresistance by targeting the ING5 expression, J Ovarian Res 10 (1) (2017) 1.

[191] S.Y. Wu, R. Rupaimoole, F. Shen et al., A miR-192-EGR1-HOXB9 regulatory network controls the angiogenic switch in cancer, Nat Commun 7 (2016) 11,169.

[192] Y. Sun, L. Hu, H. Zheng et al., miR-506 inhibits multiple targets in the epithelial-to-mesenchymal transition network and is associated with good prognosis in epithelial ovarian cancer, J Pathol 235 (1) (2015) 25–36.

[193] G. Liu, D. Yang, R. Rupaimoole et al., Augmentation of response to chemotherapy by microRNA-506 through regulation of RAD51 in serous ovarian cancers, J Natl Cancer Inst 107 (7) (2015).

[194] P. Chaluvally-Raghavan, K.J. Jeong, S. Pradeep et al., Direct upregulation of STAT3 by microRNA-551b-3p deregulates growth and metastasis of ovarian cancer, Cell Rep 15 (7) (2016) 1493–1504.

[195] P. Kanlikilicer, M.H. Rashed, R. Bayraktar et al., Ubiquitous release of exosomal tumor suppressor miR-6126 from ovarian cancer cells, Cancer Res 76 (24) (2016) 7194–7207.

[196] S.E. Bulun, Endometriosis, N Engl J Med 360 (3) (2009) 268–279.

[197] C.L. Pearce, C. Templeman, M.A. Rossing et al., Association between endometriosis and risk of histological subtypes of ovarian cancer: a pooled analysis of case-control studies, Lancet Oncol 13 (4) (2012) 385–394.

[198] N. Yanaihara, Y. Noguchi, M. Saito et al., MicroRNA gene expression signature driven by miR-9 overexpression in ovarian clear cell carcinoma, PLoS One 11 (9) (2016), e0162584.

[199] X. Zhang, G. Guo, G. Wang et al., Profile of differentially expressed miRNAs in high-grade serous carcinoma and clear cell ovarian carcinoma, and the expression of miR-510 in ovarian carcinoma, Mol Med Rep 12 (6) (2015) 8021–8031.

[200] H. Zhao, Y. Ding, B. Tie et al., miRNA expression pattern associated with prognosis in elderly patients with advanced OPSC and OCC, Int J Oncol 43 (3) (2013) 839–849.

[201] C. Chang, T. Liu, Y. Huang et al., MicroRNA-134-3p is a novel potential inhibitor of human ovarian cancer stem cells by targeting RAB27A, Gene 605 (2017) 99–107.

[202] Y. Hirata, N. Murai, N. Yanaihara et al., MicroRNA-21 is a candidate driver gene for 17q23–25 amplification in ovarian clear cell carcinoma, BMC Cancer 14 (2014) 799.

[203] A. Chao, C.H. Lai, H.C. Chen et al., Serum microRNAs in clear cell carcinoma of the ovary. J Obstet Gynecol 53 (4) (2014) 536–541.

[204] S. Suryawanshi, A.M. Vlad, H.M. Lin et al., Plasma microRNAs as novel biomarkers for endometriosis and endometriosis-associated ovarian cancer, Clin Cancer Res 19 (5) (2013) 1213–1224.

[205] X. Tian, L. Xu, P. Wang, miR-191 inhibits TNF-alpha induced apoptosis of ovarian endometriosis and endometrioid carcinoma cells by targeting DAPK1, Int J Clin Exp Pathol 8 (5) (2015) 4933–4942.

[206] M. Dong, P. Yang, F. Hua, miR-191 modulates malignant transformation of endometriosis through regulating TIMP3, Med Sci Monit 21 (2015) 915–920.

[207] X.P. Chen, Y.G. Chen, J.Y. Lan et al., MicroRNA-370 suppresses proliferation and promotes endometrioid ovarian cancer chemosensitivity to cDDP by negatively regulating ENG, Cancer Lett 353 (2) (2014) 201–210.

[208] J. Eisermann, M.J. Gast, J. Pineda et al., Tumor necrosis factor in peritoneal fluid of women undergoing laparoscopic surgery, Fertil Steril 50 (4) (1988) 573–579.

[209] R. Langhe, L. Norris, F.A. Saadeh et al., A novel serum microRNA panel to discriminate benign from malignant ovarian disease, Cancer Lett 356 (2 Pt B) (2015) 628–636.

[210] A.R. Halvorsen, G. Kristensen, A. Embleton et al., Evaluation of prognostic and predictive significance of circulating MicroRNAs in ovarian cancer patients, Dis Markers 2017 (2017) 3,098,542.

[211] M. Petrillo, G.F. Zannoni, L. Beltrame et al., Identification of high-grade serous ovarian cancer miRNA species

associated with survival and drug response in patients receiving neoadjuvant chemotherapy: a retrospective longitudinal analysis using matched tumor biopsies, Ann Oncol 27 (4) (2016) 625–634.

[212] Y.E. Choi, K. Meghani, M.E. Brault et al., Platinum and PARP inhibitor resistance due to overexpression ofmicroRNA-622 in BRCA1-mutant ovarian cancer, Cell Rep 14 (3) (2016) 429–439.

[213] C. Catalanotto, M. Pallotta, P. ReFalo et al., Redundancy of the two dicer genes in transgene-induced posttranscriptional gene silencing in Neurospora crassa, Mol Cell Biol 24 (6) (2004) 2536–2545.

[214] C. Devlin, S. Greco, F. Martelli et al., miR-210: more than a silent player in hypoxia, IUBMB Life 63 (2) (2011) 94–100.

[215] M.A. Cortez, C. Ivan, P. Zhou et al., microRNAs in cancer: from bench to bedside, Adv Cancer Res 108 (2010) 113–157.

[216] R. Rupaimoole, S.Y. Wu, S. Pradeep et al., Hypoxia-mediated downregulation of miRNA biogenesis promotes tumour progression, Nat Commun 5 (2014) 5202.

[217] Y. Kinose, K. Sawada, K. Nakamura et al., The hypoxia-related microRNA miR-199a-3p displays tumor suppressor functions in ovarian carcinoma, Oncotarget 6 (13) (2015) 11,342–56.

[218] Y. Jin, C. Wang, X. Liu et al., Molecular characterization of the microRNA-138-Fos-like antigen 1 (FOSL1) regulatory module in squamous cell carcinoma, J Biol Chem 286 (46) (2011) 40,104–9.

[219] S. Obad, C.O. dos Santos, A. Petri et al., Silencing of microRNA families by seed-targeting tiny LNAs, Nat Genet 43 (2011) 371–378.

[220] J. Guo, J.C. Evans, C.M. O'Driscoll, Delivering RNAi therapeutics with non-viral technology: a promising strategy for prostate cancer? Trends Mol Med 19 (2013) 250–261.

[221] A. Bouchie, First microRNA mimic enters clinic, Nat Biotechnol 31 (2013) 577.

[222] S. Khan, Ansarullah, D. Kumar et al., Targeting microRNAs in pancreatic cancer: microplayers in the big game, Cancer Res 73 (2013) 6541–6547.

[223] L.H. Madkour, Reactive Oxygen Species (ROS), Nanoparticles, and Endoplasmic Reticulum (ER) Stress-Induced Cell Death Mechanisms. Academic Press, 1st August 2020. https://www.elsevier.com/books/reactive-oxygen-species-ros-nanoparticles-and-endoplasmic-reticulum-er-stress-induced-cell-death-mechanisms/madkour/978-0-12-822481-6

[224] L.H. Madkour, Nanoparticles Induce Oxidative and Endoplasmic Reticulum Antioxidant Therapeutic Defenses, 1st Edition, 2020. Springer International Publishing, Switzerland. https://www.springer.com/gp/book/9783030372965?utm_campaign=3_pier05_buy_print&utm_content=en_0808 2017&utm_medium=referral&utm_source=google_books #otherversion=9783030372972

[225] L.H. Madkour, Nucleic Acids as Gene Anticancer Drug Delivery Therapy, 1st Edition, 2020. Elsevier, Academic Press. https://www.elsevier.com/books/nucleic-acids-as-gene-anticancer-drug-deliverytherapy/madkour/978-0-12-819777-6

[226] L.H. Madkour, Nanoelectronic Materials: Fundamentals and Applications (Advanced Structured Materials), 1st ed. 2019. Springer International, Switzerland. https://books.google.com.eg/books/about/Nanoelectronic_Materials.html?id=YQXCxAEACAAJ&source=kp_book_description& redir_esc=y https://www.springer.com/gp/book/9783030 216207

[227] E. van Rooij, A.L. Purcell, A.A. Levin, Developing microRNA therapeutics, Circ Res 110 (2012) 496–507.

# 10 DNA-Damaging Cancer Therapies and FDA Novel Drug Approvals

In this chapter, we will review DNA damage surveillance networks, which maintain the stability of our genome, and discuss the efforts underway to identify chemotherapeutic compounds targeting the core components of DNA double-strand break (DSB) response pathways. The theory relies primarily on the observation that the majority of tumors are deficient in the G1-DNA damage checkpoint pathway, resulting in reliance on S and $G_2$ phase checkpoints for DNA repair and cell survival. Normal tissues, however, have a functioning G1checkpoint signaling pathway that allows for DNA repair and cell survival. There is now a large body of preclinical evidence showing that checkpoint kinase inhibitors do indeed enhance the efficacy of both conventional chemotherapy and radiotherapy, and several agents have recently entered clinical trials. In addition, the treatment of BRCA1- or BRCA2-deficient tumor cells with poly(ADP-ribose) polymerase (PARP) inhibitors also leads to specific tumor killing. Due to the numerous roles of p53 in genomic stability and its defects in many human cancers, therapeutic agents that restore p53 activity in tumors are the subject of multiple clinical trials. Excitingly, additional therapeutic opportunities for checkpoint kinase inhibitors continue to emerge as biology outside their pivotal role in cell cycle arrest is further elucidated.

## 10.1 DEFECTS IN SIGNALING AND REPAIR OF DNA DAMAGE

The importance of a robust DNA damage surveillance network is underscored by the fact that defects in signaling and repair of DNA damage are causally linked with the development of genomic instability and human cancer. For example, mutation of the ATM kinase, central to cell cycle checkpoint activation after DSBs, leads to the cancer-prone syndrome ataxia telangiectasia [1]. In addition, mutation of meiotic recombination 11 (Mre11) or NBS1, components of the DSB-sensing MRN (comprised of Mre11, Rad50, and NBS1) complex, leads to the genomic instability syndromes ataxia-telangiectasia-like-disorder (AT-LD) and Nijmegen Breakage syndrome (NBS), respectively [2].

Detailed description of the DNA damage response is beyond the scope of this chapter, and the reader is referred to several excellent treatises on DNA damage responses [3–6]. Only the features of DNA DSB signaling and repair, which are potential targets for therapeutic applications, will be highlighted here.

A wide variety of DNA lesions elicit the activation of cell cycle checkpoints, controlled by the ATM and ATR kinases [7]. The sensors of DNA damage signaling that activate ATM and ATR by recruiting them to sites of DNA damage include the MRN complex and Replication Protein A (RPA)-coated single-stranded DNA (ssDNA), respectively. ATM activation is indicated by its autophosphorylation, its recruitment to chromatin, and conversion from a dimer to a monomeric form [8]. A number of proteins are required for maximal activation of ATM; these include hSSB1 and the MRN complex [9, 10]. Once activated, ATM phosphorylates downstream effector proteins to initiate cell cycle checkpoints at the G1/S, intra-S, and G2/M boundaries. The activation of these checkpoints allows repair of DNA damage before it is replicated and passed on to daughter cells and therefore preserves the genomic integrity. Checkpoint proteins 1 and 2 (Chk1 and Chk2) are key downstream checkpoint substrates of ATM and ATR (Figure 10.1). During checkpoint activation, Chk1 and Chk2 phosphorylation is necessary for the activation of the DNA damage checkpoints [11, 12]. In order to activate the checkpoints and stall the cell cycle, Chk1 or Chk2 phosphorylate several downstream substrates (see review article [13] for further details).

Activated ATM also contributes to the phosphorylation of the tumor suppressor protein p53 on serine 15, which in turn transactivates the kinase inhibitor p21, leading to the inhibition of two cyclin-dependent kinases, Cyclin E/A-Cdk2, which stall the cell cycle at the G1/S boundary (Figure 10.1). Protein p53 also plays a role in maintaining the G2/M checkpoint via transactivation of p21. Other substrates involved in checkpoint activation that are phosphorylated by ATM include Mre11, NBS1, RPA34, mediator of the damage checkpoint 1 (MDC1), structural maintenance of chromatin 1 (SMC1), and BRCA1 [14].

DNA DSBs are perhaps the biggest threat to genomic stability. Unrepaired DSBs can lead to cell death, whereas incorrectly repaired DSBs have the potential to produce chromosomal translocation and genomic instability. There are two main pathways of DSB repair, nonhomologous end joining (NHEJ) and homology-directed repair (HR).

In mammalian cells, the majority of ionizing radiation (IR)-induced DSBs are repaired by NHEJ. NHEJ occurs mainly in the G0 and G1 phase of the cycle (Figure 10.1). It is responsible for the repair of spontaneous DSBs induced by agents such as IR and programmed DSBs generated during generation of T-cell receptors and immunoglobulin molecules via V(D)J recombination. NHEJ involves rejoining of the two broken ends of the DNA in a sequence-independent manner. Consequently NHEJ is sometimes viewed

DOI: 10.1201/9781003229650-10

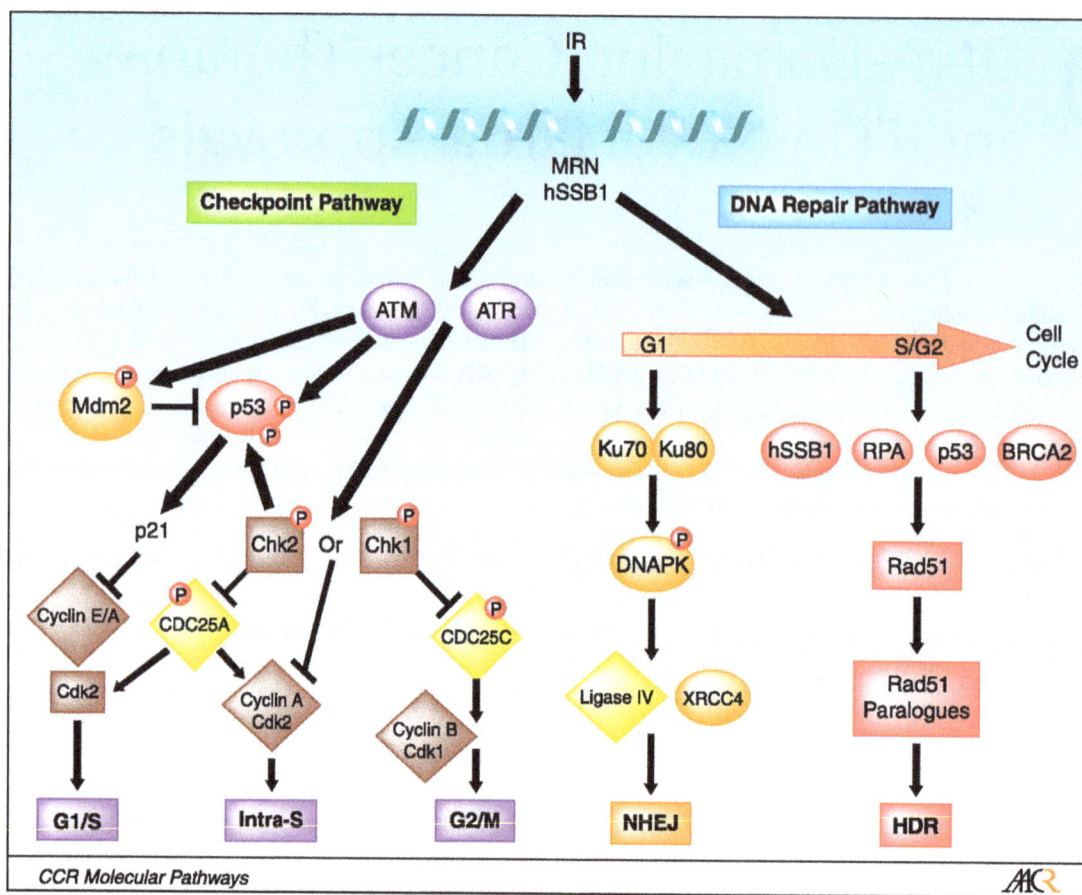

**FIGURE 10.1** DNA DSB-induced activation of checkpoint and repair pathways. Activation of the upstream kinases ATM and ATR signal cell cycle checkpoint signaling cascades through the effector kinases CHK1 and CHK2. These kinases are responsible for inhibitory phosphorylation of Cdc25, a phosphatase required for activation of Cdks, resulting in cell cycle arrest. IR-induced DSBs are detected and repaired by NHEJ or HR. NHEJ functions preferentially during G0 and G1, whereas HR is prevalent during late S and G2 phase.

as a more error prone method of DSB repair as genetic information can be lost in the repair of staggered breaks. NHEJ requires coordination of many proteins and signaling pathways. One of the main proteins involved in this repair pathway is DNA-dependent protein kinase (DNA-PK). The Ku70/80 heterodimer component of DNA-PK binds to the two DNA ends in a ring conformation. The DNA binding of Ku70/80 and aligning of the two DNA ends subsequently activates the catalytic activity of DNA-PK, which promotes the ligation of DNA ends by the XRCC4-Ligase IV complex. The recently identified NHEJ component, XRCC4-like factor (XLF), stimulates the activity of the XRCC4-DNA ligase IV complex toward noncompatible DNA ends [15]. DNA end-processing enzymes such as Artemis are also required for the processing of a subset of IR-induced DSBs in vivo [16].

HR is the second major pathway for the repair of DSBs in mammalian cells. It functions only during the late S and G2 phases of the cell cycle, when a homologous region of DNA is available (Figure 10.1). HR uses the homologous template to faithfully repair the DNA, and, consequently,

it is a more accurate form of DSB repair. HR is the predominant pathway of repair of endogenous DSBs that are produced when replication fork collapse occurs (see several reviews on this topic; refs. [4, 17, 18]).

Following induction of a DSB, one of the first events to occur is resection of the break in a 5′- to 3′-dependent manner to produce long stretches of ssDNA, which acts as a signal to recruit other DNA damage repair proteins. The MRN complex, one of the central components of HR, is stimulated to initiate resection in the early stages of HR by an interaction with CtIP [19]. A ssDNA-binding heterotrimeric complex known as RPA then binds to the exposed ssDNA. Another DNA repair protein BRCA1 is required to retain RPA at the sites of DSBs [20]. The bound RPA serves to protect the DNA from degradation by nucleases and formation of secondary structures. RPA is then removed from the DNA to allow the binding of the Rad51 recombinase, which forms a nucleoprotein filament along the DNA to facilitate completion of HR. Rad51 loading onto the DNA is dependent upon the presence of the BRCA2 protein [21, 22]. Additional members of the Rad51 family, including

Rad52 and the Rad51 paralogues, are also required to complete HR. Upon completion of HR, the DNA ends are ligated together [23].

DNA damage, if unrepaired, may result in mutation or cell death. Therefore, cells have evolved complex signaling networks to carefully monitor the integrity of the genome during DNA replication, and to initiate cell cycle arrest, repair, or apoptotic responses if errors are detected. Cancer cells, on the other hand, undergo an array of genetic changes including mutations in the DNA repair pathways that impair these controls and barriers. Over the last decade, there has been a tremendous increase in the understanding of the mechanisms of DNA damage detection, signaling, and repair, and these findings have suggested a variety of therapeutic opportunities for agents that modulate these pathways [24, 25]. As members of this network, checkpoint kinases have emerged as exciting targets, and inhibitors from AstraZeneca, Exelixis, Lilly, and Pfizer have recently entered early clinical development. In addition, there are many other checkpoint kinase (Chk)1 inhibitors in the preclinical phase, including compounds from Millenium, Merck, Abbott, Vernalis, and Chiron, so it is highly likely that in the near future, additional compounds will join those already in clinical trials.

Although it is only part of the cellular machinery that recognizes and responds to DNA damage, the signaling cascade that specifically regulates cell cycle arrest after DNA damage can itself be thought of as a complex network of interconnected pathways consisting of three main components—sensors, signal transducers, and effectors [26–28], as shown in Figure 10.2.

DNA damage triggers recruitment of multiprotein complexes (sensors) that then activate the transducers ataxia telangiectasia mutated (ATM) and ATR (ATM and Rad3 related), which belong to the phosphoinositide 3-kinase–like kinase family. It is generally accepted that ATR activation is driven by single-strand breaks formed as a result of stalled replication forks, whereas ATM is the main initiator of response to double-strand breaks resulting from ionizing radiation and other types of DNA damage [29].

Once activated, ATM and ATR phosphorylate a host of substrates, initiating a cascade that results in cell cycle arrest and DNA repair [30, 31].

Chk1 and Chk2 are checkpoint kinases downstream of ATM and ATR and play a critical role in determining cellular responses to DNA damage. Chk1 is a serine/threonine kinase and is primarily responsible for initiating cell cycle arrest, allowing time for DNA repair and cell survival [32, 33].

After its activation, Chk1 phosphorylates many serine residues on the protein phosphatase Cdc25a, facilitating recognition by ubiquitin ligases. Ubiquitination leads to its proteolysis and, thus, limits its ability to drive progression through the S phase [34–37]. Chk1 also phosphorylates Cdc25c, preventing dephosphorylation and activation of CDK1, resulting in cell cycle arrest in the G2 phase [38–40]. Further support for the pivotal role of Chk1 in cell

cycle checkpoint control has come from siRNA studies that have shown the importance of Chk1 in regulation of the S, intra-S, and the G2-M phase checkpoints [34, 41–43].

Activation of Chk2 is initiated by factors that induce single-strand breaks, such as replication stress or chemotherapeutic agents. Once activated, the effects of Chk2 on the effector proteins Cdc25a, Cdc25c, and p53 are similar to those mediated by Chk1 [35, 44–46].

Although their effects on downstream pathways share some similarities, inhibition of Chk1 or Chk2 may have profoundly different outcomes. For example, knockout animal studies revealed drastic differences in phenotype between Chk1 and Chk2 null mice. Chk1 (–/–) mice are embryonic lethal, whereas Chk2 (–/–) mice are viable and seem normal. However, tissues from Chk2–/– mice do show significant defects in G1-S checkpoint and IR-induced apoptosis [47].

Recently, further insight was gained into the interrelationship of Chk1 and Chk2 by the generation of conditional mutant mice in which Chk1 was only deleted in the T lineage [48]. It was found that in the absence of Chk1, the transition of CD4-CD8- double-negative thymocytes to CD4+CD8+ double-positive cells was blocked by an increase in apoptosis. The loss of Chk1 resulted in the activation of checkpoint kinase Chk2 in these thymocytes and conversely the loss of Chk2 resulted in the activation of Chk1. Interestingly, the double knockout mitigated some of the effects of Chk1 deletion. These data suggest that with Chk1 deletion, crosstalk between the pathways leads to Chk2-induced apoptosis [48]. Therefore, a dual inhibitor of both Chk1 and Chk2 might provide some protective effects on normal tissues such as thymocytes.

Chk2 has also been shown to be modified in Chk1lox/-ES cells expressing an inactive form of Chk1, again supporting crosstalk between the pathways [49]. Based on the crosstalk between these pathways, it could be hypothesized that it may be beneficial to target both Chk1 and Chk2 simultaneously to avoid compensatory mechanisms. Any such benefit, however, is likely to be dependent on the genetic background of the tumor because there are conflicting studies demonstrating inhibition of Chk1 and Chk2 has no benefit over inhibition of Chk1 alone [50, 51].

The phenotypic effects of the pharmacologic modulation of checkpoint pathways were first recognized in early work with caffeine (an inhibitor of ATR and ATM), and with the staurosporine analogue, UCN-01 (KW-2401, NSC 638850). It was shown that these agents could abrogate DNA damage–induced G2 arrest and selectively sensitize p53 mutant cells to radiation [52–57]. When one of the targets of UCN-01 was identified as Chk1 ($IC_{50}$, 10 nmol/L), it was recognized that Chk1 could be a useful therapeutic target to induce enhanced cytotoxicity in tumor cells in response to DNA damage [58, 59].

It should be noted, however, that UCN-01 also potently inhibits a number of other kinases including Chk2 ($IC_{50}$, 10 nmol/L) and a number of the cyclin-dependent kinases; thus, the clinical effects of this agent cannot be presumed

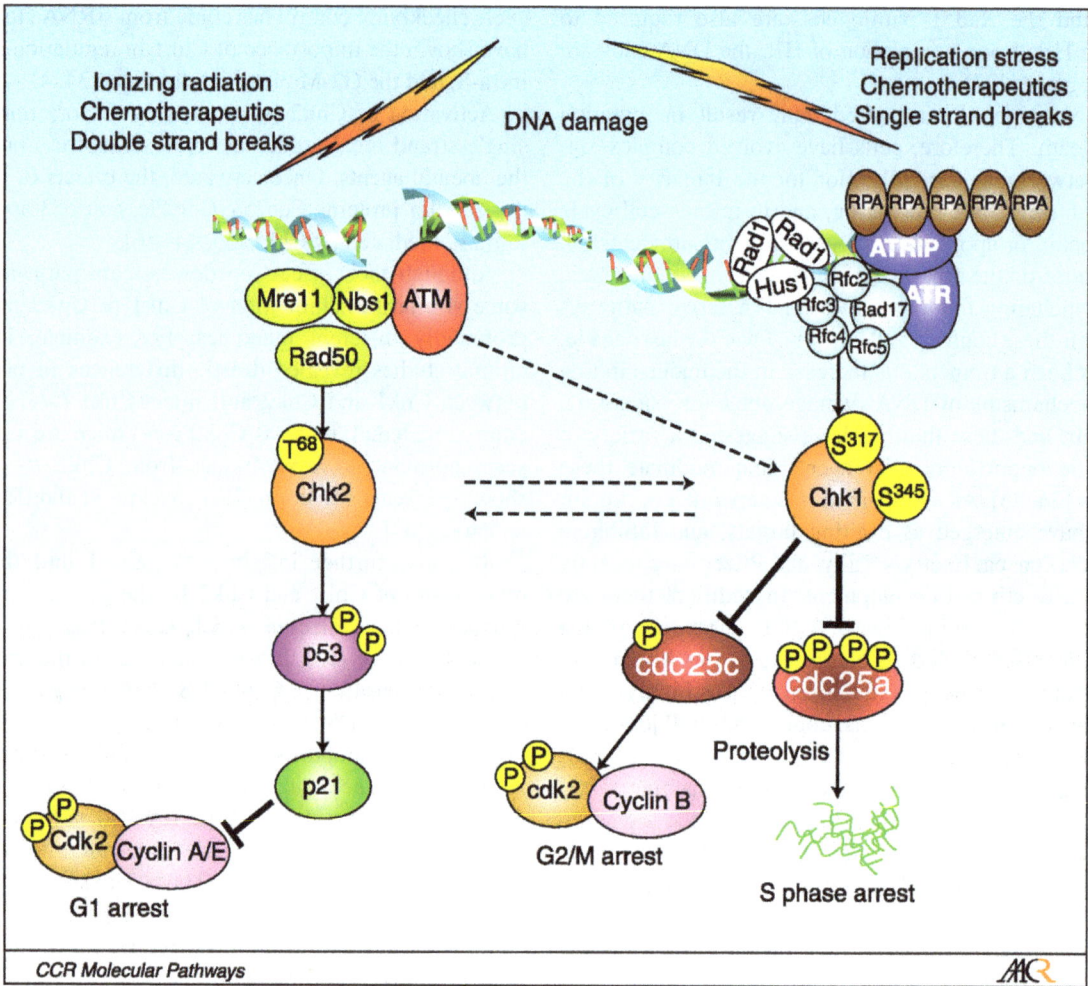

**FIGURE 10.2** Chk1 and Chk2 kinases are serine/threonine kinases that are activated by the ATM and ATR kinases in response to DNA damage. The checkpoint kinases are transducers of the DNA damage signal and both phosphorylate a number of substrates involved in the DNA damage response. Chk1 and Chk2 share a number of overlapping substrates, although it is clear that they have distinct roles in directing the response of the cell to DNA damage. Current understanding is that the checkpoint kinases are involved not only in cell cycle regulation but also in other aspects of the cellular response to DNA damage. The G1checkpoint is modulated primarily by theATM-Chk2-p53 pathway, as expression of ATR, Chk1, and Cdc25a is limited until the cell passes this restriction point. At this point, levels of ATR, Chk1, and Cdc25a all increase. If DNA damage is detected, Chk1/Chk2 are activated, Cdc25a is phosphorylated, and thus, destabilized, resulting in a p53-independent S arrest. In S phase, the same cascade can result in an intra-S arrest in response to stalled replication forks. The G2-M checkpoint prevents entry into mitosis with unrepaired DNA lesions. Initiation of this checkpoint is mediated by the ATM/ATR/Chk1/Chk2 cascades as shown, which ultimately suppresses the promitotic activity of cyclin B/cdc2. Along with their pivotal roles in the modulation of the cell cycle checkpoints, Chk1 and Chk2 are also involved in other aspects of the DNA damage response, including DNA repair, induction of apoptosis, and chromatin remodeling.

to predict the effects that will be seen with more specific inhibitors, although it is believed that the potentiation of DNA-damaging agents is primarily driven through the inhibition of Chk1 and, hence, abrogation of cell cycle arrest [55]. A number of clinical trials with UCN-01 in combination with a variety of DNA-damaging therapies are still ongoing.

The tumor cell specificity of the sensitization of the effects of DNA damage relies on the observation that the majority of tumors are deficient in the G1 DNA damage checkpoint pathway or other components of checkpoint

signaling and response. For example, high p53 mutation rates result in reliance on S and G2 phase checkpoints to repair DNA damage and promote cell survival. Therefore, abrogation of these remaining intact checkpoints should lead to enhanced tumor cell death compared with normal tissue. Inhibition of Chk1 signaling using small-molecule inhibitors, dominant negative enzymes, interference RNA, and ribozymes abrogated the S and $G_2$ checkpoints, impaired DNA repair, and selectively increased tumor cell death. This has provided strong preclinical support for this approach [25, 34, 55–62].

Recent evaluation of checkpoints activated by stalled replication forks showed that, in contrast, nontransformed cells have a more robust DNA replication checkpoint than tumor cells. In nontumor cells, both Chk1-dependent and independent pathways are activated in response to stalled replication forks. Tumor cells, on the other hand, rely entirely on Chk1 activity for a proper response. Thus, Chk1 inhibition enhanced the toxicity of hydroxyurea treatment in tumor cells but not in nontransformed cells as determined by clonogenic assays. This study therefore supports a rationale for tumor-selective effects of combined therapies such as Chk1 inhibition and agents that induce stalled replication forks such as gemcitabine, 5-fluorouracil, and hydroxyurea [63].

## 10.2 CLINICAL-TRANSLATIONAL ADVANCES

DNA damage remains the mainstay of cancer treatment; however, the toxic side effects of agents that induce DNA damage limit the dose that can be tolerated and the degree of tumor death that can be achieved. Mutations in genes required for the detection, signaling, and repair of DNA damage can lead to increased DNA damage, incorrect repair of DNA damage, and cancer [3]. To compensate for the loss of specific DNA repair pathways, different or faulty DNA repair pathways may be induced to enable tumor cells to survive. The activation of other repair pathways has been suggested to be responsible for the limited response of tumors to radio- and chemotherapy. If the DNA repair pathways essential for a tumor's survival can be identified and disrupted, this will allow chemotherapy to be much more efficient. Thus, it can be argued that targeting both checkpoint and repair pathways in combination may selectively kill tumor cells over healthy cells. Below we catalog several players in the DNA-damage response pathway that might be amenable to pharmacological interventions.

ATM is the predominant kinase responsible for the activation of multiple cell cycle checkpoints following DSB induction. The intricacies of the ATM-mediated signaling network were highlighted by a recent study showing that hundreds of phosphorylation events are dependent upon the ATM kinase [14]. The first suggestion that ATM may be an attractive target for chemotherapy was that cells from patients with the genomic instability disorder ataxia telangiectasia, resulting from a mutation in the ATM gene, were exquisitely sensitive to radiation [1]. In addition, caffeine (a known inhibitor of ATM) could increase cellular sensitivity to radiation and chemotherapeutic drugs [64, 65], although its lack of specificity and potency makes it unsuitable as a clinical agent. LY294002 is another ATM inhibitor that broadly inhibits the kinase activity of PIKKs. Like caffeine, the widespread use of LY294002 has been restricted by its lack of specificity; however, it has been used as a research tool for the design of more specific PIKK inhibitors. A highly specific small-molecule ATP

competitive inhibitor of ATM named KU-55933 was identified via screening of a drug library based on LY294002 [66]. This compound can efficiently sensitize tumor cells to radiation and DSB-inducing chemotherapeutic agents, such as camptothecin and etoposide, and there are suggestions that this compound may be used as a potential clinical treatment. There is also an indication that inhibitors of the ATM-mediated pathway may also sensitize cells deficient in other DNA repair pathways to cancer treatments. Validation of this idea was initially borne out when a high-throughput siRNA screen identified ATM as a target that disrupted the growth of cells deficient in the Fanconi anemia (FA) pathway [67]. Inhibition of ATM via siRNA was lethal to cells with a defective FA pathway. Because FA genes are disrupted in a range of cancers, the inhibition of ATM may provide a successful treatment in this subset of cancers.

In addition to ATM, the checkpoint kinases Chk1 and Chk2 are critical for cell cycle checkpoint activation following induction of DSBs. Several lines of evidence suggest that the combination treatment of tumors with genotoxic agents and Chk1 or Chk2 inhibitors may have high therapeutic value. The first inhibitor of Chk1 and Chk2 was the staurosporine inhibitor UCN-01; treatment with this compound led to G2/M checkpoint defects in IR-treated p53-deficient tumor cells [68], suggesting that inhibition of the checkpoint kinases may preferentially sensitize p53-deficient tumor cells to radio- and chemotherapy. This finding stimulated interest in the development and testing of Chk1 and Chk2 inhibitors that may show antitumor properties when combined with other chemotherapeutics. One potential clinical challenge for the use of UCN-01 as a tumor treatment is the sensitization of normal tissues due to the broad spectrum of its targets. Increased myelosuppression was seen in Phase 1 clinical trials of UCN01 with topotecan at lower doses than when topotecan was used alone [69, 70]. With increasingly selective inhibitors and biomarkers, it is likely that better therapeutic indexes can be achieved in clinical settings.

Several specific inhibitors of the Chk1 and Chk2 kinases have now emerged, including AZD7762 (AstraZeneca), XL844 (Exelixis), and PF-00477736 (Pfizer). Although these compounds all inhibit Chk1 and Chk2 at various concentrations, PF-00477736 displays the greatest specificity for Chk1. Notably, treatment with PF-00477736 also leads to strong inhibition of Chk1 auto phosphorylation, in contrast to the other checkpoint kinase inhibitors. It is unclear what effect these differences will have in a clinical setting, but it has been previously shown that inhibition of Chk1 may increase the toxicity of DNA-damaging agents. These inhibitors have been shown to sensitize cancer cells and xenografts to anticancer damaging agents [71–74]. Several clinical trials using Chk1 and Chk2 inhibitors as antitumor agents in combination with genotoxic agents, including gemcitabine, irinotecan, and cisplatin, targeting many tumor

**TABLE 10.1**

**Cancer Therapies Targeting Components of Damage-Response Pathway**

| Compound | Class | Phase | Combination | Tumor Type | Status | Ref. |
|---|---|---|---|---|---|---|
| UCN-01 | Chk1 and Chk2 inhibitor | 1/2 | Irinotecan, cisplatin, topotecan | Breast, gastric, head and neck, non–small cell lung, lung, small cell, ovarian, pancreatic, prostate, bladder, melanoma, thyroid, lung | Ongoing | [80] |
| AZD-7762 | Chk1 and Chk2 inhibitor | 1 | Irinotecan, gemcitabine | Solid tumors | Ongoing | [71] |
| XL884 | Chk1 and Chk2 inhibitor | 1 | Gemcitabine | Chronic lymphocytic leukemia | Terminated | [71] |
| PF-00477736 | Chk1 and Chk2 inhibitor | 1 | Gemcitabine | Solid tumors | Ongoing | [71] |
| Advexin | Recombinant adenovirus encoding p53 | 1–3 | — | Head and neck, breast, esophageal, prostate, ovarian, bladder, bronchoalveolar, glioblastoma | Ongoing | [78] |
| Gendicine | Recombinant adenovirus encoding p53 | Approved for use in China | — | Head and neck | — | [79] |
| SCH 58500 | Recombinant adenovirus encoding p53 | 1–2 | — | Ovarian, lung, bladder, liver | Ongoing | [79] |
| CP-31398, PRIMA-1, CDB3 | Mutant p53 reactivator | Preclinical | — | — | — | [81–83] |
| Nutlins, benzodiazepines, RITA, spiro-oxindoles, and quinolinols | MDM2-p53 interaction inhibitor | Preclinical | — | — | — | [84–88] |
| BSI-201 | PARP inhibitor | 1–3 | — | Glioma, epithelial ovarian, fallopian tube, or primary peritoneal cancer | Ongoing | [89, 90] |
| AG-014699 | PARP inhibitor | 2 | — | Breast, ovarian | Ongoing | [90] |
| Olaparib (AZD2281) | PARP inhibitor | 1–2 | — | BRCA1- and BRCA2-mutated breast and ovarian cancer | Ongoing | [90–93] |

subsets are currently underway (see Table 10.1 for further information). Together, these studies further support the potential of checkpoint kinase inhibitors to enhance the efficacy of both conventional chemotherapy and radiotherapy and increase tumor regression and tumor cell cytotoxicity in a subset of cancers.

Recent advances have also implicated checkpoint pathway activation as a major mechanism driving radioresistance in cancer stem cells. Glioblastoma is one of the most lethal human malignancies, but its treatment remains restricted because of the radioresistance found in this subset of tumors. Strikingly, the radioresistance of glioma stem cells could be reversed by the addition of a specific Chk1 and Chk2 inhibitor (debromohymenialdisine [75]), highlighting the potential of inhibiting checkpoint responses to overcome the resistance of cancer stem cells to radiotherapy.

Inactivation of p53 is a crucial event during the onset of tumorigenesis with approximately 50% of all human cancers containing somatic mutations in the p53 gene [76]. A major breakthrough in molecular-based therapy of cancer

would be to restore the function of mutated p53 in cancer cells. Because the function of p53 is reduced in 50% of cancers, efforts to use it as a chemotherapeutic target have concentrated on agents that can reactivate p53 activity in tumors [77]. At present there are two main mechanisms of increasing p53 activity in tumors, the first mechanism involves restoring wild-type p53 function via a recombinant adenovirus encoding p53 (Advexin, Phase 1–3 clinical trials [78]), and the second is Gendicine, approved for use in China and SCH 58500, Phase 1–3 clinical trials [79]. In addition, the tumor suppressor function of mutant p53 may be restored by the use of small compounds (CP-31398 and PRIMA-1) or short peptides (CDB3 and peptide 46 [77]). Although preliminary studies were encouraging, insufficient bioavailability of the small compounds has prevented their development as pharmacological therapeutics. Furthermore, the exact mechanism of rescue by these small compounds is unknown and the restricted range of p53 mutants that they can rescue limits their therapeutic potential. The second mechanism of action involves the stabilization of p53 via small molecules that target the interaction between p53 and MDM2 (the E3 ligase responsible for ubiquitin-dependent degradation of p53). These include nutlins, benzodiazepines, RITA, spiro-oxindoles, and quinolinols, all of which are at the preclinical development stages [77]. Overexpression of MDM2 is also a hallmark of the misregulation of the p53 pathway in human cancers [63]. MDM2 is a negative regulator of p53; therefore overexpression leads to the inhibition of p53 activity by altering the subcellular localization of p53, which in turn decreases its stability. Thus, in tumors that retain wild-type p53, inhibition of MDM2 can be used to reactivate p53 activity. Table 10.1 provides further information on compounds and clinical trials involving compounds targeting p53.

The MRN complex is a key component of DSB repair pathways and is also required for cell cycle checkpoint activation. Given the radiosensitive human genomic instability syndromes (discussed above) that arise from deficiencies in the Mre11 and NBS1 components of the MRN complex, this complex is an attractive target for anticancer treatments [2, 94]. In the laboratory, the use of a small peptide containing the evolutionary conserved region of NBS1, which binds to ATM, was found to disrupt the interaction between NBS1 and ATM [95]. In addition, this peptide also prevented DNA damage signaling and radiosensitized cells, suggesting that the NBS1:ATM interaction may be a successful target for anticancer drugs. A recent study used a forward genetic screen to identify a specific inhibitor of the MRN complex, designated as Mirin. Mirin inhibits MRN-dependent ATM activation and abolishes the endonuclease activity of Mre11. The consequence of the Mirin-dependent MRN inhibition is the prevention of G2/M checkpoint activation and HR repair [96]. There is still more work to be done on demonstration of its antitumor activity in xenograft models.

Rad51 is required to protect cells from tumor development via its DSB repair capacity; however, once tumors form, Rad51 can also promote the resistance of tumors to chemotherapy [97, 98]. Depleting the levels of Rad51 via antisense RNA or RNAi has been shown to sensitize tumor cells to chemotherapy agents, such as cisplatin [99–101]. The above studies implicate Rad51 as a potential target for antitumor drugs. A recent study used short peptides encompassing the Rad51 interacting domain of the BRCA2 protein to examine their inhibitory effect on Rad51 activity [101]. A peptide of 28 amino acids was found to bind Rad51, prevent its DNA binding activity, and therefore specifically inhibit the formation of the Rad51 nuclear foci and therefore the completion of HR in human cells. Although thus far this peptide has only been used as a research tool, further study of this peptide or peptidomimetics may form the basis for antitumor drugs targeting Rad51 and the HR pathway in tumors.

The ssDNA binding (SSB) proteins play essential roles in the repair of many types of DNA damage including DSBs. As well as being involved in the repair of DNA damage, SSBs are also central to other processes in which ssDNA is exposed, such as DNA replication. Until recently it was believed that RPA was the exclusive SSB involved in DNA repair in humans; however, two simple SSBs have been now identified, hSSB1 and hSSB2, with a crucial role in DNA repair [9, 102]. The cellular function of RPA and hSSB1 in DNA replication and DNA damage repair and signaling makes them a very attractive cancer therapeutic target. Indeed, one study sought to identify potential chemicals that would inhibit the interaction of RPA with DNA. Using a fluorescent-based reporter assay, they identified a number of inhibitory chemicals with possible therapeutic potential [103]. hSSB1 is also the center of a current drug discovery program that aims to find potential therapeutic inhibitors of its function.

In addition to targeting cell cycle checkpoint activation and HR, there is also evidence that targeting the NHEJ pathway may also be an effective cancer treatment. Cells deficient in Ku70/80 or the catalytic subunit of DNA-PK (DNA-PKcs) are sensitive to DSBs induced by IR or chemotherapeutic agents [104, 105], suggesting that DNA-PK may constitute a good target for chemotherapy. In addition, DNA-PK is upregulated in some cancers, implicating it as a factor required for tumor growth and survival. Indeed upregulation of DNA-PK activity was shown to impair apoptosis in B-cell chronic lymphocytic leukemia [106]. Preliminary investigations into inhibition of DNA-PK used the broad-spectrum PIKK inhibitors wortmannin and LY294002 [107]. These nonselective inhibitors were shown to sensitize tumor cells to chemotherapeutic agents and were used as a basis to develop a more specific DNA-PK inhibitor. Treatment with a flavone-based DNA-PK inhibitor IC87361 led to tumor radiosensitization in vitro and in vivo without causing toxicity [108]. In subsequent studies, a highly potent and selective DNA-PK inhibitor NU7441 with an IC50 of 13nM was identified [109]. The use of this agent led to radiosensitivity and chemosensitivity of tumor cell lines and in xenograft models in preclinical trials. A number of other agents targeting DNA-PK are currently in clinical trials [110].

Recently, the fact that HR is essential for DNA repair in the absence of Poly(ADP-ribose) polymerase (PARP-1) has been exploited for cancer treatment. PARP-1 is a member of the base excision repair pathway that repairs ssDNA breaks. Inhibition of PARP-1 leads to accumulation of single-strand breaks, which are converted to DSBs during replication. The inhibition of PARP has been shown to be synthetically lethal with loss of BRCA1 and BRCA2 [111, 112]. As a proof of concept, exposure of BRCA2-deficient murine tumors to PARP-1 inhibitors (KU 0059436/AZD2281) resulted in a profound decrease in tumor growth and survival [113]. In addition, this inhibitor has shown a low level of toxicity and was found to stabilize or regress chemotherapy-refractory ovarian, breast, or prostate cancers with BRCA1 or BRCA2 mutations, and no activity was seen outside BRCA1- and BRCA2-defective tumors.

In addition, a recent Phase 1 clinical trial also found that AZD2281 inhibited PARP-1 and exhibited antitumor activity in ovarian, breast, or prostate cancers with BRCA1 or BRCA2 mutations [91–93]. A number of PARP-1 inhibitors are currently in clinical trials as monotherapy for BRCA1- and BRCA2-deficient breast and ovarian cancers (see Table 10.1 for further detail of specific inhibitors [90]). The recent clinical finding validates synthetic lethality as a new concept in cancer drug development, and encourages the characterization and targeting of other synthetic lethality relationships in DNA damage response and repair pathways.

DNA-damaging therapies are among the most common cancer treatments and have produced significant increases in the survival of patients, particularly when used in combination with drugs with different mechanisms of action. Due to the efficacy of these anticancer treatments, DNA-damaging agents are likely to remain a standard of care for the treatment of many cancers for the foreseeable future.

Although the induction of DNA damage is an effective approach to tumor control, this mechanism also leads to significant side effects as the majority of these agents are used at the maximum tolerated dose. Toxicities to the hematologic, gastrointestinal, and other organ systems are commonly observed and limit the dose that can be tolerated and the degree of tumor control that can be achieved. Another limitation of DNA-damaging agents is that many patients develop resistance and therefore become refractory to treatment.

The hypothesis that modulation of the DNA checkpoint pathways offers the potential to sensitize cancer cells to DNA damage induced by chemotherapeutics or radiotherapy while sparing normal cells was first proposed in the 1980s and has continued to gather support. More specifically, abrogation of cell cycle checkpoint pathways via inhibition of Chk1 and Chk2 has become an increasingly appealing approach to broaden the therapeutic window of commonly used anticancer therapies including alkylating agents, topoisomerase inhibitors, antimetabolites, and radiotherapy.

Recent advances in checkpoint biology have also suggested the potential utility of combining Chk1 inhibitors with antimitotic agents. Chk1 inhibition sensitized tumor cells to paclitaxel [50]. Mechanistic studies followed, which showed that Chk1 plays a critical role in the spindle checkpoint. Chk1-deficient DT40 avian B-lymphoma cells or human BE colon cells were found to fail to undergo mitotic arrest in the presence of paclitaxel. Mechanistic studies showed Chk1 co-localization with kinetochore proteins, and in the Chk1-depleted cells, failure of critical spindle checkpoints. The spindle checkpoint failure is suggested to be due to lack of activation of aurora B kinase. Proper spindle checkpoint activation was also found to be dependent on the catalytic activity of Chk1 [114]. In contrast to studies suggesting that Chk1 is a positive regulator of the spindle checkpoint, there is also a finding that it is a negative regulator of the spindle checkpoint through

regulation of polo-like kinase 1 [115]. It is not understood whether these differences are due to the different models or techniques to deplete Chk1 that were used, but both studies clearly indicate that Chk1 plays a role in the spindle checkpoint.

As well as a way of potentiating the efficacy of DNA-damaging therapies in sensitive tumors, several recent advances have implicated checkpoint pathway activation as a major mechanism driving both chemoresistance and radioresistance. This may further broaden the clinical utility of such agents. For example, the ATR-Chk1 pathway is strongly activated in BCR/ABL-positive cells, and this was shown to contribute to the resistance of these cells to treatment with DNA crosslinking agents [116]. Increased Chk1 activity has also been found to be associated with cellular resistance to doxorubicin in K562/A02 human erythroleukemic cell lines [117]. In addition, glioma stem cells, the cell population representing radio-resistant tumor cells, were shown to promote radioresistance through activation of the checkpoint pathway; thus targeting the checkpoint response in this setting may overcome radioresistance and improve the therapeutic outcome in malignant brain cancer [118].

New findings have also led to the hypothesis that there may be certain tumor types that are particularly sensitive to Chk1 inhibition. These include triple-negative breast cancer where Chk1 has been found to be significantly upregulated [119], and colorectal cancer where Chk1 was identified as a protein that discriminated between normal and tumor mucosa [120].

In parallel with recent biological advances, there have been many recent developments in the medicinal chemistry of checkpoint kinase inhibitors, and the area has been the subject of a number of excellent reviews that give a very good overview of the design and preclinical activity of these compounds [121–129]. Overall, the data from this class of agents are compelling and consistently support the hypothesis that abrogation of DNA damage–induced checkpoints will potentiate the effects of radiotherapy and chemotherapy as exemplified below using the three Chk1/Chk2 inhibitors that have recently entered clinical trials (XL-844, AZD7762, and PF00477736). All three agents are in Phase 1 trials, and pursuing combination approach (Table 10.2). All three agents have been extensively profiled preclinically, and the reported profiles unsurprisingly show a number of similarities. XL-844, AZD7762, and PF-00477736 represent different chemical classes from both UCN-01 and each other. They are all potent inhibitors of both Chk1 and Chk2 (Table 10.2). They have all been shown to abrogate DNA damage–induced cell cycle arrest and to potentiate the effects of DNA damage–induced therapies, both in vitro and in vivo [127, 130–136]. The degree of potentiation observed is dependent on the cell line and DNA damage agent used, but in all cases, a robust response is seen in combination with gemcitabine (cell lines and xenograft models), and this is the initial focus for testing the hypothesis clinically.

**TABLE 10.2**

**Checkpoint Kinase Inhibitors in Clinical Trials**

| Agent | Structure | Chk1 IC$_{50}$ (nmol/L) | Chk2 IC$_{50}$ (nmol/L) | Combination | Tumor Setting | Phase | Status |
|---|---|---|---|---|---|---|---|
| UCN-01[a] | | 10 | 10 | Irinotecan | Breast, gastric, head/neck, | 1 | Ongoing |
| | | | | Cisplatin | NSCL, | 1 | |
| | | | | Topotecan | small cell, ovarian, | 2 | |
| | | | | | pancreas, prostate | | |
| | | | | | Bladder, melanoma, | | |
| | | | | | thyroid | | |
| | | | | | Lung | | |
| XL844 | Structure undisclosed | 2.2 | 0.2 | — | CLL | 1 | Terminated |
| | | | | Gemcitabine | Aggressive or indolent | 1 | Ongoing |
| | | | | | NHL, solid tumors | | |
| PF-00477736 | | 0.49 (Ki) | 47 (Ki) | Gemcitabine | Solid tumors | 1 | Ongoing |
| AZD7762 | | 5 | <10 | Gemcitabine | Solid tumors | 1 | Ongoing |
| | | | | Irinotecan | Solid tumors | 1 | |

*Abbreviations:* CLL, chronic lymphocytic leukemia; NHL, non–Hodgkin's lymphoma; NSCL, non–small cell lung.

[a] Nonselective kinase inhibitor with known activity against checkpoint kinases—only ongoing trials in combination with DNA-damaging agents listed.

Although they all inhibit Chk1 and Chk2, AZD7762, PF-00477736, and XL844 show a range of potencies against Chk2, with PF-00477736 demonstrating the most selectivity for Chk1 (~100-fold). Interestingly, PF-00477736 has been shown to have somewhat different effects on downstream protein phosphorylations than either XL-844 or AZD7762. For example, both in vitro and in vivo PF-00477736 combination studies have shown a decrease in Chk1 phosphorylation, in sharp contrast to the increase in Chk phosphorylation observed with other checkpoint kinase inhibitors. However, in all cases, an increase in the levels of phosphohistone 2AX (p-H2AX), a marker of double-stranded DNA damage, is seen [130, 131, 137]. It is not yet clear how these differences will affect the clinical profile of these agents, but recent findings using molecular tools such as shRNA and siRNA suggest that potentiation of the effects of DNA damage is driven primarily through Chk1 inhibition rather than Chk2 inhibition. It has also been hypothesized that Chk2 inhibition may lead to the enhanced effect of sensitizing p53-null cancer cells while protecting normal cells [121]. In support of this argument, it has been shown that an ATP-competitive inhibitor of Chk2 protected human CD4(+) and CD8(+) T cells from apoptosis after IR [138]. More recently, the selective

Chk2 inhibitor VRX0466617 (Chk1:IC$_{50}$, >10,000 nmol/L; Chk2:Ki, 11 nmol/L) has been described [139], and this together with Chk1-selective compounds such as EXEL-3611 (Chk1:IC$_{50}$, 2.4 nmol/L; Chk2:IC$_{50}$, 2400 nmol/L [140]) may shed further light on the most desirable balance of Chk1/Chk2 inhibition required to achieve the greatest therapeutic effect while minimizing detrimental effects to normal tissue.

## 10.3 SONODYNAMIC ACTIVITIES OF PORPHYRINS

Chemical agents such as porphyrins were found to be activated by ultrasound, producing significant antitumor effects. Hematoporphyrin (Hp) enhanced ultrasonically induced damage on sarcoma cells and showed a synergistic inhibitory effect on the tumor growth in combination with ultrasound at 2 MHz. Recently, other types of porphyrins such as protoporphyrin were also found to have such sonodynamic activities. Furthermore, it was found that sonochemical reactions can be greatly accelerated by superimposing the second harmonic onto the fundamental. The highest rate of iodine release from aqueous iodide was obtained at an acoustic intensity ratio between 1 MHz and 2 MHz of 1:1 while either one of the frequency components alone could not induce significant iodine release at the same total acoustic intensity. Second-harmonic superimposition in combination with sonodynamically active antitumor agents may have the potential for selective tumor treatment.

Porphyrins such as hematoporphyrin derivative are known to have antitumor effects when photochemically activated. They are used in photodynamic therapy, which is a modality of tumor treatment based on photochemical activation of drugs by high-intensity optical beams such as laser beams [141, 142]. Their photochemically induced antitumor effects have been explained in terms of singlet oxygen mechanisms. Photodynamic treatment is useful in treating surface regions, but its application to nonsuperficial tumors is quite limited due to the poor penetration of optical beams into tissues.

Ultrasound, on the other hand, can penetrate deeply into tissues while maintaining its ability to focus. This is a unique advantage when compared to electromagnetic modalities such as optical beams or microwaves in the applications for noninvasive treatment of nonsuperficial tumors. Use of ultrasound for tumor treatment has been relatively well investigated with respect to thermal effects due to ultrasound absorption [143], but few studies have been reported with respect to the effects arising from combination with drugs. Furthermore, because the drugs used in those studies had antitumor effects by themselves, it was controversial to identify the effects as being either synergistic or merely additive [144–146].

Porphyrins such as hematoporphyrin (Hp) and a gallium-deuteroporphyrin complex, ATX-70, were then found to be activated by ultrasound, producing significant antitumor effects [147–151]. Suspended tumor cells were lysed by ultrasound at a much higher rate in the presence of these porphyrins than in their absence at concentrations for which the porphyrins by themselves show no cytotoxicity [147, 150]. Experimental tumors implanted into mice have been treated with ultrasound in combination with the administration of such porphyrins and the tumor growth was significantly inhibited at an ultrasound intensity for which ultrasound alone showed no significant effect [148]. These findings allowed us to propose a new modality of cancer treatment in which an administered antitumor agent is activated selectively by ultrasound aimed at the tumor [152].

The effect of active oxygen scavengers on the ultrasonically induced in vitro cell damage and also the effect of deuterium oxide substitution of hydrogen oxide in the suspension medium were both consistent with a hypothesis that active-oxygen generation by ultrasonically activated porphyrin was the most important mediator for the porphyrin-enhanced cell damage [149]. Enhancement in ultrasonic generation of active oxygen by such aqueous porphyrins was confirmed by detecting nitroxide formation from 2,2,6,6-tetramethyl-4-piperidone by electron spin resonance spectroscopy [151].

The above-described in vitro and in vivo experiments were performed under standing wave conditions because acoustic cavitation can more easily be induced with reproducibility by standing waves than by progressive waves. In a standing wave field, microbubbles smaller than the resonant size receive ultrasonic radiation force toward a pressure antinode, where they can efficiently grow and collapse [153]. However, insonation with standing waves does not seem to be widely applicable to various therapeutic settings. We found recently that acoustic cavitation and sonochemical reactions can be efficiently induced even in a progressive wave field by superimposing the second harmonic onto the fundamental [154–157]. The intensity threshold for producing in vivo tissue damage paired with fractional harmonic emission was dramatically decreased by the second-harmonic superimposition, especially with administration of the above-described sonodynamically active agents [158–160]. The structural formula of Protoporphyrin IX disodium salt (Pp) is shown in Figure 10.3.

Synergy between the fundamental and the second harmonic was quite effective on inducing sonochemical reaction. This effect is thought to be due to the enhancement of sonochemically active cavitation by second-harmonic superimposition because fractional harmonic emission was markedly enhanced at the same time. Since the synergistic effect was so sensitive to the second-harmonic phase, the effect may be explained based on the behavior of cavitated bubbles under acoustic pressure with second-harmonic superimposition [156, 160]. In vivo effects of second-harmonic superimposition with focused transducer configuration on inducing sonodynamic reactions [158] have been reported in published papers [159, 160].

In vitro experiments in a standing wave mode demonstrated that protoporphyrin significantly enhances ultrasonically induced cell damage by a factor in the same order of magnitude as hematoporphyrin and a gallium–deuteroporphyrin complex, ATX-70, which have shown in vivo antitumor effects as well. It was also demonstrated that sonochemically effective cavitation can be induced at a high efficiency even in a progressive wave mode when the second harmonic was superimposed to the fundamental at a proper acoustic phase. Second-harmonic superimposition in combination with the sonodynamically active antitumor agents that has been described [160] may have the potential for selective tumor treatment, which may be referred to as "sonodynamic therapy."

## 10.4  FDA NOVEL DRUG APPROVALS FOR 2019

Novel drugs approved by the US Food and Drug Administration (FDA) within 2019 are outlined in Table 10.3.

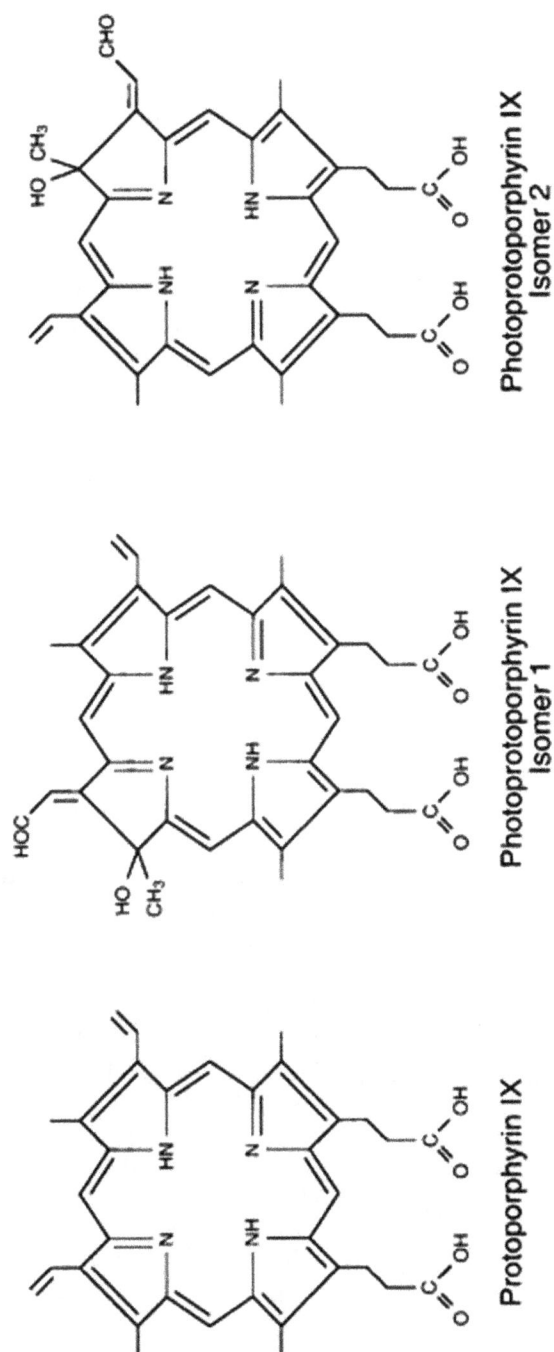

**FIGURE 10.3** Structural formulae of protoporphyrin IX and photoprotoporphyrin IX isomers.

**TABLE 10.3**
**FDA Novel Drug Approvals for 2019**

| Active Ingredient | Drug Name | Approval Date | FDA-Approved Use on Approval Date |
|---|---|---|---|
| Trastuzumab | Enhertu | 12/20/2019 | To treat metastatic breast cancer |
| Voxelotor | Oxbryta | 11/25/2019 | To treat sickle cell disease |
| Zanubrutinib | Brukinsa | 11/14/2019 | To treat certain patients with mantle cell lymphoma, a form of blood cancer |
| Elexacaftor Ivacaftor Tezacaftor | Trikafta | 10/21/2019 | To treat patients 12 years of age and older with the most common gene mutation that causes cystic fibrosis, 2014–2017 |
| Lasmiditan | Reyvow | 10/11/2019 | For the acute treatment of migraine with or without aura, in adults |
| Istradefylline | Nourianz | 8/27/2019 | To treat adult patients with Parkinson's disease experiencing "off" episodes |
| Upadacitinib | Rinvoq | 8/16/2019 | To treat adults with moderately to severely active rheumatoid arthritis |
| Fedratinib | Inrebic | 8/16/2019 | To treat adult patients with intermediate-2 or high-risk myelofibrosis |
| Entrectinib | Rozlytrek | 8/15/2019 | To treat adult patients with metastatic non–small cell lung cancer (NSCLC) whose tumors are ROS1-positive |
| Pitolisant | Wakix | 8/14/2019 | To treat excessive daytime sleepiness (EDS) in adult patients with narcolepsy |
| Pyrazinamide | | 8/14/2019 | For treatment-resistant forms of tuberculosis that affects the lungs |
| Pexidartinib | Turalio | 8/2/2019 | To treat adult patients with symptomatic tenosynovial giant cell tumor |
| Darolutamide | Nubeqa | 7/30/2019 | To treat adult patients with nonmetastatic castration-resistant prostate cancer |
| Selinexor | Xpovio | 7/3/2019 | To treat adult patients with relapsed or refractory multiple myeloma (RRMM) |
| Alpelisib | Piqray | 5/24/2019 | To treat breast cancer |
| Tafamidis Meglumine | Vyndaqel | 5/3/2019 | To treat heart disease (cardiomyopathy) caused by transthyretin-mediated amyloidosis (ATTR-CM) in adults |
| Erdafitinib | Balversa | 4/12/2019 | To treat adult patients with locally advanced or metastatic bladder cancer |
| Siponimod | Mayzent | 3/26/2019 | To treat adults with relapsing forms of multiple sclerosis |
| Triclabendazole | Egaten | 2/13/2019 | To treat fascioliasis, a parasitic infestation caused by two species of flatworms or trematodes |

For research use only. Not for human uses.

## 10.5   CONCLUSIONS

As our understanding of the mechanism and biochemical details of the DNA damage response increases [161–164], the potential ways to manipulate this pathway for the development of novel therapeutics will emerge and will have an enormous impact on future medical science. Success will rely on the availability of biomarkers to detect patients with particular checkpoint or DNA repair defects including defects in p53, HR, or FA pathways to stratify patients that should be trialed with a novel, targeted therapy.

It is likely that the response to cancer therapeutics will also become more predictable with the identification of tumor biomarkers, thus allowing for targeted, more efficient cancer treatments. Further examination of the therapeutic and oncogenic effects of checkpoint and repair inhibitors is warranted to identify the most appropriate targets. The importance of DSB signaling and repair pathways in chemo- and radiosensitization of cancer cells in vitro has been well characterized; however, the correlation between the in vivo clinical response to chemo- and radiotherapy and disruption of the DSB response pathway remains to be elucidated.

This is a complicated but promising area of research that extends far beyond the originally described roles of Chk1

and Chk2 in cell cycle arrest. Although our understanding is expanding rapidly, there are many issues and challenges still to be overcome before the potential of such agents can be fully realized. Combination dose scheduling, biomarker, and patient selection strategies are complex. The p53 hypothesis may provide guidance, but loss of p53 is not the only mechanism by which checkpoint pathways may be compromised in tumor cells. Another area with many unanswered questions is the potential for side effects from checkpoint kinase inhibition in normal tissues, as well as the possibility of exaggerated toxicity from the combination of these agents with existing chemotherapeutics. Chk1 inhibition is known to cause some genetic instability and so, long term, the risk of secondary cancer formation remains to be determined. In addition, as mentioned briefly above, emerging checkpoint biology already suggests that there may be additional opportunities for these agents in specific patient populations as in the case of drug resistance. Identification of the optimal clinical settings and those patients most likely to respond are key issues that remain to be addressed.

Overall, the discovery and development of checkpoint kinase inhibitor are areas of intense interest, and are likely to continue to grow. The preclinical data thus far provide confidence that Chk inhibitors will have broad utility in combination with many established DNA-damaging therapies across a wide range of tumor types. Hopefully, clinical validation of the potentiation hypothesis will soon be achieved. Ultimately, this approach could achieve equal or greater efficacy than is possible today, and possibly even with a lower dose of chemotherapeutic agent, thereby greatly improving patient well-being and providing a significant step forward in oncology treatment. Exciting additional opportunities, such as those outlined above in triple-negative breast and colorectal cancer, also continue to emerge and offer even further opportunities for development with this class of agent.

# REFERENCES

[1] Savitsky K, Bar-Shira A, Gilad S, et al. A single ataxia telangiectasia gene with a product similar to PI-3 kinase. Science 1995; 268:1749–53.

[2] Stewart GS, Maser RS, Stankovic T, et al. The DNA double-strand break repair gene hMRE11 is mutated in individuals with an ataxia-telangiectasia- like disorder. Cell 1999; 99:577–87.

[3] Harper JW, Elledge SJ. The DNA damage response: ten years after. Mol Cell 2007; 28:739–45.

[4] Hakem R. DNA-damage repair; the good, the bad, and the ugly. EMBO J 2008; 27:589–605.

[5] Weterings E, Chen DJ. The endless tale of non-homologous end-joining. Cell Res 2008; 18:114–24.

[6] van Gent DC, van der Burg M. Non-homologous end-joining, a sticky affair. Oncogene 2007; 26:7731–40.

[7] Abraham RT. Cell cycle checkpoint signaling through the ATM and ATR kinases. Genes Dev 2001; 15:2177–96.

[8] Bakkenist CJ, Kastan MB. DNA damage activates ATM through intermolecular autophosphorylation and dimer dissociation. Nature 2003; 421:499–506.

[9] Richard DJ, Bolderson E, Cubeddu L, et al. Single-stranded DNA-binding protein hSSB1 is critical for genomic stability. Nature 2008; 453: 677–81.

[10] Uziel T, Lerenthal Y, Moyal L, et al. Requirement of the MRN complex for ATM activation by DNA damage. EMBO J 2003; 22:5612–21.

[11] Gatei M, Sloper K, Sorensen C, et al. Ataxia-telangiectasia-mutated (ATM) and NBS1-dependent phosphorylation of Chk1 on Ser-317 in response to ionizing radiation. J Biol Chem 2003; 278:14806–11.

[12] Matsuoka S, Rotman G, Ogawa A, et al. Ataxia telangiectasiamutated phosphorylates Chk2 in vivo and in vitro. Proc Natl Acad Sci USA 2000; 97:10389–94.

[13] Branzei D, Foiani M. Regulation of DNA repair throughout the cell cycle. Nat Rev Mol Cell Biol 2008; 9:297–308.

[14] Matsuoka S, Ballif BA, Smogorzewska A, et al. ATM and ATR substrate analysis reveals extensive protein networks responsive to DNA damage. Science 2007; 316:1160–6.

[15] Ahnesorg P, Smith P, Jackson SP. XLF interacts with the XRCC4-DNA ligase IV complex to promote DNA nonhomologous end-joining. Cell 2006; 124:301–13.

[16] Ma Y, Pannicke U, Schwarz K, et al. Hairpin opening and overhang processing by an Artemis/DNA-dependent protein kinase complex in nonhomologous end joining and V(D)J recombination. Cell 2002; 108:781–94.

[17] Helleday T. Pathways for mitotic homologous recombination in mammalian cells. Mutat Res 2003; 532:103–15.

[18] Pardo B, Gomez-Gonzalez B, Aguilera A. DNA repair in mammalian cells: DNA double-strand break repair: how to fix a broken relationship. Cell Mol Life Sci 2009; 66:1039–56.

[19] Sartori AA, Lukas C, Coates J, et al. Human CtIP promotes DNA end resection. Nature 2007; 450:509–14.

[20] Chen L, Nievera CJ, Lee AY, et al. Cell cycle dependent complex formation of BRCA1.CtIP. MRN is important for DNA double-strand break repair. J Biol Chem 2008; 283:7713-20.

[21] Godthelp BC, Artwert F, Joenje H, et al. Impaired DNA damage-induced nuclear Rad51 foci formation uniquely characterizes Fanconi anemia group D1. Oncogene 2002; 21:5002–5.

[22] Yuan SS, Lee SY, Chen G, et al. BRCA2 is required for ionizing radiation-induced assembly of Rad51 complex in vivo. Cancer Res 1999; 59:3547–51.

[23] Helleday T, Lo J, van Gent DC, et al. DNA double-strand break repair: from mechanistic understanding to cancer treatment. DNA Repair (Amst) 2007; 6:923–35.

[24] Wade Harper J, Elledge SJ. The DNA damage response: ten years after. Mol Cell 2007; 28:739-45.

[25] O'Connor MJ, Martin NMB, Smith GCM. Targeted cancer therapies based on the inhibition of DNA strand break repair. Oncogene 2007; 26:7816-24.

[26] Sancar A, Lindsey-Boltz LA, Unsal-Kacmaz K, et al. Molecular mechanisms of mammalian DNA repair and the DNA damage checkpoints. Annu Rev Biochem 2004; 73:39-85.

[27] Zhou BB, Elledge SJ. The DNA damage response: putting checkpoints in perspective. Nature 2000; 408:433-9.

[28] Elledge SJ. Cell cycle checkpoints: preventing an identity crisis. Science 1996; 274:1664-72.

[29] O'Driscoll M, Jeggo PA. The role of double-strand break repair: insights from human genetics. Nat Rev Genet 2006; 7:45-54.

[30] Andreassen P, Ho G, D'Andrea A. DNA damage responses and their many interactions with the replication fork. Carcinogenesis 2006; 27:883-92.

[31] Zhou BB, Bartek J. Targeting the checkpoint kinases: chemosensitization versus chemoprotection. Nat Rev Cancer 2004; 4:216-25.

[32] Zhou BB, Sausville EA. Drug discovery targeting Chk1and Chk2 kinases. Prog Cell Cycle Res 2003; 5: 413-21.

[33] Bartek J, Lukas J. Mammalian G1- and S-phase checkpoints in response to DNA damage. Curr Opin Cell Biol 2001; 13:738-47.

[34] Zhao H, Watkins JL, Piwnica-Worms H. Disruption of the checkpoint kinase 1/cell division cycle 25A pathway abrogates ionizing radiation-induced S and G2 checkpoints. Proc Natl Acad Sci USA 2002; 99: 14795-800.

[35] Sorensen CS, Syljuasen RG, Falck J, et al. Chk1 regulates the S phase checkpoint by coupling the physiological turnover and ionizing radiation-induced accelerated proteolysis of Cdc25A. Cancer Cell 2003; 3:247-58.

[36] Mailand N, Falck J, Lukas C, et al. Rapid destruction of human Cdc25A in response to DNA damage. Science 2000; 288:1425-9.

[37] Mailand N, Podtelejnikov AV, Groth A, et al. Regulation of G(2)/M events by Cdc25A through phosphorylation-dependent modulation of its stability. EMBO J 2002; 21:5911-20.

[38] Peng CY, Graves PR, Thoma RS, et al. Piwnica-Worms H. Mitotic and G2 checkpoint control: regulation of 14 -3-3 protein binding by phosphorylation of Cdc25C on serine-216. Science 1997; 277:1501-5.

[39] Dalal SN, Schweitzer CM, Gan J, et al. Cytoplasmic localization of human cdc25C during interphase requires an intact 14 ^ 3-3 binding site. Mol Cell Biol 1999; 19:4465-79.

[40] Yang J, Winkler K, Yoshida M, et al. Maintenance of G2 arrest in the Xenopus oocyte: a role for 14-3-3-mediated inhibition of Cdc25 nuclear import. EMBOJ 1999; 18:2174-83.

[41] Zachos G, Rainey MD, Gillespie DA. Chk1-deficient tumour cells are viable but exhibit multiple checkpoint and survival defects. EMBO J 2003; 22:713-23.

[42] Gatei M, Sloper K, Soerensen C, et al. Ataxiatelangiectasia-mutated (ATM) and NBS1-dependent phosphorylation of Chk1 on Ser-317 in response to ionizing radiation. J Biol Chem 2003; 278:14806-11.

[43] Xiao Z, Chen Z, Gunasekera AH, et al. Chk1 mediates S and G2 arrests through Cdc25A degradation in response to DNA-damaging agents. J Biol Chem 2003; 278:21767-73.

[44] Bartek J, Lukas J. Chk1 and Chk2 kinases in checkpoint control and cancer. Cancer Cell 2003; 3: 421-9.

[45] Ahn J, Urist M, Prives C. The Chk2 protein kinase. DNA Repair (Amst) 2004; 3:1039-47.

[46] Falck J, Mailand N, Syljuasen RG, et al. The ATM-Chk2 - 25A checkpoint pathway guards against radioresistant DNA synthesis. Nature 2001; 410:842-7.

[47] Takai H, Naka K, Okada Y, et al. Chk2-deficient mice exhibit radio resistance and defective p53-mediated transcription. EMBO J 2002; 21:5195-205.

[48] Zaugg K, Su Y-W, Reilly PT, et al. Cross-talk between Chk1and Chk2 in double-mutant thymocytes. Proc Natl Acad Sci USA 2007; 104:3805-10.

[49] Niida H, Katsuno Y, Banerjee B, et al. Specific role of Chk1phosphorylations in cell survival and checkpoint activation. Mol Cell Biol 2007; 27:2572-81.

[50] Morgan MA, Parsels LA, Parsels JD, et al. The relationship of premature mitosis to cytotoxicity in response to checkpoint abrogation and antimetabolite treatment. Cell Cycle 2006; 5:1983-8.

[51] Xiao Z, Xue S, Sowin TJ, et al. Differential roles of checkpoint kinase1, checkpoint kinase 2, andmitogen-activated protein kinase-activated protein kinase 2 in mediating DNA damage-induced cell cycle arrest: implications for cancer therapy. Mol Cancer Ther 2006; 5:1935-43.

[52] Zhou BB, Chaturvedi P, Spring K, et al. Caffeine abolishes the mammalian G(2)/M DNA damage checkpoint by inhibiting ataxia-telangiectasiamutated kinase activity. J Biol Chem 2000; 275:10342-48.

[53] Graves PR, Yu L, Schwarz JK, et al. The Chk1protein kinase and the Cdc25C regulatory pathways are targets of the anticancer agent UCN-01. J Biol Chem 2000; 275:5600-5.

[54] Busby EC, Leistritz, DF, Abraham RT, et al. The radiosensitizing agent 7-hydroxystaurosporine (UCN-01) inhibits the DNA damage checkpoint kinase hChk1. Cancer Res 2000; 60:2108-12.

[55] Wang Q, Fan S, Eastman A, et al. UCN-01: a potent abrogator of G2 checkpont function in cancer cells disrupted with p53. J Natl Cancer Inst 1996; 88:956-65.

[56] Russell KJ, Wiens LW, Demers GW, et al. Abrogation of theG2 checkpoint results in differential radiosensitization of G1 checkpoint-deficient and G1 checkpoint-competent cells. Cancer Res 1995; 55:1639-42.

[57] Sarkaria JN, Busby EC, Tibbetts RS, et al. Inhibition of ATM and ATR kinase activities by the radiosensitizing agent, caffeine. Cancer Res 1999; 59:4375-82.

[58] Tibbetts RS, Brumbaugh KM, Williams JM, et al. A role for ATR in the DNA damage-induced phosphorylation of p53. Genes Dev 1999; 13:152-7.

[59] Koniaras K, Cuddihy AR, Christopoulos H, et al. Inhibition of Chk1-dependent G2 DNA damage checkpoint radiosensitizes p53 mutant human cells. Oncogene 2001; 20:7453-63.

[60] Luo Y, Rockow-Magnone SK, Joseph MK, et al. Abrogation of G2 checkpoint specifically sensitizes p53 defective cells to cancer chemotherapeutic agents. Anticancer Res 2001; 21:23-8.

[61] Luo Y, Rockow-Magnone SK, Kroeger PE, et al. Blocking CHK1 expression induces apoptosis and abrogates the G2 checkpoint mechanism. Neoplasia 2001; 3:411-9.

[62] Chen Z, Xiao Z, Chen J, et al. Human Chk1 expression is dispensable for somatic cell death and critical for sustaining G2 DNA damage checkpoint. Mol Cancer Ther 2003; 2:543-8.

[63] Rodriguez-Bravo V, Guaita-Esteruelas S, Salvador N, et al. Different S/M checkpoint responses of tumor and non tumor cell lines to DNA replication inhibition. Cancer Res 2007; 67:11648-56.

[64] Sarkaria JN, Busby EC, Tibbetts RS, et al. Inhibition of ATM and ATR kinase activities by the radiosensitizing agent, caffeine. Cancer Res 1999; 59:4375-82.

[65] Sabisz M, Skladanowski A. Modulation of cellular response to anticancer treatment by caffeine: inhibition of cell cycle checkpoints, DNA repair and more. Curr Pharm Biotechnol 2008; 9:325-36.

[66] Hickson I, Zhao Y, Richardson CJ, et al. Identification and characterization of a novel and specific inhibitor of the ataxia-telangiectasia mutated kinase ATM. Cancer Res 2004; 64:9152-9.

[67] Kennedy RD, Chen CC, Stuckert P, et al. Fanconi anemia pathway-deficient tumor cells are hypersensitive to inhibition of ataxia telangiectasia mutated. J Clin Invest 2007; 117:1440-9.

[68] Graves PR, Yu L, Schwarz JK, et al. The Chk1 protein kinase and the Cdc25C regulatory pathways are targets of the anticancer agent UCN-01. J Biol Chem 2000; 275:5600-5.

[69] Welch S, Hirte HW, Carey MS, et al. UCN-01 in combination with topotecan in patients with advanced recurrent ovarian cancer: a study of the Princess Margaret Hospital Phase II consortium. Gynecol Oncol 2007; 106:305-10.

[70] Tse AN, Carvajal R, Schwartz GK. Targeting checkpoint kinase 1 in cancer therapeutics. Clin Cancer Res 2007; 13:1955-60.

[71] Ashwell S, Janetka JW, Zabludoff S. Keeping checkpoint kinases in line: new selective inhibitors in clinical trials. Expert Opin Investig Drugs 2008; 17:1331-40.

[72] Zabludoff SD, Deng C, Grondine MR, et al. AZD7762, a novel checkpoint kinase inhibitor, drives checkpoint abrogation and potentiates DNA-targeted therapies. Mol Cancer Ther 2008; 7:2955-66.

[73] Matthews DJ, Yakes FM, Chen J, et al. Pharmacological abrogation of S-phase checkpoint enhances the anti-tumor activity of gemcitabine in vivo. Cell Cycle 2007; 6:104-10.

[74] Blasina A, Hallin J, Chen E, et al. Breaching the DNA damage checkpoint via PF-00477736, a novel small-molecule inhibitor of checkpoint kinase 1. Mol Cancer Ther 2008; 7:2394-404.

[75] Bao S, Wu Q, McLendon RE, et al. Glioma stem cells promote radio resistance by preferential activation of the DNA damage response. Nature 2006; 444:756-60.

[76] Hollstein M, Sidransky D, Vogelstein B, et al. p53 mutations in human cancers. Science 1991; 253:49-53.

[77] Vazquez A, Bond EE, Levine AJ, et al. The genetics of the p53 pathway, apoptosis and cancer therapy. Nat Rev Drug Discov 2008; 7:979-87.

[78] Vousden KH, Lu X. Live or let die: the cell's response to p53. Nat Rev Cancer 2002; 2:594-604.

[79] Snyder EL, Meade BR, Saenz CC, et al. Treatment of terminal peritoneal carcinomatosis by a transducible p53-activating peptide. PLoS Biol 2004; 2:E36.

[80] Ashwell S, Zabludoff S. DNA damage detection and repair pathways-recent advances with inhibitors of checkpoint kinases in cancer therapy. Clin Cancer Res 2008; 14:4032-7.

[81] Onel K, Cordon-Cardo C. MDM2 and prognosis. Mol Cancer Res 2004; 2:1-8.

[82] Vassilev LT, Vu BT, Graves B, et al. In vivo activation of the p53 pathway by small-molecule antagonists of MDM2. Science 2004; 303:844-8.

[83] Freedman DA, Wu L, Levine AJ. Functions of the MDM2 oncoprotein. Cell Mol Life Sci 1999; 55:96-107.

[84] Roth JA. Adenovirus p53 gene therapy. Expert Opin Biol Ther 2006; 6:55-61.

[85] Peng Z. Current status of gendicine in China: recombinant human Ad-p53 agent for treatment of cancers. Hum Gene Ther 2005; 16:1016-27.

[86] Harris N, Brill E, Shohat O, et al. Molecular basis for heterogeneity of the human p53 protein. Mol Cell Biol 1986; 6:4650-6.

[87] Matlashewski GJ, Tuck S, Pim D, et al. Primary structure polymorphism at amino acid residue 72 of human p53. Mol Cell Biol 1987; 7:961-3.

[88] Bond GL, Hu W, Bond EE, et al. A single nucleotide polymorphism in the MDM2 promoter attenuates the p53 tumor suppressor pathway and accelerates tumor formation in humans. Cell 2004; 119:591-602.

[89] BiPar Sciences presents interim phase 2 results for PARP inhibitor BSI-201 at San Antonio Breast Cancer Symposium. Cancer Biol Ther 2009; 8:2-3.

[90] Martin SA, Lord CJ, Ashworth A. DNA repair deficiency as a therapeutic target in cancer. Curr Opin Genet Dev 2008; 18:80-6.

[91] Fong PC, Boss DS, Yap TA, et al. Inhibition of poly(ADP-ribose) polymerase in tumors from BRCA mutation carriers. N Engl J Med 2009; 361:123-34.

[92] Tutt A, Robson M, Garber JE, et al. Phase II trial of the oral PARP inhibitor olaparib in BRCA deficient advanced breast cancer. J Clin Oncol 2009; 27:CRA501.

[93] Audeh MW, Penson RT, Friedlander M, et al. Phase II trial of the oral PARP inhibitor olaparib (AZD2281) in BRCA-deficient advanced ovarian cancer. J Clin Oncol 2009; 27:5500.

[94] Girard PM, Foray N, Stumm M, et al. Radiosensitivity in Nijmegen Breakage Syndrome cells is attributable to a repair defect and not cell cycle checkpoint defects. Cancer Res 2000; 60:4881-8.

[95] Cariveau MJ, Tang X, Cui XL, et al. Characterization of an NBS1 C-terminal peptide that can inhibit ataxia telangiectasia mutated (ATM)- mediated DNA damage responses and enhance radiosensitivity. Mol Pharmacol 2007; 72:320-6.

[96] Dupre A, Boyer-Chatenet L, Sattler RM, et al. A forward chemical genetic screen reveals an inhibitor of the Mre11-50-Nbs1 complex. Nat Chem Biol 2008; 4:119-25.

[97] Christodoulopoulos G, Malapetsa A, Schipper H, et al. Chlorambucil induction of HsRad51 in B-cell chronic lymphocytic leukemia. Clin Cancer Res 1999; 5:2178-84.

[98] Henning W, Sturzbecher HW. Homologous recombination and cell cycle checkpoints: Rad51 in tumour progression and therapy resistance. Toxicology 2003; 193:91-109.

[99] Ohnishi T, Taki T, Hiraga S, et al. In vitro and in vivo potentiation of radiosensitivity of malignant gliomas by antisense inhibition of the RAD51 gene. Biochem Biophys Res Commun 1998; 245:319-24.

[100] Collis SJ, Tighe A, Scott SD, et al. Ribozyme minigene-mediated RAD51 down-regulation increases radiosensitivity of human prostate cancer cells. Nucleic Acids Res 2001; 29:1534-8.

[101] Ito M, Yamamoto S, Nimura K, et al. Rad51 siRNA delivered by HVJ envelope vector enhances the anti-cancer effect of cisplatin. J Gene Med 2005; 7:1044-52.

[102] Richard DJ, Bolderson E, Khanna KK. Multiple human single-stranded DNA binding proteins function in genome maintenance: structural, biochemical and functional analysis. Crit Rev Biochem Mol Biol 2009:1-19, Epub 2009 Apr 14.

[103] Andrews BJ, Turchi JJ. Development of a high-throughput screen for inhibitors of replication protein A and its role in nucleotide excision repair. Mol Cancer Ther 2004; 3:385-91.

[104] Lees-Miller SP, Godbout R, Chan DW, et al. Absence of p350 subunit of DNA-activated protein kinase from a radiosensitive human cell line. Science 1995; 267:1183-5.

[105] Ouyang H, Nussenzweig A, Kurimasa A, et al. Ku70 is required for DNA repair but not for T cell antigen receptor gene recombination In vivo. J Exp Med 1997; 186:921-9.

[106] Deriano L, Guipaud O, Merle-Beral H, et al. Human chronic lymphocytic leukemia B cells can escape DNA damage-induced apoptosis through the nonhomologous end-joining DNA repair pathway. Blood 2005; 105:4776-83.

[107] Rosenzweig KE, Youmell MB, Palayoor ST, et al. Radiosensitization of human tumor cells by the phosphatidylinositol3-kinase inhibitors wortmannin and LY294002 correlates with inhibition of DNA-dependent protein kinase and prolonged G2-M delay. Clin Cancer Res 1997; 3:1149-56.

[108] Shinohara ET, Geng L, Tan J, et al. DNA dependent protein kinase is a molecular target for the development of noncytotoxic radiation-sensitizing drugs. Cancer Res 2005; 65:4987-92.

[109] Zhao Y, Thomas HD, Batey MA, et al. Preclinical evaluation of a potent novel DNA-dependent protein kinase inhibitor NU7441. Cancer Res 2006; 66:5354-62.

[110] Martin SA, Lord CJ, Ashworth A. DNA repair deficiency as a therapeutic target in cancer. Curr Opin Genet Dev 2008; 18:80-6.

[111] Bryant HE, Schultz N, Thomas HD, et al. Specific killing of BRCA2-deficient tumours with inhibitors of poly(ADP-ribose) polymerase. Nature 2005; 434:913-7.

[112] Farmer H, McCabe N, Lord CJ, et al. Targeting the DNA repair defect in BRCA mutant cells as a therapeutic strategy. Nature 2005; 434:917-21.

[113] Hay T, Matthews JR, Pietzka L, et al. Poly (ADP-ribose) polymerase-1 inhibitor treatment regresses autochthonous Brca2/p53-mutant mammary tumors in vivo and delays tumor relapse in combination with carboplatin. Cancer Res 2009; 69:3850-5.

[114] Zachos G, Black EJ, Walker M, et al. Chk1is required for spindle checkpoint function. Dev Cell 2007; 12:247-60.

[115] Tang J, Erikson RL, Liu X. Checkpoint kinase 1 (Chk1) is required for mitotic progression through negative regulation of polo-like kinase 1 (Plk1). Proc Natl Acad Sci USA 2006; 103:11964-9.

[116] Nieborowska-Skorska M, Stoklosa T, Datta M, et al. ATR-Chk1axis protects BCR/ABL leukemia cells from the lethal effect of DNA double-strand breaks. Cell Cycle 2006; 5:994-1000.

[117] Wang HY, Zhang M, Zou P, et al. Mechanism of G2/M blockage triggered by activated-Chk1in regulation of drug-resistance in K562/A02 cell line. Zhongguo Shi Yan Xue Ye Xue Za Zhi 2006; 14:1105-9.

[118] Bao S, Wu Q, McLendon RE, et al. Glioma stemcells promote radioresistance by preferential activation of the DNA damage response. Nature 2006; 444:756-60.

[119] Verlinden L, Bempt IV, Eelen G, et al. The E2Fregulated gene Chk1 is highly expressed in triple-negative estrogen receptor-/HER-2-breast carcinomas. Cancer Res 2007; 67:6574-81.

[120] Madoz-Gurpide J, Canamero M, Sanchez L, et al. A proteomics analysis of cell signaling alterations in colorectal cancer. Mol Cell Proteomics 2007; 6:2150-64.

[121] Tenzer A, Pruschy M. Potentiation of DNA-damage induced cytotoxicity by G2 checkpoint abrogators. Curr Med Chem Anti-Cancer Agents 2003; 3:35-46.

[122] Kawabe T. G2 checkpoint abrogators as anticancer drugs. Mol Cancer Ther 2004; 3:513-19.

[123] Zhou BB, Anderson HJ, Roberge M. Targeting DNA checkpoint kinases in cancer therapy. Cancer Biol Ther 2003; 2:S16-22.

[124] Prudhomme M. Combining DNA damaging agents and checkpoint 1 inhibitors. Curr Med Chem Anti-Canc Agents 2004; 4:435-8.

[125] Prudhomme M, Novel checkpoint 1 inhibitors. Rec Patents Anticancer Drug Disc 2006; 1:55-68.

[126] Tao Z-F, Lin N-H. CHK1 inhibitors for novel cancer treatment. Anticancer Agents Med Chem 2006; 6:377-88.

[127] Janetka JW, Ashwell S, Zabludoff S, et al. Inhibitors of checkpoint kinases: From discovery to the clinic. Curr Opin Drug Discov Dev 2007; 10:473-86.

[128] Arrington KL, Dudkin VY. Novel inhibitors of checkpoint kinase1. ChemMed Chem 2007; 2:1571-85.

[129] Tse AN, Carvajal R, Schwartz GK. Targeting checkpoint kinase 1in cancer therapeutics. Clin Cancer Res 2007; 13:1955-60.

[130] Matthews DJ, Yakes FM, Chen J, et al. Pharmacological abrogation of S-phase checkpoint enhances the anti-tumor activity of gemcitabine in vivo. Cell Cycle 2007; 6:104-10.

[131] Ashwell S, Caleb BL, Green S, et al. Preclinical identification of AZD7762, a Novel, Potent and Selective Inhibitor of Checkpoint Kinases. Abstracts of Papers. 2007 EORTC-NCI-AACR International Meeting on Molecular Targets and Therapeutics, San Francisco, USA (2007): A232.

[132] Blasina A, Kornmann JF, Chen E, et al. A novel inhibitor of the protein kinase CHK1: Studies on the mechanism of action [abstract 4416]. Proc Am Assoc Cancer Res 2005; 46. https://cancerres.aacrjournals.org/content/65/9_Supplement/1045.3

[133] Ninkovic S. The discovery and design of CHK kinase inhibitors. First RSC-SCI Symposium on Kinase Inhibitor Design (Part I). London (UK), 2005.

[134] McArthur GA, Raleigh J, Blasina A, et al. Imaging with FLT-PET demonstrates that PF-477736, an inhibitor of CHK1 kinase, overcomes a cell cycle checkpoint induced by gemcitabine in PC-3 xenografts [abstract 3045]. Proc Am Soc Clin Oncol 2006; 25. Journal of Clinical Oncology 24 (18_suppl):3045. https://ascopubs.org/doi/abs/10.1200/jco.2006.24.18_suppl.3045

[135] Raza Dewji M. Beyond VEGF, targeting tumor growth and angiogenesis via alternative mechanisms. First International Meeting, Targeted Therapies in Cancer: Myth or Reality? Milan (Italy), 2006.

[136] Anderes K, Blasina A, Chen E, et al. Characterization of a novel and selective inhibitor of checkpoint kinase 1: Breaching the tumor's last checkpoint defense against chemotherapeutic agents [abstract 373]. 18th EORTC-NCI-AACR Symposium on Molecular Targets and Cancer Therapeutics, Prague, Czech Republic, 2006.

[137] Li G, Elder RT, Qin K, et al. Mechanisms of radiation enhancement by the CHK1 inhibitor PF-477736. J Biol Chem 2007; 282:7287-98.

[138] Arienti KL, Brunmark A, Axe FU, et al. Checkpoint kinase inhibitors: SAR and radioprotective properties of a series of 2-arylbenzimidazoles. J Med Chem 2005; 48:1872-85.

[139] Carlessi L, Buscemi G, Larson G, et al. Biochemical and cellular characterization of VRX0466617, a novel and selective inhibitor for the checkpoint kinase Chk2. Mol Cancer Ther 2007;6: 935-44.

[140] Matthews DJ. Dissecting the roles of CHK1 and CHK2 in mitotic catrastophe using chemical genetics [abstract 344]. 18th EORTC-NCI-AACR Symposium on Molecular Targets and Cancer Therapeutics, Prague, Czech Republic, 2006.

[141] Dougherty TJ, Grindery GB, Weishaupt KR, et al. Photoradiation therapy. II. Cure of animal tumors with hematoporphyrin and light. J Nat Cancer Inst 1975; 55:115.

[142] Hayata Y, Kato H, Konaka C, et al. Haematoporphyrin in derivative and laser photoradiation in the treatment of lung cancer. Chest 1982; 81:269.

[143] Lele PP, Physical Aspect of Hyperthermia, Eds. G.H. Nussbaum. Am. Inst. Phys., New York, 1982. pp. 393-440.

[144] Kremkau FW, Kaufmann JS, Walker MM, et al. Ultrasonic enhancement of nitrogen mustard cytotoxicity in mouse leukemia. Cancer 1976; 47:1643.

[145] Akimoto R. An experimental study on enhancement of the effect of anti-cancer drug by ultrasound. J. Jpn. Soc. Cancer Therapy 1985; 3:562.

[146] Yumita N, Okumura A, Nishigaki R, et al. The combination treatment of ultrasound and antitumor drugs on Yoshida sarcoma. Jpn. J. Hyperthermia Oncol. 1987; 3:175.

[147] Yumita N, Nishigaki R, Umemura K, et al. Hematoporphyrin as a sensitizer of cell-damaging effect of ultrasound. Jpn. J. Cancer Res. 1989; 80:219.

[148] Yumita N, Nishigaki R, Umemura K, et al. Synergistic effect of ultrasound and hematoporphyrin on sarcoma 180. Jpn. J. Cancer Res. 1990; 81:304.

[149] Umemura S, Yumita N, Nishigaki R, et al. Mechanism of cell damage by ultrasound in combination with hematoporphyrin. Jpn. J. Cancer Res. 1990; 81:955.

[150] Umemura S, Yumita N, Nishigaki R. Enhancement of ultrasonically induced cell damage by a gallium-porphyrin complex, ATX-70. Jpn. J. Cancer Res. 1993; 84:582.

[151] Yumita N, Nishigaki R, Umemura K, et al. Sonochemical Activation of hematoporphyrin: an ESR study. Radiation Res. 1994; 138:171.

[152] Umemura S, Yumita N, Nishigaki R, et al. IEEE Ultrason. Symp. Proc. 1989:955-60.

[153] Henglein A. Sonochemistry: historical developments and modern aspects. Ultrasonics 1987; 25:6.

[154] Kawabata K, Umemura S. IEEE Ultrason. Syrup. Proc. 1992:1281-5.

[155] Umemura S, Kawabata K. IEEE Ultrason. Syrup. Proc. 1993:917-20.

[156] Umemura S, Kawabata K, Sasaki K. IEEE Ultrason. Syrup. Proc. 1994:1843-6.

[157] Kawabata K, Umemura S. Effect of second-harmonic superimposition on efficient induction of sonochemical effect. Ultrasonics Sonochemistry 1996; 3:1.

[158] Umemura S, Kawabata K, Sasaki K. IEEE Ultrason. Symp. Proc. 1995:1567-70.

[159] Umemura S-I, Kawabata K-I, Sasaki K. In vitro and in vivo enhancement of sonodynamically active cavitation by second-harmonic superimposition. J Acoust Soc Am 1997; 101:569. https://doi.org/10.1121/1.418120

[160] Umemura S, Kawabata K, Sasaki K, et al. Recent advances in sonodynamic approach to cancer therapy. Ultrason Sonochem 1996; 3(3):S187-91.

[161] Madkour LH. Reactive Oxygen Species (ROS), Nanoparticles, and Endoplasmic Reticulum (ER) Stress-Induced Cell Death Mechanisms. Academic Press, 2020. https://www.elsevier. com/books/reactive-oxygen-species-ros-nanoparticles-and-endoplasmic-reticulum-er-stress-induced-cell-death-mechanisms/madkour/978-0-12-822481-6

[162] Madkour LH. Nanoparticles Induce Oxidative and Endoplasmic Reticulum Antioxidant Therapeutic Defenses. 1st Edition. Springer, Switzerland, 2020. https://www. springer.com/gp/book/9783030372965?utm_campaign=3_ pier05_buy_print&utm_content=en_08082017&utm_ medium=referral&utm_source=google_books#othervers ion=9783030372972

[163] Madkour LH. Nucleic Acids as Gene Anticancer Drug Delivery Therapy. 1st Edition. Academic Press, Elsevier, 2020. https:// www.elsevier.com/books/nucleic-acids-as-gene-anticancer-drug-deliverytherapy/madkour/978-0-12-819777-6

[164] Madkour LH. Nanoelectronic Materials: Fundamentals and Applications (Advanced Structured Materials). 1st Edition. Springer, Switzerland, 2019. https://books.google. com.eg/books/about/Nanoelectronic_Materials.html?id= YQXCxAEACAAJ&source=kp_book_description&redir_ esc=y https://www.springer.com/gp/book/9783030216207

# 11 Therapeutic Potential Role of miRNAs in Pancreatic and Prostate Cancer Cells

Despite considerable progress being made in understanding pancreatic cancer (PC) pathogenesis, it still remains the 10th most often diagnosed malignancy in the world and 4th leading cause of cancer-related death in the United States with a 5-year survival rate of only 6%.

Epigenetic modifiers play important roles in fine-tuning the cellular transcriptome. Any imbalance in these processes may lead to abnormal transcriptional activity and thus result in disease state. Distortions of the epigenome have been reported in cancer initiation and progression. DNA methylation and histone modifications are principal components of this epigenome, but more recently it has become clear that microRNAs (miRNAs) are another major component of the epigenome. Interactions of these components are apparent in prostate cancer (CaP), which is the most common noncutaneous cancer and second leading cause of death from cancer in the United States.

This chapter summarizes key aspects of these mechanistic interactions within the epigenome and highlights their translational potential as functional biomarkers. To this end, exploration of The Cancer Genome Atlas (TCGA) prostate cancer data revealed that expression of key CaP miRNAs inversely associate with DNA methylation. In this chapter, we summarize the role of several miRNAs that regulate various oncogenes (KRAS) and tumor suppressor genes (p53, p16, SMAD4, etc.) involved in PC development, and their prospective roles as diagnostic and prognostic markers and as therapeutic targets.

## 11.1 EPIGENETIC CONTRIBUTIONS TO CANCER

### 11.1.1 An Epigenetic Basis for Prostate Cancer

Prostate cancer (CaP) is the most common noncutaneous cancer diagnosed and second leading cause of death in men from cancer in the United States [1, 2]. This cancer is highly heterogeneous, ranging from asymptomatic to a rapidly fatal systemic malignancy. Several genetic and epigenetic alterations are highly prevalent and appear to be important factors in the tumorigenesis and progression of CaP. The most common genetic changes associated with CaP include deletions involving NK3 homeobox 1 (NKX3.1) [3] and phosphatase and tensin homologue tumor suppressor genes (PTEN) [4, 5], amplifications of the androgen receptor (AR) [6] and MYC [7] genes, and more recently described translocations that lead to TMPRSS2:ERG and SLC45A3:ERG gene fusions [8, 9]. Recently, tumor analyses with next-generation sequencing

approaches have identified several additional novel somatic mutations, including MED12, FOXA1, and SPOP. The E3 ubiquitin ligase adaptor speckle-type poxvirus and zinc finger (POZ) domain protein (SPOP) showed recurrent mutations in 6–15% of tumors across multiple independent cohorts [10].

The complex nature of cancer phenotypes, however, cannot be explained by genetic components alone [11]. Epigenetic modifications appear to contribute significantly to transformation. These events can be defined as heritable changes to the expression and regulation of gene expression that are not associated with alterations in DNA sequences [12]. These epigenetic events including DNA methylation and histone modifications appear to play distinct yet complementary roles to genetic events and add to a fuller explanation for the basis of cancer. For example, many cancers show gene-specific and global changes in DNA CpG methylation and/or altered histone modification patterns [13, 14].

Among the epigenetic modifications, altered DNA methylation and histone modification patterns have been frequently reported in CaP [15–17]. These epigenetic modifications govern tissue-specific gene regulation by a complex machinery of several DNA- and histone-modifying enzymes including DNA methyltransferases (DNMTs) [18], histone acetyltransferases (HATs) [19], histone deacetylases (HDACs) [20], histone methyltransferases (HMTs) [21], histone demethylases (HDMTs) [22], and chromatin remodeling enzymes (reviewed in [13]).

One of the most prominent epigenetic modifications reported in CaP is hypermethylation of glutathione S-transferase (GSTP1) gene. GSTP1 is involved in detoxification of genotoxic and electrophilic compounds. Hypermethylation of GSTP1 is reported in 70–90% of patients with CaP but not in benign hyperplastic prostate tissue [23]. Similarly, upregulation of HMT enhancer of zeste homolog 2 (EZH2) is reported in metastatic CaP, and localized CaP with higher expression of EZH2 shows poorer prognosis [24]. Also, high expression of EZH2 has been reported in hormone-refractory CaP compared with benign hyperplastic prostate tissue [24, 25]. EZH2 catalyzes the trimethylation of histone H3 on Lys27 (H3K27), is involved in gene repression, and contributes to CaP tumorigenesis through silencing of tumor suppressor genes.

Recently published ENCODE project data have greatly enhanced the understanding of the transcriptional landscape of the genome. Analysis across multiple cell lines suggested that as much as 75% of the genome is transcribed into different types of RNA molecules, e.g., protein coding, long noncoding, pseudogenes, and small RNA genes [26]. Although this percentage is the subject of significant debate [27, 28], it is clear that the rate, magnitude, and diversity of RNA transcription is greater than previously suspected. Of these RNA species, the microRNAs (miRNAs) have emerged as major biological regulators. MiRNAs are small (about 19–22 nucleotides [nt]) noncoding RNA molecules and modulate the expression of many, if not all, genes in the human genome. They act by inhibition of protein translation and/or degradation of mRNA transcripts [29]. Understanding their function has contributed significantly to the knowledge of how genomic information is interpreted in physiology and corrupted in malignancy [30].

It is important to understand the interplay of DNA methylation, histone modifications, and miRNA and their cumulative effect on CaP initiation and progression. In the last 5 years, considerable work has been undertaken and has begun to reveal the true extent of epigenetic dysregulation of the genome in cancer. In this chapter we particularly focus on how miRNAs and DNA/histone modifications interact with and regulate each other and how they can be used in CaP diagnosis and treatment.

### 11.1.2 AN EMERGENT UNDERSTANDING ON THE ROLE OF MIRNA IN PROSTATE CANCER

In parallel to the well-established understanding of the role of classical epigenetic events, the role of noncoding miRNA in epigenetic regulation of gene expression has also emerged. The four types of small RNA genes, namely, small nuclear (sn)RNAs, small nucleolar (sno)RNAs, miR-NAs, and transfer (t)RNAs, correspond to 85% of total small RNAs (7053) annotated by GENCODE. Importantly, ENCODE data revealed the wide range of expression of miRNAs in different normal and cancer cell lines [26]. At any given time point 28% of miRNAs (497 of 1756) were expressed in at least one cell line. Interestingly 30% (147 of 497) of these miRNAs were detected in all 12 cell lines analyzed [26]. This observation strengthens the idea that miRNAs are important components of gene regulation networks and expressed differentially in different disease models. Furthermore, deregulation of miRNAs has been shown for a variety of solid tumors, including breast, colon, and prostate [31–33]. Thus, these studies have established an important role of miRNAs in the control of gene regulation and they are frequently deregulated in many cancers [34, 35].

Although miRNAs represent only ~1% of the genome, they are estimated to target 30% of genes [36]. The relatively small number of miRNAs and their wide coverage of protein-coding genes make them very attractive to be used as disease biomarkers. Compared with mRNA signatures, miRNAs have better and strong biomarker properties [37]. It has been estimated that miRNAs provide 20 times more power in biomarker studies as compared with mRNAs (when comparing 20, 000 mRNAs to ~1, 000 miRNAs),

which makes them ideal candidates for studies with small sample size [37].

In CaP, several miRNAs have been reported to be differentially regulated and act as both tumor suppressor and onco-miRs. For instance, miR-221 and miR-222 have been shown to be upregulated in castration-resistant prostate cancer (CRPC) cells compared with androgen-dependent CaP cell line and significantly affect the response to DHT [38]. In general, oncogenic properties of miR-221 and miR-222 are attributed to their control of the cyclin dependent kinase (Cdk) inhibitors $p27^{KIP1}$ and $p57^{KIP2}$, and thus control of G1 to S phase transition [39, 40], PI3K and PTEN signaling [41], and other targets implicated in malignant transformation, including CX43 [42], RECK [43], and ERα [44, 45].

Lin et al. observed that aggressive (Gleason grade ≥7) tumors had elevated expression of miR-221 and miR-222 compared with less aggressive (Gleason grade <7) tumor tissues, further suggesting an oncogenic role of these miR-NAs in CaP development [46]. Conversely, others have shown the loss of miR221/222 in early stage [47] as well as in aggressive CaP [48, 49]. Most recently, a miRnome-wide scan also found miR-221 and miR-222 to be downregulated when comparing 20 matched pairs of microdissected tissue samples of prostate cancer and nontumor tissue [50]. These conflicting results perhaps indicate that expression levels are differentially modulated at divergent disease stages.

Similarly, miR-125b has been identified in CaP along with miR-143 to be upregulated in metastatic CaP serum samples as compared with normal individuals [51]. Expression of miR-125b in serum of CaP patients is reported to be upregulated as compared with normal controls [51], whereas other studies reported it to be downregulated in CaP as compared with normal or benign prostatic hyperplastic (BPH) samples [47, 52, 53]. MiR-125 regulates cell proliferation in prostate cancer cell lines [54], and it has been suggested to be upregulated by androgen signaling [55]. Functionally in CaP, miR-125b has been reported to target BAK1 [56] (a proapoptotic member of the BCL-2 gene family) and EIF4EBP1 (eukaryotic translation initiation factor 4E-binding protein 1), a gene that encodes one member of a family of translation repressor proteins [52].

Other important miRNAs involved in CaP include miR-21, which is an AR-regulated miRNA and overexpressed in CRPC compared to adjacent normal tissue [56]. Overexpression of miR-21 was also observed in docetaxel-resistant variant of PC3 cells; additionally, ectopic expression of miR-21 in wild-type PC3 cells increased the resistance to docetaxel [57]. The oncogenic and drug-resistance properties of miR-21 were attributed to its control of downstream target, programmed cell death 4 (PDCD4) [57, 58].

Several miRNAs have also been identified as tumor suppressors in CaP, including miR-143, miR-145, and the miR-200 family. Expression of miR-143 and miR-145 is significantly suppressed in CaP and negatively associated with metastasis [59]. These miRs contribute to CaP progression through epithelial-mesenchymal transition (EMT) [59] and loss of their repressive effect on the EGFR/RAS/MAPK pathway [60]. MiR-205 and miR-200 family miRNAs are also downregulated in CaP and have been shown to regulate EMT by targeting ZEB1 and ZEB2 in CaP [43, 60, 61].

## 11.2 INTERPLAY BETWEEN HISTONE MODIFICATIONS, DNA METHYLATION, AND miRNA

Critically, the three different epigenetic modifications (i.e., DNA methylation, histone modifications, and miRNAs) interact with and influence one another [62]. For example, expression of miRNAs can be regulated by DNA methylation in their respective promoter regions [63]. Similarly, miRNAs can directly target chromatin-remodeling enzymes [64] and DNMTs [65] and in turn modulate the downstream effects.

### 11.2.1 REGULATION OF DNA AND HISTONE MODIFICATIONS UNDER THE CONTROL OF miRNAs IN PROSTATE CANCER

MiRNAs control gene expression posttranscriptionally by translational inhibition when base pairing between miRNA and target sequence is imperfect, while perfect or near-perfect complementarity can induce the degradation of the target mRNA [66]. One miRNA can target several genes and one gene can similarly be targeted by several miRNAs [66]. The latest release of miRBase (miRBase 19) has 1595 primary transcripts and 2038 mature miRNA sequences. The wide range of miRNA targets makes them essential in the majority of biological processes, including cell cycle regulation, differentiation, development, metabolism, and aging.

Although the biogenesis of miRNA has been well described, the regulation of miRNA expression remains less well explored. Several miRNAs have well-defined promoters, but for many the regulatory elements are not as clear.

They are often present in introns of protein-coding genes, and the promoters for the genes are considered to be promoters for the intragenic miRNAs [66, 67]. Interestingly, there is evidence for differential regulation between the host mRNA and miRNA and this may suggest an emergent role for alternative regulation processes.

Like their target genes, miRNAs can also directly or indirectly target key effectors of the epigenetic machinery (Figure 11.1). For example, miR-29 family members (29a, 29b, and 29c) are downregulated in several cancers and they are predicted to target DNA methyltransferase DNMT3A, DNMT3B, and DNMT1. Indeed, enforced expression of miR-29 family members resulted in the loss of DNA methylation in lung cancer and acute myeloid leukemia and re-expression of methylation silenced genes [68, 69]. Recently, miR-34b has been shown to target DNMT1, HDAC1, HDAC2, and HDAC4 in CaP cell lines [70]. Interestingly, miR-34b was also epigenetically silenced by DNA methylation and ectopic expression of miR-34b in CaP cells resulted in partial demethylation of 5' upstream sequence of the miR-34b gene and also showed enrichment of tri-methylated histone H3 lysine 4 (H3K4me3), a mark for an active chromatin [70]. These data provide insight into the interplay between miRNA and DNA/histone marks in CaP.

Epigenetic regulatory mechanisms such as DNA methylation (controlled by DNMTs) and histone modifications modulate epigenetic states and transcriptional responsiveness of both mRNA and miRNA target genes. These actions are controlled by antagonistic classes of enzymes, for example by histone deacetylases (HDACs) and histone acetyltransferases (HATs). There is good evidence for control of their expression by miRNA. For example, miR-101 targets include EZH2 [71] and miR-499a targets include HDAC1 [72], whereas the HAT (p300/CBP-associated factor [PCAF]) is targeted by miR-17-5p [73].

A similar miRNA-mediated epigenetic cross-talk occurs with key histone modifications. Polycomb proteins form chromatin-modifying complexes that induce gene silencing by catalyzing trimethylation of histone H3K27. There

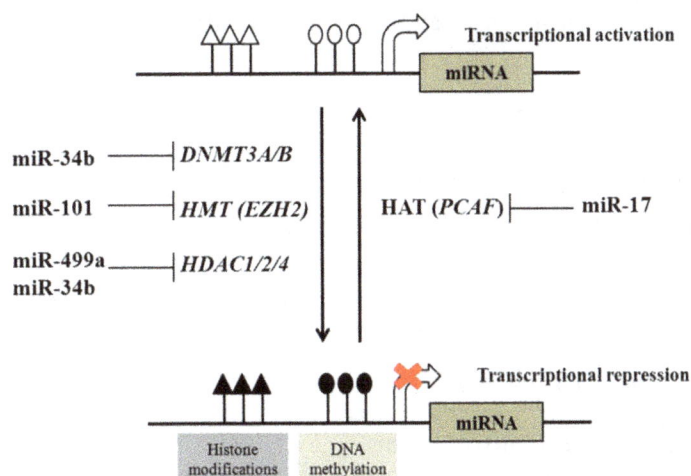

**FIGURE 11.1** Interactions of DNA/histone modification and miRNA.

are two polycomb repressor complexes (PRCs), PRC1 and PRC2, and deregulation of these complexes has been reported to associate with many cancers [74, 75]. For example, EZH2, is a highly conserved catalytic subunit within PRC2, and has been reported to be elevated in CaP and correlate with metastatic CaP and poor prognosis in localized CaP [17]. One of the important mechanisms for altered PRC2 activity arises due to elevated EZH2 level that occurs through downregulation of miR-101 [76]; expression of miR-101 is frequently lost in CaP [34, 35]. Increased EZH2 expression is associated with CaP progression and loss of miR-101 may be responsible for increased EZH2 expression in CaP [71]. MiR-101 also plays important roles in control of cell proliferation, and its expression correlates inversely with aggressiveness in CaP cell line models [71]. Re-expression of miR-101 inhibits the expression of EZH2 and attenuates cell proliferation and tumor invasiveness [71]. More recently, knockdown of Dicer, a key protein required for miRNA processing, resulted in increased expression of other PRC components, e.g., EZH2, EED, and SUZ12 (PRC2 components), and BMI1 and RING2 (PRC1 components). Together these data suggest miRNA control of the PRC complex occurs through multiple targets [77].

Upon H3K27 methylation by PRC2 proteins, PRC1 proteins are recruited to chromatin to maintain stable gene silencing by catalyzing ubiquitinylation of histone H2A. BMI1 is negatively regulated by miR-203, 200b/c while RING2 is negatively regulated by miR-181a/b and miR-200b/c [77]. Downregulation of BMI1 and RING2 by these miRNAs also resulted in decreased global ubiquityl-H2A. Interestingly, expression of these miRNAs was in turn regulated by EZH2 and silencing of EZH2 or expression of miR-101 resulted in overexpression of these miRs, and reflects the intricate co-regulation of miRNA and mRNA in feed-forward loop structures [67, 77–83]. These findings illustrate the complex interplay of epigenetic machinery and miRNA in controlling gene silencing in CaP.

MiRNA also regulates the expression of histone deacetylases (HDACs) that control gene silencing by catalyzing deacetylation of key histone residues. The class I HDACs are the most frequently overexpressed in cancers, particularly HDAC1, which is found to be overexpressed in CaP [84]. HDAC1 is a direct target of miR-499a in PC-3 cells [72], and in turn miR-449a is involved in cell-cycle regulation and is frequently downregulated in CaP. Overexpression of miR-449a in CaP cell lines leads to downregulation of HDAC1 and cell cycle arrest. The locus-specific versus global effects of HDACs are emerging and interestingly the growth arrest induced by miR-449a mediates repression of HDAC1 leading to the induction of p27$^{(kip1)}$ [72], again suggesting the combinatorial epigenetic regulation of key growth regulators.

Another class of important histone modifiers is the histone acetyltransferases (HATs). They acetylate core histones, which is associated with easily accessible chromatin for the transcriptional machinery. In CaP, p300/CBP-associated factor (PCAF), which has HAT activity, has been shown to act as a co-activator for androgen receptor (AR) [85]. Gong et al. showed that PCAF was upregulated in CaP cell lines compared with normal prostate epithelium cells and promoted DHT-stimulated AR transcriptional activity. Interestingly, they also showed that PCAF is a direct target of miR-17-5p, and it attenuated DHT-induced expression of the PSA gene and inhibited DHT-induced cell growth in LNCaP cells [73]. Taken together, these observations suggest that miRNAs play an important role in fine-tuning of both silencing and activating epigenetic marks in CaP (Table 11.1).

### 11.2.2 Regulation of miRNAs by DNA and Histone Modification in Prostate Cancer

Genes encoding miRNAs are as tightly regulated as any other gene in the genome and thus miRNA expression is also subjected to similar epigenetic regulation (Figure 11.1). Many miRNAs are present in introns of protein-coding genes and they are regulated by the promoter of host gene. In this manner the epigenetic regulation of protein-coding genes can also affect the expression of miRNA. Also, many

---

**TABLE 11.1**

**Interaction of miRNA and Epigenetic Modifications**

**Examples of miRNAs Regulated by DNA Methylation**

| DNA methylation | miR-145, miR-193b, miR-34a, miR-34b, miR-200c, miR-141, miR-205, miR-196b, miR-21, miR-615 |
|---|---|

**Examples of Epigenetic Modifiers Regulated by miRNAs**

| Epigenetic Modifiers | | Targeting miRNA |
|---|---|---|
| DNA methyltransferases (DNMTs) | DNMT3A and DNMT3B | miR-29 family |
|  | DNMT1 | miR-34b |
| Histone methyltransferases (HMTs) | EZH2 | miR-101 |
| Histone deacetylases (HDACs) | HDAC1, HDAC2, and HDAC4 | miR-34b, miR-499a |
| Histone acetyltransferases (HATs) | PCAF | miR-17-5p |

miRNAs are located in clusters in the genome and therefore more than one miRNA can be regulated in parallel by a single transcriptional event. In the genome, distribution of CpG islands (CGIs) around miRNA genes is similar to protein-coding genes [86]. For example, mapping to the genome of differentially regulated miRNAs after DNMT inhibition with 5-azacytidine (5-aza) showed that 13–24% of human miRNA genes are located within 3–10 kb of a CGI, respectively [86].

Several miRNAs are frequently deregulated in different cancers and many of them are epigenetically regulated. In CaP, miR-200 family members and miR-205 specifically are frequently altered and regulate EMT and cell migration/invasion [60, 87]. Members of the miR-200 family include five miRNAs transcribed from two clusters, one on chromosome 1 encoding miR-200b, miR-200a, and miR-429, and another on chromosome 12 encoding miR-200c and miR-141. Members of the miR-200 family combined with miR-205 regulate EMT and cell migration/invasion by directly targeting ZEB1 and ZEB2 [61]. MiR-205 is downregulated in the majority of cancers including CaP [88]. One of the proposed mechanisms for downregulation of miR-205 is through methylation of its promoter, which has been shown to be associated with chemotherapy drug resistance [89]. Specifically, downregulation of miR-205 arises due to a hypermethylated promoter, and the expression of miR-205 increased when cells were treated with 5-aza. Similarly, miR-200c and miR-141 are downregulated in PC-3 cells as compared to LNCaP, and showed hypermethylation of the promoter CpG region in PC-3 and unmethylated promoter CpG in LNCaP cells. Indeed treatment with 5-aza increased the expression of miR-200c and miR-141 in PC-3 cells [90]. These findings suggest repression of miR-200c/miR-141 is epigenetically controlled in CaP.

Another miRNA that is frequently downregulated in cancers is miR-34a [91, 92]. The expression of miR-34a is regulated by p53, and perhaps reflecting this, it induces apoptosis and cell cycle arrest [93, 94]. Again, miR-34a is downregulated in many CaP cell lines and the promoter region of miR-34a is hypermethylated in malignant CaP cell lines as compared with nonmalignant or benign prostate hyperplasia [92]. Interestingly, similar to hypermethylation observed in CaP cell lines, the promoter of miR-34a has also been demonstrated to be methylated in the majority (79%) of primary CaP tumors as compared with normal prostate stroma [92]. Additionally, miR-34a is inactivated by aberrant CpG methylation in many other types of cancer cell lines including lung (29.1%), breast (25%), colon (13%), kidney (21.4%), bladder (33.3%), and pancreatic carcinoma (15.7%) [92]. Intriguingly, miR-34a has been shown to regulate cancer stem cell (CSC) properties in CaP, and downregulation of miR-34a in prostate cancer stem cells may contribute to metastasis by regulating migratory, invasive, and metastatic properties of CSCs [95].

Other workers have investigated repressed miRNA in CaP. For example, Rauhala et al. searched for miRNAs silenced by DNA methylation in CaP cell lines and identified miR-193b to be silenced in 22Rv1 and VCaP cell lines. Expression of miR-193b was increased after treatment with 5-aza and trichostatin A (TSA) [96]. Also, miR-193b expression was significantly reduced in CaP tumors compared with adjacent normal tissue. Methylation levels at a CGI located ~1 kb upstream of the mature miR-193b locus were higher in CaP cell lines compared with a normal prostate epithelial cell line. The methylation was partially removed after treatment with 5-aza in combination with TSA, and expression of miR-193b was restored in 22Rv1 cells [96]. Overexpression of miR-193b caused reduction in cell growth and an inhibition of anchorage-independent growth, suggesting that miR-193b is an epigenetically silenced putative tumor suppressor in CaP [96].

Similarly, miR-145 has been reported as a tumor suppressor and is frequently downregulated in many cancers including CaP [97]. MiR-145 is also regulated by p53 and inhibits tumor cell growth and invasion by targeting several genes such as c-MYC [98, 99]. In CaP, miR-145 is downregulated in cell lines as well as 81% of primary tumors compared to adjacent normal tissue [100, 101]. The promoter of miR-145 was hypermethylated in CaP cell lines and proximal CpG sites were completely methylated in all three cell lines tested (PC3, LNCaP, and DU145), whereas distal CGIs showed complete methylation in PC3 and partial methylation in the other two cell lines [100]. The miR-145 promoter was most methylated in PC3 cells and treatment with 5-aza increased the expression of miR-145 in all three cell lines [100]. Another study from the same group revealed that the promoter of miR-145 was methylated in CaP tumors and the expression of miR-145 also correlated with mutation status of p53 [101]. Of all the CaP tumor samples tested, 81% showed downregulation of miR-145, of which 35% had both methylation of miR-145 promoter and p53 mutation, 29% had p53 mutation only, and 18% had methylation only. Interestingly, four CaP samples with no change in miR-145 expression had neither p53 mutation nor hypermethylation of miR-145 and 9 of 10 BPH samples with high expression of miR-145 showed no hypermethylation of miR-145 [101]. They also showed that methylation of miR-145 promoter inhibits the binding of p53 and regulation of miR-145 by p53. Renewal of miR-145 promoter activity and p53 binding led to increased apoptosis [101].

Whilst these candidate approaches are illustrative of concepts, genome-wide approaches are required to reveal the full significance of interaction between miRNA and epigenetic modifiers. Hulf and co-workers performed integrative analysis combining primary transcription, genome-wide DNA methylation, and H3K9Ac marks with miRNA expression to identify epigenetically regulated miRNA in cancer [102]. They identified three epigenetically silenced miRNAs and one activated miRNA in CaP by using the normal prostate epithelial (PrEC) and LNCaP cells with three criterion—primary transcript expression, DNA methylation loss, and H3K9Ac gain—and identified miR-205, miR-196b, and miR-21 to be epigenetically repressed and miR-615 to be epigenetically activated. As discussed above, miR-205 has been shown to be epigenetically silenced in CaP [89]. Also, miR-21 has previously been shown to be epigenetically regulated in ovarian cancer [103] and may be similarly regulated in CaP. The role of miR-196b and miR-615 is not very clear in CaP but it is interesting to note that in LNCaP cells, the HOXC cluster on chromosome 13, from which miR-615 is transcribed, is epigenetically regulated [104].

## 11.3  PANCREATIC CANCER

Pancreatic cancer (PC) is a lethal malignancy and remains a major clinical challenge. Due to its early metastatic nature, more than 80% of PC patients have invasive disease at the time of diagnosis, which makes surgical and medical intrusions mostly unsuccessful, resulting in high mortality and poor prognosis [105, 106]. With equal incidence-to-mortality ratio and a 5-year survival rate of less than 6%, PC is considered to be the most deadly and aggressive cancer compared with other malignancies [107]. Limitations of the current multimodality therapeutic regimens for PC highlight the urgent need to understand the molecular mechanism/pathway(s) governing initiation, progression, and metastasis for discovering novel diagnostic, prognostic, and therapeutic targets for better management of this lethal disease. Although several research efforts are directed to discover specific biomolecules with their utility in early diagnosis, prognosis, and therapy for PC, these currently used biomolecules do not have adequate sensitivity and specificity to detect PC in its early stages, nor can they be used to target PC.

The advent of miRNA research has opened new avenues to understand the gene regulatory mechanisms at the post-transcriptional level. MiRNAs are double-stranded, small, noncoding RNA molecules of 19–24 nt in length, regulating gene expression at the posttranscriptional level either by degradation or translational inhibition of their target mRNA [108, 109]. Details about the miRNA biogenesis, role in physiology, development, and miRNA-mediated pathological conditions have already been discussed elsewhere [109–117]. The association of miRNA's role in cancer was first described in chronic lymphocytic leukemia (CLL) [118], and subsequently several miRNAs were reported to be aberrantly expressed in many other cancers including PC [116, 119–126]. They play significant roles in initiation, progression, metastasis, and therapeutic resistance [127]. Although miRNAs represent only 3% of the human genome, they regulate 20–30% of the protein-coding genes [128, 129]. Recent studies have shown that 50% of miRNA genes are located in genomic instability regions that are generally associated with neoplastic transformation [130]. Therefore, miRNA signatures can provide better pathognomonic information about tumors than their transcriptome profiles [131]. Moreover, expression patterns of miRNAs are tissue specific [132, 133] and unique to tumor type and tissue of origin [116, 131, 134], thus making miRNAs valuable tools for diagnosis, prognosis, and therapy. Pancreatic cancer pathogenesis is a multistep process that involves a compendium of sequential genetic alterations in oncogenes like KRAS and several tumor suppressors including p53, p16, and SMAD4 [135]. In this chapter, we have summarized the role of miRNAs in PC pathogenesis with specific emphasis on their role in regulating KRAS, p53, p16, and SMAD4 expression, along with their utility as early detection/prognostic markers and as therapeutic targets.

### 11.3.1  DEREGULATED MIRNAs IN PANCREATIC CANCER

Several research studies have proposed an association between altered miRNA expression and PC. Identification of miRNA(s) at various stages of PC progression is critical to understand as their biological functions will help us develop unique diagnostic/prognostic markers and therapeutic targets. In this context, miRNA profiling of tumor samples, pancreatic juice, serum, and cyst fluid engenders a potential avenue for miRNA-based biomarker development. The first miRNA expression study was done by Poy et al. where the expression of miR-375 and miR-376 was observed in mouse pancreas (pancreas-specific miRNAs) but not in the brain, heart, and liver tissues [136]. Later, high-throughput analysis identified several miRNA signatures that were able to precisely classify tumors and differentiate PC from normal pancreas and pancreatitis [137]. In addition, profiling of PC tissue or desmoplasia revealed several precursor miRNAs that were aberrantly expressed and localized to tumor cells including miR-221, miR-376a, and miR-301 [138]. Further, extensive miRNA expression profiling in healthy individuals, pancreatitis, precursor pancreatic intraepithelial neoplasm (PanIN) lesions, and pancreatic ductal adenocarcinoma (PDAC) patients has revealed their diagnostic and prognostic utility [115, 139–141]. Subsequently, overexpression of miR-376 was reported in human Panc-1 PC cells compared with lung, breast, head and neck, colorectal, prostate, and hematopoietic cancer cells [142], although no difference in expression of miR-375 was observed [116, 142]. While expression of miR-216 and miR-217 (pancreas-specific miRNAs) and the lack of miR-133a are characteristic of normal pancreatic tissues [137], overexpression of miR-103 and miR-107, and downregulation of miR-155 are characteristic of PC specimens [143]. Furthermore, upregulation of miR-21 and downregulation of miR-150 and miR-30d in PC tissues compared with normal pancreas was reported [144]. The miR-221, miR-376a, and miR-301 are present specifically only in the tumor cells of PC and are among the top differentially expressed miRNAs [138]. Recently, miRNA microarray analyses in a panel of 15 PC cell lines revealed overexpression of miR-10a, miR-92, and miR-17-5p [145]. In addition to cell lines, several studies have reported differential expression of miRNAs that can distinguish between PDAC from chronic pancreatitis and normal pancreas [137, 146–148], and a cluster of miR30a-3p, miR-105, miR-127, miR-187, miR-452, and miR-518a-2 was also associated with better prognosis of lymph node positive patients [137]. The miR-203 was overexpressed in PC and associated with poor prognosis of patients with no residual tumors [149]. Yu et al. profiled 700 miRNAs in PanIN lesions and observed an overexpression of let-7f/g, miR-18a, -15b, -21, -29a/b/c, -31, -93, -95, -101, -103, -106b, -146a, -155, -182, -190, -193b, -194, -196b, -200a/b, -203, -222, 338-3p, -429, and 486-3p, and no or weak expression of miR-107, -139-3p/5p, -216a/b, -217, -218, and -483-5p in PanIN-3 lesions [150]. A similar miRNA expression profile using 10 PC cell lines and 17 pairs of PC/normal tissues has revealed significant upregulation of miR-196a, miR-190, miR-186, miR-221, miR-222, miR-200b, miR-15b, and miR-95 [151]. In addition, miR-376a, miR-301, miR-132, and miR-212 were reported to be overexpressed specifically in PC compared with normal or benign adjacent pancreas [138, 152]. On the contrary, miR-132 was also found to be downregulated in PC tissues compared to normal and benign tissues [153].

The miRNA deregulation is not only observed in human PC cell lines and tissues but is also evident in the circulation of PC patients. Many studies have identified differential upregulation of miRNAs including miR-642b, miR-885-5p, miR-22, miR-20a, miR-21, miR-24, miR-25, miR-99a, miR-185, miR-191, miR-155, miR-196a, and miR-18a in blood of PC patients and these miRNAs could differentiate PC patients from healthy individuals [154]. In addition, miR-2 was overexpressed in serum [155], whereas overexpression of miR-196a, miR-217, miR-451, and miR-486-5p [156] and downregulation of let-7c, let-7d, let-7f, and miR-200c were observed in the fine-needle aspiration samples (FNA) of PC patients [156].

## 11.4 MiRNA AND KRAS IN PANCREATIC CANCER

KRAS is a membrane-bound guanosine triphosphate (GTP)-binding protein from the RAS family of proteins that regulates several signal transduction pathways like RAF, MAP2K, MAPK, Tiam1, Ral-GEF, and PI3K-AKT upon receiving extracellular stimuli. Usually, RAS-associated signaling activation is short lived due to its intrinsic GTPase activity, switching off the RAS signaling cascades. However, KRAS is mostly mutated (KRASG12D) in 90% of PC [157] that results in gain of function and hyperactivation of downstream signaling cascades involved in the change of transcriptional dynamics, cell survival, and proliferation facilitating tumor progression and metastasis [158]. In spite of establishing KRASG12D as a potent oncogene, attempts to inhibit the hyperactivity or to identify various factors that induce this mutation have failed. Recent discoveries have identified several miRNAs that target KRASG12D by binding to the 3′UTR of the KRAS mRNA transcript and lead to its degradation or translational inhibition. In this section we have attempted to elucidate the role of miR-96, miR-126, miR-143/145, miR-217, and let-7 mediated regulation of oncogenic KRAS in PC.

Recently, Yu et al. have observed that miR-96, which is usually lost in PC tissues compared with normal pancreas, targets KRAS by perfect base pairing with its 3′UTR [159]. Overexpression of miR-96 in PC cell lines resulted in decreased cell migration, invasion, and proliferation, indicating its tumor suppressive potential [159]. Similarly, miR-126 was also found to be downregulated in PDAC compared with benign cystic tumors and was identified to inhibit translation of KRAS by binding to its two putative seed regions on the 3′UTR suggesting that re-expression of miR-126 can be a powerful therapeutic strategy for KRAS-mediated oncogenesis [160]. Not only do miRNAs target wild-type/mutated KRAS gene expression, KRASG12D also modulates miRNA expression during cellular transformation. RAS-responsive element-binding protein (RREB1) binds to the promoter of miR-143/145 clusters and negatively regulates their expression. Surprisingly, oncogenic KRASG12D induces RREB1 expression in PC as a means to check expression of the miR-143/145 cluster for enhanced tumorigenesis [161]. While constitutive miR-143/145 cluster expression targets RREB1 protein to inhibit a feed-forward circuit of KRAS signals via RREB1, the KRASG12D-mediated overexpression of RREB1 in turn represses the miR-143/145 cluster expression and therefore facilitates further KRAS-mediated signaling [161]. Besides the miR-143/145 cluster, another study using in situ hybridization technique revealed downregulation of miR-217 in 76% of PC tissues [162]. The same study also reported that overexpression of this miR-217 in PC cells decreased KRAS expression and phosphorylated AKT levels, suggesting that miR-217 not only downregulates KRAS expression but also affects the downstream signaling molecules involved in cell survival and proliferation [162]. Let-7 family miRNAs have multiple target sites on the 3′UTR of KRAS and downregulate its expression [163]. Several studies have shown significant downregulation of let-7 expression in PC samples compared with cancer-adjacent normal tissues, and PC patients with loss of let-7 expression were ineligible for surgery due to aggressiveness of the disease [164]. They also observed that ectopic overexpression of let-7 in PC cells significantly downregulates KRAS expression and inhibits MAPK with concomitant decrease in cell proliferation. However, let-7 overexpression fails to inhibit tumor growth and progression [164]. In addition to affecting KRAS expression in PC, miRNAs also modulate KRAS expression in breast and lung cancer. Kopp et al. observed an inverse correlation between KRAS expression and miR-200c expression in a panel of breast cancer cell lines [165]. Furthermore, they also observed that miR-200c directly regulates KRAS activity and additionally affects other tumorigenic pathways inhibiting tumor progression and resistance to therapy. Therefore, miR-NAs definitely play a key role in modulating PC as they regulate the expression of key oncogenes such as KRAS, which has been shown to be the initiating event for PC and important for tumor maintenance (Figure 11.2).

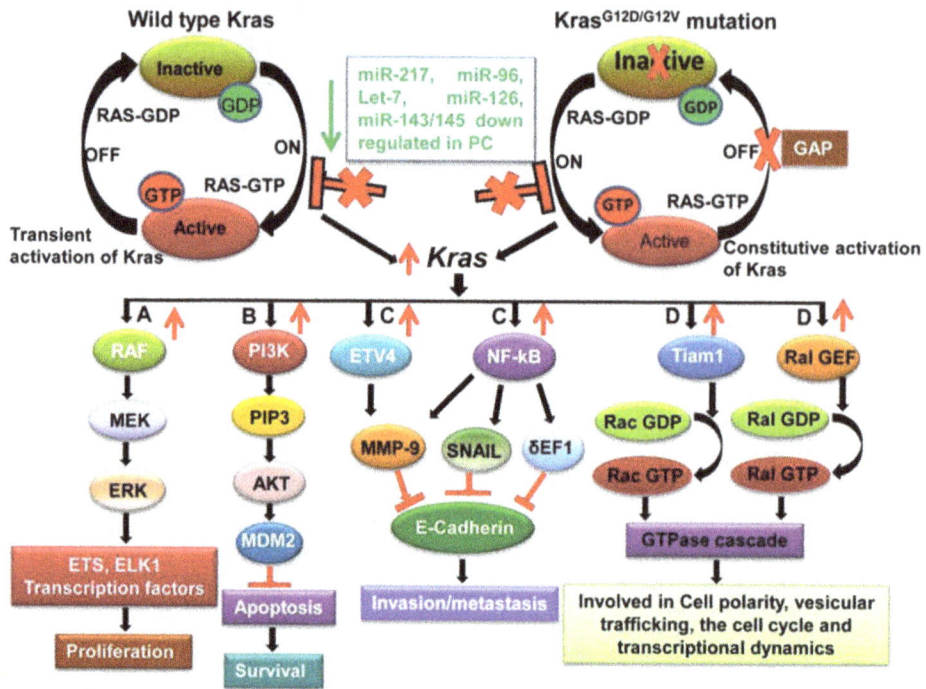

**FIGURE 11.2**  Therapeutic potential of miRNA-mediated KRAS downregulation in pancreatic cancer. More than 90% of pancreatic cancer (PC) patients have mutations (G12D/G12V) in codon 12 of KRAS that results in its constitutive activation and hyperactivation of its downstream signaling pathways accompanied by increased proliferation, motility, and survival of PC cells. KRAS-targeting miRNAs like miR-217, -96, -126, -143/145, and let-7 are significantly downregulated in PC patients, which results in increased KRAS expression. Restoration of these miRNA expressions may result in abrogation of KRAS-mediated signaling pathways such as (A) RAF/MEK/ERK will result in decreased PC cell proliferation, (B) PI3K/AKT pathway abrogation will decrease PC cell survival, (C) inhibition of NF-κB and ETV4 will result in decreased cell invasion and metastasis by upregulating E-Cadherin expression, and (D) downregulation of Tiam1 and Ral-GEF will affect cell polarity, vesicular trafficking, cell cycle and transcriptional dynamics in pancreatic cancer (miRNAs in *color green* indicate downregulated in PC). *Abbreviations:* AKT, v-akt murine thymoma viral oncogene homolog; δ-EF1, translational elongation factor 1 delta; ELK1, ELK1, member of ETS oncogene family; ERK, extracellular signal-regulated kinase; ETS, E-twenty-six family transcription factors; ETV4, human ETS translocation variant 4; MDM2, mouse double minute 2 homolog; MEK, mitogen-activated protein kinase; MMP9, matrix metalloproteinase 9; NF-κB, nuclear factor Kappa B; PI3K, phosphatidylinositol-4, 5-bisphosphate 3-kinase; PIP3, phosphatidylinositol (3, 4, 5)-trisphosphate; RAF, Raf/mil family of serine/threonine protein kinases; Ral-GEFs, Ral guanine nucleotide exchange factors; SNAIL, Snail family zinc finger 1; TIAM1, T-cell lymphoma invasion and metastasis 1.

## 11.5  MiRNA AND THE p53 PATHWAY

Inactivation of tumor suppressor genes is commonly observed in the multistep progression of various cancers. The Trp53 (p53), referred to as the guardian of the genome, is located on chromosome 17p and plays an important role in controlling cellular responses involving cell cycle arrest and programmed cell death due to various physiological stress [166]. Wild-type p53 binds to DNA and induces p21 expression to negatively regulate cell division–stimulating proteins including cyclin D1 and cyclin-dependent kinase-2 resulting in growth arrest [167]. However, more than 50% of adult human tumors have inactivating mutations or deletions in the p53 gene leading to loss of its function without affecting its translation [168]. Inactivation of p53 due to somatic mutations is observed in 50–75% of PDAC [169]. Gain-of-function mutations (R175, G245, R248, R249, R273, and R282) in the DNA-binding domain of p53 leads to loss of its function associated with uncontrolled cell

growth, cell survival, and genetic instability [168]. More precisely, an R175/R248/R273 mutation of p53 is mainly observed at late PanINs with dysplasia and occurs mostly in single allelic loss coupled with an intragenic mutation of the second allele, leading to uncontrolled cell proliferation and inhibition of apoptosis [170]. These mutations may be structural, affecting folding of the protein, or they can affect p53–DNA interactions, thereby altering transcriptional regulation of its target genes [171–173]. However, mutant p53 recognizes and binds to transcription factors/co-factors/p63/p73 proteins, which leads to transcriptional upregulation of genes involved in uncontrolled growth [174–177]. Besides in vitro studies, the transforming capabilities of mutant p53 were successfully demonstrated using genetically engineered mouse (GEM) models for various mutant p53 [169, 178, 179]. Recently, a cooperative association of mutant p53 (p53$^{R172H}$) and other genes was observed in cellular signaling pathways and cell-fate decisions by regulating cell proliferation, motility and invasion, apoptosis, inflammation, and angiogenesis either in the cancer cell itself or in the surrounding microenvironment during tumorigenesis and metastasis [180]. In addition, the cooperative role of mutant p53$^{R172H}$ and KRAS$^{G12D}$ has also been investigated during PC pathogenesis using GEM models [181].

### 11.5.1 P53-Mediated Regulation of miRNAs

Besides regulating proteins involved in cell cycle and apoptosis, wild-type p53 has recently been shown to modulate expression of several miRNAs in cancer cells, while siRNA-mediated knockdown of p53 abolished effects on miRNA expression [182]. Surprisingly, global sequence analysis has revealed that N46% of miRNA promoters have p53 binding sites suggesting that p53 can be a master regulator of miRNA expression [182] and subsequently other studies also showed transactivation of several miRNAs by p53 [183]. MiR-34 family members are evolutionarily conserved between species and are transcribed from two different gene loci [184]. While miR-34a is located at chromosome 1p36 and is expressed in a wide variety of tissues, miR-34b and miR-34c are localized to chromosome 11q23 and mainly expressed in lung tissues [185]. Recently p53 has been shown to bind to p53-responsive elements on the promoters of the miR-34 family (miR-34a/b/c) of miRNAs, thereby inducing cell cycle arrest and apoptosis [186–191]. Ectopic overexpression of miR-34 leads to G1 phase cell cycle arrest [186, 188, 191], inhibition of proliferation, colony formation [187], increased apoptosis [190–192], and induced cellular senescence in human diploid fibroblasts by targeting several of its putative target genes like CDK4, BCL-2, cyclin E, CDK6, c-MET, SIRT1, c-MYC, N-MYC, Axl, SNAIL, and LDH [188]. However, several cancers including PC harbor CpG methylation on promoters of miR-34a/b/c, and this significantly inhibits their transcription by p53 [193, 194]. PC cells have a low or undetectable level of miR-34a [195]; however, their low/loss of expression was neither associated with loss of heterozygosity (LOH), p53 mutation, nor with promoter CpG methylation [186].

Several reports have shown that mutant p53 also regulates transcription of miR-130b, miR-155, and miR-205 and thus alters the expression of their target genes like ZEB1 and ZNF652 either by inhibiting/degrading translation of their mRNA transcripts, thereby influencing invasive and metastatic potential of cells [196–198]. The mutant p53 regulates these miRNAs in either a p63-dependent (miR-155 and miR-205) or p63-independent manner (miR-130b) [196–198]. The miR-15a/16-1 family is another transcriptional target of p53 that is significantly downregulated in PC [119]. Ectopic expression of miR-15a was shown to downregulate WNT3A and FGF7 and results in reduced survival and proliferation of PC cells [199]. Furthermore, the tumor suppressor functions of this cluster were confirmed both in vitro and in vivo, whereby it downregulated ALDOA and TPI1 expression, affecting glycolysis and cell metabolism [119]. By inducing miR-145 expression, p53 downregulates c-MYC expression and impedes tumor growth both in vitro

and in vivo [200]. However, miR-145 expression was significantly downregulated in PC tissues as compared with healthy control groups [147]. The p53-regulated miR-107 expression inhibits hypoxia inducible factor-1β (HIF-1β) expression, that on its interaction with HIF-1α forms HIF-1 complex and induces genes that are involved in the pathogenesis of cancer [201]. However, miR-107 is upregulated in many solid tumors including PC [116, 201]. Besides PC, overexpressed miR-107 in breast cancer negatively regulates expression of tumor suppressor let-7 miRNA [201]. Ectopic expression of miR-107 destabilizes let-7, and thereby upregulates let-7 target genes that are involved in progression of breast cancer [201]. The miR-200 family members (miR-200c/141, miR-200a/b/429) inhibit expression of Zeb1 and Zeb2 transcription factors favoring mesenchymal to epithelial transition and reduced tumor growth [202, 203]. Transcription factor Zeb1, by upregulating Jag1, mastermind-like co-activators Maml2 and Maml3 promotes pancreatic cancer progression through notch signaling. p53 has been shown to induce expression of miR-200 family members accompanied by downregulation of Zeb1 and Zeb2 expression that resulted in decreased tumor progression. Furthermore, knockdown of p53 expression resulted in abrogation of miR-200 family members associated with upregulation of Zeb1 and Zeb2. However, Soubani et al. have recently shown that chemoresistant PC cell lines have downregulated expression of miR-200 family members including miR-200a/b/c, suggesting deregulated p53 signaling in these cells [204]. In addition, upregulated expression of Zeb1 and Jag1 was also associated with decreased miR-200 family members in PC [205]. Similarly Kim et al. have shown that p53 also induces the expression of miR-192 family members (miR-192/194/215) that target Zeb1 expression and inhibit epithelial to mesenchymal transition and induce cell cycle arrest by targeting MAD2L1, CUL5, and CDC7 genes [202, 203, 206]. Contrarily, expression of miR-192 was high in PDAC tissues as compared with normal pancreas [207]. Besides transcriptional regulation, p53 also regulates miRNA maturation posttranscriptionally by modulating the function of Drosha that is involved in the processing of pri-miRNA to pre-miRNA in the nucleus. To achieve this process, it forms a complex with DEAD box RNA helicases p68 and p72 and interacts with Drosha to process pri-miRNA to pre-miRNA (miR-16-1, miR-143, and miR-145) [208]. Significant downregulation of miR-145 and miR-16-1 as seen in PC [147, 199] speculates that point mutations like R175H/R273H in p53 may decrease the maturation of these miRNAs. Further, Dnmt3a directly interacts with p53 and inhibits p53-mediated transcription of p21 and other tumor suppressor genes in the genome [209].

### 11.5.2 miRNAs Regulating p53 Expression

Several in silico analysis tools have predicted potential binding sites on the 3′UTR of p53 mRNA for multiple miRNAs. Recently, a study revealed that ectopic expression of miR-491-5p in SW-1990 PC cells downregulates p53 expression suggesting that p53 not only has the potential to modulate miRNA expression, but in turn miRNAs can also regulate p53 expression for cooperative normal functioning of cells [210]. Besides affecting p53 expression, miR-491-5p overexpression also abrogated several signaling pathways like Jak/Stat3, PI3K/Akt accompanied by decreased cell proliferation and induction of apoptosis [210]. Tumor suppressor miR-34 cluster is also involved in inhibiting cell proliferation, migration, invasion, and epithelial to mesenchymal transition [193, 211]. It has been shown that p53, by upregulating miR-34a expression, represses NAD-dependent deacetylases that in turn increases acetylated p53 (lysine 382) and transcriptionally upregulates p21 and PUMA expression [212, 213]. Furthermore, miR-34a also represses HDM4, a negative regulator of p53, and creates a positive feedback loop on p53 expression [214]. Therefore, any imbalance in this positive feedback mechanism decreases p53 activity accompanied by enhanced cell proliferation and cell survival. Unfortunately, due to CGI promoter hypermethylation, miR-34s are frequently silenced in a variety of cancers including PC [211]. But ectopic expression of miR-34a in PC cells has been shown to induce cellular senescence and cell cycle arrest by targeting CDK6. Another miRNA, miR-214, is overexpressed in PC [116] and is involved in cancer progression, metastasis, and chemoresistance [215, 216]. This overexpressed miR-214 targets p53 and relieves negative inhibition on Nanog expression, thereby enriching the stem cell population in ovarian cancer cells [217]. While miR-125b is overexpressed in PC [137], it negatively regulates p53 expression during stress; its ectopic expression also downregulates p53 expression and inhibits apoptosis in lung fibroblast and human neuroblastoma cells [218].

The p85α is a regulatory subunit of the PI3K complex that negatively regulates p53 expression and plays a critical role in cell survival and apoptosis [219]. However, miR-29-mediated p85α downregulation has been shown to activate p53 signaling pathway and induce apoptosis in Hela cells [220]. Interestingly, overexpression of miR-29a/b/c was observed in PC tissue and the PC cell lines MiaPaCa-2, PANC-1, and BxPC-3 as compared with normal tissues and HPDE cells [151]. In contrast, miR-504 downregulates p53 expression and inhibits p53-mediated apoptosis, cell cycle arrest that leads to tumor promotion in vivo [221]. By modulating cyclin G1/PP2A and MDM2 expression, miR-122 activates and stabilizes p53 protein expression in hepatocellular carcinoma (HCC) cells [222], but it is significantly downregulated in PC [147]. Other miRNAs, such as miR-33, miR-1285, miR-380-5p, miR-25, miR-133, and miR-30d, directly bind to the 3′UTR of p53 and repress its expression [223], out of which miR-133b is downregulated, whereas miR-25 is overexpressed in mouse PC [224]. p53 also modulates MDM2 by negatively regulating miRNAs such as miR-192/194/215 and miR-605 as a feedback-loop mechanism to check p53 activity physiologically [225, 226]; however, miR-192 is significantly

upregulated in PC patients [207, 227]. The overall role of p53-mediated miRNA transcription and its regulation by miRNAs is shown in Figure 11.3.

## 11.6 p16 AND MiRNA IN PANCREATIC CANCER

The INK4A/p16, also known as cyclin dependent kinase inhibitor 2A (CDKN2A), is a tumor suppressor gene involved in cell cycle regulation and cellular senescence [228, 229]. However, promoter methylation, missense mutation, and deletions of the p16 gene lead to its inactivation in most of the cancers including PC where its function is lost in around 95% of cases [230]. In addition, miRNA-mediated regulation of p16 has been documented in many cancers including PC. Recent studies have shown that miR-24 was upregulated in PC [115, 231], and other studies have shown that it binds to both the coding and 3′UTR regions of p16 mRNA that results in its translation inhibition associated with decreased cell proliferation [232]. Similarly, miR-10b, also overexpressed in PC [233], posttranscriptionally regulates p16 expression by binding to its 3′UTR region, whereas inhibition of this miRNA in glioblastoma resulted in inhibition of tumor growth through activation of cell cycle arrest and apoptotic pathway [234]. In addition, several other miRNAs like miR-128a upregulate p16 expression by targeting its negative regulator Bmi-1, resulting in decreased growth of medulla blastoma cells [235]. Similarly, miR-20a also upregulates p16 expression and inhibits tumorigenesis [236]. DNA methyltransferase (DNMT1) has been shown to negatively regulate p16 expression, and silencing of its regulators like miR-148a and miR-152 leads to decreased p16 expression and enhanced cholangiocarcinoma growth [237]. Not only do miRNAs target p16 expression, indeed p16 signaling has been shown to regulate expression of miRNAs. In this context, ectopic overexpression of p16 in breast and glioma cells increased miR-410 and miR-650 expression and reduced CDK1 protein levels [238]. In addition, UV-induced DNA damage initiates association of p16 with sp1 transcription factor and CDK4 (cyclin-dependent kinase) causing subsequent upregulation of miR-141 and miR-146b-5p [239].

## 11.7 TGF-β/SMAD SIGNALING REGULATES MiRNAs IN PC

SMAD proteins are the backbone of the transforming growth factor beta (TGF-β) family of cytokines [240]. Binding of ligand to TGF-β receptor II (TGFβRII) results in its heterodimerization with TGF-β receptor I (TGFβRI) with their subsequent transphosphorylation and activation of substrate SMADs. The human genome encodes eight different SMAD proteins among which SMAD-1, SMAD-2, SMAD-3, SMAD-5, and SMAD-8 are regulated by the TGF-β family of receptors commonly referred to as R-SMADs. A common SMAD protein, SMAD-4 is a co-mediator that translates extracellular signals into response.

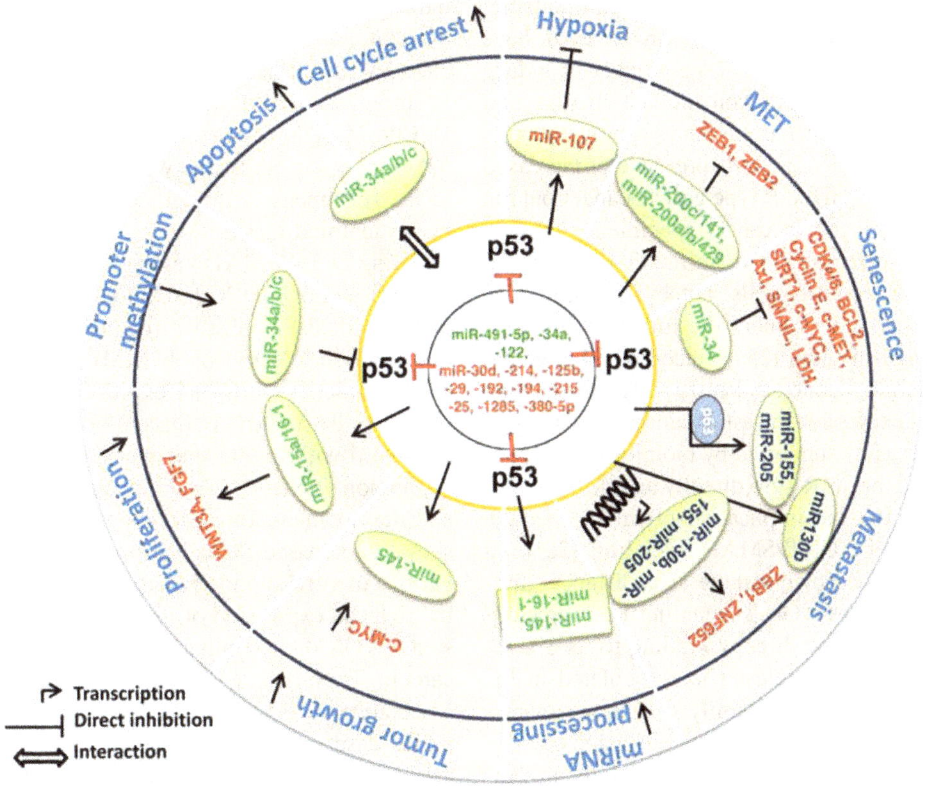

**FIGURE 11.3** MiRNA-mediated signaling of p53 in pancreatic cancer. DNA damage and stress signals like starvation, hypoxia, onco-gene hyperactivation, etc., by upregulating p53 expression transcriptionally modulate several miRNAs (miR-107, miR-34a/b/c, miR-34) that induce cell cycle arrest, apoptosis cellular senescence, and inhibits hypoxia. p53-mediated overexpression of miR-15a/16-1 cluster inhibits proliferation of pancreatic cancer cells by downregulating WNT3A and FGF7 proteins. By upregulating miR-145 expression, p53 downregulates c-Myc expression thereby decreasing cell survival in vitro and impedes tumor growth in vivo. Similarly, p53 by regulating miR-200a/b/c, miR-141, and miR-429 inhibits ZEB1 and ZEB2 transcription factors that favor mesenchymal to epithelial transition thereby impeding the tumor growth. However, mutated p53 negatively regulates miR-155, miR-205 (p63 dependent manner), and miR-130b (p63 independent) that targets ZEB1 and ZNF652 expression thereby enhancing the invasion, migration, and meta-static potential of pancreatic cancer cells. Missense mutations in p53 also interfere with miRNA (miR-145 and miR-16-1) maturation resulting in enhanced cell proliferation, invasion, and migration. Certain miRNAs like miR-491-5p, -34a, -122, -30d, -214, -125b, -29, -192, -194, -215, -25, -1285, and -380-5p regulate p53 expression by binding to its 3'UTR. Through a feedback mechanism, CpG pro-moter hypermethylation of miR- 34a/b/c could also influence p53 transcription thereby hindering p53-mediated biological functions. *Abbreviations:* AXL, AXL receptor tyrosine kinase; BCL2, B-cell leukemia/lymphoma 2; c-MET, met proto-oncogene; c-MYC, v-Myc avian myelocytomatosis viral oncogene homolog; CDK4/6, cyclin-dependent kinase 4/6; FGF7, fibroblast growth factor 7; LDH, lactate dehydrogenase; P53, tumor protein p53; SIRT-1, Sirtuin1; SNAIL, Snail family zinc finger 1; WNT3A, wingless-type MMTV integration site family, member 3A; ZEB1/2, zinc finger e-box binding homeobox 1/2; ZNF652, zinc finger protein 652.

Activated SMAD-2 and/or SMAD-3 heterodimerizes with SMAD-4 protein and translocates into the nucleus, along with other co-regulators to engage in transcriptional regu-lation of various target genes. On the other hand, SMAD-6 and SMAD-7 are inhibitory SMADs (I-SMADs), which act as a negative regulator of R-SMADs [241]. Previous studies have shown the existence of signaling cross-talk between SMAD proteins and other non-SMAD pathways such as extracellular signal-related kinase (ERK), mito-gen-activated protein kinases (MAPK), and protein kinase C (PKC). These intracellular kinases can phosphorylate R-SMADs at potential serine residues preventing accumu-lation of R-SMADs in the nucleus [242]. In normal epithe-lial cells, TGF-β acts as a potent tumor suppressor eliciting growth inhibition through downregulating transcription

factor c-Myc and upregulating cell cycle inhibitor proteins such as p15 and p21 [243, 244]. In cancer cells, the TGF-β/SMAD signaling is modulated and loss of some compo-nents of the TGF-β/SMAD pathway leads to impairment of TGF-β/SMAD-mediated growth arrest [245].

Several in vitro and in vivo findings have revealed that a variety of miRNAs are controlled and modulated under the influence of R-SMAD proteins [246–249] and the TGF-β signaling pathway. For example, TGF-β signaling induces maturation of miR-21 through various posttran-scriptional modifications of primary miR-21 into mature miR-21 through a microprocessor complex consisting of Drosha and RNA helicase p68. Thus ligand-specific activa-tion of SMAD protein may be involved in the biosynthesis of miRNAs such as miR-21, which in turn downregulate the tumor suppressor genes PDCD4 and PTEN, enhancing tumor invasion and metastasis [248]. Kong et al. have demonstrated that TGF-β/SMAD-4 signaling can induce miR-155 expression and its restoration inhibits cell migration and invasion in normal murine mammary epithelial cells [249]. Naito et al. have identified the presence of increased miR-143 expression in scirrhous type gastric cancer but not in non-scirrhous type gastric cancer specimens. Further, several studies have revealed that miR-143 is expressed mainly by stromal fibroblast cells, not by gastric cancer cells, and this expression is under the influence of TGF-β/SMAD signaling activation [250]. Recently, a functional screening approach was done to identify miRNAs that can regulate SMAD-4 expression in gastric cancer cells. Using a luciferase reporter assay supported by bioinformatics analysis, they identified that miR-199a directly targets SMAD-4 expression, suggesting the implication of miR-199a as a negative regulator of TGF-β/SMAD signaling [251]. In addition, Liu et al. have shown that the 130a/301a/454 family of miRNAs regulates TGF-β signaling by repressing the SMAD-4 expression by directly binding to its 3'UTR sequence [252]; further, this cluster is upregulated in PC [253, 254]. Among the SMAD family of tumor suppressor genes, SMAD-4 (DPC-4) is frequently inactivated in pancreatic, bile duct adenocarcinoma, and colorectal cancer. Studies have explored the fact that SMAD-4 gene is inactivated in 55% of PC patients either by deletion in both the alleles (35%) or by intragenic mutation in one allele coupled with LOH (20%) [255]. In the classical PC progression model, SMAD-4 gene inactivation occurs at the later stages of PanINs, thus loss of SMAD tumor suppressor was correlated with disease advancement or metastasis [256]. A recent study demonstrated that co-expression of SMAD-6 and/or SMAD-7 along with SMAD-4 in patients with PDAC could be one possible mechanism of SMAD-4 inactivation [257].

Traditional small molecular inhibitors like SD-208, SB431542, and LY-2157299, specific for TGF-β receptor, abrogated SMAD-mediated responses like its anti-invasive effect in PC cells [258] and inhibited cancer advancement and metastasis in xenograft mouse models [259]. Treatment of mouse preantral granulosa cells with recombinant TGF-β leads to upregulation of miR-224. Further, this upregulated miR-224 was diminished upon treatment of the above cells with SB-431542 (a TGF-β type I receptor inhibitor), indicating that anti-TGF-β therapies could block the canonical TGF-β/SMAD signaling pathway and their respective miRNA biogenesis and regulation [260]. MiRNAs play a critical role in cell-mediated immune response by controlling the dendritic cell function in cancer patients. Recently, Du et al. identified that miRNA-146a is aberrantly expressed in human CD14+ monocyte-derived dendritic cells from PC patients in vitro. Further, miR-146a expression is correlated with impairment of differentiation and inhibition of the antigen presentation function of dendritic cells through mechanism of SMAD-4 repression. Thus, miRNAs are one of the factors involved in the regulation of immune response

indirectly by modulating dendritic cells [261]. Interestingly, overexpression of other miRNAs can also directly influence SMAD-4. MiR-483-3p and miR-421 are involved in PC progression by directly regulating the tumor suppressor DPC4/SMAD4. Specifically, the expression level of miR-421 is inversely correlated with DPC4/SMAD-4 expression in PC specimens. Through various in vitro studies, it has been identified that miR-421 can also play a dual role by targeting DPC4/SMAD4 along with its downstream effectors p21 and p15 in PC [262, 263]. More recently, Li et al. had identified the negative regulatory role of miR-494 on transcriptional activator FOXM1 on pancreatic cancer cells. Clinically, miR-494 expression levels were demonstrated to be higher in adjacent normal pancreatic tissues compared with PDAC specimens, suggesting that reduced expression of miR 494 is a critical factor in PC progression and development. By restoration of miR-494, the authors showed decreased PC cell growth and metastasis through various in vitro and in vivo assays. Finally, they correlated the reduced expression of miR-494 with increased expression of FOXM1 and enhanced nuclear translocation of beta catenin. They also identified the molecular mechanism and association of SMAD4/miR-494/FOXM1/beta catenin signaling cascade and its biological impact on PC pathogenesis. In particular, miR-494 is the novel miRNA identified with tumor suppression function in PDAC [264]. Along similar lines, miR-182-5p was reported to be overexpressed in bladder cancer, which could potentially target SMAD-4 expression. The repressed SMAD-4 can influence nuclear transport of β-catenin thereby resulting in activation of WNT signaling cascade in cancer cells [123]. Also, another study by Geraldo et al. has shown that ectopic overexpression of miR-146b-5p in rat follicular thyroid carcinoma cells resulted in significant increase in cell proliferation and development of resistance to TGF-β mediated cell cycle arrest. Thus, miR-146b-5p regulates the TGF-β mediated signaling pathway by repressing SMAD4 in thyroid cancer cells [265]. Overall these studies provide evidence that SMAD proteins are (1) involved in miRNA biogenesis, (2) lost in expression in various cancers, and (3) downregulated by miRNAs (Figure 11.4).

## 11.8 SONIC HEDGEHOG SIGNALING AND MiRNA IN PANCREATIC CANCER

The Sonic Hedgehog (Shh) signal transduction pathway is critically important during embryonic development and is activated by binding of Shh to its cognate Patched-1 (PTCH1) receptor. Following PTCH1 activation, transmembrane G-protein coupled receptor (GPCR)-like protein Smoothened (SMO) is freed and activated resulting in nuclear translocation of the GLI1 family of transcription factors for target gene activation [266]. In addition, through GLI1, SHH-mediated cellular responses are also coordinated through MAPK and the phosphatidylinositol-3 kinase (PI3K) signaling cascades for enhanced proliferation and cell survival [267]. Besides being reactivated in injured

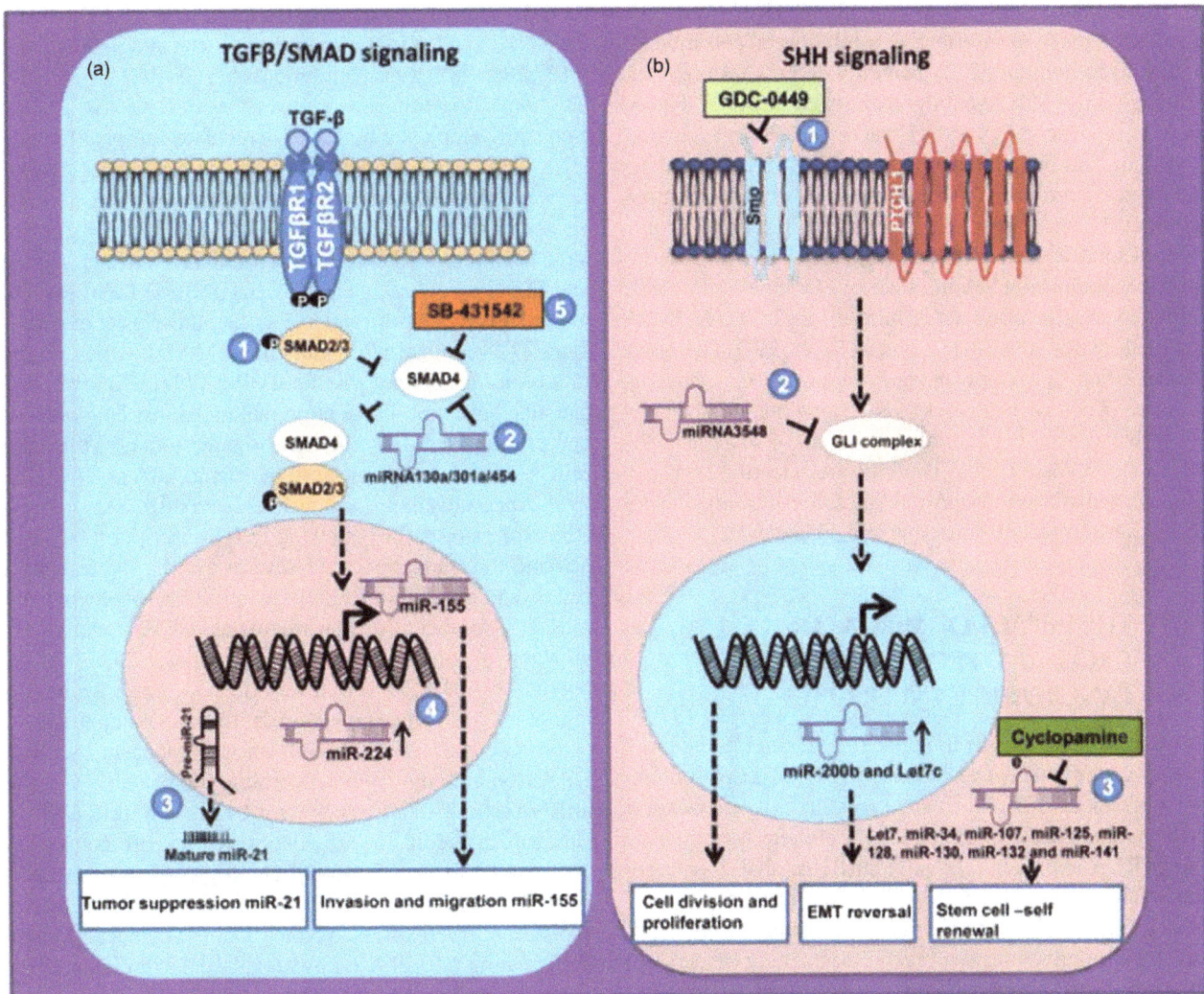

**FIGURE 11.4** **MiRNAs involved in TGF-β/SMAD and SHH signaling and their targeting approaches for pancreatic cancer pathogenesis.** (a) Binding of TGF-β to its cognate receptor leads to phosphorylation and therefore activation of SMAD signaling. (1) Phosphorylated SMADs (SMAD2 and SMAD3) form a complex with co-mediator SMAD4 and translocate to the nucleus for active transcription of the target genes. (2) Many of the miRNAs are involved in the positive and negative regulation of TGF-β/SMAD signaling cascade. MiRNA130a/301a/454 abrogate TGF-β/SMAD mediated signaling by blocking SMAD4 activation. (3) TGF-β/ SMAD signaling is also involved in miRNA biogenesis, e.g., miR-21. (4) Certain other miRNAs like miR-155 and miR-224 can also be influenced by TGF-β/SMAD mediated signaling cascade. (5) Pharmacological inhibition of SMAD4 protein by small-molecule inhibitor SB-431542 leads to modulation of miRNAs downstream of TGF-β/SMAD signaling. (b) In the presence of SHH ligand, the PTCH1–SMO association gets remitted, resulting in enhanced downstream transcriptional activation of GLI family of proteins leading to target gene expression. (1) Use of small-molecule inhibitor Vismodegib (GDC-0449) inhibits hedgehog signaling pathway thereby interfering with regulation of miR-200b and let-7c, which are critical regulators of EMT and drug resistance. (2) Indirect inhibition of SHH activation by synthetic miRNA (GLI-1-miR-3548) targets the GLI family of transcription factors that resulted in inhibition of pancreatic cancer cell division and proliferation. (3) Cyclopamine also appears to interrupt SHH signaling through regulating miRNAs (let-7, miR-34, miR-107, miR-125, miR-128, miR-130, miR-132, and miR-141) responsible for stem cell self-renewal process.

adult tissues, Shh signaling is also aberrantly activated in several malignancies promoting resistance to chemotherapeutics. While there is no expression in normal pancreas, Thayer et al. have reported aberrant expression of Shh in 70% of PC specimens and in precursor PanIN lesions (PanIN1 to PanIN3), suggesting its importance in the initiation and maintenance of PC pathogenesis [268]. Shh also contributes to the acceleration of desmoplastic response in PC [269]. Earlier studies have shown that miRNAs regulate

Shh signaling either by targeting proteins upstream of Shh or by directly targeting the components of Shh pathway members. For example, miR-196 and miR-452, by regulating Shh signaling, affect limb development [112] and epithelial to mesenchymal transition in enriched neural crest cells, respectively [270]. In addition, the miR-17–92 cluster (miR-19a, miR-20, and miR-92) was overexpressed and coordinated with the Shh signaling activation in murine and human brain cancer (medulloblastoma) [271]. Specific to PC, synthetic miR-3548 that targets GLI1 transcription factor resulted in effective inhibition of cell division and cell proliferation [272]. Not only do miRNAs regulate Shh signaling, Shh signaling also affects miRNA expression, as evidenced by the fact that use of pharmacological hedgehog inhibitor vismodegib (GDC-0449) upregulates the expression of miR-200b and let-7c that results in reversal of EMT phenotype and attenuation of drug resistance in non–small cell lung cancer (NSCLC) cells [273]; however, cyclopamine, a Shh inhibitor in PC cells, significantly inhibited expression of let-7, miR-34, miR-107, miR-125, miR-128, miR-130, miR-132, and miR-141 [274]. Therefore, identification of a complete miRNA profile that exists to regulate various components of Shh signaling, will be helpful in identifying the central mechanism of sustained Shh activation in cancer cells. The miRNAs identified through this mechanistic approach would be suitable candidates for therapeutic targets in pancreatic and other cancers.

## 11.9 THE IMPACT OF MiRNAs ON CELL CYCLE AND PROLIFERATION OF PANCREATIC CANCER CELLS

Deregulated cell proliferation due to aberrant regulation of cyclins and cyclin dependent kinases (CDKs) is a major hallmark of cancer [275]. An efficient way through which cancer cells manipulate cell cycle regulatory proteins is by altering the miRNA expression. Usually, miRNAs that downregulate cyclins, CDKs, and other cell cycle progression proteins are silenced, whereas miRNAs targeting inhibitors of cell cycle progression are overexpressed in lung cancers and PC [276–278]. Sequence homology analysis of seed regions of several differentially regulated miRNAs and 3′UTRs of mRNA indicate that the above strategy is employed by many cancers [279]. In PC, proliferation and cell cycle regulatory proteins like CCND1, CDKN1A (p21 WAF1), CDKN1B (p27 Kip1), E2F1, Rb1, CDK4, CDK6, CDKN1C (p57 Kip2), ABL-1, EZH2, c-Myc, KRAS, CDC25B, PTEN, LATS2, Spry2, RREB1, AKT, and STAT3 are all affected by aberrantly regulated miRNA [152, 159, 164, 194, 276–295]. Deregulation of these proteins leads to more rapid progression through the cell cycle and intern accelerated cell proliferation through various signaling pathways. For example miR-34a that downregulates CCND1, CDK4, and CDK6 is frequently lost in PC, resulting in increased cell proliferation [277, 278]. CDK6 is also targeted by miR-107, which like several other miRNAs is epigenetically silenced in PC cells [276]. MiR-148a is also frequently downregulated in PC and targets the CDC25B phosphatase that otherwise activates CDK1/cyclin B complex for increased cell proliferation [289]. Conversely PC cells aberrantly upregulate miRNAs that target inhibitors of cyclins and CDKs [279]. Overexpression of miR-106 targets p21, thereby preventing the G1 cell cycle block for uninterrupted cell cycle progression [287]. Similarly miR-221/222 and miR-106b family miRNAs allows cancer cells to progress through

the cell cycle by downregulating CDKN1B (p27 Kip1) and CDKN1C (p57 Kip2) circumventing the restriction point [283, 285, 287, 293]. Inhibition of TGF-β/SMAD4, p16, p53, and other tumor suppressors by miRNAs is a tactic frequently employed by PC to deregulate cell proliferation [152, 279, 292]. Overexpression of the miR-106b-25 cluster was reported in various cancers including PC, and this cluster impairs the TGF-β signaling by targeting its downstream effectors like cell cycle inhibitor CDKN1A (p21) and the proapoptotic gene BCL2L11 (BIM) [296] leading to increased cell proliferation. Further, this cluster also activates TGF-β signaling by targeting SMAD7, which results in a tumor promoting role for TGF-β [297]. However, p53 can also have gain-of-function mutations that can increase cell proliferation and other tumorigenic traits; in this case miRNA against p53 can have beneficial effects [169, 298, 299]. Aberrantly expressed miR-132 and miR-212 suppress Rb1 expression in PC, allowing E2F transcription factor to activate the transcription of several cyclins for enhanced cell cycle progression [152]. In addition, overexpressed miR-21 significantly downregulates PTEN expression allowing uncontrolled PC cell proliferation. This downregulation of PTEN was reversed upon antisense inhibition of miR-21, suggesting a useful therapeutic approach to combat PC [152, 279, 292]. Overactive oncogenes such as KRAS commonly drive cell proliferation in PC, and the miRNAs (miR-217, miR-96, miR-126, and miR-143/145 cluster) capable of downregulating KRAS, are commonly downregulated in PC [159–162]. Key players from aberrantly overexpressed oncogenic pathways alter miRNA expression. Overexpressed Notch-1 in PC induces EMT by upregulating miR-21 and downregulating expression of let-7a, let-7b, let7c, miR-200b, and miR-200c [280].

## 11.10 POTENTIAL ROLE OF MiRNAs IN PANCREATIC CANCER DIAGNOSIS

Traditionally, PC diagnosis and staging is performed by using tissue biopsies and ultrasound-guided FNAs. Serum CA19.9 (carbohydrate antigen 19.9) is the only gold standard biomarker approved by the FDA for PC, but it lacks specificity and sensitivity. Although widespread research has been focused to identify early detection markers, none has yielded a biomarker specific to PC. Because of high stability, tissue specificity, and ease of availability of miRNAs, recent studies are focused on performing miRNA expression profiling of PC tissues, serum/plasma, and other bodily fluids to identify suitable biomarker(s) for early diagnosis of this deadly disease. Several studies have revealed altered miRNA expression profiles in various cancers, and their association with various clinical parameters including disease progression, response to therapy, survival, and lymph node metastasis and their utility as prognostic and diagnostic markers [137, 150, 151, 207]. Therefore, it is very important to accurately determine the miRNA expression profile with high specificity and sensitivity in a given tissue or specific cell type. But accurate profiling of individual miRNAs with high specificity and sensitivity is technically challenging as mature miRNAs are very short (20–25 bp), differ in their GC content, lack mutual sequence features in mature miRNAs like poly-A tails, and single-nucleotide difference within miRNA families. However, recent rapid technological advancements like deep sequencing/parallel sequencing [300], oligonucleotide microarrays [301–304], northern blotting with radiolabelled probes [305], qPCR-based detection of mature miRNAs (TaqMan assays) [306–308], single-molecule detection in liquid phase [309], and in situ hybridization [310, 311] have made it feasible to profile miRNAs in various tissue/cell types with high accuracy. MiRNA microarray platform is a high-throughput method to compare differential miRNA expression profiles between the disease and the control states, whereas deep-sequencing methods involve generation of cDNA libraries from small RNA populations followed by deep sequencing using different methods [312, 313]. The advantages of deep sequencing include the following: (1) it is inexpensive and helps in the discovery of novel miRNAs, (2) it detects mature miRNAs with variation in length as well as modifications [314], and (3) it helps to identify low-abundance miRNAs with modest expression differences [312, 315–317]. Differentially expressed miRNAs are identified by either microarray or deep sequencing and are further validated by real-time qPCR using TaqMan-probed assays with high sensitivity and specificity. In addition, in situ hybridization is used to view and localize miRNAs within cells and also help us to pinpoint their distribution in tumor/stromal cell compartment/cell type-specific expression [308]. The advantage of using northern blot techniques is that it does not require technical knowledge and special equipment, but the disadvantages of this technique are poor sensitivity and more time consumption. But less time consumption and high specificity of miRNA detection were reported with the use of LNA-modified oligonucleotides [318].

Several studies have shown differential expression of miRNAs in bodily fluids of PC patients compared with normal healthy controls [154, 319, 320]. Using miRNA microarray analysis, a cohort of three miRNAs (miR-642b, miR-885-5p, and miR-22) was identified from blood that can identify early-stage PC patients from healthy control group with a sensitivity and specificity of 91% (AUC=0.97 [p<0.001]) [321]. Similarly, Frampton et al. also identified a signature of seven upregulated (miR-21, -23a, -31, -100, -143, -155, and miR-221) and three downregulated (miR-148a, -217, and miR-375) miRNAs that can identify PC patients from healthy controls [322]. Furthermore, overexpression of miR-21 and miR-31 and downregulation of miR-375 in tumor tissues were associated with poor PC patient survival [322]. Recently, Danish BIOPAC (Biomarkers in Patients with Pancreatic Adenocarcinoma) identified two diagnostic panels of circulating miRNAs (panel I includes miR-145, miR-150, miR-223, miR-636 with sensitivity of 85% and specificity of 64% with AUC of 0.86; while panel II includes miR-636, miR-26b, miR-223, miR-122, miR-150,

miR-145, miR-505, miR-34a, miR-885.5p, and miR-126*) having sensitivity and specificity of 85% with AUC of 0.93. Both these panels can significantly distinguish PC patients from healthy individuals [323]. Li et al. analyzed expression profiles of 735 miRNAs in the serum and identified miR-1290 that distinguished PC patients from healthy controls with sensitivity of 81% and specificity of 80% with AUC of 0.96 [324]. Furthermore, they observed that increased serum miR-1290 was better than CA19.9 in distinguishing early-stage PC patients from controls [324]. Additionally, combination of CA19.9 with miR-27a-3p [325] and miR-16 and miR-196a [326] expression can also precisely discriminate PC patients from healthy controls. In a similar study, a panel of seven miRNAs (miR-24, miR-20a, miR-25, miR-21, miR-185, miR-99a, and miR-191) in serum was used to accurately diagnose PC patients from healthy individuals, and can differentiate various stages of PDAC patients (Stage I patients with 96.2% and Stage II with 91.7) with high sensitivity and specificity [327]. Similarly, increased levels of serum miR-192 can distinguish PC from healthy controls with a sensitivity of 76% and specificity of 55% [207]. In addition, overexpression of miR-192 in PANC-1 PC cells results in an increase in cell proliferation, migration, and inhibition of apoptosis along with induced progression of cell cycle from the $G_0/G1$ to the S phase [207]. Plasma miR-221 level was significantly higher in PC patients and was used as an indicator of nonresectable status of the PC patients and distant metastasis [328]. In a similar study, a panel of seven miRNAs including miR-20a, miR-21, miR-24, miR-25, miR-99a, miR-185, and miR-191 in serum can accurately (83.6%) diagnose PC patients [327]. In addition to bodily fluids, Bauer et al. reported that deregulated miRNAs from tissues can discriminate between PC and pancreatitis [329]. Analysis of FNAs revealed that increased expression of miR-196a and miR-217 can accurately discriminate PC from benign pancreatic lesions with a sensitivity of 90% and specificity of 100% [330]. Similarly, downregulation of let-7 was observed in the FNA from PC patients [164]. In addition, many other studies have reported significantly higher serum concentration of miR-210, miR-18a, and miR-155 in PC patients compared with healthy controls [154, 319, 320], and concentration of blood-circulating miR-155, miR-21, and miR-210 was also significantly higher in rats with PC [331]. Although differentially expressed miR-27a-3p in peripheral blood mononucleated cell (PBMC) can sufficiently discriminate PC from benign pancreatic/peri-pancreatic diseases (BPD) with a sensitivity of 82.2% and specificity of 76.7% (AUC=0.840), combination of miR-27a-3p expression in PBMC and CA19.9 serum levels had significantly higher diagnostic accuracy than either of them alone with a sensitivity of 85.3% and specificity of 81.6% (AUC=0.886) [325]. However, it has been shown that blood-based miRNA expression profiles could not distinguish pancreatitis from PC [329].

Recently, a global miRNA expression study in pancreatic juice (exocrine pancreatic secretions) from six PC and six nonpancreatic, nonhealthy (NPNH) patients was performed using miRNA microarray. The top differentially expressed miRNAs (miR-205, miR-210, miR-492, and miR-1247) were further confirmed in pancreatic juice from 50 PC, 19 CP patients, and 19 NPNH controls, which showed marked difference in their expression from PDAC patients compared to those without pancreatic disease. These four miRNAs can diagnose PDAC with a sensitivity and specificity of 87% and 88% respectively, but inclusion of CA19.9 enhanced the sensitivity and the specificity to 91% and 100% respectively. Overexpression of the above-mentioned four miRNAs in pancreatic juice was further associated with reduced overall survival (OS) and lymph node metastasis [332]. Not only in humans, PC-bearing rats also showed increased blood levels of miR-155, miR-21, and miR-210 compared with normal control [331]. Similarly, high levels of circulating miR-196a and miR-196b were observed in KPC (KRAS$^{G12D}$; Trp53$^{R172H}$; Pdx1-Cre) mice with PanIN2/3 lesions as well as PDAC as compared to contemporary littermate control or KPC mice with PanIN1 lesions. Further, expression of these miRNAs was significantly higher in the serum of PC patients with sporadic/hereditary or individuals at risk (IAR) with multifocal PanIN2/3 lesions compared to patients with neuroendocrine pancreatic tumors or chronic pancreatitis, IAR with PanIN1, or no PanIN lesions and healthy controls with a sensitivity of 100% and specificity of 90% (AUC=0.99). However, it was observed only in few samples; therefore, further validation of miR-196a and miR-196b should be carried out to test their utility in PC diagnosis [333]. A meta-analysis study was carried out on 18 articles with a total of 2036 patients and 1444 controls to determine the role of miRNAs in early diagnosis and reported pooled sensitivity of 82% (95% CI, 78–86%), and specificity of 77% (95% CI, 73–81%) with AUC of 0.86 (95% CI, 0.83–0.89), suggesting the potential diagnostic value of miRNAs (inclusion of multiple miRNAs for diagnosis) to discriminate PC patients from healthy controls with high sensitivity and specificity [334].

## 11.11 MiRNAs AS THERAPEUTIC AGENTS IN PANCREATIC CANCER

Contributions of miRNAs in the initiation and progression of PC suggest that miRNAs can be exploited for the development of novel therapeutic strategies. The strategies followed to develop miRNA-based therapies are similar to that of other gene-targeted therapies. The main rationale for developing miRNA-based therapies over targeting a single oncogene is due to the ability of a single miRNA to target many genes and signaling pathways at once. Therefore, targeting of a single miRNA can produce dramatic results due to its ability to affect numerous cellular processes. Primarily, miRNA-targeted therapeutics can be classified into miRNA mimics and miRNA antagonists. The miRNA mimics are double-stranded RNAs that are chemically modified in vitro and used to restore the expression of miRNAs that are lost during the disease process and functionally mimic the endogenous miRNAs. On the other hand, miRNA antagonists are developed to target endogenous miRNAs that have shown a gain of function in diseased tissues. Previously, we have extensively discussed various methods that are used to deliver miRNA-based therapeutics [335, 336]; however, here we will briefly describe some currently used methods to deliver miRNA therapeutics. The miRNA antagonists are chemically synthesized single-stranded oligonucleotides complementary to a specific miRNA, with specific chemical modification (2′-O-methoxyethyl [2′-MOE] and 2′-fluoro [2′-F]) and locked nucleic acid (LNA) chemistries [337–342]. These modifications or backbone chemistries provide extreme stability, high binding affinity to target miRNAs, resistance to nuclease degradation, and lead to the sequestration of target miRNA within the RISC, inhibiting it from being processed or degraded. These modified antisense oligonucleotides (2′-MOE, LNA, and morpholinos) are utilized to antagonize miRNA activity and its processing. These oligos are frequently complementary to the miRNA guide strand, blocking its activity. They can often produce similar effects by sterically blocking only the seed region due to increased binding strength. These oligos can also antagonize miRNAs by interfering with miRNA processing by binding to the Drosha and Dicer guide sequences.

A previous study has shown that successful targeting of miR-122 in liver, by intravenous injection of antagomir, regulates genes essential for cholesterol biosynthesis, leading to a decline in concentration of miR-122 and reducing circulating levels of cholesterol [343]. Similarly, cardiac fibrosis and hypertrophy were significantly reduced after targeting miR-21 and mir-133, respectively [344, 345]. Further, chimpanzees chronically infected with hepatitis C were protected against viremia and liver pathology after targeting miR-122 with the LNA anti-miR [346]. The major drawbacks of anti-miR oligonucleotides are off-target effects that affect endogenous RNA species other than the intended target miRNAs. However, these off-target effects may be minimized by conjugating antagomir with ligands that can specifically bind to upregulated surface receptors on the target cancer cells. Besides antagomirs, several studies have successfully employed small-molecule (low-molecular-weight compound) inhibitors to target miRNAs such as miR-21, miR-122, and miR-27a [335, 347, 348] in vitro. Recent study has shown that morpholinos are capable of antagonizing miRNA by binding to the miRNA response element. They are totally immune to nucleases and at appropriate concentrations can only interfere with translation by binding to the 5′ end of mRNA near the start codon. This approach is useful when a miRNA has both procancer and anticancer targets simultaneously, because at appropriate lengths it can protect key anticancer miRNA targets while not interfering with the downregulation of the procancer targets. In order to reach effective concentrations of synthetic oligos (in vivo) at the target cells, it is beneficial to conjugate them to cholesterol or arginine-rich peptides to increase cellular uptake and diffusion out of endosomes [349]. A preclinical study involving administration of anti-miR-221 modified with cholesterol significantly enhanced liver tissue distribution and inhibited tumor growth, leading to increased survival compared with unmodified anti-miR-221 in a mouse model of HCC [350]. Similarly, substantial inhibition of tumor metastasis was achieved after therapeutic targeting of miR-10b with cholesterol-modified anti-miR-10b in a mouse mammary tumor model [351]. Another alternative method of antagonizing miRNA is the miRNA sponge, produced by a transgene. MiRNA sponges have 4–10 miRNA binding sites with small spacer nucleotides in between [352]. Efficient miRNA sponges are designed with imperfect complementarity; this hinders sponge degradation and results in longer miRNA sequestration. Sponges are typically complementary only to the seed region allowing targeting of an entire family of miRNAs. Silencing the miRNAs in the miR-17–92 cluster by employing miRNA sponges harboring multiple binding sites for these family members demonstrates the proof of principle for this approach [353]. Additional methods are being investigated, such as small molecules that interfere with Dicer function [335].

MiRNA mimics are identical, or near identical to the guide strand of endogenous miRNA. Double-stranded oligos are 100- to 1000-fold more effective than single-stranded ones at mimicking miRNA, as they are present as duplexes of the guide and the passenger strand [354]. Although the passenger strand is usually modified with cholesterol at the 3′ end to increase cellular uptake [355], miRNA mimics are still susceptible to nuclease degradation, can be targeted by the innate immune system, and also affect nontarget tissue. In light of the above, miRNA mimics have limitations with regard to therapeutic applications. To further enhance therapeutic delivery, lipid-based delivery systems have been developed. Liposomes of smaller diameter (<100 nm) allow for a high drug-to-lipid ratio. Liposomes have been successfully used to deliver miR-34a mimic intravenously in a mouse model of NSCLC resulting in reduction of tumor growth, while not having significant renal or hepatic toxicity or triggering an immune response [356]. In order to develop precise targeting of anticancer miRNAs, tumor-specific promoters can be utilized to specifically express the reintroduced miRNA in the tumor cells [357]. An example of this is the T-VISA-miR-34a construct that utilizes the human telomerase reverse transcriptase promoter, which is activated only in cancerous tissue, to express miR-34a [357]. As with other gene therapies, adeno-associated viruses can be used to enhance miRNA construct delivery [358]. Adeno-associated viruses have different serotypes that have differing affinity for various organs aiding in the targeting of miRNA, and have already proven effective in various murine cancer models [358].

### 11.11.1 Targeting of miRNAs in Pancreatic Cancer

Some preclinical studies have shown that targeting mature miRNAs or their precursors by using synthetic or chemically modified antisense oligonucleotides, and/or overexpression of some miRNAs in PC results in decreased tumor burden [311]. The oncomir miR-21 is overexpressed in the early stages of PDAC [359], and it has been shown that PC cells are addicted to miR-21 [360]. Lentiviral-based targeting of miR-21 by its antagonists resulted in increased angiogenesis, strong inhibition of PDAC-derived cell lines, and proliferation both in vitro and in vivo. Further, combinational treatment involving miR-21 antagonists and gemcitabine (125 mg/kg) leads to significant regression of tumor growth in vivo. Therefore, delivery of miR-21 antagonists may be a useful therapeutic intervention in several cancers including PC because of its upregulation in multiple cancer types [361]. The Phase 1 clinical trial (NCT01274455) involving 24 PDA patients with advanced disease treated with intratumoral injection of CYL-02 (gene-therapy product) using nonviral vector-mediated delivery such as endoscopic ultrasound followed by gemcitabine treatment showed the feasibility and safety of intratumoral injection. A recent study has shown that co-delivery of a nanosystem comprising anti-miR oligonucleotide and human serum albumin-1-palmitoyl-2-oleoyl-sn-glycero-3-ethylphosphocholine:cholesterol resulted in effective silencing of individually overexpressed miRNAs (miR-21, miR-221, miR-222, and miR-10) in PC cells leading to upregulation of their target genes. Further, treatment of PC cells with the above nanosystem containing anti-miR-21 and the chemotherapeutic drug sunitinib has produced a strong synergistic antitumor effect; therefore, the above combination therapy may have potential therapeutic value in PDACs [362]. Arora et al. developed a poly(D, L-lactide-co-glycolide) (PLGA)-based nanoformulation of miR-150 (miR-150-NF) having high encapsulation efficiency (~78%) and sustained release capacity. When PC cells were treated with miR-150-NF, it resulted in effective intracellular concentration of miR-150 mimics leading to downregulation of its target gene (MUC4), which resulted in the inhibition of cell growth, clonogenicity, motility, and invasion [363]. Further, intraperitoneal injection of LNA-based miR-21 antagonist resulted in downregulation of oncomir miR-21 leading to decreased splenomegaly in mice with systemic lupus erythematosus [364]. In contrast, overexpression of oncogenic miR-21 and miR-221 enhanced the malignant phenotype of PC cells [291]. However, inhibition of miR-21 and miR-221 with the use of antisense oligonucleotides resulted in decreased proliferation and increased apoptosis of PC cells by increasing expression of PTEN, RECK, and p27 [291]. In combination with gemcitabine, use of antisense oligonucleotides also synergistically killed PC cells [291]. In addition to oligonucleotides, targeting miR-10a expression by using retinoic acid receptor (RAR) antagonist repressed PC metastasis [313].

Overexpression of miR-143 [312] and the viral-mediated delivery of miR-143 or miR-145 resulted in decreased PC metastasis and reduced tumor formation in PC cells [161]. Restoration of tumor suppressor miRNAs such as let-7, miR-96, and miR-150 in PC cells led to decreased tumorigenesis and cell proliferation [164]. Furthermore, restoration of let-7 led to KRAS downregulation and decreased mitogen-activated protein kinase activation [164]. The use of oncogenic miR-34 mimics or its virus-mediated infection led to apoptosis, cell cycle arrest, decreased clonogenicity, invasion, and increased chemo/radiation sensitivity of PC cells by targeting Bcl-2 and Notch1/2 [314]. Several studies have shown that let-7 and miR-34 are significantly downregulated in various solid tumors as well as in cancer stem cells. Administration of let-7 and miR-34 resulted in changes in cancer stem cell phenotypic properties. Therefore, therapeutic delivery of let-7 and miR-34 may result in decreased tumor growth as well as number of viable cancer stem cells [365, 366]. Further, restoration of let-7 either in the form of a let-7 mimic or a viral-mediated therapeutic delivery resulted in significant reduction of tumor growth xenograft model for human NSCLC as well as in the KrasG12D transgenic mouse model [367]. Similarly, restoration of let-7b by intranasal inhalation of adenovirus carrying let-7b in a spontaneous Kras model for lung cancer led to a significant decrease in tumor growth compared with mice injected with only the vector [368], whereas intratumor injection of let-7 carrying plasmid into xenograft tumors produced by orthotopic implantation of Capan1 cells did not show any change in tumor growth/proliferation compared with control mice [164]. Growth of prostate and lung cancer was significantly inhibited after systemic delivery of a miR-34 mimic in lung and prostate cancer mouse models [369, 370]. Several studies have also revealed that administration of tumor suppressor miRNA mimics in mouse models revealed no toxicity in normal tissues, and therefore is well endured [358, 368, 371]. Further, delivery of miR-34a mimics in combination with lipid-containing formulation also did not elicit any nonspecific immune response [370]. All of these studies provide a rationale to use miRNAs for therapeutic intervention of lethal PC.

Not only for PC, but also in the area of cardiovascular disease, several miRNAs including miR-208/499/195 are currently under development as therapeutic agents [372]. MiR-208 is a heart muscle–specific miRNA and it stimulates hypertrophy of cardiomyocyte, fibrosis, and β-MHC expression under stress and hypoxia conditions. The miR-208 knockout mice were shown to be highly resilient to cardiomyocyte hypertrophy and fibrosis due to stress or hypoxia; therefore, targeting of miR-208 may significantly ameliorate chronic heart disease [373]. Recently, miRagen Therapeutics in collaboration with Danish Santaris Pharma developed a major LNA-based anti-miR-208 therapeutic (MGN-9103 antimir) to target miR-208 expression for treatment of chronic heart failure. In addition, they also developed two more LNA-based antagomirs to target miR-15, miR-195 (MGN-1374 antimir), and miR-451 (MGN-4893 antimir) for the treatment of myocardial infarction [374, 375] and polycythemia vera [376], respectively. MiR-195 is overexpressed in cardiac hypertrophy, and the transgenic mice overexpressing this miRNA resulted in cardiac failure and death of the animal [373]; therefore, anti-miR-195 can be used to treat cardiac disease.

Santaris Pharma has recently developed an LNA-modified antagomir Miravirsen (or SPC3649) against miR-122 that is important for induction of viral transcription of hepatitis C virus (HCV) for hepatitis C treatment [346, 377]. Treatment of chimpanzees having chronic HCV with SPC3649 has shown decreased viral RNA in serum without any adverse effects [346]. Currently, Miravirsen is under Phase 2a clinical trials and has provided a significant protection to patients with HCV [378]. Regulus Therapeutics (San Diego, United States) has generated several antagomirs (miR-10b, miR-21, miR-103/107, miR-182, miR-380-5p) to target miRNAs overexpressed in different conditions (glioblastoma, HCC, atherosclerosis, kidney fibrosis, and HCV infection), but none of the above therapeutics has reached clinical phase. Later on, it employed anti-miR-21/anti-miR-380-5p/anti-miR-182 to prevent metastasis and invasion of glioma [379], neuroblastoma [380], and melanoma [381], respectively. MiRNA Therapeutics has developed miRNA mimics to restore the expression of specific miRNAs that are lost during the course of tumorigenesis. It developed MRX34, the first miR-34a mimic compound that entered into Phase 1 clinical trials (NCT01829971). Administration of MRX34 (restoration of miR-34) produced potent antitumor effects in several mice cancer models [369]. Further, preclinical testing of the combination of MRX34 and liposomes in mouse models of HCC has shown very promising outcomes with minimal toxicity (NCT01829971) [382].

## 11.12 CLINICAL EXPLOITATION OF EPIGENETIC STATES IN PROSTATE CANCER

### 11.12.1 Epigenetic Modifications as Biomarkers in Prostate Cancer

Although the commonly used prostate-specific antigen (PSA) test has significantly increased the detection of clinically localized tumors, there are many concerns regarding sensitivity and specificity of the PSA test [383]. Therefore, there is an urgent need to identify either more accurate biomarkers for CaP or to establish markers that can be used in combination with PSA for more accurate CaP diagnosis. Understanding the molecular changes during CaP development and progression may help identify accurate biomarkers.

Accumulating evidence for the role of epigenetic modifications and miRNAs during CaP development suggests that they can be used in more accurate diagnosis of CaP initiation and progression and in turn be translated as a functional biomarker. DNA methylation has been frequently studied in CaP and contributes to both the disease initiation and progression [384, 385]. As discussed previously, methylation of GSTP1 promoter is the most frequent epigenetic modification in CaP. Hypermethylation of GSTP1 promoter is present in 90% of adenocarcinomas and 70% of high-grade prostatic intraepithelial neoplasia (high-grade PIN) lesions but not in normal prostate epithelium or hyperplastic epithelium [23]. Analysis of methylation in tumor samples is highly invasive, which is not an ideal biomarker property. However, GSTP1 methylation can be detected in bodily fluids (i.e., urine and serum) with high specificity (86.8–100%) but low sensitivity in both urine (18.8–38.9%) and serum/plasma (13.0–72.5%) [386–388]. Hence, analysis of multiple gene methylation patterns along with GSTP1 may provide increased specificity. Hoque et al. used promoter methylation patterns of nine genes in urine sediment DNA to distinguish normal and CaP individuals [386]. In general, methylation patterns in urine samples showed correlation with the methylation status in the primary tumor. A combination of only four genes (p16, ARF, MGMT, and GSTP1) was able to detect 87% of prostate cancers with 100% specificity [386]. Similarly, Rouprêt et al. showed that the promoter methylation pattern of four genes, GSTP1, RASSF1a, RARbeta2, and APC, was able to differentiate malignant from nonmalignant cases with 86% sensitivity and 89% specificity [389]. A similar combination of four genes GSTP1, PTGS2, RPRM, and TIG1 was reported using serum samples [390]. Individually, methylation status of GSTP1 was most specific (98%) in discriminating CaP and BPH with low sensitivity (42–47%). A combination of methylation patterns of four genes provides slightly high specificity (AUC=0.699) as biomarker [390].

Another class of epigenetic modifications that has been utilized to separate CaP from normal tissue is histone modifications. In a tissue microarray analysis of 113 CaP and 23 nonmalignant prostate tissues, significant reduction of H3K4me1, H3K9me2, H3K9me3, H3Ac, and H4Ac was observed [391]. Also, levels of H3Ac and H3K9me2 were able to separate CaP and nonmalignant prostate tissue with high specificity (>91%) and sensitivity (>78%) [391]. Histone modifications in bodily fluids have not been studied in detail but one study used ELISA to study global levels of H3K27Me3. This study showed significant decrease in H3K27Me3 in metastatic disease (n=28) compared with localized disease (n=33) [392].

## 11.12.2 Diagnostic and Prognostic miRNA Expression Patterns

Tumor-specific miRNA patterns are emerging as highly attractive biomarkers of cancer risk and progression. Given miRNAs are secreted into bodily fluids [393] and can be reliably extracted and measured [394], miRNAs offer significant clinical potential as highly sensitive serum-borne prognostic indicators [51, 395]. There have been several studies for miRNAs as biomarkers using serum, plasma, and urine samples (as reviewed in [396]). Mitchell et al. for the first time reported the serum expression of miR-141 was highly elevated in advanced CaP (n=25) as compared with healthy men (n=25) [51]. Expression of miR-141 was also correlated with serum PSA levels and could detect individuals with advanced CaP with 60% sensitivity at 100% specificity [51]. Expression of miR-21 has been shown

to be elevated in hormone-refractory CaP especially in those resistant to docetaxel-based chemotherapy [397]. In another approach Selth et al. identified miRNA signatures in TRAMP model and then validated four miRNAs in CaP serum. They identified miR-141, miR-298, and miR-375 to be upregulated in advanced CaP [398]. In a candidate approach, Yaman Agaoglu et al. identified miR-21 (AUC, 88%) and miR-221 (AUC, 83%) as markers to differentiate normal and CaP individuals while miR-141 was able to distinguish localized/local advanced disease (AUC=0.755) [399]. In a similar candidate approach, four miRs (miR-26a, miR-32, miR-195, miR-let7i) were tested in CaP serum samples and showed increased diagnostic accuracy in combination (sensitivity: 78.4%; specificity: 66.7%; AUC: 0.758) [400].

More recently, analysis of 742 miRNAs in plasma-derived circulating microvesicles of 78 CaP patients and 28 normal control individuals identified 12 miRNAs differentially expressed in CaP patients compared to controls. They also confirmed the association of miR-141 and miR-375 with metastatic CaP in a separate cohort of patients. Interestingly, five miRNAs identified in serum samples were also detected in urine samples. MiR-107 and miR-574-3p were measured at significantly higher concentrations in the urine of men with CaP compared to controls [401]. MiRNA expression has also been reported to correlate with other clinical parameters. MiR-141, miR-200b, and miR-375 showed upregulation in CaP and correlate with increasing tumor stage and Gleason score [402]. All together these studies show high potential of miRNAs to be noninvasive biomarkers of CaP. Additionally miRNAs can also help predict disease progression and drug response.

## 11.12.3 Insights from The Cancer Genome Atlas (TCGA) and ENCODE

We explored TCGA data for associations between DNA methylation and miRNA expression in CaP tumors. An integrated tool of all the data types generated by TCGA has been developed to understand the systems-level interaction between different data features. Statistically significant association can be identified and visualized using Regulome Explorer [403]. Regulome Explorer has been used for integrated analysis in colon, rectal, and breast cancers [404, 405]. We used Regulome Explorer to identify statistically significant correlations between miRNA expression (RNA-seq) and status of DNA methylation (Illumina 450k methylation arrays). We specifically searched for CpG sites showing negative correlation with miRNA expression on the same chromosome (using "cis" setting in distance filter and p-value cutoff of $-\log10$ (p) $\geq 6$). This analysis identified a total of 27 CpGs negatively correlated with expression of 17 miRNAs from 5 different chromosomes (Figure 11.5). Notably corroborating the discussion above, all five miRNAs of the miR-200 family (miR-200a/b/429, miR-200c/141) and miR-205 showed strong negative correlation between miRNA expression and CpG

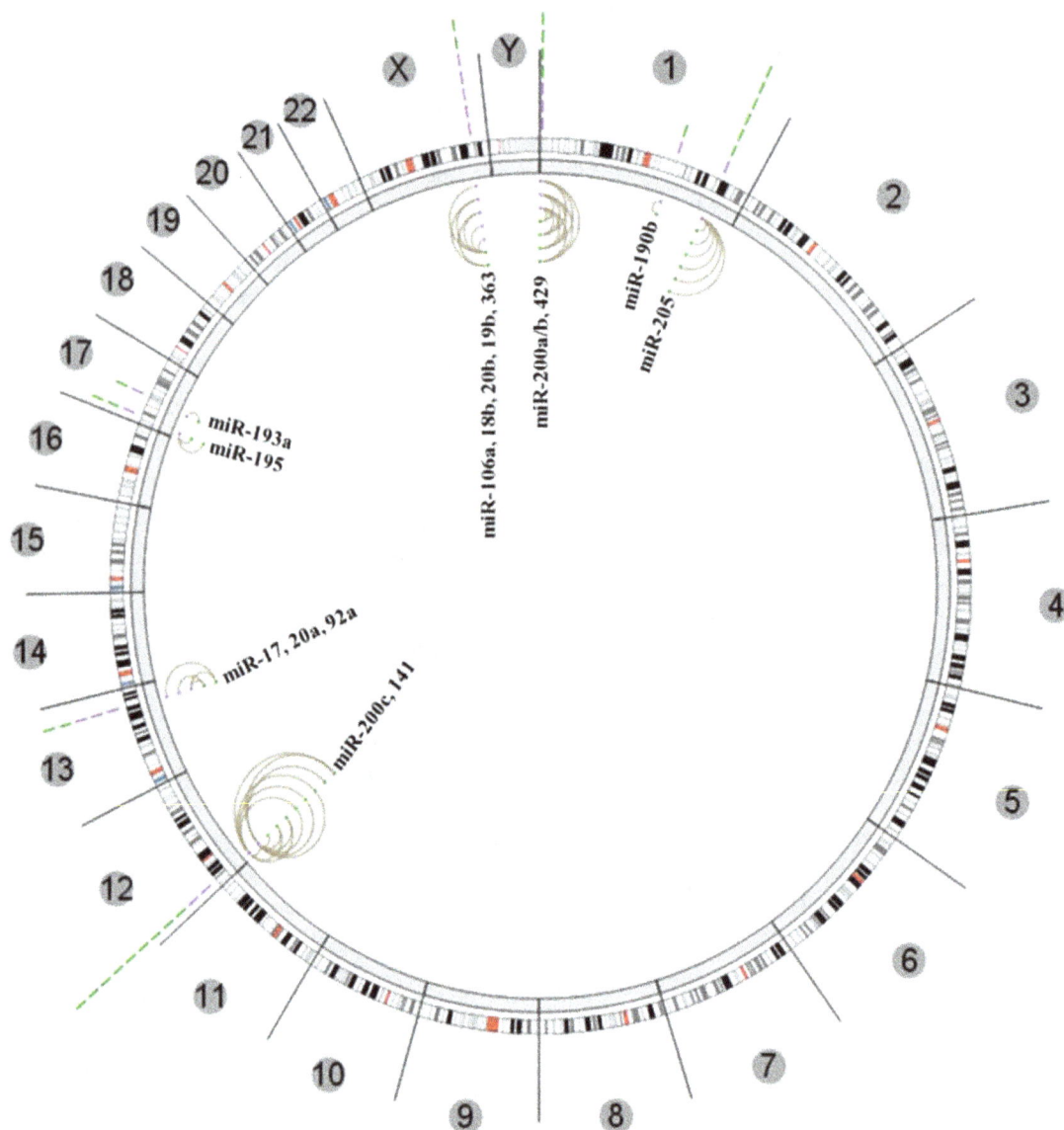

**FIGURE 11.5** Significant negative correlation between altered miRNA expression and DNA methylation pattern in primary prostate cancer tumors from the TCGA dataset. Circos plot showing miRNAs where there is a significant correlation with CpGs near miRNA gene locations. The outer ring represents all the chromosomes. MiRNA genes are represented by *purple lines* near their chromosomal locations. *Green lines* represent CpGs in the region. *Lines connecting two dots* represent the statistically significant correlation between two features selected, miRNA expression and CpG methylation in this case. [Figure generated from [403]: Regulome Explorer. Available online: http://explorer.cancerregulome.org/ (accessed on 16 July 2013), using latest version of data release for Prostate cancer —PRAD-13-March-2013].

methylation [89, 90]. Many of the miRNAs with important function in CaP did not show significant correlation in this analysis, which can be partially because of the allocation of CpGs to miRNAs in Illumina 450k array annotation. This preliminary exploration of TCGA data strengthens the observation that miRNAs are epigenetically silenced in CaP. It will be interesting to interrogate this resource in further detail to identify how other genomic data from TCGA correlate with miRNA expression and DNA methylation.

Recently published large-scale ENCODE ChIP-seq data for various epigenetic features (e.g., histone marks, chromatin accessibility, and DNA methylation) provide a key opportunity to study these associations in various cancer models including CaP [26, 406]. Additionally, as an example, we also examined the chromatin accessibility around the miRNAs identified from regulome analysis in LNCaP and PrEC cells from ENCODE. As shown in Figure 11.6, miRNAs with higher expression in CaP tumors (miR-17, miR-20a, and miR-92a), which were associated with low methylation, also showed low methylation and open chromatin in LNCaP as compared with PrEC cells. Similarly, miRNA with low expression and high methylation in CaP tumors showed more methylation and closed chromatin in LNCaP (e.g., miR-205).

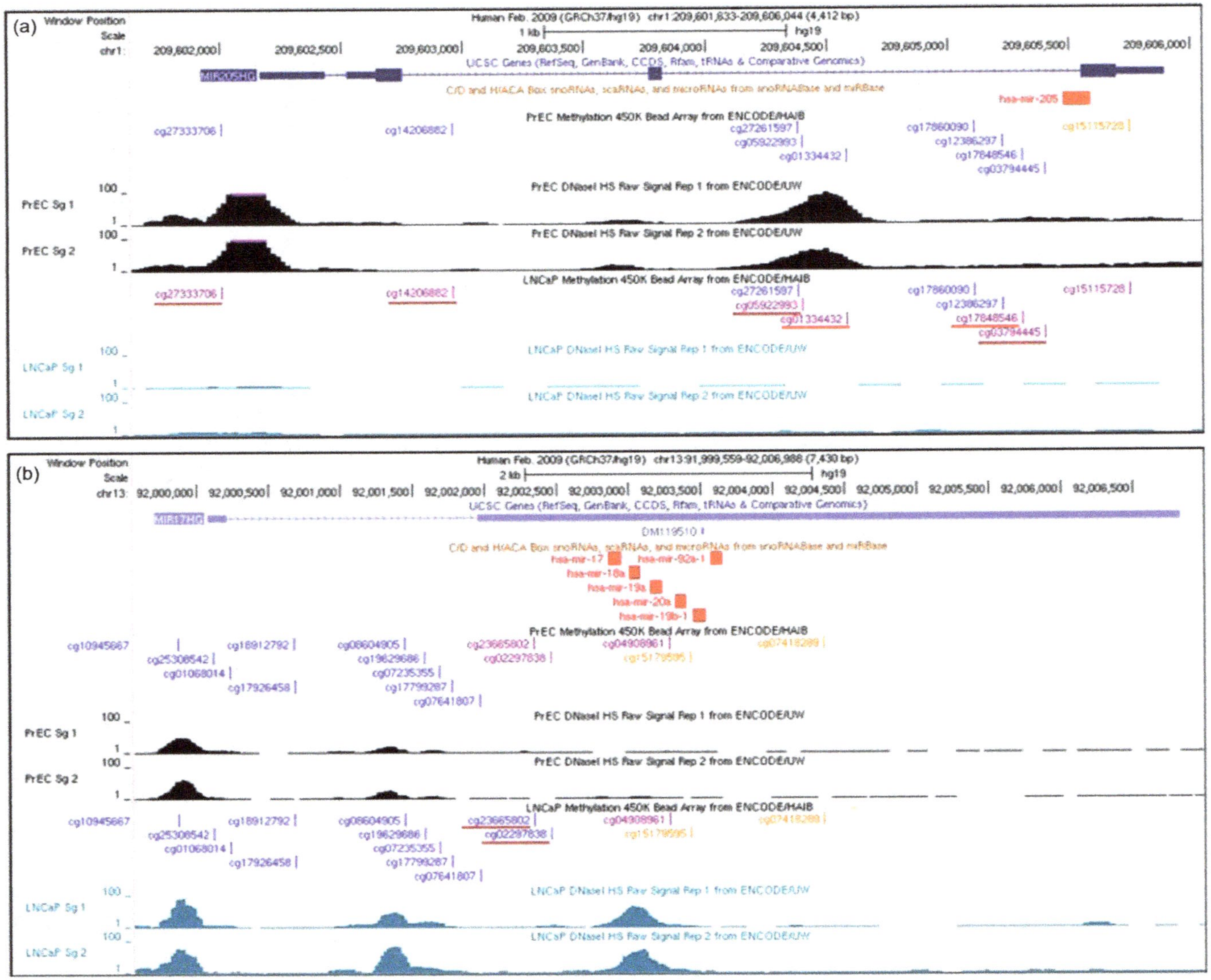


**FIGURE 11.6** UCSC genome browser tracks showing CpG methylation and DNaseI cleavage status around miRNAs. Top panel track shows the genomic location of the miRNA. Middle track shows the CpGs in genomic fragment and color of each CpG represents the beta values (ENCODE description—*orange,* beta value equal to or greater than 0.6 was considered fully methylated; *purple,* beta values between 0.2 and 0.6 were considered to be partially methylated; *blue,* beta value equal to or less than 0.2 was considered to be fully unmethylated). Bottom panel shows the DNaseI sensitivity measured in PrEC (*black*) and LNCaP (*blue*). DNaseI hypersensitive sites shown as peaks suggest accessible chromatin; conversely no DNaseI hypersensitive sites suggest closed chromatin. (a) Shows the status of miR-205. Most of the CpGs around miR-205 are partially methylated in LNCaP and unmethylated in PrEC. Supportively, DNaseI sensitivity shows open areas in PrEC cells and closed chromatin in LNCaP. (b) Shows the status of a miR-17-92 cluster. Two CpGs around this cluster are partially methylated in PrEC cells and unmethylated in LNCaP. Supportively, DNaseI sensitivity shows open chromatin in LNCaP and closed in PrEC cells. Differentially methylated CpGs are *underlined*.

## 11.13 CONCLUSIONS

Although the involvement of miRNAs in disease pathogenesis is still emerging, deregulated miRNA expression in PC suggests their critical roles in its development and progression. Many studies [407–410] have shown their involvement in a wide variety of biological processes including cell survival, proliferation, invasion/metastasis, apoptosis, and drug resistance of PC [159, 187, 195, 217, 262, 278, 285, 411]. Furthermore, differential expression profiles of circulating miRNAs among PC patients and healthy individuals render them potential biomarkers for diagnosis and prognosis. Besides the fact that a single miRNA targets multiple genes, several miRNAs frequently share the same target transcript, which makes miRNA actions more additive resulting in large changes in disease state. In light of the above, it is apparent that cancer cells frequently deregulate miRNA expression to shortcut the selection process during the evolution of benign to malignant state. In this process these tumor cells may develop addiction to either aberrantly overexpressed oncogenic or downregulated tumor suppressor miRNAs. Therefore, it seems logical that restoring tumor suppressor miRNAs to the precancer levels [277, 292] or use of miRNA inhibitors may hold promise for therapeutic intervention against cancers. Tumor promoting effects have been directly countered in experimental models primarily in two ways: (a) by nanoparticle/virally delivered miRNA, used to restore tumor suppressive miRNAs [277], and (b) by antisense blocking oligos against oncogenic miRNAs. PC cancer is composed of dense stroma, also called the desmoplastic reaction that results in poor penetration of therapeutic agents resulting in reduced efficacy due to lower available drug concentration in the center of the tumor. Use of higher amounts of miRNAs is associated with off-target effects that results in higher toxicity and thus miRNA-based therapy is a major challenge for PC therapy. For considering miRNAs as therapeutic targets in PC, several challenges need to be addressed. First specific delivery of the miRNA to PC tumors in vivo presents a significant challenge. In order to address this, nanoparticles carrying the miRNAs for delivery can be coated with ligands or antibody fragments specific to upregulated cell surface receptors on the cancer cells to increase tumor-specific miRNA delivery that will enhance efficacy and reduce targeting to normal cells. Furthermore, antisense molecules do not readily cross cell membranes and this therefore limits their use. However, recently intravenous injection of morpholino-oligos against miRNAs has shown some promise [412]. These morpholinos are absorbed by endocytosis and can diffuse through endosomes with the aid of either an octaguanidinium dendrimer or arginine-rich cell-penetrating peptides (CPP) linked to their 3′ or 5′ ends [412]. Furthermore, target-specific delivery of these morpholino-oligos may be enhanced by either altering the CPP structure or by attaching a ligand specific to an upregulated receptor on the surface of PC cells.

Most PC patients have multiple mutations that either inactivate tumor suppressors such as p16 and p53, or activate oncogenes, most importantly KRAS. Therefore, for better management of PC, using inhibition of miR-491-5p, -34a, -214, -125b, -29, -504, -122, -33, -1285, -380-5p, -25, -133, and -30d to increase p53 expression; miR-24 and miR-10b to increase p16 expression; miR-130a/301a/454 to increase Smad4 expression; miR-106 to increase p21; miR-221/222 for p27 upregulation; and miR-132 and miR-212 expression to increase Rb1 expression seems reasonable to combat this deadly disease. In addition, reintroduction of miRNA that suppresses oncogenes and inhibits cell cycle progression such as miR-217, miR-96, let-7, miR-126, and miR-143/145 to decrease KRAS expression; miR-192, -194, -215, -15a/16-1, -34a/b/c, -145, -200a/b/c, -141, -429, and miR-107 to downregulate c-Myc and cyclin expression can have significant effects on inhibiting PC growth. MiRNAs that target KRAS expression are not specific to the mutant form and therefore use of a single miRNA may not have sufficient downregulation of mutant KRAS to be effectual in vivo. It may however be of interest to utilize a cocktail of several miRNAs to downregulate mutated KRAS expression in PC. However, use of a cocktail of miRNAs at high doses to counteract the effects of oncogenic KRAS may have undesirable effects on healthy adult cells. To this end, improvements in delivery techniques and targeted delivery of miRNA cocktails may provide a larger window before toxicity becomes a problem. In addition, it will be worthless to use antisense oligos against miRNAs targeting mutated p16 and/or p53 in a subset of PC patients. It may even be counterproductive in the cases where p53 has gain-of-function mutations. Therefore, it will be of interest to genotype PC tumors to identify aberrantly expressed miRNAs, mutations of key tumor suppressors, and oncogenes for effective decision-making with regard to which miRNAs should be blocked or reintroduced.

The deregulated cross-talk between epigenetic states and miRNA transcription and action is functionally important in the progression of CaP. Studies utilizing components of epigenetic machinery may offer highly accurate, and accessible, information about CaP diagnosis and prognosis. For example, urine-borne DNA methylation patterns in combination with serum miRNA patterns offer the opportunity to be exploited as highly accurate functional biomarkers. Also, they can be used to understand the underlying epigenetic status to be prognostic of CaP disease progression risks. These components have a strong likelihood to be exploited as integrated noninvasive biomarkers of CaP progression and drug resistance.

## REFERENCES

[1] Schroder, F.H., Hugosson, J., Roobol, M.J., Tammela, T.L., Ciatto, S., Nelen, V. et al. Screening and prostate-cancer mortality in a randomized European study. N. Engl. J. Med. 360 (2009) 1320–1328.

[2] Andriole, G.L., Crawford, E.D., Grubb, R.L., 3rd, Buys, S.S., Chia, D., Church, T.R. *et al.* Mortality results from a randomized prostate-cancer screening trial. *N. Engl. J. Med. 360* (**2009**) 1310–1319.

[3] He, W.W., Sciavolino, P.J., Wing, J., Augustus, M., Hudson, P., Meissner, P.S. *et al.* A novel human prostate-specific, androgenregulated homeobox gene (nkx3.1) that maps to 8p21, a region frequently deleted in prostate cancer. *Genomics 43* (**1997**) 69–77.

[4] Cairns, P., Okami, K., Halachmi, S., Halachmi, N., Esteller, M., Herman, J.G. *et al.* Frequent inactivation of pten/mmac1 in primary prostate cancer. *Cancer Res. 57* (**1997**) 4997–5000.

[5] Li, J., Yen, C., Liaw, D., Podsypanina, K., Bose, S., Wang, S.I. *et al.* Pten, a putative protein tyrosine phosphatase gene mutated in human brain, breast, and prostate cancer. *Science 275* (**1997**) 1943–1947.

[6] Visakorpi, T., Hyytinen, E., Koivisto, P., Tanner, M., Keinanen, R., Palmberg, C. *et al. In vivo* amplification of the androgen receptor gene and progression of human prostate cancer. *Nat. Genet. 9* (**1995**) 401–406.

[7] Taylor, B.S., Schultz, N., Hieronymus, H., Gopalan, A., Xiao, Y., Carver, B.S. *et al.* Integrative genomic profiling of human prostate cancer. *Cancer Cell 18* (**2010**) 11–22.

[8] Perner, S., Demichelis, F., Beroukhim, R., Schmidt, F.H., Mosquera, J.M., Setlur, S. *et al.* Tmprss2:Erg fusion-associated deletions provide insight into the heterogeneity of prostate cancer. *Cancer Res. 66* (**2006**) 8337–8341.

[9] Tomlins, S.A., Laxman, B., Dhanasekaran, S.M., Helgeson, B.E., Cao, X., Morris, D.S. *et al.* Distinct classes of chromosomal rearrangements create oncogenic ETS gene fusions in prostate cancer. *Nature 448* (**2007**) 595–599.

[10] Barbieri, C.E., Baca, S.C., Lawrence, M.S., Demichelis, F., Blattner, M., Theurillat, J.P. *et al.* Exome sequencing identifies recurrent spop, foxa1 and med12 mutations in prostate cancer. *Nat. Genet. 44* (**2012**) 685–689.

[11] Sandoval, J., Esteller, M. Cancer epigenomics: Beyond genomics. *Curr. Opin. Genet. Dev. 22* (**2012**) 50–55.

[12] Holliday, R. The inheritance of epigenetic defects. *Science 238* (**1987**) 163–170.

[13] Esteller, M. Cancer epigenomics: DNA methylomes and histone-modification maps. *Nat. Rev. Genet. 8* (**2007**) 286–298.

[14] Campbell, M.J., Turner, B.M. Altered histone modifications in cancer. *Adv. Exp. Med. Biol. 754* (**2013**) 81–107.

[15] Chiam, K., Ricciardelli, C., Bianco-Miotto, T. Epigenetic biomarkers in prostate cancer: Current and future uses. *Cancer Lett.* (**2012**), doi:10.1016/j.canlet.2012.02.011

[16] Jeronimo, C., Bastian, P.J., Bjartell, A., Carbone, G.M., Catto, J.W., Clark, S.J. *et al.* Epigenetics in prostate cancer: Biologic and clinical relevance. *Eur. Urol. 60* (**2011**) 753–766.

[17] Shen, M.M., Abate-Shen, C. Molecular genetics of prostate cancer: New prospects for old challenges. *Genes Dev. 24* (**2010**) 1967–2000.

[18] Denis, H., Ndlovu, M.N., Fuks, F. Regulation of mammalian DNA methyltransferases: A route to new mechanisms. *EMBO Rep. 12* (**2011**) 647–656.

[19] Lee, K.K., Workman, J.L. Histone acetyltransferase complexes: One size doesn't fit all. *Nat. Rev. Mol. Cell Biol. 8* (**2007**) 284–295.

[20] Kim, H.J., Bae, S.C. Histone deacetylase inhibitors: Molecular mechanisms of action and clinical trials as anticancer drugs. *Am. J. Transl. Res. 3* (**2011**) 166–179.

[21] Albert, M., Helin, K. Histone methyltransferases in cancer. *Semin. Cell Dev. Biol. 21* (**2010**) 209–220.

[22] Kooistra, S.M., Helin, K. Molecular mechanisms and potential functions of histone demethylases. *Nat. Rev. Mol. Cell Biol. 13* (**2012**) 297–311.

[23] Nakayama, M., Bennett, C.J., Hicks, J.L., Epstein, J.I., Platz, E.A., Nelson, W.G. *et al.* Hypermethylation of the human glutathione s-transferase-pi gene (GSTP1) CpG island is present in a subset of proliferative inflammatory atrophy lesions but not in normal or hyperplastic epithelium of the prostate: A detailed study using laser-capture microdissection. *Am. J. Pathol. 163* (**2003**) 923–933.

[24] Varambally, S., Dhanasekaran, S.M., Zhou, M., Barrette, T.R., Kumar-Sinha, C., Sanda, M.G. *et al.* The polycomb group protein EZH2 is involved in progression of prostate cancer. *Nature 419* (**2002**) 624–629.

[25] Saramaki, O.R., Tammela, T.L., Martikainen, P.M., Vessella, R.L., Visakorpi, T. The gene for polycomb group protein enhancer of zeste homolog 2 (EZH2) is amplified in late-stage prostate cancer. *Genes Chromosomes Cancer 45* (**2006**) 639–645.

[26] Djebali, S., Davis, C.A., Merkel, A., Dobin, A., Lassmann, T., Mortazavi, A. *et al.* Landscape of transcription in human cells. *Nature 489* (**2012**) 101–108.

[27] Graur, D., Zheng, Y., Price, N., Azevedo, R.B., Zufall, R.A., Elhaik, E. On the immortality of television sets: —Function‖ in the human genome according to the evolution-free gospel of encode. *Genome Biol. Evol. 5* (**2013**) 578–590.

[28] Doolittle, W.F. Is junk DNA bunk? A critique of encode. *Proc. Natl. Acad. Sci. USA 110* (**2013**) 5294–5300.

[29] Bartel, D.P. MicroRNAs: Genomics, biogenesis, mechanism, and function. *Cell 116* (**2004**) 281–297.

[30] Calin, G.A., Croce, C.M. Microrna signatures in human cancers. *Nat. Rev. Cancer 6* (**2006**) 857–866.

[31] Iorio, M.V., Croce, C.M. Microrna dysregulation in cancer: Diagnostics, monitoring and therapeutics. A comprehensive review. *EMBO Mol. Med. 4* (**2012**) 143–159.

[32] Manikandan, J., Aarthi, J.J., Kumar, S.D., Pushparaj, P.N. Oncomirs: The potential role of non-coding microRNAs in understanding cancer. *Bioinformation 2* (**2008**) 330–334.

[33] Esquela-Kerscher, A., Slack, F.J. Oncomirs—MicroRNAs with a role in cancer. *Nat. Rev. Cancer 6* (**2006**) 259–269.

[34] Volinia, S., Calin, G.A., Liu, C.G., Ambs, S., Cimmino, A., Petrocca, F. *et al.* A microRNA expression signature of human solid tumors defines cancer gene targets. *Proc. Natl. Acad. Sci. USA 103* (**2006**) 2257–2261.

[35] Lu, J., Getz, G., Miska, E.A., Alvarez-Saavedra, E., Lamb, J., Peck, D. *et al.* MicroRNA expression profiles classify human cancers. *Nature 435* (**2005**) 834–838.

[36] Lewis, B.P., Burge, C.B., Bartel, D.P. Conserved seed pairing, often flanked by adenosines, indicates that thousands of human genes are microRNA targets. *Cell 120* (**2005**) 15–20.

[37] Lussier, Y.A., Stadler, W.M., Chen, J.L. Advantages of genomic complexity: Bioinformatics opportunities in microRNA cancer signatures. *J. Am. Med. Inform. Assoc. 19* (**2012**) 156–160.

[38] Sun, T., Wang, Q., Balk, S., Brown, M., Lee, G.S., Kantoff, P. The role of microrna-221 and microrna-222 in androgen-independent prostate cancer cell lines. *Cancer Res. 69* (**2009**) 3356–3363.

[39] Medina, R., Zaidi, S.K., Liu, C.G., Stein, J.L., van Wijnen, A.J., Croce, C.M. *et al.* MicroRNAs 221 and 222 bypass quiescence and compromise cell survival. *Cancer Res. 68* (**2008**) 2773–2780.

[40] Mercatelli, N., Coppola, V., Bonci, D., Miele, F., Costantini, A., Guadagnoli, M. *et al.* The inhibition of the highly expressed mir-221 and mir-222 impairs the growth of prostate carcinoma xenografts in mice. *PLoS One 3* (**2008**) e4029.

[41] Wang, L., Tang, H., Thayanithy, V., Subramanian, S., Oberg, A.L., Cunningham, J.M. *et al.* Gene networks and microRNAs implicated in aggressive prostate cancer. *Cancer Res. 69* (**2009**) 9490–9497.

[42] Hao, J., Zhang, C., Zhang, A., Wang, K., Jia, Z., Wang, G. *et al.* Mir-221/222 is the regulator of cx43 expression in human glioblastoma cells. *Oncol. Rep. 27* (**2012**) 1504–1510.

[43] Tucci, P., Agostini, M., Grespi, F., Markert, E.K., Terrinoni, A., Vousden, K.H. *et al.* Loss of p63 and its microrna-205 target results in enhanced cell migration and metastasis in prostate cancer. *Proc. Natl. Acad. Sci. USA 109* (**2012**) 15312–15317.

[44] Zhao, J.J., Lin, J., Yang, H., Kong, W., He, L., Ma, X. *et al.* Microrna-221/222 negatively regulates estrogen receptor alpha and is associated with tamoxifen resistance in breast cancer. *J. Biol. Chem. 283* (**2008**) 31079–31086.

[45] Miller, T.E., Ghoshal, K., Ramaswamy, B., Roy, S., Datta, J., Shapiro, C.L. *et al.* Microrna-221/222 confers tamoxifen resistance in breast cancer by targeting p27kip1. *J. Biol. Chem. 283* (**2008**) 29897–29903.

[46] Lin, D., Cui, F., Bu, Q., Yan, C. The expression and clinical significance of GTP-binding RAS-like 3 (arhi) and microrna 221 and 222 in prostate cancer. *J. Int. Med. Res. 39* (**2011**) 1870–1875.

[47] Tong, A.W., Fulgham, P., Jay, C., Chen, P., Khalil, I., Liu, S. *et al.* Microrna profile analysis of human prostate cancers. *Cancer Gene Ther. 16* (**2009**) 206–216.

[48] Spahn, M., Kneitz, S., Scholz, C.J., Stenger, N., Rudiger, T., Strobel, P. *et al.* Expression of microrna-221 is progressively reduced in aggressive prostate cancer and metastasis and predicts clinical recurrence. *Int. J. Cancer 127* (**2010**) 394–403.

[49] Martens-Uzunova, E.S., Jalava, S.E., Dits, N.F., van Leenders, G.J., Moller, S., Trapman, J. *et al.* Diagnostic and prognostic signatures from the small non-coding RNA transcriptome in prostate cancer. *Oncogene 31* (**2012**) 978–991.

[50] Wach, S., Nolte, E., Szczyrba, J., Stohr, R., Hartmann, A., Orntoft, T. *et al.* Microrna profiles of prostate carcinoma detected by multiplatform microrna screening. *Int. J. Cancer 130* (**2012**) 611–621.

[51] Mitchell, P.S., Parkin, R.K., Kroh, E.M., Fritz, B.R., Wyman, S.K., Pogosova-Agadjanyan, E.L. *et al.* Circulating microRNAs as stable blood-based markers for cancer detection. *Proc. Natl. Acad. Sci. USA 105* (**2008**) 10513–10518.

[52] Ozen, M., Creighton, C.J., Ozdemir, M., Ittmann, M. Widespread deregulation of microrna expression in human prostate cancer. *Oncogene 27* (**2008**) 1788–1793.

[53] Porkka, K.P., Pfeiffer, M.J., Waltering, K.K., Vessella, R.L., Tammela, T.L., Visakorpi, T. Microrna expression profiling in prostate cancer. *Cancer Res. 67* (**2007**) 6130–6135.

[54] Lee, Y.S., Kim, H.K., Chung, S., Kim, K.S., Dutta, A. Depletion of human microRNA mir-125b reveals that it is critical for the proliferation of differentiated cells but not for the down-regulation of putative targets during differentiation. *J. Biol. Chem. 280* (**2005**) 16635–16641.

[55] Ribas, J., Ni, X., Haffner, M., Wentzel, E.A., Salmasi, A.H., Chowdhury, W.H. *et al.* Mir-21: An androgen receptor-regulated microRNA that promotes hormone-dependent and hormone-independent prostate cancer growth. *Cancer Res. 69* (**2009**) 7165–7169.

[56] Shi, X.B., Xue, L., Yang, J., Ma, A.H., Zhao, J., Xu, M. *et al.* An androgen-regulated miRNA suppresses bak1 expression and induces androgen-independent growth of prostate cancer cells. *Proc. Natl. Acad. Sci. USA 104* (**2007**) 19983–19988.

[57] Shi, G.H., Ye, D.W., Yao, X.D., Zhang, S.L., Dai, B., Zhang, H.L. *et al.* Involvement of microrna-21 in mediating chemo-resistance to docetaxel in androgen-independent prostate cancer pc3 cells. *Acta Pharmacol. Sin. 31* (**2010**) 867–873.

[58] Lu, Z., Liu, M., Stribinskis, V., Klinge, C.M., Ramos, K.S., Colburn, N.H. *et al.* Microrna-21 promotes cell transformation by targeting the programmed cell death 4 gene. *Oncogene 27* (**2008**) 4373–4379.

[59] Peng, X., Guo, W., Liu, T., Wang, X., Tu, X., Xiong, D. *et al.* Identification of mirs-143 and -145 that is associated with bone metastasis of prostate cancer and involved in the regulation of emt. *PLoS One 6* (**2011**) e20341.

[60] Kong, D., Li, Y., Wang, Z., Banerjee, S., Ahmad, A., Kim, H.R. *et al.* Mir-200 regulates pdgf-d-mediated epithelial-mesenchymal transition, adhesion, and invasion of prostate cancer cells. *Stem Cells 27* (**2009**) 1712–1721.

[61] Gregory, P.A., Bert, A.G., Paterson, E.L., Barry, S.C., Tsykin, A., Farshid, G. *et al.* The mir-200 family and mir-205 regulate epithelial to mesenchymal transition by targeting zeb1 and sip1. *Nat. Cell Biol. 10* (**2008**) 593–601.

[62] Liep, J., Rabien, A., Jung, K. Feedback networks between microRNAs and epigenetic modifications in urological tumors. *Epigenetics 7* (**2012**) 315–325.

[63] Saito, Y., Jones, P.A. Epigenetic activation of tumor suppressor microRNAs in human cancer cells. *Cell Cycle 5* (**2006**) 2220–2222.

[64] Chen, J.F., Mandel, E.M., Thomson, J.M., Wu, Q., Callis, T.E., Hammond, S.M. *et al.* The role of microRNA-1 and microRNA-133 in skeletal muscle proliferation and differentiation. *Nat. Genet. 38* (**2006**) 228–233.

[65] Duursma, A.M., Kedde, M., Schrier, M., le Sage, C., Agami, R. Mir-148 targets human dnmt3b protein coding region. *RNA 14* (**2008**) 872–877.

[66] Filipowicz, W., Bhattacharyya, S.N., Sonenberg, N. Mechanisms of post-transcriptional regulation by micrornas: Are the answers in sight? *Nat. Rev. Genet. 9* (**2008**) 102–114.

[67] Thorne, J.L., Maguire, O., Doig, C.L., Battaglia, S., Fehr, L., Sucheston, L.E. *et al.* Epigenetic control of a vdr-governed feed-forward loop that regulates p21(waf1/cip1) expression and function in non-malignant prostate cells. *Nucleic Acids Res. 39* (**2011**) 2045–2056.

[68] Garzon, R., Liu, S., Fabbri, M., Liu, Z., Heaphy, C.E., Callegari, E. *et al.* Microrna-29b induces global DNA hypomethylation and tumor suppressor gene reexpression in acute myeloid leukemia by targeting directly dnmt3a and 3b and indirectly dnmt1. *Blood 113* (**2009**) 6411–6418.

[69] Fabbri, M., Garzon, R., Cimmino, A., Liu, Z., Zanesi, N., Callegari, E. *et al.* Microrna-29 family reverts aberrant methylation in lung cancer by targeting DNA methyltransferases 3a and 3b. *Proc. Natl. Acad. Sci. USA 104* (**2007**) 15805–15810.

[70] Majid, S., Dar, A.A., Saini, S., Shahryari, V., Arora, S., Zaman, M.S. *et al.* Mirna-34b inhibits prostate cancer through demethylation, active chromatin modifications, and akt pathways. *Clin. Cancer Res. 19* (**2013**) 73–84.

[71] Cao, P., Deng, Z., Wan, M., Huang, W., Cramer, S.D., Xu, J. *et al.* Microrna-101 negatively regulates ezh2 and its expression is modulated by androgen receptor and hif-1alpha/hif-1beta. *Mol. Cancer 9* (**2010**) 108.

[72] Noonan, E.J., Place, R.F., Pookot, D., Basak, S., Whitson, J.M., Hirata, H. *et al.* Mir-449a targets hdac-1 and induces growth arrest in prostate cancer. *Oncogene 28* (**2009**) 1714–1724.

[73] Gong, A.Y., Eischeid, A.N., Xiao, J., Zhao, J., Chen, D., Wang, Z.Y. *et al.* Mir-17-5p targets the p300/cbp-associated factor and modulates androgen receptor transcriptional activity in cultured prostate cancer cells. *BMC Cancer 12* (**2012**) 492.

[74] Margueron, R., Reinberg, D. The polycomb complex prc2 and its mark in life. *Nature 469* (**2011**) 343–349.

[75] Bracken, A.P., Helin, K. Polycomb group proteins: Navigators of lineage pathways led astray in cancer. *Nat. Rev. Cancer 9* (**2009**) 773–784.

[76] Varambally, S., Cao, Q., Mani, R.S., Shankar, S., Wang, X., Ateeq, B. *et al.* Genomic loss of microrna-101 leads to overexpression of histone methyltransferase ezh2 in cancer. *Science 322* (**2008**) 1695–1699.

[77] Cao, Q., Mani, R.S., Ateeq, B., Dhanasekaran, S.M., Asangani, I.A., Prensner, J.R. *et al.* Coordinated regulation of polycomb group complexes through micrornas in cancer. *Cancer Cell 20* (**2011**) 187–199.

[78] Cohen, E.E., Zhu, H., Lingen, M.W., Martin, L.E., Kuo, W.L., Choi, E.A. *et al.* A feed-forward loop involving protein kinase calpha and micrornas regulates tumor cell cycle. *Cancer Res. 69* (**2009**) 65–74.

[79] Brosh, R., Shalgi, R., Liran, A., Landan, G., Korotayev, K., Nguyen, G.H. *et al.* P53-repressed mirnas are involved with e2f in a feed forward loop promoting proliferation. *Mol. Syst. Biol. 4* (**2008**) 229.

[80] Polioudakis, D., Bhinge, A.A., Killion, P.J., Lee, B.K., Abell, N.S., Iyer, V.R. A myc-microrna network promotes exit from quiescence by suppressing the interferon response and cell-cycle arrest genes. *Nucleic Acids Res. 41* (**2013**) 2239–2254.

[81] Hassan, M.Q., Maeda, Y., Taipaleenmaki, H., Zhang, W., Jafferji, M., Gordon, J.A. *et al.* Mir-218 directs a wnt signaling circuit to promote differentiation of osteoblasts and osteomimicry of metastatic cancer cells. *J. Biol. Chem. 287* (**2012**) 42084–42092.

[82] McInnes, N., Sadlon, T.J., Brown, C.Y., Pederson, S., Beyer, M., Schultze, J.L. *et al.* Foxp3 and foxp3-regulated micrornas suppress satb1 in breast cancer cells. *Oncogene 31* (**2012**) 1045–1054.

[83] Da Costa Martins, P.A., Salic, K., Gladka, M.M., Armand, A.S., Leptidis, S., el Azzouzi, H. *et al.* Microrna-199b targets the nuclear kinase dyrk1a in an auto-amplification loop promoting calcineurin/nfat signalling. *Nat. Cell Biol. 12* (**2010**) 1220–1227.

[84] Weichert, W., Roske, A., Gekeler, V., Beckers, T., Stephan, C., Jung, K. *et al.* Histone deacetylases 1, 2 and 3 are highly expressed in prostate cancer and hdac2 expression is associated with shorter psa relapse time after radical prostatectomy. *Br. J. Cancer 98* (**2008**) 604–610.

[85] Nagy, Z., Tora, L. Distinct gcn5/pcaf-containing complexes function as co-activators and are involved in transcription factor and global histone acetylation. *Oncogene 26* (**2007**) 5341–5357.

[86] Dudziec, E., Miah, S., Choudhry, H.M., Owen, H.C., Blizard, S., Glover, M. *et al.* Hypermethylation of cpg islands and shores around specific micrornas and mirtrons is associated with the phenotype and presence of bladder cancer. *Clin. Cancer Res. 17* (**2011**) 1287–1296.

[87] Gandellini, P., Folini, M., Longoni, N., Pennati, M., Binda, M., Colecchia, M. *et al.* Mir-205 exerts tumor-suppressive functions in human prostate through down-regulation of protein kinase cepsilon. *Cancer Res. 69* (**2009**) 2287–2295.

[88] Iorio, M.V., Croce, C.M. Micrornas in cancer: Small molecules with a huge impact. *J. Clin. Oncol. 27* (**2009**) 5848–5856.

[89] Bhatnagar, N., Li, X., Padi, S.K., Zhang, Q., Tang, M.S., Guo, B. Downregulation of mir-205 and mir-31 confers resistance to chemotherapy-induced apoptosis in prostate cancer cells. *Cell Death Dis. 1* (**2010**) e105.

[90] Vrba, L., Jensen, T.J., Garbe, J.C., Heimark, R.L., Cress, A.E., Dickinson, S. *et al.* Role for DNA methylation in the regulation of mir-200c and mir-141 expression in normal and cancer cells. *PLoS One 5* (**2010**) e8697.

[91] Hermeking, H. The mir-34 family in cancer and apoptosis. *Cell Death Differ. 17* (**2010**) 193–199.

[92] Lodygin, D., Tarasov, V., Epanchintsev, A., Berking, C., Knyazeva, T., Korner, H. *et al.* Inactivation of mir-34a by aberrant cpg methylation in multiple types of cancer. *Cell Cycle 7* (**2008**) 2591–2600.

[93] He, L., He, X., Lim, L.P., de Stanchina, E., Xuan, Z., Liang, Y. *et al.* A microrna component of the p53 tumour suppressor network. *Nature 447* (**2007**) 1130–11, 34.

[94] Chang, T.C., Wentzel, E.A., Kent, O.A., Ramachandran, K., Mullendore, M., Lee, K.H. *et al.* Transactivation of mir-34a by p53 broadly influences gene expression and promotes apoptosis. *Mol. Cell 26* (**2007**) 745–752.

[95] Liu, C., Kelnar, K., Liu, B., Chen, X., Calhoun-Davis, T., Li, H. *et al.* The microrna mir-34a inhibits prostate cancer stem cells and metastasis by directly repressing cd44. *Nat. Med. 17* (**2011**) 211–215.

[96] Rauhala, H.E., Jalava, S.E., Isotalo, J., Bracken, H., Lehmusvaara, S., Tammela, T.L. *et al.* Mir-193b is an epigenetically regulated putative tumor suppressor in prostate cancer. *Int. J. Cancer 127* (**2010**) 1363–1372.

[97] Sachdeva, M., Mo, Y.Y. Mir-145-mediated suppression of cell growth, invasion and metastasis. *Am. J. Transl. Res. 2* (**2010**) 170–180.

[98] Sachdeva, M., Zhu, S., Wu, F., Wu, H., Walia, V., Kumar, S. *et al.* P53 represses c-myc through induction of the tumor suppressor mir-145. *Proc. Natl. Acad. Sci. USA 106* (**2009**) 3207–3212.

[99] Sachdeva, M., Mo, Y.Y. Microrna-145 suppresses cell invasion and metastasis by directly targeting mucin 1. *Cancer Res. 70* (**2010**) 378–387.

[100] Zaman, M.S., Chen, Y., Deng, G., Shahryari, V., Suh, S.O., Saini, S. *et al.* The functional significance of microrna-145 in prostate cancer. *Br. J. Cancer 103* (**2010**) 256–264.

[101] Suh, S.O., Chen, Y., Zaman, M.S., Hirata, H., Yamamura, S., Shahryari, V. *et al.* Microrna-145 is regulated by DNA methylation and p53 gene mutation in prostate cancer. *Carcinogenesis 32* (**2011**) 772–778.

[102] Hulf, T., Sibbritt, T., Wiklund, E.D., Bert, S., Strbenac, D., Statham, A.L. *et al.* Discovery pipeline for epigenetically deregulated mirnas in cancer: Integration of primary mirna transcription. *BMC Genomics 12* (**2011**) 54.

[103] Iorio, M.V., Visone, R., di Leva, G., Donati, V., Petrocca, F., Casalini, P. *et al.* Microrna signatures in human ovarian cancer. *Cancer Res. 67* (**2007**) 8699–8707.

[104] Coolen, M.W., Stirzaker, C., Song, J.Z., Statham, A.L., Kassir, Z., Moreno, C.S. *et al.* Consolidation of the cancer genome into domains of repressive chromatin by long-range epigenetic silencing (lres) reduces transcriptional plasticity. *Nat. Cell Biol. 12* (**2010**) 235–246.

[105] Matsuno, S., Egawa, S., Fukuyama, S., Motoi, F., Sunamura, M., Isaji, S. *et al.* Pancreatic cancer registry in Japan: 20 years of experience, *Pancreas 28* (**2004**) 219–230.

[106] Sultana, A., Tudur, S.C., Cunningham, D., Starling, N., Neoptolemos, J.P., Ghaneh, P. Meta-analyses of chemotherapy for locally advanced and metastatic pancreatic cancer: Results of secondary end points analyses, *Br. J. Cancer 99* (**2008**) 6–13.

[107] Siegel, R., Naishadham, D., Jemal, A. Cancer statistics, 2013, *CA Cancer J. Clin. 63* (**2013**) 11–30.

[108] Ambros, V. The functions of animal microRNAs, *Nature 431* (**2004**) 350–355.

[109] Bartel, D.P. MicroRNAs: Genomics, biogenesis, mechanism, and function, *Cell 116* (**2004**) 281–297.

[110] Chen, C.Z., Li, L., Lodish, H.F., Bartel, D.P. MicroRNAs modulate hematopoietic lineage differentiation, *Science 303* (**2004**) 83–86.

[111] Chen, J.F., Mandel, E.M., Thomson, J.M., Wu, Q., Callis, T.E., Hammond, S.M. *et al.* The role of microRNA-1 and microRNA-133 in skeletal muscle proliferation and differentiation, *Nat. Genet. 38* (**2006**) 228–233.

[112] Hornstein, E., Mansfield, J.H., Yekta, S., Hu, J.K., Harfe, B.D., McManus, M.T. *et al.* The microRNA miR-196 acts upstream of Hoxb8 and Shh in limb development, *Nature 438* (**2005**) 671–674.

[113] Rachagani, S., Kumar, S., Batra, S.K. MicroRNA in pancreatic cancer: Pathological, diagnostic and therapeutic implications, *Cancer Lett. 292* (**2010**) 8–16.

[114] Sempere, L.F., Freemantle, S., Pitha-Rowe, I., Moss, E., Dmitrovsky, E., Ambros, V. Expression profiling of mammalian microRNAs uncovers a subset of brain expressed microRNAs with possible roles in murine and human neuronal differentiation, *Genome Biol. 5* (**2004**) R13.

[115] Szafranska, A.E., Davison, T.S., John, J., Cannon, T., Sipos, B., Maghnouj, A. *et al.* MicroRNA expression alterations are linked to tumorigenesis and nonneoplastic processes in pancreatic ductal adenocarcinoma, *Oncogene 26* (**2007**) 4442–4452.

[116] Volinia, S., Calin, G.A., Liu, C.G., Ambs, S., Cimmino, A., Petrocca, F. *et al.* A microRNA expression signature of human solid tumors defines cancer gene targets, *Proc. Natl. Acad. Sci. USA 103* (**2006**) 2257–2261.

[117] Momi, N., Kaur, S., Rachagani, S., Ganti, A.K., Batra S.K. Smoking and microRNA dysregulation: A cancerous combination, *Trends Mol. Med. 20* (**2014**) 36–47.

[118] Calin, G.A., Dumitru, C.D., Shimizu, M., Bichi, R., Zupo, S., Noch, E. *et al.* Frequent deletions and down-regulation of micro-RNA genes miR15 and miR16 at 13q14 in chronic lymphocytic leukemia, *Proc. Natl. Acad. Sci. USA 99* (**2002**) 15524–15529.

[119] Calin, G.A., Cimmino, A., Fabbri, M., Ferracin, M., Wojcik, S.E., Shimizu, M. *et al.* MiR-15a and miR-16-1 cluster functions in human leukemia, *Proc. Natl. Acad. Sci. USA. 105* (**2008**) 5166–5171.

[120] Deng, M., Tang, H., Zhou, Y., Zhou, M., Xiong, W., Zheng, Y. *et al.* miR-216b suppresses tumor growth and invasion by targeting KRAS in nasopharyngeal carcinoma, *J. Cell Sci. 124* (**2011**) 2997–3005.

[121] Diaz, R., Silva, J., Garcia, J.M., Lorenzo, Y., Garcia, V., Pena, C. *et al.* Deregulated expression of miR-106a

[122] Drakaki, A., Iliopoulos, D. MicroRNA-gene signaling pathways in pancreatic cancer, *Biomed. J. 36* (**2013**) 200–208.

[123] Hirata, H., Ueno, K., Shahryari, V., Tanaka, Y., Tabatabai, Z.L., Hinoda, Y. *et al.* Oncogenic miRNA-182-5p targets Smad4 and RECK in human bladder cancer, *PLoS ONE 7* (**2012**) e51056.

[124] Takamizawa, J., Konishi, H., Yanagisawa, K., Tomida, S., Osada, H., Endoh, H. *et al.* Reduced expression of the let-7 microRNAs in human lung cancers in association with shortened postoperative survival, *Cancer Res. 64* (**2004**) 3753–3756.

[125] Macha, M.A., Seshacharyulu, P., Krishn, S.R., Pai, P., Rachagani, S., Jain, M. *et al.* MicroRNAs (miRNA) as Biomarker(s) for Prognosis and Diagnosis of Gastrointestinal (GI) Cancers, *Curr. Pharm. Des.* **2014**.

[126] Radhakrishnan, P., Mohr, A.M., Grandgenett, P.M., Steele, M.M., Batra, S.K., Hollingsworth, M.A. MicroRNA-200c modulates the expression of MUC4 and MUC16 by directly targeting their coding sequences in human pancreatic cancer, *PLoS ONE 8* (**2013**) e73356.

[127] Calin, G.A., Croce, C.M. MicroRNA signatures in human cancers, *Nat. Rev. Cancer 6* (**2006**) 857–866.

[128] Bentwich, I., Avniel, A., Karov, Y., Aharonov, R., Gilad, S., Barad, O. *et al.* Identification of hundreds of conserved and nonconserved human microRNAs, *Nat. Genet. 37* (**2005**) 766–770.

[129] Carthew, R.W. Gene regulation by microRNAs, *Curr. Opin. Genet. Dev. 16* (**2006**) 203–208.

[130] Calin, G.A., Sevignani, C., Dumitru, C.D., Hyslop, T., Noch, E., Yendamuri, S. *et al.* Human microRNA genes are frequently located at fragile sites and genomic regions involved in cancers, *Proc. Natl. Acad. Sci. USA 101* (**2004**) 2999–3004.

[131] Lu, J., Getz, G., Miska, E.A., Alvarez-Saavedra, E., Lamb, J., Peck, D. *et al.* MicroRNA expression profiles classify human cancers, *Nature 435* (**2005**) 834–838.

[132] Babak, T., Zhang, W., Morris, Q., Blencowe, B.J., Hughes, T.R. Probing microRNAs with microarrays: Tissue specificity and functional inference, *RNA 10* (**2004**) 1813–1819.

[133] Sood, P., Krek, A., Zavolan, M., Macino, G., Rajewsky, N. Cell-type-specific signatures of microRNAs on target mRNA expression, *Proc. Natl. Acad. Sci. USA 103* (**2006**) 2746–2751.

[134] Yanaihara, N., Caplen, N., Bowman, E., Seike, M., Kumamoto, K., Yi, M. *et al.* Unique microRNA molecular profiles in lung cancer diagnosis and prognosis, *Cancer Cell 9* (**2006**) 189–198.

[135] Hezel, A.F., Kimmelman, A.C., Stanger, B.Z., Bardeesy, N., DePinho, R.A. Genetics and biology of pancreatic ductal adenocarcinoma, *Genes Dev. 20* (**2006**) 1218–1249.

[136] Poy, M.N., Eliasson, L., Krutzfeldt, J., Kuwajima, S., Ma, X., Macdonald, P.E. *et al.* A pancreatic islet-specific microRNA regulates insulin secretion, *Nature 432* (**2004**) 226–230.

[137] Bloomston, M., Frankel, W.L., Petrocca, F., Volinia, S., Alder, H., Hagan, J.P. *et al.* MicroRNA expression patterns to differentiate pancreatic adenocarcinoma from normal pancreas and chronic pancreatitis, *JAMA 297* (**2007**) 1901–1908.

[138] Lee, E.J., Gusev, Y., Jiang, J., Nuovo, G.J., Lerner, M.R., Frankel, W.L. *et al.* Expression profiling identifies microRNA signature in pancreatic cancer, *Int. J. Cancer 120* (**2007**) 1046–1054.

predicts survival in human colon cancer patients, *Genes Chromosomes Cancer 47* (**2008**) 794–802.

[139] Fang, Y., Yao, Q., Chen, Z., Xiang, J., William, F.E., Gibbs, R.A. *et al.* Genetic and molecular alterations in pancreatic cancer: Implications for personalized medicine, *Med. Sci. Monit. 19*:916–26.

[140] Khan, S., Ansarullah, Kumar, D., Jaggi, M., Chauhan, S.C. Targeting microRNAs in pancreatic cancer: Microplayers in the big game, *Cancer Res. 73* (**2013**) 6541–6547.

[141] Yu, J., Wang, F. Recent progress in microRNA study: Benefits from technique advance, *Sci. China Life Sci. 55* (**2012**) 649–650.

[142] Jiang, J., Lee, E.J., Gusev, Y., Schmittgen, T.D. Real-time expression profiling of microRNA precursors in human cancer cell lines, *Nucleic Acids Res. 33* (**2005**) 5394–5403.

[143] Roldo, C., Missiaglia, E., Hagan, J.P., Falconi, M., Capelli, P., Bersani, S. *et al.* MicroRNA expression abnormalities in pancreatic endocrine and acinar tumors are associated with distinctive pathologic features and clinical behavior, *J. Clin. Oncol. 24* (**2006**) 4677–4684.

[144] Srivastava, S.K., Bhardwaj, A., Singh, S., Arora, S., Wang, B., Grizzle, W.E. *et al.* MicroRNA-150 directly targets MUC4 and suppresses growth and malignant behavior of pancreatic cancer cells, *Carcinogenesis 32* (**2011**) 1832–1839.

[145] Ohuchida, K., Mizumoto, K., Lin, C., Yamaguchi, H., Ohtsuka, T., Sato, N. *et al.* MicroRNA-10a is overexpressed in human pancreatic cancer and involved in its invasiveness partially via suppression of the HOXA1 gene, *Ann. Surg. Oncol. 19* (**2012**) 2394–2402.

[146] Munding, J.B., Adai, A.T., Maghnouj, A., Urbanik, A., Zollner, H., Liffers, S.T. *et al.* Global microRNA expression profiling of microdissected tissues identifies miR-135b as a novel biomarker for pancreatic ductal adenocarcinoma, *Int. J. Cancer 131* (**2012**) E86–E95.

[147] Papaconstantinou, I.G., Manta, A., Gazouli, M., Lyberopoulou, A., Lykoudis, P.M., Polymeneas, G. *et al.* Expression of microRNAs in patients with pancreatic cancer and its prognostic significance, *Pancreas 42* (**2013**) 67–71.

[148] Schultz, N.A., Andersen, K.K., Roslind, A., Willenbrock, H., Wojdemann, M., Johansen, J.S. Prognostic microRNAs in cancer tissue from patients operated for pancreatic cancer—five microRNAs in a prognostic index, *World J. Surg. 36* (**2012**) 2699–2707.

[149] Ikenaga, N., Ohuchida, K., Mizumoto, K., Yu, J., Kayashima, T., Sakai, H. *et al.* MicroRNA-203 expression as a new prognostic marker of pancreatic adenocarcinoma, *Ann. Surg. Oncol. 17* (**2010**) 3120–3128.

[150] Yu, J., Li, A., Hong, S.M., Hruban, R.H., Goggins, M. MicroRNA alterations of pancreatic intraepithelial neoplasias, *Clin. Cancer Res. 18* (**2012**) 981–992.

[151] Zhang, Y., Li, M., Wang, H., Fisher, W.E., Lin, P.H., Yao, Q. *et al.* Profiling of 95 microRNAs in pancreatic cancer cell lines and surgical specimens by real-time PCR analysis, *World J. Surg. 33* (**2009**) 698–709.

[152] Park, J.K., Henry, J.C., Jiang, J., Esau, C., Gusev, Y., Lerner, M.R. *et al.* miR-132 and miR-212 are increased in pancreatic cancer and target the retinoblastoma tumor suppressor, *Biochem. Biophys. Res. Commun. 406* (**2011**) 518–523.

[153] Zhang, S., Hao, J., Xie, F., Hu, X., Liu, C., Tong, J. *et al.* Downregulation of miR-132 by promoter methylation contributes to pancreatic cancer development, *Carcinogenesis 32* (**2011**) 1183–1189.

[154] Morimura, R., Komatsu, S., Ichikawa, D., Takeshita, H., Tsujiura, M., Nagata, H. *et al.* Novel diagnostic value of circulating miR-18a in plasma of patients with pancreatic cancer, *Br. J. Cancer 105* (**2011**) 1733–1740.

[155] Ali, S., Almhanna, K., Chen, W., Philip, P.A., Sarkar, F.H. Differentially expressed miRNAs in the plasma may provide a molecular signature for aggressive pancreatic cancer, *Am. J. Transl. Res. 3* (**2010**) 28–47.

[156] Ali, S., Saleh, H., Sethi, S., Sarkar, F.H., Philip, P.A. MicroRNA profiling of diagnostic needle aspirates from patients with pancreatic cancer, *Br. J. Cancer 107* (**2012**) 1354–1360.

[157] di Magliano, M.P., Logsdon, C.D. Roles for KRAS in pancreatic tumor development and progression, *Gastroenterology 144* (**2013**) 1220–1229.

[158] Spaargaren, M., Bischoff, J.R., McCormick, F. Signal transduction by Ras-like GTPases: A potential target for anticancer drugs, *Gene Expr. 4* (**1995**) 345–356.

[159] Yu, S., Lu, Z., Liu, C., Meng, Y., Ma, Y., Zhao, W. *et al.* miRNA-96 suppresses KRAS and functions as a tumor suppressor gene in pancreatic cancer, *Cancer Res. 70* (**2010**) 6015–6025.

[160] Jiao, L.R., Frampton, A.E., Jacob, J., Pellegrino, L., Krell, J., Giamas, G. *et al.* MicroRNAs targeting oncogenes are down-regulated in pancreatic malignant transformation from benign tumors, *PLoS ONE 7* (**2012**) e32068.

[161] Kent, O.A., Chivukula, R.R., Mullendore, M., Wentzel, E.A., Feldmann, G., Lee, K.H. *et al.* Repression of the miR-143/145 cluster by oncogenic Ras initiates a tumor-promoting feed-forward pathway, *Genes Dev. 24* (**2010**) 2754–2759.

[162] Zhao, W.G., Yu, S.N., Lu, Z.H., Ma, Y.H., Gu, Y.M., Chen, J. The miR-217 microRNA functions as a potential tumor suppressor in pancreatic ductal adenocarcinoma by targeting KRAS, *Carcinogenesis 31* (**2010**) 1726–1733.

[163] Johnson, S.M., Grosshans, H., Shingara, J., Byrom, M., Jarvis, R., Cheng, A. *et al.* RAS is regulated by the let-7 microRNA family, *Cell 120* (**2005**) 635–647.

[164] Torrisani, J., Bournet, B., du Rieu, M.C., Bouisson, M., Souque, A., Escourrou, J. *et al.* Let-7 microRNA transfer in pancreatic cancer-derived cells inhibits in vitro cell proliferation but fails to alter tumor progression, *Hum. Gene Ther. 20* (**2009**) 831–844.

[165] Kopp, F., Wagner, E., Roidl, A. The proto-oncogene KRAS is targeted by miR-200c, *Oncotarget 5* (**2014**) 185–195.

[166] Vousden, K.H., Prives, C. Blinded by the light: The growing complexity of p53, *Cell 137* (**2009**) 413–431.

[167] Levine, A.J., Oren, M. The first 30 years of p53: Growing ever more complex, *Nat. Rev. Cancer 9* (**2009**) 749–758.

[168] Muller, P.A., Vousden, K.H. p53 mutations in cancer, *Nat. Cell Biol. 15* (**2013**) 2–8.

[169] Morton, J.P., Timpson, P., Karim, S.A., Ridgway, R.A., Athineos, D., Doyle, B. *et al.* Mutant p53 drives metastasis and overcomes growth arrest/senescence in pancreatic cancer, *Proc. Natl. Acad. Sci. USA 107* (**2010**) 246–251.

[170] Redston, M.S., Caldas, C., Seymour, A.B., Hruban, R.H., Da, C.L., Yeo, C.J. *et al.* p53 mutations in pancreatic carcinoma and evidence of common involvement of homocopolymer tracts in DNA microdeletions, *Cancer Res. 54* (**1994**) 3025–3033.

[171] Cho, Y., Gorina, S., Jeffrey, P.D., Pavletich, N.P. Crystal structure of a p53 tumor suppressor–DNA complex: Understanding tumorigenic mutations, *Science 265* (**1994**) 346–355.

[172] Sigal, A., Rotter, V. Oncogenic mutations of the p53 tumor suppressor: The demons of the guardian of the genome, *Cancer Res. 60* (**2000**) 6788–6793.

[173] Thukral, S.K., Lu, Y., Blain, G.C., Harvey, T.S., Jacobsen, V.L. Discrimination of DNA binding sites by mutant p53 proteins, *Mol. Cell. Biol. 15* (**1995**) 5196–5202.

[174] Sampath, J., Sun, D., Kidd, V.J., Grenet, J., Gandhi, A., Shapiro, L.H. *et al.* Mutant p53 cooperates with ETS and selectively upregulates human MDR1 not MRP1, *J. Biol. Chem. 276* (**2001**) 39359–39367.

[175] Stambolsky, P., Tabach, Y., Fontemaggi, G., Weisz, L., Maor-Aloni, R., Siegfried, Z. *et al.* Modulation of the vitamin D3 response by cancer-associated mutant p53, *Cancer Cell 17* (**2010**) 273–285.

[176] Strano, S., Dell'Orso, S., Di, A.S., Fontemaggi, G., Sacchi, A., Blandino, G. Mutant p53: An oncogenic transcription factor, *Oncogene 26* (**2007**) 2212–2219.

[177] Weisz, L., Oren, M., Rotter, V. Transcription regulation by mutant p53, *Oncogene 26* (**2007**) 2202–2211.

[178] Dittmer, D., Pati, S., Zambetti, G., Chu, S., Teresky, A.K., Moore, M. *et al.* Gain of function mutations in p53, *Nat. Genet. 4* (**1993**) 42–46.

[179] Lang, G.A., Iwakuma, T., Suh, Y.A., Liu, G., Rao, V.A., Parant, J.M. *et al.* Gain of function of a p53 hot spot mutation in a mouse model of Li–Fraumeni syndrome, *Cell 119* (**2004**) 861–872.

[180] Hanahan, D., Weinberg, R.A. The hallmarks of cancer, *Cell 100* (**2000**) 57–70.

[181] Hingorani, S.R., Wang, L., Multani, A.S., Combs, C., Deramaudt, T.B., Hruban, R.H. *et al.* Trp53R172H and KrasG12D cooperate to promote chromosomal instability and widely metastatic pancreatic ductal adenocarcinoma in mice, *Cancer Cell 7* (**2005**) 469–483.

[182] Xi, Y., Shalgi, R., Fodstad, O., Pilpel, Y., Ju, J. Differentially regulated micro-RNAs and actively translated messenger RNA transcripts by tumor suppressor p53 in colon cancer, *Clin. Cancer Res. 12* (**2006**) 2014–2024.

[183] Hunten, S., Siemens, H., Kaller, M., Hermeking, H. The p53/microRNA network in cancer: Experimental and bioinformatics approaches, *Adv. Exp. Med. Biol. 774* (**2013**) 77–101.

[184] Lin, C.P., Choi, Y.J., Hicks, G.G., He, L. The emerging functions of the p53–miRNA network in stem cell biology, *Cell Cycle 11* (**2012**) 2063–2072.

[185] Hermeking, H. The miR-34 family in cancer and apoptosis, *Cell Death Differ. 17* (**2010**) 193–199.

[186] Bommer, G.T., Gerin, I., Feng, Y., Kaczorowski, A.J., Kuick, R., Love, R.E. *et al.* p53- mediated activation of miRNA34 candidate tumor-suppressor genes, *Curr. Biol. 17* (**2007**) 1298–1307.

[187] Corney, D.C., Flesken-Nikitin, A., Godwin, A.K., Wang, W., Nikitin, A.Y. MicroRNA-34b and microRNA-34c are targets of p53 and cooperate in control of cell proliferation and adhesion-independent growth, *Cancer Res. 67* (**2007**) 8433–8438.

[188] He, L., He, X., Lim, L.P., de Stachina, E., Xuan, Z., Liang, Y. *et al.* A microRNA component of the p53 tumour suppressor network, *Nature 447* (**2007**) 1130–1134.

[189] Hermeking, H. p53 enters the microRNA world, *Cancer Cell 12* (**2007**) 414–418.

[190] Raver-Shapira, N., Marciano, E., Meiri, E., Spector, Y., Rosenfeld, N., Moskovits, N. *et al.* Transcriptional activation of miR-34a contributes to p53- mediated apoptosis, *Mol. Cell 26* (**2007**) 731–743.

[191] Tarasov, V., Jung, P., Verdoodt, B., Lodygin, D., Epanchintsev, A., Menssen, A. *et al.* Differential regulation of microRNAs by p53 revealed by massively parallel sequencing: miR-34a is a p53 target that induces apoptosis and G1-arrest, *Cell Cycle 6* (**2007**) 1586–1593.

[192] Welch, C., Chen, Y., Stallings, R.L. MicroRNA-34a functions as a potential tumor suppressor by inducing apoptosis in neuroblastoma cells, *Oncogene 26* (**2007**) 5017–5022.

[193] Lodygin, D., Tarasov, V., Epanchintsev, A., Berking, C., Knyazeva, T., Korner, H. *et al.* Inactivation of miR-34a by aberrant CpG methylation in multiple types of cancer, *Cell Cycle 7* (**2008**) 2591–2600.

[194] Vogt, M., Munding, J., Gruner, M., Liffers, S.T., Verdoodt, B., Hauk, J. *et al.* Frequent concomitant inactivation of miR-34a and miR- 34b/c by CpG methylation in colorectal, pancreatic, mammary, ovarian, urothelial, and renal cell carcinomas and soft tissue sarcomas, *Virchows Arch. 458* (**2011**) 313–322.

[195] Chang, T.C., Wentzel, E.A., Kent, O.A., Ramachandran, K., Mullendore, M., Lee, K.H. *et al.* Transactivation of miR-34a by p53 broadly influences gene expression and promotes apoptosis, *Mol. Cell 26* (**2007**) 745–752.

[196] Dong, P., Karaayvaz, M., Jia, N., Kaneuchi, M., Hamada, J., Watari, H. *et al.* Mutant p53 gain-of-function induces epithelial–mesenchymal transition through modulation of the miR-130b-ZEB1 axis, *Oncogene 32* (**2013**) 3286–3295.

[197] Neilsen, P.M., Noll, J.E., Mattiske, S., Bracken, C.P., Gregory, P.A., Schulz, R.B. *et al.* Mutant p53 drives invasion in breast tumors through up-regulation of miR-155, *Oncogene 32* (**2013**) 2992–3000.

[198] Tucci, P., Agostini, M., Grespi, F., Markert, E.K., Terrinoni, A., Vousden, K.H. *et al.* Loss of p63 and its microRNA-205 target results in enhanced cell migration and metastasis in prostate cancer, *Proc. Natl. Acad. Sci. USA 109* (**2012**) 15312–15317.

[199] Zhang, X.J., Ye, H., Zeng, C.W., He, B., Zhang, H., Chen, Y.Q. Dysregulation of miR-15a and miR-214 in human pancreatic cancer, *J. Hematol. Oncol. 3* (**2010**) 46.

[200] Sachdeva, M., Zhu, S., Wu, F., Wu, H., Walia, V., Kumar, S. *et al.* p53 represses c-Myc through induction of the tumor suppressor miR-145, *Proc. Natl. Acad. Sci. USA 106* (**2009**) 3207–3212.

[201] Chen, P.S., Su, J.L., Cha, S.T., Tarn, W.Y., Wang, M.Y., Hsu, H.C. *et al.* miR-107 promotes tumor progression by targeting the let-7 microRNA in mice and humans, *J. Clin. Invest. 121* (**2011**) 3442–3455.

[202] Gregory, P.A., Bert, A.G., Paterson, E.L., Barry, S.C., Tsykin, A., Farshid, G. *et al.* The miR-200 family and-miR-205 regulate epithelial to mesenchymal transition by targeting ZEB1 and SIP1, *Nat. Cell Biol. 10* (**2008**) 593–601.

[203] Kim, T., Veronese, A., Pichiorri, F., Lee, T.J., Jeon, Y.J., Volinia, S. *et al.* p53 regulates epithelial– mesenchymal transition through microRNAs targeting ZEB1 and ZEB2, *J. Exp. Med. 208* (**2011**) 875–883.

[204] Soubani, O., Ali, A.S., Logna, F., Ali, S., Philip, P.A., Sarkar, F.H. Re-expression of miR-200 by novel approaches regulates the expression of PTEN and MT1-MMP in pancreatic cancer, *Carcinogenesis 33* (**2012**) 1563–1571.

[205] Brabletz, S., Bajdak, K., Meidhof, S., Burk, U., Niedermann, G., First, E. *et al.* The ZEB1/miR-200 feedback loop controls Notch signalling in cancer cells, *EMBO J. 30* (**2011**) 770–782.

[206] Braun, C.J., Zhang, X., Savelyeva, I., Wolff, S., Moll, U.M., Schepeler, T. *et al.* p53-Responsive micrornas 192 and 215 are capable of inducing cell cycle arrest, *Cancer Res. 68* (**2008**) 10094–10104.

[207] Zhao, C., Zhang, J., Zhang, S., Yu, D., Chen, Y., Liu, Q. *et al.* Diagnostic and biological significance of microRNA-192 in pancreatic ductal adenocarcinoma, *Oncol. Rep. 30* (**2013**) 276–284.

[208] Suzuki, H.I., Yamagata, K., Sugimoto, K., Iwamoto, T., Kato, S., Miyazono, K. Modulation of microRNA processing by p53, *Nature 460* (**2009**) 529–533.

[209] Wang, Y.A., Kamarova, Y., Shen, K.C., Jiang, Z., Hahn, M.J., Wang, Y. *et al.* DNA methyltransferase-3a interacts with p53 and represses p53-mediated gene expression, *Cancer Biol. Ther. 4* (**2005**) 1138–1143.

[210] Guo, R., Wang, Y., Shi, W.Y., Liu, B., Hou, S.Q., Liu, L. MicroRNA miR-491-5p targeting both TP53 and Bcl-XL induces cell apoptosis in SW1990 pancreatic cancer cells through mitochondria mediated pathway, *Molecules 17* (**2012**) 14733–14747.

[211] Chim, C.S., Wong, K.Y., Qi, Y., Loong, F., Lam, W.L., Wong, L.G. *et al.* Epigenetic inactivation of the miR-34a in hematological malignancies, *Carcinogenesis 31* (**2010**) 745–750.

[212] Brooks, C.L., Gu, W. How does SIRT1 affect metabolism, senescence and cancer? *Nat. Rev. Cancer 9* (**2009**) 123–128.

[213] Yamakuchi, M., Ferlito, M., Lowenstein, C.J. miR-34a repression of SIRT1 regulates apoptosis, *Proc. Natl. Acad. Sci. USA 105* (**2008**) 13421–13426.

[214] Okada, N., Lin, C.P., Ribeiro, M.C., Biton, A., Lai, G., He, X. *et al.* A positive feedback between p53 and miR-34 miRNAs mediates tumor suppression, *Genes Dev. 28* (**2014**) 438–450.

[215] Penna, E., Orso, F., Cimino, D., Tenaglia, E., Lembo, A., Quaglino, E. *et al.* MicroRNA-214 contributes to melanoma tumour progression through suppression of TFAP2C, *EMBO J. 30* (**2011**) 1990–2007.

[216] Yang, H., Kong, W., He, L., Zhao, J.J., O'Donnell, J.D., Wang, J. *et al.* MicroRNA expression profiling in human ovarian cancer: miR-214 induces cell survival and cisplatin resistance by targeting PTEN, *Cancer Res. 68* (**2008**) 425–433.

[217] Xu, C.X., Xu, M., Tan, L., Yang, H., Permuth-Wey, J., Kruk, P.A. *et al.* MicroRNA miR-214 regulates ovarian cancer cell stemness by targeting p53/Nanog, *J. Biol. Chem. 287* (**2012**) 34970–34978.

[218] Minh T.N. Le, Cathleen Teh, Ng Shyh-Chang, Huangming Xie, Beiyan Zhou, Vladimir Korzh, Harvey F. Lodish, and Bing Lim. MicroRNA-125b is a novel negative regulator of p53, Genes Dev. 23(7) (2009) 862–876.

[219] Reif, K., Gout, I., Waterfield, M.D., Cantrell, D.A. Divergent regulation of phosphatidylinositol 3-kinase P85 alpha and P85 beta isoforms upon T cell activation, *J. Biol. Chem. 268* (**1993**) 10780–10788.

[220] Park, S.Y., Lee, J.H., Ha, M., Nam, J.W., Kim, V.N. miR-29 miRNAs activate p53 by targeting p85 alpha and CDC42, *Nat. Struct. Mol. Biol. 16* (**2009**) 23–29.

[221] Hu, W., Chan, C.S., Wu, R., Zhang, C., Sun, Y., Song, J.S. *et al.* Negative regulation of tumor suppressor p53 by microRNA miR-504, *Mol. Cell 38* (**2010**) 689–699.

[222] Fornari, F., Gramantieri, L., Giovannini, C., Veronese, A., Ferracin, M., Sabbioni, S. *et al.* MiR-122/cyclin G1 interaction modulates p53 activity and affects doxorubicin sensitivity of human hepatocarcinoma cells, *Cancer Res. 69* (**2009**) 5761–5767.

[223] Herrera-Merchan, A., Cerrato, C., Luengo, G., Dominguez, O., Piris, M.A., Serrano, M. *et al.* miR-33-mediated downregulation of p53 controls hematopoietic stem cell self-renewal, *Cell Cycle 9* (**2010**) 3277–3285.

[224] Janakiram, N.B., Mohammed, A., Qian, L., Steele, V.E., Rao, C.V. Cancer miRNA profiling and target identification in PanINs and pancreatic cancer in LSLKras$^{G12D/+}$ mice: Implication for human pancreatic cancer chemoprevention and treatment, **2014**.

[225] Pichiorri, F., Suh, S.S., Rocci, A., De, L.L., Taccioli, C., Santhanam, R. *et al.* Downregulation of p53-inducible microRNAs 192, 194, and 215 impairs the p53/MDM2 autoregulatory loop in multiple myeloma development, *Cancer Cell 18* (**2010**) 367–381.

[226] Xiao, J., Lin, H., Luo, X., Luo, X., Wang, Z. miR-605 joins p53 network to form a p53: miR-605:Mdm2 positive feedback loop in response to stress, *EMBO J. 30* (**2011**) 524–532.

[227] Mohr, A.M., Bailey, J.M., Lewallen, M.E., Liu, X., Radhakrishnan, P., Yu, F. *et al.* MUC1 regulates expression of multiple microRNAs involved in pancreatic tumor progression, including the miR-200c/141 cluster, *PLoS ONE 8* (**2013**) e73306.

[228] Hara, E., Smith, R., Parry, D., Tahara, H., Stone, S., Peters, G. Regulation of p16CDKN2 expression and its implications for cell immortalization and senescence, *Mol. Cell. Biol. 16* (**1996**) 859–867.

[229] Serrano, M., Lee, H., Chin, L., Cordon-Cardo, C., Beach, D., DePinho, R.A. Role of the INK4a locus in tumor suppression and cell mortality, *Cell 85* (**1996**) 27–37.

[230] Okamoto, A., Demetrick, D.J., Spillare, E.A., Hagiwara, K., Hussain, S.P., Bennett, W.P. *et al.* Mutations and altered expression of p16INK4 in human cancer, *Proc. Natl. Acad. Sci. USA 91* (**1994**) 11045–11049.

[231] Zhang, L., Jamaluddin, M.S., Weakley, S.M., Yao, Q., Chen, C. Roles and mechanisms of microRNAs in pancreatic cancer, *World J. Surg. 35* (**2011**) 1725–1731.

[232] Lal, A., Kim, H.H., Abdelmohsen, K., Kuwano, Y., Pullmann Jr., R., Srikantan, S. et al. p16(INK4a) translation suppressed by miR-24, *PLoS ONE 3* (**2008**) e1864.

[233] Nakata, K., Ohuchida, K., Mizumoto, K., Kayashima, T., Ikenaga, N., Sakai, H. *et al.* MicroRNA-10b is overexpressed in pancreatic cancer, promotes its invasiveness, and correlates with a poor prognosis, *Surgery 150* (**2011**) 916–922.

[234] Gabriely, G., Yi, M., Narayan, R.S., Niers, J.M., Wurdinger, T., Imitola, J. *et al.* Human glioma growth is controlled by microRNA-10b, *Cancer Res. 71* (**2011**) 3563–3572.

[235] Venkataraman, S., Alimova, I., Fan, R., Harris, P., Foreman, N., Vibhakar, R. MicroRNA 128a increases intracellular ROS level by targeting Bmi-1 and inhibits medulloblastoma cancer cell growth by promoting senescence, *PLoS ONE 5* (**2010**) e10748.

[236] Poliseno, L., Pitto, L., Simili, M., Mariani, L., Riccardi, L., Ciucci, A. *et al.* The proto-oncogene LRF is under posttranscriptional control of MiR-20a: Implications for senescence, *PLoS ONE 3* (**2008**) e2542.

[237] Braconi, C., Huang, N., Patel, T. MicroRNA-dependent regulation of DNA methyltransferase-1 and tumor suppressor gene expression by interleukin-6 in human malignant cholangiocytes, *Hepatology 51* (**2010**) 881–890.

[238] Chien, W.W., Domenech, C., Catallo, R., Kaddar, T., Magaud, J.P., Salles, G. *et al.* Cyclin-dependent kinase 1 expression is inhibited by p16(INK4a) at the posttranscriptional level through the microRNA pathway, *Oncogene 30* (**2011**) 1880–1891.

[239] Al-Khalaf, H.H., Mohideen, P., Nallar, S.C., Kalvakolanu, D.V., Aboussekhra, A. The cyclin-dependent kinase

inhibitor p16INK4a physically interacts with transcription factor Sp1 and cyclin-dependent kinase 4 to transactivate microRNA-141 and microRNA-146b-5p spontaneously and in response to ultraviolet light induced DNA damage, *J. Biol. Chem. 288* (**2013**) 35511–35525.

[240] Whitman, M. Smads and early developmental signaling by the TGFbeta superfamily, *Genes Dev. 12* (**1998**) 2445–2462.

[241] Massague, J., Seoane, J., Wotton, D. Smad transcription factors, *Genes Dev. 19* (**2005**) 2783–2810.

[242] Hayashida, T., Decaestecker, M., Schnaper, H.W. Crosstalk between ERKMAP kinase and Smad signaling pathways enhances TGF-beta-dependent responses in human mesangial cells, *FASEB J. 17* (**2003**) 1576–1578.

[243] Adhikary, S., Eilers, M. Transcriptional regulation and transformation by Myc proteins, *Nat. Rev. Mol. Cell Biol. 6* (**2005**) 635–645.

[244] Donovan, J., Slingerland, J. Transforming growth factor-beta and breast cancer: Cell cycle arrest by transforming growth factor-beta and its disruption in cancer, *Breast Cancer Res. 2* (**2000**) 116–124.

[245] Nicolas, F.J., Hill, C.S. Attenuation of the TGF-beta-Smad signaling pathway in pancreatic tumor cells confers resistance to TGF-beta-induced growth arrest, *Oncogene 22* (**2003**) 3698–3711.

[246] Blahna, M.T., Hata, A. Smad-mediated regulation of microRNA biosynthesis, *FEBS Lett. 586* (**2012**) 1906–1912.

[247] Chung, A.C., Huang, X.R., Meng, X., Lan, H.Y. miR-192 mediates TGF-beta/Smad3- driven renal fibrosis, *J. Am. Soc. Nephrol. 21* (**2010**) 1317–1325.

[248] Davis, B.N., Hilyard, A.C., Lagna, G., Hata, A. SMAD proteins control DROSHA mediated microRNA maturation, *Nature 454* (**2008**) 56–61.

[249] Kong, W., Yang, H., He, L., Zhao, J.J., Coppola, D., Dalton, W.S. *et al.* MicroRNA- 155 is regulated by the transforming growth factor beta/Smad pathway and contributes to epithelial cell plasticity by targeting RhoA, *Mol. Cell. Biol. 28* (**2008**) 6773–6784.

[250] Naito, Y., Sakamoto, N., Oue, N., Yashiro, M., Sentani, K., Yanagihara, K. *et al.* MicroRNA-143 regulates collagen type III expression in stromal fibroblasts of scirrhous type gastric cancer, *Cancer Sci. 105* (**2014**) 228–235.

[251] Zhang, Y., Fan, K.J., Sun, Q., Chen, A.Z., Shen, W.L., Zhao, Z.H. *et al.* Functional screening for miRNAs targeting Smad4 identified miR-199a as a negative regulator of TGF-beta signalling pathway, *Nucleic Acids Res. 40* (**2012**) 9286–9297.

[252] Liu, L., Nie, J., Chen, L., Dong, G., Du, X., Wu, X. *et al.* The oncogenic role of microRNA-130a/301a/454 in human colorectal cancer via targeting Smad4 expression, *PLoS ONE 8* (**2013**) e55532.

[253] Chen, Z., Chen, L.Y., Dai, H.Y., Wang, P., Gao, S., Wang, K. miR-301a promotes pancreatic cancer cell proliferation by directly inhibiting Bim expression, *J. Cell. Biochem. 113* (**2012**) 3229–3235.

[254] Lu, Z., Li, Y., Takwi, A., Li, B., Zhang, J., Conklin, D.J. *et al.* miR- 301a as an NF-kappaB activator in pancreatic cancer cells, *EMBO J. 30* (**2011**) 57–67.

[255] Hruban, R.H., Offerhaus, G.J., Kern, S.E., Goggins, M., Wilentz, R.E., Yeo, C.J. Tumor suppressor genes in pancreatic cancer, *J. Hepatobiliary. Pancreat. Surg. 5* (**1998**) 383–391.

[256] Singh, P., Srinivasan, R., Wig, J.D. Major molecular markers in pancreatic ductal adenocarcinoma and their roles in screening, diagnosis, prognosis, and treatment, *Pancreas 40* (**2011**) 644–652.

[257] Singh, P., Srinivasan, R., Wig, J.D., Radotra, B.D. A study of Smad4, Smad6 and Smad7 in surgically resected samples of pancreatic ductal adenocarcinoma and their correlation with clinicopathological parameters and patient survival, *BMC. Res. Notes 4* (**2011**) 560–564.

[258] Gaspar, N.J., Li, L., Kapoun, A.M., Medicherla, S., Reddy, M., Li, G. *et al.* Inhibition of transforming growth factor beta signaling reduces pancreatic adenocarcinoma growth and invasiveness, *Mol. Pharmacol. 72* (**2007**) 152–161.

[259] Smith, A.L., Robin, T.P., Ford, H.L. Molecular pathways: Targeting the TGF-beta pathway for cancer therapy, *Clin. Cancer Res. 18* (**2012**) 4514–4521.

[260] Yao, G., Yin, M., Lian, J., Tian, H., Liu, L., Li, X. *et al.* MicroRNA-224 is involved in transforming factor-beta-mediated mouse granulosa cell proliferation and granulosa cell function by targeting Smad4, *Mol. Endocrinol. 24* (**2010**) 540–551.

[261] Du, J., Wang, J., Tan, G., Cai, Z., Zhang, L., Tang, B. *et al.* Aberrant elevated microRNA-146a in dendritic cells (DC) induced by human pancreatic cancer cell line BxPC-3-conditioned medium inhibits DC maturation and activation, *Med. Oncol. 29* (**2012**) 2814–2823.

[262] Hao, J., Zhang, S., Zhou, Y., Hu, X., Shao, C. MicroRNA 483-3p suppresses the expression of DPC4/Smad4 in pancreatic cancer, *FEBS Lett. 585* (**2011**) 207–213.

[263] Hao, J., Zhang, S., Zhou, Y., Liu, C., Hu, X., Shao, C. MicroRNA 421 suppresses DPC4/Smad4 in pancreatic cancer, *Biochem. Biophys. Res. Commun. 406* (**2011**) 552–557.

[264] Li, L., Li, Z., Kong, X., Xie, D., Jia, Z., Jiang, W. *et al.* Downregulation of microRNA-494 via loss of SMAD4 increases FOXM1 and beta-catenin signaling in pancreatic ductal adenocarcinoma cells, *Gastroenterology 147* (**2014**) 485–497.

[265] Geraldo, M.V., Yamashita, A.S., Kimura, E.T. MicroRNAmiR-146b-5p regulates signal transduction of TGF-beta by repressing SMAD4 in thyroid cancer, *Oncogene 31* (**2012**) 1910–1922.

[266] Hui, C.C., Angers, S. Gli proteins in development and disease, *Annu. Rev. Cell Dev. Biol. 27* (**2011**) 513–537.

[267] Dosch, J.S., Pasca di Magliano, M., Simeone, D.M. Pancreatic cancer and hedgehog pathway signaling: New insights, *Pancreatology 10* (**2010**) 151–157.

[268] Thayer, S.P., di Magliano, M.P., Heiser, P.W., Nielsen, C.M., Roberts, D.J., Lauwers, G.Y. *et al.* Hedgehog is an early and late mediator of pancreatic cancer tumorigenesis, *Nature 425* (**2003**) 851–856.

[269] Bailey, J.M., Swanson, B.J., Hamada, T., Eggers, J.P., Singh, P.K., Caffery, T. *et al.* Sonic hedgehog promotes desmoplasia in pancreatic cancer, *Clin. Cancer Res. 14* (**2008**) 5995–6004.

[270] Sheehy, N.T., Cordes, K.R., White, M.P., Ivey, K.N., Srivastava, D. The neural crest enriched microRNA miR-452 regulates epithelial–mesenchymal signaling in the first pharyngeal arch, *Development 137* (**2010**) 4307–4316.

[271] Uziel, T., Karginov, F.V., Xie, S., Parker, J.S., Wang, Y.D., Gajjar, A. *et al.* The miR-17 ~ 92 cluster collaborates with the Sonic Hedgehog pathway in medulloblastoma, *Proc. Natl. Acad. Sci. USA 106* (**2009**) 2812–2817.

[272] Tsuda, N., Ishiyama, S., Li, Y., Ioannides, C.G., Abbruzzese, J.L., Chang, D.Z. Synthetic microRNA designed to target glioma-associated antigen 1 transcription factor inhibits division and induces late apoptosis in pancreatic tumor cells, *Clin. Cancer Res. 12* (**2006**) 6557–6564.

[273] Ahmad, A., Maitah, M.Y., Ginnebaugh, K.R., Li, Y., Bao, B., Gadgeel, S.M. *et al.* Inhibition of Hedgehog signaling sensitizes NSCLC cells to standard therapies through

modulation of EMT-regulating miRNAs, *J. Hematol. Oncol. 6* (6) (**2013**) 77.

[274] Luo, G., Long, J., Cui, X., Xiao, Z., Liu, Z., Shi, S. *et al.* Highly lymphatic metastatic pancreatic cancer cells possess stem cell-like properties, *Int. J. Oncol. 42* (**2013**) 979–984.

[275] Hanahan, D., Weinberg, R.A. Hallmarks of cancer: The next generation, *Cell 144* (**2011**) 646–674.

[276] Lee, K.H., Lotterman, C., Karikari, C., Omura, N., Feldmann, G., Habbe, N. *et al.* Epigenetic silencing of MicroRNA miR-107 regulates cyclin dependent kinase 6 expression in pancreatic cancer, *Pancreatology 9* (**2009**) 293–301.

[277] Pramanik, D., Campbell, N.R., Karikari, C., Chivukula, R., Kent, O.A., Mendell, J.T. *et al.* Restitution of tumor suppressor microRNAs using a systemic nanovector inhibits pancreatic cancer growth in mice, *Mol. Cancer Ther. 10* (**2011**) 1470–1480.

[278] Sun, F., Fu, H., Liu, Q., Tie, Y., Zhu, J., Xing, R. *et al.* Downregulation of CCND1 and CDK6 by miR-34a induces cell cycle arrest, *FEBS Lett. 582* (**2008**) 1564–1568.

[279] Sanchez-Diaz, P.C., Hsiao, T.H., Chang, J.C., Yue, D., Tan, M.C., Chen, H.I. *et al.* De-regulated microRNAs in pediatric cancer stem cells target pathways involved in cell proliferation, cell cycle and development, *PLoS ONE 8* (**2013**) e61622.

[280] Bao, B., Wang, Z., Ali, S., Kong, D., Li, Y., Ahmad, A. *et al.* Notch-1 induces epithelial–mesenchymal transition consistent with cancer stem cell phenotype in pancreatic cancer cells, *Cancer Lett. 307* (**2011**) 26–36.

[281] Bao, B., Ali, S., Banerjee, S., Wang, Z., Logna, F., Azmi, A.S. *et al.* Curcumin analogue CDF inhibits pancreatic tumor growth by switching on suppressor microRNAs and attenuating EZH2 expression, *Cancer Res. 72* (**2012**) 335–345.

[282] Bao, B., Wang, Z., Ali, S., Ahmad, A., Azmi, A.S., Sarkar, S.H. *et al.* Metformin inhibits cell proliferation, migration and invasion by attenuating CSC function mediated by deregulating miRNAs in pancreatic cancer cells, *Cancer Prev. Res. (Phila.) 5* (**2012**) 355–364.

[283] Basu, A., Alder, H., Khiyami, A., Leahy, P., Croce, C.M., Haldar, S. MicroRNA-375 and microRNA-221: Potential noncoding RNAs associated with antiproliferative activity of benzyl isothiocyanate in pancreatic cancer, *Genes Cancer 2* (**2011**) 108–119.

[284] Bhutia, Y.D., Hung, S.W., Patel, B., Lovin, D., Govindarajan, R. CNT1 expression influences proliferation and chemosensitivity in drug-resistant pancreatic cancer cells, *Cancer Res. 71* (**2011**) 1825–1835.

[285] Greither, T., Grochola, L.F., Udelnow, A., Lautenschlager, C., Wurl, P., Taubert, H. Elevated expression of microRNAs 155, 203, 210 and 222 in pancreatic tumors is associated with poorer survival, *Int. J. Cancer 126* (**2010**) 73–80.

[286] Heyn, H., Schreek, S., Buurman, R., Focken, T., Schlegelberger, B., Beger, C. MicroRNA miR-548d is a superior regulator in pancreatic cancer, *Pancreas 41* (**2012**) 218–221.

[287] Ivanovska, I., Ball, A.S., Diaz, R.L., Magnus, J.F., Kibukawa, M., Schelter, J.M. *et al.* MicroRNAs in the miR 106b family regulate p21/CDKN1A and promote cell cycle progression, *Mol. Cell. Biol. 28* (**2008**) 2167–2174.

[288] Li, Y., VandenBoom, T.G., Kong, D., Wang, Z., Ali, S., Philip, P.A. *et al.* Upregulation of miR-200 and let-7 by natural agents leads to the reversal of epithelial-to-mesenchymal transition in gemcitabine-resistant pancreatic cancer cells, *Cancer Res. 69* (**2009**) 6704–6712.

[289] Liffers, S.T., Munding, J.B., Vogt, M., Kuhlmann, J.D., Verdoodt, B., Nambiar, S. *et al.* MicroRNA-148a is downregulated in human pancreatic ductal adenocarcinomas and regulates cell survival by targeting CDC25B, *Lab. Invest. 91* (**2011**) 1472–1479.

[290] Ma, Y., Yu, S., Zhao, W., Lu, Z., Chen, J. miR-27a regulates the growth, colony formation and migration of pancreatic cancer cells by targeting Sprouty2, *Cancer Lett. 298* (**2010**) 150–158.

[291] Nalls, D., Tang, S.N., Rodova, M., Srivastava, R.K., Shankar, S. Targeting epigenetic regulation of miR-34a for treatment of pancreatic cancer by inhibition of pancreatic cancer stem cells, *PLoS ONE 6* (**2011**) e24099.

[292] Park, J.K., Lee, E.J., Esau, C., Schmittgen, T.D. Antisense inhibition of microRNA-21 or-221 arrests cell cycle, induces apoptosis, and sensitizes the effects of gemcitabine in pancreatic adenocarcinoma, *Pancreas 38* (**2009**) e190–e199.

[293] Su, A., He, S., Tian, B., Hu, W., Zhang, Z. MicroRNA-221 mediates the effects of PDGFBB on migration, proliferation, and the epithelial–mesenchymal transition in pancreatic cancer cells, *PLoS ONE 8* (**2013**) e71309.

[294] Sun, M., Estrov, Z., Ji, Y., Coombes, K.R., Harris, D.H., Kurzrock, R. Curcumin (diferuloylmethane) alters the expression profiles of microRNAs in human pancreatic cancer cells, *Mol. Cancer Ther. 7* (**2008**) 464–473.

[295] Yu, J., Ohuchida, K., Mizumoto, K., Sato, N., Kayashima, T., Fujita, H. *et al.* MicroRNA, hsa-miR-200c, is an independent prognostic factor in pancreatic cancer and its upregulation inhibits pancreatic cancer invasion but increases cell proliferation, *Mol. Cancer 9* (**2010**) 169.

[296] Petrocca, F., Vecchione, A., Croce, C.M. Emerging role of miR-106b-25/miR-17-92 clusters in the control of transforming growth factor beta signaling, *Cancer Res. 68* (**2008**) 8191–8194.

[297] Smith, A.L., Iwanaga, R., Drasin, D.J., Micalizzi, D.S., Vartuli, R.L., Tan, A.C. *et al.* The miR-106b-25 cluster targets Smad7, activates TGF-beta signaling, and induces EMT and tumor initiating cell characteristics downstream of Six1 in human breast cancer, *Oncogene 31* (**2012**) 5162–5171.

[298] Freed-Pastor, W.A., Prives, C. Mutant p53: One name, many proteins, *Genes Dev. 26* (**2012**) 1268–1286.

[299] Weissmueller, S., Manchado, E., Saborowski, M., Morris, J.P., Wagenblast, E., Davis, C.A. *et al.* Mutant p53 drives pancreatic cancer metastasis through cell-autonomous PDGF receptor beta signaling, *Cell 157* (**2014**) 382–394.

[300] Morin, R.D., O'Connor, M.D., Griffith, M., Kuchenbauer, F., Delaney, A., Prabhu, A.L. *et al.* Application of massively parallel sequencing to microRNA profiling and discovery in human embryonic stem cells, *Genome Res. 18* (**2008**) 610–621.

[301] Castoldi, M., Schmidt, S., Benes, V., Noerholm, M., Kulozik, A.E., Hentze, M.W. *et al.* A sensitive array for microRNA expression profiling (miChip) based on locked nucleic acids (LNA), *RNA 12* (**2006**) 913–920.

[302] Davison, T.S., Johnson, C.D., Andruss, B.F. Analyzing micro-RNA expression using microarrays, *Methods Enzymol. 411* (**2006**) 14–34.

[303] Liang, R.Q., Li, W., Li, Y., Tan, C.Y., Li, J.X., Jin, Y.X. *et al.* An oligonucleotide microarray for microRNA expression analysis based on labeling RNA with quantum dot and nanogold probe, *Nucleic Acids Res. 33* (**2005**) e17.

[304] Thomson, J.M., Parker, J., Perou, C.M., Hammond, S.M. A custom microarray platform for analysis of microRNA gene expression, *Nat. Methods 1* (**2004**) 47–53.

[305] Valoczi, A., Hornyik, C., Varga, N., Burgyan, J., Kauppinen, S., Havelda, Z. Sensitive and specific detection of microRNAs by northern blot analysis using LNA-modified oligonucleotide probes, *Nucleic Acids Res. 32* (**2004**) e175.

[306] Duncan, D.D., Eshoo, M., Esau, C., Freier, S.M., Lollo, B.A. Absolute quantitation of microRNAs with a PCR-based assay, *Anal. Biochem. 359* (**2006**) 268–270.

[307] Raymond, C.K., Roberts, B.S., Garrett-Engele, P., Lim, L.P., Johnson, J.M. Simple, quantitative primer-extension PCR assay for direct monitoring of microRNAs and short interfering RNAs, *RNA 11* (**2005**) 1737–1744.

[308] Shi, Z., Johnson, J.J., Stack, M.S. Fluorescence in situ hybridization for microRNA detection in archived oral cancer tissues, *J. Oncol.* 903581 (**2012**).

[309] Neely, L.A., Patel, S., Garver, J., Gallo, M., Hackett, M., McLaughlin, S. *et al.* A single-molecule method for the quantitation of microRNA gene expression, *Nat. Methods 3* (**2006**) 41–46.

[310] Deo, M., Yu, J.Y., Chung, K.H., Tippens, M., Turner, D.L. Detection of mammalian microRNA expression by in situ hybridization with RNA oligonucleotides, *Dev. Dyn. 235* (**2006**) 2538–2548.

[311] Kloosterman, W.P., Wienholds, E., de, B.E., Kauppinen, S., Plasterk, R.H. In situ detection of miRNAs in animal embryos using LNA-modified oligonucleotide probes, *Nat. Methods 3* (**2006**) 27–29.

[312] Pritchard, C.C., Cheng, H.H., Tewari, M. MicroRNA profiling: Approaches and considerations, *Nat. Rev. Genet. 13* (**2012**) 358–369.

[313] Lu, C., Meyers, B.C., Green, P.J. Construction of small RNA cDNA libraries for deep sequencing, *Methods 43* (**2007**) 110–117.

[314] Kawahara, Y., Zinshteyn, B., Sethupathy, P., Iizasa, H., Hatzigeorgiou, A.G., Nishikura, K. Redirection of silencing targets by adenosine-to-inosine editing of miRNAs, *Science 315* (**2007**) 1137–1140.

[315] Margulies, M., Egholm, M., Altman, W.E., Attiya, S., Bader, J.S., Bemben, L.A. *et al.* Genome sequencing in microfabricated high-density picolitre reactors, *Nature 437* (**2005**) 376–380.

[316] Berezikov, E., Cuppen, E., Plasterk, R.H. Approaches to microRNA discovery, *Nat. Genet. 38* (Suppl.:S2-7) (**2006**) S2–S7.

[317] Berezikov, E., Thuemmler, F., van Laake, L.W., Kondova, I., Bontrop, R., Cuppen, E. *et al.* Diversity of microRNAs in human and chimpanzee brain, *Nat. Genet. 38* (**2006**) 1375–1377.

[318] Varallyay, E., Burgyan, J., Havelda, Z. MicroRNA detection by northern blotting using locked nucleic acid probes, *Nat. Protoc. 3* (**2008**) 190–196.

[319] Habbe, N., Koorstra, J.B., Mendell, J.T., Offerhaus, G.J., Ryu, J.K., Feldmann, G. *et al.* MicroRNA miR-155 is a biomarker of early pancreatic neoplasia, *Cancer Biol. Ther. 8* (**2009**) 340–346.

[320] Ho, A.S., Huang, X., Cao, H., Christman-Skieller, C., Bennewith, K., Le, Q.T. *et al.* Circulating miR-210 as a novel hypoxia marker in pancreatic cancer, *Transl. Oncol. 3* (**2010**) 109–113.

[321] Ganepola, G.A., Rutledge, J.R., Suman, P., Yiengpruksawan, A., Chang, D.H. Novel blood-based microRNA biomarker panel for early diagnosis of pancreatic cancer, *World J. Gastrointest. Oncol. 6* (**2014**) 22–33.

[322] Frampton, A.E., Giovannetti, E., Jamieson, N.B., Krell, J., Gall, T.M., Stebbing, J. *et al.* A microRNA meta-signature for pancreatic ductal adenocarcinoma, *Expert. Rev. Mol. Diagn. 14* (**2014**) 267–271.

[323] Schultz, N.A., Dehlendorff, C., Jensen, B.V., Bjerregaard, J.K., Nielsen, K.R., Bojesen, S.E. *et al.* MicroRNA biomarkers in whole blood for detection of pancreatic cancer, *JAMA 311* (**2014**) 392–404.

[324] Li, A., Yu, J., Kim, H., Wolfgang, C.L., Canto, M.I., Hruban, R.H. *et al.* MicroRNA array analysis finds elevated serum miR-1290 accurately distinguishes patients with low-stage pancreatic cancer from healthy and disease controls, *Clin. Cancer Res. 19* (**2013**) 3600–3610.

[325] Wang, W.S., Liu, L.X., Li, G.P., Chen, Y., Li, C.Y., Jin, D.Y. *et al.* Combined serum CA19-9 and miR-27a-3p in peripheral blood mononuclear cells to diagnose pancreatic cancer, *Cancer Prev. Res. (Phila.) 6* (**2013**) 331–338.

[326] Liu, J., Gao, J., Du, Y., Li, Z., Ren, Y., Gu, J. *et al.* Combination of plasma microRNAs with serum CA19-9 for early detection of pancreatic cancer, *Int. J. Cancer 131* (**2012**) 683–691.

[327] Liu, R., Chen, X., Du, Y., Yao, W., Shen, L., Wang, C. *et al.* Serum microRNA expression profile as a biomarker in the diagnosis and prognosis of pancreatic cancer, *Clin. Chem. 58* (**2012**) 610–618.

[328] Kawaguchi, T., Komatsu, S., Ichikawa, D., Morimura, R., Tsujiura, M., Konishi, H. *et al.* Clinical impact of circulating miR-221 in plasma of patients with pancreatic cancer, *Br. J. Cancer 108* (**2013**) 361–369.

[329] Bauer, A.S., Keller, A., Costello, E., Greenhalf, W., Bier, M., Borries, A. *et al.* Diagnosis of pancreatic ductal adenocarcinoma and chronic pancreatitis by measurement of microRNA abundance in blood and tissue, *PLoS ONE 7* (**2012**) e34151.

[330] Szafranska, A.E., Doleshal, M., Edmunds, H.S., Gordon, S., Luttges, J., Munding, J.B. *et al.* Analysis of microRNAs in pancreatic fine-needle aspirates can classify benign and malignant tissues, *Clin. Chem. 54* (**2008**) 1716–1724.

[331] Yabushita, S., Fukamachi, K., Tanaka, H., Sumida, K., Deguchi, Y., Sukata, T. *et al.* Circulating microRNAs in serum of human K-ras oncogene transgenic rats with pancreatic ductal adenocarcinomas, *Pancreas 41* (**2012**) 1013–1018.

[332] Wang, J., Raimondo, M., Guha, S., Chen, J., Diao, L., Dong, X. *et al.* Circulating microRNAs in pancreatic juice as candidate biomarkers of pancreatic cancer, *J. Cancer. 5* (**2014**) 696–705.

[333] Slater, E.P., Strauch, K., Rospleszcz, S., Ramaswamy, A., Esposito, I., Kloppel, G. *et al.* MicroRNA-196a and-196b as potential biomarkers for the early detection of familial pancreatic cancer, *Transl. Oncol. 7* (**2014**) 464–471.

[334] Ding, Z., Wu, H., Zhang, J., Huang, G., Ji, D. MicroRNAs as novel biomarkers for pancreatic cancer diagnosis: A meta-analysis based on 18 articles, *Tumour Biol. 35* (**2014**) 8837–8848.

[335] Bose, D., Jayaraj, G.G., Kumar, S., Maiti, S. A molecular-beacon-based screen for small molecule inhibitors of miRNA maturation, *ACS Chem. Biol. 8* (**2013**) 930–938.

[336] Pai, P., Rachagani, S., Are, C., Batra, S.K. Prospects of miRNA-based therapy for pancreatic cancer, *Curr. Drug Targets 14* (**2013**) 1101–1109.

[337] Davis, S., Lollo, B., Freier, S., Esau, C. Improved targeting of miRNA with antisense oligonucleotides, *Nucleic Acids Res. 34* (**2006**) 2294–2304.

[338] Davis, S., Propp, S., Freier, S.M., Jones, L.E., Serra, M.J., Kinberger, G. *et al.* Potent inhibition of microRNA in vivo without degradation, *Nucleic Acids Res. 37* (**2009**) 70–77.

[339] Esau, C.C., Monia, B.P. Therapeutic potential for microR-NAs, *Adv. Drug Deliv. Rev. 59* (**2007**) 101–114.

[340] Esau, C.C. Inhibition of microRNA with antisense oligo-nucleotides, *Methods 44* (**2008**) 55–60.

[341] Petersen, M., Wengel, J. LNA: A versatile tool for therapeu-tics and genomics, *Trends Biotechnol. 21* (**2003**) 74–81.

[342] Stenvang, J., Kauppinen, S. MicroRNAs as targets for antisense-based therapeutics, *Expert. Opin. Biol. Ther. 8* (**2008**) 59–81.

[343] Krutzfeldt, J., Rajewsky, N., Braich, R., Rajeev, K.G., Tuschl, T., Manoharan, M. *et al.* Silencing of microRNAs in vivo with 'antagomirs', *Nature 438* (**2005**) 685–689.

[344] Care, A., Catalucci, D., Felicetti, F., Bonci, D., Addario, A., Gallo, P. *et al.* MicroRNA-133 controls cardiac hypertro-phy, *Nat. Med. 13* (**2007**) 613–618.

[345] Thum, T., Gross, C., Fiedler, J., Fischer, T., Kissler, S., Bussen, M. *et al.* MicroRNA-21 contributes to myocardial disease by stimulating MAP kinase signalling in fibro-blasts, *Nature 456* (**2008**) 980–984.

[346] Lanford, R.E., Hildebrandt-Eriksen, E.S., Petri, A., Persson, R., Lindow, M., Munk, M.E. *et al.* Therapeutic silencing of microRNA-122 in primates with chronic hepatitis C virus infection, *Science 327* (**2010**) 198–201.

[347] Gumireddy, K., Young, D.D., Xiong, X., Hogenesch, J.B., Huang, Q., Deiters, A. Small molecule inhibitors of microrna miR-21 function, *Angew. Chem. Int. Ed. Engl. 47* (**2008**) 7482–7484.

[348] Young, D.D., Connelly, C.M., Grohmann, C., Deiters, A. Small molecule modifiers of microRNAmiR-122 func-tion for the treatment of hepatitis C virus infection and hepatocellular carcinoma, *J. Am. Chem. Soc. 132* (**2010**) 7976–7981.

[349] Staton, A.A., Giraldez, A.J. Use of target protector morpho-linos to analyze the physiological roles of specific miRNA–mRNA pairs in vivo, *Nat. Protoc. 6* (**2011**) 2035–2049.

[350] Park, J.K., Kogure, T., Nuovo, G.J., Jiang, J., He, L., Kim, J.H. *et al.* miR-221 silencing blocks hepatocellular car-cinoma and promotes survival, *Cancer Res. 71* (**2011**) 7608–7616.

[351] Ma, L., Reinhardt, F., Pan, E., Soutschek, J., Bhat, B., Marcusson, E.G. *et al.* Therapeutic silencing of miR-10b inhibits metastasis in a mouse mammary tumor model, *Nat. Biotechnol. 28* (**2010**) 341–347.

[352] Ebert, M.S., Sharp, P.A. MicroRNA sponges: Progress and possibilities, *RNA 16* (**2010**) 2043–2050.

[353] Kluiver, J., Gibcus, J.H., Hettinga, C., Adema, A., Richter, M.K., Halsema, N. *et al.* Rapid generation of microRNA sponges for microRNA inhibition, *PLoS ONE 7* (**2012**) e29275.

[354] Bader, A.G., Brown, D., Stoudemire, J., Lammers, P. Developing therapeutic microRNAs for cancer, *Gene Ther. 18* (**2011**) 1121–1126.

[355] Montgomery, R.L., van Rooij, E. Therapeutic advances in microRNA targeting, *J. Cardiovasc. Pharmacol. 57* (**2011**) 1–7.

[356] Trang, P., Wiggins, J.F., Daige, C.L., Cho, C., Omotola, M., Brown, D. *et al.* Systemic delivery of tumor suppressor microRNA mimics using a neutral lipid emulsion inhibits lung tumors in mice, *Mol. Ther. 19* (**2011**) 1116–1122.

[357] Li, L., Xie, X., Luo, J., Liu, M., Xi, S., Guo, J. *et al.* Targeted expression of miR-34a using the T-VISA system suppresses breast cancer cell growth and invasion, *Mol. Ther. 20* (**2012**) 2326–2334.

[358] Kota, J., Chivukula, R.R., O'Donnell, K.A., Wentzel, E.A., Montgomery, C.L., Hwang, H.W. *et al.* Therapeutic

[359] Nair, V.S., Maeda, L.S., Ioannidis, J.P. Clinical outcome prediction by microRNAs in human cancer: A systematic review, *J. Natl. Cancer Inst. 104* (**2012**) 528–540.

[360] Medina, P.P., Nolde, M., Slack, F.J. OncomiR addiction in an in vivo model of microRNA-21-induced pre-B-cell lym-phoma, *Nature 467* (**2010**) 86–90.

[361] Sicard, F., Gayral, M., Lulka, H., Buscail, L., Cordelier, P. Targeting miR-21 for the therapy of pancreatic cancer, *Mol. Ther. 21* (**2013**) 986–994.

[362] Passadouro, M., Pedroso de Lima, M.C., Faneca, H. MicroRNA modulation combined with sunitinib as a novel therapeutic strategy for pancreatic cancer, *Int. J. Nanomedicine 9* (**2014**) 3203–3217.

[363] Arora, S., Swaminathan, S.K., Kirtane, A., Srivastava, S.K., Bhardwaj, A., Singh, S. *et al.* Synthesis, characterization, and evaluation of poly (D, L-lactide-coglycolide)- based nanoformulation of miRNA-150: Potential implications for pancreatic cancer therapy, *Int. J. Nanomedicine 9* (**2014**) 2933–2942.

[364] Garchow, B.G., Bartulos, E.O., Leung, Y.T., Tsao, P.Y., Eisenberg, R.A., Caricchio, R. *et al.* Silencing of microRNA-21 in vivo ameliorates autoimmune splenomeg-aly in lupus mice, *EMBO Mol. Med. 3* (**2011**) 605–615.

[365] Ji, Q., Hao, X., Zhang, M., Tang, W., Yang, M., Li, L. *et al.* MicroRNA miR-34 inhibits human pancreatic cancer tumor-initiating cells, *PLoS ONE 4* (**2009**) e6816.

[366] Yu, F., Yao, H., Zhu, P., Zhang, X., Pan, Q., Gong, C. *et al.* let-7 regulates self renewal and tumorigenicity of breast cancer cells, *Cell 131* (**2007**) 1109–1123.

[367] Trang, P., Medina, P.P., Wiggins, J.F., Ruffino, L., Kelnar, K., Omotola, M. *et al.* Regression of murine lung tumors by the let-7 microRNA, *Oncogene 29* (**2010**) 1580–1587.

[368] Esquela-Kerscher, A., Trang, P., Wiggins, J.F., Patrawala, L., Cheng, A., Ford, L. *et al.* The let-7 microRNA reduces tumor growth in mouse models of lung cancer, *Cell Cycle 7* (**2008**) 759–764.

[369] Liu, C., Kelnar, K., Liu, B., Chen, X., Calhoun-Davis, T., Li, H. *et al.* The microRNA miR-34a inhibits prostate can-cer stem cells and metastasis by directly repressing CD44, *Nat. Med. 17* (**2011**) 211–215.

[370] Wiggins, J.F., Ruffino, L., Kelnar, K., Omotola, M., Patrawala, L., Brown, D. *et al.* Development of a lung cancer therapeutic based on the tumor suppressor microRNA-34, *Cancer Res. 70* (**2010**) 5923–5930.

[371] Takeshita, F., Patrawala, L., Osaki, M., Takahashi, R.U., Yamamoto, Y., Kosaka, N. *et al.* Systemic delivery of syn-thetic microRNA-16 inhibits the growth of metastatic pros-tate tumors via downregulation of multiple cell-cycle genes, *Mol. Ther. 18* (**2010**) 181–187.

[372] Wahid, F., Shehzad, A., Khan, T., Kim, Y.Y. MicroRNAs: Synthesis, mechanism, function, and recent clinical trials, *Biochim. Biophys. Acta 1803* (**2010**) 1231–1243.

[373] van, R.E., Sutherland, L.B., Liu, N., Williams, A.H., McAnally, J., Gerard, R.D. *et al.* A signature pattern of stress-responsive microRNAs that can evoke cardiac hyper-trophy and heart failure, *Proc. Natl. Acad. Sci. USA 103* (**2006**) 18255–18260.

[374] Hullinger, T.G., Montgomery, R.L., Seto, A.G., Dickinson, B.A., Semus, H.M., Lynch, J.M. *et al.* Inhibition of miR-15 protects against cardiac ischemic injury, *Circ. Res. 110* (**2012**) 71–81.

[375] Porrello, E.R., Mahmoud, A.I., Simpson, E., Johnson, B.A., Grinsfelder, D., Canseco, D. *et al.* Regulation of neonatal

and adult mammalian heart regeneration by the miR-15 family, *Proc. Natl. Acad. Sci. USA 110* (**2013**) 187–192.

[376] Patrick, D.M., Zhang, C.C., Tao, Y., Yao, H., Qi, X., Schwartz, R.J. *et al.* Defective erythroid differentiation in miR-451 mutant mice mediated by 14-3-3zeta, *Genes Dev. 24* (**2010**) 1614–1619.

[377] Jopling, C.L., Yi, M., Lancaster, A.M., Lemon, S.M., Sarnow, P. Modulation of hepatitis C virus RNA abundance by a liver-specific MicroRNA, *Science 309* (**2005**) 1577–1581.

[378] Janssen, H.L., Reesink, H.W., Lawitz, E.J., Zeuzem, S., Rodriguez-Torres, M., Patel, K. *et al.* Treatment of HCV infection by targeting microRNA, *N. Engl. J. Med. 368* (**2013**) 1685–1694.

[379] Gabriely, G., Wurdinger, T., Kesari, S., Esau, C.C., Burchard, J., Linsley, P.S. *et al.* MicroRNA 21 promotes glioma invasion by targeting matrix metalloproteinase regulators, *Mol. Cell. Biol. 28* (**2008**) 5369–5380.

[380] Swarbrick, A., Woods, S.L., Shaw, A., Balakrishnan, A., Phua, Y., Nguyen, A. *et al.* miR-380-5p represses p53 to control cellular survival and is associated with poor outcome in MYCN-amplified neuroblastoma, *Nat. Med. 16* (**2010**) 1134–1140.

[381] Huynh, C., Segura, M.F., Gaziel-Sovran, A., Menendez, S., Darvishian, F., Chiriboga, L. *et al.* Efficient in vivo microRNA targeting of liver metastasis, *Oncogene 30* (**2011**) 1481–1488.

[382] Bader, A.G. miR-34 — a microRNA replacement therapy is headed to the clinic, *Front. Genet. 3* (**2012**) 120.

[383] Cookson, M.S., Aus, G., Burnett, A.L., Canby-Hagino, E.D., D'Amico, A.V., Dmochowski, R.R. *et al.* Variation in the definition of biochemical recurrence in patients treated for localized prostate cancer: The American urological association prostate guidelines for localized prostate cancer update panel report and recommendations for a standard in the reporting of surgical outcomes. *J. Urol. 177* (**2007**) 540–545.

[384] Goering, W., Kloth, M., Schulz, W.A. DNA methylation changes in prostate cancer. *Methods Mol. Biol. 863* (**2012**) 47–66.

[385] Perry, A.S., Watson, R.W., Lawler, M., Hollywood, D. The epigenome as a therapeutic target in prostate cancer. *Nat. Rev. Urol. 7* (**2010**) 668–680.

[386] Hoque, M.O., Topaloglu, O., Begum, S., Henrique, R., Rosenbaum, E., van Criekinge, W. *et al.* Quantitative methylation-specific polymerase chain reaction gene patterns in urine sediment distinguish prostate cancer patients from control subjects. *J. Clin. Oncol.* (**2005**) *23* 6569–6575.

[387] Gonzalgo, M.L., Pavlovich, C.P., Lee, S.M., Nelson, W.G. Prostate cancer detection by gstp1 methylation analysis of postbiopsy urine specimens. *Clin. Cancer Res. 9* (**2003**) 2673–2677.

[388] Jeronimo, C., Usadel, H., Henrique, R., Silva, C., Oliveira, J., Lopes, C. *et al.* Quantitative gstp1 hypermethylation in bodily fluids of patients with prostate cancer. *Urology 60* (**2002**) 1131–1135.

[389] Roupret, M., Hupertan, V., Yates, D.R., Catto, J.W., Rehman, I., Meuth, M. *et al.* Molecular detection of localized prostate cancer using quantitative methylation-specific pcr on urinary cells obtained following prostate massage. *Clin. Cancer Res. 13* (**2007**) 1720–1725.

[390] Ellinger, J., Haan, K., Heukamp, L.C., Kahl, P., Buttner, R., Muller, S.C. *et al.* Cpg island hypermethylation in cell-free serum DNA identifies patients with localized prostate cancer. *Prostate 68* (**2008**) 42–49.

[391] Ellinger, J., Kahl, P., von der Gathen, J., Rogenhofer, S., Heukamp, L.C., Gutgemann, I. *et al.* Global levels of histone modifications predict prostate cancer recurrence. *Prostate 70* (**2010**) 61–69.

[392] Deligezer, U., Yaman, F., Darendeliler, E., Dizdar, Y., Holdenrieder, S., Kovancilar, M. *et al.* Post-treatment circulating plasma bmp6 mrna and h3k27 methylation levels discriminate metastatic prostate cancer from localized disease. *Clin. Chim. Acta 411* (**2010**) 1452–1456.

[393] Cortez, M.A., Bueso-Ramos, C., Ferdin, J., Lopez-Berestein, G., Sood, A.K., Calin, G.A. Micrornas in body fluids-the mix of hormones and biomarkers. *Nat. Rev. Clin. Oncol. 8* (**2011**) 467–477.

[394] Chen, X., Ba, Y., Ma, L., Cai, X., Yin, Y., Wang, K. *et al.* Characterization of micrornas in serum: A novel class of biomarkers for diagnosis of cancer and other diseases. *Cell Res. 18* (**2008**) 997–1006.

[395] El-Hefnawy, T., Raja, S., Kelly, L., Bigbee, W.L., Kirkwood, J.M., Luketich, J.D. *et al.* Characterization of amplifiable, circulating rna in plasma and its potential as a tool for cancer diagnostics. *Clin. Chem. 50* (**2004**) 564–573.

[396] Selth, L.A., Tilley, W.D., Butler, L.M. Circulating micrornas: Macro-utility as markers of prostate cancer? *Endocr. Relat. Cancer 19* (**2012**) R99–R113.

[397] Zhang, H.L., Yang, L.F., Zhu, Y., Yao, X.D., Zhang, S.L., Dai, B. *et al.* Serum mirna-21: Elevated levels in patients with metastatic hormone-refractory prostate cancer and potential predictive factor for the efficacy of docetaxel-based chemotherapy. *Prostate 71* (**2011**) 326–331.

[398] Selth, L.A., Townley, S., Gillis, J.L., Ochnik, A.M., Murti, K., Macfarlane, R.J. *et al.* Discovery of circulating micrornas associated with human prostate cancer using a mouse model of disease. *Int. J. Cancer 131* (**2012**) 652–661.

[399] Yaman Agaoglu, F., Kovancilar, M., Dizdar, Y., Darendeliler, E., Holdenrieder, S., Dalay, N. *et al.* Investigation of mir-21, mir-141, and mir-221 in blood circulation of patients with prostate cancer. *Tumour Biol. 32* (**2011**) 583–588.

[400] Mahn, R., Heukamp, L.C., Rogenhofer, S., von Ruecker, A., Muller, S.C., Ellinger, J. Circulating micrornas (mirna) in serum of patients with prostate cancer. *Urology 77* (**2011**) 1265.e9–1265.e16.

[401] Bryant, R.J., Pawlowski, T., Catto, J.W., Marsden, G., Vessella, R.L., Rhees, B. *et al.* Changes in circulating microrna levels associated with prostate cancer. *Br. J. Cancer 106* (**2012**) 768–774.

[402] Brase, J.C., Johannes, M., Schlomm, T., Falth, M., Haese, A., Steuber, T. *et al.* Circulating mirnas are correlated with tumor progression in prostate cancer. *Int. J. Cancer 128* (**2011**) 608–616.

[403] Regulome Explorer. Available online: http://explorer.cancerregulome.org/ (accessed on 16 July **2013**).

[404] Cancer Genome Atlas Network. Comprehensive molecular portraits of human breast tumours. *Nature 490* (**2012**) 61–70.

[405] Cancer Genome Atlas Network. Comprehensive molecular characterization of human colon and rectal cancer. *Nature 487* (**2012**) 330–337.

[406] Thurman, R.E., Rynes, E., Humbert, R., Vierstra, J., Maurano, M.T., Haugen, E. *et al.* The accessible chromatin landscape of the human genome. *Nature 489* (**2012**) 75–82.

[407] Madkour, L.H. *Reactive Oxygen Species (ROS), Nanoparticles, and Endoplasmic Reticulum (ER) Stress-Induced Cell Death Mechanisms.* Academic Press, **2020**. https://www.elsevier.com/books/reactive-oxygen-species-ros-nanoparticles-

and-endoplasmic-reticulum-er-stress-induced-cell-death-mechanisms/madkour/978-0-12-822481-6

[408] Madkour, L.H. *Nanoparticles Induce Oxidative and Endoplasmic Reticulum Antioxidant Therapeutic Defenses.* 1st ed. Springer: Switzerland, **2020**. https://www.springer.com/gp/book/9783030372965?utm_campaign=3_pier05_buy_print&utm_content=en_08082017&utm_medium=referral&utm_source=google_books#otherversion=9783030372972

[409] Madkour, L.H. *Nucleic Acids as Gene Anticancer Drug Delivery Therapy.* 1st ed. Elsevier, **2020**. https://www.elsevier.com/books/nucleic-acids-as-gene-anticancer-drug-deliverytherapy/madkour/978-0-12-819777-6

[410] Madkour, L.H. *Nanoelectronic Materials: Fundamentals and Applications (Advanced Structured Materials).* 1st ed. Springer: Switzerland, **2019**. https://books.google.com.eg/books/about/Nanoelectronic_Materials.html?id=YQXCxAEACAAJ&source=kp_book_description&redir_esc=y https://www.springer.com/gp/book/9783030216207

[411] Hou, B., Jian, Z., Chen, S., Ou, Y., Li, S., Ou, J. Expression of miR-216a in pancreatic cancer and its clinical significance, *Nan. Fang Yi. Ke. Da. Xue. Xue. Bao 32* (**2012**) 1628–1631.

[412] Moulton, J.D., Jiang, S. Gene knockdowns in adult animals: PPMOs and vivomorpholinos, *Molecules 14* (**2009**) 1304–1323.

# 12 Regulation of miRNAs and Their Role in Regeneration and Cancer Diseases

The key miRNAs identified in liver development and chronic liver disease will be discussed together with, where possible, the target messenger RNAs that these miRNAs regulate to profoundly alter these processes. Hepatotropic viruses such as hepatitis B virus (HBV) and hepatitis C virus (HCV) are the major etiological agents associated with development of HCC. Progression of HCC is a multi-step process that requires sequential or parallel deregulation of oncogenic and tumor suppressive pathways leading to chromosomal instability and neoplastic phenotype. Not surprisingly, miRNAs are fast emerging as central players in myriads of malignancies including HCC. MiRNAs are reported to participate in the initiation and progression of HCC and have also been clinically correlated with risk assessment, disease grade, aggressiveness, and prognosis. Despite extensive data available on the role of miRNAs in HCC, there is a pressing need to integrate and evaluate these datasets to find their correlation, if any, with causal agents in order to devise novel interventional modalities.

## 12.1 INTRODUCTION

It has been estimated that approximately 97% of the human genome consists of non–protein coding DNA [1]. Although originally considered to be "junk DNA," the Encyclopedia of DNA Elements (ENCODE) project, a consortium with the central aim to define all functional elements encoded in the human genome, concluded that the vast majority (~80%) of the human genome is not only transcribed, but has biochemical function [2]. More recently, the role of noncoding RNAs has become a major focus of research efforts, effectively laying to rest the dogma that RNA functions only in a structural capacity or as an intermediary between DNA sequence and translated protein.

Many types of noncoding RNAs have now been described including infrastructural RNAs (ribosomal (rRNA), transfer (tRNA), small nuclear (snRNA), and small nucleolar (snoRNA) RNAs, as well as regulatory RNAs. Regulatory noncoding RNAs can be further divided into long noncoding (lncRNAs) and small noncoding RNAs, which include small interfering (siRNA), piwi-associated (piRNA), and micro (miRNA) RNA. Many excellent reviews cover the multitude of noncoding RNAs in detail [3–6]. The purpose of this review is to examine the function of miRNAs, and in particular their role in liver development, regeneration, and disease. To assist the reader to identify miRNAs involved in these processes we have summarized the miRNAs that were reported in at least two or more studies in Figure 12.1.

The liver is an excellent system in which to study miRNAs in development, as liver organogenesis has been well characterized, and extensive analyses have defined the transcriptional changes that occur during cellular differentiation of hepatic progenitors and through different stages of embryogenesis to form the functional adult liver (reviewed in [7]). The liver is also remarkable in regard to its ability to regenerate following surgical resection of up to 70% of the liver mass [8], allowing the study of in vivo organ growth and cell repopulation in the adult. Furthermore, recent research has ascribed a role for miRNAs in all aspects of liver disease, from initiation, progression, diagnosis, and treatment. Advantageously, there are many established models of acute and chronic liver injury that are routinely used to study a host of liver pathologies, and the role that miRNAs may play. Whilst numerous etiologies may underlie liver disease, this chapter will cover four of the most chronic liver diseases, namely alcoholic liver disease (ALD), nonalcoholic fatty liver disease (NAFLD), viral hepatitis, and primary liver cancer.

As interest in noncoding RNAs intensifies, particularly in the miRNA field, there is an increasing number of publications citing large-scale array analyses with little or no target gene validation or link to biological significance. For miRNA data to be of value for future investigators, a thorough understanding of the function of dysregulated miRNAs and the target messenger RNAs (mRNAs) they regulate is essential. Ideally, where possible, studies should demonstrate a cause and effect of regulated target mRNAs, for example siRNA silencing of target genes, to validate the biological relevance of changes in miRNAs and the targets they control. Therefore to assist the reader to identify meaningful data amongst a vast repository of published literature, this chapter will primarily focus on papers for which validation of potential targets of differentially regulated miRNAs has been performed.

According to a WHO estimate, hepatitis B virus (HBV) and hepatitis C virus (HCV) together account for ~78% of HCC incidence worldwide and are the second leading cause of cancer mortality [9]. High morbidity observed in HCC is majorly attributed to lack of early detection markers and poor prognosis, which limit the options for chemotherapy, adjuvant therapies, or surgical procedures [10]. Hence, explorations of novel frontiers in HCC diagnosis and therapeutics remain high-priority research areas. Recent studies suggest an indispensable role played by miRNAs in tumor

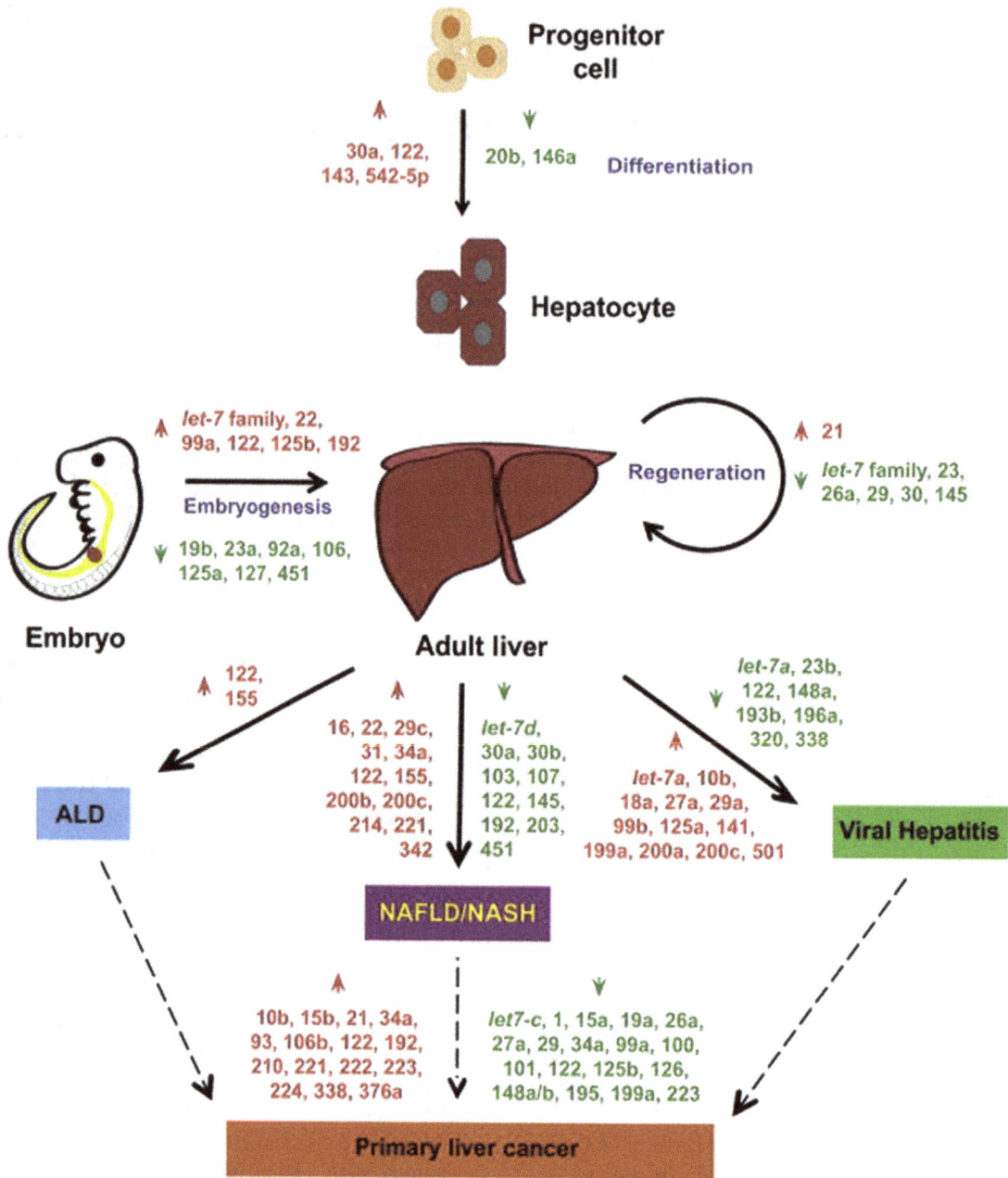

**FIGURE 12.1** MiRNA regulation during liver development, regeneration, and disease. Schematic illustration of the changes in miRNA abundance during liver development, regeneration, and disease. MiRNAs cited as being upregulated (*red*) or downregulated (*green*) during differentiation of hepatic progenitors, embryogenesis, and regeneration, as well as in four of the most common types of chronic liver disease (alcoholic liver disease [ALD], nonalcoholic fatty liver disease/nonalcoholic steatohepatitis [NAFLD/NASH], viral hepatitis, and primary liver cancer) are represented. A minimum of two reports showing dysregulation of a miRNA in the same direction was the criterion for inclusion.

growth and immune evasion. MiRNAs constitute a major class of well-conserved, small, noncoding RNAs that can up- or downregulate gene expression [10, 11]. Accordingly, miRNAs can function both as tumor suppressor (TS-miR) and oncogene (onco-miR) impacting both pro- and antiproliferative cascades [12]. To date, nearly 2000 different miRNAs have been identified in humans. The stability of miRNAs in blood circulation makes them ideal candidates for use in diagnosis and treatment of cancers. Further,

tumor miRNA profiles can be helpful in defining relevant subtypes, patient survival, treatment response, and risk prediction [13]. Given the vast repertoire of cancer-related pathways regulated by miRNAs, this class of biomolecules appears to be a driving force for oncogenesis as well as an "Achilles' heel" for therapeutic targeting. The intricate processes of miRNA biogenesis and maturation are easy targets of stealth hepatitis viruses [14]. In the next sections, we overview how these processes are modulated in HCC.

## 12.2 MiRNA

The first miRNA (lin-4) was identified in 1993 in *Caenorhabditis elegans* as a regulator of developmental timing [15, 16]. However, it was not until 2000 with the discovery that let-7 miRNA was highly conserved between invertebrates and vertebrates [17, 18] that research into miRNAs really took off. MiRNAs represent the most abundant class of small endogenous non–protein coding RNAs, and function as posttranscriptional regulators of gene expression.

The biogenesis of miRNAs is a multistep process that was first elucidated in *C. elegans* by several research groups (for review see [19]). Briefly, miRNA genes are transcribed into hairpin stem-loop structures known as primary miRNA (pri-miRNA) by RNA polymerase II. Pri-miRNAs are then processed within the nucleus by the Drosha–DGCR8 complex to form precursor miRNAs, which are transported to the cytoplasm by Exportin 5. Further processing by the Dicer1-TARBP2 complex generates the mature miRNA duplexes that are typically 18–25 nucleotides (nt) in length. Finally, the single active strand of the mature miRNA is loaded into the RNA-induced silencing complex (RISC), which delivers the miRNA to its target mRNA where it binds with imperfect complementarity, typically in the 3′untranslated region (UTR) [20].

The binding of miRNAs to their target mRNAs induces gene silencing through destabilization and degradation of the mRNA and/or via translational repression (reviewed in [21]). Due to the imperfect nature of miRNA–mRNA binding, a single miRNA species is able to target multiple mRNAs (target multiplicity). Furthermore, a single mRNA can be targeted and regulated by several different miRNAs (miRNA redundancy). It has been estimated that approximately 30% of all human genes are under the control of miRNAs [22]. It is therefore not surprising that miRNAs are now implicated in the control of almost all known biological processes.

## 12.3 LIVER DEVELOPMENT

The liver develops from the definitive endodermal epithelium of the embryonic foregut. While the liver serves as the initial site of hematopoiesis, hepatic specification only commences at mouse gestation day E8.5, corresponding to approximately 3 weeks of human gestation. Hepatic progenitor cells are hepatoblasts, bipotential cells that are able to differentiate to form cholangiocytes (bile duct epithelial cells) and hepatocytes, the two main cell types of the liver. Differentiation of hepatoblasts begins at approximately E13.5 in the mouse (~6 weeks in humans) and continues until birth. Within the adult liver, a small population of cells resembling embryonic hepatoblasts persist within the canals of Hering—the terminal bile ductules. Known as liver progenitor cells (LPCs), these cells play an important role in the liver's response to chronic injury.

### 12.3.1 MiRNAs in Liver Development

There have been three independent studies concerned with identifying miRNAs that are involved in liver development [23] comparing embryonic and adult livers, and an embryo with its liver removed to identify panels of miRNAs specifically expressed at different developmental stages. MiRNAs highly expressed in embryonic compared to adult liver included miR-18a, -92a, -409-3p, -451, and -483-3p, and miRNAs enriched in adult compared to embryonic liver included miR-22, -23b, -99a, -125b, -192, and let-7a, b and c (Figure 12.1). Additionally the authors linked miRNA profiles with gene expression data and found an inverse correlation between the expression of selected miRNAs and their predicted targets. Furthermore they showed that the type I receptor gene *TGFβR1* is negatively regulated by let-7c during development [23]. To assist the reader, we have summarized the miRNAs with validated targets that are included in this chapter.

Liu et al. [24] took a different approach to profile miRNAs in fetal and adult liver tissue as well as a chorionic villus. Expression profiles of the 42 miRNAs identified revealed a number of miRNAs differentially expressed between the fetal and adult liver samples, in particular miR-122, -148a, -192, -194, -451, -21, and let-7a were all reduced in the adult liver compared to fetal liver. As expected, bioinformatic analyses revealed numerous potential targets for identified miRNAs, including multiple genes involved in proliferation and growth such as PinX1 [24].

It is difficult to compare these studies directly because they assessed liver tissues from markedly different developmental stages (7–10 weeks [23] compared with 18–35 weeks [24]). Furthermore, the studies used different techniques to assess the miRNA profile. Nevertheless, both human studies identified several miRNAs, namely miR-122, -192, -194, -451, and to a lesser extent miR-483-3p, which were detected in embryonic/fetal liver. Of these, however, only miR-451 was highly expressed in the embryonic/fetal liver compared with adult liver in both studies.

More recently, using next-generation sequencing to compare miRNA gene expression profiles of E8.5 foregut endoderm, E14.5 hepatoblasts, and adult rodent liver [25], Wei et al. identified two miRNAs, miR-302b and -20a, which were both highly enriched in the endoderm yet rapidly downregulated during liver development. Interestingly, these miRNAs share a common target (Tgfbr2), which could regulate transforming growth factor beta (TGFβ) signaling.

Many miRNAs exhibit tissue and developmental timing specificity. MiR-122 is a liver-specific miRNA accounting for ~70% of all expressed miRNAs in the liver whilst undetectable in other tissue types assessed [26]. Furthermore, miR-122 is specifically switched on during embryogenesis, and is maintained in adulthood [27]. Although many studies have investigated the role of miR-122 in different liver pathologies, in particular viral hepatitis, little has been published on the role of this miRNA in liver development.

One study by Xu et al. (2010) [28] identified a link between transcription factors enriched in the liver and the expression of miR-122 in developing embryos. Furthermore, a number of miR-122 targets were identified that are known to be involved in regulating cellular proliferation and differentiation during development. Using miRNA overexpression and knockdown experiments, the authors demonstrated that one of these targets, Cutl1, is negatively regulated by miR-122 during liver development [28].

Besides tissue specificity, miRNAs also exhibit cell-specific expression. In one study, in situ hybridization on frozen liver sections was performed to identify miRNAs expressed in particular cell populations within the developing liver including hepatocytes, hepatoblasts of the developing ductal plate, and mature cholangiocytes, as well as cells within the periportal space and scattered blood cells throughout the lobule [29]. With a specific focus on miRNAs that regulate development of the ductal plate, which gives rise to the mature bile ducts, it was revealed that miR-30a and -30c are biliary specific in both mouse and human [29]. In support of this, miR-30a silencing in zebrafish resulted in an impairment of biliary function. Although the authors identified potential targets of miR-30a in a mouse embryonic LPC line, including Ak1, an enzyme involved in cellular energy homeostasis, and Tnrc6a, which encodes a component of the RISC complex, neither has been specifically linked to biliary function.

The importance of miRNAs in development has been emphasized by studies showing that animals devoid of Dicer1, resulting in loss of mature miRNAs, die early during embryogenesis [30]. However, a separate study [31] found that transgenic mice specifically lacking Dicer1 in alpha-fetoprotein (Afp)-expressing hepatoblast-derived hepatocytes develop normally. This was despite large-scale changes in liver gene expression, including the increased expression of selected miRNA-targeted mRNAs that were experimentally validated. Although liver damage accompanied by hepatic apoptosis and inflammation was eventually observed in older animals (2–4 months), it was concluded by the authors that mature miRNAs are not required for hepatic function [31].

As noted by the authors [32], a limitation of this study for assessing the role of miRNAs in liver development is that the Dicer1 transcript was not substantially depleted until birth; therefore the processing of some mature miRNAs would likely have continued throughout development. Furthermore, due to the transgenic model used (AlfpCre), Dicer1 was deleted only in hepatocytes derived from Afp-expressing hepatoblasts and not in other cell types within the liver including the hepatoblasts themselves or cholangiocytes, which is expected to be significant. Therefore studies in which Dicer1 is deleted earlier in development and/or in other hepatic cell types need to be performed before the role of miRNAs in hepatic development can be clearly defined. For instance, one study that targeted Dicer in vitro using shRNA in

hepatic stellate cells observed a reduced proliferation rate as well as downregulation of fibrosis-related genes including type I collagen (Col1A1), α-smooth muscle actin (αSMA), and tissue inhibitor of metalloproteinases (TIMP) [33]; however, these changes have yet to be documented in vivo.

Whilst in vivo studies enable the most relevant assessment of biological processes with regard to translation as an endpoint, in vitro models can make significant contributions, especially where information on direct cellular and/or molecular interactions is sought. Often these studies involve the differentiation of cultured cell lines. LPC lines can be derived from normal fetal and adult liver, as well as regenerating liver, and these can be successfully differentiated in vitro to form hepatocytes and cholangiocytes [34]. Identifying miRNA expression changes that occur during the differentiation of progenitor cells into hepatocytes provides an insight into the molecular mechanisms controlling this developmental event. Using this approach, researchers identified several miRNAs [35] in the differentiated cells that showed similar expression patterns to freshly isolated hepatocytes, yet were significantly different from the undifferentiated LPCs. These miRNAs included miR-122 and -101 that increased, and miR-146a, -146b, and -142-5p that decreased, upon differentiation (Figure 12.1).

Studies have also investigated changes in miRNAs during differentiation of other progenitor cell types into hepatocytes, including human embryonic stem cells [36] and human mesenchymal stem cells derived from the umbilical cord [37]. Upregulation of miR-10a, -122, and -143 and downregulation of miR-20b, -30a, and -146a were observed in two out of the three studies; however, no miRNA was consistently altered in all three studies, suggesting that the cell type from which they were derived may influence the miRNA expression profile of the differentiated cells. A more recent study has shown that the overexpression of miR-122 could indirectly promote the hepatic differentiation of mouse embryonic stem cells (ESCs) via indirect induction of two liver-specific transcription factors, FoxA1 and HNF4α [38].

## 12.4  MiRNAs IN LIVER REGENERATION

The liver is one of the few organs with the ability to regenerate following loss through trauma or surgical resection. Partial hepatectomy (PH), removing up to two-thirds of the liver mass in rodents, is commonly used to study liver regeneration. Following two-thirds PH, replacement of liver mass is achieved by proliferation of mature hepatocytes that each undergo an average of 1.4 rounds of replication to reestablish normal liver weight within 5–7 days (8–15 days in humans) (reviewed in [39]). Numerous publications since 2010 have focused on the role of individual miRNAs in regulating hepatocyte proliferation during liver regeneration (Figure 12.1), including miR-21, -127, and -26a [40–43], or during the termination of liver regrowth in which miR-34a increases and miR-23b decreases [44, 45], following PH.

As reported previously [46], one of the complicating factors in identifying consistent changes in miRNAs during liver regeneration has been that individual groups use different models, e.g., mouse, rat, or human, and look at different time points during the regenerative process. Despite this, several reports have provided valuable insight into the changes in individual miRNAs during liver regeneration. Several groups have performed array-based studies to assess global miRNA changes during the highly proliferative phase of liver regeneration. All demonstrated that miRNA expression patterns changed considerably during liver regeneration, and putative targets for some of these regulated miRNAs have been identified [47–50]. Shu et al. (2011) observed that miRNA regulation post PH followed a biphasic pattern where miRNAs were upregulated within the first 3–18 h following PH before a general downregulation of as much as 70% of miRNAs by 24 h post PH. Additionally, increased expression of genes associated with miRNA biogenesis such as Tarbp2 and Dicer1 correlated with the decreased expression of some miRNAs, indicating that a negative feedback loop exists between miRNAs and their regulatory genes.

Several studies have reported the induction of miR-21 during the first 24 h of liver regeneration following PH thus making miR-21 the most consistently altered miRNA during the early stages of regeneration [40, 41, 47, 50–53]. Consistent with this, several miR-21 targets have been reported that are involved in hepatocyte proliferation. These include Btg2, which inhibits FoxM1b, that is required for hepatocyte proliferation [47]; RhoB leading to increases in Cyclin D1 [41]; and Peli1, a ubiquitin ligase involved in NFκB signaling, which may be part of a negative feedback loop required to inactivate NFκB in latter stages of liver regeneration [40]. Interestingly, miR-21 abundance is further enhanced by the bile acid, ursodeoxycholic acid (UDCA), and by ethanol following PH in rats [51, 52]; however, despite elevated miR-21 levels, the presence of ethanol blocked hepatocyte proliferation [52].

Several publications have reported increases in other miRNAs during liver regeneration but none as consistently altered as miR-21. These include miR-34a, which targets Met and Inhbb, possibly for termination of regeneration [44] and miR-221, which targets the cyclin-dependent kinase (CDK) inhibitors Cdkn1b (p27), and Cdkn1c (p57), and Aryl hydrocarbon nuclear translocator (Arnt) [54]. Many miRNAs are also repressed during liver regeneration. Several papers have reported the repression of miR-26a that targets the cell cycle genes, Cyclin D2 (Ccnd2) and Cyclin E2 (Ccne2), which are involved in proliferation in the remnant liver following PH [43, 48, 49]. Indeed a nice correlation between downregulated miRNAs and highly upregulated genes was identified, for example decreased miR-26a correlated well with elevated Ccne2 [48]. Interestingly, as noted below, increased expression of miR-26a can suppress tumor growth arising from c-MYC overexpression [55], highlighting miR-26a as an important regulator of hepatocyte proliferation.

Repression of miR-378 was also seen during the first 18 h after PH and miR-378 inhibited Odc1, which controls DNA synthesis in hepatocytes [47]. Other studies have reported decreases in miR-22a, -30b, let-7 family members, and, miR-122a and -150 within 12–48 h after PH [48, 56]. Of note miR-150 represses the hepatocyte growth factor, VEGF-A; thus increases in VEGF-A via reduced miR-150 levels could be important for the regenerative response [56]. Finally, in accordance with the other studies, [49] confirmed that changes in miRNA expression were most evident during the peak of DNA replication, occurring approximately 24 h following PH. By coupling miRNA expression data with proteomic analysis 23 putative targets of 13 downregulated miRNAs (including miR-26a and let-7e) were identified, of which many were associated with regulation of cell metabolism including pyruvate carboxylase and glutamate dehydrogenase as well as genes involved in protein translation (e.g., eukaryotic translation elongation factor 1 alpha) and cell shape (e.g., keratins) [49].

It is worth noting that some caution is required while interpreting these types of studies. As mentioned above, miR-21 is consistently increased during liver regeneration; however, a recent elegant study that examined the miRNA load on RISC complexes found that while miR-21 was increased following PH, its load on RISC in fact decreased [53]. Thus simply determining the abundance of a particular miRNA and a correlated change of a potential target may not be sufficient to account for the changes seen in liver regeneration. Furthermore, other miRNAs with an altered RISC load may be more biologically relevant.

## 12.5 TRANSCRIPTIONAL REGULATION OF MiRNA

The transcription of miRNA is guided by RNA polymerase (pol II) regulators, which are often deregulated in case of liver pathologies. In HBV- and HCV-associated pathogenesis, c-Myc along with viral oncoproteins modulates miRNA expression, to create an oncogenic milieu [57] facilitating the binding of transcriptional repressor complexes to miRNA promoters to allow its sustained expression [58, 59]. For example, c-Myc binding to miR-122 promoter (a liver-specific TS-miR) prevents RNA pol II recruitment and H3K9 acetylation [60]. c-Myc also downregulates HNF3b, a liver-specific transcription factor, involved in transcription of miR-122. Because c-Myc itself is a target of certain miRNAs, the transcriptional control of its regulators sets on a positive feedback loop for c-Myc expression in HCC [58, 61].

c-Myc is also reported to stimulate Drosha promoter and, thus, indirectly regulate the stability of DiGeorge syndrome critical region gene 8 mRNA [62]. Although c-Myc normally functions as a negative regulator of miRNAs, the levels of onco-miRs like miR-21 and miR-17-92 polycistron are elevated in the c-Myc microenvironment, which downregulate tumor suppressors [63]. HCV infection per se can cause downregulation of TS-miR-181c by promoting the recruitment of CCAAT/enhancer binding protein b (C/EBP-b) [64]. In contrast, HBV positively regulates miR-181a transcription—an index of poor survival, whereas it negatively regulates tumor suppressors, WIF1 and DKK3 (controllers of Wnt signaling pathway and miR181a targets) possibly via miR181a [65–67]. Inactivation of p53 in HCC leads to downregulation of its transcriptional targets such as miR-34, miR-200, and miR-15/16, which allows cell proliferation and metastasis [68]. p53 inactivation also alters miRNA biogenesis either by directly binding to Drosha or via downregulation of DICER-1 [69], which may define miRNA target genes in HCC.

## 12.6 POSTTRANSCRIPTIONAL REGULATION OF MiRNA

Hepatitis viruses often deregulate pro-proliferative pathways in HCC affecting phase-specific control of cyclins by miRNAs [14]. Impairment of DICER-1 is also frequently observed in HCC combined with tumor stemness [70]. Posttranscriptional regulators of Argonaute (Ago-1 and Ago-2) such as lin-41 are overexpressed in HCC in a c-Myc dependent fashion, which downregulates Ago protein [71]. Low levels of miR-99 and miR-199a induce Ago-2 expression with consequent increase in miR-21 [72, 73]. Modulation of TGFβ signaling in HCC also involves miRNAs to promote tumorigenesis [74, 75]. Interestingly, TGFβ effector SMAD1/5 along with RNA helicase p68 increases the maturation rate of miR-21 and miR-199 contributing to vascularization [75].

## 12.7 EPIGENETIC ALTERATIONS OF MiRNA

Concordant hypermethylation of miRNA genes is frequently seen in HCC [76, 77]. The master controllers of proliferation, such as c-Met, are epigenetically silenced by TS-miRs [78]. An auto-feedback loop of hypermethylation and gene suppression is suggested for miR-148a and miR-152 in HCC [79, 80]. Acetylation status at the pri-miRNA promoters can also suppress miRNA and help relieve the suppressive effect of TS-miRs [81]. Besides, sulfated glycolipids can facilitate intrahepatic metastasis [82].

HBV and HCV oncoproteins may also regulate miRNA expression [83] by engaging DNA methyltransferases (DNMT), which cause global hypermethylation [84, 85]. The HBx oncoprotein is known to sequester the epigenetic modifier PPARγ in order to downregulate miR-122 in HCC [86]. The HBx-mediated epigenetic silencing of miR-205 stabilizes HBx mRNA, and aggravates disease [87]. Similarly, HCV core protein can downregulate miR-124 and miR-345 levels by inducing DNMT expression and abrogating apoptosis [88].

## 12.8 SINGLE-NUCLEOTIDE POLYMORPHISM AND GENETIC ALTERATIONS

Single-nucleotide polymorphism (SNP) in the regulatory or coding regions of miRNA is essential for onco-miR expression or silencing of TS-miRs in HCC. TS-miRs such as miR-34b/c are poorly expressed in HCC owing to SNP RS4938723, which inhibits the recruitment of transcription factor GATA [89]. Alternately, SNPs in the coding region of miR-196a2 and promoter of miR-106b-25 upregulate their expression and contribute toward HCC [90, 91]. SNP in the stem-loop of pre-miR-146a affects its processing efficiency and increases the risk of HCC [92]. As miRNA genes are often located close to fragile sites [93], their translocation is frequently observed in many cancers. The translocation of 5′end of the hcr gene (encodingmiR-122) to c-myc locus can cause a massive 50-fold increase in c-Myc expression and a consequent downregulation of miR-122 in woodchuck hepatitis virus–related HCC [74, 94].

Thus, miRNAs have emerged as crucial players in virus–host interactions, where hepatitis viruses can alter miRNA biogenesis, which in turn control key cellular pathways to establish a successful infection as discussed in the next section.

## 12.9 DEREGULATION OF MiRNA IN HCC

Deregulated miRNAs play a pivotal role in supporting viral replication and perturbing key cellular processes in HBV- and HCV-associated HCC as depicted in Figure 12.2.

The downregulation of TS-miRs and concomitant elevation of onco-miRs also affect some key cellular processes that support HCC.

## 12.10 SUPPRESSION OF APOPTOSIS

Cancer cells evade apoptosis by perturbing the balance between pro- and antiapoptotic factors. In aggressive HCC cases, miR-25 and miR-221 are overexpressed, which target proapoptotic proteins Bim and Bmf, respectively [95, 96]. Downregulated miR-29, miR-101, and miR-122 relieve suppression of their antiapoptotic targets Bcl-2, Mcl-1, and Bcl-w, respectively, leading to increased survival in HCC [97]. HBx oncoprotein downregulates let-7 and miR-15a/16 leading to increased antiapoptotic activity [98]. Similarly, low levels of TS-miR-29c in HCC target tumor necrosis factor alpha-induced protein 3 to prevent apoptosis [99].

## 12.11 ALTERATION OF SIGNALING PATHWAYS

Tumor suppressor PTEN, a negative regulator of PI3K/AKT signaling is frequently targeted by elevated levels of miR-21, miR-222, and miR-29a in HCC [97, 100]. MiR-199-3p targets mTOR and c-Met, a HGF receptor, which controls downstream PAK4/Raf/MEK/ERK pathway [101]. Restoration of miR-199-3p levels in hepatoma cells leads to G1 arrest, enhanced susceptibility to hypoxia, and drug [101].

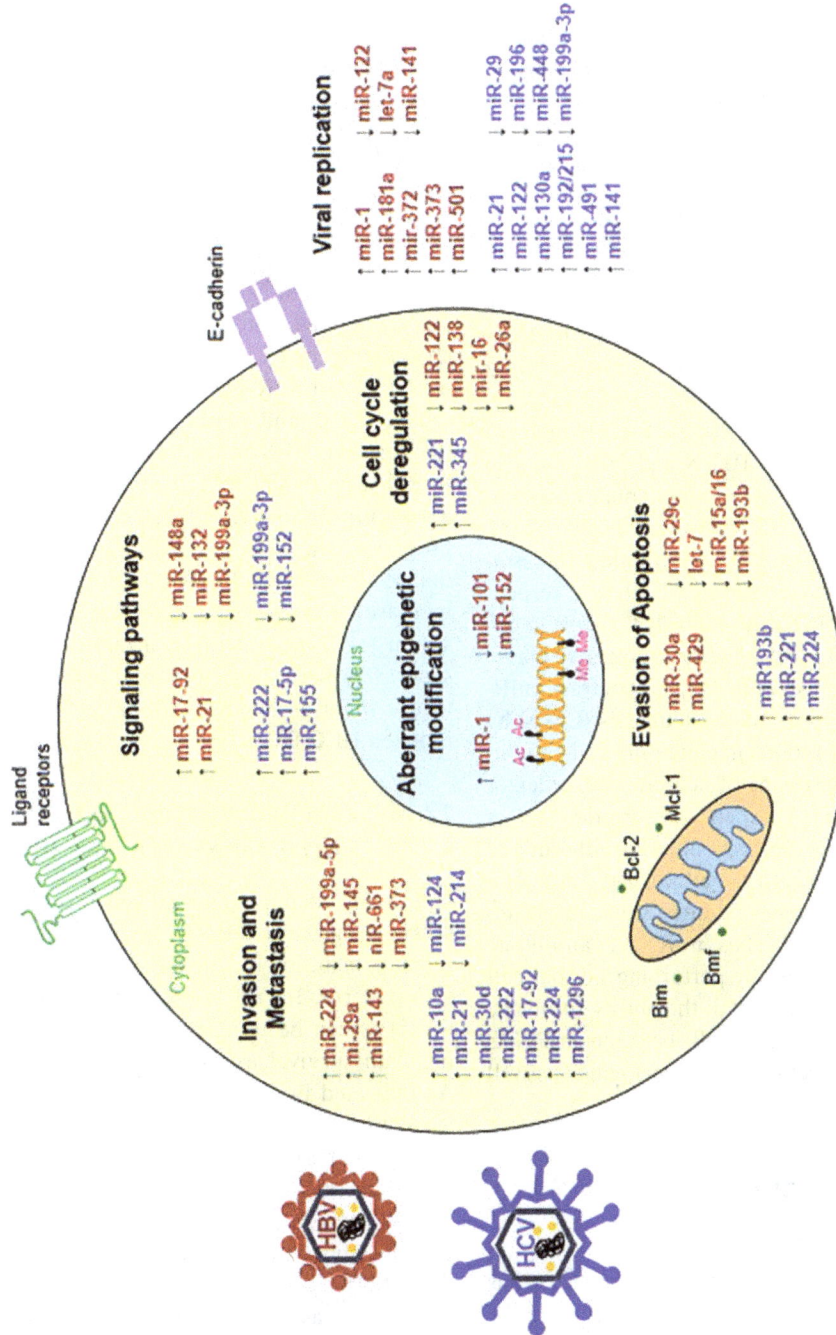

**FIGURE 12.2** Status of miRNAs during hepatitis virus infection. MiRNAs are upregulated (↑) or downregulated (↓) in HBV- (*red*) or HCV-associated HCC (*blue*) and interfere with key cellular pathways such as cell signaling, apoptosis, cell cycle, and metastasis as well as viral replication to promote hepatocarcinogenesis.

HBx suppresses the p53-dependent activation of miR-148a, resulting in upregulation of AKT, ERK, and activation of mTOR pathway, which promotes tumor growth and metastasis in a HCC mouse model [102]. MiR-17-5p, a member of the onco-miR-17-92 cluster, activates p38-MAPK and Hsp27 pathway to promote cell migration and proliferation [103]. Further, miR-222 overexpression confers metastatic potential on cancer cells by activating Akt pathway [104]. Besides its pro-inflammatory role, miR-155 upregulation by HCV results in increased nuclear accumulation of signal transducer β-catenin to promote HCC [105].

## 12.12 EPITHELIAL–MESENCHYMAL TRANSITION AND METASTASIS

Prior to metastasis, tumor cells undergo epithelial–mesenchymal transition (EMT) involving loss of E-cadherin, and gain of vimentin, collagen I, and fibronectin. MiR-224, a highly expressed miRNA in HCC, promotes tumor growth and metastasis by silencing its target genes Cdc42, CDH1, PAK2, and BCL-2 [106]. Elevated levels of miR-29a and miR-148a, which target PTEN and miR-143, also downregulate fibronectin type III domain-containing protein 3B to enhance hepatoma cell migration [107, 108]. HBx engages miR-661 and miR-373 to stimulate metastasis associated-1 protein and suppress E-cadherin, respectively [109]. TS-mir-34a and miR-125b that, respectively, target c-Met and LIN28B2 oncogenes are inversely correlated with metastasis in HCC [110, 111]. In contrast, miR-200c suppresses EMT of liver cancer stem cells (CSCs) by upregulating E-cadherin, reducing vimentin, and inhibiting metastasis through repression of neurotrophic receptor tyrosine kinase 2 [112]. Zeb1 and Zeb2 transcription factors are also targeted by miR-141/200c to alter E-cadherin and related gene expression involved in cell polarity [113]. Not surprisingly, miR-141/200c cluster is often silenced in cancer by DNA methylation [114]. Thus, miRNAs can initiate the multistep process of cancer by perturbing normal cell homeostasis and endowing cells with the ability to invade and metastasize. Further, the crosstalk between miRNAs could have a multiplier effect on their target genes as well as drug resistance cancer phenotype.

## 12.13 MiRNAs AS DIAGNOSTIC MARKERS IN HCC

The lack of effective diagnostic methods for early HCC has rendered the overall survival rate to a low 0–14% from the time of clinical diagnosis [115]. Modest precision of currently used diagnostic markers such as alpha-fetoprotein (AFP) calls for testing the prospect of miRNAs as HCC biomarkers. Circulating miRNAs are highly stable in serum owed to their resistance to RNAse, extreme pH, and temperature, hence are perfect as biomarkers for detecting early-stage, presymptomatic diseases such as HCC, as depicted in Figure 12.3.

The serum levels of miRNAs undergo alterations in HCC patients as evident from low levels of miR-16 and miR-199a, and high levels of miR-21, miR-221, miR-222, miR-223, and miR-224 in serum samples [13]. MiR-125b is downregulated in ~70% of primary HCC samples, thus could be a good candidate for diagnosis [111]. Further, to expand the repertoire of prospective miRNAs in early diagnosis of HCC, downregulated levels of TS-miR-129 in HCC can be detected in plasma samples from 85% of stage I HCC patients as compared to AFP in just 10% of stage I cases [116]. Detection of HBV- or HCV-positive HCC cases, especially those independent of cirrhosis etiology, poses a greater challenge due to lack of biomarkers. In light of this, the unique expression profile of serum miRNAs in HBV- and HCV-positive HCC patients can serve as a fingerprint for distinguishing between HBV and HCV cases. Not surprisingly, miRNAs such as miR-1269, miR-224, and miR-224-3p are significantly altered specifically in HCV-associated HCCs [117]. In contrast, miR-152 and miR-143 are aberrantly regulated in HBV-related HCC and hence constitute potential diagnostic markers for HBV-related HCC cases [118, 119]. Serum miRNAs enriched in exosomes can also serve as valuable noninvasive HCC biomarkers for both diagnostic and prognostic purposes. Indeed, recent findings have shown that serum exosomal miR-21 from HCC patients provides increased sensitivity of detection compared with whole serum [120]. Interestingly, a report on urinary miRNAs such as miR-618 and miR-650 has opened the prospect to use them as biomarkers for early detection of HCV-induced HCC [121].

## 12.14 PREDICTIVE PROGNOSTIC VALUE OF MiRNAs IN HCC

It is being increasingly realized that miRNAs may possess an edge over mRNAs as prognostic indicators owed to their stability in clinical samples and robust expression patterns. In fact, expression profiles of a panel of 20 miRNAs can be used as a metastatic predictor, correlating with survival as well as relapse rates in HCC [122]. As indicated in Figure 12.3, hypermethylation of miRNA promoters seems to correlate with poor prognosis, exemplified by aberrant methylation of miR-9, and corroborates with clinical outcomes [123]. Further, miR-199a/b and miR-139, which are frequently downregulated in most HCC patients, show a significant correlation with poor survival [123, 124]. Likewise, low expression of miR-124 in HCC seems to be associated with more aggressive behavior and shorter survival [123]. Also, high levels of miR-22, miR-221, and C19MC miRNA correlate with increased risk of tumor recurrence and shorter survival [100]. Interestingly, the patterns of miRNA expression as well as SNPs in some miRNAs seem potent as predictors of patient response to various therapeutic strategies as well as disease risk [96]. In Chinese and Turkish populations, a positive association has been noticed between rs11614913 (C→T) SNP in

**DIAGNOSIS**

MiR-125b: Down-regulation detected in 70% HCC primary samples

MiR-129: Hyper-methylation detected in 85% of stage I HCC patients, early diagnosis

MiRs specific to HBV-related HCC: miR-152, miR-143

MiRs specific to HCV-related HCC: miR -1269, miR-224, miR-224-3p

Novel miRNA-based diagnostic approaches: Exosomal (miR-21) and urinary (miR-618, miR-650) miRs

**PROGNOSIS**

MiR-26a, miR-9, miR-199a/b, miR-139, miR124: low levels as predictors of aggressive phenotype, survival, tumor recurrence

MiR-9: hyper-methylation correlated with poor prognosis

MiR-22, miR-221, C19MC cluster: high levels correlated with increased risk of recurrence, short survival.

SNPs- risk predictors for HCC

**HCC THERAPY**

**PRE-CLINICAL STUDIES**

Silencing onco-miRs-

LNA based miR-122 treatment in non-human primates: suppression of HCV viral load.

miR-221 targeting oligos in orthotopic HCC mouse model: decrease in cell transformation and improved survival.

Reintroduction of TS-miRs-

AAV-mediated delivery of miR-122 and miR-26a: suppressed tumorigenesis in mouse models of HCC.

Restoration of 2'O-methyl-modified and cholesterol conjugated miR-375: repression of metastasis in heptoma xenograft mice.

**CLINICAL TRIALS**

Cancer targeting miRNA mimic miR-34 (MRX34)- Phase 1 trial in HCC patients.

Modified antisense oligo antagonist of miR-122, miravirsen: Phase 2a trials in HCV-infected patients

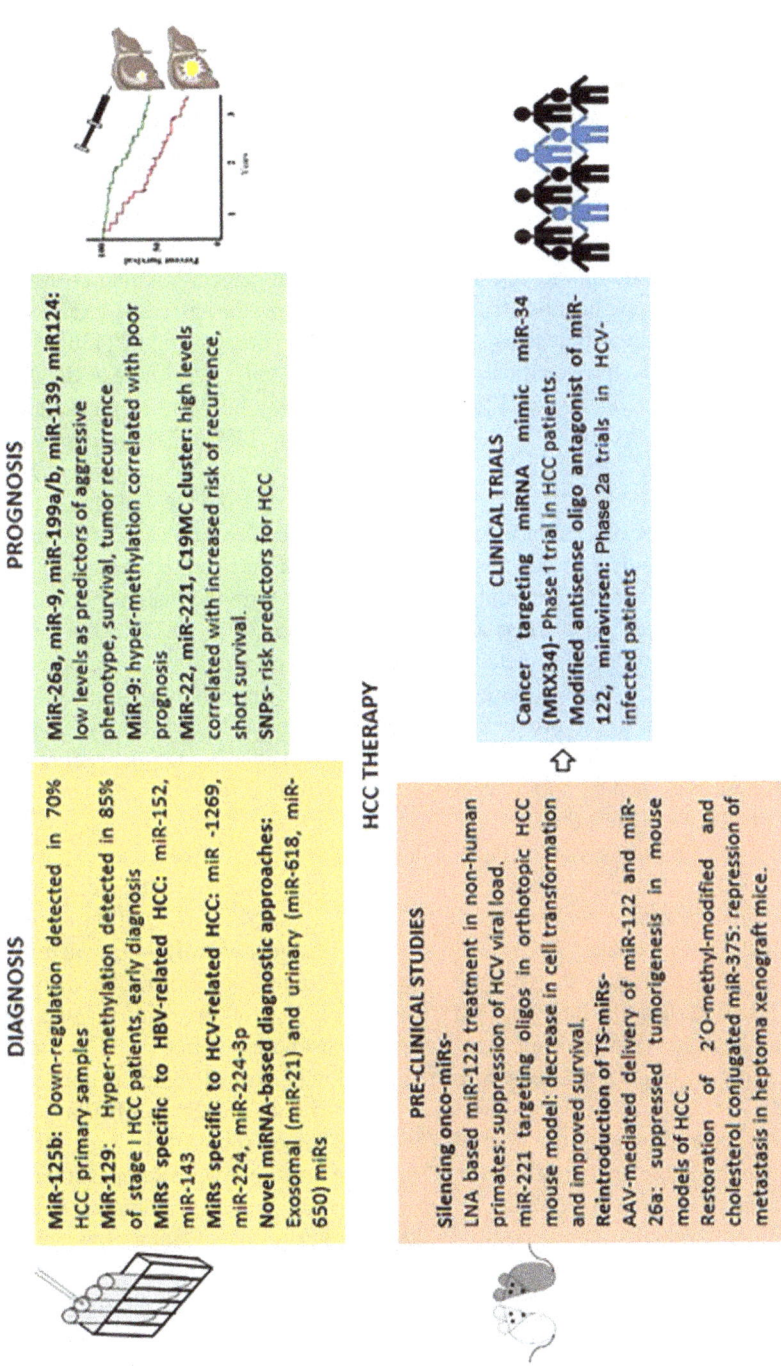

**FIGURE 12.3** MiRNAs in clinical management of HCC. MiRNAs with potential in HCC diagnosis, prognosis, and therapy (preclinical studies and clinical trials).

miR-196a-2 and HCC susceptibility [125, 126]. Similarly, "TTCA" insertion (rs3783553) in 3′-UTR of IL-α gene, which ablates the binding site for miR-122 and miR-378 leading to upregulation of IL1-α, correlates well with HCC development [127].

## 12.15   MiRNAs AND LIVER DISEASE

Together with generation of bile and secreted blood proteins including albumin and coagulation factors, the liver is responsible for removal of toxic substances from the circulation. Loss of liver function leads to hepatic encephalopathy followed by progression to coma, seizures, cerebral edema, and death [128]. Encephalopathy is frequently observed in chronic liver disease, which results from a number of etiologies including excessive alcohol consumption, obesity, or viral hepatitis. Characterized initially by hepatic steatosis (fatty liver), chronic liver disease may progress to steatohepatitis, liver fibrosis, and eventually liver cirrhosis, a major risk factor for the development of liver cancer [129].

### 12.15.1   ALCOHOLIC LIVER DISEASE

Excessive alcohol consumption is a major risk factor for the development of ALD. Whilst the mechanism of how alcohol induces liver damage is complex, chronic inflammation of the liver, most likely a result of oxidative stress and reactive oxygen species produced during alcohol metabolism by specific enzymes in the liver, appears to be key. Direct damage of liver cells, in addition to the production of an inflammatory cascade involving many inflammatory mediators, may contribute to alcoholic hepatitis and eventual ALD [130]. Epidemiological data suggest that 10–35% of alcoholics will develop alcoholic hepatitis, regardless of the amount of alcohol consumed (reviewed in [131]). This suggests that there are additional genetic, epigenetic, and/or environmental factors that can modulate ALD risk. Published data of miRNA expression profiles in ALD are limited; however, there are a few studies that have focused on the expression of liver miRNAs following chronic ethanol feeding of rodents, or treatment of cultured cells with ethanol.

The Lieber–DeCarli diet is a commonly used model of chronic ethanol feeding in rodents. There are two studies in which the authors [32] subjected mice to an ethanol diet for 4–5 weeks before profiling miRNA expression in the liver [132, 133]. Dolganiuc et al. (2009) found 2% of all liver miRNAs were altered following ethanol feeding; upregulated miRNAs included miR-320, -486, -705 and -1224, and downregulated miRNAs included miR-27b, -182, -183, -200a, -214, -199a-3p, and -322. Whilst the authors did not include any target prediction or validation experiments, some of the dysregulated miRNAs have published targets involved in proliferation and modulation of alcohol-induced liver injury. In the second study, miR-155 and miR-132 were upregulated in the liver of alcohol-fed mice [133]. MiR-155 was consistently increased in both isolated hepatocytes and

Kupffer cells (liver macrophages). Furthermore, isolated hepatocytes also showed decreased miR-125b expression, whereas an increase in miR-132 was only seen in isolated Kupffer cells. This finding was complemented by a subsequent study [134] in which a correlation was identified between increased miR-155 expression and increased TNFα production in macrophages treated with alcohol. The authors [32] also suggested increased NF-κB activation as a mechanism for miR-155 upregulation following ethanol exposure.

A third study did not use the Lieber–DeCarli diet but administered ethanol to mice by gastric intubation and profiled plasma miRNAs to evaluate the possibility of using miRNAs as noninvasive biomarkers of liver injury [135]. Plasma miR-122 was significantly elevated within 0.5 h of ethanol administration, and correlated with an increase in alanine aminotransferase (ALT) activity, a traditional marker of hepatocyte injury [136]. Most notably, miR-122 plasma abundance was increased following alcohol exposure before any histopathological findings of liver damage were apparent [135]. This suggests that circulating miRNAs could be developed as early diagnostic markers of liver injury resulting from alcohol abuse.

Comprehensive studies published by [137, 138] specifically focused on potential targets of regulated miRNAs in an attempt to provide insight into how miRNAs may contribute to toxicity in ALD. In the first study, miR-217 was increased in both cultured hepatocytes exposed to ethanol, and in livers from ethanol-fed mice, where it caused excess fat accumulation [137]. Overexpressed miR-217 resulted in a decrease in the NAD⁺-dependent deacetylase Sirtuin-1 (Sirt1), as well as downstream effectors encoding lipogenic or fatty acid oxidation enzymes [137]. This study provides a crucial link between ethanol exposure and fat accumulation that may lead to hepatic steatosis. In the second study, Caspase-2 (CASP2) in addition to SIRT1 were identified as direct targets of miR-34a, which increased following 4 weeks of ethanol administration to mice [138]. Furthermore, the abundance of two matrix metalloproteases (MMP-2 and 9) was increased in ethanol-exposed mouse liver tissue, and following transfection of normal human hepatocytes with a miR-34a precursor. MMPs are proteolytic enzymes involved in cell remodeling, suggesting their regulation by miR-34a may be involved in hepatic remodeling in ALD.

A recently published study on human ALD patients identified six miRNAs that could regulate as many as 79 downstream targets [139]. While the authors did not provide any validation in this study, six miRNAs, namely miR-570, -122, -34b, -29c, -922, and -185, are implicated to regulate genes with diverse biological functions such as the immune response, the inflammatory response, and glutathione metabolism. Intriguingly, with the exception of miR-122, there appears to be little overlap between the miRNAs identified in these human samples with those from rodent experiments. However as this study had a sample size of only ten subjects per group, additional human studies may yield more consistent results.

## 12.15.2 Nonalcoholic Fatty Liver Disease

Like ALD, NAFLD is a chronic liver disease characterized by hepatic steatosis; however, this occurs in the absence of excessive alcohol consumption. Obesity, insulin resistance, and type 2 diabetes have been recognized as major risk factors for NAFLD, with prevalence increasing from ~30% in the general population to ~74% amongst diabetics, and up to 90% in the morbidly obese (reviewed in [140]). NAFLD patients are at risk of developing nonalcoholic steatohepatitis (NASH), which increases their risk of progression through hepatic fibrosis and cirrhosis to liver cancer.

As with other conditions, changes in the abundance of miRNAs have been linked to NAFLD. Comprehensive studies by several groups have focused on the identification of miRNA target genes involved in the pathogenesis of NAFLD/NASH in human patients. MiR-34a, which was dysregulated in some cases of ALD [138], was also increased in the liver of patients with NASH (Figure 12.1), and correlated with a decrease in its target SIRT1 [141–143]. Decreased SIRT1 results in dephosphorylation of AMP kinase and increases the active form of HMG-CoA reductase, a key regulator of cholesterol synthesis that has been linked to NAFLD [144]. A miR-34a associated reduction in SIRT1 is also linked to reduced hepatic malonyl-coA levels, which promotes β-oxidation of fatty acids to reduce hepatic steatosis [143]. Finally, in morbidly obese patients increases in miR-34a, acetylated-p53 levels, and apoptosis correlated with disease severity, while SIRT1 abundance was diminished in the NAFLD liver [142]. Interestingly, feeding rats with UDCA resulted in inhibition of the miR-34a/SIRT1/p53 pathway suggesting p53 increases miR-34a abundance to repress SIRT1 in a positive feedback loop [142]. In addition to miR-34a, miR-296-5p was also decreased in NASH patients [145], and studies in a human hepatoma cell line identified PUMA as a direct target. PUMA is a protein that is likely involved in apoptotic damage in fatty liver disease, evidenced by miR-296-5p inhibition experiments in which cells became sensitized to lipoapoptosis. A recent study by [146], which assessed RNA expression in the visceral adipose tissue of morbidly obese patients, found that several enzymes involved in miRNA processing namely, Dicer1, Drosha, and DGCR8, are elevated in patients with NASH compared with the non-NASH NAFLD cohort. This highlights a novel mechanism of miRNA regulation that may contribute to pathogenesis in patients with NAFLD.

Genetic and dietary rodent models of obesity have been developed to mimic the development of NAFLD. These models have been used to profile changes in miRNA expression, and identify associated gene targets that may play a role in the progression of obesity to NAFLD, and subsequent development of NASH. Additionally, cellular models of steatosis provide an excellent system in which to study the effects of single miRNAs on gene expression.

In an extensive study by [147], miRNAs were profiled in the livers of mice fed a choline-deficient, low methionine, amino acid–defined diet that is known to promote NASH that will progress to hepatocarcinogenesis. Many significantly altered miRNAs were identified, including upregulation of miR-155, -221, -223, -342-3p, and -34a and downregulation of miR-323-5p and let-7a. This study also characterized the regulation of miR-155, including induction of miR-155 expression by the transcription factor NF-κB, as mentioned previously, and subsequent downregulation of the transcription factor C/EBPβ by miR-155 binding to its 3′UTR. Lipoprotein lipase (Lpl), a regulator of lipid metabolism, was identified as a target of miR-467b, which was decreased in the livers of mice fed a high-fat diet and in cultured hepatocytes [148].

Development of an in vitro steatotic cell model by overloading a human hepatocyte cell line with fat resulting in a high concentration of free fatty acids was utilized by one group to identify miRNAs dysregulated under conditions mimicking NAFLD [149]. Numerous miRNAs were altered under conditions of fat-overload, with miR-10b, -22, -29b-1, and -107 showing the greatest changes. PPARα, a transcription factor and lipid regulator in the liver, was verified as a direct target of miR-10b using miRNA overexpression and knockdown in steatotic L02 cells [149]. Consistently, a recent paper by [150] confirmed the regulation of PPARα by miR-10b, which lies downstream of Kruppel-like factor 6 (KLF6). Interestingly, however, using in vivo studies the authors found that mice lacking KLF6 were protected from hepatic steatosis when fed a high-fat diet for 16 weeks, which was accompanied by a decrease in PPARα activity. Furthermore, miR-10b was significantly decreased and negatively correlated with KLF6 expression in NAFLD patients with concomitant inflammation [150].

There is evidence to suggest that NAFLD, like ALD, is a complex disease regulated by a number of genetic and environmental factors. This was highlighted by two studies, the first of which showed variation in the miRNA expression profiles of rats fed three different diets (high fat, high sugar, or both) to induce NAFLD [151]. In particular, miR-21, -24, -99a, and -140-3p exhibit markedly different expression patterns in each diet. In the second study, C57BL/6J and DBA/2J mice fed the same lipogenic methyl-deficient diet capable of inducing a NASH-like pathology exhibited marked strain-specific differences in the severity of NASH symptoms [152]. Disease severity correlated directly with a greater upregulation of fibrosis-related genes. Furthermore, whilst similar miRNAs (including miR-29c, -34a, -122, -155, and -200b) were dysregulated in the livers of both mouse strains, the magnitude of the changes was often greater in the strain with the more severe phenotype [152]. The authors [32] also correlated expression patterns of miRNAs with their published targets in mouse liver and in cultured hepatocytes, including miR-155 with Cebpβ and Socs1, and miR-200b with Zeb1 and its downstream target E-cadherin.

Despite the complexity of NAFLD/NASH development, a "molecular signature" of dysregulated miRNAs was identified that could distinguish steatosis from steatohepatitis in a rat model [153]. This signature included 14 upregulated miRNAs such as miR-29c, -34a, -146, and -210, and six downregulated miRNAs including miR-33, -145, and -196b. It was hypothesized that these miRNAs play a pivotal role in the transition from one disease state to the next, and thus may be potential targets for preventing this form of chronic liver disease progression.

Several studies have also investigated the potential use of miRNAs as biomarkers for detecting early-stage cases of NAFLD or their use as therapeutic targets to treat the disease (reviewed in [154]). MiR-15b was upregulated in the livers of a rat model of NAFLD and also in the sera of a cohort of 69 patients with fatty liver compared with control subjects [155]. In a more comprehensive study with 403 participants, several miRNAs, namely miR-21, -34a, -122, and -451, were higher in patients with NAFLD than those without [156]. Not surprisingly because it is a liver-specific miRNA, the serum level of miR-122 positively correlated with the severity of liver steatosis [156].

### 12.15.3 VIRAL HEPATITIS

Chronic viral hepatitis remains one of the most significant causes of chronic liver disease and primary liver cancer, with approximately 80% of HCC cases associated with chronic HBV or HCV infections [157]. Treatment options for chronic HBV/HCV infection include interferon (IFN) therapy and administration of drugs such as Ribavirin; however successful clearance of virus is typically observed in only 40% of cases [158]. Alternative therapies are thus needed to decrease the burden of HBV and HCV infection worldwide. Recent reports have shown antiviral effects of some mammalian miRNAs by direct targeting of HBV and HCV viral genes [159]. MiRNAs therefore offer great potential for the development of novel therapeutics for viral hepatitis; however, a comprehensive understanding of how host miRNA expression is affected by viral infection is first required.

In a study comparing miRNA expression levels in livers of control subjects and patients with chronic HBV or HCV infection, it was found that eight downregulated miRNAs, including miR-26a, -29c, -219, and 320 (Figure 12.1), could distinguish normal liver from virally infected liver (HBV or HCV) using hierarchical cluster analysis [160]. Despite having distinct virological features, the clinical manifestation of chronic HBV and HCV infections is indistinguishable [161]. Interestingly, in the above-mentioned study the authors were able to show that miRNA profiles could differentiate HBV- and HCV-infected liver. In this comparison, 13 miRNAs showed decreased expression in the HCV liver whereas six miRNAs showed decreased expression in the HBV liver. Furthermore, global gene expression analysis in each sample was used to correlate altered miRNAs with their potential targets. This revealed antigen presentation, lipid metabolism, immune response, and cell cycle

pathway activation in HCV infection, and cell death, DNA damage, and signal transduction pathway activation in HBV infection [160]. Recently, in a study of ten patients chronically infected with HCV, it was found that 22 miRNAs were increased, including miR-144, -486-3p, and -200c, and 35 miRNAs were decreased, including miR-802, -556-3p, and -615-5p, more than two-fold [162]. Gene ontology (GO) analysis identified several dysregulated miRNA clusters affecting diverse pathways such as phospholipase C signaling (miR-200c, -20b, and -31), response to growth factors and hormones (miR-141, -107, and -200c), and cell proliferation (miR-20b, -10b, and -141). Specifically, increased expression of miR-200c in HCV liver can lead to downregulation of its target FAP-1, and consequent increase in c-Src to promote fibrosis [162].

Differential miRNA profiles between HBV and HCV infections can be explained by the fact that viral proteins can directly alter the expression of host miRNAs. In separate studies, Dicer and Drosha, two core elements of the miRNA biogenesis pathway, were negatively regulated by the HCV core protein and HBV protein X (HBx), respectively [163, 164]. Inhibition of Dicer or Drosha would decrease the expression of mature miRNAs, significantly impacting gene regulation. The expression of the HBx has also been inversely correlated with let-7a expression in a human hepatoma cell line, resulting in upregulation of its target, STAT3, and increased cell proliferation [165]. Furthermore, HBx upregulated miR-29a, which induced hepatoma cell migration via decreased expression of the tumor suppressor PTEN and downstream phosphorylation of AKT, a regulator of cell growth, survival, and migration [166]. HBx was also shown to decrease miR-132 abundance in HCC cells via methylation of the miR-132 promoter, leading to inhibition of AKT signaling [167]. Reduced miR-132 abundance was also observed in HBV-related HCC tumor tissue compared with adjacent nontumor tissue, and the levels of miR-132 in patient serum and tumor tissue were significantly correlated [167]. Taken together, these results indicate an oncogenic role for dysregulated miRNA in viral infection.

Interestingly, HCV has recently been reported to promote hepatic steatosis via induction of miR-27. By comparing HBV- and HCV-infected hepatoma cells, it was demonstrated that HCV infection produced higher levels of miR-27a, and this promoted an increase in cellular lipid [168]. MiR-27a was shown to target several genes involved in lipid metabolism including the transcription factor RXRα and the lipid transporter ATP-binding cassette subfamily A member 1 (ABCA1). Furthermore, miR-27a enhanced IFN signaling in vitro, and patients who expressed high levels of miR-27a in the liver showed a more favorable response to IFN-based therapy [168]. Similarly, [169] showed that HCV infection of cells both in vitro and in vivo increased cellular abundance of miR-27b and decreased expression of its target PPARα, resulting in lipid accumulation and hepatic steatosis. These studies indicate that miRNAs may be mediators of HCV-induced hepatic steatosis.

In addition to modulation of host cell properties, host miRNAs can directly regulate viral replication of both HBV and HCV. For example, miR-372 and miR-373 that are upregulated in HBV-infected liver can stimulate the production of HBV DNA and proteins via downregulation of the transcription factor nuclear factor I/B (NFIB) [170]. NFIB has previously been implicated in the regulation of several viruses such as feline leukemia, human papilloma, and mouse mammary tumor viruses. In another study that assessed the effect of miRNA mimics on HBV replication, miR-1 was shown to indirectly increase HBV transcription, gene expression, and progeny secretion through modulation of several host genes including farnesoid X receptor α (FXRA), which enhanced HBV promoter activity [171]. It has also been shown that HBV significantly increases miR-501 abundance in hepatoma cells, and suppression of miR-501 reduced HBV replication via repression of HBXIP, a negative regulator of HBV replication, identifying miR-501 as a promoter of viral replication [172]. Likewise, HBx expression in similar cells reduced miR-15 and miR-16 abundance, which could prevent HBV replication by targeting viral HBx and HBV polymerase protein (HBp) [173]. Conversely, HBV infection of hepatoma cells can also induce the miR-17-92 cluster (miR-17-5p, -18a, -19a, -19b, -20a, and -92a-1) in a c-MYC dependent manner to restrict HBV replication [174]. Similarly for HCV, miR-141 could downregulate expression of the tumor suppressor DLC-1, increasing HCV replication and cell proliferation in primary human hepatocytes [175]. Furthermore, using miRNA mimics and inhibitors, four miRNAs (miR-24, -149-3p, -638, and -1181) dysregulated in a HCV infected human hepatoma cell line were able to influence HCV entry or HCV RNA abundance, indicative of viral replication [176]. Collectively, these studies highlight the intricate interplay between HBV and HCV and the host cellular miRNAs and the regulatory effects they exert on viral replication.

As a liver-specific miRNA, considerable attention has focused on the role of miR-122 in regulating chronic liver disease, particularly in HBV/HCV infection. A search of the literature revealed more than 100 publications linked to the search terms "miR-122 and HBV or HCV." Subsequent to the discovery that HBV and HCV can induce markedly different host miRNA expression profiles, a striking difference in the role of miR-122 in HBV and HCV infection has been observed. It has been shown that cellular HCV RNA decreased [177] when miR-122 was inactivated. Since then, several published studies have revealed a requirement of miR-122 and/or its precursor molecule for HCV replication and stability [178–180] and successfully validated an anti-miR-122 therapy for suppression of viremia in chronically HCV-infected primates [181]. Moreover, a study in which the serum from 102 HCV patients was compared with healthy controls identified a 23-fold increase of miR-122, indicating

that it may be a useful biomarker to detect chronic HCV infections [182]. Conversely, miR-122 was found to be significantly downregulated in the livers of chronic HBV-infected patients, inversely correlating with viral load [183]. Further investigation disclosed that loss of miR-122 enhances HBV transcription through a complex signaling pathway involving upregulation of cyclin G1 (CCNG1), which binds p53, thus preventing p53-mediated repression of a HBV enhancer element [183]. Consistent with these findings, miR-122 is decreased in HBV-associated but not HCV-associated HCC [184]. The mechanism of miR-122 downregulation is mediated, at least in part, by the binding of HBx to PPARγ leading to inhibition of miR-122 transcription [86].

At first glance miR-122 appears to be a good candidate for development of a miRNA-based therapy for modulating viral replication in both HCV- and HBV-infected patients. However, a separate study revealed a subset of chronic hepatitis C individuals with decreased levels of miR-122 that would be poor candidates for miR-122-based therapy [185]. Thus a more thorough understanding of the role of miR-NAs in viral infection is required before successful miRNA therapies can be developed.

### 12.15.4 PRIMARY LIVER CANCER

Liver cancer is the sixth most commonly diagnosed cancer, and the third leading cause of cancer-related death worldwide [186]. HCC accounts for the majority (85–90%) of primary liver cancers in adults, with the remainder attributed to cholangiocarcinoma (CC) as well as other rare entities [129]. Chronic liver diseases, such as chronic viral hepatitis as well as alcohol abuse, metabolic disorders, and aflatoxin B exposure, are major etiologic risk factors for the development of primary liver cancer, in particular HCC. Notably, in the Western world the sharp increase in the incidence of NAFLD and NASH is predicted to supersede alcohol as the most prevalent risk factor for liver cancer within the next decade. Curative treatment of liver cancer remains challenging as traditional chemotherapies and radiation therapies are largely ineffective against advanced HCC. Surgical intervention by means of liver resection or transplantation as well as radiofrequency ablation for eligible patients therefore offers the most effective treatment option for liver cancer. However, in the majority of patients, diagnosis occurs at advanced stages, leading to ineligibility for liver transplant. Furthermore, high recurrence rates and frequent metastases result in poor patient outcomes (reviewed in [187]). These observations clearly indicate that liver cancer is a substantial healthcare problem and the development of novel molecular approaches to treat HCC is therefore a growing field of interest.

Hepatocarcinogenesis is a multistage process that results from sequential accumulation of (epi)-genetic events in three core cellular processes, i.e., cell fate, cell survival, and genome maintenance [188]. MiRNAs are implicated in all of these processes and, like protein-coding genes, can function as oncogenes or tumor suppressors depending on the mRNAs they target. MiRNAs are frequently located in cancer-associated genomic regions or fragile sites, and dysregulated miRNA expression has been reported in many cancers, including liver cancer (reviewed in [189, 190]). A number of studies have assessed the role of miRNAs with regulatory effects on key pro-oncogenic processes in the context of primary liver cancer. Downregulation of several miRNAs in HCC cells including miR-26a, -101, -22, -99a, and -129-5p (Figure 12.1) has been reported with subsequent upregulation of their respective targets, Ccnd2 and Ccne2, MCL1, HDAC4, IGF-1R and mTOR, and Valosin-containing protein (VCP) [55, 191–194]. Each of the targets has been reported [32] to regulate different aspects of tumor growth including cell cycle progression (e.g., Cyclin D2) and apoptotic resistance (e.g., Mcl-1). Similarly, increased miR-25 abundance was observed in CC lines and patient samples, and correlated with apoptotic resistance via downregulation of the TRAIL receptor, death receptor-4 (TRAIL-DR4) [195]. Additionally, many recent studies have compared miRNA expression profiles in liver cancer to normal liver, and there are several excellent reviews that describe these data comprehensively [196–198]. Therefore this chapter will focus mainly on more recent studies that have analyzed miRNAs, in particular those with translational implications such as predictors of patient outcome. These include miRNAs involved in early diagnosis and tumor classification, disease progression, chemotherapy resistance, tumor metastasis, and recurrence of HCC and CC.

Recently, global analyses of miRNAs by next-generation sequencing identified nine miRNAs (miR-122, -99a, -101, -192, -199a/b-3p, and several let-7 family members) with differential expression between normal human liver, hepatitis, and HCC accounting for ~88.2% of the "miRNome" in the liver [199]. The study also demonstrated that low miR-199a/b-3p expression was associated with poorer outcome for HCC patients. Moreover, subsequent targeting of miR-199a/b-3p using adeno-associated virus (AAV) 8 inhibited tumor growth via interacting with PAK4/Raf/MEK/ERK pathway.

Given the prominent role of miR-122 in liver homeostasis, several studies have demonstrated the relevance of this miRNA in liver cancer development [200, 201]. The deletion of murine miR-122 not only leads to hepatic steatosis and inflammation but also to HCC development by activating oncogenic pathways, e.g., TGFβ, MAPK, and PTEN, as well as increased infiltration of inflammatory cells that produce proinflammatory cytokines, including IL-6 and TNF [202, 203]. Furthermore, in a clinical context, HCCs with low miR-122 displayed a particularly poor prognosis and showed enrichment of gene sets commonly associated with cancer progression [204]. Functionally, loss of miR-122 resulted in an increase of cell migration and invasion indicating that miR-122 is a marker of hepatocyte-specific differentiation and an important determinant in the control of metastasis.

Another global miRNA analysis performed on a large cohort of 104 HCC, 90 adjacent cirrhotic livers, 21 normal livers, and 35 HCC cell lines identified a set of 12 dysregulated miRNAs (including miR-21, -221, -222, -34a, -519a, -93, -96, and let-7c) that were associated with liver cancer progression, where miR-221/222 were the most upregulated [205]. Moreover, miR-221/222 enhanced cell proliferation via targeting CDKN1B (p27) [205], and induced TRAIL resistance and enhanced cellular migration through the activation of the AKT pathway as well as metalloproteinases by targeting the PTEN and TIMP3 tumor suppressors [206]. Interestingly, as in liver regeneration, miR-21 is also highly increased in HCC and affected cell proliferation, migration, and invasion by targeting the PTEN/mTOR pathway [207]. Besides the above-mentioned studies, several other miRNAs have been demonstrated to harbor both tumor suppressive (e.g., miR-1, -26, -29, -34a, -195, and -223) and oncogenic activity (e.g., miR-9, -181, and -224) in HCC underlining the relevance of miRNAs as targets for diagnostic or therapeutic interventions in liver cancer (see review by [208]).

MiRNAs show great potential as biomarkers since they are secreted into the circulation through vesicles (microvesicles or exosomes) and can be detected in almost all bodily fluids [209]. Consequently, several studies have assessed the potential of miRNAs to aid diagnosis of primary liver cancer, including improving the sensitivity and specificity of HCC detection by evaluating tissue and plasma miRNA expression [210]. Using a cohort of 96 HCCs [211] demonstrated that the combined use of miR-15b and miR-130 yielded 98.2% sensitivity and 91.5% specificity, and detection sensitivity remained high (96.7%) in a subgroup of HCCs with low circulating Afp, a commonly used serum biomarker for HCC (<20 ng/mL). Furthermore, these miRNAs successfully identified Afp-negative, early-stage HCC cases indicating their potential as complementary diagnostic tools. In another screen of sera from more than 500 patients, six miRNAs were identified that were significantly increased in HCC vs control samples [118]. Among these, miR-375 and miR-92a were also enriched in patients infected with HBV. Moreover, a combination of three of these miRNAs (miR-25, -375, and let-7f) accurately distinguished HCC from control patients.

Similarly, using a cohort of 934 patient plasma samples [212] identified a miRNA panel (miR-122, -192, -21, -223, -26a, -27a, and -801) with high diagnostic accuracy for HCC irrespective of disease status. However, it was particularly useful for the diagnosis of early-stage HBV-related HCCs, and could discriminate HCC from healthy, chronic HBV and cirrhotic patients. Other miRNAs associated with HCC included miR-222, -223, -21, and -122 [213, 214]. Another recent study showed that circulating miR-21 was significantly increased in patients with HCC compared to patients with chronic hepatitis or healthy individuals [210], and miR-21 expression was reduced following surgical resection of HCC. Finally, combination of miR-21 expression levels with traditional markers of HCC such as Afp increased the sensitivity and specificity of detection thus improving the diagnostic power. Together, these data provide compelling evidence for the feasibility of utilizing circulating miRNAs as biomarkers for HCC diagnosis.

Prognostic classification using expression profiles has a long-standing history in HCC [215]. In recent years, the power of miRNAs for classifying liver cancers has been repeatedly demonstrated. Profiling miRNA by microarray has revealed subclasses associated with clinicopathological features as well as mutations in several oncogenic pathways such as β-Catenin and HNF1A [216]. A recent study that investigated the heterogeneity of miRNA profiles of 89 HCCs identified three clinically distinct subclasses of HCCs that showed enrichment of specific miRNAs [217]. The same study also identified an oncogenic role for miR-517a in vitro and in vivo, potentially via activation of NF-κB or MAPK/ERK signaling. Other miRNA profiling of paired tumor and nontumorous adjacent liver tissue from 73 HCC patients identified 13 and 56 recurrence-related miRNAs, respectively [218]. Importantly, while the number of miRNAs associated with recurrence was significantly higher in nontumorous tissue and predicted late recurrence, the tumor-derived miRNAs gave superior accuracy in prediction of early recurrence.

Given the described changes of several miRNAs during hepatocarcinogenesis, it is not surprising that miRNAs have emerged as attractive druggable targets in liver cancer [219]. Kota et al. [55] demonstrated that induction of miR-26a expression in hepatoma cells leads to a cell cycle arrest by directly targeting Cyclin D2 and E2. Subsequently, systemic in vivo delivery of this miRNA by AAVs in a mouse model of HCC resulted in reduced tumorigenicity without overt cellular toxicity in normal cells. Recent studies by two groups established the potential of targeting miR-221 as a cancer therapeutic [220, 221]. First, Park et al. [220] demonstrated that anti-miR-221 could effectively reduce in vivo miR-221 levels and lead to subsequent inhibition of tumor cell proliferation as well as increased apoptosis. Callegari et al. [221] then utilized a transgenic mouse model with liver-specific constitutive overexpression of miR-221 and resultant concomitant inhibition of its target protein-coding genes (i.e., CDK inhibitors Cdkn1b [p27], and Cdkn1c [p57], and B-cell lymphoma 2-modifying factor [Bmf]). The miR-221 transgenic mouse develops spontaneous nodular liver lesions, and consistent with the previous study, in vivo delivery of anti-miR-221 oligonucleotides led to a significant reduction of tumorigenicity in the same model.

Chemotherapeutic resistance is another major challenge in the treatment of primary liver cancer and contributes to poor patient outcome. Resistance of liver cancer cell lines to cytotoxic compounds including IFN-α, 5-fluorouracil, and cisplatin has been associated with changes in miR-146a, -193a-3p, and -199a-5p abundance, respectively [222–224]. However, the proposed mechanism for how each miRNA confers resistance differs. For example, miR-193a-3p and miR-146a suppress apoptosis via modulation

of proapoptotic CASP2L through SRSF2, and SMAD4, respectively, whereas decreased miR-199a-5p in resistant cells increases cell proliferation by activation of autophagy via ATG7. Importantly, in all three studies, the sensitivity to the respective chemotherapeutics could be altered in vitro by modulating miRNA abundance [222–224]. Ji et al. (2009) [225] recently dissected the miRNA contribution for the response to adjuvant therapy with IFN-α, thereby confirming the therapeutic potential of miRNA-guided treatment modalities in HCC. The authors profiled miRNAs from 445 HCC patients comprising several cohorts, and demonstrated that the miRNA profile in liver tissue is vastly different between men and women with HCC. Furthermore, the authors identified miR-26 as a good predictor for sensitivity to IFN-α treatment and, more recently, developed a simple and reliable diagnostic test (MIR26-DX) to select HCC patients for adjuvant IFN-α therapy, thereby setting a benchmark first step to successfully utilize miRNAs in the clinic [226].

Liver cancer has a strong tendency for metastasis and early tumor recurrence following surgical intervention, contributing significantly to poor patient outcome. Besides malignant transformation and promotion of HCC, several miRNAs have been implicated in promotion or repression of this process (reviewed in [227, 228]). Several groups have investigated the role of miRNAs in metastasis of primary liver cancer. Not surprisingly, a prominent role for miR-122 was also confirmed in this process. Loss of miR-122 induced the generation of intrahepatic metastasis by promoting angiogenesis via regulation of ADAM17 [229, 230]. Li et al. (2009) correlated decreased miR-34a abundance with metastasis and invasion of HCC tumors and identified c-MET as a direct target of miR-34a, which can modulate cell migration and invasion through ERK1/2 signaling. A recent study further demonstrated that reduction of miR-26a was associated with HCC recurrence and metastasis [231]. Xu et al. (2013) [102] showed that HBx directly repressed miR-148a expression in a p53-dependent manner thereby promoting cancer growth and metastasis through targeting of hematopoietic pre-B-cell leukemia transcription factor-interacting protein (HPIP). Accordingly, inhibition of HPIP expression by miR-148a, reduced AKT and ERK levels and subsequently mTOR through AKT/ERK signaling. The authors concluded that miR-148a activation or HPIP inhibition might be a useful strategy for cancer treatment. Another recently described tumor suppressive miRNA with antimetastatic properties is miR-612 [232]. MiR-612 exerted its function by regulating AKT2 during epithelial–mesenchymal transition (EMT) and metastasis and showed an inverse correlation with tumor size, stage, EMT, and metastasis in HCC patients. Furthermore, miR-612 not only affected local invasion but also intravasation at distant sites indicating that this miRNA is involved throughout the entire metastatic process.

Budhu et al. [233] generated a 20-miRNA signature from a cohort of 241 HCC patients that efficiently predicted the occurrence of venous metastases and this correlated with patient outcome. This included four increased miR-NAs including miR-219-1, and 16 decreased miRNAs including miR-30c-1, -148a, and -34a (Figure 12.1). A separate study also compared the miRNA profiles of primary HCC and venous metastasis for 20 matched patients [234]. Interestingly, although nontumorous livers exhibited distinct miRNA profiles compared with primary HCCs and venous metastases, no apparent difference in the profiles of primary HCCs and venous metastases was revealed. However, miRNA levels were markedly reduced in venous metastases compared with primary HCCs, suggesting that miRNA deregulation occurs early in hepatocarcinogenesis and that the generation of metastasis is aggravated by a stepwise disruption of the deregulated miRNAs. Few studies have been performed that compared metastatic and nonmetastatic CC; however, in one study, miR-214 was decreased in intrahepatic metastatic CC tissues compared to nonmetastatic tissues, and inhibition of miR-214 promoted metastasis of a CC cell line in vitro [235]. Subsequent upregulation of TWIST1, a miR-214 target gene associated with EMT, and consequent decrease in E-cadherin, could explain the EMT phenotype and may be crucial for CC metastasis in vivo.

The ability to predict an individual's risk of liver tumor recurrence would be highly beneficial to aid candidate selection for transplant and in tailoring appropriate postoperative care for the individual. Altered expression of certain miRNAs in the cancerous liver has been linked to an increased risk of disease recurrence following resection. Specifically, three studies defined miRNA expression signatures, consisting of 7, 18, and 67 miRNAs, respectively, from HCC tissue samples to distinguish patients with a high risk of recurrence from those with a low risk [236–238]. From these studies miR-10b, -21, -34c, -155, and let-7d positively correlated with recurrence, whilst miR-15b, -20a, -24, -122, -145, and -182 negatively correlated with recurrence. Further investigation revealed a role for miR-15b in cell proliferation and inhibition of apoptosis, with the antiapoptotic gene BCL-W identified as a direct target of miR-15b [236]. Finally, the gene ontology (GO) and Kyoto Encyclopedia of Genes and Genomes (KEGG) [238] performed a bioinformatics pathway analysis, which identified signal transduction, cell cycle, cell differentiation, cell proliferation, and apoptosis as the most significant GO categories for predicted miRNA targets, with MAPK signaling, pathways in cancer, Wnt signaling, and insulin signaling as the most significant KEGG pathways for dysregulated miR-NAs in HCC recurrence.

## 12.16 MiRNAs IN HCC THERAPY

Alterations in miRNA expression are frequently associated with HCC disease physiology; hence miRNAs could be used as potential druggable targets in HCC management. One of the approaches in miRNA-based HCC therapy involves using antagomirs against onco-miRs. Nonhuman primates

chronically infected with HCV when treated with locked nucleic acid (LNA) specific for miR-122 (a positive regulator of HCV replication), exhibited long-term suppression of HCV viral load, supporting therapeutic use of miRNA in HCC [10]. In another preclinical study, involving an orthotopic HCC mouse model, oligonucleotides targeting miR-221 inhibited cell transformation and improved survival, underscoring its potential in HCC therapeutics [13]. Another potential therapeutic approach for HCC treatment entails restoration of TS-miRs, thus serving as anticancer agents. For instance, AAV-mediated delivery of miR-122 and miR-26a and systemic restoration of miR-124 can suppress tumorigenesis in animal models of HCC. Similarly, restoration of miR-375 ($2'O$-methyl-modified and cholesterol conjugated form) and miR-29 could inhibit tumorigenesis in preclinical HCC models [10, 13].

Delivery of miRNA mimics can also be used in HCC therapy. Indeed, a cancer-targeting miRNA mimic of miR-34 (MRX34) is in Phase 1 clinical trial performed by Mirna Therapeutics Inc., in HCC patients [13, 239]. Interestingly, the mimic miRNAs are delivered using Smarticles, which are anionic at neural pH, but attain cationic charge in acidic tumor environment thus minimizing off-target effects [13]. Importantly, the therapeutic potential of miR-122 antagonist Miravirsen was evident from a multicentric Phase 2A trial, which showed sequestration of mature miR-122 and reduction of viral load [240].

Though miRNAs possess tremendous therapeutic potential for HCC, a major concern remains their delivery system. MiRNAs can be incorporated into PEGylated stable nucleic acid lipid particles (SNALPS), to extend the circulation time. Further, virus-like particle (VLP)–dependent delivery gives the leverage of natural tissue tropism, albeit with risk of eliciting immune response [13].

The emergence of miRNAs as novel clinical biomarkers is set to change the face of HCC diagnosis and therapeutic procedures. The proposition of miRNA profiles serving as signatures to distinguish HCV from HBV cases, though awaiting clinical evaluation, offers the advantage of accurate diagnosis and appropriate therapeutic course.

## 12.17 CONCLUSIONS AND FUTURE DIRECTIONS

The literature [241–244] covered by this chapter clearly demonstrates that miRNAs play an important and extensive role in liver development and regeneration as well as contributing to or preventing chronic liver disease. Furthermore, it appears possible to use miRNA expression data to distinguish different types of liver pathologies emanating from different etiologies. This strongly advocates miRNA profiling as a viable alternate means to diagnose liver disease and predict patient outcome, and further suggests that miRNA manipulation may be the basis of effective therapy for the treatment of liver disease. The use of anti-miR-122 to suppress HCV viremia [181] and the observation that adeno-associated viral delivery of miR-26a to the livers of mice can suppress tumor growth arising from c-MYC overexpression [55] are two encouraging examples that highlight the potential of miRNA-based therapy to treat liver disease. Indeed the efficacy of an anti-miR-122 therapeutic (Miravirsen) for the treatment of genotype 1 HCV is currently in Phase 2 clinical trials [245].

Whilst significant progress has been made in our understanding of the biological role of miRNAs, the complexity of miRNA regulation, including specific cell/tissue type and temporal expression, target multiplicity, and miRNA redundancy, presents many challenges to the miRNA researcher. The purpose of this chapter was to examine the role of miRNAs in liver development, regeneration, and disease. Remarkably, only a handful of miRNAs are consistently reported in these processes, including miR-23, -29, -30, -122, and -192 and members of the let-7 family (Figure 12.1). Not surprisingly, as a liver-specific miRNA, miR-122 featured prominently in all processes discussed except liver regeneration. Furthermore, it was the only miRNA to be consistently dysregulated in all four chronic liver diseases covered by this review. It is noteworthy that while some miRNAs featured prominently in one disease or process, e.g., miR-34a in NAFLD/NASH or miR-21 in primary liver cancer, they were not consistently reported in other conditions (Figure 12.1). Our criterion for inclusion in Figure 12.1 was that a miRNA be reported at least twice for a particular process/condition and the majority of miRNAs did not meet this criterion. Thus the conclusions that can be drawn from these types of studies are two-fold: miRNAs regulating these processes are in fact quite unique with little overlap, or, alternatively, that the full complement of miRNAs participating in these processes is not yet known. It is likely however, that the above-mentioned miRNAs are critical in liver development and disease.

As miRNAs exert their effect by targeting protein-coding mRNAs, verifying targets of dysregulated miRNAs is of utmost importance. Whilst a large body of research has been undertaken thus far, much of it highly descriptive, detailing numerous dysregulated miRNAs, many studies lack definitive evidence of functional relevance by showing that appropriate target mRNAs are affected.

Drawing meaningful conclusions from this body of work has also been complicated by the fact that research groups often use different model systems, time points, and methodologies, and currently, the field lacks studies that verify the findings of other groups. Looking ahead, miRNA research in liver development and disease could be significantly enhanced by encouraging comprehensive target validation for all miRNA studies. Importantly, this should be performed under well-defined conditions, for example cell/tissue type or disease model, as it is clear that these variables can affect outcome in regard to miRNA expression and target regulation.

The heterogeneity in individual cases of cancer including HCC demands development of personalized medicine to ensure the most effective treatment with minimal side effects. In such a scenario, unique tumor-specific miRNA signatures, as reviewed here, will help design accurate diagnostics and therapeutics tailored to individual needs. The ease of delivering oligonucleotides to liver and high tolerance of normal liver cells to supplements of deficient miRNAs make HCC an ideal model to test miRNA therapy [10]. Preliminary evidence in nonhuman primates indicates that miRNA-based treatment of chronic HCV infection provides prolonged alleviation of virus-induced liver pathology with a high barrier to viral resistance, suggesting miRNAs may function as better antivirals than conventional drugs [10]. CSCs, resistant to conventional chemotherapy and cause of cancer recurrence, form the major obstacle in successful treatment of the disease. The discovery of stemness-associated miRNAs in liver cancer such as miR-181, miR-150, and miR-548c-5p warrants further evaluation of their potential as druggable targets in therapies against liver CSCs [246]. The gender-specific discrepancy in HCC incidence can be in part explained by miRNA signatures such as higher levels of TS miR-26a and miR-26b in females that may provide them a protective edge over males [247]. Elevated levels of miR-18a in HCC female patients, which suppresses protective effects of estrogen by targeting estrogen receptor-alpha, can be used as a risk predictor of HCC development in female populations [248].

Finally, the quest for yet elusive HBV- and HCV-encoded miRNAs poses a challenge, which when overcome may help in devising novel strategies to silence viral miRNA and cure liver cancer.

## REFERENCES

[1] Frith MC, Pheasant M, Mattick JS. The amazing complexity of the human transcriptome. Eur J Hum Genet (2005) **13**:894–7.

[2] Bernstein BE, Birney E, Dunham I, Green ED, Gunter C, Snyder M, et al. An integrated encyclopedia of DNA elements in the human genome. Nature (2012) **489**:57–74.

[3] Mattick JS, Makunin IV. Non-coding RNA. Hum Mol Genet (2006) **15**:R17–29.

[4] Ponting CP, Oliver PL, Reik W. Evolution and functions of long noncoding RNAs. Cell (2009) **136**:629–41.

[5] Siomi H, Siomi MC. On the road to reading the RNA-interference code. Nature (2009) **457**:396–404.

[6] Costa FF. Non-coding RNAs: meet thy masters. Bioessays (2010) **32**:599–608.

[7] Si-Tayeb K, Lemaigre FP, Duncan SA. Organogenesis and development of the liver. Dev Cell (2010) **18**:175–89.

[8] Steer CJ. Liver regeneration. FASEB J (1995) **9**:1396–400.

[9] Stewart BW, Wild CP. World Cancer Report 2014. Lyon: International Agency for Research on Cancer, World Health Organization (2014).

[10] Callegari E, Gramantieri L, Domenicali M, D'Abundo L, Sabbioni S, Negrini M. microRNAs in liver cancer: a model for investigating pathogenesis and novel therapeutic approaches. Cell Death Differ (2015) **22**:46–57. doi:10.1038/cdd.2014.136

[11] Vasudevan S, Tong Y, Steitz JA. Switching from repression to activation: microRNAs can up-regulate translation. Science (2007) **318**:1931–4. doi:10.1126/science.1149460

[12] Fabbri M, Garzon R, Cimmino A, Liu Z, Zanesi N, Callegari E, et al. microRNA- 29 family reverts aberrant methylation in lung cancer by targeting DNA methyl-transferases 3A and 3B. Proc Natl Acad Sci USA (2007) 104:15805–10. doi:10.1073/pnas.0707628104

[13] Yang N, Ekanem NR, Sakyi CA, Ray SD. Hepatocellular carcinoma and microRNA: new perspectives on therapeutics and diagnostics. Adv Drug Deliv Rev (2014) 81C: 62–74. doi:10.1016/j.addr.2014.10.029

[14] Chu R, Mo G, Duan Z, Huang M, Chang J, Li X, et al. miRNAs affect the development of hepatocellular carcinoma via dysregulation of their biogenesis and expression. Cell Commun Signal (2014) 12:45. doi:10.1186/s12964-014-0045-y

[15] Lee RC, Feinbaum RL, Ambros V. The C. elegans heterochronic gene lin-4 encodes small RNAs with antisense complementarity to lin-14. Cell (1993) 75:843–54.

[16] Wightman B, Ha I, Ruvkun G. Posttranscriptional regulation of the heterochronic gene lin-14 by lin-4 mediates temporal pattern formation in C. elegans. Cell (1993) 75:855–62.

[17] Pasquinelli AE, Reinhart BJ, Slack F, Martindale MQ, Kuroda MI, Maller B, et al. Conservation of the sequence and temporal expression of let-7 heterochronic regulatory RNA. Nature (2000) 408:86–9.

[18] Reinhart BJ, Slack FJ, Basson M, Pasquinelli AE, Bettinger JC, Rougvie AE, et al. The 21- nucleotide let-7 RNA regulates developmental timing in Caenorhabditis elegans. Nature (2000) 403:901–6.

[19] Bartel DP. MicroRNAs: genomics, biogenesis, mechanism, and function. Cell (2004) 116:281–97.

[20] Bartel DP. MicroRNAs: target recognition and regulatory functions. Cell (2009) 136:215–33.

[21] Valencia-Sanchez MA, Liu J, Hannon GJ, Parker R. Control of translation and mRNA degradation by miRNAs and siRNAs. Gene Dev (2006) 20:515–24.

[22] Lewis BP, Burge CB, Bartel DP. Conserved seed pairing, often flanked by adenosines, indicates that thousands of human genes are microRNA targets. Cell (2005) 120:15–20.

[23] Tzur G, Israel A, Levy A, Benjamin H, Meiri E, Shufaro Y, et al. Comprehensive gene and microRNA expression profiling reveals a role for microRNAs in human liver development. PLoS One (2009) 4:e7511.

[24] Liu D, Fan J, Zeng W, Zhou Y, Ingvarsson S, Chen H. Quantitative analysis of miRNA expression in several developmental stages of human livers. Hepatol Res (2010a) 40:813–22.

[25] Wei W, Hou J, Alder O, Ye X, Lee S, Cullum R, et al. Genome-wide microRNA and messenger RNA profiling in rodent liver development implicates mir302b and mir20a in repressing transforming growth factor-beta signaling. Hepatology (2013a) 57:2491–501.

[26] Lagos-Quintana M, Rauhut R, Yalcin A, Meyer J, Lendeckel W, Tuschl T. Identification of tissue-specific microRNAs from mouse. Curr Biol (2002) 12:735–9.

[27] Chang J, Nicolas E, Marks D, Sander C, Lerro A, Buendia MA, et al. miR-122, a mammalian liver-specific microRNA, is processed from hcr mRNA and may downregulate the high affinity cationic amino acid transporter CAT-1. RNA Biol (2004) 1:106–13.

[28] Xu H, He JH, Xiao ZD, Zhang QQ, Chen YQ, Zhou H, et al. Liver-enriched transcription factors regulate microRNA-122 that targets CUTL1 during liver development. Hepatology (2010) 52:1431–2.

[29] Hand NJ, Master ZR, EauClaire SF, Weinblatt DE, Matthews RP, Friedman JR. The microRNA-30 family is required for vertebrate hepatobiliary development. Gastroenterology (2009a) 136:1081–90.

[30] Bernstein E, Kim SY, Carmell MA, Murchison EP, Alcorn H, Li MZ, et al. Dicer is essential for mouse development. Nat Genet (2003) 35:215–7.

[31] Hand NJ, Master ZR, Lay JL, Friedman JR. Hepatic function is preserved in the absence of mature microRNAs. Hepatology (2009b) 49:618–26.

[32] Finch ML, Marquardt JU, Yeoh GC, Callus BA. Regulation of microRNAs and their role in liver development, regeneration and disease. Int J Biochem Cell Biol (September 2014) 54:288-303. https://doi.org/10.1016/j.biocel.2014.04.002

[33] Yu F, Lin Z, Zheng J, Gao S, Lu Z, Dong P. Suppression of collagen synthesis by Dicer gene silencing in hepatic stellate cells. Mol Med Rep (2014) 9:707–14.

[34] Fougère-Deschatrette C, Imaizumi-Scherrer T, Strick-Marchand H, Morosan S, Charneau P, Kremsdorf D, et al. Plasticity of hepatic cell differentiation: bipotential adult mouse liver clonal cell lines competent to differentiate in vitro and in vivo. Stem Cells (2006) 24:2098–109.

[35] Chen Y, Zhou H, Sarver AL, Zeng Y, Roy-Chowdhury J, Steer CJ, et al. Hepatic differentiation of liver-derived progenitor cells and their characterization by microRNA analysis. Liver Transplant (2010) 16:1086–97.

[36] Kim N, Kim H, Jung I, Kim Y, Han DKYM. Expression profiles of miRNAs in human embryonic stem cells during hepatocyte differentiation. Hepatol Res (2011) 41:170–83.

[37] Cui L, Zhou X, Li J, Wang L, Wang J, Li Q, et al. Dynamic microRNA profiles of hepatic differentiated human umbilical cord lining-derived mesenchymal stem cells. PLoS One (2012) 7:e44737.

[38] Deng XG, Qiu RL, Wu YH, Li ZX, Xie P, Zhang J, et al. Overexpression of miR-122 promotes the hepatic differentiation and maturation of mouse ESCs through a miR-122/FoxA1/HNF4a-positive feedback loop. Liver Int (2014) 34:281–95.

[39] Michalopoulos GK. Liver regeneration. J Cell Physiol (2007) 213:286–300.

[40] Marquez RT, Wendlandt E, Galle CS, Keck K, McCaffrey AP. MicroRNA-21 is upregulated during the proliferative phase of liver regeneration, targets Pellino-1, and inhibits NF-κB signaling. Am J Physiol Gastrointest Liver Physiol (2010) 298:G41–535.

[41] Ng R, Song G, Roll GR, Frandsen NM, Willenbring H. A microRNA-21 surge facilitates rapid cyclin D1 translation and cell cycle progression in mouse liver regeneration. J Clin Invest (2012) 122:1097–108.

[42] Pan C, Chen H, Wang L, Yang S, Fu H, Zheng Y, et al. Down-regulation of MiR- 127 facilitates hepatocyte proliferation during rat liver regeneration. PLoS One (2012) 7:e39151.

[43] Zhou J, Ju W, Wang D, Wu L, Zhu X, Guo Z, et al. Down-regulation of microRNA-26a promotes mouse hepatocyte proliferation during liver regeneration. PLoS One (2012) 7:e33577.

[44] Chen H, Sun Y, Dong R, Yang S, Pan C, Xiang D, et al. Mir-34a is upregulated during liver regeneration in rats and is associated with the suppression of hepatocyte proliferation. PLoS One (2011a) 6:e20238.

[45] Yuan B, Dong R, Shi D, Zhou Y, Zhao Y, Miao M, et al. Down-regulation of miR-23b may contribute to activation of the TGF-b1/Smad3 signalling pathway during the

termination stage of liver regeneration. FEBS Lett (2011) **585**:927–34.

[46] Raschzok N, Sallmon H, Dame C, Sauer IM. Liver regeneration after partial hepatectomy: inconsistent results of expression screenings for human, mouse, and rat microRNAs. Am J Physiol Gastrointest Liver Physiol (2012) **302**:G470–1.

[47] Song G, Sharma AD, Roll GR, Ng R, Lee AY, Blelloch RH, et al. MicroRNAs control hepatocyte proliferation during liver regeneration. Hepatology (2010) **51**:1735–43.

[48] Chen X, Murad M, Cui YY, Yao LJ, Venugopal SK, Dawson K, et al. miRNA regulation of liver growth after 50% partial hepatectomy and small size grafts in rats. Transplantation (2011b) **91**:293–9.

[49] Raschzok N, Werner W, Sallmon H, Billecke N, Dame C, Neuhaus P, et al. Temporal expression profiles indicate a primary function for microRNA during the peak of DNA replication after rat partial hepatectomy. Am J Physiol Regul Integr Comp Physiol (2011) **300**:R1363–72.

[50] Shu J, Kren BT, Xia Z, Wong PYP, Li L, Hanse EA, et al. Genomewide microRNA down-regulation as a negative feedback mechanism in the early phases of liver regeneration. Hepatology (2011) **54**:609–19.

[51] Castro RE, Ferreira DM, Zhang X, Borralho PM, Sarver AL, Zeng Y, et al. Identification of microRNAs during rat liver regeneration after partial hepatectomy and modulation by ursodeoxycholic acid. Am J Physiol Gastrointest Liver Physiol (2010) **299**:G887–97.

[52] Dippold RP, Vadigepalli R, Gonye GE, Hoek JB. Chronic ethanol feeding enhances miR-21 induction during liver regeneration while inhibiting proliferation in rats. Am J Physiol Gastrointest Liver Physiol (2012) **303**:G733–43.

[53] Schug J, McKenna LB, Walton G, Hand N, Mukherjee S, Essuman K, et al. Dynamic recruitment of microRNAs to their mRNA targets in the regenerating liver. BMC Genomics (2013) **14**:264.

[54] Yuan Q, Loya K, Rani B, Mobus S, Balakrishnan A, Lamle J, et al. MicroRNA-221 overexpression accelerates hepatocyte proliferation during liver regeneration. Hepatology (2013) **57**:299–310.

[55] Kota J, Chivukula RR, O'Donnell KA, Wentzel EA, Montgomery CL, Hwang H-W, et al. Therapeutic microRNA delivery suppresses tumorigenesis in a murine liver cancer model. Cell (2009) **137**:1005–17.

[56] Yu ZY, Bai YN, Luo LX, Wu H, Zeng Y. Expression of microRNA-150 targeting vascular endothelial growth factor-A is downregulated under hypoxia during liver regeneration. Mol Med Rep (2013) **8**:287–93.

[57] Bui TV, Mendell JT. Myc: maestro of microRNAs. Genes Cancer (2010) **1**:568–75. doi:10.1177/1947601910377491

[58] Han H, Sun D, Li W, Shen H, Zhu Y, Li C, et al. Ac-Myc-microRNA functional feedback loop affects hepatocarcinogenesis. Hepatology (2013) **57**:2378–89. doi:10.1002/hep.26302

[59] Wang L, Zhang X, Jia LT, Hu SJ, Zhao J, Yang JD, et al. c-Myc-mediated epigenetic silencing of microRNA-101 contributes to dysregulation of multiple pathways in hepatocellular carcinoma. Hepatology (2014) **59**:1850–63. doi:10.1002/hep.26720

[60] Wang B, Hsu SH, Wang X, Kutay H, Bid HK, Yu J, et al. Reciprocal regulation of microRNA-122 and c-Myc in hepatocellular cancer: role of E2F1 and transcription factor dimerization partner 2. Hepatology (2014) **59**:555–66. doi:10.1002/hep.26712

[61] O'Donnell KA, Wentzel EA, Zeller KI, Dang CV, Mendell JT. c-Myc-regulated microRNAs modulate E2F1expression.Nature (2005) **435**:839–43. doi:10.1038/nature03677

[62] Wang X, Zhao X, Gao P, Wu M. c-Myc modulates microRNA processing via the transcriptional regulation of Drosha. SciRep (2013) **3**:1942. doi:10.1038/srep01942

[63] Connolly E, Melegari M, Landgraf P, Tchaikovskaya T, Tennant BC, Slagle BL, et al. Elevated expression of the miR-17-92 polycistron and miR-21in hepadnavirus-associated hepatocellular carcinoma contributes to the malignant phenotype. Am J Pathol (2008) **173**:856–64. doi:10.2353/ajpath.2008.080096

[64] Mukherjee A, Shrivastava S, Bhanja Chowdhury J, Ray R, Ray RB. Transcriptional suppression of miR-181c by hepatitis C virus enhances homeobox A1 expression. J Virol (2014) **88**:7929–40. doi:10.1128/JVI.00787-14

[65] Zou C, Li Y, Cao Y, Zhang J, Jiang J, Sheng Y, et al. Up-regulated microRNA-181a induces carcinogenesis in hepatitis B virus-related hepatocellular carcinoma by targeting E2F5. BMC Cancer (2014) **17**(14):97. doi:10.1186/1471-2407-14-97.

[66] Ji D, Chen Z, Li M, Zhan T, Yao Y, Zhang Z, et al. microRNA-181a promotes tumor growth and liver metastasis in colorectal cancer by targeting the tumor suppressor WIF-1. Mol Cancer (2014) **13**:86. doi:10.1186/1476-4598-13-86

[67] Ding Z, Qian YB, Zhu LX, Xiong QR. Promoter methylation and mRNA expression of DKK-3 and WIF-1 in hepatocellular carcinoma. World J. Gastroenterol (2009) **15**:2595 601. doi:10.3748/wjg.15.2595

[68] Feng Z, Zhang C, Wu R, Hu W. Tumor suppressor p53 meets microRNAs. J Mol Cell Biol (2011) **3**:44–50. doi:10.1093/jmcb/mjq040

[69] Hermeking H. microRNAs in the p53 network: micromanagement of tumour suppression. Nat Rev Cancer (2012) **12**:613–26. doi:10.1038/nrc3318

[70] Iliou MS, da Silva-Diz V, Carmona FJ, Ramalho-Carvalho J, Heyn H, Villanueva A, et al. Impaired DICER1 function promotes stemness and metastasis in colon cancer. Oncogene (2014) **33**:4003–15. doi:10.1038/onc.2013.398

[71] Chen YL, Yuan RH, Yang WC, Hsu HC, Jeng YM. The stem cellE3-ligase Lin-41 promotes liver cancer progression through inhibition of microRNA-mediated gene silencing. J Pathol (2013) **229**:486–96. doi:10.1002/path.4130

[72] Zhang J, Jin H, Liu H, Lv S, Wang B, Wang R, et al. MiRNA-99a directly regulates Ago2 through translational repression in hepatocellular carcinoma. Oncogenesis (2014) **3**:e97. doi:10.1038/oncsis.2014.11

[73] Hou J, Lin L, Zhou W, Wang Z, Ding G, Dong Q, et al. Identification of miRNomes in human liver and hepatocellular carcinoma reveals miR-199a/b-3p as therapeutic target for hepatocellular carcinoma. Cancer Cell (2011) **19**:232–43. doi:10.1016/j.ccr.2011.01.001

[74] Mott JL. microRNAs involved in tumor suppressor and oncogene pathways: implications for hepatobiliary neoplasia. Hepatology (2009) **50**:630–7. doi:10.1002/hep.23010

[75] Davis BN, Hilyard AC, Lagna G, Hata A. SMAD proteins control DROSHA-mediated microRNA maturation. Nature (2008) **454**:56–61. doi:10.1038/nature07086

[76] Pogribny IP, Rusyn I. Role of epigenetic aberrations in the development and progression of human hepato cellular carcinoma. Cancer Lett (2014) **342**:223–30. doi:10.1016/j.canlet.2012.01.038

[77] Anwar SL, Albat C, Krech T, Hasemeier B, Schipper E, Schweitzer N, et al. Concordant hypermethylation of intergenic microRNA genes in human hepatocellular carcinoma as new diagnostic and prognostic marker. Int J Cancer (2013) 133:660–70. doi:10.1002/ijc.28068

[78] Buurman R, Gürlevik E, Schäffer V, Eilers M, Sandbothe M, Kreipe H, et al. Histone deacetylases activate hepatocyte growth factor signaling by repressing microRNA-449 in hepato cellular carcinoma cells. Gastroenterology (2012) 143:811–20. doi:10.1053/j.gastro.2012.05.033

[79] Long XR, He Y, Huang C, Li J. microRNA-148a is silenced by hypermethylation and interacts with DNA methyltransferase1 in hepatocellular carcinogenesis. Int J Oncol (2014) 44:1915–22. doi:10.3892/ijo.2014.2373

[80] Huang J, Wang Y, Guo Y, Sun S. Down-regulated microRNA-152 induces aberrant DNA methylation in hepatitis B virus-related hepatocellular carcinoma by targeting DNA methyltransferase1. Hepatology (2010) 52:60–70. doi:10.1002/hep.23660

[81] Yuan JH, Yang F, Chen BF, Lu Z, Huo XS, Zhou WP, et al. The histone deacetylase 4/SP1/microrna-200a regulatory network contributes to aberrant histone acetylation in hepatocellular carcinoma. Hepatology (2011) 54:2025–35. doi:10.1002/hep.24606

[82] Dong YW, Wang R, Cai QQ, Qi B, Wu W, Zhang YH, et al. Sulfatide epigenetically regulates miR-223 and promotes the migration of human hepatocellular carcinoma cells. J Hepatol (2013) 60:792–801. doi:10.1016/j.jhep.2013.12.004

[83] Okamoto Y, Shinjo K, Shimizu Y, Sano T, Yamao K, Gao W, et al. Hepatitis virus infection affects DNA methylationin mice with humanized livers. Gastroenterology (2013) 146:562 72. doi:10.1053/j.gastro.2013.10.056

[84] Wei X, Tan C, Tang C, Ren G, Xiang T, Qiu Z, et al. Epigenetic represssion of miR-132 expression by the hepatitis B virus x protein in hepatitis B virus-related hepatocellular carcinoma. Cell Signal (2013) 25:1037–43. doi:10.1016/j.cellsig.2013.01.019

[85] Wei X, Xiang T, Ren G, Tan C, Liu R, Xu X, et al. miR-101 is down regulated by the hepatitis B virus x protein and induces aberrant DNA methylation by targeting DNA methyltransferase 3A. Cell Signal (2013) 25:439–46. doi:10.1016/j.cellsig.2012.10.013

[86] Song K, Han C, Zhang J, Lu D, Dash S, Feitelson M, etal. Epigenetic regulation of microRNA-122 by peroxisome proliferator activated receptor-gamma and hepatitis b virus X protein in hepatocellular carcinoma cells. Hepatology (2013) 58:1681–92. doi:10.1002/hep.26514

[87] Zhang T, Zhang J, Cui M, Liu F, You X, Du Y, et al. Hepatitis B virus X protein inhibits tumor suppressor miR-205 through inducing hypermethylation of miR-205 promoter to enhance carcinogenesis. Neoplasia (2013) 15:1282–91.

[88] Shiu TY, Huang SM, Shih YL, Chu HC, Chang WK, Hsieh TY. Hepatitis C virus core protein down-regulates p21(Waf1/Cip1) and inhibits curcumin-induced apoptosis through microRNA-345 targeting in human hepatoma cells. PLoS One (2013) 8:e61089. doi:10.1371/journal.pone.0061089

[89] Xu Y, Liu L, Liu J, Zhang Y, Zhu J, Chen J, et al. A potentially functional poly- morphism in the promoter region ofmiR-34b/c is associated with an increased risk for primary hepatocellular carcinoma. Int J Cancer (2011) 128:412–7. doi:10.1002/ijc.25342

[90] Li XD, Li ZG, Song XX, Liu CF. A variant in microRNA-196a2 is associated with susceptibility to hepatocellular carcinoma in Chinese patients with cirrhosis. Pathology (2010) 42:669–73. doi:10.3109/00313025.2010.522175

[91] Liu Y, Zhang Y, Wen J, Liu L, Zhai X, Liu J, et al. A genetic variant in the promoter region of miR-106b-25 cluster and risk of HBV infection and hepatocellular carcinoma. PLoS One (2012) 7:e32230. doi:10.1371/journal.pone.0032230

[92] Xu T, Zhu Y, Wei QK, Yuan Y, Zhou F, Ge YY, et al. A functional polymorphism in the miR-146a gene is associated with the risk for hepatocellular carcinoma. Carcinogenesis (2008) 29:2126–31. doi:10.1093/carcin/bgn195

[93] Calin GA, Sevignani C, Dumitru CD, Hyslop T, Noch E, Yendamuri S, et al. Human microRNA genes are frequently located at fragile sites and genomic regions involved in cancers. Proc Natl Acad Sci USA (2004) 101:2999–3004. doi:10.1073/pnas.0307323101

[94] Möröy T, Marchio A, Etiemble J, Trépo C, Tiollais P, Buendia MA. Rearrangement and enhanced expression of c-myc in hepatocellular carcinoma of hepatitis virus infected woodchucks. Nature (1986) 324:276–9. doi:10.1038/324276a0.

[95] Li Y, Tan W, Neo TW, Aung MO, Wasser S, Lim SG, et al. Role of the miR-106b-25 microRNA cluster in hepatocellular carcinoma. Cancer Sci (2009) 100:1234–42. doi:10.1111/j.1349-7006.2009.01164.x

[96] Gramantieri L, Fornari F, Ferracin M, Veronese A, Sabbioni S, Calin GA, et al. microRNA-221 targets Bmf in hepatocellular carcinoma and correlates with tumor multifocality. Clin Cancer Res (2009) 15:5073–81. doi:10.1158/1078-0432.CCR-09-0092

[97] Law PT, Wong N. Emerging roles of microRNA in the intracellular signaling networks of hepatocellular carcinoma. J Gastroenterol Hepatol (2011) 26:437–49. doi:10.1111/j.1440-1746.2010.06512.x

[98] Liu N, Zhang J, Jiao T, Li Z, Peng J, Cui Z, et al. Hepatitis B virus inhibits apoptosis of hepatoma cells by sponging the microRNA15a/16 cluster. J Virol (2013) 87:13370–8. doi:10.1128/JVI.02130-13

[99] Wang CM, Wang Y, Fan CG, Xu FF, Sun WS, Liu YG, et al. MiR-29c tar gets TNFAIP3, inhibits cell proliferation and induces apoptosis in hepatitis B virus-related hepatocellular carcinoma. Biochem Biophys Res Commun (2011) 411:586–92. doi:10.1016/j.bbrc.2011.06.191

[100] Sun J, Lu H, Wang X, Jin H. microRNAs in hepatocellular carcinoma: regulation, function, and clinical implications. Sci World J (2013) 2013:924206. doi:10.1155/2013/924206

[101] Fornari F, Milazzo M, Chieco P, Negrini M, Calin GA, Grazi GL, et al. MiR-199a-3p regulates mTOR and c-Met to influence the doxorubicin sensitivity of human hepatocarcinoma cells. Cancer Res (2010) 70:5184–93. doi:10.1158/0008-5472.CAN-10-0145

[102] Xu X, Fan Z, Kang L, Han J, Jiang C, Zheng X, et al. Hepatitis B virus X protein represses miRNA-148a to enhance tumorigenesis. J Clin Invest (2013) 123:630–45. doi:10.1172/JCI64265

[103] Yang F, Yin Y, Wang F, Wang Y, Zhang L, Tang Y, et al. miR-17-5p Promotes migration of human hepatocellular carcinoma cells through the p38mitogen-activated protein kinase-heat shock protein 27 pathway. Hepatology (2010) 51:1614–23. doi:10.1002/hep.23566

[104] Wong QW, Ching AK, Chan AW, Choy KW, To KF, Lai PB, et al. MiR-222 overexpression confers cell migratory advantages in hepatocellular carcinoma through enhancing AKT signaling. Clin Cancer Res (2010) 16:867–75. doi:10.1158/1078-0432.CCR-09-1840

[105] Zhang Y, Wei W, Cheng N, Wang K, Li B, Jiang X, et al. Hepatitis C virus-induced up-regulation of microRNA-155 promotes hepatocarcinogenesis by activating Wnt signaling. Hepatology (2012) **56**:1631–40. doi:10.1002/hep.25849

[106] Zhang Y, Takahashi S, Tasaka A, Yoshima T, Ochi H, Chayama K. Involvement of microRNA-224 in cell proliferation, migration, invasion, and anti-apoptosis in hepatocellular carcinoma. J Gastroenterol Hepatol (2013) **28**:565–75. doi:10. 1111/j.1440-1746.2012.07271.x

[107] Kong G, Zhang J, Zhang S, Shan C, Ye L, Zhang X. Upregulated microRNA-29a by hepatitis B virus X protein enhances hepatoma cell migration by targeting PTEN in cell culture model. PLoS One (2011) **6**:e19518. doi:10.1371/journal.pone.0019518

[108] Yuan K, Lian Z, Sun B, Clayton MM, Ng IO, Feitelson MA. Role of miR-148a in hepatitis B associated hepatocellular carcinoma. PLoS One (2012) **7**:e35331. doi:10.1371/journal.pone.0035331

[109] Tian Y, Yang W, Song J, Wu Y, Ni B. Hepatitis B virus X protein-induced aberrant epigenetic modifications contributing to human hepatocellular carcinoma pathogenesis. Mol Cell Biol (2013) **33**:2810–6. doi:10.1128/MCB.00205-13

[110] Li N, Fu H, Tie Y, Hu Z, Kong W, Wu Y, et al. miR-34a inhibits migration and invasion by down-regulation of c-Met expression in human hepatocellular carcinoma cells. Cancer Lett (2009) **275**:44–53. doi:10.1016/j.canlet.2008.09.035

[111] Liang L, Wong CM, Ying Q, Fan DN, Huang S, Ding J, et al. microRNA-125b suppressed human liver cancer cell proliferation and metastasis by directly targeting oncogene LIN28B2. Hepatology (2010) **52**:1731–40. doi:10.1002/hep.23904

[112] Hung CH, Chiu YC, Chen CH, Hu TH. microRNAs in hepatocellular carcinoma: carcinogenesis, progression, and therapeutic target. Biomed Res Int (2014) **2014**:486407. doi:10.1155/2014/486407

[113] Park SM, Gaur AB, Lengyel E, Peter ME. The miR-200 family determines the epithelial phenol type of cancer cells by targeting the E-cadherin repressors ZEB1 and ZEB2. Genes Dev (2008) **22**:894–907. doi:10.1101/gad.1640608

[114] Vrba L, Jensen TJ, Garbe JC, Heimark RL, Cress AE, Dickinson S, et al. Role for DNA methylation in the regulation of miR-200c and miR-141 expression in normal and cancer cells. PLoS One (2010) **5**:e8697. doi:10.1371/journal.pone.0008697

[115] D'Anzeo M, Faloppi L, Scartozzi M, Giampieri R, Bianconi M, Del Prete M, et al. The role of micro-RNAs in hepatocellular carcinoma: from molecular biology to treatment. Molecules (2014) **19**(5):6393–406. doi:10.3390/molecules19056393

[116] Lu CY, Lin KY, Tien MT, Wu CT, Uen YH, Tseng TL. Frequent DNA methylation of MiR-129-2 and its potential clinical implication in hepatocellular carcinoma. Genes Chromosomes Cancer (2013) **52**(7):636–43. doi:10.1002/gcc.22059

[117] Hou W, Bonkovsky HL. Non-coding RNAs in hepatitis C-induced hepatocellular carcinoma: dysregulation and implications for early detection, diagnosis and therapy. World J Gastroenterol (2013) **19**(44):7836–45. doi:10.3748/wjg.v19.i44.7836

[118] Li LM, Hu ZB, Zhou ZX, Chen X, Liu FY, Zhang JF, et al. Serum microRNA profiles serve as novel biomarkers for HBV infection and diagnosis of HBV-positive hepatocarcinoma. Cancer Res (2010) **70** (23):9798–807. doi:10.1158/0008-5472.CAN-10-1001

[119] Wei YF, Cui GY, Ye P, Chen JN, Diao HY. microRNAs may solve the mystery of chronic hepatitis B virus infection. World J Gastroenterol (2013) **19** (30):4867–76. doi:10.3748/wjg.v19.i30.4867

[120] Wang H, Hou L, Li A, Duan Y, Gao H, Song X. Expression of serum exosomal microRNA-21 in human hepatocellular carcinoma. Biomed Res Int (2014) **2014**:864-894. doi:10.1155/2014/864894

[121] Abdalla MA, Haj-Ahmad Y. Promising candidate urinary microRNA biomarkers for the early detection of hepatocellular carcinoma among high-risk hepatitis C virus Egyptian patients. J Cancer (2011) **3**:19–31. doi:10.7150/jca.3.19

[122] Budhu A, Jia HL, Forgues M, Liu CG, Goldstein D, Lam A, et al. Identification of metastasis-related microRNAs in hepatocellular carcinoma. Hepatology (2008) **47**(3):897–907. doi:10.1002/hep.22160

[123] Anwar SL, Lehmann U. DNA methylation, microRNAs, and their crosstalk as potential biomarkers in hepatocellular carcinoma. World J Gastroenterol (2014) **20**:7894–913. doi:10.3748/wjg.v20.i24.7894

[124] Wang C, Song B, Song W, Liu J, Sun A, Wu D, et al. Under expressed microRNA-199b-5p targets hypoxia-inducible factor-1α in hepatocellular carcinoma and predicts prognosis of hepatocellular carcinoma patients. J Gastroenterol Hepatol (2011) **26** (11):1630–7. doi:10.1111/j.1440-1746.2011.06758.x

[125] Guo J, Jin M, Zhang M, Chen K. A genetic variant in miR-196a2 increased digestive system cancer risks: a meta-analysis of 15 case-control studies. PLoS One (2012) **7**(1):e30585. doi:10.1371/journal.pone.0030585

[126] Akkız H, Bayram S, Bekar A, Akgöllü E, Ulger Y. A functional polymorphism in pre-microRNA-196a-2 contributes to the susceptibility of hepatocellular carcinoma in a Turkish population: a case-control study. J Viral Hepat (2011) **18**(7):e399–407. doi:10.1111/j.1365-2893.2010.01414.x

[127] Gao Y, He Y, Ding J, Wu K, Hu B, Liu Y, et al. An insertion/deletion polymorphism at miRNA-122-binding site in the interleukin-1 alpha3' untranslated region confers risk for hepatocellular carcinoma. Carcinogenesis (2009) **30**(12):2064–9. doi:10.1093/carcin/bgp283.

[128] Butterworth R. Hepatic encephalopathy. Alcohol Res Health (2003) **27**:240–6.

[129] El-Serag HB, Rudolph L. Hepatocellular carcinoma: epidemiology and molecular carcinogenesis. Gastroenterology (2007) **132**:2557–76.

[130] Szabo G. Moderate drinking, inflammation and liver disease. Ann Epidemiol (2007) **17**:S49–54.

[131] Mandayam S, Jamal MM, Morgan TR. Epidemiology of alcoholic liver disease. Semin Liver Dis (2004) **24**:217–32.

[132] Dolganiuc A, Petrasek J, Kodys K, Catalano D, Mandrekar P, Velayudham A, et al. MicroRNA expression profile in Lieber–DeCarli diet-induced alcoholic and methionine choline deficient diet-induced nonalcoholic steatohepatitis models in mice. Alcohol Clin Exp Res (2009) **33**:1704–10.

[133] Bala S, Szabo G. MicroRNA signature in alcoholic liver disease. Int J Hepatol (2012):498232.

[134] Bala S, Marcos M, Kodys K, Csak T, Catalano D, Mandrekar P, et al. Up-regulation of microRNA-155 in macrophages contributes to increased tumor necrosis factor α (TNFα) production via increased mRNA half-life in alcoholic liver disease. J Biol Chem (2011) **286**:1436–44.

[135] Zhang Y, Jia Y, Zheng R, Guo Y, Wang Y, Guo H, et al. Plasma microRNA-122 as a biomarker for viral-, alcohol-,

and chemical-related hepatic diseases. Clin Chem (2010c) **56**:1830–8.

[136] Ozer J, Ratner M, Shaw M, Bailey W, Schomaker S. The current state of serum biomarkers of hepatotoxicity. Toxicology (2008) **245**:194–205.

[137] Yin H, Hu M, Zhang R, Shen Z, Flatow L, You M. MicroRNA-217 promotes ethanol induced fat accumulation in hepatocytes by down-regulating SIRT1. J Biol Chem (2012) **287**:9817–26.

[138] Meng F, Glaser SS, Francis H, Yang F, Han Y, Stokes A, et al. Epigenetic regulation of miR-34a expression in alcoholic liver injury. Am J Pathol (2012) **181**:804–17.

[139] Liu Y, Chen SH, Jin X, Li YM. Analysis of differentially expressed genes and microRNAs in alcoholic liver disease. Int J Mol Med (2013) **31**:547–54.

[140] Baffy G, Brunt EM, Caldwell SH. Hepatocellular carcinoma in non-alcoholic fatty liver disease: an emerging menace. J Hepatol (2012) **56**:1384–91.

[141] Min HK, Kapoor A, Fuchs M, Mirshahi F, Zhou H, Maher J, et al. Increased hepatic synthesis and dysregulation of cholesterol metabolism is associated with the severity of nonalcoholic fatty liver disease. Cell Metab (2012) **15**:665–74.

[142] Castro RE, Ferreira DM, Afonso MB, Borralho PM, Machado MV, Cortez-Pinto H, et al. miR-34a/SIRT1/p53 is suppressed by ursodeoxycholic acid in the rat liver and activated by disease severity in human non-alcoholic fatty liver disease. J Hepatol (2013) **58**:119–25.

[143] Derdak Z, Villegas KA, Harb R, Wu AM, Sousa A, Wands JR. Inhibition of p53 attenuates steatosis and liver injury in a mouse model of non-alcoholic fatty liver disease. J Hepatol (2013) **58**:785–91.

[144] Caballero F, Fernández A, De Lacy AM, Fernández-Checa JC, Caballería J, García-Ruiz C. Enhanced free cholesterol, SREBP-2 and StAR expression in human NASH. J Hepatol (2009) **50**:789–96.

[145] Cazanave SC, Mott JL, Elmi NA, Bronk SF, Masuoka HC, Charlton MR, et al. A role for miR-296 in the regulation of lipoapoptosis by targeting PUMA. J Lipid Res (2011) **52**:1517–25.

[146] Sharma H, Estep M, Birerdinc A, Afendy A, Moazzez A, Elairny H, et al. Expression of genes for microRNA-processing enzymes is altered in advanced non-alcoholic fatty liver disease. J Gastroen Hepatol (2013) **28**:1410–5.

[147] Wang B, Majumder S, Nuovo G, Kutay H, Volinia S, Patel T, et al. Role of microRNA-155 at early stages of hepatocarcinogenesis induced by choline-deficient and amino acid—defined diet in C57BL/6 mice. Hepatology (2009) **50**:1152–61.

[148] Ahn J, Lee H, Chung CH, Ha T. High fat diet induced downregulation of microRNA-467b increased lipoprotein lipase in hepatic steatosis. Biochem Biophys Res Commun (2011) **414**:664–9.

[149] Zheng L, Lv GC, Sheng J, Yang YD. Effect of miRNA-10b in regulating cellular steatosis level by targeting PPAR-α expression, a novel mechanism for the pathogenesis of NAFLD. J Gastroen Hepatol (2010) **25**:156–63.

[150] Bechmann LP, Vetter D, Ishida J, Hannivoort RA, Lang UE, Kocabayoglu P, et al. Posttranscriptional activation of PPAR alpha by KLF6 in hepatic steatosis. J Hepatol (2013) **58**:1000–6.

[151] Alisi A, Da Sacco L, Bruscalupi G, Piemonte F, Panera N, De Vito R, et al. Mirnome analysis reveals novel molecular determinants in the pathogenesis of diet-induced nonalcoholic fatty liver disease. Lab Invest (2011) **91**:283–93.

[152] Pogribny IP, Starlard-Davenport A, Tryndyak VP, Han T, Ross SA, Rusyn I, et al. Difference in expression of hepatic microRNAs miR-29c, miR-34a, miR-155, and miR-200b is associated with strain-specific susceptibility to dietary nonalcoholic steatohepatitis in mice. Lab Invest (2010) **90**:1437–46.

[153] Jin X, Chen YP, Kong M, Zheng L, Yang YD, Li YM. Transition from hepatic steatosis to steatohepatitis: unique microRNA patterns and potential downstream functions and pathways. J Gastroen Hepatol (2012) **27**:331–40.

[154] Ceccarelli S, Panera N, Gnani D, Nobili V. Dual role of microRNAs in NAFLD. Int J Mol Sci (2013) **14**:8437–55.

[155] Zhang Y, Cheng X, Lu Z, Wang J, Chen H, Fan W, et al. Upregulation of miR-15b in NAFLD models and in the serum of patients with fatty liver disease. Diabetes Res Clin Pract (2013) **99**:327–34.

[156] Yamada H, Suzuki K, Ichino N, Ando Y, Sawada A, Osakabe K, et al. Associations between circulating microRNAs (miR-21, miR-34a, miR-122 and miR-451) and non-alcoholic fatty liver. Clin Chim Acta (2013) **424**:99–103.

[157] El-Serag HB. Epidemiology of viral hepatitis and hepatocellular carcinoma. Gastroenterology (2012) **142**:1264–73.

[158] Walsh K, Alexander GJM. Update on chronic viral hepatitis. Postgrad Med J (2001) **77**:498–505.

[159] Russo A, Potenza N. Antiviral effects of human microRNAs and conservation of their target sites. FEBS Lett (2011) **585**:2551–5.

[160] Ura S, Honda M, Yamashita T, Ueda T, Takatori H, Nishino R, et al. Differential microRNA expression between hepatitis B and hepatitis C leading disease progression to hepatocellular carcinoma. Hepatology (2009) **49**:1098–112.

[161] Geller SA. Hepatitis B and hepatitis C. Clin Liver Dis (2002) **6**:317–34.

[162] Ramachandran S, Ilias Basha H, Sarma NJ, Lin Y, Crippin JS, Chapman WC, et al. Hepatitis C virus induced miR200c down modulates FAP-1, a negative regulator of Src signaling and promotes hepatic fibrosis. PLoS One (2013) **8**:e70744.

[163] Chen W, Zhang Z, Chen J, Zhang J, Zhang J, Wu Y, et al. HCV core protein interacts with Dicer to antagonize RNA silencing. Virus Res (2008) **133**:250–8.

[164] Ren M, Qin D, Li K, Qu J, Wang L, Wang Z, et al. Correlation between hepatitis B virus protein and microRNA processor Drosha in cells expressing HBV. Antivir Res (2012) **94**:225–31.

[165] Wang Y, Lu Y, Toh ST, Sung WK, Tan P, Chow P, et al. Lethal-7 is down-regulated by the hepatitis B virus x protein and targets signal transducer and activator of transcription 3. J Hepatol (2010) **53**:57–66.

[166] Kong G, Zhang J, Zhang S, Shan C, Ye L, Zhang X. Upregulated microRNA-29a by hepatitis B virus X protein enhances hepatoma cell migration by targeting PTEN in cell culture model. PLoS One (2011) **6**:e19518.

[167] Wei X, Tan C, Tang C, Ren G, Xiang T, Qiu Z, et al. Epigenetic repression of miR-132 expression by the hepatitis B virus x protein in hepatitis B virus-related hepatocellular carcinoma. Cell Signal (2013b) **25**:1037–43.

[168] Shirasaki T, Honda M, Shimakami T, Horii R, Yamashita T, Sakai Y, et al. MicroRNA-27a regulates lipid metabolism and inhibits hepatitis C virus replication in human hepatoma cells. J Virol (2013) **87**:5270–86.

[169] Singaravelu R, Chen R, Lyn RK, Jones DM, O'Hara S, Rouleau Y, et al. Hepatitis C virus induced up-regulation

of microRNA-27: a novel mechanism for hepatic steatosis. Hepatology (2014) **59**:98–108.

[170] Guo H, Liu H, Mitchelson K, Rao H, Luo M, Xie L, et al. MicroRNAs-372/373 promote the expression of hepatitis B virus through the targeting of nuclear factor I/B. Hepatology (2011) **54**:808–19.

[171] Zhang X, Zhang E, Ma Z, Pei R, Jiang M, Schlaak JF, et al. Modulation of hepatitis B virus replication and hepatocyte differentiation by microRNA-1. Hepatology (2011) **53**:1476–85.

[172] Jin J, Tang S, Xia L, Du R, Xie H, Song J, et al. MicroRNA-501 promotes HBV replication by targeting HBXIP. Biochem Biophys Res Commun (2013) **430**:1228–33.

[173] Wang Y, Jiang L, Ji X, Yang B, Zhang Y, Fu XD. Hepatitis B viral RNA directly mediates down-regulation of the tumor suppressor microRNA miR-15a/miR-16-1 in hepatocytes. J Biol Chem (2013) **288**:18484–93.

[174] Jung YJ, Kim JW, Park SJ, Min BY, Jang ES, Kim NY, et al. c-Myc-mediated overexpression of miR-17-92 suppresses replication of hepatitis B virus in human hepatoma cells. J Med Virol (2013) **85**:969–78.

[175] Banaudha K, Kaliszewski M, Korolnek T, Florea L, Yeung ML, Jeang KT, et al. MicroRNA silencing of tumor suppressor DLC-1 promotes efficient hepatitis C virus replication in primary human hepatocytes. Hepatology (2011) **53**:53–61.

[176] Liu X, Wang T, Wakita T, Yang W. Systematic identification of microRNA and messenger RNA profiles in hepatitis C virus-infected human hepatoma cells. Virology (2010b) **398**:57–67.

[177] Jopling CL, Yi M, Lancaster AM, Lemon SM, Sarnow P. Modulation of hepatitis C virus RNA abundance by a liver-specific microRNA. Science (2005) **309**:1577–81.

[178] Jangra RK, Yi M, Lemon SM. Regulation of hepatitis C virus translation and infectious virus production by the MicroRNA miR-122. J Virol (2010) **84**:6615–25.

[179] Cox EM, Sagan SM, Mortimer SA, Doudna JA, Sarnow P. Enhancement of hepatitis C viral RNA abundance by precursor miR-122 molecules. RNA (2013) **19**:1825–32.

[180] Mortimer SA, Doudna JA. Unconventional miR-122 binding stabilizes the HCV genome by forming a trimolecular RNA structure. Nucleic Acids Res (2013) **41**:4230–40.

[181] Lanford RE, Hildebrandt-Eriksen ES, Petri A, Persson R, Lindow M, Munk ME, et al. Therapeutic silencing of microRNA-122 in primates with chronic hepatitis C virus infection. Science (2010) **327**:198–201.

[182] van der Meer AJ, Farid WR, Sonneveld MJ, de Ruiter PE, Boonstra A, van Vuuren AJ, et al. Sensitive detection of hepatocellular injury in chronic hepatitis C patients with circulating hepatocyte-derived microRNA-122. J Viral Hepat (2013) **20**:158–66.

[183] Wang S, Qiu L, Yan X, Jin W, Wang Y, Chen L, et al. Loss of microRNA 122 expression in patients with hepatitis B enhances hepatitis B virus replication through cyclin G1-modulated P53 activity. Hepatology (2012) **55**:730–41.

[184] Spaniel C, Honda M, Selitsky SR, Yamane D, Shimakami T, Kaneko S, et al. MicroRNA-122 abundance in hepatocellular carcinoma and non-tumor liver tissue from Japanese patients with persistent HCV versus HBV infection. PLoS One (2013) **8**:e76867.

[185] Sarasin-Filipowicz M, Krol J, Markiewicz I, Heim MH, Filipowicz W. Decreased levels of microRNA miR-122 in individuals with hepatitis C responding poorly to interferon therapy. Nat Med (2009) **15**:31–3.

[186] Jemal A, Bray F, Center MM, Ferlay J, Ward E, Forman D. Global cancer statistics. CA-Cancer J Clin (2011) **61**:69–90.

[187] Rahbari NN, Mehrabi A, Mollberg NM, Müller SA, Koch M, Büchler MW, et al. Hepatocellular carcinoma current management and perspectives for the future. Ann Surg (2011) **253**:453–69.

[188] Vogelstein B, Papadopoulos N, Velculescu VE, Zhou S, Diaz LA Jr, Kinzler KW. Cancer genome landscapes. Science (2013) **339**:1546–58.

[189] Calin GA, Croce CM. MicroRNA signatures in human cancers. Nat Rev Cancer (2006) **6**:857–66.

[190] Zhang B, Pan X, Cob GP, Anderson TA. MicroRNAs as oncogenes and tumor suppressors. Dev Biol (2007) **302**:1–12.

[191] Su H, Yang J-R, Xu T, Huang J, Xu L, Yuan Y, et al. MicroRNA-101, down-regulated in hepatocellular carcinoma, promotes apoptosis and suppresses tumorigenicity. Cancer Res (2009) **69**:1135–42.

[192] Zhang J, Yang Y, Yang T, Liu Y, Li A, Fu S, et al. MicroRNA-22, downregulated in hepatocellular carcinoma and correlated with prognosis, suppresses cell proliferation and tumourigenicity. Brit J Cancer (2010b) **103**:1215–20.

[193] Li D, Liu X, Lin L, Hou J, Li N, Wang C, et al. MicroRNA-99a inhibits hepatocellular carcinoma growth and correlates with prognosis of patients with hepatocellular carcinoma. J Biol Chem (2011) **286**:36677–85.

[194] Liu Y, Hei Y, Shu Q, Dong J, Gao Y, Fu H, et al. VCP/p97, down-regulated by microRNA-129-5p, could regulate the progression of hepatocellular carcinoma. PLoS One (2012b) **7**:e35800.

[195] Razumilava N, Bronk SF, Smoot RL, Fingas CD, Werneburg NW, Roberts LR, et al. miR-25 targets TNF-related apoptosis inducing ligand (TRAIL) death receptor-4 and promotes apoptosis resistance in cholangiocarcinoma. Hepatology (2012) **55**:465–75.

[196] Gramantieri L, Fornari F, Callegari E, Sabbioni S, Lanza G, Croce CM, et al. MicroRNA involvement in hepatocellular carcinoma. J Cell Mol Med (2008) **12**:2189–204.

[197] Varnholt H. The role of microRNAs in primary liver cancer. Ann Hepatol (2008) **7**:104–13.

[198] Isomoto H. Epigenetic alterations associated with cholangiocarcinoma (Review). Oncol Rep (2009) **22**:227–32.

[199] Hou J, Lin L, Zhou W, Wang Z, Ding G, Dong Q, et al. Identification of miRNomes in human liver and hepatocellular carcinoma reveals miR-199a/b-3p as therapeutic target for hepatocellular carcinoma. Cancer Cell (2011) **19**:232–43.

[200] Gramantieri L, Ferracin M, Fornari F, Veronese A, Sabbioni S, Liu CG, et al. Cyclin G1 is a target of miR-122a, a microRNA frequently down-regulated in human hepatocellular carcinoma. Cancer Res (2007) **67**:6092–9.

[201] Zhang R, Wang L, Yu GR, Zhang X, Yao LB, Yang AG. MicroRNA-122 might be a double-edged sword in hepatocellular carcinoma. Hepatology (2009) **50**:1322–3.

[202] Hsu SH, Wang B, Kota J, Yu J, Costinean S, Kutay H, et al. Essential metabolic, antiinflammatory, and anti-tumorigenic functions of miR-122 in liver. J Clin Invest (2012) **122**:2871–83.

[203] Tsai WC, Hsu SD, Hsu CS, Lai TC, Chen SJ, Shen R, et al. MicroRNA-122 plays a critical role in liver homeostasis and hepatocarcinogenesis. J Clin Invest (2012) **122**:2884–97.

[204] Coulouarn C, Factor VM, Andersen JB, Durkin ME, Thorgeirsson SS. Loss of miR-122 expression in liver

cancer correlates with suppression of the hepatic phenotype and gain of metastatic properties. Oncogene (2009) **28**:3526–36.

[205] Pineau P, Volinia S, McJunkin K, Marchio A, Battiston C, Terris B, et al. miR-221 overexpression contributes to liver tumorigenesis. Proc Nat Acad Sci USA (2010) **107**:264–9.

[206] Garofalo M, Di Leva G, Romano G, Nuovo G, Suh SS, Ngankeu A, et al. miR-221&222 regulate TRAIL resistance and enhance tumorigenicity through PTEN and TIMP3 downregulation. Cancer Cell (2009) **16**:498–509.

[207] Meng F, Henson R, Wehbe-Janek H, Ghoshal K, Jacob ST, Patel T. MicroRNA-21 regulates expression of the PTEN tumor suppressor gene in human hepatocellular cancer. Gastroenterology (2007) **133**:647–58.

[208] Huang S, He X. The role of microRNAs in liver cancer progression. Br J Cancer (2011) **104**:235–40.

[209] Borel F, Konstantinova P, Jansen PL. Diagnostic and therapeutic potential of miRNA signatures in patients with hepatocellular carcinoma. J Hepatol (2012) **56**:1371–83.

[210] Tomimaru Y, Eguchi H, Nagano H, Wada H, Kobayashi S, Marubashi S, et al. Circulating microRNA-21 as a novel biomarker for hepatocellular carcinoma. J Hepatol (2012) **56**:167–75.

[211] Liu AM, Yao TJ, Wang W, Wong KF, Lee NP, Fan ST, et al. Circulating miR-15b and miR-130b in serum as potential markers for detecting hepatocellular carcinoma: a retrospective cohort study. BMJ Open (2012a) **2**:e000825.

[212] Zhou J, Yu L, Gao X, Hu J, Wang J, Dai Z, et al. Plasma microRNA panel to diagnose hepatitis B virus-related hepatocellular carcinoma. J Clin Oncol (2011) **29**:4781–8.

[213] Qi P, Cheng SQ, Wang H, Li N, Chen YF, Gao CF. Serum microRNAs as biomarkers for hepatocellular carcinoma in Chinese patients with chronic hepatitis B virus infection. PLoS One (2011) **6**:e28486.

[214] Xu J, Wu C, Che X, Wang L, Yu D, Zhang T, et al. Circulating microRNAs, miR-21, miR-122, and miR-223, in patients with hepatocellular carcinoma or chronic hepatitis. Mol Carcinog (2011) **50**:136–42.

[215] Marquardt JU, Galle PR, Teufel A. Molecular diagnosis and therapy of hepatocellular carcinoma (HCC): an emerging field for advanced technologies. J Hepatol (2012) **56**:267–75.

[216] Ladeiro Y, Couchy G, Balabaud C, Bioulac-Sage P, Pelletier L, Rebouissou S, et al. MicroRNA profiling in hepatocellular tumors is associated with clinical features and oncogene/tumor suppressor gene mutations. Hepatology (2008) **47**:1955–63.

[217] Toffanin S, Hoshida Y, Lachenmayer A, Villanueva A, Cabellos L, Minguez B, et al. MicroRNA-based classification of hepatocellular carcinoma and oncogenic role of miR-517a. Gastroenterology (2011) **140**:1618–28.

[218] Sato F, Hatano E, Kitamura K, Myomoto A, Fujiwara T, Takizawa S, et al. MicroRNA profile predicts recurrence after resection in patients with hepatocellular carcinoma within the Milan criteria. PLoS One (2011) **6**:e16435.

[219] Szabo G, Sarnow P, Bala S. MicroRNA silencing and the development of novel therapies for liver disease. J Hepatol (2012) **57**:462–6.

[220] Park JK, Kogure T, Nuovo GJ, Jiang J, He L, Kim JH, et al. miR-221 silencing blocks hepatocellular carcinoma and promotes survival. Cancer Res (2011) **71**:7608–16.

[221] Callegari E, Elamin BK, Giannone F, Milazzo M, Altavilla G, Fornari F, et al. Liver tumorigenicity promoted by microRNA-221 in a mouse transgenic model. Hepatology (2012) **56**:1025–33.

[222] Tomokuni A, Eguchi H, Tomimaru Y, Wada H, Kawamoto K, Kobayashi S, et al. miR-146a suppresses the sensitivity to interferon-α in hepatocellular carcinoma cells. Biochem Biophys Res Commun (2011) **414**:675–80.

[223] Ma K, He Y, Zhang H, Fei Q, Niu D, Wang D, et al. DNA methylation-regulated miR-193a-3p dictates resistance of hepatocellular carcinoma to 5-fluorouracil via repression of SRSF2 expression. J Biol Chem (2012) **287**:5639–49.

[224] Xu N, Zhang J, Shen C, Luo Y, Xia L, Xue F, et al. Cisplatin-induced downregulation of miR-199a-5p increases drug resistance by activating autophagy in HCC cell. Biochem Biophys Res Commun (2012) **423**:826–31.

[225] Ji J, Shi J, Budhu A, Yu Z, Forgues M, Roessler S, et al. MicroRNA expression, survival, and response to interferon in liver cancer. N Engl J Med (2009) **361**:1437–47.

[226] Ji J, Yu L, Yu Z, Forgues M, Uenishi T, Kubo S, et al. Development of a miR-26 companion diagnostic test for adjuvant interferon-alpha therapy in hepatocellular carcinoma. Int J Biol Sci (2013) **9**:303–12.

[227] Zhang H, Li Y, Lai M. The microRNA network and tumor metastasis. Oncogene (2010a) **29**:937–48.

[228] Giordano S, Columbano A. MicroRNAs: new tools for diagnosis, prognosis, and therapy in hepatocellular carcinoma. Hepatology (2013) **57**:840–7.

[229] Tsai WC, Hsu PW, Lai TC, Chau GY, Lin CW, Chen CM, et al. MicroRNA-122, a tumor suppressor microRNA that regulates intrahepatic metastasis of hepatocellular carcinoma. Hepatology (2009) **49**:1571–82.

[230] Li N, Fu H, Tie Y, Hu Z, Kong W, Wu Y, et al. miR-34a inhibits migration and invasion by down-regulation of c-Met expression in human hepatocellular carcinoma cells. Cancer Lett 2009; **275**:44–53.

[231] Yang X, Liang L, Zhang XF, Jia HL, Qin Y, Zhu XC, et al. MicroRNA-26a suppresses tumor growth and metastasis of human hepatocellular carcinoma by targeting interleukin-6-Stat3 pathway. Hepatology (2013) **58**:158–70.

[232] Tao ZH, Wan JL, Zeng LY, Xie L, Sun HC, Qin LX, et al. miR-612 suppresses the invasive-metastatic cascade in hepatocellular carcinoma. J Exp Med (2013) **210**:789–803.

[233] Budhu A, Jia H-L, Forgues M, Liu C-G, Goldsteir D, Lam A, et al. Identification of metastasis-related microRNAs in hepatocellular carcinoma. Hepatology (2008) **47**:897–907.

[234] Wong CM, Wong CC, Lee JM, Fan DN, Au SL, Ng IO. Sequential alterations of microRNA expression in hepatocellular carcinoma development and venous metastasis. Hepatology (2012) **55**:1453–61.

[235] Li B, Han Q, Zhu Y, Yu Y, Wang J, Jiang X. Down-regulation of miR-214 contributes to intrahepatic cholangiocarcinoma metastasis by targeting twist. FEBS J (2012) **279**:2393–8.

[236] Chung GE, Yoon J-H, Myung SJ, Lee J-H, Lee S-H, Lee S-M, et al. High expression of microRNA-15b predicts a low risk of tumor recurrence following curative resection of hepatocellular carcinoma. Oncol Rep (2010) **23**:113–9.

[237] Barry CT, D'Souza M, McCall M, Safadjou S, Ryan C, Kashyap R, et al. Micro RNA expression profiles as adjunctive data to assess the risk of hepatocellular carcinoma recurrence after liver transplantation. Am J Transplant (2012) **12**:428–37.

[238] Han Z-B, Zhong L, Teng M-J, Fan J-W, Tang H-M, Wu J-Y, et al. Identification of recurrence-related microRNAs in hepatocellular carcinoma following liver transplantation. Mol Oncol (2012) **6**:445–57.

[239] Shaikh F, Goff LW. Decoding hepatocellular carcinoma: the promise of microRNAs. Hepatobiliary Surg Nutr (2014) **3**(2):93–4. doi:10.3978/j.issn.2304-3881.2014.02.08

[240] Janssen HL, Reesink HW, Lawitz EJ, Zeuzem S, Rodriguez-Torres M, Patel K, et al. Treatment of HCV infection by targeting microRNA. N Engl J Med (2013) **368**:1685–94. doi:10.1056/NEJMoa1209026

[241] Madkour LH. Reactive Oxygen Species (ROS), Nanoparticles, and Endoplasmic Reticulum (ER) Stress-Induced Cell Death Mechanisms. Academic Press (2020). https://www.elsevier.com/books/reactive-oxygen-species-ros-nanoparticles-and-endoplasmic-reticulum-er-stress-induced-cell-death-mechanisms/madkour/978-0-12-822481-6

[242] Madkour LH. Nanoparticles Induce Oxidative and Endoplasmic Reticulum Antioxidant Therapeutic Defenses. 1st Edition. Switzerland: Springer (2020). https://www.springer.com/gp/book/9783030372965?utm_campaign=3_pier05_buy_print&utm_content=en_08082017&utm_medium=referral&utm_source=google_books#otherversion=9783030372972

[243] Madkour LH. Nucleic Acids as Gene Anticancer Drug Delivery Therapy. 1st Edition. Elsevier (2020). https://www.elsevier.com/books/nucleic-acids-as-gene-anticancer-drug-deliverytherapy/madkour/978-0-12-819777-6

[244] Madkour LH. Nanoelectronic Materials: Fundamentals and Applications (Advanced Structured Materials). 1st edition. Switzerland: Springer (2019). https://books.google.com.eg/books/about/Nanoelectronic_Materials.html?id=YQXCxAEACAAJ&source=kp_book_description&redir_esc=y https://www.springer.com/gp/book/9783030216207

[245] Lindow M, Kauppinen S. Discovering the first microRNA-targeted drug. J Cell Biol (2012) **199**:407–12.

[246] Ma L, Chua MS, Andrisani O, So S. Epigenetics in hepatocellular carcinoma: an update and future therapy perspectives. World J Gastroenterol (2014) **20** (2):333–45. doi:10.3748/wjg.v20.i2.333

[247] Gao J, Liu QG. The role of miR-26 in tumors and normal tissues (Review). Oncol Lett (2011) **2**(6):1019–23.

[248] Liu WH, Yeh SH, Lu CC, Yu SL, Chen HY, Lin CY, et al. microRNA-18a prevents estrogen receptor-alpha expression, promoting proliferation of hepatocellular carcinoma cells. Gastroenterology (2009) **136**(2):683–93. doi:10.1053/j.gastro.2008.10.029

# 13 Novel Classes of Noncoding RNAs and Cancer Biology Therapeutic Targets

For many years, the central dogma of molecular biology has been that RNA functions mainly as an informational intermediate between a DNA sequence and its encoded protein. But one of the great surprises of modern biology was the discovery that protein-coding genes represent less than 2% of the total genome sequence, and subsequently the fact that at least 90% of the human genome is actively transcribed. Thus, the human transcriptome was found to be more complex than a collection of protein-coding genes and their splice variants. Although initially argued to be spurious transcriptional noise or accumulated evolutionary debris arising from the early assembly of genes and/or the insertion of mobile genetic elements, recent evidence suggests that the noncoding RNAs (ncRNAs) may play major biological roles in cellular development, physiology, and pathologies. NcRNAs could be grouped into two major classes based on the transcript size; small ncRNAs and long ncRNAs. Each of these classes can be further divided, whereas novel subclasses are still being discovered and characterized. In recent years, small ncRNAs called microRNAs (miRNAs) have been studied most frequently; with more than 10,000 hits at the PubMed database, evidence has begun to accumulate describing the molecular mechanisms by which a wide range of novel RNA species function, providing insight into their functional roles in cellular biology and in human disease.

An improved understanding of the role of ncRNAs in multiple myeloma would provide valuable information about key cancer-promoting pathways and might be highly useful for diagnostic and prognostic assessments. This knowledge might also lead to advancement in the management of multiple myeloma through the development of novel personalized ncRNA-based therapies.

We review the recent remarkable advancement in the understanding of the biological functions of human ncRNAs in multiple myeloma, including the biogenesis, the mechanisms of expression, the relevance as biomarkers, and, mostly, the therapeutic potential. Special emphasis is given to miRNAs, the best-characterized class of ncRNA.

## 13.1 INTRODUCTION

The abundance of nontranslated functional RNAs in the cell has been a textbook truth for decades. Most of these noncoding RNAs (ncRNAs) fulfill essential functions, such as ribosomal RNAs (rRNAs) and transfer RNAs (tRNAs) involved in mRNA translation, small nuclear RNAs (snRNAs) involved in splicing and small nucleolar RNAs (snoRNAs) involved in the modification of rRNAs. The central dogma of molecular biology, developed from the study of simple organisms like *Escherichia coli*, has been that RNA functions mainly as an informational intermediate between a DNA sequence ("gene") and its encoded protein. The presumption was that most genetic information that specifies biological form and phenotype is expressed as proteins, which have not only diverse catalytic and structural functions, but also regulate the activity of the system in various ways. This is largely true in prokaryotes and presumed also to be true in eukaryotes [1]. But one of the great surprises of modern biology was definitely the discovery that the human genome encodes only ~20,000 protein-coding genes, representing less than 2% of the total genome sequence (Figure 13.1). Subsequently, with the advent of tiling resolution genomic microarrays and whole-genome and transcriptome-sequencing technologies (ENCODE project), it was determined that at least 90% of the genome is actively transcribed. The human transcriptome was found to be more complex than a collection of protein-coding genes and their splice variants, showing extensive antisense, overlapping, and ncRNA expression [1, 2]. Although initially argued to be spurious transcriptional noise or accumulated evolutionary debris arising from the early assembly of genes and/or the insertion of mobile genetic elements, recent evidence suggests that the proverbial "dark matter" of the genome may play a major biological role in cellular development, physiology, and pathologies. In general, the more complex an organism, the greater is its number of ncRNAs. The enticing possibility that although the number of protein-coding transcripts between organisms is similar, the ultimate control of cellular function may be through interactions between proteins and ncRNA, is corroborated by the fact that the majority of chromatin-modifying complexes do not have DNA-binding capacity and, therefore, must utilize a third party in binding to DNA. It has been largely demonstrated that this third party may be represented by transcription factors as well as by ncRNAs [2, 3].

The beginnings of the present-day understanding on regulatory ncRNAs were inspired mainly by the pioneering ideas of John S. Mattick, who has long argued that proteins comprise only a minority of the eukaryotic genome's information output. Considering the unique ability of RNA to both fold in three-dimensional space and hybridize in a sequence-specific manner to other nucleic acids, ncRNAs are proposed to behave as a digital-to-analogue processing network, allowing the expansion of complexity in biological systems, well beyond purely protein-based regulatory networks [4].

DOI: 10.1201/9781003229650-13

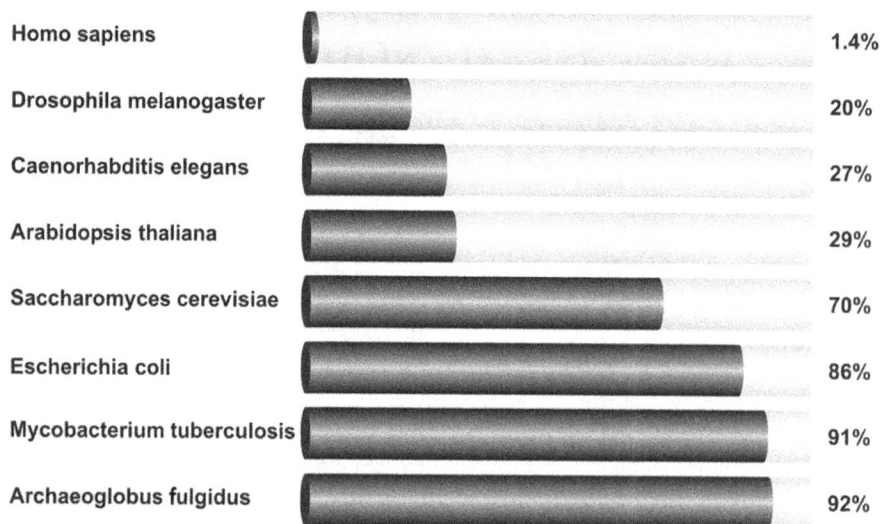

**FIGURE 13.1**   The percentage of protein-coding gene sequences in several eukaryotic and bacterial genomes.

NcRNAs are grouped into two major classes based on transcript size; small ncRNAs and long ncRNAs (lncRNAs) (classification of recently discovered ncRNAs is summarized in Table 13.1). Small ncRNAs are represented by a broad range of known and newly discovered RNA species, with many being associated with 5′ or 3′ regions of protein-coding genes. This class includes the well-documented miRNAs, siRNAs, piRNAs, etc. Most of them significantly extended our view of molecular carcinogenesis, and at present they are subject of intensive translational research in this field. In contrast to miRNAs, lncRNAs are mRNA-like transcripts ranging in length from 200 nucleotides (nt) to ~100 kilobases (kb) and lacking significant open reading frames. LncRNAs' expression levels appear to be lower than protein-coding genes, and some lncRNAs are preferentially expressed in specific tissues. The small numbers of characterized human lncRNAs have been associated with a spectrum of biological processes including alternative splicing or nuclear import. Moreover they can serve as structural components, precursors to small RNAs and even as regulators of mRNA decay. Furthermore, accumulating reports of misregulated lncRNA (HOTAIR, MALAT1, HULC, T-UCRs, etc.) expressions across numerous cancer types suggest that aberrant lncRNA expression may be an important contributor to tumorigenesis. In this chapter, we summarize recent knowledge of novel classes of ncRNAs, their biology, and function, with special focus on their significance in cancer biology and oncology translational research, which is the field where the number of publications focusing on this topic is rapidly growing [5–7].

## TABLE 13.1
## Types of Recently Discovered Human Noncoding RNAs

| | Class | Symbol | Characteristics | Disease/Biological Function Associations |
|---|---|---|---|---|
| Small noncoding RNAs | MicroRNAs | miRNAs | 18≠5 nt; account for 1–2% of the human genome; control the 50% of protein-coding genes; guide suppression of translation; Drosha- and Dicer-dependent small ncRNAs | Initiation of various disorders including many, if not all, cancers/regulation of proliferation, differentiation, and apoptosis involved in human development |
| | Small interfering RNAs | siRNAs | 19–23 nt; made by Dicer processing; guide sequence-specific degradation of target mRNA | Great potential in diseases treatment/posttranscriptional gene silencing mainly through RISC degradation mechanism; defense against pathogenic nucleic acids |
| | Piwi-interacting RNAs | piRNAs | 26–30 nt; bind Piwi proteins; Dicer independent; exist in genome clusters; principally restricted to the germline and somatic cells bordering the germline | Relationship between piRNAs and diseases has not yet been discovered/involved in germ cell development, stem self-renewal, and retrotransposon silencing |
| | Small nucleolar RNAs | snoRNAs | 60–300 nt; enriched in the nucleolus; in vertebrate are excised from pre-mRNA introns; bind snoRNP proteins | Association with development of some cancers/important function in the maturation of other noncoding RNAs, above all, rRNAs and snRNAs; miRNA-like snoRNAs regulate mRNAs |

*(Continued)*

**TABLE 13.1** *(Continued)*

**Types of Recently Discovered Human Noncoding RNAs**

| Class | | Symbol | Characteristics | Disease/Biological Function Associations |
|---|---|---|---|---|
| | Promoter-associated small RNAs | PASRs | 20–200 nt; modified 5′ (capped) ends; coincide with the transcriptional start sites of protein- and noncoding genes; made from transcription of short-capped transcripts | |
| | Transcription initiation RNAs | tiRNAs | ~18 nt; have the highest density just downstream of transcriptional start sites; show patterns of positional conservation; preferentially located in GC-rich promoters | Relationship with diseases has not yet been discovered/involved in the regulation of the transcription of protein-coding genes by targeting epigenetic silencing complexes |
| | Centromere repeat–associated small interacting RNAs | crasiRNAs | 34–42 nt; processed from long dsRNAs | Relationship between crasiRNAs and diseases has not yet been discovered/involved in the recruitment of heterochromatin and/or centromeric proteins |
| | Telomere-specific small RNAs | tel-sRNAs | ~24 nt; Dicer independent; 2′-O-methylated at the 3′ terminus; evolutionarily conserved from protozoa to mammals; have not been described in human up to now | Relationship between tel-sRNAs and diseases has not yet been discovered/epigenetic regulation |
| | Pyknons | | Subset of patterns of variable length; form mosaics in untranslated and protein-coding regions; more frequently in 3′ UTR | Expected association with cancer biology/possible link with posttranscriptional silencing of genes, mainly involved in cell communication, regulation of transcription, signaling, transport, etc. |
| Long noncoding RNAs | Long intergenic noncoding RNAs | lincRNAs | Ranging from several hundreds to tens of thousands nt; lie within the genomic intervals between two genes; transcriptional cis-regulation of neighboring genes | Involved in tumorigenesis and cancer metastasis/involved in diverse biological processes such as dosage compensation and/or imprinting |
| | Long intronic noncoding RNAs | | Lie within the introns; evolutionary conserved; tissue and subcellular expression specified | Aberrantly expressed in human cancers/possible link with posttranscriptional gene silencing |
| | Telomere-associated ncRNAs | TERRAs | 100 bp to >9 kb; conserved among eukaryotes; synthesized from C-rich strand; polyadenylated; form intermolecular G-quadruplex structure with single-stranded telomeric DNA | Possible impact on telomere-associated diseases including many cancers/negative regulation of telomere length and activity through inhibition of telomerase |
| | Long noncoding RNAs with dual functions | | Both protein-coding and functionally regulatory RNA capacity | Deregulation has been described in breast and ovarian tumors/modulate gene expression through diverse mechanisms |
| | Pseudogene RNAs | | Gene copies that have lost the ability to code for a protein; potential to regulate their protein-coding cousin; made through retrotransposition; tissue specific | Often deregulated during tumorigenesis and cancer progression/regulation of tumor suppressors and oncogenes by acting as miRNA decoys |
| | Transcribed-ultraconserved regions | T-UCRs | Longer than 200 bp; absolutely conserved between orthologous regions of human, rat, and mouse; located in both intra- and intergenic regions | Expression is often altered in some cancers; possible involvement in tumorigenesis/antisense inhibitors for protein-coding genes or other ncRNAs |

Multiple myeloma (MM) is a hematologic malignancy characterized by the accumulation of neoplastic plasma cells (PCs) in the bone marrow (BM), which causes bone destruction and marrow failure [8]. MM is characterized by marked genetic heterogeneity that has important implications for tumor pathogenesis, disease outcome, and response to treatment [9–11]. Recently, the introduction of drugs such as thalidomide, bortezomib, and lenalidomide in the current therapeutic approach to MM, has witnessed a dramatic improvement of disease control and long-term patient survival. However, while new research platforms are emerging [12–18] and many other novel agents are presently in an advanced phase of preclinical and clinical investigation [19–26], currently available experience reveals that a substantial portion of high-risk patients do not benefit even from double or triple combinations of novel agents with or without conventional cytotoxic agents. Therefore, MM is still an incurable and highly therapeutically challenging disease.

Recently, the discovery of ncRNAs has changed the landscape of cancer biology [27]. Early studies have shown that the expression of miRNAs, the best-characterized class of ncRNAs, is deregulated in cancer and experimental data indicate that targeting miRNA expression may produce important phenotypical changes in cancer cells. Based on these observations, miRNA-based anticancer therapies are being developed, either alone or in combination with current targeted therapies, with the goal to improve disease response and increase cure rates. The advantage of miRNA-based approaches relies on the ability to concurrently target, at multiple effector levels, major pathways involved in cell differentiation, proliferation, and survival [27]. This chapter discusses the current challenges, as well as the potential addressing strategies, for the development of a novel ncRNA-based personalized medicine for MM.

## 13.2   NONCODING RNAs: CLASSIFICATION

In recent years, it has become evident that the non-protein-coding component, which represents approximately 98.5% of the whole human genome, has a pivotal role in regulating physiologic cellular processes and contributes to molecular alterations in pathologic conditions [28]. The functional relevance of the non-protein-coding genome especially emerges from a class of small ncRNAs called miRNAs, whose aberrant expression and/or function is now widely recognized as a hallmark of human cancer. However, miRNAs are just the tip of the iceberg and other recently discovered ncRNAs might contribute to the pathogenesis of human cancer [29].

### 13.2.1   Small Noncoding RNAs

Posttranscriptional RNA silencing or RNA interference (RNAi) is a naturally conserved mechanism of regulation of gene expression described in almost all eukaryotic species including humans [30, 31]. It is mostly triggered by dsRNA precursors that vary in length and origin. These dsRNAs are rapidly processed into short RNA duplexes subsequently generating small ncRNAs, which are associated with Argonaute family proteins and guide the recognition and ultimately the cleavage or translational repression of complementary single-stranded RNAs, such as messenger RNAs or viral genomic/antigenomic RNAs. Moreover, the small ncRNAs have also been implicated in guiding chromatin modifications [31, 32]. Since the discovery of the first small ncRNA, various classes of small ncRNAs have been identified. Based on whether their biogenesis is dependent on Dicer, the dsRNA-specific RNA III ribonuclease, all the known eukaryotic small ncRNAs can be classified into two groups: Dicer-dependent, such as miRNAs, small interfering RNAs (siRNAs), and in some cases small nucleolar RNAs (snoRNAs); and Dicer-independent

small ncRNAs, such as PIWI-interacting RNAs (piRNAs) (Figure 13.2) [33]. Moreover, phylogenetic analysis indicates that known Argonaute family proteins can be divided into two subgroups namely AGO based on AGO1 and PIWI based on PIWI. Interestingly, Ago proteins interact with miRNAs and siRNAs while the Piwi subgroup is characterized by interaction with piRNAs [34]. Biogenesis of other small ncRNAs is less known or completely undescribed yet. These RNAs are generally classified according to their genome and function localization. Among them belong promoter-associated small RNAs (PASRs), transcription initiation RNAs´(tiRNAs), centromere repeat–associated small interacting RNAs (crasiRNAs), and telomere-specific small RNAs (tel-sRNAs). To the class of small ncRNAs also belong the recently discovered pyknons that, as suggested by current findings, are involved in many biological functions. It has been many times described that some of above-mentioned small ncRNAs play important roles in the pathogenesis of various diseases including tumors. In this respect, the most-studied ncRNAs are miRNAs, which have been described in many, if not all, cancers [35–38].

### 13.2.2   MiRNAs

The most frequently studied subclass of small ncRNAs are miRNAs, originally discovered by Victor Ambros in *Caenorhabditis elegans*. They are 18–25 nt long, evolutionarily conserved, single-stranded RNA molecules involved in specific regulation of gene expression in eukaryotes [39]. It is predicted that miRNA genes account for 1–2% of the human genome and control the activity of ~50% of all protein-coding genes [40, 41]. Early annotation for the genomic position of miRNAs indicated that most miRNAs are located in intergenic regions (>1 kb away from annotated or predicted genes), although a sizeable minority was found in the intronic regions of known genes in the sense or antisense orientation. This led to the postulation that most miRNA genes are transcribed as autonomous transcription units [41]. A detailed analysis of miRNA gene expression showed that miRNA genes can be transcribed from their own promoters and that miRNAs are generated by RNA polymerase II (RNAPII) as primary transcripts (pri-miRNAs). These are processed to short, 70-nt, stem-loop structures known as pre-miRNAs by the ribonuclease called Drosha and the double-stranded-RNA-binding protein known as Pasha (or DGCR8—DiGeorge critical region 8), which together compose a multiprotein complex termed a *microprocessor*. The pre-miRNAs are transported to cytoplasm by the RAN GTP-dependent transporter exportin 5 (XPO5). In the cytoplasm, the pre-miRNAs are processed to mature miRNA duplexes by their interaction with the endonuclease enzyme Dicer in complex with dsRNA binding protein TRBP [41, 42]. One strand (the "guide strand") of the resulting 18- to 25-nt mature miRNA duplex ultimately gets integrated into the miRNA-induced silencing complex (miRISC) with the central part formed by proteins of the Argonaute family, whereas the other strand (passenger or

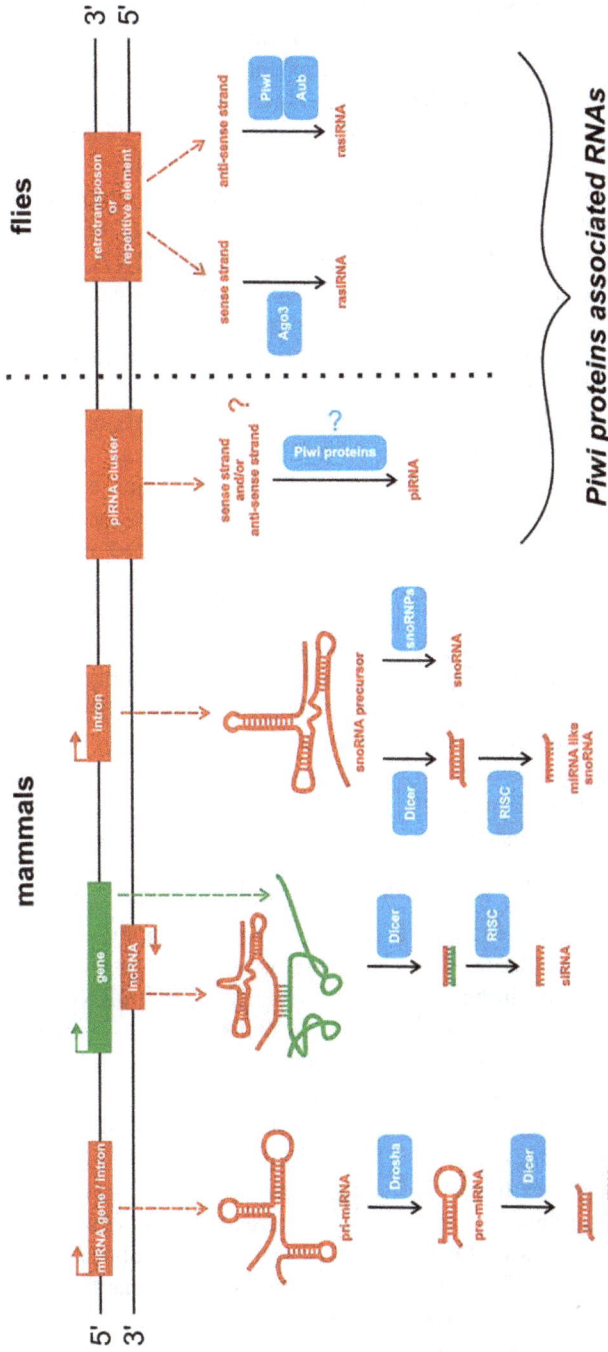

**FIGURE 13.2** Short ncRNA biogenesis pathways.

miRNA*) is released and degraded. The retained ("guide") strand is the one that has the less stably base-paired 5' end in the miRNA/miRNA* duplex. Generally, most miRNA genes produce one dominant miRNA species. However, the ratio of miRNA to miRNA* can vary in different tissues or developmental stages, which probably depends on specific properties of the pre-miRNA or miRNA duplex, or on the activity of different accessory processing factors [41]. Moreover, the ratio might be modulated by the availability of mRNA targets as a result of enhanced destabilization of either miRNA or miRNA* occurring in the absence of respective complementary mRNAs [42]. Mature miRNAs in miRISC exert their regulatory effects by binding to imperfect complementary sites. MiRNAs repress target-gene expression posttranscriptionally, apparently at the level of translation, through a miRISC complex that is similar to, or possibly identical with, that used for the RNAi pathway discussed later. Perfect complementarity of mRNA-miRNA allows Ago-catalyzed cleavage of the mRNA strand, whereas central mismatches exclude cleavage and promote repression of mRNA translation. Consistent with translational control, miRNAs that use this mechanism reduce the protein levels of their target genes, but the mRNA levels of these genes are barely affected [43–45]. Current studies indicate that miRNA targeting in mammalian cells occurs predominantly through binding to sequences within 3'UTRs [46, 47]; however, inhibition of gene expression through targeting the 5'UTR has been also demonstrated [48]. Nevertheless, statistical analyses of conserved miRNA target sequences proved that mammalian miRNA target sites rarely occur within 5'UTRs [46, 47, 49]. Moreover, it was found out that miR-10a induces, rather than inhibits, protein expression through binding to 5'UTRs of cellular transcripts [45]. It is therefore supposed that binding to 5'UTR results in mechanistic effects divergent from 3'UTR binding.

Most of the miRNAs described to date regulate crucial cell processes such as proliferation, differentiation, and apoptosis. Therefore, these RNAs are involved in human development as well as in initiation of various disorders including many, if not all, cancers where miRNAs have been found to be also significant prognostic and predictive markers [35, 50–57]. Examples of miRNAs with significant functional effects in cancer are mentioned below.

Bloomston et al. [58] identified six miRNAs linked to long-term survival in pancreatic adenocarcinoma. They found also that expression level of miR-196a-2 was able to predict patients' survival, because higher miRNA levels marked the poor survivors group. In hepatocellular carcinoma (HCC), upregulation of miR-221 and downregulation of miR-122 were associated with shorter time to recurrence [59, 60]. MiR-21 is upregulated in many solid tumors, including CRC. Slabý et al. [61] proved that miR-21 overexpression shows a strong correlation with the established prognostic factors as nodal stage, metastatic disease, and UICC stage. Moreover, Kulda et al. [62] correlated miR-21 expression to disease-free interval (DFI). There was shorter DFI in patients with a higher expression of miR-21. Several studies proved that downregulated expression of miR-221, miR-137, miR-372, miR-182*, let-7, and miR-34a is associated with shorter survival in patients with lung cancer [63–65]. Breast cancer metastatic process has been connected with upregulation of miR-10b [66] and with loss of expression of miR-126 and miR-335 [67]. Finally, higher levels of miR-15b were associated with poor survival and recurrence in melanoma [68]. Another important question for management of cancer patients is the possibility of predicting therapy response. Nakajima et al. [69] identified let-7g and miR-181b as significant indicators for chemoresponse to S-1-based chemotherapy. The same year, Markou et al. [70] demonstrated that inhibition of miR-21 and miR-200b increases the sensitivity of cholangiocarcinoma cells to gemcitabine. Yang et al. [71] identified miR-214, a miRNA upregulated in ovarian cancer, as responsible for cisplatin resistance through its action on PTEN/AKT pathway. Subsequently, a large number of publications have confirmed many 3'UTRs of oncogenes and tumor suppressor genes to be direct targets of selected miRNAs. According to a recent study by Nagel et al. [72], miR-135a and miR-135b decrease translation of the APC transcript in vitro. Concerning CRC, KRAS oncogene has been reported to be a direct target of the let-7 miRNA [73]. Another miRNA associated with KRAS regulation is miR-143 [74]. MiRNA array-based studies revealed the p85β regulatory subunit of PI3K as a direct target of miR-126 [75]. Moreover, another important regulatory component of PI3K pathway, the tumor suppressor gene PTEN, is strongly repressed by miR-21 in HCC [76]. MiR-17-5p belongs to a highly conserved, polycistronic miRNA cluster miR-17-92. Yu et al. [77] described the function of this cluster as a negative regulator of cell cycle and proliferation of human breast cancer cells, which directly regulates cyclin D1 (CCND1). The same cluster is also involved in malignancies of B-cell origin [78] and a direct regulation by c-MYC has been reported [79, 80]. Some of the most often deregulated miRNAs with their experimentally proved mRNA targets are summarized in Table 13.2; however, the number of described miRNAs and putative targets is much higher, and it is not possible to mention all of them.

MiRNAs belong to the most abundant class of small RNAs in animals. They represent approximately 1% of the genome of different species and each has hundreds of different targets: it has been estimated that approximately 30% of genes are regulated by at least one microRNA [81]. MiRNAs can be classified as intronic or intergenic: intronic miRNAs, which are more common, share the same promoter with the corresponding encoded genes and are then spliced out of the transcript and processed into mature miRNAs; intergenic miRNAs are transcribed by their own promoters and locate in genomic regions apart from the gene cluster regions [82]. Biogenesis of miRNAs takes place in the nucleus, where a pri-miRNA hairpin is transcribed by a RNA polymerase II and is subsequently cleaved by Drosha, a member of the RNA polymerase III

**TABLE 13.2**

**Gene Targets of the Most Common Described Human Cancer–Associated MiRNAs**

| MiRNA | Associated Cancers | In Vitro Confirmed Gene Targets |
|---|---|---|
| **MiR-21** | CRC, PC, RCC, GBM, BrC, NSCLC, BCL, PTC, HCC, HNSCC, ESCC, GC, CML, CCC, MM, OC, M, LC, PDA | PDCD4, TIMP3, RhoB, Spry1∅, PTEN, TM1, CDK2AP1, ANP32A, SMARCA4, ANKRD46, THRB, Cdc25A, BMPRRII, LRRFIP1, BTG2, MARCKS, TPM1 |
| **MiR-155** | NSCLC, SCLC, HCC, BrC, M, CCC, HL, PDA, RCC, GBM, PTC, CML, CRC, SPA, AML, NPC, CLL | FOXO3A, SOX6, SATB1, SKI, Wee1, SOCS1, SHIP1, S/EBPβ, IFN-γRα, AGTR1, FGF7, ZNF537, ZIC3, IKBKE, RhoA, BACH1, ZIC3, HIVEP2, CEBPB, ZNF652, ARID2, SMAD5, TP53INP1 |
| **MiR-145** | BrC, CRC, ESCC, NSCLC, PC, BCL, OC, GC, BlC, NPC, HCC | c-Myc, ERK5, FSCN1, SMAD2/3, IGF-1R, FLI1, DFF45, mucin 1, MYO6, CBFB, PPP3CA, CLINT1, ICP4, RTKN |
| **MiR-221** **MiR-222** | BrC, PC, CRC, M, GBM, ALL, HCC, PTC, PDA, GC, CML, NSCLC, AML, OC | DVL2, KIT, CDKN1B, Bmf, p27, HOXB5, CDKN1C/p57, CDKN1B/p27, MMP1, SOD2, TIMP3, Dicer1, ERα, ARHI, PUMA, p27Kip1, p57 |
| **Let-7a** | M, HL, nHL, CRC, SLC, NSCLC, GC, HNSCC, ESCC, OC, CLL, HCC | PRDM-1, STAT3, Caspase-3, Integrin β3, PRDM1/blimp-1 |
| **MiR-16** | LC, OC, NPC, GC, PC, BrC, HCC, MM, CLL, HL | VEGFR2, FGFR1, Zyxin, Cyclin, E1, Bmi-1, BRCA-1, BCL2 |
| **MiR-200** | BrC, PDA, GC, HNSCC, M, OC, PC | FN1, MSN, NTRK2, LEPR, ARHGAP19, ZEB1/2, Flt1/VEGFR1, FAP-1, FOG2, ERRFI-1 |
| **MiR-205** | M, BrC, PC, ESCC, HNSCC | Runx2, E2F1, ErbB3, Zeb1 |
| **MiR-31** | PTC, CRC, BrC, LC, GC, HCC | LATS2, WAVE3, SATB2, ITGA5, RDX, RhoA, FIH |
| **MiR-126** | CRC, GC, BrC, SCLC, AML, NSCLC, HCC | SLC7A5, SOX2, PLAC1, VEGFA, PIK3R2, Crk, EGFL7, p85beta |
| **MiR-210** | PDA, RCC, BrC, PC, GBM, NSCLC, OC, GC, HNSCC | FGFRL1, SDHD, MNT |
| **MiR-9** | GBM, PC, nHL, EC, OC | CAMTA1, PDGFR-β, CDX2, PRDM-1, E-cadherin, NFkappaB1 |
| **MiR-141** | PC, EC, CRC, HNSCC, LC, BrC, ESCC, OC, RCC | SIP1, YAP1 |
| **MiR-122** | HCC, RCC | Bcl-w, ADAM17 |

*Abbreviations:* ALL, acute lymphocytic leukemia; AML, acute myeloid leukemia; BCL, B-cell lymphoma; BlC, bladder cancer; BrC, breast cancer; CCC, cervical cell carcinoma; CLL, chronic lymphocytic leukemia; CML, chronic myelogenous leukemia; CRC, colorectal cancer; EC endometrial cancer; ESCC, esophagus squamous cell carcinoma; GBM, glioblastoma multiforme; GC, gastric cancer; HCC, hepatocellular carcinoma; HL, Hodgkin lymphoma; HNSCC, head and neck squamous cell carcinoma; LC, lung cancer; LCM, laryngeal carcinoma; M, melanoma; MM, multiple myeloma; nHL, non-Hodgkin lymphoma; NPC, nasopharyngeal carcinoma; NSCLC, non–small cell lung cancer; OC, ovarian cancer; PC, prostate cancer; PDA, pancreatic ductal adenocarcinoma; PTC, papillary thyroid carcinoma; RCC, renal cell carcinoma; SCLC, small cell lung cancer; SPA, sporadic pituitary adenoma.

family, in a 70–100 bp pre-miRNA that translocates in the cytoplasm by exportin 5; here, Dicer, which also belongs to the RNA polymerase III family, cleaves pre-miRNAs leading to 20–22 bp miRNA/miRNA* duplexes. After cleavage, the miRNA duplex is unwound by an as yet unknown RNA helicase and the mature miRNA strands bind to an Argonaute protein into a RNP complex, known as miRNA-containing ribonucleoprotein complex, Argonaute or miRNA-containing RNA-induced silencing complex, that drives the mature miRNA strand to the 3′-UTR mRNA target sequence. 3′-UTR binding represses translation or induces deadenylation, the first step of mRNA decay, depending on the degree of complementarity between miRNA and its target sequence (Figure 13.3a) [83, 84].

### 13.2.3 piRNAs

PIWI-interacting RNA (piRNAs) are 25–31 nt in length and bind PIWI proteins, the other subclass of the Argonaute family whose expression is restricted to germline cells [85]. Molecular mechanisms underlying piRNA biogenesis are still elusive. PiRNAs are a highly complex mix of sequences, with thousands of distinct piRNA sequences, derived from defined genomic regions called piRNA clusters. The main function of piRNAs is to silence expression and block mobilization of transposable elements [86], as identified in *Drosophila melanogaster*: PIWI proteins mediated cleavage of transposable element transcripts [87] and heterochromatin gene silencing [88]. Four PIWI proteins are expressed in humans: PIWIL1/HIWI, PIWIL2/HILI, PIWIL3, and PIWIL4/HIWI2 [89]. Unlike the other classes of small regulatory RNAs, piRNAs are generated from single-stranded RNA transcripts by a Dicer-independent mechanism. PiRNAs have a preference for uridine at their 5′-ends, and have a HEN1-methyltransferase-catalyzed 2′-O-methylribose modification at their 3′-ends [90]. Three major classes of PIWI proteins (i.e., PIWIL1, PIWIL2, and PIWIL4) are involved in a "ping-pong" amplification process, which creates antisense piRNAs that repress the transcript of origin (Figure 13.3b) [91]. While miRNA profiles in cancer have been widely characterized, the study of PIWI and/or piRNAs' role in pathogenesis of human cancer is still in its infancy.

### 13.2.4 snoRNAs

SnoRNAs are molecules of 60–300 nt in length bound to small nucleolar ribonucleoproteins (snoRNPs), which are complexes responsible for sequence-specific-2′-O-methylation and pseudouridylation of ribosomal RNA (rRNA). Two main groups of snoRNAs have been reported: (1) the box H/ACA snoRNAs (SNORAs), which bind conserved core box H/ACA snoRNP proteins, specifically acting as guide RNA in pseudouridylation of target rRNAs, and (2) the box C/D snoRNAs (SNORDs), which bind conserved core box C/D snoRNP proteins involved in 2′-O-ribose methylation. rRNA posttranscriptional

modifications take place in the nucleolus and facilitate rRNA folding and stability [92]. SnoRNA coding sequences are predominantly located in introns. The mature snoRNAs originating after splicing, debranching, and trimming of the original transcript, are either exported—in this case, working in the rRNA processing—or remain in the nucleus, where they regulate alternative splicing or are involved in as yet unknown activities (Figure 13.3c) [93].

### 13.2.5 lncRNAs

This class of ncRNA represents the largest portion of the mammalian noncoding transcriptome [28]; they act by positively or negatively modifying the expression or processing of their target genes, which can be either coding or noncoding. Many identified long noncoding RNAs (lncRNAs) are transcribed by RNA polymerase II, spliced, and often contain canonical polyadenylation signals. On the other hand, it has been also found that some of these lncRNAs are transcribed by polymerase III [94]. Moreover, lncRNAs promoters are bound and regulated by transcriptional factors and epigenetically marked with specific histone modifications [95]. LncRNAs are developmentally regulated and tissue specific, and have been associated with a spectrum of biological processes, for example, alternative splicing, modulation of protein activity, alternation of protein localization, and epigenetic regulation. LncRNAs can be also precursors of small RNAs and even tools for miRNA silencing. Long intergenic noncoding RNAs (lincRNAs) belong to lncRNA family, which locate within the genomic regions intervalling two genes. More than 3000 human lincRNAs have been identified, but <1% have been characterized. LincRNAs are involved in diverse biological processes, such as imprinting or cancer metastasis [95]. Ultraconserved regions (UCRs) are a subset of conserved sequences located in both intragenic and intergenic regions. There are 481 described UCRs, some of which overlap with coding exons [96]. Surprisingly, 68% of UCRs are transcribed, constituting a new category of ncRNAs, the T-UCRs [97].

Other classes of ncRNA have been described that are associated with the transcriptional start sites (TSSs) of genes: for example, promoter-associated small RNAs [98], TSS-associated RNAs (TSSa-RNAs) [99], promoter upstream transcripts, transcription initiation RNAs [100], and telomeric repeat-containing RNAs, which are transcribed from telomeres and cooperate to maintain the integrity of telomeric heterochromatin through regulating telomerase activity [101].

### 13.2.6 Small Interfering RNAs

Another class of small ncRNAs involved in posttranscriptional RNA silencing is so-called small interfering RNA (siRNA). They are produced from long dsRNAs of exogenous or endogenous origin [102]. These short helical RNA molecules are formed by two at least partially

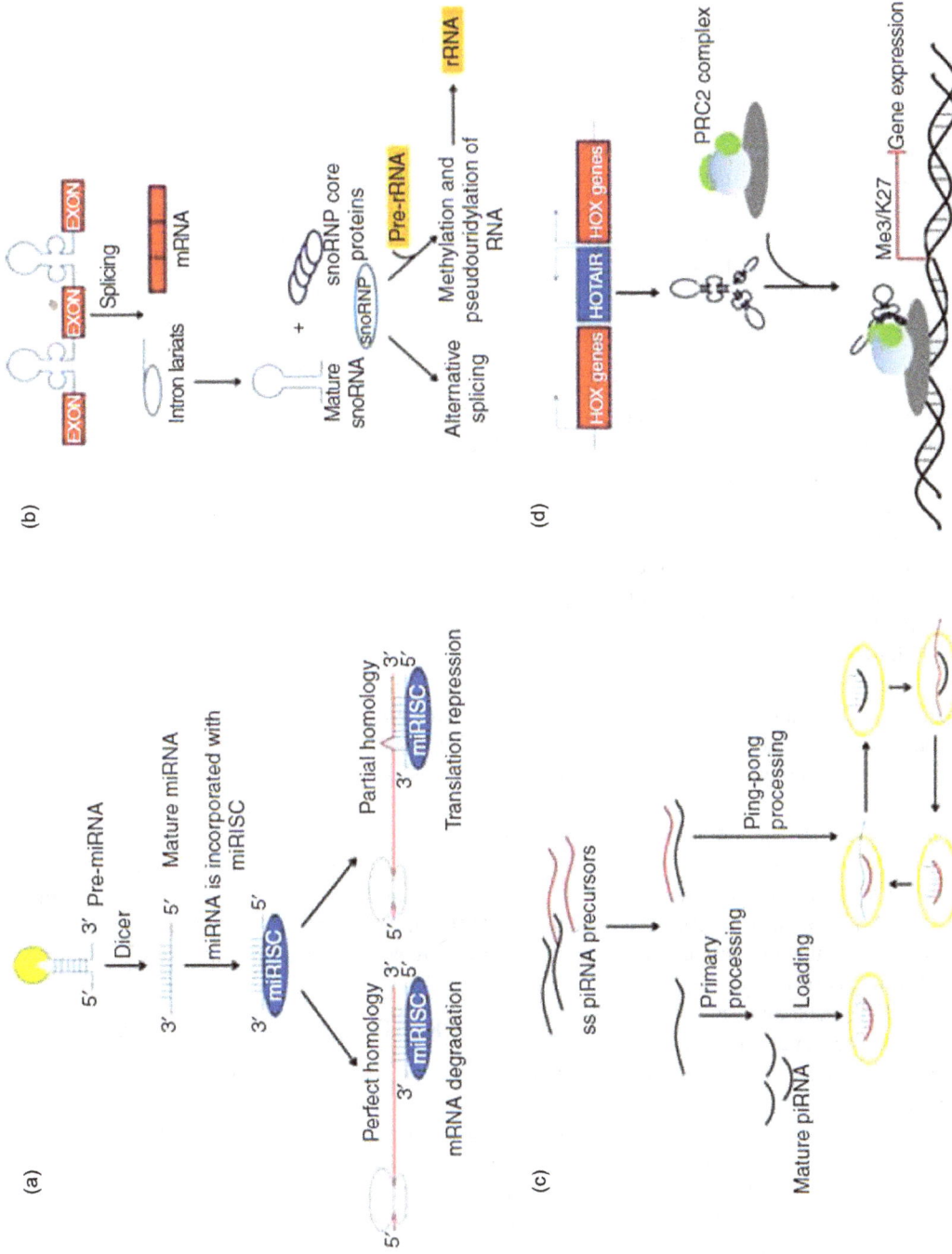

**FIGURE 13.3** **Schematic diagram of ncRNA biogenesis and mode of action.** (a) After being transcribed in the nucleus from a primary-miRNA, precursor miRNAs (pre-miRNAs) are exported from the nucleus by exportin 5. Further processing by Dicer and TAR RNA-binding protein 2 (TARBP2) generates mature miRNAs, which are loaded into the RNA-induced silencing complex (RISC). MiRNAs function through degradation of protein-coding transcripts or translational repression. (b) Mature small nucleolar RNAs (snoRNAs) generated by splicing, debranching, and trimming are either exported from the nucleus, in which case they function in ribosomal RNA (rRNA) processing, or remain in the nucleus, where they mainly regulate alternative splicing. (c) PIWI-interacting RNAs (piRNAs) are expressed as single-stranded RNAs (ss piRNAs) or produced through a secondary amplification loop via sense and antisense intermediates. The PIWI ribonucleoprotein (piRNP) complex functions in transposon repression through target degradation and epigenetic silencing. (d) The lincRNA HOX transcript antisense RNA (HOTAIR) is transcribed from the HOXC locus and functions in the binding and recruitment of the polycomb-repressive complexes (PRCs), like PRC2 and LSD1. Through an undetermined mechanism, the HOTAIR-PRC2-LSD1 complex is redirected to the HOXD locus on chromosome 2 where genes involved in metastasis suppression are silenced through H3K27 methylation.

**TABLE 13.3**

**Small RNA-Based Therapeutics in Clinical Trials (Adapted from [114])**

| Gene Target | Drug Type | Drug Name | Clinical Phase | Notes |
|---|---|---|---|---|
| Bcl-2 | LNA-oligo | SPC2996 | 1/2 | CLL |
| Immunoproteasome β-subunits LMP2, LMP7 and MECL1 | siRNA | Proteasome siRNA | 1 | Metastatic lymphoma |
| PLK1 | siRNA | PLK SNALP | Preclinical | |
| M2 subunit of ribonucleotide reductase | siRNA | CALAA-01 | 1 | Solid tumors |
| PKN3 | siRNA | Atu027 | 1 | Solid tumors |
| KSP and VEGF | siRNA | ALN-VSP | 1 | Solid tumors |
| Survivin | LNA-oligo | EZN3042 | 1/2 | Solid tumors |
| HIF-1α | LNA-oligo | EZN2968 | 1/2 | Solid tumors |
| Furin | shRNA | FANG vaccine | 1 | Solid tumors |
| eiF-4E | LNA-oligo | eIF-4E ASO | 1 | Solid tumors |
| Survivin | LNA-oligo | Survivin ASO | 2 | Solid tumors |

complementary RNA single strands, namely the passenger strand and the guide strand. Typical strand lengths of these dsRNAs are 19–23 nt and they are made by Dicer processing as miRNAs [103]. One of the arisen single strands is subsequently incorporated into RISC (RNA-induced silencing complex) which guides sequence-specific degradation of complementary target mRNAs unlike miRNA that rather suppresses translation and does not lead to degradation of the mRNA target [31, 104, 105]. SiRNAs are used in gene-silencing experiments worldwide and have become a specific and powerful tool to turn off the expression of target genes, and also are a promising experimental tool in molecular oncology. SiRNAs could be used in cancer therapy by several strategies. These include the suppression of over-expressed oncogenes, retarding cell division by interfering with cyclins and related genes, or enhancing apoptosis by inhibiting antiapoptotic genes.

For example, Vassilev et al. [106] developed new siRNA-based inhibitors of the p53-MDM2 protein interaction. A year later, Wu et al. [52] demonstrated that downregulation of RPL6 (ribosomal protein L6) in gastric cancer SGC7901 and AGS cell lines by siRNA reduced colony forming ability and cell growth. Moreover, the cell cycle of these cells was suppressed in G1 phase. Similarly, CDK8-specific siRNA transfection downregulated the expression of CDK8 in colon cancer cells, which was also associated with a decrease in the expression of β-catenin, inhibition of proliferation, increased apoptosis, and G0/G1 cell cycle arrest [107]. Dufort et al. [108] described that cell transfection of IGF-IR siRNAs decreased proliferation, diminished phosphorylation of downstream signaling pathway proteins, AKT and ERK, and caused a G0/G1 cell cycle block in two murine breast cancer cell lines, EMT6 and C4HD. The IGF-IR silencing also induced secretion of two proinflammatory cytokines, TNF-α and IFN-γ. Another study showed that mTOR-siRNA transfection significantly inhibits cell proliferation, increases the level of apoptosis, and decreases

migration of NSCLC cells, and could be used as an alternative therapy targeting mTOR with fewer side effects [109]. RNAi against multidrug resistance genes or chemoradioresistance and angiogenesis targets may also provide beneficial cancer treatments. He et al. [110] proved that silencing of MDR1 by siRNA led to decreased P-glycoprotein activity and lower drug resistance of L2-RAC cells, which could be used as a novel approach of combined gene and chemotherapy for yolk sac carcinoma. Another study showed that combination of proteasome inhibitors with Mcl-1 siRNA enhances the ultimate anticancer effect in DLD-1, LOVO, SW620, HCT-116, SKOV3, and H1299 cell lines [111]. Bansal et al. [112] state that selective siRNA depletion of CDK1 increases sensitivity of patients with ovarian cancer to cisplatin-induced apoptosis. The number of publications dealing with siRNAs is rapidly growing, and successful cancer therapy by siRNA in vitro and in vivo provides the enthusiasm for potential therapeutic applications of this technique [113]. Some examples of siRNA cancer therapies in clinical trials are summarized in Table 13.3.

## 13.3 PIWI PROTEIN–ASSOCIATED RNAs

Extensive research in the past few years has revealed that members of the Argonaute protein family are key players in gene-silencing pathways guided by small RNAs. This family is further divided into AGO and PIWI subfamilies [115]. It was proved that the AGO proteins are present in diverse tissues and bind to miRNAs and siRNAs, whereas PIWI proteins are especially present in germline, and associate with a new class of small ncRNAs termed PIWI-interaction RNAs (piRNAs). PiRNAs are typically 24–32 nt long RNAs that are generated by a Dicer-independent mechanism. It was thought that they are derived only from transposons and other repeated sequence elements [116], and, therefore, they were alternatively designated as repeat-associated small interfering RNAs (rasiRNAs) [117]. But it is now clear that piRNAs can be also derived from complex DNA sequence elements [118] and that rasiRNAs are a subset of piRNAs.

The precise mechanism of piRNA biogenesis is not clear, but in 2007 Brennecke et al. [116] described a new mechanism similar to secondary siRNA generation, called a ping-pong model. They observed that antisense piRNAs associate with PIWI/AUB complex while sense piRNAs associate with AGO3 protein. This information led to the suggestion that PIWI and AUB proteins bind to maternally deposited piRNA (primary piRNA) and this complex is subsequently bound to the transcripts produced by retrotransposons and cleaves a transcript generating sense piRNAs (secondary piRNAs) that bind to AGO3. Finally, piRNA-AGO3 complex binds to the retrotransposon transcript, creating another set of antisense piRNAs. However, the model of piRNA biogenesis is still incomplete and precise mechanisms of action remain poorly characterized (for a review, see [119–121]).

The PIWI subfamily as well as piRNAs have been implicated in germ cell development, stem cell self-renewal, and retrotransposon silencing. Recently, several studies were published describing the association between HIWI (the human ortholog of PIWI) expression and a diverse group of cancers including pancreatic [122], gastric [123], adenocarcinomas, sarcomas [124], HCC [125], colorectal [126], gliomas [127], and esophageal squamous cell carcinomas [128]. It was proved that higher levels of HIWI mRNA are connected with worse clinical outcome. Moreover, the expression pattern of HIWI in gastric cancer tissues was similar to that of Ki67 and suppression of HIWI-induced cell cycle arrest in the G2/M phase [123]. Lee et al. [129] described that PIWIL2 (PIWI-like 2) protein is widely expressed in tumors and inhibits apoptosis through activation of STAT3/BCL-X(L) signaling pathway. Similarly, the newest study of Lu et al. [130] shows that this protein forms a PIWIL2/STAT3/c-Src complex, where STAT3 is phosphorylated by c-Src and translocated to nucleus. Subsequently, STAT3 binds to p53 promoter and represses its transcription. These findings indicate that PIWI proteins may be involved in the development of different types of cancer and could be a potential target for cancer therapy. Recently, it was also proved, that not only PIWI proteins, but also piRNAs can play an important role in carcinogenesis. It was discovered that expression of piR-823 in gastric cancer tissues was significantly lower than in noncancerous tissues. Artificial increase of the piR-823 levels in gastric cancer cells inhibited their growth. Moreover, the observations from the xenograft nude mice model confirmed its tumor-suppressive properties [131]. On the contrary, levels of piR-651 were upregulated in gastric, colon, lung, and breast cancer tissues compared with the paired noncancerous tissues. The growth of gastric cancer cells was efficiently inhibited by a piR-651 inhibitor and the cells were arrested at the G2/M phase [132]. Interestingly, the peripheral blood levels of piR-651 and piR-823 in the patients with gastric cancer were significantly lower than those from controls. Thus,

piRNAs may be valuable biomarkers for detecting circulating gastric cancer cells [133]. Resolving the function of PIWI proteins and piRNAs has broad implications not only in understanding their essential role in fertility, germline, stem cell development, and basic control and evolution of animal genomes, but also in the biology of cancers [34].

## 13.4 SMALL NUCLEOLAR RNAs

Small nucleolar RNAs (snoRNAs), 60–300 nt long, represent one of the abundant groups of small ncRNAs characterized in eukaryotes. SnoRNAs are enriched in the nucleolus, which is the most prominent organelle in the interphase nucleus providing the cellular locale for the synthesis and processing of cytoplasmic ribosomal RNAs (rRNAs) [134]. Most snoRNAs are located within introns of protein-coding genes and are transcribed by RNA polymerase II; however, they can also be processed from introns of longer ncRNA precursors [135]. Nevertheless, while vertebrate snoRNAs are prevalently excised from pre-mRNA introns, in plant and yeast these RNAs are mainly generated from independent transcription units, as either monocistronic or (especially in plants) polycistronic snoRNA transcripts [136].

All snoRNAs fall into two major classes based on the presence of short consensus sequence motifs. The first group contains the box C (RUGAUGA) and D (CUGA) motifs, whereas members of the second group are characterized by the box H (ANANNA) and ACA elements [137]. In both classes of snoRNAs, short stems bring the conserved boxes close to one another to constitute the structural core motifs of the snoRNAs, which coordinate the binding of specific proteins to form small nucleolar RNPs (snoRNPs) distinct for both groups [134, 138]. SnoRNAs have important functions in the maturation of other ncRNAs. Above all, they manage posttranscriptional modification of rRNA and snRNA by 2′-O-methylation and pseudouridylation (for a review, see [134]). Interestingly, a number of human snoRNAs with miRNA-like function have been identified. These snoRNAs are processed to small 20–25 nt long RNAs that stably associate with Ago proteins. Processing is independent of Drosha but requires Dicer. Moreover, cellular target mRNA, whose activity is regulated by snoRNA, was identified [139].

Several studies have indicated that alterations of snoRNAs play important functions in cancer development and progression. The first report linking snoRNAs to cancer was published in 2002 by Chang et al. [140]. They proved that h5sn2, a box H/ACA snoRNA, was significantly downregulated in human meningiomas compared with normal brain tissues. Subsequently, Dong et al. [141] identified snoRNA U50 as a reasonable candidate for the 6q tumor suppressor gene in prostate cancer and this statement was confirmed in another study describing involvement of snoRNA U50 in the development and/or progression of breast cancer [142]. Interestingly, chromosome 6q14-15 is a breakpoint of chromosomal translocation t(3;6)(q27;q15) for human B-cell lymphoma [143]. The same year, the

GAS5 (growth arrest-specific transcript 5) was identified to control mammalian apoptosis and cell growth. GAS5 transcript levels were found to be significantly lower in breast cancer samples relative to adjacent unaffected normal breast epithelial tissues. Despite the fact that this gene has no significant protein-coding potential, it was proved that several snoRNAs are encoded in its introns [144]. By profiling ncRNA signatures in NSCLC tissues and matched noncancerous lung tissues, four snoRNAs (snoRD33, snoRD66, snoRD76 [145], and snoRA42 [146]) were found to be overexpressed in lung tumor tissues and it is supposed that they could be used as potential markers for early detection of NSCLC [145]. Moreover, snoRD33 is located at chromosome 19q13.3 that contains oncogenes involved in different malignances including lung cancer, whereas snoRD66 and snoRD76 are located at chromosomal regions 3q27.1 and 1q25.1, respectively. These two chromosomal segments are the most frequently amplified in human solid tumors [50, 147, 148]. Recently, low levels of four snoRNAs (RNU44, RNU48, RNU43, RNU6B), commonly used for normalization of miRNA expression, were associated with a poor prognosis of the cancer patients [149]. Martens-Uzunova et al. [150] analyzed the composition of the entire small transcriptome by Illumina/Solexa deep sequencing and revealed several snoRNAs with deregulated expression in samples of patients with prostate cancer. The newest publication concerning snoRNAs proved that snoRD112-114 at the DLK1-DIO3 locus is ectopically expressed in acute promyelocytic leukemia (APL), which shows that a relationship exists between a chromosomal translocation and expression of snoRNA loci. Moreover, in vitro experiments revealed that the snoRD114-1 [14q (II-1)] variant promotes cell growth through G0/G1 to S phase transition mediated by the Rb/p16 pathways [151]. Finally, it was also published that snoRNAs are present in stable form in plasma and serum samples [145, 149] and therefore could be used as fluid-based biomarkers for cancers. These facts indicate that snoRNAs are critically associated with the development and progression of cancer; however, further research for comprehensively understanding their role in carcinogenesis is required.

## 13.5   PROMOTER-ASSOCIATED RNAs

Recently, a new class of ncRNAs known as promoter-associated RNAs (paRNAs) (sometimes termed as promoter-upstream transcripts, PROMPTs [152]; transcription start site-associated RNAs [153]; or promoter-proximal transcription start site RNAs [154]), was discovered. These ncRNAs are derived from eukaryotic promoters and have the potential to regulate the transcription of protein-coding genes by targeting epigenetic silencing complexes [114, 155, 156]. Their size ranges from 18–200 nt, and they include long, small, and tiny RNAs.

The short paRNAs (PASRs) were identified in 2007 [157] using RNA maps. They are located near the promoter or transcription start site (TSS), but they are not associated with known protein-coding genes. These transcripts are 20–90 nt long and it was proved that they are not Dicer product [153]. Human PASRs are expressed at low levels and their number per gene is positively correlated with promoter activity and mRNA level [152]. The tiny paRNAs or transcription initiation RNAs (tiRNAs) are shorter than 23 nt and they are transcribed in both sense and antisense directions around the promoter [158]. Furthermore, they are closely associated with highly expressed promoters and are preferentially located in GC-rich promoters [114, 158]. It is still unclear how these two classes of small RNAs are related to one another, or if they share common biogenesis pathways [158]. Recently, a long paRNA (PALR, 100–200 nt) has been identified at a single-gene level, and it was associated with regulatory functions (for a review, see [155, 156, 159, 160]), especially with modification of DNA methylation [161].

It is supposed, that because of the potential of paRNAs to regulate transcription, their deregulation could be associated with different types of diseases, including cancer. It was proved that transfection of mimetic paRNAs into HeLa and HepG2 cells resulted in the transcriptional repression of human C-MYC and connective tissue growth factor (CTGF) [162]. Hawkins et al. [163] described that targeting of the human ubiquitin C gene (UbC) with a small paRNA led to long-term silencing, which correlated with an early increase in histone methylation and a later increase in DNA methylation at the targeted locus. Furthermore, it was shown that PASRs play an important role in maintaining accessible chromatin architecture for transcription and releasing negative supercoils during transcription [153]. Concerning tiRNAs, they may have similar functions like PASRs; moreover they are usually found at CTCF-binding sites. Taft et al. [164] proved that overexpression of tiRNAs decreased CTCF binding and associated gene expression, whereas inhibition of tiRNAs resulted in increased CTCF localization and associated gene expression. Wang et al. [165] described that an RNA-binding protein TLS (for translocated in liposarcoma) can specifically bind to CREB-binding protein (CBP) and p300 histone acetyltransferase depending on its allosteric modulation by PALRs, and so repress gene target CCND1 in human cell lines. Finally, it was shown that paRNAs have the potential to form double-stranded RNAs and to be processed into endogenous siRNAs [166]. These facts indicate that this novel class of ncRNAs has a great potential to regulate expression of various tumor suppressors and oncogenes on a transcriptional level and therefore be involved in human cancerogenesis.

## 13.6   CENTROMERE REPEAT ASSOCIATED SMALL INTERACTING RNAs

Cell stresses can induce incorrect centromere function manifesting in loss of sister chromatid cohesion, abnormal chromosome segregation, and aneuploidy, which have been observed in many human diseases including cancers [167]. These defects are often correlated with the aberrant accumulation of centromere satellite transcripts [168]. Moreover, it was observed that human cells under stress accumulate large transcripts of SatIII satellites [169]. The accumulation of similar transcripts in vertebrate cells is thought to result from defective RNA processing of larger transcripts that leads to a reduction of the small RNAs that participate in the recruitment of specific histones critical for centromere function [168, 170]. Research on mammalian models uncovered the strong bidirectional promoter capability of the kangaroo endogenous retrovirus (KERV-1) LTR to produce long double-stranded RNAs for both KERV-1 and surrounding sequences, including sat23. These long dsRNAs are then processed into centromere repeat associated small interacting RNAs (crasiRNAs), 34–42 nt in length. Unfortunately, the mechanism by which full-length KERV-1 and sat23 transcripts are processed into crasiRNAs remains unknown. The crasiRNAs are involved in the recruitment of heterochromatin and/or centromeric proteins. These findings have profound implications for understanding of centromere function and epigenetic identity by suggesting that a retrovirus, KERV-1, may participate in the organization of centromere chromatin structures indispensable to chromosome segregation in vertebrates [167]. These small centromere-associated ncRNAs occur conserved among eukaryotes suggesting their impact also in human.

## 13.7 TELOMERE-SPECIFIC SMALL RNAs

Another group of recently described short ncRNAs is telomere-specific small RNAs (tel-sRNAs). Tel-sRNAs are ~24 nt long, Dicer-independent, and 2′-O-methylated at the 3′ terminus. They are asymmetric with specificity toward telomere G-rich strand, and evolutionarily conserved from protozoan to mammalian cells. Interestingly, tel-sRNAs are upregulated in cells that carry null mutation of H3K4 methyltransferase MLL and downregulated in cells that carry null mutations of histone H3K9 methyltransferase SUV39H, suggesting that they are subject to epigenetic regulation. These results support that tel-sRNAs are heterochromatin-associated pi-like small RNAs [171]. Recently, it was also reported that an 18-mer RNA oligo of (UUAGGG)3 has potential to inhibit telomerase TERT activity in vitro by RNA duplex formation in the template region of the telomerase RNA component [172]. Therefore, it is supposed that tel-sRNAs containing UUAGGG repeats could act as sensors of chromatin status and create a feedback loop between the telomeric heterochromatic regulation and telomere length control. Although tel-sRNAs have not been described in humans to date, they could play an important role in carcinogenesis and contribute to unlimited replicative potential of cancer cells.

## 13.8 PYKNONS

Pyknons are a subset of 127,998 patterns of variable length, which form mosaics in untranslated as well as protein-coding regions of human genes. Nevertheless, they are found more frequently in the 3′UTR of genes than in other regions of the human genome [173, 174]. Pyknons are present in a statistically significant manner in genes that are involved in specific processes such as cell communication, transcription, regulation of transcription, signaling, transport, etc. Pyknons involve ~40% of the known miRNA sequences, thus suggesting a possible link with posttranscriptional gene silencing and RNA interference [174]. Different sets of pyknons are connected to allele-specific sequence variations of disease-associated SNPs and miRNAs, suggesting that increased susceptibility to multiple common human disorders is associated with global alterations in genome-wide regulatory templates affecting the biogenesis and functions of ncRNAs [175].

In the time since their discovery, evidence has been slowly accumulating that these pyknon motifs mark transcribed, ncRNA sequences with potential functional relevance in human disease. Tsirigos et al. [176] described two GO terms (GO:0006281/DNA repair, GO:0006298/mismatch repair) that were significantly enriched in pyknon-containing regions of the human introns. They pointed out that these two terms are uniquely associated with pyknons, and a search of the ENSEMBL database [177] for human genes labeled with these two GO terms identified a MLH1 gene that has been associated with hereditary nonpolyposis colorectal cancer and other types of carcinomas and microsatellite instabilities. The human MLH1 transcript has 17 introns and the authors proved that these introns contain more than ten different pyknons. Nevertheless, further research for comprehensively understanding their role in carcinogenesis is necessary.

## 13.9 LONG NONCODING RNAs

Long noncoding RNAs (lncRNAs) are the broadest class encompassing all non-protein-coding RNA species with length more than 200 nt, however, frequently ranging up to 100 kb. Many identified lncRNAs are transcribed by RNA polymerase II (RNAPII), spliced, and usually contain canonical polyadenylation signals, but this is not a fast rule [2]. On the other hand, Pagano et al. [178] found out that some of these lncRNAs are due to their promoter structure likely to be transcribed by polymerase III (RNAPIII) and they marked them as co-genes since they could specifically co-act with a protein-coding pol II gene. There is substantial evidence to suggest that lncRNAs mirror protein-coding genes. Additionally, lncRNAs' promoters are bound and regulated by transcriptional factors and epigenetically marked with specific histone modifications [179]. LncRNAs are developmentally and tissue specific, and have been associated with a spectrum of biological processes, for example, alternative splicing, modulation of protein activity, alternation of protein localization, and epigenetic regulation. LncRNAs can also be precursors of small RNAs and even tools for miRNAs silencing [114, 180–184]. However, one of their primary tasks appears to be regulators of protein-coding gene expression (Figure 13.4) [185].

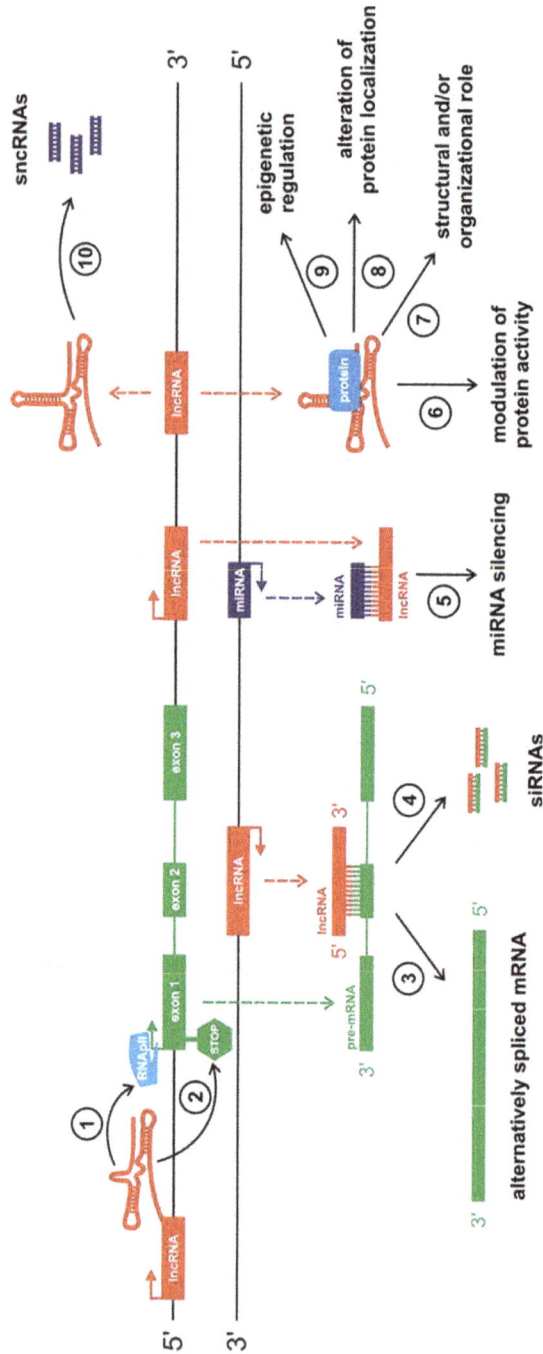

**FIGURE 13.4  Schematic illustration of lncRNAs functioning.** LncRNA transcribed from an upstream noncoding promoter can negatively (**1**) or positively (**2**) affect expression of the downstream gene by inhibiting RNA polymerase II recruitment and/or inducing chromatin remodeling, respectively. LncRNA is able to hybridize to the pre-mRNA and block recognition of the splice sites by the spliceosome, thus resulting in an alternatively spliced transcript (**3**). Alternatively, hybridization of the sense and antisense transcripts can allow Dicer to generate endogenous siRNAs (**4**). The binding of lncRNA to the miRNA function silencing (**5**). The complex of lncRNA and specific protein partners can modulate the activity of the protein (**6**), is involved in structural and organization roles of the cell (**7**), alters the protein localizes in the cell (**8**), and affects epigenetic processes (**9**). Finally, long ncRNAs can be processed to the small RNAs (**10**).

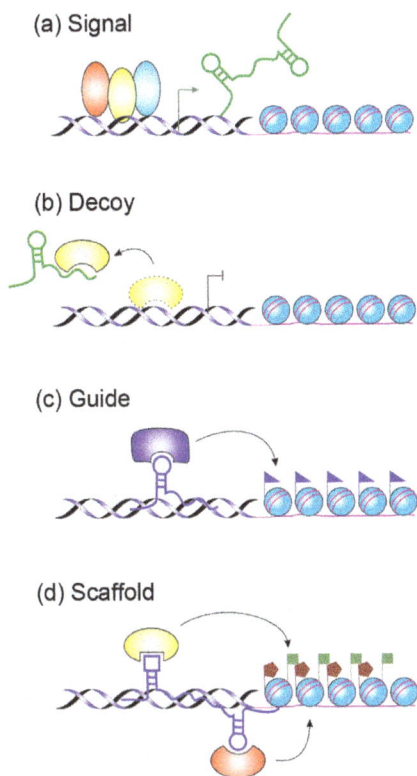

**FIGURE 13.5 Schematic diagram of the four mechanisms of lncRNA functioning.** (a) LncRNAs can function as signals and regulate gene expression. (b) LncRNAs can titrate transcription factors and other proteins away from chromatin or they can function as decoy for miRNA target sites. (c) LncRNAs can recruit chromatin-modifying enzymes to target genes and therefore function as guides. (d) LncRNAs can bring together multiple proteins to form ribonucleoprotein complexes (modified according to [186]).

Recently, Wang et al. [186] described four different mechanisms of lncRNAs action. Their research supposes that these molecules can function as signals, decoys, and guides, or as scaffolds (Figure 13.5).

It is not surprising, then, that dysregulation of lncRNAs seems to be an important feature of many complex human diseases, including cancer (Table 13.4), ischemic heart disease [187], and Alzheimer's disease [188]. Dysregulation of lncRNAs that function as regulators of the expression of tumor suppressors or oncogenes, and not the protein-coding sequence itself, may be one of the "hits" that leads to oncogenesis [2]. That is why they might be suitable as potential biomarkers and targets for novel therapeutic approaches in the future.

## 13.10 LONG INTERGENIC NONCODING RNAs

Long intergenic noncoding RNAs (lincRNAs) are newly discovered ncRNAs belonging to lncRNAs. RNAs of this subclass range in length from several hundred to tens of thousands of bases and they lie within the genomic intervals between two genes. More than 3000 human lincRNAs

**TABLE 13.4**

**Human Cancer–Associated lncRNAs (Adapted from [4])**

| LncRNA | Size | Cytoband | Cancer Types | References |
|---|---|---|---|---|
| HOTAIR | 2158 nt | 12q13.13 | Breast | [7, 183] |
| MALAT1/α/ NEAT2 | 7.5 kb | 11q13.1 | Breast, lung, uterus, pancreas, colon, prostate, liver, osteosarcoma, neuroblastoma, cervix | [189–194] |
| HULC | 500 nt | 6p24.3 | Liver | [195, 196] |
| BC200 | 200 nt | 2p21 | Breast, cervix, esophagus, lung, ovary, parotid, tongue | [197, 198] |
| H19 | 2.3 kb | 11p15.5 | Bladder, lung, liver, breast, endometrial, cervix esophagus, ovary, prostate, colorectal | [199–202] |
| BIC/ MIRHG155/ MIRHG2 | 1.6 kb | 21q11.2 | B-cell lymphoma | [203] |
| PRNCR1 | 13 kb | 8q24.2 | Prostate | [204] |
| LOC285194 | 2105 nt | 3q13.31 | Osteosarcoma | [205] |
| PCGEM1 | 1643 nt | 2g32.2 | Prostate | [206–208] |
| UCA1/CUDR | 1.4–2.7 kb | 19p13.12 | Bladder, colon, cervix, lung, thyroid, liver, breast, esophagus, stomach | [209] |
| DD3/PCA3 | 0.6–4 kb | 9q21.22 | Prostate | [210, 211] |
| anti-NOS2A | 1.9 kb | 17q23.2 | Brain | [212] |
| uc.73A | 201 nt | 2q22.3 | Colon | [213] |
| uc.338 | 590 nt | 12q13.13 | Liver | [214] |
| ANRIL/ p15AS/ CDK2BAS | 34.8 kb | 9p21.3 | Prostate, leukemia | [215–218] |
| MEG3 | 1.6 kb | 14q32.2 | Brain | [219–221] |
| GAS5/ SNHG2 | isoforms | 1q25.1 | Breast | [144] |
| SRA-1/SRA | 1965 nt | 5q31.3 | Breast, uterus, ovary | [222, 223] |
| PTENP1 | 3.9 kb | 9p13.3 | Prostate | [224, 225] |
| ncRAN | 2186–2087 nt | 17q25.1 | Bladder, neuroblastoma | [226, 227] |
| LSINCT5 | 2.6 kb | 5p15.33 | Breast, ovary | [228] |

have been identified, but less than 1% have been characterized [179, 229]. It was shown that distinct lincRNAs are involved in diverse biological processes such as imprinting or cancer metastasis [7, 183, 229]. Moreover, recent studies proved that lincRNAs are exquisitely regulated during development and in response to diverse signaling cues, and exhibit distinct gene expression patterns in primary tumors and metastases [179]. Therefore, these lncRNAs could be utilized for cancer diagnosis, prognosis, and serve as potential therapeutic targets.

Recently it has been demonstrated that lncRNAs can act as natural "miRNA sponges" to reduce miRNA levels [198]. The most highly upregulated transcript found in a microarray-based study of gene expression in HCC was determined to be the ncRNA HULC, or highly upregulated in liver cancer. Transcribed from chromosome 6p24.3, this lncRNA demonstrates the hallmarks of a typical mRNA molecule, including a single spliced GT-AG intron, canonical polyadenylation signals upstream of the poly(A) tail, and nuclear export demonstrating strong localization to the cytoplasm. Although HULC was found to co-purify with ribosomes, no translation product for this lncRNA has been detected, supporting its classification as a noncoding transcript [199]. In addition to liver cancer, HULC was found to be highly upregulated in hepatic colorectal cancer metastasis and in HCC cell lines producing hepatitis B virus (HBV) [200]. HULC exists as part of an intricate autoregulatory network, which when perturbed, resulted in increased HULC expression (Figure 13.6a). The HULC RNA appeared to function as a "molecular decoy" or "miRNA sponge" sequestering miR-372, of which one function is the translational repression of PRKACB, a kinase targeting cAMP response element binding protein (CREB). Once activated, the CREB protein was able to promote HULC transcription by maintaining an open chromatin structure at the HULC promoter resulting in increased HULC transcription [201].

Another well-known RNA that belongs to lncRNA subclass described in the previous paragraph is HOX antisense intergenic RNA (HOTAIR) (see Figure 13.6b). HOTAIR is a 2.2 kb gene localized within the human HOXC gene cluster on the long arm of chromosome 2. It has been shown that this lincRNA has a potential to regulate HOXD genes in trans via the recruitment of polycomb repressive complex 2 (PRC2), followed by the trimethylation of lysine 27 of histone H3 [7]. In general, the 5′ region of the RNA binds the PRC2 complex responsible for H3K27 methylation, while the 3′ region of HOTAIR binds LSD1 (flavin-dependent monoamine oxidase), a histone lysine demethylase that mediates enzymatic demethylation of H3K4Me2. HOTAIR exists in mammals, has poorly conserved sequences and considerably conserved structures, and has evolved faster than nearby HOXC genes [230]. HOTAIR was one of the first metastasis-associated lncRNAs, described to have a fundamental role in cancer. This lncRNA was found to be highly upregulated in both primary and metastatic breast tumors, showing up to 2000-fold increased transcription over normal breast tissue. This phenotype seems to be closely linked with PRC2-dependent gene repression induced by HOTAIR. High levels of HOTAIR expression correlate with both metastasis and poor survival rate,

connecting lncRNAs with tumor invasiveness and patient prognosis [183]. In addition, it was observed that the high expression level of HOTAIR in HCC could be a candidate biomarker for predicting tumor recurrence in HCC patients who have undergone liver transplant therapy and might be a potential therapeutic target [231]. Huarte et al. [232] identified several lincRNAs that are regulated by p53. Furthermore, they proved that lincRNAs-p21 serves as a repressor in p53-dependent transcriptional responses, since inhibition of this lincRNA affected the expression of hundreds of gene targets enriched for genes normally repressed by p53.

While targeting cancer-specific miRNAs has proven to be successful, it will be necessary to design molecules with potential to inhibit lincRNAs. Gupta et al. [183] proved that these molecules can be depleted by siRNAs, but this possibility is quite complicated because of extensive secondary structures in lincRNAs [230]. Nevertheless, it is evident that cancer-associated lincRNAs may provide new approaches to the diagnosis and treatment of cancer.

## 13.11   LONG INTRONIC NONCODING RNAs

The biogenesis of long intronic ncRNAs is poorly understood at this time. Nevertheless, there are some indirect evidences that indicate an involvement of RNA polymerase II (RNAPII). Among such evidences belongs concordant and co-regulated expression profiles of many intronic ncRNAs and their corresponding protein-coding genes, the broad contribution of RNAPII-associated transcription factors, and physiological stimuli in the transcription of intronic ncRNAs as well the presence of poly(A+) tail [233–237]. Nonetheless, it is described that over 10% of long intronic poly(A+) ncRNAs are upregulated compared with only 4% of protein-coding transcripts after treatment with the RNAPII specific inhibitor α-amanitin [233, 236, 238]. These findings suggest that some intronic ncRNA and peculiar protein-coding RNAs could be transcribed by another RNA polymerase such as the recently described spRNAP-IV, whose transcriptional output seems to be enhanced by α-amanitin, or also could be transcribed by RNAP III [233, 238–242].

Similarly to lincRNAs, there are also described evolutionarily conserved long intronic ncRNA sequences from mouse and human [243, 244]. When the introns of a larger selection of vertebrates were aligned, the length of the conserved region became only 100 bp, while in the alignment of a smaller group of closely related species (human-mouse-cow-dog) the evolutionary conservation of the region extended to as much as 750 bp [244].

The widespread occurrence, tissue and subcellular expression specificity, evolutionary conservation, environment alteration responsiveness, and aberrant expression in human cancers are features that accredit intronic ncRNAs to be mediators of gene expression regulation. A few sets of intronic ncRNAs have the same tissue expression pattern

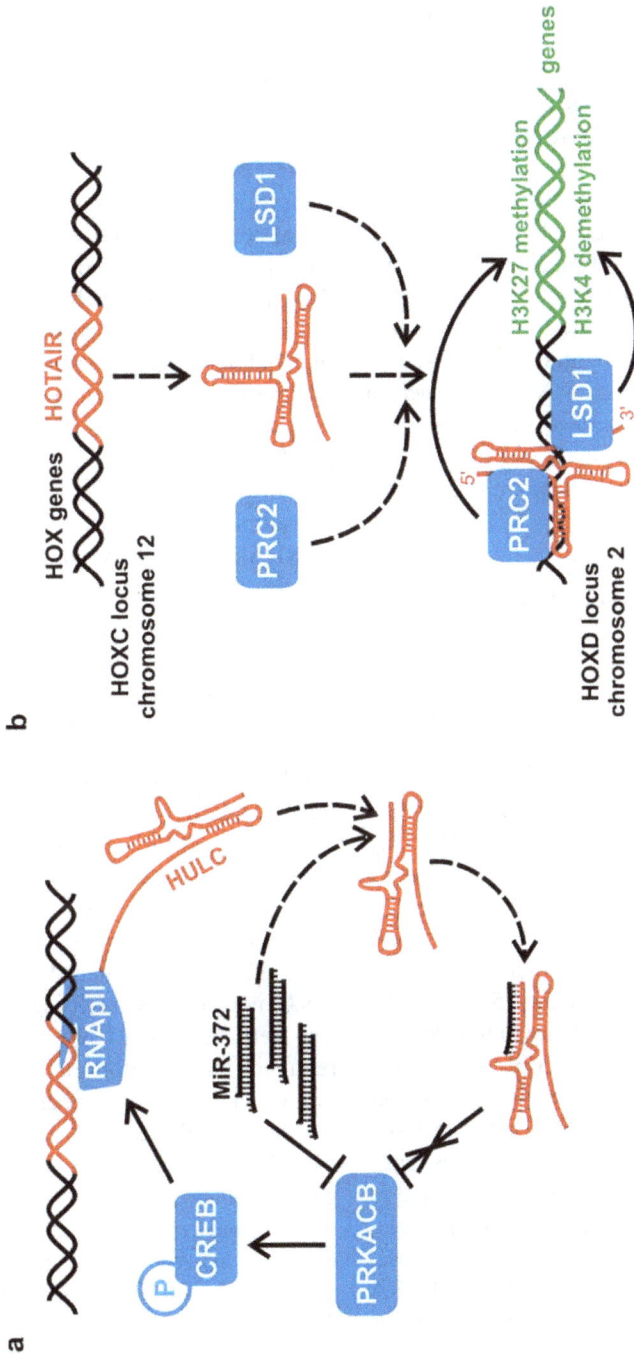

**FIGURE 13.6** Proposed mechanism of HULC upregulation in hepatocellular carcinoma (a) and HOTAIR-mediated gene silencing of 40 kb of the HOXD locus (b).

as the corresponding protein-coding genes, whereas others are inversely correlated. These findings point to complex regulatory relationships between intronic ncRNAs and their host loci [233, 236, 245, 246]. Some small ncRNAs are encoded within intronic regions; moreover, intronic miRNAs tend to be present in large introns with 5′-biased position distribution, what correlates with the previous observation that most long intronic transcripts are expressed within first introns of the host genes. Thus, it is expected that a number of long intronic ncRNAs are processed into smaller ncRNAs [111, 233, 247, 248]. Similar to lincRNA HOTAIR, Heo et al. [249] described a long intronic ncRNA termed as cold-assisted intronic ncRNA (COLDAIR), which is required for the vernalization-mediated epigenetic repression of FLC mediated by PRC2. Interestingly, the newest study of Tahira et al. [250] shows that long intronic ncRNAs are differentially expressed in primary and metastatic pancreatic cancer. Moreover, loci harboring intronic lncRNAs differentially expressed in pancreatic ductal carcinoma metastases were enriched in genes associated to the MAPK pathway. These findings indicate potential relevance of this class of transcripts in biological processes related to malignant transformation and metastasis.

## 13.12  TELOMERE-ASSOCIATED ncRNAs

Telomeres protect linear chromosome ends from being recognized and processed as double-strand breaks by DNA repair activities. This protective function of telomeres is essential for chromosome stability. Until recently, the heavily methylated state of subtelomeric regions, the geneless nature of telomeres, and the observed telomere position effect led to the notion that telomeres are transcriptionally silent [251]. This hypothesis was recently challenged when several groups independently demonstrated that subtelomeric and telomeric regions, although devoid of genes, have the potential to be transcribed into telomeric UUAGGG-repeat containing ncRNAs (TERRA) [252–254]. TERRA molecules are conserved among eukaryotes and have been identified also in human. TERRA transcripts are synthesized from the C-rich strand and polyadenylated, and their synthesis is α-amanitin-sensitive, suggesting that they are transcripts of RNAPII [251, 255]. TERRA molecules range between 100 bp and >9 kb in length and were reported to form intermolecular G-quadruplex structure with single-stranded telomeric DNA, but can also fold into a compact repeated structure containing G-quartets [254]. TERRA transcripts can be found throughout the different stages of the cell cycle, and their levels are affected by several factors that include telomere length, tumor stage, cellular stress, developmental stage, and telomeric chromatin structure [251].

TERRA most likely negatively regulates telomere length [254]. Increased TERRA levels by interfering with TERRA decay, such as the impairment of non-sense-mediated RNA decay in human cells or by deletion of the 5′–3′exonuclease Rat1p in *Saccharomyces cerevisiae*, are associated with a loss of telomere reserve [252, 255]. Current models propose a role for TERRA in controlling telomerase activity. In yeast, the formation of a DNA/RNA hybrid between TERRA and telomeres is thought to inhibit elongation by telomerase, whereas in mammals, TERRA was shown to efficiently inhibit telomerase activity in vitro, presumably by base pairing with the template region of the RNA component of telomerase [251, 253, 255]. Caslini et al. [256] described that telomere uncapping through either TRF2 shelterin protein knockdown or exposure to telomere G-strand DNA oligonucleotides significantly increases the transcription of TERRA, an effect mediated by the functional cooperation between transcriptional regulator MLL and the tumor suppressor p53. Sampl et al. [257] found that the expression of TERRA in patients with glioblastoma multiforme negatively correlates with the grade. Moreover, this finding of a diagnostic value of TERRA levels in astrocytoma WHO grade 2 to 4 corresponded with preliminary data in advanced stages of human tumors of larynx, colon, and lymph node [253]. Unfortunately, it is largely unclear how the expression of TERRA and the amount of TERRA transcripts are regulated in the cell [251]. Nevertheless, TERRA opens new avenues for telomere research that will impact on telomere-associated diseases including many cancers [258].

## 13.13  LONG NcRNAs WITH DUAL FUNCTIONS

Until not long ago, ncRNAs were strictly considered as RNA molecules with regulatory functions but not associated with the protein-coding capacity typical of mRNAs. However, the recent identification and characterization of bifunctional RNAs, i.e., RNAs for which coding capacity and activity as functional regulatory RNAs have been reported, suggest that a definite categorization of some RNA molecules is far from being straightforward [259]. The steroid receptor RNA activator (SRA) is a unique co-regulator that functions as a ncRNA, although incorporation of an additional 5′ region can result in translation of an SRA protein (SRAP) that also has co-activator activity [223, 260, 261]. SRA was initially shown to enhance gene expression through a ribonucleoprotein complex with steroid receptors and SRC-1 [260]. Currently, SRA is known as an RNA co-activator for many other nuclear receptors. In addition, SRA may act as an RNA scaffold for co-repressor complexes [259, 262]. SRA transcripts have been identified in normal human tissues, with a higher expression in liver, skeletal muscle, adrenal and pituitary glands, whereas intermediate expression levels were observed in the placenta, lung, kidney, and pancreas [260]. In some pathological cases, increased RNA levels of SRA were reported, like in breast and ovarian tumors [222, 263, 264]. Interestingly, levels of SRA expression could be characteristic of tumor grade or particular sub-types of lesions among different tumors. Indeed, serous ovarian tumors showed higher levels of SRA than granulosa tumor cells [259, 263].

## 13.14 PSEUDOGENE RNAs

Pseudogenes are gene copies that have lost the ability to code for a protein; they are typically identified through annotation of disabled, decayed, or incomplete protein-coding sequences. These molecules have long been labeled as "junk" DNA, failed copies of genes that arise during the evolution of genomes. However, recent results showed that some pseudogenes appear to harbor the potential to regulate their protein-coding cousins [265, 266]. Processed pseudogenes are made through retrotransposition of mRNAs, especially as a possible byproduct of LINE-1 (Long INterspersed Elements) retrotransposition. Thus, these mRNAs are reverse transcribed and re-integrated into the genomic DNA [267, 268]. The parent gene of the mRNA does not need to be on the same chromosome as the retrotransposed copy. Retrotransposed mRNAs have three possible fates in the genome: formation of processed genes, formation of nontranscribed pseudogenes, or formation of pseudogenes transcribed into RNAs [265]. Interestingly, some of these RNAs exhibit a tissue-specific pattern of activation. Pseudogene transcripts can be processed into short interfering RNAs that regulate coding genes through the RNAi pathway. In another remarkable discovery, it has been shown that pseudogene RNAs are capable of regulating tumor suppressors and oncogenes by acting as microRNA decoys [266, 268]. Moreover, Devor et al. [269] found that primate-specific miRNAs, miR-220 and miR-492, each lie within a processed pseudogene. Several studies also show deregulated expression of these molecules during cancer progression, which provides evidence for the functional involvement of pseudogene RNAs in carcinogenesis and suggests these molecules as a potential novel diagnostic or therapeutic target in human cancers. One of these pseudogenes is myosin light chain kinase pseudogene (MYLK). MYLKP1 is partially duplicated from the original MYLK gene that encodes nonmuscle and smooth muscle myosin light chain kinase (smMLCK) isoforms and regulates cell contractility and cytokinesis. Despite strong homology with the smMLCK promoter (~90%), the MYLKP1 promoter is minimally active in normal bronchial epithelial cells, but highly active in lung adenocarcinoma cells. Moreover, MYLKP1 and smMLCK exhibit negatively correlated transcriptional patterns in normal and cancer cells with MYLKP1 strongly expressed in cancer cells and smMLCK highly expressed in nonneoplastic cells. For instance, expression of smMLCK decreased in colon carcinoma tissues compared with normal colon tissues. Mechanistically, MYLKP1 overexpression inhibits smMLCK expression in cancer cells by decreasing RNA stability, leading to increased cell proliferation.

These findings provide strong evidence for the functional involvement of pseudogenes in carcinogenesis and suggest MYLKP1 as a potential novel diagnostic or therapeutic target in human cancers [270]. Using massively parallel signature sequencing (MPSS) technology, RT-PCR, and 5′ rapid amplification of cDNA ends (RACE) a novel androgen regulated and transcribed pseudogene of kallikreins termed as KLK31P was discovered. It was further proved that this pseudogene may play an important role in prostate carcinogenesis [271]. He et al. [272] found that pseudogene RNAs are also able to regulate a dosage of PTEN tumor suppressor during tumor development. Pseudogene RNAs, however, warrant further investigation into the true extent of their function [266, 270].

## 13.15 TRANSCRIBED ULTRACONSERVED REGIONS

Ultraconserved regions (UCRs) are a subset of conserved sequences that are located in both intra- and intergenic regions. They are 481 sequences, longer than 200 bp, that are absolutely conserved between orthologous regions of human, rat, and mouse genomes [273]. Calin et al. [213] have proved in cancer systems that differentially expressed UCR could alter the functional characteristics of malignant cells. The link between genomic location of UCRs and analyzed cancer-related genomic elements is highly statistically significant and comparable to that reported for miRNAs. UCRs are frequently located at fragile sites and genomic regions involved in cancers. Using northern blot, qRT-PCR, and microarray analysis, it was revealed that UCRs have distinct signatures in human leukemias and carcinomas [213].

The majority of UCRs are transcribed (T-UCRs) in normal human tissues, both ubiquitously and tissue specifically. From the molecular point of view, untranscribed UCRs might have regulatory functions as enhancers [274], while many functions can be assigned for T-UCRs, such as antisense inhibitors for protein-coding genes or other ncRNAs, including miRNAs. On the other hand, instead of T-UCRs interacting with protein-coding genes and miRNAs, it is possible that miRNAs control T-UCRs. Evidence supporting this prediction is that many T-UCRs have significant antisense complementarity with particular miRNAs and negative correlation between expression of specific T-UCRs and predicted interactor miRNAs [213, 275].

The expression of many T-UCRs is significantly altered in cancer, especially in adult chronic lymphocytic leukemias, colorectal and HCCs, and neuroblastomas [213]. Their aberrant transcription profiles can be used to distinguish types of human cancers and have been linked to patient outcome [276]. Especially in neuroblastoma, functional T-UCR annotations, inferred through a functional genomics approach and validated using cellular models, reveal associations with several cancer-related cellular processes such as apoptosis and differentiation [277]. Further, DNA hypomethylation induces release of T-UCR silencing in cancer cells. Studies of primary human tumors have shown that hypermethylation of T-UCR CpG islands is a common event among the various tumor types. Thus in addition to miRNAs, another class of ncRNAs (T-UCRs) undergoes DNA methylation-associated inactivation in transformed cells, and so supports the model that both epigenetic and genetic alterations in coding and noncoding sequences cooperate in human tumorigenesis. Most importantly, restoration of T-UCR expression was observed upon treatment with the DNA-demethylating agent [275]. Another study proved that SNPs (single-nucleotide polymorphisms) rs9572903 and rs2056116 in ultraconserved regions were associated with increased familial breast cancer risk [278]. Because of the increasing number of studies concerning T-UCRs being published, it is supposed that the more specific roles of these molecules in cancer will be known in a short time.

## 13.16   DEREGULATED EXPRESSION OF ncRNAs IN MM

### 13.16.1   MiRNAs

The discovery of miRNAs has been followed by a worldwide research effort to discover their roles in cancer. So far, results have been impressive, and it is now evident that miRNAs form central nodal points in cancer developmental pathways [279, 280]. A groundbreaking report demonstrated that >50% of miRNA genes are located in cancer-associated genomic regions or in fragile sites, suggesting that miRNAs are critical players in the pathogenesis of human cancers [281].

There is emerging evidence that some miRNAs can function as either oncogenes or tumor suppressors [279]. Those miRNAs whose expression is increased in tumors may be considered as oncogenes (oncomiRs) that promote tumor development by inhibiting tumor suppressor genes and/or genes controlling cell differentiation or apoptosis. A prototypical example is the miR-17-92 cluster, a miRNA polycistron located at chromosome 13q31, a genomic locus amplified in lung cancer and in several kinds of lymphoma, including diffuse large B-cell lymphoma. This cluster, which is regulated by c-MYC, an important transcription factor that is overexpressed in many human cancers [282], was found specifically upregulated in MM as compared with monoclonal gammopathy of undetermined significance (MGUS) or normal PCs and able to promote proliferation of MM cells in vitro [282]. Conversely, underexpressed miRNAs in cancers, such as some members of the let-7 family, may function as tumor suppressor genes mainly by downregulating oncogenes [279]. By the introduction of high-throughput screenings, miRNA expression patterns have been explored in several hematological malignancies, such as chronic lymphocytic leukemia [283] and acute myeloid leukemia [284]. During the last 5 years,

available information about miRNA expression in MM has grown rapidly [285, 286]. Overall, these studies unveiled a number of miRNAs that control critical genes in MM development and, more importantly, revealed that miRNA pattern of expression in MM is associated with specific genetic abnormalities. However, results from different studies appear somehow incongruous, and this caveat relies upon differences on platforms for miRNA profiling, the number of miRNAs analyzed, and the statistical study design. An initial study aimed at profiling miRNA expression by using both miRNA microarrays and quantitative real-time PCR, in a panel of 49 MM cell lines, 16 bone marrow CD138+ plasma cells isolated from MM and 6 from MGUS patients [282]. In this work, the authors tried to overcome the limitation arising from the limited number of primary MGUS and MM samples by validating their results in MM cell lines, and found a common miRNA signature likely associated to the multistep transformation process of MM. Of note, they found miR-21, miR-106b-25 cluster, miR-181a, and miR-181b upregulated in MGUS patients; moreover, by comparing MGUS and MM samples with normal plasma cells, they found some miRNAs, including miR-32 and miR-17-92 cluster, upregulated only in MM cells.

With the aim of identifying miRNAs associated with the major recurrent cytogenetic abnormalities associated to MM, other authors performed high-density miRNA profiling by using highly purified malignant plasma cells from 40 primary tumors [287]. Among 74 miRNAs, 26 miRNAs were identified by multiclass analysis as highly differentially expressed across the five translocation and cyclin D (TC) molecular classification of MM groups. Specific signatures were found to be associated with t(4;14) or translocated MAF genes and, to a lesser extent, with t(11;14) and TC2 group. Other chromosomal abnormalities recurrently observed in MM samples (i.e., 13q deletion, 1q gain/amplification, 17p deletion, and the HD status) appeared to be associated with less strong miRNA signatures. Most of the 26 miRNAs significantly discriminating the TC groups, that is, miR-155, miR-221, miR-222, and the let-7 family, have previously been found to be involved in solid and hematologic tumors, where they are predominantly overexpressed and readily act as oncomiRs. Notably, miR-155 has been found deregulated in Waldenstrom macroglobulinemia, suggesting a role in the proliferation and growth of WM cells acting on signaling cascades, including MAPK/ERK, PI3K/AKT, and nuclear factor kB pathways [288]. Interestingly, the same authors evaluated the impact of allelic imbalances commonly occurring in MM on miRNA expression; some of the miRNA coding sequences map to hotspot-altered regions in MM: for example, miR-17 and miR-20a, belonging to a cluster at 13q31, which was deleted in almost 40% of patients. Furthermore, a number of them are located in odd-numbered chromosomes involved in hyperdiploidy, and three (miR-1231, miR-205, and miR-215) belong to the long arm of chromosome 1, which was gained at a frequency of >30%. It has been also identified a significant correlation between miR-140-3p expression and the occurrence of LOH at 16q22.1-q23.1, which has been described as recurrent in MM [289]. Most importantly, they adopted an integrative analysis in this study based on computational target prediction and miRNA/mRNA profiling for each patient. This strategy disclosed a network of putative functional miRNA-target interactions with a potential role in MM pathogenesis [287]. Another study indeed supports the notion that deregulated expression of miRNAs is associated with different genetic subtypes of MM [290], although no clearly separated clusters containing specific cytogenetic abnormalities could be observed.

Similarly, other authors performed integrative miRNA/gene expression profile analyses of MM cells from 52 newly diagnosed patients, and correlated miRNA expression with a validated miRNA-based risk stratification score and proliferation index [291]. Among 464 human miRNAs analyzed, the authors found that mean expression levels of 95 miRNAs were higher in MM samples as compared with healthy donors. Unsupervised hierarchical clustering divided the 95 expressed miRNAs in two differentially expressed clusters, where the cluster 1 patients had significantly higher gene expression profile-defined risk score, suggesting a positive correlation between overall miRNA expression and risk scores. Dysregulation of miRNAs in cancer can occur through genetic alterations and/or epigenetic changes, which can affect the production of the primary miRNA transcript, their processing to mature miRNAs, and/or interactions with mRNA targets.

### 13.16.1.1 Genomic Alterations

One of the first associations observed between miRNAs and cancer development was miR-15 and miR-16 dysregulation in most B-cell chronic lymphocytic leukemias as a result of chromosome 13q14 deletion [292]. MiRNA profiling of CD138+-selected MM bone marrow–derived cells from 15 patients with relapsed/refractory MM in comparison with CD138+ cells isolated from bone marrow of healthy donors and three MM cell lines revealed complete lack of expression of miR-15a and miR-16-1 in those patients with chromosome 13 deletion [del(13)], together with a significantly decreased expression in the remaining patients without del(13) [293]. Conversely, others [294] failed in detecting any significant correlation between chromosome 13 status and miR-15a and miR-16 expression. Transfection of synthetic pre–miR-15a and pre–miR-16-1 in MM cell lines resulted in reduced proliferation and cell cycle inhibition through the targeting of NF-κB, even in the context of the BM milieu, and reduced angiogenesis [293].

Endothelial cells and angiogenesis regulate MM proliferation in the context of the BM milieu. It is well known that several cytokines secreted by MM cells and BM endothelial cells support the growth of MM cells. As a direct consequence, MM cell-supported endothelial cell growth is characterized by increased BM angiogenesis, which further promotes MM progression and acquired MM cell resistance to conventional and novel therapies. Interestingly, multiple

myeloma-bone marrow stromal cells (MM-BMSCs) can induce changes in the miRNome: downregulation of tumor suppressor miR-15a [295] or miR-29b [296], or upregulation of the oncomiR-21 [297] that results in drug resistance, are observed when MM cells are grown in the presence of BMSCs, although mechanisms underlying this effect remain to be determined.

The recent discovery of tumor-specific genetic defects in the miRNA-processing machinery, such as in the genes that encode TARBP2 [298], DICER1 [299], and exportin 5 [298], has strengthened the relevance of these pathways in cellular transformation. Some authors found that MGUS expresses higher levels of Dicer, but not of Drosha, as compared to both smoldering and symptomatic MM, and in close similarity to normal plasma cells. A specific group of 32 miRNAs, including, among others, miR-7 family, miR-125a, and miR-30c, were specifically upregulated in the group of patients with high Dicer levels [300].

Changes in the expression of miRNA coding sequences may also be linked to amplification of genomic regions. This is exemplified in the 17q23 region: this region hosts the gene coding for miR-21, which is usually upregulated in tumors [301]. Moreover, analysis of global miRNA expression in a panel of molecularly well-characterized human MM cell lines (HMCLs) identified 16 miRNAs mapped to chromosomal regions frequently involved in numerical imbalances in MM, whose expression significantly correlated with the copy number of the corresponding miRNA genes; among these, miR-22 expression was also affected by chromosome arm 17p loss in a representative panel of primary MM tumors [302].

The number of point mutations or single-nucleotide polymorphisms (SNPs) in human pri-miRNA or pre-miRNA sequences (also referred to as miRSNPs) has been usually reported to be rare [303]. Nevertheless, several studies have indicated that SNPs may explain changes in miRNA expression between tumor and normal tissues.

In a recent study, the prognostic impact and the functional consequences of six miRSNPs in 137 patients with MM undergoing autologous stem cell transplantation (ASCT) were evaluated. In detail, the authors analyzed the occurrence of SNP rs3660, located in the 3' UTR of KRT81, a predicted binding site for several miRNAs, miR-17 among others [304]. This gene was previously observed to be upregulated in MM and to inhibit expression of SOCS-1, a negative regulator of interleukin-6 (IL-6) growth signaling [282]. Notably, the presence of the SNP could affect the protein levels and functionally resulted in lower proliferation and improved patient survival. The authors also analyzed rs11077, a polymorphism found in the 3' UTR region of the XPO5 gene, demonstrating a significantly longer progression-free survival and long-term survival in carrier patients after ASCT [304].

### 13.16.1.2 Transcriptional Regulation

Transcription of miRNA-coding sequences by RNA polymerase II is under the control of several transcription factors [305]. The tumor suppressor p53 has been shown as a transcriptional regulator of several miRNAs. It acts as a transcription activator of the miR-34a gene, as it can bind to the promoter region of this gene. In a panel of 11 MM human cell lines, we have recently shown that those carrying a wt p53 gene express higher levels of miR-34a; moreover, treatment of these cell lines with the MDM2-inhibitor nutlin-3 leads to upregulated miR-34a levels, thus indicating a p53-mediated transcriptional control of miR-34a gene and decreased translation of miR-34a targets BCL2 and CDK6 [306].

Using the same experimental approach, it has been shown that nutlin-3-mediated p53 activation results in miR-34, miR-192, miR-194, and miR-215 upregulation, which mediates p53 anti-proliferative activity in MM. These miRNAs are recognized as critical regions in MM pathogenesis since they target MDM2 and IGF-1, critical regulators of MM development and progression [307]. The expression of miR-214, a tumor suppressor miRNA that inhibits MM cell proliferation and induces apoptosis, is also activated by p53; of note, miR-214 inhibits gankyrin protein by directly targeting PSMD10 gene, which in turn, activates p53 expression, thus establishing a positive feedback loop [308].

The oncogenic transcription factor c-Myc has been described as a modulator of the expression of a number of miRNAs. Indeed, by binding to promoter regions, Myc has been shown to act as expression activator of the miR-17-92 cluster in MM cells. Interestingly, overexpression of miR-17 or miR-18 could partly abrogate Myc-knockdown-induced MM cell apoptosis, thus demonstrating a significant role for this cluster as effector of c-Myc oncogenic activity [309].

Increased activity of the transcription factor Sp1 indeed occurs in MM [310]. We have recently shown that Sp1 inhibits the transcriptional efficiency of miR-29b promoter; on the other hand, miR-29b targets Sp1 and reduces its mRNA and protein levels, thus resulting in the occurrence of a negative feedback loop. Interestingly, such a molecular loop is amenable to pharmacologic intervention, since bortezomib, Sp1, or PI3K-inhibitors anti-MM activity is paralleled by Sp1 downregulation and miR-29b upregulation [296]. Notably, enforced expression of miR-29b seems also effective against MM-related bone disease since it antagonizes molecular pathways triggering osteoclast differentiation and blocks osteoclast activation induced by MM cells [311].

### 13.16.1.3 Epigenetic Regulation

Epigenetic therapy has recently emerged as a state-of-the-art strategy in cancer treatment, and clinical-grade DNA methyltransferase inhibitors have been approved for the treatment of myelodysplastic syndrome. This appears relevant since methylation of tumor suppressor miRNAs can be reversed by hypomethylating agents, leading to restoration of expression and tumor suppressor function of these miRNAs. MiR-34b/c was found methylated in 75% of MM cell lines but not in normal marrow controls.

Five-Aza-2′-deoxycytidine led to miR-34b/c promoter demethylation and re-expression; moreover, restoration of miR-34b/c reduced cellular proliferation and enhanced apoptosis of MM cells. Importantly, the authors showed that methylation of miR-34b/c occurred in 5.3% of cases at diagnosis and 52.2% at relapse/disease progression, thus supporting a role of hypermethylated miR-34b/c as biomarker of advanced disease [312]. Downregulation of tumor suppressor miR-214 in MM can be also ascribed to promoter hypermethylation [308]; likewise, tumor suppressor miR-203, localized to 14q32, was upregulated by demethylating agents and its promoter methylation appears an early event in myelomagenesis since comparable methylation of miR-203 is observed in MGUS and MM at diagnosis or relapse [313].

A specific subgroup of miRNAs, called epi-miRNAs, is able to directly target key enzymatic effectors of the epigenetic machinery (such as DNA methyltransferases, histone deacetylases, and polycomb genes), therefore indirectly affecting the expression of epigenetically regulated oncogenes and tumor suppressor genes. Our group has recently demonstrated that miR-29b is an epi-miRNA that targets de novo DNMT3A/B in MM cells, thus reducing global DNA methylation levels. Interestingly, treatment of MM cells with 5-aza-2′-deoxycytidine does not affect miR-29b levels, but significantly potentiates the growth-inhibitory effects exerted by synthetic miR-29b oligonucleotides in MM cells [314]. In Table 13.5 the most relevant MM-associated miRNAs are reported.

### 13.16.2 Other ncRNAs

Mounting evidence has recently shown that sno/scaRNAs may be actively involved in tumorigenesis [317]. Overexpression of SNORD25, SNORD27, SNORD30, and SNORD31 in smoldering MM patients showing a rapid progression to symptomatic disease has been reported [318]. Interestingly, upregulation of small cajal body-specific RNA 22 (SCARNA22) was reported in MMs harboring the t(4;14); of note, this SCARNA is localized within intron 18 of the WHSC1/MMSET gene involved in the translocation and constitutively expressed in these patients. Interestingly, SCARNA22 can suppress oxidative stress both in vitro and in vivo, facilitates cell proliferation, and exerts chemoprotective effect, suggesting an oncogenic role for this ncRNA in MM [319].

Recently, we have carried out the first study of global expression profiles of sno/scaRNAs in a panel of 55 MMs, 8 sPCLs, and 4 normal tonsil PCs. Unsupervised analysis of sno/scaRNA expression profiles using Human Gene 1.0 ST arrays revealed a general pattern of downregulation of sno/scaRNA expression in MM and sPCL patients compared with normal cells [320]. Specifically, sPCLs show a highly significant downregulation of SNORD32A, which is known to have a noncanonical role in the mediation of oxidative and endoplasmic reticulum stress-induced response pathways in vitro and in mouse models [321].

**TABLE 13.5**

**Overview of the Most Frequently Deregulated miRNAs in MM**

| miRNA | MM-Target | Function | References |
|---|---|---|---|
| miR-15a | MAP3KIP3, cyclin D1, cyclin D2, | Tumor suppressor | [293] |
| miR-16-1 | CDC25A, BCL2, AKT3, rS6, MAPKs | | |
| miR-34a | BCL2, CDK6, NOTCH1 | Tumor suppressor | [306] |
| miR-192-2-194 cluster | MDM2, IGF-1 | Tumor suppressor | [307] |
| miR-194-1-215 cluster | IGF-1R | | |
| miR-19a | SOCS1 | Oncogene | [309] |
| miR-19b | BIM | | |
| miR-181a | PCAF | Oncogene | [282] |
| miR-181b | PCAF | | |
| miR-214 | PSMD10, ASF1B | Tumor suppressor | [308] |
| miR-29b | MCL-1, CDK6, SP1, DNMT3A | Tumor suppressor | [296, 311, 314] |
| | DNMT3B, c-FOS, MMP2 | | |
| miR-21 | PTEN, Rhob, BTG2, PIAS3 | Oncogene | [315] |
| miR-221/222 | p27, p57, PTEN | Oncogene | [316] |
| miR-203 | CREB-1 | Tumor suppressor | [313] |

Despite the general pattern of downregulation observed in myeloma cells, hyperdiploid MMs upregulate specific snoRNAs. Interestingly, the integrated analysis of sno/scaRNA expression profiles with genome-wide copy number alterations and occurrence of loss of heterozygosity allowed the identification of SNOR116-11, -23, -25, -26, -28, and -29, and SNORD115-8, belonging to the SNORD115 and SNORD116 cluster; SNORA68, SNORD105, SNORD41, SNORD32A, and SNORD15B are mapped in regions generally gained in MM.

Recent studies proved that lincRNAs exhibit distinct gene expression patterns in primary tumors and metastases. For instance, it has been proved that differentially expressed UCR could alter the functional characteristics of malignant cells [97]. The link between genomic location of UCRs and analyzed cancer-related genomic elements is highly statistically significant and comparable to that reported for miRNAs.

HOX antisense intergenic RNA is a 2.2 kb sequence located within the human HOXC gene cluster on the long arm of chromosome 2. Interestingly, this lincRNA has a potential to regulate HOXD genes in trans via the recruitment of polycomb repressive complex 2, followed by the trimethylation of lysine 27 of histone H3 (Figure 13.3d) [322]. Since experimental evidence clarifying lncRNA roles in MM is presently missing, future studies in this field are eagerly warranted.

## 13.17 ncRNAs AS POTENTIAL CLINICAL BIOMARKERS IN MM

Quantitative analyses of miRNA expression in tumor tissues indeed indicate that miRNAs can be considered as appropriate biomarkers that might help to discriminate tumor types. NcRNA expression levels can be easily determined by in situ hybridization, for example, on a tumor section

and its normal adjacent counterparts. Mature-miRNA-specific stem-loop RT-PCR is an alternative system for very short ncRNA detection; moreover, deep-sequencing technologies, such as SOLiD (Applied Biosystems, Foster City, USA) or Genome Analyzer (Illumina, San Diego, USA), have become available for high-throughput detection of ncRNAs. Recently, a possible innovative approach for early detection has been represented by the determination of circulating miRNAs in plasma. MiRNAs can be found in detectable amounts in small-membrane vesicles (about 100 nm in diameter) called exosomes, which protect miRNAs against RNases and are secreted by a wide range of cell types such as hematopoietic cells, dendritic cells, mast cells, neurons cells, and tumor cells [323]. In such exosomes, miRNAs are bound by RNA-binding proteins such as Ago2 and nucleophosmin [324]. Recent studies demonstrate that circulating miRNAs provide a basis for a convenient and noninvasive diagnostic tool, in some cases, endowed with prognostic value, in MM. Microarray miRNA profiling of plasma samples of 12 MM patients compared with 8 healthy controls revealed that the total miRNA expression level was higher in samples from MM patients than in controls, with six miRNAs (miR-148a, miR-181a, miR-20a, miR-221, miR-625, and miR-99b) specifically upregulated in MM. Interestingly, the correlation of these six miRNAs with patients' clinicopathological data and survival disclosed a prognostic value for miR-148a and miR-20a [325]. Importantly, differences in circulating miRNAs might provide a powerful diagnostic tool for distinguishing MM patients from those affected by MGUS. For instance, circulating miR-1308 and miR-720 in combination showed excellent specificity and selectivity in distinguishing MGUS/MM patients from healthy controls, whereas the combination miR-1246/miR-1208 can distinguish MM from MGUS patients [326]. Furthermore, plasma miR-92a levels were found significantly reduced in newly diagnosed patients as compared with normal healthy controls, whereas no difference could be detected between MGUS and smoldering MM patients. Notably, circulating miRNA expression levels do not seem to overlap with other established biomarkers: in fact, increased blood levels of miR-720 and miR-1308 in MM patients did not correlate with the levels of M-paraprotein, thus providing alternative information on the progression of the disease [326].

## 13.18  ncRNA-BASED THERAPEUTIC STRATEGIES IN MM

So far, functional studies aimed at understanding the role of ncRNAs in normal and cancer cells have been based on strategies relying upon overexpression or inhibition of deregulated miRNAs. Despite the possibility of expressing exogenous small RNAs in cells through transient or stable transfection, or viral transduction of a pri-miRNA transgene, pre-miRNA, mature miRNA/miRNA*, or miRNA inhibitors (i-miRNA), the investigational use of ncRNAs for therapeutic purposes is mostly based on the use of short oligonucleotides for the achievement of replacement and reactivation or inhibition and silencing. The in vivo delivery of RNA oligonucleotides as therapeutics is restricted because of charge density, molecular weight, and their intrinsic instability in the presence of nucleases. Additionally, intracellular accumulation and endosomal escape are relevant barriers in the delivery of these macromolecules. Many viral and nonviral delivery vectors have been investigated in detail to overcome these barriers and several studies report applications for oligonucleotides as drugs in different diseases together with the investigation of various delivery systems. Both local and systemic strategies have been utilized to replace or inhibit ncRNAs' effect in different diseases including cancer [327].

The replacement approaches by use of miRNA mimic oligonucleotides seem particularly appealing. In comparison with single-gene targeting agents as siRNA or antisense oligonucleotides, miRNA mimics act not only on their target genes but they exert a fine-tuning of the gene expression by modulating multiple genes, and consequently, relevant athwart-related pathways to integrate them in meaningful networks in relation to different biological cell status [328]. Through the delivery of a +mature or engineered miRNA precursor to specific biologic compartments, the so-called miRNA replacement therapy aims to restore, to reinforce, or to functionally redirect common miRNA-target networks.

The antitumor activity of in vivo exogenously delivered let-7 and miR-34a mimics has been recently reported. The therapeutic potential of let-7 was assessed in mouse models of NSCLC, where the treatment of xenografted tumors significantly reduced the tumor burden [329]. In the same experimental system, therapeutic delivery was also successfully demonstrated using mimics of the tumor suppressor miR-34a, where the mice treated displayed a 60% tumor reduction compared with mice treated with a miRNA

control [330]. Recently, our group has demonstrated tumor growth inhibition and survival improvement in different MM mouse models with TP53 impairment by miR-34a replacement [306]. In all these studies, miRNAs were delivered by a neutral lipid emulsion (NLE) that showed an optimal biodistribution in different organs in vivo [331]. By this approach, our group also demonstrated tumor bioavailability by the detection of high levels of miR-34a in MM xenografts together with significant downregulation of miR-34a canonical targets [306]. The alternative delivery of miRNA mimics by adenoviral vectors has been reported in a HCC mouse model. Adenovirus-mediated overexpression of miR-26a suppressed tumorigenesis after one dose of viral injection with a good safety profile [332]. Despite the growing evidence of the safety of adenovirus and different ongoing clinical trials using this delivery platform, the viral vectors face several impediments to translation into the clinical practice.

The anti-miRNA therapeutic strategies involve inhibition of endogenous functions of well-characterized miRNAs. The use of antisense oligonucleotides is supported by the base-pairing to the miRNA, specifically to its seed sequence complementary to mRNA target 3'UTR. This approach is based on base-pairing specificity and the oligonucleotide design requires profound chemical modifications in order to stabilize the molecules for in vivo delivery. Chemical modifications are not indicated for miRNA mimics where the adaption to the RISC pocket requires specific spatial structure. The most advanced engineered nucleotides include a class of bicyclic RNA analogues in which the furanose ring in the sugar-phosphate backbone is chemically locked in a RNA mimicking N-type (C3'-endo) conformation by the introduction of a 2'-O,4'-C methylene bridge, named locked nucleic acid (LNA) [333]. LNA inhibitors containing one or more LNA monomer(s) exhibit nuclease resistance and increased binding affinity to the target miRNAs. Each LNA monomer increases the duplex melting temperature (Tm) by 2–8°C with respect to unmodified duplexes. Moreover, power LNA inhibitors (LNA Exiqon) have improved nuclease resistance through phosphorothioate linkage in which a sulfur atom replaces one of the nonbridging oxygen atoms in the phosphate group, producing a more effective silencing.

Recently, a 15-mer LNA-modified miR-122 inhibitor harboring a complete phosphorothioate backbone has been successfully investigated in vivo in nonhuman primates, opening the opportunity to treat chronically infected HCV patients taking benefit from high-specificity miR-122 targeting and low toxicity [334].

Although the use of miRNA inhibitors is still in its infancy, preclinical evidence strongly supports the potential for future applications [288]. Recently, our group demonstrated that enforced expression or lentivirus-based constitutive expression of miR-21 inhibitors triggered significant growth inhibition of primary patient MM cells or IL-6 dependent/independent MM cell lines and overcame the protective activity of the human bone marrow milieu in vitro. Importantly, this effect was confirmed in vivo in SCID mice bearing human MM tumors, together with upregulation of PTEN and downregulation of p-AKT in retrieved xenografts following treatment with miR-21 inhibitors, providing a rationale for a clinical study of miR-21 inhibitors in MM [315]. Moreover, the same group reported that miR-221/222 antagonism elicited strong anti-MM activity in vitro and in vivo. Most importantly, in the aim of clinical translation of these findings, the activity of miR-221/222 inhibitors was achieved in a subgroup of patients classified as TC2 and TC4, suggesting the MM genotyping may predict the therapeutic response of the disease to miR-221/222 inhibitor therapy [316].

Interestingly, the miRNA antagonistic potential has been recently enhanced using seed-targeting 8-mer LNA oligonucleotides, named tiny LNAs [335]. Transfection of miR-21 tiny LNA inhibitors into prostate cancer and HCC cells resulted in inhibition of miR-21 with consequent upregulation of direct targets. The biodistribution analysis suggested that LNA oligonucleotides are taken up by a broad range of tissues and excreted by both urine and bile in mice. High serum stability and nuclease resistance of LNA miRNA inhibitors, together with efficient on-target and low off-target activity and low toxicity in primates, indicate the remarkable potential of this approach.

## 13.19 EXPERT OPINION

A plethora of investigational drugs against MM unfortunately failed to translate into clinical effective agents. Many challenges still face molecularly targeted therapy of MM: molecular targets with differential expression in MM cells as compared to their normal counterpart may be definitively useless if they are functionally redundant; moreover, interaction of MM cells with tumor microenvironment confers drug resistance through the activation of multiple oncogenic pathways. The recent discovery of ncRNAs has dramatically improved the understanding of gene regulatory networks and has provided a new basis for development of new diagnostics and therapeutics for MM. Major expectations in the therapeutic arena have been raised following the identification of aberrant miRNA expression in cancer cells, including MM cells, which are likely associated to specific patterns of genetic abnormalities [280]. The validation in prospective studies of already available data on prognostic and diagnostic miRNA profiles in tumor tissues or as circulating biomarkers would indeed enable, in the near future, their introduction as diagnostic, prognostic, or predictive tools. Moreover, a deeper knowledge about "driver miRNA lesions" for specific MM subtypes will be of paramount importance for the design of personalized miRNA-based approaches, which, in turn, could harness such miRNA abnormalities. Indeed, many opportunities are presently offered by miRNAs as therapeutic tools in MM [285, 286, 296, 306, 311, 314–316].

Our thought is that an individualized miRNA-based strategy might have a high chance of providing an effective therapeutic tool. In fact, this approach might be useful to overcome one of the major limitations of traditional target therapy, that is, the development of drug resistance: theoretically, the ability of miRNAs to target several oncogenic-related mRNAs at the same time would make a cell much more sensitive to agents targeting the miRNAome "as a whole," thus producing relevant molecular perturbations, which contribute to kill MM cells or revert the malignant phenotype. Moreover, recent findings allow predicting that manipulating miRNA expression by using synthetic mimics or anti-miRNA molecules, could also integrate currently used anti-MM agents, through the potentiation of their therapeutic activity at a molecular level [296, 314] and/or the decrease of undesired effects.

Although the initial proof-of-principle of miRNAs as therapeutics exploited viral-based delivery methods, the translation into clinics further requires the development of safe delivery vehicles. These include packaging of miRNAs into lipid-based nanoparticles that can be locally or systemically delivered to tumor tissues [306]. Expansions to the available knowledge on the mechanisms that facilitate their distribution in tissues, their stability in biological fluids, and their selective uptake by MM cells represent important issues to be addressed in future. In addition, the analysis of delivered miRNAs treatments in the context of powerful predictive preclinical animal models recapitulating the clinical disease scenario [12, 14, 15], would provide valuable information allowing miRNA modulation to become a central feature of MM treatment. While targeting cancer-specific miRNAs has proven to be successful in MM, it will be necessary to increase knowledge of the whole ncRNA transcriptome, in order to design molecules with potential to restore expression or inhibit the function of still-unexplored ncRNAs; however, this latter point appears highly challenging because of the complex secondary structure of lncRNAs.

## 13.20 CONCLUSIONS AND FUTURE PERSPECTIVES

For a long time, the central dogma of molecular biology proposed RNA molecules primarily to be informational "messengers" between DNA and protein [336–339]. But, surprisingly, only 2% of the human genome sequence encodes proteins, while a large part of it is devoted to the expression of ncRNAs, which are divided into two main groups according to their nucleotide length—small and long ncRNAs. These molecules are suggested to be important regulators of gene expression. Nevertheless, the two groups of ncRNAs are distinct in their biological functions and mechanisms of gene regulations. Small ncRNAs are involved mainly in posttranscriptional gene regulation using translational repression or RNAi pathway, while long ncRNAs are much more involved in epigenetic regulation.

In many cases, differential expression of ncRNAs is becoming recognized as one of the hallmarks of cancer cells, indicating their potential usage as novel diagnostic, prognostic, or predictive biomarkers. Growing evidence also suggests that ncRNAs have promising potential in targeted regulation of gene expression and, therefore, in cancer-targeted therapy. However, the function of many ncRNAs remains unknown and it will be necessary to discover the precise mechanisms by which these molecules are involved in carcinogenesis.

The discovery of ncRNAs has significantly changed the landscape of cancer biology. Several classes of ncRNAs have been identified, and their biological role and mode of action are still under deep investigation. At present, most of the studies have focused on miRNAs, whereas data about long ncRNAs' role in human cancer are slowly emerging because of technical issues related to their complex structure. Cumulative evidence supports the notion that miRNAs function as critical regulators of cell commitment and differentiation during human normal hematopoiesis and their dysregulation contributes to the pathogenesis of common malignancies like MM. Of note, abnormal miRNA expression, mostly due to genetic or epigenetic mechanisms, correlates with MM-specific cytogenetic abnormalities and is now regarded as an intrinsic feature of MM growth and progression. Early studies have also established that miRNA signatures are endowed with prognostic power and could predict outcome and treatment response in MM. Most importantly, recent findings indicate that targeting aberrant miRNA networks in MM using miRNA mimics or inhibitors may result in significant antitumor activity in vitro and in vivo in different robust preclinical models, thus highlighting the therapeutic potential of such agents as novel powerful tools for a personalized miRNA-based therapy of MM.

In conclusion, information obtained in MM primary samples, cell lines, and mouse models is booming in terms of diagnostic/prognostic and therapeutic potential of ncRNAs in MM. It is hoped that this will be translated into clinical practice in the future.

## REFERENCES

Papers of special note have been highlighted as either of interest (•) or of considerable interest (••) to readers.

[1] Stein LD. Human genome: end of the beginning. Nature 2004; 431:915–916.

[2] Taft RJ, Pang KC, Mercer TR, et al. Non-coding RNAs: regulators of disease. J Pathol 2010; 220:126–139.

[3] Knowling S, Morris KV. Non-coding RNA and antisense RNA. Nature's trash or treasure? Biochimie 2011; 93:1922–1927.

[4] Mattick JS. Non-coding RNAs: the architects of eukaryotic complexity. EMBO Rep 2001; 2:986–91.

[5] Costa FF. Non-coding RNAs: new players in eukaryotic biology. Gene 2005; 357:83–94.

[6] Okamura K, Chung W-J, Ruby JG, et al. The Drosophila hairpin RNA pathway generates endogenous short interfering RNAs. Nature 2008; 453:803–6.

[7] Rinn JL, Kertesz M, Wang JK, et al. Functional demarcation of active and silent chromatin domains in human HOX loci by noncoding RNAs. Cell 2007; 129:1311–23.

[8] Palumbo A, Anderson K. Multiple myeloma. N Engl J Med 2011; 364:1046–60.

[9] Richardson PG, Delforge M, Beksac M, et al. Management of treatment-emergent peripheral neuropathy in multiple myeloma. Leukemia 2012; 26:595–608.

[10] Tassone P, Tagliaferri P, Rossi M, et al. Genetics and molecular profiling of multiple myeloma: novel tools for clinical management? Eur J Cancer 2006; 42:1530–8.

[11] Morabito F, Recchia AG, Mazzone C, et al. Targeted therapy of multiple myeloma: the changing paradigm at the beginning of the new millennium. Curr Cancer Drug Targets 2012; 12:743–56.

[12•] Calimeri T, Battista E, Conforti F, et al. A unique three-dimensional SCID polymeric scaffold (SCID-synth-hu) models for in vivo expansion of human primary multiple myeloma cells. Leukemia 2011; 25:707–11.
  • This publication reports on the first murine bio-synthetic platform of human multiple myeloma recapitulated within a human bone marrow milieu in vivo.

[13] Tassone P, Neri P, Burger R, et al. Mouse models as a translational platform for the development of new therapeutic agents in multiple myeloma. Curr Cancer Drug Targets 2012; 12:814–22.

[14•] Tassone P, Neri P, Carrasco DR, et al. A clinically relevant SCID-hu in vivo models of human multiple myeloma. Blood 2005; 106:713–16.
  • This publication reports on a novel model of human multiple myeloma recapitulated within a human bone marrow milieu.

[15•] Tassone P, Neri P, Kutok JL, et al. A SCID-hu in vivo model of human Waldenstrom macroglobulinemia. Blood 2005; 106:1341–5.
  • This publication reports on a novel model of human Waldenstrom's macroglobulinemia recapitulated within a human bone marrow milieu in vivo.

[16] Tassone P, Tagliaferri P, Rossi M, et al. challenging the current approaches to multiple myeloma-related bone disease: from bisphosphonates to target therapy. Curr Cancer Drug Targets 2009; 9:854–70.

[17] Tassone P, Tagliaferri P, Fulciniti MT, et al. Novel therapeutic approaches based on the targeting of microenvironment-derived survival pathways in human cancer: experimental models and translational issues. Curr Pharm Des 2007; 13:487–96.

[18] Ditzel Santos D, Ho AW, Tournilhac O, et al. Establishment of BCWM.1 cell line for Waldenstrom's macroglobulinemia with productive in vivo engraftment in SCID-hu mice. Exp Hematol 2007; 35:1366–75.

[19] Rossi M, Di Martino MT, Morelli E, et al. Molecular targets for the treatment of multiple myeloma. Curr Cancer Drug Targets 2012; 12:757–67.

[20] Neri P, Kumar S, Fulciniti MT, et al. Neutralizing B-cell activating factor antibody improves survival and inhibits osteoclastogenesis in a severe combined immunodeficient human multiple myeloma model. Clin Cancer Res 2007; 13:5903–9.

[21] Tassone P, Galea E, Forciniti S, et al. The IL-6 receptor super-antagonist Sant7 enhances antiproliferative and apoptotic effects induced by dexamethasone and zoledronic acid on multiple myeloma cells. Int J Oncol 2002; 21:867–73.

[22] Tassone P, Forciniti S, Galea E, et al. Synergistic induction of growth arrest and apoptosis of human myeloma cells by

the IL-6 super-antagonist Sant7 and Dexamethasone. Cell Death Differ 2000; 7:327–8.

[23] Tassone P, Gozzini A, Goldmacher V, et al. In vitro and in vivo activity of the maytansinoid immunoconjugate huN901-N2'-deacetyl-N2'-(3-mercapto-1-oxopropyl)-maytansine against CD56+ multiple myeloma cells. Cancer Res 2004; 64:4629–36.

[24] Neri P, Ren L, Gratton K, et al. Bortezomib-induced "BRCAness" sensitizes multiple myeloma cells to PARP inhibitors. Blood 2011; 118:6368–79.

[25] Neri P, Ren L, Azab AK, et al. Integrin beta7-mediated regulation of multiple myeloma cell adhesion, migration, and invasion. Blood 2011; 117:6202–13.

[26] Neri P, Tagliaferri P, Di Martino MT, et al. In vivo antimyeloma activity and modulation of gene expression profile induced by valproic acid, a histone deacetylase inhibitor. Br J Haematol 2008; 143:520–31.

[27] Garzon R, Marcucci G, Croce CM. Targeting microRNAs in cancer: rationale, strategies and challenges. Nat Rev Drug Discov 2010; 9:775–89.

[28] Mercer TR, Dinger ME, Mattick JS. Long non-coding RNAs: insights into functions. Nat Rev Genet 2009; 10:155–9.

[29] Esteller M. Non-coding RNAs in human disease. Nat Rev Genet 2011; 12:861–74.

[30] Elbashir SM, Harborth J, Lendeckel W, et al. Duplexes of 21-nucleotide RNAs mediate RNA interference in cultured mammalian cells. Nature 2001; 411:494–8.

[31] Meister G, Tuschl T. Mechanisms of gene silencing by double-stranded RNA. Nature 2004; 431:343–9.

[32] Lippman Z, Martienssen R. The role of RNA interference in heterochromatic silencing. Nature 2004, 431:364–70.

[33] Houwing S, Kamminga LM, Berezikov E, et al. A role for Piwi and piRNAs in germ cell maintenance and transposon silencing in Zebrafish. Cell 2007; 129:69–82.

[34] Seto AG, Kingston RE, Lau NC. The coming of age for Piwi proteins. Mol Cell 2007; 26:603–9.

[35] Sana J, Hajduch M, Michalek J, et al. MicroRNAs and glioblastoma: roles in core signalling pathways and potential clinical implications. J Cell Mol Med 2011; 15:1636–44.

[36] Slaby O, Bienertova-Vasku J, Svoboda M, et al. Genetic polymorphisms and microRNAs: new direction in molecular epidemiology of solid cancer. J Cell Mol Med 2012; 16:8–21.

[37] Slaby O, Svoboda M, Michalek J, et al. MicroRNAs in colorectal cancer: translation of molecular biology into clinical application. Mol Cancer 2009; 8:102.

[38] Redova M, Svoboda M, Slaby O. MicroRNAs and their target gene networks in renal cell carcinoma. Biochem Biophys Res Commun 2011; 405:153–6.

[39] Lee RC, Feinbaum RL, Ambros V. The C. elegans heterochronic gene lin-4 encodes small RNAs with antisense complementarity to lin-14. Cell 1993; 75:843–54.

[40] Griffiths-Jones S. miRBase: the microRNA sequence database. Methods Mol Biol 2006; 342:129–38.

[41] Krol J, Loedige I, Filipowicz W. The widespread regulation of microRNA biogenesis, function and decay. Nat Rev Genet 2010; 11:597–610.

[42] Roberts APE, Lewis AP, Jopling CL. miR-122 activates hepatitis C virus translation by a specialized mechanism requiring particular RNA components. Nucleic Acids Res 2011; 39:7716–7729.

[43] Grey F, Tirabassi R, Meyers H, et al. A viral microRNA down-regulates multiple cell cycle genes through mRNA 5'UTRs. PLoS Pathog 2010; 6:e1000967.

[44] Tsai N-P, Lin Y-L, Wei L-N. MicroRNA mir-346 targets the 5'-untranslated region of receptor-interacting protein 140 (RIP140) mRNA and upregulates its protein expression. Biochem J 2009; 424:411–8.

[45] Ørom UA, Nielsen FC, Lund AH. MicroRNA-10a binds the 5'UTR of ribosomal protein mRNAs and enhances their translation. Mol Cell 2008; 30:460–71.

[46] Farh KK-H, Grimson A, Jan C, et al. The widespread impact of mammalian MicroRNAs on mRNA repression and evolution. Science 2005; 310:1817–21.

[47] Lim LP, Lau NC, Garrett-Engele P, et al. Microarray analysis shows that some microRNAs downregulate large numbers of target mRNAs. Nature 2005; 433:769–73.

[48] Lee I, Ajay SS, Yook JI, et al. New class of microRNA targets containing simultaneous 5'-UTR and 3'-UTR interaction sites. Genome Res 2009; 19:1175–83.

[49] Chi SW, Zang JB, Mele A, et al. Argonaute HITS-CLIP decodes microRNA-mRNA interaction maps. Nature 2009; 460:479–86.

[50] Li M, Li J, Ding X, et al. microRNA and cancer. AAPS J 2010; 12:309–17.

[51] Kwak PB, Iwasaki S, Tomari Y. The microRNA pathway and cancer. Cancer Sci 2010; 101:2309–15.

[52] Wu WKK, Law PTY, Lee CW, et al. MicroRNA in colorectal cancer: from benchtop to bedside. Carcinogenesis 2011; 32:247–53.

[53] Lin P-Y, Yu S-L, Yang P-C. MicroRNA in lung cancer. Br J Cancer 2010; 103:1144–48.

[54] Lakomy R, Sana J, Hankeova S, et al. MiR-195, miR-196b, miR-181c, miR-21 expression levels and O-6-methylguanine-DNA methyltransferase methylation status are associated with clinical outcome in glioblastoma patients. Cancer Sci 2011; 102:2186–90.

[55] Satoh J. MicroRNAs and their therapeutic potential for human diseases: aberrant microRNA expression in Alzheimer's disease brains. J Pharmacol Sci 2010; 114: 269–75.

[56] Slaby O, Jancovicova J, Lakomy R, et al. Expression of miRNA-106b in conventional renal cell carcinoma is a potential marker for prediction of early metastasis after nephrectomy. J Exp Clin Cancer Res 2010; 29:90.

[57] Slaby O, Lakomy R, Fadrus P, et al. MicroRNA-181 family predicts response to concomitant chemoradiotherapy with temozolomide in glioblastoma patients. Neoplasma 2010; 57:264–9.

[58] Bloomston M, Frankel WL, Petrocca F, et al. MicroRNA expression patterns to differentiate pancreatic adenocarcinoma from normal pancreas and chronic pancreatitis. JAMA 2007; 297:1901–8.

[59] Fornari F, Gramantieri L, Giovannini C, et al. MiR-122/cyclin G1 interaction modulates p53 activity and affects doxorubicin sensitivity of human hepatocarcinoma cells. Cancer Res 2009; 69:5761–7.

[60] Gramantieri L, Fornari F, Ferracin M, et al. MicroRNA-221 targets Bmf in hepatocellular carcinoma and correlates with tumor multifocality. Clin Cancer Res 2009; 15:5073–81.

[61] Slaby O, Svoboda M, Fabian P, et al. Altered expression of miR-21, miR-31, miR-143 and miR-145 is related to clinicopathologic features of colorectal cancer. Oncology 2007; 72:397–402.

[62] Kulda V, Pesta M, Topolcan O, et al. Relevance of miR-21 and miR-143 expression in tissue samples of colorectal carcinoma and its liver metastases. Cancer Genet Cytogenet 2010; 200:154–60.

[63] Takamizawa J, Konishi H, Yanagisawa K, et al. Reduced expression of the let-7 microRNAs in human lung cancers in association with shortened postoperative survival. Cancer Res 2004; 64:3753–6.

[64] Gallardo E, Navarro A, Viñolas N, et al. miR-34a as a prognostic marker of relapse in surgically resected non-small-cell lung cancer. Carcinogenesis 2009; 30:1903–9.

[65] Yu S-L, Chen H-Y, Chang G-C, et al. MicroRNA signature predicts survival and relapse in lung cancer. Cancer Cell 2008; 13:48–57.

[66] Ma L, Teruya-Feldstein J, Weinberg RA. Tumour invasion and metastasis initiated by microRNA-10b in breast cancer. Nature 2007; 449:682–8.

[67] Tavazoie SF, Alarcón C, Oskarsson T, et al. Endogenous human microRNAs that suppress breast cancer metastasis. Nature 2008; 451:147–52.

[68] Satzger I, Mattern A, Kuettler U, et al. MicroRNA-15b represents an independent prognostic parameter and is correlated with tumor cell proliferation and apoptosis in malignant melanoma. Int J Cancer 2010; 126:2553–62.

[69] Mohri T, Nakajima M, Fukami T, et al. Human CYP2E1 is regulated by miR-378. Biochem Pharmacol 2010; 79:1045–52.

[70] Markou A, Tsaroucha EG, Kaklamanis L, et al. Prognostic value of mature microRNA-21 and microRNA-205 overexpression in non-small cell lung cancer by quantitative real-time RT-PCR. Clin Chem 2008; 54:1696–704.

[71] Yang H, Kong W, He L, et al. MicroRNA expression profiling in human ovarian cancer: miR-214 induces cell survival and cisplatin resistance by targeting PTEN. Cancer Res 2008; 68:425–33.

[72] Nagel R, le Sage C, Diosdado B, et al. Regulation of the adenomatous polyposis coli gene by the miR-135 family in colorectal cancer. Cancer Res 2008; 68:5795–802.

[73] Johnson SM, Grosshans H, Shingara J, et al. RAS is regulated by the let-7 microRNA family. Cell 2005; 120: 635–47.

[74] Chen X, Guo X, Zhang H, et al. Role of miR-143 targeting KRAS in colorectal tumorigenesis. Oncogene 2009; 28:1385–92.

[75] Guo C, Sah JF, Beard L, et al. The noncoding RNA, miR-126, suppresses the growth of neoplastic cells by targeting phosphatidylinositol 3-kinase signaling and is frequently lost in colon cancers. Genes Chromosomes Cancer 2008; 47:939–46.

[76] Meng F, Henson R, Wehbe-Janek H, et al. MicroRNA-21 regulates expression of the PTEN tumor suppressor gene in human hepatocellular cancer. Gastroenterology 2007; 133:647–58.

[77] Yu Z, Wang C, Wang M, et al. A cyclin D1/microRNA 17/20 regulatory feedback loop in control of breast cancer cell proliferation. J Cell Biol 2008; 182:509–17.

[78] Ota A, Tagawa H, Karnan S, et al. Identification and characterization of a novel gene, C13orf25, as a target for 13q31-q32 amplification in malignant lymphoma. Cancer Res 2004; 64:3087–95.

[79] O'Donnell KA, Wentzel EA, Zeller KI, et al. c-Myc-regulated microRNAs modulate E2F1 expression. Nature 2005; 435:839–43.

[80] He L, Thomson JM, Hemann MT, et al. A microRNA polycistron as a potential human oncogene. Nature 2005; 435:828–33.

[81] Bartel DP. MicroRNAs: genomics, biogenesis, mechanism, and function. Cell 2004; 116:281–97.

[82] Ying SY, Lin SL. Current perspectives in intronic micro RNAs (miRNAs). J Biomed Sci 2006; 13:5–15.

[83] Denli AM, Tops BB, Plasterk RH, et al. Processing of primary microRNAs by the Microprocessor complex. Nature 2004; 432:231–5.

[84] Eulalio A, Huntzinger E, Nishihara T, et al. Deadenylation is a widespread effect of miRNA regulation. RNA 2009; 15:21–32.

[85] Aravin AA, Hannon GJ, Brennecke J. The Piwi-piRNA pathway provides an adaptive defense in the transposon arms race. Science 2007; 318:761–4.

[86] Klattenhoff C, Theurkauf W. Biogenesis and germline functions of piRNAs. Development 2008; 135:3–9.

[87] Gunawardane LS, Saito K, Nishida KM, et al. A slicer-mediated mechanism for repeat-associated siRNA 5' end formation in Drosophila. Science 2007; 315:1587–90.

[88] Pal-Bhadra M, Leibovitch BA, Gandhi SG, et al. Heterochromatic silencing and HP1 localization in Drosophila are dependent on the RNAi machinery. Science 2004; 303:669–72.

[89] Sasaki T, Shiohama A, Minoshima S, et al. Identification of eight members of the Argonaute family in the human genome small star, filled. Genomics 2003; 82:323–30.

[90] Horwich MD, Li C, Matranga C, et al. The Drosophila RNA methyltransferase, DmHen1, modifies germline piRNAs and single-stranded siRNAs in RISC. Curr Biol 2007; 17:1265–72.

[91] Carmell MA, Girard A, van de Kant HJ, et al. MIWI2 is essential for spermatogenesis and repression of transposons in the mouse male germline. Dev Cell 2007; 12:503–14.

[92] King TH, Liu B, McCully RR, et al. Ribosome structure and activity are altered in cells lacking snoRNPs that form pseudouridines in the peptidyl transferase center. Mol Cell 2003; 11:425–35.

[93] Ni J, Tien AL, Fournier MJ. Small nucleolar RNAs direct site-specific synthesis of pseudouridine in ribosomal RNA. Cell 1997; 89:565–73.

[94] Pagano A, Castelnuovo M, Tortelli F, et al. New small nuclear RNA gene-like transcriptional units as sources of regulatory transcripts. PLoS Genet 2007; 3:e1.

[95] Guttman M, Amit I, Garber M, et al. Chromatin signature reveals over a thousand highly conserved large non-coding RNAs in mammals. Nature 2009; 458:223–7.

[96] Bejerano G, Lowe CB, Ahituv N, et al. A distal enhancer and an ultraconserved exon are derived from a novel retroposon. Nature 2006; 441:87–90.

[97] Calin GA, Liu CG, Ferracin M, et al. Ultraconserved regions encoding ncRNAs are altered in human leukemias and carcinomas. Cancer Cell 2007; 12:215–29.

[98] Kapranov P, Cheng J, Dike S, et al. RNA maps reveal new RNA classes and a possible function for pervasive transcription. Science 2007; 316:1484–8.

[99] Seila AC, Calabrese JM, Levine SS, et al. Divergent transcription from active promoters. Science 2008; 322: 1849–51.

[100] Taft RJ, Glazov EA, Cloonan N, et al. Tiny RNAs associated with transcription start sites in animals. Nat Genet 2009; 41:572–8.

[101] Feuerhahn S, Iglesias N, Panza A, et al. TERRA biogenesis, turnover and implications for function. FEBS Lett 2010; 584:3812–8.

[102] Sakamoto KM. Knocking down human disease: potential uses of RNA interference in research and gene therapy. Pediatr Res 2004; 55:912–3.

[103] Noll B, Seiffert S, Vornlocher H-P, et al. Characterization of small interfering RNA by non-denaturing ion-pair reversed-phase liquid chromatography. J Chromatogr A 2011; 1218:5609–17.

[104] Martinez J, Patkaniowska A, Urlaub H, et al. Single stranded antisense siRNAs guide target RNA cleavage in RNAi. Cell 2002; 110:563–74.

[105] Martinez J, Tuschl T. RISC is a 5′ phosphomonoester-producing RNA endonuclease. Genes Dev 2004; 18: 975–80.

[106] Vassilev LT. Small-molecule antagonists of p53-MDM2 binding: research tools and potential therapeutics. Cell Cycle 2004; 3:419–21.

[107] He S-B, Yuan Y, Wang L, et al. Effects of cyclin dependent kinase 8 specific siRNA on the proliferation and apoptosis of colon cancer cells. J Exp Clin Cancer Res 2011; 30:109.

[108] Durfort T, Tkach M, Meschaninova MI, et al. Small interfering RNA targeted to IGF-IR delays tumor growth and induces proinflammatory cytokines in a mouse breast cancer model. PLoS ONE 2012; 7:e29213.

[109] Matsubara H, Sakakibara K, Kunimitsu T, et al. Non-small cell lung carcinoma therapy using mTOR-siRNA. Int J Clin Exp Pathol 2012; 5:119–25.

[110] He Y, Bi Y, Hua Y, et al. Ultrasound microbubble-mediated delivery of the siRNAs targeting MDR1 reduces drug resistance of yolk sac carcinoma L2 cells. J Exp Clin Cancer Res 2011; 30:104.

[111] Zhou W, Hu J, Tang H, et al. Small interfering RNA targeting mcl-1 enhances proteasome inhibitor-induced apoptosis in various solid malignant tumors. BMC Cancer 2011; 11:485.

[112] Bansal N, Marchion DC, Bicaku E, et al. BCL2 antagonist of cell death kinases, phosphatases, and ovarian cancer sensitivity to cisplatin. J Gynecol Oncol 2012; 23:35–42.

[113] Wang X, Chen Y, Ren J, et al. Small interfering RNA for effective cancer therapies. Mini Rev Med Chem 2011; 11:114–24.

[114] Costa FF. Non-coding RNAs: Meet thy masters. Bioessays 2010; 32:599–608.

[115] Carmell MA, Xuan Z, Zhang MQ, et al. The argonaute family: tentacles that reach into RNAi, developmental control, stem cell maintenance, and tumorigenesis. Genes Dev 2002; 16:2733–42.

[116] Brennecke J, Aravin AA, Stark A, et al. Discrete small RNA-generating loci as master regulators of transposon activity in Drosophila. Cell 2007; 128:1089–103.

[117] Aravin AA, Lagos-Quintana M, Yalcin A, et al. The small RNA profile during Drosophila melanogaster development. Dev Cell 2003; 5:337–50.

[118] Aravin AA, Sachidanandam R, Girard A, et al. Developmentally regulated piRNA clusters implicate MILI in transposon control. Science 2007; 316:744–7.

[119] Samji T. PIWI, piRNAs, and germline stem cells: what's the link? Yale J Biol Med 2009; 82:121–4.

[120] Thomson T, Lin H. The biogenesis and function of PIWI proteins and piRNAs: progress and prospect. Annu Rev Cell Dev Biol 2009; 25:355–76.

[121] Klattenhoff C, Theurkauf W. Biogenesis and germline functions of piRNAs. Development 2008; 135:3–9.

[122] Grochola LF, Greither T, Taubert H, et al. The stem cell-associated Hiwi gene in human adenocarcinoma of the pancreas: expression and risk of tumour-related death. Br J Cancer 2008; 99:1083–8.

[123] Liu X, Sun Y, Guo J, et al. Expression of Hiwi gene in human gastric cancer was associated with proliferation of cancer cells. Int J Cancer 2006; 118:1922–9.

[124] Taubert H, Greither T, Kaushal D, et al. Expression of the stem cell self-renewal gene Hiwi and risk of tumour-related death in patients with soft-tissue sarcoma. Oncogene 2007; 26:1098–100.

[125] Zhao Y-M, Zhou J-M, Wang L-R, et al. HIWI is associated with prognosis in patients with hepatocellular carcinoma after curative resection. Cancer 2012; 118(10):2708–17.

[126] Zeng Y, Qu L, Meng L, et al. HIWI expression profile in cancer cells and its prognostic value for patients with colorectal cancer. Chin Med J 2011; 124:2144–9.

[127] Sun G, Wang Y, Sun L, et al. Clinical significance of Hiwi gene expression in gliomas. Brain Res 2011; 1373:183–8.

[128] He W, Wang Z, Wang Q, et al. Expression of HIWI in human esophageal squamous cell carcinoma is significantly associated with poorer prognosis. BMC Cancer 2009; 9:426.

[129] Lee JH, Schütte D, Wulf G, et al. Stem-cell protein Piwil2 is widely expressed in tumors and inhibits apoptosis through activation of Stat3/Bcl-XL pathway. Hum Mol Genet 2006; 15:201–11.

[130] Lu Y, Zhang K, Li C, et al. Piwil2 suppresses p53 by inducing phosphorylation of signal transducer and activator of transcription 3 in tumor cells. PLoS ONE 2012; 7:e30999.

[131] Cheng J, Deng H, Xiao B, et al. piR-823, a novel noncoding small RNA, demonstrates in vitro and in vivo tumor suppressive activity in human gastric cancer cells. Cancer Lett 2012; 315:12–7.

[132] Cheng J, Guo J-M, Xiao B-X, et al. piRNA, the new noncoding RNA, is aberrantly expressed in human cancer cells. Clin Chim Acta 2011; 412:1621–5.

[133] Cui L, Lou Y, Zhang X, et al. Detection of circulating tumor cells in peripheral blood from patients with gastric cancer using piRNAs as markers. Clin Biochem 2011; 44:1050–7.

[134] Kiss T. Small nucleolar RNAs: an abundant group of noncoding RNAs with diverse cellular functions. Cell 2002; 109:145–8.

[135] Bortolin M-L, Kiss T. Human U19 intron-encoded snoRNA is processed from a long primary transcript that possesses little potential for protein coding. RNA 1998; 4:445–54.

[136] Weinstein LB, Steitz JA. Guided tours: from precursor snoRNA to functional snoRNP. Curr Opin Cell Biol 1999; 11:378–84.

[137] Ganot P, Caizergues-Ferrer M, Kiss T. The family of box ACA small nucleolar RNAs is defined by an evolutionarily conserved secondary structure and ubiquitous sequence elements essential for RNA accumulation. Genes Dev 1997; 11:941–56.

[138] Vidovic I, Nottrott S, Hartmuth K, et al. Crystal structure of the spliceosomal 15.5kD protein bound to a U4 snRNA fragment. Mol Cell 2000; 6:1331–42.

[139] Ender C, Krek A, Friedländer MR, et al. A human snoRNA with microRNA-like functions. Mol Cell 2008; 32:519–28.

[140] Chang L-S, Lin S-Y, Lieu A-S, et al. Differential expression of human 5 S snoRNA genes. Biochem Biophys Res Commun 2002; 299:196–200.

[141] Dong X-Y, Rodriguez C, Guo P, et al. SnoRNA U50 is a candidate tumor-suppressor gene at 6q14.3 with a mutation associated with clinically significant prostate cancer. Hum Mol Genet 2008; 17:1031–42.

[142] Dong X-Y, Guo P, Boyd J, et al. Implication of snoRNA U50 in human breast cancer. J Genet Genomics 2009; 36:447–54.

[143] Tanaka R, Satoh H, Moriyama M, et al. Intronic U50 small-nucleolar-RNA (snoRNA) host gene of no protein-coding potential is mapped at the chromosome breakpoint t(3;6) (q27;q15) of human B- cell lymphoma. Genes Cells 2000; 5:277–87.

[144] Mourtada-Maarabouni M, Pickard MR, Hedge VL, et al. GAS5, a non-protein-coding RNA, controls apoptosis and is downregulated in breast cancer. Oncogene 2009; 28:195–208.

[145] Liao J, Yu L, Mei Y, et al. Small nucleolar RNA signatures as biomarkers for non-small-cell lung cancer. Mol Cancer 2010; 9:198.

[146] Mei Y-P, Liao J-P, Shen J-P, et al. Small nucleolar RNA 42 acts as an oncogene in lung tumorigenesis. Oncogene 2012; 31(22):2794–804.

[147] Jiang F, Yin Z, Caraway NP, et al. Genomic profiles in stage I primary non small cell lung cancer using comparative genomic hybridization analysis of cDNA microarrays. Neoplasia 2004; 6:623–35.

[148] Gebhart E. Double minutes, cytogenetic equivalents of gene amplification, in human neoplasia - a review. Clin Transl Oncol 2005; 7:477–85.

[149] Gee HE, Buffa FM, Camps C, et al. The small nucleolar RNAs commonly used for microRNA normalisation correlate with tumour pathology and prognosis. Br J Cancer 2011; 104:1168–77.

[150] Martens-Uzunova ES, Jalava SE, Dits NF, et al. Diagnostic and prognostic signatures from the small non-coding RNA transcriptome in prostate cancer. Oncogene 2012; 31:978–91.

[151] Valleron W, Laprevotte E, Gautier E-F, et al. Specific small nucleolar RNA expression profiles in acute leukemia. UK: Leukemia: Official Journal of the Leukemia Society of America, Leukemia Research Fund, 2012.

[152] Preker P, Nielsen J, Kammler S, et al. RNA exosome depletion reveals transcription upstream of active human promoters. Science 2008; 322:1851–4.

[153] Seila AC, Calabrese JM, Levine SS, et al. Divergent transcription from active promoters. Science 2008; 322:1849–51.

[154] Core LJ, Lis JT. Transcription regulation through promoter-proximal pausing of RNA polymerase II. Science 2008; 319:1791–2.

[155] Morris KV, Santoso S, Turner A-M, et al. Bidirectional transcription directs both transcriptional gene activation and suppression in human cells. PLoS Genet 2008; l(4):e1000258.

[156] Schwartz JC, Younger ST, Nguyen N-B, et al. Antisense transcripts are targets for activating small RNAs. Nat Struct Mol Biol 2008; 15:842–8.

[157] Kapranov P, Cheng J, Dike S, et al. RNA maps reveal new RNA classes and a possible function for pervasive transcription. Science 2007; 316:1484–8.

[158] Taft RJ, Kaplan CD, Simons C, et al. Evolution, biogenesis and function of promoter-associated RNAs. Cell Cycle 2009; 8:2332–8.

[159] Martianov I, Ramadass A, Serra Barros A, et al. Repression of the human dihydrofolate reductase gene by a non-coding interfering transcript. Nature 2007; 445:666–70.

[160] Han J, Kim D, Morris KV. Promoter-associated RNA is required for RNA directed transcriptional gene silencing in human cells. Proc Natl Acad Sci USA 2007; 104:12422–7.

[161] Imamura T, Yamamoto S, Ohgane J, et al. Noncoding RNA directed DNA demethylation of Sphk1 CpG island. Biochem Biophys Res Commun 2004; 322:593–600.

[162] Fejes-Toth Kata, Sotirova Vihra, Sachidanandam Ravi, et al. Post-transcriptional processing generates a diversity of 5′-modified long and short RNAs. Nature 2009; 457:1028–32.

[163] Hawkins PG, Santoso S, Adams C, et al. Promoter targeted small RNAs induce long-term transcriptional gene silencing in human cells. Nucleic Acids Res 2009; 37:2984–95.

[164] Taft RJ, Hawkins PG, Mattick JS, et al. The relationship between transcription initiation RNAs and CCCTC-binding factor (CTCF) localization. Epigenetics Chromatin 2011; 4:13.

[165] Wang X, Arai S, Song X, et al. Induced ncRNAs allosterically modify RNA binding proteins in cis to inhibit transcription. Nature 2008; 454:126–30.

[166] Watanabe T, Totoki Y, Toyoda A, et al. Endogenous siRNAs from naturally formed dsRNAs regulate transcripts in mouse oocytes. Nature 2008; 453:539–43.

[167] Carone DM, Longo MS, Ferreri GC, et al. A new class of retroviral and satellite encoded small RNAs emanates from mammalian centromeres. Chromosoma 2009; 118:113–25.

[168] Bouzinba-Segard H, Guais A, Francastel C. Accumulation of small murine minor satellite transcripts leads to impaired centromeric architecture and function. Proc Natl Acad Sci USA 2006; 103:8709–14.

[169] Valgardsdottir R, Chiodi I, Giordano M, et al. Structural and functional characterization of noncoding repetitive RNAs transcribed in stressed human cells. Mol Biol Cell 2005; 16:2597–604.

[170] Fukagawa T, Nogami M, Yoshikawa M, et al. Dicer is essential for formation of the heterochromatin structure in vertebrate cells. Nat Cell Biol 2004; 6:784–91.

[171] Cao F, Li X, Hiew S, et al. Dicer independent small RNAs associate with telomeric heterochromatin. RNA 2009; 15:1274–81.

[172] Horard B, Gilson E. Telomeric RNA enters the game. Nat Cell Biol 2008; 10:113–5.

[173] Meynert A, Birney E. Picking pyknons out of the human genome. Cell 2006; 125:836–8.

[174] Rigoutsos I, Huynh T, Miranda K, et al. Short blocks from the noncoding parts of the human genome have instances within nearly all known genes and relate to biological processes. Proc Natl Acad Sci USA 2006; 103:6605–10.

[175] Glinsky GV. Human genome connectivity code links disease-associated SNPs, microRNAs and pyknons. Cell Cycle 2009; 8:925–30.

[176] Tsirigos A, Rigoutsos I. Human and mouse introns are linked to the same processes and functions through each genome's most frequent nonconserved motifs. Nucleic Acids Res 2008; 36:3484–93.

[177] Stabenau A, McVicker G, Melsopp C, et al. The Ensembl core software libraries. Genome Res 2004; 14:929–33.

[178] Pagano A, Castelnuovo M, Tortelli F, et al. New small nuclear RNA gene-like transcriptional units as sources of regulatory transcripts. PLoS Genet 2007; 3:174–84.

[179] Guttman M, Amit I, Garber M, et al. Chromatin signature reveals over a thousand highly conserved large non-coding RNAs in mammals. Nature 2009; 458:223–7.

[180] Chen Y, Song Y, Wang Z, et al. Altered expression of MiR-148a and MiR-152 in gastrointestinal cancers and its clinical significance. J Gastrointest Surg 2010; 14:1170–9.

[181] Lipovich L, Johnson R, Lin C-Y. MacroRNA underdogs in a microRNA world: evolutionary, regulatory, and biomedical significance of mammalian long non-protein-coding RNA. Biochim Biophys Acta 2010; 1799:597–615.

[182] Tripathi V, Ellis JD, Shen Z, et al. The nuclearretained non-coding RNA MALAT1 regulates alternative splicing by modulating SR splicing factor phosphorylation. Mol Cell 2010; 39:925–38.

[183] Gupta RA, Shah N, Wang KC, et al. Long non-coding RNA HOTAIR reprograms chromatin state to promote cancer metastasis. Nature 2010; 464:1071–6.

[184] Ørom UA, Derrien T, Beringer M, et al. Long noncoding RNAs with enhancer-like function in human cells. Cell 2010; 143:46–58.

[185] Mattick JS, Amaral PP, Dinger ME, et al. RNA regulation of epigenetic processes. Bioessays 2009; 31:51–9.

[186] Wang KC, Chang HY. Molecular mechanisms of long non-coding RNAs. Mol Cell 2011; 43:904–14.

[187] Ishii N, Ozaki K, Sato H, et al. Identification of a novel non-coding RNA, MIAT, that confers risk of myocardial infarction. J Hum Genet 2006; 51:1087–99.

[188] Faghihi MA, Modarresi F, Khalil AM, et al. Expression of a noncoding RNA is elevated in Alzheimer's disease and drives rapid feedforward regulation of beta-secretase. Nat Med 2008; 14:723–30.

[189] Ji P, Diederichs S, Wang W, et al. MALAT-1, a novel noncoding RNA, and thymosin beta4 predict metastasis and survival in early-stage non-small cell lung cancer. Oncogene 2003; 22:8031–41.

[190] Lin R, Maeda S, Liu C, et al. A large noncoding RNA is a marker for murine hepatocellular carcinomas and a spectrum of human carcinomas. Oncogene 2007; 26:851–8.

[191] Davis IJ, Hsi B-L, Arroyo JD, et al. Cloning of an Alpha-TFEB fusion in renal tumors harboring the t(6;11)(p21;q13) chromosome translocation. Proc Natl Acad Sci USA 2003; 100:6051–6.

[192] Guo F, Li Y, Liu Y, et al. Inhibition of metastasis-associated lung adenocarcinoma transcript 1 in CaSki human cervical cancer cells suppresses cell proliferation and invasion. Acta Biochim Biophys Sin (Shanghai) 2010; 42:224–9.

[193] Fellenberg J, Bernd L, Delling G, et al. Prognostic significance of drug-regulated genes in high-grade osteosarcoma. Mod Pathol 2007; 20:1085–94.

[194] Koshimizu T, Fujiwara Y, Sakai N, et al. Oxytocin stimulates expression of a noncoding RNA tumor marker in a human neuroblastoma cell line. Life Sci 2010; 86: 455–60.

[195] Panzitt K, Tschernatsch MMO, Guelly C, et al. Characterization of HULC, a novel gene with striking up-regulation in hepatocellular carcinoma, as noncoding RNA. Gastroenterology 2007; 132:330–42.

[196] Matouk IJ, Abbasi I, Hochberg A, et al. Highly upregulated in liver cancer noncoding RNA is overexpressed in hepatic colorectal metastasis. Eur J Gastroenterol Hepatol 2009; 21:688–92.

[197] Chen W, Böcker W, Brosius J, et al. Expression of neural BC200 RNA in human tumours. J Pathol 1997; 183:345–51.

[198] Iacoangeli A, Lin Y, Morley EJ, et al. BC200 RNA in invasive and preinvasive breast cancer. Carcinogenesis 2004; 25:2125–33.

[199] Brannan CI, Dees EC, Ingram RS, et al. The product of the H19 gene may function as an RNA. Mol Cell Biol 1990; 10:28–36.

[200] Gabory A, Jammes H, Dandolo L. The H19 locus: role of an imprinted noncoding RNA in growth and development. Bioessays 2010; 32:473–80.

[201] Hibi K, Nakamura H, Hirai A, et al. Loss of H19 imprinting in esophageal cancer. Cancer Res 1996; 56:480–2.

[202] Berteaux N, Lottin S, Adriaenssens E, et al. Hormonal regulation of H19 gene expression in prostate epithelial cells. J Endocrinol 2004; 183:69–78.

[203] Eis PS, Tam W, Sun L, et al. Accumulation of miR-155 and BIC RNA in human B cell lymphomas. Proc Natl Acad Sci USA 2005; 102:3627–32.

[204] Chung S, Nakagawa H, Uemura M, et al. Association of a novel long non-coding RNA in 8q24 with prostate cancer susceptibility. Cancer Sci 2011; 102:245–52.

[205] Pasic I, Shlien A, Durbin AD, et al. Recurrent focal copy-number changes and loss of heterozygosity implicate two noncoding RNAs and one tumor suppressor gene at chromosome 3q13.31 in osteosarcoma. Cancer Res 2010; 70:160–71.

[206] Petrovics G, Zhang W, Makarem M, et al. Elevated expression of PCGEM1, a prostate-specific gene with cell growth-promoting function, is associated with high-risk prostate cancer patients. Oncogene 2004; 23:605–11.

[207] Srikantan V, Zou Z, Petrovics G, et al. PCGEM1, a prostate-specific gene, is overexpressed in prostate cancer. Proc Natl Acad Sci USA 2000; 97:12216–21.

[208] Fu X, Ravindranath L, Tran N, et al. Regulation of apoptosis by a prostate-specific and prostate cancer-associated noncoding gene, PCGEM1. DNA Cell Biol 2006; 25:135–41.

[209] Wang X-S, Zhang Z, Wang H-C, et al. Rapid identification of UCA1 as a very sensitive and specific unique marker for human bladder carcinoma. Clin Cancer Res 2006; 12:4851–8.

[210] Bussemakers MJ, van Bokhoven A, Verhaegh GW, et al. DD3: a new prostate-specific gene, highly overexpressed in prostate cancer. Cancer Res 1999; 59:5975–9.

[211] de Kok JB, Verhaegh GW, Roelofs RW, et al. DD3(PCA3), a very sensitive and specific marker to detect prostate tumors. Cancer Res 2002; 62:2695–8.

[212] Korneev SA, Korneeva EI, Lagarkova MA, et al. Novel noncoding antisense RNA transcribed from human anti-NOS2A locus is differentially regulated during neuronal differentiation of embryonic stem cells. RNA 2008; 14:2030–7.

[213] Calin GA, Liu C, Ferracin M, et al. Ultraconserved regions encoding ncRNAs are altered in human leukemias and carcinomas. Cancer Cell 2007; 12:215–29.

[214] Braconi C, Valeri N, Kogure T, et al. Expression and functional role of a transcribed noncoding RNA with an ultra-conserved element in hepatocellular carcinoma. Proc Natl Acad Sci USA 2011; 108:786–91.

[215] Yu W, Gius D, Onyango P, et al. Epigenetic silencing of tumour suppressor gene p15 by its antisense RNA. Nature 2008; 451:202–6.

[216] Folkersen L, Kyriakou T, Goel A, et al. Relationship between CAD risk genotype in the chromosome 9p21 locus and gene expression. Identification of eight new ANRIL splice variants. PLoS ONE 2009; 4:e7677.

[217] Yap KL, Li S, Muñoz-Cabello AM, et al. Molecular interplay of the noncoding RNA ANRIL and methylated histone H3 lysine 27 by polycomb CBX7 in transcriptional silencing of INK4a. Mol Cell 2010; 38:662–74.

[218] Pasmant E, Laurendeau I, Héron D, et al. Characterization of a germ-line deletion, including the entire INK4/ARF locus, in a melanoma-neural system tumor family: identification of ANRIL, an antisense noncoding RNA whose expression coclusters with ARF. Cancer Res 2007; 67:3963–9.

[219] Miyoshi N, Wagatsuma H, Wakana S, et al. Identification of an imprinted gene, Meg3/Gtl2 and its human homologue

MEG3, first mapped on mouse distal chromosome 12 and human chromosome 14q. Genes Cells 2000; 5:211–20.

[220] Zhang X, Zhou Y, Mehta KR, et al. A pituitary-derived MEG3 isoform functions as a growth suppressor in tumor cells. J Clin Endocrinol Metab 2003; 88:5119–26.

[221] Zhang X, Rice K, Wang Y, et al. Maternally expressed gene 3 (MEG3) noncoding ribonucleic acid: isoform structure, expression, and functions. Endocrinology 2010; 151:939–47.

[222] Leygue E, Dotzlaw H, Watson PH, et al. Expression of the steroid receptor RNA activator in human breast tumors. Cancer Res 1999; 59:4190–3.

[223] Chooniedass-Kothari S, Emberley E, Hamedani MK, et al. The steroid receptor RNA activator is the first functional RNA encoding a protein. FEBS Lett 2004; 566:43–7.

[224] Poliseno L, Salmena L, Zhang J, et al. A coding-independent function of gene and pseudogene mRNAs regulates tumour biology. Nature 2010; 465:1033–8.

[225] Alimonti A, Carracedo A, Clohessy JG, et al. Subtle variations in Pten dose determine cancer susceptibility. Nat Genet 2010; 42:454–8.

[226] Yu M, Ohira M, Li Y, et al. High expression of ncRAN, a novel non-coding RNA mapped to chromosome 17q25.1, is associated with poor prognosis in neuroblastoma. Int J Oncol 2009; 34:931–8.

[227] Zhu Y, Yu M, Li Z, et al. ncRAN, a newly identified long noncoding RNA, enhances human bladder tumor growth, invasion, and survival. Urology 2011; 77:510. e1–5.

[228] Silva JM, Boczek NJ, Berres MW, et al. LSINCT5 is over expressed in breast and ovarian cancer and affects cellular proliferation. RNA Biol 2011; 8:496–505.

[229] Khalil AM, Guttman M, Huarte M, et al. Many human large intergenic noncoding RNAs associate with chromatin-modifying complexes and affect gene expression. Proc Natl Acad Sci USA 2009; 106:11667–72.

[230] Tsai M-C, Manor O, Wan Y, et al. Long noncoding RNA as modular scaffold of histone modification complexes. Science 2010; 329:689–93.

[231] Yang Z, Zhou L, Wu L-M, et al. Overexpression of long non-coding RNA HOTAIR predicts tumor recurrence in hepatocellular carcinoma patients following liver transplantation. Ann Surg Oncol 2011; 18:1243–50.

[232] Huarte M, Guttman M, Feldser D, et al. A large intergenic noncoding RNA induced by p53 mediates global gene repression in the p53 response. Cell 2010; 142:409–19.

[233] Louro R, Smirnova AS, Verjovski-Almeida S. Long intronic noncoding RNA transcription: expression noise or expression choice? Genomics 2009; 93:291–8.

[234] Louro R, Nakaya HI, Amaral PP, et al. Androgen responsive intronic non-coding RNAs. BMC Biol 2007; 5:4.

[235] Cawley S, Bekiranov S, Ng HH, et al. Unbiased mapping of transcription factor binding sites along human chromosomes 21 and 22 points to widespread regulation of noncoding RNAs. Cell 2004; 116:499–509.

[236] Nakaya HI, Amaral PP, Louro R, et al. Genome mapping and expression analyses of human intronic noncoding RNAs reveal tissue specific patterns and enrichment in genes related to regulation of transcription. Genome Biol 2007; 8:R43.

[237] Dinger ME, Amaral PP, Mercer TR, et al. Long noncoding RNAs in mouse embryonic stem cell pluripotency and differentiation. Genome Res 2008; 18:1433–45.

[238] Kravchenko JE, Rogozin IB, Koonin EV, et al. Transcription of mammalian messenger RNAs by a nuclear RNA polymerase of mitochondrial origin. Nature 2005; 436:735–9.

[239] Li S-C, Tang P, Lin W-C. Intronic microRNA: discovery and biological implications. DNA Cell Biol 2007; 26:195–207.

[240] Borchert GM, Lanier W, Davidson BL. RNA polymerase III transcribes human microRNAs. Nat Struct Mol Biol 2006; 13:1097–101.

[241] Massone S, Vassallo I, Castelnuovo M, et al. RNA polymerase III drives alternative splicing of the potassium channel-interacting protein contributing to brain complexity and neurodegeneration. J Cell Biol 2011; 193:851–66.

[242] Massone S, Vassallo I, Fiorino G, et al. 17A, a novel noncoding RNA, regulates GABA B alternative splicing and signaling in response to inflammatory stimuli and in Alzheimer disease. Neurobiol Dis 2011; 41:308–17.

[243] Louro R, El-Jundi T, Nakaya HI, et al. Conserved tissue expression signatures of intronic noncoding RNAs transcribed from human and mouse loci. Genomics 2008; 92:18–25.

[244] Rearick D, Prakash A, McSweeny A, et al. Critical association of ncRNA with introns. Nucleic Acids Res 2011; 39:2357–66.

[245] Mercer TR, Dinger ME, Sunkin SM, et al. Specific expression of long noncoding RNAs in the mouse brain. Proc Natl Acad Sci USA 2008; 105:716–21.

[246] Katayama S, Tomaru Y, Kasukawa T, et al. Antisense transcription in the mammalian transcriptome. Science 2005; 309:1564–6.

[247] Hirose T, Ideue T, Nagai M, et al. A spliceosomal intron binding protein, IBP160, links position-dependent assembly of intronencoded box C/D snoRNP to pre-mRNA splicing. Mol Cell 2006; 23:673–84.

[248] Filipowicz W, Pogačić V. Biogenesis of small nucleolar ribonucleoproteins. Curr Opin Cell Biol 2002; 14:319–27.

[249] Heo JB, Sung S. Vernalization-mediated epigenetic silencing by a long intronic noncoding RNA. Science 2011; 331:76–9.

[250] Tahira AC, Kubrusly MS, Faria MF, et al. Long noncoding intronic RNAs are differentially expressed in primary and metastatic pancreatic cancer. Mol Cancer 2011; 10:141.

[251] Isken O, Maquat LE. Telomeric RNAs as a novel player in telomeric integrity. F1000 Biol Rep 2009; 1:90.

[252] Azzalin CM, Reichenbach P, Khoriauli L, et al. Telomeric repeat containing RNA and RNA surveillance factors at mammalian chromosome ends. Science 2007; 318: 798–801.

[253] Schoeftner S, Blasco MA. Developmentally regulated transcription of mammalian telomeres by DNA-dependent RNA polymerase II. Nat Cell Biol 2008; 10:228–36.

[254] Schoeftner S, Blasco MA. A "higher order" of telomere regulation: telomere heterochromatin and telomeric RNAs. EMBO J 2009; 28:2323–36.

[255] Luke B, Panza A, Redon S, et al. The Rat1p 5′ to 3′ exonuclease degrades telomeric repeat-containing RNA and promotes telomere elongation in Saccharomyces cerevisiae. Mol Cell 2008; 32:465–77.

[256] Caslini C, Connelly JA, Serna A, et al. MLL associates with telomeres and regulates telomeric repeat-containing RNA transcription. Mol Cell Biol 2009; 29:4519–26.

[257] Sampl S, Pramhas S, Stern C, et al. Expression of telomeres in astrocytoma WHO grade 2 to 4: TERRA level correlates with telomere length, telomerase activity, and advanced clinical Ggade. Transl Oncol 2012; 5:56–65.

[258] Schoeftner S, Blasco MA. Chromatin regulation and non-coding RNAs at mammalian telomeres. Semin Cell Dev Biol 2010; 21:186–93.

[259] Ulveling D, Francastel C, Hubé F. When one is better than two: RNA with dual functions. Biochimie 2011; 93:633–44.

[260] Lanz RB, McKenna NJ, Onate SA, et al. A steroid receptor coactivator, SRA, functions as an RNA and is present in an SRC-1 complex. Cell 1999; 97:17–27.

[261] Kawashima H, Takano H, Sugita S, et al. A novel steroid receptor co-activator protein (SRAP) as an alternative form of steroid receptor RNA-activator gene: expression in prostate cancer cells and enhancement of androgen receptor activity. Biochem J 2003; 369:163–71.

[262] Xu K, Liang X, Cui D, et al. miR-1915 inhibits Bcl-2 to modulate multidrug resistance by increasing drug-sensitivity in human colorectal carcinoma cells. Mol Carcinog 2011; doi:10.1002/mc.21832

[263] Hussein-Fikret S, Fuller PJ. Expression of nuclear receptor coregulators in ovarian stromal and epithelial tumours. Mol Cell Endocrinol 2005; 229:149–60.

[264] Lanz RB, Chua SS, Barron N, et al. Steroid receptor RNA activator stimulates proliferation as well as apoptosis in vivo. Mol Cell Biol 2003; 23:7163–76.

[265] Harrison PM, Zheng D, Zhang Z, et al. Transcribed processed pseudogenes in the human genome: an intermediate form of expressed retrosequence lacking protein-coding ability. Nucleic Acids Res 2005; 33:2374–83.

[266] Pink RC, Wicks K, Caley DP, et al. Pseudogenes: pseudo-functional or key regulators in health and disease? RNA 2011; 17:792–8.

[267] Esnault C, Maestre J, Heidmann T. Human LINE ret-rotransposons generate processed pseudogenes. Nat Genet 2000; 24:363–7.

[268] Terai G, Yoshizawa A, Okida H, et al. Discovery of short pseudogenes derived from messenger RNAs. Nucleic Acids Res 2010; 38:1163–71.

[269] Devor EJ. Primate microRNAs miR-220 and miR-492 lie within processed pseudogenes. J Hered 2006; 97:186–90.

[270] Han YJ, Ma SF, Yourek G, et al. A transcribed pseudo-gene of MYLK promotes cell proliferation. FASEB J 2011; 25:2305–12.

[271] Lu W, Zhou D, Glusman G, et al. KLK31P is a novel andro-gen regulated and transcribed pseudogene of kallikreins that is expressed at lower levels in prostate cancer cells than in normal prostate cells. Prostate 2006; 66:936–44.

[272] He L. Posttranscriptional regulation of PTEN dosage by noncoding RNAs. Sci Signal 2010; 3:pe39.

[273] Bejerano G, Pheasant M, Makunin I, et al. Ultraconserved elements in the human genome. Science 2004; 304:1321–5.

[274] Nobrega MA, Ovcharenko I, Afzal V, et al. Scanning human gene deserts for long-range enhancers. Science 2003; 302:413.

[275] Lujambio A, Portela A, Liz J, et al. CpG island hyper-methylation-associated silencing of noncoding RNAs transcribed from ultraconserved regions in human cancer. Oncogene 2010; 29:6390–401.

[276] Scaruffi P, Stigliani S, Moretti S, et al. Transcribed-ultra conserved region expression is associated with outcome in high-risk neuroblastoma. BMC Cancer 2009; 9:441.

[277] Mestdagh P, Fredlund E, Pattyn F, et al. An integrative genomics screen uncovers ncRNA T-UCR functions in neu-roblastoma tumours. Oncogene 2010; 29:3583–92.

[278] Yang R, Frank B, Hemminki K, et al. SNPs in ultra-conserved elements and familial breast cancer risk. Carcinogenesis 2008; 29:351–5.

[279] Zhang B, Pan X, Cobb GP, et al. microRNAs as oncogenes and tumor suppressors. Dev Biol 2007; 302:1–12.

[280] Lionetti M, Agnelli L, Lombardi L, et al. MicroRNAs in the pathobiology of multiple myeloma. Curr Cancer Drug Targets 2012; 12:823–37.

[281••] Calin GA, Sevignani C, Dumitru CD, et al. Human microRNA genes are frequently located at fragile sites and genomic regions involved in cancers. Proc Natl Acad Sci USA 2004; 101:2999–3004.

•• This is one of the first reports demonstrating the associa-tion of miRNA loci with genomic regions altered in cancer.

[282] Pichiorri F, Suh SS, Ladetto M, et al. MicroRNAs regulate critical genes associated with multiple myeloma pathogen-esis. Proc Natl Acad Sci USA 2008; 105:12885–90.

[283] Calin GA, Liu CG, Sevignani C, et al. MicroRNA profil-ing reveals distinct signatures in B cell chronic lymphocytic leukemias. Proc Natl Acad Sci USA 2004; 101:11755–60.

[284] Jongen-Lavrencic M, Sun SM, Dijkstra MK, et al. MicroRNA expression profiling in relation to the genetic heterogeneity of acute myeloid leukemia. Blood 2008; 111:5078–85.

[285] Tassone P, Tagliaferri P. Editorial: new approaches in the treatment of multiple myeloma: from target-based agents to the new era of microRNAs (dedicated to the memory of Prof. Salvatore Venuta). Curr Cancer Drug Targets 2012; 12:741–2.

[286] Tagliaferri P, Rossi M, Di Martino MT, et al. Promises and challenges of MicroRNA-based treatment of multiple myeloma. Curr Cancer Drug Targets 2012; 12:838–46.

[287•] Lionetti M, Biasiolo M, Agnelli L, et al. Identification of microRNA expression patterns and definition of a microRNA/mRNA regulatory network in distinct molecular groups of multiple myeloma. Blood 2009; 114:e20–6.

• This paper demonstrates that miRNAs associate to specific cytogenetic abnormalities in MM.

[288] Zhang Y, Roccaro AM, Rombaoa C, et al. LNA-mediated anti-miR-155 silencing in low-grade Bcell lymphomas. Blood 2012; 120:1678–86.

[289] Agnelli L, Mosca L, Fabris S, et al. A SNP microarray and FISH-based procedure to detect allelic imbalances in mul-tiple myeloma: an integrated genomics approach reveals a wide gene dosage effect. Genes Chromosomes Cancer 2009; 48:603–14.

[290] Gutierrez NC, Sarasquete ME, Misiewicz-Krzeminska I, et al. Deregulation of microRNA expression in the different genetic subtypes of multiple myeloma and correlation with gene expression profiling. Leukemia 2010; 24:629–37.

[291] Zhou Y, Chen L, Barlogie B, et al. High-risk myeloma is associated with global elevation of miRNAs and overex-pression of EIF2C2/AGO2. Proc Natl Acad Sci USA 2010; 107:7904–9.

[292] Calin GA, Dumitru CD, Shimizu M, et al. Frequent dele-tions and down-regulation of micro- RNA genes miR15 and miR16 at 13q14 in chronic lymphocytic leukemia. Proc Natl Acad Sci USA 2002; 99:15524–9.

[293] Roccaro AM, Sacco A, Thompson B, et al. MicroRNAs 15a and 16 regulate tumor proliferation in multiple myeloma. Blood 2009; 113:6669–80.

[294] Corthals SL, Jongen-Lavrencic M, de Knegt Y, et al. Micro-RNA-15a and micro-RNA-16 expression and chro-mosome 13 deletions in multiple myeloma. Leuk Res 2010; 34:677–81.

[295] Hao M, Zhang L, An G, et al. Bone marrow stromal cells protect myeloma cells from bortezomib induced

apoptosis by suppressing microRNA-15a expression. Leuk Lymphoma 2011; 52:1787–94.

[296••] Amodio N, Di Martino MT, Foresta U, et al. miR-29b sensitizes multiple myeloma cells to bortezomib-induced apoptosis through the activation of a feedback loop with the transcription factor Sp1. Cell Death Dis 2012; 3:e436.
•• This publication describes the anti-tumor activity of miR-29b and its ability to potentiate bortezomib effects against multiple myeloma cells.

[297] Wang X, Li C, Ju S, et al. Myeloma cell adhesion to bone marrow stromal cells confers drug resistance by microRNA-21 up-regulation. Leuk Lymphoma 2011; 52:1991–8.

[298] Melo SA, Moutinho C, Ropero S, et al. A genetic defect in exportin-5 traps precursor microRNAs in the nucleus of cancer cells. Cancer Cell 2010; 18:303–15.

[299] Hill DA, Ivanovich J, Priest JR, et al. DICER1 mutations in familial pleuropulmonary blastoma. Science 2009; 325:965.

[300] Sarasquete ME, Gutierrez NC, Misiewicz-Krzeminska I, et al. Upregulation of Dicer is more frequent in monoclonal gammopathies of undetermined significance than in multiple myeloma patients and is associated with longer survival in symptomatic myeloma patients. Haematologica 2011; 96:468–71.

[301] Bonci D. MicroRNA-21 as therapeutic target in cancer and cardiovascular disease. Recent pat Cardiovasc Drug Discov 2010; 5:156–61.

[302] Lionetti M, Agnelli L, Mosca L, et al. Integrative high-resolution microarray analysis of human myeloma cell lines reveals deregulated miRNA expression associated with allelic imbalances and gene expression profiles. Genes Chromosomes Cancer 2009; 48:521–31.

[303] Saunders MA, Liang H, Li WH. Human polymorphism at microRNAs and microRNA target sites. Proc Natl Acad Sci USA 2007; 104:3300–5.

[304] de Larrea CF, Navarro A, Tejero R, et al. Impact of MiRSNPs on survival and progression in patients with multiple myeloma undergoing autologous stem cell transplantation. Clin Cancer Res 2012; 18:3697–704.

[305] Ohler U, Yekta S, Lim LP, et al. Patterns of flanking sequence conservation and a characteristic upstream motif for microRNA gene identification. RNA 2004; 10:1309–22.

[306••] Di Martino MT, Leone E, Amodio N, et al. Synthetic miR-34a mimics as a novel therapeutic agent for multiple myeloma: in vitro and in vivo evidence. Clin Cancer Res 2012; 18:6260–70.
•• This is the first publication demonstrating the in vivo anti-multiple myeloma activity of miR-34a mimics delivered through a neutral lipid emulsion.

[307] Pichiorri F, Suh SS, Rocci A, et al. Downregulation of p53-inducible microRNAs 192, 194, and 215 impairs the p53/MDM2 autoregulatory loop in multiple myeloma development. Cancer Cell 2010; 18:367–81.

[308] Misiewicz-Krzeminska I, Sarasquete ME, Quwaider D, et al. Restoration of miR-214 expression reduces growth of myeloma cells through a positive regulation of P53 and inhibition of DNA replication. Haematologica 2013; 98(4):640–8.

[309] Chen L, Li C, Zhang R, et al. miR-17-92 cluster microRNAs confers tumorigenicity in multiple myeloma. Cancer Lett 2011; 309:62–70.

[310] Fulciniti M, Amin S, Nanjappa P, et al. Significant biological role of sp1 transactivation in multiple myeloma. Clin Cancer Res 2011; 17:6500–9.

[311••] Rossi M, Pitari MR, Amodio N, et al. miR-29b negatively regulates human osteoclastic cell differentiation and function: implications for the treatment of multiple myeloma-related bone disease. J Cell Physiol 2013 Jul; 228(7):1506–15. DOI: 10.1002/jcp.24306
•• This paper is the first demonstration of the anti-osteoclastogenic activity of miR-29b in the context of multiple myeloma-related bone disease.

[312] Wong KY, Yim RL, So CC, et al. Epigenetic inactivation of the MIR34B/C in multiple myeloma. Blood 2011; 118:5901–4.

[313] Wong KY, Liang R, So CC, et al. Epigenetic silencing of MIR203 in multiple myeloma. Br J Haematol 2011; 154:569–78.

[314••] Amodio N, Leotta M, Bellizzi D, et al. DNA-demethylating and anti-tumor activity of synthetic miR-29b mimics in multiple myeloma. Oncotarget 2012; 3:1246–58.
•• This publication demonstrates demethylating activity of synthetic miR-29b mimics in multiple myeloma cells.

[315••] Leone E, Morelli E, Di Martino MT, et al. Targeting miR-21 inhibits in vitro and in vivo multiple myeloma cell growth. Clin Cancer Res 2013; 8:2096–106.
•• This is the first publication demonstrating the in vivo anti-multiple myeloma activity of miR-21 inhibitors.

[316••] Di Martino MT, Gulla A, Cantafio ME, et al. In vitro and in vivo anti-tumor activity of miR-221/222 inhibitors in multiple myeloma. Oncotarget 2013; 4:242–55.
•• This is the first publication demonstrating the in vivo activity of mir-221/222 inhibitors against TC2 and TC4 multiple myeloma cells.

[317] Williams GT, Farzaneh F. Are snoRNAs and snoRNA host genes new players in cancer? Nat Rev Cancer 2012; 12:84–8.

[318] Lopez-Corral L, Mateos MV, Corchete LA, et al. Genomic analysis of high-risk smoldering multiple myeloma. Haematologica 2012; 97:1439–43.

[319••] Chu L, Su MY, Maggi LB Jr, et al. Multiple myeloma-associated chromosomal translocation activates orphan snoRNA ACA11 to suppress oxidative stress. J Clin Invest 2012; 122:2793–806.
•• This paper describes the identification of a snoRNA involved in multiple myelomas carrying the t(4;14) translocation.

[320•] Ronchetti D, Todoerti K, Tuana G, et al. The expression pattern of small nucleolar and small Cajal body-specific RNAs characterizes distinct molecular subtypes of multiple myeloma. Blood Cancer J 2012; 2:e96.
• This is the first snoRNA expression profile study carried out in multiple myeloma.

[321] Michel CI, Holley CL, Scruggs BS, et al. Small nucleolar RNAs U32a, U33, and U35a are critical mediators of metabolic stress. Cell Metabol 2011; 14:33–44.

[322] Rinn JL, Kertesz M, Wang JK, et al. Functional demarcation of active and silent chromatin domains in human HOX loci by noncoding RNAs. Cell 2007; 129:1311–23.

[323] Keller S, Sanderson MP, Stoeck A, et al. Exosomes: from biogenesis and secretion to biological function. Immunol Lett 2006; 107:102–8.

[324] Reid G, Kirschner MB, van Zandwijk N. Circulating microRNAs: association with disease and potential use as biomarkers. Crit Rev Oncol Hematol 2011; 80:193–208.

[325] Huang JJ, Yu J, Li JY, et al. Circulating microRNA expression is associated with genetic subtype and survival of multiple myeloma. Med Oncol 2012; 29:2402–8.

[326] Jones CI, Zabolotskaya MV, King AJ, et al. Identification of circulating microRNAs as diagnostic biomarkers for use in multiple myeloma. Br J Cancer 2012; 107:1987–96.

[327] Burnett JC, Rossi JJ, Tiemann K. Current progress of siRNA/shRNA therapeutics in clinical trials. Biotechnol J 2011; 6:1130–46.

[328] Lal A, Navarro F, Maher CA, et al. miR-24 Inhibits cell proliferation by targeting E2F2, MYC, and other cell-cycle genes via binding to "seedless" 3'UTR microRNA recognition elements. Mol Cell 2009; 35:610–25.

[329] Trang P, Medina PP, Wiggins JF, et al. Regression of murine lung tumors by the let-7 microRNA. Oncogene 2010; 29:1580–7.

[330] Wiggins JF, Ruffino L, Kelnar K, et al. Development of a lung cancer therapeutic based on the tumor suppressor microRNA-34. Cancer Res 2010; 70:5923–30.

[331•] Trang P, Wiggins JF, Daige CL, et al. Systemic delivery of tumor suppressor microRNA mimics using a neutral lipid emulsion inhibits lung tumors in mice. Mol Ther 2011; 19:1116–22.
   • This is one of the first reports showing therapeutic activity of systemically delivered miRNAs against cancer cells.

[332] Kota J, Chivukula RR, O'Donnell KA, et al. Therapeutic microRNA delivery suppresses tumorigenesis in a murine liver cancer model. Cell 2009; 137:1005–17.

[333] Petersen M, Wengel J. LNA: a versatile tool for therapeutics and genomics. Trends Biotechnol 2003; 21:74–81.

[334••] Elmen J, Lindow M, Schutz S, et al. LNA-mediated microRNA silencing in non-human primates. Nature 2008; 452:896–9. This paper demonstrates the therapeutic efficacy and lack of toxicity of LNAs in non-human primates. 100. Obad S, dos Santos CO, Petri A, et al. Silencing of microRNA families by seed-targeting tiny LNAs. Nat Genet 2011; 43:371-8.
   •• This publication describes the efficiency of tiny LNAs in silencing miRNAs.

[335••] Obad S, dos Santos CO, Petri A, et al. Silencing of microRNA families by seed-targeting tiny LNAs. Nat Genet 2011; 43:371-8.
   •• This publication describes the efficiency of tiny LNAs in silencing miRNAs.

[336] Madkour LH. Reactive Oxygen Species (ROS), Nanoparticles, and Endoplasmic Reticulum (ER) Stress-Induced Cell Death Mechanisms. Academic Press, 2020. https://www.elsevier.com/books/reactive-oxygen-species-ros-nanoparticles-and-endoplasmic-reticulum-er-stress-induced-cell-death-mechanisms/madkour/978-0-12-822481-6

[337] Madkour LH. Nanoparticles Induce Oxidative and Endoplasmic Reticulum Antioxidant Therapeutic Defenses. 1st Edition. Switzerland: Springer, 2020. https://www.springer.com/gp/book/9783030372965?utm_campaign=3_pier05_buy_print&utm_content=en_08082017&utm_medium=referral&utm_source=google_books#otherversion=9783030372972

[338] Madkour LH. Nucleic Acids as Gene Anticancer Drug Delivery Therapy. 1st Edition. Elsevier, 2020. https://www.elsevier.com/books/nucleic-acids-as-gene-anticancer-drug-deliverytherapy/madkour/978-0-12-819777-6

[339] Madkour LH. Nanoelectronic Materials: Fundamentals and Applications (Advanced Structured Materials). 1st Edition. Switzerland: Springer, 2019. https://books.google.com.eg/books/about/Nanoelectronic_Materials.html?id=YQXCxAEACAAJ&source=kp_book_description&redir_esc=y https://www.springer.com/gp/book/9783030216207

# 14 Advances in the Inhibition and Optimization of Checkpoint Kinases by Small Molecules for the Treatment of Cancer

## 14.1 DNA-TARGETING THERAPIES

Conventional DNA-targeting therapies, such as chemotherapy and radiation, are among the most commonly used cancer treatments. As a class they have produced significant increases in the survival of patients, particularly when used in combination with drugs that function via different mechanisms of action. Because of the efficacy of these anticancer treatments, DNA-targeting agents are likely to remain a standard-of-care for the treatment of many cancers for the foreseeable future. Although DNA targeting has proved an effective approach to tumor control, this mechanism also leads to significant side effects, as the majority of these agents are used at the maximum tolerated dose. Toxicities to the hematological, gastrointestinal, and other organ systems are commonly observed, resulting in a limit to the degree of tumor control that can be achieved. Another limitation of DNA-damaging agents is that many patients develop resistance and therefore become refractory to treatment.

Resistance can arise through multiple mechanisms, including modulation of cellular levels of the drug, and defects in apoptosis or DNA repair.

Results from cell cycle research have led to the hypothesis that tumors may be selectively sensitized to DNA-damaging agents, leading to improved antitumor activity and a wider therapeutic margin for such agents. This approach relies on the observation that the majority of tumors are deficient in the G1 DNA damage checkpoint pathway, resulting in a reliance on the S- and G2-phase checkpoints to repair DNA damage and survive. The S- and G2-phase checkpoints are regulated by checkpoint kinase 1 (CHK1), a serine/threonine kinase that is activated by ataxia telangiectasia–mutated protein kinase (ATM) and ATM and Rad3-related protein kinase (ATR) phosphorylation in response to DNA damage (Figure 14.1).

Inhibition of CHK1 signaling in tumor cells has been shown to abrogate the S and G2 checkpoints, impair DNA repair, and result in increased tumor cell death. However, noncancerous tissue has a functioning G1 checkpoint signaling pathway, allowing for DNA repair and cell survival.

The hypothesis that abrogation of the G2 DNA damage checkpoint could lead to potentiation of the effects of DNA-damaging agents was first proposed in the 1980s, and research conducted in the 1990s provided further support to the theory [1–4]. Lau and Pardee presented results that caffeine [1, 2], later identified as an inhibitor of ATR and ATM [5–6], potentiates the lethal effects of nitrogen mustard by abrogation of G2 arrest, thus inducing cells to undergo mitosis prior to DNA repair. As a result of this early research using caffeine, additional studies demonstrated selective radio sensitization of p53 mutant cells following abrogation of the G2 checkpoint by caffeine or the staurosporine analog UCN-01 (1, KW-2401, NSC-638850, Kyowa Hakko Kogyo Co Ltd; Figure 14.2) [7–9]. This result suggested that p53 mutant tumors might be selectively killed through inhibition of G2 checkpoint activity. One of the targets of UCN-01 was subsequently identified as CHK1, thereby promoting CHK1 as a useful therapeutic target that could lead to enhanced cytotoxicity in tumor cells [10, 11]. Another staurosporine analog, SB-218078 (2; Figure 14.2), demonstrated similar activities [12]. More precise targeting of CHK1 function using dominant negative enzymes, interference RNA, and ribozymes, further supported the hypothesis that CHK1 activity is critical for maintaining cell cycle arrest, and that inhibition of CHK1 function can potentiate tumor cells to multiple DNA-damaging modalities [13–19].

This chapter focuses on advances made in checkpoint kinase inhibitor design and development from 2005 to the present, and which have not fully been covered by prior reviews [20–25]. Over the past few years, significant advances have been made by many investigators in this area of research. There are currently three agents undergoing Phase 1 clinical trials, with additional candidates undoubtedly to follow. In the same time period, there has also been an almost exponential increase in the number of new publications and patent applications in this field. In addition to medicinal chemistry advances, significant advances made in the understanding of checkpoint biology have led to new therapeutic opportunities for checkpoint kinase inhibitors and novel approaches to facilitate clinical trials.

DOI: 10.1201/9781003229650-14

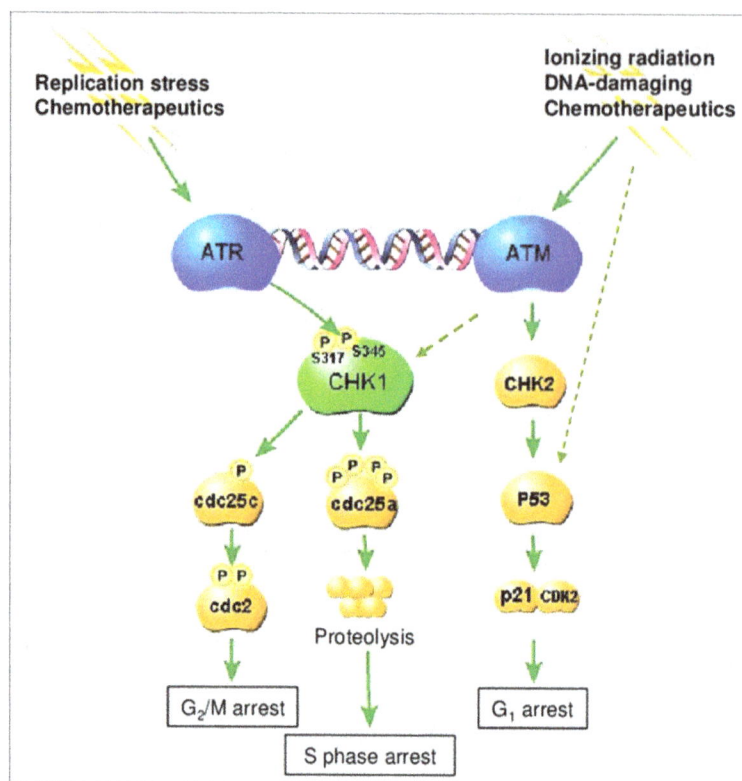

**FIGURE 14.1**  The DNA-damage response pathway. CHK1 is activated by ATR phosphorylation of Ser[317] and/or Ser[345] after DNA damage is detected by ATM/ATR kinases. Downstream phosphorylation events result in G2/M- and S-phase cell cycle arrest.

## 14.2   OVERVIEW OF CHK1 INHIBITORS IN THE CLINIC

The first inhibitor in this class of sensitizing agents to enter the clinic was the nonselective kinase inhibitor UCN-01. This compound is an inhibitor of CHK1 ($IC_{50}$=10 nM) plus many other kinases, including CHK2 ($IC_{50}$=10 nM) and several of the cyclin-dependent kinases. Therefore, the clinical effects of this agent may not be associated solely with CHK1 inhibition [26–28]. Since late 2005, three novel small-molecule CHK1 inhibitors have begun clinical evaluation: XL-844 (Exelixis Inc.), AZD-7762 (AstraZeneca plc), and PF-477736 (Pfizer Inc.).

### 14.2.1   XL-844

In September 2005, the first Phase 1 clinical study of a small-molecule checkpoint kinase inhibitor, XL-844 (EXEL-9844; Exelixis Inc.), was initiated for the treatment of chronic lymphocytic leukemia. XL-844 is an

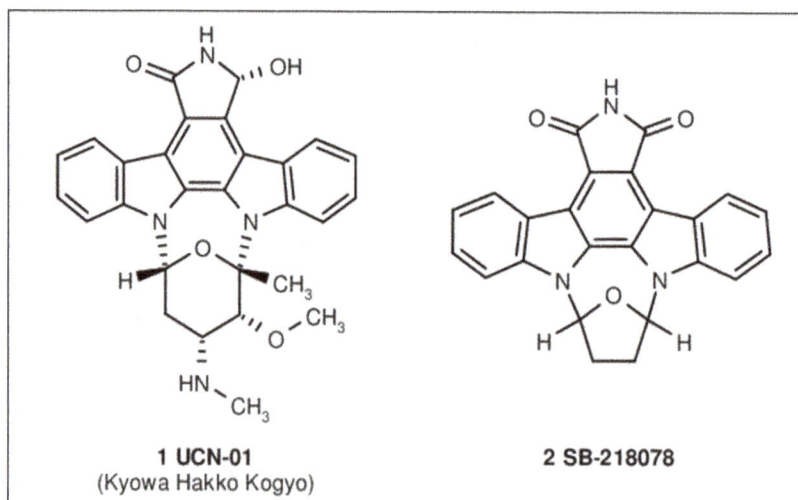

**FIGURE 14.2**   The structures of UCN-01 and SB-218078.

orally bioavailable aminopyrazine kinase inhibitor of both CHK1 ($IC_{50}$=2.2 nM) and CHK2 ($IC_{50}$=0.2 nM). XL-844 has been profiled against 86 other kinases, four of which have $IC_{50}$ values of <30 nM (i.e., Flt-4 [6 nM], KDR [12 nM], PDGF [25 nM], and Flt-3 [28 nM]) [29•]. In combination with gemcitabine, XL-844 has been demonstrated preclinically to abrogate the S-phase checkpoint, potentiate the effects of gemcitabine, and cause the predicted protein changes downstream of CHK1. In addition, combination treatment with gemcitabine led to an increase in activated pCHK-S317 levels and also an increase in the levels of phosphohistone 2AX (p-H2AX), a marker of double-stranded DNA damage. A PANC-1 xenograft study demonstrated that XL-844 potentiates the effects of gemcitabine in vivo, while showing no single-agent activity. In this study, the tumor growth inhibition increased from 70% in a group of mice administered gemcitabine alone, to 91% in the group receiving combination therapy. Hematological assessment indicated that the combination treatment did not lead to enhanced hematological toxicity. In a similar manner, XL-844 was evaluated in combination with daunorubicin. In a chronic myeloid leukemia efficacy model (K562), XL-844 in combination with daunorubicin resulted in a 93% increase in mean survival time as compared to daunorubicin alone. Again, there was neither evidence of single-agent XL-844 activity nor any change in daunorubicin plasma concentration in the combination treatment group. In addition to XL-844, another Exelixis compound, EXEL-3611, has been highlighted as a checkpoint kinase inhibitor [30]. EXEL-3611 is 1000-fold selective for CHK1 ($IC_{50}$=2.4 nM) over CHK2 ($IC_{50}$=2400 nM). The author of this study [31] demonstrated through fluorescence-activated cell sorting (FACS) analysis that XL-844 was more effective than EXEL-3611 at abrogating daunorubicin-induced cell cycle arrest. The dual CHK1/CHK2 inhibitor XL-844 also caused >4N polyploidy cell content and mitotic catastrophe. Evaluation of the structure-activity relationship (SAR) for this series [32] demonstrated that, of the specific examples having an $IC_{50}$ value <50 nM for CHK1, the majority contained basic amide substituents with a preference for 3S-aminopiperidine. It is also notable that most of the compounds are metasubstituted, in particular with a benzylic amide or amine group, predominantly indane or hydroxyindane. The structure of XL-844 has not been disclosed [33]. It is worth noting that chiral piperidine 3 (Figure 14.3) and methyl amide 4 (Figure 14.3) exhibit similar potency against CHK1, but it is surmised that they may have differing activities against CHK2, analogous to XL-844 and EXEL-3611.

## 14.2.2 AZD-7762

Since 2004, researchers at AstraZeneca plc have published on several structural classes of checkpoint kinase inhibitors [34, 35], and have recently advanced the small-molecule compound AZD-7762 into the clinic for use in combination with gemcitabine in the treatment of solid tumors. As with

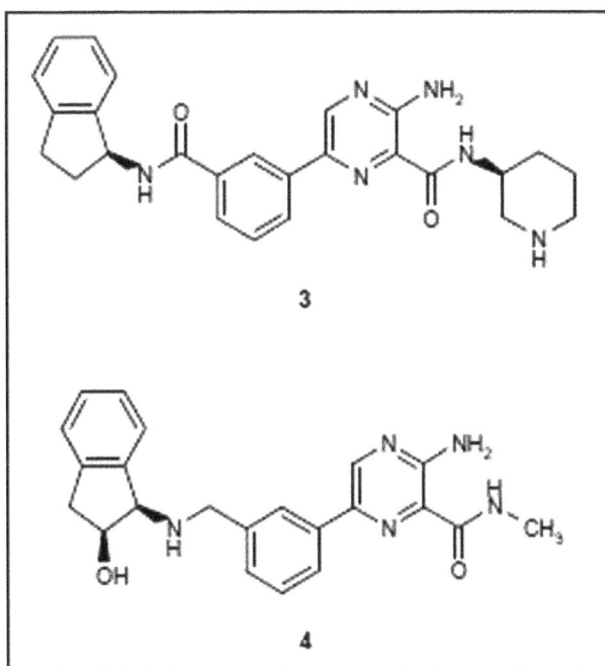

**FIGURE 14.3** The structures of selected chiral piperidine and methyl amide CHK1 inhibitors from Exelixis.

XL-844, AZD-7762 demonstrates potent and dual inhibition of both CHK1 ($IC_{50}$=5 nM, Ki=3.6 nM) and CHK2 ($IC_{50}$ <10 nM) [36]. Researchers at AstraZeneca have identified CHK1 inhibitors from diverse structural classes. There are currently two patent filings from AstraZeneca claiming thiophene carboxamide ureas as CHK1 inhibitors in both the 3-ureido series (e.g., compound 5, $IC_{50}$=124 nM; Figure 14.4) and the isomeric 2-ureido series (e.g., compound 6, $IC_{50}$=2 nM; Figure 14.4) [37, 38]. Structures containing a wide variety of substitutions on the core ring, urea, and amide groups are exemplified, as well as inhibitors containing other five-membered ring heterocyclic cores.

Additional CHK1 inhibitor patent applications from AstraZeneca include triazolones (e.g., compound 7, $IC_{50}$=0.9 µM; Figure 14.5) [39], thienopyridine carboxamides (e.g., compound 8, $IC_{50}$=0.55 µM; Figure 14.5), and other novel five- to six-membered fused system scaffolds, such as thiazolopyridines and indoles (e.g., compound 9; Figure 14.5) [40], although only limited biological data have been disclosed.

## 14.2.3 PF-477736

In 2005, researchers at Pfizer Inc. disclosed a "prototype" diazapinoindolone CHK1 inhibitor, PF-394691 (10, Ki=0.75 nM, $EC_{50}$=40 nM; Figure 14.6) [41, 42]. More recently, preclinical data for CHK1 inhibitor PF-477736 (11, Ki=0.49 nM, $EC_{50}$=45 nM; Figure 14.6) have been presented [43–49], and in early 2007, a Phase 1 clinical trial of PF-477736 in combination with gemcitabine was announced. The in vitro potency demonstrated by both PF-394691 and PF-477736 is similar, suggesting that PF-477736 may have been based

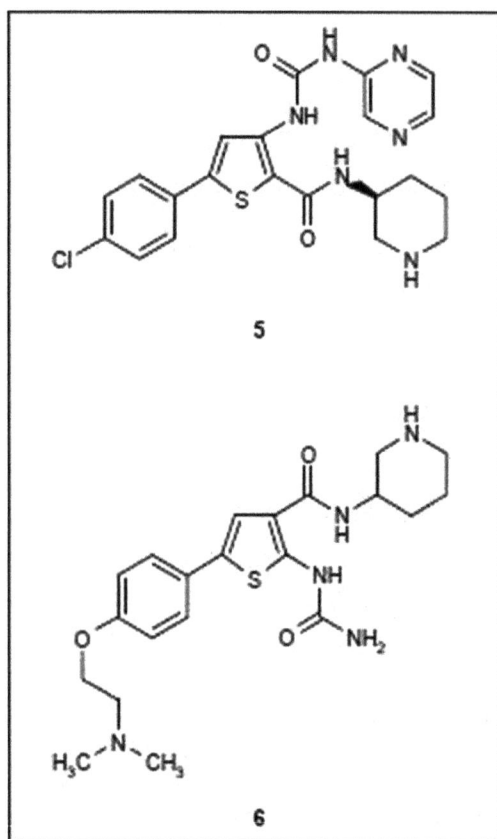

**FIGURE 14.4** The structures of selected thiophene carbox-amide urea CHK1 inhibitors from AstraZeneca.

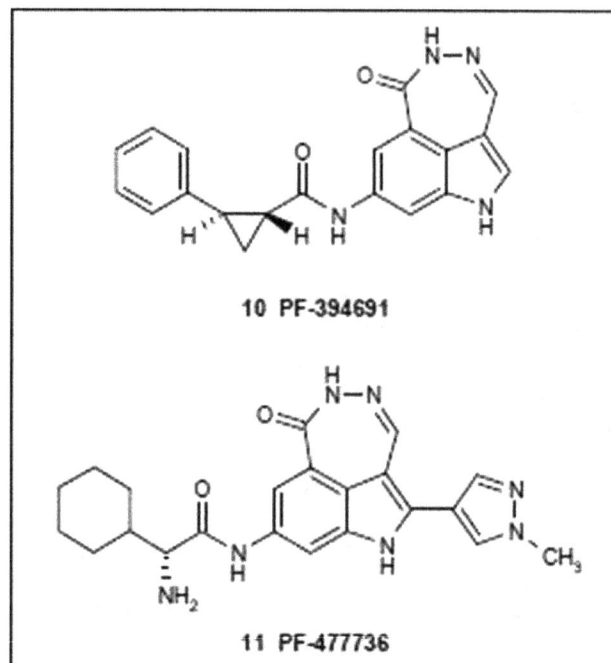

**FIGURE 14.6** The structures of PF-394691 and PF-477736 from Pfizer.

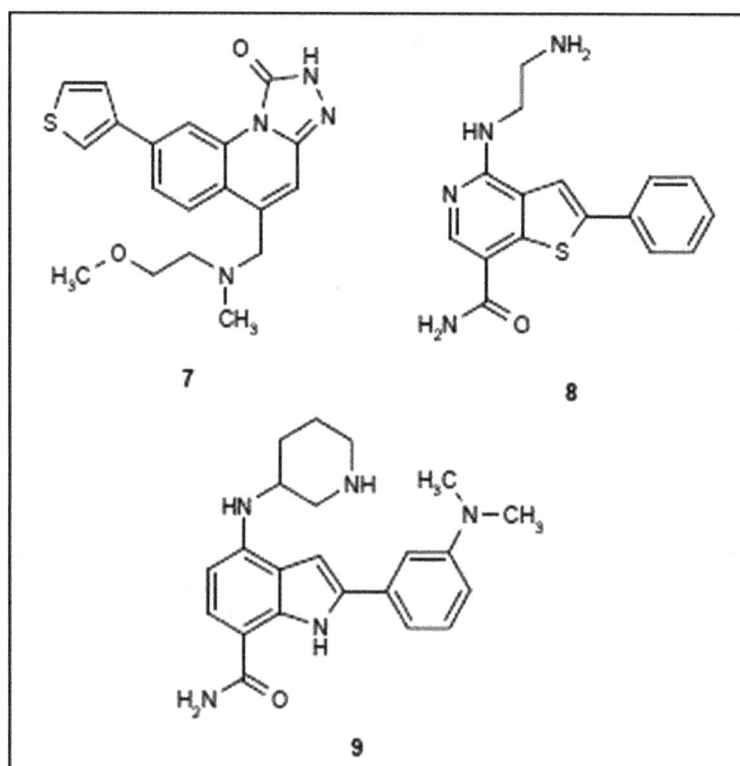

**FIGURE 14.5** The structures of selected triazolone, thienopyridine carboxamide, and indole CHK1 inhibitors from AstraZeneca.

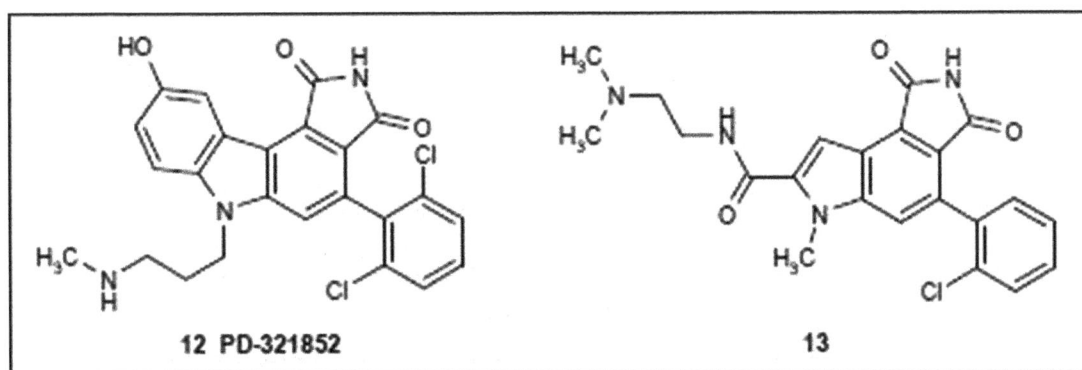

**FIGURE 14.7**    The structures of PD-321852 and an indacene indolocarbazole CHK1 inhibitor from Pfizer.

on optimization of PF-394691 for improved solubility and physical properties.

Unlike the CHK1 inhibitors from Exelixis and AstraZeneca, PF-477736 is 100-fold selective for CHK1 over CHK2. In vitro, PF-477736 potentiated the activity of a number of DNA-damaging agents across several cell lines. The potentiation factor was dependent on the specific cell line and DNA-damaging agent used, with the most chemopotentiation observed with gemcitabine. When dosed intravenously (bolus), PF-477736 provided a consistent half-life of ~3 h in rat, dog, and monkey. PF-477736, dosed in combination with gemcitabine in a Colo205 mouse xenograft model, resulted in a shift in log cell kill (LCK) of 0.75. In a second efficacy study using an HT29 mouse xenograft model and PF-477736 in combination with irinotecan, an LCK shift of 0.92 was observed for the most efficacious regimen. Results from both in vitro and in vivo PF-477736 combination studies have shown a decrease in CHK1 phosphorylation on both Ser317 and Ser345 [50]; this is in sharp contrast to the increase in CHK1 phosphorylation on Ser317 observed with other checkpoint kinase inhibitors. In addition, there was a marked increase in caspase-3 activation and p-H2AX levels. It is interesting to note that similar analogs of the diazapinoindolone scaffold encompassed by PF-477736 are also reported to be inhibitors of poly(ADP-ribose) polymerase (PARP), another enzyme involved in the DNA damage response.

Researchers at Pfizer have also published details of CHK1 inhibitors from an indolocarbazole series, such as PD-321852 (**12**, $IC_{50}$=5 nM; Figure 14.7) [51]. The SAR of this series of compounds as CHK1 and Wee1 kinase inhibitors has been described [52, 53], and in a recent patent publication indacene variants of the indolocarbazoles (e.g., compound **13**; Figure 14.7) as CHK1 and Wee1 kinase inhibitors were disclosed [54].

Researchers at Pfizer have reported two additional series of CHK1 inhibitor compounds since 2005, including benzimidazole carbamates (e.g., compound **14**, Ki <1 μM; Figure 14.8) [55] and pyrazolo catechols (e.g., compound **15**, Ki <1 nM, $EC_{50}$=60 nM; Figure 14.8) [56].

## 14.3    DRUG DESIGN AND SAR OF CHK1 INHIBITORS FROM PRECLINICAL RESEARCH PROGRAMS

In addition to the research programs described above, a number of other pharmaceutical companies have been actively pursuing checkpoint kinase inhibitors.

### 14.3.1    ICOS CORP CHK1 INHIBITORS

ICOS Corp. (a subsidiary of Eli Lilly and Co. Ltd.) was one of the first companies to disclose checkpoint kinase inhibitors, and has continued its interest in bis-aryl urea-based

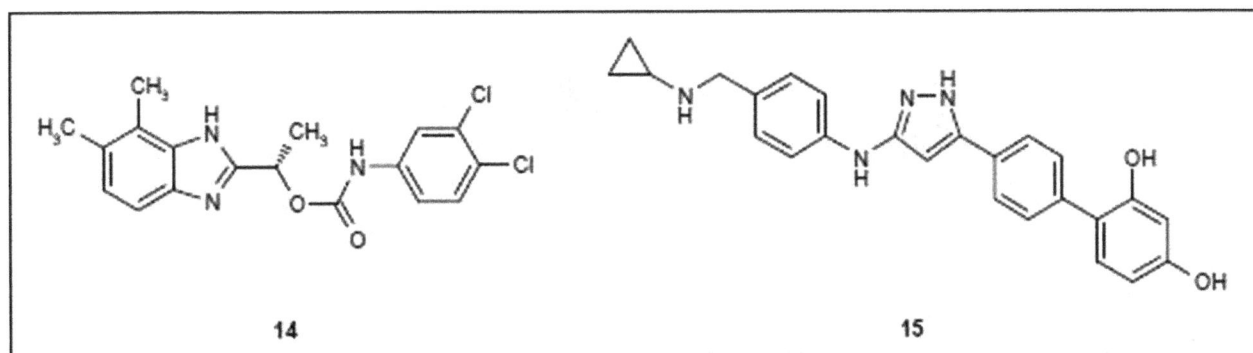

**FIGURE 14.8**    The structures of selected benzimidazole carbamate and pyrazolo catechol CHK1 inhibitors from Pfizer.

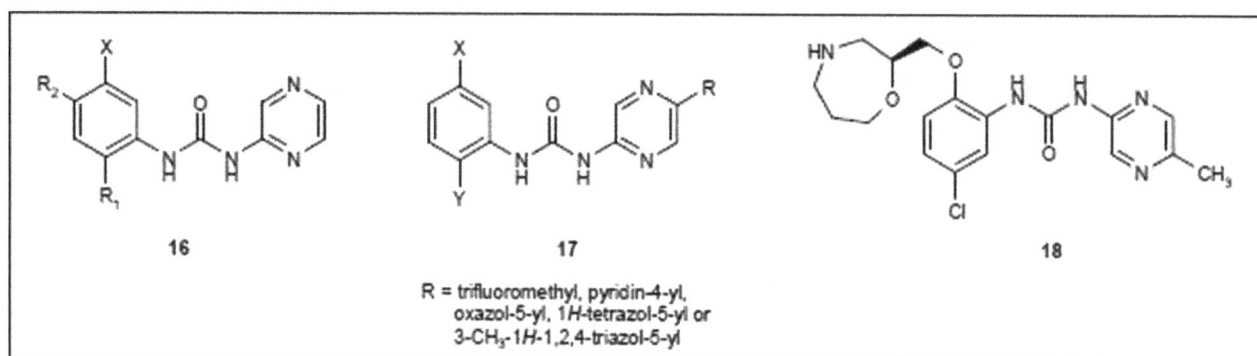

R = trifluoromethyl, pyridin-4-yl,
oxazol-5-yl, 1H-tetrazol-5-yl or
3-CH₃-1H-1,2,4-triazol-5-yl

**FIGURE 14.9**   The structures of selected pyrazinyl urea checkpoint kinase inhibitors from ICOS.

CHK1 inhibitors, with a number of patent applications published since 2005 [57–60]. Initial SAR studies from ICOS showed a clear preference for pyrazine as one of the aryl groups, and that increased potency and solubility could be achieved with relatively simple substituents [61, 62]. More recent patent applications continue to build upon this initial research (e.g., compounds **16** to **18**; Figure 14.9); however, some of the examples provided, while still maintaining the pyrazinyl urea core, include a more diverse set of 5-position substituents, as exemplified by general structure **17** and compound **18**. Although no detailed biological data were included in the applications, the compounds were stated to inhibit isolated human CHK1 and to have activity in a cell-based abrogation assay, as well as leading to increased cell death associated with DNA damage.

### 14.3.2   MILLENNIUM PHARMACEUTICALS INC. CHK1 INHIBITORS

Initial research from Millennium Pharmaceuticals Inc. also described bis-aryl ureas as potent inhibitors of CHK1 [63]. For these compounds, one of the aryl groups was either a substituted pyrazinyl moiety or a substituted pyridyl moiety. This area of research has continued to be of interest to Millennium Pharmaceuticals, as demonstrated by the publication of a further application on related compounds, such as those of general structure **19** or compound **20** (both Figure 14.10) [64].

Since 2005, patent applications from Millennium Pharmaceuticals have also claimed checkpoint kinase inhibitors based on different structural scaffolds, namely 2,5-dihydropyrazolo[4,3-c]quinolin-4-ones (e.g., compound **21**; Figure 14.11) [65(a)] and N-substituted 2-amino-5H,7H-benzo[b]pyrimido[4,5,d]azepin-6-ones (e.g., compound **22**; Figure 14. 11) [66, 67]. The latter series of compounds were claimed not only as CHK1 inhibitors but also as inhibitors of CHK2, Aurora, and PLK, with the potency and selectivity profile depending upon the specific example.

The synthesis of **21** is shown in Figure 14.12 [65(b)]. The route began with the preparation of 6-bromoindazole by the method of Bartsch and Yang [65(c)]. Iodination followed by a regioselective Suzuki cross-coupling reaction with the depicted indole boronic acid proceeded smoothly to form the 3-(indol-2-yl)indazole core structure. Palladium-catalyzed coupling of the product bromide with propargyl alcohol in pyrrolidine [65(d)] and subsequent oxidation to the aldehyde set the stage for the construction of the triazole by way of a 3+2 cyclization with sodium azide [65(e)]. Sodium borohydride reduction of the resulting aldehyde then provided **21**.

Insight into the superior binding affinity of **21** was gained through analysis of an inhibitor-bound X-ray structure (Figure 14.13) [65(f)]. The structure revealed that the hydroxyl group of the triazole had displaced one of three conserved water molecules that make up a key hydrogen bonding network in HI. While desolvation of the hydroxyl

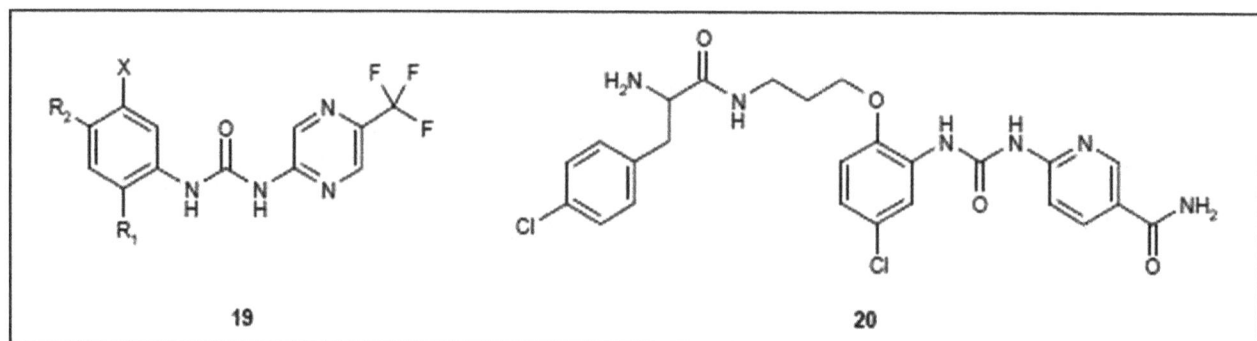

**FIGURE 14.10**   The structures of selected bis-aryl urea CHK1 inhibitors from Millennium Pharmaceuticals.

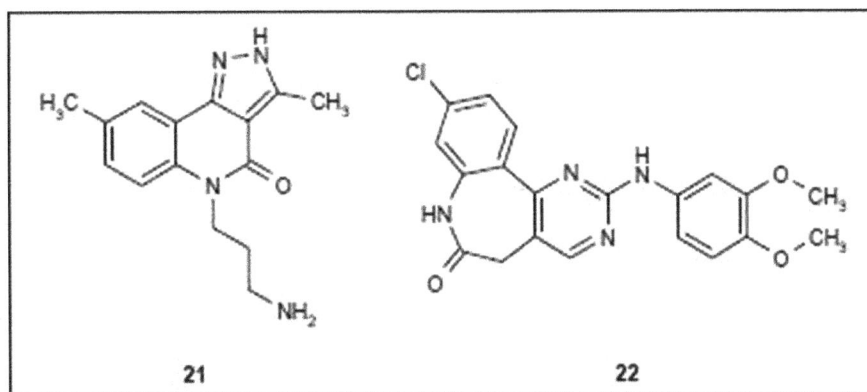

**FIGURE 14.11** The structures of selected quinolinone and azepinone CHK1 inhibitors from Millennium Pharmaceuticals.

group may be an unfavorable thermodynamic event, the entropic gain of resolution of the ordered water molecule released from HI compensates for it [65(g)]. Moreover, the X-ray structure of **21** showed that the hydroxyl group made an additional hydrogen bond to the protein backbone at Asp148. Taken together, we concluded that the gain in binding affinity with **21** was largely due to the "anchoring" of the hydroxyl group within the depicted hydrogen bonding array.

Millennium Pharmaceuticals has not reported comprehensive preclinical profiles of these newer inhibitors, although there are some details provided in the patent applications where a number of isolated enzyme assays are described, and an indication of the potency range is given for various examples. Results from an in vitro antiproliferation assay demonstrated an increased level of cell death when several DNA-damaging agents were combined with CHK1 inhibitors.

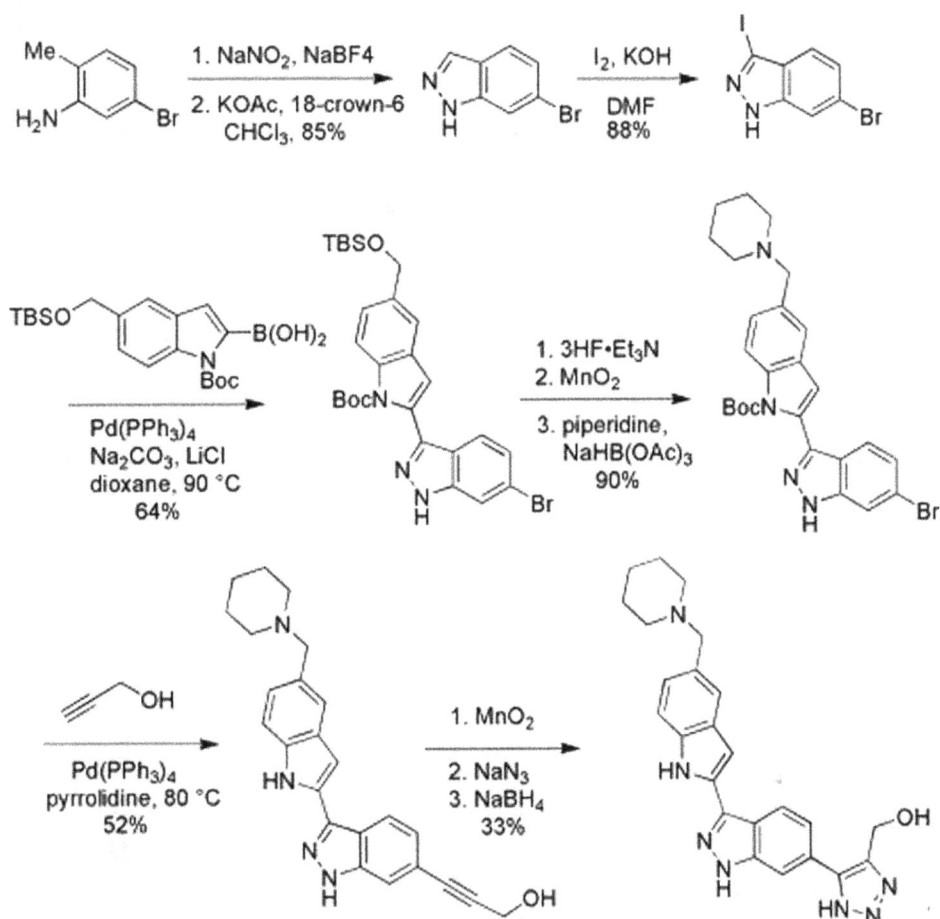

**FIGURE 14.12** Synthesis of **21**.

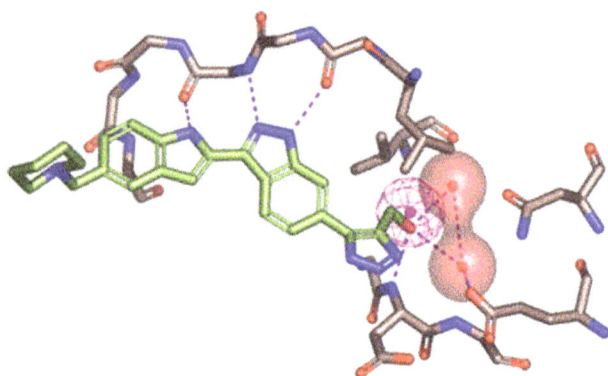

FIGURE 14.13 X-ray crystallographic structure of **21** (*green*) bound to Chek1 (*light brown*) with water molecules shown (*red spheres and surface*). Water molecule displaced by **21**, but present in X-ray structure of **1** shown as a mesh sphere (*magenta*). (Ref. [65])

### 14.3.3 ABBOTT LABORATORIES CHK1 INHIBITORS

There have been a number of reports originating from Abbott Laboratories since mid-2005, describing the medicinal chemistry and preclinical biology of CHK1 inhibitors based on four structurally diverse scaffolds. Similar to ICOS and Millennium Pharmaceuticals, researchers at Abbott have been actively pursuing pyrazine urea inhibitors. In particular, the Abbott compounds utilize a cyano-substituted pyrazine ring. Two compounds, A-641397 (**23**, $IC_{50}=8.3$ nM; Figure 14.14) and A-690002 (**24**, $IC_{50}=6.5$

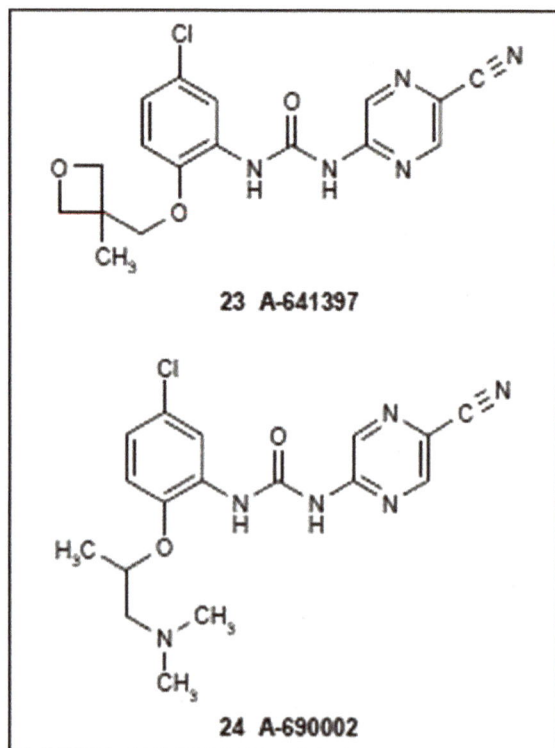

**23  A-641397**

**24  A-690002**

FIGURE 14.14 The structures of A-641397 and A-690002.

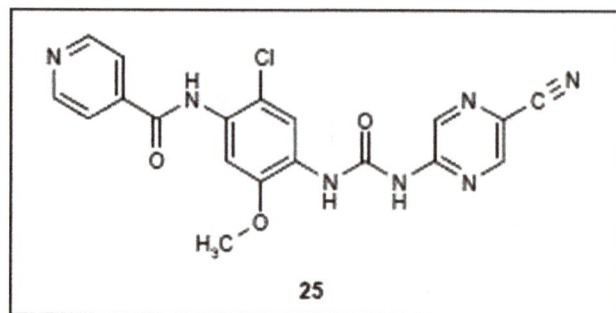

**25**

FIGURE 14.15 The structure of a selected pyrazinylurea CHK1 inhibitor from Abbott Laboratories.

nM; Figure 14.14) have been reported [68•]. Studies of the antiproliferative activity of A-641397 in combination with doxorubicin and A-641397 in combination with camptothecin, display potentiation factors of >22 and 14–20, respectively. An increase in caspase activation and abrogation of cell cycle arrest were demonstrated, as well as an anticipated decrease in phosphorylated cyclin-dependent kinase 2 (CDK2) and cell division cycle 2 (CDC2). In the same assays, A-690002 in combination with doxorubicin and camptothecin was approximately 10-fold more potent than A-641397 [68•].

Publications have described the SAR of this series of compounds with various substituents at the 4-position and to a lesser extent the 2-position of the phenyl ring [69–71].

A wide range of substitutions and functional groups were tolerated at these positions, with only small changes in CHK1 enzyme potency. This may, in part, be due to the fact that these positions are orientated toward solvent and thus could have a minimal effect on binding. Several analogs from this series were screened in an MTS and FACS assay with doxorubicin; one analog showed a potentiation factor of >136 (compound **25**, $IC_{50}=8$ nM; Figure 14.15).

In two 2005 patent applications [72, 73] and other reports from researchers at Abbott [74, 75], cyclic ether CHK1 inhibitors that encompass the biaryl pyrazine urea framework were described. The optimal cyclic ether size was stated to be a 15-membered ring consisting of five methylene units. Keeping the 5-position chlorine substituent constant, the authors described the SARs for various O-, N- and C-linked 4-position substituents. Although all three of these substitutions resulted in potent inhibitors of CHK1, the C-linked compound **26** (Figure 14.16) demonstrated a Ki value of 4.3 nM against CHK1 and excellent selectivity against a panel of 70 kinases. Submicromolar potency was only observed for five other kinases (IKKβ, PDK1, EphA2, FGFR1, and FGFR3). Compound **26** (MTS $EC_{50}=0.86$ μM; FACS $EC_{50}=0.19$ μM) in combination with doxorubicin provided potentiation factors of 33 and >53 in the MTS and FACS cellular assays, respectively. Similarly to the acyclic ureas, the pyrazine nitrogen moiety forms a water-mediated hydrogen bond with the side chain of Asn59, while the nitrile group forms a second hydrogen bond with the Lys38 side chain.

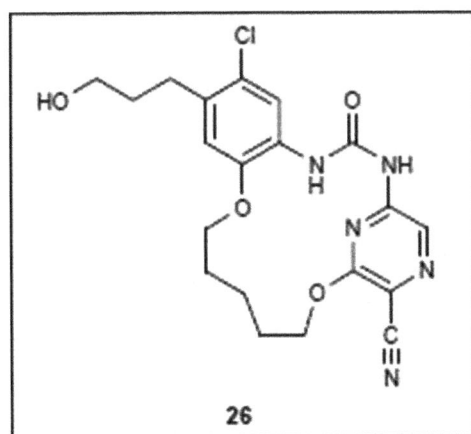

**FIGURE 14.16**  The structure of a selected cyclic ether CHK1 inhibitor from Abbott.

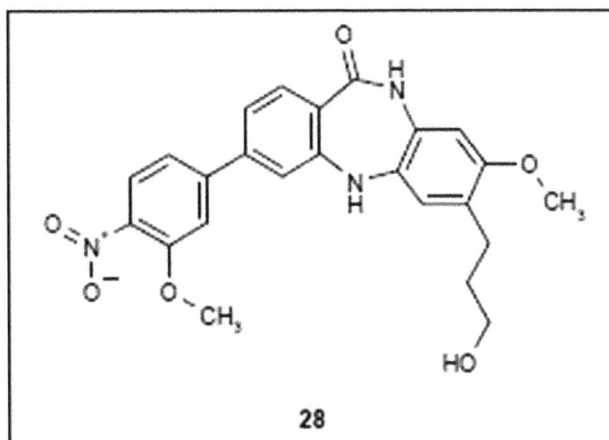

**FIGURE 14.18**  The structure of a selected dibenzodiazepinone CHK1 inhibitor from Abbott.

In separate publications, researchers from Abbott reported on the biological evaluation of indolinone CHK1 inhibitors [76, 77]. The most potent analog (compound **27**; Figure 14.17) had an $IC_{50}$ value of 4 nM, and because of its reported kinase selectivity profile exhibited antiproliferative activity as a single agent ($EC_{50}$=10 nM). The X-ray crystal structure of one of these analogs shows the phenol substituent making an interaction with Glu[55] near the ribose pocket.

Researchers from Abbott recently presented data relating to the SAR of their dibenzodiazepinone series of CHK1 inhibitors (e.g., compound **28**, $IC_{50}$=2.6 nM; Figure 14.18) [78–80]. In combination with doxorubicin, compound **28** (MTS $EC_{50}$=1.2 μM, FACS $EC_{50}$=0.45 μM) showed potentiation factors of >50 and >22 in the MTS and FACS cellular assays, respectively.

The most recent series of compounds identified by researchers at Abbott are tricyclic pyrazoles (e.g., compound **29**; Figure 14.19). The authors demonstrated a clear SAR, indicating that acidic substituents placed on an aryl ring of the core scaffold led to improved potency [81, 82]. X-ray

crystal structures showed water-mediated hydrogen bonding to Glu[55] and Asn[59]. A superior replacement for the acid was found to be a phenolic ring, which can be rationalized by displacement of water by the aryl ring while retaining the key hydrogen bonding to Glu[55] and Asn[59]. This CHK1-binding mode is reminiscent of the Pfizer pyrazolo catechol inhibitors described above. The SAR clearly indicates that this interaction is important for potent CHK1 inhibition in both series. The most promising analog in the Abbott pyrazole series, compound **29**, demonstrated a CHK1 $IC_{50}$ value of 6.2 nM, along with potentiation in vitro. In this series, changing the amide to an amine yielded increased CHK1 activity, but also resulted in the introduction of single-agent cellular activity, probably indicating a switch in kinase selectivity profiles.

### 14.3.4  Chiron Corp. CHK1 Inhibitors

There have been many reports published by researchers at Chiron Corp. covering two preclinical candidate CHK1 inhibitors, CHIR-124 (**30**; Figure 14.20) and CHIR-600 (**31**; Figure 14.20) [83–85]. Medicinal chemistry and biological evaluation of the benzimidazole quinolinone series, exemplified by CHIR-124, have recently been presented in two papers [86, 87]. A basic amine group, such as a quinuclidine, was determined to be important for potent CHK1 inhibition (e.g., CHIR-124, $IC_{50}$=0.32 nM), and the orientation and stereochemistry of the amine were critical. For example, changing from S-quinuclidine to R-quinuclidine resulted in a >60-fold decrease in potency. Furthermore, it was found that, while switching from S-quinuclidine to pyrrolidine or piperidine did not affect potency, moving the ring nitrogen by one atom resulted in a three-fold decrease in CHK1 inhibition. The addition of a chlorine or methyl group at the 6-position resulted in a 20-fold increase in potency. This series of inhibitors binds to the CHK1 hinge region with three hydrogen bonds through the benzimidazole nitrogen and the quinolinone amide, and the 6-position substituent lies in a small hydrophobic pocket. Rationalization for the stereochemical preference of the amine can be seen from

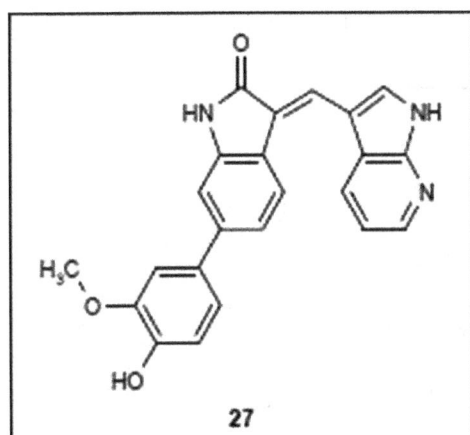

**FIGURE 14.17**  The structure of a selected indolinone CHK1 inhibitor from Abbott.

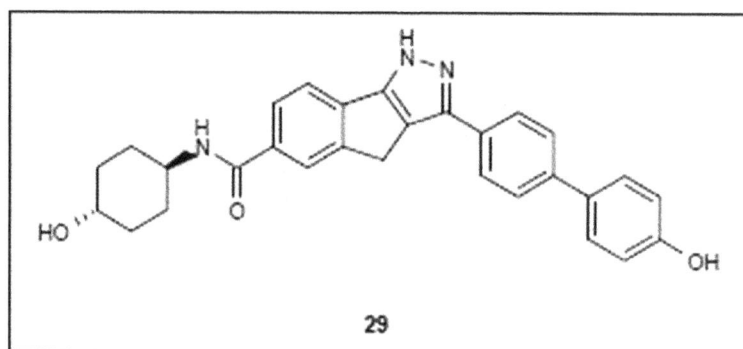

**FIGURE 14.19**    The structure of a selected tricyclic pyrazole CHK1 inhibitor from Abbott.

the X-ray crystal structure, which shows a tridentate hydrogen bond from the amine nitrogen to Glu[134] and Glu[91] near the ribose sugar pocket. This series has excellent potency for CHK1, but also inhibits several other kinases with submicromolar potency (e.g., PDGFR $IC_{50}$=6.6 nM and Flt3 $IC_{50}$=5.8 nM). Nonetheless, the fold selectivities observed are good relative to CHK1 and synergistic effects of CHIR-124 with various topoisomerase poisons observed both in vitro and in vivo. The original patent application for the benzimidazole series claimed the inhibition of many other Ser/Thr and Tyr kinases in addition to CHK1 [88], but in late 2005, a further patent application from Chiron claimed examples from this series specifically as CHK1 inhibitors [89].

### 14.3.5  MERCK AND CO. INC. CHK1 INHIBITORS

The most recent company to publish in the field of checkpoint kinases is Merck & Co. Inc. In addition to six patent applications since July 2006 [90–95], two medicinal chemistry papers have also been published describing indolylquinolinone [96, 97] and indolylindazole derivatives [98, 99] as CHK1 inhibitors. Using high-throughput screening, researchers at Merck identified indazole and quinolinone leads (e.g., compounds **32** [$IC_{50}$=12 nM] and **33** [$IC_{50}$=144 nM]; Figure 14.21). The latter series has a similar structure to the Chiron quinolinone series, except that it has

an indole in place of the benzimidazole group, and binds to the CHK1 active site in a similar manner. Utilizing structure-based design, both of the Merck lead compounds were optimized for potency via C6-substitution. After exploration of a variety of heterocycles and functional groups, the addition of a pyrazole group was demonstrated to be optimal in the quinolinone series (e.g., compound **34**, $IC_{50}$=0.65 nM; Figure 14.21). The potency improvement derives from complex interactions with Glu[55], Asp[148], and conserved water molecules in the hydrophobic region near the ATP sugar pocket. Although potent at inhibiting the CHK1 enzyme, cellular potency in a CHK1 abrogation assay was dramatically reduced in this series, which was attributed to a lack of cellular permeability. The authors improved cellular activity by removing the dibasic amine of compound **33**, resulting in compound **34**, which had an $EC_{50}$ value of 97 nM, but this is still a >140-fold differential between enzyme and cell $EC_{50}$ values.

There are currently no patent filings covering this particular quinolinone series, although four recent patent publications from Merck have claimed a quinolinone core structure fused to one or more additional five- or six-membered rings (e.g., compounds **35** to **38**; Figure 14.22), although no biological activity has been disclosed [91–94].

Similar to the quinolinone series, the indazole series (e.g., compound **32**) suffers from poor enzyme-to-cell correlation [98, 99]. Along with excellent CHK1 enzyme

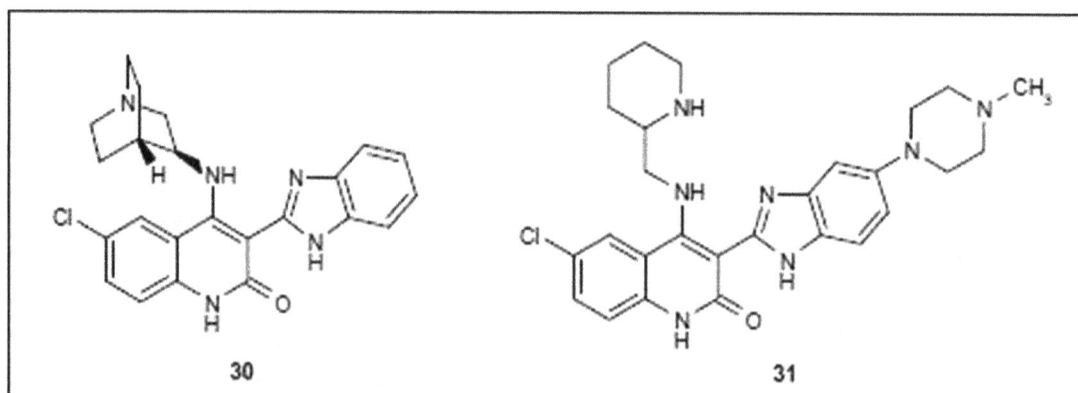

**FIGURE 14.20**    The structures of CHIR-124 and CHIR-600 from Chiron.

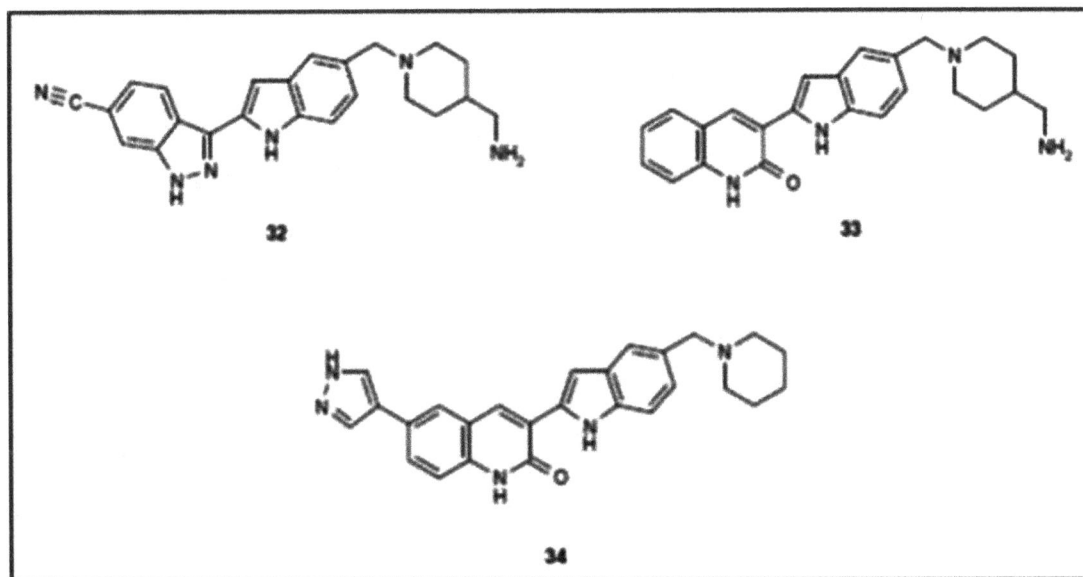

**FIGURE 14.21** The structures of selected indazole and quinolinone CHK1 inhibitors from Merck.

potency, this series demonstrated good potency toward CDK7, which could mask any cell cycle abrogation resulting from CHK1 inhibition. Researchers from Merck describe an optimization of this series by "designing-out" CDK7 binding utilizing structure-based design. The original lead compound showed a lack of cell activity in both abrogation and CHK1 autophosphorylation assays. Substitution at the C6-position of the indazole was explored in an attempt to reduce CDK7 activity. This position is directed toward the gatekeeper residue, which is much larger in CDK7 (Phe) than in CHK1 (Leu). Results indicated that larger substituents reduce CDK7 activity, while retaining CHK1 potency. More importantly, these modifications led to cellular activity suggesting that CDK7 inhibition did indeed play a role in masking the cellular effects of CHK1 inhibition, as seen with the early examples in the series. However, there are some inconsistencies in the results, which suggest the interplay of other factors, such as cellular permeability, being partly responsible for the lack of cellular CHK1 activity. For example, the most potent compound (**39**; Figure 14.23) had a 600-fold reduction in potency moving from enzyme to cell,

but still showed 3000-fold selectivity for CHK1 over CDK7, while the most promising compound (**40**; Figure 14.23) had a better $IC_{50}/EC_{50}$ ratio (370-fold) but decreased selectivity against CDK7. Interestingly, compound **40** exploits a phenol interaction akin to the corresponding Merck quinolinone series and also to the Abbott pyrazole inhibitors, as described above.

The final series of CHK1 inhibitors claimed by Merck are aminothiazoles [95]. There are many examples included in the patent application, but no biological data were reported. The majority of the examples provided contain 1,3-diamino pyridine as the amino substituent and an additional pyridine carboxamide substituent attached to the thiazole (e.g., compound **41**; Figure 14.24). Although there are many CHK1 binding modes possible, it appears that the 2-pyridylaminothiazole may be the ATP site hinge binder.

### 14.3.6 Vernalis plc CHK1 Inhibitors

Researchers at Vernalis plc have published two structure-based papers describing the identification of chemically

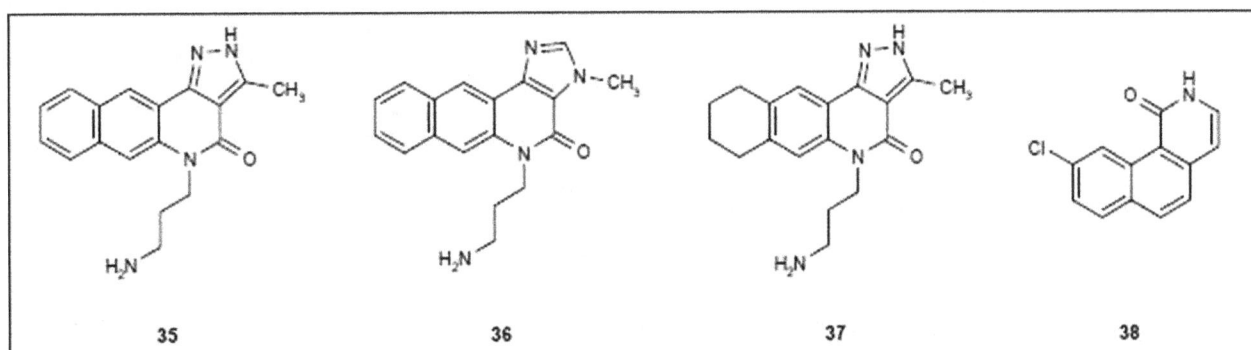

**FIGURE 14.22** The structures of selected quinolinone CHK1 inhibitors from Merck.

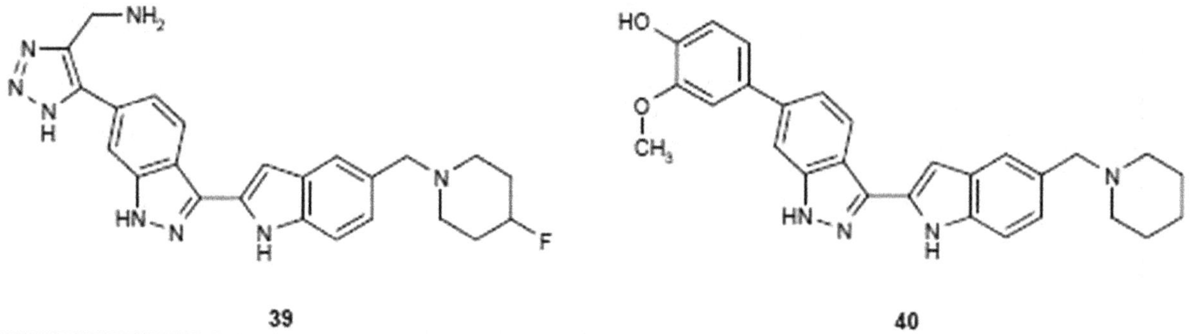

**FIGURE 14.23** The structures of selected indolyl indazole CHK1 inhibitors from Merck.

diverse CHK1 inhibitors [100, 101]. Although these papers provide insight into crystallographic binding motifs to CHK1, the most potent compound (**42**; Figure 14.25) was of limited potency ($IC_{50}$=1.4 μM).

Another paper described the discovery of a "buried" pocket in the CHK1 ATP binding site, as exemplified with indazole **43** (Figures 14.24 and 14.25) [102•]. Interestingly, this pocket is the same as the one described by researchers at Abbott and Merck (discussed earlier). It appears that these three research groups independently discovered this buried hydrophobic pocket adjacent to the sugar-binding pocket for ATP, and it is the region where CHK1 inhibitors can exploit interactions with Glu[55], Asn[59], and Asp[148] (Figure 14.26).

More recently, Vernalis researchers have described pyrazole benzimidazole amides as CHK1 and PDK1 inhibitors [103]. A majority of the examples provided have a basic amine substituent similar to that of other CHK1 inhibitors described above. Biological activities are disclosed for compounds in this series against CHK1, PDK1, AKT1, PKA, and CDK2. Compound **44** (Figure 14.27) had $IC_{50}$ values of 2 and 73 nM against CHK1 and PDK, respectively, but >10 μM for the other kinases. Many of the examples described are equipotent for CHK1 and PDK1. Improvements in selectivity for CHK1 were realized from various substitutions on the benzimidazole, and it appears that the basic amine of the piperidine could be making an interaction with the Asp[148] adjacent to the ATP ribose-binding pocket, similarly to other CHK1 inhibitors described above.

### 14.3.7 OTHER CHK1 INHIBITORS

Scientists at BioFocus DPI have reported a series of indazolylpyrimidines as CHK1 inhibitors, but only three of the examples disclosed showed an $IC_{50}$ value of <1 μM; for example, compound **45** (Figure 14.28), which had a >50-fold selectivity for CHK1 over CDK1 ($IC_{50}$=37.2 μM), giving an $IC_{50}$ value for CHK1 of <0.74 μM [104].

Merck KGaA researchers have claimed diamino squaric acid derivatives (e.g., compounds **46** and **47**; Figure 14.29) as CHK1, CHK2, and SGK inhibitors [105–108], representing another distinct class of kinase inhibitors.

In addition to pharmaceutical company interest, various research groups in academia have been involved in checkpoint kinase inhibitor design and synthesis [109–122]. One of the research groups most active in the area is the Prudhomme group, with many publications relating to CHK1 inhibitors [23, 24, 109–120]. Its major focus of research is the design of more potent and selective indolocarbazole CHK1 inhibitors, related to UCN-01.

## 14.4 PRECLINICAL OVERVIEW OF SUGGESTED siRNA DELIVERY SYSTEMS AND PROS/CONCERN OF DNA-BASED GENE DELIVERY CARRIERS

RNAi therapeutics have unique advantages over conventional pharmaceutical drugs. RNA interference is an endogenous gene regulation process, thus almost all genes can be modulated by siRNAs. The identification and

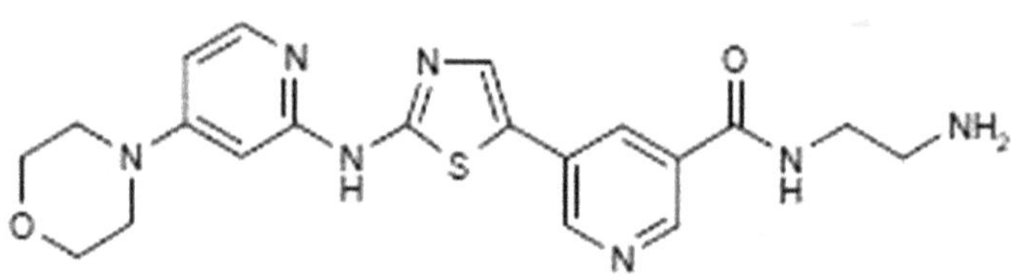

**FIGURE 14.24** The structures of selected aminothiazole CHK1 inhibitors from Merck.

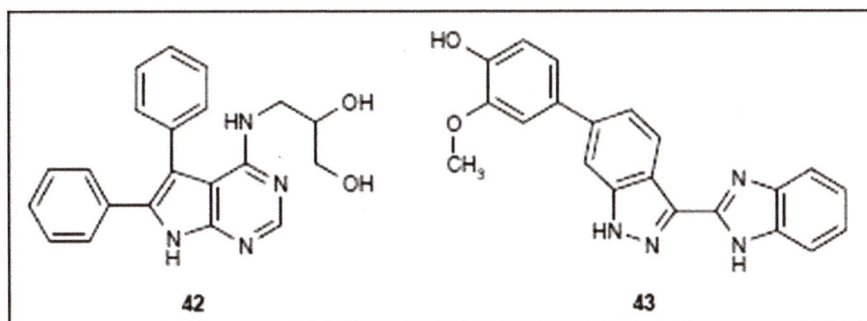

**FIGURE 14.25** The structures of selected CHK1 inhibitors from Vernalis.

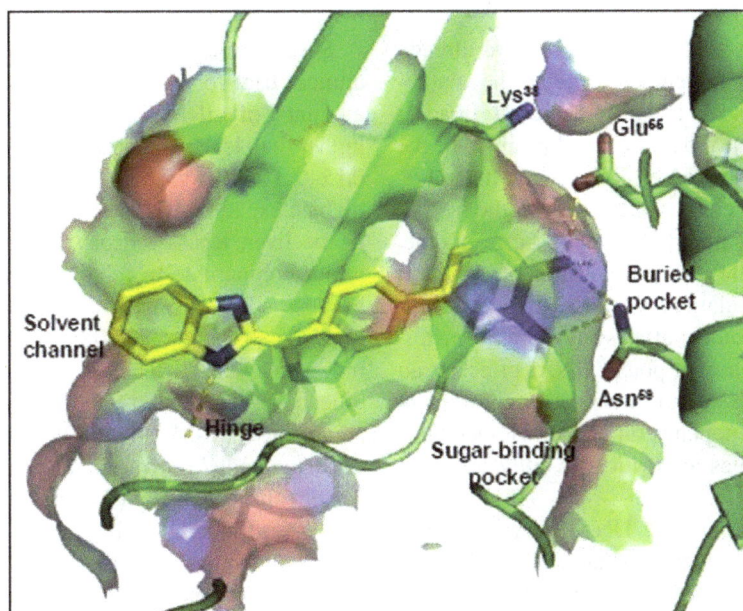

**FIGURE 14.26** Illustration of the buried pocket identified by researchers at Vernalis. The figure shows compound **43** bound to CHK1. Interactions of the inhibitor with the ATP hinge region and the buried pocket encompassing Glu[55] and Asn[59] are highlighted (PDB ID: 2C3K).

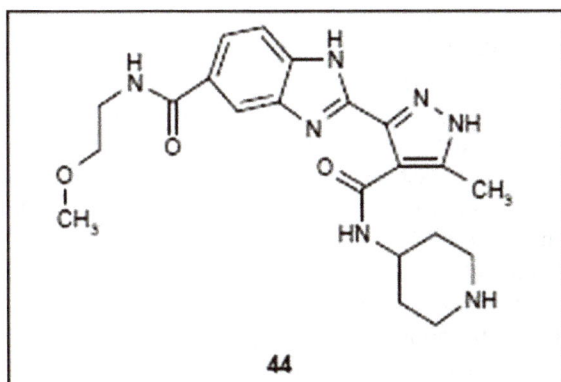

**FIGURE 14.27** The structure of a selected pyrazolo benzimidazole amide CHK1 inhibitor from Vernalis.

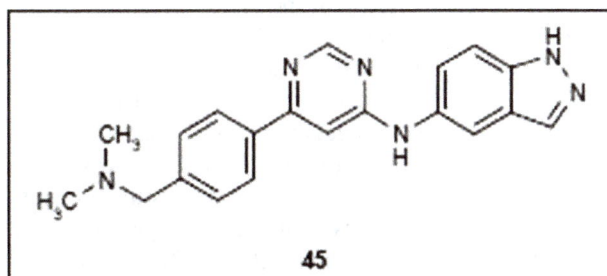

**FIGURE 14.28** The structure of a selected indazolylpyrimidine CHK1 inhibitor from BioFocus.

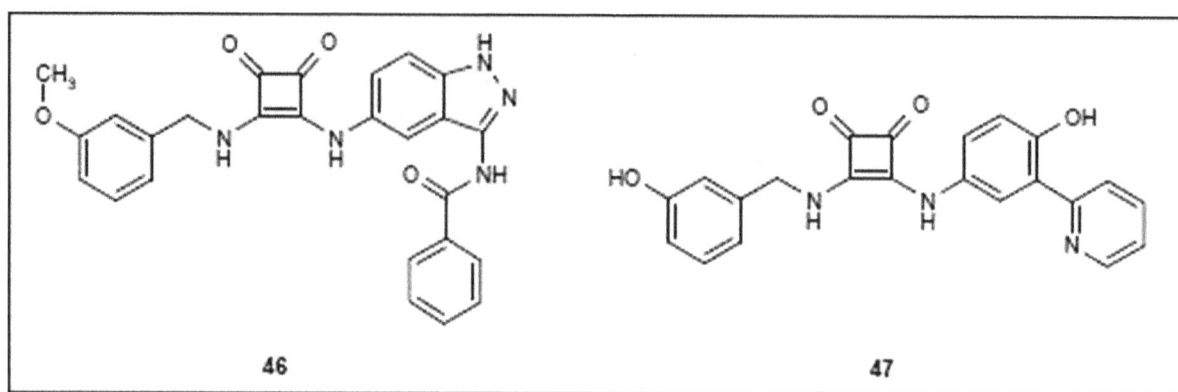

**FIGURE 14.29** The structure of selected diamino squaric acid CHK1, CHK2, and SGK inhibitors from Merck KGaA.

selection of highly potent siRNA sequences have already been accomplished for many gene targets, and the synthesis of siRNAs on a large scale has been achieved. In addition, RNAi therapeutics have demonstrated promise in the treatment of cancers, viral diseases, and genetic disorders. Although significant progress has been made in the field of siRNA delivery, there remain challenges to be overcome. These challenges include: (1) the minimization of off-target effects and immune stimulation, (2) target-specific accumulation of RNAi therapeutics after systemic administration, and (3) the induction of a potent RNAi effect at an acceptable dose level. The key to therapeutic achievement using an RNAi approach depends on delivery issues, thus advanced delivery strategies are critical to fully optimize the power of siRNAs.

Engineered design of synthetic DNA/RNA molecules can generate predefined structures that can easily self-assemble to form nanoparticles with multiple functionalities. The field of oligonucleotide-based nanotechnology for biomedical applications is just emerging, but will play an important role in the delivery of siRNA. In particular, oligonucleotide-based structural RNAi systems described in this chapter are promising as a new generation of gene delivery carriers for cancer therapy. To realize clinical application of structural RNAi systems, the potency of the delivery systems needs to be optimized. One of the solutions may be the incorporation of highly specific ligands within the system. Preclinical data from various biopharmaceutical companies have suggested that the delivery of ligand-conjugated siRNA can be highly improved by the utilization of the engineered design of structural RNAi systems [123]. Another considerable issue in the delivery of structural RNAi systems is the facilitated endosomal release of these materials. It is important to understand the endosomal escape mechanism of structural RNAi systems and endeavor to use the endolytic properties to accelerate the transfer of active siRNAs into the cytoplasm. Future prospects of multimerized/branched siRNA structures and oligonucleotide-based structural RNAi systems with defined size and functionality will continue to improve the precision and efficacy of siRNA delivery.

Several multimolecular delivery vehicles are under clinical trial for RNAi-based cancer therapy but the dose amounts of siRNA (0.1–1.5 mg/kg) are comparatively higher than levels observed in diseases of other organs (e.g., 0.15–0.3 mg/kg in the liver). This may indicate that the highest expected RNAi efficacy in tumor is similar to that in the liver. RNAi efficacy in rapid-growing cancer cells is not comparable to relatively slow-growing hepatocytes because siRNA concentration in cytoplasm will dilute in divided cells. But more efficient delivery vehicles for tumor may contribute to increase RNAi rather than the present efficacies. Compared to a clinically approved trastuzumab emtansine (half-life 1–4 days) and clinically tested anticancer drug-loaded polymeric micelles (half-life 16–80 h), the current clinically tested vehicles showed shorter circulation properties (half-life <2 h) (Table 14.1).

This indicated that the current delivery vehicle needed better performance. The clinical trial results and new biological evidences provide the clues for development of the next vehicle design. (i) The vehicle should exhibit long blood circulation properties (half-life ≥2 h). The higher amounts of circulating delivery vehicle (containing siRNA) will increase the possibility that the vehicle diffuses/accumulates into tumor microenvironment. Some vehicles introduced in this book showed half-life longer than 2 h, but their doses for tumor growth inhibition in animal model were not significantly lower than other vehicles. These results indicate that other aspects in vehicle design should be considered. (ii) The vehicle should be <30 nm diameter size to enhance diffusion/accumulation in tumor because the nanoparticle with this diameter size penetrated in thick fibrotic stroma and hypovascular tumor in animal models. This book indicates that fewer numbers examples of Doxil (diameter size 90 nm) in clinical tumor accumulation may be hampered by this size limitation. Eventually, the behavior of the delivery vehicle inside tumor is governed by diffusion, implicating that smaller particles with <30 nm size are also preferred to reach cancer cells. Fabrication of these small nanoparticles has gradually been realized by various materials and techniques, e.g., unimer polyion complex/gold nanoparticles and polymers [139–142]. Repeatedly, we

**TABLE 14.1**

**Summary of Blood Circulation and Size of Delivery Vehicles in this Section**

| Delivery Formulation | Half-Life in Mouse or Patient | Hydrodynamic Diameter (nm) |
|---|---|---|
| Naked siRNA [124, 125] | 3–10 min | 7 (length) × 2 (diameter) |
| ALN-VSP02 [126] | ≤2 h in patient | 80–100 |
| Atu027 [127, 128] | ≤2 h in patient | 120 |
| CALAA-01 [129, 130] | ≤30 min in patient | 60–150 |
| Hydrophobic interaction [131] | 10 min | 140 |
| Hydrophobic interaction [132] | 18 h | 100 |
| Hydrophobic interaction [133] | 3–4 h | 130 |
| Redox potential responsiveness [134] | 20 min | 40 |
| Extracellular pH responsiveness [135] | 5 h | 150 |
| MMP responsiveness [136] | ≤1 h | 80 |
| High quantity of siRNA [137] | 4 min in the first phase, 27 h in the second phase | 300 |
| Gold nanoparticle [138] | 1 min in the first phase, 8.5 h in the second phase | 31–34 |
| Gold nanoparticle [139] | 30 min | 40 |

emphasize that size distribution of vehicle in buffer or fetal bovine serum does not guarantee the same size distribution in bloodstream. (iii) Other functionalities (e.g., selective release of siRNA, high cell-specific recognition, and high endosome escapability) must endow the delivery vehicle, which simultaneously satisfies both (i) and (ii). To date, it is not clear which functionality is the most critical factor to enhance RNAi in patients. Furthermore, the delivery vehicle that satisfies (i) and (ii) but not (iii) does not expect to exhibit superior RNAi than the current vehicles in clinical trials. Ultimately, simpler formulation of delivery vehicles can be more easily translated to their clinical use because of better quality control as well as lower possibility of unexpected adverse effects. The success of RNAi-based cancer therapy is closely associated with tumor biology as well as architecture of delivery vehicles. Tumor cell plasticity evokes a resistance mechanism against clinical treatments, and cancer stem cells are gradually being identified as the root of cancer recurrence. New target RNA genes should be discovered to increase apoptosis in cancer cells and simultaneously reduce side effects in normal and healthy cells. Multidisciplinary research studies will guide the development of highly effective and safer RNAi-based drugs in clinical trials.

## 14.5 CONCLUSION AND FUTURE PERSPECTIVES

The high incidence of cancers has triggered a pressing need to develop new cancer therapeutics. MiRNAs are associated with cell proliferation, metastasis, tumor progression, invasion, sustained angiogenesis, apoptosis, and drug- and radio-resistance. In the last 10 years there has been an intense focus on miRNA research including miRNA targets, their impact on the up- or downregulation of various genes, and on pathological processes. For many years, studies of siRNA have progressively advanced toward novel treatment strategies against cancer. Recently the clinical success rate of RNAi therapeutics is increasing and now some are on the right track to gain FDA approval in the next year or two. Small interfering RNA (siRNA) has gained attention as a potential therapeutic reagent due to its ability to inhibit specific genes in many genetic diseases.

Cancer is a multifactorial and epigenetic disease. MiRNAs, frequently deregulated in cancer, can be controlled by epigenetic alterations but can also function as epigenetic players, suggesting that epigenetic mechanisms and miRNAs can interact on a bidirectional level. Interestingly, epigenetic changes can be reversed by certain drugs, and because there is a tight link between epigenetics and miRNAs, it is conceivable to assess the therapeutic targeting of epigenetic miRNA regulation mechanisms in cancer. The studies described here provide an update regarding miRNA and epigenetics in cancer.

According to the functions of piRNAs such as transcriptional and posttranscriptional regulatory, which are certainly not restricted to silencing transposable elements, they seem to present great potential for future interventions in the course of diseases, including cancer, and also provide new insights into cancer epigenetics.

Furthermore, many miRNAs are remarkably changed in almost all human cancers [143]. Therefore, miRNAs have been used as biomarkers and targets for treatment. For example, tumor suppressive miRNAs, including miR-21, miR-34, let-7, miR-145, miR-122, and miR-10b, have been widely investigated for their therapeutic potential. However, the delivery of miRNAs remains a critical issue for effective clinical cancer treatment. It is worth mentioning that the role of miRNAs in the modulation of MDR transporters renders them a useful avenue for overcoming resistance to chemotherapy and radiotherapy by serving as anti-MDR sensitizers.

Thus the conclusions that can be drawn from these types of studies are two-fold; that miRNAs regulating these processes are in fact quite unique with little overlap, or alternatively that the full complement of miRNAs participating in these processes is not yet known. However, more studies are needed to characterize all the interactions between miRNAs and epigenetics to better develop "ad hoc" cancer biomarkers and/or to identify new therapeutic targets.

Despite recent advances, there is significant additional work needed to realize the therapeutic potential of miRNAs. In part, this is because there are no perfect delivery approaches either in vitro or in vivo. Each system we have reviewed has its own advantages and disadvantages. Adenoviral- and lentiviral-based delivery carriers for miRNA are limited by immune responses [144], and non-viral-based vectors are challenged by their unsatisfying efficiency, toxicity, and lack of specificity [145]. The delivery of miRNAs in vivo is usually hindered by biological barriers such as RES clearance, poor targeting efficiency, and low tissue/cell penetration. Considering the disadvantages of the present miRNA treatments, some future directions that should be studied include: (1) The low toxicity of delivery vectors should be further improved. The overall design philosophy of delivery carriers has been fixed in that almost all delivery vectors have a cationic surface for packing miRNAs via electrostatic interactions. However, while the increased number and density of the amines used to achieve a cationic surface generally promote the transfection efficiency, particularly for polymeric and dendrimer-based vectors, increasing the positive charge density is associated with higher cytotoxicity. Certainly, the side effects of the toxic polymers (e.g., PEI and PAMAM) must be minimized prior to human trials. The cationic lipid- and polymer-based nanoparticles can be used to deliver miRNAs locally to the tumor for reducing unwanted drug exposure. However, when given systemically, surface charge–dependent toxicities have been found. More work is needed to identify new biocompatible materials. For example, researchers have conjugated PEG and other chemical moieties to the periphery of cationic vectors. This has reduced the side effect of the vectors by masking the charge of the primary amine groups. It is also possible that a biodegradable system (e.g., reactive oxygen sensitive, enzyme degradable, pH responsive) will provide more efficient delivery of miRNAs with lower toxicity [146–148]. (2) The specificity of vectors is an important issue. The targeting ability of miRNAs represents their special characteristic compared with common drugs. In this regard, an individualized treatment system based on genomic differences between cancers will be an important future design feature. In addition, an improved organ specificity of the treatment systems will be beneficial. Vectors will change with the target tissues, different cell lines, and cancers. External moieties, such as targeting peptides (e.g., cRGD, DS4-3, and RVG), ligands (e.g., hyaluronic acid and folate acid), antibodies (e.g., scFv, GD2, and ch14.18), aptamers (e.g., A10-3.2 and AS1411), and other molecules that enhance active targeted delivery are being studied to better direct the NPs to particular organs or cells. Furthermore, these moieties will help reduce doses, making the treatment more acceptable to patients and thereby enhancing compliance and comfort. (3) Another challenge is to improve the ability of miRNAs to cross physiological barriers, particularly the blood–brain barrier (BBB) in the therapy of brain tumors. There has been progress in vitro but the BBB remains a significant problem in vivo although some studies show that complexes with penetrating functional peptides, Pluronic, and transferrin have improved the transfection efficiency [149–155]. Such applications in brain delivery of miRNAs are still rare.

In the long term, future work should focus on identifying effective concentrations, dosage regimens, and combination with chemotherapy and radiotherapy, apart from the vectors. Once in use, the long-term results following miRNA therapy will need to be assessed. A better understanding of the chromatin structure around individual miRNA genes in normal and cancer cells will help us understand the mechanisms of deregulation that occur during tumorigenesis. Furthermore, the regulation of miRNA expression by epigenetic treatment reveals promising new avenues for the design of innovative strategies in the fight against human cancer [156]. Evidence presented here provides a strong rationale for developing therapies that combine chromatin modifier inhibitors (such as DNA demethylating agents or histone deacetylase inhibitors) and depletion/overexpression of particular miRNAs with the goal of reactivating tumor suppressors and normalizing the aberrant patterns of methylation that occur in cancer. This is illustrated by the recent finding that miR-203 (which targets ABL1 and BCR-ABL1 oncogenes) is silenced in hematopoietic malignancies by genetic loss and also epigenetically by promoter CpG hypermethylation [157]. Restoration of miR-203 expression (either directly or through the action of demethylating drugs) suppresses proliferation of the tumor cells.

The possible presence and function of other types of small noncoding RNAs (ncRNAs) in the nucleus of mammalian cells remains a largely obscure area. Whereas in plants or fission yeast the role of small RNAs in sequence-specific targeting of chromatin is clear, there is as yet only limited evidence of small RNAs or, in general, ncRNAs functioning in roles other than posttranscriptional regulation in mammals. Further investigations will undoubtedly clarify the details of the mechanisms by which ncRNAs in mammals may control chromatin modifications and ultimately determine their role not only in normal cell functioning but also in the onset of disease.

To date, different clinical trials have demonstrated the use of miRNA-based therapy as a promising strategy for the treatment of different diseases, making miRNA highly relevant for clinical use [158]. In this regard, the delivery of miR-193a-3p mimics by nano-sized particles could represent a novel therapeutic tool for the treatment of cancer because it may hamper tumor-aggressive properties in tumor xenograft models by restoring the miR original levels. On the other hand, the local delivery of anti-miR-193a-3p molecules could be an effective intervention for local ischemic diseases. These findings may pave the way to further studies aimed to elucidate the possible use of miR-193a-3p for experimental therapeutic procedures.

The ability of siRNA to orchestrate coordinated cellular responses via potent and specific gene silencing makes it a unique and powerful therapeutic option. The potential to tailor siRNA to silence any gene of interest endows this strategy with broad applicability. The promise of therapeutic RNAi in local delivery has already been adapted for clinical trials in a variety of indications, including respiratory syncytial virus infection, macular degeneration, pancreatic ductal adenocarcinoma, and pachyonychia congenita [159–164]. A comprehensive list of local siRNA therapeutics in varying stages of clinical trials is provided in [165]. Of note is that only one of these technologies incorporates a mechanism for controlled siRNA release, the siG12D LODER system from Silenseed Ltd. Despite the pioneering progress of siG12D LODER in the clinical pipeline, there remains a strong need to further translate in vitro silencing from local delivery depots to functional improvements in relevant preclinical models in vivo. Many systems showing promising silencing in vitro have not been explored in medically relevant animal disease models. In addition, of the multitude of pathologically relevant genes identified with in vitro RNAi studies, only a small portion have been examined in conjunction with established local delivery systems. Further investigation of potential synergism between the impact of the delivery depots themselves and the silencing of genetic targets is an exciting avenue to explore, especially for tissue-regenerative applications. Another complicating factor relevant to the clinical adoption of local siRNA therapeutics is the plethora of polymers, siRNA carriers, and construct types pursued in basic research. Among local siRNA therapeutics in the clinical pipeline, systems achieving sustained delivery are notably absent. Instead, simple topical treatments or local injection strategies predominate due to the far greater ease with which these treatments proceed through the regulatory pathway. In order to reap the therapeutic benefits of controlled-release delivery systems that maintain strong gene silencing activity without repeated doses, a balance must be struck between complicated fabrication processes and therapeutic efficacy. Additionally, obfuscating comparison of strategies for therapeutic delivery of siRNA is the growing use of complex and costly modified siRNA chemistries. Emerging clinical studies have departed from use of unmodified siRNA due to recognized immunogenic effects, and recent industry-sponsored strategies employ heavily modified siRNA molecules in conjunction with the company's patent-protected delivery strategy [166, 167]. Due to the variety of modified siRNA used and because widespread use of the highly modified siRNA is often not financially feasible in early-stage or academic research applications, variation in silencing efficacy due to potency of the siRNA molecules must be carefully assessed when evaluating and comparing the published results of different delivery technologies. Despite these challenges, the ability to localize the siRNA therapeutic to a site of interest, minimizing physiological clearance and off-site effects, endows local delivery methods with an enormous advantage over systemic delivery techniques.

The hierarchical nano-assemblies were capable of simultaneously transporting siRNA and DOX into Bel-7402 cells, and FA-directed internalization significantly increased the delivery efficiency. Systematic biological experiments revealed that DOX-inducible upregulation of antiapoptotic BCL-2 gene in Bel-7402 cells was significantly suppressed by co-delivered BCL-2 siRNA. Consequently, cell apoptosis was enhanced and potency of DOX in inducing cell death was greatly potentiated through synergistic effect of the two therapeutic agents. A series of experiments were carried out to optimize nano-complex formulation for maximized performance. Our results showed potential of the hierarchical nano-assembly as a facile nano-platform for siRNA and hydrophobic drug co-delivery in biomedical applications. Recent advancements in wound-healing applications have been particularly promising. As research and understanding of the variety of delivery systems and genetic targets continue to expand, functional improvements following sustained, localized siRNA silencing will increase rapidly. With properly designed material delivery platforms and new molecular targets identified, powerful RNAi-based local therapies will soon be clinically realized.

The applications of SiNPs in in vitro and in vivo tumor imaging demonstrated the potential for further developing the SiNPs into clinically useful reagents. Fluorescence imaging and MRI are currently the main strategies for the applications of SiNPs in cancer imaging. X-ray/CT may also be used along with SiNPs as imaging agents. In conclusion, the studies of theranostic micellar drug and gene delivery systems not only synergistically combined gene silencing and chemotherapy but also served as a negative MRI contrast agent and therefore might be a candidate novel nanomedicine for colorectal cancer therapy.

The results of these studies demonstrate that for the efficient siRNA delivery through the skin layers (i.e., stratum corneum and epidermis) and deposition in the upper dermis, a correct balance of lipoplex size, charge, and edge activator content is required. The developed lipoplexes were able to not only permeate through the skin layers but also effectively internalize into the viable cells of basal epidermis and knock down the expression of target proteins. Given that no active-delivery approach such as ultrasound was used to enhance skin permeation, the developed liposomal system can be considered a major step forward toward a simple and efficient drug delivery of macromolecules via the topical route for therapy of various skin diseases. In the near future, it is our intention to develop a lipoplex-embedded hydrogel system to enhance local retention time of the lipoplexes on the skin and make a suitable formulation for in vivo studies. It is also noteworthy that in this study we used skin from one donor to ensure that the observed differences are due to formulation variables and not the donors. To demonstrate the broad application of the developed formulation, in the future the efficacy of the system needs to be tested in different skin types considering race, age, and gender.

This book will offer perspectives on future applications of siRNA therapeutics. Since RNAi was discovered, various nonviral vector delivery systems for siRNA delivery have been explored extensively. Although significant advances have been made in the development of efficient in vivo siRNA delivery, there are still many challenges and barriers that must be overcome to achieve the ideal formulation in terms of selectivity, efficacy, and safety. Only a few nanoparticle-based siRNA delivery systems have been approved by the FDA and are in clinical trials for cancer therapy. Delivery systems can improve specificity of cancer cell targeting, prevent nonspecific delivery of siRNA, and may also protect the siRNA during transport. Nanoparticles conjugated to the targeting ligand for effective siRNA delivery increase the chance of binding the tumor surface receptor; however, the process also increases the overall size of the nanoparticle. The PEG coating of nanoparticles reduces uptake by RES, resulting in enhanced circulatory half-life, but reduces targeting specificity because PEG molecules sterically disrupt selective conjugation. Thus, the selection of appropriate cell-specific targeting moieties and careful design of stable and potent nanoparticle delivery systems is required for future development.

Other major challenges for RNAi-based cancer therapeutics include controlling the specificity of the siRNA, minimizing off-target effects, increasing resistance to nuclease degradation, and avoiding immune responses such as α/β interferons, RNA-dependent kinase effects, and toll-like immunity. Chemical modification of the siRNA, such as inserting a $2'$-$O$-methyl ribose in the nucleotide in the second position of the guide strand, could reduce silencing of most off-target transcripts with complementarity to the siRNA guide [168]. Dual-targeted siRNA drugs, such as ALN-VSP02, which targets VEGF and KSP, may reduce the potential for off-target gene silencing and increase the chances of knocking down the desired target.

Tremendous advances in the inhibition of checkpoint kinases by small molecules have occurred over the past 2 years. In addition, novel chemical inhibitors have enhanced the understanding of key hydrogen bonding, electrostatic, and hydrophobic interactions, leading to the rational design of potent and selective CHK1 inhibitors. The majority of potent CHK1 inhibitors contain a hydrophilic moiety that forms interactions with the ATP-binding pocket in either the "buried" hydrophobic pocket or near the adjacent ribose sugar-binding pocket. The compelling therapeutic rationale for inhibiting CHK1 to improve standard chemotherapy and radiation therapy for cancer patients is finally being put to the test with three structurally diverse, ATP-competitive, small molecules now in the clinic [169•]. Existing checkpoint inhibitors in clinical development, XL-9844, AZD-7762, and PF-477736, are all potent inhibitors of CHK1 but have differing selectivities against other kinases, including CHK2. It is currently unclear whether these different kinase selectivity profiles will lead to different clinical outcomes. Furthermore, recent developments in checkpoint kinase biology have the potential to impact future development of checkpoint kinase inhibitors and clinical trials [170–179, 180••, 181, 182, 183••]. In addition to CHK1 and CHK2, other emerging therapeutic targets for potentiating the effects of DNA damage are PARP, ATM, ATR, CDC25, DNA-PK, Wee1 kinase, and hsp90. Judging from the intense efforts being made in the field of sensitization approaches, we can expect additional clinical candidates to be brought forward for evaluation in the near future.

In conclusion, information obtained in MM primary samples, cell lines, and mouse models is booming in terms of diagnostic/prognostic and therapeutic potential of ncRNAs in MM. It is hoped that this will be translated into clinical practice in the future.

Various nanoparticle-based delivery systems [184–187] such as cationic lipids, polymers, dendrimers, and inorganic nanoparticles [188–206] have been demonstrated to provide effective and efficient siRNA delivery in vitro and in vivo. Future studies must focus on the in vivo safety profiles of the various delivery systems, including undesirable immune stimulation and cytotoxicity. It is critical to develop safe, biocompatible, and biodegradable nanoparticle delivery systems for the clinical application of RNAi-based cancer therapeutics. However, a multisensitization individualized therapy [207–216] designed specifically for diseases and each type of cancer cell may be more likely to prove viable to overcome cancer cell resistance.

Recently, a novel algorithm including CA-125, HE4, and body mass index in the diagnosis of endometrial cancer has been studied [217] as shown in Figure 14.30: (1) SerumCA-125 and HE4 levels are significantly higher in endometrial cancer patients. (2) Model to distinguish endometrial cancer from benign gynecological diseases is proposed. (3) Diagnostic algorithm includes serum CA-125 and HE4 levels and BMI (AUC=0.80). (4) Serum HE4 levels differentiate patients with lymphovascular invasion (AUC=0.81). (5) Serum HE4 levels stratify patients with deep myometrial invasion (AUC=0.78) [217].

Recent study on uterine lymphatic anatomy confirmed previous publications' description of two consistent lymphatic pathways with pelvic SLNs in women with EC; an upper paracervical pathway (UPP) with draining medial external and/or obturator lymph nodes and a lower paracervical pathway (LPP) with draining internal iliac and/or presacral lymph nodes (Figure 14.31) [218–222]. These pathways communicate with finer lymphatics at the level

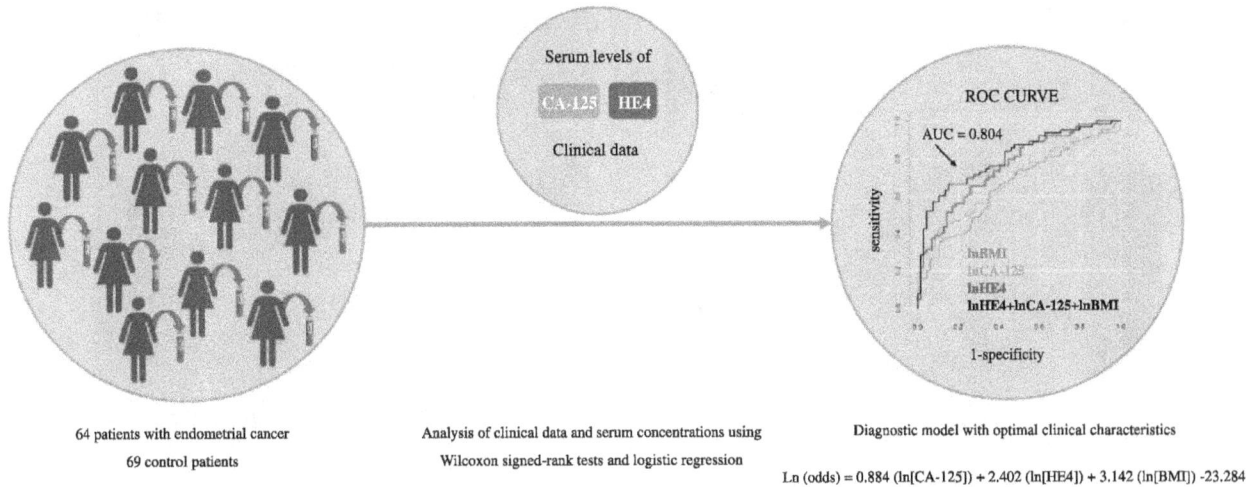

**FIGURE 14.30** Novel algorithm includes CA-125, HE4, and body mass index in the diagnosis of endometrial cancer.

of the cardinal ligaments, thereafter dividing into separate noncommunicating courses lateral and medial to the common iliac arteries with further drainage to the para-aortic area.

According to most publications, lymph node dissection (LND) along the UPP constitutes a full pelvic lymphadenectomy [223–225].

In addition, 87 SLNs were detected in the internal iliac vein, parametrial, and presacral areas in 57 patients, and accounted for 3.5%, 1.2%, and 3.2% of all SLN nodes, respectively (Figure 14.32) [226].

In conclusion, (1) SLN mapping has a high sensitivity and NVP in high-risk endometrial cancer, (2) SLN mapping may be an appropriate staging procedure in high-risk endometrial cancer, and (3) Cervical injection for SLN mapping appears to adequately represent uterine drainage. Accurate surgical staging is an important prognostic predictor of survival in patients with endometrial cancer (EC). Given the

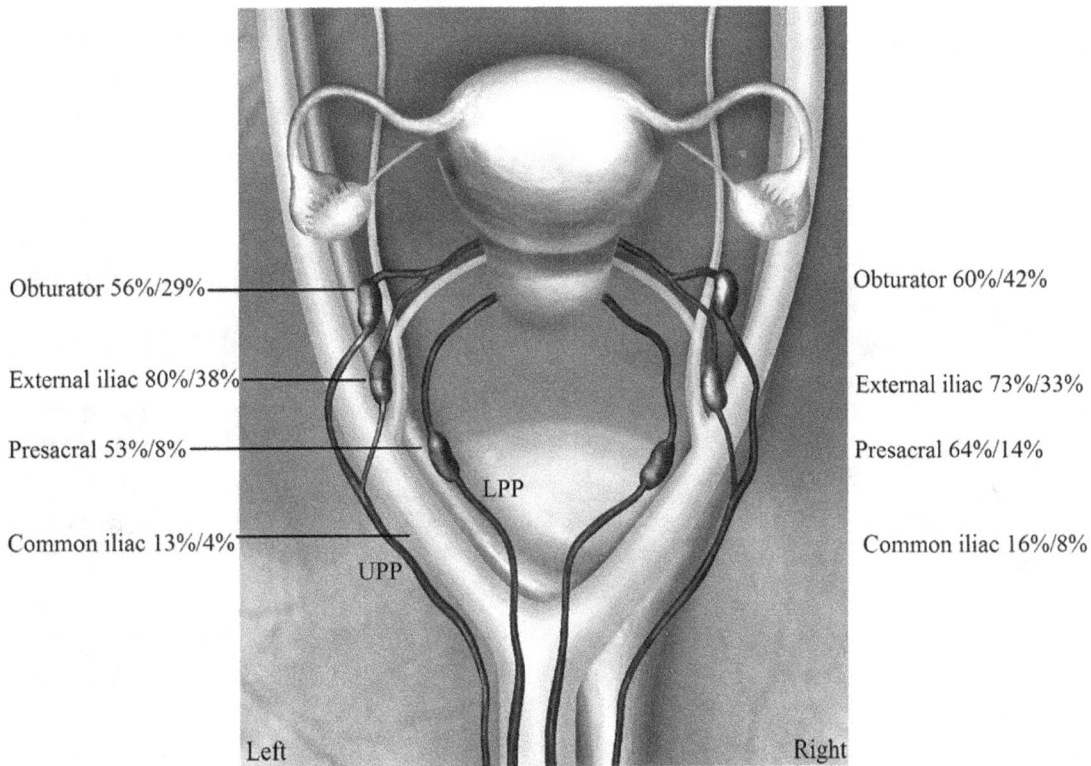

**FIGURE 14.31** Schematic overview of the pelvic uterine lymphatic pathways with typical localization of sentinel lymph nodes and percentage of nonmetastatic/metastatic sentinel lymph nodes per lymph compartment in endometrial cancer (percentages refer to the total number of patients/node positive patients, patients can have more than one nonmetastatic/metastatic sentinel lymph node). *Abbreviations:* LPP, lower paracervical pathway; UPP, upper paracervical pathway.

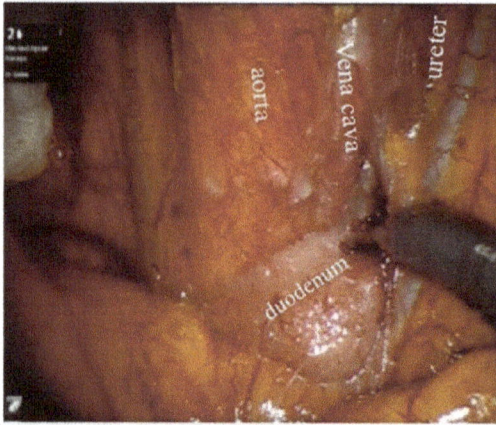

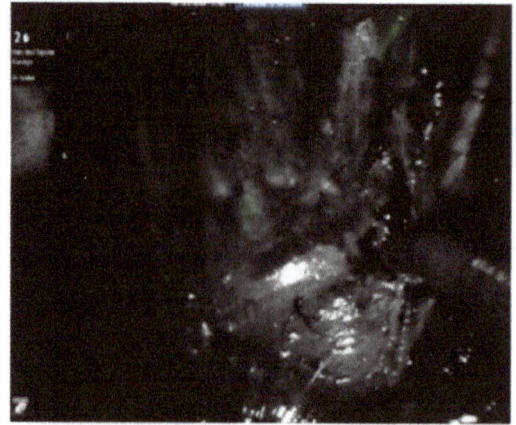

SLN detection that is high in para-aortic region just inferior to duodenum with lymphatic pathway tracking upwards after passing the midline

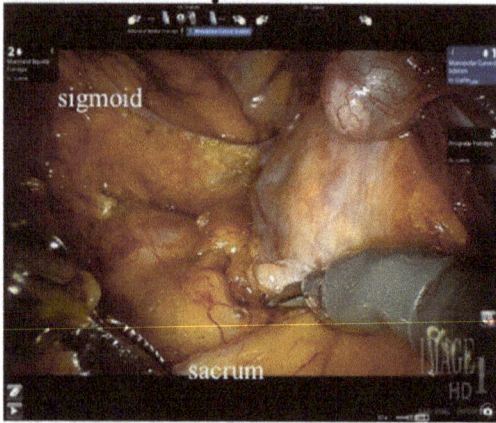

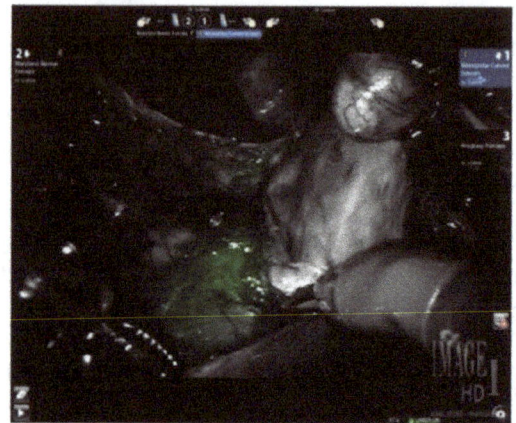

SLN detection in right presacral area

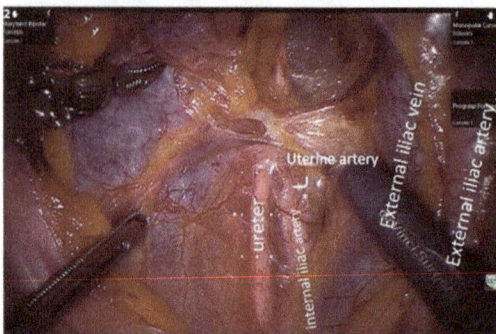

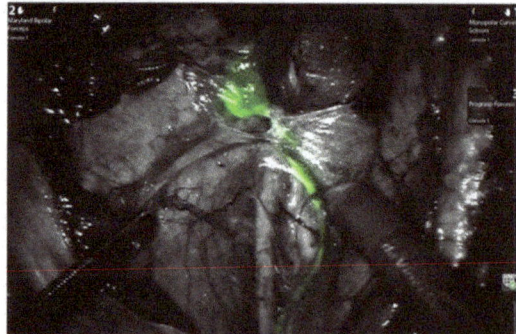

SLN detection in right parametrial area

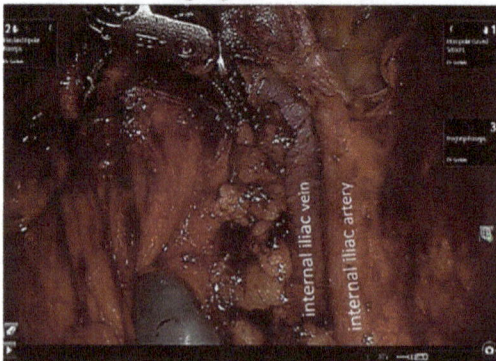

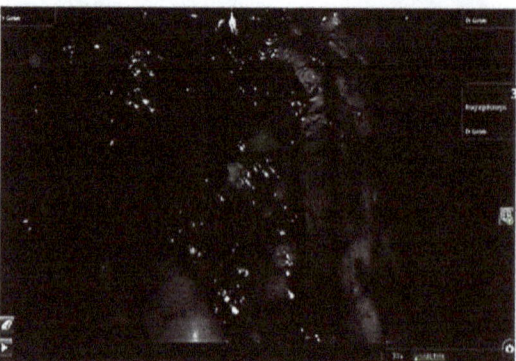

SLN detection adjacent to right hypogastric vein

**FIGURE 14.32** Direct (*left panel*) and immunofluorescent (*right panel*) visualization following intracervical injection of ICG. SLN detection that is high in para-aortic region just inferior to duodenum with lymphatic pathway tracking upwards after passing the midline. SLN detection in right presacral area. SLN detection in right parametrial area. SLN detection adjacent to right hypogastric vein.

high sensitivity and high negative predictive value found in the study [227], we believe that the use of SLN mapping appears to be an appropriate staging procedure in high-risk endometrial cancer.

It has been reported [228] in a large French cohort of patients with endometrial carcinoma that isolated local recurrences are rare. Vaginal site of recurrence is probably the main parameter associated with prolonged survival. Recurrence's treatment remains unclear and better understanding of the prognostic factors associated with survival in these patients is determinant.

Recent developments in structural siRNA and RNAi nanotechnology have enabled more refined and reliable in vivo gene silencing with multiple advantages over naked siRNAs. Recent progress in RNA biology has broadened the scope of therapeutic targets of RNA drugs for cancer therapy. Thus, siRNAs have been considered one of the most noteworthy developments that are able to regulate gene expression following a process known as RNA interference (RNAi).

## REFERENCES

[1] Lau CC, Pardee AB: Mechanism by which caffeine potentiates lethality of nitrogen mustard. *Proc Natl Acad Sci USA* (1982) **79**(9):2942–2946.

[2] Das SK, Lau CC, Pardee AB: Abolition by cycloheximide of caffeine-enhanced lethality of alkylating agents in hamster cells. *Cancer Res* (1982) **42**(11):4499–4504.

[3] Musk SR, Steel GG: Override of the radiation-induced mitotic block in human tumor cells by methylxanthines and its relationship to the potentiation of cytotoxicity. *Int J Radiat Biol* (1990) **57**(6):1105–1112.

[4] Teicher BA, Holden SA, Herman TS et al.: Efficacy of pentoxyfylline as a modulator of alkylating agent activity *in vitro* and *in vivo*. *Anticancer Res* (1991) **11**(4):1555–1560.

[5] Sarkaria JN, Busby EC, Tibbetts RS et al.: Inhibition of ATM and ATR kinase activities by the radiosensitizing agent, caffeine. *Cancer Res* (1999) **59**(17):4375–4382.

[6] Tibbetts RS, Brumbaugh KM, Williams JM et al.: A role for ATR in the DNA damage-induced phosphorylation of p53. *Genes Dev* (1999) **13**(2):152–157.

[7] Ribeiro JC, Barnetson AR, Jackson P et al.: Caffeine-increased radiosensitivity is not dependent on a loss of G2/M arrest or apoptosis in bladder cancer cell lines. *Int J Radiat Biol* (1999) **75**(4):481–492.

[8] Moser BA, Brondello J-M, Baber-Furnari B et al.: Mechanism of caffeine-induced checkpoint override in fission yeast. *Mol Cell Biol* (2000) **20**(12):4288–4294.

[9] Wang Q, Fan S, Eastman A et al.: UCN-01: A potent abrogator of G2 checkpoint function in cancer cells with disrupted p53. *J Natl Cancer Inst* (1996) **88**(14):956–965.

[10] Graves PR, Yu L, Schwarz JK et al.: The CHK1 protein kinase and the Cdc25C regulatory pathways are targets of the anticancer agent UCN-01. *J Biol Chem* (2000) **275**(8):5600–5605.

[11] Busby EC, Leistritz DF, Abraham RT et al.: The radiosensitizing agent 7-hydroxystaurosporine (UCN-01) inhibits the DNA damage checkpoint kinase hCHK1. *Cancer Res* (2000) **60**(8):2108–2112.

[12] Jackson JR, Gilmartin A, Imburgia C et al.: An indolo-carbazole inhibitor of human checkpoint kinase (CHK1) abrogates cell cycle arrest caused by DNA damage. *Cancer Res* (2000) **60**(3):566–572.

[13] Koniaras K, Cuddihy AR, Christopoulos H et al.: Inhibition of CHK1-dependent G2 DNA damage checkpoint radiosensitizes p53 mutant human cells. *Oncogene* (2001) **20**(51):7453–7463.

[14] Luo Y, Rockow-Magnone SK, Joseph MK et al.: Abrogation of G2 checkpoint specifically sensitize p53 defective cells to cancer chemotherapeutic agents. *Anticancer Res* (2001) **21**(1A):23–28.

[15] Luo Y, Rockow-Magnone SK, Kroeger PE et al.: Blocking CHK1 expression induces apoptosis and abrogates the G2 checkpoint mechanism. *Neoplasia* (2001) **3**(5):411–419.

[16] Zhao H, Watkins JL, Piwnica-Worms H: Disruption of the checkpoint kinase 1/cell division cycle 25A pathway abrogates ionizing radiation-induced S and G2 checkpoints. *Proc Natl Acad Sci USA* (2002) **99**(23):14795–14800.

[17] Kaufmann WK, Campbell CB, Simpson DA et al.: Degradation of ATM-independent decatenation checkpoint function in human cells is secondary to inactivation of p53 and correlated with chromosomal destabilization. *Cell Cycle* (2002) **1**(3):210–219.

[18] Chen Z, Xiao Z, Chen J et al.: Human CHK1 expression is dispensable for somatic cell death and critical for sustaining G2 DNA damage checkpoint. *Mol Cancer Therap* (2003) **2**(6):543–548.

[19] Luo Y, Leverson JD: New opportunities in chemosensitization and radiosensitization: Modulating the DNA-damage response. *Expert Rev Anticancer Ther* (2005) **5**(2):333–342.

[20] Tenzer A, Pruschy M: Potentiation of DNA-damage-induced cytotoxicity by G2 checkpoint abrogators. *Curr Med Chem Anticancer Agents* (2003) **3**(1):35–46.

[21] Kawabe T: G2 checkpoint abrogators as anticancer drugs. *Mol Cancer Ther* (2004) **3**(4):513–519.

[22] Zhou B-B, Anderson HJ, Roberge M: Targeting DNA checkpoint kinases in cancer therapy. *Cancer Biol Ther* (2003) **2**(4 Suppl 1): S16-S22.

[23] Prudhomme M: Combining DNA damaging agents and checkpoint 1 inhibitors. *Curr Med Chem Anticancer Agents* (2004) **4**(5):435–438.

[24] Prudhomme M: Novel checkpoint 1 inhibitors. *Rec Patents Anticancer Drug Disc* (2006) **1**(1):55–68.

[25] Tao Z-F, Lin N-H: CHK1 inhibitors for novel cancer treatment. *Anticancer Agents Med Chem* (2006) **6**(4):377–388.

[26] Hotte SJ, Oza A, Winquist EW et al.: Phase I trial of UCN-01 in combination with topotecan in patients with advanced solid cancers: a Princess Margaret hospital phase II consortium study. *Ann Oncol* (2006) **17**(2):334–340.

[27] Perez RP, Lewis LD, Beelen AP et al.: Modulation of cell cycle progression in human tumors: A pharmacokinetic and tumor molecular pharmacodynamic study of cisplatin plus the CHK1 inhibitor UCN-01 (NSC 638850). *Clin Cancer Res* (2006) **12**(23):7079–7085.

[28] Sampath D, Cortes J, Estrov Z et al.: Pharmacodynamics of cytarabine alone and in combination with 7-hydroxystaurosporine (UCN-01) in AML blasts *in vitro* and during a clinical trial. *Blood* (2006) **107**(6):2517–2524.

[29] Matthews DJ, Yakes FM, Chen J et al.: Pharmacological abrogation of S-phase checkpoint enhances the antitumor activity of gemcitabine *in vivo*. *Cell Cycle* (2007) **6**(1):104–110.

[30] Matthews DJ: Dissecting the roles of CHK1 and CHK2 in mitotic catastrophe using chemical genetics. *18th EORTC-NCI-AACR Symposium on Molecular Targets and Cancer Therapeutics*, Prague, Czech Republic (2006): Abs 344.

[31] Janetka JW, Ashwell S, Zabludoff S et al.: Inhibitors of checkpoint kinases: From discovery to the clinic. *Curr Opin Drug Discovery Dev* (2007) **10**(4):473–486. © The Thomson Corporation ISSN 1367–6733

[32] Exelixis Inc, Buhr CA, Baik T-G et al.: Preparation of substituted pyrazines as protein kinase modulators. WO-2003093297 (2003).

[33] Clary DO: Inhibition of CHKs in a leukemia model abrogates DNA damage checkpoints and promotes mitotic catastrophe. *Proc Am Assoc Cancer Res* (2007) 48:Abs 5385.

[34] Lyne PD, Kenny PW, Cosgrove DA et al.: Identification of compounds with nanomolar binding affinity for checkpoint kinase-1 using knowledge-based virtual screening. *J Med Chem* (2004) **47**(8):1962–1968.

[35] Lyne PD: Virtual screening for kinase inhibitors. *228th ACS National Meeting* (2004):CINF-073.

[36] Ashwell S, Caleb BL, Green S et al.: Preclinical identification of AZD7762, a novel potent and selective inhibitor of checkpoint kinases. Abstracts of Papers. 2007 EORTC-NCI-AACR International Meeting on Molecular Targets and Therapeutics, San Francisco, USA (2007): A232.

[37] Astrazeneca PLC, Ashwell S, Gero T et al.: Preparation of thiophene compounds as CHK1 inhibitors. WO-2005016909 (2005).

[38] Astrazeneca UK Ltd, Ashwell S, Gero T et al.: Preparation of substituted heterocycles, particularly ureidothiophenes, as CHK1 kinase inhibitors for treating neoplasm. WO-2005066163 (2005).

[39] Astrazeneca AB, Ashwell S, Ezhuthachan J et al.: Preparation of novel fused triazolones as antitumor agents. WO 2004081008 (2004).

[40] Astrazeneca AB, Daly K, Heron N et al.: Substituted thienopyridines and related compounds and their preparation, pharmaceutical compositions, and use as CHK1, PDK1 and PAK inhibitors in the treatment of cancer. WO-2006106326 (2006).

[41] Pfizer Inc, Ninkovic S, Bennett MJ et al.: Preparation of diazepinoindolones as CHK-1 kinase inhibitors. WO-2004063198 (2004).

[42] Anderes KL, Blasina A, Castillo R et al.: Small molecule CHK1 inhibitor potentiates antitumor activity of chemotherapeutic agents *in vivo*. *Proc Am Assoc Cancer Res* (2005) **46**:Abs 4417.

[43] Blasina A, Kornmann JF, Chen E et al.: A novel inhibitor of the protein kinase CHK1: Studies on the mechanism of action. *Proc Am Assoc Cancer Res* (2005) **46**:Abs 4416.

[44] Ninkovic S: The discovery and design of CHK kinase inhibitors. *First RSC-SCI Symposium on Kinase Inhibitor Design (Part I)*, London, UK (2005).

[45] McArthur GA: Imaging with FLT-PET demonstrates that PF-477736, an inhibitor of CHK1 kinase, overcomes a cell cycle checkpoint induced by gemcitabine in PC-3 xenografts. *Proc Am Soc Clin Oncol* (2006) **25**:Abs 3045.

[46] Raza Dewji M: Beyond VEGF, targeting tumor growth and angiogenesis via alternative mechanisms. *First International Meeting, Targeted Therapies in Cancer: Myth or Reality?*, Milan, Italy (2006).

[47] Anderes K, Blasina A, Chen E et al.: Characterization of a novel and selective inhibitor of checkpoint kinase 1: Breaching the tumor's last checkpoint defense against chemotherapeutic agents. *18th EORTC-NCI-AACR Symposium on Molecular Targets and Cancer Therapeutics*, Prague, Czech Republic (2006):Abs 373.

[48] Hallin M, Zhang C, Yan Z et al.: PF-00477736 an inhibitor of CHK1 enhances the antitumor activity of docetaxel indicating a role for CHK1 in the mitotic spindle checkpoint. *Proc Am Assoc Cancer Res* (2007) **48**:Abs 4373.

[49] Cullinane C, Raleigh J, Anderes K et al.: Mechanisms of radiation enhancement by the CHK1 inhibitor PF-477736. *Proc Am Assoc Cancer Res* (2007) **48**:Abs 5386.

[50] Li G, Elder RT, Qin K et al.: Mechanisms of radiation enhancement by the CHK1 inhibitor PF-477736. *J Biol Chem* (2007) **282**(10):7287–7298.

[51] Parsels LA, Parsels JD, Booth RJ et al.: The small-molecule CHK1 inhibitor, PD-321852, causes synergistic depletion of CHK1 protein and clonogenic death when combined with gemcitabine in colorectal and pancreatic tumor cells. *Proc Am Assoc Cancer Res* (2006) **47**:Abs 4911.

[52] Booth JR, Sheehan DJ, Ortwine DF et al.: The small-molecule CHK1 inhibitor, PD-321852, causes synergistic depletion of CHK1 protein and clonogenic death when combined with gemcitabine in colorectal and pancreatic tumor cells. *232nd ACS National Meeting* (2006):MEDI-131.

[53] Palmer BD, Thompson AM, Booth RJ et al.: 4-Phenylpyrrolo[3,4-c]carbazole-1,3(2H,6H)-dione inhibitors of the checkpoint kinase Wee1. Structure-activity relationships for chromophore modification and phenyl ring substitution. *J Med Chem* (2006) **49**(16):4896–4911.

[54] Pfizer Inc, Booth RJ, Lee HH et al.: Inhibitors of checkpoint kinases (Wee1 and CHK1). US-2005250836 (2005).

[55] Agouron Pharmaceuticals Inc, Rui EY, Johnson TO Jr: Preparation of benzimidazolyl alkyl carbamates for modulating the activity of CHK1 kinase. US-2005148643 (2005).

[56] PFIZER INC, Johnson MD, Teng M et al.: Preparation of aminopyrazoles as CHK1 checkpoint protein kinase inhibitors. WO-2005009435 (2005).

[57] ICOS CORP, Diaz F, Farouz FS et al.: Heteroaryl urea derivatives useful for inhibiting CHK1 and their preparation, pharmaceutical compositions, and use in the treatment of diseases related to DNA damage. WO-2006105262 (2006).

[58] ICOS CORP, Farouz FS, Holcomb R et al.: Aryl- and heteroaryl-substituted urea derivatives as CHK1 inhibitors, their preparation, pharmaceutical compositions, and use in therapy. WO-2006021002 (2006).

[59] ICOS CORP, Gaudino JJ, Cook AW: Preparation of aryl heteroaryl urea derivatives useful for inhibiting CHK1. WO-2006014359 (2006).

[60] ICOS CORP, Burgess LE, Cook AW et al.: Preparation of bisarylurea derivatives useful for inhibiting CHK1. WO-2006012308 (2006).

[61] Kesicki EA, Gaudino JJ, Cook AW et al.: Discovery of pyrazinyl ureas as inhibitors of the cell cycle checkpoint kinase CHK1. *228th ACS National Meeting* (2004):MEDI-225.

[62] Diaz FA, Holcomb R, Farouz F et al.: Development of highly selective CHK-1 inhibitors with low nanomolar cellular activity. *233rd ACS National Meeting* (2007): MEDI-121.

[63] Millennium Pharmaceuticals Inc, Boyle RG, Imogai HJ et al.: Preparation of diarylureas as CHK-1 kinase inhibitors for the treatment of cancer. WO-2003101444 (2003).

[64] Millennium Pharmaceuticals Inc, Boyle RG, Imogal HJ et al.: Preparation of diarylureas as CHK1 kinase inhibitors for treating cancer. WO-2005072733 (2005).

[65] (a) Millennium Pharmaceuticals Inc, Boyle RG, Imogai HJ et al.: Preparation of 2,5-dihydropyrazolo[4,3-c]quinolin-4-ones as CHK-1 inhibitors for treating cancer]. WO-2005118583 (2005). (a) and (b) For full

experimental detail on compounds described in this paper, see WO2006086255. Google Scholar (c) Bartsch RA, Yang IW: J Heterocycl Chem, (1984) 21:1063. Cross Ref View Record in Scopus (d) Alami M, Ferri F, Linstrunelle G: Tetrahedron Lett (1993) 34:6403. Article Download PDF View Record in Scopus (e) Journet M, Cai, D, Kowal JJ et al.: Tetrahedron Lett (2001) 42: 9117. Article Download PDF View Record in Scopus (f) PDB coordinates for 21: 2HOG. Google Scholar (g) Dunitz JD: Science (1994) 264:670. View Record in Scopus.

[66] Millennium Pharmaceuticals Inc, Boyle RG, Imogai HJ et al.: Preparation of pyrazoloquinolinone derivatives as CHK-1 inhibitors. WO-2005028474 (2005).

[67] Millennium Pharmaceuticals Inc, Blackburn C, Claiborne CF et al.: Preparation of lactam compounds useful as protein kinase inhibitors. WO-2006041773 (2006).

[68] Chen Z, Xiao Z, Gu WZ et al.: Selective CHK1 inhibitors differentially sensitize p53-deficient cancer cells to cancer therapeutics. Int J Cancer (2006) 119(12):2784–2794.

[69] Tao Z-F, Li G, Wang GT et al.: Pyrazyl phenyl ureas as potent and selective CHK 1 inhibitors: The exploration of C6-position of pyrazyl ring and SAR studies at C4-position of phenyl ring. 229th ACS National Meeting (2005):MEDI-145.

[70] Wang GT, Li G, Mantei RA et al.: 1-(5-Chloro-2-alkoxyphenyl)-3-(5-cyanopyrazin-2-yl)ureas [correction of cyanopyrazi] as potent and selective inhibitors of CHK1 kinase: synthesis, preliminary SAR, and biological activities. J Med Chem (2005) 48(9):3118–3121.

[71] Li G, Hasvold LA, Tao ZF et al.: Synthesis and biological evaluation of 1-(2,4,5-trisubstituted phenyl)-3-(5-cyano-2-pyrazinyl)-ureas as potent CHK1 kinase inhibitors. Bioorg Med Chem Lett (2006) 16(8):2293–2298.

[72] Abbott Laboratories, Lin N-H, LI G et al.: Preparation of substituted macrocyclic diaryl urea kinase inhibitors as potential anticancer agents. US-2005215556 (2005).

[73] Abbott Laboratories, Tao Z-F, Lin N-H et al.: A preparation of macrocyclic N-(heteroaryl)urea derivatives, useful as kinase inhibitors. US-2005096324 (2005).

[74] Tao Z-F, Wang L, Tong Y et al.: Synthesis and biological evaluation of 1-(2,4,5-trisubstituted phenyl)-3-(5-cyano-2-pyrazinyl)ureas as potent CHK1 kinase inhibitors. 232nd ACS National Meeting (2006):MEDI-128.

[75] Tao Z-F, Wang L, Stewart KD et al.: Structure-based design, synthesis, and biological evaluation of potent and selective macrocyclic checkpoint kinase 1 inhibitors. J Med Chem (2007) 50(7):1514–1527.

[76] Lin N-H, Xia P, Kovar P et al.: Synthesis and biological evaluation of 3-ethylidene-1,3-dihydro-indol-2-one as novel checkpoint 1 kinase inhibitors. 229th ACS National Meeting (2005):MEDI-146.

[77] Lin NH, Xia P, Kovar P et al.: Synthesis and biological evaluation of 3-ethylidene-1,3-dihydro-indol-2-ones as novel checkpoint 1 inhibitors. Bioorg Med Chem Lett (2006) 16(2):421–426.

[78] Wang L, Sullivan GM, Hexamer L et al.: Design, synthesis, and biological activity of 5,10-dihydro-dibenzo[b,e][1,4]diazepin-11-one based potent, and selective CHK-1 inhibitors. 232nd ACS National Meeting (2006):MEDI-125.

[79] Hexamer L, Wang L, Sullivan GM et al.: Investigation of novel 7-substituted and 8-substituted-5,10-dihydro-dibenzo[b,e][1,4]-diazepin-11-one based potent, and selective CHK-1 inhibitors. 233rd ACS National Meeting (2007):MEDI-123.

[80] Hasvold LA, Wang L, Sullivan GM et al.: Investigation of novel 7,8-disubstituted-5,10-dihydro-dibenzo[b,e][1,4] diazepin-11-ones as potent and selective CHK1 inhibitors. 233rd ACS National Meeting (2007):MEDI-122.

[81] Tong Y, Claiborne A, Stewart KD et al.: Discovery of 1,4-dihydroindeno[1,2-c]pyrazoles as a novel class of potent and selective checkpoint kinase 1 inhibitors. Bioorg Med Chem (2007) 15(7):2759–2767.

[82] Tong Y, Claiborne A, Tao Z-F et al.: Discovery of tricyclic pyrazoles as potent checkpoint-1 kinase inhibitors. 232nd ACS National Meeting (2006):MEDI-126.

[83] Tse AN, Embry M, Aardalen K et al.: CHIR 124, a novel and potent inhibitor of CHK1, potentiates the anti-tumor activity of topoisomerase I poisons in vitro and in vivo. Proc Am Assoc Cancer Res (2004) 45:Abs 2324.

[84] Hibner B: Small molecule inhibitors of CHK1 cause abrogation of the G2/M cell cycle checkpoint in vitro and tumor regression in a murine xenograft. Proc Am Assoc Cancer Res (2004) 45:Abs 2325.

[85] Liu X, Sampath D, Tseng J-C et al.: Abrogation of S-phase and G2 cell cycle checkpoints by small molecule inhibitors of the DNA damage kinase, CHK1. Proc Am Assoc Cancer Res (2005) 46:Abs 1682.

[86] Ni ZJ, Barsanti P, Brammeier N et al.: 4-(Aminoalkylamino)-3-benzimidazole-quinolinones as potent CHK-1 inhibitors. Bioorg Med Chem Lett (2006) 16(12):3121–3124.

[87] Tse AN, Rendahl KG, Sheikh T et al.: CHIR-124, a novel potent inhibitor of CHK1, potentiates the cytotoxicity of topoisomerase I poisons in vitro and in vivo. Clin Cancer Res (2007) 13(2 Pt 1):591–602.

[88] Chiron Corporation, Barsanti PA, Bussiere D et al.: Preparation of benzimidazole quinolinones for inhibiting a serine/threonine kinase. WO-2004018419 (2004).

[89] Chiron Corporation, Gesner TG, Barsanti PA et al.: Preparation of benzimidazole quinolinones for inhibiting a checkpoint kinase 1 and their use in combination therapy for cancer. US-2005256157 (2005).

[90] Merck & Co Inc, Brnardic EJ, Fraley ME et al.: New 2,5-dihydro-pyrazolo[4,3-c]quinolin-4-one derivatives are checkpoint kinase 1 inhibitors – useful for the treatment of cancer. WO-2006074207 (2006).

[91] Merck & Co Inc, Fraley ME, Steen JT: Imidazole derivatives as antitumor inhibitors of human checkpoint kinase CHK1. WO-2007015837 (2007).

[92] Merck & Co Inc, Brnardic EJ, Fraley ME et al.: Pyrazoloquinolinone inhibitors of checkpoint kinase CHK1 for use in cancer treatment. WO-2006074281 (2006).

[93] Merck & Co Inc, Arrington KL, Dudkin VY et al.: Benzoisoquinolines and aza derivatives as antitumor inhibitors of human checkpoint kinase CHK1. WO-2007008502 (2007).

[94] Merck & Co Inc, Arrington KL, Fraley ME et al.: Preparation of indolylindazole derivatives as inhibitors of checkpoint kinases. WO-2006086255 (2006).

[95] Merck & Co Inc, Arrington KL, Dudkin VY et al.: Preparation of (pyridyl)(amino)thiazole derivatives as checkpoint kinase inhibitors for treatment of cancer. WO-2006135604 (2006).

[96] Huang S, Garbaccio RM, Fraley ME et al.: Development of 6-substituted indolylquinolinones as potent Chek1 kinase inhibitors. Bioorg Med Chem Lett (2006) 16(22):5907–5912.

[97] Huang S, Garbaccio R, Fraley M et al.: Development of substituted indolylquinolinones as potent Chek1 kinase inhibitors. 232nd ACS National Meeting (2006):MEDI-129.

[98] Fraley ME, Steen JT, Brnardic EJ et al.: 3-(Indol-2-yl)inda-zoles as Chek1 kinase inhibitors: Optimization of potency and selectivity via substitution at C6. *Bioorg Med Chem Lett* (2006) **16**(23):6049–6053.

[99] Steen JT, Fraley ME, Brnardic E et al.: Development of substituted indolylindazoles as potent and selective Chek1 kinase inhibitors. *232nd ACS National Meeting* (2006):MEDI-127.

[100] Foloppe N, Fisher LM, Howes R et al.: Structure-based design of novel CHK1 inhibitors: Insights into hydrogen bonding and protein-ligand affinity. *J Med Chem* (2005) **48**(13):4332–4345.

[101] Foloppe N, Fisher LM, Howes R et al.: Identification of chemically diverse CHK1 inhibitors by receptorbased virtual screening. *Bioorg Med Chem* (2006) **14**(14):4792–4802.

[102] Foloppe N, Fisher LM, Francis G et al.: Identification of a buried pocket for potent and selective inhibition of CHK1: Prediction and verification. *Bioorg Med Chem* (2006) **14**(6):1792–1804.

[103] Vernalis Research Ltd, Walmsley DL, Drysdale MJ et al.: Preparation of pyrazolyl benzimidazoles as PDK1 and CHK1 kinase inhibitors for the treatment of cancer and autoimmune disorders. WO-2006134318 (2006).

[104] Biofocus DPI, Birault V, Woodland CA: Pyrimidin-4-yl-1*H*-indazol-5-yl-amines as CHK-1 kinase inhibitors, their preparation, pharmaceutical compositions, and use in therapy. WO-2005103036 (2005).

[105] Merck Patent GmbH, Mederski W, Gericke R et al.: Preparation of 3-amino-4-[(phenylmethyl)-amino]-3-cyclobutene-1,2-diones as CHK1, CHK2 and SGK kinase inhibitors. WO-2006105865 (2006).

[106] Merck Patent GmbH, Mederski W, Gericke R et al.: 3-Oxo-indazole-squaric acid derivatives. WO-2007022858 (2007).

[107] Merck Patent GmbH, Mederski W, Emde U et al.: Preparation of *N*-pyridinylphenyl-3-4-diaminocyclobut-3-ene-1,2-diones as CHK1, CHK2 and/or SGK kinase inhibitors for treating cancer. WO-2007014607 (2007).

[108] Merck Patent GmbH, Mederski W, Emde U et al.: New quadratic acid derivatives are kinase signal transduction modulators – useful for treating eg cancer, diabetes and obesity. WO-2007014608 (2007).

[109] Prudhomme M: Staurosporines and structurally-related indolocarbazoles as antitumor agents. In: *Anticancer Agents from Natural Products*. Cragg GM, Kingston DGI, Newman DJ (Eds), CRC, London, UK (2005):499–517.

[110] Prudhomme M: Biological targets of antitumor indolocar-bazoles bearing a sugar moiety. *Curr Med Chem Anticancer Agents* (2004) **4**(6):509–521.

[111] Henon H, Anizon F, Pfeiffer B et al.: Synthesis of dipyrrolo[3,4-*a*:3,4-*c*]carbazole-1,3,4,6-tetraones bearing a sugar moiety. *Tetrahedron* (2006) **62**(6):1116–1123.

[112] Anizon F, Pfeiffer B, Prudhomme M: Synthesis of pyridine-[3',2':4,5]pyrrolo[3,2-*g*]pyrrolo[3,4-*e*]indolizine-1,3-dione and pyrrolo-[3,2-*c*]pyrazole skeletons. *Tetrahedron Lett* (2006) **47**(4):433–436.

[113] Conchon E, Aboab B, Golsteyn RM et al.: Synthesis, *in vitro* antiproliferative activities, and CHK1 inhibitory properties of indolylpyrazolones and indolylpyridazinedione. *Eur J Med Chem* (2006) **41**(12):1470–1477.

[114] Hénon H, Messaoudi S, Anizon F et al.: Bis-imide granulatimide analogues as potent checkpoint 1 kinase inhibitors. *Eur J Pharmacol* (2007) **554**(2–3):106–112.

[115] Conchon E, Anizon F, Golsteyn RM et al.: Synthesis, *in vitro* antiproliferative activities, and CHK1 inhibitory

properties of dipyrrolo[3,4-*a*:3,4-*c*]carbazole-triones. *Tetrahedron* (2006) **62**(48):11136–11144.

[116] Messaoudi S, Anizon F, Peixoto P et al.: Synthesis and biological activities of 7-aza rebeccamycin analogues bearing the sugar moiety on the nitrogen of the pyridine ring. *Bioorg Med Chem* (2006) **14**(22):7551–7562.

[117] Hénon H, Anizon F, Golsteyn RM et al.: Synthesis and biological evaluation of new dipyrrolo[3,4-*a*:3,4-*c*]carbazole-1,3,4,6-tetraones, substituted with various saturated and unsaturated side chains via palladium catalyzed cross-coupling reactions. *Bioorg Med Chem* (2006) **14**(11):3825–3834.

[118] Henon H, Anizon F, Kucharczyk N et al.: Expedited synthesis of substituted dipyrrolo[3,4-*a*:3,4-*c*]carbazole-1,3,4,6-tetraones structurally related to granulatimide. *Synthesis* (2006) **4**:711–715.

[119] Prudhomme M, Messaoudi S, Anizon F et al.: Aza-rebeccamycin analogs as potent checkpoint kinase 1 inhibitors. *233rd ACS National Meeting* (2007):MEDI-124.

[120] Prudhomme M, Hénon H, Hugon B et al.: Granulatimide analogues as potent CHK1 inhibitors. *Proc Am Assoc Cancer Res* (2006) **47**:Abs134.

[121] Roy S, Eastman A, Gribble GW: Synthesis of bisindolylmaleimides related to GF109203x and their efficient conversion to the bioactive indolocarbazoles. *Org Biomol Chem* (2006) **4**(17):3228–3234.

[122] Roy S, Eastman A, Gribble GW: Synthesis of novel bioactive indolocarbazoles. *229th ACS National Meeting* (2005):ORGN-131.

[123] Nair JK, Willoughby JL, Chan A et al.: Multivalent N-acetylgalactosamine-conjugated siRNA localizes in hepatocytes and elicits robust RNAi-mediated gene silencing. *J Am Chem Soc* (2014) **136**:16958–16961.

[124] Gao S, Dagnaes-Hansen F, Nielsen EJB et al.: The effect of chemical modification and nanoparticle formulation on stability and biodistribution of siRNA in mice. *Mol Ther* (2009) **17**:1225–1233.

[125] Turner JJ, Jones SW, Moschos SA et al.: MALDI-TOF mass spectral analysis of siRNA degradation in serum confirms an RNAse A-like activity. *Mol Bio Syst* (2007) **3**:43–50.

[126] Tabernero J, Shapiro GI, LoRusso PM et al.: First-in-humans trial of an RNA interference therapeutic targeting VEGF and KSP in cancer patients with liver involvement. *Cancer Discov* (2013) **3**:406–417.

[127] Schultheis B, Strumberg D, Vank AC et al.: First-in-human phase I study of the liposomal RNA interference therapeutic Atu027 in patients with advanced solid tumors. *J Clin Oncol* (2014) **32**:4141–4148.

[128] Santel A, Aleku M, Keil O et al.: A novel siRNA-lipoplex technology for RNA interference in the mouse vascular endothelium. *Gene Ther* (2006) **13**:1222–1234.

[129] Bartlett DW, Davis ME: Physiochemical and biological characterization of targeted, nucleic acid-containing nanoparticles. *Bioconjug Chem* (2007) **18**:456–468.

[130] Zuckerman JE, Gritli I, Tolcher A et al.: Correlating animal and human phase Ia/Ib clinical data with CALAA-01, a targeted, polymer-based nanoparticle containing siRNA. *Proc Natl Acad Sci USA* (2014) **111**:11449–11454.

[131] Kim HJ, Oba M, Pittella F et al.: PEG-detachable cationic polyaspartamide derivatives bearing stearoyl moieties for systemic siRNA delivery toward subcutaneous BxPC3 pancreatic tumor. *J Drug Target* (2012) **20**:33–42.

[132] Yagi N, Manabe I, Tottori T et al.: A nanoparticle system specifically designed to deliver short interfering

RNA inhibits tumor growth in vivo. *Cancer Res* (2009) **69**:6531–6538.

[133] Sun Q, Kang Z, Xue L et al.: A collaborative assembly strategy for tumor-targeted siRNA delivery. *J Am Chem Soc* (2015) **137**:6000–6010.

[134] Oe Y, Christie RJ, Naito M et al.: Actively-targeted polyion complex micelles stabilized by cholesterol and disulfide cross-linking for systemic delivery of siRNA to solid tumors. *Biomaterials* (2014) **35**:7887–7895.

[135] Li J, Yu X, Wang Y et al.: A reduction and pH dualsensitive polymeric vector for long-circulating and tumor-targeted siRNA delivery. *Adv. Mater.* (2014) **26**:8217–8224.

[136] Wang H-X, Yang X-Z, Sun C-Y et al.: Matrix metalloproteinase 2-responsive micelle for siRNA delivery. *Biomaterials* (2014) **35**:7622–7634.

[137] Deng ZJ, Morton SW, Ben-Akiva E et al.: Layer-by-layer nanoparticles for systemic codelivery of an anticancer drug and siRNA for potential triple-negative breast cancer treatment. *ACS Nano* (2013) **7**:9571–9584.

[138] Jensen SA, Day ES, Ko CH et al.: Spherical nucleic acid nanoparticle conjugates as an RNAi-based therapy for glioblastoma. *Sci Transl Med* (2013) **5**:209ra152.

[139] Kim HJ, Takemoto H, Yi Y et al.: Precise engineering of siRNA delivery vehicles to tumors using polyion complexes and gold nanoparticles. *ACS Nano* (2014) **8**:8979–8991.

[140] Shimizu H, Hori Y, Kaname S et al.: siRNA-based therapy ameliorates glomerulonephritis. *J Am Soc Nephrol* (2010) **21**:622–633.

[141] Sizovs A, Song X, Waxham MN et al.: Precisely tunable engineering of sub-30 nm monodisperse oligonucleotide nanoparticles. *J Am Chem Soc* (2014) **136**:234–240.

[142] Dohmen C, Edinger D, Fröhlich T et al.: Nanosized multifunctional polyplexes for receptor-mediated siRNA delivery. *ACS Nano* (2012) **6**:5198–5208.

[143] Li F, Mahato RI: MicroRNAs and drug resistance in prostate cancers. *Mol Pharm* (2014) **11**:2539–2552.

[144] Choi KM, Choi SH, Jeon H et al.: Chimeric capsid protein as a nanocarrier for siRNA delivery: stability and cellular uptake of encapsulated siRNA. *ACS Nano* (2011) **5**:8690–8699.

[145] Ledford H: Drug giants turn their backs on RNA interference. *Nature* (2010) **468**:487.

[146] Zeng X, Sun YX, Qu W et al.: Biotinylated transferrin/avidin/biotinylated disulfide containing PEI bioconjugates mediated p53 gene delivery system for tumor targeted transfection. *Biomaterials* (2010) **31**:4771–4780.

[147] Batrakova EV, Kabanov AV: Pluronic block copolymers: evolution of drug delivery concept from inert nanocarriers to biological response modifiers. *J Control Release* (2008) **130**:98–106.

[148] Panyam J, Labhasetwar V: Biodegradable nanoparticles for drug and gene delivery to cells and tissue. *Adv Drug Deliv Rev* (2012) **64**:61–71.

[149] Hu QL, Jiang QY, Jin X et al.: Cationic microRNA-delivering nanovectors with bifunctional peptides for efficient treatment of PANC-1 xenograft model. *Biomaterials* (2013) **34**:2265–2276.

[150] Hwang DW, Son S, Jang J et al.: A brain-targeted rabies virus glycoprotein-disulfide linked PEI nanocarrier for delivery of neurogenic microRNA. *Biomaterials* (2011) **32**:4968–4975.

[151] Price TO, Farr SA, Yi X et al.: Transport across the blood–brain barrier of pluronic leptin. *J Pharmacol Exp Ther* (2010) **333**:253–263.

[152] Zhang X, Koh CG, Yu B et al.: Transferrin receptor targeted lipopolyplexes for delivery of antisense oligonucleotide g3139 in a murine k562 xenograft model. *Pharm Res* (2009) **26**:1516–1524.

[153] Pardridge WM: Blood–brain barrier drug delivery of IgG fusion proteins with a transferrin receptor monoclonal antibody. *Expert Opin Drug Deliv* (2014):1–16.

[154] Zhou QH, Boado RJ, Hui EK et al.: Chronic dosing of mice with a transferrin receptor monoclonal antibody-glial-derived neurotrophic factor fusion protein. *Drug Metab Dispos* (2011) **39**:1149–1154.

[155] Huang X, Schwind S, Yu B et al.: Targeted delivery of microRNA-29b by transferrin-conjugated anionic lipopolyplex nanoparticles: a novel therapeutic strategy in acute myeloid leukemia. *Clin Cancer Res* (2013) **19**:2355–2367.

[156] Yang N, Coukos G, Zhang L: MicroRNA epigenetic alterations in human cancer: one step forward in diagnosis and treatment. *Int J Cancer* (2008) **122**:963–8.

[157] Bueno MJ, Pérez de Castro I, Gómez de Cedrón M et al.: Genetic and epigenetic silencing of microRNA-203 enhances ABL1 and BCR-ABL1 oncogene expresión. *Cancer Cell* (2008) **13**:496–506.

[158] Rupaimoole R, Slack FJ: MicroRNA therapeutics: towards a new era for the management of cancer and other diseases. *Nat Rev Drug Discovery* (2017) **16**(3):203–222.

[159] Whitehead KA, Langer R, Anderson DG: Knocking down barriers: advances in siRNA delivery. *Nat Rev Drug Discovery* (2009) **8**:129–138.

[160] Zamora MR, Budev M, Rolfe M et al.: RNA interference therapy in lung transplant patients infected with respiratory syncytial virus. *Am J Respir Crit Care* (2011) **183**:531–538.

[161] Kaiser PK, Symons RC, Shah SM et al.: SiRNA-027 study I. RNAi-based treatment for neovascular age-related macular degeneration by siRNA-027. *Am J Opthalmol* (2010) **150**:33–39. e32.

[162] Leachman SA, Hickerson RP, Schwartz ME et al.: First-in-human mutation-targeted siRNA phase ib trial of an inherited skin disorder. *Mol Ther* (2010) **18**:442–446.

[163] Thompson JD: Clinical development of synthetic siRNA therapeutics. *Drug Discov Today* (2013) **10**: e133–e138.

[164] Kanasty R, Dorkin JR, Vegas A et al.: Delivery materials for siRNA therapeutics. *Nat Mater* (2013) **12**:967–977.

[165] Sarett SM, Nelson CE, Duvall CL: Technologies for controlled, local delivery of siRNA. *J Controlled Release* (2015) **218**:94–113.

[166] Thompson JD: Clinical development of synthetic siRNA therapeutics. *Drug Discov Today* **10** (2013) e133–e138.

[167] Snead NM, Rossi JJ: RNA interference trigger variants: getting the most out of RNA for RNA interference-based therapeutics. *Nucleic Acid Ther* (2012) **22**:139–146.

[168] Jackson AL, Bartz SR, Schelter J et al.: Expression profiling reveals off-target gene regulation by RNAi. *Nature Biotechnol* (2003) **21**(6):635–637.

[169] Tse AN, Carvajal R, Schwartz GK: Targeting checkpoint kinase 1 in cancer therapeutics. *Clin Cancer Res* (2007) **13**(7):1955–1960.

[170] Ho CC, Siu WY, Chow JP et al.: The relative contribution of CHK1 and CHK2 to adriamycin-induced checkpoint. *Exp Cell Res* (2005) **304**(1):1–15.

[171] Morgan MA, Parsels LA, Parsels JD et al.: The relationship of premature mitosis to cytotoxicity in response to checkpoint abrogation and antimetabolite treatment. *Cell Cycle* (2006) **5**(17):1983–1988.

[172] Xiao Z, Xue J, Sowin TJ et al.: Differential roles of check-point kinase 1, checkpoint kinase 2, and mitogen-activated protein kinase-activated protein kinase 2 in mediating DNA damage induced cell cycle arrest: Implications for cancer therapy. *Mol Cancer Ther* (2006) **5**(8):1935–1943.

[173] Xiao Z, Xue J, Sowin TJ et al.: A novel mechanism of checkpoint abrogation conferred by CHK1 downregulation. *Oncogene* (2005) **24**(8):1403–1411.

[174] Morgan MA, Parsels LA, Parsels JD et al.: Role of check-point kinase 1 in preventing premature mitosis in response to gemcitabine. *Cancer Res* (2005) **65**(15):6835–6842.

[175] Cho SH, Toouli CD, Fujii GH et al.: CHK1 is essential for tumor cell viability following activation of the replication checkpoint. *Cell Cycle* (2005) **4**(1):131–139.

[176] Karnitz LM, Flatten KS, Wagner JM et al.: Gemcitabine-induced activation of checkpoint signaling pathways that affect tumor cell survival. *Mol Pharmacol* (2005) **68**(6):1636–1644.

[177] Sampath D, Cortes J, Estrov Z et al.: Pharmacodynamics of cytarabine alone and in combination with 7-hydroxys-taurosporine (UCN-01) in AML blasts *in vitro* and during a clinical trial. *Blood* (2006) **107**(6):2517–2524.

[178] Xiao Z, Xue J, Semizarov D et al.: Novel indication for cancer therapy: CHK1 inhibition sensitizes tumor cells to antimitotics. *Int J Cancer* (2005) **115**(4):528–538.

[179] Ren Q, Liu R, Dicker A et al.: CHK1 affects cell sensitiv-ity to microtubule-targeted drugs. *J Cell Physiol* (2005) **203**(1):273–276.

[180] Zachos G, Black EJ, Walker M et al.: CHK1 is required for spindle checkpoint function. *Dev Cell* (2007) **12**(2):247–260.

[181] Duensing A, Teng X, Liu Y et al.: A role of the mitotic spin-dle checkpoint in the cellular response to DNA replication stress. *J Cell Biochem* (2006) **99**(3):759–769.

[182] Wang H-Y, Zhang M, Zou P et al.: Mechanism of G2/M blockage triggered by activated-CHK1 in regulation of drug-resistance in K562/A02 cell line. *J Exp Hematol* (2006) **14**(6):1105–1109.

[183] Bao S, Wu Q, McLendon RE et al.: Glioma stem cells pro-mote radioresistance by preferential activation of the DNA damage response. *Nature* (2006) **444**(7120):756–760.

[184] Madkour LH: *Reactive Oxygen Species (ROS), Nanoparticles, and Endoplasmic Reticulum (ER) Stress-Induced Cell Death Mechanisms.* Academic Press, 2020. https://www.elsevier.com/books/reactive-oxygen-species-ros-nanoparticles-and-endoplasmic-reticulum-er-stress-induced-cell-death-mechanisms/madkour/978-0-12-822481-6

[185] Madkour LH: *Nanoparticles Induce Oxidative and Endoplasmic Reticulum Antioxidant Therapeutic Defenses.* 1st ed. Springer International Publishing: Switzerland, 2020. https://www.springer.com/gp/book/9783030372965?utm_campaign=3_pier05_buy_print&utm_content=en_08082017&utm_medium=referral&utm_source=google_books#otherversion=9783030372972

[186] Madkour LH: *Nucleic Acids as Gene Anticancer Drug Delivery Therapy.* 1st ed. Elsevier, 2020. https://www.elsevier.com/books/nucleic-acids-as-gene-anticancer-drug-deliverytherapy/madkour/978-0-12-819777-6

[187] Madkour LH: *Nanoelectronic Materials: Fundamentals and Applications (Advanced Structured Materials).* 1st ed. Springer International Publishing: Switzerland, 2019. https://books.google.com.eg/books/about/Nanoelectronic_Materials.html?id=YQXCxAEACAAJ&source=kp_book_description&redir_esc=y https://www.springer.com/gp/book/9783030216207

[188] Madkour LH: *Nanoelectronic Materials Fundamentals and Applications*, Front Matter. Advanced Structured Materials Book Series. STRUCTMAT. Springer International Publishing, 2019, **116**, i–xlv. https://link.springer.com/content/pdf/bfm%3A978-3-030-21621-4%2F1.pdf

[189] Madkour LH: *Introduction to Nanotechnology (NT) and Nanomaterials (NMs).* Advanced Structured Materials Book Series. STRUCTMAT. Springer International Publishing, 2019, **116**, 1–47. https://link.springer.com/chapter/10.1007%2F978-3-030-21621-4_1

[190] Madkour LH: *Principles of Computational Simulations Devices and Characterization of Nanoelectronic Materials.* Advanced Structured Materials Book Series. STRUCTMAT. Springer International Publishing, 2019, **116**, 49–89. https://link.springer.com/chapter/10.1007%2F978-3-030-21621-4_2

[191] Madkour LH. *Where Are Nanomaterials (Nms) Found?* Advanced Structured Materials Book Series. STRUCTMAT. Springer International Publishing, 2019, **116**, 91–100. https://link.springer.com/chapter/10.1007%2F978-3-030-21621-4_3

[192] Madkour LH: *Benefits of Nanomaterials and Nanowire Geometry.* Advanced Structured Materials Book Series. STRUCTMAT. Springer International Publishing, 2019, **116**, 101–121. https://link.springer.com/chapter/10.1007%2F978-3-030-21621-4_4

[193] Madkour LH: *Why So Much Interest in Nanomaterials (NMs)?* Advanced Structured Materials Book Series. STRUCTMAT. Springer International Publishing, 2019, **116**, 123–140. https://link.springer.com/chapter/10.1007%2F978-3-030-21621-4_5

[194] Madkour LH: *Examples of Nanomaterials with Various Morphologies.* Advanced Structured Materials Book Series. STRUCTMAT. Springer International Publishing, 2019, **116**, 141–164. https://link.springer.com/chapter/10.1007%2F978-3-030-21621-4_6

[195] Madkour LH: *Carbon Nanomaterials and Two-Dimensional Transition Metal Dichalcogenides (2D TMDCs).* Advanced Structured Materials book Series. STRUCTMAT. Springer International Publishing, 2019, **116**, 165–245. https://link.springer.com/chapter/10.1007%2F978-3-030-21621-4_7

[196] Madkour LH: *Nanoelectronics and Role of Surfaces Interfaces.* Advanced Structured Materials Book Series. STRUCTMAT. Springer International Publishing, 2019, **116**, 247–267. https://link.springer.com/chapter/10.1007%2F978-3-030-21621-4_8

[197] Madkour LH: *Classification of Nanostructured Materials.* Advanced Structured Materials Book Series. STRUCTMAT. Springer International Publishing, 2019, **116**, 269–307. https://link.springer.com/chapter/10.1007%2F978-3-030-21621-4_9

[198] Madkour LH: *Processing of Nanomaterials (NMs).* Advanced Structured Materials Book Series. STRUCTMAT. Springer International Publishing, 2019, **116**, 309–353. https://link.springer.com/chapter/10.1007%2F978-3-030-21621-4_10

[199] Madkour LH: *Techniques for Elaboration of Nanomaterials.* Advanced Structured Materials Book Series. STRUCTMAT. Springer International Publishing, 2019, **116**, 355–391. https://link.springer.com/chapter/10.1007%2F978-3-030-21621-4_11

[200] Madkour LH: *Synthesis Methods For 2D Nanostructured Materials, Nanoparticles (NPs), Nanotubes (NTs) and*

*Nanowires (NWs)*. Advanced Structured Materials Book Series. STRUCTMAT. Springer International Publishing, 2019, **116**, 393–456. https://link.springer.com/chapter/10.1007%2F978-3-030-21621-4_12

[201] Madkour LH: *Chemistry and Physics for Nanostructures Semiconductivity*. Advanced Structured Materials Book Series. STRUCTMAT. Springer International Publishing, 2019, **116**, 457–478. https://link.springer.com/chapter/10.1007%2F978-3-030-21621-4_13

[202] Madkour LH: *Properties of Nanostructured Materials (NSMs) and Physicochemical Properties of (NPs)*. Advanced Structured Materials Book Series. STRUCTMAT. Springer International Publishing, 2019, **116**, 479–564. https://link.springer.com/chapter/10.1007%2F978-3-030-21621-4_14

[203] Madkour LH: *Applications of Nanomaterials and Nanoparticles*. Advanced Structured Materials Book Series. STRUCTMAT. Springer International Publishing, 2019, **116**, 565–603. https://link.springer.com/chapter/10.1007%2F978-3-030-21621-4_15

[204] Madkour LH: *Environmental Impact of Nanotechnology and Novel Applications of Nano Materials and Nano Devices*. Advanced Structured Materials Book Series. STRUCTMAT. Springer International Publishing, 2019, **116**, 605–699. https://link.springer.com/chapter/10.1007%2F978-3-030-21621-4_16

[205] Madkour LH: *Interfacing Biology Systems with Nanoelectronics for Nanodevices*. Advanced Structured Materials Book Series. STRUCTMAT. Springer International Publishing, 2019, **116**, 701–759. https://link.springer.com/chapter/10.1007%2F978-3-030-21621-4_17

[206] Madkour LH: Nanoelectronic materials fundamentals and applications, Back Matter. *Advanced Structured Materials Book Series*. STRUCTMAT. Springer International Publishing, 2019, **116**, 761–783. https://link.springer.com/content/pdf/bbm%3A978-3-030-21621-4%2F1.pdf

[207] Madkour LH: Ecotoxicology-nanotoxicology and reactive oxygen species (ROS) stress combination of free radicals and nanoparticles towards antioxidant defense therapeutics. *J Targeted Drug Delivery* (2019) **3**(1):1–58.

[208] Madkour LH: Nanoparticles as targeted drug co-delivery in cancer therapeutics. *Chron Pharm Sci J* (2019) **3**(2):800–804. https://scientiaricerca.com/srcops/SRCOPS-03-00081.php

[209] Madkour LH: Applications of gold nanoparticles in medicine and therapy. *Pharm Pharmacol Int J* (2018) **6**(3):157–174. DOI:10.15406/ppij.2018.06.00172. http://medcraveonline.com/PPIJ/PPIJ-06-00172.pdf

[210] Madkour LH: Biogenic–biosynthesis metallic nanoparticles (MNPs) for pharmacological. *Chron Pharm Sci J* (2018) **2**(1):384–444. https://scientiaricerca.com/srcops/SRCOPS-02-00038.php

[211] Madkour LH: Ecofriendly green biosynthesized of metallic nanoparticles: bio-reduction mechanism, characterization and pharmaceutical applications in biotechnology industry. *Global Drugs Therap* (2018) **3**(1):1–11 http://www.oatext.com/ecofriendly-green-biosynthesized-of-metallic-nanoparticles-bio-reduction-mechanism-characterization-and-pharmaceutical-applications-in-biotechnology-industry.php

[212] Madkour LH: Toxic effects of environmental heavy metals on cardiovascular pathophysiology and heart health function: chelation therapeutics. *UPI J Pharm, Med Health Sci* (2018) **1**(1):19–62. https://uniquepubinternational.com/wp-content/uploads/2018/03/UPI-JPMHS-2018-7.pdf

[213] Madkour LH: Review Article: Advanced AuNMs as nanomedicine's central goals capable of active targeting in both imaging and therapy in biomolecules. *Global Drugs Therap* (2017) **2**(6):1–12. http://www.oatext.com/advanced-aunms-as-nanomedicines-central-goals-capable-of-active-targeting-in-both-imaging-and-therapy-in-biomolecules.php

[214] Madkour LH: Biotechnology of nucleic acids medicines as gene therapeutics and their drug complexes. *Chron Pharm Sci J* (2017) **1**(4):204–253. https://scientiaricerca.com/srcops/pdf/SRCOPS-01-00023.pdf

[215] Madkour LH: Advanced AuNMs as nanomedicine's central goals capable of active targeting in both imaging and therapy in biomolecules. *Glob Drugs Therap* (2017) **2**(6):1–12. https://www.oatext.com/pdf/GDT-2-136.pdf

[216] Madkour LH: Vision for life sciences: interfaces between nanoelectronic and biological systems. *Global Drugs Therap* (2017) **2**(4):1–4. DOI: 10.15761/GDT.1000126. https://oatext.com/Vision-for-life-sciences-interfaces-between-nanoelectronic-and-biological-systems.php

[217] Knific T, Osredkar J, Smrkolj Š et al.: Novel algorithm including CA-125, HE4 and body mass index in the diagnosis of endometrial cancer. *Gynecol Oncol* (2017) **147**:126–132.

[218] Persson J, Geppert B, Lönnerfors C et al.: Description of a reproducible anatomically based surgical algorithm for detection of pelvic sentinel lymph nodes in endometrial cancer. *Gynecol Oncol* (2017) **147**:120–125.

[219] Geppert B, Lonnerfors C, Bollino M et al.: A study on uterine lymphatic anatomy for standardization of pelvic sentinel lymph node detection in endometrial cancer. *Gynecol Oncol* (2017) **145** (2):256–261.

[220] Delamere G: *The Lymphatics*. W.T. Keener and Co: Chicago, 1904.

[221] Leveuf J, Godard H: Les lymphatiques de l'utérus. *Rev Chir* (1923):219–224.

[222] Ackerman LV, Regato J: Cancer. *Diagn, Treat Prognosis*. The CV Mosby Company: St Louis, 1947.

[223] Favero G, Pfiffer T, Ribeiro A et al.: Laparoscopic sentinel lymph node detection after hysteroscopic injection of technetium-99 in patients with endometrial cancer. *Int J Gynecol Cancer* (2015) **25**(3):423–430.

[224] Solima E, Martinelli F, Ditto A et al.: Diagnostic accuracy of sentinel node in endometrial cancer by using hysteroscopic injection of radiolabeled tracer. *Gynecol Oncol* (2012) **126**(3):419–423.

[225] Mariani A, Webb MJ, Keeney GL et al.: Routes of lymphatic spread: a study of 112 consecutive patients with endometrial cancer. *Gynecol Oncol* (2001) **81**(1):100–104.

[226] How J, Boldeanu I, Lau S et al.: Unexpected locations of sentinel lymph nodes in endometrial cancer. *Gynecol Oncol* (2017) **147**:18–23.

[227] Touhami O, Grégoire J, Renaud M-C et al.: Performance of sentinel lymph node (SLN) mapping in high-risk endometrial cancer. *Gynecol Oncol* (2017) **147**:549–553.

[228] Dabi Y, Uzan J, Bendifallah S et al.: For the Groupe de Recherche FRANCOGYN. Prognostic value of local relapse for patients with endometrial cancer. *EJSO* (2017) **43**:2143–2149

•• of outstanding interest
• of special interest

# 15 Recent Therapeutic Prospects of miRNAs and siRNA Delivery Systems in Cancer Treatment Nanobiotechnology

## 15.1 ROLE OF MiRNAs IN REGULATING DISEASE

MicroRNA (miRNA) molecules are small, single-stranded RNA molecules that function to regulate networks of genes. They play important roles in normal female reproductive tract biology, as well as in the pathogenesis and progression of epithelial ovarian cancer (EOC). Drosha, Dicer, and Argonaute proteins are components of the miRNA regulatory machinery and mediate miRNA production and function. Understanding the regulation of miRNA molecule production and function may facilitate the development of novel diagnostic and therapeutic strategies to improve the prognosis of women with EOC. Additionally, understanding miRNA molecules and miRNA regulatory machinery associations with clinical features may influence prevention and early-detection efforts.

MiRNAs were initially identified in *Caenorhabditis elegans* in 1993, as small noncoding RNAs, which modulate eukaryotic gene expression at posttranscriptional levels. Lee et al. first reported that the lin-4 gene was involved in larval development of *C. elegans* [1]. MiRNAs play an important role in various physiological and pathological conditions including embryonic differentiation, viral infection, cardiac hypertrophy, hematopoiesis, and oncogenesis. MiRNAs control a wide variety of cellular processes, such as proliferation, cell death, differentiation, motility, and invasiveness [2, 3].

Different biological and physicochemical factors govern the effective delivery of miRNAs (Figure 15.1). Delivery can be improved through various approaches, such as [4, 5]:

1. Minimization of degradation and elimination of miRNAs by optimization of particle size, surface charge, and chemical modification;
2. Use of targeting ligands and cell-penetrating moieties to improve the tissue permeation; and
3. Use of fusogenic peptides to avoid intracellular disposition of the miRNAs.

MiRNA-based therapeutics represent one of the major commercial areas of interest in today's biotechnology market [6]. MiRNAs are well-identified and associated in various disease pathologies such as is the case for miR-208 in

heart failure [7], miR-15/195 in post-myocardial infarction remodeling [8], miR-145 in vascular disease [9], miR-451 in myeloproliferative disease [10], miR-29 in pathological fibrosis [11], miR34 and let-7 in cancer [12], and miR122 in liver transplant [13].

Roy and Sen [6] first demonstrated that miRNAs regulate the cellular redox environment via a NADPH oxidase-dependent mechanism in human microvascular endothelial cells. They demonstrated that the hypoxia-sensitive miR-200b was involved in the induction of angiogenesis by directly targeting erythroblastosis virus E26 oncogene homolog 1 (Ets-1) in human microvascular endothelial cells. These studies lend support to the potential role of miRNAs in wound healing and angiogenesis. MiRNAs block translation of messenger RNAs (mRNAs) or repress the synthesis of protein via mRNA destabilization [5, 6]. They also serve a fundamental role during the development of the organism, through effects on cell differentiation and metabolism [14]. In 2000, a second miRNA (let-7) was identified from *C. elegans* and characterized as a 21-nucleotide (nt) small RNA [15]. Since these early studies, it is now known that more than 60% of human protein-coding genes are regulated by miRNAs [16]. The miRBase database provides a searchable online repository for published miRNA sequences and associated annotation [14]. The miRBase database contains more than 28,645 hairpin precursors and more than 35,828 mature miRNA sequences in 223 species (Release 21) [17].

Hebert et al. observed that a small number of miRNAs have altered expression levels in patients with Alzheimer's disease [18]. Caporali and Emanueli reviewed the role of miRNAs in angiogenesis focusing on post-ischemic neovascularization [19]. Using locked nucleic acid, anti-miR21 oligonucleotides, bimodal imaging vectors, and neural precursor cells (NPCs) expressing a secretable variant of the cytotoxic agent tumor necrosis factor-related apoptosis-inducing ligand (STRAIL), the combined suppression of miR-21 and NPCS-TRAIL leads to a synergistic increase in caspase activity and significantly decreased cell viability in human glioma cells in vitro [20]. In a recent report, Hsu et al. found that influenza A virus infections led to increased inflammatory and antiviral responses in primary bronchial epithelial cells from healthy nonsmoking and smoking subjects. The authors reported increased expression of

DOI: 10.1201/9781003229650-15

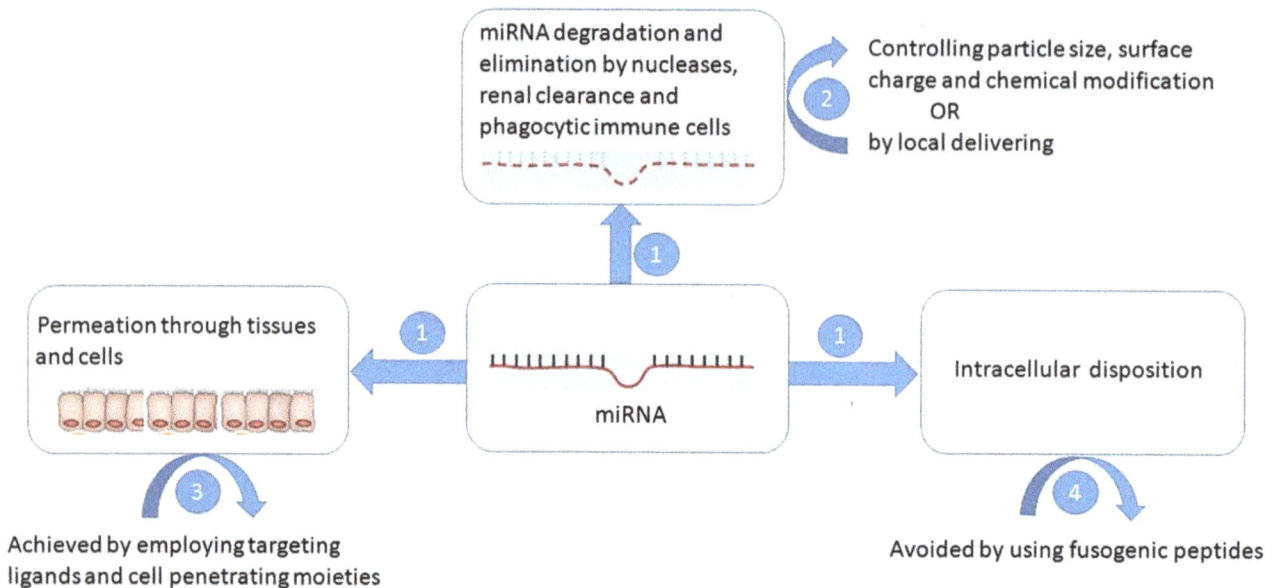

**FIGURE 15.1** Challenges (1) and solutions (2–4) for effective delivery of miRNAs.

miR-125a or miR-125b due to influenza A virus infection, which reduced the expression of A20 (TNFAIP3), a negative regulator of NF-κB-mediated inflammatory responses, and mitochondrial antiviral signaling. This leads to the exaggeration of inflammation and impaired antiviral responses [21]. MiRNAs also play an important role in autism spectrum disorder, and recent findings associate the condition with genetic variants in miRNA genes, miRNA biogenesis genes, and miRNA targets [22].

Calin et al. evaluated the miRNA expression profiles in chronic lymphocytic leukemia samples. They identified germline mutations in the miR-16-1-miR-15a primary precursor, which caused low levels of miRNA expression in vitro and in vivo and was associated with the deletion of the normal allele [23]. In a recent study, Zhao et al. found that the downregulation of miR-493-5p in hepatocellular carcinoma (HCC) was correlated with tumor size, tumor differentiation, grade, and tumor/node/metastasis stage of HCC patients. The authors [24] reported that the miR-493-5p was downregulated in HCC. This miRNA acts to suppress the proliferation of HCC cells via targeting Golgiprotein73 [25]. Sun et al. observed the upregulation of miR-223 in colorectal cancer tissues and the downregulation of RAS p21GTPase-activating protein 1 (RASA1) in colorectal cancer tissues. The in vivo xenograft model of colorectal cancer suggested that the upregulation of miR-223 could promote tumor growth and that the inhibition of miR-223 might prevent solid tumor growth [24, 26].

## 15.2 DESCRIPTION OF MiRNA BIOGENESIS AND REGULATION

RNA polymerase II transcribes miRNA molecules from genomic DNA into a primary miRNA molecule (pri-miRNA). Pri-miRNA molecules are typically >200 nt in length with a characteristic stem-loop structure. Furthermore, miRNA clusters containing multiple stem-loop structures [27], each coding for a mature miRNA molecule, can be in the kilobase size range. Pri-miRNA molecules are recognized by Drosha, an RNAse III, which cuts the double-stranded RNA into ~70-nt precursor miRNA (pre-miRNA) in the nucleus. Pre-miRNA molecules are exported to the cytoplasm and are processed by Dicer, an RNAse III, into two unique single-stranded mature miRNA molecules representing each side of the stem-loop structure. Mature miRNA molecules are loaded onto the Argonaute-containing RNA-induced silencing complex (RISC). Within this structure, mature miRNA molecules function to repress gene expression by complementary binding of the 3′ untranslated region (UTR) of the target gene to the miRNA seed sequence, nucleotides 2–8 of the mature miRNA molecule, leading to transcript degradation, and subsequent gene product loss [28]. Studies have shown that miRNA target genes play an important role in EOC cancer biology [29]. Thus, miRNA molecules and their biogenesis regulation as mediated by miRNA machinery are clinically important.

The regulation of eukaryotic gene expression is based on the cytoplasmic control of mRNA translation and degradation. The regulation of miRNAs is referred to as RNA interference (RNAi) or RNA silencing [30]. MiRNAs are generated through the cleavage of primary miRNA (pri-miRNA), which incorporates into the effector complex RISC [31]. The presence of a single (or multiple) imperfect hairpin structure(s) with a stem of approximately 33 bp is the characteristic feature of a pri-miRNA [19].

In brief, the miRNAs are first transcribed into pri-miRNAs in the presence of polymerase II or polymerase III.

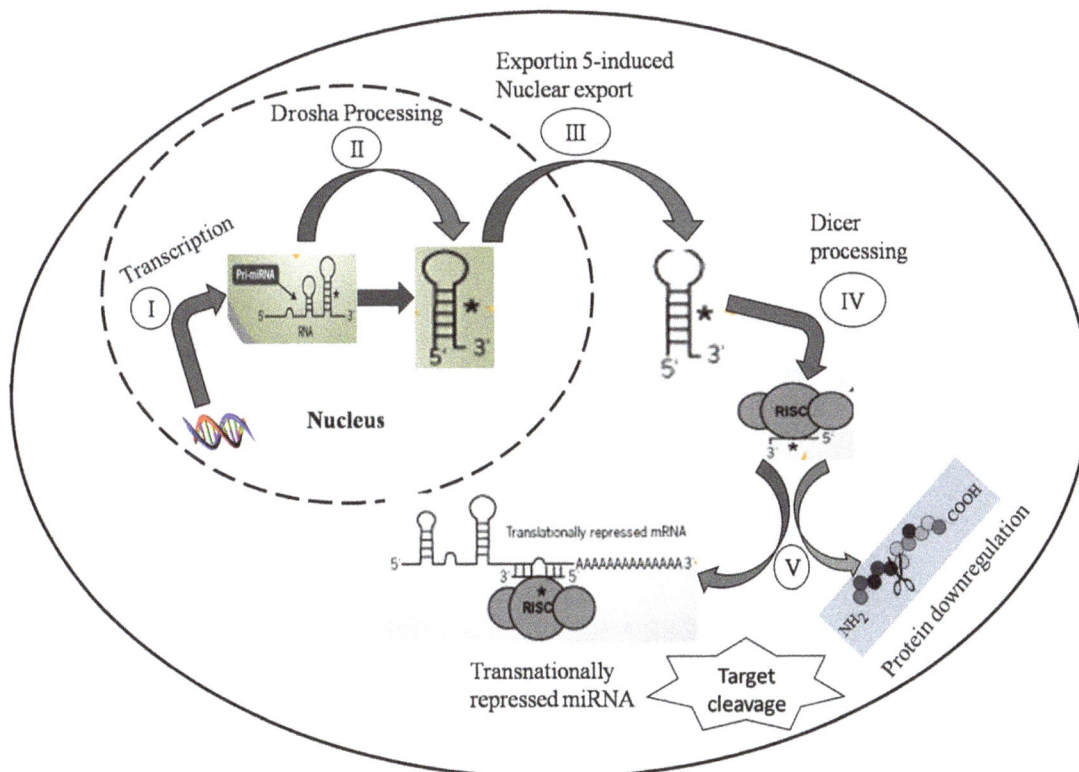

**FIGURE 15.2** The process of miRNA biogenesis. MiRNA genesis initiates in the nucleus. (I) In the presence of RNA polymerase II, miRNA genes are transcribed and produce pri-miRs. (II) Pri-miRNAs are catalyzed by Drosha to produce pre-miRNAs. (III) After nuclear export to the cytoplasm, Dicer processes pre-miRNAs into 20-bp miR:miR duplexes. (IV) After miRNA duplex unwinding initiated by Dicer, one strand is selected to function as mature miR and loaded into the RNA-induced silencing complex (RISC). (V) The mature miRNA leads to mRNA translational repression, as well as protein downregulation. (Adapted from [5, 33, 34].)

In the presence of the RNAse III enzyme, Drosha, the synthesis of pri-miRNAs begins in the nucleus. In complex with other proteins (double-stranded RNA-binding domain protein called Pasha in *Drosophila* or DGCR8 in mammals), the pri-miRNAs convert into precursor miRNAs (pre-miRNAs). Pre-miRNAs are transported into the cytoplasm by exportin-5, a RanGTP-dependent dsRNA-binding protein, and are subsequently processed by Dicer (a cytoplasmic endonuclease RNAse III enzyme) that generates a miRNA duplex [26].

The mature duplex miRNA is then incorporated into the RISC, a ribonucleoprotein effector containing a catalytic endonuclease core (Argonaute 2); Dicer, a dsRNA-binding protein-transactivating response RNA-binding protein; and a protein activator kinase R [32]. The process of miRNA biogenesis is presented in Figure 15.2.

Circulating miRNAs are released into the blood, and their expression level is specifically related to disease stage. Circulating miRNAs are considerably stable and can be easily evaluated through blood sampling and following molecular analysis [35]. A single miRNA molecule can target multiple mRNAs [36]. MiRNAs have attracted wide attention by both the biological and formulation scientists due to their unique functional significance and mode of action [37].

## 15.3 MiRNA AND SiRNA DELIVERY SYSTEMS FOR APPLICATIONS IN CANCER THERAPY

In the last decades, alternative delivery systems functionalized with RNA (miRNA and siRNA) have gained increasing attention. Here, we summarize the mechanism of action and the benefits of such delivery systems in cancer therapy. At the moment, these delivery systems are intensively studied in in vitro/in vivo experiments and also in clinical trials.

MiRNAs are short molecules of noncoding RNAs consisting of 20–24 nt that exhibit important roles in all biological pathways in multicellular organisms, including mammals [38–40], and are responsible for the regulation of posttranscriptional gene expression [41, 42]. MiRNAs are necessary for normal cellular processes, being deregulated in almost all diseases, including cancer [40, 43]. Mature miRNAs are processed from hairpin structures into pre-miRNA precursors (60–100 nt). Mature miRNAs are obtained after RNase III Dicer recognition and cleavage into small dsRNA duplexes [44]. MiRNAs are involved in many processes, such as development, cell proliferation, apoptosis, tumorigenesis [41], cell cycle control, differentiation, migration, and metabolism [38]. MiRNAs can act similar to oncogenes or tumor suppressor genes [41, 44–48].

In mammals, miRNAs binding to the 3′-UTR halt translation and result in mRNA degradation [42]. Dysregulation of miRNAs causes gene expression imbalance that is related to the dysregulation of key cellular pathways [49]. In recent published papers, different types of miRNAs were used in the medical field as therapeutic agents loaded on the surface of nanoparticles [50].

A great applicability of miRNAs is provided by the use of artificial structures able to restore the normal expression level of a gene, called miRNA mimics and anti-miRNAs (anti-miRs) [48]. MiRNA mimics are RNA duplexes that are identical to the mature miRNA sequence. An miRNA mimic is designed to have the function of the endogenous miRNA, attempting to restore its loss of function as a tumor suppressor. Anti-miRs are synthetic-modified oligonucleotides that help to explain the function of miRNAs and their targets [48, 51]. They are complementary to the mature miRNAs or their precursors, designed to block their function in RISC (Figure 15.3) [52].

SiRNAs are short RNA duplexes that gained attention due to their role in gene regulation, making them targets for drug discovery and development. SiRNAs act by their ability to specifically inhibit a target gene [53]. Delivery systems are designed to protect the cargo from premature nuclease degradation [53]. MiRNA and siRNA have an intracellular site of action, but due to their intrinsic properties such as hydrophilic nature, high molecular weight, and negative charge, the permeability for the cellular membrane is decreased. Both types of RNA molecules are used for many disorders, and due to their similar physicochemical properties and intracellular site of action, similar delivery technologies can be used for both transcripts [54]. Therapeutic applications of siRNAs rely on their local delivery to the specific tissue or tumor site. In cancer therapy, the systemic delivery of siRNA has become a major topic and faces many challenges such as interaction with specific gene targets, obtaining pharmacologically effective levels, stability in circulation, improved cellular uptake, monitoring the distribution, and therapeutic efficacies. Due to these challenges, novel delivery vehicles such as lipid, polymer, and nanoparticles were developed for the systemic delivery of siRNAs [50]. These delivery systems have been tested on animal models, and some disadvantages regarding toxicity, immune and inflammatory responses, gene-control, and gene-targeting issues were observed [55].

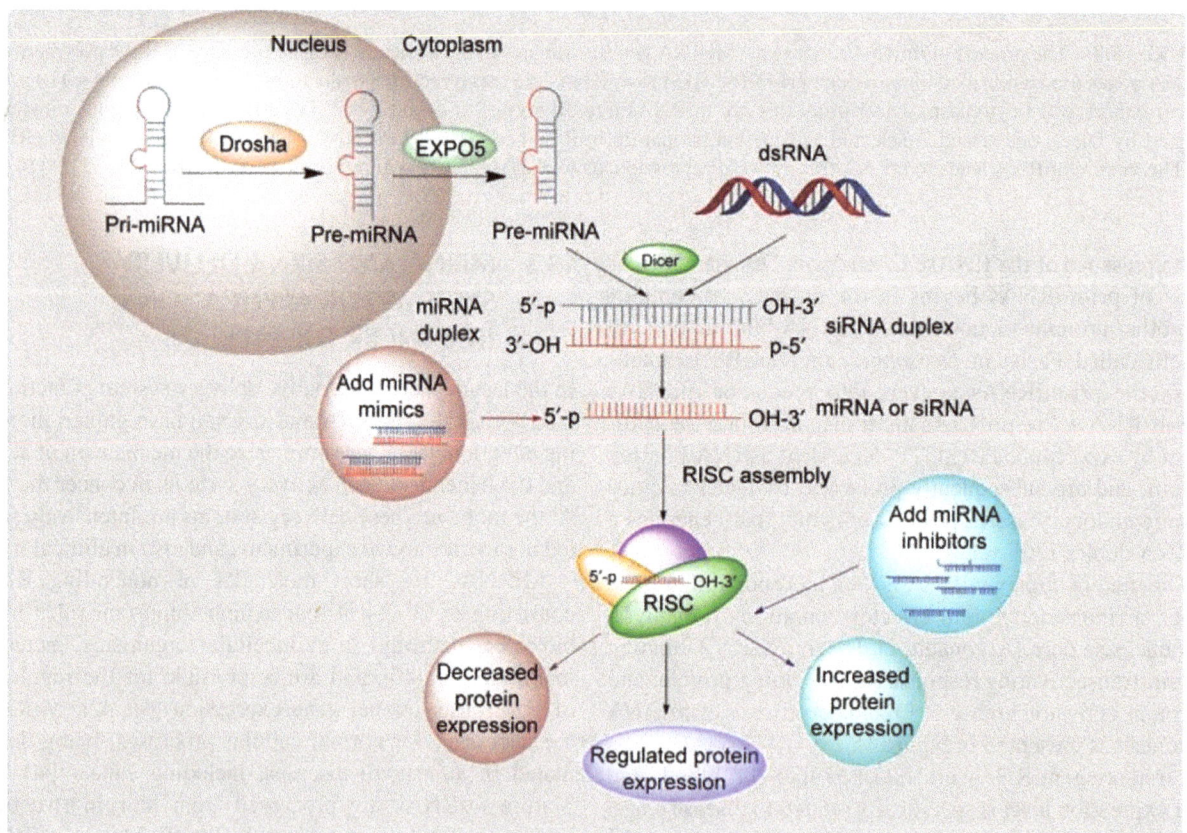

FIGURE 15.3   MiRNA and siRNA mechanism. MiRNA is first transcribed in the nucleus as primary miRNA and then is activated by the RNase III Drosha to create precursor miRNA. The siRNA mechanism starts from dsRNA being transferred into cytoplasm. MiRNA mimic involves the reintroduction of a tumor suppressor miRNA to restore a loss of function. Anti-miRNA traps the endogenous miRNA in a configuration that is unable to be processed by RISC. *Abbreviations:* dsRNA, double-stranded RNA; expo5, exportin-5; miRNA, microRNA; pre-miRNA, precursor microRNA; pri-miRNA, primary microRNA; RISC, RNA-induced silencing complex; siRNA, small interfering RNA.

SiRNA and miRNA present similar physicochemical properties but distinct functions [50]. They are short RNA duplexes that produce gene-silencing effects by targeting mRNA, but their mechanism of action and requirements for sequence design and therapeutic applications are different. For clinical development, they present some disadvantages, such as poor stability in vivo, delivery challenges, and off-target effects [54].

## 15.4 MiRNA DELIVERY THROUGH NANOPARTICLES IN CANCER THERAPY

Nanotechnology is one of the principal areas that can be used to effectively formulate the delivery of miRNAs due to their small size and low molecular weight [56]. When miRNAs are attached to the surface of nanoparticles, the delivery efficacy of miRNA-conjugated nanoparticles increases relative to free miRNA [57].

In recent years, extensive efforts have been focused on the development of nanocarriers that exhibit high delivery efficiency for miRNAs in various diseased conditions [58]. The first polymer-drug nanoconjugate system was reported by Jatzkewitz in the 1950s [59]. In contribution to the field, Bangham discovered the first liposomal formulation in the mid-1960s [60, 61] and Scheffel and coworkers developed the first albumin-based paclitaxel nanoparticles (Abraxane) in 1972 [62]. Abraxane was approved by the US Food and Drug Administration (FDA) in 2005 for the treatment of breast cancer [63]. A brief description of the various nanoformulations and their approval status is provided in [24].

An ideal nanoparticulate drug carrier system is one that displays prolonged systemic circulation time, is present at appropriate concentrations at the target site, retains its therapeutic efficacy against physiological barriers, and finally is metabolized in the body. To fulfill all of these properties, an ideal drug carrier system should possess physical properties of small size, high drug encapsulation efficacy, efficient localization of the carriers by tumor cells (effective binding to the specific targets through the ligands), prolonged circulation time, and finally controlled release of drug at the target site [64, 65].

Systemic circulation time, biodistribution, and cellular internalization of nanoparticles depend on particle size and surface properties such as surface charge. Nanoparticles accumulate in the spleen due to filtration and are removed by the reticuloendothelial system. The optimum particle size of nanocarriers is in the range of 100–200 nm, as this is ideal for enhanced permeability and retention in tumor cells [65]. This size range avoids filtration in the spleen and is large enough to avoid the uptake in the liver [65]. Nanoparticles may be opsonized and thus be recognized and eliminated by macrophages. Opsonization is the key factor that determines the fate of nanoparticles to a large extent in blood circulation. The circulation time of nanoparticles can be altered by modifying their surface charge [65]. An increased cellular uptake of nanoparticles with cationic surfaces has been reported by Chen et al. [66].

It has been established that due to the large effective surface area–volume ratio and higher particle concentrations, the nanoparticulate drug carrier systems can maintain the drug concentration at the target site within the desired level [64, 65]. Accumulation of nanoparticles at target sites can be increased selectively by engineering the particle surface with polymer selection and coupling of targeting ligands. Cellular uptake and drug-release profiles from nanocarriers are mainly dependent on the particle shape. Champion et al. reported a zero-order release profile from hemispherical nanoparticles [67]. Figure 15.4 represents a schematic presentation of a multifunctional nanocarrier. Various nanotechnology-based drug products for cancer therapy under different phases of clinical investigations are presented in [24].

Recently, Kato et al. reported bone morphogenetic protein 2 displays a higher osteogenic effect on MC3T3-E1 cells grown on titanium with nanotopography compared with control titanium [68]. Antonellis et al. developed miRNA-199b-5p-encapsulated stable nucleic acid lipid particles, and a significant impairment of Hes-1 protein levels and cancer stem cell markers in a range of different tumorigenic cell lines (colon, breast, prostate, glioblastoma, and medulloblastoma) was observed [69]. Yoo et al. demonstrated the inhibition of mature miRNA in a metastatic breast cancer cell line using novel layered gadolinium hydroxychloride nanoparticles. Specifically, anti-miRNA oligonucleotides delivered with layered gadolinium hydroxychloride nanoparticles remained functional by inducing changes in the expression of their downstream effect and/or by curbing the invasive properties. Layered gadolinium hydroxychloride nanoparticles provide a promising multifunctional platform for miRNA therapeutics by virtue of their diagnostic, imaging, and therapeutic capabilities. These nanoparticles had shown good cellular uptake profile [70]. Chen et al. developed polycation-hyaluronic acid-loaded nanoparticles containing single-chain antibody fragments for systemic delivery of miRNAs into the

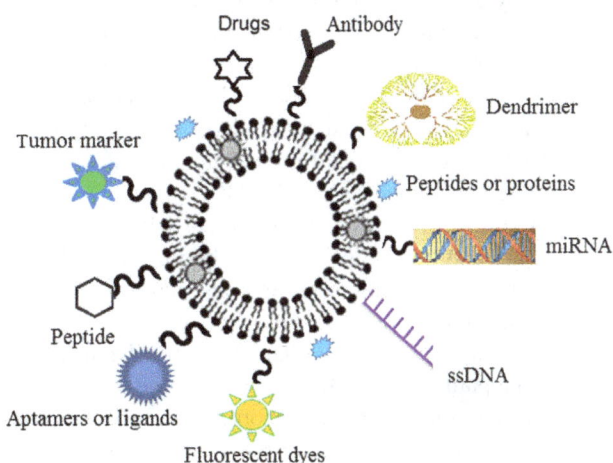

**FIGURE 15.4** Schematic representation of a multifunctional nanocarrier.

lung metastasis of murine B16F10 melanoma. MiRNAs delivered by fabricated nanoparticles significantly downregulated the survivin expression in metastatic tumors and reduced tumor load in the lung [71].

A robust method for delivering miRNAs into cells using cysteamine-functionalized gold nanoparticles was developed by Ghosh et al. The preparation was validated on the basis of the highest payload, the lowest toxicity, efficient uptake, the fastest endosomal escape, and increased half-lives using two different tumor models [72]. The first report on a silica nanoparticle-based delivery system for miRNA delivery to neuroblastoma tumors in a murine orthotopic xenograft model was demonstrated by Tivnan et al. These carriers resulted in a decrease in tumor growth, increase in apoptosis, and a reduction in tumor vascularization [73]. Polycationic liposome-hyaluronic acid nanoparticles for the delivery of miRNA have also been reported by several investigators [74]. Systemic delivery of anti-miR-155 peptide nucleic acids using polylactic-co-glycolic acid (PLGA) polymer nanoparticles showed enhanced delivery efficiency and achieved therapeutic effects. The surface of the nanoparticles was modified with penetration, a cell-penetrating peptide [75]. Liu et al. [76] reported PEGylated liposome-polycation-hyaluronic acid nanoparticle formulation modified with cyclic arginine-glycine-aspartic acid (RGD) peptide for specific delivery of anti-miRNA antisense oligonucleotides to target $\alpha v \beta 3$ integrin present on the tumor neovasculature. Anti-miR-296 antisense oligonucleotides delivered by nanoparticles decreased microvessel formation within Matrigel by suppressing the invasion of cluster of differentiation 31 (CD31)-positive cells and prompting hepatocyte growth factor-regulated tyrosine kinase substrate expression in angiogenic endothelial cells [76].

MiRNA gold nanoparticle conjugates have been shown to control cellular processes by supplementing the endogenous miRNA levels in human prostate cancer cells. The miRNA gold nanoparticles mimicked human miR-205 and decreased the expression of protein kinase C epsilon (PRKCε) by 52% compared with cells treated with control particles functionalized with nontargeting sequences [77]. Valadi et al. reported miRNA delivery from exosomes to human and mouse mast cells. The in vitro translation study proved the functioning of exosome mRNAs. They reported that the RNA from mast cell exosomes is transferable to other mouse and human mast cells [78].

Saraiva et al. developed nanoparticles to deliver miR-124 into neural stem/progenitor cells and boost neuronal differentiation and maturation in vitro. The intracerebroventricular injection of miR-124 nanoparticles increased the number of new neurons in the olfactory bulb of healthy and 6- hydroxidopamine-lesioned mice, a model for Parkinson's disease. Importantly, miR-124 nanoparticles enhanced the migration of new neurons into the 6-hydroxidopamine-lesioned striatum, culminating in motor function improvement [79].

Cai et al. developed monomethoxy(polyethylene glycol)- poly(D,L-lactide-co-glycolide)-poly(L-lysine)-lactobionic acid-antivascular endothelial growth factor antibody (mPEGPLGA-PLL-LA/VEGFab or PEAL-LA/VEGFab) nanoparticles to restore the expression of miR-99a both in vitro and in vivo, to inhibit hepatic carcinoma progression. The nanoparticles selectively and effectively delivered miR-99a to hepatic carcinoma cells based on the double-targeting character of these nanoparticles, thereby offering potential for translation into effective clinical therapies for hepatic carcinoma [80]. Mandli et al. reported novel electrochemical biosensor gold nanoparticles for miRNA-21 determination. The developed biosensor exhibited selective and sensitive detection with a linear range from 200 pM to 388 nM, and the detection limit was 100 pM [81].

MiRNA-145-based magnetic nanoparticles have shown promising anticancer efficacy in pancreatic cancer cells. The miR-NA145 re-expression resulted in the downregulation of mucin MUC13, pAKT, and HER2 and inhibition of cell proliferation, migration, clonogenicity, and invasion of cancer cells [82]. Nanoparticles prepared using biodegradable polycationic prodrug synthesized from a polyamine analog N1,N11-bisethylnorspermine have shown promising results in simultaneous regulation of polyamine metabolism and miRNA delivery for combination cancer therapy [83]. Chen et al. developed liposome-polycation-hyaluronic acid nanoparticles modified with a tumor-targeting single-chain antibody fragment for systemic co-delivery of miRNA and siRNA into experimental lung metastasis of murine B16F10 melanoma. These nanoparticles significantly reduced tumor load and downregulated the survivin expression in the metastatic tumor [71]. Babar and coworkers fabricated surface-modified anti-miRNA-loaded PLGA nanoparticles using double-emulsion solvent evaporation technique. The study claimed systemic delivery of antisense peptide nucleic acid, which inhibited miRNA-155 and suggested promising therapeutic benefits for lymphoma [84]. All the above-stated attempts clearly emphasize and highlight the application and benefits of nanotechnology as a carrier for miRNAs in cancer treatment. A summary of current commercial investments in miRNA therapeutics is found in [24].

## 15.5 ROLE OF NANOPARTICLES IN MiRNA BIOSENSOR CHEMOTHERAPEUTIC DELIVERY TECHNOLOGY

Various electrochemical nanobiosensors have been fabricated for detection or quantification of valuable miRNAs. The first electrochemical miRNA biosensor was described in 2006 [85]. Azimzadeh et al. reviewed the role of nanotechnology in nanobiosensor development for application in miRNA detection [35]. Biomarkers are divided into two categories based on their application, that is, diagnostic biomarkers and screening biomarkers. MiRNAs are one of the most reliable biomarkers reported for the early detection, diagnosis, metastasis, prognosis, and assessment of treatment [86, 87]. Northern blotting, microarray, and polymerase chain reaction are the most common methods used for the detection of miRNAs [88].

Wang et al. reported that alpha-hemolysin protein-based nanopore miRNA sensors detected miRNAs at the single molecular level in plasma samples from lung cancer patients without amplification of the miRNA [89]. In a recent report, Wei et al. developed a magnetic fluorescent miRNA-sensing system for the rapid and sensitive detection of miRNAs from cell lysates and serum samples [90]. Albumin nanoparticles were prepared from inherently biocompatible bovine serum albumin. The results suggested a broad linear detection range of 10 fM–10 nM and a low detection limit of 9 fM within 100 min by detecting a model target miRNA-21. Based on gold nanoparticle-decorated molybdenum sulfide ($MoS_2$) nanosheet, a dual-mode electronic biosensor was developed for miRNA-21 detection. The proposed biosensor displayed high selectivity and stability to determine miRNA-21 in human serum samples with satisfactory results [91].

The development of miRNA biosensors has attracted great attention. MiRNA is considered an ideal biomarker for cancer detection in early stages. Electrochemical biosensors have been used widely due to the fact that they can provide simple, rapid, and reliable detection [92].

Gold nanoparticles are conjugated with different types of agents such as drugs, monoclonal antibodies, aptamers, peptides, RNA transcripts (siRNA and miRNA), fluorophores, PEG, and 1,4,7,10-tetraazacyclododecane-1,4,7,10-tetraacetic acid for chelating [93] Cu, or small natural or synthetic molecules [94]. Dendrimers can be functionalized with various targeting moieties such as folic acid, peptides, monoclonal antibodies, and sugar groups [95]. Anthracyclines (ANTs) are cancer drugs effectively used for treating malignant neoplasms [96]. Doxorubicin, an amphiphilic drug with small molecular weight [97, 98], is an effective ANT antibiotic [99, 100] used for the treatment of multiple types of cancers. Doxorubicin's advantages include high response rate, increased time to disease progression, and low therapeutic index. It is not recommended for patients with high risks of developing cardiac toxicity [101–104] with adjacent congestive heart failure, arrhythmias, or conductivity dysfunction [98]. The high rates of adverse drug reactions limit the use of this drug. The adverse effects are dose dependent and are characterized by injuries in the liver, brain, kidney [105], and gastrointestinal tract [106]. Enhancing strategies for protection from these severe side effects [107] by reducing the cumulative dose of cardiotoxic agents [108], while increasing the cytotoxic effects on malignant cells [107], has become a high priority. Drugs affect DNA intercalation and helicase inhibition causing a cytotoxic mechanism [99], DNA cross-linking, and free-radical formation [109]. Tumor growth is inhibited by triggering cell apoptosis or inhibition of angiogenesis [110]. Doxorubicin interacts with dsDNA and nucleic acids, causing single- and double-strand breaks, but at the same time, it increases cell membrane permeability and inactivates membrane receptors [96]. Damage induced by free radicals affects the heart through high oxidative metabolism and low levels of antioxidant defenses, and regarding the liver, free radicals induce cell death and tissue damage [105]. Through active internalization, doxorubicin exhibits accelerated intracellular trafficking in vivo. Greater accumulation of doxorubicin relies on internalization of carriers, which triggers a mechanism for fast release of drugs in cancer cells.

The delivery of chemotherapeutics into cancer cells is based on liposomal penetration into the tumor interstitia and release of therapeutic agents. Poor penetration leads to low antitumor efficacy. Through nanoparticles and targeting ligands with low affinity, intratumor distributions can be improved. Intracellular trafficking of the released agent is affected by the pH values and size of the agent. For doxorubicin, a weak base that limits the permeability of the cell membranes leads to decreased endosomal pH values and allows diffusion through the endosomal membrane to the cytoplasm [111]. Another problem during therapy is the development of resistance to treatment. Doxorubicin is a substrate for various ATP-binding cassette membrane pumps, such as P-glycoprotein and multidrug resistance-related proteins. As a result of the overexpression of these transporters, doxorubicin can be removed from cancer cells, reducing its intracellular accumulation [112].

Doxorubicinol, a doxorubicin metabolite, is cytotoxic, being responsible for the adverse effects. To alleviate acute toxicities [113] and improve antitumor effect, the chemotherapeutic agent is encapsulated into a PEGylated liposome. This clinical product, Doxil (liposomal doxorubicin), is an anticancer drug [114] that shows increased microvascular permeability of the tumor with a prolonged duration of circulation in plasma and decreased distribution volume [115], increasing tumor uptake through EPR effect [98, 113]. To avoid drug interaction with plasma proteins such as opsonins, high-density lipoproteins, and low-density lipoproteins, or elimination by macrophages, liposomes' surface is conjugated with PEG. The external part of Doxil is formed from a lipid bilayer coated with PEG, and the molecules of doxorubicin are encapsulated into the aqueous core [98, 116] with better pharmacokinetic profile [117]. Doxil is the first nanodrug approved by the FDA, and is extensively used in the clinic for cancer patients [114]. Through lipid bilayer stabilization and steric hindrance, the interaction with the plasma proteins is decreased, and the recognition by the macrophages is diminished [118]. Doxil is used to treat AIDS-related Kaposi's sarcoma, metastatic breast cancer, and ovarian cancer [97, 113, 119] with a better toxicity profile and reduced incidence of cardiotoxicity [120, 121]. At high doses or short dosing intervals, some forms of toxicity, including acute infusion reaction, mucositis, palmarplantar erythrodysesthesia [101], nausea, or vomiting, are present [116]. Doxorubicin is deposited in the liver, spleen, and tumor [120, 121]. An attractive strategy to decrease tumor growth via modulating different signaling pathways is to combine multiple drugs with different mechanisms of action [110].

## 15.6 MiRNAs ARE IMPORTANT REGULATORS OF CANCER MULTIDRUG RESISTANCE AND METASTATIC CAPACITY

Extracellular vesicles serve as important intercellular vectors in the regulation of many biological processes. Microparticles (MPs) are a subtype of extracellular vesicle, typically defined as having a size of 0.1–1 μm in diameter, exposure of phosphatidylserine, and the expression of surface antigens originating from their donor cells [122]. MPs also contain functional proteins, second messengers, growth factors, and genetic material from the cell of origin and confer onto recipient cells biological effects [123–132]. MPs together with exosomes are important cancer biomarkers through their discrete protein and nucleic acid signatures [133].

MPs are important mediators in the dissemination of functional multidrug resistance (MDR) in cancer cell populations. Cancer MDR is a significant cause of treatment failure and disease relapse and is attributed to the overexpression of drug efflux transporters belonging to the ATP-binding cassette (ABC) superfamily. MPs can confer MDR through the packaging and intercellular transfer of these functional resistance proteins and nucleic acids from MDR cells to drug-responsive cells [123, 125, 129, 130].

In addition to the presence of proteins and transcripts encoding resistance proteins, these MPs also carry miRNAs, which are involved in the acquisition of MDR and other deleterious traits associated with this complex phenotype. MPs comprise the major source of systemic RNA, including miRNA, the aberrant expression of which appears to be associated with stage, progression, and spread of many cancers [124, 125].

In a study conducted by Bebawy and coworkers in 2012, Affymetrix miRNA microarray was used to explore the miRNA expression profiles of MPs shed from MDR cells as well as cells that acquired MDR following MP transfer [124]. The analysis showed the selective packaging of seven miRNAs including miR-1228*, miR-1246, miR-1308, miR-149*, miR-455-3p, miR-638, and miR 923 within the MP cargo upon release from MDR cells. The pathway analysis of the predicted targets for these miRNAs showed target genes to be significantly related to pathways in cancer and at least seven other pathways that were cancer related. The selective packaging of miRNAs was also reported by the same team for miR-451 and miR-326 in MPs shed from MDR breast and leukemia cells in

earlier studies, supporting the presence of a mechanism for the selective dissemination and transfer of miRNAs within cancer cell populations [125]. Following coculture with recipient cells, there was again an increase in the levels of miR-1246, miR-1308, miR-1228*, miR-149*, miR-638, and miR-923 in recipient cells [124]. Certainly, this was attributed to the unloading of MP cargo to the recipient following binding and uptake; however, MP effects on the transcriptional regulation of these miRNAs in recipient cells also cannot be excluded. Indeed, the acquired cell population demonstrated miRNA expression trends reflective of the donor cells, a finding consistent with previous reports demonstrating a BRE templating of the transcriptional landscape of recipient cells to reflect that of the donor cells [125].

The above-mentioned miRNAs play important roles in cancer cell biology. For instance, the NF-κB inhibitor, NKIRAS1, is targeted by miR-1308. MiR-1308 is upregulated in many cancers, including in aggressive inflammatory breast cancer [134, 135]. Likewise, miR-1228* is highly expressed in malignant mesothelioma [136].

The role of miRNAs in the regulation of MDR is an emerging area [124, 126, 131]. MiR-455-3p, among others selectively packaged in the MP cargo shed from MDR cells, targets the MDR protein, P-glycoprotein (P-gp), and HIF1AN (hypoxia-inducible factor 1, alpha subunit inhibitor). HIF-1 alpha has been shown to induce MDR in HCC [136]. This is consistent with miRNA profiling studies conducted by Bebawy and coworkers [124], which showed resistant leukemia cells to have lower levels of this miRNA compared to drug-sensitive cells, a finding consistent with the resistant state. In the acquired cell population following MP transfer, miR-455-3p levels were suppressed supporting the acquisition of increased P-gp levels.

The role of miRNAs in the regulation of the functional redundancy that resides among members of the ABC transporters, of which the prototypical members include the MDR proteins P-gp (ABCB1) and MDR protein (ABCC1), has also recently been demonstrated [137]. This significant functional redundancy that exists between members of this superfamily of drug transporters is attributed to significant sequence homology, broad and overlapping substrate specificities, and significant tissue co-localization. This ensures a fail-proof mechanism for cell survival, more so in the context of malignancy. This was first attributed to the suppression of endogenous ABCC1 transcript levels in recipient CEM cells by MPs shed from ABCB1-overexpressing VLB100 cells to the presence of miR-326 in the MP cargo and to its subsequent transfer to recipient cells [125]. Indeed, the suppressive effect of miR-326 on ABCC1 has been shown previously and MRP1 expression is inversely correlated with miR-326 in advanced breast cancer [138].

The transfer of miR-326 from P-gp-mediated MDR breast cancer and leukemia cells to recipient cells was previously shown [125]. Relative to breast cancer MPs, leukemic MPs were shown to package significantly greater amounts of the ABCB1 transcripts. Despite the miR-326 levels in both breast cancer and leukemia MPs being comparable, suppression of ABCC1 was only observed in recipient cells following coculture with the latter [125]. The molecular basis for the differential effect of miR-326 on ABCC1 suppression in breast cancer and leukemia cells has been shown to be regulated by the presence of ABCB1 transcript demonstrating a novel mechanism regulating the expression of ABC transporters in cancer. This work positions ABCB1 mRNA as a transcriptional regulator of ABCC1 through its actions on miR-326 of which there are no known putative binding sites [131].

Other miRNAs associated with cancer MDR include miR-27a and miR-451. These miRNAs have been detected in resistant breast cancer and leukemia cells as well as in their MP cargo [92] with the former possibly playing a role in the upregulation of ABCB1 transcript in recipient breast cancer cells. MiR-345 and miR-7 have also been previously shown to target ABCC1 in MDR breast cancer cells relative to parental cells [139].

Although the development of MDR and metastases are both major considerations in the clinical treatment of cancer, their significance in the context of one another has only recently been studied. Microparticles shed from MDR breast cancer cells also mediate the intercellular transfer of miRNA-503 to alter the migration and invasion capacities of recipient breast cancer cell populations [128]. Microarray analysis identified miRNAs common to the resistant state and which contribute to the dissemination of metastatic traits. Among the miRNAs identified, miR-503 was downregulated in recipient cells following coculture with MPs isolated from drug-resistant cells. MiR-503 was shown to be inversely associated with metastatic and invasive capacities, as demonstrated using wound-healing/scratch migration assays and Matrigel-coated transwell invasion assays. This is consistent with earlier reports whereby miR-503 was previously shown to be involved in the development of drug resistance and metastatic traits, with reduced levels of miR-503 being identified in drug-resistant cells and highly metastatic cells [140, 141]. Activation of the NF-κB pathway has been shown to suppress the expression of miR-503 in epithelial cells and may also be responsible for the miRNA-503 suppression following MP transfer [142]. Reduced levels of miR-503 have been observed in cisplatin-resistant non–small cell lung cancer (NSCLC) cells, while its overexpression resensitizes cells to cisplatin via modulation of the apoptosis regulator Bcl-2. MiR-503 has also been shown to directly target and repress the Fanconi anemia complementation group A protein (FANCA) gene to sensitize NSCLC to cisplatin treatment [143].

MiR-503 is also an important tumor suppressor, the overexpression of which inhibits the migration and invasion of highly invasive HCC cells. MiR-503 also acts to induce G1 cell cycle arrest and reduce cell proliferation [140, 144, 145]. Its tumor suppressive activity has been shown to be via regulation of PI3K/AKT signaling by its effect on inhibiting AKT activation [141, 146].

MiRNA-494 is another miRNA found in abundance in recipient cells following coculture with MPs isolated from MDR cells [128]. MiR-494 is predicted to target the focal adhesion kinase family-interacting protein of 200-kDa (FIP200) gene. FIP200 binds directly to the kinase domain of proline-rich tyrosine kinase 2 (PYK-2) to inhibit its activity [147]. MiR-494 downregulation of the FIP200 gene may be responsible for enhanced PYK2-dependent phosphorylation of AKT and activation of PI3K/AKT pathway resulting in increased metastatic capacity in recipient cells following MP coculture [128]. MPs shed from resistant cells not only mediate the intercellular transfer of MDR but are also implicated in promoting migration and invasion in recipient cells, potentially providing a link between these two deleterious traits.

## 15.7 MiRNA DETECTION

MiRNA is associated with cancer diagnosis. Jou et al. reported a two-step sensing platform for sensitive detection of miR-141, a promising biomarker for prostate cancer. The first step of the sensing platform used CdSe/ZnS QDs modified with FRET quencher-functionalized nucleic acids, which contained a telomerase primer sequence together with a recognition sequence for the miR-141 recognition sequence. The FRET quencher exhibited covalent binding to the nucleic acid-functionalized CdSe/ZnS QDs. When miR-141 hybridized with the probe, a duplex was formed, which would be cleaved by duplex-specific nuclease (DSN). The cleavage released the quencher unit and activated the fluorescence of the QDs. This cleavage also led to exposure of the telomerase primer sequence. The second step involved the primer unit elongation stimulated by telomerase/dNTPs, incorporation of hemin, and chemiluminescence generated with the help of luminol/$H_2O_2$. This platform helped detect miR-141 in a serum sample and discriminated healthy individuals from prostate cancer carriers [148].

In addition to the detection of extracellular nucleic acids, nanoparticles have also been developed as intracellular nucleic acid sensors. Seferos et al. [149] demonstrated that it is possible to use novel gold nanoparticle probes modified by oligonucleotides hybridized to complements labeled with a fluorophore as transfection agents and cellular "nanoflares" to detect mRNA in living cells. Nanoflares overcome many challenges in the creation of effective and sensitive intracellular probes and show a large signal-to-noise ratio and sensitivity to changes in the number of RNA transcripts in cells. Nanoflares, which show high orientation, dense oligonucleotide coating, and can enter cells without the need for cytotoxic transfection agents [150], are useful for detecting intracellular mRNA.

Meanwhile, researchers have developed nanoflares for simultaneous intracellular detection of various mRNA transcripts. In these multiplexed nanoflare studies, AuNPs functionalized with two to three DNA recognition strands and later hybridized with short complementary reporter strands were generated as nanoflares. For example, the use of multiplexed nanoflares to detect survivin in addition to actin has been investigated for normalizing nanoflare fluorescence differences in cellular uptake. Therefore, the technique is comparable with conventional qRT-PCR for quantification of intracellular mRNA but can be performed at the single live cell level.

In some cases, the nanoflare platform was expanded to quantify intracellular RNA and detect spatiotemporal localization in living cells [151]. In this work, β-actin targeting nanoflares were incubated with HeLa cells and presented an obviously different intracellular distribution, exhibiting strong colocalization with mitochondria, which has not been previously demonstrated. Further, Smart-Flares were utilized for studying melanoma tumor cell heterogeneity [152]. These Smart-Flares were able to quantify genomic expression at the single-cell level, thus expanding our knowledge of cancer and metastasis. Investigating the heterogeneity of cancer cells is crucial for identifying novel biomarkers for early cancer diagnosis.

Halo et al. [153] reported nanoflares, which were applied to capture live circulating breast cancer cells. These nanoflares could detect target mRNA in model metastatic breast cancer (MBC) cell lines in human blood and exhibited high recovery and fidelity reaching 99%. They also used nanoflares together with later cultured mammospheres to reimplant the retrieved live recurrent breast cancer cells into whole human blood. Only 100 live cancer cells could be detected per mL of blood. Relying on the NanoFlare technology, it was possible to simultaneously isolate and characterize intracellular live cancer cells from whole blood. The authors [154] demonstrated the ability of nanoflares to collect CTCs for future culture and study. In addition, nanoflares contribute to the technology of combining intracellular markers with cell-surface markers for dually identifying putative CTCs. The combined method is likely to enhance the function of more platforms to specifically identify CTCs and subpopulations of CTCs. The authors think that nanoflares provide the first gene-based approach to detect, isolate, and characterize live cancer cells in the blood and are likely to contribute to cancer diagnosis, prognosis, and prediction, as well as personalized treatment.

Lee et al. reported an approach based on an elegant plasmonic nanoparticle network structure, generating a plasmon-coupled dimer able to detect single mRNA variants [155]. They applied the method to the detection and quantification of BRCA1 mRNA splice variants in vitro and in vivo. Two probes conjugated to nanoparticles were connected to the BRCA1 mRNA target in a sequence-specific manner, and as a result, the signal exhibited a spectral shift due to dimer formation. They demonstrated that their method is powerful and can successfully detect, quantify, and differentiate between different BRCA1 splice variants with single-copy sensitivity, thereby laying a foundation for quantitative, single-cell genetic profiling in the future.

## 15.8 DELIVERY SYSTEMS USED IN CANCER RESEARCH

This section highlights the ideal delivery systems used in cancer therapy, such as vectors, gold nanoparticles, liposomes, hybrid systems, dendrimers, and carbon nanotubes (CNTs). The main purpose of this section is to describe delivery systems' designs, and to present their benefits and effects in the medical field. Novel nanoparticles that can target multiple altered mechanisms represent an important tool in cancer treatment and can be employed in viral and nonviral delivery systems [156], each of which has its specific advantages and disadvantages [50].

The optimal way to deliver gene therapy is by direct administration of the therapeutic gene to the target site. However, this is extremely inefficient, unreliable, and feasible only in tumors. Generally, in gene-therapy approaches, the genetic material is delivered via the intravenous route; as nucleic acids are susceptible to degradation by nucleases and rapid clearance in systemic circulation [114], a vector is required to pack, protect, and transport the genetic material to its site of action.

### 15.8.1 VIRAL DELIVERY SYSTEMS

Viral vectors are viruses capable of delivering genetic material into specific cells with the purpose of increasing gene expression or inhibiting the production of a target protein [157, 158]. Among the viral vectors used for gene delivery are adenoviruses, retroviruses and lentiviruses [156]. Viral vectors are efficient in gene delivery and expression, but their drawbacks, such as low transgenic size, high cost [156], immunogenicity, oncogenicity [157], and toxicity [158], limit their use. Retroviruses can be used for miRNA delivery inside somatic and germline cells. These types of viral vectors belong to the RNA virus family, and their size is between 7–11 kb. The cargo is delivered and integrated inside the target cell's genomic DNA during the mitotic phase of the cell cycle, infecting just the dividing cell. Lentiviruses belong to the retrovirus family and incorporate the foreign genetic material inside the host genome. These viruses are able to affect both dividing and nondividing cells through infecting postmitotic and terminal differentiated cells. Lentiviruses exhibit a high transfection efficiency and long-term stable expression. Adenoviruses contain double-stranded DNA (dsDNA) and are specific for

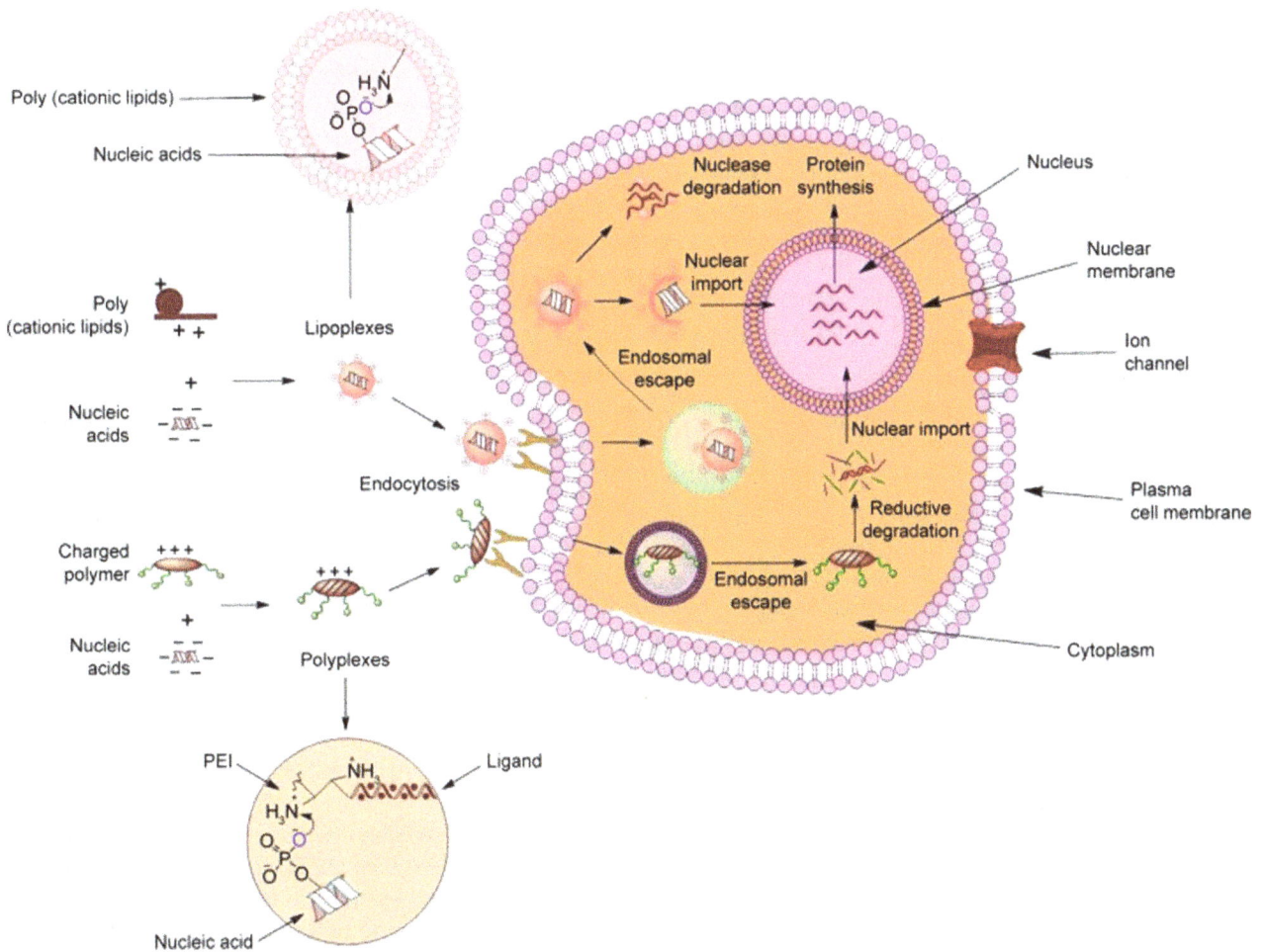

**FIGURE 15.5** Nonviral gene delivery using lipoplexes and polyplexes. Nucleic acid is complexed with these two types of nonviral delivery systems, and it is internalized through receptor-mediated endocytosis. A large amount of complexes are degraded after their internalization in the endosomal compartments. Only a small fraction enters the nucleus and elicits desired gene expression. *Abbreviation:* PEI, polyethylenimine.

miRNA gene delivery. Similar to lentiviruses, adenoviruses infect both dividing and nondividing cells [159]. In addition, several tests employ adeno-associated virus-mediated gene delivery, which has the advantage to overcome resistance to conventional anticancer therapies [160–162] and lead to a cell differentiation inhibition [161].

### 15.8.2 Nonviral Delivery Systems

The alternative vectors available for drug delivery are nonviral vectors. Due to their cationic charge, these nanostructures interact with negatively charged DNA or RNA structures through electrostatic interactions obtaining cationic polymers (polyplexes) and cationic lipids (lipoplexes) (Figure 15.5) [157].

Cationic polymers are completely soluble in water and do not contain a hydrophobic moiety. They can be synthesized with different functional groups that are attached by substitution or addition, in different lengths and with different geometry. Cationic lipids are amphiphilic molecules that contain positive charges. Through the positive charge, lipoplexes are bound to a hydrophobic domain including two alkyl chains. This charge is associated with an amine group with different degrees of substitution, that is, amidine, guanidium, and pyridinium. Cationic polymers differ from cationic lipids in some properties such as chemical structures, nucleic acid interactions, and their behavior inside the cell [161]. Nonviral vectors can be delivered through physical as well as chemical methods. When delivered through physical methods, the vector delivers the gene to the target by applying a physical force that increases cell membrane permeability. These methods cause cell injuries and increase the apoptotic rate, but are unable to prevent nuclease cleavage [159]. The physical methods used are electroporation, ultrasound, microinjection, and hydrodynamic applications. Meanwhile, natural and synthetic viral vectors can be delivered through chemical methods to deliver the gene to the target. These delivery systems were developed to improve the ligand's ability to attach on the surface, or to be able to encapsulate and deliver foreign genetic materials into the specific cell type. The main advantages are represented by the low immunogenic response, capability to carry large inserts, selected modifications [156], easy synthesis, and cell-/tissue-specific targeting [157]. Viral and nonviral delivery systems have different features, which are detailed further. Due to their ability to transfer their genetic material into host cells, viral vectors present higher transfection efficiency [159, 162]. However, there are difficulties in large-scale production mainly due to the size of the carried DNA, mutagenesis [163], toxicity, and immunogenicity [159], which limit the viral vectors' progression. Nonviral vectors have the ability to deliver nucleic acids into cells, with lower transfection efficiency than viral vectors [163], but are safer [159], protect the cargo from the immune system, and can manage larger DNA fragments [162]. In cancer therapy, it is important to use efficient vectors that can surpass different natural barriers such as extracellular and intracellular membranes, and deliver the genetic material to its target site [163]. In addition, the side effects such as toxicity, mutagenesis, and immunogenicity must be avoided by using materials that are biodegradable and compatible with the systems.

### 15.8.3  Gold Nanoparticles

Gold nanoparticles are the key focus of biomedical research due to their physical–chemical properties such as shape, surface area, amphiphilicity, carrier capabilities, and biocompatibility. However, because of various drawbacks such as low encapsulation efficiency, poor storage stability, and slow endosomal escape, the use of these nanoparticles is limited [164]. Gold nanoparticles can be conjugated with small strands of RNA such as miRNA and siRNA, or a wide range of small molecules or monoclonal antibodies through physical and chemical bonds, leading to various sizes. By loading nanoparticles with RNA, the genetic sequence becomes a target for the cancer cell. SiRNA and miRNA perform gene silencing, with siRNA inhibiting the translation of mRNA and miRNA cleaving the mRNA [165]. Nanoparticles are conjugated with monoclonal antibodies to target a variety of mechanisms, from the simple blocking of the antigen receptor in the effector cells, to cytotoxic action of the cells that express the corresponding antigen. Therapeutic applications include the development of targeted drug delivery systems, hyperthermia, regenerative medicine, or radiotherapy. These nanostructures gain attention in various medical fields due to their advantages. Moreover, gold nanoparticles exhibit a low cytotoxicity to the normal cells [166], increase the lifespan of the cargo in the bloodstream [165], enable easy size control, improve surface chemistry [167], increase therapeutic effects, increase accumulation of drug into the

cancer cells, and improve pharmacokinetic effects and biodistribution [168, 169].

### 15.8.4  Liposomes

Liposomes were the first colloidal drug carriers [170] used in gene therapy [171] and are used for targeted delivery of natural or synthetic chemotherapeutics. They consist of a phospholipid bilayer surface enclosing an aqueous core [172–174]. Liposomes are closed spherical vesicles [175] and can encapsulate both hydrophilic and hydrophobic drugs [176], which can be released through diffusion or cell internalization. Liposomes have a structure of hydrophilic heads stabilized by surfactants, and multiple hydrophobic tails (Figure 15.6) [174].

Due to this structure, aqueous hydrophilic components can be entrapped in the interior, while the lipophilic components can be incorporated between the lipid bilayers. These carriers are attractive for drug designing due to their biodegradability, biocompatibility, low toxicity, ability to encapsulate multicomponents [177], and ease of surface manipulation, and have been approved for multiple clinical trials [174]. However, liposomes have several disadvantages such as low encapsulation efficiency, poor storage stability, easy oxidation of liposomal phospholipids, and short release time [177].

Liposomes are widely used to carry the drug inside the lipid bilayer and transfer the contents to plasma proteins by diffusion [178]. Liposomes have been clinically used to improve drug delivery to tumor sites and diminish the side effects of chemotherapy or antimicrobial therapies [179], as well as to enhance specificity to injurious sites [170]. The stability of liposomes is influenced by the lipid composition and structure, and this contributes to the optimization of liposomal product design [180]. The stability of liposomal nanostructures includes multiple aspects, such as colloidal and biological stability. Should colloidal stability lack, liposomes form larger-sized particles, and their efficiency as delivery systems is reduced [170].

Encapsulation of drugs into liposomes has allowed the delivery of therapeutic agents to the target [181] and also avoided their uptake by the reticuloendothelial system [182, 183]. Due to specific stimuli present at the tumor site, the liposomes are able to target the tumor cells and release the chemotherapeutic agents, which are encapsulated into the nanoparticles [184]. Liposomal encapsulation of doxorubicin, a topoisomerase II inhibitor [182], changes its pharmacology and pharmacokinetics and leads to enhanced drug delivery into tumor sites [185] and reduced toxic effects, in comparison to classical treatments [183]. Liposomes release their load in the tumor vessels, which then diffuse through the vascular sites into the distal tumor areas [186]. Tissue distribution and pharmacokinetic profile of doxorubicin are altered by the PEGylated liposomal doxorubicin, which decreases the rate of left ventricular cardiac dysfunction and symptomatic congestive heart failure [182, 187].

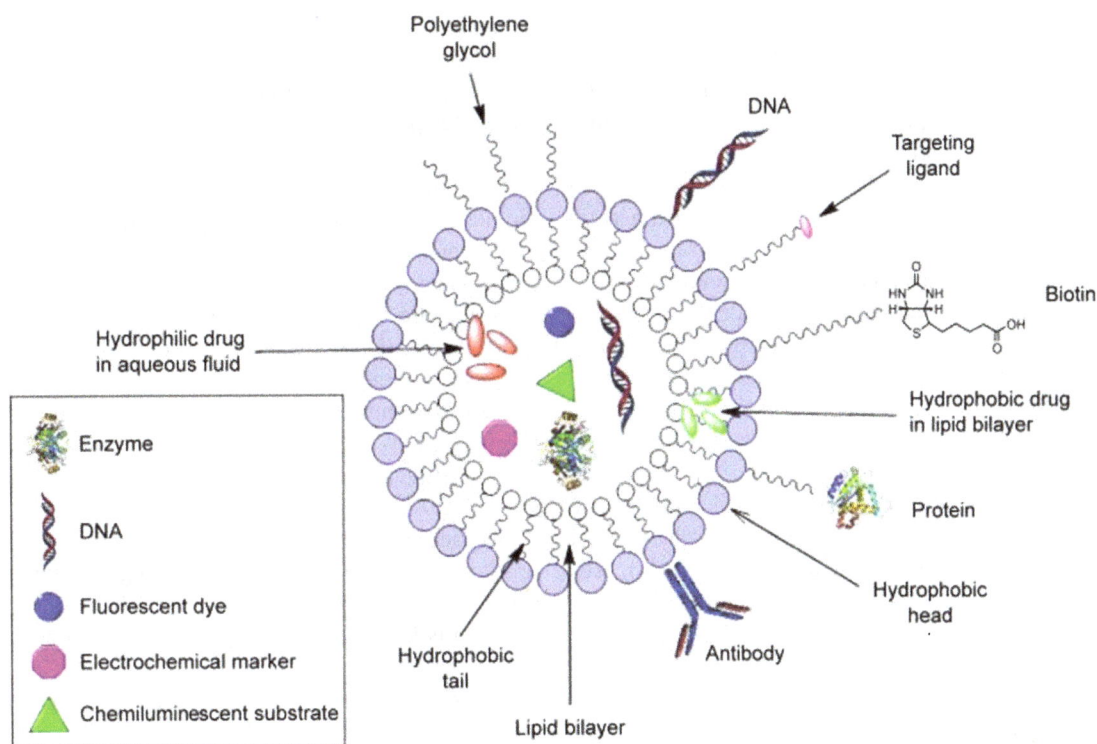

**FIGURE 15.6** Structure of liposomes. Liposomes are colloidal drug carriers consisting of a phospholipid bilayer surface enclosing an aqueous core. Hydrophilic components can be entrapped inside the aqueous core, while the lipophilic components can be incorporated between the lipid bilayers. On the liposomes surface, different particles that target the interest cells can be attached. To avoid the immune system response, the liposomes surface is loaded with a polymer called polyethylene glycol. Thus, the cargo is protected and is discharged into the target cells.

### 15.8.5 HYBRID SYSTEMS

The polymer–lipid hybrid system is a mixture of polymeric nanoparticles and liposomes. The components involved in the hybrid system design present interesting features for potential use in cancer therapy. The core of the hybrid system consists of a biodegradable hydrophobic polymer that allows the encapsulation of water-soluble drugs and thus assures a continuous release. To increase the circulation time in the bloodstream and avoid immune system response, the hybrid system is coated with a hydrophilic shell. Between the hydrophobic core and the hydrophilic shell, the system has a lipid monolayer that prevents diffusion of encapsulated drugs and reduces water penetration inside the nanoparticles [188]. Chavanpatil et al. developed a polymer–surfactant hybrid system for encapsulation of water-soluble drugs and enhancing their release, consisting of polymer (sodium alginate) and dioctyl sodium sulfosuccinate (Aerosol OT [AOT], which is an anionic surfactant) forming AOT-alginate nanoparticles [189]. Bellocq et al. developed a polymer–cyclodextrin hybrid system for siRNA delivery. This system was made by condensation of a polycation cyclodextrin with nucleic acid and coating with polyethylene glycol (PEG) to enhance the stability in biological fluids [190]. Wong et al. presented a polymer–lipid hybrid nanoparticle containing cationic doxorubicin, anionic soybean oil–based polymer, and stearic acid for water dispensing of the nanoparticle [191]. Hybrid systems designed from noble metals are promising anticancer agents used in diagnostics and anticancer therapy. Shmarakov et al. have tested bimetallic silver–gold (Ag–Au) nanoparticles on lung cancer cells and observed that Ag–Au NPs may serve as a suitable prototype to develop anticancer agents and drug vehicles [192]. Fakhri et al. used in their experiments Ag–Au bimetallic nanoparticles and demonstrated that they have antiproliferative effects on the human lung cancer cell line A549 and the human breast cancer cell line MCF-7 [193]. Mittal et al. synthesized silver–selenium (Ag–Se) bimetallic nanoparticles functionalized with quercetin and gallic acid and showed their potential antitumor activity against Dalton's lymphoma cells [194]. Wu et al. showed that Ag–Au bimetallic nanostructures present significant cytotoxic effects against the breast cancer cell line MCF-7 [195]. Another study presented the effects of gold–platinum (Au–Pt) bimetallic nanoparticles on cervical cancer, and Alshatwi et al. studied the potential cytotoxic effects of bimetallic nanoparticles [196].

### 15.8.6 DENDRIMERS

Dendrimers are well-organized nanoscopic macromolecules and have an essential role in the emerging field of nanomedicine. Due to their high water solubility, biocompatibility, polyvalence, and precise molecular weight, dendrimers are gaining considerable attention in modern biomedicine. These characteristics make them ideal carriers for drug delivery and targeting applications. They have the ability to interact with cell membranes, cell organelles, and proteins. In addition, dendrimers with cationic surface tend to interact with the lipid bilayer, facilitating increased permeability and decreased integrity of biological membranes. Interaction between dendrimers and cell membrane determines the mechanism that causes the leakage of cytosol proteins [197]. Cell toxicity is determined by the number of end groups and surface charges. It was shown that cationic dendrimers such as polyamidoamine, polypropylenimine, and poly-L-lysine expose toxicity in a dose-dependent manner. To prevent the toxicity of dendrimers, the surface groups of cationic dendrimers are modified with neutral molecules. The positive charge of end groups may interact with the negative charge of the membrane, which increases the permeability and facilitates the intracellular delivery of agents. In the case of cationic dendrimers with high charge density, the interaction with the membrane may result in the disruption of membrane integrity and the leakage of important intracellular components, which finally causes cell death and toxicity [198]. Through physical and chemical bonds, dendrimers interact with different classes of drugs, and they can be used for the incorporation of hydrophobic/hydrophilic molecules inside their empty cavities through nonbonding interactions. Another alternative is to attach the drug molecule to its periphery, thus obtaining a complex. The complex is formed due to the electrostatic interactions or conjugation between the drug and the dendrimers. Moreover, the covalent conjugation of drugs to dendrimers may include PEG, p-amino benzoic acid, p-amino hippuric acid, and lauryl chains or biodegradable linkages including amide or ester bonds. These conjugates have been found to increase the stability of drugs and blood resistance time, and cause enhanced therapeutic action [197].

### 15.8.7 Carbon Nanotubes

CNTs have cylindrical shape and belong to the fullerene family of carbon allotropes [199], consisting of a hexagonal arrangement of sp2-hybridized carbon atoms. The wall of CNTs is formed from single or multiple layers of graphene sheets. When a single sheet is rolled up, it forms single-walled carbon nanotubes, and multiwalled carbon nanotubes are obtained by rolling up more than one sheet [200]. CNTs display abilities for drug loading on the surface or in the inner core through covalent and noncovalent interactions. These nanoparticles are able to immobilize therapeutic agents such as drugs, proteins, DNA, and antibodies on the outer wall, or encapsulate them inside the nanotubes, decreasing the cytotoxicity for healthy tissues. Due to their nanoneedle-like structure, carbon nanoparticles are efficiently taken up and translocated into the cytoplasm of target cells without causing cell death. Their applications are limited due to the fact that CNTs are hydrophobic in

nature and insoluble in water, and are accumulated in internal organs, having a low degradation rate [201]. To eliminate the undesirable effects and to facilitate their use in medical applications, various methods of functionalization of CNTs, such as adsorption and electrostatic and covalent interaction, are used. To increase the systemic retention, circulation time, and the solubility of CNTs, a hydrophilic biocompatible polymer with neutral charge such as PEG or polyethylene oxide is used [202, 203].

Schematic illustrations of the working mechanisms of miRNA and siRNA are shown in Figure 15.7.

Tissues targeted by siRNA and miRNA therapeutics are currently being investigated at the clinical stage (Figure 15.8).

A chemical modification of siRNA is shown in Figure 15.9.

Surface modification of siRNA carriers to decrease opsonization, renal clearance, and degradation has been shown in Figure 15.10.

Schematic illustration of various nanocarriers is shown in Figure 15.11.

Figure 15.12 represents a schematic image of LNPs siRNA showing a nanostructured core.

A schematic representation of several strategies for encapsulating siRNA in liposomes is shown in Figure 15.13.

### 15.9 NANOSCALE IMMUNOTHERAPY

Tumor immunotherapy has been shown to complement available treatment modalities to fight severe cases of relapsed disease, to provide so far unseen clinical results. This novel treatment method is based on the use of in vitro–modified adoptively transferred T lymphocytes to express an artificial signaling molecule named CAR, which would specifically redirect the modified lymphocytes to the surface antigens expressed only by malignant cells [204]. CARs typically encode an extracellular antibody-derived domain to bind a surface antigen linked to an intracellular signaling domain that mediates T-cell activation such as TCR $\zeta$ chain, and various co-stimulatory domains such as CD28 or 4-1BB intracellular chains [205]. In principle, any surface antigen can be targeted with CAR in a non-HLA-restricted manner. A large number of CARs targeting diverse tumors have been recently developed, and there are ongoing multiple clinical trials. One of the most successful examples of CAR-based immunotherapy is the treatment of B-cell acute and chronic leukemia by targeting the cell surface antigen CD19 [206]. However, in spite of these highly encouraging results, major questions remain unresolved. The major problem with CAR-based therapy is the unpredictable treatment responses. A subset of patients encounter limited or missing expansion of infused CAR T cells that might be caused by inefficient activation of CAR T cell in combination with immunosuppressive environment within the tumor stroma. In other patients, there is a complete depletion of all B-type lymphocytes also including the nonmalignant ones.

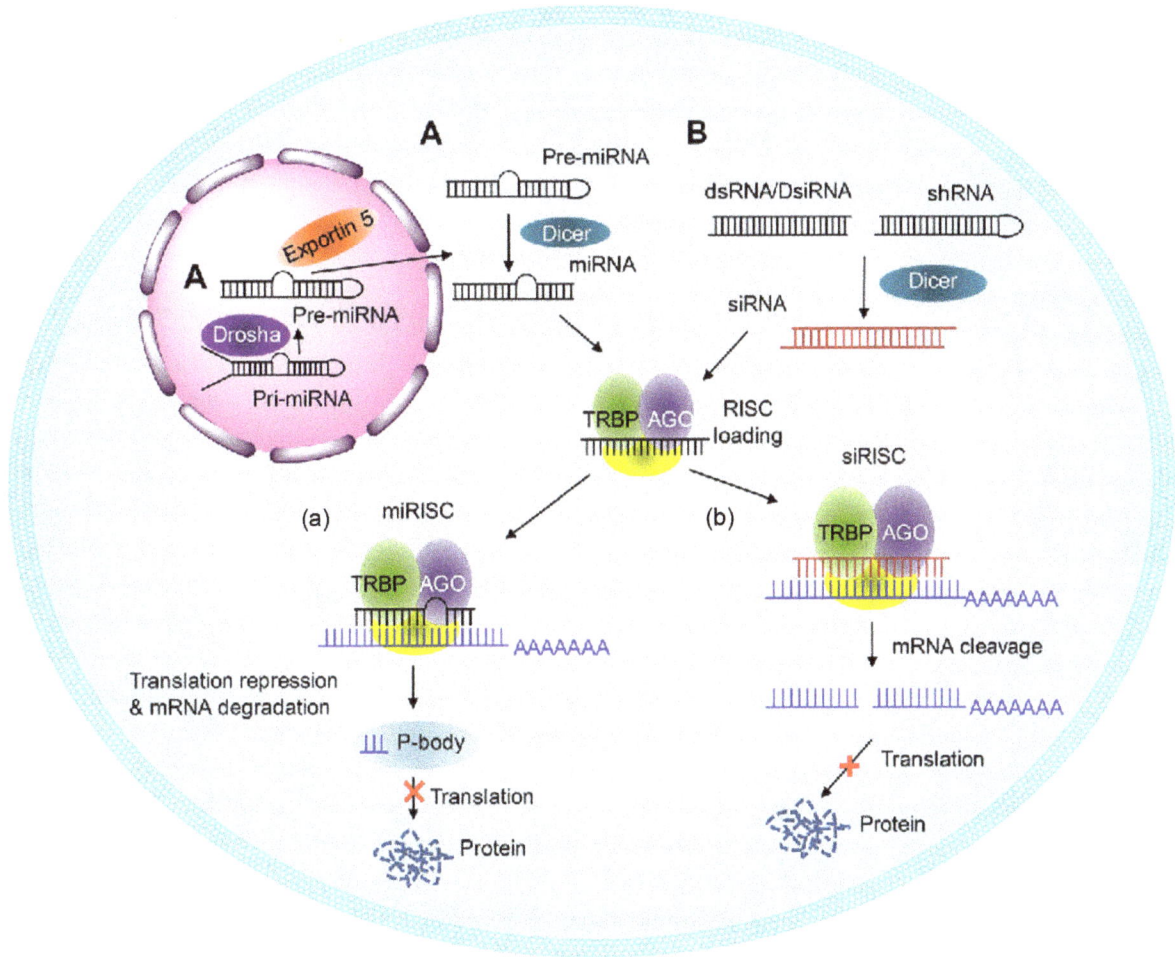

**FIGURE 15.7** Schematic illustration of the working mechanisms of miRNA (a) and siRNA (b).

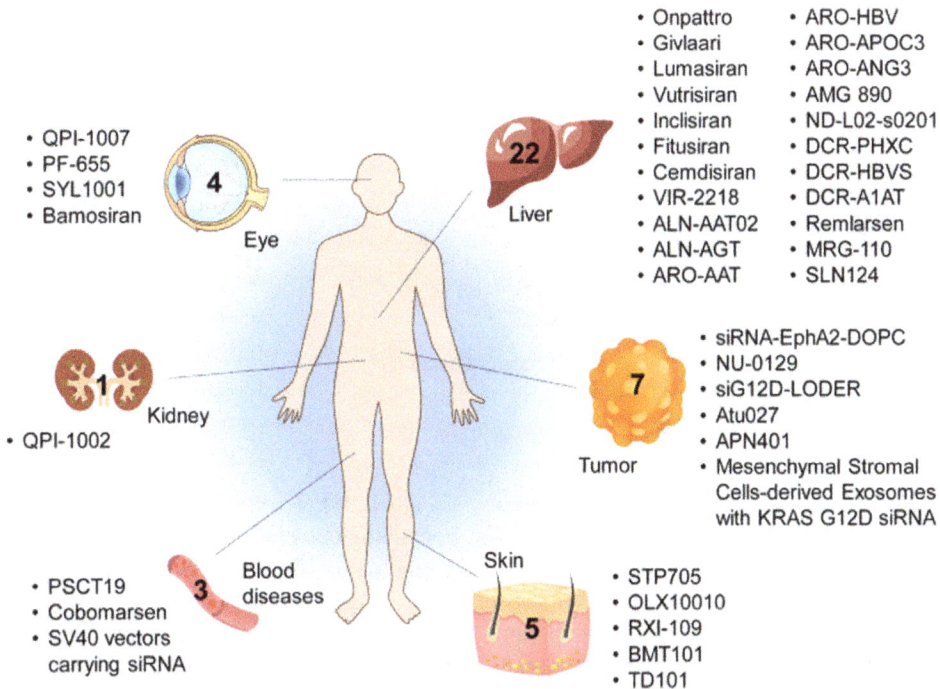

**FIGURE 15.8** Schematic illustration of the tissues targeted by siRNA and miRNA therapeutics currently being investigated at the clinical stage.

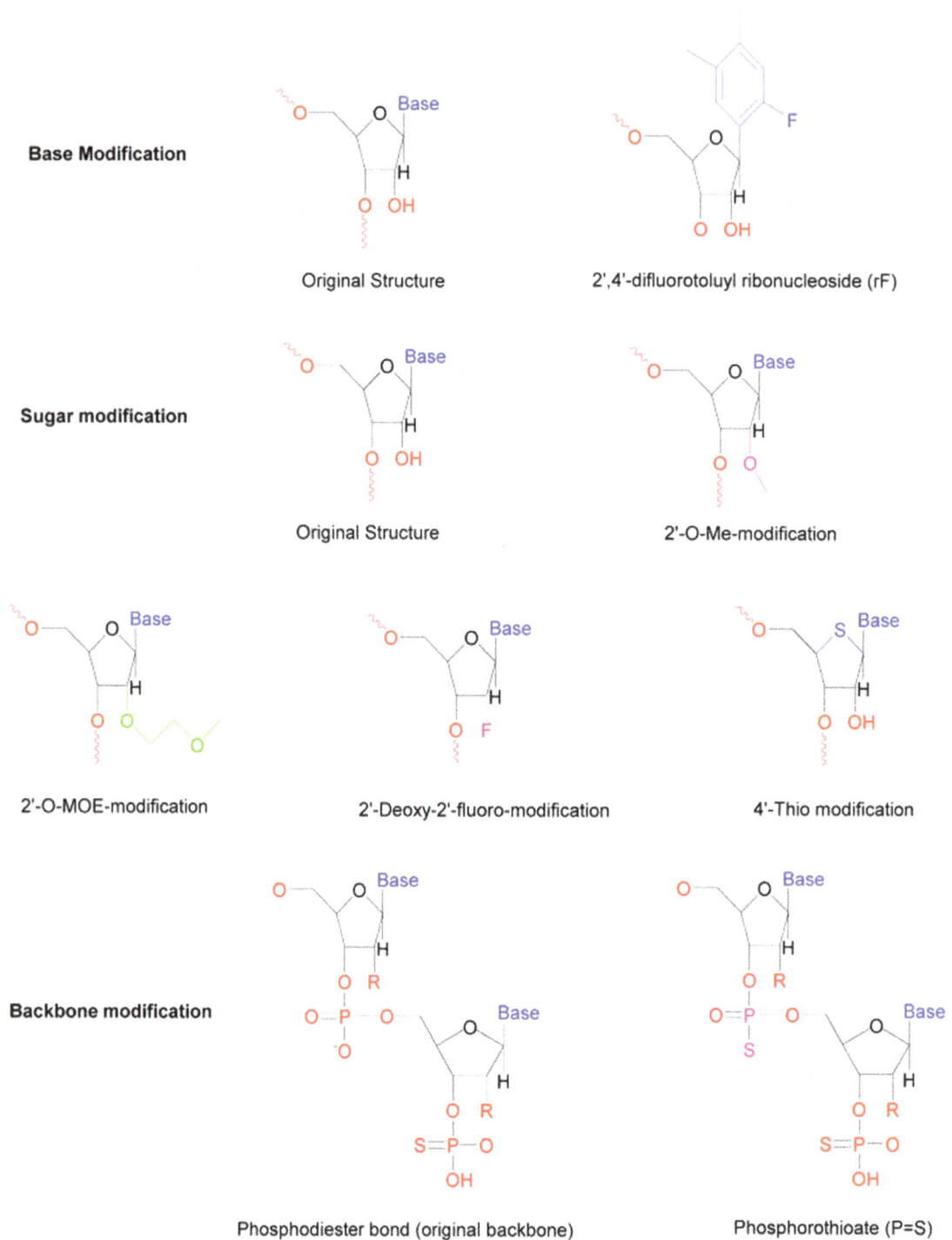

**FIGURE 15.9**  Chemical modifications of siRNA.

Advanced medicinal products constitute a group of novel treatment modalities and are rapidly being developed by scientific communities worldwide [207]. Recently, CAR-based cancer immunotherapy was listed as one of the breakthroughs of the year 2016, and this method is being tested in many clinical studies in the United States and European Union for the therapy of many types of cancer. Adoptive T-cell therapy is an important approach to cancer treatment that assumes the infusion of tumor-specific T cells [208]. T cells for adoptive therapy of patients with hematological malignancies can originate from an allogeneic donor. Several techniques have been developed to remove alloreactive T cells from grafts of stem cells to reduce the risk of graft-versus-host disease (GVHD). It has been shown recently that alloreactivity stems mainly from naïve T cells (CD45RA+) and not from memory T cells. Depletion of graft cells from naïve T cells could be a new approach for GVHD prophylaxis. Moreover, the depletion

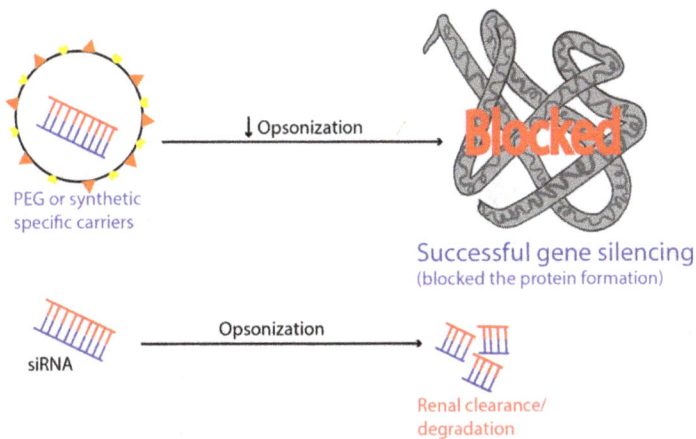

**FIGURE 15.10** Surface modification of siRNA carriers to decrease opsonization, renal clearance, and degradation.

**FIGURE 15.11** Schematic illustration of various nanocarriers.

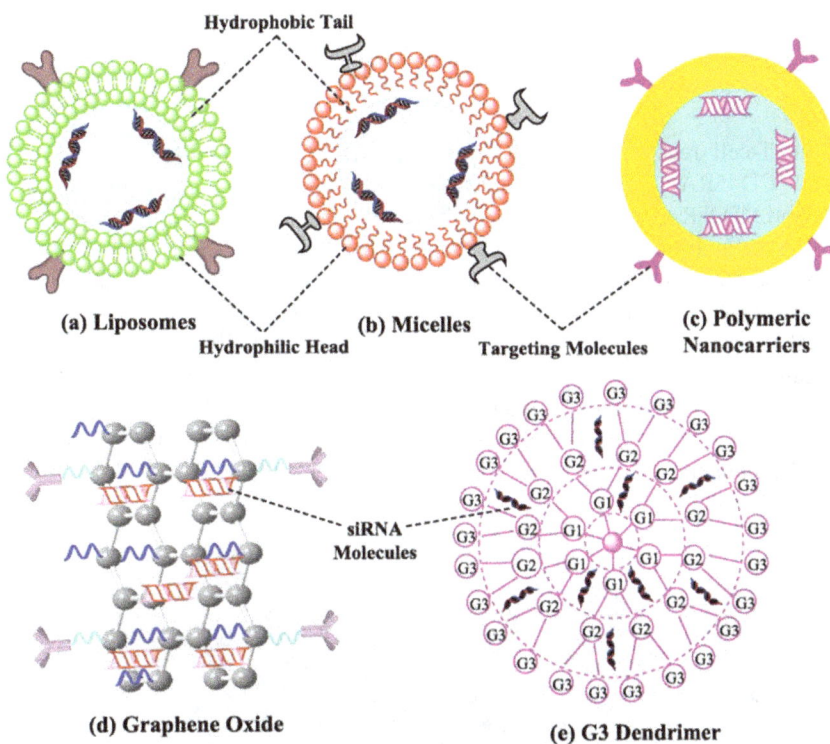

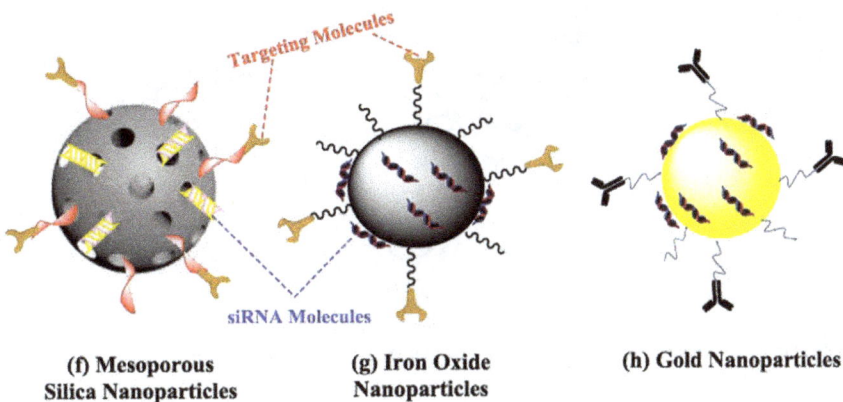

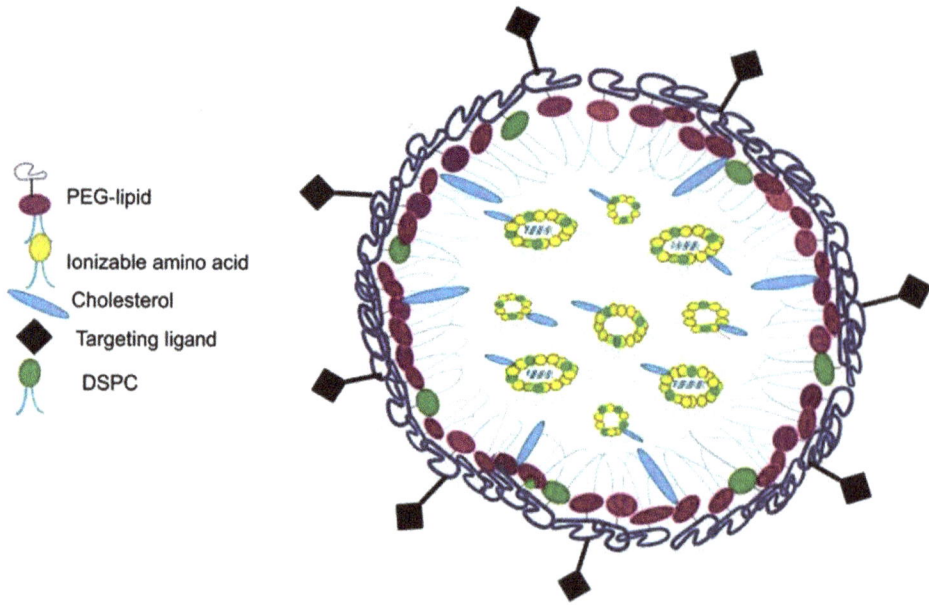

**FIGURE 15.12**   Schematic image of LNPs siRNA showing a nanostructured core.

procedure sustained the T-cell reactivity to pathogens. The elimination of all CD45RA+ cells also depletes naïve CD4+ Treg cells and CD45RA+ effector memory T cells (TEMRA). TEMRA cells are expanded in subjects chronically infected with HCMV [209], and they represent senescent and terminally differentiated (CD27-, CD28-, CD57+) cytotoxic T cells, which produce large amounts of granzyme and IFN gamma and have a low proliferative potential, and thus, the presence of such cells in expanded antiviral T cells is disadvantageous. The nanoscale immunotherapy represents a fundamental starting point for all forms of cancer.

## 15.10   CONCLUSIONS AND FUTURE PROSPECTS

MiRNA therapeutics provide a promising strategy for the treatment of disease. The present miRNA research is mainly centered on posttranscriptional gene silencing induced by RISC binding to the 3′-UTR of the mRNA [210]. Extensive research is required to develop rapid and sensitive analytical methods for the identification of miRNAs present in a particular cell, tissue, or fluids (such as serum and plasma) [32]. Gene silencing with miRNAs using nanoparticles is another area of investigation [211]. During the past decade,

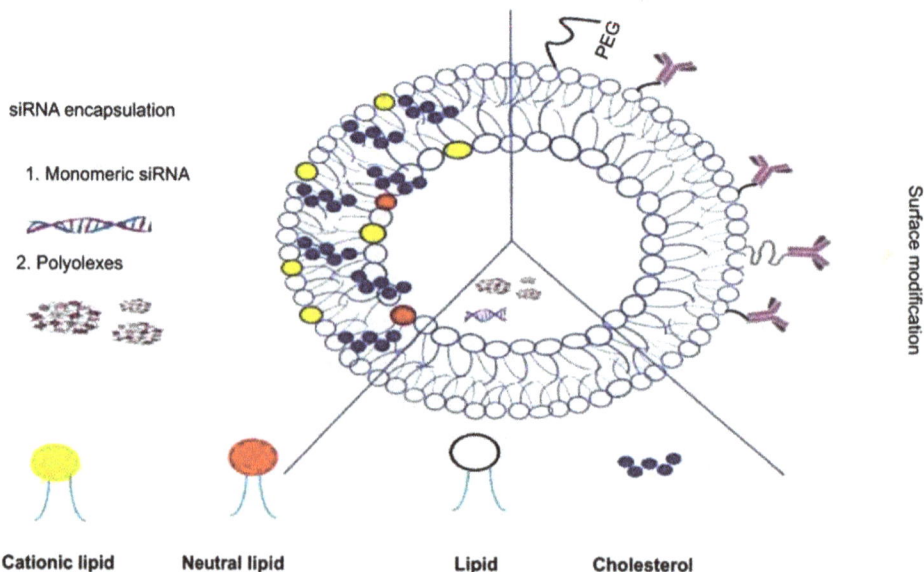

**FIGURE 15.13**   Schematic representation of several strategies for encapsulating siRNA in liposomes.

a strong research focus has been on the biology of miRNAs with special attention to miRNA regulation and miRNAs as biological targets in human disease.

The first miRNA-targeted drug, LNA-antimir-122, is under Phase 2 clinical trial (Miravirsen, Santaris Pharma). Over the last several years, extensive efforts have been given for the development of liposome- and nanoparticle-associated and naked oligonucleotides for targeted delivery of miRNA [210]. Various therapeutic miRNAs have been studied in the context of exosome delivery vehicles [212]. Unfortunately, only a few miRNA therapeutics have reached clinical trials [24].

An ideal carrier system should protect the therapeutic agent from the circulatory nucleases and deliver it intact to the target site. Despite the advancement in miRNA-based therapies to clinical trials, there remain many hurdles that need to be overcome for the use of the novel nanocarrier-based delivery technologies. Irrespective of their clinical significances, nanocarriers are also not devoid of limitations. These delivery vehicles are reported for intravenous or subcutaneous administration. The development of oral delivery vehicles is needed in advancing miRNA delivery through clinical development and commercial application [213].

A successful nanocarrier-based miRNA therapeutic must be safe and composed of biocompatible active/effector ingredients together with an organic or inorganic core;

it must have a sustainable half-life (PEGylation is the most used method to achieving this) and should have target specificity and outstanding pharmacokinetic profile; the manufacturing process must be robust and feasible. The product needs FDA approval [36, 214–217]. There is significant risk for investment in miRNA therapeutics due to the biological challenges, the cost of production and scale-up, and clinical approval challenges [216].

Overall, the application of nanotechnology is paving a new path in the development of effective drug delivery systems containing miRNAs. This will introduce new vistas in clinically considering their various merits like maximum efficacy, targeted effects, and improved patient compliance.

Nanotechnology is a field with high applicability in basic and translational medicine. Nanomedicine relies on various nanostructure designs, which are conjugated with a wide range of specific targeting agents used for clinical applications, like early diagnosis or disease treatment. The specific agents are attached to the nanoparticles' surface, which promote the accumulation and delivery of those agents in the neoplastic tissue. Nanomedicine's aim is to replace the chemotherapeutic drugs that are highly invasive or nonspecific with particular targeting agents with potential in detection, diagnosis, imaging, targeted delivery, and controlled release of therapeutic cargo. In clinical trials, delivery

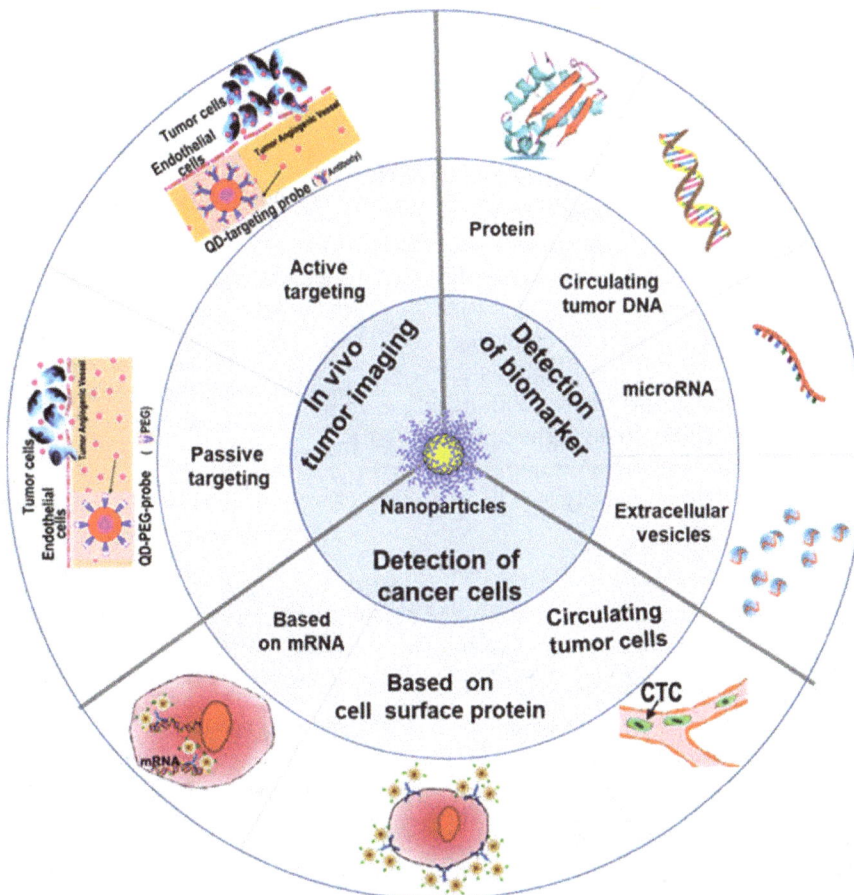

**FIGURE 15.14** Schematic illustration of nanotechnology applications in cancer diagnosis.

system–based therapies meet many obstacles such as therapeutic molecule stability, decreased nonspecific inflammation, controlled release of therapeutic molecules, specificity, and efficiency of the delivery systems. To enhance the stability of therapeutic molecules and to decrease the nonspecific immunogenicity, chemical modifications are required, while for efficient and specific delivery, tailored carriers are needed. RNA therapeutics exhibit great potential in clinical trials, while needing advanced delivery strategies to perform and play their roles in cancer therapy. There are some important component designs, such as PEGylated, tumor-specific ligand that coats the nanoparticles in combination with other light-, thermal-, pH-, or magnetic-sensitive components, that enhance the precision, specificity, and efficiency of therapeutic molecules to act on tumor sites and tumor cells. Also, a great potential is offered by biochemical modifications that increase the potency and decrease the off-target effects and other side effects of therapeutic drugs, allowing the implementation of new personalized drugs in clinical use.

The recent progress in nanotechnology-based application in cancer diagnosis has been summarized in this review (Figure 15.14) [154].

In the past 10 years, many efforts have been made to develop assays for cancer diagnosis based on nanotechnology. Compared with the currently available cancer diagnostics in the clinic, a variety of NP-based assays showed improvement in terms of selectivity and sensitivity or offered entirely new capacities that could not be achieved with traditional approaches. These advances will improve the survival rate of cancer patients by enabling early detection. In addition, these advances could be used to monitor cancer progress in response to treatment, which may contribute to the development of better strategies for cancer treatment.

Over the last decade, great progress has been made in the field of nanotechnology-based cancer diagnosis, and our understanding in this field has greatly improved. Although only a few NP-based assays have advanced to clinical trials, with close collaboration among researchers, engineers, and clinicians, nanotechnology-based cancer diagnosis is poised to move into the clinic in the near future. With its high sensitivity, specificity, and multiplexed measurement capacity, nanotechnology provides great opportunities to improve cancer diagnosis, which will ultimately lead to an improved cancer patient survival rate.

## REFERENCES

[1] Lee RC, Feinbaum RL, Ambros V. The C. elegans heterochronic gene LIN-4 encodes small RNAs with antisense complementarity to LIN-14. *Cell.* 1993; 75:843–54.

[2] Giza DE, Vasilescu C, Calin GA. Key principles of miRNA involvement in human diseases. *Discoveries (Craiova).* 2014; 2(4):e34. doi:10.15190/d.2014.26

[3] Ha TY. MicroRNAs in human diseases: from cancer to cardiovascular disease. *Immune Netw.* 2011; 11(3):135–54.

[4] Huiyuan W, Yifan J, Huige P, et al. Recent progress in microRNA delivery for cancer therapy by nonviral synthetic vectors. *Adv Drug Deliv Rev.* 2015; 81:142–60.

[5] Dua K, Hansbro NG, Foster PS, et al. MicroRNAs as therapeutics for future drug delivery systems in treatment of lung diseases. *Drug Deliv and Transl Res.* 2017; 7(1):168–78.

[6] Roy S, Sen CK. miRNA in wound inflammation and angiogenesis. *Microcirculation.* 2012; 19(3):224–32.

[7] Montgomery RL, Hullinger TG, Semus HM, et al. Therapeutic inhibition of miR-208a improves cardiac function and survival during heart failure. *Circulation.* 2011; 124(14):1537–47.

[8] Wong L, Wang J, Liew OW, et al. MicroRNA and heart failure. *Int J Mol Sci.* 2016; 17:502. https://doi.org/10.3390/ijms17040502

[9] Ji LY, Jiang DQ, Dong NN. The role of miR-145 in microvasculature. *Pharmazie.* 2013; 68(6):387–91.

[10] Alizadeh S, Azizi SG, Soleimani M, et al. The role of microRNAs in myeloproliferative neoplasia. *Int J Hematol Oncol Stem Cell Res.* 2016; 10(3):172–85.

[11] Deng Z, He Y, Yang X, et al. MicroRNA-29: a crucial player in fibrotic disease. *Mol Diagn Ther.* 2017; 21(3): 285–94.

[12] Stahlhut C, Slack FJ. Combinatorial action of microRNAs let-7 and miR-34 effectively synergizes with erlotinib to suppress nonsmall cell lung cancer cell proliferation. *Cell Cycle.* 2015; 14(13): 2171–80.

[13] Degliangeli F, Pompa PP, Fiammengo R. Nanotechnology-based strategies for the detection and quantification of microRNA. *Chem Eur J.* 2014; 20:9476–92.

[14] Kong W, Zhao JJ, He L, et al. Strategies for profiling microRNA expression. *J Cell Physiol.* 2009; 218:22–5.

[15] He L, Hannon GJ. MicroRNAs: small RNAs with a big role in gene regulation. *Nat Rev.* 2004; 5:522–31.

[16] Davis BN, Hata A. Regulation of microRNA biogenesis: amiRiad of mechanisms. *Cell Commun Signal.* 2009; 7:18. https://doi.org/10.1186/1478-811X-7-18

[17] The miRBase Sequence Database - Release 21 June 2014. http://www.mirbase.org/index.shtml. Accessed 14 April 2017.

[18] Hebert SS, Horre K, Nicolai L, et al. Loss of microRNA cluster miR-29a/b-1 in sporadic Alzheimer's disease correlates with increased BACE1/β-secretase expression. *Proc Natl Acad Sci USA.* 2008; 105:6415–20.

[19] Caporali A, Emanueli C. MicroRNA regulation in angiogenesis. *Vasc Pharmacol.* 2011; 55:79–86.

[20] Corsten MF, Miranda R, Kasmieh R, et al. MicroRNA-21 knockdown disrupts glioma growth in vivo and displays synergistic cytotoxicity with neural precursor cell–delivered S-TRAIL in human gliomas. *Cancer Res.* 2007; 67(19):8994–9000.

[21] Hsu AC, Dua K, Starkey MR, et al. MicroRNA-125a and -b inhibit A20 and MAVS to promote inflammation and impair antiviral response in COPD. *JCI Insight.* 2017; 2(7):e90443. https://doi.org/10.1172/jci.insight.90443

[22] Hu Y, Ehli EA, Boomsma DI. MicroRNAs as biomarkers for psychiatric disorders with a focus on autism spectrum disorder: current progress in genetic association studies, expression profiling, and translational research. *Autism Res* 2017; 10(7):1184–203.

[23] Calin GA, Ferracin M, Cimmino A, et al. A microRNA signature associated with prognosis and progression in chronic lymphocytic leukemia. *N Engl J Med.* 2005; 353(17):1793–801.

[24] Awasthi R, Rathbone MJ, Hansbro PM, et al. Therapeutic prospects of microRNAs in cancer treatment through nanotechnology. *Drug Deliv Transl Res.* 2018; 8:97–110 https://doi.org/10.1007/s13346-017-0440-1

[25] Zhao J, Xu T, Wang F, et al. miR-493-5p suppresses hepatocellular carcinoma cell proliferation through targeting GP73. *Biomed Pharmacother.* 2017; 90:744–51.

[26] Sun D, Wang C, Long S, et al. EBP-bactivated microRNA-223 promotes tumour growth through targeting RASA1 in human colorectal cancer. *Br J Cancer.* 2015; 112:1491–500.

[27] Wang X, Ivan M, Hawkins SM. The role of MicroRNA molecules and MicroRNA-regulating machinery in the pathogenesis and progression of epithelial ovarian cancer. *Gynecol Oncol* 2017; 147:481–7.

[28] Bartel DP. MicroRNAs: genomics, biogenesis, mechanism, and function. *Cell* 2004; 116 (2):281–97.

[29] Creighton CJ, Hernandez-Herrera A, Jacobsen A, et al. Integrated analyses of microRNAs demonstrate their widespread influence on gene expression in high-grade serous ovarian carcinoma. *PLoS One* 2012; 7 (3):e34546.

[30] Valencia-Sanchez MA, Liu J, Hannon GJ, et al. Control of translation and mRNA degradation by miRNAs and siRNAs. *Genes Dev.* 2006; 20:515–24.

[31] MacFarlane LA, Murphy PR. MicroRNA: biogenesis, function and role in cancer. *Curr Genomics.* 2010; 11(7):537–61.

[32] Catuogno S, Esposito CL, Quintavalle C, et al. Recent advance in biosensors for micrornas detection in cancer. *Cancers.* 2011; 3:1877–98.

[33] Conde J, Edelman ER, Artzi N. Target-responsive DNA/RNA nanomaterials for microRNA sensing and inhibition: the jack-of-all-trades in cancer nanotheranostics? *Adv Drug Deliv Rev.* 2015; 81:169–83.

[34] Broderick JA, Zamore PD. MicroRNA therapeutics. *Gene Ther.* 2011; 18:1104–10.

[35] Azimzadeh M, Rahaie M, Nasirizadeh N, et al. Electrochemical miRNA biosensors: the benefits of nanotechnology. *Nanomed Res J.* 2017; 2(1):36–48.

[36] Pasquinelli AE. MicroRNAs and their targets: recognition, regulation and an emerging reciprocal relationship. *Nat Rev Genet.* 2012; 13(4):271–82.

[37] Li M, Li J, Ding X, et al. MicroRNA and cancer. *AAPS J.* 2010; 12(3):309–17.

[38] Jansson MD, Lund AH. MicroRNA and cancer. *Mol Oncol.* 2012; 6(6):590–610.

[39] Reddy KB. MicroRNA (miRNA) in cancer. *Cancer Cell Int.* 2015; 15:38.

[40] Braicu C, Cojocneanu-Petric R, Chira S, et al. Clinical and pathological implications of miRNA in bladder cancer. *Int J Nanomedicine.* 2015; 10:791–800.

[41] Ekin A, Karatas OF, Culha M, et al. Designing a gold nanoparticle-based nanocarrier for microRNA transfection into the prostate and breast cancer cells. *J Gene Med.* 2014; 16(11–12):331–335.

[42] Berindan-Neagoe I, Calin GA. Molecular pathways: microRNAs, cancer cells, and microenvironment. *Clin Cancer Res.* 2014; 20(24):6247–53.

[43] Ghosh R, Singh LC, Shohet JM, et al. A gold nanoparticle platform for the delivery of functional microRNAs into cancer cells. *Biomaterials.* 2013; 34(3):807–16.

[44] Geng Y, Lin D, Shao L, et al. Cellular delivery of quantum dot-bound hybridization probe for detection of intracellular pre-microRNA using chitosan/poly(γ-glutamic acid) complex as a carrier. *PLoS One.* 2013; 8(6):e65540.

[45] Zaharie F, Muresan MS, Petrushev B, et al. Exosome-carried microRNA-375 inhibits cell progression and dissemination via Bcl-2 blocking in colon cancer. *J Gastrointestin Liver Dis.* 2015; 24(4): 435–43.

[46] Grewal R, Cucuianu A, Swanepoel C, et al. The role of microRNAs in the pathogenesis of HIV-related lymphomas. *Crit Rev Clin Lab Sci.* 2015; 52(5):232–41.

[47] Muresan M, Zaharie F, Bojan A, et al. MicroRNAs in liver malignancies. Basic science applied in surgery. *J BUON.* 2015; 20(2):361–75.

[48] Braicu C, Tomuleasa C, Monroig, P et al. Exosomes as divine messengers: are they the Hermes of modern molecular oncology? *Cell Death Differ.* 2015; 22(1):34–45.

[49] Kim JH, Yeom JH, Ko JJ, et al. Effective delivery of anti-miRNA DNA oligonucleotides by functionalized gold nanoparticles. *J Biotechnol.* 2011; 155(3):287–92.

[50] Jurj A, Braicu C, Pop L-A, et al. The new era of nanotechnology, an alternative to change cancer treatment. *Drug Des Dev Ther.* 2017; 11:2871–90.

[51] Li L, Zhou L, Li Y, et al. MicroRNA-21 stimulates gastric cancer growth and invasion by inhibiting the tumor suppressor effects of programmed cell death protein 4 and phosphatase and tensin homolog. *J BUON.* 2014; 19(1):228–36.

[52] Ben-Shushan D, Markovsky E, Gibori H, et al. Overcoming obstacles in microRNA delivery towards improved cancer therapy. *Drug Deliv Transl Res.* 2014; 4(1):38–49.

[53] Irimie AI, Braicu C, Pileczki V, et al. Knocking down of p53 triggers apoptosis and autophagy, concomitantly with inhibition of migration on SSC-4 oral squamous carcinoma cells. *Mol Cell Biochem.* 2016; 419(1–2):75–82.

[54] Lam JK, Chow MY, Zhang Y, et al. siRNA versus miRNA as therapeutics for gene silencing. *Mol Ther Nucleic Acids.* 2015; 4:e252.

[55] Guo W, Chen W, Yu W, et al. Small interfering RNA-based molecular therapy of cancers. *Chin J Cancer.* 2013; 32(9):488–493.

[56] Gandhi NS, Tekade RK, Chougule MB. Nanocarrier mediated delivery of siRNA/miRNAin combination with chemotherapeutic agents for cancer therapy: current progress and advances. *J Control Release.* 2014; 194:238–56.

[57] Yalcin S, Gunduz U. Nanoparticle based delivery of miRNAs to overcome drug resistance in breast cancer. *J Nanomed Nanotechnol.* 2016; 7:414. https://doi.org/10.4172/2157-7439.1000414

[58] Zhou J, Shum KT, Burnett JC, et al. Nanoparticle-based delivery of RNAi therapeutics: progress and challenges. *Pharmaceuticals.* 2013; 6:85–107.

[59] Jatzkewitz H. Incorporation of physiologically-active substances into a colloidal blood plasma substitute. *I Inc Mescaline Peptide Polyvinylpyrrolidone Hoppe-Seyler's Zeitschrift fur Physiol Chemie.* 1954; 297:149.

[60] Bangham A, Horne R. Negative staining of phospholipids and their structural modification by surface-active agents as observed in the electron microscope. *J Mol Biol.* 1964; 8:660–8.

[61] Bangham A, Standish M, Watkins J. Diffusion of univalent ions across the lamellae of swollen phospholipids. *J Mol Biol.* 1965; 13:238–52.

[62] Scheffel U, Wagner HN, Rhodes BA, et al. Albumin microspheres for study of reticuloendothelial system. *J Nucl Med.* 1972; 13:498.

[63] Gradishar WJ, Tjulandin S, Davidson N, et al. Phase III trial of nanoparticle albumin-bound paclitaxel compared with polyethylated castor oil-based paclitaxel in women with breast cancer. *J Clin Oncol.* 2005; 23:7794–803.

[64] Awasthi R, Pant I, Kulkarni GT, et al. Nano-structure mediated drug delivery: opportunities and challenges. *Curr Nanomed.* 2016; 6(2):78–104.

[65] Vasir JK, Reddy MK, Labhasetwar V. Nanosystems in drug targeting: opportunities and challenges. *Curr Nanosci.* 2005; 1:47–64.

[66] Chen J, Hessler JA, Putchakayala K, et al. Cationic nanoparticles induce nanoscale disruption in living cell plasma membranes. *J Phys Chem B.* 2009; 113:11179–85.

[67] Champion JA, Katare YK, Mitragotri S. Particle shape: a new design parameter for micro and nanoscale drug delivery carriers. *J Control Release.* 2007; 121:3–9.

[68] Kato RB, Roy B, De Oliveira FS, et al. Nanotopography directs mesenchymal stem cells to osteoblast lineage through regulation of microRNASMAD-BMP-2 circuit. *J Cell Physiol.* 2014; 229(11):1690–6.

[69] De Antonellis P, Liguori L, Falanga A, et al. MicroRNA 199b-5p delivery through stable nucleic acid lipid particles (SNALPs) in tumorigenic cell lines. *Naunyn Schmiedeberg's Arch Pharmacol.* 2013; 386(4):287–302.

[70] Yoo SS, Razzak R, Bedard E, et al. Layered gadolinium-based nanoparticle as a novel delivery platform for microRNA therapeutics. *Nanotechnology.* 2014; 25:425102.

[71] Chen Y, Zhu X, Zhang X, et al. Nanoparticles modified with tumor-targeting scFv deliver siRNA and miRNA for cancer therapy. *Mol Ther.* 2010; 18(9):1650–6.

[72] Ghosh R, Singh LC, Shohet JM, et al. A gold nanoparticle platform for the delivery of functional microRNAs into cancer cells. *Biomaterials.* 2013; 34:807–16.

[73] Tivnan A, Orr WS, Gubala V, et al. Inhibition of neuroblastoma tumor growth by targeted delivery of microRNA-34a using antidisialoganglioside GD2 coated nanoparticles. *PLoS One.* 2012; 7(5):e38129. https://doi.org/10.1371/journal.pone.0038129

[74] Medina OP, Zhu Y, Kairemo K. Targeted liposomal drug delivery in cancer. *Curr Pharm Des.* 2004; 10:2981–9.

[75] Babar IA, Cheng CJ, Booth CJ, et al. Nanoparticle-based therapy in an in vivo microRNA-155 (miR-155)-dependent mouse model of lymphoma. *Proc Natl Acad Sci USA.* 2012; 109(26):E1695–704. https://doi.org/10.1073/pnas.1201516109

[76] Liu XQ, Song WJ, Sun TM, et al. Targeted delivery of antisense inhibitor of miRNA for antiangiogenesis therapy using cRGD-functionalized nanoparticles. *Mol Pharm.* 2011; 8(1):250–9.

[77] Hao L, Patel PC, Alhasan AH, et al. Nucleic acid-gold nanoparticle conjugates as mimics of microRNA. *Small.* 2011; 7(22):3158–62.

[78] Valadi H, Ekstrom K, Bossios A, et al. Exosome mediated transfer of mRNAs and microRNAs is a novel mechanism of genetic exchange between cells. *Nat Cell Biol.* 2007; 9:654–9.

[79] Saraiva C, Ferreira L, Bernardino L. Traceable microRNA-124 loaded nanoparticles as a new promising therapeutic tool for Parkinson's disease. *Neurogenesis (Austin).* 2016; 3(1):e1256855. https://doi.org/10.1080/23262133.2016.1256855

[80] Cai C, Xie Y, Wu L, et al. PLGA-based dual targeted nanoparticles enhance miRNA transfection efficiency in hepatic carcinoma. *Sci Rep.* 2017; 7:46250. https://doi.org/10.1038/srep46250

[81] Mandli J, Mohammadi H, Amine A. Electrochemical DNA sandwich biosensor based on enzyme amplifiedmicroRNA-21 detection and gold nanoparticles. *Bioelectrochemistry.* 2017; 116:17–23.

[82] Setua S, Khan S, Yallapu MM, et al. Restitution of tumor suppressor microRNA-145 using magnetic nanoformulation for pancreatic cancer therapy. *J Gastrointest Surg.* 2017; 21(1):94–105.

[83] Xie Y, Murray-Stewart T, Wang Y, et al. Self-immolative nanoparticles for simultaneous delivery of microRNA and targeting of polyamine metabolism in combination cancer therapy. *J Control Release.* 2017; 246:110–9.

[84] Babar IA, Cheng CJ, Booth CJ, et al. Nanoparticle-based therapy in an in vivo microRNA-155 (miR-155)-dependent mouse model of lymphoma. *Proc Natl Acad Sci USA.* 2012; 109(26):E1695–704.

[85] Xia N, Zhang L. Nanomaterials-based sensing strategies for electrochemical detection of microRNAs. *Materials.* 2014; 7:5366–84.

[86] Wu L, Qu X. Electrochemical DNA biomarker detection: recent achievements and challenges. *Chem Soc Rev.* 2015; 44(10):2963–97.

[87] Li J, Tan S, Kooger R, et al. MicroRNAs as novel biological targets for detection and regulation. *Chem Soc Rev.* 2014; 43(2):506–17.

[88] Vigneshvar S, Sudhakumari CC, Senthilkumaran B, et al. Recent advances in biosensor technology for potential applications—an overview. *Frontiers in Bioengineering and Biotechnology.* 2016; 4:Article 11. https://doi.org/10.3389/fbioe.2016.00011

[89] Wang Y, Zheng D, Tan Q, et al. Nanopore-based detection of circulating microRNAs in lung cancer patients. *Nat Nanotechnol.* 2011; 6(10):668–74.

[90] Wei T, Du D, Wang Z, et al. Rapid and sensitive detection of microRNA via the capture of fluorescent dyes-loaded albumin nanoparticles around functionalized magnetic beads. *Biosens Bioelectron.* 2017; 94:56–62.

[91] Su S, Cao W, Liu W, et al. Dual-mode electrochemical analysis of microRNA-21 using gold nanoparticle-decorated $MoS_2$ nanosheet. *Biosens Bioelectron.* 2017; 94:552–9.

[92] Keshavarz M, Behpour M, Rafiee-pour MA. Recent trends in electrochemical microRNA biosensors for early detection of cancer. *RSC Adv.* 2015; 5:35651–60.

[93] Prabhu RH, Patravale VB, Joshi MD. Polymeric nanoparticles for targeted treatment in oncology: current insights. *Int J Nanomed.* 2015; 10:1001–18.

[94] Zhang Z, Liu Y, Jarreau C, et al. Nucleic acid-directed self-assembly of multifunctional gold nanoparticle imaging agents. *Biomater Sci.* 2013; 1(10):1055–64.

[95] Madaan K, Kumar S, Poonia N, et al. Dendrimers in drug delivery and targeting: drug-dendrimer interactions and toxicity issues. *J Pharm Bioallied Sci.* 2014; 6(3):139–50.

[96] Kamendi H, Zhou Y, Crosby M, et al. Doxorubicin: comparison between 3-h continuous and bolus intravenous administration paradigms on cardio-renal axis, mitochondrial sphingolipids and pathology. *Toxicol Appl Pharmacol.* 2015; 289(3):560–72.

[97] Soininen SK, Repo JK, Karttunen V, et al. Human placental cell and tissue uptake of doxorubicin and its liposomal formulations. *Toxicol Lett.* 2015; 239(2):108–14.

[98] Zakaria S, Gamal-Eldeen AM, El-Daly SM, et al. Synergistic apoptotic effect of Doxil® and aminolevulinic acid-based photodynamic therapy on human breast adenocarcinoma cells. *Photodiagnosis Photodyn Ther.* 2014; 11(2):227–38.

[99] Rose PG, Blessing JA, Lele S, et al. Evaluation of pegylated liposomal doxorubicin (Doxil) as second-line chemotherapy of squamous cell carcinoma of the cervix: a phase II study of the Gynecologic Oncology Group. *Gynecol Oncol.* 2006; 102(2):210–3.

[100] Ma Y, Fan X, Li L. pH-sensitive polymeric micelles formed by doxorubicin conjugated prodrugs for co-delivery of doxorubicin and paclitaxel. *Carbohydr Polym*. 2016; 137:19–29.

[101] Dellapasqua S, Mazza M, Rosa D, et al. Pegylated liposomal doxorubicin in combination with low-dose metronomic cyclophosphamide as preoperative treatment for patients with locally advanced breast cancer. *Breast*. 2011; 20(4):319–23.

[102] Yang X, Gao H, Qian F, et al. Internal standard method for the measurement of doxorubicin and daunorubicin by capillary electrophoresis with in-column double optical-fiber LED-induced fluorescence detection. *J Pharm Biomed Anal*. 2016; 117:118–24.

[103] Koh JS, Yi CO, Heo RW, et al. Protective effect of cilostazol against doxorubicin-induced cardiomyopathy in mice. *Free Radic Biol Med*. 2015; 89:54–61.

[104] Niu J, Xue A, Chi Y, et al. Induction of miRNA-181a by genotoxic treatments promotes chemotherapeutic resistance and metastasis in breast cancer. *Oncogene*. 2016; 35(10):1302–13.

[105] Wang Y, Mei X, Yuan J, et al. Taurine zinc solid dispersions attenuate doxorubicin-induced hepatotoxicity and cardiotoxicity in rats. *Toxicol Appl Pharmacol*. 2015; 289(1):1–11.

[106] Mohammadi ZA, Aghamiri SF, Zarrabi A, et al. A comparative study on non-covalent functionalization of carbon nanotubes by chitosan and its derivatives for delivery of doxorubicin. *Chem Phys Lett*. 2015; 642:22–8.

[107] Adwas AA, Elkhoely AA, Kabel AM, et al. Anti-cancer and cardioprotective effects of indol-3-carbinol in doxorubicin-treated mice. *J Infect Chemother*. 2016; 22(1):36–43.

[108] Oki Y, Ewer MS, Lenihan DJ, et al. Pegylated liposomal doxorubicin replacing conventional doxorubicin in standard R-CHOP chemotherapy for elderly patients with diffuse large B-cell lymphoma: an open label, single arm, phase II trial. *Clin Lymphoma Myeloma Leuk*. 2015; 15(3):152–8.

[109] Rios-Doria J, Durham N, Wetzel L, et al. Doxil synergizes with cancer immunotherapies to enhance antitumor responses in syngeneic mouse models. *Neoplasia*. 2015; 17(8):661–70.

[110] Zhang P, Li J, Ghazwani M, et al. Effective co-delivery of doxorubicin and dasatinib using a PEG-Fmoc nanocarrier for combination cancer chemotherapy. *Biomaterials*. 2015; 67:104–14.

[111] Bandekar A, Karve S, Chang MY, et al. Antitumor efficacy following the intracellular and interstitial release of liposomal doxorubicin. *Biomaterials*. 2012; 33(17):4345–52.

[112] Kopecka J, Campia I, Olivero P, et al. A LDL-masked liposomal-doxorubicin reverses drug resistance in human cancer cells. *J Control Release*. 2011; 149(2):196–205.

[113] Soundararajan A, Bao A, Phillips WT, et al. [(186)Re] Liposomal doxorubicin (Doxil): in vitro stability, pharmacokinetics, imaging and biodistribution in a head and neck squamous cell carcinoma xenograft model. *Nucl Med Biol*. 2009; 36(5):515–24.

[114] Schilt Y, Berman T, Wei X, et al. Using solution X-ray scattering to determine the high-resolution structure and morphology of PEGylated liposomal doxorubicin nanodrugs. *Biochim Biophys Acta*. 2016; 1860(1 Pt A):108–19.

[115] Yokoi K, Chan D, Kojic M, et al. Liposomal doxorubicin extravasation controlled by phenotype-specific transport properties of tumor microenvironment and vascular barrier. *J Control Release*. 2015; 217:293–9.

[116] Ko EM, Lippmann Q, Caron WP, et al. Clinical risk factors of PEGylated liposomal doxorubicin induced palmar plantar erythrodysesthesia in recurrent ovarian cancer patients. *Gynecol Oncol*. 2013; 131(3):683–8.

[117] Biswas S, Deshpande PP, Perche F, et al. Octa-arginine-modified pegylated liposomal doxorubicin: an effective treatment strategy for non-small cell lung cancer. *Cancer Lett*. 2013; 335(1):191–200.

[118] Tomuleasa C, Braicu C, Irimie A, et al. Nanopharmacology in translational hematology and oncology. *Int J Nanomedicine*. 2014; 9:3465–79.

[119] Lee BS, Yip AT, Thach AV, et al. The targeted delivery of doxorubicin with transferrin-conjugated block copolypeptide vesicles. *Int J Pharm*. 2015; 496(2):903–11.

[120] Kushnir CL, Angarita AM, Havrilesky LJ, et al. Selective cardiac surveillance in patients with gynecologic cancer undergoing treatment with pegylated liposomal doxorubicin (PLD). *Gynecol Oncol*. 2015; 137(3):503–7.

[121] Gill SE, Savage K, Wysham WZ, et al. Continuing routine cardiac surveillance in long-term use of pegylated liposomal doxorubicin: is it necessary? *Gynecol Oncol*. 2013; 129(3):544–7.

[122] Nowotnik DP, Cvitkovic E. ProLindac™ (AP5346): a review of the development of an HPMADACH platinumpolymer therapeutic. *Adv Drug Deliv Rev*. 2009; 61:1214–9.

[123] Bebawy M, Combes V, Lee E, et al. Membrane microparticles mediate transfer of P-glycoprotein to drug sensitive cancer cells. *Leukemia*. 2009; 23(9):1643–9.

[124] Jaiswal R, Luk F, Gong J, et al. Microparticle conferred microRNA profiles—implications in the transfer and dominance of cancer traits. *Mol Cancer*. 2012a; 11:37. https://doi.org/10.1186/1476-4598-11-37

[125] Jaiswal R, Gong J, Sambasivam S, et al. Microparticle-associated nucleic acids mediate trait dominance in cancer. *FASEB J*. 2012b; 26(1):420–9.

[126] Lu JF, Luk F, Gong J, et al. Microparticles mediate MRP1 intercellular transfer and the retemplating of intrinsic resistance pathways. *Pharmacol Res*. 2013; 76:77–83.

[127] Pokharel D, Padula MP, Lu JF, et al. Proteome analysis of multidrug-resistant, breast cancerderived microparticles. *J Extracell Vesicles*. 2014; 3:24384. https://doi.org/10.3402/jev.v3.24384

[128] Gong J, Luk F, Jaiswal R, et al. Microparticles mediate the intercellular regulation of microRNA-503 and proline-rich tyrosine kinase 2 to alter the migration and invasion capacity of breast cancer cells. *Front Oncol*. 2014; 4(220):1–11. https://doi.org/10.3389/fonc.2014.00220

[129] Pokharel D, Wijesinghe P, Oenarto V, et al. Deciphering cell-to-cell communication in acquisition of cancer traits: extracellular membrane vesicles are regulators of tissue biomechanics. *OMICS*. 2016; 20(8):462–9.

[130] Lu JF, Pokharel D, Padula MP, et al. A novel method to detect translation of membrane proteins following microvesicle intercellular transfer of nucleic acids. *J Biochem*. 2016; 160(5):281–9.

[131] Lu JF, Pokharel D, Bebawy M. A novel mechanism governing the transcriptional regulation of ABC transporters in MDR cancer cells. *Drug Deliv Transl Res*. 2017; 7(2):276–85.

[132] Jaiswal R, Johnson MS, Pokharel D, et al. Microparticles shed from multidrug resistant breast cancer cells provide a parallel survival pathway through immune evasion. *BMC Cancer*. 2017; 17:104. https://doi.org/10.1186/s12885-017-3102-2

[133] Krishnan SR, Luk F, Brown RD, et al. Isolation of human CD138(+) microparticles from the plasma of patients with multiple myeloma. *Neoplasia*. 2016; 18(1):25–32.

[134] Wu Q, Lu Z, Li H, et al. Next-generation sequencing of microRNAs for breast cancer detection. *J Biomed Biotechnol*. 2011; 2011:597145. https://doi.org/10.1155/2011/597145

[135] Lerebours GCF, Tozlu-Kara S, Vacher S, et al. MicroRNA expression profiling of inflammatory breast cancer. In *Thirty-Second Annual CTRC-AACR San Antonio Breast Cancer Symposium*, San Antonio, TX. Cancer Research, 2009: Abstract nr 6118; 2009. December 15, 2009.

[136] Guled M, Lahti L, Lindholm PM, et al. CDKN2A, NF2, and JUN are dysregulated among other genes by miRNAs in malignant mesothelioma—a miRNA microarray analysis. *Genes Chromosom Cancer*. 2009; 48(7):615–23.

[137] Zhu H, Chen XP, Luo SF, et al. Involvement of hypoxia-inducible factor-1-alpha in multidrug resistance induced by hypoxia in HepG2 cells. *J Exp Clin Cancer Res*. 2005; 24(4):565–74.

[138] Liang Z, Wu H, Xia J, et al. Involvement of miR-326 in chemotherapy resistance of breast cancer through modulating expression of multidrug resistance-associated protein 1. *Biochem Pharmacol*. 2010; 79:817–24.

[139] Pogribny IP, Filkowski JN, Tryndyak VP, et al. Alterations of microRNAs and their targets are associated with acquired resistance ofMCF-7 breast cancer cells to cisplatin. *Int J Cancer*. 2010; 127(8):1785–94.

[140] Qiu T, Zhou L, Wang T, et al. miR-503 regulates the resistance of non-small cell lung cancer cells to cisplatin by targeting Bcl-2. *Int J Mol Med*. 2013; 32:593–8.

[141] Zhang Y, Chen X, Lian H, et al. MicroRNA-503 acts as a tumor suppressor in glioblastoma for multiple anti-tumor effects by targeting IGF-1R. *Oncol Rep*. 2014; 31(3):1445–52.

[142] Zhou R, Gong A-Y, Chen D, et al. Histone deacetylases and NF-kB signaling coordinate expression of CX3CL1 in epithelial cells in response to microbial challenge by suppressing miR-424 and miR-503. *PLoS One*. 2013; 8(5): e65153. https://doi.org/10.1371/journal.pone.0065153

[143] Li N, Zhang F, Li S, et al. Epigenetic silencing of MicroRNA- 503 regulates FANCA expression in non-small cell lung cancer cell. *Biochem Biophys Res Commun*. 2014; 444(4):611–6.

[144] Xiao F, Zhang W, Chen L, et al. MicroRNA-503 inhibits the G1/S transition by downregulating cyclin D3 and E2F3 in hepatocellular carcinoma. *J Transl Med*. 2013; 11:195. https://doi.org/10.1186/1479-5876-11-195

[145] Forrest AR, Kanamori-Katayama M, Tomaru Y, et al. Induction of microRNAs, mir-155, mir-222, mir-424 and mir-503, promotes monocytic differentiation through combinatorial regulation. *Leukemia*. 2010; 24(2):460–6.

[146] Yang Y, Luo J, Zhai X, et al. Prognostic value of phospho-Akt in patients with non-small cell lung carcinoma: a meta-analysis. *Int J Cancer*. 2014; 135(6):1417–24.

[147] Ueda H, Abbi S, Zheng C, et al. Suppression of Pyk2 kinase and cellular activities by Fip200. *J Cell Biol*. 2000; 149(2):423–30.

[148] Jou AF, Lu CH, Ou YC, et al. Diagnosing the miR-141 prostate cancer biomarker using nucleic acid-functionalized CdSe/ZnS QDs and telomerase. *Chem Sci*. 2015; 6:659.

[149] Seferos DS, Giljohann DA, Hill HD, et al. Nano-flares: probes for transfection and mRNA detection in living cells. *J Am Chem Soc*. 2007; 129:15477.

[150] Choi CH, Hao L, Narayan SP, et al. Mechanism for the endocytosis of spherical nucleic acid nanoparticle conjugates. *Proc Natl Acad Sci USA*. 2013; 110:7625.

[151] Briley WE, Bondy MH, Randeria PS, et al. Quantification and real-time tracking of RNA in live cells using Sticky-flares. *Proc Natl Acad Sci USA*. 2015; 112:9591.

[152] Seftor EA, Seftor R, Weldon D, et al. Melanoma tumor cell heterogeneity: a molecular approach to study subpopulations expressing the embryonic morphogen nodal. *Semin Oncol*. 2014; 41:259.

[153] Halo TL, McMahon KM, Angeloni NL, et al. NanoFlares for the detection, isolation, and culture of live tumor cells from human blood. *Proc Natl Acad Sci USA*. 2014; 111:17104.

[154] Zhang Y, Li M, Gao X, et al. Nanotechnology in cancer diagnosis: progress, challenges and opportunities. *J Hematol Oncol*. 2019; 12:137. https://doi.org/10.1186/s13045-019-0833-3

[155] Lee K, Cui Y, Lee LP, et al. *Nat nanotechnol*. 2014; 9:474.

[156] Morille M, Passirani C, Vonarbourg A, et al. Progress in developing cationic vectors for non-viral systemic gene therapy against cancer. *Biomaterials*. 2008; 29(24–25):3477–96.

[157] He CX, Tabata Y, Gao JQ. Non-viral gene delivery carrier and its three-dimensional transfection system. *Int J Pharm*. 2010; 386(1–2):232–42.

[158] Itaka K, Kataoka K. Recent development of nonviral gene delivery systems with virus-like structures and mechanisms. *Eur J Pharm Biopharm*. 2009; 71(3):475–83.

[159] Yang N. An overview of viral and nonviral delivery systems for microRNA. *Int J Pharm Investig*. 2015; 5(4):179–81.

[160] Lee JH, Kim Y, Yoon YE, et al. Development of efficient adeno-associated virus (AAV)-mediated gene delivery system with a phytoactive material for targeting human melanoma cells. *N Biotechnol*. 2017; 37(Pt B):194–9.

[161] Elouahabi A, Ruysschaert JM. Formation and intracellular trafficking of lipoplexes and polyplexes. *Mol Ther*. 2005; 11(3):336–47.

[162] Riley MK, Vermerris W. Recent advances in nanomaterials for gene delivery-a review. *Nanomaterials (Basel)*. 2017; 7(5): 1–19.

[163] McErlean EM, McCrudden CM, McCarthy HO. Chapter 03: Multifunctional delivery systems for cancer gene therapy. In: Hashad D, editor. *Gene Therapy – Principles and Challenges*. Rijeka: InTech, 2015.

[164] Ghosh R, Singh LC, Shohet JM, et al. A gold nanoparticle platform for the delivery of functional microRNAs into cancer cells. *Biomaterials*. 2013; 34(3):807–16.

[165] Crew E, Tessel MA, Rahman S, et al. MicroRNA conjugated gold nanoparticles and cell transfection. *Anal Chem*. 2012; 84(1):26–9.

[166] Berce C, Lucan C, Petrushev B, et al. In vivo assessment of bone marrow toxicity by gold nanoparticle-based bioconjugates in Crl:CD1(ICR) mice. *Int J Nanomedicine*. 2016; 11:4261–73.

[167] Ekin A, Karatas OF, Culha M, et al. Designing a gold nanoparticle-based nanocarrier for microRNA transfection into the prostate and breast cancer cells. *J Gene Med*. 2014; 16(11–12):331–5.

[168] Orza A, Soriţău O, Tomuleasa C, et al. Reversing chemoresistance of malignant glioma stem cells using gold nanoparticles. *Int J Nanomedicine*. 2013; 8:689–702.

[169] Tomuleasa C, Soritau O, Orza A, et al. Gold nanoparticles conjugated with cisplatin/doxorubicin/capecitabine lower

the chemoresistance of hepatocellular carcinoma-derived cancer cells. *J Gastrointestin Liver Dis.* 2012; 21(2):187–96.

[170] Tao L, Faig A, Uhrich KE. Liposomal stabilization using a sugar-based, PEGylated amphiphilic macromolecule. *J Colloid Interface Sci.* 2014; 431:112–6.

[171] Sugiyama I, Sadzuka Y. Enhanced antitumor activity of different double arms polyethyleneglycol-modified liposomal doxorubicin. *Int J Pharm.* 2013; 441(1–2):279–84.

[172] Beloglazova NV, Goryacheva OA, Speranskaya ES, et al. Silica-coated liposomes loaded with quantum dots as labels for multiplex fluorescent immunoassay. *Talanta.* 2015; 134:120–5.

[173] Bunker A. Poly(ethylene glycol) in drug delivery, why does it work, and can we do better? All atom molecular dynamics simulation provides some answers. *Phys Procedia.* 2012; 34:24–33.

[174] Kleinstreuer C, Childress E, Kennedy A. Chapter 10 – Targeted drug delivery: multifunctional nanoparticles and direct micro-drug delivery to tumors. In: Becker SM, Kuznetsov AV, editors. *Transport in Biological Media.* Boston: Elsevier; 2013:391–416.

[175] Danhier F, Feron O, Préat V. To exploit the tumor microenvironment: passive and active tumor targeting of nanocarriers for anti-cancer drug delivery. *J Control Release.* 2010; 148(2):135–46.

[176] Seguin J, Brullé L, Boyer R, et al. Liposomal encapsulation of the natural flavonoid fisetin improves bioavailability and antitumor efficacy. *Int J Pharm.* 2013; 444(1–2):146–54.

[177] Shin GH, Kim JT, Park HJ. Recent developments in nanoformulations of lipophilic functional foods. *Trends Food Sci Technol.* 2015; 46(1):144–57.

[178] Lee BK, Yun YH, Park K. Smart nanoparticles for drug delivery: boundaries and opportunities. *Chem Eng Sci.* 2015; 125:158–64.

[179] Refuerzo JS, Alexander JF, Leonard F, et al. Liposomes: a nanoscale drug carrying system to prevent indomethacin passage to the fetus in a pregnant mouse model. *Am J Obstet Gynecol.* 2015; 212(4):508.e1–e7.

[180] Shibata H, Yoshida H, Izutsu K, et al. Interaction kinetics of serum proteins with liposomes and their effect on phospholipase-induced liposomal drug release. *Int J Pharm.* 2015; 495(2):827–39.

[181] Wang TW, Yeh CW, Kuan CH, et al. Tailored design of multifunctional and programmable pH-responsive self-assembling polypeptides as drug delivery nanocarrier for cancer therapy. *Acta Biomater.* 2017; 58:54–66.

[182] Constantinidou A, Jones RL, Scurr M, et al. Pegylated liposomal doxorubicin, an effective, well-tolerated treatment for refractory aggressive fibromatosis. *Eur J Cancer.* 2009; 45(17): 2930–4.

[183] Ananda S, Nowak AK, Cher L, et al. Phase 2 trial of temozolomide and pegylated liposomal doxorubicin in the treatment of patients with glioblastoma multiforme following concurrent radiotherapy and chemotherapy. *J Clin Neurosci.* 2011; 18(11):1444–8.

[184] Liu C, Xu XY. A systematic study of temperature sensitive liposomal delivery of doxorubicin using a mathematical model. *Comput Biol Med.* 2015; 60:107–16.

[185] Rohlfing S, Aurich M, Schöning T, et al. Nonpegylated liposomal doxorubicin as a component of R-CHOP is an effective and safe alternative to conventional doxorubicin in the treatment of patients with diffuse large B-cell lymphoma and preexisting cardiac diseases. *Clin Lymphoma Myeloma Leuk.* 2015; 15(8):458–63.

[186] Zhao Y, Alakhova DY, Kim JO, et al. A simple way to enhance Doxil® therapy: drug release from liposomes at the tumor site by amphiphilic block copolymer. *J Control Release.* 2013; 168(1):61–9.

[187] Reynolds JG, Geretti E, Hendriks BS, et al. HER2-targeted liposomal doxorubicin displays enhanced anti-tumorigenic effects without associated cardiotoxicity. *Toxicol Appl Pharmacol.* 2012; 262(1):1–10.

[188] Prabhu RH, Patravale VB, Joshi MD. Polymeric nanoparticles for targeted treatment in oncology: current insights. *Int J Nanomedicine.* 2015; 10:1001–18.

[189] Chavanpatil MD, Khdair A, Gerard B, et al. Surfactant-polymer nanoparticles overcome P-glycoprotein-mediated drug efflux. *Mol Pharm.* 2007; 4(5):730–8.

[190] Bellocq NC, Pun SH, Jensen GS, et al. Transferrin-containing, cyclodextrin polymer-based particles for tumor-targeted gene delivery. *Bioconjug Chem.* 2003; 14(6):1122–32.

[191] Wong HL, Rauth AM, Bendayan R, et al. A new polymer-lipid hybrid nanoparticle system increases cytotoxicity of doxorubicin against multidrug-resistant human breast cancer cells. *Pharm Res.* 2006; 23(7):1574–85.

[192] Shmarakov I, Mukha I, Vityuk N, et al. Antitumor activity of alloy and core-shell-type bimetallic AgAu nanoparticles. *Nanoscale Res Lett.* 2017; 12(1):333.

[193] Fakhri A, Tahami S, Naji M. Synthesis and characterization of core-shell bimetallic nanoparticles for synergistic antimicrobial effect studies in combination with doxycycline on burn specific pathogens. *J Photochem Photobiol B.* 2017; 169:21–6.

[194] Mittal AK, Kumar S, Banerjee UC. Quercetin and gallic acid mediated synthesis of bimetallic (silver and selenium) nanoparticles and their antitumor and antimicrobial potential. *J Colloid Interface Sci.* 2014; 431:194–9.

[195] Wu P, Gao Y, Zhang H, et al. Aptamer-guided silver-gold bimetallic nanostructures with highly active surface-enhanced Raman scattering for specific detection and near-infrared photothermal therapy of human breast cancer cells. *Anal Chem.* 2012; 84(18):7692–9.

[196] Alshatwi AA, Athinarayanan J, Periasamy VS. Green synthesis of bimetallic Au@Pt nanostructures and their application for proliferation inhibition and apoptosis induction in human cervical cancer cell. *J Mater Sci Mater Med.* 2015; 26(3):148.

[197] Madaan K, Kumar S, Poonia N, et al. Dendrimers in drug delivery and targeting: drug-dendrimer interactions and toxicity issues. *J Pharm Bioallied Sci.* 2014; 6(3):139–50.

[198] Kesharwani P, Jain K, Jain NK. Dendrimer as nanocarrier for drug delivery. *Prog Polym Sci.* 2014; 39(2):268–307.

[199] Zhang W, Zhang Z, Zhang Y. The application of carbon nanotubes in target drug delivery systems for cancer therapies. *Nanoscale Res Lett.* 2011; 6:555.

[200] Kushwaha SKS, Ghoshal S, Rai AK, et al. Carbon nanotubes as a novel drug delivery system for anticancer therapy: a review. *Braz J Pharm Sci.* 2013; 49(4):629–43.

[201] Gherman C, Tudor MC, Constantin B, et al. Pharmacokinetics evaluation of carbon nanotubes using FTIR analysis and histological analysis. *J Nanosci Nanotechnol.* 2015; 15(4):2865–9.

[202] Farahani BV, Behbahani GR, Javadi N. Functionalized multi walled carbon nanotubes as a carrier for doxorubicin: drug adsorption study and statistical optimization of drug loading by factorial design methodology. *J Braz Chem Soc.* 2016; 27(4):694–705.

[203] Fletcher CDM, Bridge JA, Hogendoorn P, et al. *WHO Classification of Tumours of Soft Tissue and Bone*. 4th ed. Lyon: IARC Press; 2013.

[204] Kalos M, Levine BL, Porter DL, et al. T cells with chimeric antigen receptors have potent antitumor effects and can establish memory in patients with advanced leukemia. *Sci Transl Med*. 2011; 3(95):95ra73.

[205] Imai C, Mihara K, Andreansky M, et al. Chimeric receptors with 4-1BB signaling capacity provoke potent cytotoxicity against acute lymphoblastic leukemia. *Leukemia*. 2004; 18(4):676–84.

[206] Porter DL, Levine BL, Kalos M, et al. Chimeric antigen receptor-modified T cells in chronic lymphoid leukemia. *N Engl J Med*. 2011; 365(8):725–33.

[207] Tomuleasa C, Fuji S, Cucuianu A, et al. MicroRNAs as biomarkers for graft-versus-host disease following allogeneic stem cell transplantation. *Ann Hematol*. 2015; 94(7):1081–92.

[208] Perica K, Varela JC, Oelke M, et al. Adoptive T cell immunotherapy for cancer. *Rambam Maimonides Med J*. 2015; 6(1):e0004.

[209] Pardoll DM. The blockade of immune checkpoints in cancer immunotherapy. *Nat Rev Cancer*. 2012; 12(4): 252–64.

[210] He C, Hu Y, Yin L, et al. Effects of particle size and surface charge on cellular uptake and biodistribution of polymeric nanoparticles. *Biomaterials*. 2010; 31(13):3657–66.

[211] Xiao K, Li Y, Luo J, et al. The effect of surface charge on in vivo biodistribution of PEG-oligocholic acid based micellar nanoparticles. *Biomaterials*. 2011; 32(13):3435–46.

[212] Goodman CM, McCusker CD, Yilmaz T, et al. Toxicity of gold nanoparticles functionalized with cationic and anionic side chains. *Bioconjug Chem*. 2004; 15(4):897–900.

[213] Natarajan JV, Nugraha C, Ng XW, et al. Sustained-release from nanocarriers: a review. *J Control Release*. 2014; 193:122–38.

[214] Smith BR, Kempen P, Bouley D, et al. Shape matters: intravital microscopy reveals surprising geometrical dependence for nanoparticles in tumor models of extravasation. *Nano Lett*. 2012; 12(7): 3369–77.

[215] Yin PT, Shah BP, Lee KB. Combined magnetic nanoparticle-based microRNA and hyperthermia therapy to enhance apoptosis in brain cancer cells. *Small*. 2014; 10(20):4106–12.

[216] Schwerdt A, Zintchenko A, Concia M, et al. Hyperthermia-induced targeting of thermosensitive gene carriers to tumors. *Hum Gene Ther*. 2008; 19(11):1283–92.

[217] Zhang Z, Liu Y, Jarreau C, et al. Nucleic acid-directed self-assembly of multifunctional gold nanoparticle imaging agents. *Biomater Sci*. 2013; 1(10):1055–64.

# List of Abbreviations

| | |
|---|---|
| α SMA | α-smooth muscle actin |
| 2'O-Methyl | the methylation of the oxygen at position 2 in the riboses |
| 3′-ends of siRNA | sticky siRNA |
| 5-aza-dC | 5-azacytidine |
| AASLD | American Association for the Study of Liver Diseases |
| AATD | alpha-1 antitrypsin deficiency |
| AAV | adeno-associated virus |
| ABC | ATP-binding cassette |
| ABCA1 | ATP-binding cassette subfamily A member 1 |
| Abraxane | albumin-based paclitaxel nanoparticles |
| AD | adenocarcinoma |
| AD | Alzheimer's disease |
| ADRB2 β2 | adrenergic receptor |
| AFAP | actin filament-associated protein |
| Afp | alpha-fetoprotein |
| Ag–Au NP | silver–gold bimetallic nanoparticle |
| Ago | Argonaute protein subfamily |
| AGO1 | Argonaute 1 |
| Ag–Se NP | silver–selenium bimetallic nanoparticle |
| aiRNA | asymmetric interfering RNA |
| AL | acute leukemia |
| ALD | alcoholic liver disease |
| ALL | acute lymphoblastic leukemia |
| AML | acute myeloid leukemia |
| ANT | anthracycline |
| AOT | aerosol OT |
| APC | adenomatous polyposis coli |
| APL | acute promyelotic leukemia |
| apoB | apolipoprotein B |
| AR | androgen receptor |
| ARE | adenylate uridylate (AU)-rich element |
| ARF | alterative reading frame |
| ARNT | aryl hydrocarbon nuclear translocator |
| ASCT | autologous stem cell transplantation |
| ASGP-R | asialoglycoprotein receptor |
| ASO | antisense oligonucleotide |
| AST | aminotransferase |
| AT | antithrombin |
| AT-LD | ataxia-telangiectasia-like-disorder |
| ATM | ataxia-telangiectasia mutated |
| ATP | adenosine triphosphate |
| ATR | atom transfer radical |
| ATRP | atom transfer radical polymerization |
| AU | adenylate uridylate |
| AUC | area under the curve |
| Au NP | gold nanoparticle |
| Au–Pt NP | gold–platinum bimetallic nanoparticle |
| BBB | blood–brain barrier |
| BCL | B-cell lymphoma |

| | |
|---|---|
| B-CLL | B-cell lymphocytic leukemia |
| BIC | B-cell integration cluster |
| BLC | bladder cancer |
| BM | basal membrane |
| BM | bone marrow |
| BM(PEG)₂ | 1,8-bis(maleimidodiethylene) glycol |
| BPD | benign pancreatic/peri-pancreatic disease |
| bPEI | branched PEI |
| BPH | benign prostate hyperplasia |
| BrC | breast cancer |
| BRCA1 | breast cancer 1 |
| BRCA2 | breast cancer 2 |
| BSP | bisulfite genomic sequencing PCR |
| BTC | biliary tract cancer |
| CA19.9 | carbohydrate antigen 19.9 |
| CaP | prostate cancer |
| CAF | cancer-associated fibroblast |
| CA | contrast agent |
| CB | cord blood |
| CBC | complete differential blood count |
| CBP | CREB-binding protein |
| CC | cholangiocarcinoma |
| CCC | cervical cell carcinoma |
| CCND2 | cyclin D2 |
| CCNE2 | cyclin E2 |
| CCND1 | cyclin D1 |
| CCNG1 | cyclin G1 |
| CD | cycle day |
| CD | Crohn's disease |
| CD31 | cluster of differentiation 31 |
| CD44 | cyclin D44 |
| CDA | cytidine deaminase |
| CDC2 | cell division cycle 2 |
| CDH13 | cadherin 13, H-cadherin (heart) |
| CDK | cyclin-dependent kinase |
| CDK1 | cyclin-dependent kinase 1 |
| CDK2 | cyclin-dependent kinase 2 |
| CDKN2A | cyclin-dependent kinase inhibitor 2A |
| CEA | carcinoembryonic antigen |
| ceRNA | competing endogenous RNA |
| CGI | CpG island |
| CHC | chronic hepatitis C infection |
| ChIP | chromatin immunoprecipitation |
| Chk1 | checkpoint proteins 1 |
| Chk2 | checkpoint proteins 2 |
| Chol-siRNA | cholesterol-conjugated siRNA |
| CIMP | CpG island methylator phenotype |
| CK | creatine kinase |
| CK2 | casein kinase II |
| CLL | chronic lymphocytic leukemia |
| c-Myc | oncogenic downregulating transcription factor |

| | |
|---|---|
| CML | chronic myelogenous leukemia |
| CNT | carbon nanotube |
| COBRA | combined bisulfite restriction analysis |
| CP | chronic pancreatitis |
| CPP | cell penetrating peptide |
| CR | complete response |
| crasiRNA | centromere repeat associated small interacting RNA |
| CRC | colorectal cancer |
| CREB | cAMP response element binding protein |
| cRGD | cyclic Arg-Gly-Asp peptides |
| cRGD-LPHNP | nanoparticles modified with cRGD |
| CRPC | castration-resistant prostate cancer |
| CSC | cancer stem cell |
| CSN | COP9 signalosome complex |
| CT | computed tomography |
| CTCF | transcriptional repressor also known as 11-zinc finger protein |
| CTGF | connective tissue growth factor |
| CTC | circulating tumor cell |
| DAPK | death-associated protein kinase |
| DCK | deoxycytidine kinase |
| DDAB | dimethyldioctadecylammonium bromide |
| del(13) | 13 deletion |
| DFI | disease-free interval |
| DFS | disease-free survival |
| DGCR8 | DiGeorge syndrome critical region 8 |
| DKO | double knockout |
| DLBCL | diffuse large B-cell lymphoma |
| DM2 | dominant multisystemic |
| DMD | Duchenne muscular dystrophy |
| DMMAn | dimethylmaleic anhydride |
| DMSO | dimethyl sulfoxide |
| DNA | deoxyribonucleic acid |
| DNA NF | DNA nanoflower |
| DNA-PK | DNA-dependent protein kinase |
| DNA-PKcs | catalytic subunit of DNA-PK |
| DNMT | DNA methyltransferase |
| DNMT1 | DNA methyltransferase 1 |
| DNR | DNA nanoribbon |
| DNR-T | DNR with three staple stands |
| DOPC | 1,2-dioleoyl-sn-glycero-3-phosphocholine |
| DOX | doxorubicin |
| DPC | dynamic polyconjugates |
| ds | double-stranded |
| DSB | DNA double-strand break |
| DSB | double-strand break |
| dsDNA | double-stranded DNA |
| DSN | duplex-specific nuclease |
| dsRBP | double-stranded RNA-binding protein |
| dsRNA | double-stranded RNA molecules |
| DTME | dithiobismaleimidoethane |
| EC | endometrial cancer |
| ECFC | endothelial colony-forming cells |
| ECM | extracellular matrix |
| EGFR | epidermal growth factor receptor |

| | |
|---|---|
| EGFRvIII | epidermal growth factor receptor variant III |
| eIF4E | essential translation initiation factor |
| EMT | epithelial-mesenchymal transition |
| Eo | eosinophils |
| EOC | epithelial ovarian cancer |
| EphA2 | ephrin type-A receptor 2 |
| EPR | enhanced permeability and retention |
| ER | estrogen receptor |
| ER-α | estrogen receptor alpha |
| ERK | extracellular signal related kinase |
| ESC | embryonic stem cell |
| ESC | enhanced stabilization chemistry |
| ESCC | esophageal squamous cell carcinoma |
| esiRNA | endogenous small interfering RNA (siRNA) |
| EST | expressed sequence tag |
| Ets-1 | erythroblastosis virus E26 oncogene homolog 1 |
| EV | extracellular vesicle |
| expo5 | exportin-5 |
| EZH2 | enhancer of zeste homolog 2 |
| F7 | factor VII |
| FA | folate acid |
| FA | Fanconi anemia |
| FACS | fluorescence-activated cell sorting |
| FANCA | Fanconi anemia complementation group A protein |
| FA-pRNA-AmiRs | FA-conjugated bacteriophage Phi29 pRNA |
| FB4 | transferrin receptor |
| FC | fold-change |
| FDA | US Food and Drug Administration |
| FFPE | formalin-fixed paraffin-embedded |
| FIP200 | kinase family-interacting protein of 200-kDa |
| FLT3 | fms-related tyrosine kinase 3 |
| FNA | fine-needle aspiration |
| FR | folate receptor |
| FXRA | farnesoid X receptor α |
| GalNAc | N-acetylgalactosamine |
| GalNAc–siRNA | N-acetylgalactosamine-siRNA conjugates |
| GBM | glioblastoma multiforme |
| GC | gastric cancer |
| G-CIMP | glioma CpG island methylator phenotype |
| GD$_2$ | disialoganglioside |
| GEM | genetically engineered mouse |
| GFR | glomerular filtration rate |
| GnRH | gonadotropin releasing hormone |
| GO | gene ontology |
| GPCR | G-protein coupled receptor |
| GPT | glycerol propoxylate triacrylate |
| GRP | gastrin-releasing peptide |
| GSTP1 | glutathione S-transferase-π |
| GTEx | genotype-tissue expression |
| GTP | guanosine triphosphate |
| GVHD | graft-versus-host disease |
| H2A | histone 2A |

| | | | | |
|---|---|---|---|---|
| **H3K4me3** | trimethylated histone H3 lysine 4 | | **IHC** | immunohistochemical |
| **H3K27** | histone H3 on Lys 27 | | **IL-6** | interleukin-6 |
| **HA** | hyaluronic acid | | **IPMN** | intraductal papillary mucinous neoplasms |
| **HA-CS** | NPs chitosan NPs modified with hyaluronic acid | | **IR** | ionizing radiation |
| **H2B** | histone 2B | | **IRS1** | insulin receptor substrate 1 |
| **H3K9me3** | trimethylation of H3K9 | | **i.t.** | intratumoral |
| **HAT** | histone acetyltransferase | | **JNK** | C-Jun-NH$_2$-kinase |
| **HB** | hypocrellin B | | **kb** | kilobase |
| **HBEC** | human bronchial epithelial cell | | **KEGG** | Kyoto Encyclopedia of Genes and Genomes |
| **HBp** | HBx and HBV polymerase protein | | **KERV-1** | kangaroo endogenous retrovirus |
| **HBV** | hepatitis B virus | | **KLF6** | Kruppel-like factor 6 |
| **HBx** | HBV protein X | | **KLK4** | Kallikrein-related peptidase 4 |
| **HCC** | hepatocellular carcinoma | | **KRAS** | Ras family of oncogenes, includes two other genes: HRAS and NRAS |
| **HCV** | hepatitis C virus | | **KSHV** | Kaposi's-sarcoma-associated herpes virus |
| **HDAC** | histone deacetylase | | **KSP** | kinesin spindle protein |
| **HDL** | high-density lipoprotein | | **LAC** | lung adenocarcinoma |
| **HDM** | histone demethylase | | **LATS2** | large tumor suppressor homologue 2 |
| **hENT1** | human equilibrative nucleoside transporter 1 | | **LC** | lung cancer |
| | | | **LC** | laryngeal carcinoma |
| **HER2** | human epidermal growth factor receptor 2 | | **LCC** | large-cell carcinoma |
| **HI** | healthy individual | | **LCK** | log cell kill |
| **HIF-1α** | hypoxia-inducible factor 1 alpha | | **LCM** | laryngeal carcinoma |
| **HIF-1β** | hypoxia-inducible factor 1 beta | | **LCST** | lower critical solution temperature |
| **HIWI** | human ortholog of PIWI | | **LDH** | lactate dehydrogenase |
| **HL** | Hodgkin lymphoma | | **LDH** | layered double hydroxide |
| **HLH** | helix-loop-helix | | **LDL** | low-density lipoprotein |
| **HMCL** | human MM cell line | | **LEU2** | leukemia-associated gene 2 |
| **HMT** | histone methyltransferase | | **LHRH** | luteinizing hormone releasing hormone |
| **HMW-PEI** | high-molecular-weight PEI | | **lincRNA** | long intergenic noncoding RNA |
| **HNPCC** | hereditary nonpolyposis colorectal cancer | | **LINE 1** | long interspersed nuclear element 1 |
| **HNSCC** | head and neck squamous cell carcinoma | | **LMP2** | low molecular mass polypeptides 2 |
| **HOTAIR** | HOX transcript antisense RNA | | **LMP7** | low molecular mass polypeptides 7 |
| **HOX** | homeobox | | **LMW-PEI** | low-molecular-weight PEI |
| **Hp** | hematoporphyrin | | **LNA** | locked nucleic acid |
| **HPIP** | hematopoietic pre-B cell leukemia transcription factor-interacting protein | | **lncRNA** | long noncoding RNA |
| | | | **LND** | lymph node dissection |
| **HR** | hazard ratio | | **LNM** | lymph node metastasis |
| **HR** | homology-directed repair | | **LNP** | lipid nanoparticle |
| **HSC** | hematopoietic stem cell | | **LOH** | loss of heterozygosity |
| **hSSB** | human single-stranded DNA binding protein | | **LOI** | loss of imprinting |
| | | | **LPC** | liver progenitor cell |
| **hTERT** | human telomerase reverse transcriptase | | **LPEI** | linear PEI |
| **HU** | Hounsfield unit | | **LPH** | liposome-polycation hyaluronic acid |
| **HULC** | highly upregulated in liver cancer | | **LPL** | lipoprotein lipase |
| **HUVEC** | human umbilical vein endothelial cell | | **LSC** | leukemia stem cell |
| **IAR** | individuals at risk | | **LSCC** | lung squamous cell carcinoma |
| **IC$_{50}$** | half maximal (50%) inhibitory concentration | | **LSD1** | lysine-specific demethylase 1 |
| | | | **M** | melanoma |
| **ICAM-1** | intercellular adhesion molecule-1 | | **MAPK** | mitogen-activated protein kinase |
| **ID** | injected dose | | **MB** | molecular beacon |
| **IFN** | interferon | | **MBC** | metastatic breast cancer cell |
| **IFN-α** | interferon alpha | | **MBD** | methyl-CpG binding protein |
| **IGF II** | insulin-like growth factor II | | **MBDs** | methyl-binding domain proteins |
| **IGFBP-3** | insulin-like growth factor-binding protein-3 | | **MCF-7** | minichromosome factor 7 |
| **IgG** | immunoglobulin G | | **MCL-1** | myeloid leukemia cell sequence-1 |
| **IgY** | immunoglobulin Y | | **mCRPC** | metastatic castration-resistant prostate cancer |

| | | | |
|---|---|---|---|
| **MDC1** | mediator of the damage checkpoint 1 | **NHEJ** | nonhomologous end joining |
| **MDM2** | mouse double minute 2 | **NHL** | non–Hodgkin's lymphoma |
| **MDR** | multidrug resistance | **NK** | natural killer (cells) |
| **MDS** | myelodysplastic syndrome | **NKX3.1** | NK3 homeobox 1 |
| **MECL1** | multicatalytic endopeptidase complex subunit | **NLE** | neutral lipid emulsion |
| | | **NP** | nanoparticle |
| **MGMT** | methylguanine-DNA methyltransferase | **NPC** | nasopharyngeal carcinoma |
| **MGUS** | monoclonal gammopathy of undetermined significance | **NPNH** | nonpancreatic, nonhealthy |
| | | **NRP1** | neuropilin 1 |
| **miR** | microRNA | **NSCL** | non–small cell lung |
| **miR-21i** | miR-21i inhibitor | **NSCLC** | non–small cell lung cancer |
| **miRBase** | miRNA-based investigations | **(nt)** | nucleotide |
| **miRISC** | miRNA-induced silencing complex | **OC** | ovarian cancer |
| **miRNA** | microRNA | **OPN** | osteopontin protein |
| **miRNP** | miRNA-containing ribonucleoprotein complex | **OR** | odds ratio |
| | | **ORF** | open reading frame |
| **miRSP** | miRNA sponge | **OS** | overall survival |
| **MLH1** | MutL homolog | **OSCC** | oral squamous cell carcinoma |
| **MLL1** | mixed lineage leukemia | **P** | gene promoter |
| **MLP** | melittin-like peptide | **p53** | Trp53 |
| **MM** | multiple myeloma | **PAMAM** | poly(amidoamine) dendrimer |
| **MM-BMSC** | multiple myeloma-bone marrow stromal cell | **PanIN** | pancreatic intraepithelial neoplasm |
| **MMSET** | multiple myeloma SET domain | **paRNA** | promoter-associated RNA |
| **MNP** | magnetic nanoparticle | **PARP** | poly(ADP-ribose) polymerase |
| **Mo** | monocytes | **PASR** | promoter-associated small RNA |
| **MoS₂** | molybdenum sulfide | **PAX5α** | paired box gene 5α |
| **MPM** | malignant pleural mesothelioma | **PB** | peripheral blood |
| **MP** | microparticle | **PBAVE** | poly(butyl) amino vinyl ether |
| **MPSS** | massively parallel signature sequencing | **PBMC** | peripheral blood mononucleated cell |
| **Mre11** | meiotic recombination 11 | **P bodies** | processing bodies |
| **MRI** | magnetic resonance imaging | **PC** | prostate cancer |
| **mRNA** | messenger RNA | **PCA** | pancreatic cancer |
| **MRP1** | multidrug resistance-associated protein 1 | **PCAF** | p300/CBP-associated factor |
| **MRP2** | multidrug resistance-associated protein 2 | **PcG** | Polycomb group protein |
| **MRX34** | cancer-targeting miRNA mimic of miR-34 | **PCNA** | proliferating cell nuclear antigen |
| **MSNP** | mesoporous silica nanoparticle | **PCR** | polymerase chain reaction |
| **MSP** | methylation-specific PCR | **PC** | plasma cell |
| **MSH2** | human mutS homolog 2 | **PCSK9** | proprotein convertase subtilisin/kexin type 9 |
| **MSI** | microsatellite instability | **PD** | Parkinson's disease |
| **MSKCC** | Memorial Sloan Kettering Cancer Center | **PD** | progressive disease |
| **MSP** | methylation-specific PCR | **PD1** | programmed cell death protein 1 |
| **Ms-SNuPE** | methylation-sensitive single-nucleotide primer extension | **PDA** | pancreatic ductal adenocarcinoma |
| | | **PDAC** | pancreatic ductal adenocarcinoma |
| **MSV** | silicon-based multistage vector | **PDCD4** | downstream target, programmed cell death 4 |
| **multi-siRNA** | multimerized siRNA | **PDGFC** | polymorphism in platelet-derived growth factor C |
| **MVP** | major vault protein | | |
| **Myc** | transcription factor | **PDI** | low polydispersity |
| **MYLK** | myosin light chain kinase pseudogene | **PDL1** | programmed cell death ligand 1 |
| **N** | neutrophil | **PDMAEMA** | poly-dimethylaminoethyl methacrylate |
| **NACT** | neoadjuvant chemotherapy | **PDP–PE** | pyridyldithio propionate–PE |
| **NAFLD** | nonalcoholic fatty-liver disease | **PEC** | polyelectrolyte complex |
| **NBS** | Nijmegen breakage syndrome | **PEG** | polyethylene glycol |
| **nc** | noncoding | **PEG–PE** | PEG–phosphoethanolamine |
| **ncRNA** | non-protein-coding RNA | **PEI** | polyethylene amine |
| **NFκB** | nuclear factor kappa B | **PE-SH** | phosphothioethanol |
| **NFIB** | transcription factor nuclear factor I/B | **P-gp** | P-glycoprotein |
| **NGS** | next-generation sequencing | **PH** | partial hepatectomy |

| | |
|---|---|
| PI3K | phosphatidylinositol-3 kinase |
| PIKK | phosphatidylinositol-3 kinase-related kinase |
| piRNA | PIWI-interacting RNA |
| piRNA | PIWI-associated RNA |
| piRNP | PIWI ribonucleoprotein |
| PIWIL2 | PIWI-like 2 |
| PKC | protein kinase C |
| PKN3 | protein kinase N3 |
| PKR | protein kinase R |
| PLA | polylactic acid |
| PLGA | poly(lactic-co-glycolic acid) |
| PLK1 | Polo-like kinase 1 |
| PLL | poly-L-lysine |
| PLT | platelets |
| PML-RARα | promyelocytic leukemia-retinoic acid receptor α |
| PMS2 | postmeiotic segregation increased 2 |
| PRKCε | protein kinase C epsilon |
| PRKDC | protein kinase, DNA-activated, catalytic polypeptide |
| PROMPT | promoter-upstream transcript |
| pol II | RNA polymerase |
| poly-siRNA | polymerized siRNA |
| POZ | poxvirus and zinc finger |
| Pp | protoporphyrin IX disodium salt |
| PPAR | peroxisome proliferator-activated receptor alpha |
| PRC2 | polycomb repressive complex 2 |
| PRC | polycomb repressive complex like PRC2 and LSD1 |
| PrEC | prostate epithelial cancer |
| pre-miRNA | precursor microRNA |
| pri-miRNA | primary transcript microRNA molecule |
| PS | phosphorothioate |
| PS2 | phosphodithioate |
| PSA | prostate-specific antigen |
| PSMA | prostate-specific membrane antigen |
| PTC | papillary thyroid carcinoma |
| PTCH1 | Patched-1 |
| PTEN | phosphatase and tensin homolog |
| PYK-2 | proline-rich tyrosine kinase 2 |
| QD | quantum dot |
| QPCR | quantitative PCR |
| qRT-PCR | quantitative reverse transcription-polymerase chain reaction |
| RACE | rapid amplification of cDNA ends |
| RAR | retinoic acid receptor |
| RA | retinoic acid |
| RASA1 | RAS p21GTPase-activating protein 1 |
| rasiRNA | repeat-associated small interfering RNA |
| RASSF1A | RAS association family 1A |
| Rb | retinoblastoma |
| RCC | renal cell carcinoma |
| RCT | rolling circle transcription |
| RES | reticuloendothelial system |
| RGD | Arg-Gly-Asp |
| rHDL | reconstituted high-density lipoprotein |
| RIG-1 | retinoic acid inducible protein |
| RIP | RNA immunoprecipitation |
| RISC | RNA-induced silencing complex |
| RNA | ribonucleic acid |
| RNAi | RNA interference |
| RNAPII | RNA polymerase II |
| RNAPIII | RNA polymerase III |
| RNase | ribonuclease |
| ROC | receiver-operating characteristic |
| RPA | replication protein A |
| RPL6 | ribosomal protein L6 |
| RPN2 | human-ribophorin II |
| RRM2 | ribonucleotide reductase M2 |
| rRNA | ribosomal RNA |
| RSV | N respiratory syncytial virus nucleocapsid gene |
| RTP801 | Redd1 (for regulated in development and DNA damage responses) |
| RT-PCR | real-time polymerase chain reaction |
| RT-PCR | reverse transcription-polymerase chain reaction |
| RT-qPCR | real-time quantitative reverse transcription polymerase chain reaction |
| RVG | rabies virus glycoprotein |
| SAGE | serial analysis of gene expression |
| SCARNA22 | small cajal body-specific RNA 22 |
| SCC | squamous cell carcinoma |
| scFv | single-chain variable fragment |
| SCLC | small cell lung cancer |
| sd-rxRNA | lipophilically-enriched RNA-antisense hybrid |
| SELEX | selection of aptamers in DNA/RNA libraries technology |
| SFRP | secreted frizzled-related protein |
| SHH | Sonic Hedgehog |
| shRNA | short hairpin RNA |
| SINE | short interspersed nuclear element |
| siRNA | small interfering RNA |
| siRNN | short interfering ribonucleic neutral |
| sisiRNA | small internally segmented interfering RNA |
| SLE | systemic lupus erythematosus |
| SLEDAI | SLE disease activity index |
| SLN | solid lipid nanoparticle |
| SMAD | similar mothers against decapentaplegic of genes |
| SMC1 | structural maintenance of chromatin 1 |
| SMO | smoothened |
| SMPB | sulfo-succinimidyl 4-[p-maleimidophenyl] butyrate |
| smMLCK | smooth-muscle myosin light chain kinase |
| SNALP | stable nucleic acid lipid nanoparticle |
| SNALP | stable nucleic acid lipid particle |
| sncRNA | small noncoding RNA |
| SND1 | staphylococcal nuclease homology domain containing 1 |
| SNORA | the box H/ACA snoRNA |

| | | | |
|---|---|---|---|
| **SNORD** | the box C/D snoRNA | $T_m$ | melting point |
| **snRNA** | small nuclear RNA | **TMEA** | tris-(2-maleimidoethyl) amine |
| **(sn)RNA** | small nuclear RNA | **TNBC** | triple-negative breast cancer |
| **(sno)RNA** | small nucleolar RNA | **TNFα** | tumor necrosis factor alpha |
| **snoRNP** | small nucleolar ribonucleoprotein | **TNFSF4** | tumor necrosis factor superfamily member 4 |
| **SNP** | single-nucleotide polymorphism | **TNM** | tumor node metastasis |
| **SPA** | sporadic pituitary adenoma | **TocsiRNA** | α-tocopherol-conjugated siRNA |
| **SPE** | spermine | **TP53** | tumor protein 53 |
| **SPN** | solitary pulmonary nodule | **TP53INP1** | tumor protein 53-induced nuclear protein 1 |
| **SPOP** | speckle-type poxvirus and zinc finger (POZ) domain protein | **TPA** | tissue polypeptide antigen |
| | | **TPN** | tumor penetrating nanocomplex |
| **SRA** | steroid receptor RNA activator | **TR** | tandem repeat |
| **SRAP** | SRA protein | **TRAMP** | transgenic adenocarcinoma of mouse prostate |
| **SR-BI** | scavenger receptor class B type I (SR-BI) | | |
| **–S–S–** | disulfide linkage | **TRBP** | trans-activator RNA binding protein |
| **ss** | single-stranded | **(t)RNA** | transfer RNA |
| **SSB** | single-stranded DNA binding | **TrpV1** | transient receptor potential cation channel subfamily V member 1 |
| **ssDNA** | single-stranded DNA | | |
| **ssRNA** | single-stranded (DNA or RNA) | **TrxG** | trithorax group |
| **STAT3** | signal transducer and activator of transcription 3 | **TrxGbound** | transcriptionally competent euchromatin |
| | | **TSA** | tricostatin A |
| **STC** | standard template chemistry | **TSGs** | tumor-suppressor genes |
| **TACSTD1** | tumor-associated calcium signal transducer 1 | **ts-miRNA** | tumor suppressor miRNA |
| | | **TSSa-RNA** | TSS-associated RNA |
| **TAM** | tumor-associated macrophage | **TSS** | transcriptional start site |
| **T-ALL** | T-cell acute lymphoblastic leukemia | **TTR** | transthyretin |
| **TARBP2** | TAR RNA-binding protein 2 | **T-UCR** | UCR is transcribed |
| **TBH bilayer-to-hexagonal** | HII phase transition | **TUTase** | terminal uridylyl transferase |
| | | **UbC** | human ubiquitin C gene |
| **TC** | translocation and cyclin | **UC** | ulcerative colitis |
| **TCGA** | The Cancer Genome Atlas | **UCK** | uridine-cytidine kinase |
| **TD** | tardive dyskinesia | **UCR** | ultraconserved region |
| **tel-sRNA** | telomere-specific small RNA | **UDCA** | ursodeoxycholic acid |
| **TET1-3** | ten-eleven translocation 1–three protein | **uPA** | urokinase-type plasminogen activator |
| **TF** | transcription factor | **UTR** | untranslated region |
| **Tf-NP** | NPs with transferring | **VEGF** | vascular endothelial growth factor |
| **TfR** | transferring receptor | **VEGF-A** | vascular endothelial growth factor A |
| **TGF-β** | transforming growth factor beta | **vitamin E** | alpha-tocopherol |
| **TGFβRI** | TGF-β receptor I | **VLP** | virus-like particle |
| **TGFβRII** | TGF-β receptor II | **WHO** | World Health Organization |
| **TIC** | tumor-initiating cell | **XIAP** | X-linked inhibitor of apoptosis protein |
| **TIMP** | tissue inhibitor of metalloproteinase | **XLF** | XRCC4-like factor |
| **TIMP3** | tissue inhibitor of metalloproteinase 3 | **XPO5** | nuclear export protein exportin-5 |
| **tiRNA** | transcription initiation RNA | **ZAP** | zeta-associated protein |
| **TLDA** | TaqMan low-density array | | |
| **TLR** | Toll-like receptor | | |

# Index

Note: Locators in *italics* represent figures and **bold** indicate tables in the text.

For Product Safety Concerns and Information please contact our EU
representative  GPSR@taylorandfrancis.com
Taylor & Francis Verlag GmbH, Kaufingerstraße 24, 80331 München, Germany